AF361759

Emerging Trends in TiO$_2$ Photocatalysis and Applications

Emerging Trends in TiO$_2$ Photocatalysis and Applications

Editors

Trong-On Do
Sakar Mohan

MDPI • Basel • Beijing • Wuhan • Barcelona • Belgrade • Manchester • Tokyo • Cluj • Tianjin

Editors
Trong-On Do
Laval University
Canada

Sakar Mohan
Laval University
Canada

Editorial Office
MDPI
St. Alban-Anlage 66
4052 Basel, Switzerland

This is a reprint of articles from the Special Issue published online in the open access journal *Catalysts* (ISSN 2073-4344) (available at: https://www.mdpi.com/journal/catalysts/special_issues/TiO2_photocatal).

For citation purposes, cite each article independently as indicated on the article page online and as indicated below:

LastName, A.A.; LastName, B.B.; LastName, C.C. Article Title. *Journal Name* **Year**, *Article Number*, Page Range.

ISBN 978-3-03936-706-1 (Pbk)
ISBN 978-3-03936-707-8 (PDF)

Contents

About the Editors

Trong-On Do is a full professor in the Department of Chemical Engineering at Laval University, Canada. He received his MSc and PhD from University of P. and M. Curie (France) and carried out postdoctoral research in Prof. G. Bond's group at Brunel University (UK), and then the French Catalysis Institute (France). He spent two years (1997–1999) in Profs. Hashimoto/Fujishima's group at KAST under the Japanese STA Fellowship Award. His current research focuses on the design and synthesis of innovative and smart nanomaterials as photocatalysts and their applications in renewable energy and environmental remediation.

Sakar Mohan received his PhD from the University of Madras, India in 2015. Later, he joined in the research group of Prof. Trong-On Do, Laval University, Canada for his postdoc. Currently, he is working as an Assistant Professor in the Centre for Nano and Material Sciences, Jain University, India since 2017. His research interest includes photocatalysis, membranes, biodiesel, sensors and biomaterials.

Editorial

Editorial: Special Issue on "Emerging Trends in TiO$_2$ Photocatalysis and Applications"

Trong-On Do [1,*] and Sakar Mohan [1,2]

[1] Department of Chemical Engineering, Laval University, Quebec G1V 0A6, Canada; sakar.mohan.1@ulaval.ca
[2] Centre for Nano and Material Sciences, Jain University, Bangalore 562112, India
[*] Correspondence: Trong-On.Do@gch.ulaval.ca

Received: 4 June 2020; Accepted: 8 June 2020; Published: 13 June 2020

It is not an exaggerated fact that the semiconductor titanium dioxide (TiO$_2$) has been evolved as a prototypical material to understand the photocatalytic process and has been demonstrated for various photocatalytic applications such as pollutants degradation, water splitting, heavy metal reduction, CO$_2$ conversion, N$_2$ fixation, bacterial disinfection, etc., as depicted in Figure 1. [1,2] The rigorous photocatalytic studies over TiO$_2$ have paved ways to understand the various chemical processes involved and physical parameters (optical and electrical) required to design and construct diverse photocatalytic systems. [3,4] Accordingly, it has been realized that an effective photocatalyst should have ideal band edge potential, narrow band gap energy, reduced charge recombination, enhanced charge separation, improved interfacial charge transfer, surface-rich catalytic sites, etc. These studies further highlighted that single component catalysts may not be good enough to achieve the required/enhanced photocatalytic process. As a result, many strategies have been developed to design a variety of photocatalytic systems, which include doping, composite formation, sensitization, co-catalyst loading, etc. [5] The doping strategy includes cationic and anionic doping, where it is found that the essential purposeof doping is to tune the band gap energy of the photocatalyst by introducing the new energy levels of the doped elements underneath the conduction band (CB) and above the valence band (VB) of the semiconductor photocatalyst, respectively. On the other hand, the composite formation serves in multiple ways to almost meet all the requirements to achieve a quantum efficient photocatalytic process. The basis of composite formation is found to redesign the charge transport kinetics in the bulk and surface/interface of the integrated photocatalyst systems. These composite systems generally include p-n heterojunction, Z-scheme, etc. Similarly, the mechanism of sensitizing the photocatalysts includes the integration of plasmonic metal nanoparticles, carbon-based materials, 2D materials, quantum dots, and metal organic frameworks to enhance their optical absorption, electrical transportation properties, etc. [6] Interestingly, the co-catalyst loading serves as an *'engineered-catalytic-site'* for the specific redox process to achieve the selective photocatalytic reactions. Furthermore, the unique systems, such as ferroelectric-based photocatalysts, are found to be more interesting as they are governed by their inherent internal electrical field and surface polarization properties. For instance, the ferroelectric properties intrinsically facilitate the adsorption of the surrounding molecules, carrier separation, and interfacial charge transfer via band bending phenomenon, etc. Similarly, the influence of defects in photocatalysis has been well studied over TiO$_2$, where the concepts of "self-doping", "oxygen vacancy", "colored TiO$_2$", etc. have been well addressed in TiO$_2$photocatalysts.

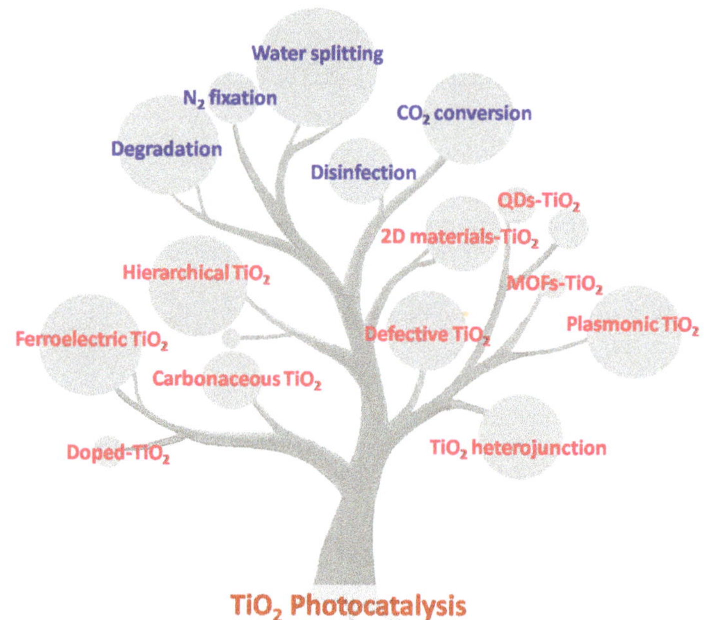

Figure 1. Overview of TiO$_2$-based various photocatalytic systems and their applications.

Towards highlighting the above mentioned diversities in TiO$_2$ photocatalysis, there have been many interesting research works on TiO$_2$, involving material designs for various photocatalytic applications published in this Special Issue. These material systems include TiO$_2$ QDs@g-C$_3$N$_4$ p-n junction, [7] oxygen defective TiO$_2$ nanorod array, [8] TiO$_2$/N-doped graphene QDs, [9] TiO$_2$/HKUST-1, [10] TiO$_2$-Carbon composite, [11] Ru-Ti oxide, [12] TiO$_2$ coated porous glass fiber cloth, [13] Ag/Fe$_3$O$_4$/TiO$_2$ nanofibers, [14] Pd-doped TiO$_2$, [15] N-doped TiO$_2$, [16] C/N/S-doped TiO$_2$, [17] Mo/W co-doped TiO$_2$, [18] Fe-doped TiO$_2$, [19] N-doped graphene QDs-TiO$_2$, [20] Nd-doped TiO$_2$, [21] Cu-doped TiO$_2$ thin film, [22] surface engineered TiO$_2$, [23] etc., for various photocatalytic applications, such as the degradations of a variety of pollutants, [24–30] biomass reforming, [10] heavy metal reduction, [14] and bacterial disinfections, [22] etc. In addition to these original research papers, some excellent review papers have also been published in this Special Issue, focusing on the various TiO$_2$-based photocatalytic systems and their mechanisms and applications. [1–6] To this end, it is highlighted that future works in TiO$_2$ should involve developing new material systems based on TiO$_2$. For instance, instead of doping N into TiO$_2$, the composition/phase tunable Ti oxy-nitride systems should be developed and so should the Ti oxy-phosphates, oxy-sulfurs, oxy-carbons, etc. From application perspectives, TiO$_2$ should be investigated for its photocatalytic efficiencies towards the production of H$_2$/O$_2$ from atmospheric vapor, dark-photocatalytic activities, hydrogen storage, biodiesel productions, etc. However, the research should also be continued on bare TiO$_2$ to achieve an in depth understanding of the photocatalytic mechanisms towards finding new photocatalytic applications.

Conflicts of Interest: The authors declare no conflict of interest.

References

1. Sakar, M.; Mithun Prakash, R.; Do, T.O. Insights into the TiO$_2$-based photocatalytic systems and their mechanisms. *Catalysts* **2019**, *9*, 680. [CrossRef]
2. Serpone, N. Heterogeneous photocatalysis and prospects of TiO$_2$-based photocatalyticDeNOxing the atmospheric environment. *Catalysts* **2018**, *8*, 553. [CrossRef]

3. Higashimoto, S. Titanium-dioxide-based visible-light-sensitive photocatalysis: Mechanistic insight and applications. *Catalysts* **2019**, *9*, 201. [CrossRef]

4. Zhang, B.; Cao, S.; Du, M.; Ye, X.; Wang, Y.; Ye, J. Titanium dioxide (TiO_2) mesocrystals: Synthesis, growth mechanisms and photocatalytic properties. *Catalysts* **2019**, *9*, 91. [CrossRef]

5. Kang, X.; Liu, S.; Dai, Z.; He, Y.; Song, X.; Tan, Z. Titanium dioxide: From engineering to applications. *Catalysts* **2019**, *9*, 191. [CrossRef]

6. Ren, Y.; Dong, Y.; Feng, Y.; Xu, J. Compositing two-dimensional materials with TiO_2 for photocatalysis. *Catalysts* **2018**, *8*, 590. [CrossRef]

7. Wang, S.; Wang, F.; Su, Z.; Wang, X.; Han, Y.; Zhang, L.; Xiang, J.; Du, W.; Tang, N. Controllable fabrication of heterogeneous p-TiO_2 QDs@g-C_3N_4 p-n junction for efficient photocatalysis. *Catalysts* **2019**, *9*, 439. [CrossRef]

8. Yang, B.; Chen, G.; Tian, H.; Wen, L. Improvement of the photoelectrochemical performance of TiO_2nanorod array by PEDOT and oxygen vacancy Co-modification. *Catalysts* **2019**, *9*, 407. [CrossRef]

9. Ou, N.Q.; Li, H.J.; Lyu, B.W.; Gui, B.J.; Sun, X.; Qian, D.J.; Jia, Y.; Wang, X.; Yang, J. Facet-dependent interfacial charge transfer in TiO2/nitrogen-doped graphene quantum dots heterojunctions for visible-light driven photocatalysis. *Catalysts* **2019**, *9*, 345. [CrossRef]

10. Martínez, F.M.; Albiter, E.; Alfaro, S.; Luna, A.L.; Justin, C.C.; Barrera-Andrade, J.M.; Remita, H.; Valenzuela, M.A. Hydrogen production from glycerol photoreforming on TiO_2/HKUST-1 composites: Effect of preparation method. *Catalysts* **2019**, *9*, 338. [CrossRef]

11. Songo, M.M.; Moutloali, R.; Suprakas Sinha, R. Development of TiO2-carbon composite acid catalyst for dehydration of fructose to 5-hydroxymethylfurfural. *Catalysts* **2019**, *9*, 126. [CrossRef]

12. Shi, J.; Hui, F.; Yuan, J.; Yu, Q.; Mei, S.; Zhang, Q.; Li, J.; Wang, W.; Yang, J.; Lu, J. Ru-Ti oxide based catalysts for HCl oxidation: The favorable oxygen species and influence of Ce additive. *Catalysts* **2019**, *9*, 108. [CrossRef]

13. Yanagida, S.; Hirayama, K.; Iwasaki, K.; Yasumori, A. Adsorption and photocatalytic decomposition of gaseous 2-propanol using TiO_2-coated porous glass fiber cloth. *Catalysts* **2019**, *9*, 82. [CrossRef]

14. Chang, Y.H.; Wu, M.C. Enhanced photocatalytic reduction of Cr(VI) by combined magnetic tio2-based nfs and ammonium oxalate hole scavengers. *Catalysts* **2019**, *9*, 72. [CrossRef]

15. Rusinque, B.; Escobedo, S.; de Lasa, H. Photocatalytic hydrogen production under near-uv using Pd-doped mesoporous TiO_2 and ethanol as organic scavenger. *Catalysts* **2019**, *9*, 33. [CrossRef]

16. Rangel, R.; Cedeño, V.J.; Espino, J.; Bartolo-Pérez, P.; Rodríguez-Gattorno, G.; Alvarado-Gil, J.J. Comparing the efficiency of N-doped TiO2 and N-doped Bi2MoO6 photo catalysts for MB and lignin photodegradation. *Catalysts* **2018**, *8*, 668. [CrossRef]

17. El-Hosainy, H.M.; El-Sheikh, S.M.; Ismail, A.A.; Hakki, A.; Dillert, R.; Killa, H.M.; Ibrahim, I.A.; Bahnemann, D.W. Highly selective photocatalytic reduction of o-dinitrobenzene to o-phenylenediamine over non-metal-doped TiO_2 under simulated solar light irradiation. *Catalysts* **2018**, *8*, 641. [CrossRef]

18. Avilés-García, O.; Espino-Valencia, J.; Romero-Romero, R.; Rico-Cerda, J.L.; Arroyo-Albiter, M.; Solís-Casados, D.A.; Natividad-Rangel, R. Enhanced photocatalytic activity of titania by co-doping with Mo and W. *Catalysts* **2018**, *8*, 631. [CrossRef]

19. Ramírez-Sánchez, I.M.; Bandala, E.R. Photocatalytic degradation of estriol using iron-doped TiO_2 under high and low UV irradiation. *Catalysts* **2018**, *8*, 625. [CrossRef]

20. Li, F.; Li, M.; Luo, Y.; Li, M.; Li, X.; Zhang, J.; Wang, L. The synergistic effect of pyridinic nitrogen and graphitic nitrogen of nitrogen-doped graphene quantum dots for enhanced TiO_2nanocomposites' photocatalytic performance. *Catalysts* **2018**, *8*, 438. [CrossRef]

21. Eguchi, R.; Takekuma, Y.; Ochiai, T.; Nagata, M. Improving interfacial charge-transfer transitions in Nb-doped TiO_2 electrodes with 7,7,8,8-tetracyanoquinodimethane. *Catalysts* **2018**, *8*, 367. [CrossRef]

22. Moongraksathum, B.; Shang, J.Y.; Chen, Y.W. Photocatalytic antibacterial effectiveness of Cu-doped TiO_2 thin film prepared via the peroxo sol-gel method. *Catalysts* **2018**, *8*, 352. [CrossRef]

23. Kobayashi, K.; Takashima, M.; Takase, M.; Ohtani, B. Mechanistic study on facet-dependent deposition of metal nanoparticles on decahedral-shaped anatasetitaniaphotocatalyst particles. *Catalysts* **2018**, *8*, 542. [CrossRef]

24. Mortazavian, S.; Saber, A.; James, D.E. Optimization of photocatalytic degradation of Acid Blue 113 and Acid Red 88 textile dyes in a UV-C/TiO_2 suspension system: Application of response surface methodology (RSM). *Catalysts* **2019**, *9*, 360. [CrossRef]

25. Li, Q.; Wang, L.; Fang, X.; Zhang, L.; Li, J.; Xie, H. Synergistic effect of photocatalytic degradation of hexabromocyclododecane in water by UV/TiO_2/persulfate. *Catalysts* **2019**, *9*, 189. [CrossRef]

26. Schneider, O.M.; Liang, R.; Bragg, L.; Jaciw-Zurakowsky, I.; Fattahi, A.; Rathod, S.; Peng, P.; Servos, M.R.; Norman Zhou, Y. Photocatalytic degradation of microcystins by TiO_2 using UV-led controlled periodic illumination. *Catalysts* **2019**, *9*, 181. [CrossRef]

27. He, Y.; Li, H.; Guo, X.; Zheng, R. Bleached wood supports for floatable, recyclable, and efficient three dimensional photocatalyst. *Catalysts* **2019**, *9*, 115. [CrossRef]

28. Koysuren, H.N. Solid-phase photocatalytic degradation of polyvinyl borate. *Catalysts* **2018**, *8*, 499. [CrossRef]

29. Sun, P.; Zhang, J.; Liu, W.; Wang, Q.; Cao, W. Modification to L-H kinetics model and its application in the investigation on photodegradation of gaseous benzene by nitrogen-doped TiO_2. *Catalysts* **2018**, *8*, 326. [CrossRef]

30. Fu, W.; Shi, Z.; Bai, H.; Dai, J.; Lu, Z.; Lei, F.; Zhang, D.; Zhao, L.; Zhang, Z. Facile formation of anatase nanoparticles on H-titanate nanotubes at low temperature for efficient visible light-driven degradation of organic pollutants. *Catalysts* **2020**, *10*, 695. [CrossRef]

 *catalysts*

Communication

Facile Formation of Anatase Nanoparticles on H-Titanate Nanotubes at Low Temperature for Efficient Visible Light-Driven Degradation of Organic Pollutants

Weiwei Fu [1],*, Zhiqiang Shi [2], Helong Bai [1], Jinyu Dai [2], Zhiming Lu [2], Feifei Lei [2], Deguang Zhang [2], Lun Zhao [1],* and Zongtao Zhang [2],*

[1] College of Chemistry, Changchun Normal University, Changchun 130032, China; baihelong@ccsfu.edu.cn
[2] State Key Laboratory of Inorganic Synthesis & Preparative Chemistry, Jilin University, Changchun 130012, China; shizq17@jlu.edu.cn (Z.S.); daijy16@mails.jlu.edu.cn (J.D.); zmlu18@mails.jlu.edu.cn (Z.L.); leiff18@mails.jlu.edu.cn (F.L.); zhangdg17@mails.jlu.edu.cn (D.Z.)
* Correspondence: fuweiwei@ccsfu.edu.cn (W.F.); zhaolun@mail.cncnc.edu.cn (L.Z.); zzhang@jlu.edu.cn (Z.Z.); Tel.: +86-431-8616-8099 (W.F.)

Received: 13 May 2020; Accepted: 17 June 2020; Published: 19 June 2020

Abstract: Anatase nanoparticles (5–10 nm) generated on H-titanate nanotube surface (H-titanate/anatase) were prepared by an ingenious and simple method. H-titanate tubes were prepared by a hydrothermal reaction of Ti powder in concentrated NaOH solution and an ion exchange process with HNO_3 solution. After that, at a relatively low drying temperature (100 °C), a small quantity of anatase nanoparticles were in-situ formed on the H-titanate tubes surface by a surface dehydration reaction. In-situ transformation can form a strong interface coupling between H-titanate and anatase, which is conducive to accelerating charge transfer and improving its photocatalytic activity. In addition, the smaller average crystal size, the large specific surface areas (BET), the nanotubed and layered structure and the synergistic effect of dual phases would be beneficial to improving the photocatalytic efficiency.

Keywords: in-situ formation; anatase nanoparticles; H-titanate nanotubes; dual-phase; low temperature

---

1. Introduction

Water contamination resulting from the rapid development of industrialization has attracted worldwide attention. Photocatalytic degradation is the most promising strategy to completely solve the organic pollutants problem [1]. Among all kinds of conversion systems, TiO_2-based materials for water pollution are considered to be an environmentally friendly and promising way to efficiently utilize solar energy [2,3]. Under UV irradiation, anatase can degrade a broad range of tenacious and toxic organic contaminants in water, and it is nontoxic, relatively cheap and chemically stable. However, its photocatalytic application is limited owing to the rapid recombination of the excited electron–hole pairs, the low visible light activity and low surface area [4,5]. Coupling anatase with another semiconductor favors a narrow band gap and the electron–hole separation [6–9], so as to improve the quantum efficiency. Moreover, the synergistic effect between two different phases can also enhance the photocatalytic activity [6,10].

It was reported that H-titanate was formed from TiO_2 reacting with a concentrated NaOH solution to form titanate and then the ion exchange reacted with a dilute acid solution [11–13]. After calcination at a high temperature, H-titanate can transform into anatase TiO_2 [13]. Based on the above formation mechanism of TiO_2-H-titanate-TiO_2, it is estimated that, at low calcination temperature, a small amount of anatase TiO_2 will be formed in situ on the H-titanate nanotube surface. Several papers have reported the synthesis of titanate/anatase composites; however, most preparation processes were carried out under high pressure and high temperature conditions, which consumed more energy [14–16]. For example, Xiong et al. reported a nitrogen-doped titanate–anatase core–shell nanobelts. In this paper, the titanate–anatase was obtained by calcination at 500 °C [14]. Yan et al. demonstrated that the titanate nanotube/anatase nanoparticle composites could be prepared by the hydrothermal method using as-obtained titanate tube dispersed into a HNO_3 solution [15]. Herein, we report that a dual-phase photocatalyst (anatase nanoparticles (5–10 nm) was generated in situ on a H-titanate nanotube surface) was obtained via a controllable surface dehydration reaction at low temperature (100 °C) and atmospheric pressure, which exhibited a higher visible light photocatalytic activity than P25; pure H-titanate nanotubes and pure anatase. The efficient visible light photocatalytic activity can be attributed to: (1) In situ transformation can form strong interfacial coupling between H-titanate and anatase, which is favorable to accelerating charge transfer [17]; (2) H-titanate has a layered structure, which is composed of TiO_6 octahedra sheets sharing four edges, similar to that of anatase crystals, which is easy to form a heterostructure between anatase and H-titanate [14,18,19]. In the meantime, at a low drying temperature, dual-phase catalyst retained the nanotubed and layered structures, which was beneficial for the high BET surfaces to adsorb organic pollutants and promote the diffusion of organic molecules inside the pores. The smaller average crystal size of anatase nanoparticles means a stronger redox ability in the photocatalytic process. Therefore, in the presence of the dual-phase catalyst, rhodamine B (RhB) and methylene blue (MB) can be completely decomposed in a very short time under visible light irradiation. In addition, the synergistic effect would be beneficial to improve the photocatalytic efficiency.

2. Results and Discussion

Figure 1 shows the X-ray diffraction (XRD) patterns of the as-synthesized material. As shown in Figure 1a, there are several diffraction peaks located at $2\theta = 9.8°$, $24.4°$, $28.4°$, $48.4°$ suggesting that the as-prepared sample is layered titanate with a component of $H_2Ti_2O_4(OH)_2$ [20] and a 0.9 nm interlayer distance, which is further confirmed by TEM observation. After drying at 100 °C, the crystal structure of H-titanate was well maintained; the diffraction peaks of anatase-type TiO_2 (JCPDS no. 21-1272) [21] are clearly observed (Figure 1b), showing that dual-phase H-titanate/anatase was obtained at a low temperature, which may be due to the dehydration of the H-titanate. The broad peaks indicate its low crystallinity and nanosized crystallites. With increasing calcination temperature, the peaks of anatase become narrower and sharper (Figure 1c,d). After calcination at 500 °C for 5 h, all diffraction peaks of H-titanate disappeared (Figure 1e), which suggested that the layered H-titanate was completely transformed into anatase. The apparent sharpening of peaks suggests its high crystallinity. To further confirm the coexistence of these two TiO_2 phases in the as-synthesized products, Raman spectroscopy was tested for the H-titanate and dual-phase catalysts (Figure 2). In Figure 2b, the peaks that center at 143 (E_g), 514 (A_{1g}) and 636 cm^{-1} (E_g) belong to the anatase phase [22], while others match well with the hydrogen titanate phase (shown in Figure 2a). The Raman spectrum in Figure 2b is composed of the characteristic peaks of the H-titanate and anatase phase, which are in good agreement with the XRD analysis.

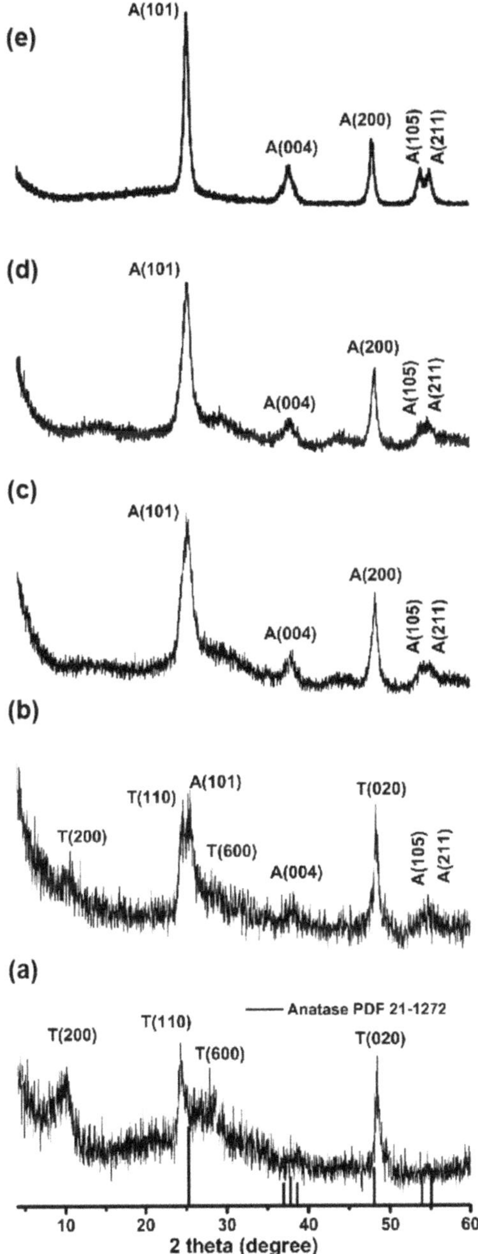

Figure 1. XRD patterns of (**a**) TiO$_2$-60 (the as-prepared H-titanate tubes), (**b**) TiO$_2$-100, (**c**) TiO$_2$-200, (**d**) TiO$_2$-300, (**e**) TiO$_2$-500, T is the H-titanate phase and A is the anatase phase.

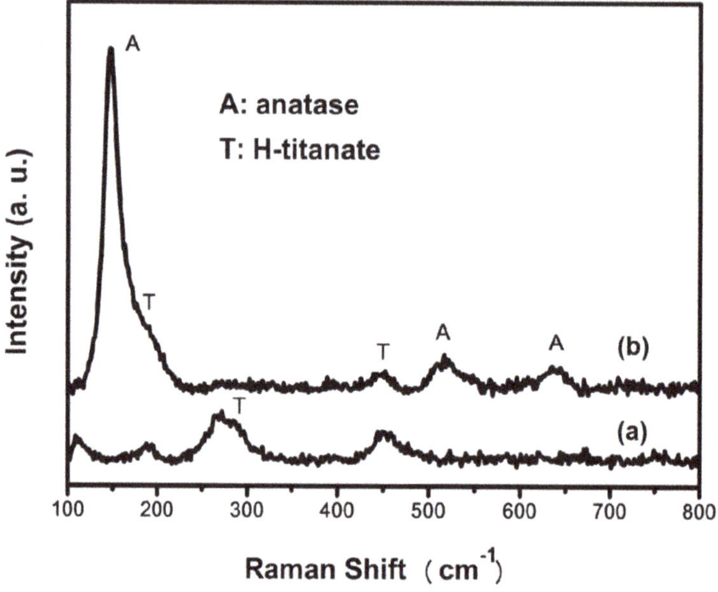

Figure 2. Raman spectra of (**a**) TiO$_2$-60 and (**b**) TiO$_2$-100.

The detailed characterization and crystal structure of the H-titanate/anatase composite were investigated via transmission electron microscopy (TEM), as shown in Figure 3. H-titanate was formed via the hydrothermal reaction of titanium powder with a concentrated NaOH solution to form sodium titanate and then a subsequent ion exchange reaction with HNO$_3$ solution at room temperature. The dual-phase H-titanate/anatase catalysts were prepared by in situ generation of anatase nanoparticles on the H-titanate tube surfaces by a controllable surface dehydration reaction at a low drying temperature (100 °C). Figure S1a shows images of only the nanotubes, whereas Figure 3b shows that some nanoparticles adhered to the surface of nanotubes, which indicates that a new phase was obtained after drying at 100 °C. The average diameter of the as-synthesized nanoparticles is about 5–10 nm. Figure 3b1 (high-resolution transmission electron microscopy (HRTEM) images) distinctly reveals a lattice fringe spacing of 0.35 nm, which corresponds well with the (101) anatase. Figure 3b2 shows that the nanotubes are crystallized of layered H-titanate. It also indicates that the H-titanate retains its morphology of nanotubes, and the anatase reveals the morphology of nanoparticles. There are only particles in Figure S1b, which indicates that when the calcination temperature increased to 500 °C, H-titanate nanotubes completely transformed into anatase nanoparticles. The above results coincide with the XRD analysis. At a low drying temperature (100 °C), the photocatalyst retained its original nanotube structure and a small amount of anatase nanoparticles were formed on the H-titanate nanotube surfaces. Specific surface area (BET) is a key factor in photodegradation. High specific surface area offers more reaction sites for dye molecules and hydroxyl groups. Therefore, the porosities of the samples were determined by N$_2$ sorption. Figure 3c exhibits the N$_2$ sorption isotherms of all as-synthesized products and the DFT (Density Functional Theory) pore size distributions of the dual-phase photocatalyst. The obtained isotherms of the samples are the typical IUPAC type-IV isotherm. We can see that the isotherms decreased with the decrease in the amount of H-titanate nanotubes. The corresponding DFT pore size distribution curve (inset of Figure 3c) of the dual-phase catalyst shows two pore sizes centered at ~1.0 nm and ~12 nm, respectively, which directly proves the layered and nanotubed structure of the dual phase catalyst. BET measurements show that the as-synthesized H-titanate tubes have a surface area of 245 m^2·g^{-1}, and dual-phase photocatalyst has a surface area of 174 m^2·g^{-1}. As the calcination temperature raised to 500 °C, the surface area of the product decreased to 95 m^2·g^{-1} (Table S1). The decrease in surface area may be due to the decreased proportion of layered H-titanate nanotubes. In other words, the specific surface area of the catalysts is enormously increased by the H-titanate nanotubes, which is hoping to enhance the photocatalytic property of the nanomaterials.

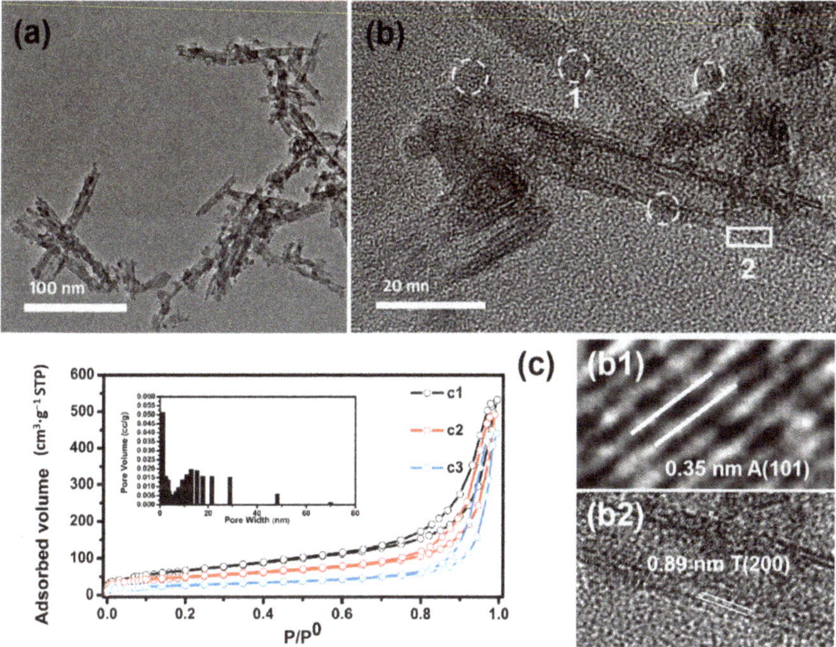

Figure 3. (**a**,**b**) are the TEM images of TiO$_2$-100. (**b1**) and (**b2**) are the HRTEM images of TiO$_2$-100 for the enlarged view of the circle and rectangle areas in (**b**). (**c**) Reversible nitrogen gas adsorption isotherm for the (**c1**) TiO$_2$-60, (**c2**) TiO$_2$-100, (**c3**) TiO$_2$-500 measured at 77 K. The inset is the pore size distribution of the TiO$_2$-100 photocatalyst calculated via DFT method.

UV-Vis diffuse reflectance absorption spectra of as-synthesized H-titanate nanotubes (TiO$_2$-60), dual-phase H-titanate/anatase (TiO$_2$-100) and anatase nanoparticles (TiO$_2$-500) are shown in the Figure 4a. As indicated in Figure 4a, in the range of 200-800 nm, the H-titanate nanotubes, dual-phase H-titanate/anatase and anatase samples demonstrate similar absorption. Compared with that of pure H-titanate and anatase, the absorption edge of dual-phase H-titanate/anatase moved toward a longer wavelength. It may be owing to the synergistic effect of H-titanate and anatase, resulting in a narrow band gap, which is a crucial role to the realization of solar energy conversion. The Kubelka-Munk method is often used to estimate the band gap energy (E$_g$) of as-prepared products [23]. Figure 4b reveals the Tauc plots of $(\alpha h\nu)^2$ vs. photon energy ($h\nu$) of H-titanate tubes, dual-phase H-titanate/anatase and anatase photocatalyst. The band gap (E$_g$) can be acquired by extending the vertical segment to the $h\nu$ axis. As shown in Figure 4b, the E$_g$ of the H-titanate, H-titanate/anatase, and anatase is 3.48, 3.30, and 3.36 eV, respectively, which reveals the same results with the ultraviolet-visible diffffuse reflectance spectra (UV-Vis DRS) analysis. The photoluminescence (PL) technique is usually used to investigate the charge carrier separation and transfer processes because PL emission results from the free photogenerated carriers recombination [24]. Figure 4c shows the PL spectra of dual-phase H-titanate/anatase, H-titanate and anatase excited at 315 nm. The emission intensity of dual-phase H-titanate/anatase is much weaker than that of pure H-titanate and anatase, which can be attributed to the formation of H-titanate/anatase heterojunction in two semiconductor interfaces. Different band edge positions of H-titanate and anatase can reduce the recombination of the carriers [19,25]. That enables more free electrons and holes to participate in the photocatalytic reactions and accelerates the photocatalytic process.

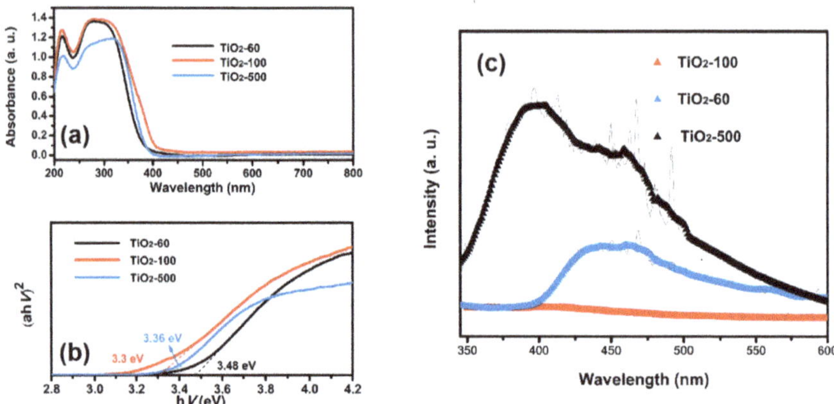

Figure 4. (**a**) UV/Vis diffuse reflectance absorption spectra, (**b**) plots of $(\alpha h v)^2$ versus hv, and (**c**) PL spectra of TiO$_2$-60, TiO$_2$-100 and TiO$_2$-500.

The photocatalytic activities of as-prepared samples were studied by degradation of methyl orange (MO), rhodamine B (RhB) and methylene blue (MB) in aqueous solution under visible light at room temperature. In order to evaluate the photocatalytic efficiency, P25 was chosen as the photocatalytic reference material. In general, high specific surface area can improve the adsorption performance of the materials. Therefore, prior to the photocatalytic degradation studies, the adsorption properties of the materials were investigated. In general, a high degree of surface adsorption was observed within the first 30 min of stirring time in dark before attaining the saturation level. In Figure 5a and Figure S2A, P25 shows no obvious adsorption capacity, whereas the as-received dual-phase H-titanate/anatase catalyst shows 90% MB, 25% RhB, 8.9% MO dye adsorption, the H-titanate shows 92% MB, 59% RhB, 9.1% MO adsorption due to their nanotubed and layered structures, larger specific surface areas, which are beneficial for enhancing the photocatalytic activity. In addition, the electrostatic attraction between the catalysts and the dye molecules plays an important role in the different adsorption behavior of several dyes [26]. MB, RhB is cationic and MO is anionic in the aqueous solution, while the nanotubes possess negative surface charge, which is more favorable to absorb MB and RhB. Although both RhB and MB are cationic dyes, the adsorption capacity of nanotubes for MB is much higher than that of RhB, which is due to the different molecular structures of dyes. The more linear shape and smaller size of MB molecule mean a weaker steric hindrance during the adsorption process [26]. In order to verify the adsorption of dye on H-titanate/anatase, FTIR analysis was carried out after a MB adsorption and degradation test (Figure S3). According to the previous report [27], in adsorption test, the peaks at 2926 cm^{-1} and 666 cm^{-1} indicated that the MB was adsorbed onto the H-titanate/anatase nanotubes surfaces. After the degradation process, the peaks disappeared, suggesting that MB degraded completely.

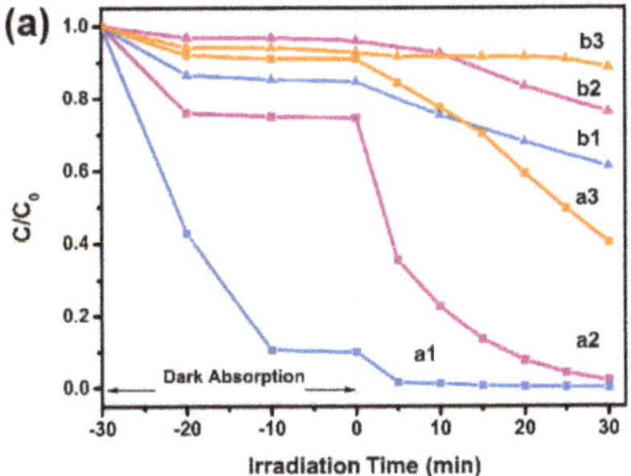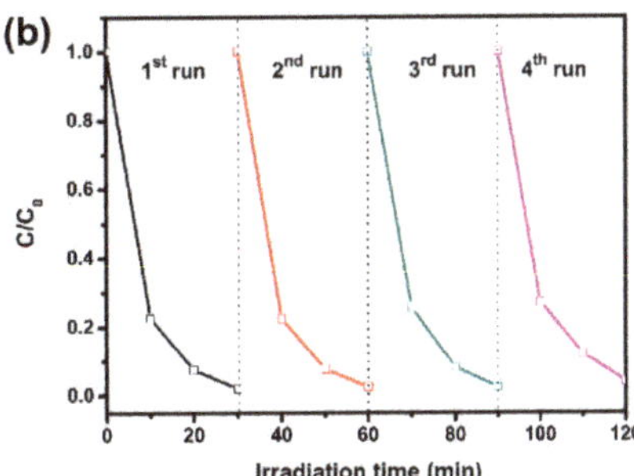

Figure 5. (**a**) Photocatalytic degradation of methylene blue (MB) (**a1**), rhodamine B (RhB) (**a2**), methyl orange (MO) (**a3**) over the dual-phase H-titanate/anatase catalysts and photocatalytic degradation of MB (**b1**), RhB (**b2**), MO (**b3**) over P25 under visible light irradiation, (**b**) cycling experiments of H-titanate/anatase catalysts for RhB degradation under visible light irradiation.

When visible light was turned on, as shown in Figure 5a, RhB was almost decomposed within only 30 min (Figure 5a2 and Figure S4), MO was removed by about 60% (Figure 5a3) within 30 min illumination, and the MB completely decomposed in just 5 min (Figure 5a1) in the test of the dual-phase H-titanate/anatase photocatalyst, whereas P25 showed no significant degradation for the dyes in 30 min. Additionally, without any photocatalyst, RhB, MB and MO were hardly degraded by visible light [28–30], which demonstrated the high reactivity of as-synthesized dual-phase nanotubes catalysts. Compared with many reported TiO_2-based materials, it also shows higher photocatalytic activity for the degradation of organic dyes under visible light. Xiong et al. reported that under the visible light irradiation, MB degraded completely in about 175 min in the presence of the nitrogen-doped titanate-anatase core-shell nanobelts catalyst [14]. S-doped $Na_2Ti_6O_{13}$@TiO_2 core-shell nanorods can completely decompose MB in 100 min under visible light [31]. Li et al. demonstrated that within 60 min of UV light irradiation, RhB was completely decomposed in the presence of the double-shell anatase-rutile TiO_2 spheres [32]. The report of Pan et al. showed that within 40 min visible light irradiation, MO can be degraded by about 50% in the presence of GQD-TiO_2 heterojunctions [33]. Figure S2A shows the photocatalytic activity of all as-synthesized catalysts for MB, RhB and MO under visible light irradiation. Compared with pure H-titanate and anatase catalysts, the dual-phase catalyst exhibited the best photoactivity. Figure S2B displays the degradation rate of the materials (dual phase H-titanate/anatase and P25) for MO, RhB under visible light irradiation. It is noted that the dual-phase catalyst displays better degradation efficiency than P25. H-titanate/anatase photocatalyst shows 15 times higher efficiency for the photodegradation of RhB and 17 times higher efficiency for the photodegradation of MO compared to P25 under visible light radiation, which demonstrated the high photocatalytic activity of the dual-phase H-titanate/anatase. In order to evaluate the stability of the dual-phase photocatalyst, several photodegradation tests of RhB under visible light were carried out (Figure 5b). After four cycles of photocatalytic degradation, within about 30 min, the RhB could be decomposed completely, which indicates that the dual-phase H-titanate/anatase has excellent photocatalytic stability.

Based on above results, the dual-phase (H-titanate/anatase) catalyst reveals a high visible light photo-degradation ability. It can be expounded by the following involved reasons: (1) In-Situ generation can form a strong interfacial coupling between H-titanate and anatase, which is useful for accelerating charge transfer and improving the photocatalytic activity [17]; (2) The dual-phase catalyst retains the nanotubed and layered structures, and possesses a high BET surface area. Large specific surface area and pore structure can provide more active sites to adsorb organic pollutants and promote the diffusion of organic molecules inside the pores. The layered titanate product has been considered as an excellent adsorbent [34,35]. It would be beneficial to improve the photocatalytic activity. In addition, the meso-nanotubes structure is also conducive to the rapid diffusion of quantum, which is formed in the photocatalytic process, further promoting photogenerated charge transport to improve the separation rate [36,37]; (3) The synergetic effect between H-titanate and anatase is one of the major ingredients for its enhanced visible light photocatalytic activity. When two phases combined, a staggered band gap was formed, which lead to the efficient charge separation of the cross phase junction [38]. As shown in Figure 6, under visible light irradiation, anatase in dual-phase can be excited to generate electron-hole pairs. Electrons are excited from the valence band (VB) to the conduction band (CB). According to the energy band data, the CB and the VB energy potentials in anatase are −0.26 and 2.94 eV [39], while those of titanate are −0.50 and 3.03 eV [31]. The photogenerated electrons accumulated on anatase will migrate from the CB of anatase to that of titanate due to the potential difference. In this way, titanate can effectively collect photogenerated electrons and anatase collect holes. These electrons react with the surface adsorbed O_2 to form $O_2^-\cdot$, because the CB edge potential of titanate (−0.50 eV) is more negative than the standard redox potential of $O_2/O_2^-\cdot$ (−0.33 eV) [40]. In the mean time, the VB potential of anatase (2.94 eV) is more positive than the standard redox potential of $\cdot OH/OH^-$ (1.99 eV) [41,42], the holes react with OH^- to generate $\cdot OH$ radicals. Then the organic pollutants could be mineralized by the produced $\cdot OH$ and $O_2^-\cdot$ radicals. Hence, the above efficient separation of photogenerated electron-hole pairs process improves the photodegradation rate of dyes; (4) The new anatase phase with a small average crystal size means a stronger redox ability because of the quantum size effect [43]. Combining all above factors, the dual-phase H-titanate/anatase photocatalysts displayed high visible light activity.

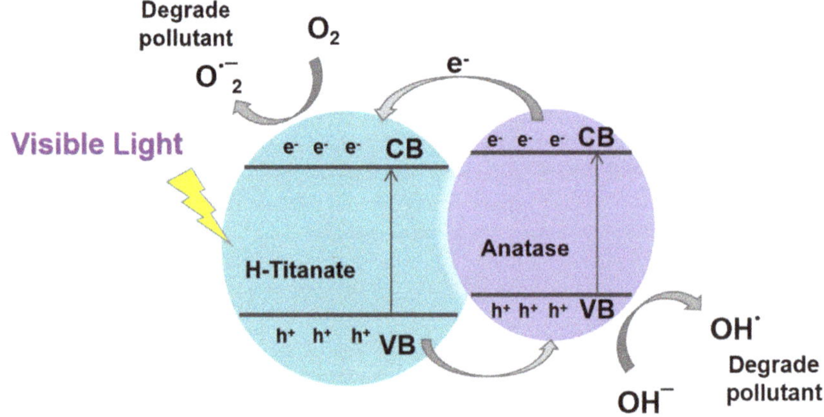

Figure 6. The proposed photocatalytic mechanism of H-titanate/anatase composite.

3. Materials and Methods

3.1. Materials

The titanium powder was purchased from Aladdin, Tianjin, China. The NaOH and the HNO_3 were purchased from Sinopharm Chemical Reagent Co., Ltd., Shanghai, China.

3.2. Catalyst Preparation

The photocatalysts were prepared by the following synthetic route. Firstly, 1.5 g titanium powder was mixed with 35 mL NaOH solution (10 M) and stirred at room temperature for 5 h. The mixture was transferred into a Teflon-lined autoclave and kept at 150 °C for 72 h. The obtained precipitates were washed with deionized water until neutral and dried in an oven (60 °C) overnight. Afterwards, the ion-exchange reaction was followed with 0.5 M HNO_3 solution for 3 h at room temperature. There were three times ion-exchange reactions for the product. The final product was dried at 60 °C for 24 h to produce the hydrogen titanate (H-titanate) tubes and then dried at 100 °C for 24 h to yield hydrogen titanate/anatase nanotubes. The as-prepared H-titanate was calcined in air at 200, 300 and 500 °C for 5 h to get TiO_2 products. Henceforth, these samples are referred to as TiO_2-60, TiO_2-100, TiO_2-200, TiO_2-300 and TiO_2-500, respectively.

3.3. Catalyst Characterization

X-ray powder diffraction (XRD) analysis was carried out using a D/Max-2550 X-ray powder diffractometer (Tokyo, Japan) with Cu Kα radiation. The ultraviolet-visible diffuse reflectance spectra of the samples were measured on a UV-Vis-NIR spectrophotometer (Shimadzu U-4100, Shanghai, China) detecting absorption over the range of 200–800 nm. The morphologies of samples were measured on a Tecnai G2 S-Twin F20 transmission electron microscopy (TEM, FEI, Hillsboro, FL, USA). N_2 adsorption–desorption isotherms were obtained at 77 K on a Micromeritics ASAP 2020 sorptometer (Norcross, GA, USA). Raman spectra were recorded using a Renishaw InVia Raman spectrometer (London, UK) with a wavelength of 532 nm. Room temperature photoluminescence (PL) spectra with an excitation wavelength of 315 nm were measured on a FLUOROMAX-4 (Beijing, China).

3.4. Photocatalytic Activity Test

The photocatalytic activities of the photocatalysts were performed at room temperature in a glass reactor fitted with a Xe lamp (300 W). A 420 nm cut-on filter was used to ensure that only visible light illuminated the photocatalyst. The reaction liquid was prepared by mixing 0.25 g photocatalysts and 100 mL rhodamine B (or 10 mg/L MB; 10 mg/L MO) aqueous solution (10 mg/L). Then, the suspension was stirred in dark for 30 min to reach adsorption-desorption equilibrium before irradiation. Then it was irradiated under visible light. The suspension (8 mL) was withdrawn from the irradiated solution at preset time intervals and centrifuged to separate the photocatalyst particles, and then the supernatants were analyzed by UV-Vis spectrophotometer (UV-2450, Shanghai, China).

4. Conclusions

In conclusion, anatase nanoparticles generated on H-titanate nanotubes surfaces were successfully synthesized by an ingenious method. H-titanate tubes were prepared by a hydrothermal synthesis of Ti powder in concentrated NaOH solution and an ion exchange process with HNO_3 solution. After that, at a relatively low drying temperature, a small amount of anatase nanoparticles were in-situ formed on the surface of the H-titanate tubes by the surface dehydration reaction. It showed higher photocatalytic activity than pure H-titanate nanotube, anatase, and P25 under visible light (100% rhodamine B (RhB) and methylene blue (MB), 60% methyl orange (MO) degraded in 30 min under visible light irradiation) due to the in-situ transformation, the smaller average crystal size, the nanotubed and layered structure, the large BET surface areas and the synergistic effect of the H-titanate/anatase dual phases, which can accelerate the transfer of electron-hole pairs and inhibit their recombination. This work provides an ingenious and simple method to prepare an efficient visible-light-responsive TiO_2-based photocatalyst for solving environment problems.

Supplementary Materials: The following are available online at http://www.mdpi.com/2073-4344/10/6/695/s1, Figure S1: TEM images of (a) TiO_2-60, (b) TiO_2-500, Figure S2: (A) Photocatalytic degradation of RhB over the (a_1) dual-phase H-titanate/anatase, (a_2) H-titanate, (a_3) anatase catalysts; photocatalytic degradation of MB over the (b_1) dual-phase H-titanate/anatase, (b_2) H-titanate, (b_3) anatase catalysts; photocatalytic degradation of MO over the (c_1) dual-phase H-titanate/anatase, (c_2) H-titanate, (c_3) anatase catalysts under visible light irradiation. (B) Photocatalytic kinetic plot of the (a_1) P25, (a_2) dual-phase H-titanate/anatase for degradation of RhB; photocatalytic kinetic plot of the (b_1) P25, (b_2) dual-phase H-titanate/anatase for degradation of MO under visible light irradiation.

Author Contributions: Z.Z. and W.F. conceived and designed the experiments; Funding acquisition, L.Z.; Z.S., H.B., J.D., Z.L., F.L. and D.Z. performed the experiments and analyzed the data; W.F. wrote the paper. All authors have read and agreed to the published version of the manuscript.

Funding: This work was supported by the Science and Technology Development Planning of Jilin Province, China (No. 20170101098JC).

Conflicts of Interest: The authors declare no conflict of interest.

References

1. Cao, S.W.; Low, J.X.; Yu, J.G.; Jaroniec, M. Polymeric Photocatalysts Based on Graphitic Carbon Nitride. *Adv. Mater.* **2015**, *27*, 2150–2176. [CrossRef]

2. Zhang, X.; Chen, Y.; Xiao, Y.; Zhou, W.; Tian, G.; Fu, H. Enhanced charge transfer and separation of hierarchical hydrogenated TiO_2 nanothorns/carbon nanofibers composites decorated by NiS quantum dots for remarkable photocatalytic H_2 production activity. *Nanoscale* **2018**, *10*, 4041–4050. [CrossRef]

3. Dai, G.; Qin, H.; Zhou, H.; Wang, W.; Luo, T. Template-free fabrication of hierarchical macro/mesoporous SnS_2/TiO_2 composite with enhanced photocatalytic degradation of Methyl Orange (MO). *Appl. Surf. Sci.* **2018**, *430*, 488–495. [CrossRef]

4. Lyu, J.; Gao, J.; Zhang, M.; Fu, Q.; Sun, L.; Hu, S.; Zhong, J.; Wang, S.; Li, J. Construction of homojunction-adsorption layer on anatase TiO_2 to improve photocatalytic mineralization of volatile organic compounds. *Appl. Catal. B* **2017**, *202*, 664–670. [CrossRef]

5. Meng, A.; Zhang, L.; Cheng, B.; Yu, J. TiO_2–MnOx–Pt Hybrid Multiheterojunction Film Photocatalyst with Enhanced Photocatalytic CO_2-Reduction Activity. *ACS Appl. Mater. Interfaces* **2018**, *11*, 5581–5589. [CrossRef] [PubMed]

6. Shi, Y.; Li, H.; Wang, L.; Shen, W.; Chen, H. Novel α-Fe2O3/CdS Cornlike Nanorods with Enhanced Photocatalytic Performance. *ACS Appl. Mater. Interfaces* **2012**, *4*, 4800–4806. [CrossRef] [PubMed]

7. Ran, J.; Jaroniec, M.; Qiao, S.Z. Cocatalysts in Semiconductor-based Photocatalytic CO_2 Reduction: Achievements, Challenges, and Opportunities. *Adv. Mater.* **2018**, *30*, 1704649(1)–1704649(31). [CrossRef]

8. Fu, W.W.; Li, G.D.; Wang, Y.; Zeng, S.J.; Yan, Z.J.; Wang, J.W.; Xin, S.G.; Zhang, L.; Wu, S.W.; Zhang, Z.T. Facile formation of mesoporous structured mixed-phase (anatase/rutile) TiO_2 with enhanced visible light photocatalytic activity. *Chem. Commun.* **2018**, *54*, 58–61. [CrossRef]

9. Meng, A.; Zhang, L.; Cheng, B.; Yu, J. Dual Cocatalysts in TiO_2 Photocatalysis. *Adv. Mater.* **2019**, *31*, 1807660(1)–1807660(31).

10. Niu, S.; Zhang, R.; Zhang, Z.; Zheng, J.; Jiao, Y.; Guo, C. In situ construction of the $BiOCl/Bi_2Ti_2O_7$ heterojunction with enhanced visible-light photocatalytic activity. *Inorg. Chem. Front.* **2019**, *6*, 791–798. [CrossRef]

11. Bavykin, D.V.; Friedrich, J.M.; Walsh, F.C. Protonated Titanates and TiO_2 Nanostructured Materials: Synthesis, Properties, and Applications. *Adv. Mater.* **2006**, *18*, 2807–2824. [CrossRef]

12. Mao, Y.; Wong, S.S. Size- and Shape-Dependent Transformation of Nanosized Titanate into Analogous Anatase Titania Nanostructures. *J. Am. Chem. Soc.* **2006**, *128*, 8217–8226. [CrossRef] [PubMed]

13. Buchholcz, B.; Haspel, H.; Kukovecz, Á.; Kónya, Z. Low-temperature conversion of titanate nanotubes into nitrogen-doped TiO_2 nanoparticles. *CrystEngComm* **2014**, *16*, 7486–7492. [CrossRef]

14. Xiong, Z.; Zhao, X. Nitrogen-doped titanate-anatase core-shell nanobelts with exposed {101} anatase facets and enhanced visible light photocatalytic activity. *J. Am. Chem. Soc.* **2012**, *134*, 5754–5757. [CrossRef]

15. Yan, Y.; Qiu, X.; Wang, H.; Li, L.; Fu, X.; Wu, L.; Li, G. Synthesis of titanate/anatase composites with highly photocatalytic decolorization of dye under visible light irradiation. *J. Alloys Compd.* **2008**, *460*, 491–495. [CrossRef]

16. Harsha, N.; Ranya, K.R.; Babitha, K.B.; Shukla, S.; Biju, S.; Reddy, M.L.P.; Warrier, K.G.K. Hydrothermal Processing of Hydrogen Titanate/Anatase-Titania Nanotubes and Their Application as Strong Dye-Adsorbents. *J. Nanosci. Nanotechnol.* **2010**, *10*, 1–13. [CrossRef]

17. Wang, P.; Yi, X.; Lu, Y.; Yu, H.; Yu, J. In-situ synthesis of amorphous H_2TiO_3-modified TiO_2 and its improved photocatalytic H_2-evolution performance. *J. Colloid Interface Sci.* **2018**, *532*, 272–279. [CrossRef]

18. Cheng, Y.H.; Huang, Y.Z.; Kanhere, P.D.; Subramaniam, V.P.; Gong, D.G.; Zhang, S.; Highfield, J.; Schreyer, M.K.; Chen, Z. Dual-Phase Titanate/Anatase with Nitrogen Doping for Enhanced Degradation of Organic Dye under Visible Light. *Chem. Eur. J.* **2011**, *17*, 2575–2578. [CrossRef]

19. Cai, J.; Zhu, Y.; Liu, D.; Meng, M.; Hu, Z.; Jiang, Z. Synergistic Effect of Titanate-Anatase Heterostructure and Hydrogenation-Induced Surface Disorder on Photocatalytic Water Splitting. *ACS Catal.* **2015**, *5*, 1708–1716. [CrossRef]

20. Zhang, L.; Lin, H.; Wang, N.; Lin, C.; Li, J. The evolution of morphology and crystal form of titanate nanotubes under calcination and its mechanism. *J. Alloys Compd.* **2007**, *431*, 230–235. [CrossRef]

21. Xiong, Z.; Dou, H.; Pan, J.; Ma, J.; Xu, C.; Zhao, X.S. Synthesis of mesoporous anatase TiO_2 with a combined template method and photocatalysis. *CrystEngComm* **2010**, *12*, 3455–3457. [CrossRef]

22. Pant, B.; Park, M.; Park, S.-J. TiO2 NPs Assembled into a Carbon Nanofififiber Composite Electrode by a One-Step Electrospinning Process for Supercapacitor Applications. *Polymers* **2019**, *11*, 899. [CrossRef] [PubMed]

23. Nowak, M.; Kauch, B.; Szperlich, P. Determination of energy band gap of nanocrystalline SbSI using diffuse reflectance spectroscopy. *Rev. Sci. Instrum.* **2009**, *80*, 046107(1)–046107(3). [CrossRef] [PubMed]

24. Zhang, W.F.; Zhang, M.S.; Yin, Z.; Chen, Q. Photoluminescence in anatase titanium dioxide nanocrystals. *Appl. Phys. B Laser Opt.* **2000**, *70*, 261–265. [CrossRef]

25. Ng, J.; Xu, S.; Zhang, X.; Yang, H.Y.; Sun, D.D. Hybridized Nanowires and Cubes: A Novel Architecture of a Heterojunctioned $TiO_2/SrTiO_3$ Thin Film for Efficient Water Splitting. *Adv. Funct. Mater.* **2010**, *20*, 4287–4294. [CrossRef]

26. Wang, Q.; Zhang, B.; Lu, X.; Zhang, X.; Zhu, H.; Li, B. Multifunctional 3D $K_2Ti_6O_{13}$ nanobelt-built architectures towards wastewater remediation: Selective adsorption, photodegradation, mechanism insight and photoelectrochemical investigation. *Catal. Sci. Technol.* **2018**, *8*, 6180–6195. [CrossRef]

27. Pant, B.; Ojha, G.P.; Kim, H.-Y.; Park, M.; Park, S.-J. Fly-ash-incorporated electrospun zinc oxide nanofififibers: Potential material for environmental remediation. *Environ. Pollut.* **2019**, *245*, 163–172. [CrossRef]

28. Fu, W.; Ding, S.; Wang, Y.; Wu, L.; Zhang, D.; Pan, Z.; Wang, R.; Zhang, Z.; Qiu, S. F, Ca co-doped TiO_2 nanocrystals with enhanced photocatalytic activity. *Dalton Trans.* **2014**, *43*, 16160–16163. [CrossRef]

29. Nguyen-Le, M.-T.; Lee, B.-K. Novel fabrication of a nitrogen-doped mesoporous TiO_2-nanorod titanate heterojunction to enhance the photocatalytic degradation of dyes under visible light. *RSC Adv.* **2016**, *6*, 31347–31350.

30. Niu, F.; Chen, D.; Qin, L.; Zhang, N.; Wang, J.; Chen, Z.; Huang, Y. Facile Synthesis of Highly Efficient p–n Heterojunction CuO/BiFeO3 Composite Photocatalysts with Enhanced Visible-Light Photocatalytic Activity. *ChemCatChem* **2015**, *7*, 3279–3289. [CrossRef]

31. Liu, C.; Liang, J.-Y.; Han, R.-R.; Wang, Y.-Z.; Zhao, J.; Huang, Q.-J.; Chen, J.; Hou, W.-H. S-doped $Na_2Ti_6O_{13}$@TiO_2 core-shell nanorods with enhanced visible light photocatalytic performance. *Phys. Chem. Chem. Phys.* **2015**, *17*, 15165–15172. [CrossRef] [PubMed]

32. Li, S.; Chen, J.; Zheng, F.; Li, Y.; Huang, F. Synthesis of the double-shell anatase-rutile TiO_2 hollow spheres with enhanced photocatalytic activity. *Nanoscale* **2013**, *5*, 12150–12155. [CrossRef] [PubMed]

33. Pan, D.; Jiao, J.; Li, Z.; Guo, Y.; Feng, C.; Liu, Y.; Wang, L.; Wu, M. Efficient Separation of Electron–Hole Pairs in Graphene Quantum Dots by TiO_2 Heterojunctions for Dye Degradation. *ACS Sustain. Chem. Eng.* **2015**, *3*, 2405–2413. [CrossRef]

34. Tang, Y.; Lai, Y.; Gong, D.; Goh, K.-H.; Lim, T.-T.; Dong, Z.; Chen, Z. Ultrafast Synthesis of Layered Titanate Microspherulite Particles by Electrochemical Spark Discharge Spallation. *Chem. Eur. J.* **2010**, *16*, 7704–7708. [CrossRef]

35. Lim, Y.W.L.; Tang, Y.X.; Cheng, Y.H.; Chen, Z. Morphology, crystal structure and adsorption performance of hydrothermally synthesized titania and titanate nanostructures. *Nanoscale* **2010**, *2*, 2751–2757. [CrossRef] [PubMed]

36. Luo, Z.; Poyraz, A.S.; Kuo, C.-H.; Miao, R.; Meng, Y.; Chen, S.-Y.; Jiang, T.; Wenos, C.; Suib, S.L. Crystalline Mixed Phase (Anatase/Rutile) Mesoporous Titanium Dioxides for Visible Light Photocatalytic Activity. *Chem. Mater.* **2015**, *27*, 6–17. [CrossRef]

37. Gurulakshmi, M.; Selvaraj, M.; Selvamani, A.; Vijayan, P.; Sasi Rekha, N.R.; Shanthi, K. Enhanced visible-light photocatalytic activity of V_2O_5/S-TiO_2 nanocomposites. *Appl. Catal. A* **2012**, *449*, 31–46. [CrossRef]

38. Kho, Y.K.; Iwase, A.; Teoh, W.Y.; Mädler, L.; Kudo, A.; Amal, R. Photocatalytic H_2 Evolution over TiO_2 Nanoparticles. The Synergistic Effect of Anatase and Rutile. *J. Phys. Chem. C* **2010**, *114*, 2821–2829. [CrossRef]

39. Meng, A.; Zhang, J.; Xu, D.; Cheng, B.; Yu, J. Enhanced photocatalytic H_2-production activity of anatase TiO_2 nanosheet by selectively depositing dual cocatalysts on {101} and {001} facets. *Appl. Catal. B* **2016**, *198*, 286–294. [CrossRef]

40. Zhang, D.; Wang, Q.; Wang, L.; Zhang, L. Magnetically separable $CdFe_2O_4$/graphene catalyst and its enhanced photocatalytic properties. *J. Mater. Chem. A* **2015**, *3*, 3576–3585. [CrossRef]

41. Li, K.; Gao, S.; Wang, Q.; Xu, H.; Wang, Z.; Huang, B.; Dai, Y.; Lu, J. In-Situ-Reduced Synthesis of Ti^{3+} Self-Doped TiO_2/g-C_3N_4 Heterojunctions with High Photocatalytic Performance under LED Light Irradiation. *ACS Appl. Mater. Interfaces* **2015**, *7*, 9023–9030. [CrossRef] [PubMed]

42. Yu, J.; Wang, S.; Low, J.; Xiao, W. Enhanced photocatalytic performance of direct Z-scheme g-C_3N_4-TiO_2 photocatalysts for the decomposition of formaldehyde in air. *Phys. Chem. Chem. Phys.* **2013**, *15*, 16883–16890. [CrossRef] [PubMed]

43. Liu, Z.Y.; Sun, D.D.; Guo, P.; Leckie, J.O. One-Step Fabrication and High Photocatalytic Activity of Porous TiO_2 Hollow Aggregates by Using a Low-Temperature Hydrothermal Method Without Templates. *Chem. Eur. J.* **2007**, *13*, 1851–1855. [CrossRef] [PubMed]

 *catalysts*

Review

Insights into the TiO$_2$-Based Photocatalytic Systems and Their Mechanisms

Mohan Sakar [1,2], **Ravikumar Mithun Prakash** [2] **and Trong-On Do** [1,*]

[1]   Department of Chemical Engineering, Laval University, Quebec, QC G1V 0A8, Canada
[2]   Centre for Nano and Material Sciences, Jain University, Bangalore 562 112, India
*   Correspondence: Trong-On.Do@gch.ulaval.ca

Received: 22 July 2019; Accepted: 7 August 2019; Published: 9 August 2019

Abstract: Photocatalysis is a multifunctional phenomenon that can be employed for energy applications such as H$_2$ production, CO$_2$ reduction into fuels, and environmental applications such as pollutant degradations, antibacterial disinfection, etc. In this direction, it is not an exaggerated fact that TiO$_2$ is blooming in the field of photocatalysis, which is largely explored for various photocatalytic applications. The deeper understanding of TiO$_2$ photocatalysis has led to the design of new photocatalytic materials with multiple functionalities. Accordingly, this paper exclusively reviews the recent developments in the modification of TiO$_2$ photocatalyst towards the understanding of its photocatalytic mechanisms. These modifications generally involve the physical and chemical changes in TiO$_2$ such as anisotropic structuring and integration with other metal oxides, plasmonic materials, carbon-based materials, etc. Such modifications essentially lead to the changes in the energy structure of TiO$_2$ that largely boosts up the photocatalytic process via enhancing the band structure alignments, visible light absorption, carrier separation, and transportation in the system. For instance, the ability to align the band structure in TiO$_2$ makes it suitable for multiple photocatalytic processes such as degradation of various pollutants, H$_2$ production, CO$_2$ conversion, etc. For these reasons, TiO$_2$ can be realized as a prototypical photocatalyst, which paves ways to develop new photocatalytic materials in the field. In this context, this review paper sheds light into the emerging trends in TiO$_2$ in terms of its modifications towards multifunctional photocatalytic applications.

Keywords: TiO$_2$; semiconductors; photocatalysis; redox reactions; band gap engineering; nanostructures

---

1. Introduction

Since the observation of an enhanced electrolysis of water (H$_2$O) molecules into H$_2$ and O$_2$ using TiO$_2$ as photo-anode and Pt as cathode under UV light irradiation, [1] the research on TiO$_2$ is gaining significant momentum towards its 'photocatalytic' process, which is coined later on. In 1977, Schrauzer and Guth reported the Pt/Rh metal modified-TiO$_2$ powders for the photocatalytic splitting of water molecules [2]. Followed by such pioneering work in the field, a range of semiconducting materials have been explored for the photocatalytic properties towards various photocatalytic applications [3–12]. Accordingly, there has been prompt progress in developing various photocatalytic systems to convert the chemical energy through water splitting [13–16] into H$_2$ and O$_2$ and other associated reactions [17,18]. Specifically, diverse binary oxide-based photocatalysts have been developed and demonstrated as reliable photocatalysts [19–21].

Despite the emergence of various binary oxide photocatalytic systems, TiO$_2$ is considered as the most promising material due to its unprecedented stability, excellent physiochemical properties with ease of synthesis, availability, and relatively lower cost [22–24]. In addition to this, TiO$_2$ exhibits three polymorphs, namely anatase, rutile, and brookite [25], in which the anatase phase is widely used because of its photocatalytic efficiency as its conduction band has been positioned in the appropriate negative potential, which is the favorable band edge position for redox reactions [26]. Despite such merits and reliable properties, TiO_2 lacks in some of the other specific crucial properties for photocatalysis, such as wide bang gap energy, rapid charge recombination, insufficient transportation, etc. [27]. To surpass such limitations, TiO_2 has been modified in many different ways through chemical and physical modifications, where the former involves doping, composite formation, defects creation, functionalization, plasmonic sensitization, co-catalyst loading, etc., and the other involves size, morphology, and shape modifications, etc. [28].

In this review, we have essentially focused on the versatile modifications of TiO_2 such as morphology modifications, doped TiO_2, hetero-junctions, Z-scheme, plasmonic, ferroelectric/perovskite, chalcogenides, metal–organic frameworks, carbon-based TiO_2, defective TiO_2, etc. TiO_2 may be the only material that has been used to construct the any given aforementioned photocatalytic systems and investigated for almost all the photocatalytic applications such as dye degradations, pharmaceutical degradations, H_2 evolution, O_2 evolution, CO_2 reduction, heavy metal reduction, N_2 fixation, organic synthesis, antimicrobial disinfection, etc. Unlike other existing reviews, which merely provides TiO_2 modifications such as doping, etc., this review paper gives insights into the modifications of TiO_2 towards developing various photocatalytic systems as a whole, which can be prototyped using other materials.

2. Mechanics of TiO_2 Photocatalysis

Photocatalysis (PC) is the process of performing a chemical reaction in the presence of light and a photoactive catalyst, where the charge carriers (electron hole) get separated by the incident photons with sufficient energy and transferred to the respective bands and involved in the redox reactions. The following equations show the reaction mechanism of the photocatalytic process [29,30].

$$\text{Incident photon: Photocatalyst} + h\nu \rightarrow e^- + h^+ \tag{1}$$

$$\text{Reduction: } 2H^+ + 2e^- \rightarrow H_2 \; \Delta E = 0 \text{ V} \tag{2}$$

$$\text{Oxidation: } 2H_2O + 4h^+ \rightarrow O_2 + 4H^+ \; \Delta E = 1.23 \text{ V} \tag{3}$$

$$\text{Overall: } 2H_2O \rightarrow 2H_2 + O_2 \; (\Delta G = +237.2 \text{ kJmol}^{-1}) \tag{4}$$

As mentioned in the reaction equations, the incident photons generate the photo-induced electron hole (e^-/h^+) pairs in the semiconductor and the electron involved in the reduction reactions, while the holes are involved in the oxidation reactions. The first and foremost prerequisite for a photocatalyst is to have an appropriate band edge potential (valence band/VB, conduction band/CB) to induce the required redox species. Considering the PC process in TiO_2, the VB and CB level of TiO_2 lies at +2.9 and −0.3 eV, respectively, which leads to the band gap energy of 3.2 eV. It should be noted that the VB and CB level of TiO_2 lies at more positive and more negative values in comparison with the standard redox potential of O_2/H_2O (1.23 eV) and H+/H_2 (0 eV) vs. normal hydrogen electrode (NHE), which is one of the more favorable conditions for the photocatalytic redox reactions [31,32].

Apart from the band edge positions, the photocatalytic process also requires enhanced surface reactivity, charge separation, and transportations mechanisms [33]. Upon excitation, the photocatalyst should facilitate the transportation of electrons to the surface, which essentially determines the surface chemistry and reactivity of the photocatalyst. The surface of TiO_2 typically contains more defects, which are often found to be oxygen vacancies; the unpaired electrons in such defects are transferred to the conduction band of TiO_2 and facilitate the catalytic reactions in the system [34]. Interestingly, the accumulation of electrons leads to the band bending phenomenon in TiO_2 that considerably redesigns the transportation of charges or energy to the surrounding molecules [35]. Charge recombination dynamics is one of the serious issues in a photocatalyst. Regarding TiO_2, with its indirect band gap, it is proposed that the recombination process occurs via non-radiative pathways and, thus, the lifetime of charge carriers in TiO_2 varies from picoseconds to milliseconds [36,37]. In addition, the observed relatively enhanced PC efficiency of TiO_2 can also be ascribed to its electron and hole trapping [38]. Generally, the photo-induced charge carriers do not tend to recombine directly due to the factors such as carrier trapping, band bending, etc. Accordingly, it is predicted that the holes in TiO_2 can be trapped either at the "bridging" O^{2-} or "surface bound" OH^- anions, which results in the generation of $O^{-\bullet}$ and/or $OH^\bullet$ centers, respectively. Similarly, the photo-induced electrons can be forced to migrate into the bulk from surface, where they can be delocalized in possible Ti sites. Furthermore, it is also predicted that in TiO_2 it is of more possible for bulk trapping rather than surface trapping and thereby TiO_2 shows relatively enhanced photocatalytic activities as compared to the other semiconducting oxide-based photocatalysts [38–40].

3. Versatile Modifications of TiO_2 and Their Photocatalytic Mechanisms

TiO_2 as a photocatalyst has been modified in a variety of ways that generally includes (i) morphological, (ii) defective, (iii) elemental doping (cationic/anionic), (iv) plasmonic metal-loading, composites with (v) binary oxides, (vi) perovskite systems, (vii) metal–organic frameworks, (viii) carbon materials, (ix) chalcogenides, etc. These modifications essentially lead to development of new photocatalytic systems, enhancing (i) the overall visible light/full-sunlight absorption, (ii) charge separation, (iii) recombination resistance, (iv) charge transportations, and (v) tuning of the band edge potential of the system. Accordingly, the following section presents some of the recent studies that mainly highlight the photocatalytic mechanism/functions in such chemically and physically modified TiO_2.

3.1. Morphology-Dependent Photocatalytic Properties of TiO_2

Photocatalysis can be influenced by the size, shape, and morphology of the photocatalyst due to the spatial confinements of electrons in the system [41,42]. For instance, compared to bulk, the surface reactivity is higher for the nanoparticles, where their high surface area/energy facilitates the enhanced (i) catalytic activity on the surface, (ii) surface adsorption of the molecules, and (iii) promotion of charge carriers to surface. The size parameter also considerably influences the band-gap energy as well as band-edge position in a photocatalyst. Similarly, the geometrics of photocatalyst also influences the PC process. For instance, compared to the particles, the one-dimensional nanostructures show improved activity due to the enhanced "delocalization of electrons" in the conduction band of the photocatalyst [43,44]. Further, photocatalysts also demonstrate the crystal-facet-dependent efficiencies towards various photocatalytic applications. TiO_2 nanocrystals with different shapes, as shown in Figure 1a–f, have been synthesized and demonstrated for photo-reforming of methanol into hydrogen under UV light [45].

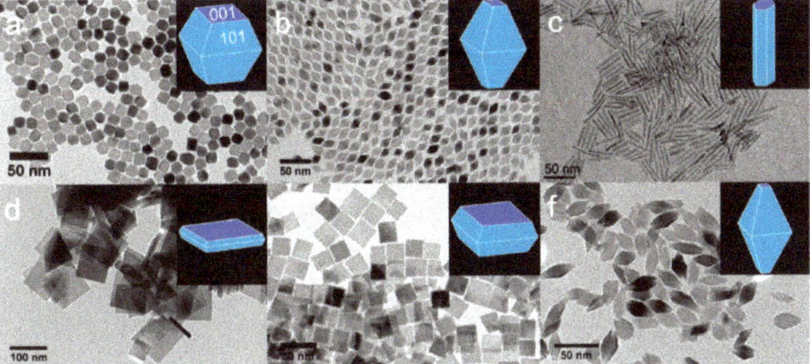

Figure 1. TEM images of TiO_2 NCs synthesized using the precursor TiF_4 (**a,d**), a mixed precursor of TiF_4 and $TiCl_4$ (**b,e**), and $TiCl_4$ (**c,f**). Those depicted in a–c and d–f are synthesized in the presence of OLAM and 1-ODOL, respectively. (reproduced with permission from ref. [45]).

In another study, the synthesis of TiO_2 solid and hollow nanocubes have been demonstrated, as shown in Figure 2, and applied for the photocatalytic-mediated synthesis of benzimidazole under UV and visible conditions [46]. Similarly, TiO_2 with different morphologies such as nanospheres, nanocubes, nanotubes, nanorods, nanoflowers, nanosheets, and nanofibers have been synthesized and studied for their photocatalytic applications [47–53]. The size and morphology control over TiO_2 photocatalyst exhibit significant influences over their (i) optical properties such as tunable band-gap energy, repositioning of band edge positions, visible light absorption, etc., (ii) electronic properties such as increased carrier lifetime, enhanced photocurrent conduction, reduced recombination, and (iii) surface properties such as enhanced surface energy, porous structures, enhanced surface adsorption, etc. Realizing the photocatalytic phenomenon, these properties are very much important to achieve the enhanced efficiencies in the photocatalytic materials.

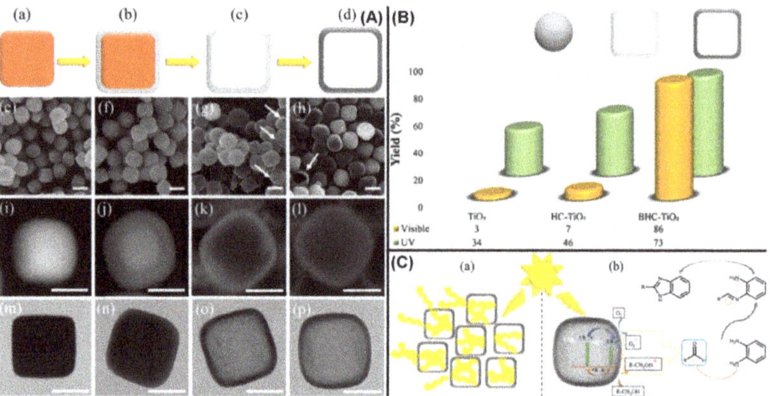

Figure 2. (**A**) Overall flowchart for fabrication of black hollow nanocubic (BHC)-TiO_2 (**a–p**), (**B**) Comparison photocatalytic activity of different TiO_2 nanostructures in the synthesis of benzimidazole under UV and visible conditions; (**C**) Schematic diagram of the light scattering effect caused by BHC-TiO_2 nanocubes (**a**) and schematic of the proposed mechanism for benzimidazole preparation by BHC-TiO_2 architecture (**b**) (reproduced with permission from ref. [46]).

3.2. Doped TiO_2

Doping can be essentially classified into two categories, (i) cationic and (ii) anionic doping. Accordingly, TiO_2 has been widely modified through doping under both categories. The cationic and anionic doping in TiO_2 leads to the formation of new energy levels underneath the conduction band and above the valence band [54]. The former doping has often been found to reduce the band gap energy and facilitates the visible light absorption and charge separation in TiO_2, whereas the latter often helps in shifting of the VB position, mitigates the defects, and enhances the chemical stability of TiO_2 [55]. The anionic dopants such as N, C, S, and P have been largely doped in TiO_2. Among them, the N doping showed relatively enhanced photocatalytic activity due to the increased stability in the system. Similarly, there are variety of elements doped at the cationic site of TiO_2 and explored for their photocatalytic activities under UV-visible light.

3.2.1. Anionic Doping inTiO_2

Chen et al. reported the origin of visible-light absorption characteristics of C-, N-, and S-doped TiO_2 nanomaterials [56]. In their studies, the TiO_2-P25 showed the typical band-edge absorption around 390 nm with band gap energy of 3.2 eV, while the C and S doping also showed the same values, however the N-doping showed an absorption around 415 nm with band gap energy of 3.0 eV. Further, their valence band-X ray photoelectron spectra revealed an interesting feature that the doping of C, S, and N created additional states in the TiO_2 system, as shown in Figure 3A [56]. These additional states were attributed to the C 2p, S 3p, and N 2p orbitals and they were found to add deeper states into the band gap of TiO_2 in the order of C > N > S. Emy et al. reported the band gap engineering in the anionic co-doped TiO_2 [57]. According to their investigations, they have explained that in F-doped TiO_2, the band gap reduction is mediated by the presence of surface Ti^{3+} defects underneath the CB, while in N-doped TiO_2, the mid-band states have been formed as the N species fill voids as impurities above the VB. On the other hand, the co-doping of N and F into TiO_2 leads to the biggest band gap reduction to 2.24 eV from 3.19 eV, where it is attributed to the doping induced creation of defects and shifting of the VB tail towards Fermi level as shown in Figure 3B [57].

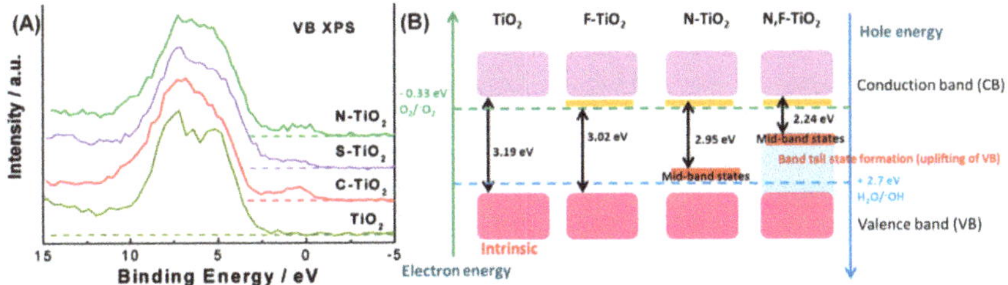

Figure 3. (**A**) Valence band (VB) XPS spectra of pure and (C, S, N)-doped TiO_2; (**B**) proposed band gap engineering structure for all (F, N) doped TiO_2 (reproduced with permission from refs. [56,57], respectively).

Based on the available experimental evidences and theoretical results obtained by Wang et al. [58], we have concluded that both the bang gap narrowing and the overlapping of O 2p state with the dopant-induced states strongly affect the photocatalytic activities of anion-doped TiO_2. However, Kuznetsov et al. [59] have reported that the visible light absorption happening in these doped-TiO_2 may be due to the formation of color centers and may not be due to the band gap narrowing. Further, they have also argued that the red shift in the absorption edge could be due to the emergence of color centers and the doping (heavily) may completely lead to the formation of material with completely different chemical composition from TiO_2 with different electronic band structures. However, it should be noted that the anion-doped TiO_2 is considered as the second-generation photocatalysts [60].

3.2.2. Cationic Doping in TiO_2

As described, the cationic doping essentially introduces the intra-band energy levels close to the CB of TiO_2, which leads to the red shift in the optical property of the system and it is also observed in various cations such as transition metal, [61–63], rare-earth [64–66], and other metals [67–69] doped TiO_2. However, the main drawback of the cation doping is the creation of more trapping sites for charge carriers (both electrons and holes) that considerably reduces the efficiency of the photocatalyst. This is because the trapped carriers tend to recombine with the respective mobile carriers in the system. The mechanism of cation doping is essentially to tune the Fermi level and electronic structure of d-electron configuration in TiO_2, thereby to tune the energy levels to absorb the visible light energy and to enhance the overall photocatalytic efficiency of the system as shown in Figure 4a–c [70–72].

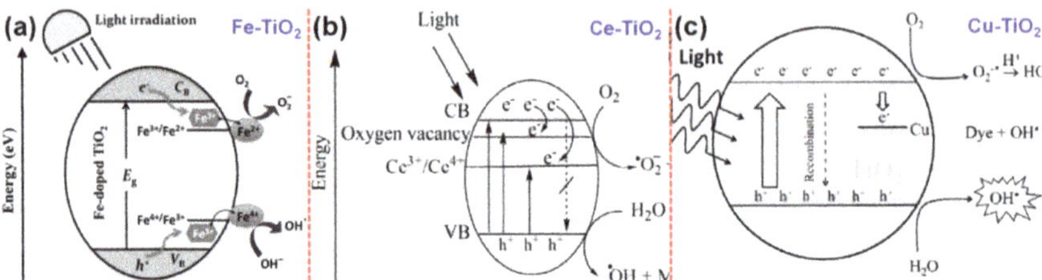

Figure 4. Band gap engineering in TiO$_2$ via (**a**) Fe, (**b**) Ce, (**c**) Cu doping, showing the formation of dopant energy states underneath the conduction band of TiO$_2$ and associated carrier dynamics (reproduced with permission from refs. [70–72], respectively).

Consequently, there have been many cations doped in TiO$_2$ towards enhancing its PC activities. In such cation doping, TiO$_2$ has been doped with the (i) transition metals such as Sc, V, Cr, Mn, Fe, Co, Ni, Cu, Zn, Y, Zr, Nb, Mo, Cd, and W [73–84]; (ii) rare-earth metals such as Ce, Pr, Nd, Sm, Eu, Gd, Tb, Dy, Er, Yb, and La [85–89]; and (iii) other metals such as Li, Mg, Ca, Se, Sr, Al, Sn, and Bi [90–97]. In the case of rare earth elements doping, the electronic configurations such as 4f, 5d, and 6s are found to be favorable to tune the band edge positions, density of states, and width of VB and CB via altering the crystal, electronic, and optical structures in TiO$_2$ [98–100]. In addition, the rare earth elements tend to form complexes through their f-orbital and form various Lewis-based organic compounds, thereby improving the photocatalytic activities of TiO$_2$ [101,102]. For instance, lanthanum (La) leads to the NIR absorption in TiO$_2$ [103], cerium (Ce) owing to its tunable electronic configuration of 4f states, such as $4f^05d^0$ (Ce^{4+}) and $4f^5d^0$ (Ce^{3+}), where it leads to the formation of mid-band gap in TiO$_2$ that facilitates the absorption of in the visible region 400–500 nm [104,105].

3.3. Hetero-Junction TiO$_2$

Coupling of TiO$_2$ with other semiconductors, especially narrow band gap semiconductors to form a heterojunction, is considered to be one of the promising strategies to improve the photocatalytic efficiencies of the system [106,107]. The selection of semiconductors towards forming the heterojunction should be made in such a way that they have different band edge potential and conducting types. For instance, Figure 5a,b depicts the charge transfer mechanisms in the p-n and non p-n junctions between the semiconductors [107]. Such configuration provides several features to the system, such as it helps improve the (i) charge separation, (ii) life time of the charge carriers, (iii) recombination resistance, and (iv) interfacial charge transportations towards the adsorbed molecules [106,107]. The semiconductor that coupled with the host-semiconductor would typically act as a sensitizer. In such cases, it is the sensitizers that get excited and transfer/inject the carriers into the host-semiconductor and, therefore, the VB of the sensitizer should be more cathodic than the VB of TiO$_2$, so that the holes cannot migrate to the TiO$_2$; thereby, the charge separation remains in the system [108]. These kinetics facilitate the phenomenon of electron injections into TiO$_2$ as demonstrated in Figure 5c,d [109]. Based on such thermodynamics of heterojunction formations, Bessekhouad et al. developed Cu$_2$O/TiO$_2$, Bi$_2$O$_3$/TiO$_2$, and ZnMn$_2$O$_4$/TiO$_2$ heterojunctions towards the photocatalytic degradation of multiple organic pollutants Orange II, benzamide, and 4-hydroxybenzoic under UV-visible light [109]. In this study, they have discussed that the CB of Cu$_2$O is positioned at -1.54 eV, which is more negative than the CB of TiO$_2$ (-0.41 eV) that favored the transfer of electrons to TiO$_2$ from Cu$_2$O. Importantly, such electrons-transfer kinetics led to the faster degradation of Orange II molecules as compared to benzamide and 4-hydroxybenzoic molecules as they require more holes oxidation. The same results were also observed in the case of Bi$_2$O$_3$/TiO$_2$ heterojunction. In the case of ZnMn$_2$O$_4$/TiO$_2$ heterojunction, the CB position of ZnMn$_2$O$_4$ is estimated to be $+0.062$ eV, which is greater than the CB of TiO$_2$. Under such circumstances, the electrons excited to the CB of ZnMn$_2$O$_4$ could not be transferred to TiO$_2$, but the opposite would happen when the TiO$_2$ is excited. However, ZnMn$_2$O$_4$/TiO$_2$

heterojunction was not found to be effective and, in fact, it had a tendency to decrease the efficiency of TiO$_2$. From their results, they finally concluded that the band edge positions of the semiconductors involved should be compatible for an effective inter-particle electron injection to happen in the system and, more importantly, the generated holes must be promoted and react highly at the surface to have an improved carrier separation process.

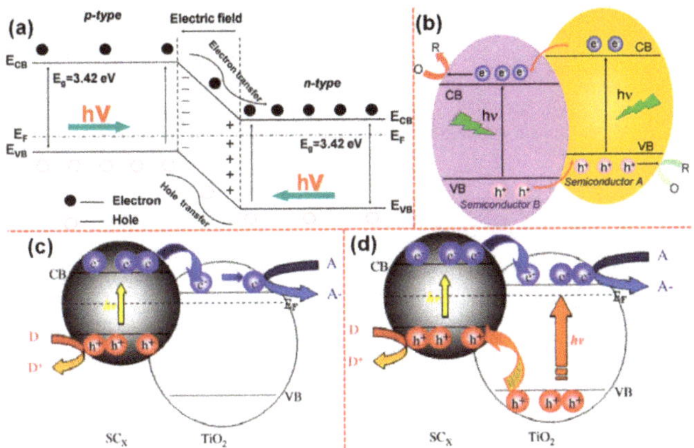

Figure 5. Schematic diagram showing the energy band structure and electron-hole pair separation in the (**a**) p-n heterojunction; (**b**) non p-n heterojunction; (**c**) energy diagram illustrating the coupling of two SC in which vectoral electron transfer occurs from the light-activated SC to the non-activated TiO$_2$; (**d**) diagram depicting the coupling of SC in which vectoral movement of electrons and holes is possible (reproduced with permission from refs. [107,109]).

As aforementioned, the charge transportation mechanism in heterojunction structure is dependent upon the band-edge levels of the semiconductors forming the heterojunction. For instance, the Fe$_3$O$_4$/TiO$_2$ has been widely studied in this direction. Liu et al. [110] reported the 3D flower-like α-Fe$_2$O$_3$@TiO$_2$ core-shell nanostructures, in which the observed photocatalytic efficiency was attributed to the interfacial charge transportation.

As shown in Figure 6a, where they have irradiated the photocatalyst under UV-visible light, it will excite both the semiconductors. Upon the contact of α-Fe$_2$O$_3$ with TiO$_2$ system, the excited electrons in α-Fe$_2$O$_3$ get injected into the CB of TiO$_2$ due to the relative work function of α-Fe$_2$O$_3$ (5.88 eV) and TiO$_2$ (4.308 eV) system as it leads to the positioning of CB of TiO$_2$ to be positioned below the CB α-Fe$_2$O$_3$. The study by Xia et al. [111] proposed the charge transfer kinetics in α-Fe$_2$O$_3$@TiO$_2$ system under UV and visible light irradiation separately, as shown in Figure 6b. They explained that under visible light irradiation, the carriers get excited in α-Fe$_2$O$_3$ and transferred to TiO$_2$, whereas no excitation would happen in TiO$_2$ as the system is irradiated by visible light and, subsequently, the charges carrier would be promoted to the surface and perform the photocatalytic redox reaction. On the other hand, it was observed that the system was irradiated under UV light, carriers in TiO$_2$ get excited, and the α-Fe$_2$O$_3$ becomes recombination center of the photo-induced carriers; as a result, α-Fe$_2$O$_3$@TiO$_2$ exhibits relatively poor photocatalytic activity. To address such issues and towards making the α-Fe$_2$O$_3$@TiO$_2$ to work efficiently, Lin et al. [112] developed TiO$_2$ with abundant oxygen vacancies via self-doping, which greatly shifted the VB edge position to 2.50 eV (vs. NHE), which is very close to that of α-Fe$_2$O$_3$ (2.48 eV) and unaltered CB position with respect to the CB position of α-Fe$_2$O$_3$, as shown in Figure 6c. However, despite the considerable amount of research that has been done on TiO$_2$-based heterojunction photocatalyst, the carrier dynamics and their transportation, and thereby the photocatalytic process, should be studied in detail [113,114].

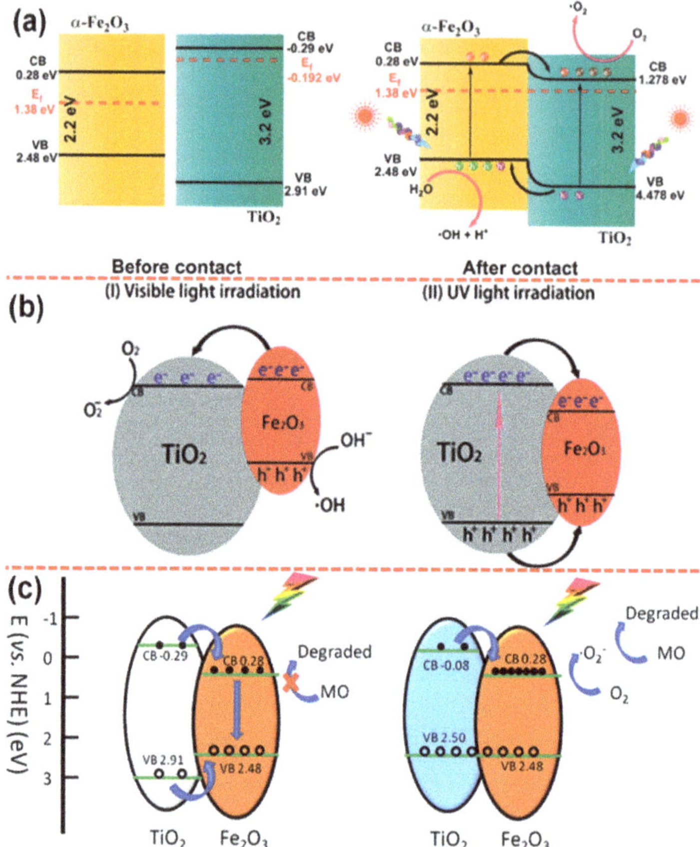

Figure 6. (**a**,**b**) Schematic diagram of the band edge positions and charge transfer mechanism in various α-Fe₂O₃@TiO₂ photocatalytic systems under UV and visible light irradiation. (**c**) The presence of abundant oxygen vacancies in TiO₂ shifts its VB edge position and aligns it to the VB of Fe₂O₃ (reproduced with permission from refs. [110–112], respectively).

3.4. Z-Scheme-Based TiO₂

The concept of Z-scheme photocatalytic process is essentially derived from the natural photosynthesis process, which demonstrated a significantly enhanced potential towards accomplishing high photocatalytic efficiencies [115]. The Z-scheme photocatalyst is typically constructed by coupling two photocatalytic semiconductors, which is likely similar to the conventional heterojunction photocatalyst [116]. However, Z-scheme has a unique mechanism for the injection/transfer of charge carrier into the adjacent semiconductor, as shown in Figure 7a,b [117]. Notably, among the two coupled photocatalysts in Z-scheme, one will be an oxidation and the other will be a reduction photocatalyst. The selection of such oxidation and reduction photocatalyst will be based on the VB and CB edge position, which is dependent upon the specific applications [118]. As a result of such meticulous construction, Z-scheme systems demonstrate exotic features such as (i) simultaneous strong reduction-oxidation abilities, (ii) spatial separation of reduction and oxidation active sites, (iii) enhanced carrier-separation efficiency with high redox abilities, and (iv) extended light absorption range [119,120].

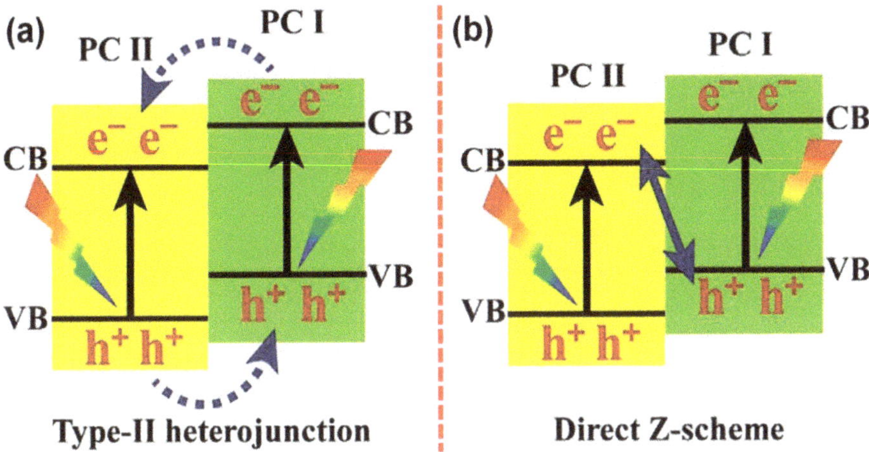

Figure 7. Schematic illustration of the (**a**) typical heterojunction and (**b**) Z-scheme photocatalysts (reproduced with permission from ref. [117]).

In the Z-scheme-based systems, TiO_2 has been largely used as oxidation photocatalyst owing to their low VB position and accordingly, it has been coupled with the other photocatalytic systems such as CdS [121,122], g-C_3N_4 [123–125], NiS [126], $ZnIn_2S_4$ [127], Cu_2O [128], and WO_{3-x} [129] owing to their high CB position that act as the reduction photocatalysts. As shown in Figure 7a [117], in the typical heterojunction photocatalyst, the separated electron holes in PCI will be injected into the respective CB and VB of the PCII. In contrast, the charge transfer mechanism in Z-scheme always follows a signature pathway in which the electrons excited to the CB of low VB photocatalyst will be injected into the VB of the high CB photocatalyst (Figure 7b) [117]. As listed above, Figure 8a–d shows the mechanism of various TiO_2-based Z-scheme photocatalysts. Interestingly, Fu et al. [128] proposed a Z-scheme system mediated by Ag located at the interface of the TiO_2 and Cu_2O. They observed that the TiO_2 and Cu_2O coupled photocatalyst demonstrated a relatively poor photocatalytic performance; as a result, they proposed that upon the irradiation of TiO_2 and Cu_2O, the electrons in the CB of Cu_2O get transferred into the TiO_2 and meanwhile, the holes in VB of TiO_2 get transferred to Cu_2O. Such a process essentially led to the depletion of hole density in the VB of TiO_2 and it increased in the VB of Cu_2O. Under such circumstances, due to the low positive VB edge position of Cu_2O, it has insufficient energy to oxidize the OH or H_2O molecules. To address such an issue, they introduced Ag into the interfacial contact of TiO_2 and Cu_2O, as shown in Figure 8e [128].

In this TiO_2–Ag–Cu_2O system, firstly, the equilibrium in Fermi levels has been established; thereby, upon irradiation, the excited electrons in the TiO_2 CB get injected into Ag and due to the localized electric field created by Ag, these electrons are further injected into the Cu_2O and enhanced the photocatalytic efficiency of the system. Further, they proposed that this system keeps the photo-induced holes on more positive potential (VB of TiO_2) and electrons on more negative (CB of Cu_2O), which essentially enhance the redox ability as well as the charge separation efficiencies of the system as a whole.

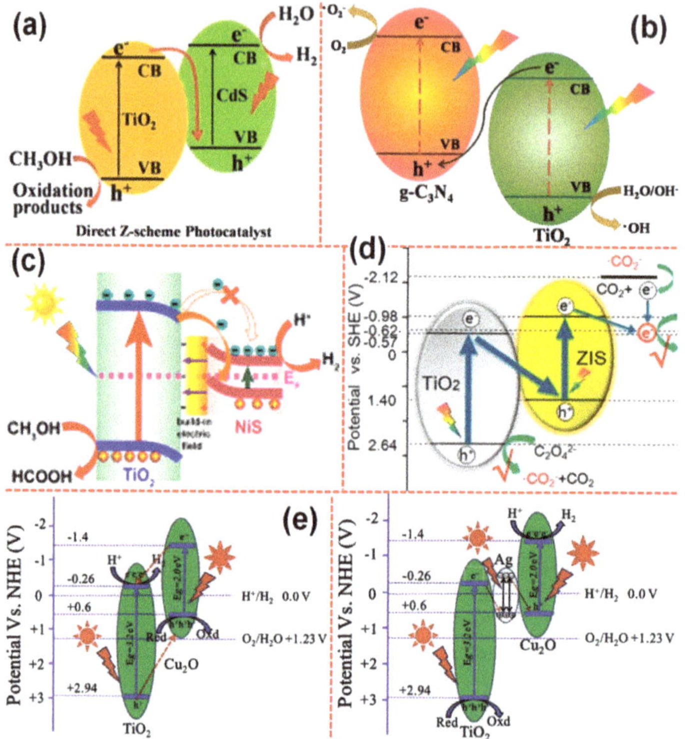

Figure 8. Charge transfer mechanism in various Z-scheme-based TiO_2 photocatalysts, (**a**) CdS/TiO_2, (**b**) g-C_3N_4/TiO_2, (**c**) NiS/TiO_2, (**d**) $ZnIn_2S_4$/TiO_2, and (**e**) TiO_2–Ag–Cu_2O (reproduced with permission from refs. [121,124,126–128], respectively).

3.5. Plasmonic TiO_2

Plasmonic photocatalysis is one of the emerging and interesting concepts in this field [130]. These types of photocatalysts make use of the plasmonic nanoparticles to harvest energy in the visible region [131]. It extends the absorption range of the photocatalyst in UV-visible-IR region [132]. The plasmonic nanoparticles also play an important role in alerting the charge transfer mechanism in the host photocatalysts. The plasmon-mediated process in photocatalysts can occur in four different ways, (i) direct migration of carriers from the plasmonic particles to photocatalyst, (ii) indirect migration of carriers between the plasmonic particles and photocatalyst via the localized surface plasmon resonance (LSPR), (iii) localized plasmonic heating, and (iv) radiative transfer of photons from the plasmonic particles to the photocatalyst, where these photons will excite the photocatalyst to generate the electron hole pairs in the system [131–133]. However, the origins and functions of plasmonic photocatalysts are under hot debate.

Noble metals such as Ag, Au, Pd, and Pt have been integrated with TiO_2 to produce the TiO_2-based plasmonic photocatalysts. Among them, Ag–TiO_2 has been relatively largely studied with different configurations [134–137]. Plasmonic sensitization conventionally happens by the deposition of plasmonic nanoparticles (NPs) onto the surface of the host photocatalyst. However, there have been other configurations such as core-shell structuring [137], filling up the plasmonic NPs into the pores of the host photocatalyst, and composite-like formation [135]. As aforementioned, the plasmonic nanoparticles can extend the light absorption in the visible region and they can also substantially influence the charge transfer kinetics the photocatalyst. However, there are essentially two pathways proposed regarding their charge transfer, which is either from the (i) plasmonic NPs to photocatalyst or (ii) photocatalyst to plasmonic NPs [130]. As a result, it has also been proposed that the scheme of such charge transfer is also determined by the relative band edge potential, conducting type (n/p-type), and work function of the photocatalyst and plasmonic metal, respectively, and also determined by the light source that is used to excite the plasmonic photocatalyst system, as shown in Figure 9a–b [134,138].

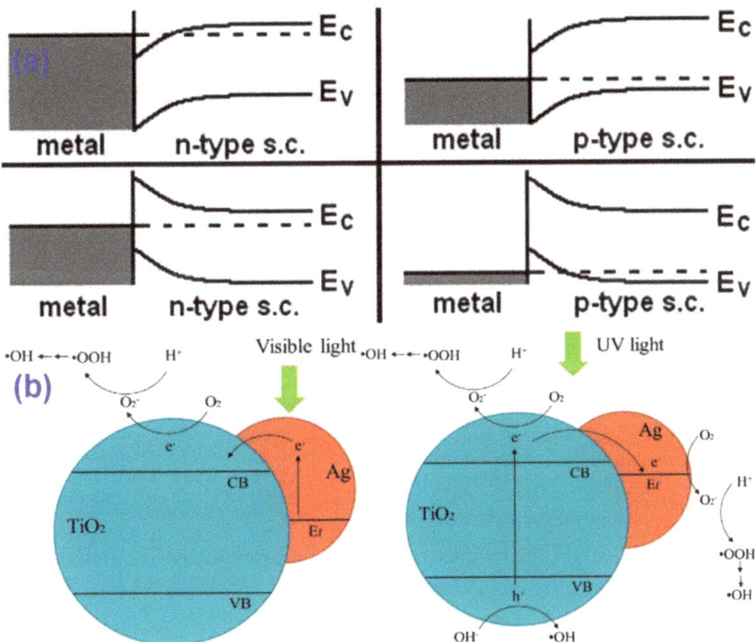

Figure 9. (a) Band bending occurs in the metal-semiconductor junction and (b) charge transfers in plasmonic photocatalyst, depending upon the light source irradiated (reproduced with permission from refs. [134,138], respectively).

As depicted in Figure 9a [138], the work function of the metal nanoparticle with respect to the host semiconductor also directs the course of charge transfer in the plasmonic photocatalyst. For instance, the work function of Au, Ag, and anatase TiO_2 has the work function of 5.23, 4.25–4.37, and 5.10 eV, respectively, where the Au–TiO_2 and Ag–TiO_2 follow the Schottky-junction and Ohmic-junction, respectively, for the charge transfer in the system, as shown in Figure 10a,b [139]. Compared to the Ag and Au, the surface plasmon resonance (SPR) properties of Pt/Pd-deposited TiO_2 has been less explored [140]. However, these metal NPs have been explored as a co-catalyst for various photocatalyst systems [141–143]. This is because the plasmonic peak of Pt NPs appears below 450 nm, while the SPR properties of Ag and Au can be well tuned in visible to IR region, and therefore, the Pt and Pd NPs have not been typically used for developing the plasmonic photocatalysts [144–146].

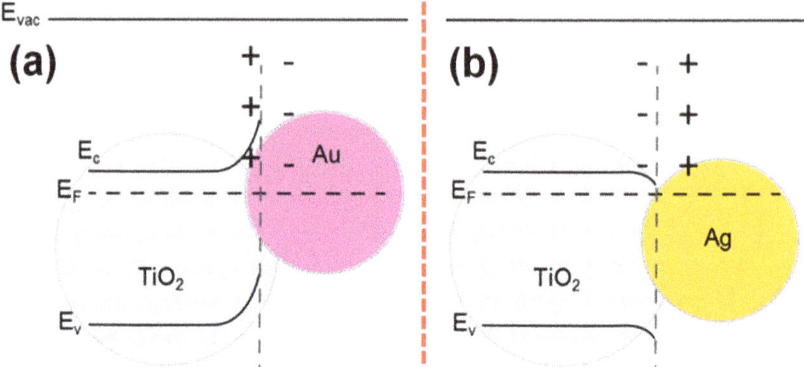

Figure 10. Work function dependent band-bending in (**a**) Au/TiO$_2$, (**b**) Ag/TiO$_2$ plasmonic systems (reproduced with permission from ref. [139]).

3.6. Ferroelectrics Modified TiO$_2$

Ferroelectrics are defined by the spontaneous electric polarization that can be induced by an external electric field, where the induced spontaneous polarization will be permanent in the material and it essentially originates from the off-center displacements of ions in a non-centrosymmetric crystal system [147]. In ferroelectric materials, the internal screening induced by the free carriers and the bulk defects lead to the distribution of charge carriers in the near surface of the material, which essentially creates a space-charge region and band bending in the system [148]. These features greatly help in the photocatalytic process. The bands of ferroelectrics bend at the near the surface or interface region, depending upon the positive or negative spontaneous polarizations, as shown in Figure 11a,b [149].

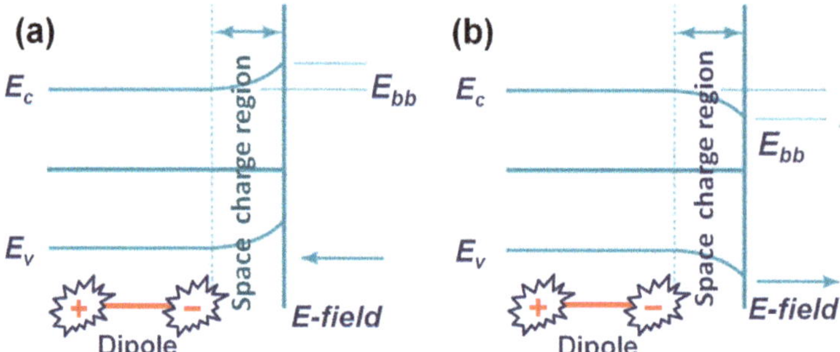

Figure 11. Schematic diagram of band bending in a ferroelectric material; (**a**) a surface with negative polarity and (**b**) a surface with positive polarity.

For instance, in a negatively polarized surface, the electrons will be depleted from the surface, which leads to a creation of a spatial-charge layer (depletion layer) with "upward" band-bending. On the other hand, in a positively polarized surface, the electrons will be accumulated for screening, which leads to a "downward" band bending in the system along with formation of a spatial accumulation charge layer. Thereby, these interesting features in ferroelectric, along with such deformed migration of charge carriers, largely helpful to exhibit exotic photo-active chemical properties [150,151]. The features such as the spontaneous polarization, deformed migration of carriers, surface charges, band bending process, and the external and/or internal screening effects altogether direct the photo-induced charge carriers in a ferroelectric toward an effective oxidation and reduction reaction for various photocatalytic applications [152–158].

Ferroelectric materials such as $BaTiO_3$ [159–161], $BiFeO_3$ [162,163], $PbTiO_3$ [164] have been successfully integrated with TiO_2 to produce ferroelectric-TiO_2 photocatalysts. Zhang et al. have explained how the ferroelectric phenomenon influences the photocatalytic activity of the system, where they demonstrated it using $BiFeO_3/TiO_2$ system [162]. They proposed a plausible energy level for the $BiFeO_3/TiO_2$ system, as shown in Figure 12a,b. According to this diagram, the energy levels at the interface of $BiFeO_3$ (BFO) and TiO_2 are strongly influenced by the induced polarization in $BiFeO_3$, where it bends the band of BFO upward when the polarization in negative (i.e., away from the surface) and downward when the polarization is positive (i.e., towards the surface). Under such circumstances, the photo-induced electrons in negative domains are impeded by the energy barrier at the interface; meanwhile, in positive domains, the electrons are moved to the interface, in such a way that it facilitates the photocatalytic activity with enough redox abilities of the excited charge carriers in the system [163].

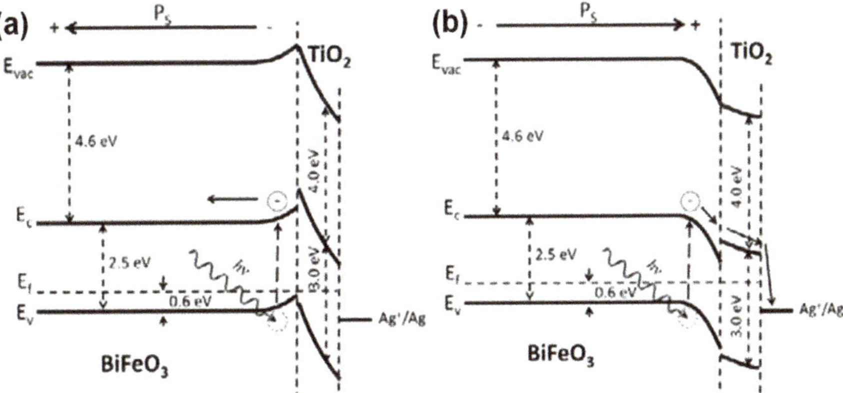

Figure 12. The energy bands at the $BiFeO_3/TiO_2$ interface bend (**a**) upward and (**b**) downward corresponding to the applied polarization (reproduced with permission from ref. [162]).

3.7. Carbon-Based TiO_2 Composites

Carbon-based materials-modified TiO_2 photocatalysts demonstrate significant enhancements in the photocatalytic process due to various reasons such as (i) high surface area, (ii) enhanced electrical conductivity, (iii) tunable optical properties, (iv) improved surface adsorption efficiency, and (v) controllable structural features [165–167]. These properties essentially help improve the overall properties of the photocatalysts. For instance, the enhanced surface area populates more catalytic-sites on the surface of the catalysts. The enhanced electrical conductivity improves the charge separation and transportation characteristics of the system. The tunable optical properties help activate the photocatalyst under a desirable light source such as visible light and/or sunlight. The improved surface adsorption essentially paves the way for the adsorption of surrounding molecules onto the surface of the photocatalyst that eventually enhances the interfacial interaction of the photocatalyst and molecules. Finally, the controllable structural features of carbon materials such as quantum dots (fullerenes) [168–170], 2D materials (graphene, g-C_3N_4) [171,172], 1D materials (carbon nanotubes (CNTs), carbon fibers) [173–176], and 3D materials (carbon spheres, flowers) [177,178] offer unique charge transportations and improve the overall efficiency of the carbon-based photocatalytic materials.

TiO_2 has been modified by the variety of carbon-based materials such as carbon doping, carbon coating, composites with activated carbon, graphene/graphene oxide/reduced-graphene oxide, g-C_3N_4, CNTs, carbon fibers, anisotropic carbon structures, etc. [165–178]. The general photocatalytic mechanisms of these carbon-based TiO_2 systems are summarized in Figure 13a–d [168,179–181] Yu et al. [168], have reported the mechanism of carbon quantum dots (CQDs)-integrated TiO_2 towards photocatalytic H_2 production. The CQDs play a dual vital role in the improved photocatalytic properties.

During the photocatalytic excitation under UV light, the CQDs act as (i) electron reservoirs and (ii) photo-sensitizers. The former role of CQDs essentially plays a role in trapping the photo-generated electrons from the conduction band of TiO_2 and facilitates the enhanced process of electrons-holes separation. On the other hand, the latter characteristics of π-conjugated CQDs is to sensitize the TiO_2 as similar to the organic dyes, towards making it a visible light active "dyade"-like structure, where it gives the electrons to the CB of TiO_2 and leads to the visible light-driven hydrogen production (Figure 13a) [168].

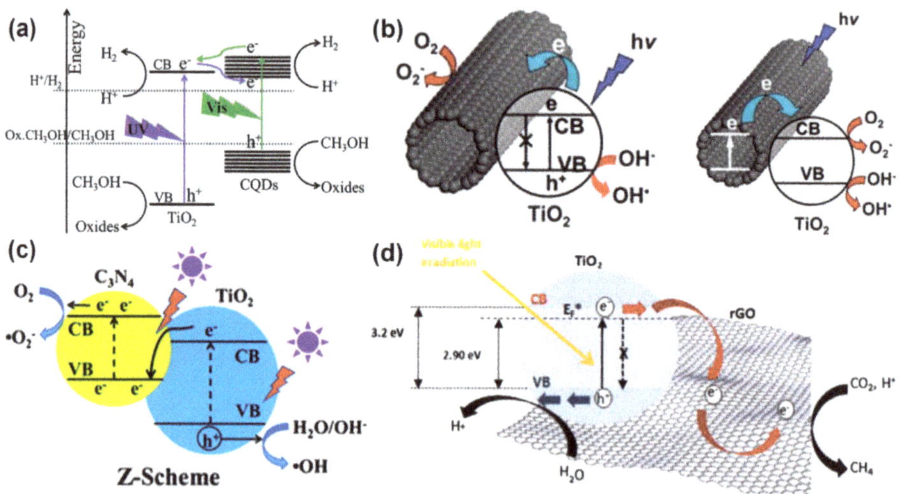

Figure 13. Photocatalytic mechanism in various carbon-TiO_2 systems, (**a**) carbon QD-TiO_2, (**b**) carbon nanotubes (CNT)-TiO_2, (**c**) g-C_3N_4-rGo-TiO_2, (**d**) rGO-TiO_2 (reproduced with permission from refs. [168,179–181], respectively).

The carbon nanotubes (CNTs), owing to their large electron-storage capacity (per electron for every 32 C-atoms), accept the photo-induced electrons from the supported semiconductor and, thereby, they largely hinder the recombination of charge carriers [179]. It is believed that the excellent conductive nature of the CNTs promotes the electron-hole separation via the formation of a heterojunction between CNTs and semiconductors. For instance, as similar to the carbon QDs, the CNTs also play a dual role in the photocatalytic process. Accordingly, the freely moving electrons in the excited TiO_2 get transferred into the CNTs scaffolds, where the excess holes in the VB in TiO_2 are set to reach and react with the H_2O and OH^- to generate radicals such $OH^\bullet$ as shown in Figure 13b [179]. On the other hand, it is known that TiO_2 is UV-driven, but it is observed that the CNTs-TiO_2 nanocomposites have become visible light driven, which is attributed to the photo-sensitizing effect of CNTs. In this scenario, the photo-induced electrons in CNTs (sensitizers) get injected into the CB of TiO_2 and lead to reducing the adsorbed molecular oxygen to form the superoxide species. In parallel, the holes in these positively charged CNTs react with H_2O and form $OH^\bullet$ radicals, as shown in Figure 13b.

Yu et al. [180] have demonstrated that the coupling between TiO_2 and g-C_3N_4 cannot lead to the formation of heterojunction; rather, it always tends to form the Z-scheme-based photocatalyst system. Based on their experiments, they have explained the phenomenon that if TiO_2 and g-C_3N_4 form a heterojunction, then the following scenario will emerge. Under the UV exposure, the photo-induced holes will get transferred from the VB of TiO_2 to that of the g-C_3N_4 and the electrons will get transferred from CB of g-C_3N_4 to that of the TiO_2. As a result, the holes of g-C_3N_4 cannot oxidize the adsorbed H_2O or OH^- to form the $OH^\bullet$ radicals due to the higher potential of VB of g-C_3N_4 with respect to the H_2O/OH^- couple. Such a process eventually leads to the lower oxidation, thereby the photocatalytic efficiency of the system is much lower than the TiO_2. However, the observed photocatalytic efficiency of TiO_2/g-C_3N_4 is higher than the individual counterparts, which essentially means that this system forms a direct Z-scheme system without the electron mediator, as shown in Figure 13c [180].

The photocatalytic mechanism in the reduced graphene oxide (rGO)-TiO_2 composite has been proposed by Tan et al. [181] as shown in Figure 13d. In the rGO-TiO_2 composite, the d and π orbital of TiO_2 and rGO, respectively, matches well in their energy levels and they overlap each other well (d-π). As a result, rGO is bound to serve as an electron-collector as well as a transporter towards effectively separating the photo-induced electron-hole pairs, which eventually enhances the lifetime of the charge carriers as well, and thereby the photocatalytic efficiency of the rGO-TiO_2 system [182–185].

3.8. 2D-Transition Metal Chalcogenides Modified TiO_2

It is well established that the large surface-to-volume ratio of 2D nanostructures can provide more surface-active sites for the photocatalytic reactions. The planar structure of 2D materials essentially favors the charge transportations across the interfaces of the catalyst and surrounding phases and thereby it drastically improves the photocatalytic efficiencies [186]. Moreover, as compared to other nanostructures, the 2D nanostructures exhibit exotic properties owing to the atomic arrangements with surface atomic elongation and structural-disorder characteristics [187]. These interesting physical structure-induced properties of 2D materials largely contribute in enhancing the photo-stability and chemical durability of the photocatalyst. Furthermore, 2D materials, due to their flat band potential and effective band bending at the interface, help tune the band gap energies and band-edge positions of the photocatalysts [188]. Specifically, when these 2D materials couple with the other metal and metal oxides, their unique 2D structures serve as a matrix for those integrated materials and enhance the optical and electrical properties of the system as a whole [189–191]. In this direction, the 2D transition metal chalcogenides (2D TMC) with general chemical formula of MX_2, M = Mo, or W and X = S, Se, or Te serve as both the independent or composite photocatalytic materials [191]. Accordingly, TiO_2 has been modified with these 2D TMC materials to avail their structural features and unique properties towards various photocatalytic applications.

Among the listed 2D TMC materials, the MoS_2/TiO_2 system has been largely explored for the photocatalytic applications [192–197]. Interestingly, the charge transfer in this system depends upon the photon energy used to excite the system. The Figure 14a,b shows the charge transfer in a MoS_2/TiO_2 system that irradiated under UV and visible light, respectively [198,199]. When the MoS_2/TiO_2 system irradiated under UV light, the electrons that were excited in TiO_2 will be transferred to the attached MoS_2 nanosheets; thereby, this process significantly limits the electron hole recombination and promotes carrier separation by effectively transporting to the adsorbed H^+ ions to reduce them to produce molecular hydrogen. On the other hand, when the MoS_2/TiO_2 system is irradiated under visible light, the electron transfer occurs from the MoS_2 to TiO_2, as shown in Figure 14b [199]. It should be noted that the TiO_2 used in this study is doped with N species that facilitates visible light absorption in TiO_2 as well. Therefore, the coupling of MoS_2 with TiO_2 promotes the excited electrons to the CB of TiO_2 from the CB of MoS_2. The further photocatalytic reactions essentially occur via the conventional redox reactions on the surface of the photocatalyst.

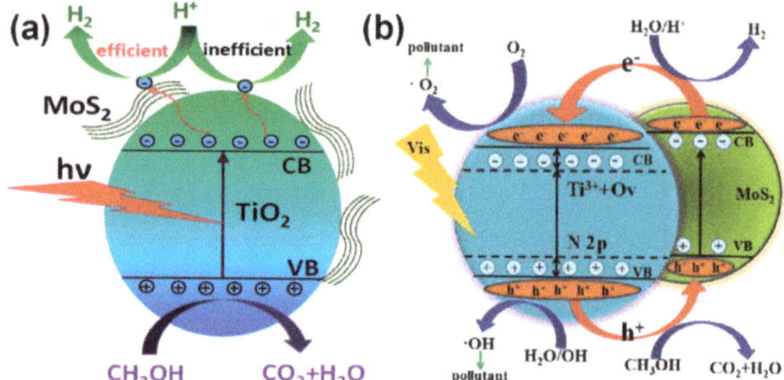

Figure 14. Photocatalytic charge transfer process in MoS$_2$/TiO$_2$ under the irradiation of (**a**) UV light and (**b**) visible light (reproduced with permission from refs. [198,199]).

Zhang et al., have reported the possible charge transfer mechanism in P25-TiO$_2$/MoS$_2$ and P25-TiO$_2$/WS$_2$ systems under UV-visible irradiation [200]. Accordingly, the excited electrons in P25-TiO$_2$/MoS$_2$ migrate from the CB of TiO$_2$ to the CB of MoS$_2$, while it occurs vice versa in the P25-TiO$_2$/WS$_2$ system, as shown in Figure 15a,b [200]. The observed charge transfer mechanism is essentially due to the relative band-edge potentials of the semiconductors involved in the composite.

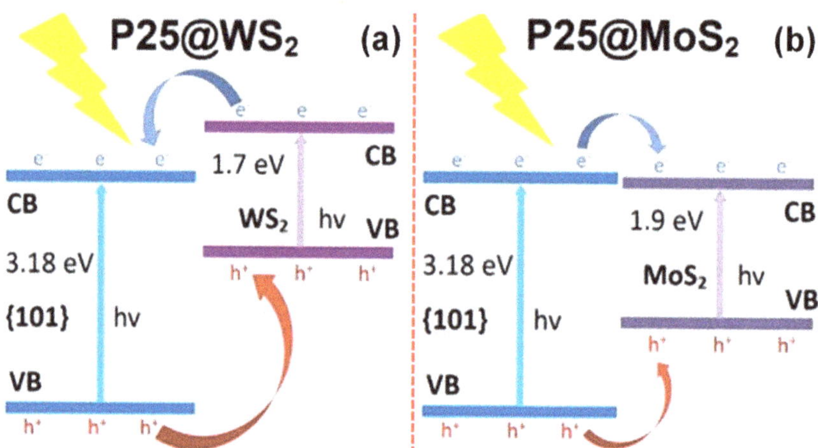

Figure 15. Photocatalytic charge transfer mechanism in (**a**) P25-TiO$_2$/WS$_2$ and (**b**) P25-TiO$_2$/MoS$_2$ (reproduced with permission from ref. [200]).

Similar to the aforementioned systems, there are alternative hypotheses to explain the charge transfer mechanism in MoSe$_2$/TiO$_2$ system. Chu et al. [201] and Shen et al. [202] have proposed that the MoSe$_2$/TiO$_2$ follows the heterojunction mechanism towards the charge transfer process in the system, as shown in Figure 16a [201]. Accordingly, the type-II heterostructure, which formed between MoSe$_2$ and TiO$_2$, facilitates the electron transfer from the CB of MoSe$_2$ to that of TiO$_2$ and reduces the recombination process, prolongs the lifetime of the carriers, and provides an enhanced conductivity in the system towards transporting the carriers to the surrounding for the effective photocatalytic process. On the other hand, Zheng et al. proposed that this system follows the Z-scheme to transfer the charges from the TiO$_2$ to MoSe$_2$ [203]. According to their hypothesis, the MoSe$_2$/TiO$_2$ (nanotubes) photocatalyst could not form a type-II heterojunction. This may be because of the reason that the holes in TiO_2 VB are likely to migrate into the $MoSe_2$ VB if type-II has been formed. However, their experimental investigations using ESR and PL demonstrated that the proposed charge is not possible, owing to the low potential of 0.98 V that cannot effectively oxidize the adsorbed surface H_2O to produce $OH^\bullet$ radicals. Therefore, the photo-generated electrons in the CB of TiO_2 might have been transferred and recombined with the holes in $MoSe_2$ VB, leaving the holes in the VB of TiO_2 and electrons in the CB of $MoSe_2$ via constructing a 'direct Z-Scheme' to augment the photocatalytic redox reactions in the system, as shown in Figure 16b [203]. Similarly, the WS_2/TiO_2 system has also been explored for various photocatalytic applications and their mechanisms [204–210].

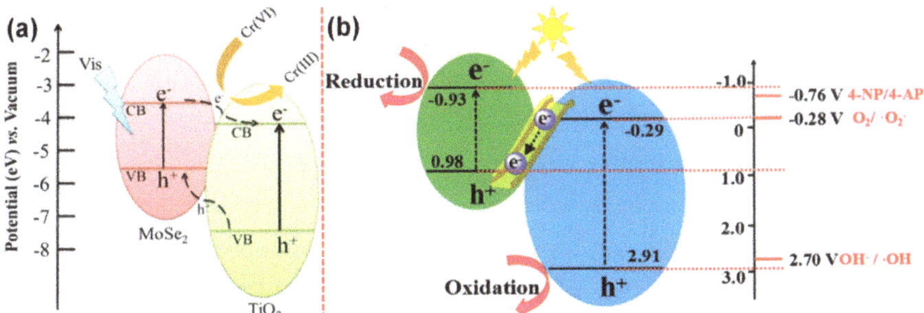

Figure 16. Charge transfer mechanism in $MoSe_2/TiO_2$ (**a**) heterojunction and (**b**) Z-scheme (reproduced with permission from refs. [201,203]).

3.9. Metal-Organic Framework-TiO$_2$ Composites

Metal–organic frameworks (MOFs) are an exotic class of crystalline materials with inherent porous structures. MOFs are constructed using the metal clusters that interconnected by organic ligands built into a 3D networked structure. Their unique properties, such as the well-ordered porosity, very high specific surface area, and tunable surface chemistry, have made them a promising material for various applications, including photocatalysis. MOFs can be reliable photocatalytic materials due to semiconductor-like properties. In addition, they possess high surface area that largely facilitates enhanced surface catalytic activities; the metal clusters play a role in the effective absorption of incident photons and charge separation, while the ligands favor the charge transportations in the system. However, the major issue in MOFs is the moderate charge separation that considerably reduces the overall photocatalytic efficiency of the MOFs [211–216].

Yao et al. proposed the observed superior photocatalytic efficiency of TiO_2@-NH_2-UiO-66 composites towards the degradation of styrene [217]. According to their findings, (i) the plenty of available interconnected nanopore facilitated the enhanced and rapid diffusion of the surrounding styrene molecules into the pores of MOFs, where the encapsulated TiO_2 effectively oxidized the molecules with the produced oxidation radical species, and (ii) the linkers in MOFs acted as antenna to augment the light absorption and sensitize the TiO_2 and led to the effective absorption of light towards the transportation of charge carriers in the system; thereby, it demonstrated excellent photocatalytic activity [217].

Similarly, the photocatalytic efficiency of TiO_2/NH_2-UiO-66 nanocomposites towards CO_2 reduction has been demonstrated by Crake et al. [218]. Based on their observations, the composite of NH_2-UiO-66 and TiO_2 can lead to the formation of type-II heterojunction. This could essentially be because of the factor that the CB position of NH_2-UiO-66 lies at −0.6 eV, while the TiO_2 CB lies at a more negative potential at −0.28 eV, as shown in Figure 17a [218]. They have further proposed that the photocatalytic activity of TiO_2/NH_2-UiO-66 nanocomposites was mainly ruled out by (i) the concentration of TiO_2, (ii) the effective charge separation characteristics of NH_2-UiO-66, and (iii) the enhanced availability of charge carriers at the interface of the TiO_2/NH_2-UiO-66 system. Ling et al. have synthesized a ternary nanocomposite composed of TiO_2/UiO-66-NH_2/graphene oxide and studied towards the photocatalytic dye (RhB) degradation and H_2 evolution [219]. They have reported that, under the visible excitation, the electrons tend to transfer from RhB* to CB of MOFs to CB of TiO_2 due to the cascading potential of these systems. Under such circumstances, the integrated GO captures the electrons from the CB of TiO_2 that eventually enhances charge separation, thereby accelerating the dye removal. On the other hand, the electrons from GO further migrate to the Pt and lead to the H_2 production. It is also possible that the electrons from RhB* can get directly injected into Pt and produce H_2, as shown in Figure 17b [219]. Similarly, there have been other TiO_2/MOFs-based photocatalytic systems reported [220–225].

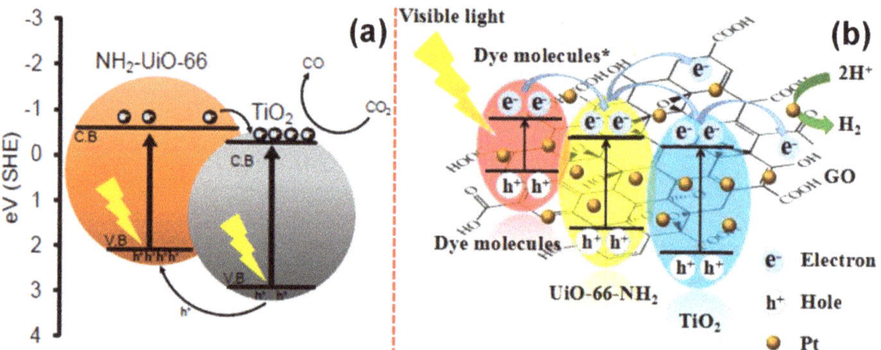

Figure 17. Photocatalytic charge transfer process in (a) TiO_2/NH_2-UiO-66 for CO_2 reduction and (b) TiO_2/NH_2-UiO-66/GO/Pt for dye removal and H_2 production (reproduced with permission from refs. [218,219]).

3.10. Reduced/Defective/Colored TiO_{2-x} Photocatalysts

The off-stoichiometricity in TiO_2, which is induced by processes such as self-doping by Ti^{3+} ions and oxygen vacancy creations (V_o), plays an important role in enhancing the visible light absorption and photocatalytic efficiency of the TiO_2 materials [226–230]. Based on such a modification approach, TiO_2 has been synthesized in a variety of "colors" such as black, blue, red, and yellow. The reduced band gap energy in off-stoichiometric TiO_2 essentially originates due to the formation of localized energy states (0.75–1.18eV) underneath the CB minimum of the TiO_2 [231]. As compared to any other modification strategies, the self-doping and/or oxygen vacancy creation is more favorable for maintaining the intrinsic properties of the TiO_2 as well as to introduce the visible light absorption characteristics and enhance the photocatalytic efficiencies of TiO_2 [232–234].

The first black-TiO_2 was produced by Chen et al. with band gap energy of around 1.0 eV via high-pressure hydrogenation process in the crystalline TiO_2 [235]. The general mechanism for the formation of black TiO_2 is broadly attributed to the presence of Ti^{3+} by self-doping, formation of hydroxyl groups on the surface, oxygen vacancies, Ti-H bonds, and the formation of H-energy states in the mid-gap of the TiO_2 band structure, which eventually dispersed the VB in TiO_2, as shown in Figure 18a [235]. Zhu et al. synthesized the stable blue TiO_2 nanoparticles [236] and proposed the origin that the observed blue color could be due to the high concentration of Ti^{3+} defects in the bulk and the formation of mid-gap electronic energy states beneath the band gap of TiO_2. As a result, the observed enhanced photocatalytic properties were attributed to their unique structural features, which is the disordered-core/ordered-shell-like structure. This essentially means that the TiO_2 was stoichiometric at the surface while it was off-stoichiometric in the core. These features collectively improved the overall photocatalytic efficiencies of the blue TiO_2 by enhancing the charge separation and transportation, as shown in Figure 18b [236].

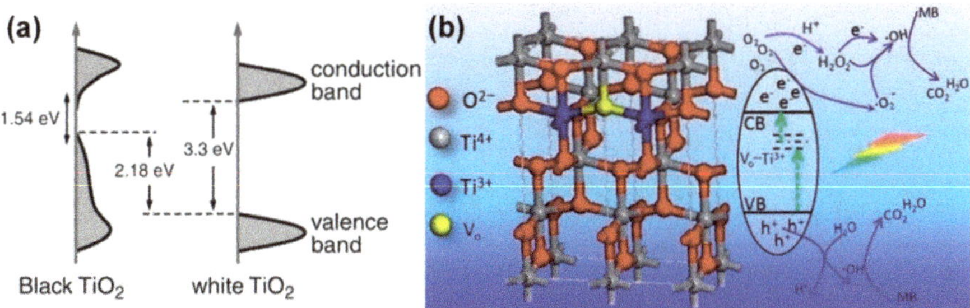

Figure 18. Band gap structure of (**a**) black-TiO_2 and (**b**) blue-TiO_2 (reproduced with permission from refs. [235,236]).

Wu et al. developed ultra-small yellow TiO_2 nanoparticles via simple sol-gel process with UV treatment technique. Based on their experimental findings, the origin of the observed yellow color of TiO_2 could be due to titanium vacancies (V_{Ti}) and titanium interstitials (Ti_i) as shown in Figure 19a [237]. Interestingly, Liu et al. prepared the red anatase TiO_2 via a gradient co-doping of B-N into the system. It was observed that the band gap energy varied from 1.94 eV on the surface to 3.22 eV in the core, as shown in Figure 19b [238].

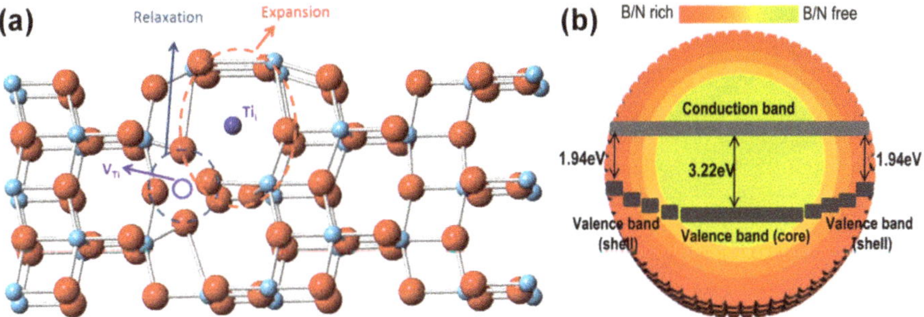

Figure 19. (**a**) Structure of yellow-TiO_2 and (**b**) anatase red TiO_2 via gradient B-N co-doping (reproduced with permission from refs. [237,238]).

Ren et al. reported that the $NaBH_4$ reduced TiO_2 photocatalysts with a range of colors such as white, light-yellow, light-grey, and dark-grey, which were prepared by varying the concentration of the reducing agent $NaBH_4$, as shown in Figure 20a [239]. The observed color variation was attributed to the self-doping of Ti^{3+} ions into the TiO_2. Similarly, Fan et al. reported the synthesis of TiO_2 with white, dark brown, light brown, yellow, light yellow, gray, yellowish gray, and yellowish white color (Figure 20b) that were derived from the amorphous hydrated TiO_2 through hydroxylated and N-doping process with a controlled degree of disorders using a heating treatment technique [240]. In this study, the observed color variation was attributed to the heating process that turned the Ti–OH bonds in amorphous TiO_2 into the Ti–O bonds that transformed the disordered TiO_6 octahedron into a regular 3D structure. As a result, the formed hydroxylated anatase TiO_2 with enhanced degree of disorder strongly influenced the optical transition in TiO_2 and narrowed down the band gap energy. Further, these colored TiO_2 materials have also demonstrated enhanced photocatalytic efficiencies towards the degradation of acid fuchsin under visible light.

Figure 20. Photographic images of the (**a**) chemically reduced TiO_2 with increasing concentration of $NaBH_4$ and (**b**) hydroxylated and N-doped anatase TiO_2 that were derived from the amorphous hydrate TiO_2 at the increasing processing temperature (reproduced with permission from refs. [239,240]).

4. Summary and Outlook

Undoubtedly, TiO_2 is indeed an interesting material for various photocatalytic applications. As described, the fundamental photocatalytic process involves the excitation of photo-induced carriers and their successful transfer to the surface to produce the desired redox species towards the designated photocatalytic application. The versatile applications emerge essentially due to the produced redox species with appropriate energy, which is dictated by the band edge potential of the photocatalyst. Since TiO_2 inherently meets such requirements, it has been successfully used for various photocatalytic applications. However, TiO_2 has limitations such as its wide-band gap, moderate charge separation efficiency, etc. To overcome such limitations, TiO_2 has been both physically and chemically modified. Accordingly, herein we provided a glimpse on the various modifications that were performed on TiO_2 towards enhancing its photocatalytic efficiencies. These modifications include morphological modifications, anionic-cationic doping, heterojunction formations, Z-scheme formations, plasmonic integrations, ferroelectric integrations, carbon-based materials integrations, 2D transition metal chalcogenide integrations, metal–organic framework integrations, and defects inducements in TiO_2. We also have discussed the charge transfer mechanism that manifests in these various modified-TiO_2 photocatalytic systems.

TiO_2 can be a prototype photocatalyst, which can be used to design new photocatalytic materials. The meticulous investigations on TiO_2 for their photocatalytic mechanism can be better applied towards its effective applications in photocatalysis. In this direction, the further improvement in TiO_2 could be the establishment of techniques to intrinsically modify the TiO_2 towards their photocatalytic enhancements. Such known techniques are the inducement of defective structures in TiO_2 through self-doping, atoms in interstitial positions, oxygen-, and Ti-vacancies. For instance, instead of doping the N atoms into TiO_2, the O atoms can be partially replaced by N atoms to form oxy-nitrides and so the oxy-phosphates, oxy-sulfur, oxy-carbon, etc., can be formed by partially replacing the O atoms with P, S, and C, respectively. These modifications may lead to the formation of entirely different TiO_2-based materials with possibly new crystal phase and structure and can exhibit enhanced photocatalytic efficiencies. Towards applications, TiO_2 can be explored for new photocatalytic processes such as the production of H_2/O_2 from the atmospheric vapor, dark-photocatalysis, hydrogen storage, biodiesel productions, etc. TiO_2 should be consistently explored towards further understanding of their photocatalytic mechanisms and finding new photocatalytic applications.

Funding: This work was supported by the Natural Science and Engineering Research Council of Canada (NSERC) through the Collaborative Research and Development (CRD), Strategic Project (SP), and Discovery Grants (DG). MS gratefully acknowledges the Department of Science and Technology, Govt. of India for the funding support through the DST-INSPIRE Faculty Award [DST/INSPIRE/04/2016/002227, 14-02-2017].

Acknowledgments: We would also like to thank EXP Inc. and SiliCycle Inc. for their support.

Conflicts of Interest: The authors declare no conflict of interest. The funders had no role in the design of the study; in the collection, analyses, or interpretation of data; in the writing of the manuscript, or in the decision to publish the results.

References

1. Fujishima, A.; Honda, K. Electrochemical photolysis of water at a semiconductor electrode. *Nature* **1972**, *238*, 37–38. [CrossRef]
2. Schrauzer, G.N.; Guth, T.D. Photocatalytic reactions. 1. Photolysis of water and photoreduction of nitrogen on titanium dioxide. *J. Am. Chem. Soc.* **1977**, *99*, 7189–7193. [CrossRef]
3. Kiwi, J.; Gratzel, M. Projection, size factors, and reaction dynamics of colloidal redox catalysts mediating light induced hydrogen evolution from water. *J. Am. Chem. Soc.* **1979**, *101*, 7214–7217. [CrossRef]
4. Kawai, T.; Sakata, T. Conversion of carbohydrate into hydrogen fuel by a photocatalytic process. *Nature* **1980**, *286*, 474–476. [CrossRef]
5. Sato, S.; White, J.M. Photoassisted water-gas shift reaction over platinized titanium dioxide catalysts. *J. Am. Chem. Soc.* **1980**, *102*, 7206–7210. [CrossRef]
6. Bagheri, S.; Yousefi, A.T.; Do, T.O. Photocatalytic pathway toward degradation of environmental pharmaceutical pollutants: Structure, kinetics and mechanism approach. *Catal. Sci. Technol.* **2017**, *7*, 4548–4569. [CrossRef]
7. Moser, J.; Gratzel, M. Light-induced electron transfer in colloidal semiconductor dispersions: Single vs. dielectronic reduction of acceptors by conduction-band electrons. *J. Am. Chem. Soc.* **1983**, *105*, 6547–6555. [CrossRef]
8. Bahnemann, D.; Henglein, A.; Lilie, J.; Spanhel, L. Flash photolysis observation of the absorption spectra of trapped positive holes and electrons in colloidal titanium dioxide. *J. Phys. Chem.* **1984**, *88*, 709–711. [CrossRef]
9. Nozik, A.J.; Williams, F.; Nenadovic, M.T.; Rajh, T.; Micic, O.I. Size quantization in small semiconductor particles. *J. Phys. Chem.* **1985**, *89*, 397–399.
10. Anpo, M.; Shima, T.; Kodama, S.; Kubokawa, Y. Photocatalytic hydrogenation of propyne with water on small-particle titania: Size quantization effects and reaction intermediates. *J. Phys. Chem.* **1987**, *91*, 4305–4310. [CrossRef]
11. Hong, A.P.; Bahnemann, D.W.; Hoffmann, M.R. Cobalt (II) tetrasulfophthalocyanine on titanium dioxide. 2. Kinetics and mechanisms of the photocatalytic oxidation of aqueous sulfur dioxide. *J. Phys. Chem.* **1987**, *91*, 6245–6251. [CrossRef]
12. Zhang, J.L.; Minagawa, M.; Matsuoka, M.; Yamashita, H.; Anpo, M. Photocatalytic decomposition of NO on Ti-HMS mesoporous zeolite catalysts. *Catal. Lett.* **2000**, *66*, 241–243. [CrossRef]
13. Frank, S.N.; Bard, A.J. Heterogeneous photocatalytic oxidation of cyanide ion in aqueous solutions at titanium dioxide powder. *J. Am. Chem. Soc.* **1977**, *99*, 303–304. [CrossRef]
14. Halmann, M. Photoelectrochemical reduction of aqueous carbon dioxide on p-type gallium phosphide in liquid junction solar cells. *Nature* **1978**, *275*, 115–116. [CrossRef]
15. Anpo, M.; Chiba, K.; Tomonari, M.; Coluccia, S.; Che, M.; Fox, M.A. Photocatalysis on Native and Platinum-Loaded TiO_2 and ZnO Catalysts-Origin of Different Reactivities on Wet and Dry Metal Oxides. *Bull. Chem. Soc. Jpn.* **1991**, *64*, 543–551. [CrossRef]
16. Domen, K.; Naito, S.; Soma, M.; Onishi, T.; Tamaru, K. Photocatalytic decomposition of water vapour on an $NiO\text{-}SrTiO_3$ catalyst. *J. Chem. Soc. Chem. Commun.* **1980**, *12*, 543–544. [CrossRef]
17. Boonstra, A.H.; Mutsaers, C. Relation between the photoadsorption of oxygen and the number of hydroxyl groups on a titanium dioxide surface. *J. Phys. Chem.* **1975**, *79*, 1694–1698. [CrossRef]
18. Yun, C.; Anpo, M.; Kubokawa, Y. UV irradiation-induced fission of a C=C or C≡C bond adsorbed on TiO_2. *J. Chem. Soc. Chem. Commun.* **1980**, 609. [CrossRef]
19. Anpo, M.; Nakaya, H.; Kodama, S.; Kubokawa, Y.; Domen, K.; Onishi, T. Photocatalysis over binary metal oxides. Enhancement of the photocatalytic activity of titanium dioxide in titanium-silicon oxides. *J. Phys. Chem.* **1986**, *90*, 1633–1636. [CrossRef]

20. Anpo, M.; Kawamura, T.; Kodama, S.; Maruya, K.; Onishi, T. Photocatalysis on titanium-aluminum binary metal oxides: Enhancement of the photocatalytic activity of titania species. *J. Phys. Chem.* **1988**, *92*, 438–440. [CrossRef]

21. Dohshi, S.; Takeuchi, M.; Anpo, M. Photoinduced superhydrophilic properties of Ti-B binary oxide thin films and their photocatalytic reactivity for the decomposition of NO. *J. Nanosci. Nanotechnol.* **2001**, *1*, 337–342. [CrossRef] [PubMed]

22. Liu, Y.; Li, Z.; Green, M.; Just, M.; Li, Y.Y.; Chen, X. Titanium dioxide nanomaterials for photocatalysis. *J. Phys. D: Appl. Phys.* **2017**, *50*, 193003. [CrossRef]

23. Jenny, S.; Matsuoka, M.; Takeuchi, M.; Zhang, J.; Horiuchi, Y.; Anpo, M.; Detlef, W. Bahnemann. Understanding TiO_2 photocatalysis: Mechanisms and materials. *Chem. Rev.* **2014**, *114*, 9919–9986.

24. Hashimoto, K.; Irie, H.; Fujishima, A. TiO_2 photocatalysis: A historical overview and future prospects. *Jpn. J. Appl. Phys.* **2005**, *44*, 8269. [CrossRef]

25. Coronado, D.R.; Gattorno, G.R.; Pesqueira, M.E.E.; Cab, C.; de Coss, R.; Oskam, G. Phase-pure TiO_2 nanoparticles: Anatase, brookite and rutile. *Nanotechnology* **2008**, *19*, 145605. [CrossRef] [PubMed]

26. Zhang, J.; Zhou, P.; Liu, J.; Yu, J. New understanding of the difference of photocatalytic activity among anatase, rutile and brookite TiO_2. *Phys. Chem. Chem. Phys.* **2014**, *16*, 20382–20386.

27. Linsebigler, A.L.; Lu, G.; Yates, J.T., Jr. Photocatalysis on TiO_2 surfaces: Principles, mechanisms, and selected results. *Chem. Rev.* **1995**, *95*, 735–758. [CrossRef]

28. Girish Kumar, S.; Gomathi Devi, L. Review on modified TiO_2 photocatalysis under UV/visible light: Selected results and related mechanisms on interfacial charge carrier transfer dynamics. *J. Phys. Chem. A.* **2011**, *115*, 13211–13241. [CrossRef]

29. de Lasa, H.; Serrano, B.; Salaices, M. Establishing Photocatalytic Kinetic Rate Equations: Basic Principles and Parameters. In *Photocatalytic Reaction Engineering*; Springer: Boston, MA, USA, 2005.

30. Ravelli, D.; Dondi, D.; Fagnonia, M.; Albini, A. Photocatalysis. A multi-faceted concept for green chemistry. *Chem. Soc. Rev.* **2009**, *38*, 1999–2011. [CrossRef]

31. Nakata, K.; Fujishima, A. TiO_2 photocatalysis: Design and applications. *J. Photochem. Photobio. C Photochem. Rev.* **2012**, *13*, 169–189. [CrossRef]

32. Parrino, F.; De Pasquale, C.; Palmisano, L. Influence of surface-related phenomena on mechanism, selectivity, and conversion of TiO_2-induced photocatalytic reactions. *ChemSusChem* **2019**, *12*, 589–602. [CrossRef] [PubMed]

33. Fujishima, A.; Zhang, X.; Tryk, D.A. TiO_2 photocatalysis and related surface phenomena. *Surf. Sci. Rep.* **2008**, *63*, 515–582. [CrossRef]

34. Pan, X.; Yang, M.Q.; Fu, X.; Zhang, N.; Xu, Y.-J. Defective TiO_2 with oxygen vacancies: Synthesis, properties and photocatalytic applications. *Nanoscale* **2013**, *5*, 3601–3614. [CrossRef] [PubMed]

35. Scanlon, D.O.; Dunnill, C.W.; Buckeridge, J.; Shevlin, S.A.; Logsdail, A.J.; Woodley, S.M.; Catlow, C.R.; Powell, M.J.; Palgrave, R.G.; Parkin, I.P.; et al. Band alignment of rutile and anatase TiO_2. *Nat. Mater.* **2013**, *12*, 798–801. [CrossRef] [PubMed]

36. Guo, Q.; Zhou, C.; Ma, Z.; Ren, Z.; Fan, H.; Yang, X. Fundamental Processes in Surface Photocatalysis on TiO_2. In *Heterogeneous Photocatalysis*; Green Chemistry and Sustainable Technology; Colmenares, J., Xu, Y.J., Eds.; Springer: Berlin/Heidelberg, Germany, 2016.

37. Yalavarthi, R.; Naldoni, A.; Kment, S.; Mascaretti, L.; Kmentová, H.; Tomanec, O.; Schmuki, P.; Zboril, R. Radiative and non-radiative recombination pathways in mixed-phase TiO_2 nanotubes for PEC water-splitting. *Catalysts* **2019**, *9*, 204. [CrossRef]

38. Nguyen, C.C.; Vu, N.N.; Do, T.O. Efficient hollow double-shell photocatalysts for the degradation of organic pollutants under visible light and in darkness. *J. Mater. Chem. A* **2016**, *4*, 4413–4419. [CrossRef]

39. Sakar, M.; Nguyen, C.C.; Vu, M.H.; Do, T.O. Materials and mechanisms of photo-assisted chemical reactions under light and dark: Can day-night photocatalysis be achieved? *ChemSusChem* **2018**, *11*, 809–820. [CrossRef]

40. Dinh, C.T.; Pham, M.H.; Seo, Y.; Kleitz, F.; Do, T.O. Design of multicomponent photocatalysts for hydrogen production under visible light using water-soluble titanate nanodisks. *Nanoscale* **2014**, *6*, 4819–4829. [CrossRef]

41. Dinh, C.T.; Seo, Y.; Nguyen, T.D.; Kleitz, F.; Do, T.O. Controlled synthesis of titanate nanodisks as versatile building blocks for the design of hybrid nanostructures. *Angew. Chem. Int. Ed.* **2012**, *51*, 6608–6612. [CrossRef]

42. Dinh, C.T.; Nguyen, T.D.; Kleitz, F.; Do, T.O. Shape-controlled synthesis of highly crystalline titania nanocrystals. *ACS Nano* **2009**, *11*, 3737–3743. [CrossRef]

43. Ng, J.; Pan, J.H.; Sun, D.D. Hierarchical assembly of anatase nanowhiskers and evaluation of their photocatalytic efficiency in comparison to various one-dimensional TiO_2 nanostructures. *J. Mater. Chem.* **2011**, *21*, 11844–11853. [CrossRef]

44. Conceição, D.S.; Ferreira, D.P.; Graça, C.A.L.; Julio, M.F.; Ilharco, L.M.; Velosa, A.C.; Santos, P.F.; Vieira Ferreira, L.F. Photochemical and photocatalytic evaluation of 1D titanate/TiO_2 based nanomaterials. *Appl. Surf. Sci.* **2017**, *392*, 418–429. [CrossRef]

45. Gordon, T.R.; Cargnello, M.; Paik, T.; Mangolini, F.; Weber, R.T.; Fornasiero, P.; Murray, C.B. Nonaqueous synthesis of TiO_2 nanocrystals using TiF_4 to engineer morphology, oxygen vacancy concentration, and photocatalytic activity. *J. Am. Chem. Soc.* **2012**, *134*, 6751–6761. [CrossRef] [PubMed]

46. Ziarati, A.; Badiei, A.; Luque, R. Black Hollow TiO_2 nanocubes: Advanced nanoarchitectures for efficient visible light photocatalytic applications. *Appl. Catal. B Environ.* **2018**, *238*, 177–183. [CrossRef]

47. Dinh, C.T.; Nguyen, T.D.; Kleitz, F.; Do, T.O. A novel single-step route based on solvothermal technique to shape-controlled titanium dioxide nanocrystals. *Can. J. Chem. Eng.* **2012**, *90*, 8–17. [CrossRef]

48. Kang, X.; Song, X.Z.; Han, Y.; Cao, J.; Tan, Z. Defect-engineered TiO_2 hollow spiny nanocubes for phenol degradation under visible light irradiation. *Sci. Rep.* **2018**, *8*, 5904. [CrossRef] [PubMed]

49. Zhou, X.; Liu, N.; Schmuki, P. Photocatalysis with TiO_2 nanotubes: "Colorful" reactivity and designing site-specific photocatalytic centers into TiO_2 nanotubes. *ACS Catal.* **2017**, *7*, 3210–3235. [CrossRef]

50. Zhao, Q.E.; Wen, W.; Xia, Y.; Wu, J.M. Photocatalytic activity of TiO_2 nanorods, nanowires and nanoflowers filled with TiO_2 nanoparticles. *Thin Solid Films* **2018**, *648*, 103–107. [CrossRef]

51. Nguyen, C.C.; Vu, N.N.; Do, T.O. Recent advances in the development of sunlight-driven hollow structure photocatalysts and their applications. *J. Mater. Chem. A* **2015**, *7*, 8187–8208. [CrossRef]

52. Li, M.; Chen, Y.; Li, W.; Li, X.; Tian, H.; Wei, X.; Ren, Z.; Han, G. Ultrathin anatase TiO_2 nanosheets for high-performance photocatalytic hydrogen production. *Small* **2017**, *13*, 1604115. [CrossRef]

53. Choi, S.K.; Kim, S.; Lim, S.K.; Park, H. Photocatalytic comparison of TiO_2 Nanoparticles and Electrospun TiO_2 nanofibers: Effects of mesoporosity and interparticle charge transfer. *J. Phys. Chem. C* **2010**, *114*, 16475–16480. [CrossRef]

54. Ribeiro, R.A.P.; de Lazaro, S.R.; de Oliveira, C.R. Band-Gap engineering for photocatalytic applications: Anionic and cationic doping of TiO_2 anatase. *Curr. Phys. Chem.* **2016**, *6*, 22–27. [CrossRef]

55. Yalçın, Y.; Kılıç, M.; Çınar, Z. The role of non-metal doping in TiO_2 photocatalysis. *J. Adv. Oxid. Technol.* **2016**, *13*, 281–296. [CrossRef]

56. Chen, X.; Burda, C. The electronic origin of the visible-light absorption properties of C-, N- and S-doped TiO_2 nanomaterials. *J. Am. Chem. Soc.* **2008**, *130*, 5018–5019. [CrossRef] [PubMed]

57. Emy, M.S.; Sharifah, B.A.H. Effect of band gap engineering in anionic-doped TiO_2 photocatalyst. *Appl. Surf. Sci.* **2017**, *391*, 326–336.

58. Wang, H.; Lewis, J.P. Second-generation photocatalytic materials: Anion-doped TiO_2. *J. Phys. Condens. Matter* **2006**, *18*, 421–434. [CrossRef]

59. Kuznetsov, V.N.; Serpone, N. on the origin of the spectral bands in the visible absorption spectra of visible-light-active TiO_2 specimens analysis and assignments. *J. Phys. Chem. C* **2009**, *113*, 15110–15123. [CrossRef]

60. Serpone, N. Is the band gap of pristine TiO_2 narrowed by anion- and cation-doping of titanium dioxide in second-generation photocatalysts? *J. Phys. Chem. B* **2006**, *110*, 24287–24293. [CrossRef]

61. Fan, W.Q.; Bai, H.Y.; Zhang, G.H.; Yan, Y.S.; Liu, C.B.; Shi, W.D. Titanium dioxide macroporous materials doped with iron: Synthesis and photo-catalytic properties. *CrystEngComm* **2014**, *16*, 116–122. [CrossRef]

62. Chang, S.M.; Liu, W.S. The roles of surface-doped metal ions (V, Mn, Fe, Cu, Ce, and W) in the interfacial behavior of TiO_2 photocatalysts. *Appl. Catal. B Environ.* **2014**, *156*, 466–475. [CrossRef]

63. Hahn, R.; Stark, M.; Killian, M.S.; Schmuki, P. Photocatalytic properties of in situ doped TiO_2-nanotubes grown by rapid breakdown anodization. *Catal. Sci. Technol.* **2013**, *3*, 1765–1770. [CrossRef]

64. Ishii, M.; Towlson, B.; Harako, S.; Zhao, X.W.; Komuro, S.; Hamilton, B. Roles of electrons and holes in the luminescence of rare-earth-doped semiconductors. *Electr. Commun. Jpn.* **2013**, *96*, 1–7. [CrossRef]

65. Tobaldi, D.M.; Pullar, R.C.; Gualtieri, A.F.; Seabra, M.P.; Labrincha, J.A. Sol-gel synthesis, characterisation and photocatalytic activity of pure, W-, Ag- and W/Ag codoped TiO_2 nanopowders. *Chem. Eng. J.* **2013**, *214*, 364–375. [CrossRef]

66. de Lima, J.F.; Harunsani, M.H.; Martin, D.J.; Kong, D.; Dunne, P.W.; Gianolio, D.; Kashtiban, R.J.; Sloan, J.; Serra, O.A.; Tang, J.; et al. Control of chemical state of cerium in doped anatase TiO_2 by solvothermal synthesis and its application in photocatalytic water reduction. *J. Mater. Chem. A* **2015**, *3*, 9890–9898. [CrossRef]

67. Li, H.; Liu, J.; Qian, J.; Li, Q.; Yang, J. Preparation of Bi-doped TiO_2 nanoparticles and their visible light photocatalytic performance. *Chin. J. Catal.* **2014**, *35*, 1578–1589. [CrossRef]

68. Klaysri, R.; Wichaidit, S.; Tubchareon, T.; Nokjan, S.; Piticharoenphun, S.; Mekasuwandumrong, O.; Praserthdam, P. Impact of calcination atmospheres on the physiochemical and photocatalytic properties of nanocrystalline TiO_2 and Si-doped TiO_2. *Ceram. Int.* **2015**, *41*, 11409–11417. [CrossRef]

69. Zhao, Y.; Liu, J.; Shi, L.; Yuan, S.; Fang, J.; Wang, Z.; Zhang, M. Solvothermal preparation of Sn^{4+} doped anatase TiO_2 nanocrystals from peroxo-metal-complex and their photocatalytic activity. *Appl. Catal. B Environ.* **2011**, *103*, 436–443. [CrossRef]

70. Carneiro, J.O.; Azevedo, S.; Fernandes, F.; Freitas, E.; Pereira, M.; Tavares, C.J.; Lanceros-Mendez, S.; Teixeira, V. Synthesis of iron-doped TiO_2 nanoparticles by ball-milling process: The influence of process parameters on the structural, optical, magnetic, and photocatalytic properties. *J. Mater. Sci.* **2014**, *49*, 7476–7488. [CrossRef]

71. Makdee, A.; Unwiset, P.; Chanapattharapol, K.C.; Kidkhunthod, P. Effects of Ce addition on the properties and photocatalytic activity of TiO_2, investigated by X-ray absorption spectroscopy. *Mater. Chem. Phys.* **2018**, *213*, 431–443. [CrossRef]

72. Unwiset, P.; Makdee, A.; Chanapattharapol, K.C.; Kidkhunthod, P. Effect of Cu addition on TiO_2 surface properties and photocatalytic performance: X-ray absorption spectroscopy analysis. *J. Phys. Chem. Solids* **2018**, *120*, 231–240. [CrossRef]

73. Zhang, D.R.; Liu, H.L.; Han, S.Y.; Piao, W.X. Synthesis of Sc and V-doped TiO_2 nanoparticles and photodegradation of rhodamine-B. *J. Indus. Eng. Chem.* **2013**, *19*, 1838–1844. [CrossRef]

74. Ould-Chikh, S.; Proux, O.; Afanasiev, P.; Khrouz, L.; Hedhili, M.N.; Anjum, D.H.; Harb, M.; Geantet, C.; Basset, J.; Puzenat, E. Photocatalysis with chromium-doped TiO_2: Bulk and surface doping. *ChemSusChem* **2014**, *7*, 1361–1371. [CrossRef] [PubMed]

75. Deng, Q.R.; Xia, X.H.; Guo, M.L.; Gao, Y.; Shao, G. Mn-doped TiO_2 nanopowders with remarkable visible light photocatalytic activity. *Mater. Lett.* **2011**, *65*, 2051–2054. [CrossRef]

76. Sun, L.; Zhai, J.; Li, H.; Zhao, Y.; Yang, H.; Yu, H. Study of homologous elements: Fe, Co, and Ni dopant effects on the photoreactivity of TiO_2 nanosheets. *ChemCatChem* **2014**, *6*, 339–347. [CrossRef]

77. Karunakaran, C.; Abiramasundari, G.; Gomathisankar, P.; Manikandan, G.; Anandi, V. Cu-doped TiO_2 nanoparticles for photocatalytic disinfection of bacteria under visible light. *J. Colloid Interface Sci.* **2010**, *352*, 68–74. [CrossRef] [PubMed]

78. Aware, D.V.; Jadhav, S.S. Synthesis, characterization and photocatalytic applications of Zn-doped TiO_2 nanoparticles by sol–gel method. *Appl. Nanosci.* **2016**, *6*, 965–972. [CrossRef]

79. Jiang, X.; Gao, Y.; Li, C.; You, F.; Yao, J.; Ji, Y. Preparation of hollow yttrium-doped TiO_2 microspheres with enhanced visible-light photocatalytic activity. *Mater. Res. Express* **2019**, *6*, 065510. [CrossRef]

80. Gao, B.; Lim, T.M.; Subagio, D.P.; Lim, T.T. Zr-doped TiO_2 for enhanced photocatalytic degradation of bisphenol A. *Appl. Catal. A Gen.* **2010**, *375*, 107–115. [CrossRef]

81. Kou, Y.; Yang, J.; Li, B.; Fu, S. Solar photocatalytic activities of porous Nb-doped TiO_2 microspheres by coupling with tungsten oxide. *Mater. Res. Bull.* **2015**, *63*, 105–111. [CrossRef]

82. Avilés-García, O.; Espino-Valencia, J.; Romero, R.; Rico-Cerda, J.L.; Arroyo-Albiter, M.; Natividad, R. W and Mo doped TiO_2: Synthesis, characterization and photocatalytic activity. *Fuel* **2017**, *198*, 31–41. [CrossRef]

83. Hao, H.Y.; He, C.X.; Tian, B.Z.; Zhang, J.L. Study of photocatalytic activity of Cd-doped mesoporous nanocrystalline TiO_2 prepared at low temperature. *Res. Chem. Intermed.* **2009**, *35*, 705. [CrossRef]

84. Chandan, H.R.; Sakar, M.; Ashesh, M.; Ravishankar, T.N.; Ramakrishnappa, T.; Sergio, R.T.; Geetha Balakrishna, R. Observation of oxo-bridged yttrium in TiO_2 nanostructures and their enhanced photocatalytic hydrogen generation under UV/Visible light irradiations. *Mater. Res. Bull.* **2018**, *104*, 212–219.

85. Štengl, V.; Bakardjieva, S.; Murafa, N. Preparation and photocatalytic activity of rare earth doped TiO$_2$ nanoparticles. *Mater. Chem. Phys.* **2009**, *114*, 217–226. [CrossRef]

86. Al-Maliki, F.J.; Al-Lamey, N.H. Synthesis of Tb-doped titanium dioxide nanostructures by sol–gel method for environmental photocatalysis applications. *J. Sol-Gel Sci. Technol.* **2017**, *81*, 276–283. [CrossRef]

87. Singh, K.; Harish, S.; Kristya, A.P.; Shivani, V.; Archana, J.; Navaneethan, M.; Shimomura, M.; Hayakawa, Y. Erbium doped TiO$_2$ interconnected mesoporous spheres as an efficient visible light catalyst for photocatalytic applications. *Appl. Surf. Sci.* **2018**, *449*, 755–763. [CrossRef]

88. Jiang, X.; Li, C.; Liu, S.; Zhang, F.; You, F.; Yao, C. The synthesis and characterization of ytterbium-doped TiO$_2$ hollow spheres with enhanced visible-light photocatalytic activity. *RSC Adv.* **2017**, *7*, 24598–24606. [CrossRef]

89. Shwetharani, R.; Sakar, M.; Chandan, H.R.; Geetha Balakrishna, R. Observation of simultaneous photocatalytic degradation and hydrogen evolution on the lanthanum modified TiO$_2$ nanostructures. *Mater. Lett.* **2018**, *218*, 262–265. [CrossRef]

90. Ravishankar, T.N.; Nagaraju, G.; Dupont, J. Photocatalytic activity of Li-doped TiO$_2$ nanoparticles: Synthesis via ionic liquid-assisted hydrothermal route. *Mater. Res. Bull.* **2016**, *78*, 103–111. [CrossRef]

91. Shivaraju, H.P.; Midhun, G.; Anil Kumar, K.M.; Pallavi, S.; Pallavi, N.; Behzad, S. Degradation of selected industrial dyes using Mg-doped TiO$_2$ polyscales under natural sun light as an alternative driving energy. *Appl. Water Sci.* **2017**, *7*, 3937–3948. [CrossRef]

92. Fu, W.; Ding, S.; Wang, Y.; Wu, L.; Zhang, D.; Pan, Z.; Wang, R.; Zhang, Z.; Qiu, S. F, Ca co-doped TiO$_2$ nanocrystals with enhanced photocatalytic activity. *Dalton Trans.* **2014**, *43*, 16160–16163. [CrossRef] [PubMed]

93. Xie, W.; Li, R.; Xu, Q. Enhanced photocatalytic activity of Se-doped TiO$_2$ under visible light irradiation. *Sci. Rep.* **2018**, *8*, 8752. [CrossRef] [PubMed]

94. Nguyen, C.C.; Dinh, C.T.; Do, T.O. Hollow Rh/Sr-codoped TiO$_2$ photocatalyst for efficient sunlight-driven organic compound degradation. *RSC Adv.* **2017**, *7*, 3480–3487. [CrossRef]

95. Murashkina, A.A.; Murzin, P.D.; Rudakova, A.V.; Ryabchuk, V.K.; Emeline, A.V.; Detlef, W. Bahnemann. Influence of the dopant concentration on the photocatalytic activity: Al-doped TiO$_2$. *J. Phys. Chem. C* **2015**, *119*, 24695–24703. [CrossRef]

96. Liqiang, J.; Honggang, F.; Baiqi, W.; Dejun, W.; Baifu, X.; Shudan, L.; Jiazhong, S. Effects of Sn dopant on the photoinduced charge property and photocatalytic activity of TiO$_2$ nanoparticles. *Appl. Catal. B Environ.* **2006**, *62*, 282–291. [CrossRef]

97. Nan, W.; Xing, L.; Yanling, Y.; Tingting, G.; Xiaoxuan, Z.; Siyang, J.; Tingting, Z.; Yi, S.; Zhiwei, Z. Enhanced photocatalytic degradation of sulfamethazine by Bi-doped TiO$_2$ nano-composites supported by powdered activated carbon under visible light irradiation. *Sep. Purif. Technol.* **2019**, *211*, 673–683.

98. Li, W. Influence of electronic structures of doped TiO$_2$ on their photocatalysis. *Phys. Status Solidi R* **2015**, *9*, 10–27. [CrossRef]

99. Boulbar, E.L.; Millon, E.; Leborgne, C.B.; Cachoncinlle, C.; Hakim, B.; Ntsoenzok, E. Optical properties of rare earth-doped TiO$_2$ anatase and rutile thin films grown by pulsed-laser deposition. *Thin Solid Films* **2014**, *553*, 13–16. [CrossRef]

100. Shwetharani, R.; Sakar, M.; Fernando, C.A.N.; Binas, V.; Geetha Balakrishna, R. Recent advances and strategies applied to tailor energy levels, active sites and electron mobility in titania and its doped/composite analogues for hydrogen evolution in sunlight. *Catal. Sci. Technol.* **2019**, *9*, 12–46. [CrossRef]

101. Ma, Y.T.; Li, S.D. Photocatalytic activity of TiO$_2$ nanofibers with doped La prepared by electrospinning method. *J. Chin. Chem. Soc.* **2015**, *62*, 380–384. [CrossRef]

102. Borlaf, M.; Colomer, M.T.; de Andrés, A.; Cabello, F.; Serna, R.; Moreno, R. TiO$_2$/Eu^{3+} thin films with high photoluminescence emission prepared by electrophoretic deposition from nanoparticulate sols. *Eur. J. Inorg. Chem.* **2014**, *30*, 5152–5159. [CrossRef]

103. Du, J.; Li, B.; Huang, J.; Zhang, W.; Peng, H.; Zou, J. Hydrophilic and photocatalytic performances of lanthanum doped titanium dioxide thin films. *J. Rare Earth* **2013**, *31*, 992–996. [CrossRef]

104. Maddila, S.; Oseghe, E.O.; Jonnalagadda, S.B. Photocatalyzed ozonation by Ce doped TiO$_2$ catalyst degradation of pesticide Dicamba in water. *J. Chem. Technol. Biotechnol.* **2016**, *91*, 385–393. [CrossRef]

105. Choudhury, B.; Borah, B.; Choudhury, A. Extending photocatalytic activity of TiO$_2$ nanoparticles to visible region of illumination by doping of cerium. *Photochem. Photobiol.* **2012**, *88*, 257–264. [CrossRef] [PubMed]

106. Vu, T.T.D.; Mighri, F.; Ajji, A.; Do, T.O. Synthesis of titanium dioxide/cadmium sulfide nanosphere particles for photocatalyst applications. *Ind. Eng. Chem. Res.* **2014**, *53*, 3888–3897. [CrossRef]

107. Wang, H.; Zhang, L.; Chen, Z.; Hu, J.; Li, S.; Wang, Z.; Liu, J.; Wang, X. Semiconductor heterojunction photocatalysts: Design, construction, and photocatalytic performances. *Chem. Soc. Rev.* **2014**, *43*, 5234–5244. [CrossRef] [PubMed]

108. Afroz, K.; Moniruddin, M.; Bakranov, N.; Kudaibergenov, S.; Nuraje, N. A heterojunction strategy to improve the visible light sensitive water splitting performance of photocatalytic materials. *J. Mater. Chem. A* **2018**, *6*, 21696–21718. [CrossRef]

109. Bessekhouad, Y.; Robert, D.; Weber, J.V. Photocatalytic activity of Cu_2O/TiO_2, Bi_2O_3/TiO_2 and $ZnMn_2O_4/TiO_2$ heterojunctions. *Catal. Today* **2005**, *101*, 315–321. [CrossRef]

110. Liu, J.; Yang, S.; Wu, W.; Tian, Q.; Cui, S.; Dai, Z.; Ren, F.; Xiao, X.; Jiang, C. 3D Flowerlike α-Fe_2O_3@TiO_2 core-shell nanostructures: General synthesis and enhanced photocatalytic performance. *ACS Sustain. Chem. Eng.* **2015**, *3*, 2975–2984. [CrossRef]

111. Xia, Y.; Yin, L. Core-shell structured α-Fe_2O_3@TiO_2 nanocomposites with improved photocatalytic activity in visible light region. *Phys. Chem. Chem. Phys.* **2013**, *15*, 18627–18634. [CrossRef]

112. Lin, Z.; Liu, P.; Yan, J.; Yang, G. Matching energy levels between TiO_2 and α-Fe_2O_3 in a core-shell nanoparticle for the visible-light photocatalysis. *J. Mater. Chem. A* **2015**, *3*, 14853–14863. [CrossRef]

113. Moniz, S.J.A.; Shevlin, S.A.; Martin, D.J.; Guo, Z.X.; Tang, J. Visible-light driven heterojunction photocatalysts for water splitting—A critical review. *Energy Environ. Sci.* **2015**, *8*, 731–759. [CrossRef]

114. Ge, J.; Zhang, Y.; Heo, Y.J.; Park, S.J. Advanced design and synthesis of composite photocatalysts for the remediation of wastewater: A review. *Catalysts* **2019**, *9*, 122. [CrossRef]

115. 115. Danlian, H.; Sha, C.; Zeng, G.; Gong, X.; Zhou, C.; Cheng, M.; Xue, W.; Yan, X.; Li, J. Artificial Z-scheme photocatalytic system: What have been done and where to go? *Coordination Chem. Rev.* **2019**, *385*, 44–80.

116. Bard, A.J. Photoelectrochemistry and heterogeneous photo-catalysis at semiconductors. *J. Photochem.* **1979**, *10*, 59–75. [CrossRef]

117. Jiang, W.; Zong, X.; An, L.; Hua, S.; Miao, X.; Luan, S.; Wen, Y.; Tao, F.F.; Sun, Z. Consciously constructing heterojunction or direct z-scheme photocatalysts by regulating electron flow direction. *ACS Catal.* **2018**, *8*, 2209–2217. [CrossRef]

118. Xu, Q.; Zhang, L.; Yu, J.; Wageh, S.; Al-Ghamdi, A.A.; Jaroniec, M. Direct Z-scheme photocatalysts: Principles, synthesis, and applications. *Mater. Today* **2018**, *21*, 1042–1063. [CrossRef]

119. Qi, K.; Cheng, B.; Yu, J.; Ho, W. A review on TiO_2-based Z-scheme photocatalysts. *Chin. J. Catal.* **2017**, *38*, 1936–1955. [CrossRef]

120. Li, H.; Tu, W.; Zhou, Y.; Zou, Z. Z-scheme photocatalytic systems for promoting photocatalytic performance: Recent progress and future challenges. *Adv. Sci.* **2016**, *3*, 1500389. [CrossRef]

121. Meng, A.; Zhu, B.; Zhong, B.; Zhang, L.; Cheng, B. Direct Z-scheme TiO_2/CdS hierarchical photocatalyst for enhanced photocatalytic H_2-production activity. *Appl. Surface Sci.* **2017**, *422*, 518–527. [CrossRef]

122. Dinh, C.T.; Pham, M.H.; Kleitz, F.; Do, T.O. Design of water-soluble CdS-titanate-nickel nanocomposites for photocatalytic hydrogen production under sunlight. *J. Mater. Chem. A* **2013**, *1*, 13308–13313. [CrossRef]

123. Jo, W.K.; Natarajan, T.S. Influence of TiO_2 morphology on the photocatalytic efficiency of direct Z-scheme g-C_3N_4/TiO_2 photocatalysts for isoniazid degradation. *Appl. Surf. Sci.* **2017**, *422*, 518–527.

124. Zhou, D.; Chen, Z.; Yang, Q.; Dong, X.; Zhang, J.; Qin, L. In-situ construction of all-solid-state Z-scheme g-C_3N_4/TiO_2 nanotube arrays photocatalyst with enhanced visible-light-induced properties. *Solar Energy Mater. Solar Cells* **2016**, *157*, 399–405. [CrossRef]

125. Liao, W.; Murugananthan, M.; Zhang, Y. Synthesis of Z-scheme g-C_3N_4-Ti^{3+}/TiO_2 material: An efficient visible light photoelectrocatalyst for degradation of phenol. *Phys. Chem. Chem. Phys.* **2015**, *17*, 8877–8884. [CrossRef] [PubMed]

126. Xu, F.; Zhang, L.; Cheng, B.; Yu, J. Direct Z-scheme TiO_2/NiS core-shell hybrid nanofibers with enhanced photocatalytic H_2-production activity. *ACS Sustain. Chem. Eng.* **2018**, *6*, 12291–12298. [CrossRef]

127. Li, Q.; Xia, Y.; Yang, C.; Lv, K.; Lei, M.; Li, M. Building a direct Z-scheme heterojunction photocatalyst by $ZnIn_2S_4$ nanosheets and TiO_2 hollowspheres for highly-efficient artificial photosynthesis. *Chem. Eng. J.* **2018**, *349*, 287–296. [CrossRef]

128. Fu, J.; Cao, S.; Yu, J. Dual Z-scheme charge transfer in TiO_2-Ag-Cu_2O composite for enhanced photocatalytic hydrogen generation. *J. Materiomics* **2015**, *1*, 124–133. [CrossRef]

129. Pan, L.; Zhang, J.; Jia, X.; Ma, Y.H.; Zhang, X.; Wang, L.; Zou, J.J. Highly efficient Z-scheme WO_{3-x} quantum dots/TiO_2 for photocatalytic hydrogen generation. *Chin. J. Catal.* **2017**, *38*, 253–259. [CrossRef]

130. Zhang, X.; Chen, Y.L.; Liu, R.S.; Tsai, D.P. Plasmonic photocatalysis. *Rep. Prog. Phys.* **2013**, *76*, 046401. [CrossRef] [PubMed]

131. Wang, P.; Huang, B.; Dai, Y.; Whangbo, M.H. Plasmonic photocatalysts: Harvesting visible light with noble metal nanoparticles. *Phys. Chem. Chem. Phys.* **2012**, *14*, 9813–9825. [CrossRef]

132. Wu, J.; Zhang, Z.; Liu, B.; Fang, Y.; Wang, L.; Dong, B. UV-Vis-NIR-driven plasmonic photocatalysts with dual-resonance modes for synergistically enhancing H_2 generation. *Sol. RRL* **2018**, *2*, 1800039. [CrossRef]

133. Dinh, C.T.; Hoang, Y.; Kleitz, F.; Do, T.O. Three-dimensional ordered assembly of thin-shell Au/TiO_2 hollow nanospheres for enhanced visible-light-driven photocatalysis. *Angew. Chem. Int. Ed.* **2014**, *53*, 6618–6623. [CrossRef] [PubMed]

134. He, Y.; Basnet, P.; Hunyadi Murph, S.E.; Zhao, Y. Ag nanoparticle embedded TiO_2 composite nanorod arrays fabricated by oblique angle deposition: Toward plasmonic photocatalysis. *ACS Appl. Mater. Interfaces* **2013**, *5*, 11818–11827. [CrossRef] [PubMed]

135. Chen, Z.; Fang, L.; Dong, W.; Zheng, F.; Shena, M.; Wang, J. Inverse opal structured Ag/TiO_2 plasmonic photocatalyst prepared by pulsed current deposition and its enhanced visible light photocatalytic activity. *J. Mater. Chem. A* **2014**, *2*, 824–832. [CrossRef]

136. Dinh, C.T.; Nguyen, T.D.; Kleitz, F.; Do, T.O. A new route to size and population control of silver clusters on colloidal TiO_2 nanocrystals. *ACS Appl. Mater. Interfaces* **2011**, *3*, 2228–2234. [CrossRef] [PubMed]

137. Hirakawa, T.; Kamat, P.V. Charge separation and catalytic activity of Ag@TiO_2 core-shell composite clusters under UV-irradiation. *J. Am. Chem. Soc.* **2005**, *127*, 3928–3934. [CrossRef] [PubMed]

138. Luth, H. *Solid Surfaces, Interfaces, and Films*; Springer: Berlin, Germany, 2001.

139. Kozlov, D.A.; Lebedev, V.A.; Polyakov, A.Y.; Khazova, K.M.; Garshev, A.V. The microstructure effect on the Au/TiO_2 and Ag/TiO_2 nanocomposites photocatalytic activity. *Nanosyst. Phys. Chem. Math.* **2018**, *9*, 266–278. [CrossRef]

140. Leong, K.H.; Chu, H.Y.; Ibrahim, S.; Saravanan, P. Palladium nanoparticles anchored to anatase TiO_2 for enhanced surface plasmon resonance-stimulated, visible-light-driven photocatalytic activity. *Beilstein J. Nanotechnol.* **2015**, *6*, 428–437. [CrossRef]

141. Tapin, B.; Epron, F.; Especel, C.; Ly, B.K.; Pinel, C.; Besson, M. Study of monometallic Pd/TiO_2 catalysts for the hydrogenation of succinic acid in aqueous phase. *ACS Catal.* **2013**, *3*, 2327–2335. [CrossRef]

142. Keihan, A.H.; Rasoulnezhad, H.; Mohammadgholi, A.; Sajjadi, A.; Hosseinzadeh, R.; Farhadian, M.; Hosseinzadeh, G. Pd nanoparticle loaded TiO_2 semiconductor for photocatalytic degradation of Paraoxon pesticide under visible-light irradiation. *J. Mater. Sci. Mater. Electron.* **2017**, *28*, 16718–16727. [CrossRef]

143. Pham, M.H.; Dinh, C.T.; Vuong, G.T.; Do, T.O. General route toward hollow photocatalyst with two cocatalysts separated on two surface sides for hydrogen generation. *Phys. Chem. Chem. Phys.* **2014**, *16*, 5937–5941. [CrossRef]

144. Galinska, A.; Walendziewski, J. Photocatalytic water splitting over Pt-TiO_2 in the presence of sacrificial reagents. *Energy Fuels* **2005**, *19*, 1143–1147. [CrossRef]

145. Nguyen, C.C.; Nguyen, D.T.; Do, T.O. A novel route to synthesize C/Pt/TiO_2 phase tunable anatase-rutile TiO_2 for efficient sunlight-driven photocatalytic applications. *Appl. Catal. B Environ.* **2018**, *226*, 46–52. [CrossRef]

146. Liu, K.; Litke, A.; Su, Y.; van Campenhout, B.G.; Pidko, E.A.; Hensen, E.J.M. Photocatalytic decarboxylation of lactic acid by Pt/TiO_2. *Chem. Commun.* **2016**, *52*, 11634–11637. [CrossRef] [PubMed]

147. Fang, L.; You, L.; Liu, J. Ferroelectrics in Photocatalysis. In *Ferroelectric Materials for Energy Applications*; Huang, H., Scott, J.F., Eds.; Wiley-VCH Verlag GmbH & Co. KGaA: Weinheim, Germany, 2019.

148. Cui, Y.; Briscoe, J.; Dunn, S. Effect of ferroelectricity on solar-light-driven photocatalytic activity of $BaTiO_3$-influence on the carrier separation and stern layer formation. *Chem. Mater.* **2013**, *25*, 4215–4223. [CrossRef]

149. Jones, P.M.; Dunn, S. Photo-reduction of silver salts on highly heterogeneous lead zirconate titanate. *Nanotechnology* **2007**, *18*, 185702. [CrossRef]

150. Park, S.; Lee, C.W.; Kang, M.G.; Kim, S.; Kim, H.J.; Kwon, J.E.; Park, S.Y.; Kang, C.Y.; Hong, K.S.; Nam, K.T. A ferroelectric photocatalyst for enhancing hydrogen evolution: Polarized particulate suspension. *Phys. Chem. Chem. Phys.* **2014**, *16*, 10408–10413. [CrossRef] [PubMed]

151. Li, L.; Salvador, P.A.; Rohrer, G.S. Photocatalysts with internal electric fields. *Nanoscale* **2014**, *6*, 24–42. [CrossRef] [PubMed]

152. Huang, H.; Tu, S.; Du, X.; Zhang, Y. Ferroelectric spontaneous polarization steering charge carriers migration for promoting photocatalysis and molecular oxygen activation. *J. Colloid Interface Sci.* **2018**, *509*, 113–122. [CrossRef]

153. Cui, Y.; Briscoe, J.; Wang, Y.; Tarakina, N.V.; Dunn, S. Enhanced photocatalytic activity of heterostructured ferroelectric $BaTiO_3/\alpha$-Fe_2O_3 and the significance of interface morphology control. *ACS Appl. Mater. Interfaces* **2017**, *9*, 24518–24526. [CrossRef]

154. Fu, Q.; Wang, X.; Li, C.; Sui, Y.; Han, Y.; Lv, Z.; Song, B.; Xu, P. Enhanced photocatalytic activity on polarized ferroelectric $KNbO_3$. *RSC Adv.* **2016**, *6*, 108883–108887. [CrossRef]

155. Al-keisy, A.; Ren, L.; Cui, D.; Xu, Z.; Xu, X.; Su, X.; Hao, W.; Doua, S.X.; Du, Y. A ferroelectric photocatalyst Ag10Si4O13 with visible-light photooxidation properties. *J. Mater. Chem. A* **2016**, *4*, 10992–10999. [CrossRef]

156. Sakar, M.; Balakumar, S.; Saravanan, P.; Bharathkumar, S. Compliments of confinements: Substitution and dimension induced magnetic origin and band-bending mediated photocatalytic enhancements in $Bi_{1-x}Dy_xFeO_3$ particulate and fiber nanostructures. *Nanoscale* **2015**, *7*, 10667–10679. [CrossRef] [PubMed]

157. Sakar, M.; Balakumar, S.; Saravanan, P.; Bharathkumar, S. Particulates vs. fibers: Dimension featured magnetic and visible light driven photocatalytic properties of Sc modified multiferroic bismuth ferrite nanostructures. *Nanoscale* **2016**, *8*, 1147–1160. [CrossRef] [PubMed]

158. Sakar, M.; Balakumar, S.; Bhaumik, I.; Gupta, P.K.; Jaisankar, S.N. Nanostructured $Bi_{(1-x)}Gd_{(x)}FeO_3$-A multiferroic photocatalyst on its sunlight driven photocatalytic activity. *RSC Adv.* **2014**, *4*, 16871–16878.

159. Li, R.; Li, Q.; Zong, L.; Wang, X.; Yang, J. $BaTiO_3/TiO_2$ heterostructure nanotube arrays for improved photoelectrochemical and photocatalytic activity. *Electrochim. Acta* **2013**, *91*, 30–35. [CrossRef]

160. Küçük, O.; Teber, S.; Kaya, I.C.; Akyildiz, H.; Kalem, V. Photocatalytic activity and dielectric properties of hydrothermally derived tetragonal $BaTiO_3$ nanoparticles using TiO_2 nanofibers. *J. Alloys Compd.* **2018**, *765*, 82–91. [CrossRef]

161. Li, Q.; Li, R.; Zong, L.; He, J.; Wang, X.; Yang, J. Photoelectrochemical and photocatalytic properties of Ag-loaded $BaTiO_3/TiO_2$ heterostructure nanotube arrays. *Int. J. Hydrog. Energy* **2013**, *38*, 12977–12983. [CrossRef]

162. Zhang, Y.; Salvador, P.A.; Rohrer, G.S. Ferroelectric-enhanced photocatalysis with $TiO_2/BiFeO_3$. In *Energy Technology 2014*; Wang, C., de Bakker, J., Belt, C.K., Jha, A., Neelameggham, N.R., Pati, S., Prentice, L.H., Tranell, G., Brinkman, K.S., Eds.; The Minerals, Metals & Materials Society: Pittsburgh, PA, USA, 2014.

163. Li, S.; Lin, Y.H.; Zhang, B.P.; Li, J.F.; Nan, C.W. $BiFeO_3/TiO_2$ core-shell structured nanocomposites as visible-active photocatalysts and their optical response mechanism. *J. Appl. Phys.* **2009**, *105*, 054310. [CrossRef]

164. Liu, G.; Ma, L.; Yin, L.C.; Wan, G.; Zhu, H.; Zhen, C.; Yang, Y.; Liang, Y.; Tan, J.; Cheng, H.M. Selective chemical epitaxial growth of TiO_2 islands on ferroelectric $PbTiO_3$ crystals to boost photocatalytic activity. *Joule* **2018**, *2*, 1–13. [CrossRef]

165. Leary, R.; Westwood, A. Carbonaceous nanomaterials for the enhancement of TiO_2 photocatalysis. *Carbon* **2011**, *49*, 741–772. [CrossRef]

166. Khalid, N.R.; Majid, A.; Tahir, M.B.; Niaz, N.A.; Khalid, S. Carbonaceous-TiO_2 nanomaterials for photocatalytic degradation of pollutants: A review. *Ceram. Int.* **2017**, *43*, 14552–14571. [CrossRef]

167. Sakthivel, S.; Kisch, H. Daylight photocatalysis by carbon-modified titanium dioxide. *Angew. Chem. Int. Ed.* **2003**, *42*, 4908–4911. [CrossRef] [PubMed]

168. Yu, H.; Zhao, Y.; Zhou, C.; Shang, L.; Peng, Y.; Cao, Y.; Wu, L.Z.; Tunga, C.H.; Zhang, T. Carbon quantum dots/TiO_2 composites for efficient photocatalytic hydrogen evolution. *J. Mater. Chem. A* **2014**, *2*, 3344–3351. [CrossRef]

169. Yu, X.; Liu, J.; Yu, Y.; Zuo, S.; Li, S. Preparation and visible light photocatalytic activity of carbon quantum dots/TiO_2 nanosheet composites. *Carbon* **2014**, *68*, 718–724. [CrossRef]

170. Sathish Kumar, M.; Yamini Yasoda, K.; Kumaresan, D.; Kothurkar, N.K.; Batabyal, S.K. TiO_2-carbon quantum dots (CQD) nanohybrid: Enhanced photocatalytic activity. *Mater. Res. Express* **2018**, *5*, 075502. [CrossRef]

171. Ren, Y.; Dong, Y.; Feng, Y.; Xu, J. Compositing two-dimensional materials with TiO_2 for photocatalysis. *Catalysts* **2018**, *8*, 590. [CrossRef]

172. Baca, M.; Kukułka, W.; Cendrowski, K.; Mijowska, E.; Kaleńczuk, R.J.; Zielińska, B. Graphitic carbon nitride and titanium dioxide modified with 1 D and 2 D carbon structures for photocatalysis. *ChemSusChem* **2019**, *12*, 612–620. [CrossRef] [PubMed]

173. Olowoyo, J.O.; Kumar, M.; Jain, S.L.; Babalola, J.O.; Vorontsov, A.V.; Kumar, U. Insights into reinforced photocatalytic activity of the CNT-TiO$_2$ nanocomposite for co2 reduction and water splitting. *J. Phys. Chem. C* **2019**, *123*, 367–378. [CrossRef]

174. Shaban, M.; Ashraf, A.M.; Abukhadra, M.R. TiO$_2$ nanoribbons/carbon nanotubes composite with enhanced photocatalytic activity; fabrication, characterization, and application. *Sci. Rep.* **2018**, *8*, 781. [CrossRef]

175. Li, X.; Lin, H.; Chen, X.; Niu, H.; Zhang, T.; Liua, J.; Qu, F. Fabrication of TiO$_2$/porous carbon nanofibers with superior visible photocatalytic activity. *New J. Chem.* **2015**, *39*, 7863–7872. [CrossRef]

176. Kim, S.; Lim, S.K. Preparation of TiO$_2$-embedded carbon nanofibers and their photocatalytic activity in the oxidation of gaseous acetaldehyde. *Appl. Catal. B Environ.* **2008**, *84*, 16–20. [CrossRef]

177. Cao, L.; Sahu, S.; Anilkumar, P.; Bunker, C.E.; Xu, J.; Shiral Fernando, K.A.; Wang, P.; Guliants, E.A.; Tackett, K.N.; Sun, Y.P. Carbon nanoparticles as visible-light photocatalysts for efficient CO$_2$ conversion and beyond. *J. Am. Chem. Soc.* **2011**, *133*, 4754–4757. [CrossRef] [PubMed]

178. Zhong, J.; Chen, F.; Zhang, J. Carbon-deposited TiO$_2$: Synthesis, characterization, and visible photocatalytic performance. *J. Phys. Chem. C* **2010**, *114*, 933–939. [CrossRef]

179. Zhang, W.D.; Bin Xu, B.; Jiang, L.C. Functional hybrid materials based on carbon nanotubes and metal oxides. *J. Mater. Chem.* **2010**, *20*, 6383–6391. [CrossRef]

180. Yu, J.; Wang, S.; Lowa, J.; Xiao, W. Enhanced photocatalytic performance of direct Z-scheme g-C$_3$N$_4$-TiO$_2$ photocatalysts for the decomposition of formaldehyde in air. *Phys. Chem. Chem. Phys.* **2013**, *15*, 16883–16890. [CrossRef] [PubMed]

181. Tan, L.L.; Ong, W.J.; Chai, S.P.; Mohamed, A.R. Reduced graphene oxide-TiO$_2$ nanocomposite as a promising visible-light-active photocatalyst for the conversion of carbon dioxide. *Nanoscale Res. Lett.* **2013**, *8*, 465. [CrossRef] [PubMed]

182. Tan, L.; Chai, S.; Mohamed, A.R. Synthesis and applications of graphene-based TiO$_2$ photocatalysts. *ChemSusChem* **2012**, *5*, 1868–1882. [CrossRef] [PubMed]

183. Nasr, M.; Balme, S.; Eid, C.; Habchi, R.; Miele, P.; Bechelany, M. Enhanced visible-light photocatalytic performance of electrospun rGO/TiO$_2$ composite nanofibers. *J. Phys. Chem. C* **2017**, *121*, 261–269. [CrossRef]

184. Pan, X.; Zhao, Y.; Liu, S.; Korzeniewski, C.L.; Wang, S.; Fan, Z. Comparing graphene-TiO$_2$ nanowire and graphene-TiO$_2$ nanoparticle composite photocatalysts. *ACS Appl. Mater. Interfaces* **2012**, *4*, 3944–3950. [CrossRef] [PubMed]

185. Yu, H.; Xiao, P.; Tian, J.; Wang, F.; Yu, J. Phenylamine-functionalized rGO/TiO$_2$ photocatalysts: Spatially separated adsorption sites and tunable photocatalytic selectivity. *ACS Appl. Mater. Interfaces* **2016**, *8*, 29470–29477. [CrossRef] [PubMed]

186. Luo, B.; Liu, G.; Wang, L. Recent advances in 2D materials for photocatalysis. *Nanoscale* **2016**, *8*, 6904–6920. [CrossRef] [PubMed]

187. Singh, A.K.; Mathew, K.; Zhuang, H.L.; Hennig, R.G. Computational screening of 2D materials for photocatalysis. *J. Phys. Chem. Lett.* **2015**, *6*, 1087–1098. [CrossRef] [PubMed]

188. Zhang, C.; Huang, H.; Ni, X.; Zhou, Y.; Kang, L.; Jiang, W.; Chen, H.; Zhong, J.; Liu, F. Band gap reduction in van der Waals layered 2D materials via a de-charge transfer mechanism. *Nanoscale* **2018**, *10*, 16759–16764. [CrossRef] [PubMed]

189. Ida, S.; Ishihara, T. Recent progress in two-dimensional oxide photocatalysts for water splitting. *J. Phys. Chem. Lett.* **2014**, *5*, 2533–2542. [CrossRef] [PubMed]

190. Nguyen, C.C.; Vu, N.N.; Chabot, S.; Kaliaguine, S.; Do, T.O. Role of C$_x$N$_y$-triazine in photocatalysis for efficient hydrogen generation and organic pollutant degradation under solar light irradiation. *Solar RRL* **2017**, *1*, 1700012. [CrossRef]

191. Haque, F.; Daeneke, T.; Kalantar-zadeh, K.; Ou, J.Z. Two-dimensional transition metal oxide and chalcogenide-based photocatalysts. *Nano-Micro Lett.* **2018**, *10*, 23. [CrossRef] [PubMed]

192. Guo, L.; Yang, Z.; Marcus, K.; Li, Z.; Luo, B.; Zhou, L.; Wang, X.; Du, Y.; Yang, Y. MoS$_2$/TiO$_2$ heterostructures as nonmetal plasmonic photocatalysts for highly efficient hydrogen evolution. *Energy Environ. Sci.* **2018**, *11*, 106–114. [CrossRef]

193. Sun, Y.; Lin, H.; Wang, C.; Wu, Q.; Wang, X.; Yang, M. Morphology-controlled synthesis of TiO_2/MoS_2 nanocomposites with enhanced visible-light photocatalytic activity. *Inorg. Chem. Front.* **2018**, *5*, 145–152. [CrossRef]

194. Dong, Y.; Chen, S.Y.; Lu, Y.; Xiao, Y.X.; Hu, J.; Wu, S.M.; Deng, Z.; Tian, G.; Chang, G.G.; Li, J.; et al. Hierarchical $MoS_2@TiO_2$ heterojunctions for enhanced photocatalytic performance and electrocatalytic hydrogen evolution. *Chem. Asian J.* **2018**, *13*, 1609. [CrossRef]

195. Zhang, W.; Xiao, X.; Zheng, L.; Wan, C. Fabrication of TiO_2/MoS_2 composite photocatalyst and its photocatalytic mechanism for degradation of methyl orange under visible light. *Can. J. Chem. Eng.* **2015**, *93*, 1594–1602. [CrossRef]

196. Zhao, F.; Rong, Y.; Wan, J.; Hu, Z.; Peng, Z.; Wang, B. MoS_2 quantum dots@TiO_2 nanotube composites with enhanced photoexcited charges separation and high-efficiency visible-light driven photocatalysis. *Nanotechnology* **2018**, *29*, 105403. [CrossRef] [PubMed]

197. Liu, X.; Xing, Z.; Zhang, H.; Wang, W.; Zhang, Y.; Li, Z.; Wu, X.; Yu, X.; Zhou, W. Fabrication of 3D mesoporous black $TiO_2/MoS_2/TiO_2$ nanosheets for visible-light-driven photocatalysis. *ChemSusChem* **2016**, *9*, 1118–1124. [CrossRef] [PubMed]

198. Zhu, Y.; Ling, Q.; Liu, Y.; Wang, H.; Zhu, Y. Photocatalytic H_2 evolution on MoS_2-TiO_2 catalysts synthesized via mechanochemistry. *Phys. Chem. Chem. Phys.* **2015**, *17*, 933–940. [CrossRef] [PubMed]

199. Liu, X.; Xing, Z.; Zhang, Y.; Li, Z.; Wu, X.; Tan, S.; Yu, X.; Zhu, Q.; Zhou, W. Fabrication of 3D flower-like black N-$TiO_{2-x}@MoS_2$ for unprecedented-high visible-light-driven photocatalytic performance. *Appl. Catal. B Environ.* **2017**, *201*, 119–127. [CrossRef]

200. Zhang, J.; Zhang, L.; Ma, X.; Ji, Z. A study of constructing heterojunction between two-dimensional transition metal sulfides (MoS_2 and WS_2) and (101),(001) faces of TiO_2. *Appl. Surf. Sci.* **2018**, *430*, 424–437. [CrossRef]

201. Chu, H.; Lei, W.; Liua, X.; Lib, J.; Zheng, W.; Zhu, G.; Li, C.; Pan, L.; Sun, C. Synergetic effect of TiO_2 as co-catalyst for enhanced visible light photocatalytic reduction of Cr (VI) on $MoSe_2$. *Appl. Catal. A Gen.* **2016**, *521*, 19–25. [CrossRef]

202. Shen, Y.; Ren, X.; Xu, G.; Huang, Z.; Qi, X. Mixed-dimensional TiO_2 nanoparticles with $MoSe_2$ nanosheets for photochemical hydrogen generation. *J. Mater. Sci. Mater. Electron.* **2017**, *28*, 2023–2028. [CrossRef]

203. Zheng, X.; Yang, L.; Li, Y.; Yang, L.; Luo, S. Direct Z-scheme $MoSe_2$ decorating TiO_2 nanotube arrays photocatalyst for water decontamination. *Electrochim. Acta* **2019**, *298*, 663–669. [CrossRef]

204. Cao, S.; Liu, T.; Hussain, S.; Zeng, W.; Peng, X.; Pan, F. Hydrothermal synthesis, characterization and optical absorption property of nanoscale WS_2/TiO_2 composites. *Phys. E Low-Dimens. Syst. Nanostruct.* **2015**, *68*, 171–175. [CrossRef]

205. Ren, X.; Qiao, H.; Huang, Z.; Tang, P.; Liu, S.; Luo, S.; Yao, H.; Qi, X.; Zhong, J. Investigating the photocurrent generation and optoelectronic responsivity of WS_2-TiO_2 heterostructure. *Optics Commun.* **2018**, *406*, 118–123. [CrossRef]

206. Cho, E.C.; Chang-Jian, C.W.; Zheng, J.H.; Huang, J.H.; Lee, K.C.; Ho, B.C.; Hsiao, Y.S. Microwave-assisted synthesis of TiO_2/WS_2 heterojunctions with enhanced photocatalytic activity. *J. Taiwan Ins. Chem. Eng.* **2018**, *91*, 489–498. [CrossRef]

207. Wu, Y.; Liu, Z.; Li, Y.; Chen, J.; Zhu, X.; Na, P. Construction of 2D-2D TiO_2 nanosheet/layered WS_2 heterojunctions with enhanced visible-light-responsive photocatalytic activity. *Chin. J. Catal.* **2019**, *40*, 60–69. [CrossRef]

208. Wu, Y.; Liu, Z.; Li, Y.; Chen, J.; Zhu, X.; Na, P. WS_2 nanodots-modified TiO_2 nanotubes to enhance visible-light photocatalytic activity. *Mater. Lett.* **2019**, *240*, 47–50. [CrossRef]

209. Ho, W.; Yu, J.C.; Lin, J.; Yu, J.; Li, P. Preparation and photocatalytic behavior of MoS_2 and WS_2 nanocluster sensitized TiO_2. *Langmuir* **2004**, *20*, 5865–5869. [CrossRef] [PubMed]

210. Dowla Biswas, M.R.U.; Ali, A.; Cho, K.Y.; Oh, W.C. Novel synthesis of WSe_2-Graphene-TiO_2 ternary nanocomposite via ultrasonic technics for high photocatalytic reduction of CO_2 into CH_3OH. *Ultrason. Sonochem.* **2018**, *42*, 738–746. [CrossRef]

211. Nasalevich, M.A.; van der Veen, M.; Kapteijn, F.; Gascon, J. Metal-organic frameworks as heterogeneous photocatalysts: Advantages and challenges. *CrystEngComm* **2014**, *16*, 4919–4926. [CrossRef]

212. Zhang, T.; Lin, W. Metal-organic frameworks for artificial photosynthesis and photocatalysis. *Chem. Soc. Rev.* **2014**, *43*, 5982–5993. [CrossRef]

213. Li, Y.; Xu, H.; Ouyang, S.; Ye, J. Metal-organic frameworks for photocatalysis. *Phys. Chem. Chem. Phys.* **2016**, *18*, 7563–7572. [CrossRef]

214. Dhakshinamoorthy, A.; Li, Z.; Garcia, H. Catalysis and photocatalysis by metal organic frameworks. *Chem. Soc. Rev.* **2018**, *47*, 8134–8172. [CrossRef]

215. Xiao, J.D.; Jiang, H.L. Metal-organic frameworks for photocatalysis and photothermal catalysis. *Acc. Chem. Res.* **2019**, *52*, 356–366. [CrossRef]

216. Qiu, J.; Zhang, X.; Feng, Y.; Zhang, X.; Wang, H.; Yao, J. Modified metal-organic frameworks as photocatalysts. *Appl. Catal. B Environ.* **2018**, *231*, 317–342. [CrossRef]

217. Yao, P.; Liu, H.; Wang, D.; Chen, J.; Li, G.; An, T. Enhanced visible-light photocatalytic activity to volatile organic compounds degradation and deactivation resistance mechanism of titania confined inside a metal-organic framework. *J. Colloid Interface Sci.* **2018**, *522*, 174–182. [CrossRef] [PubMed]

218. Crake, A.; Christoforidis, K.C.; Kafizas, A.; Zafeiratos, S.; Petit, C. CO_2 capture and photocatalytic reduction using bifunctional TiO_2/MOF nanocomposites under UV-vis irradiation. *Appl. Catal. B Environ.* **2017**, *210*, 131–140. [CrossRef]

219. Ling, L.; Wang, Y.; Zhang, W.; Ge, Z.; Duan, W.; Liu, B. Preparation of a novel ternary composite of TiO_2/UiO-66-NH_2/Graphene Oxide with enhanced photocatalytic activities. *Catal. Lett.* **2018**, *148*, 1978–1984. [CrossRef]

220. Cardoso, J.C.; Stulp, S.; de Brito, J.F.; Flor, J.B.S.; Frem, R.C.G.; Zanoni, M.V.B. MOFs based on ZIF-8 deposited on TiO_2 nanotubes increase the surface adsorption of CO_2 and its photoelectrocatalytic reduction to alcohols in aqueous media. *Appl. Catal. B Environ.* **2018**, *225*, 563–573. [CrossRef]

221. Wang, M.; Wang, D.; Li, Z. Self-assembly of CPO-27-Mg/TiO_2 nanocomposite with enhanced performance for photocatalytic CO_2 reduction. *Appl. Catal. B Environ.* **2016**, *183*, 47–52. [CrossRef]

222. Zhang, B.; Zhang, J.; Tan, X.; Shao, D.; Shi, J.; Zheng, L.; Zhang, J.; Yang, G.; Han, B. MIL-125-NH_2@TiO_2 Core-shell particles produced by a post-solvothermal route for high-performance photocatalytic H_2 production. *ACS Appl. Mater. Interfaces* **2018**, *10*, 16418–16423. [CrossRef]

223. Zhao, C.W.; Li, Y.A.; Wang, X.R.; Chen, G.J.; Liu, Q.K.; Ma, J.P.; Dong, Y.B. Fabrication of Cd (II)-MOF-based ternary photocatalytic composite materials for H_2 production via gel-to-crystal approach. *Chem. Commun.* **2015**, *51*, 15906–15909. [CrossRef]

224. Li, X.; Pi, Y.; Hou, Q.; Yu, H.; Li, Z.; Li, Y.; Xiao, J. Amorphous TiO_2@NH_2-MIL-125(Ti) homologous MOF encapsulated heterostructures with enhanced photocatalytic activity. *Chem. Commun.* **2018**, *54*, 1917–1920. [CrossRef]

225. Xue, C.; Zhang, F.; Chang, Q.; Dong, Y.; Wang, Y.; Hu, S.; Yang, J. MIL-125 and NH_2-MIL-125 modified TiO_2 nanotube array as efficient photocatalysts for pollute degradation. *Chem. Lett.* **2018**, *47*, 711–714. [CrossRef]

226. Xing, M.; Fang, W.; Nasir, M.; Ma, Y.; Zhang, J.; Anpo, M. Self-doped Ti^{3+}-enhanced TiO_2 nanoparticles with a high-performance photocatalysis. *J. Catal.* **2013**, *297*, 236–243. [CrossRef]

227. Zhang, R.; Yang, Y.; Leng, S.; Wang, Q. Photocatalytic activity of Ti^{3+} self-doped dark TiO_2 ultrafine nanorods, grey SiO_2 nanotwin crystalline, and their composite under visible light. *Mater. Res. Express* **2018**, *5*, 045044. [CrossRef]

228. Liu, X.; Gao, S.; Xu, H.; Lou, Z.; Wang, W.; Huang, B.; Dai, Y. Green synthetic approach for Ti^{3+} self-doped TiO_{2-x} nanoparticles with efficient visible light photocatalytic activity. *Nanoscale* **2013**, *5*, 1870–1875. [CrossRef] [PubMed]

229. Hao, W.; Li, X.; Qin, L.; Han, S.; Kang, S.Z. Facile preparation of Ti^{3+} self-doped TiO_2 nanoparticles and their dramatic visible photocatalytic activity for fast restoration of highly concentrated Cr (VI) effluent. *Catal. Sci. Technol.* **2019**, *9*, 2523–2531. [CrossRef]

230. Zheng, Z.; Huang, B.; Meng, X.; Wang, J.; Wang, S.; Lou, Z.; Wang, Z.; Qin, X.; Zhang, X.; Dai, Y. Metallic zinc-assisted synthesis of Ti^{3+} self-doped TiO_2 with tunable phase composition and visible-light photocatalytic activity. *Chem. Commun.* **2013**, *49*, 868–870. [CrossRef] [PubMed]

231. Naldoni, A.; Allieta, M.; Santangelo, S.; Marelli, M.; Fabbri, F.; Cappelli, S.; Bianchi, C.L.; Psaro, R.; Santo, V.D. Effect of nature and location of defects on bandgap narrowing in black TiO_2 nanoparticles. *J. Am. Chem. Soc.* **2012**, *134*, 7600–7603. [CrossRef] [PubMed]

232. Ullattil, S.G.; Narendranath, S.B.; Pillai, S.C.; Periyat, P. Black TiO_2 nanomaterials: A review of recent advances. *Chem. Eng. J.* **2018**, *343*, 708–736. [CrossRef]

233. Wang, B.; Shen, S.; Mao, S.S. Black TiO$_2$ for solar hydrogen conversion. *J. Materiomics* **2017**, *3*, 96–111. [CrossRef]

234. Chen, X.; Liu, L.; Huang, F. Black titanium dioxide (TiO$_2$) nanomaterials. *Chem. Soc. Rev.* **2015**, *44*, 1861–1885. [CrossRef]

235. Chen, X.B.; Liu, L.; Yu, P.Y.; Mao, S.S. Increasing solar absorption for photocatalysis with black hydrogenated titanium dioxide nanocrystals. *Science* **2011**, *331*, 746–750. [CrossRef]

236. Zhu, Q.; Peng, Y.; Lin, L.; Fan, C.M.; Gao, G.Q.; Wang, R.X.; Xu, A.W. Stable blue TiO$_{2-x}$ nanoparticles for efficient visible light photocatalysts. *J. Mater. Chem. A* **2014**, *2*, 4429–4437. [CrossRef]

237. Wu, Q.; Huang, F.; Zhao, M.; Xu, J.; Zhou, J.; Wang, Y. Ultra-small yellow defective TiO$_2$ nanoparticles for co-catalyst free photocatalytic hydrogen production. *Nano Energy* **2016**, *24*, 63–71. [CrossRef]

238. Liu, G.; Yin, L.C.; Wang, J.; Niu, P.; Zhen, C.; Xie, Y.; Cheng, H.M. A red anatase TiO$_2$ photocatalyst for solar energy conversion. *Energy Environ. Sci.* **2012**, *5*, 9603–9610. [CrossRef]

239. Ren, R.; Wen, Z.; Cui, S.; Hou, Y.; Guo, X.; Chen, J. Controllable synthesis and tunable photocatalytic properties of Ti^{3+}-doped TiO$_2$. *Sci. Rep.* **2015**, *5*, 10714. [CrossRef] [PubMed]

240. Fan, C.; Chen, C.; Wang, J.; Fu, X.; Ren, Z.; Qian, G.; Wang, Z. Enhanced photocatalytic activity of hydroxylated and N-doped anatase derived from amorphous hydrate. *J. Mater. Chem. A* **2014**, *2*, 16242–16249. [CrossRef]

 *catalysts* 

Article

Controllable Fabrication of Heterogeneous p-TiO$_2$ QDs@g-C$_3$N$_4$ p-n Junction for Efficient Photocatalysis

Songbo Wang [1,2], Feifan Wang [1], Zhiming Su [1], Xiaoning Wang [1], Yicheng Han [1], Lei Zhang [1], Jun Xiang [1], Wei Du [1] and Na Tang [1,*]

[1] Tianjin Key Laboratory of Brine Chemical Engineering and Resource Eco-utilization, College of Chemical Engineering and Materials Science, Tianjin University of Science & Technology, Tianjin 300457, China; wangsongbo@tust.edu.cn (S.W.); wangfeifan4500@sina.com (F.W.); suzhiming0321@sina.com (Z.S.); wangxiaoning@sina.com (X.W.); hanyicheng1998@sina.com (Y.H.); Leizhang@tust.edu.cn (L.Z.); jxiang@tust.edu.cn (J.X.); duwei@tust.edu.cn (W.D.)

[2] Tianjin Key Laboratory of Marine Resources and Chemistry, Tianjin University of Science & Technology, Tianjin 300457, China

* Correspondence: tjtangna@tust.edu.cn; Tel./Fax: +86-22-60602745

Received: 4 April 2019; Accepted: 6 May 2019; Published: 10 May 2019

Abstract: Photocatalytic technology has been considered to be an ideal approach to solve the energy and environmental crises, and TiO$_2$ is regarded as the most promising photocatalyst. Compared with bare TiO$_2$, TiO$_2$ based p-n heterojunction exhibits a much better performance in charge separation, light absorption and photocatalytic activity. Herein, we developed an efficient method to prepare p-type TiO$_2$ quantum dots (QDs) and decorated graphitic carbonitrile (g-C$_3$N$_4$) nanocomposites, while the composition and structure of the TiO$_2$@g-C$_3$N$_4$ were analyzed by X-ray diffraction, Fourier transform infrared spectroscopy, thermogravimetric analysis, transmission electron microscopy, X-ray photoelectron spectroscopy and UV-visible diffuse reflectance spectroscopy characterizations. The characterization results reveal the surface decorated TiO$_2$ quantum dots is decomposed by titanium glycerolate, which exhibits p-type conductivity. The presence of p-n heterojunction over interface is confirmed, and photoluminescence results indicate a better performance in transfer and separation of photo-generated charge carriers than pure semiconductors and type-II heterojunction. Moreover, the synergy of p-n heterojunction over interface, strong interface interaction, and quantum-size effect significantly contributes to the promoted performance of TiO$_2$ QDs@g-C$_3$N$_4$ composites. As a result, the as-fabricated TiO$_2$ QDs@g-C$_3$N$_4$ composite with a p/n mass ratio of 0.15 exhibits improved photo-reactivity of 4.3-fold and 5.4-fold compared to pure g-C$_3$N$_4$ in degradation of organic pollutant under full solar spectrum and visible light irradiation, respectively.

Keywords: p-n heterojunction; g-C$_3$N$_4$; TiO$_2$; charge separation; photocatalysis

---

1. Introduction

With the increasing concerns regarding the global environmental and energy-related crises over the past decades, photocatalytic technology has been considered to be an effective approach since the foundation of Fujishima-Honda effect in 1972 [1,2]. The key for an efficient photocatalytic process lies in the design and construction of highly active photocatalyst, which requires a wide light absorption edge, fast transfer and separation of photo-generated charge carriers, and quick surface redox reaction [3–7]. TiO$_2$ is regarded as the most promising photocatalyst due to the advantages such as earth abundance, low price, excellent thermal and chemical stability, and being environmentally friendly [8,9]. Unfortunately, the broad band gap (i.e., 3.2 eV for anatase) means TiO$_2$ can only be excited by ultraviolet (UV) light, which is less than 5% in the solar spectrum. Meanwhile, the fast charge recombination rate in single TiO$_2$ also results in a low quantum efficiency. Therefore, it is an urgently necessary to promote the light absorption and charge separation efficiency of TiO_2 to meet the requirements of industrial applications [10,11].

Recently, many researchers have focused on the modification of TiO_2 towards an improved photocatalytic performance, including for: morphology modulation [12–14], metal or nonmetal doping [15–18], defect engineering [19–22], and fabrication of hetero/homojunction [2,23–27]. In our previous reports, we found that the introduction of titanium vacancies into TiO_2 will widen the valence band, which controls the mobility of holes inherently and thus increase charge separation efficiency. Moreover, the introduction of metal vacancies can alter TiO_2 from an n-type semiconductor to a p-type semiconductor [28]. Pan et-al. modified p-type TiO_2 with n-type TiO_2 quantum dots (QDs) to construct p-n homojunction, and the resulted p-n homojunction exhibits significantly high photo-activity compared with pure type TiO_2, which is attributed to the formation of large electronic field over the interface [29]. Moreover, besides the influence of p-n homojunction, quantum sized TiO_2 can also enhance the charge separation due to the quantum size effect [13,30–32]. However, in these reports, both the titanium defected TiO_2 and TiO_2 QDs still absorb only UV light, which limits their applications under sunlight.

Graphic carbon nitride (g-C_3N_4) is a promising metal-free photocatalyst in the field of H_2 production, organic pollutant degradation, CO_2 reduction, and artificial photosynthesize due to the suitable band structure (2.7~2.8 eV), excellent chemical and thermal stability [33–37]. Usually, g-C_3N_4 is synthesized by thermal condensation of melamine, urea or other triazine derivatives, but the resultant product exhibits an irregular 2D aggregation structure, leading to a lower surface area and electrical conductivity, as well as an increased charge carrier recombination [38,39]. Both theoretical and experimental results suggest that nanosheet structured g-C_3N_4 will exhibit a larger surface area and lower charge transfer resistance, so the exfoliation of bulk g-C_3N_4 into nanosheets has been regarded as an effective approach to promote the photocatalytic performance of g-C_3N_4 [40,41]. Wang et-al. applied a liquid exfoliation method on bulk g-C_3N_4 to prepare layer g-C_3N_4, and the charge transfer resistance decreased by 75% according to the electrochemical impedance spectroscopy (EIS) results, indicating a lower charge recombination [42]. Cheng et-al. conducted an exfoliation treatment on bulk g-C_3N_4 using a simple thermal oxidation etch method, and g-C_3N_4 nanosheets with layer thickness of 2 nm and surface area of 306 m^2/g were obtained. Compared with bulk g-C_3N_4, the charge carrier life in the as-prepared g-C_3N_4 nanosheets is prolonged because of the quantum confinement effect [43]. Li and coworkers also reported the synthesizing of 6–9 atomic thick g-C_3N_4 nanosheets by a thermal exfoliation approach, the photocatalytic degradation rate is 2.9 times higher than that of bulk g-C_3N_4 [44]. Therefore, the exfoliation of bulk g-C_3N_4 to nanosheets can effectively increase the surface area, and thus improve charge transfer and separation.

Construction of heterojunction is an effective approach to improve the photocatalytic performance by combining both the advantage of two semiconductors, and the electronic field will enhance the charge transfer and separation across the interface. Type II heterojunction with staggered band alignment is the most widely studied structure. Jiang et-al. placed 5.5 nm sized TiO_2 on g-C_3N_4 nanosheets. Due to the formation of type-II heterojunction, the resultant composites exhibited a Rhodamine B (RhB) degradation rate of 2.5-fold compared to pure g-C_3N_4 [25]. However, this is due to the fact that the work function of the same types semiconductor is closed to each other, which limits the driving force in type-II heterojunction. Instead, with different conductivity types semiconductor contacts, p-n heterojunction will be formed and the difference of the work function is large enough to realize a more efficient charge carrier separation [26,29]. For example, Wang et-al. reported that Cu_2O/TiO_2 p-n junction exhibits a much better photoelectrochemical activity than that of pure TiO_2 and pure Cu_2O [27].

In this work, based on the improvement of our previous work [26,28,29], nanosized titanium glycerolate (TiGly), precursors of p-type TiO_2 quantum dots, were successfully synthesized and in-situ deposited on the surface of g-C_3N_4, TiO_2@g-C_3N_4 nanosheet p-n heterojunctions were then obtained after calcination in air. This p-n junction can achieve the function of killing three birds with one stone: p-type TiO_2 QDs for promoted charge separation, g-C_3N_4 nanosheets for lower charge resistance and p-n junction for enhanced charge transfer over interface. Therefore, compared with pure g-C_3N_4 and TiO_2/g-C_3N_4 type-II heterojunction, TiO_2 QDs@g-C_3N_4 p-n heterojunctions exhibit promoted electron-hole separation efficiency and excellent photocatalytic performance in degradation of organic pollutant and hydrogen evolution.

2. Results and Discussion

The synthesizing process of bulk g-C$_3$N$_4$ and TiO$_2$@g-C$_3$N$_4$ p-n heterojunctions were diagrammatically presented in Scheme 1. Consistent with the literature reported, melamine was calcined in air at 550 °C and then bulk g-C$_3$N$_4$ were obtained. Previously, we have demonstrated that titanium glycerolate (TiGly) is the precursor of p-type TiO$_2$, the organic groups will be gradually removed after calcination, and titanium vacancies (V$_{Ti}$) will be introduced into TiO$_2$ during the conjunction process of the remaining Ti-O-Ti parallel lattice chains. Herein, in order to fabricate TiO$_2$@g-C$_3$N$_4$ p-n heterojunction, bulk g-C$_3$N$_4$ is synthesized firstly, and then TiGly nanoparticles were in-situ deposited on the surface of g-C$_3$N$_4$ and labeled as TGC-*x* according to the added g-C$_3$N$_4$. After calcination in air, g-C$_3$N$_4$ was exfoliated and TiGly nanoparticles were decomposed to p-type TiO$_2$ QDs (labeled as PTC-*x*). The sharp peak at about 27.4° in the X-ray diffraction (XRD) pattern of TGC-*x* (Figure 1a) can be attributed to the periodic accumulation of layers of conjugated aromatic systems in bulk g-C$_3$N$_4$ [45]. In the sample TGC-20 and TGC-40, a weak peak appearing at about 10.2°, which is the characteristic peaks corresponding to TiGly [28], indicating the presence of TiGly nanoparticles. As for TGC-60, the content of TiGly is too low to be detected.

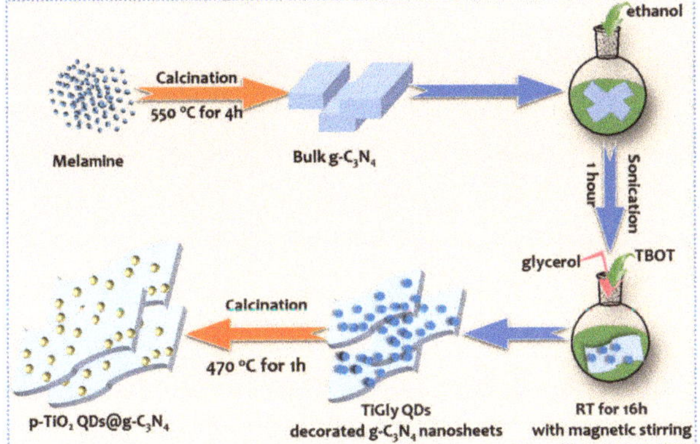

Scheme 1. Schematic fabrication procedures of g-C$_3$N$_4$ and TiO$_2$@g-C$_3$N$_4$ p-n heterojunctions.

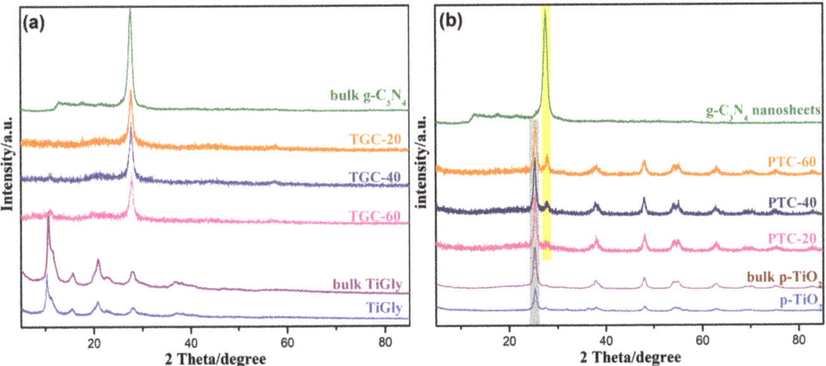

Figure 1. (**a**) XRD pattern of bulk g-C$_3$N$_4$, bulk and nanoparticle TiGly and TGC-*x* before calcination; (**b**) XRD pattern of g-C$_3$N$_4$ nanosheets, bulk and nanoparticle p-TiO$_2$ and PTC-*x* after calcination.

After being calcined at 470 °C for 1 h, TiGly nanoparticles will be decomposed into anatase p-type TiO$_2$. As shown in Figure 1b, no diffraction peaks of TiGly can be observed, only anatase TiO$_2$ (JCPDS No. 21-1272) and g-C$_3$N$_4$ (JCPDS No. 87-1526) can be observed in PTC-x, with no other visible phases or impurities. Notably, the full-width at half maxima (FWHM) of peaks at 25.3° corresponding to (101) planes of p-type TiO$_2$ nanoparticles is broadened from 0.57 to 0.65 compared with bulk p-TiO$_2$, so the average particle size of the TiO$_2$ nanoparticles is smaller than the bulk p-TiO$_2$ according to Scherer Equation. The same phenomenon can also be observed in bulk g-C$_3$N$_4$ and g-C$_3$N$_4$ nanosheets, the peak at 27.4° corresponding to (002) plane of g-C$_3$N$_4$ nanosheets is broadened and drastically weakened from PTC-60 to PTC-20, which is due to the decreased thickness of nanosheets during the thermal exfoliation process [39]. Moreover, we also calculated the lattice constant of the p-type TiO2 nanoparticles based on XRD patterns, the results indicate that a = b = 3.793 Å, slightly larger than normal anatase TiO2 (a = b = 3.785 Å for JCPDS No. 21-1272), whereas the c axis shrinks from 9.514 Å to 9.488 Å, which is identical with that of titanium defected TiO$_2$ [28].

Thermogravimetric (TG) tests were conducted to quantify the relative content of TiO$_2$ QDs in the as-prepared samples. As shown in Figure 2, the weight loss of the samples below 80 °C is due to the removal of surface absorbed water, whereas the weight loss between 80 °C and 400 °C is attributed to the decomposition of titanium glycerolate as shown in Equation (1) [28,29]. Since bulk g-C$_3$N$_4$ exhibits almost no weight loss in this range, the weight loss of TGC-x can all assigned to the decomposition of TiGly. As shown in Figure 2, the weight loss is 20.03%, 9.77% and 7.98% for TGC-20, TGC-40 and TGC-60, respectively. Therefore, according to Equations (2) and (3), the mass content of TiGly in TGC-x can be calculated as 43.7%, 21.3% and 17.4%, respectively, corresponding to a p-type TiO$_2$ QDs mass content of 29.6%, 12.8% and 10.2% in PTC-20, PTC-40 and PTC-60, respectively. Additionally, there is a sight weight increase after 80 °C for all the samples, which may be due to the impurity in the sample gas used in TG tests, but this deviation will not affect the calculated QDs content in the samples.

$$TiGly\ (Ti_3(C_3H_4O_3)_4) + 14O_2 \rightarrow 3Ti_{1-x}O_2 + 12CO_2 + 10H_2O \tag{1}$$

$$y_{TG} = \frac{W_s}{W_{TG}} \tag{2}$$

$$y_{TiO2} = \frac{m_{TiO2}}{m_{TiO2} + m_{g-C3N4}} = \frac{y_{TG}(1 - W_{TG})}{y_{TG}(1 - W_{TG}) + 1 - y_{TG}} = \frac{y_{TG} - y_{TG}W_{TG}}{1 - y_{TG}W_{TG}} \tag{3}$$

Among these figures: y_{TG} is the calculated content of TiGly, %; W_s is the weight loss of TGC-x, %; W_{TG} is the weight loss of pure TiGly, %; m_{TiO2} is the calculated mass of TiO$_2$ in unit mass of PTC-x, g; m_{g-C3N4} is the calculated mass of g-C$_3$N$_4$ in unit mass of PTC-x, g; y_{TiO2} is the calculated mass content of p-type TiO$_2$ QDs in PTC-x, %.

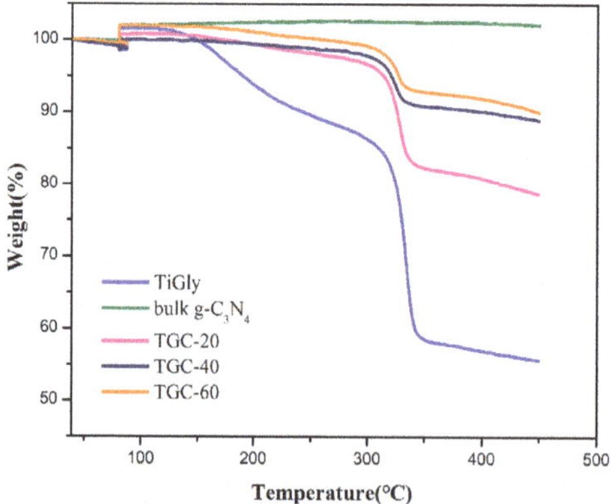

Figure 2. TG profiles of bulk g-C$_3$N$_4$, TiGly and TGC-x.

High resolution transmission electron microscopy (HRTEM) analyses were conducted to reveal the morphology and composition of the samples. Unlike the densely packed bulk g-C$_3$N$_4$ reported in literatures [39], the as-prepared g-C$_3$N$_4$ exhibits a nanosheet structure, the edges are curled and rough due to the minimizing surface energy (Figure 3a), which provides a lower resistance pathway for charge transfer. As shown in Figure 3b–d, the light contrast nanosheets are g-C$_3$N$_4$ nanosheets whereas the dark contrast are the p-type TiO$_2$ nanoparticles decomposed from TiGly. Figure 3e shows the lattice fingers of 0.35 nm and 0.48 nm in the dark contrast, corresponding to the (101) and (002) planes of anatase TiO$_2$, which reveal the exposure of (010) facet of anatase, consistent with the XRD results [31]. The typical size of TiO$_2$ nanoparticles is 4–5 nm according to the particle size distribution result shown in the inset of Figure 3e, which is accordance with the characteristic size of quantum confinement effect. It is noteworthy that when the p-type TiO$_2$ QDs are in-situ grown on the surface of g-C$_3$N$_4$, there is a strong interaction that exists between them so that the TiO$_2$ QDs cannot be peeled off even after a 30 min ultra-sonication process. Moreover, the strong interaction implies an obvious interface heterojunction between TiO$_2$ QDs and g-C$_3$N$_4$, which will enhance the transfer of photo generated electrons and holes [25]. In addition, in agreement with the TG results, the relative content of TiO$_2$ QDs decreases from PTC-20 to PTC-60.

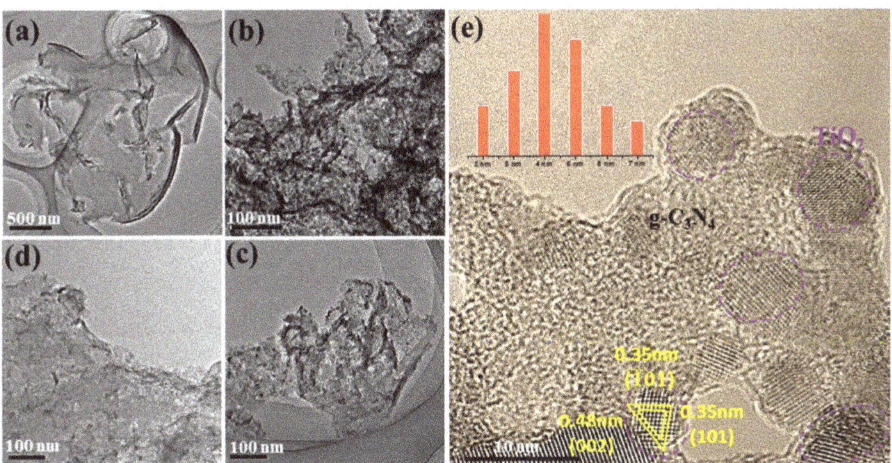

Figure 3. TEM images of (**a**) g-C$_3$N$_4$ nanosheets, (**b**) PTC-20, (**c**) PTC-40 and (**d**) PTC-60, (**e**) is the enlarged image of PTC-40, inset: size distribution of TiO$_2$ QDs in PTC-40.

In order to further investigate the composition and interaction between p-type TiO$_2$ QDs and g-C$_3$N$_4$ in the as-prepared composites, Fourier transform infrared spectroscopy (FT-IR) and X-ray photoelectron spectroscopy (XPS) characterizations were conducted. As shown in Figure 4a, for the TGC-*x*, there are three characteristic bands corresponding to g-C$_3$N$_4$, namely the broad peaks located at 3000–3400 cm^{-1} assigned to the stretching of N–H bonds, the strong peaks at 1250–1650 cm^{-1} due to the stretching vibration of C=N heterocycles and C–N bonds, and the band around 808 cm^{-1} related to the ring vibration of s-triazine [45–48]. Moreover, besides the above three peaks, another three peaks related to TiGly appearing in TGC-*x*, the peaks located at around 1000–1150 cm^{-1} were attributed to the alcoholic Ti–O–C stretching mode, the apparent peak at 611 cm^{-1} was indexed to the stretching mode of Ti–O bonds, and the bands located at 2855–2927 cm^{-1} were assigned to the glycerol C-H stretching vibration [49]. Besides, the broad peak over 3000–3600 cm^{-1} was attributed to the presence of physically adsorbed water and glycerol O-H stretching mode. Therefore, by combining with the XRD, TG and FT-IR results, it is proved that the as-prepared TGC-*x* samples is composed by TiGly and g-C$_3$N$_4$.

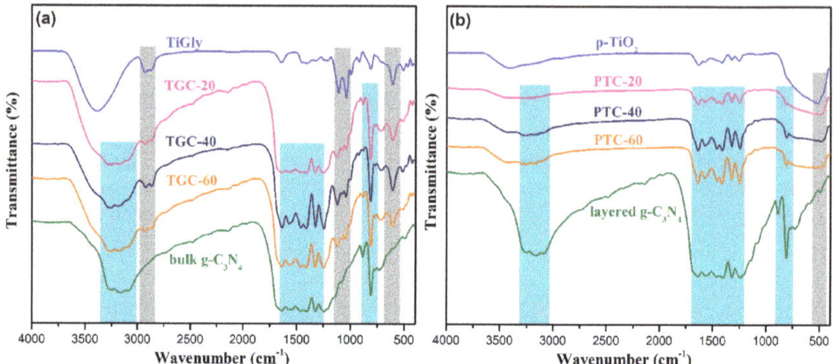

Figure 4. (**a**) FT-IR patterns of bulk g-C$_3$N$_4$, TiGly and TGC-*x* before calcination; (**b**) FT-IR patterns of layered g-C$_3$N$_4$, p-TiO$_2$ and PTC-*x*.

Upon calcination, the major FT-IR peaks of g-C$_3$N$_4$ almost all remained in PTC-*x* samples (Figure 4b). However, the bands assigned to C–H and O–H in TiGly disappeared, while only the stretching band of Ti–O remained, indicating the transformation of TiGly to TiO$_2$, further confirming the formation of TiO$_2$ QDs@g-C$_3$N$_4$ heterojunction. Moreover, with the decrement of g-C$_3$N$_4$ content from PTC-60 to PTC-20, the intensity of s-triazine ring vibration 808 cm^{-1} also decreases, which is in agreement with the above TG results. It is noteworthy that the stretching vibration of Ti–O–Ti shifted significantly towards a lower wavenumber in PTC-*x* composites, suggesting a strong interaction exists between p-TiO$_2$ QDs and g-C$_3$N$_4$ [25], which is in favor of charge transfer across the interface and thus promotes the photocatalytic performance of the heterojunction.

XPS spectrum were recorded to study the status of the C, N, Ti and O elements in the composites. Figure 5a shows the C1s XPS spectra of the samples, there are two main peaks located at 284.8 eV and 288.5 eV, respectively. The peak located at higher binding energy is attributed to the sp3-bonded C of N-C=N$_2$ in g-C$_3$N$_4$, and the peak located at 284.8 eV is due to the surface contaminated carbon during XPS test and sp2-hybridized carbon atoms presented in graphic domains [29]. In the N1s XPS spectra (Figure 5b), a asymmetric profiles can be observed in all the samples, with a main peak at 401 eV and a shoulder peak at lower binding energy, the main peak is due to the sp2-hybridized nitrogen (C–N–C), while the shoulder peak is usually attributed to amino functional groups with a hydrogen atom (C–NH) and sp3-hybridized nitrogen (N-[C]3) [25,39]. For the Ti 2p XPS spectra of the PTC-*x* (Figure 5c), the binding energy of Ti 2p$_{3/2}$ and Ti 2p$_{1/2}$ are observed at 458.9 eV and 464.4 eV, respectively, suggests the existence of TiO$_2$ in the samples. Notably, compared with the Ti 2p binding energy of 458.4 eV in n-type TiO$_2$ [29], the Ti 2p binding energy in the as-prepared samples shifted towards a higher binding energy, indicating the existence of titanium vacancies (V$_{Ti}$) and p-type properties of the TiO$_2$ QDs [28].

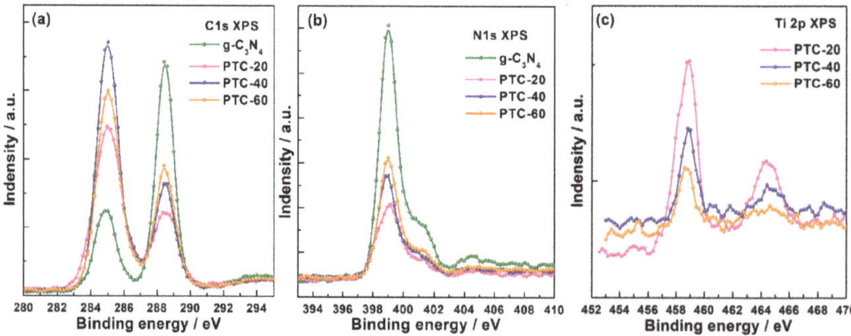

Figure 5. (**a**) C1s, (**b**) N1s and (**c**) Ti 2p XPS spectra of layered g-C$_3$N$_4$, p-TiO$_2$ and PTC-*x*, respectively.

The O1s XPS spectrum were shown in Figure 6, there is only one symmetrical peak at around 532.7 eV appearing in pure g-C_3N_4, which is attributed to the loosely bonded oxygen species on the surface (O_2, H_2O or OH groups), no peaks corresponding to C–O and N–C–O appeared at 531.4 eV, indicating that no O doping process occurred in g-C_3N_4 during the calcination process [38]. As for the TiO_2@g-C_3N_4 composites, the O1s peaks split into two peaks located at 532.7 eV and 529.8 eV, respectively. The lower binding energy is assigned to the oxygen anions (O^{2-}) in the crystal lattice of anatase [28], and the concentration of TiO_2 QDs is proportional to the intensity of this peak. As shown in Figure 6b–d, from PTC-20 to PTC-60, the intensity of this peak becomes obviously weak, confirming the gradually decreased content of TiO_2 QDs in the composites. These results are in good agreement with the TEM, TG and FT-IR results, indicating that the composites are composed of g-C_3N_4 and p-type TiO_2 QDs, and that the content of TiO_2 QDs decreases from PTC-20 to PTC-60.

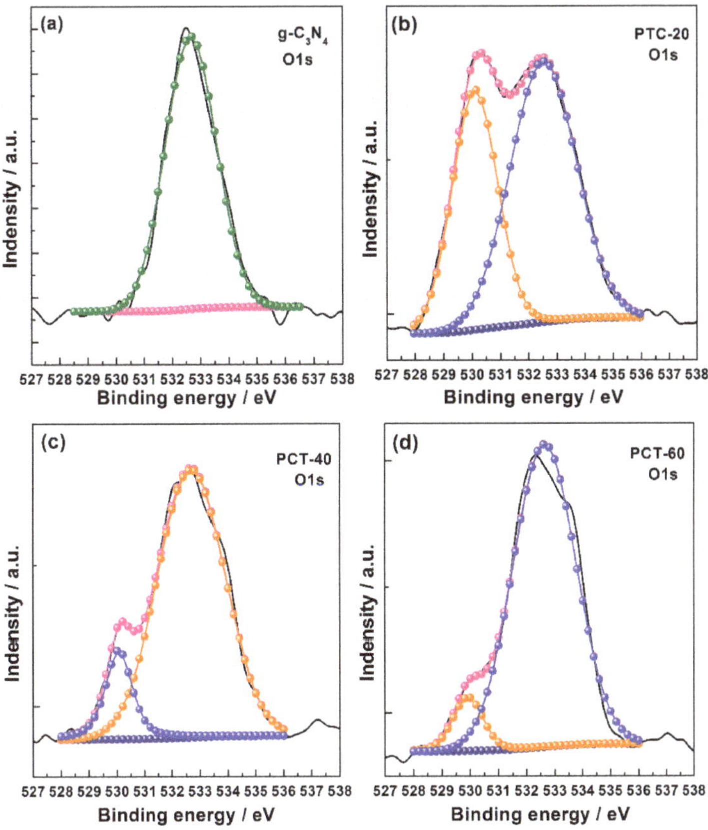

Figure 6. Fitted O1s XPS spectra of (**a**) layered g-C_3N_4, (**b**) PTC-20, (**c**) PTC-40 and (**d**) PTC-60, respectively.

The light absorption properties of the as-prepared samples were characterized by UV-visible diffuse reflection spectrum (UV-vis DRS). As illustrated in Figure 7a, pure g-C_3N_4 can absorb light from UV region to visible light up to 465 nm, while p-type TiO_2 only absorb UV light up to 400 nm. As for the TiO_2 QDs@g-C_3N_4 composites, there is a remarkable absorption edge extension to the visible region compared with single TiO_2 [39]. Moreover, with the increasing content of g-C_3N_4 form PTC-20 to PTC-60, the optical absorption edge of p-n heterojunction shifts towards a longer wavelength, indicating a stronger light absorption in the visible light region, which is beneficial for the improvement of photocatalytic performance. Meanwhile, the quantum-size effect of the p-type TiO_2

QDs is illustrated by VB XPS spectra. As shown in Figure 7b, a blue shift of VB edge is observed in the as-prepared heterojunctions due to the existence of TiO_2 nanoparticles, and this tendency becomes more and more obvious with the increment of TiO_2, confirming the quantum dots nature of the surface decorated p-type TiO_2 nanoparticles [31].

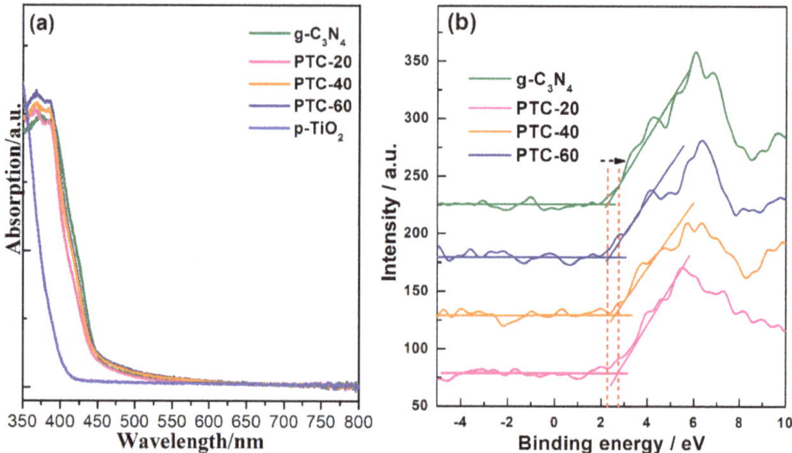

Figure 7. (**a**) UV-vis DRS spectra and (**b**) valence band density of states (DOS) of layered g-C_3N_4, PTC-20, PTC-40 and PTC-60.

As mentioned above, we have demonstrated that the surface deposited TiO_2 QDs is abundant in titanium vacancies, while both experimental and density functional theoretical (DFT) results have indicated that metal defected TiO_2 is p-type semiconductor [28]. In general, g-C_3N_4 exhibits n-type conductivity, therefore, p-n heterojunctions will be formed across the interface, which can afford a large electrical filed and more efficient charge separation. Therefore, according to the band gap and valence band (VB) position of pure g-C_3N_4 and TiO_2, the band alignment of the p-n heterojunction (PTC-40) is diagrammed in Scheme 2. Both TiO_2 and g-C_3N_4 can be excited under $\lambda > 365$ nm, charge redistribution process will occur to equilibrate the Fermi level (E_f) due to the large difference of E_f between p-type TiO_2 and n-type g-C_3N_4, the consequence is that the photogenerated electrons migrate to the conduction band (CB) of TiO_2 and holes to the g-C_3N_4 VB. However, only g-C_3N_4 can be excited under visible light irradiation ($\lambda > 400$ nm), and the photogenerated electrons will still transfer to the CB of TiO_2, while the holes tend to keep stay in the VB of g-C_3N_4. In both cases, an efficient electron-hole spatial separation can be realized and the lifetime of charge carriers can also be prolonged, these separated electrons can react with O_2 or H_2O to form reactive oxygen species ($\cdot O_2^-$) or H_2, while the holes will oxidize an organic pollutant or sacrificial agent directly. In addition, the presence of metal vacancies will enlarge the width of VB, which controls the mobility of holes inherently [28], and thus the synergy influence of metal vacancies, quantum confinement effect, and nanosheet structure can effectively promote the charge transfer and separation across the interface.

The facilitated charge separation by the p-n heterojunction was confirmed by photoluminescence (PL) spectra. As shown in Figure 8, pure g-C_3N_4 has an emission peak around 450 nm, whereas p-TiO_2 exhibits a signal at around 400nm, which is due to the band-band transition, namely the energy corresponding to the emission is close to the excitation energy of g-C_3N_4 and TiO_2, respectively [25,29]. The PL intensity of PTC-*x* is much lower than pure p-type TiO_2 and g-C_3N_4, indicating a promoted electron-hole separation by the p-n heterojunction. Notably, PTC-40 exhibits the lowest PL intensity among the samples, suggesting a higher charge separation efficiency and a better photocatalytic performance, which also indicates that there exists an approximate p/n ratio between the p-type TiO_2 and n-type g-C_3N_4 in the heterojunction.

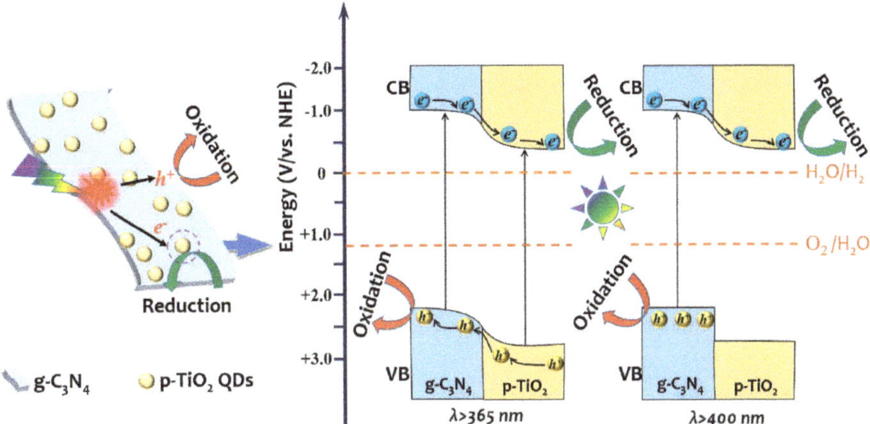

Scheme 2. Schematic illustration of TiO$_2$ QDs@g-C$_3$N$_4$ p-n heterojunction and the pathway for electron-hole transfer and separation over the junction under simulated solar irradiation (λ > 365 nm) and visible light irradiation (λ > 400 nm).

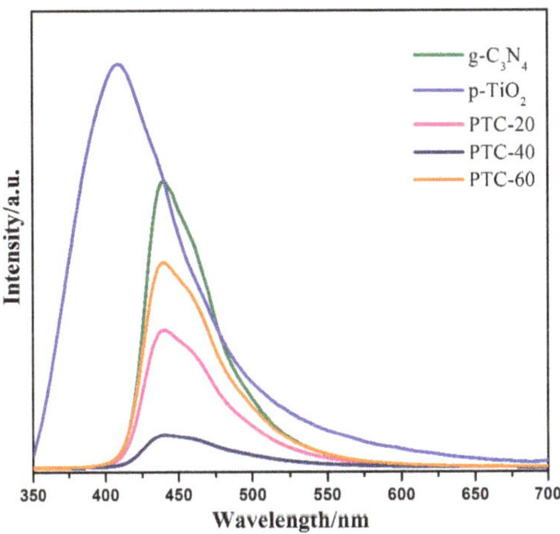

Figure 8. Steady state PL spectra of layered g-C$_3$N$_4$, PTC-20, PTC-40 and PTC-60.

Photocatalytic degradation of organic pollutants and photocatalytic water splitting were conducted to evaluate the performance of as-prepared TiO$_2$@g-C$_3$N$_4$ p-n heterojunctions. Figure 9a shows the photodegradation rate of the samples based on pseudo-first-order reaction mode. Under simulated sunlight irradiation (λ > 365nm), the degradation rate of MO for PTC-40 with a TiO$_2$ (p) to g-C$_3$N$_4$ (n) mass ratio of 0.15 is 0.52 min^{-1}·g^{-1}, which is the highest among the PTC-*x* and is 4.3-fold higher than that of pure g-C$_3$N$_4$. Moreover, we also tested the photoactivity with a cut-off filter of λ > 400 nm to evaluate the visible light photocatalytic performance, the reaction rate exhibits the same trends with that under simulated solar spectrum (Figure 9b), and the reaction rate for MO degradation of PTC-40 is 0.125 min^{-1}·g^{-1}, 5.4-fold higher than pure g-C$_3$N$_4$. This result confirms that p-n heterojunction exhibits a better photocatalytic performance than single photocatalyst and that the as-prepared TiO$_2$@g-C$_3$N$_4$ is active under visible light.

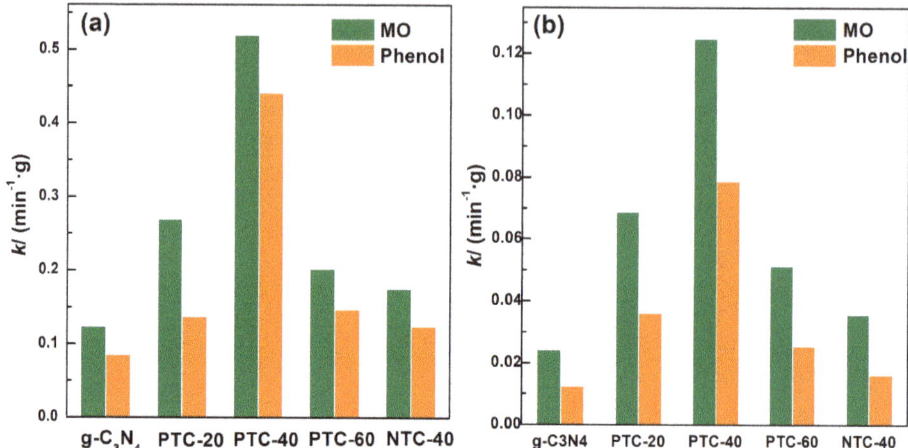

Figure 9. Pseudo-first-order reaction rate in degradation of methyl orange (MO) and phenol under: (**a**) full solar spectrum and (**b**) visible light ($\lambda > 400$ nm).

The photocatalytic H_2 evolution reaction rate is shown in Figure 10a, the H_2 release rate for g-C_3N_4, PTC-20, PTC-40 and PTC-60 is 186 $\mu mol \cdot g^{-1} \cdot h^{-1}$, 712 $\mu mol \cdot g^{-1} \cdot h^{-1}$, 1072 $\mu mol \cdot g^{-1} \cdot h^{-1}$ and 838 $\mu mol \cdot g^{-1} \cdot h^{-1}$, respectively. It is clearly that the H_2 evolution rate of PTC-40 is the fastest, which is 5.8-fold of pure g-C_3N_4, 1.5-fold of PTC-20 and 1.3-fold of PTC-60. Combining with the PL results and photoactivity, we can get the conclusion that PTC-40 has the most efficient electron-hole separation, and thus exhibits the best photocatalytic performance among the as-fabricated samples. Meanwhile, this phenomenon also indicates that a suitable p/n ratio is required to construct the best p-n heterojunction. In comparison, we also loaded n-type TiO_2 QDs on the surface of g-C_3N_4 to construct a type-II heterojunction and found that the promotion in photoactivity is far from that by p-n heterojunction (1.44-fold vs. 5.8-fold in H_2 evolution), illustrating that p-n heterojunctions are more effective in accelerating photogenerated charge carrier separation and promoting the photocatalytic performance.

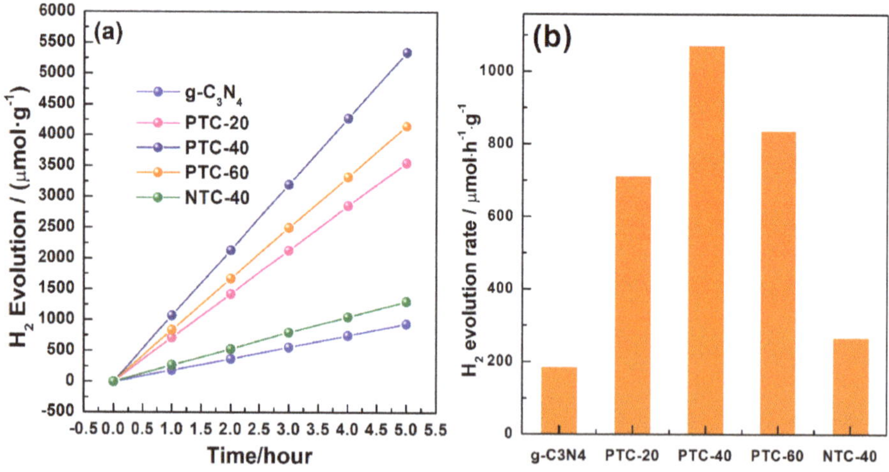

Figure 10. (**a**) Time course of hydrogen evolution and (**b**) hydrogen generation rate of the samples.

3. Experimental

3.1. Materials

Ethanol, glycerol, melamine and methyl orange (MO) were all purchased from Tianjin Jiangtian Fine Chemical Research Institute. Titanium butoxide ($C_{16}H_{36}O_4Ti$, TBOT), triethanolamine (TEOA) and phenol were purchased from J&K chemical. Milli-Q ultra-pure water with a resistivity larger than 18.2 MΩ·cm was used in all experiments. All the chemicals were reagent grade and used as received.

3.2. Preparation of Bulk g-C₃N₄

Bulk g-C_3N_4 was synthesized by thermal annealing melamine under air atmosphere. Typically, 5 g of melamine was put into an airtight crucible, then the crucible was placed into a muffle furnace and calcined at 550 °C for 4 h with a ramping rate of 5 °C/min. After being cooled down to room temperature naturally, the obtained yellow powder is bulk g-C_3N_4.

3.3. Preparation of TiO₂ QDs@g-C₃N₄ p-n Heterojunction

As shown in Scheme 1, for the fabrication of TiO$_2$ QDs@g-C_3N_4 p-n heterojunction, x (x = 20, 40, 60) mg g-C_3N_4 was dispersed in 148 mL ethanol and sonicated for one hour. After that, 5 mL glycerol was added into the solution. After being stirred for another 20 min, 400 µL TBOT was dropwised into the solution and then titanium glycerolate (TiGly) was in-situ grown on the surface of g-C_3N_4, the solution was stirred at room temperature for 16 h. The resulted powders (labeled as TGC-x) were washed with water and absolute ethanol for several times, and dried at 70 °C overnight. After calcination of TGC-x in air at 470 °C for 1 h, TiO$_2$ QDs@g-C_3N_4 p-n heterojunctions were obtained and labeled as PTC-x (x = 20, 40, 60). As a reference, n-type TiO$_2$ QDs deposited g-C_3N_4 was synthesized with the same procedure (x = 40 mg) except that glycerol was replaced by 233 µL NH$_3$·H$_2$O for the purpose of triggering the nucleation of TiO$_2$ on g-C_3N_4 as previously reported (the reference sample was labeled as NTC-40) [29].

3.4. Characterization of Photocatalysts

The calcination temperature for the decomposition of titanium glycerolate and the content of TiO$_2$ was determined by thermogravimetric analysis (TGA Q500, TA Instruments, DE, USA) with air gas flow at 50 mL/min in a range of 30–500 °C (5 °C/min). In order to obtain the composition of the samples, Fourier transform infrared spectroscopy (FT-IR, Bruker Tensor-27 spectrum, Bavaria, Germany) was conducted before and after calcination, the FT-IR spectra were acquired in the range of 400–4000 cm^{-1} with a resolution of 1 cm^{-1}.

All the samples were characterized with an X-ray diffractometer (XRD-6100, Shimadzu, Kyoto, Japan) to determine the crystalline properties. The X-ray diffractometer was equipped with a Cu Kα radiation at 40 kV and 30 mA at a scanning rate of 5°/min. The diffraction patterns were determined over 2 theta range of 5°–90° with a resolution of 2°/min. To obtain the average size of the crystalline size, Scherrer equation was used (D = 0.9λ/βcosθ), whereas the lattice constants was calculated according to Bragg equation of 2dhklsinθ = λ, where λ is the applied wavelength, θ is the Bragg angle and β is the FWHM value.

The chemical states of the as-prepared samples were characterized by X-ray photoelectron spectroscopy (PHI-1600, ULVAC-PHI, Kanagawa, Japan) with Al Kα radiation, and the binding energy was calibrated by the C1s peak (284.8 eV) of the contamination carbon. High resolution transmission electron microscopy (HRTEM) analysis was carried out using a Tecnai G^2 F-20 transmission electron microscope (FEI, OR, USA) with a field-emission gun operation at 200 kV.

The band gap and the light absorption properties of the samples were determined with a UV-vis diffuse reflectance spectrum (U-3010, Hitachi Ltd., Lbarakiken, Japan) with a 60 mm diameter integrating sphere using BaSO$_4$ as the reflectance sample. Steady-state photoluminescence spectra (PL) spectra were measured by a Fluorolog3-21 (Horiba JobinYvon, NJ, USA) with the excitation light at 325 nm.

3.5. Photocatalytic Degradation and Hydrogen Evolution

Photodegradation of organic pollutants (phenol and MO) was conducted in an opening quartz chamber (150 mL) vertically irradiated by a 300 W high-pressure xenon lamp (PLS-SXE300, Beijing Perfect Light Co. Ltd., Beijing, China) located on the upper position. The irradiation area was about 20 cm^2. The light density was measured using a radiometer (Photoelectric Instrument Factory, Beijing Normal University, Model UV-A, Beijing, China), and the results indicate that the light density at 365 nm and 400 nm was 34.7 mW/cm^2 and 32.5 mW/cm^2, respectively. Reaction conditions included the following: a temperature of 25 °C, solution volume of 100 mL, C_0 (phenol) of 400 μmol·L^{-1}; C_0 (MO) of 120 μmol·L^{-1}; photocatalyst of 0.1 g·L^{-1}. Prior to the reaction, the suspension was stirred without irradiation for 20 min to achieve an adsorption equilibrium. Samples were withdrawn, centrifuged and analyzed using a U-3010 UV-vis spectrometer.

Photocatalytic hydrogen production was carried out in a Pyrex top-irradiation reaction vessel connected to closed glass gas system. 10 mg catalyst dispersed in 120 mL aqueous solution containing TEOA (30 vol.%). The temperature of reaction solution was maintained at 0 °C. The resultant hydrogen was analyzed using an off-line gas chromatography (Bruker 450-GC, CA, USA) equipped with a thermal conductive detector (TCD), 5 Å molecular sieve column, and N$_2$ as carrier gas.

4. Conclusions

In this work, p-TiO$_2$ QDs@g-C$_3$N$_4$ p-n heterojunctions were fabricated by in-situ decorating titanium-defected TiO$_2$ QDs on the surface of g-C$_3$N$_4$, in which TiO$_2$ QDs bring up p-type conductivity and g-C$_3$N$_4$ affords for n-type conductivity. The as-prepared p-n heterojunction exhibits higher charge separation efficiency and photocatalytic performance in H$_2$ evolution reaction and degradation of organic pollutant than pure g-C$_3$N$_4$ and TiO$_2$/g-C$_3$N$_4$ type-II heterojunction under both UV-light and visible light irradiation, which can be ascribed to the synergy of a large electrical field over interface, a strong interface interaction, and the quantum confinement effect. In all the samples, PTC-40 with a p/n mass ratio of 0.15 exhibits the best photocatalytic performance. This work demonstrates that the construction of p-n heterojunction is an effective pathway to accelerate the electron-hole separation that is the key for a highly efficient photocatalyst.

Author Contributions: Conceptualization, N.T. and S.W.; methodology, L.Z., W.D. and J.X.; formal analysis, F.W.; investigation, S.W. and F.W.; data curation, F.W., Z.S., X.W. and Y.H.; writing—original draft preparation, F.W.; writing—review and editing, S.W.

Funding: This research was funded by the National Natural Science Foundation of China (No. 21808172), Tianjin Municipal Natural Science Foundation (No. 18JCQNJC05800, 18JCZDJC37200), Innovation Fund for Young Talents of TUST and the foundation of Tianjin Key Laboratory of Marine Resources and Chemistry (TUST, No. 201707).

Conflicts of Interest: The authors declare no conflict of interest.

References

1. Fujishima, A.; Honda, K. Electrochemical Photolysis of Water at a Semiconductor Electrode. *Nature* **1972**, *238*, 37–38. [CrossRef]
2. Wu, S.-M.; Liu, X.-L.; Lian, X.-L.; Tian, G.; Janiak, C.; Zhang, Y.-X.; Lu, Y.; Yu, H.-Z.; Hu, J.; Wei, H.; et al. Homojunction of Oxygen and Titanium Vacancies and its Interfacial n–p Effect. *Adv. Mater.* **2018**, *30*, 1802173. [CrossRef]
3. Zhang, N.; Yang, M.-Q.; Liu, S.; Sun, Y.; Xu, Y.-J. Waltzing with the Versatile Platform of Graphene to Synthesize Composite Photocatalysts. *Chem. Rev.* **2015**, *115*, 10307–10377. [CrossRef] [PubMed]
4. Wondraczek, L.; Tyystjärvi, E.; Méndez-Ramos, J.; Müller, F.A.; Zhang, Q. Shifting the Sun: Solar Spectral Conversion and Extrinsic Sensitization in Natural and Artificial Photosynthesis. *Adv. Sci.* **2015**, *2*, 1500218. [CrossRef]
5. Zhang, H.; Liu, G.; Shi, L.; Liu, H.; Wang, T.; Ye, J. Engineering coordination polymers for photocatalysis. *Nano Energy* **2016**, *22*, 149–168. [CrossRef]

6.   Kapilashrami, M.; Zhang, Y.; Liu, Y.-S.; Hagfeldt, A.; Guo, J. Probing the Optical Property and Electronic Structure of TiO_2 Nanomaterials for Renewable Energy Applications. *Chem. Rev.* **2014**, *114*, 9662–9707. [CrossRef] [PubMed]

7.   Fresno, F.; Portela, R.; Suárez, S.; Coronado, J.M. Photocatalytic materials: recent achievements and near future trends. *J. Mater. Chem. A* **2014**, *2*, 2863–2884. [CrossRef]

8.   Chen, X.; Mao, S.S. Titanium Dioxide Nanomaterials: Synthesis, Properties, Modifications, and Applications. *Chem. Rev.* **2007**, *107*, 2891–2959. [CrossRef] [PubMed]

9.   Pan, X.; Yang, M.-Q.; Fu, X.; Zhang, N.; Xu, Y.-J. Defective TiO_2 with oxygen vacancies: synthesis, properties and photocatalytic applications. *Nanoscale* **2013**, *5*, 3601–3614. [CrossRef]

10. Ma, Y.; Wang, X.; Jia, Y.; Chen, X.; Han, H.; Li, C. Titanium Dioxide-Based Nanomaterials for Photocatalytic Fuel Generations. *Chem. Rev.* **2014**, *114*, 9987–10043. [CrossRef]

11. Wang, Z.; Yang, C.; Lin, T.; Yin, H.; Chen, P.; Wan, D.; Xu, F.; Huang, F.; Lin, J.; Xie, X.; et al. Visible-light photocatalytic, solar thermal and photoelectrochemical properties of aluminium-reduced black titania. *Energy Environ. Sci.* **2013**, *6*, 3007–3014. [CrossRef]

12. Yang, H.G.; Sun, C.H.; Qiao, S.Z.; Zou, J.; Liu, G.; Smith, S.C.; Cheng, H.M.; Lu, G.Q. Anatase TiO_2 single crystals with a large percentage of reactive facets. *Nature* **2008**, *453*, 638. [CrossRef]

13. Li, L.; Yan, J.; Wang, T.; Zhao, Z.J.; Zhang, J.; Gong, J.; Guan, N. Sub-10 nm rutile titanium dioxide nanoparticles for efficient visible-light-driven photocatalytic hydrogen production. *Nat. Commun.* **2015**, *6*, 5881. [CrossRef] [PubMed]

14. Liu, G.; Yang, H.G.; Pan, J.; Yang, Y.Q.; Lu, G.Q.; Cheng, H.-M. Titanium Dioxide Crystals with Tailored Facets. *Chem. Rev.* **2014**, *114*, 9559–9612. [CrossRef] [PubMed]

15. Di Valentin, C.; Pacchioni, G.; Selloni, A. Reduced and n-Type Doped TiO_2: Nature of Ti^{3+} Species. *J. Phys. Chem. C* **2009**, *113*, 20543–20552. [CrossRef]

16. Livraghi, S.; Paganini, M.C.; Giamello, E.; Selloni, A.; Di Valentin, C.; Pacchioni, G. Origin of Photoactivity of Nitrogen-Doped Titanium Dioxide under Visible Light. *J. Am. Chem. Soc.* **2006**, *128*, 15666–15671. [CrossRef]

17. Li, J.-G.; Büchel, R.; Isobe, M.; Mori, T.; Ishigaki, T. Cobalt-Doped TiO_2 Nanocrystallites: Radio-Frequency Thermal Plasma Processing, Phase Structure, and Magnetic Properties. *J. Phys. Chem. C* **2009**, *113*, 8009–8015. [CrossRef]

18. Cao, J.; Zhang, Y.; Liu, L.; Ye, J. A p-type Cr-doped TiO_2 photo-electrode for photo-reduction. *Chem. Commun.* **2013**, *49*, 3440–3442. [CrossRef]

19. Wang, Y.; Han, P.; Lv, X.; Zhang, L.; Zheng, G. Defect and Interface Engineering for Aqueous Electrocatalytic CO_2 Reduction. *Joule* **2018**, *2*, 2551–2582. [CrossRef]

20. Wang, T.; Liu, L.; Ge, G.; Liu, M.; Zhou, W.; Chang, K.; Yang, F.; Wang, D.; Ye, J. Two-dimensional titanium oxide nanosheets rich in titanium vacancies as an efficient cocatalyst for photocatalytic water oxidation. *J. Catal.* **2018**, *367*, 296–305. [CrossRef]

21. Nowotny, M.K.; Bogdanoff, P.; Dittrich, T.; Fiechter, S.; Fujishima, A.; Tributsch, H. Observations of p-type semiconductivity in titanium dioxide at room temperature. *Mater. Lett.* **2010**, *64*, 928–930. [CrossRef]

22. Wu, Q.; Huang, F.; Zhao, M.; Xu, J.; Zhou, J.; Wang, Y. Ultra-small yellow defective TiO_2 nanoparticles for co-catalyst free photocatalytic hydrogen production. *Nano Energy* **2016**, *24*, 63–71. [CrossRef]

23. Dahl, M.; Liu, Y.; Yin, Y. Composite Titanium Dioxide Nanomaterials. *Chem. Rev.* **2014**, *114*, 9853–9889. [CrossRef] [PubMed]

24. Li, H.; Zhou, Y.; Tu, W.; Ye, J.; Zou, Z. State-of-the-Art Progress in Diverse Heterostructured Photocatalysts toward Promoting Photocatalytic Performance. *Adv. Funct. Mater.* **2015**, *25*, 998–1013. [CrossRef]

25. Tong, Z.; Yang, D.; Xiao, T.; Tian, Y.; Jiang, Z. Biomimetic fabrication of g-C_3N_4/TiO_2 nanosheets with enhanced photocatalytic activity toward organic pollutant degradation. *Chem. Eng. J.* **2015**, *260*, 117–125. [CrossRef]

26. Wang, S.; Huang, C.-Y.; Pan, L.; Chen, Y.; Zhang, X.; Fazale, A.; Zou, J.-J. Controllable fabrication of homogeneous ZnO p-n junction with enhanced charge separation for efficient photocatalysis. *Catal. Today* **2018**. [CrossRef]

27. Wang, M.; Sun, L.; Lin, Z.; Cai, J.; Xie, K.; Lin, C. p–n Heterojunction photoelectrodes composed of Cu_2O-loaded TiO_2 nanotube arrays with enhanced photoelectrochemical and photoelectrocatalytic activities. *Energy Environ. Sci.* **2013**, *6*, 1211–1220. [CrossRef]

28. Wang, S.; Pan, L.; Song, J.-J.; Mi, W.; Zou, J.-J.; Wang, L.; Zhang, X. Titanium-Defected Undoped Anatase TiO$_2$ with p-Type Conductivity, Room-Temperature Ferromagnetism, and Remarkable Photocatalytic Performance. *J. Am. Chem. Soc.* **2015**, *137*, 2975–2983. [CrossRef]

29. Pan, L.; Wang, S.; Xie, J.; Wang, L.; Zhang, X.; Zou, J.-J. Constructing TiO$_2$ p-n homojunction for photoelectrochemical and photocatalytic hydrogen generation. *Nano Energy* **2016**, *28*, 296–303. [CrossRef]

30. Ma, L.; Han, H.; Pan, L.; Tahir, M.; Wang, L.; Zhang, X.; Zou, J.-J. Fabrication of TiO$_2$ nanosheets via Ti^{3+} doping and Ag$_3$PO$_4$ QD sensitization for highly efficient visible-light photocatalysis. *RSC Adv.* **2016**, *6*, 63984–63990. [CrossRef]

31. Pan, L.; Zou, J.-J.; Wang, S.; Huang, Z.-F.; Yu, A.; Wang, L.; Zhang, X. Quantum dot self-decorated TiO$_2$ nanosheets. *Chem. Commun.* **2013**, *49*, 6593–6595. [CrossRef]

32. Anpo, M.; Kawamura, T.; Kodama, S.; Maruya, K.; Onishi, T. Photocatalysis on titanium-aluminum binary metal oxides: enhancement of the photocatalytic activity of titania species. *J. Phys. Chem.* **1988**, *92*, 438–440. [CrossRef]

33. Cao, S.; Yu, J. g-C$_3$N$_4$-Based Photocatalysts for Hydrogen Generation. *J. Phys. Chem. Lett.* **2014**, *5*, 2101–2107. [CrossRef] [PubMed]

34. Wang, X.; Blechert, S.; Antonietti, M. Polymeric Graphitic Carbon Nitride for Heterogeneous Photocatalysis. *ACS Catal.* **2012**, *2*, 1596–1606. [CrossRef]

35. Low, J.; Cao, S.; Yu, J.; Wageh, S. Two-dimensional layered composite photocatalysts. *Chem. Commun.* **2014**, *50*, 10768–10777. [CrossRef]

36. Butchosa, C.; Guiglion, P.; Zwijnenburg, M.A. Carbon Nitride Photocatalysts for Water Splitting: A Computational Perspective. *J. Phys. Chem. C* **2014**, *118*, 24833–24842. [CrossRef]

37. Zhang, H.; Zuo, X.; Tang, H.; Li, G.; Zhou, Z. Origin of photoactivity in graphitic carbon nitride and strategies for enhancement of photocatalytic efficiency: insights from first-principles computations. *Phys. Chem. Chem. Phys.* **2015**, *17*, 6280–6288. [CrossRef]

38. Huang, Z.-F.; Song, J.; Pan, L.; Wang, Z.; Zhang, X.; Zou, J.-J.; Mi, W.; Zhang, X.; Wang, L. Carbon nitride with simultaneous porous network and O-doping for efficient solar-energy-driven hydrogen evolution. *Nano Energy* **2015**, *12*, 646–656. [CrossRef]

39. Zhang, J.-W.; Gong, S.; Mahmood, N.; Pan, L.; Zhang, X.; Zou, J.-J. Oxygen-doped nanoporous carbon nitride via water-based homogeneous supramolecular assembly for photocatalytic hydrogen evolution. *Appl. Catal. B-Environ.* **2018**, *221*, 9–16. [CrossRef]

40. Ye, C.; Li, J.-X.; Li, Z.-J.; Li, X.-B.; Fan, X.-B.; Zhang, L.-P.; Chen, B.; Tung, C.-H.; Wu, L.-Z. Enhanced Driving Force and Charge Separation Efficiency of Protonated g-C$_3$N$_4$ for Photocatalytic O$_2$ Evolution. *ACS Catal.* **2015**, *5*, 6973–6979. [CrossRef]

41. Zhang, S.; Li, J.; Wang, X.; Huang, Y.; Zeng, M.; Xu, J. Rationally designed 1D Ag@AgVO$_3$ nanowire/graphene/protonated g-C$_3$N$_4$ nanosheet heterojunctions for enhanced photocatalysis via electrostatic self-assembly and photochemical reduction methods. *J. Mater. Chem. A* **2015**, *3*, 10119–10126. [CrossRef]

42. Yang, S.B.; Gong, Y.J.; Zhang, J.S.; Zhan, L.; Ma, L.L.; Fang, Z.Y.; Vajtai, R.; Wang, X.C.; Ajayan, P.M. Exfoliated Graphitic Carbon Nitride Nanosheets as Efficient Catalysts for Hydrogen Evolution Under Visible Light. *Adv. Mater.* **2013**, *25*, 2452–2456. [CrossRef]

43. Niu, P.; Zhang, L.L.; Liu, G.; Cheng, H.M. Graphene-Like Carbon Nitride Nanosheets for Improved Photocatalytic Activities. *Adv. Funct. Mater.* **2012**, *22*, 4763–4770. [CrossRef]

44. Xu, H.; Yan, J.; She, X.; Xu, L.; Xia, J.; Xu, Y.; Song, Y.; Huang, L.; Li, H. Graphene-analogue carbon nitride: novel exfoliation synthesis and its application in photocatalysis and photoelectrochemical selective detection of trace amount of Cu^{2+}. *Nanoscale* **2014**, *6*, 1406–1415. [CrossRef]

45. Kang, Y.; Yang, Y.; Yin, L.-C.; Kang, X.; Liu, G.; Cheng, H.-M. An Amorphous Carbon Nitride Photocatalyst with Greatly Extended Visible-Light-Responsive Range for Photocatalytic Hydrogen Generation. *Adv. Mater.* **2015**, *27*, 4572–4577. [CrossRef]

46. Jun, Y.-S.; Lee, E.Z.; Wang, X.; Hong, W.H.; Stucky, G.D.; Thomas, A. From Melamine-Cyanuric Acid Supramolecular Aggregates to Carbon Nitride Hollow Spheres. *Adv. Funct. Mater.* **2013**, *23*, 3661–3667. [CrossRef]

47. Liang, Q.; Li, Z.; Yu, X.; Huang, Z.-H.; Kang, F.; Yang, Q.-H. Macroscopic 3D Porous Graphitic Carbon Nitride Monolith for Enhanced Photocatalytic Hydrogen Evolution. *Adv. Mater.* **2015**, *27*, 4634–4639. [CrossRef]

48. Zhang, J.; Zhang, M.; Lin, L.; Wang, X. Sol Processing of Conjugated Carbon Nitride Powders for Thin-Film Fabrication. *Angew. Chem. Int. Edit.* **2015**, *54*, 6297–6301. [CrossRef]

49. Das, J.; Freitas, F.S.; Evans, I.R.; Nogueira, A.F.; Khushalani, D. A facile nonaqueous route for fabricating titania nanorods and their viability in quasi-solid-state dye-sensitized solar cells. *J. Mater. Chem.* **2010**, *20*, 4425–4431. [CrossRef]

 *catalysts*

Article

Improvement of the Photoelectrochemical Performance of TiO₂ Nanorod Array by PEDOT and Oxygen Vacancy Co-Modification

Bin Yang [1], Guoqiang Chen [2], Huiwen Tian [3],* and Lei Wen [1],*

[1] National Center for Materials Service Safety, University of Science and Technology Beijing, Beijing 100083, China; binyang@ustb.edu.cn

[2] Beijing General Research Institute of Mining & Metallurgy Group, Beijing 100160, China; ccggqq871020@126.com

[3] Key Laboratory of Marine Environmental Corrosion and Bio-fouling, Institute of Oceanology, Chinese Academy of Sciences, Qingdao 266071, China

* Correspondence: tianhuiwen1983@foxmail.com (H.T.); wenlei@ustb.edu.cn (L.W.); Tel.: +86-010-62333132 (H.T.); +86-0532-82898832 (L.W.)

Received: 25 March 2019; Accepted: 19 April 2019; Published: 30 April 2019

Abstract: In this study, oxygen vacancy modified TiO₂ nanorod array photoelectrode was prepared by reducing hydrogen atmosphere to increase its free charge carrier density. Subsequently, a p-type conductive poly 3,4-ethylenedioxythiophene (PEDOT) layer was deposited on the surface of oxygen vacancy modified TiO₂, to inhibit the surface states. Meanwhile, a p-n heterojunction formed between PEDOT and TiO₂ to improve the separation of photo-induced carriers further. The photocurrent of TiO₂ nanorod array increased to nearly 0.9 mA/cm² after the co-modification under standard sunlight illumination, whose value is nearly nine times higher than that of pure TiO₂ nanorod array. Thus, this is a promising modification method for TiO₂ photoanode photoelectrochemical (PEC) performance improving.

Keywords: oxygen vacancy; polymeric composites; photoelectrochemistry; co-modification; solar energy conversion

---

1. Introduction

TiO₂ has been widely investigated in the past few decades since Fujishima and Honda first reported its potential in the fields of photocatalysis and photoelectrochemistry in 1972 [1]. The theoretical limited photocurrent densities of anatase and rutile TiO₂ are 1.1 mA/cm² and 1.8 mA/cm² under solar light illumination, respectively. [2] Limited by its low solar light utilization rate and high photo-generated carrier recombination rate, many modification methods have been researched, such as metal doped [3], non-metal doped [4], and construct heterojunction [5]. Several elements have been introduced into TiO₂, such as Fe [6], S [7], and N [8]. Metal and non-metal doping could narrow the bandgap, extend the light absorption range and increase the charge carrier density to improve its photocatalysis performance. However, the introduction of heterogeneous atoms is likely to cause asymmetric doping or impurities, which would serve as recombination centers for the photo-generated electrons and holes, therefore reducing the PEC performance. Many previous research works showed that the formation of surface oxygen vacancy [9–13] could increase the charge carrier density of the semiconductor to improve its PEC performance. Wang et al. [14] obtained a yellowish ZnO with a narrowing band gap by introducing the oxygen vacancies into ZnO crystal, which increased the free charge density of the ZnO, so that the transfer process of the photogenerated charges became feasible.

Polymer organic semiconductors with good film-forming properties, high conductivity, high visible light transmittance and excellent stability are widely used in the field of photoelectrode modification. Park et al. [15] used a blend of 100 nm TiO_2 scattering particles in PEDOT:PSS (poly 3,4-ethylenedioxythiophene:poly styrenesulfonate) solution to fabricate transparent electrode films. When utilized in an organic photovoltaic device, a power conversion efficiency of 7.92% was achieved. Sakai et. al. [16] assembled PEDOT and TiO_2 layer-by-layer to switch electric conductivity in response to ultraviolet and visible light. PEDOT is a promising material to modify the TiO_2 photoanode to improve its PEC performance [17–20].

Therefore, in this work, we prepared oxygen vacancy modified TiO_2 nanorod array photoanode with high charge mobility capacity. Then, a p-type PEDOT layer was covered on the surface of oxygen vacancy modified TiO_2 photoanode to inhibit the undesirable surface state and construct a p-n heterojunction to accelerate the separation capacity of photo-generated carriers [5].

2. Results and Discussion

The XRD (X-ray Diffraction) patterns of series samples were shown in Figure 1, all diffraction peaks of the prepared three photoelectrodes can be indexed as rutile-type and anatase-type TiO_2 (JCPDS No. 21-1276, JCPDS No. 21-1272) [21,22]. The characteristic diffraction peaks at $2\theta = 36.08°$, 54.32, 62.74, and 69.78° corresponded to the (101), (211), (002), and (112) crystal planes of rutile-type TiO_2, and the XRD peaks at $2\theta = 63.68°$ corresponded to the (204) crystal planes of anatase-type TiO_2. The other characteristic diffraction peaks at $2\theta = 26.57°$, 37.76°, 51.75°, and 65.74° corresponded to the (110), (200), (211), and (301) crystal planes of SnO_2 (JCPDS No. 46-1088), which caused by the fluorine doped tin oxide (FTO) conductive glass. So, the prepared TiO_2 nanorod array included rutile phases and little anatase phases. The TiO_2 nanorod array preparing method in this work was referred to in Liu's work [17]. The vanished peaks for anatase and rutile TiO_2 at 25.4° and 27.4° on the XRD curves maybe attributed to the crystal face inhibition effect of the oriented growth nanorod structure, whose results are similar to Liu's work [17].

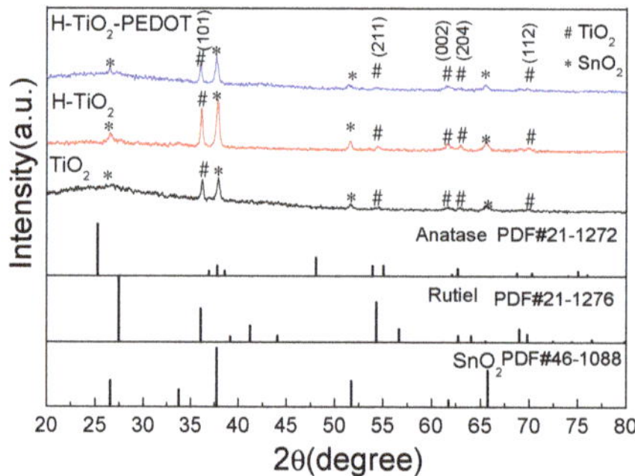

Figure 1. XRD patterns of TiO_2, H-TiO_2, and H-TiO_2-PEDOT.

The SEM technique was employed to observe the surface morphologies of the series samples, and the results are shown in Figure 2. As presented in Figure 2A–C, both TiO_2, H-TiO_2, and H-TiO_2-PEDOT appear to have a distinct nanorod structure. The cross-section image of H-TiO_2-PEDOT shown in Figure 2D reveals that the TiO_2 nanorod is growing vertically on the FTO substrate. The nanorods are tetragonal in shape with square top facets, the expected growth habit for the tetragonal crystal structure. The nanorods are nearly perpendicular to the FTO substrate. After 8 h of growth, the average diameter and length, as determined from SEM, were 90 ± 20 nm and 1 ± 0.2 μm, respectively. The peaks of (101) crystal planes for rutile and (204) for anatase TiO_2 can be clearly observed in the HRTEM image inset in Figure 2D, which is in agreement with the XRD results. Meanwhile, PEDOT layer can be observed at the edge area of TiO_2 nanorod. Elements distribution of H-TiO_2-PEDOT were tested by STEM and STEM-EDS mapping. The STEM mapping shown in Figure 2E reveals the uniform distribution of Ti, O, and S element on the surface of the nanorod, where the S element corresponding to the PEDOT deposition layer. This result indicates that PEDOT layer was successfully deposited on the surface of H-TiO$_2$ photoelectrode.

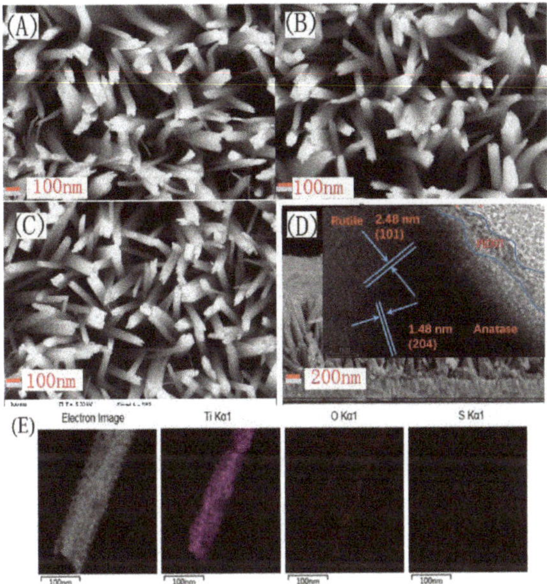

Figure 2. SEM images of (**A**) TiO$_2$, (**B**) hydrogen treated TiO$_2$ (H-TiO$_2$), (**C**) PEDOT modified hydrogen treated TiO$_2$ (H-TiO$_2$-PEDOT) and (**D**) cross-section image of H-TiO$_2$-PEDOT. Insert is the HRTEM image of H-TiO$_2$-PEDOT. (**E**) STEM mapping of H-TiO$_2$-PEDOT.

Similar results could be observed on EDS mapping (Figure 3), in which, the O, Ti, and Sn element corresponding to TiO$_2$ nanorod and FTO substrate were evenly distributed throughout all the H-TiO$_2$-PEDOT photoelectrode, besides, C and S elements could be observed simultaneously, which is corresponding to the STEM mapping showed in Figure 2E.

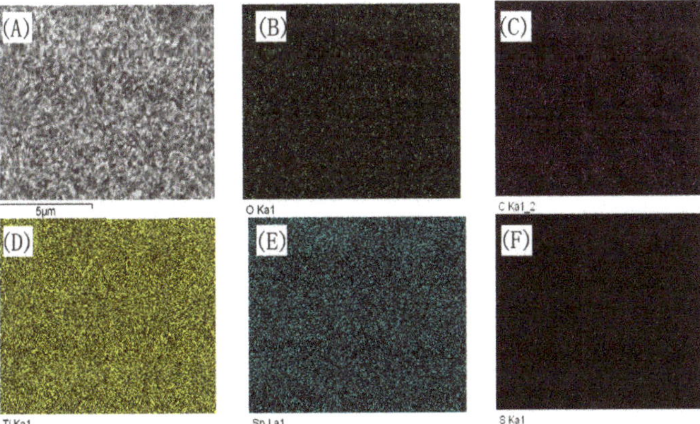

Figure 3. EDS mapping of H-TiO$_2$-PEDOT. (**A**) Scanning area, (**B**) O Element, (**C**) C Element, (**D**) Ti Element, (**E**) Sn Element, (**F**) S Element.

To determine the surface composition and chemical states of the series samples, high-resolution XPS spectra of O 1s and S 2p were used (see Figure 4). The characteristic peaks at 529.8 eV and

530.8 eV correspond to the lattice oxygen (O_{lat}) and the vacancies of O element (O_{def}). Compared to TiO_2, the peak area of O_{def} in H-TiO_2 was enhanced after the hydrogen treatment, indicating the increase of oxygen vacancies from 34.2% to 43.77%, which might improve the PEC performance [16]. Oxygen vacancy concentration refers to the proportion of oxygen vacancy peak area to the total oxygen peak area. Because XPS can only read the distribution of surface elements, the peak area of oxygen (O 1s, O_{lat}, O_{def}) becomes smaller after PEDOT loading, but the relative content is credible. Next, the peak area of O_{def} in H-TiO_2-PEDOT was reduced further after PEDOT deposition, which can be ascribed to the protection of PEDOT layer. The S was observed in the XPS spectra of H-TiO_2-PEDOT indicating PEDOT was introduced successfully, which corresponds to the result of the XPS survey spectra shown in Figure 4C. Because of the low loading amount of PEDOT, noises can be found on the XPS S2p curve. The characteristic peaks of Ti 2p did not shift after the hydrogen treatment, and the deposition of PEDOT (Figure 4D) indicated that the unique TiO_2 nanorod structure was preserved.

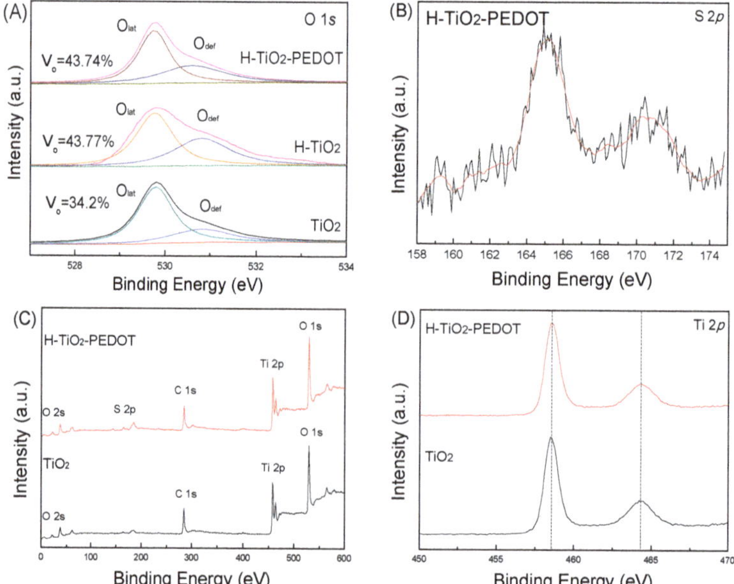

Figure 4. High-resolution XPS spectra of O 1s (**A**) and S 2p (**B**) of the TiO_2, H-TiO_2 and H-TiO_2-PEDOT. XPS survey spectra (**C**), high-resolution XPS spectra of Ti 2p (**D**) of the TiO_2 and H-TiO_2-PEDOT.

The PEC performance results of series samples are presented in Figure 5. Figure 5A is the current density-time curve of the series of electrodes, and Figure 5B is the current density-voltage curve of the series of electrodes. The current density-voltage curve shows that the current density of the TiO_2 sample at zero bias (vs. Ag/AgCl) is about 0.07 mA/cm^2. The current density of H-TiO_2 sample at zero bias is about 0.27 mA/cm^2. The current density of H-TiO_2-PEDOT sample at zero bias is about 0.33 mA/cm^2. In the voltage range from −0.5 to 0.5 V, the photocurrent density of sample H-TiO_2-PEDOT is higher than that of sample H-TiO_2, and the PEC performance of pure TiO_2 nanorod array is the worst. Figure 5C is the impedance data of each sample in the absence of light. The arc radius of pure TiO_2 is the largest, corresponding to the largest impedance. The arc radius of H-TiO_2 is the smallest, corresponding to the smallest impedance. After PEDOT deposition, the arc radius of H-TiO_2-PEDOT become larger because of the impedance of PEDOT. After oxygen vacancies modification, the arc radius and impedance of the obtained H-TiO_2 sample decreases. PEDOT conductive layer coated on the hydrogen treated TiO_2 photoelectrode make the arc radius of the obtained H-TiO_2-PEDOT further smaller, indicating a smaller impedance of this sample.

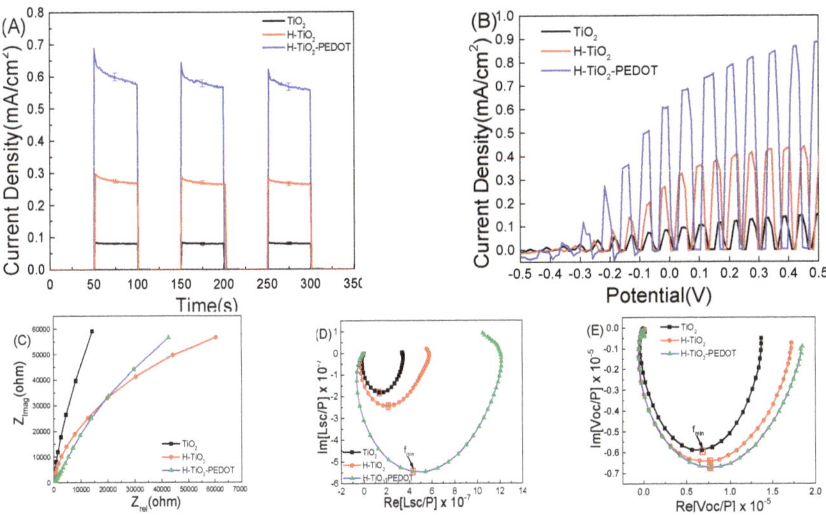

Figure 5. The current density-time curves (I-T) (**A**), the current density-potential (I-V) (**B**), AC impedance spectroscopy (EIS) (**C**), intensity modulated photocurrent spectroscopy (CIMPS) (**D**) and intensity modulated photovoltage spectroscopy (CIMVS) (**E**) of series photoelectrodes.

Figure 5D shows the CIMPS data of each sample under a monochrome light LED-365 nm with a 5% amplitude. The electron migration time of the sample can be obtained by converting the frequency of the minimum imaginary component into Equation (1), which is shown in the Experimental Section. Electron migration time represents the sum of the photogenerated electron time from excitation to the back electrode FTO and the time of photogenerated holes oxidation of the electrolytes in the electric double layer. Figure 5E is the CIMVS data in the same testing conditions. The electronic lifetime can be obtained by introducing the obtained frequency into Equation (2).

From the calculation results shown in Table 1, the electron migration time decreases obviously after hydrogen treatment. However, hydrogen treatment also introduces defects in the surface and bulk phase, which increases the probability of secondary recombination to reduce the lifetime of photogenerated electron holes. After the PEDOT conductive layer deposition, the surface state cannot be oxidized by air, meanwhile, p-n junction can be formed between TiO_2 and PEDOT thin film. The formation of p-n junction electric field accelerates the separation of photogenerated electron holes and reduces the electron migration time. The charge collection efficiency of these three samples was also calculated and the calculation process is shown in Equation (3). $H-TiO_2$-PEDOT photoanode shows a 37.71% charge collection efficiency whose value is higher than that of TiO_2 and $H-TiO_2$, indicating that more real hot carriers can be used in the PEC process.

Table 1. The calculated data through CIMPS and CIMVS results.

	f_{min} **(CIMPS)**	t_r	f_{min} **(CIMVS)**	t_{rec}	η **(%)**
TiO_2	172.24	0.924499	154.22	1.032523	10.46
$H-TiO_2$	673.58	0.236402	536.63	0.296733	20.33
$H-TiO_2$-PEDOT	845.47	0.18834	526.63	0.302367	37.71

IPCE of the series of electrodes were tested and the results are shown in Figure 6A. It can be seen that the photoelectric conversion efficiency of hydrogen-reduced TiO_2 is significantly higher than that of non-reduced TiO_2. After loading PEDOT on the photoelectrode, the $H-TiO_2$-PEDOT electrodes reducing surface state have more than 60% photoelectric conversion efficiency. Figure 6B presents the ultraviolet-visible diffuse reflectance result of the series photoanodes. It can be seen that the absorption band edge of pure TiO_2 is about 400 nm, because of anatase (band gap 3.2 eV) and rutile (band gap 3.0 eV) mix phase. After hydrogen treatment, an indicated absorption can be found from 400 nm to 600 nm, because of the oxygen vacancy energy level formed on the top of the TiO_2 valance band. The light absorption capacity of oxygen modified TiO_2 nanorod array did not change after the PEDOT outer layer loading. Comparing with Figure 6A, there is no photocurrent response of H-TiO_2-PEDOT photoanode in the wavelength area from 400 nm to 600 nm, indicating that there is no IPCE contribution from oxygen vacancy surface energy level. Figure 6C is a photocurrent-time curve measured continuously for 4 h under 0.5 V (vs. Ag/AgCl) external bias voltage. After 4 h continuous illumination, the photocurrent generating by H-TiO_2-PEDOT photoanode decays less than 10% of the initial value, showing acceptable stability. Meanwhile, the oxygen and hydrogen evolution performance were tested during the PEC stability testing for 4 h, and the result shown in Figure 6C indicate that the H-TiO_2-PEDOT photoanode can completely split pure water into hydrogen and oxygen under simulated sunlight illumination.

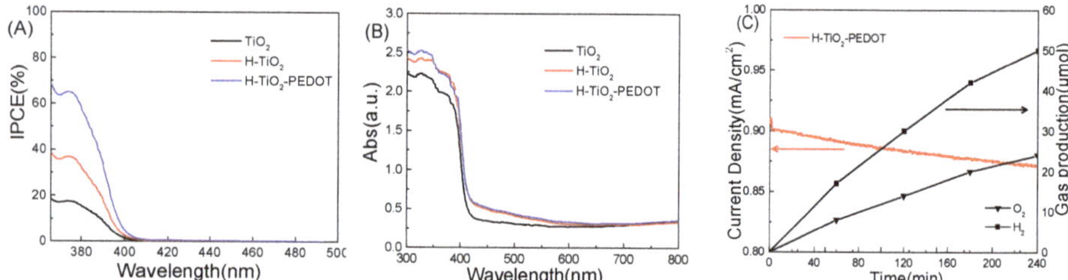

Figure 6. The IPCE curves of series photoelectrodes under 0.5 V (vs. Ag/AgCl) (**A**), UV-Vis DRS of series photoelectrodes (**B**), the stability test of H-TiO_2-PEDOT under 0.5 V (vs. Ag/AgCl) and corresponding oxygen and hydrogen evolution performance (**C**).

In Table 2, the related research on TiO_2 electrodes in recent years is listed. When comparing them, we can see that H-TiO_2-PEDOT electrodes presented in this work obtained relatively high PEC performance.

Table 2. Statistical list of references.

Electrode	Light Source	Voltage	Electrolyte	Current Density
TiO_2 B-NRs [23]	Xe lamp 88 mW cm^{-2}	1.1 V	1 M KOH	0.8 mA/cm^2
TiO_2 nanorod array [24]	AM 1.5 100 mW cm^{-2}	0.5 V	0.5 M NaClO$_4$	15 µA/cm^2
Carbon Dot/TiO_2 Nanorod [25]	Xe lamp 88 mW cm^{-2}	0 V	0.1 M NaSO$_4$ + 0.01 M Na$_2$S	0.35 mA/cm^2
H:TiO_2 nanotube arrays [26]	AM 1.5G 100 mW cm^{-2}	0 V	1 M NaOH	0.6 mA/cm^2
TiO_2 nanotubes [27]	UV light 70 mW cm^{-2}	0.2 V	1 M KOH	0.125 mA/cm^2
This Work	Simulated sunlight 100 mW cm^{-2}	0.5 V	0.1 M NaSO$_4$ + KPi	0.9 mA/cm^2

The PEC performance improving the mechanism of H-TiO_2-PEDOT nanorod photoanode is shown in Figure 7. Firstly, a nanorod array structure of TiO_2 was prepared, which provided a unique route for the photogenerated electron transfer and reduced the recombination rate. In addition, after hydrogen treatment, oxygen vacancies formed on the surface of TiO_2 nanorod, increasing the concentrations of free charge carriers. Lastly, a PEDOT layer was deposited on the surface of oxygen vacancy modified TiO_2, to inhibit the surface states and improve the separation of photo-induced carriers further by p-n heterojunction formation between PEDOT and TiO_2. Thus, more photogenerated holes were transferred to the PEDOT layer and oxidized water, whereas more photogenerated electrons were transferred to the FTO substrate through the TiO_2 nanorod to improve the PEC performance of H-TiO_2-PEDOT photoanode.

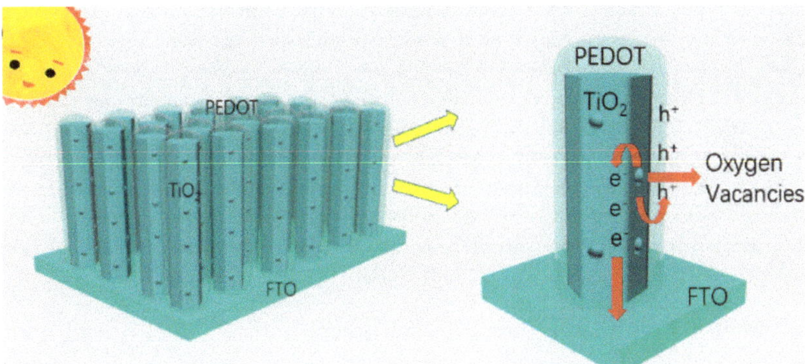

Figure 7. Schematic diagram of the mechanism of H-TiO$_2$-PEDOT nanorod thin film under simulated sunlight.

3. Materials and Methods

All reagents used in this study were purchased from Aladdin Industrial Corporation (Shanghai, China) with analytical grade. Tetrabutyl titanate, 3,4-ethylenedioxythiophene, and sodium dodecyl sulfonate were not further purified.

3.1. TiO$_2$ Nanorod Array and Oxygen Vacancy Modified TiO$_2$ Nanorod Preparation

The TiO$_2$ nanorod arrays were prepared through the solvothermal method. In a typical synthesis, 0.5 mL tetrabutyl titanate was dissolved in 15 mL of hydrochloric acid (36.5%) under continuous stirring, and then 15 mL of deionized water was added for another 5 min to obtain a homogenous solution. The mixed solution was then transferred into a 50 mL Teflon stainless steel autoclave, then two cleaned FTO substrates were immersed into the mixture and kept at 160 °C for 8 h in an oven. After that, the FTO substrates were cleaned with deionized water and then dried under ambient conditions, followed by annealing at 450 °C for 2 h with a ramping rate of 10 °C/min in air in a muffle furnace to obtain TiO$_2$ nanorod array. Then, TiO$_2$ nanorod array was reduced by annealing at 350 °C for 0.5 h with a ramping rate of 10 °C/min in hydrogen conditions, which was denoted as H-TiO$_2$.

3.2. PEDOT Preparation

The PEDOT was coated by H-TiO$_2$ nanorod array through electrodeposition method. Typically, 1 mL 3,4-ethylenedioxythiophene (EDOT) and 20 mmol of sodium dodecyl sulfonate (SDS) were dissolved into 200 mL of deionized water under continuous stirring to prepare precursor solution, the deposition process was carried out in a three-electrode system in the above solution. The as-prepared H-TiO$_2$ photoelectrodes, platinum and Ag/AgCl electrode were served as the working, counter, and reference electrodes, respectively. The electrodeposition was carried out using multi-current steps containing 0.01 s of 1 mA of anodic pulse, 0.004 s of 1 mA of cathodic pulse and 0.5 s of 0 A rest current. This above process is termed as one cycle, and 20 cycles were repeated, the obtained electrode was denoted as H-TiO$_2$-PEDOT. Three-electrode system was used to test the H-TiO$_2$-PEDOT stability with an applied bias of 0.5 V (vs. Ag/AgCl). At the same time, oxygen and hydrogen evolution performance were detected by gas chromatography (97900II) regularly.

3.3. Characterization

The micromorphology of the prepared photoelectrodes was characterized using a field emission scanning electron microscope (FE-SEM, Ultra 55, Zeiss, Oberkochen, Germany) and a field emission transmission electron microscope (FE-HRTEM, JEM-2100F, Beijing, China). TEM sample was scraping the electrode film into powder and filling the power with alcohol or acetone in a small container. Then a small amount of powder sample was put into it, next, it was placed in an ultrasonic oscillator to vibrate for more than 15 min, and then a copper mesh with supporting film was used to gently pull it out from the solution. The elemental compositions of the photoelectrodes were tested through energy dispersive spectroscopy (EDS, X-max, Oxford Instruments, Oxford, England) and scanning transmission electron microscopy (STEM, JEM-2100F, Tokyo, Japan) mapping. X-ray diffraction (XRD, D/MAX-2500/PC, Rigaku Co., Tokyo, Japan) was used to identify the crystalline structures of the prepared series photoelectrodes. The elementary composition and bonding information of the materials were analyzed using X-ray photoelectron spectroscopy (XPS; Axis Ultra, Kratos Analytical Ltd., Kratos Analytical, Manchester, England). Characterization of the optical absorption properties of a series of electrodes was done by UV-Vis diffuse reflectance (TU-1901, Persee Co., Beijing, China).

3.4. PEC Performance Testing

PEC performance measurements were performed in a traditional three-electrode experimental system using Zahner Zennium Pro Electrochemical Workstation (Zahner, Kronach, Germany). The prepared series photoelectrodes, Ag/AgCl (saturated KCl), and a piece of platinum acted as the working, reference, and counter electrodes, respectively. The series photoelectrodes were illuminated under a standard solar simulator (AM1.5G) (LSE341-Zahner, Kronach, Germany). All tests were performed in 0.1 M Na_2SO_4 electrolyte. The photocurrent test with time (I-t) curves was measured at a bias potential of 0 V (vs. Ag/AgCl). The linear sweep voltammetry (I-V) curves were measured from -0.5 to 1.5 V (vs. Ag/AgCl) at a scan rate of 0.02 V s^{-1}. The IPCE of the photoelectrodes were tested at 0.5 V (vs. Ag/AgCl) bias potential using an IPCE tester (TLS03-Zahner, Germany). Electrochemical impedance spectroscopy (EIS) tests were performed at OCP vs. Ag/AgCl (saturated KCl) over the frequency range between 10^5 and 10^{-1} Hz. Control intensity modulated photocurrent/photovoltage spectroscopy (CIMPS/CIMVS) measured series photoelectrodes with an LED white light source (LSW) from 100 K to 0.1 Hz. The electron transit time (τ_r) and electron lifetime (τ_{rec}) can be obtained by the following Equations:

$$\tau_r = 1/(2\pi\, f_{CIMPS}) \tag{1}$$

$$\tau_{rec} = 1/(2\pi\, f_{CIMVS}) \tag{2}$$

$$\eta(\%) = (1 - \tau_r/\tau_{rec}) \times 100\% \tag{3}$$

where f_{CIMPS}/f_{CIMVS} is the frequency of the minimum imaginary component.

4. Conclusions

In this study, PEDOT modified oxygen vacancy-TiO_2 nanorod was prepared, oxygen vacancy can improve the charge transfer capacity of TiO_2. Meanwhile, the PEDOT could not only serve as the protective layer to inhibit the surface states, but also to fabricate a p-n junction to increase the separation efficiency of the photo-generated electrons and holes. Thus, a near 0.9 mA/cm^2 photocurrent of TiO_2 nanorod array was achieved after oxygen vacancy and PEDOT co-modification under standard sunlight illumination. Furthermore, the PEC stability test showed that the photocurrent generating by H-TiO_2-PEDOT photoanode decays less than 10% of the initial value after 4 h of continuous illumination. Meanwhile, the H-TiO_2-PEDOT photoanode can completely split pure water into hydrogen and oxygen under simulated sunlight illumination. Thus, oxygen vacancy and PEDOT co-modification is a promising method for TiO_2 photoanode PEC performance improving.

Author Contributions: Conceptualization, B.Y.; methodology, H.T., L.W.; experiment and analysis, B.Y., G.C.

Funding: This research received no external funding.

Acknowledgments: This work was financially supported by the National Natural Science Foundation of China (Grant Nos. 51679227).

Conflicts of Interest: The authors declare no conflict of interest.

References

1. Fujishima, A.; Honda, K. Electrochemical photolysis of water at a semiconductor electrode. *Nature* **1972**, *238*, 37–38. [CrossRef] [PubMed]
2. Liu, C.; Dasgupta, N.; Yang, P. Semiconductor nanowires for artificial photosynthesis. *Chem. Mater.* **2013**, *26*, 415–422. [CrossRef]
3. Yang, L.; Luo, S.; Li, Y.; Xiao, Y.; Kang, Q.; Cai, Q. High Efficient Photocatalytic Degradation of p-Nitrophenol on a Unique Cu_2O/TiO_2 p-n Heterojunction Network Catalyst. *Environ. Sci. Technol.* **2010**, *44*, 7641. [CrossRef] [PubMed]
4. Kim, H.; Moon, G.; Monllor-Satoca, D.; Park, Y.; Choi, W. Solar Photoconversion Using Graphene/TiO_2 Composites: Nanographene Shell on TiO_2 Core versus TiO_2 Nanoparticles on Graphene Sheet. *J. Phys. Chem. C* **2012**, *116*, 1535–1543. [CrossRef]
5. Ghosh, M.; Liu, J.; Chuang, S.; Jana, S. Fabrication of hierarchical V_2O_5 nanorods on TiO_2 nanofibers and their enhanced photocatalytic activity under visible light. *ChemCatChem* **2018**, *10*, 3305–3318. [CrossRef]
6. Sonawane, R.; Kale, B.; Dongare, M. Preparation and photo-catalytic activity of Fe-TiO_2 thin films prepared by sol-gel dip coating. *Mater. Chem. Phys.* **2004**, *85*, 52–57. [CrossRef]
7. Tian, H.; Zhang, X.; Bu, Y. Sulfur-and carbon-codoped carbon nitride for photocatalytic hydrogen evolution performance improvement. *ACS Sustain. Chem. Eng.* **2018**, *6*, 7346–7354. [CrossRef]
8. Zeng, L.; Lu, Z.; Li, M.; Yang, J.; Song, W.; Zeng, D.; Xie, C. A modular calcination method to prepare modified N-doped TiO_2 nanoparticle with high photocatalytic activity. *Appl. Catal. B Environ.* **2016**, *183*, 308–316. [CrossRef]
9. Lei, F.; Sun, Y.; Liu, K.; Gao, S.; Liang, L.; Pan, B.; Xie, Y. Oxygen vacancies confined in ultrathin indium oxide porous sheets for promoted visible-light water splitting. *J. Am. Chem. Soc.* **2014**, *136*, 6826–6829. [CrossRef]
10. Bu, Y.; Ren, J.; Zhang, H.; Yang, D.; Chen, Z.; Ao, J. Photogenerated-carrier separation along edge dislocation of WO_3 single-crystal nanoflower photoanode. *J. Mater. Chem. A* **2018**, *6*, 8604–8611. [CrossRef]
11. Liu, X.; Zhou, K.; Wang, L.; Wang, B.; Li, Y. Oxygen vacancy clusters promoting reducibility and activity of ceria nanorods. *J. Am. Chem. Soc.* **2009**, *131*, 3140–3141. [CrossRef] [PubMed]
12. Bu, Y.; Tian, J.; Chen, Z.; Zhang, Q.; Li, W.; Tian, F.; Ao, J. Optimization of the Photo-Electrochemical Performance of Mo-Doped $BiVO_4$ Photoanode by Controlling the Metal-Oxygen Bond State on (020) Facet. *Adv. Mater. Interfaces* **2017**, *4*, 1601235. [CrossRef]
13. Wang, S.; Pan, L.; Song, J.; Mi, W.; Zou, J.; Wang, L.; Zhang, X. Titanium-defected undoped anatase TiO_2 with p-type conductivity, room-temperature ferromagnetism, and remarkable photocatalytic performance. *J. Am. Chem. Soc.* **2015**, *137*, 2975–2983. [CrossRef]
14. Wang, J.; Wang, Z.; Huang, B.; Ma, Y.; Liu, Y.; Qin, X.; Zhang, X.; Dai, Y. Oxygen vacancy induced band-gap narrowing and enhanced visible light photocatalytic activity of ZnO. *ACS Appl. Mater. Interfaces* **2012**, *4*, 4024–4030. [CrossRef]
15. Park, Y.; Müller-Meskamp, L.; Vandewal, K.; Leo, K. PEDOT: PSS with embedded TiO_2 nanoparticles as light trapping electrode for organic photovoltaics. *Appl. Phys. Lett.* **2016**, *108*, 253302. [CrossRef]
16. Sakai, N.; Prasad, G.; Ebina, Y.; Takada, K.; Sasaki, T. Layer-by-layer assembled TiO_2 nanoparticle/PEDOT-PSS composite films for switching of electric conductivity in response to ultraviolet and visible light. *Chem. Mater.* **2006**, *18*, 3596–3598. [CrossRef]
17. Liu, B.; Aydil, E. Growth of oriented single-crystalline rutile TiO_2 nanorods on transparent conducting substrates for dye-sensitized solar cells. *J. Am. Chem. Soc.* **2009**, *131*, 3985–3990. [CrossRef]
18. Wu, F.; Yu, Y.; Yang, H.; German, L.; Li, Z.; Chen, J.; Yang, W.; Huang, L.; Shi, W.; Wang, L.; et al. Simultaneous Enhancement of Charge Separation and Hole Transportation in a TiO_2-$SrTiO_3$ Core-Shell Nanowire Photoelectrochemical System. *Adv. Mater.* **2017**, *29*, 1701432. [CrossRef] [PubMed]
19. Fonseca, S.; Moreira, T.; Parola, A.; Pinheiro, C.; Laia, C. PEDOT electrodeposition on oriented mesoporous silica templates for electrochromic devices. *Sol. Energy Mater. Sol. Cells* **2017**, *159*, 94–101. [CrossRef]
20. Taggart, D.; Yang, Y.; Kung, S.; McIntire, T.; Penner, R. Enhanced thermoelectric metrics in ultra-long electrodeposited PEDOT nanowires. *Nano Lett.* **2010**, *11*, 125–131. [CrossRef]
21. Ghosh, M.; Lohrasbi, M.; Chuang, S.; Jana, S. Mesoporous titanium dioxide nanofibers with a significantly enhanced photocatalytic activity. *ChemCatChem* **2016**, *8*, 2525–2535. [CrossRef]

22. Nunes, D.; Pimentel, A.; Santos, L.; Barquinha, P.; Fortunato, E.; Martins, R. Photocatalytic TiO_2 nanorod spheres and arrays compatible with flexible applications. *Catalysts* **2017**, *7*, 60. [CrossRef]
23. Cho, I.; Chen, Z.; Forman, A.; Kim, D.; Rao, P.; Jaramillo, T.; Zheng, X. Branched TiO_2 nanorods for photoelectrochemical hydrogen production. *Nano Lett.* **2011**, *11*, 4978–4984. [CrossRef] [PubMed]
24. Wolcott, A.; Smith, W.; Kuykendall, T.; Zhao, Y.; Zhang, J. Photoelectrochemical water splitting using dense and aligned TiO_2 nanorod arrays. *Small* **2009**, *5*, 104–111. [CrossRef] [PubMed]
25. Bian, J.; Huang, C.; Wang, L.; Hung, T.; Daoud, W.; Zhang, R. Carbon dot loading and TiO_2 nanorod length dependence of photoelectrochemical properties in carbon dot/TiO_2 nanorod array nanocomposites. *ACS Appl. Mater. Interfaces* **2014**, *6*, 4883–4890. [CrossRef] [PubMed]
26. Wang, G.; Wang, H.; Ling, Y.; Tang, Y.; Yang, X.; Fitzmorris, R.; Li, Y. Hydrogen-treated TiO_2 nanowire arrays for photoelectrochemical water splitting. *Nano Lett.* **2011**, *11*, 3026–3033. [CrossRef] [PubMed]
27. Li, Y.; Yu, H.; Song, W.; Li, G.; Yi, B.; Shao, Z. A novel photoelectrochemical cell with self-organized TiO_2 nanotubes as photoanodes for hydrogen generation. *Int. J. Hydrogen Energy* **2011**, *36*, 14374–14380. [CrossRef]

Article

Optimization of Photocatalytic Degradation of Acid Blue 113 and Acid Red 88 Textile Dyes in a UV-C/TiO$_2$ Suspension System: Application of Response Surface Methodology (RSM)

Soroosh Mortazavian [1], Ali Saber [2,*] and David E. James [2]

[1] Department of Mechanical Engineering, University of Nevada, Las Vegas, Las Vegas, NV 89154, USA; mortazav@unlv.nevada.edu

[2] Department of Civil and Environmental Engineering and Construction, University of Nevada, Las Vegas, Las Vegas, NV 89154, USA; dave.james@unlv.edu

* Correspondence: sabersic@unlv.nevada.edu; Tel.: +1-702-285-2836

Received: 19 March 2019; Accepted: 8 April 2019; Published: 14 April 2019

Abstract: Textile industries produce copious amounts of colored wastewater some of which are toxic to humans and aquatic biota. This study investigates optimization of a bench-scale UV-C photocatalytic process using a TiO$_2$ catalyst suspension for degradation of two textile dyes, Acid Blue 113 (AB 113) and Acid Red 88 (AR 88). From preliminary experiments, appropriate ranges for experimental factors including reaction time, solution pH, initial dye concentration and catalyst dose, were determined for each dye. Response surface methodology (RSM) using a cubic IV optimal design was then used to design the experiments and optimize the process. Analysis of variance (ANOVA) was employed to determine significance of experimental factors and their interactions. Results revealed that among the studied factors, solution pH and initial dye concentration had the strongest effects on degradation rates of AB 113 and AR 88, respectively. Least-squares cubic regression models were generated by step-wise elimination of non-significant (p-value > 0.05) terms from the proposed model. Under optimum treatment conditions, removal efficiencies reached 98.7% for AB 113 and 99.6% for AR 88. Kinetic studies showed that a first-order kinetic model could best describe degradation data for both dyes, with degradation rate constants of $k_{1,\,AB\,113}$ = 0.048 min^{-1} and $k_{1,\,AR\,88}$ = 0.059 min^{-1}.

Keywords: process optimization; response surface methodology; kinetic study; Advanced oxidation processes (AOPs); TiO$_2$ catalyst; textile wastewater

---

1. Introduction

Dyes are widely used in several industries such as textile industry, paper, plastics, food, cosmetics and so forth. [1]. The textile industry has large water consumption and thereby, produces copious amounts of colored wastewater. It has been estimated that 1–20% of total dye consumption is lost during the dying process, which is subsequently introduced to the receiving water bodies [2]. Some dyes are carcinogenic and toxic to humans and aquatic biota [3], requiring appropriate treatments. Methods for color removal are generally divided into three main groups: physical, chemical and biological treatments. Physical methods, such as adsorption and screening, only transfer pollutants from one phase to another; therefore post-treatment is necessary for complete removal of contaminants [4]. The toxic nature and complex molecular structures of many dyes limit their biological degradation [5]. Hence, biological methods are usually not able to treat colored wastewaters [6]. In addition, biological methods have a disadvantage of producing large volumes of sludge [7]. Chemical methods, on the other hand, have demonstrated more promising results [4]. In chemical treatment methods, instead of transferring contaminants from one phase to another, the dyes are converted into harmless substances. Advanced oxidation processes (AOPs) are among the most powerful chemical treatment techniques used for removal of organic compounds. AOPs are characterized by in-situ generation of hydroxyl radicals ($^\bullet$OH), which are strongly oxidizing species (oxidative potential +2.8 V) [6]. Hydroxyl radicals unselectively attack organic molecules to degrade them into simpler and less harmful compounds and ultimately, convert them into CO_2, H_2O and mineral acids [4,8,9]. Photocatalytic degradation is an advanced oxidation method in which hydroxyl radicals are generated by irradiating UV light on a semiconductor catalyst [10]. In the past several decades, titanium dioxide (TiO_2) has been proved to be more efficient for the photocatalytic processes than other semiconductors. It is inexpensive, non-toxic, water-insoluble, highly reactive and photochemically stable [6].

In a photocatalytic reaction, when TiO_2 particles are illuminated with a light source having energy greater than its band gap (E_{g,TiO_2} = 3.2 eV), electrons in the valence band promote to the conduction band, creating electron-hole pairs [6,11]. Formation of electron-hole pairs is a fast-reversible reaction. To prevent this, an electron acceptor, which is dissolved in most cases, is necessary to entrap free electrons and reduce the rate of electron-hole recombination [10]. Photogenerated electrons may also react with dye molecules and reduce them [6]. Holes, on the other hand, can react with hydroxide ions (OH^-) or adsorbed H_2O on the catalyst's surface and generate hydroxyl radicals. Finally, dye molecules will react with the formed radicals. This reaction takes place on the surface of the catalyst particles and will continue until complete mineralization of the organic species [10]. The photocatalytic reactions described above can be summarized as [8]:

$$TiO_2 + h\nu \rightarrow e^- + h^+ \tag{1}$$

$$e^- + O_2 \rightarrow O{\cdot}_2^- \tag{2}$$

$$h^+ + Organic \rightarrow CO_2 \tag{3}$$

$$h^+ + H_2O \rightarrow {\cdot}OH + H^+ \tag{4}$$

$${\cdot}OH + Organic \rightarrow CO_2 + H_2O \tag{5}$$

A number of studies have investigated aqueous phase photodegradation of various dyes using TiO_2 catalyst [1,7,9,11–18]. Sohrabi and Ghavami (2008) [1] studied photocatalytic degradation of Direct Red 23 using UV/TiO_2 system. They reported an increase in dye decomposition rate with increasing TiO_2 concentration up to 4.0 g/L; the rate then decreased with further increases in catalyst dose. Juang et al. (2010) [7] investigated photodegradation and mineralization of single and binary Acid Orange 7 (AO7) and Reactive Red 2 (RR2) under UV irradiation in TiO_2 suspensions. Their results showed that after 20 min of UV irradiation with 0.5 g/L TiO_2, complete removals of single AO7 and RR2 were achieved at pH 6.8. Photocatalytic degradation of Amaranth dye was investigated in a UV-C/TiO_2 system by Gupta et al. (2012) [13]. They obtained degradation efficiencies of 17%, 26%, 38% and 64% for UV, UV + H_2O_2, UV + TiO_2 and UV + TiO_2 + H_2O_2 systems, respectively, after 100 min irradiation. Barakat (2011) [9] investigated the removal of Procion® yellow H-EXL dye over TiO_2 suspension and obtained 100% photodegradation efficiency under optimum conditions of pH = 5.0, TiO_2 dose = 1.0 g/L and dye concentration = 10 mg/L. Toor et al. (2006) [14] evaluated the photocatalytic degradation of Direct Yellow 12 in a shallow pond slurry using TiO_2 suspension under irradiation of UV light using black fluorescent lamps lies in UV-A range. After 1.5 h and under optimum conditions (TiO_2 dosage = 2.0 g/L, pH = 4.5 and initial dye concentration = 100 mg/L) complete decolorization was achieved. Khataee et al. (2009) [15] investigated degradation of three azo dyes by UV-A irradiation using immobilized TiO_2 and achieved complete decolorization after 6 h at natural pH and an initial dye concentration of 30 mg/L.

However, to the best knowledge of authors, there are limited numbers of studies [19] which have assessed interaction effects between operational factors in the photocatalytic degradation process of dyes and optimized the process.

There are several classical methods for design and optimization of experiments. For instance, the one-factor-at-a-time method does not consider the interactions among experimental factors [20,21]. The full factorial method, considers interaction effects through a great number of experiments but can be time-consuming and costly in multi-variable systems [20,21]. Response surface methodology (RSM) is a collection of statistical and mathematical methods used for development of a functional relationship between a response of interest and a number of input variables [22]. This method is applied for designing experiments, evaluating the effects of individual operational parameters and their interaction effects and optimizing the parameters, with a significant reduction in the number of experiments [23–25]. Several recent studies have optimized the response of various environmental treatment processes using models based on RSM [19,24–33]. For example, Saber et al. (2014, 2017) [26,27] used RSM to optimize Fenton and photo-Fenton processes for treatment of petroleum refinery effluents, Cifuentes et al. (2017) [34] used RSM for simulation of the ethanol's catalytic steam reforming, Li et al. (2018) [35] used RSM to investigate photocatalytic performance and degradation mechanism of Aspirin by TiO_2, Inger et al. (2019) [36] optimized ammonia oxidation using RSM and Aljuboury et al. (2016) [37] optimized TiO_2/ZnO photodegradation of petroleum refinery wastewaters by using RSM.

The current work, for the first time, optimizes experimental conditions for photocatalytic degradation of two anionic textile dyes, Acid Blue 113 (AB 113) and Acid Red 88 (AR 88), in a TiO_2 suspension system using UV-C irradiation. A simple enclosed bench-scale batch photoreactor was constructed for this study. Mercury vapor UV-C lamps were positioned over a relatively shallow free surface dye solution to provide sufficient light penetration as well as reduce costs of employing quartz tubes to immerse a UV lamp in the solution. In RSM-designed experiments, initial dye concentration, catalyst loading and solution pH were considered as independent parameters. The degradation efficiency of dyes was the target response. Modeling photocatalytic degradation efficiency, examining the influences of several variable parameters on degradation efficiency and their interactions and determining optimum conditions for dye removal were conducted using RSM with a cubic IV optimization method. Kinetic studies were also conducted to evaluate dyes' photodegradation rates under optimum conditions.

2. Results and Discussions

2.1. Stage 1: Preliminary Experiments

Figure 1 shows the effects of different experimental factors on photocatalytic degradation of AB 113 and AR 88 textile dyes in the first stage one-factor-at-a-time preliminary experiments.

Figure 1a shows that increasing TiO_2 dose up to 2.0 g/L and 1.0 g/L, enhanced removal efficiencies of AB 113 and AR 88, respectively. Increases in catalyst dose above these thresholds resulted in decreased removal efficiencies. Even though increase of TiO_2 particles in the solution might provide more active sites for the dye molecules to be adsorbed and degraded, excessive amounts of catalyst particles might aggregate leading to a decreased number of active sites. Excessive TiO_2 doses can also increase the opacity of solution and consequently reduce the penetration of UV light and thus decrease the treatment efficiency [1,38]. These results agreed with the previous studies [1,9,14,38] reporting decreased photocatalytic efficiency when applying TiO_2 dose above an optimum value.

Figure 1b shows that increasing reaction time improved removal percentages for both dyes. A higher overall removal efficiency was observed for AR 88 compared with AB 113 over the entire reaction time. As observed, only 7.4% and 8.7% increases in removal efficiencies were observed after 90 min for AB 113 and AR 88 dyes, respectively, indicating that 90 min was sufficient for most of the dye degradation reactions to occur. Hence, a fixed reaction time of 90 min was considered for both dyes in the main experiments (i.e., second stage experiments), while other experimental factors (e.g., initial dye concentration, catalyst dose and pH) were optimized using RSM.

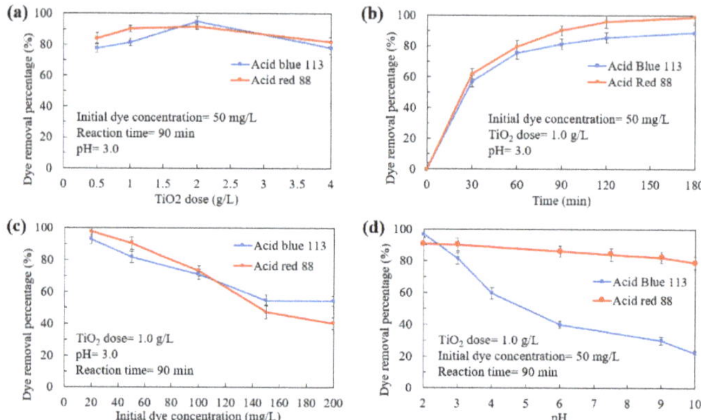

Figure 1. Results of preliminary experiments for photocatalytic degradation of Acid Blue (AB) 113 and Acid Red (AR) 88 textile dyes. Effects of changing (**a**) TiO_2 dose (g/L) (Initial dye concentration = 50 mg/L, pH = 3.0, Reaction time= 90 min), (**b**) reaction time (min) (Initial dye concentration = 50 mg/L, pH = 3.0, TiO_2 dose = 1.0 g/L), (**c**) Initial dye concentration (mg/L) (Reaction time = 90 min, pH = 3.0, TiO_2 dose = 1.0 g/L) and (**d**) pH (Initial dye concentration = 50 mg/L, Reaction time = 90 min, TiO_2 dose = 1.0 g/L). Error bars show standard deviation of duplicate runs.

Figure 1c illustrates that, as expected, increasing dye concentration from 20 mg/L to 200 mg/L decreased removal efficiencies in both dyes (from 97.6% to 40% for AR 88 and from 92.6% to 54.4% for AB 113). This was likely because increasing dye concentration in the solution while maintaining a constant catalyst dose caused the fixed number of catalysis sites to be saturated faster [39]. In addition, increased dye concentration probably decreased the light transmittance in the solution. Decreased UV penetration can reduce the activation rate of TiO_2 particles and hinder the generation of $^{\bullet}OH$ radicals, resulting in decreased photocatalytic degradation efficiencies for both dyes [39].

Figure 1d represents the effects of changing pH from 2.0 to 10.0 on removal efficiencies of AB 113 and AR 88. For AB 113, increasing pH from 2.0 to 6.0 caused a marked decrease in removal efficiency from 96.7% to 40.3% (56.4% decrease). Further increasing pH above 6.0 up to 9.0 decreased efficiency by only an additional 18%. For AR 88, increasing solution pH from 2 to 10 only resulted in 11.8% decrease in degradation efficiency. Detailed discussion about the effects of solution pH on each of dyes and possible interactions with other factors are presented next in Section 2.2.

In order to investigate the sole contribution of adsorption in removal of dyes from the aqueous solutions, experiments were conducted under dark conditions (i.e., without UV-C radiation) for both AB 113 and AR 88 dyes. After 90 min reaction under dark conditions in the closed photoreactor under the fixed experimental conditions of TiO_2 dose = 1.0 g/L, initial dye concentration = 50 mg/L and pH = 3.0, concentrations of both AB 113 and AR 88 remained unchanged when the reacted samples were analyzed by the UV-VIS spectrophotometer. This showed that adsorption onto TiO_2 particles did not by itself have a significant role in dye removal, demonstrating that the observed removal of dyes in the UV-C/TiO_2 system were due to photocatalytic process.

2.2. Stage 2: Process Optimization

2.2.1. Response Surface Plots, Fitted Models and ANOVA

Figures 2 and 3 show the results obtained from the 30 experimental runs (i.e., stage 2) for AB 113 and AR 88, respectively, after 90 min reaction time. Analysis of variance (ANOVA) for AB 113 and AR 88 are presented in Tables 1 and 2, respectively. Significance of the model terms was evaluated based on computed F-statistic values and their associated *p*-values. Least-squares cubic regression models were generated by eliminating non-significant terms (*p*-value > 0.05). Reduced cubic models for AB 113 and AR 88 are expressed in the Equations (6) and (7), respectively.

$$\text{AB 113 Removal (\%)} = 76.75 - 21.30A - 17.41B - 12.47C + 2.28AB - 2.26AC - 8.06A^2 - 11.57B^2 - 5.14A^2C - 4.44AB^2 + 5.23AC^2 + 16.52B^3 \tag{6}$$

$$\text{AR 88 removal (\%)} = 74.21 - 2.00A - 7.67B - 19.61C - 2.03AC - 1.25BC - 11.18B^2 - 1.63ABC + 3.57AB^2 - 3.99AC^2 - 3.30B^2C - 4.44A^3 + 7.18B^3 - 3.95C^3 \tag{7}$$

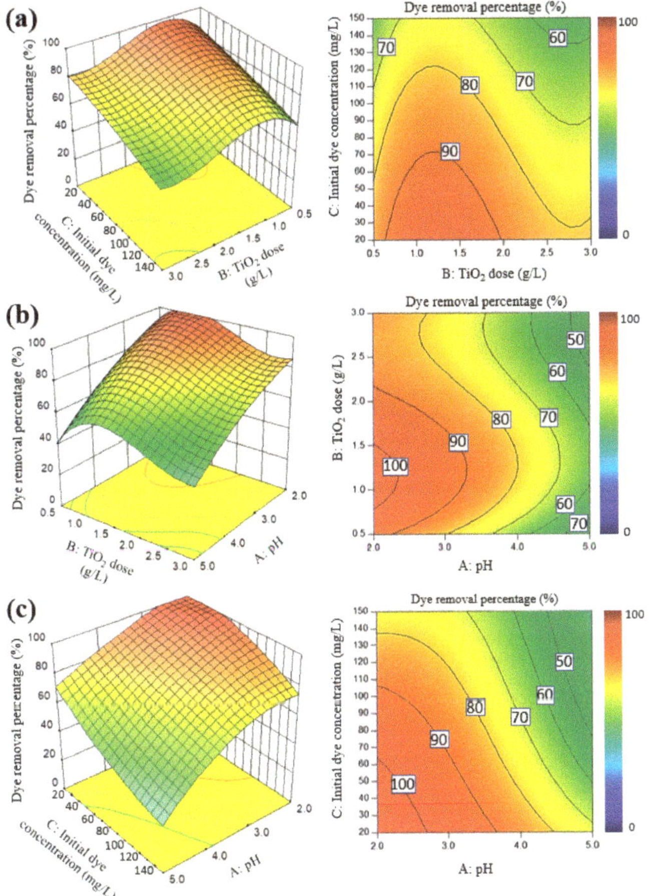

Figure 2. Response surface and contour plots for photocatalytic degradation of AB 113 as a function of (**a**) C: initial dye concentration (mg/L) and B: TiO$_2$ dose (g/L) (pH = 3.0, reaction time = 90 min), (**b**) A: pH and B: TiO$_2$ dose (g/L) (initial dye concentration = 50 mg/L, reaction time = 90 min) and (**c**) A: pH and C: initial dye concentration (mg/L) (TiO$_2$ dose = 1.0 g/L, reaction time = 90 min).

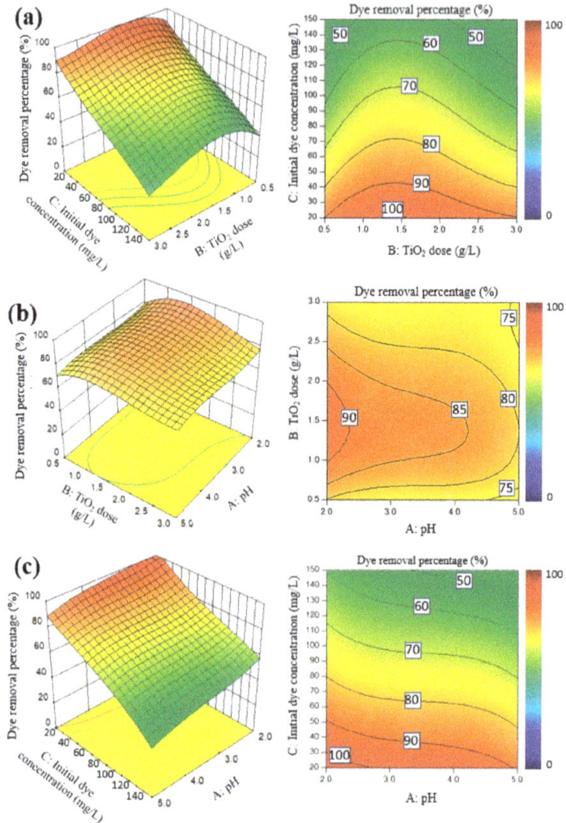

Figure 3. Response surface and contour plots for photocatalytic degradation efficiency of AR 88 as a function of (**a**) C: initial dye concentration (mg/L) and B: TiO$_2$ dose (g/L) (pH = 3.0, reaction time = 90 min), (**b**) A: pH and B: TiO$_2$ dose (g/L) (initial dye concentration = 50 mg/L, reaction time = 90 min) and (**c**) A: pH and C: initial dye concentration (mg/L) (TiO$_2$ dose = 1.0 g/L, reaction time = 90 min).

Table 1. Analysis of variance for modified cubic model obtained for photocatalytic degradation of AB 113 in the UV-C/TiO$_2$ system.

Source	F-Statistic Value	p-Value
Model	86.33	<0.0001
A-pH	129.56	<0.0001
B-TiO$_2$ dose (g/L)	18.05	0.0005
C-Initial dye concentration (mg/L)	44.81	<0.0001
AB	4.11	0.0578
AC	4.58	0.0462
A^2	20.69	0.0002
B^2	50.53	<0.0001
A^{2}C	5.13	0.0360
AB^2	4.56	0.0467
AC^2	5.87	0.0261
B^3	14.68	0.0012
Lack of Fit	0.51	0.8499

Table 2. Analysis of variance for modified cubic model obtained for photocatalytic degradation of AR 88 in the UV-C/TiO$_2$ system.

Source	F-Statistic Value	*p*-Value
Model	217.009	<0.0001
A-pH	0.773	0.392
B-TiO$_2$ dose (g/L)	14.338	0.0016
C-Initial dye concentration (mg/L)	92.905	<0.0001
AC	9.611	0.0069
BC	3.301	0.088
B^2	90.316	<0.0001
ABC	4.961	0.0406
AB^2	6.835	0.0188
AC^2	6.427	0.0221
B^{2}C	5.822	0.0282
A^3	3.379	0.0847
B^3	10.334	0.0054
C^3	3.466	0.0811
Lack of Fit	1.312	0.4046

Figure 2 shows that removal efficiencies greater than 90% for AB 113 occurred for conditions of pH between 2.0 and 3.0, TiO$_2$ dose of 0.7 to 2.0 g/L and dye concentration of 20 to 65 mg/L.

For AR 88, Figure 3 shows that removal efficiencies greater than 90% occurred for conditions of pH between 2.0 and 4.5, TiO$_2$ dose of 1.0 to 2.0 g/L and dye concentration between 20 and 30 mg/L.

As observed from Figure 2 and Equation (6), AB 113 removal efficiency decreased with increasing pH (term A) and initial dye concentration (term C), which is consistent with preliminary experiments. pH could affect the speciation of dye molecules and consequently, the electrostatic force between catalysts particles and dye molecules. Since chemical reactions associated with photocatalytic degradation take place on the surface of the catalyst particles [10,38], adsorption of the contaminants' molecules onto catalyst surface is an essential step for efficient photocatalytic degradation [40].

The point of zero charge (zpc) for TiO$_2$ Degussa P 25 is 6.5 [41]. Hence, the TiO$_2$ surface was positively charged at pH < 6.5 and negatively charged at pH > 6.5. On the other hand, AB 113 is a disulfonate acid dye (having two sulfonated ($-SO_3^-$) groups) with an acid dissociation constant (pKa) of 0.5 [38]. Therefore, AB 113 tended to be negatively charged at pH > 0.5. The increased density of positive charges on the surface of TiO$_2$ particles under acidic conditions was likely to be favorable for adsorbing AB 113 [42] and consequently improved photodegradation efficiency. The cubic regression model also showed that there were interaction effects between these two factors (i.e., A for pH and C for dye concentration) due to presence of the statistically-significant terms AC, A^2C and AC^2 in the equation. These significant interactions indicated that changing pH affected the speciation and ionization state of AB 113 dye molecules as well as TiO$_2$ particles' surface charge [43].

Effects of catalyst dose (term B) on AB 113 dye degradation were also consistent with preliminary experiments, as well as studies reported in the literature [1,9,14], which is improving removal efficiency by increasing TiO$_2$ dose to an optimum point and then reducing the efficiency at greater values. Statistically significant (*p*-value < 0.05) interactions between the catalyst dose (term B) and the solution pH (term A) were also observed in ANOVA results. These interactions could be due to the effect of pH on the surface charge of TiO$_2$, affecting the adsorption of dye molecules on its surface. Since the reaction between hydroxide ions in the solution and holes on the surface of TiO$_2$ particles could generate hydroxyl radicals, an alkaline environment could be favorable for hydroxyl radicals' generation. However, the electrostatic repulsion between the negatively charged surface of TiO$_2$ and OH$^-$ anions in an alkaline environment would hinder the formation of hydroxyl radicals leading to a reduced degradation efficiency [44]. Venkatachalam et al. [45] reported that an acidic environment is beneficial for photocatalytic degradation by TiO$_2$, since it minimizes electron-holes recombination and enhances $^\bullet$OH production. In addition, TiO$_2$ particles agglomerate in alkaline conditions, leading to a reduced exposed surface area to the energy source (UV-C light) [46]. This could be another reason for the decreased removal efficiency of both dyes at higher pH values.

Mohammadzadeh et al. (2015) [38] investigated the photodegradation of AB 113 using ZnO-Ag catalyst under UV illumination. They reported that although there is a stronger electrostatic attraction force between ZnO-Ag and AB 113 at $0.5 < pH < 9.0$, which is favorable for photocatalytic degradation reactions, since the catalyst dissolves at $pH < 3.0$, the overall decolorization was enhanced at higher pH values. Ma et al. (2011) [47] studied photooxidation of three azo dyes including AB 113 using TiO_2/H_2O_2 under vacuum ultraviolet (VUV, $\lambda < 190$ nm) irradiation. Consistent with the results of the present study, they found that lower pH values (in their experimental range of $3.0 < pH < 9.0$) enhanced AB 113 photodegradation.

Figure 3 shows that AR 88 removal efficiencies exceeding 90% occurred for dye concentrations between 20 and 53 mg/L and TiO_2 doses between 1.0 and 2.0 g/L. In the studied range, pH (term A) was not found to be a significant factor in AR 88 removal efficiency. Increasing initial AR 88 dye concentration (term C) reduced its degradation efficiency. Table 2 and Equation (7) indicate high interactions between all three factors (AC, BC, ABC, AB^2, AC^2 and B^2C). Presence of interaction terms between A and C (i.e., pH and initial dye concentration) in the regression model suggested that these two factors were not completely independent. Hence, it could be inferred that changing pH in a wider range probably would show more intense effects on the response. This could also be explained by the electrostatic repulsion between negatively charged surface of TiO_2 particles in alkaline environment and negatively charged sulfonic groups ($R\text{-}SO_3^-$) present in structure of AR 88 [3,43]. Therefore, increased negative surface charge of TiO_2 particles due to increased pH could have hindered adsorption of AR 88 onto catalyst surface and consequently, reduced the removal efficiency. However, a pK_a of 10.7 for AR 88 [48] suggests that AR 88 was not highly ionized in the pH range used in this study. Thus, pH effects on degradation of AR 88 were not very significant. Similar to AB 113, changing TiO_2 dose (term B) showed an optimum point for AR 88 photodegradation beyond which increasing catalyst dose reduced removal efficiency. For AR 88, the pH parameter (term A) was kept in the model because of the significant interaction effects between pH and other factors (i.e., AC, AC^2, ABC, AB^2).

The obtained models' p-values of < 0.0001 demonstrated significance of cubic models for both dyes. High lack-of-fit p-values of 0.8499 and 0.4046 for AB 113 and AR 88, respectively, confirmed that both reduced cubic models were statistically significant. High calculated F-statistic values and correspondingly low associated p-values for each retained parameter (Tables 1 and 2) indicated highly significant effects of each retained specific parameter or combination of parameters on removal efficiency. The potency of experimental variables on changing removal percentages could be graded as pH > initial dye concentration > TiO_2 dose for AB 113 and initial dye concentration > TiO_2 dose > pH for AR 88. In addition, interaction between pH and the second power of initial dye concentration for AB 113 (AC^2) was found to be the most significant interaction, while for AR 88 the interaction between pH and initial dye concentration for (AC) was the most significant interaction (See Figure S1).

A summary of the fitted models' statistical characteristics for two studied dyes, as well as for the cubic models before modification is shown in Table 3.

Values of R^2_{adj} of 0.9700 for AB 113 and 0.9898 for AR 88 (Table 3) indicate that both reduced models could describe a very large portion of the variance in the design space. Table 3 shows coefficients of variation (standard deviation/mean) of 5.88% and 3.47% for AB 113 and AR 88, respectively, meaning that standard deviations were 5.88% and 3.47% of the mean, respectively.

Table 3. Summary of fitted models' characteristics for photocatalytic degradation of AB 113 and AR 88 textile dyes in UV-C/TiO$_2$ system, before and after removing insignificant terms.

Item	AB 113		AR 88	
	Initial Cubic Model	Reduced Cubic Model	Initial Cubic Model	Reduced Cubic Model
Standard deviation	3.90	3.78	2.35	2.27
Mean	64.24	64.24	65.35	65.35
Coefficient of variation, %	6.06	5.88	3.60	3.47
PRESS	1205.01	664	1780.39	265.73
R^2	0.9890	0.9814	0.9962	0.994
R^2_{adj}	0.9682	0.9700	0.9890	0.990
Adequate precision	26.392	34.834	33.969	42.061

Adequate precision was obtained 34.834 and 42.061 for AB 113 and AR 88, respectively. Values greater than 4 for this factor are desirable. High values denote an adequate signal and show that the model can navigate the design space [23]. Comparing values before and after models' modification, it is observed in Table 3 that the modifications enhanced the signal-to-noise ratio for both dyes, reflected as increased adequate precision values.

For a specific model, a lower value for the Predicted Residual Error Sum of Squares (PRESS, see Equation (13) in Section 3.4) is favorable, showing that the model is not overly sensitive to any single data point [49]. Table 3 shows that PRESS values decreased by 81.5% and 567% after modification of AB 113 and AR 88 models, respectively, indicating that the cubic models were improved by removing statistically insignificant data points.

2.2.2. Optimization

In order to verify the accuracy of the reduced cubic models in predicting optimum treatment condition, a third round of experiments was carried out under optimum conditions. Predicted optimum operating conditions and removal efficiencies as well as the obtained experimental results are shown in Table 4. Removal efficiencies of 98.7% and 99.6% under optimum conditions were achieved for AB 113 and AR 88 dyes, respectively; values reasonably close to 100% removal efficiencies predicted by the reduced cubic models.

Table 4. Optimum conditions for photocatalytic degradation of AB 113 and AR 88 dyes in UV-C/TiO$_2$ suspension system.

	pH	Initial Dye Concentration (mg/L)	TiO$_2$ Dose (g/L)	Predicted Removal Efficiency (%)	Achieved Removal Efficiency (%)
AB 113	2.21	43.13	0.98	100%	98.7%
AR 88	2.36	22.40	1.22	100%	99.6%

Regarding the practical applications of the optimized conditions, it should be noted that although highly acidic conditions—causing high operational costs—were proposed for the complete degradation of both AB 113 and AR 88, Figures 2 and 3 illustrate that high removal efficiencies could be achieved in a wider range of operating conditions. For example, having an initial dye concentration of 50 mg/L and a reaction time of 90 min, removal efficiencies of almost >80% could be achieved with pH increased to 4.0 and 5.0 for AB 113 and AR 88, respectively.

2.3. Stage 3: Kinetics of Photocatalytic Degradation

In order to evaluate photocatalytic degradation rates of AB 113 and AR 88, kinetic studies were performed for each dye under the optimum experimental conditions. Results are shown in Figure 4 and reaction rate constants and model characteristics associated with fitted kinetic models are shown in Table 5.

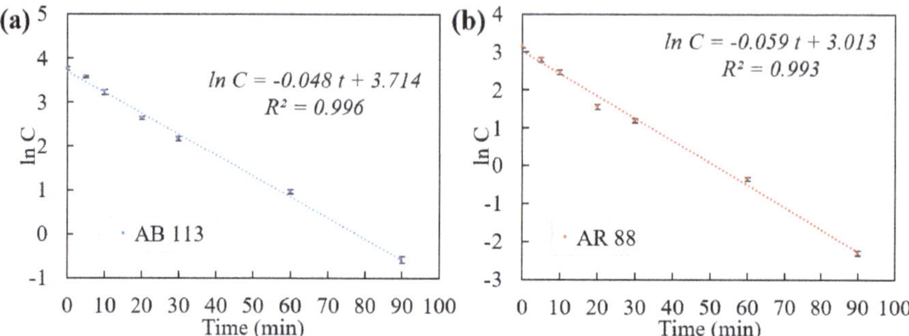

Figure 4. First-order kinetic models for photocatalytic degradation of (**a**) AB 113 and (**b**) AR 88 dyes in UV/TiO$_2$ suspension system under optimized conditions for each dye (AB 113: initial dye concentration = 43.13 mg/L, reaction time = 90 min, TiO$_2$ dose = 0.98 g/L, pH = 2.2; AR 88: initial dye concentration = 22.40 mg/L, reaction time = 90 min, TiO$_2$ dose = 1.22 g/L, pH = 2.4). Error bars show standard deviation.

Table 5. Characteristics of First order kinetic models for photocatalytic degradation of AB 113 and AR 88 in a UV-C/TiO$_2$ suspension system.

	k_1 (min^{-1})	p-Value for k_1	$C_{o,\,model}$	p-Value for ln C_o	R^2	RMSE (mg/L)
AB 113	0.048	2.13×10^{-7}	41	1.13×10^{-8}	0.996	1.72
AR 88	0.059	8.60×10^{-7}	20.4	3.70×10^{-7}	0.993	1.12

High coefficients of determination ($R^2_{AB\,113} = 0.996$ $R^2_{AR\,88} = 0.993$) and low root mean square error (*RMSE*) values ($RMSE_{C,\,AB\,113} = 1.72$ and $RMSE_{C,\,AR\,88} = 1.12$) between the first order kinetic models and experimental values shown in Table 5 demonstrated that first-order kinetic models were appropriate for observed dye degradation under optimum conditions.

AR 88 showed a higher degradation rate compared to AB 113 ($k_{1,\,AR\,88} = 0.059$ min^{-1} with a p-value of 8.60×10^{-7} compared to $k_{1,\,AB\,113} = 0.048$ min^{-1} with a p-value of 2.13×10^{-7}) under optimum conditions. This is consistent with the preliminary results shown in Figure 1b indicating a higher removal percentage for AR 88 compared to AB 113 at all reaction times.

Ma et al. (2011) [47] reported a pseudo-first order rate constant of $k = 0.2469$ min^{-1} at pH = 3.0 as the highest degradation rate for AB 113 with an initial dye concentration of 0.0523 mM in the studied pH range of 3.0 to 11.0, using a VUV/TiO$_2$ system, with the VUV lamp immersed in the dye solution and TiO$_2$ dose of 0.5 g/L. After 60 min, 60% of AB 113 was decomposed. The higher reaction rate obtained by Ma et al. (2011) [47] compared to the present study could be due to the application of VUV, with a lower wavelength and thus higher energy compared to UV-C, which potentially enhance the excitation of TiO$_2$ particles. In addition, immersing the VUV lamp inside the dye solution using a quartz tube also provides a better exposure of catalyst particles to the energy source. The present study proposed a more economical approach by using a higher wavelength UV-C (meaning a lower energy, primary emission band 254 nm) source and eliminating the use of quartz-tube through direct radiation of UV on the solution surface. Mohammadzadeh et al. (2015) [38] obtained a pseudo-first-order rate constant of 0.007 min^{-1} for AB 113 photodegradation in a ZnO-Ag/UV system with an immersed UV lamp, under the conditions of initial dye concentration = 40 mg/L, catalyst dose = 0.15 g/L and optimum pH = 8.0. After 90 min, almost 50% of AB 113 degraded. Their lower degradation rate compared to the present study might be due to the application of a different catalyst at a lower dose.

Anandan et al. (2008) [11] studied the photocatalytic degradation of AR 88 using Ag-loaded TiO$_2$ particles (Ag/TiO$_2$) under visible light and compared the photodegradation rates with using unloaded

TiO$_2$. They obtained first-order rate constants of about 0.006 min^{-1} and 0.008 min^{-1} for TiO$_2$ and Ag/TiO$_2$, respectively, using an initial AR 88 concentration of 0.034 mg/L, a TiO$_2$ dose of 0.6 g/L, with no pH adjustments. After 425 min, 55% TOC removal was observed using Ag/TiO$_2$. Konyar et al. (2017) [50] studied photocatalytic degradation of AR 88 using sintered-reticulated ZnO catalyst under UV-A and UV-C radiations, in a quartz tube reactor surrounded by a cylindrical light assembly. They obtained pseudo-first order rate constants of about 0.007 min^{-1} and 0.009 min^{-1} for photodegradation under UV-A and UV-C radiations, respectively, having initial AR 88 concentration of 50 mg/L and catalyst dose of 40 g/L, without pH adjustment. After 180 min, 60% and 80% color removal percentages were obtained under UV-A and UV-C radiation, respectively.

As shown in Figure 5, AB 113 is a diazo naphthyl dye and AR 88 is a mono-azo naphthyl dye, having conjugated chromophores responsible for their color. When TiO$_2$ is added to the dye solutions, AB 113 and AR 88 molecules are adsorbed mainly through their sulfonate groups [51]. The main degradation pathway proposed by previous researchers studying photodegradation of naphthyl azo dyes in AOP systems [51,52] is the attack of hydroxyl radicals to the naphthalene ring, forming a hydroxylated naphthyl azo dye which is subsequently cleaved. Additionally, hydroxyl radicals attack the aromatic rings with azo groups resulting in azo bond cleavage. Both these reactions result in chromophoric group destruction [52]. Mohammadzadeh et al. (2015) [38] investigated degradation pathway and reaction byproducts for AB 113 photodegradation using a ZnO-Ag nanophotocatalyst under UV radiation. They showed that cleavage of azo bond during a 90 min photocatalysis reaction resulted in the formation of 4-diazenyl-1-naphthylamine, 1-naphthyldiazene or 5-diazenyl-1-naphthol intermediate compounds, which were gradually converted to CO$_2$ and H$_2$O [38]. Madhavan et al. (2010) [53] proposed a pathway for the TiO$_2$ mediated photocatalytic degradation of AR 88 by investigating reaction intermediates using a mass spectrometer and showed formation of hydroxyamino naphthol (4) and 4-aminonaphthalene sulfonic acid (5) as the intermediate products. In the proposed pathway, they showed that AR 88 photocatalytic degradation was mainly due to the hydroxyl radical attack to the aromatic rings [53]. It is expected that the mechanism of hydroxyl radicals attack to AB 113 and AR 88 molecules in the UV-C/TiO$_2$ system to be similar to the mechanisms proposed by Mohammadzadeh et al. (2015) [38] and Madhavan et al. (2010) [53], consisting of cleavage of azo bonds.

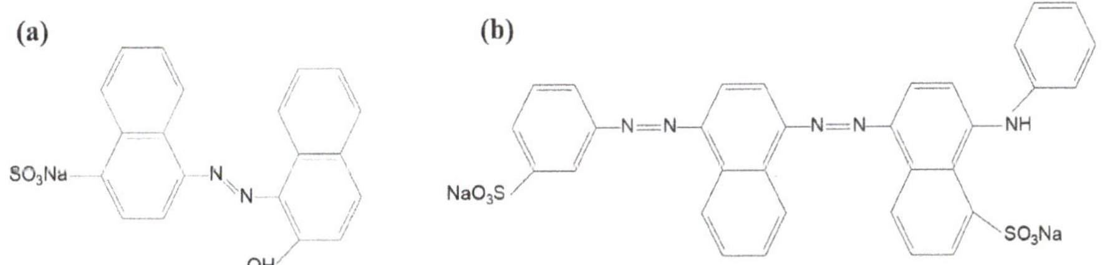

Figure 5. Molecular structure of (**a**) Acid Red 88, pKa = 10.7 and (**b**) Acid Blue 113, pKa = 0.5.

3. Materials and Methods

3.1. Materials and Equipment

Acid Red 88 (AR 88) (also known as Fast Red A or 2-Naphthol Red; CAS number 1658-56-6; molecular formula C$_{20}$H$_{13}$N$_2$NaO$_4$S; molecular weight 400.38 g/mole) and Acid Blue 113 (AB113) (Fast Navy Blue 5R; CAS number 3351-05-1; molecular formula C$_{32}$H$_{21}$N$_5$Na$_2$O$_6$S$_2$; molecular weight 681.65 g/mole) were purchased from Sigma-Aldrich (St. Louis, MO, USA) (dye content 75%). Both of these dyes are common azo acid dyes, usually applied for wool, nylon, rayon and polyester dyeing [54]. Acid dyes are negatively charged dyes [55], which are protonated in pH values below their acid dissociation constant (pKa). The chemical structure of AR 88 and AB 113 are shown in Figure 5. Titanium dioxide (TiO$_2$) Degussa P25 with an average particle size of 30 nm [14] and surface area of 57 m^2/g [11] was purchased from Merck (Kenilworth, NJ, USA) (reagent grade) and used as received.

Solutions of 1 M, 0.1 M and 0.01 M of HCl and NaOH were used to adjust solution pH to pre-determined values before initialization of photocatalytic process. Standard buffer solutions of pH of 4.0 and 7.0 were used to calibrate the pH meter (Jenway, staffordshire, UK) 3045 Ion Analyzer pH meter with a Sentek (Stepney, Australia) single-junction, glass body combination electrode filled with AgCl). To separate TiO_2 particles from treated solutions, 8 mL of treated dye solutions were poured in 15 mL-polypropylene centrifuge tubes (17 mm × 120 mm). A Sigma (St. Louis, MO, USA) 201 centrifuge machine was used at 4000× *g* rpm for 40 min to separate particles. The supernatant was then decanted and used for analysis. Laboratory scales [Sartorius-AC 121S-00MS (Göttingen, Germany) and Rad Wag-WTB 3000 (Radom, Poland)] with the resolution of 0.001 g were used to measure the mass of dyes and TiO_2 particles. All the experiments were carried out using DI water (with an electrical resistivity of 1 MOhm/cm at 25 °C). A Rayleigh (Beijing, China) UV1601 UV/VIS spectrophotometer was calibrated against standard dye solution concentrations and used to measure the dye concentration.

3.2. Photoreactor

Photocatalytic degradation experiments were conducted in a batch reactor, shown schematically in Figure 6. The reactor setup consisted of two UV-C lamps [each lamp: Philips (Somerset, NJ, USA) TUV G30T8 25PK; 30 W, 0.37 A, 102 V; primary emission 253.7 nm, UV-C radiation 12 W, 10% depreciation during 9000 h; 90 cm length, 28 mm diameter], two 4.5-volt rotary agitators, an aeration pump [Hailea (Guangdong, China) ACO 5505, 6 Watt, air output = 5.5 L/min] with two output tubes and two cylindrical dishes (Schott, Germany) for holding dye solutions, with an inside diameter of 13 cm, height of 7.5 cm and bottom thickness of 5 mm. UV lamps were placed on two concrete columns with a height of 25 cm on the top of dye solutions' containers. Distance from the UV lamps to the surface of dye solutions was 20.7 cm. To ensure a homogenous stirring of catalyst particles in dye solutions during the photocatalytic process, samples were agitated at 150 rpm using rotary agitators. Aeration pump tubes were placed in sample containers to supply the oxygen demand for photocatalytic reaction with the oxygen flow rate of 3.9×10^{-2} mole/min, as well as to achieve a uniform suspension of TiO_2 particles in the aqueous solutions. To prevent UV-C radiation leakage, the reactor setup was covered with a cardboard box of 30 cm × 40 cm × 110 cm dimensions. The internal surface of the box was completely covered with aluminum foil to prevent escape of UV radiation. By reflecting the radiation toward the samples, the removal efficiency would likely be enhanced.

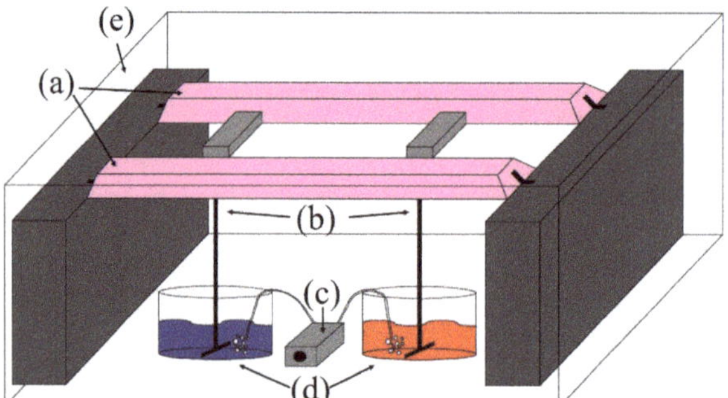

Figure 6. Schematic of the UV-C/TiO_2 photoreactor set up: (**a**) UV-C lamps, (**b**) agitators, (**c**) aeration pump, (**d**) sample containers (**e**) photoreactor cover.

3.3. Experimental Procedure and Measurements

Fresh dye solutions were prepared by adding appropriate amounts of dye powder and DI water in 500-mL volumetric flasks. The solutions were agitated for 10 min on a rotary shaker to obtain a homogenous dye solution and then poured into the cylindrical reaction dishes. Dye solution pH was adjusted to the predetermined levels and the predetermined masses of TiO_2 powder were added to the solutions. The suspensions were immediately placed in the photoreactor and the mechanical agitators and air pump were simultaneously turned on sand worked in the dark for 5 min. After 5 min, the two UV lamps were turned on to initiate the photocatalytic reactions. Reaction time was measured from the beginning of UV irradiation. All the experiments were carried out at room temperature (23 ± 0.1 °C). To monitor the effectiveness of the process, light absorbance of the samples was measured by the UV-VIS spectrophotometer at characteristic wavelengths of 505 nm and 565 nm for AR 88 and AB 113, respectively [16,56]. Dye concentrations in treated samples were determined from measured light extinction based on Beer–Lambert's law as expressed below:

$$\log_{10} (I/I_0) = A \tag{8}$$

$$A = \varepsilon \cdot L \cdot C \tag{9}$$

where I_0/I is the ratio of incident light to transmitted light, A is light absorbance, ε is the molar absorption coefficient (L mg^{-1} cm^{-1}), L (cm) is the length of solution that light passes through, which is equal to cell thickness used in spectrophotometer and C is the concentration of solution (mg/L) [57]. For dye concentrations ranging between 20 to 200 mg L^{-1}, the light absorption versus dye concentration plots at the peak of each dye's absorption spectrum were linear for both dyes (Figure S2). The extinction coefficient (ε) showed values of 0.20 L mg^{-1} cm^{-1} for AR 88 and 0.21 L mg^{-1} cm^{-1} for AB 113 (Figure S1).

Degradation efficiency, R (%), was calculated using Equation (10):

$$R(\%) = \frac{C_i - C_f}{C_i} \times 100 \tag{10}$$

where C_i and C_f are initial and final dye concentrations (mg/L).

3.4. Statistical Analysis

Apart from R^2 and standard deviation as the two well-known statistical analysis measures, analysis of variance (ANOVA) uses other standard factors to evaluate significance of a fitted regression model to a data set. These factors include coefficient of variation, adequate precision and predicted residual error sum of squares.

Coefficient of variation is the standard deviations which is expressed as the mean percentage:

$$\text{Coefficient of variation } (\%) = \frac{\text{Standard deviation}}{\text{mean}} \times 100 \tag{11}$$

Adequate precision is an indicator for measuring signal to noise ratio of the model, which is calculated as:

$$\text{Adequate precision} = \frac{\text{maximum predicted response} - \text{minimum predicted response}}{\text{Average standard deviation of all predicted response}} \tag{12}$$

Predicted residual error sum of squares (PRESS) is a measure between the fitted values and observed values. From a fitted model, each observation from the data set is removed, the model is refitted and the predicted value at that excluded point is calculated. The PRESS is calculated as: [50]

$$PRESS = \sum_{i=1}^{n} (y_i - \hat{y}_{-i})^2 \tag{13}$$

where n is the number of data points, y_i is the outcome of ith data point and the $\hat{y}_{-i}$ is the prediction of ith data point from the refitted model excluded ith data [58].

The F-test in ANOVA investigates if the variance between the means of two populations are significantly different. The F-statistic is the ratio of the "between-group variability" to the "within-group variability", or:

$$F = \frac{variation\ between\ sample\ means}{variation\ within\ the\ samples} \tag{14}$$

The p-value tests the null hypothesis which expresses that data from all groups are from populations with equal means. In other words, p-value determines that if all the populations really have the same mean, what is the chance that random sampling would result in the means as far apart as observed. The p-value is computed from a comparison of the computed F-statistic to the critical value of the F-statistic for the given number of degrees of freedom.

If the null hypothesis of no significant difference between sample manes is true, the F-statistic is expected to be close to 1. A large F-statistic means that the variation among the group means is more than is expected to occur by chance. Therefore, a large F-statistic, if it exceeds the critical F-statistic for a pre-established level of significant (typically $p < 0.05$) can lead to the rejection of the null hypothesis, meaning that the data were not likely to have been sampled from populations with the same mean.

3.5. Experimental Design and Optimization

3.5.1. Preliminary Experiments

Experiments and optimization were performed in three stages. In the first stage, preliminary experiments were conducted to determine the ranges of experimental factors to be used in the main experiments. In the second stage, the main experiments were conducted to determine optimal experimental conditions using the response surface methodology. Finally, in the third stage, reaction kinetics were investigated under optimized operational conditions.

Four independent factors including pH, initial dye concentration, TiO_2 dose and reaction time were used to evaluate dye removal efficiencies in the preliminary first-stage experiments. Adopting a one factor-at-a-time approach, three out of four variables were held constant and the fourth was varied in 4 or 5 levels. Table 6 shows variables and levels used in the preliminary experiments. Two replicate runs were conducted for each combination.

Table 6. Experimental factors and levels used in preliminary experiments for photocatalytic degradation of AB 113 and AR 88 in the UV-C/TiO_2 system.

Variables	Levels				
pH	2.0	3.0 *	6.0	9.0	10.0
Initial dye concentration (mg/L)	20	50 *	100	150	200
TiO_2 dose (g/L)	0.5	1.0 *	2.0	4.0	-
Reaction time (min)	30	60	90 *	120	180

* Fixed value of variable when other factors changed.

3.5.2. Experimental Design Using Response Surface Methodology

Experimental design, statistical analyses, mathematical modeling and optimizations were accomplished using Design Expert software (Design-Expert®, V 10, Stat-Ease, Inc., Minneapolis, MN, USA). Similar to the paper by Saber et al. [26], a cubic IV optimal design method was employed to investigate the effects of input factors and their interactions on dye removal percentage (i.e., the target response), in the second experimental stage. Cubic IV optimal design minimizes the integral of prediction variance through the design space and results in a lower prediction variance throughout an area of interest [59]. Ranges of experimental variables were considered based on the results from preliminary experiments. As Table 7 shows, six levels of pH, six levels of TiO_2 dose and six levels of initial dye concentration were considered as the independent variables in the optimal design. For each dye, 30 combinations of conditions were developed according to the cubic IV optimal design algorithm. The achieved removal efficiency for each run was considered as the target response. In order to account for experimental errors, three replicate runs were conducted for each combination of conditions and the average removal efficiency of the three runs was reported for each experimental run.

Table 7. Experimental factors and their levels used in cubic IV optimal design for photocatalytic degradation of AB 113 and AR 88.

Factors	Levels					
A: pH	2.0	2.5	3.0	3.5	4.0	5.0
B: TiO_2 dose (g/L)	0.5	1.0	1.5	2.0	2.5	3.0
C: Initial dye concentration (mg/L)	20	50	60	80	115	150

Analysis of variance (ANOVA) was performed on the fitted cubic models for each dye to evaluate the significance of the fitted models and to identify the relative significance of experimental factors and their interactions on the removal efficiency for each dye. Three-dimensional response surface and contour plots were generated based on the cubic least-squares regression models obtained from ANOVA.

3.5.3. Kinetic Studies

In the third stage, additional experiments were conducted to investigate reaction kinetics under the optimized treatment conditions determined from the cubic IV optimal experimental design.

A first-order kinetic model can be described as:

$$\ln (C) = \ln (C_o) - k_1 t \tag{15}$$

where, t is the reaction time (min), C_o and C (mg/L) are the initial dye concentration (mg/L) and dye concentration at time t, respectively and k_1 is the first-order reaction rate constant (min^{-1}). When evaluating goodness of fit for a kinetic model, root mean square error ($RMSE$) (Equation (16)) was used along with R^2 value to evaluate model validity [60].

$$RMSE = \sqrt{\frac{\sum_{i=1}^{n}\left(C_m - C_{exp}\right)^2}{n}} \tag{16}$$

where C_{exp} and C_m are the experimental and calculated values (based on the fitted kinetic models) of dye concentration and n is the number of data points.

4. Conclusions

This study optimized photodegradation of AB 113 and AR 88 dyes in a UV-C/TiO_2 suspension system using RSM, considering initial dye concentration, solution pH and catalyst dose as variant factors and the removal percentage as the target response. Under optimum conditions, kinetics of photocatalytic degradation of AB 113 and AR 88 were also investigated. Analysis of variance showed that reduced cubic models could well describe the removal of AB 113 and AR 88 dyes. The F-test showed that the solution pH and initial dye concentration were the most important parameters for removal of AB 113 and AR 88, respectively. Although pH was the most significant parameter affecting AB 113 removal efficiency, it was found to be insignificant for AR 88 removal. However, pH of the AR 88 solution showed significant interactions with the other two factors. Degradation efficiencies of 98.7% and 99.6% were achieved under optimum conditions for AB 113 and AR 88, respectively. The present study demonstrated almost complete degradation of AB 113 and AR 88 in 90 min under the optimized conditions obtained using RSM, with first-order rate constants of degradation rate constants of $k_{1,\,AB\,113} = 0.048$ min^{-1} and $k_{1,\,AR\,88} = 0.059$ min^{-1}.

Results show that a UV-C/TiO$_2$ photocatalytic degradation process can be considered as a promising and cost-effective technique for dye removal from textile industry effluents. As a proposal, a dyeing plant could adjust conditions in its effluent to the optimized values to obtain more than 98.7% destruction of waste dye and then adjust effluent pH to neutral values and filter to remove and recycle the TiO$_2$ particles prior to discharge to a receiving water. Additional batch experiments should be conducted at a bench scale in multi dye solutions to evaluate competition between various dyes, followed by pilot-scale application of this treatment method on a real textile wastewater in a flow-through reactor to evaluate limitations due to incomplete mixing and dispersion.

Supplementary Materials: The following are available online at http://www.mdpi.com/2073-4344/9/4/360/s1, Figure S1: Computed F-values for significant ($p < 0.05$) regression model terms associated with photocatalytic degradation of (a) AB 113 (a) and (b) AR 88 [*A*: pH, *B*: TiO$_2$ dose (g/L) and *C*: Initial dye concentration: (mg/L)]., Figure S2: Absorbance versus dye concentrations graphs for (a) AB 113 and (b) AR 88, showing linearity of data for both dyes in the studied dye concentration between 0 mg/L to 200 mg/L.

Author Contributions: Conceptualization, S.M. and A.S.; Data curation, S.M. and A.S.; Formal analysis, S.M., A.S. and D.E.J.; Funding acquisition, D.E.J.; Investigation, S.M.; Methodology, S.M. and A.S.; Software, S.M. and A.S.; Supervision, D.E.J.; Validation, S.M. and A.S.; Visualization, S.M.; Writing—original draft, S.M.; Writing—review & editing, A.S. and D.E.J.

Funding: The publication fees for this article were supported by the UNLV University Libraries Open Article Fund. Authors thank Isfahan University of Technology for partially funding (1391-7) this research.

Acknowledgments: Authors would like to acknowledge Amir Taebi for his professional advice and Behnaz Harandizadeh for her help in running experiments. The authors greatly appreciate the anonymous reviewers of this manuscript for their constructive suggestions which improved the quality of this study.

Conflicts of Interest: The authors declare no conflict of interest.

References

1. Sohrabi, M.R.; Ghavami, M. Photocatalytic degradation of Direct Red 23 dye using UV/TiO$_2$: Effect of operational parameters. *J. Hazard. Mater.* **2008**, *153*, 1235–1239. [CrossRef]

2. Houas, A.; Lachheb, H.; Ksibi, M.; Elaloui, E.; Guillard, C.; Herrmann, J.-M. Photocatalytic degradation pathway of methylene blue in water. *Appl. Catal. B Environ.* **2001**, *31*, 145–157. Available online: https://ac-els-cdn-com.ezproxy.library.unlv.edu/S0926337300002769/1-s2.0-S0926337300002769-main.pdf?_tid=7e7afe96-d229-11e7-85a6-00000aab0f26&acdnat=1511646312_3be9e82c1ae096d6f42aa8199f88814d (accessed on 25 November 2017). [CrossRef]

3. Konicki, W.; Sibera, D.; Mijowska, E.; Lendzion-Bieluń, Z.; Narkiewicz, U. Equilibrium and kinetic studies on acid dye Acid Red 88 adsorption by magnetic ZnFe2O4 spinel ferrite nanoparticles. *J. Colloid Interface Sci.* **2013**, *398*, 152–160. [CrossRef] [PubMed]

4. Gupta, A.K.; Pal, A.; Sahoo, C. Photocatalytic degradation of a mixture of Crystal Violet (Basic Violet 3) and Methyl Red dye in aqueous suspensions using Ag+ doped TiO$_2$. *Dyes Pigments* **2006**, *69*, 224–232. [CrossRef]

5. Han, F.; Kambala, V.S.R.; Srinivasan, M.; Rajarathnam, D.; Naidu, R. Tailored titanium dioxide photocatalysts for the degradation of organic dyes in wastewater treatment: A review. *Appl. Catal. A Gen.* **2009**, *359*, 25–40. [CrossRef]

6. Konstantinou, I.K.; Albanis, T.A. TiO$_2$-assisted photocatalytic degradation of azo dyes in aqueous solution: Kinetic and mechanistic investigations: A review. *Appl. Catal. B Environ.* **2004**, *49*, 1–14. [CrossRef]

7. Juang, R.S.; Lin, S.H.; Hsueh, P.Y. Removal of binary azo dyes from water by UV-irradiated degradation in TiO$_2$ suspensions. *J. Hazard. Mater.* **2010**, *182*, 820–826. [CrossRef]

8. Ahmed, S.; Rasul, M.G.; Martens, W.N.; Brown, R.; Hashib, M.A. Heterogeneous photocatalytic degradation of phenols in wastewater: A review on current status and developments. *Desalination* **2010**, *261*, 3–18. [CrossRef]

9. Barakat, M.A. Adsorption and photodegradation of Procion yellow H-EXL dye in textile wastewater over TiO$_2$ suspension. *J. Hydro-Environ. Res.* **2011**, *5*, 137–142. [CrossRef]

10. Soutsas, K.; Karayannis, V.; Poulios, I.; Riga, A.; Ntampegliotis, K.; Spiliotis, X.; Papapolymerou, G. Decolorization and degradation of reactive azo dyes via heterogeneous photocatalytic processes. *Desalination* **2010**, *250*, 345–350. [CrossRef]

11. Anandan, S.; Kumar, P.S.; Pugazhenthiran, N.; Madhavan, J.; Maruthamuthu, P. Effect of loaded silver nanoparticles on TiO$_2$ for photocatalytic degradation of Acid Red 88. *Sol. Energy Mater. Sol. Cells* **2008**, *92*, 929–937. [CrossRef]

12. Zayani, G.; Bousselmi, L.; Pichat, P.; Mhenni, F. Photocatalytic degradation of the Acid Blue 113 textile azo dye in aqueous suspensions of four commercialized TiO(2) samples. *J. Environ. Sci. Health A Tox. Hazard. Subst. Environ. Eng.* **2008**, *43*, 202–209. [CrossRef]

13. Gupta, V.K.; Jain, R.; Mittal, A.; Saleh, T.A.; Nayak, A.; Agarwal, S.; Sikarwar, S. Photo-catalytic degradation of toxic dye amaranth on TiO$_2$/UV in aqueous suspensions. *Mater. Sci. Eng. C* **2012**, *32*, 12–17. [CrossRef] [PubMed]

14. Toor, A.T.; Verma, A.; Jotshi, C.K.; Bajpai, P.K.; Singh, V. Photocatalytic degradation of Direct Yellow 12 dye using UV/TiO$_2$ in a shallow pond slurry reactor. *Dyes Pigments* **2006**, *68*, 53–60. [CrossRef]

15. Khataee, A.R.; Pons, M.N.; Zahraa, O. Photocatalytic degradation of three azo dyes using immobilized TiO$_2$ nanoparticles on glass plates activated by UV light irradiation: Influence of dye molecular structure. *J. Hazard. Mater.* **2009**, *168*, 451–457. [CrossRef] [PubMed]

16. Balachandran, K.; Venckatesh, R.; Sivaraj, R.; Rajiv, P. TiO$_2$ nanoparticles versus TiO$_2$-SiO$_2$ nanocomposites: A comparative study of photo catalysis on acid red 88, Spectrochim. *Acta Part A Mol. Biomol. Spectrosc.* **2014**, *128*, 468–474. [CrossRef] [PubMed]

17. Gao, B.; Yap, P.S.; Lim, T.M.; Lim, T.T. Adsorption-photocatalytic degradation of Acid Red 88 by supported TiO$_2$: Effect of activated carbon support and aqueous anions. *Chem. Eng. J.* **2011**, *171*, 1098–1107. [CrossRef]

18. Moon, J.; Yun, C.Y.; Chung, K.W.; Kang, M.S.; Yi, J. Photocatalytic activation of TiO$_2$ under visible light using Acid Red 44. *Catal. Today* **2003**, *87*, 77–86. [CrossRef]

19. Khataee, A.R.; Zarei, M. Photoelectrocatalytic decolorization of diazo dye by zinc oxide nanophotocatalyst and carbon nanotube based cathode: Determination of the degradation products. *Desalination* **2011**, *278*, 117–125. [CrossRef]

20. Cavazzuti, M. *Optimization Methods: From Theory to Design Scientific and Technological Aspects in Mechanics*; Springer: New York, NY, USA, 2013. [CrossRef]

21. Antony, J. Taguchi or classical design of experiments: A perspective from a practitioner. *Sens. Rev.* **2006**, *26*, 227–230. [CrossRef]

22. Khuri, A.I.; Mukhopadhyay, S. Response surface methodology, Wiley Interdiscip. *Rev. Comput. Stat.* **2010**, *2*, 128–149. [CrossRef]

23. Myers, R.H.; Montgomery, D.C.; Anderson-Cook, C.M. *Response Surface Methodology: Process and Product Optimization Using Designed Experiments*, 3rd ed.; John Wiley & Sons: New York, NY, USA, 2009.

24. Wei, L.; Zhu, H.; Mao, X.; Gan, F. Electrochemical oxidation process combined with UV photolysis for the mineralization of nitrophenol in saline wastewater. *Sep. Purif. Technol.* **2011**, *77*, 18–25. [CrossRef]

25. Sahu, J.N.; Acharya, J.; Meikap, B.C. Response surface modeling and optimization of chromium(VI) removal from aqueous solution using Tamarind wood activated carbon in batch process. *J. Hazard. Mater.* **2009**, *172*, 818–825. [CrossRef] [PubMed]

26. Saber, A.; Mortazavian, S.; James, D.E.; Hasheminejad, H. Optimization of Collaborative Photo-Fenton Oxidation and Coagulation for the Treatment of Petroleum Refinery Wastewater with Scrap Iron. *Water Air Soil Pollut.* **2017**, *228*. [CrossRef]

27. Saber, A.; Hasheminejad, H.; Taebi, A.; Ghaffari, G. Optimization of Fenton-based treatment of petroleum refinery wastewater with scrap iron using response surface methodology. *Appl. Water Sci.* **2014**, *4*, 283–290. [CrossRef]

28. Zhu, X.; Tian, J.; Liu, R.; Chen, L. Optimization of fenton and electro-fenton oxidation of biologically treated coking wastewater using response surface methodology. *Sep. Purif. Technol.* **2011**, *81*, 444–450. [CrossRef]

29. Benatti, C.T.; Tavares, C.R.G.; Guedes, T.A. Optimization of Fenton's oxidation of chemical laboratory wastewaters using the response surface methodology. *J. Environ. Manag.* **2006**, *80*, 66–74. [CrossRef] [PubMed]

30. Ahmadi, M.; Vahabzadeh, F.; Bonakdarpour, B.; Mofarrah, E.; Mehranian, M. Application of the central composite design and response surface methodology to the advanced treatment of olive oil processing wastewater using Fenton's peroxidation. *J. Hazard. Mater.* **2005**, *123*, 187–195. [CrossRef]

31. Bianco, B.; de Michelis, I.; Vegliò, F. Fenton treatment of complex industrial wastewater: Optimization of process conditions by surface response method. *J. Hazard. Mater.* **2011**, *186*, 1733–1738. [CrossRef] [PubMed]

32. Bagheri, A.R.; Ghaedi, M.; Asfaram, A.; Bazrafshan, A.A.; Jannesar, R. Comparative study on ultrasonic assisted adsorption of dyes from single system onto Fe_3O_4 magnetite nanoparticles loaded on activated carbon: Experimental design methodology. *Ultrason. Sonochem.* **2017**, *34*, 294–304. [CrossRef] [PubMed]

33. Mohajeri, S.; Aziz, H.A.; Isa, M.H.; Zahed, M.A.; Adlan, M.N. Statistical optimization of process parameters for landfill leachate treatment using electro-Fenton technique. *J. Hazard. Mater.* **2010**, *176*, 749–758. [CrossRef] [PubMed]

34. Cifuentes, B.; Figueredo, M.; Cobo, M. Response Surface Methodology and Aspen Plus Integration for the Simulation of the Catalytic Steam Reforming of Ethanol. *Catalysts* **2017**, *7*, 15. [CrossRef]

35. Li, L.; Ma, Q.; Wang, S.; Song, S.; Li, B.; Guo, R.; Cheng, X.; Cheng, Q. Photocatalytic Performance and Degradation Mechanism of Aspirin by TiO_2 through Response Surface Methodology. *Catalysts* **2018**, *8*, 118. [CrossRef]

36. Inger, M.; Dobrzyńska-Inger, A.; Rajewski, J.; Wilk, M. Optimization of Ammonia Oxidation Using Response Surface Methodology. *Catalysts* **2019**, *9*, 249. [CrossRef]

37. Aljuboury, D.a.A.; Palaniandy, P.; Aziz, H.B.A.; Feroz, S.; Amr, S.S.A. Evaluating photo-degradation of COD and TOC in petroleum refinery wastewater by using TiO_2/ZnO photo-catalyst. *Water Sci. Technol.* **2016**, *74*, 1312–1325. [CrossRef]

38. Mohammadzadeh, S.; Olya, M.E.; Arabi, A.M.; Shariati, A.; Nikou, M.R.K. Synthesis, characterization and application of ZnO-Ag as a nanophotocatalyst for organic compounds degradation, mechanism and economic study. *J. Environ. Sci.* **2015**, *35*, 194–207. [CrossRef] [PubMed]

39. Mohan, D.; Pittman, C.U. Activated carbons and low cost adsorbents for remediation of tri- and hexavalent chromium from water. *J. Hazard. Mater.* **2006**, *137*, 762–811. [CrossRef]

40. Alkaim, A.F.; Kandiel, T.A.; Hussein, F.H.; Dillert, R.; Bahnemann, D.W. Enhancing the photocatalytic activity of TiO_2 by pH control: A case study for the degradation of EDTA. *Catal. Sci. Technol.* **2013**, *3*, 3216. [CrossRef]

41. Hu, C.; Yu, J.C.; Hao, Z.; Wong, P.K. Effects of acidity and inorganic ions on the photocatalytic degradation of different azo dyes. *Appl. Catal. B Environ.* **2003**, *46*, 35–47. [CrossRef]

42. Shirzad-Siboni, M.; Jafari, S.J.; Giahi, O.; Kim, I.; Lee, S.M.; Yang, J.K. Removal of acid blue 113 and reactive black 5 dye from aqueous solutions by activated red mud. *J. Ind. Eng. Chem.* **2014**, *20*, 1432–1437. [CrossRef]

43. Lima, E.C.; Royer, B.; Vaghetti, J.C.P.; Simon, N.M.; da Cunha, B.M.; Pavan, F.A.; Benvenutti, E.V.; Cataluña-Veses, R.; Airoldi, C. Application of Brazilian pine-fruit shell as a biosorbent to removal of reactive red 194 textile dye from aqueous solution. Kinetics and equilibrium study. *J. Hazard. Mater.* **2008**, *155*, 536–550. [CrossRef] [PubMed]

44. Habibi, M.H.; Vosooghian, H. Photocatalytic degradation of some organic sulfides as environmental pollutants using titanium dioxide suspension. *J. Photochem. Photobiol. A Chem.* **2005**, *174*, 45–52. [CrossRef]

45. Venkatachalam, N.; Palanichamy, M.; Murugesan, V. Sol-gel preparation and characterization of alkaline earth metal doped nano TiO_2: Efficient photocatalytic degradation of 4-chlorophenol. *J. Mol. Catal. A Chem.* **2007**, *273*, 177–185. [CrossRef]

46. Fox, M.A.; Dulay, M.T. Heterogeneous photocatalysis. *Chem. Rev.* **1993**, *93*, 341–357. [CrossRef]

47. Ma, C.-M.; Hong, G.-B.; Chen, H.-W.; Hang, N.-T.; Shen, Y.-S. Photooxidation Contribution Study on the Decomposition of Azo Dyes in Aqueous Solutions by VUV-Based AOPs. *Int. J. Photoenergy* **2011**, *2011*, 1–8. [CrossRef]

48. Saharan, V.K.; Pandit, A.B.; Kumar, P.S.S.; Anandan, S. Hydrodynamic cavitation as an advanced oxidation technique for the degradation of Acid Red 88 dye. *Ind. Eng. Chem. Res.* **2012**, *51*, 1981–1989. [CrossRef]

49. Spiess, A.N. qpcR package V1.4-0, Modeling and Analysis of Real-Time PCR Data. 2018. Available online: https://www.rdocumentation.org/packages/qpcR/versions/1.4-0 (accessed on 31 May 2018).

50. Konyar, M.; Yildiz, T.; Aksoy, M.; Yatmaz, H.C.; Öztürk, K. Reticulated *ZnO* Photocatalyst: Efficiency Enhancement in Degradation of Acid Red 88 Azo Dye by Catalyst Surface Cleaning. *Chem. Eng. Commun.* **2017**, *204*, 711–716. [CrossRef]

51. Camarillo, R.; Rincón, J. Photocatalytic Discoloration of Dyes: Relation between Effect of Operating Parameters and Dye Structure. *Chem. Eng. Technol.* **2011**, *34*, 1675–1684. [CrossRef]

52. Stylidi, M.; Kondarides, D.I.; Verykios, X.E. Visible light-induced photocatalytic degradation of Acid Orange 7 in aqueous TiO_2 suspensions. *Appl. Catal. B Environ.* **2004**, *47*, 189–201. [CrossRef]

53. Madhavan, J.; Kumar, P.S.S.; Anandan, S.; Grieser, F.; Ashokkumar, M. Degradation of acid red 88 by the combination of sonolysis and photocatalysis. *Sep. Purif. Technol.* **2010**, *74*, 336–341. [CrossRef]

54. Trotma, E.R. *Dyeing and Chemical Technology of Textile Fibres*; Griffin: London, UK, 1970; Available online: https://www.scribd.com/doc/101550453/Dyeing-and-Chemical-Technology-of-Textile-Fibres (accessed on 3 May 2017).

55. Martínez-Huitle, C.A.; Brillas, E. Decontamination of wastewaters containing synthetic organic dyes by electrochemical methods: A general review. *Appl. Catal. B Environ.* **2009**, *87*, 105–145. [CrossRef]

56. Shu, H.Y.; Chang, M.C.; Chen, C.C.; Chen, P.E. Using resin supported nano zero-valent iron particles for decoloration of Acid Blue 113 azo dye solution. *J. Hazard. Mater.* **2010**, *184*, 499–505. [CrossRef] [PubMed]

57. Meyers, R. *Encyclopedia of Physical Science and Technology*; Academic Press: New York, NY, USA, 2002.

58. Zumel, N. Estimating Generalization Error with the PRESS statistic|Win-Vector Blog, (2014) 12. Available online: http://www.win-vector.com/blog/2014/09/estimating-generalization-error-with-the-press-statistic/ (accessed on 2 April 2019).

59. Anderson, M.J.; Whitcomb, P.J. *RSM Simplified: Optimizing Processes Using Response Surface Methods for Design of Experiments*; Productivity Press: New York, NY, USA, 2005. Available online: https://www.crcpress.com/RSM-Simplified-Optimizing-Processes-Using-Response-Surface-Methods-for/Whitcomb-Anderson/p/book/9781563272974 (accessed on 3 May 2017).

60. Saber, A.; Tafazzoli, M.; Mortazavian, S.; James, D.E. Investigation of kinetics and absorption isotherm models for hydroponic phytoremediation of waters contaminated with sulfate. *J. Environ. Manag.* **2018**, *207*, 276–291. [CrossRef] [PubMed]

 *catalysts*

Article

Facet-Dependent Interfacial Charge Transfer in TiO$_2$/Nitrogen-Doped Graphene Quantum Dots Heterojunctions for Visible-Light Driven Photocatalysis

Nan-Quan Ou [1,†], Hui-Jun Li [1,†], Bo-Wen Lyu [1], Bo-Jie Gui [1], Xiong Sun [1], Dong-Jin Qian [2], Yanlin Jia [3,4,*], Xianying Wang [1,5,*] and Junhe Yang [1,5]

[1] School of Materials Science and Technology, University of Shanghai for Science and Technology, Shanghai 200093, China; ounanquan@163.com (N.-Q.O.); huijunli0701@126.com (H.-J.L.); bowenlyu0324@163.com (B.-W.L.); bojie_gui@163.com (B.-J.G.); sunxiong1993@163.com (X.S.); jhyang@usst.edu.cn (J.Y.)
[2] Shanghai Key Laboratory of Molecular Catalysis and Innovative Materials, Fudan University, Shanghai 200433, China; djqian@fudan.edu.cn
[3] College of Materials Science and Engineering, Central South University, Changsha 410083, China
[4] College of Materials Science and Engineering, Beijing University of Technology, Beijing 100124, China
[5] Shanghai Innovation Institute for Materials, Shanghai 200444, China
[*] Correspondence: xianyingwang@usst.edu.cn (X.W.); jiayanlin@126.com (Y.J.)
[†] These two authors contributed equally to this work.

Received: 26 February 2019; Accepted: 8 April 2019; Published: 9 April 2019

Abstract: Interfacial charge transfer is crucial in the efficient conversion of solar energy into fuels and electricity. In this paper, heterojunction composites were fabricated, comprised of anatase TiO$_2$ with different percentages of exposed {101} and {001} facets and nitrogen-doped quantum dots (NGQDs) to enhance the transfer efficiency of photo-excited charge carriers. The photocatalytic performances of all samples were evaluated for RhB degradation under visible light irradiation, and the hybrid containing TiO$_2$ with 56% {001} facets demonstrated the best photocatalytic activity. The excellent photoactivity of TiO$_2$/NGQDs was owed to the synergistic effects of the following factors: (i) The unique chemical features of NGQDs endowed NGQDs with high electronic conductivities and provided its direct contact with the TiO$_2$ surface via forming Ti–O–C chemical bonds. (ii) The co-exposed {101} and {001} facets were beneficial for the separation and transfer of charge carriers in anatase TiO$_2$. (iii) The donor-acceptor interaction between NGQDs and electron-rich {101} facets of TiO$_2$ could remarkably enhance the photocurrent, thus hindering the charge carriers recombination rate. Extensive characterization of their physiochemical properties further showed the synergistic effect of facet-manipulated electron-hole separation in TiO$_2$ and donor-acceptor interaction in graphene quantum dots (GQDs)/TiO$_2$ on photocatalytic activity.

Keywords: electron transfer; graphene quantum dots; heterojunction; photocatalysis; TiO$_2$

---

1. Introduction

Anatase TiO$_2$ is generally considered a better photocatalyst than rutile, mainly due to its attributes of longer exciton diffusion length, higher electron mobility, and longer carrier life time [1,2]. The photocatalytic activity of anatase has been revealed to depend closely on the crystal surface [3,4]. Clear pictures have now shown that reduction and oxidation reactions would preferentially occur on {101} and {001} facets, respectively. Furthermore, it has been found that the {101} surface is attractive for electrons in aqueous solutions while excess electrons tend to strongly avoid the {001} surface via surface science experiments and first-principles simulations [5]. The substantial electrons in anatase TiO_2 generated via photoexcitation play an important part in many energy-related applications. However, due to intrinsic defects and the fast carrier recombination rate, electron trapping in anatase TiO_2 are unavoidable, which hamper the overall photocatalytic activity [6,7].

To solve this problem, fabrication of heterojunctions modified anatase TiO_2 has emerged as a promising method [8–10]. Angus and co-workers reported the development of a hetero-structured material by using pre-formed carbon nitride nanosheets (CNNS) composite with facet-controllable TiO_2. The materials possess an excellent CO_2 adsorption capacity and charge transfer rate, thus leading to the improvement of the photocatalytic activity of TiO_2 [11]. Luca Rimoldi et al. have then reported a method to combine TiO_2 with WO_3. Due to the admirable properties of WO_3, the photocatalytic activity enhanced remarkably [12]. Through a series of experiments and calculations, Latterly Olowoyo et al. have also found that carbon nanotubes (CNTs) can strongly be attached to the {101} facet of TiO_2, since the atomic orbitals of anatase overlap with the orbitals of the CNTs [13].

As a novel class of quantum dots (QDs), graphene quantum dots (GQDs) have currently attracted intensive interest in fabricating new heterojunctions, due to their large surface areas, high electron mobilities, conductivity, and adjustable band gaps [14,15]. These properties ensure discrete electronic levels, which could allow for light-induced electron injection, efficient carrier transfer, and long-lived excited states [16–20]. GQDs have also been explored as the light absorber and heteroatoms-doped GQDs are expected to realize absorption in the visible region [21]. A variety of surface functional groups on GQDs could provide better covalent chemical linking between anatase and GQDs, facilitating charge separation and transfer behaviors [22–24]. Pan et al. have found that monodispersed amine-functionalized GQDs anatase TiO_2 heterojunctions have an absorption range extended into the visible light region and a much lower carrier recombination rate. They attributed the improved performance to the proper energy position of GQDs/TiO_2 23]. Then, Yu and co-workers reported the decoration of GQDs on {001} faceted anatase TiO_2 with an exposed percentage of 65%–75%. The experiments show a promotion of photocatalytic hydrogen evolution rate of the composites compared to bare anatase, which might originate from the higher-charge separation efficiency. [25] Zheng and co-workers also utilized TiO_2 and sulfur, nitrogen co-doped GQDs (SN-GQDs) to develop an efficient photocatalyst for synthesizing H_2O_2. They testified that SN-GQDs induced visible light absorption, promoted charge transfer, and provided active sites for $\cdot OOH$ formation [26]. Recently, Prezhdo et al. built several models of donor-acceptor interaction between GQDs and TiO_2 via stacking and covalent bonding, respectively, to provide guidance for subsequent photocatalysis application [16].

These findings provide strong evidence and also motivation for deeply understanding the models of electron-hole separation dynamics at heterojunction interfaces [3,21,27,28]. There has still been no adequate investigation, considering the influence of facet-dependent photogenerated charge-carrier separation in anatase TiO_2, on the photocatalytic activity of GQDs/TiO_2. Since both theoretical and experimental studies had shown that {101} crystal facets are electron-rich while {001} crystal facets are hole-rich in anatase TiO_2, it would be desirable to elucidate the different donor-acceptor interaction between NGQDs and TiO_2 with specific facet composition, thus providing explicit guidance in constructing heterojunction structures with superior performance.

Herein, we have designed a heterojunction composite via depositing nitrogen-doped GQDs on anatase TiO_2 with different exposure percentages of {001} and {101} facets, combining the advantages of facet and interfacial modification to maximize the driving force promoting charge carrier transfer. The visible light-driven dye degradation performances on anatase TiO_2 with {101}, {001}, and {001}-{101} facets, and their corresponding heterojunctions with nitrogen-doped GQDs (NGQDs) have been systematically studied. To further understand the electron transfer mechanism, the relationships of the morphology, chemical states, optical, and electrical properties with the photocatalytic activity were intensively analyzed.

2. Results and Discussion

2.1. Structural Characterization

The crystalline phases of different samples before and after decoration with NGQDs were firstly identified. Figure 1 shows the X-ray diffraction (XRD) patterns of anatase TiO_2 with different exposed facets. The diffraction peaks appeared around 2θ values of 25.3, 38.6, 48.0, 53.9, and 62.1, assigned to the (101), (112), (200), (105), and (213) crystal planes of anatase titania, respectively [27]. For bare anatase TiO_2, with the increase of Hydrofluoric acid (HF) volume, the {004} diffraction peak was broadened, implying the thickness of the TiO_2 along the {001} direction was decreased. Meanwhile, the intensity of the {200} diffraction peaks was enhanced, indicative of the increasing side length of the nanoparticles along the {100} direction. According to the two peaks, the percentage of the exposed {001} facet could be estimated, based on the calculation method reported in the literature. Shown in Table 1, the percentage was increased when increasing the HF volume [29]. The calculation method was described in the supporting information, as shown in Figure S1. By comparison, the XRD patterns of TiO_2/NGQDs exhibited similar but much lower diffraction peaks. The XRD pattern of NGQDs showed a wide weak diffraction peak centered at 26.8°, assigned to the {002} facet [30]. Though the diffraction peak of NGQDs was not observed in the composites, which might be due to the relatively low diffraction intensity of NGQDs, the peak attributed to the {101} facet of anatase decreased obviously. The possible reason is that the {002} crystal orientation in NGQDs influenced the {101} surface of TiO_2, thus leading to decrease of the peak [31].

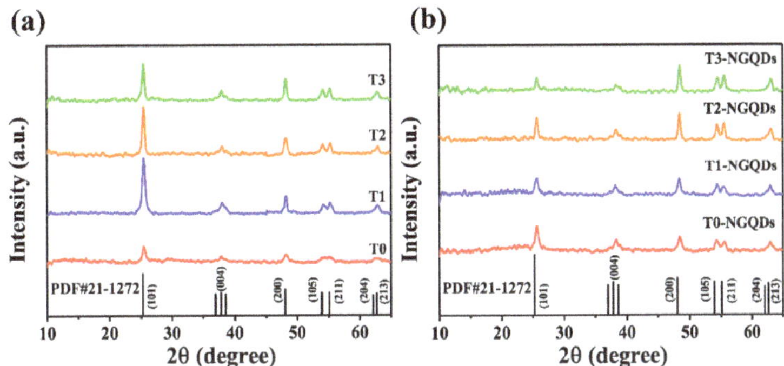

Figure 1. XRD patterns of different samples (**a**) without and (**b**) with nitrogen-doped quantum dots (NGQDs) decoration.

Table 1. Structural information of different anatase titania samples.

Samples	Average thickness (nm)	Average length (nm)	Percentage of {001}
T0	8.9	8.1	12%
T1	8.5	13.8	56%
T2	6.1	21.6	71%
T3	4.3	25.5	85%

In Raman spectroscopy, all samples show similar peaks centered at 144, 394, 514, and 636 cm^{-1} shown in Figure S2. When increasing the addition amount of HF, the intensity of the Eg peak at 144 cm^{-1} decreased simultaneously. The Eg peak is mainly attributed to the symmetric stretching vibration of O−Ti−O TiO_2 [29]. A higher percentage of exposed {001} facets generally represented fewer amount of symmetric O−Ti−O stretching vibration modes, thus leading to the decreasing intensity of Eg peak in the Raman spectra. Thus, it could be concluded from the Raman spectra that the exposure percentage of {001} facets increased with the increase of HF volume. Shown in the inset of Figure

S2b, two characteristic peaks of D and G band appeared at 1351 cm^{-1} and 1590 cm^{-1}, respectively, confirming the presence of graphite-like structure in the composites.

X-ray photoelectron spectroscopy (XPS) spectra were measured to study the bonding conditions in the heterojunctions, shown in Figure 2. The Ti 2p spectra showed two peaks with the binding energies of 458.9 eV and 464.6 eV, which are assigned to Ti 2p$^{3/2}$ and 2p$^{1/2}$ spin-orbital splitting photoelectrons, respectively. The splitting values indicated Ti^{4+} chemical states in these samples, while no Ti^{3+} forms were observed [32]. The C 1s spectra could be then fitted into three Gaussian peaks (288.8 eV, 286.1 eV, and 284.8 eV). The peak at 288.8 eV was assigned to the sp^2 hybridized carbon in the skeleton of NGQDs and also some carbon contaminants from the ambience. The other two peaks corresponded to the oxygenated carbon, representative of carboxyl carbon (288.8 eV) and hydroxyl carbon (286.1 eV) functional groups, respectively [33–36]. No Ti–C bond related peak (282 eV) was observed, implying that the NGQDs were probably anchored to the surface of TiO$_2$ via Ti–O–C bonds. In some reported studies, functional groups including C–O and COOH were evaluated to identify the existence of Ti–O–C bonding [32,37]. These groups are not that stable and might be converted to the epoxy group during the composite formation process.

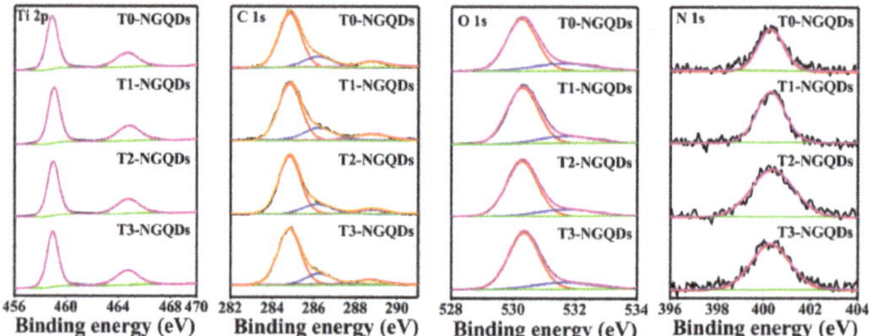

Figure 2. XPS spectra of different TiO$_2$/NGQDs composites.

The evidence of Ti–O–C bonding formation was further provided in the O 1s XPS spectra. It was fitted into two symmetric peaks. The peaks at 530.3 eV is ascribed to the oxygen in crystal lattice (Ti–O–Ti) and the other peak at 531.6 eV is believed to result from the Ti–O–C bonding, based on previously reported cases [25,32,38]. These results indicated the composite formation of TiO$_2$/NGQDs through the C–O–Ti bonds. The C–O–Ti bonds are capable of mediating the coupling between NGQDs and TiO$_2$, which could promote the interfacial electron transfer. The N 1s spectra revealed a peak centered at 400.3 eV, which could be assigned to the pyrrolic N (400.5 eV). The nitrogen atoms are mainly introduced by the NGQDs, demonstrating the successful decoration of NGQDs on the surface of TiO$_2$ [30]. Furthermore, the percentages of different bonds according to the fitting results of the XPS high-resolution spectra were calculated, shown in Table 2 and Table S1. Similar percentages implied similar bonding and chemical composition in different composites.

Table 2. Percentages of different bonds according to the fitting results of the XPS high-resolution spectra calculated from Figure 2.

Bond	T0-NGQDs	T1-NGQDs	T2-NGQDs	T3-NGQDs
% of Ti 2p	26.57	27.63	27.23	27.69
% of C 1s	21.18	19.4	20.65	18.99
O–C=O/Ti–O–C 288.8 eV	7.91	8.03	7.84	7.83
C–O 286.1 eV	19.53	20.30	19.05	19.26
C=C 284.8 eV	72.56	71.07	73.40	73.34
% of O 1s	51.69	52.33	51.53	52.85
Ti–O–Ti 530.3 eV	77.80	80.97	80.84	80.42
Ti–O–C 531.6 eV	22.20	19.03	19.16	19.58
% of N 1s	0.55	0.64	0.58	0.47

2.2. Morphology Characterization

The morphology of anatase TiO_2 with and without NGQDs decoration were characterized by Transmission Electron Microscope (TEM) and High Resolution Transmission Electron Microscope (HR-TEM). The morphology and crystal facets of TiO_2 remained similar before and after NGQDs modification, as can be seen from Figure 3 and Figure S3. Figure S3 shows that sample T0 is mostly composed of nanoparticles with a truncated octahedral bipryramid. After adding HF into the reaction system, it could be found that the anatase TiO_2 mostly consists of nanoplates and the plate size increases with the increase of HF volume, which is consistent with the XRD calculation results. Figure S4 shows the TEM image of NGQDs, which has an average size of ~2.4 nm. Clear lattice fringes demonstrative of its well-crystalline structure, and the autocorrelated HRTEM lattice image (inset in Figure S4b) show a 0.21 nm lattice fringe assigned to the {100} plane of GQDs [39,40].

The formation of NGQDs/TiO_2 heterojunctions could be obviously observed in the TEM images, shown in Figure 3, of which the NGQDs were uniformly decorated on both {001} and {101} facets of TiO_2. The NGQDs are shown with red circles. No selective deposition of NGQDs on a specific facet of anatase TiO_2 was found. The autocorrelated HRTEM lattice image (inset in Figure 3c,f) both show 0.21 nm lattice fringes assigned to the {100} plane of GQDs. Meanwhile, the autocorrelated HRTEM lattice image (inset in Figure 3c,f) also show clear lattice fringes of 0.35 nm and 0.19 nm, which could be assigned to the {101} and {001} facet of TiO_2, respectively. The HRTEM lattice images of all composites show that the lattices of both NGQDs and TiO_2 are simultaneously recognized, revealing good attachment of NGQDs over the TiO_2 surface.

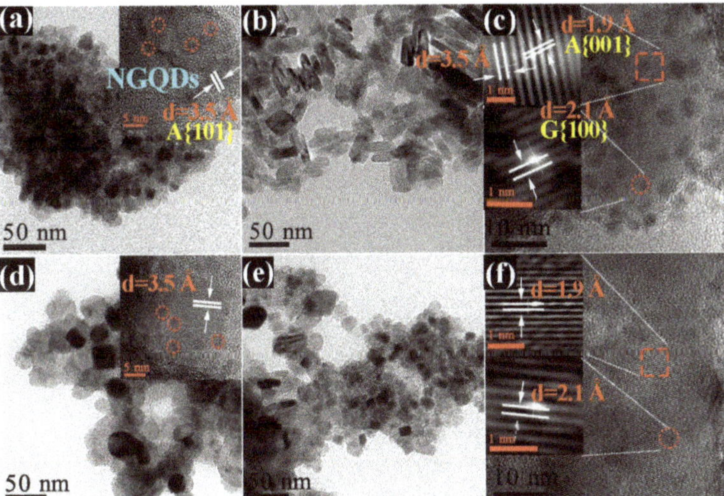

Figure 3. TEM images of (**a**) T0-NGQDs, (**b**,**c**) T1-NGQDs, (**d**) T2-NGQDs, and (**e**–**f**) T3-NGQDs. The inset in (**a**) is the HR-TEM image of T0-NGQDs. The inset in (**c**) is the autocorrelated HRTEM lattice images recorded from the corresponding selected areas. The inset in (**d**) is the HRTEM image of T2-NGQDs. The inset in (**f**) is the autocorrelated HRTEM lattice images recorded from the corresponding selected areas.

2.3. Optical and Electrical Properties

The optical properties of all samples were investigated via the Ultraviolet-Visible (UV-Vis) diffuse reflection spectroscopy (UV-DRS). All anatase TiO_2 demonstrate an absorption threshold near 400 nm in the ultraviolet region. The band gaps of the anatase could be obtained based on the Kubelka–Munk rule, seen from the inset in Figure 4a. The band gaps of T0, T1, T2, and T3 are 3.22, 3.28, 3.30, and 3.32 eV, respectively. Apparently, with the increase of the {001} facet percentage, the light absorption edge was slightly blue shifted [21]. The UV-vis spectrum of NGQDs is shown in Figure S4d, and the band gap was approximately 1.56 eV, according to our previous work [30,41]. After decorating NGQDs, it was found that there was an increasing visible light absorption for all composites. Generally, the resultant extended light absorption was due to the existence of Ti–O–C chemical bonds between GQDs and TiO_2 [32,37]. The interaction could improve the interfacial carrier transfer rate, which is beneficial to photocatalysis under visible light irradiation. Meanwhile, the presence of the energy gap of NGQDs further ensured the long-lived excited states and absorbance of solar photons in the broad solar spectrum.

The electrochemical Mott–Schottky experiments of the anatase were then measured (Figure 4c). The plots present a positive slope and the flat band potential values were recalculated vs. NHE (Normal Hydrogen Electrode). Combined with the band gaps of anatase TiO_2 and NGQDs, we proposed electronic band structures for all composites, shown in Figure 4d. It was hypothesized that the downshift of conduction band (CB) band level of the anatase might make the electrons less reductive and also weaken the dynamics of the electron transfer rate between TiO_2 and NGQDs [42–44].

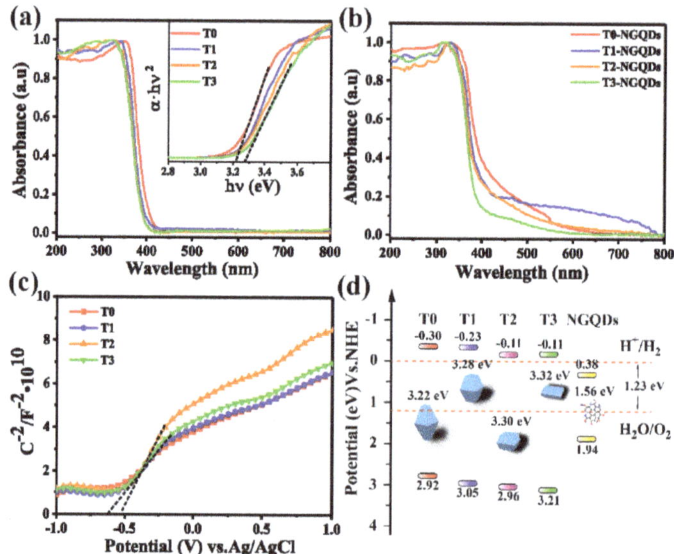

Figure 4. UV-vis diffuse reflection spectra (UV-DRS) of (**a**) bare anatase TiO_2, (**b**) TiO_2/NGQDs composites, (**c**) Mott–Schottky plots of bare anatase TiO_2, and (**d**) band structure diagram of different samples and NGQDs. The inset in (**a**) is the Tauc plot of the corresponding bare anatase TiO_2.

Moreover, the photogenerated charge carrier separation and transfer rate was determined using photocurrent responses, shown in Figure 5. Fast and uniform photocurrents with good reproducibility was demonstrated, indicative of relatively reversible photo-responses. Via three on-off cycles under visible light irradiation (>420 nm), it was found that the photocurrent density of bare anatase TiO_2 decreased in the order of T2, T3, T1, and T0. This suggests that defects (oxygen vacancies) are possibly formed on the surface of pristine anatase TiO_2 with small size under light illumination in our work, which would introduce defect energy levels in the band gap and lead to electron-hole separation under visible light irradiation [25]. Meanwhile, the synergistic effect of {001} and {101} facets would also affect the charge carrier separation efficiency, resulting in the difference of photocurrent.

After depositing NGQDs, their corresponding photocurrent responses were all enhanced obviously. The photocurrent density of TiO_2/NGQDs composites decreased in the order of T1-NGQDs, T2-NGQDs, T0-NGQDs, and T3-NGQDs. The photocurrent density of sample T1-NGQDs was the highest, about three times larger than that of sample T3-NGQDs. The photocurrent enhancement could be attributed to the promoted separation rate of photogenerated charge carriers, owing to the introduction of NGQDs. The unique chemical features of NGQD endow it with superior carrier mobility and excellent electronic conductivity [45,46]. Moreover, the extended π-electron systems of NGQD provide its sufficient contact with the surface of titania, and the formation of Ti–O–C bonding could also facilitate the donor-acceptor interaction [16]. These factors contributed to the apparent enhancement of photocurrent density. We calculated multiple times the photocurrents of the anatase TiO_2 with and without NGQDs decoration, shown in Figure 5c. Interestingly, the times decreased with the decrease of exposed {101} facet percentage in TiO_2. The phenomenon indicated a noticeable improvement of electron-hole separation efficiency between the NGQDs and the electron-rich {101} facet, compared to that between the NGQDs and the hole-rich {001} facet. The difference demonstrated that efficient electron transfer existed in the interfacial interaction between NGQDs and TiO_2 with high {101} facet exposure.

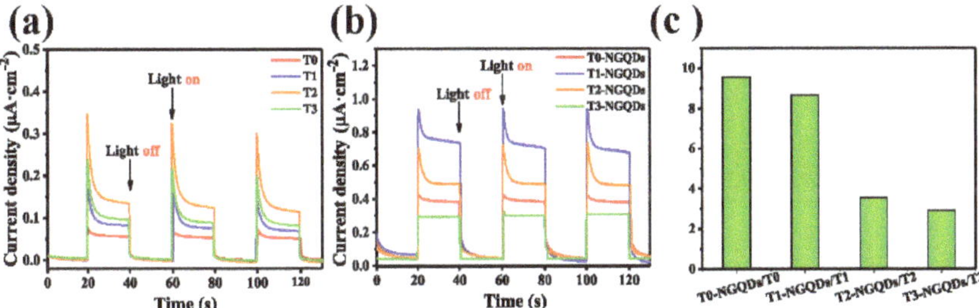

Figure 5. Periodic on-off photocurrent output of (**a**) bare anatase TiO_2, and (**b**) TiO_2/NGQDs composites. (c) The multiple times of the photocurrents of the anatase TiO_2 with and without NGQDs decoration.

Electrochemical impedance spectroscopy (EIS) measurements were utilized to investigate the mechanism of photocurrent improvement. The semicircle diameter generally indicates the carrier transfer resistance. In Figure 6, all Nyquist plots of the composites presented as semicircle, the corresponding arc radius of EIS Nyquist plots all decreased compared to that of bare anatase TiO_2, which is consistent with the photocurrent output. Among all photocatalysts, sample T1-NGQDs exhibited the smallest semicircle while sample T0 the largest. This result is demonstrative of more effective carrier separation and transfer process in the heterojunctions [47,48]. Thus, it could be concluded that the percentage of exposed {001} facet indeed had different effects on the electronic properties of the composites, which might be attributed to the different interfacial interaction between NGQDs and TiO_2 with different facet compositions.

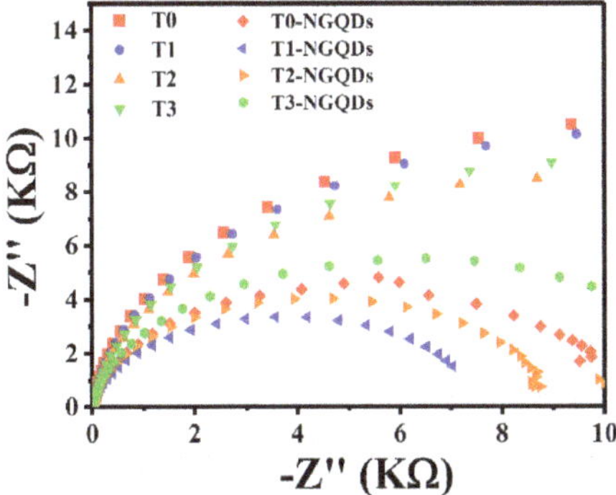

Figure 6. Electrochemical impedance spectroscopy (EIS) Nyquist plots of different TiO_2 samples without and with NGQDs decoration.

Photoluminescence (PL) emission spectrum could help to directly understand the carrier behaviors and observe the radiative recombination of charge carriers [49,50]. Generally, PL emission signals are caused by the photo-induced carrier recombination process. Lower intensity is relevant to better photocatalytic performance. All peaks shape similarly in Figure 7. After decorating NGQDs, the composites exhibited a slight decrease in the PL intensity, compared to the bare anatase, which was probably due to the efficient electron transfer from the CB band of anatase to NGQDs. Thus, the trapping and recombination of charge carriers could be hindered.

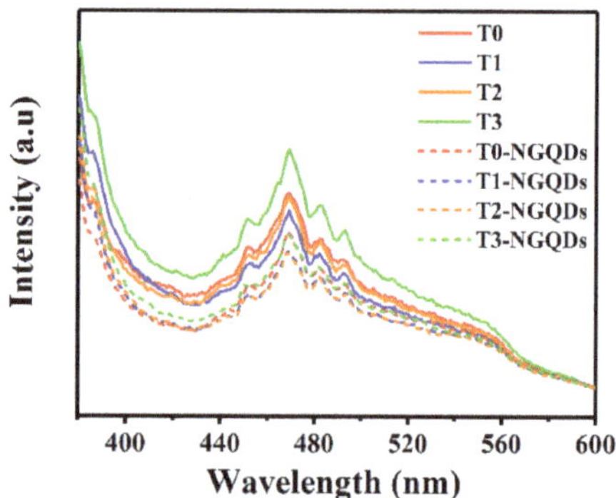

Figure 7. PL spectra of different TiO_2 samples without and with NGQDs decoration.

2.4. Photocatalytic Performance

The photocatalytic performances of all samples were evaluated for RhB photodegradation under visible light irradiation after achieving absorption-desorption balance in the dark, as seen from Figure 8. Obviously, the photocatalytic degradation efficiency of the TiO_2/NGQDs heterojunctions was greatly enhanced compared with that of the bare anatase TiO_2. Among the composites, approximately 96% of the dye was photo-degraded by the T1-NGQDs within 3 h. The degradation process followed the pseudo-first-order kinetics:

$$-\ln\left(\frac{c}{c_0}\right) = kt \tag{1}$$

where k is equal to the corresponding slope of the fitting line, representing the rate constant indicative of the photocatalytic efficiency. To prove the efficiency of sample T1-NGQDs, the photocatalytic degradation activity of other three organic pollutants including methylene blue (MB), methyl orange (MO), and phenol (Phe) were compared. It demonstrated that the T1-NGQDs exhibited high photocatalytic activity for degrading MB dye as well as other common organic species (including phenol, colorless aqueous solution), shown in Figure S6.

The calculated k values of different samples are shown in Figure 8c (the red column corresponds to the bare anatase TiO_2 and the blue column to the composites). Before decorating NGQDs, sample T2 and T3 exhibited nearly the same constant rate, higher than that of T0 and T1. RhB could be easily absorbed on the titania surface with reactive {001} facet exposure, leading to subsequent dye self-photosentization process decomposing RhB under visible light irradiation [51–53].

After decoration of NGQDs, it was found that the values were much higher, and the calculated k of T1-NGQDs reached the highest, about 0.8 h^{-1}. The multiple times of the k values of the anatase TiO_2 with and without NGQDs decoration were then calculated, shown in Figure 8d. Notably, it was shown that the times decreased with the increase of the exposed {001} facet percentage in TiO_2, which is consistent with the photocurrent variation in Figure 5c. The same law illustrates the photocatalytic degradation performance depends greatly on the interfacial charge carrier separation and transfer rate, demonstrative of the significant donor-acceptor interaction between NGQDs and {101} facets. At the same time, it was supposed that the different roles of {001} and {101} facets in separating the photogenerated electron-hole are unignorable, since the hybrid containing TiO_2 with 56% {001} facets exhibited a higher reaction rate value than that with 12% {001} facets.

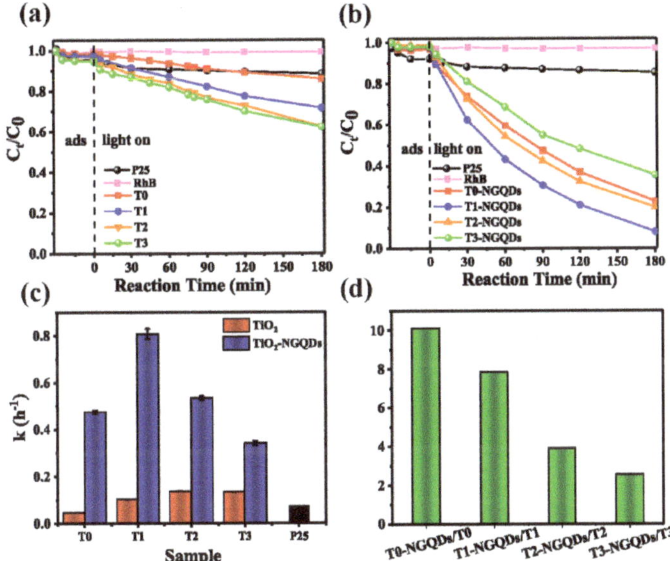

Figure 8. The photocatalytic degradation of RhB for different samples under visible light irradiation. Change of the relative concentration (C_t/C_0) of RhB in (**a**) bare anatase TiO_2, (**b**) TiO_2/NGQDs composites as a function of irradiation time up to 180 min. (**c**) Plot of k ($\ln(C_0/C_t)$) values for RhB degradation in different samples. (**d**) The times of the k values of the anatase with and without NGQDs decoration.

2.5. Photocatalytic Mechanism

The schematic representation of the electron-hole separation and transfer in TiO_2/NGQDs heterojunction composites during the photocatalytic reaction is shown in Scheme 1. Under visible light irradiation, both NGQDs and anatase TiO_2 of avoidable intrinsic defects were capable to generate photo-excited electrons. Obviously, the narrow energy gaps for NGQDs allow for rich hot electrons to produce when the excitation wavelength is larger than 420 nm. According to the measured band potential values, the energy levels in Figure 4d further demonstrate that the band configuration contributes to electron injections from the CB level of TiO_2 to the LUMO of NGQDs. The donor-acceptor interaction was greatly promoted due to the full contact and C–O–Ti formation between NGQDs and TiO_2. The appropriate band alignments explain for the reason why the photocatalytic activities were all enhanced after depositing NGQDs on the surface of TiO_2, compared to that of bare TiO_2.

On the other hand, for bare TiO_2, it was found that the anatase TiO_2 with the higher percentage of exposed {001} facet owned a much better photocatalytic performance which is due to the high activity of the {001} facet. However, the reaction rate value (k) achieved the highest for T1-NGQDs rather than T3-NGQDs. The different variation trend of k value in bare TiO_2 and TiO_2/NGQDs heterojunctions imply the possible influence of facets in the interfacial electron transfer process. It revealed that the photo-excited electron and holes behave differently in TiO_2, where electrons could be easily trapped in the {101} facet while holes tend to run to the {001} facet. Since NGQDs are a good electron transport medium, its deposition on the titania surface with more percentages of exposed {101} facets could result in a higher electron transfer efficiency, which is consistent with the increasing multiple times of photocurrent and also k values with and without NGQDs decoration. Simultaneously, the different roles of {101} and {001} facets in anatase TiO_2 on charge carrier separation is not negligible, considering that the k value of T1-NGQDs is higher than that of T0-NGQDs. By comparing the degradation activity of TiO_2 before and after NGQDs decoration, it could be concluded that there is a synergistic effect of facet-manipulated electron-hole separation in TiO_2 and donor-acceptor interaction in $GQDs/TiO_2$ on the visible light driven photocatalytic performance.

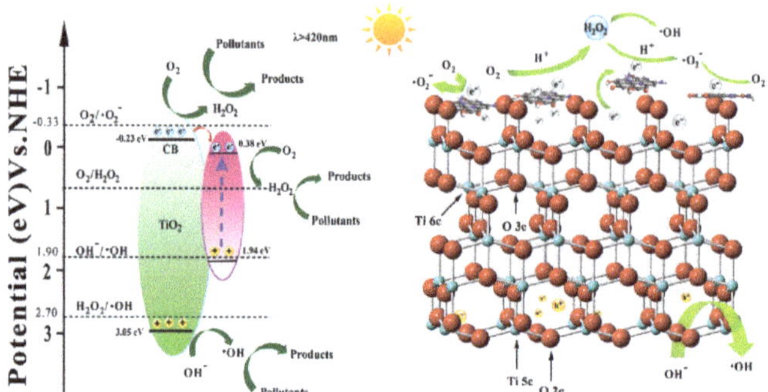

Scheme 1. Schematic illustration of the proposed band alignment and interfacial electron transfer process for the $TiO_2/NGQDs$ heterojunction composites under visible light irradiation ($\lambda > 420$ nm).

3. Materials and Methods

3.1. Synthesis of Anatase TiO_2

The anatase TiO_2 with different exposed percentages of {001} and {101} facets were prepared via the traditional hydrothermal method. Typically, different volumes (0.2, 0.4, and 0.8 mL) of hydrofluoric acid ($\geq$40.0%. Sinopharm Chemical Reagent Co.,Ltd, Shanghai, China) were added into the mixture of tetrabutyl titanate (5 mL; $\geq$99.0%, Aladdin, Shanghai, China) and ethanol (8 mL). After stirring for 30 min, the mixture was then transferred into a 50 mL Teflon-lined autoclave and heated at 180 °C for 24 h. After that, the products were collected by centrifugation, followed by being rinsed several times with absolute ethyl alcohol and dried at 60 °C overnight. The obtained anatase TiO_2 co-exposed with {001} and {101} facets were named as T1, T2, and T3 respectively. To prepare TiO_2 with the {101} dominating plane, the same procedure was conducted but with 0.4 mL of H_2O, which was named T0.

3.2. Synthesis of NGQDs

The NGQDs were prepared according to a one-step hydrothermal process reported by Sun and co-workers [54]. In a typical run, 1.44 g of urea (AR, Aladdin, Shanghai, China) and 1.68 g of citric acid (GR, Aladdin, Shanghai, China) were dissolved in 40 mL of deionized water (DI water). The solution was transferred into a 50 mL Teflon-lined stainless autoclave and heated at 180 °C for 8 h. The final product was centrifuged several times at 10,000 rpm for 5 min with absolute alcohol. The obtained NGQDs precipitate was dried at 80 °C for 1 h to obtain the NGQDs powders.

3.3. Synthesis of $TiO_2/NGQDs$ Heterojunction Composites

An ultrasonic-hydrothermal method was used to prepare the $TiO_2/NGQDs$ heterojunction photocatalysts. A certain amount of TiO_2 and GQDs at a low doping level (1.0 wt%) was added into 40 mL of DI water. The suspension solution was placed in an ultrasonic bath for 30 min and then transferred into a 50 mL Teflon-lined stainless autoclave by heating at 120 °C for another 2 h. After centrifuging several times with DI water and being dried at 60 °C overnight, the final powders were named as T0-NGQDs, T1-NGQDs, T2-NGQDs, and T3-NGQDs, respectively.

3.4. Characterization

The XRD patterns of all samples were recorded with a PAN analytical X'Pert Pro MPD diffractometer (Pananalytical, Holland) using Cu-Kα radiation (λ = 0.1541 nm), and the data was collected from 20° to 80° (2θ). UV-vis diffuse reflectance spectroscopy (DRS) were taken at room temperature measured using $BaSO_4$ as the reference on a UV-3150 spectrophotometer (Shimadzu, Kyoto, Japan). The photoluminescence (PL) spectroscopy was performed using a RF-5301pc fluorescence spectroscopy (Shimadzu, Kyoto, Japan). Band gap energies were calculated by analysis of the Tauc-plots resulting from Kubelka–Munk transformation of absorption spectra. High resolution transmission electron microscope (HRTEM) were conducted by a Phillips/FEI Tecnai F20 S-TWIN TEM (Hillsborough, OR, USA) instrument operating at 200 kv. X-ray photoelectron spectroscopy (XPS) measurements were carried out on an ESCALAB 250 Xi (Thermo Scientific, MA, USA) using non-monochromatized Mg-Kα X-ray as the excitation source. The binding energies for the samples were calibrated by setting the measured binding energy of C 1s to 284.60 eV. The Raman spectra were measured on a LabRAM HR Evolution (Horiba, Tokyo, Japan) at room temperature using the 532 nm line of an argon ion laser as the excitation source.

3.5. Photocatalytic Performance

The photocatalytic performance of the as-synthesized photocatalysts were examined under a 300 W Xe lamp (PLS-SXE 300/300 UV, Perfect Light, Shanghai, China) equipped with a 420 nm cut-off filter as the visible light irradiation source. A total of 15 mg of catalysts were added into 50 mL of a 10.0 mg·L^{-1} solution of different dyes: Rhodamin B (RhB), methylene blue (MB), methyl orange (MO), and phenol (Phe) (Sinopharm Chemical Reagent Co. Ltd., Shanghai, China). The molar concentration was 0.026, 0.031, 0.030, and 0.11 mmol/L, respectively. Before irradiation, the suspension was stirred in the dark to ensure the adsorption-desorption equilibrium of RhB on the surface of the photocatalyst.

3.6. Photoelectrochemical Measurements

The photocurrent measurements, electrochemical impedance spectroscopy (EIS) and Mott–Schottky experiments were conducted on an electrochemical analyzer (CHI 660C work station, CHI, Shanghai, China). The employed standard three-electrode configuration included a platinum plate (as the counter electrode), an Ag/AgCl electrode (as the reference electrode), and a working electrode. The working electrodes were prepared as follows: 40 mg of powders and 5 mg of $Mg(NO_3)_2 \cdot 6H_2O$ ($\geq$98%; Alfa Aesar, Shanghai, China) were dispersed in 100 mL of isopropanol. The suspension was ultrasound for 1 h. A clean SnO_2 transparent conductive glass doped with fluorine, FTO (as cathode) facing the stainless-steel anode was then immersed into this suspension. The distance between the two electrodes was fixed at about 5 cm. The Mg^{2+} adsorbed samples suspension was loaded in a quartz vessel as the electrolyte, and the electrophoresis process was performed at 60 V for 120 s. After the electrophoretic deposition (EPD) process, the prepared electrodes were washed by ethanol and deionized water several times and dried at room temperature. A 350 W xenon lamp with a cut-off filter (λ > 420 nm) was used as a light source and placed 20 cm away from the working electrode. The working electrode was immersed in 0.1 M Na_2SO_4 aqueous solution. The EIS measurements were performed over a range from 0.01 to 1000 Hz at 0.2 V, and the amplitude of the applied potential in each case was 5 mV.

4. Conclusions

In summary, nitrogen-doped graphene quantum dots were successfully deposited onto the surface of anatase TiO_2 with different percentages of exposed {001} facets to form TiO_2/GQDs heterojunction composites. The photocatalytic performances of the hybrid containing TiO_2 with 56% {001} facets exhibited the highest reaction rate value under visible light irradiation. The successful decoration of NGQDs on the TiO_2 surface extended the light absorption edge into the visible light region. Due to Ti–O–C formation and high electron conductivity of NGQDs, the photocurrent responses of the composites were all enhanced obviously, compared to that of the bare samples. Meanwhile, the different roles of {101} and {001} facets in anatase TiO_2 on charge carrier separation was also not negligible. The improved photocatalytic activity was due to the synergistic effect of facet-manipulated electron-hole separation in TiO_2 and the remarkable donor-acceptor interaction between NGQDs and the electron-rich {101} facet. The existence of the {101} facet contributed to the interfacial electron transfer that played a vital role in improving the photocatalytic activity of the NGQDs/TiO_2 heterojunctions. Furthermore, the existence of both facets in anatase assisted the further enhancement of photocatalytic performance. This work provided a new clue to improve the interfacial charge transfer in faceted semiconductor related heterojunctions.

Supplementary Materials: The following are available online at http://www.mdpi.com/2073-4344/9/4/345/s1, Figure S1: (a) Slab model of anatase TiO_2 single crystal. (b) Equilibrium model of anatase TiO_2 single crystal. (Calculation method of the percentage of {001} facets). Figure S2: (a) Raman spectra of different TiO_2 samples without and with NGQDs decoration. (b) Raman spectra of T1 before and after decoration of NGQDs. The inset in Figure S2 b is the enlargement of 1200–1700 cm^{-1} of T1-NGQDs. Figure S3: TEM images of (a) T0, (b) T1, (c) T2, and (d) T3. Figure S4: (a) TEM image, (b) HRTEM image, (c) AFM image, (d) UV-vis spectra and PL spectra of the GQDs (the excitation wavelength is 365 nm), (e) Raman spectra, and (f) XRD pattern of NGQDs. The inset in (a) is the size distribution of NGQDs. The inset in (b) is the autocorrelated HRTEM lattice images recorded. Figure S5: HRTEM images of the anatase TiO_2 decorate with NGQDs. Figure S6: (a) The photocatalytic degradation of different pollutants for T1-NGQDs, (b) plot of k values for different pollutants degradation in T1-NGQDs.

Author Contributions: The manuscript was written through contributions of all authors. N.-Q.O., H.-J.L. designed the experiments; N.-Q.O., B.-W.L., B.-J.G. and X.S. performed the experiments and analyzed the data; N.-Q.O. and H.-J.L. wrote the paper; D.-J.Q. revised the paper. X.W., Y.J. and J.Y. acted as supervisor.

Funding: This research was funded by National Natural Science Foundation of China (51572173, 51602197, 51771121 and 51702212), Shanghai Municipal Science and Technology Commission (16060502300, 16JC1402200 and 18511110600), Shanghai Academic/Technology Research Leader Program (19XD1422900), Shanghai Eastern Scholar Program (QD2016014).

Conflicts of Interest: The authors declare no conflicts of interest.

References

1. Tang, H.; Prasad, K.; Sanjinès, R.; Schmid, P.E.; Lévy, F. Electrical and optical properties of TiO_2 anatase thin films. *J. Appl. Phys.* **1994**, *75*, 2042–2047. [CrossRef]

2. Xu, M.; Gao, Y.; Moreno, E.M.; Kunst, M.; Muhler, M.; Wang, Y.; Idriss, H.; Woll, C. Photocatalytic activity of bulk TiO_2 anatase and rutile single crystals using infrared absorption spectroscopy. *Phys. Rev. Lett.* **2011**, *106*, 138302. [CrossRef] [PubMed]

3. Yu, J.; Low, J.; Xiao, W.; Zhou, P.; Jaroniec, M. Enhanced Photocatalytic CO_2-Reduction Activity of Anatase TiO_2 by Co-exposed {001} and {101} Facets. *J. Am. Chem. Soc.* **2014**, *136*, 8839–8842. [CrossRef]

4. Tachikawa, T.; Yamashita, S.; Majima, T. Evidence for crystal-face-dependent TiO_2 photocatalysis from single-molecule imaging and kinetic analysis. *J. Am. Chem. Soc.* **2011**, *133*, 7197–7204. [CrossRef] [PubMed]

5. Selcuk, S.; Selloni, A. Facet-dependent trapping and dynamics of excess electrons at anatase TiO_2 surfaces and aqueous interfaces. *Nat. Mater.* **2016**, *15*, 1107–1118. [CrossRef]

6. Chen, Q.; Ma, W.; Chen, C.; Ji, H.; Zhao, J. Anatase TiO_2 mesocrystals enclosed by (001) and (101) facets: Synergistic effects between Ti^{3+} and facets for their photocatalytic performance. *Chem. Eur. J.* **2012**, *18*, 12584–12589. [CrossRef]

7. Hyam, R.S.; Lee, J.; Cho, E.; Khim, J.; Lee, H. Effect of Annealing Environments on Self-Organized TiO_2 Nanotubes for Efficient Photocatalytic Applications. *Nanosci. Nanotechnol.* **2012**, *12*, 8908–8912. [CrossRef]

8. Wang, X.; Sun, G.; Li, N.; Chen, P. Quantum dots derived from two-dimensional materials and their applications for catalysis and energy. *Chem. Soc. Rev.* **2016**, *45*, 2239–2262. [CrossRef]

9. Wegner, K.D.; Hildebrandt, N. Quantum dots: Bright and versatile in vitro and in vivo fluorescence imaging biosensors. *Chem. Soc. Rev.* **2015**, *44*, 4792–4834. [CrossRef]

10. Yuan, F.; Li, S.; Fan, Z.; Meng, X.; Fan, L.; Yang, S. Shining carbon dots: Synthesis and biomedical and optoelectronic applications. *Nano Today* **2016**, *11*, 565–586. [CrossRef]

11. Crake, A.; Christoforidis, K.C.; Godin, R.; Moss, B.; Kafizas, A.; Zafeiratos, S.; Durrant, J.R.; Petit, C. Titanium dioxide/carbon nitride nanosheet nanocomposites for gas phase CO_2 photoreduction under UV-visible irradiation. *Appl. Catal. B* **2019**, *242*, 369–378. [CrossRef]

12. Olowoyo, J.O.; Kumar, M.; Jain, S.L.; Babalola, J.O.; Vorontsov, A.V.; Kumar, U. Insights into Reinforced Photocatalytic Activity of the CNT–TiO_2 Nanocomposite for CO_2 Reduction and Water Splitting. *J. Phys. Chem. C* **2018**, *123*, 367–378. [CrossRef]

13. Rimoldi, L.; Giordana, A.; Cerrato, G.; Falletta, E.; Meroni, D. Insights on the photocatalytic degradation processes supported by TiO_2/WO_3 systems. The case of ethanol and tetracycline. *Catal. Today* **2019**, *328*, 210–215. [CrossRef]

14. Zheng, X.T.; Ananthanarayanan, A.; Luo, K.Q.; Chen, P. Glowing graphene quantum dots and carbon dots: Properties, syntheses, and biological applications. *Small* **2015**, *11*, 1620–1655. [CrossRef]

15. Freeman, R.; Finder, T.; Bahshi, L.; Gill, R.; Willner, I. Functionalized CdSe/ZnS QDs for the detection of nitroaromatic or RDX explosives. *Adv. Mater.* **2012**, *24*, 6416–6421. [CrossRef]

16. Long, R.; Casanova, D.; Fang, W.-H.; Prezhdo, O.V. Donor-Acceptor Interaction Determines the Mechanism of Photo-Induced Electron Injection from Graphene Quantum Dots into TiO_2: Stacking Supersedes Covalent Bonding. *J. Am. Chem. Soc.* **2017**, *139*, 2619–2629. [CrossRef]

17. Wang, W.-S.; Wang, D.-H.; Qu, W.-G.; Lu, L.-Q.; Xu, A.-W. Large Ultrathin Anatase TiO_2 Nanosheets with Exposed {001} Facets on Graphene for Enhanced Visible Light Photocatalytic Activity. *J. Phys. Chem. C* **2012**, *116*, 19893–19901. [CrossRef]

18. Wang, F.; Wu, Y.; Wang, Y.; Li, J.; Jin, X.; Zhang, Q.; Li, R.; Yan, S.; Liu, H.; Feng, Y.; et al. Construction of novel Z-scheme nitrogen-doped carbon dots/{0 0 1} TiO_2 nanosheet photocatalysts for broad-spectrum-driven diclofenac degradation: Mechanism insight, products and effects of natural water matrices. *Chem. Eng. J.* **2019**, *356*, 857–868. [CrossRef]

19. Wang, W.; Ni, Y.; Xu, Z. One-step uniformly hybrid carbon quantum dots with high-reactive TiO_2 for photocatalytic application. *J. Alloys Compd.* **2015**, *622*, 303–308. [CrossRef]

20. Baker, D.R.; Kamat, P.V. Photosensitization of TiO_2 Nanostructures with CdS Quantum Dots: Particulate versus Tubular Support Architectures. *Adv. Funct. Mater.* **2009**, *19*, 805–811. [CrossRef]

21. Kenrick, J.; Williams, C.A.N.; Yan, X.; Li, L.-S.; Zhu, X. Hot Electron Injection from Graphene Quantum Dots to TiO_2. *ACS Nano* **2013**, *7*, 1388–1394.

22. Yang, N.; Liu, Y.; Wen, H.; Tang, Z.; Zhao, H.; Li, Y.; Wang, D. Photocatalytic Properties of Graphdiyne and Graphene Modified TiO_2: From Theory to Experiment. *ACS Nano* **2013**, *7*, 1504–1512. [CrossRef]

23. Pan, D.; Jiao, J.; Li, Z.; Guo, Y.; Feng, C.; Liu, Y.; Wang, L.; Wu, M. Efficient Separation of Electron–Hole Pairs in Graphene Quantum Dots by TiO_2 Heterojunctions for Dye Degradation. *ACS Sustain. Chem. Eng.* **2015**, *3*, 2405–2413. [CrossRef]

24. Ma, Y.; Chen, A.Y.; Xie, X.; Wang, X.; Wang, D.; Wang, P.; Li, H.-J.; Yang, J.; Li, Y. Doping effect and fluorescence quenching mechanism of N-doped graphene quantum dots in the detection of dopamine. *Talanta* **2018**, *196*, 563–571. [CrossRef] [PubMed]

25. Yu, S.; Zhong, Y.Q.; Yu, B.Q.; Cai, S.Y.; Wu, L.Z.; Zhou, Y. Graphene quantum dots to enhance the photocatalytic hydrogen evolution efficiency of anatase TiO_2 with exposed {001} facet. *Phys. Chem. Chem. Phys.* **2016**, *18*, 20338–20381. [CrossRef]

26. Zheng, L.; Su, H.; Zhang, J.; Walekar, L.S.; Vafaei Molamahmood, H.; Zhou, B.; Long, M.; Hu, Y.H. Highly selective photocatalytic production of H_2O_2 on sulfur and nitrogen co-doped graphene quantum dots tuned TiO_2. *Appl. Catal. B* **2018**, *239*, 475–484. [CrossRef]

27. Jiang, B.; Tian, C.; Zhou, W.; Wang, J.; Xie, Y.; Pan, Q.; Ren, Z.; Dong, Y.; Fu, D.; Han, J.; et al. In situ growth of TiO_2 in interlayers of expanded graphite for the fabrication of TiO_2-graphene with enhanced photocatalytic activity. *Chem. Eur. J.* **2011**, *17*, 8379–8387. [CrossRef]

28. Liu, L.; Jiang, Y.; Zhao, H.; Chen, J.; Cheng, J.; Yang, K.; Li, Y. Engineering Coexposed {001} and {101} Facets in Oxygen-Deficient TiO_2 Nanocrystals for Enhanced CO_2 Photoreduction under Visible Light. *ACS Catal.* **2016**, *6*, 1097–1108. [CrossRef]

29. Tian, F.; Zhang, Y.; Zhang, J.; Pan, C. Raman Spectroscopy: A New Approach to Measure the Percentage of Anatase TiO_2 Exposed (001) Facets. *J. Phys. Chem. C* **2012**, *116*, 7515–7519. [CrossRef]

30. Li, H.-J.; Sun, X.; Xue, F.; Ou, N.; Sun, B.-W.; Qian, D.-J.; Chen, M.; Wang, D.; Yang, J.; Wang, X. Redox Induced Fluorescence On–Off Switching Based on Nitrogen Enriched Graphene Quantum Dots for Formaldehyde Detection and Bioimaging. *ACS Sustain. Chem. Eng.* **2018**, *6*, 1708–1716. [CrossRef]

31. Qu, A.; Xie, H.; Xu, X.; Zhang, Y.; Wen, S.; Cui, Y. High quantum yield graphene quantum dots decorated TiO_2 nanotubes for enhancing photocatalytic activity. *Appl. Surf. Sci.* **2016**, *375*, 230–241. [CrossRef]

32. Rajender, G.; Kumar, J.; Giri, P.K. Interfacial charge transfer in oxygen deficient TiO_2-graphene quantum dot hybrid and its influence on the enhanced visible light photocatalysis. *Appl. Catal. B* **2018**, *224*, 960–972. [CrossRef]

33. Shen, K.; Xue, X.; Wang, X.; Hu, X.; Tian, H.; Zheng, W. One-step synthesis of band-tunable N, S co-doped commercial TiO_2/graphene quantum dots composites with enhanced photocatalytic activity. *RSC Adv.* **2017**, *7*, 23319–23327. [CrossRef]

34. Wang, W.; Xu, D.; Cheng, B.; Yu, J.; Jiang, C. Hybrid carbon@TiO_2 hollow spheres with enhanced photocatalytic CO_2 reduction activity. *J. Mater. Chem. A* **2017**, *5*, 5020–5029. [CrossRef]

35. Xu, C.; Han, Q.; Zhao, Y.; Wang, L.; Li, Y.; Qu, L. Sulfur-doped graphitic carbon nitride decorated with graphene quantum dots for an efficient metal-free electrocatalyst. *J. Mater. Chem. A* **2015**, *3*, 1841–1846. [CrossRef]

36. Han, S.; Hu, X.; Wang, J.; Fang, X.; Zhu, Y. Novel Route to Fe-Based Cathode as an Efficient Bifunctional Catalysts for Rechargeable Zn-Air Battery. *Adv. Energy Mater.* **2018**, *8*, 1800955. [CrossRef]

37. Li, H.-J.; Ou, N.-Q.; Sun, X.; Sun, B.-W.; Qian, D.-J.; Chen, M.; Wang, X.; Yang, J. Exploitation of the synergistic effect between surface and bulk defects in ultra-small N-doped titanium suboxides for enhancing photocatalytic hydrogen evolution. *Catal. Sci. Technol.* **2018**, *8*, 5515–5525. [CrossRef]

38. Rimoldi, L.; Pargoletti, E.; Meroni, D.; Falletta, E.; Cerrato, G.; Turco, F.; Cappelletti, G. Concurrent role of metal (Sn, Zn) and N species in enhancing the photocatalytic activity of TiO_2 under solar light. *Catal. Today* **2018**, *313*, 40–46. [CrossRef]

39. Ding, H.; Yu, S.B.; Wei, J.S.; Xiong, H.M. Full-Color Light-Emitting Carbon Dots with a Surface-State-Controlled Luminescence Mechanism. *ACS Nano* **2016**, *10*, 484–491. [CrossRef]

40. Ding, H.; Wei, J.S.; Zhang, P.; Zhou, Z.Y.; Gao, Q.Y.; Xiong, H.M. Solvent-Controlled Synthesis of Highly Luminescent Carbon Dots with a Wide Color Gamut and Narrowed Emission Peak Widths. *Small* **2018**, *14*, e1800612. [CrossRef]

41. Yang, H.; Wang, P.; Wang, D.; Zhu, Y.; Xie, K.; Zhao, X.; Yang, J.; Wang, X. New Understanding on Photocatalytic Mechanism of Nitrogen-Doped Graphene Quantum Dots-Decorated $BiVO_4$ Nanojunction Photocatalysts. *ACS Omega* **2017**, *2*, 3766–3773. [CrossRef]

42. Deming, C.P.; Mercado, R.; Gadiraju, V.; Sweeney, S.W.; Khan, M.; Chen, S. Graphene Quantum Dots-Supported Palladium Nanoparticles for Efficient Electrocatalytic Reduction of Oxygen in Alkaline Media. *ACS Sustain. Chem. Eng.* **2015**, *3*, 3315–3323. [CrossRef]

43. Safardoust-Hojaghan, H.; Salavati-Niasari, M. Degradation of methylene blue as a pollutant with N-doped graphene quantum dot/titanium dioxide nanocomposite. *J. Clean. Prod.* **2017**, *148*, 31–36. [CrossRef]

44. Wang, S.; Cole, I.S.; Li, Q. Quantum-confined bandgap narrowing of TiO_2 nanoparticles by graphene quantum dots for visible-light-driven applications. *Chem. Commun.* **2016**, *52*, 9208–9211. [CrossRef]

45. Wang, D.; Chen, J.-F.; Dai, L. Recent Advances in Graphene Quantum Dots for Fluorescence Bioimaging from Cells through Tissues to Animals. *Part. Part. Syst. Charact.* **2015**, *32*, 515–523. [CrossRef]

46. Zou, X.; Liu, M.; Wu, J.; Ajayan, P.M.; Li, J.; Liu, B.; Yakobson, B.I. How Nitrogen-Doped Graphene Quantum Dots Catalyze Electroreduction of CO_2 to Hydrocarbons and Oxygenates. *ACS Catal.* **2017**, *7*, 6245–6250. [CrossRef]

47. Li, H.-J.; Qian, D.J.; Chen, M. Templateless Infrared Heating Process for Fabricating Carbon Nitride Nanorods with Efficient Photocatalytic H2 Evolution. *ACS Appl. Mater. Interfaces* **2015**, *7*, 25162–25170. [CrossRef]

48. Yan, M.; Zhu, F.; Gu, W.; Sun, L.; Shi, W.; Hua, Y. Construction of nitrogen-doped graphene quantum dots-$BiVO_4$/g-C_3N_4 Z-scheme photocatalyst and enhanced photocatalytic degradation of antibiotics under visible light. *RSC Adv.* **2016**, *6*, 61162–61174. [CrossRef]

49. Liu, J.; Li, J.; Sedhain, A.; Lin, J.; Jiang, H. Structure and Photoluminescence Study of TiO_2 Nanoneedle Texture along Vertically Aligned Carbon Nanofiber Arrays. *J. Phys. Chem. C* **2008**, *112*, 17127–17132. [CrossRef]

50. Nadica, D.; Abazovic, M.I.C.O.; Dramicanin, M.D. Photoluminescence of Anatase and Rutile TiO_2 Particles. *J. Phys. Chem. B* **2006**, *110*, 25366–25370.

51. Hu, C.; Zhang, X.; Li, X.; Yan, Y.; Xi, G.; Yang, H.; Bai, H. Au photosensitized TiO_2 ultrathin nanosheets with {001} exposed facets. *Chem. Eur. J.* **2014**, *20*, 13557–13560. [CrossRef]
52. Li, K.; Xu, Y.; He, Y.; Yang, C.; Wang, Y.; Jia, J. Photocatalytic fuel cell (PFC) and dye self-photosensitization photocatalytic fuel cell (DSPFC) with BiOCl/Ti photoanode under UV and visible light irradiation. *Environ. Sci. Technol.* **2013**, *47*, 3490–3497. [CrossRef]
53. Wang, X.; Xia, R.; Muhire, E.; Jiang, S.; Huo, X.; Gao, M. Highly enhanced photocatalytic performance of TiO_2 nanosheets through constructing TiO_2/TiO_2 quantum dots homojunction. *Appl. Surf. Sci.* **2018**, *459*, 9–15. [CrossRef]
54. Qu, D.; Zheng, M.; Du, P.; Zhou, Y.; Zhang, L.; Li, D.; Tan, H.; Zhao, Z.; Xie, Z.; Sun, Z. Highly luminescent S, N co-doped graphene quantum dots with broad visible absorption bands for visible light photocatalysts. *Nanoscale* **2013**, *5*, 12272–12278. [CrossRef]

Article

Hydrogen Production from Glycerol Photoreforming on TiO$_2$/HKUST-1 Composites: Effect of Preparation Method

Fabián M. Martínez [1], Elim Albiter [1], Salvador Alfaro [1], Ana L. Luna [2], Christophe Colbeau-Justin [2], José M. Barrera-Andrade [1], Hynd Remita [2] and Miguel A. Valenzuela [1,*]

[1]   Laboratorio de Catálisis y Materiales, ESIQIE-Instituto Politécnico Nacional, México City 07738, Mexico; fmm003@eng.ucsd.edu (F.M.M.); ealbitere@ipn.mx (E.A.); salfaroh@ipn.mx (S.A.); jmanban@yahoo.com.mx (J.M.B.-A.)

[2]   Laboratoire de Chimie Physique, CNRS UMR 8000 Université Paris-Sud, 91405 Orsay, France; aluna0786@gmail.com (A.L.L.); christophe.colbeau-justin@u-psud.fr (C.C.-J.); hynd.remita@u-psud.fr (H.R.)

*   Correspondence: mavalenz@ipn.mx; Tel.: +52-55-5729-6000 (ext. 55112)

Received: 12 March 2019; Accepted: 31 March 2019; Published: 4 April 2019

Abstract: Coupling metal-organic frameworks (MOFs) with inorganic semiconductors has been successfully tested in a variety of photocatalytic reactions. In this work we present the synthesis of TiO$_2$/HKUST-1 composites by grinding, solvothermal, and chemical methods, using different TiO$_2$ loadings. These composites were used as photocatalysts for hydrogen production by the photoreforming of a glycerol-water mixture under simulated solar light. Several characterization techniques were employed, including X-ray diffraction (XRD), UV-Vis diffuse reflectance spectroscopy (DRS), infrared spectroscopy (FTIR), and time-resolved microwave conductivity (TRMC). A synergetic effect was observed with all TiO$_2$/HKUST-1 composites (mass ratio TiO$_2$/MOF 1:1), which presented higher photocatalytic activity than that of individual components. These results were explained in terms of an inhibition of the charge carrier (hole-electron) recombination reaction after photoexcitation, favoring the electron transfer from TiO$_2$ to the MOF and creating reversible Cu^{1+}/Cu^0 entities useful for hydrogen production.

Keywords: hydrogen production; photocatalysis; TiO$_2$-HKUST-1 composites; solar light

1. Introduction

Nowadays, one of the most important necessities of society is the use of natural renewable resources to produce energy, minimizing the use of fossil fuels and reducing the associated harmful pollution produced by their combustion. On the other hand, hydrogen is considered a good candidate as a green energy carrier because it produces null pollution during its combustion, and it can be obtained from renewable sources [1–3]. The use of hydrogen as an energy carrier has several benefits, such as the many different storage possibilities, its ability to be converted to other energy forms with ease and to be produced from water with near-zero emissions, and its high conversion efficiency [2]. However, there are also severe limitations for the widespread use of hydrogen, for example, as a fuel for transportation. If we are planning to use hydrogen-combustion and hydrogen-fuel-cell vehicles in the future, we must first resolve outstanding issues, such as the efficient and safe storage of hydrogen, creating a fueling infrastructure, and reducing its production costs [4]. Certainly, one possibility to reduce the production cost of hydrogen is the use of green energy sources. In this sense, hydrogen production using solar energy can be categorized as: (a) thermal, (b) photovoltaic, (c) bio-photolysis, and (c) photo-electrochemical [5]. Although most of the production methods involve renewable sources, they are not well understood, and their development implies an increase in production costs and low global efficiency [2–5].

Photocatalytic hydrogen generation can be obtained mostly by two different approaches: (1) photocatalytic water splitting and (2) photocatalytic reforming of organics [6]. The first method relates to the capability of water to be reduced and oxidized by reacting with photogenerated electrons and positive holes, during semiconductor irradiation, in the presence of selected co-catalysts. The second approach is based on the ability of some organic species—namely, sacrificial agents—to donate electrons to the positive holes of the illuminated photocatalyst and be oxidized, generating proton ions, while photogenerated electrons reduce the latter to produce hydrogen in the presence of proper co-catalysts.

Glycerol is a sustainable compound that can be used for hydrogen production by photocatalytic reactions (photoreforming). Although this reaction has been studied extensively, the overall performance towards hydrogen evolution is low, and in many cases, a high photocatalytic activity is only achieved with UV-light irradiation. For this reason, the search for new materials, active and stable in the presence of sunlight, is of great interest [7–9].

Metal organic frameworks (MOFs) are obtained by the self-assembly of metal ions and organic ligands through the formation of covalent bonds or the presence of inter-molecular forces between them [10]. MOFs present a long-range periodic structure with good crystallinity, and MOF-based structures take some unique properties of both organic and inorganic porous materials. They exhibit several advantages such as a high surface area, tunable pore size, easy preparation, flexibility, and structural diversity [11]. CuMOF, also known as HKUST-1, copper-benzene-1,3,5-tricarboxylate (Cu-BTC), MOF-199 or Basolite® C300, was first assembled by Chui et al. [12] through the formation of coordination bonds between trimesic acid (H_3BTC) and Cu ions [13].

MOFs have been investigated in many fields, such as sensing, drug delivery, sequestration, separation, molecular transport, electronics, bioreactors, optics, energy production, and catalysis, among others [14]. Applications in photocatalysis have been reported in the last decade, and since then, several articles and reviews have been published focusing on artificial photosynthesis (i.e., water splitting and CO_2 photoreduction) [15,16], organic photosynthesis [17], and pollutants degradation [18,19].

Specifically, in solar-driven hydrogen evolution with the presence of a sacrificial electron donor, e.g., alcohols, most of the MOFs cannot be used as a stable and efficient photocatalyst for this application individually [20]. Certain modifications of the pristine MOF, including the decoration of the organic linker or metal center, combination with semiconductors, metal nanoparticles loading, decoration with reduced graphene oxide, sensitization, pyrolyzation, and incorporation with other functional materials, have been tested to increase their activity and stability under visible light [19,21].

Hybrid nanocomposites of semiconductors with MOFs have attracted increased attention because they improve charge transfer mechanisms with a lower charge recombination and more efficient light harvesting [22]. Hybrid nanocomposites based on TiO_2 and HKUST-1 are exciting materials which could show synergic effects enhancing photocatalytic activity under visible light. Only a few investigations have reported the synthesis, structure, and properties (i.e., as photocatalysts in hydrogen production) of TiO_2/HKUST-1 nanocomposites [23–26]. Particularly, it has been reported that in these nanocomposites, HKUST-1 is transformed to $Cu-Cu_2O$ nanoparticles after calcination at 400 °C, presenting better rates of hydrogen production in comparison with Cu deposited on TiO_2 by conventional methods [24]. There are contradictory results concerning the stability of HKUST-1. For example, when this MOF was used in aqueous media, it decomposed after 24 h of reaction [27]. However, TiO_2/HKUST-1 composites synthesized using ionic liquids as solvents showed high activity and stability during photo-oxidation/photoreduction reactions [28].

In this context, it would be very useful to know the role played by TiO_2-HKUST interactions on the activity and stability in glycerol photoreforming. Therefore, TiO_2/HKUST-1 composites were synthesized by employing three methods: the first composite was prepared by grinding the commercial reagents, Aeroxide® TiO_2 P25 (Evonik, P25), and HKUST-1 (Basolite® C300). These composites were designated as TiO_2 P25/com-HKUST-1. The second one was formed by TiO_2 prepared by a solvothermal route in the presence of the commercial HKUST-1 (TiO_2-ST/com-HKUST-1), and the third composite was prepared by synthesizing HKUST-1 by a chemical route in the presence of TiO_2 P25 (TiO_2 P25/syn-HKUST-1). Furthermore, the aim of the present work was focused on the effect of the preparation method of TiO_2/HKUST-1 composites, as well as the mass ratio TiO_2:MOF employed, on their photocatalytic properties for hydrogen production, using glycerol as a sacrificial agent.

2. Results

2.1. Photocatalytic Hydrogen Evolution

Due to the lack of studies regarding the effect of the optimal amount of TiO_2 that can be deposited on the HKUST-1, Figure 1 shows the photocatalytic hydrogen evolution rates as a function of TiO_2 content. All the experiments were conducted under similar operating conditions, and the H_2 production rate was estimated after 8 h of irradiation time. As can be observed, the results demonstrate a synergic photocatalytic activity between HKUST-1 and TiO_2, and the best performance corresponds to the composites with 50 wt % TiO_2. In the case of the catalyst prepared by grinding (50TiO_2 P25/com-HKUST-1) the production rate was 2.9 mmol $\times$ g^{-1} $\times$ h^{-1}, 2.4 mmol $\times$ g^{-1} $\times$ h^{-1} for the 50TiO_2-ST/com-HKUST-1, and 4.5 mmol $\times$ g^{-1} $\times$ h^{-1} for the 50TiO_2 P25/syn-HKUST-1. Note that the photoactivity of the synthesized HKUST-1 and commercial HKUST-1 was insignificant, and as a comparison, the production rate shown by TiO_2 P25 was 1.1 mmol $\times$ g^{-1} $\times$ h^{-1}.

It is very important to highlight that during the reaction, there was a change in color in the photocatalysts from light blue (original composite color) to reddish brown (spent composite, see Figure 2b), which is indicative that the Cu^{2+} originally present in the HKUST-1 was partially reduced towards Cu^{1+} or Cu^0 [28,29]. This observation suggests that HKUST-1 assembled with Cu ions and benzene 1,3,5-tricarboxylate ligands (Cu-BTC) is an unstable material when irradiated in an aqueous medium, and functions as a precursor of Cu reduced species interacting with TiO_2, as co-catalysts in the production of hydrogen. Note that the highest amounts of CO_2 and CH_4 were obtained with the 50TiO_2 P25/syn-HKUST-1 composite, which comes from the photocatalytic oxidation of an aqueous solution of glycerol. Indeed, it has been proposed that a secondary alcohol photoreforming can produce methane via β-hydride elimination, which could explain the origin of produced methane [30]. Figure 1d compares the amount of hydrogen produced as a function of the irradiation time for the prepared 50TiO_2/HKUST-1 composites. Hydrogen production followed almost the same trend with the three photocatalysts. However, a higher photoactivity was observed with the 50TiO_2 P25/syn-HKUST-1 composite.

It is important to point out that only the 50TiO_2 P25/syn-HKUST-1 composite presented long-term activity, and it was evaluated in five cycles, under simulated solar light. Figure 2a shows the H_2 production rate reached during 8 h of reaction time in each cycle. It can be seen from the second cycle that the composite shows a reduction on the production rate, and in the fifth cycle the observed reduction was ca. 50% of the production rate observed in the first run. An explanation for this unfavorable behavior can be given in terms of a partial reduction of Cu^{2+} contained in the original HKUST-1 by the photogenerated electrons in the TiO_2 conduction band, which was clearly demonstrated by the color change of the original HKUST-1 from light blue (original composite) to reddish brown (spent composite), as shown in Figure 2b. Furthermore, the zone attributed to the d-d spin allowed the transition of the Cu^{2+} between 500–800 nm (discussed later), which was modified to a reddish-brown color, as is characteristic of Cu reduced species in the HKUST-1 structure [31]. Interestingly, after each reaction cycle and subsequent washing and purging of the reaction cell, the solid returned to the original light blue color of the composite. This means that the HKUST-1 structure was not completely destroyed, otherwise it would be forming the Cu^{1+}-Cu^{2+} MOF meta-stable phase with photocatalytic activity to reduce protons to hydrogen. This behavior was previously reported in other applications of

HKUST-1 [28,29,31,32] however, this is the first time that it has been observed in the photocatalytic hydrogen evolution reaction, which requires a systematic and thorough study.

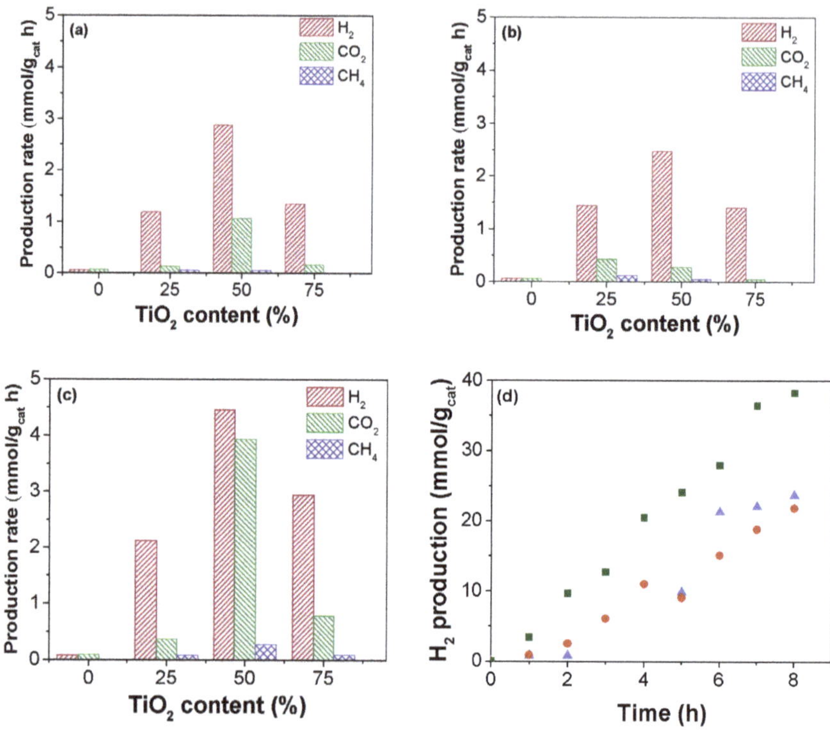

Figure 1. Effect of TiO_2 content on the H_2, CO_2, and CH_4 production rates of TiO_2/HKUST-1 composites, after 8 h irradiation with simulated solar light with (**a**) TiO_2 P25/com-HKUST-1, (**b**) TiO_2-ST/com-HKUST-1, and (**c**) TiO_2 P25/syn-HKUST-1 (**d**) Hydrogen evolution vs. time for TiO_2/HKUST-1 composites: 50TiO_2 P25/com-HKUST-1 (▲), 50TiO_2-ST/com-HKUST-1 (●), and 50TiO_2 P25/syn-HKUST-1 (■).

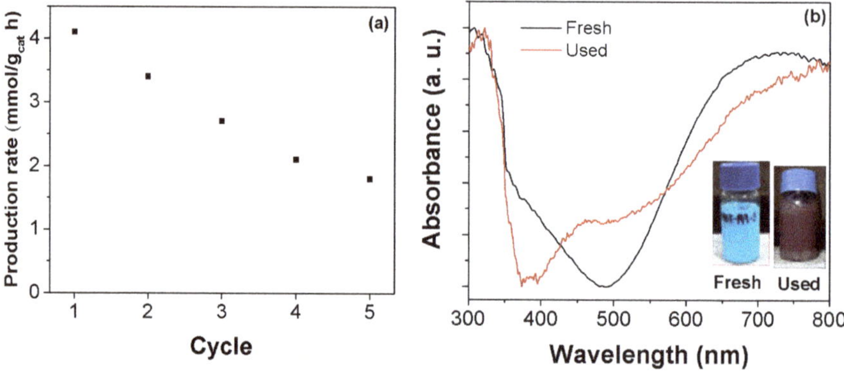

Figure 2. (**a**) Photocatalytic stability tests of 50TiO_2 P25/syn-HKUST-1, under simulated solar light, (**b**) UV-Vis diffuse reflectance spectra (DRS) of fresh and spent 50TiO_2 P25/syn-HKUST-1 composites.

2.2. Characterization

Figure 3 shows the Fourier-transform infrared (FT-IR) spectra of com-HKUST-1, syn-HKUST-1, and TiO$_2$/HKUST-1 composites. Clearly, the com-HKUST-1 and syn-HKUST-1 spectra are quite similar to those reported in previous works [33–35], which indicates that the method employed for the synthesis of syn-HKUST-1 was effective. In these spectra, several signals appeared in the range from 1300 to 1500 cm^{-1} and from 1500 to 1700 cm^{-1}, which are associated with the interactions between the carboxylate anion—in symmetric and asymmetric modes respectively—with the metal ion [34,35]. The signals indicated at 1110, 765, and 740 cm^{-1} are associated with the C-H vibration modes in the aromatic ring [34]. The band at 1060 cm^{-1} is attributed to the presence of copper coordinated *N,N*-dimethylformamide (DMF) molecules [33], and the band centered at 507 cm^{-1} is assigned to the Cu-O stretching mode [35]. On the other hand, all the bands mentioned previously appeared in the spectrum of the 50TiO$_2$ P25/com-HKUST-1 composite, indicating a weak interaction between TiO$_2$ P25 and com-HKUST-1. In the case of the 50TiO$_2$-ST/com-HKUST-1 and 50TiO$_2$ P25/syn-HKUST-1, the bands between 1300 and 1700 cm^{-1} were less defined, and the signals at 507, 740, and 765 cm^{-1} were replaced by a broad band (500–900 cm^{-1}) in the case of TiO$_2$-ST/com-HKUST-1 and two bands (500–700 and 765–830 cm^{-1}) in the spectrum of 50TiO$_2$ P25/syn-HKUST-1. These last results could indicate that there is a chemical interaction between TiO$_2$ and HKUST-1 when either component is obtained by a chemical route.

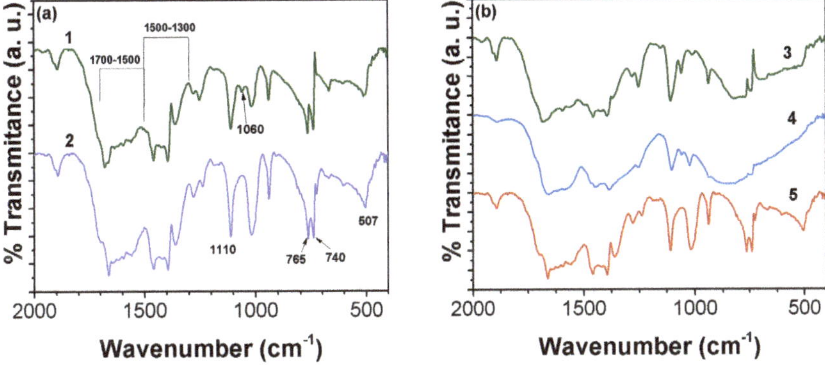

Figure 3. Fourier-transform infrared (FT-IR) spectra of (**a**) syn-HKUST-1 (1), com-HKUST-1 (2) (**b**) 50TiO$_2$ P25/syn-HKUST-1 (3), 50TiO$_2$ ST/com-HKUST-1 (4), and 50TiO$_2$ P25/com-HKUST-1 (5).

The optical properties of HKUST-1 and the TiO$_2$/Cu MOFs composites were investigated by UV-Vis diffuse reflectance spectroscopy (DRS). As can be seen in Figure 4a, the as-prepared HKUST-1 sample and the commercial HKUST-1 showed a similar spectrum, exhibiting two characteristic absorption bands centered at 300 and 700 nm, similar values to those reported in the literature [36]. Note that one shoulder can also be detected at 375 nm. The first band located in the UV region is assigned to π-π^* transitions of the ligands and the band in the visible zone is attributed to the d-d spin and allowed transition of the Cu^{2+} [37]. The shoulder at 375 nm is ascribed to the ligand-to-metal charge transfer (LMCT), and the additional broad absorption band between 500 and 800 nm is assigned to the d-d spin and allowed transition of the Cu^{2+} (d^9) ions [37].

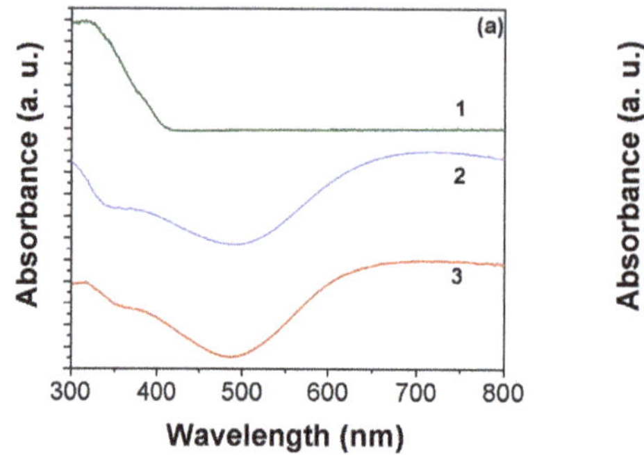
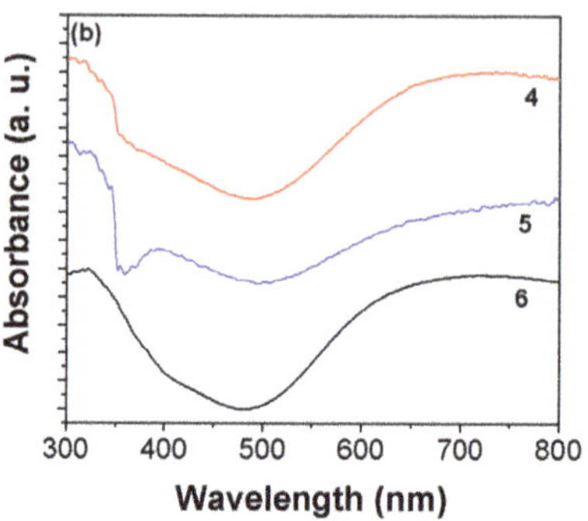

Figure 4. UV-Vis DRS spectra of (**a**) P25 (1), syn-HKUST-1 (2), com-HKUST-1 (3), and (**b**) TiO$_2$/HKUST-1 composites: 50TiO$_2$ P25/com-HKUST-1 (4), 50TiO$_2$ ST/com-HKUST-1 (5), and 50TiO$_2$ P25/syn-HKUST-1 (6).

Figure 4b corresponds to the UV-Vis DRS spectra of the TiO$_2$/HKUST-1 composites. In general, all composite photocatalysts showed similar absorption behavior to the pristine HKUST-1. However, it is worth noting a slight change of their absorption edge to the UV zone (350–400 nm), compared to those of commercial and synthesized HKUST-1 (Figure 4a), which can be related to the TiO$_2$-HKUST-1 interaction. On the other hand, the slight differences in the 50TiO$_2$ ST/com-HKUST-1 composite spectrum (e.g., a lower absorption in the visible region) could be related to a shielding effect by TiO$_2$, partially inhibiting the visible light absorption of the Cu^{2+} ions in the HKUST-1 structure, because TiO$_2$, in this particular composite, was grown in intimate contact with the commercial HKUST-1.

Figure 5a presents the X-ray diffraction (XRD) pattern of syn-HKUST-1, which is quite similar to the pattern of com-HKUST-1, showing the main reflections peaks at 11.6°, 13.4°, 17.4° and 19° [38,39]. These results prove that the HKUST-1 structure was successfully obtained using our described preparation method. XRD patterns of 50TiO$_2$ P25/com-HKUST-1, 50TiO$_2$ ST/com-HKUST-1, and 50TiO$_2$ P25/syn-HKUST-1 composites are shown in Figure 5b. These three samples displayed the same reflections described earlier, indicating that the HKUST-1 structure was preserved despite the preparation method used. Note that a higher crystallinity is observed in samples 50TiO$_2$ P25/com-HKUST-1 and 50TiO$_2$ P25/syn-HKUST-1 in comparison with sample 50TiO$_2$ ST/com-HKUST-1, which means that a poor crystallization of TiO$_2$ occurred due to the low synthesis temperature (i.e., 100 °C) compared to that reported in the literature, above 150 °C under solvothermal process [40].

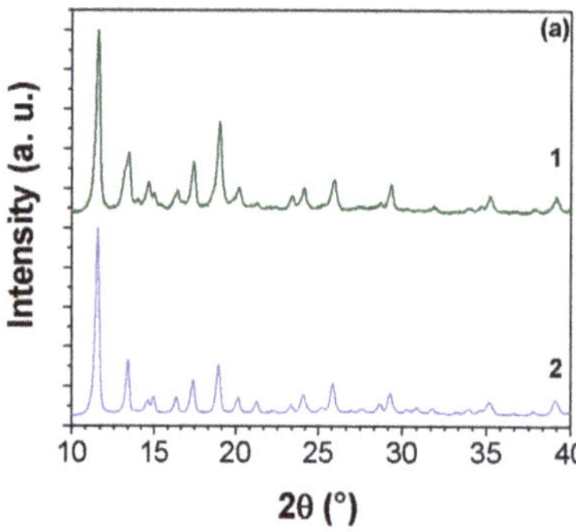
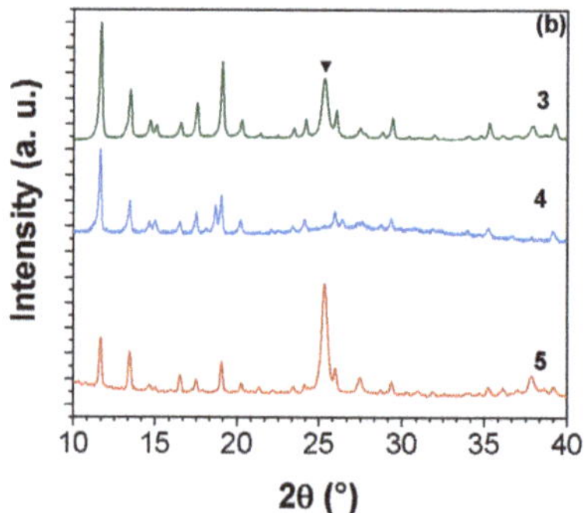

Figure 5. X-ray diffraction patterns of (**a**) as-sensitized HKUST-1 (1), commercial HKUST-1 (2), and (**b**) 50TiO$_2$ P25/com-HKUST-1 (3), 50TiO$_2$ ST/com-HKUST-1 (4), and 50TiO$_2$ P25/syn-HKUST-1 (5). ▼ denotes the peak corresponding to TiO$_2$ anatase.

Figure 6a compares the time-resolved microwave conductivity (TRMC) profiles of TiO_2 P25, syn-HKUST-1, and com-HKUST-1, obtained under a wavelength excitation of 355 nm. The highest signal was exhibited by TiO_2 P25, which also presented a long-time decay. It is worth noting that syn-HKUST-1 showed a TRMC signal because it behaves like a semiconductor material; however, it decays faster than TiO_2 P25, revealing a short lifetime of photogenerated electrons. Surprisingly, com-HKUST-1, which presented very similar structure and light absorption (see Figure 4a) to those of syn-HKUST-1, displayed a much lower TRMC signal, revealing a great difficulty in executing the charge separation after irradiation with UV light of 355 nm.

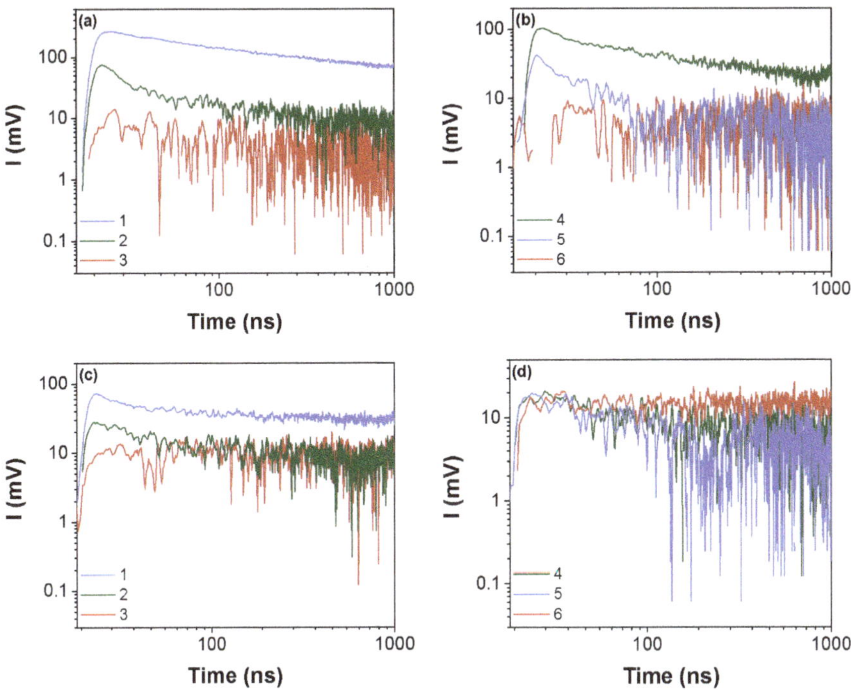

Figure 6. Time-resolved microwave conductivity (TRMC) transient signal of (**a,c**) P25 (1), syn HKUST 1 (2), com-HKUST-1 (3) (**b,d**) TiO_2/HKUST-1 composites: 50TiO_2 P25/com-HKUST-1 (4), 50TiO_2 ST/com-HKUST-1) (5), and 50TiO_2 P25/syn-HKUST-1 (6). The excitation wavelengths were 355 nm (**a,b**) and 410 nm (**c,d**).

On the other hand, by analyzing the TRMC signals of the composites in Figure 6b, the sample prepared by grinding (sample 4) displayed a decay profile quite similar than that obtained with TiO_2 P25, indicating that the charge carrier dynamics are mainly due to the TiO_2 P25 contribution. Unexpectedly, the 50TiO_2 ST/com-HKUST-1 (sample 5) presented a TRMC signal with an I_{max} value slightly smaller than that of the grinding composite (sample 4), clearly showing a charge carrier separation, but with a short time decay. The decay signal abruptly becomes highly noisy after 70 ns, denoting a charge carrier recombination or electron transfer from TiO_2 to HKUST-1. The 50TiO_2 ST/com-HKUST-1 (sample 6) did not display a clear TRMC signal, similar to that shown by the com-HKUST-1(Figure 6a), which can be connected with the XRD results, meaning that TiO_2 was poorly crystallized.

As simulated solar light is being used in the photocatalytic evaluation, it is interesting to see the TRMC signals of the materials under 410 nm excitation. As can be seen in Figure 6c, TiO_2 P25, com-HKUST-1 and syn-HKUST-1MOFs presented similar behaviors to those presented under 355 nm excitation, but all the samples showed a lower I_{max} value. The TiO_2 P25 signal can be attributed to the presence of rutile, which has a bandgap of 3.0 eV, making possible the generation of electron-hole pairs under visible-light irradiation. All the composites shown in Figure 6d exhibited a small TRMC signal, which means that they have the capacity to generate electron-hole pairs under 410 nm excitation.

3. Discussion

The above results suggest that the photocatalytic performance of TiO_2 is improved by the incorporation of HKUST-1, forming a semiconductor-MOF composite regardless the preparation method. Nonetheless, the integration of HKUST-1 with TiO_2 P25 (TiO_2 P25/syn-HKUST-1) by a chemical method showed greater photocatalytic activity and stability compared with to grinding (TiO_2 P25/com-HKUST-1) or the TiO_2-ST/com-HKUST-1. At first glance, the greater photocatalytic activity shown by the composites in comparison with TiO_2 P25 or HKUST-1 is explained by the synergy between the semiconductor and the MOF, inhibiting electron-hole recombination. This cooperative behavior implied the partial reduction of the Cu^{2+} contained in the MOF, forming Cu^{1+} species which absorb visible light and could contribute to the proton reduction, as shown in Figure 7. The creation of reversible Cu^{1+}/Cu^{2+} entities in the composite was attributed to the electron transfer from P25 to HKUST-1 generating in situ species, i.e., Cu^{1+}-Cu^{2+} MOF, giving rise to improved hydrogen production.

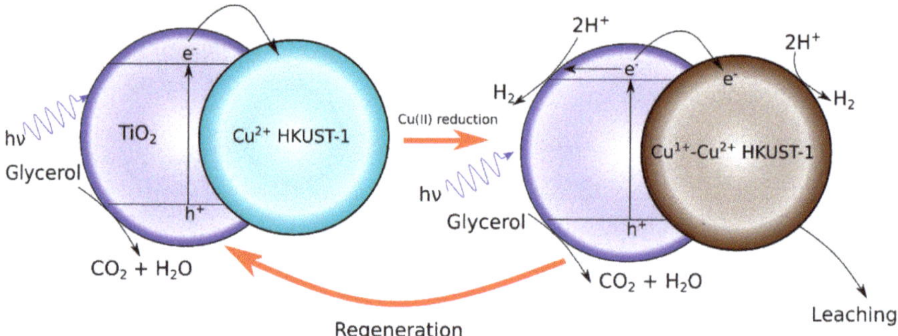

Figure 7. Proposed photocatalytic mechanism of glycerol photoreforming with the TiO_2 P25/syn-HKUST-1 under solar light irradiation.

Furthermore, it was found that the higher photoactivity and stability shown by the TiO_2 P25/syn-HKUST-1 can also be related to a strong interaction between the two components, which was not seen with the other two composites prepared by grinding or mixing poor crystallized TiO_2 with commercial HKUST-1.

4. Materials and Methods

4.1. Materials

Copper (II) acetate monohydrate ($Cu(OAc)_2H_2O$), trimesic acid (H_3BTC), *N,N*-dimethylformamide (DMF), Triethylamine (Et_3N), and Titanium (IV) isopropoxide were purchased from Sigma-Aldrich (St. Louis, MO, USA), Ethanol (EtOH) and ammonium hydroxide 30% were purchased from Panreac Chemicals (Chicago, IL, USA), and deionized water (H_2O), Aeroxide® TiO_2 P25 (Degussa), and Basolite® C300 (Sigma-Aldrich) were used as reference materials without further purification.

4.2. Preparation Methods

4.2.1. Grinding (TiO$_2$ P25/com-HKUST-1)

The composites were prepared by grinding the TiO$_2$ P25 and commercial HKUST-1 powders by hand in an agate mortar until a homogeneous light blue color was obtained. Five composites were prepared with a TiO$_2$ content of 25, 50 and 75 wt %.

4.2.2. TiO$_2$ Solvothermal Deposition on Commercial HKUST-1 (TiO$_2$-ST/com-HKUST-1)

TiO$_2$ was prepared by mixing 1.14 mL of titanium (IV) isopropoxide with 100 mL of anhydrous ethanol and sonicating this for 5 min. Then, under vigorous magnetic stirring, concentrated nitric acid (70% *v/v*) was added drop by drop to get a pH around 1. The solution was diluted with 10 mL of distilled water, and 8 mL of ammonium hydroxide was added as a precipitating agent. Subsequently, a given amount of commercial HKUST-1 was added to the suspension under vigorous stirring. The resultant suspension was placed in a homemade PTFE-lined autoclave and sealed hermetically, and then introduced into a convective furnace (Fisher Scientific, Pittsburgh PA, USA) at 100 °C for 24 h. The composite formed was recovered and washed five times with a mixture of DMF/EtOH/H$_2$O (molar ratio of 1:1:1) to eliminate the residues of any organic compound. Then, the product was dried at 50 °C for 5 h. The solid was ground in an agate mortar and sieved using a US 80 mesh to get a homogeneous particle size. Three composites were prepared by this method with a nominal TiO$_2$ content of 25, 50, and 75 wt %.

4.2.3. TiO$_2$ P25 Incorporation During HKUST-1 Synthesis (TiO$_2$ P25/syn-HKUST-1)

The synthesis route of synthesized HKUST-1 mainly followed the procedure reported by Tranchemontagne et al. [41] with some modifications, such as the integration of the TiO$_2$ P25 during the preparation of precursor solution. First, two solutions were prepared: one solution containing 100 mg of trimesic acid (H$_3$BTC) dissolved in 6 mL of a mixture of DMF/EtOH/H$_2$O with a molar ratio (1:1:1), and a second solution contained 200 mg of Cu(OAc)$_2$ × H$_2$O dissolved in 6 mL of the solvent DMF/EtOH/H$_2$O. Both solutions were mixed under magnetic stirring to get a homogeneous solution. Then, 0.2 mL of Et$_3$N was added drop by drop as an oxidant agent to the reaction mixture under magnetic stirring. After that, a given quantity of TiO$_2$ P25 (25, 50 and 75 wt %) was added to the MOFs precursor solution. This suspension was kept under magnetic stirring for 24 h, and the powder was obtained by centrifugation. The solid was washed five times with 5 mL of DMF to eliminate the residues of any organic compound, and finally, the solid was dried—at 50 °C for 5 h—milled and sieved to get a homogeneous particle size.

4.3. Characterization Techniques

All composites were characterized by several techniques. X-ray diffraction patterns were recorded on a Siemens D-5000 diffractometer (Munich, Germany), with a copper anode and Cu-Kα radiation over a 2 theta range of 10–80° using a step size of 4 °/min. FTIR and UV-Vis spectra of powder samples were respectively obtained using a Nicolet system (Nexus 470, Thermo Fisher Scientific, Waltham, MA, USA) (with KBr pellet samples) and a GBC spectrophotometer (Cintra 20, GBC Scientific, Hampshire, IL, USA), repectively.

The dynamics of the charge carriers in the photocatalysts were studied by the TRMC technique. The TRMC set-up consists of two main components: (1) a pulse light source, which has the objective to photo-excite the samples and (2) microwave source. A Gunn diode of Kα band at 30 GHz was used to generate the incident microwaves. A tunable laser in the range between 220 and 2000 nm (NT342B; EKSPLA, Vilnius, Lithuania) was used as a light source. It was equipped with an optical parametric oscillator (OPO). The laser delivered 8 ns FWMH pulses with a frequency of 10 Hz. The selected excitation wavelengths were 355 and 410 nm, with a light energy density of 747.6 µJ × cm^{-2} and 2.6 mJ × cm^{-2}, respectively.

4.4. Photocatalytic H$_2$ Evolution

The composite powders were evaluated in the hydrogen production reaction using a glycerol-water solution with a volumetric ratio glycerol/water = 1:9. The photocatalytic reaction was carried out in a 25 mL glass cell. The composites (1 g/L) and the glycerol-water mixture was placed in the cell and mixed to form a homogeneous suspension, and then purged with nitrogen to eliminate all dissolved oxygen. Before irradiation, the reaction cell was maintained under stirring for one hour for adsorption/desorption equilibration and then irradiated with a solar simulator (Model 9600, 150 W; Newport Corporation, Irvine, CA, USA) for 8 h or 24 h. The gas mixture (H$_2$, CO$_2$, CH$_4$) produced during the reaction was analyzed in a Perkin–Elmer Gas Chromatograph (Autosystem XL, Waltham, MA, USA).

5. Conclusions

A series of TiO$_2$/HKUST-1 composites were successfully prepared using grinding, solvothermal, and chemical methods. All composites showed a higher photocatalytic activity than the individual components, particularly those containing a TiO$_2$/HKUST-1 weight ratio of 1:1. These results demonstrated the effect of the synthesis method of composites on photocatalytic activity and stability. The best performance was obtained with the composite prepared by a chemical route, i.e., the synthesis of HKUST-1 in the presence of TiO$_2$ P25, leading to a strong interaction between the two components. The higher photocatalytic performance of the composites, compared with TiO$_2$ or HKUST-1, was explained regarding a synergy between the semiconductor and the HKUST-1, inhibiting electron-hole recombination. There was experimental evidence of the reversible partial reduction of Cu^{2+} towards the Cu^{1+}-Cu^{0} entities contained in HKUST-1, which could indicate the in situ formation of highly active HKUST-1 co-catalysts, improving the photocatalytic activity.

Author Contributions: Conceptualization, S.A., and M.A.V.; investigation, F.M.M., and A.L.L.; methodology, C.C.-J., M.A.V., and H.R.; writing—original draft preparation, M.A.V., and S.A.; writing—review and editing, E.A., and J.M.B.-A.

Funding: This research was sponsored by "Consejo Nacional de Ciencia y Tecnología" México (Project No. 153356), and "Instituto Politécnico Nacional" (Projects SIP 20194976 and SIP 20196347).

Acknowledgments: M.V. and H.R. acknowledge Université Paris-Saclay for the financial support through the Chaire Jean d'Alembert program and the IRS MOMENTOM (Ininiative de Recherche Stratégique).

Conflicts of Interest: The authors declare no conflict of interest.

References

1. Andrews, J.; Shabani, B. Re-envisioning the role of hydrogen in a sustainable energy economy. *Int. J. Hydrog. Energy* **2012**, *37*, 1184–1203. [CrossRef]
2. Dincer, I.; Acar, C. Review and evaluation of hydrogen production methods for better sustainability. *Int. J. Hydrog. Energy* **2015**, *40*, 11094–11111. [CrossRef]
3. Dodds, P.E.; Staffell, I.; Hawkes, A.D.; Li, F.; Grünewald, P.; McDowall, W.; Ekins, P. Hydrogen and fuel cell technologies for heating: A review. *Int. J. Hydrog. Energy* **2015**, *40*, 2065–2083. [CrossRef]
4. Liao, C.-H.; Huang, C.-W.; Wu, J.C.S. Hydrogen Production from Semiconductor-based Photocatalysis via Water Splitting. *Catalysts* **2012**, *2*, 490–516. [CrossRef]
5. Bozoglan, E.; Midilli, A.; Hepbasli, A. Sustainable assessment of solar hydrogen production techniques. *Energy* **2012**, *46*, 85–93. [CrossRef]
6. Christoforidis, K.C.; Fornasiero, P. Photocatalytic Hydrogen Production: A Rift into the Future Energy Supply. *ChemCatChem* **2017**, *9*, 1523–1544. [CrossRef]
7. Babu, V.J.; Vempati, S.; Uyar, T.; Ramakrishna, S. Review of one-dimensional and two-dimensional nanostructured materials for hydrogen generation. *Phys. Chem. Chem. Phys.* **2015**, *17*, 2960–2986. [CrossRef] [PubMed]
8. Guo, L.; Jing, D.; Liu, M.; Chen, Y.; Shen, S.; Shi, J.; Zhang, K. Functionalized nanostructures for enhanced photocatalytic performance under solar light. *Beilstein J. Nanotechnol.* **2014**, *5*, 994–1004. [CrossRef]

9. Kumar, S.; Kumar, A.; Bahuguna, A.; Sharma, V.; Krishnan, V. Two-dimensional carbon-based nanocomposites for photocatalytic energy generation and environmental remediation applications. *Beilstein J. Nanotechnol.* **2017**, *8*, 1571–1600. [CrossRef]

10. Rowsell, J.L.C.; Yaghi, O.M. Metal–organic frameworks: A new class of porous materials. *Microporous Mesoporous Mater.* **2004**, *73*, 3–14. [CrossRef]

11. Nguyen, L.T.L.; Nguyen, T.T.; Nguyen, K.D.; Phan, N.T.S. Metal–organic framework MOF-199 as an efficient heterogeneous catalyst for the aza-Michael reaction. *Appl. Catal. Gen.* **2012**, *425–426*, 44–52. [CrossRef]

12. Chui, S.S.-Y.; Lo, S.M.-F.; Charmant, J.P.H.; Orpen, A.G.; Williams, I.D. A Chemically Functionalizable Nanoporous Material [$Cu_3(TMA)_2(H2O)_3$]$_n$. *Science* **1999**, *283*, 1148–1150. [CrossRef]

13. Prestipino, C.; Regli, L.; Vitillo, J.G.; Bonino, F.; Damin, A.; Lamberti, C.; Zecchina, A.; Solari, P.L.; Kongshaug, K.O.; Bordiga, S. Local Structure of Framework Cu(II) in HKUST-1 Metallorganic Framework: Spectroscopic Characterization upon Activation and Interaction with Adsorbates. *Chem. Mater.* **2006**, *18*, 1337–1346. [CrossRef]

14. Kumar, P.; Vellingiri, K.; Kim, K.-H.; Brown, R.J.C.; Manos, M.J. Modern progress in metal-organic frameworks and their composites for diverse applications. *Microporous Mesoporous Mater.* **2017**, *253*, 251–265. [CrossRef]

15. Wen, M.; Mori, K.; Kuwahara, Y.; An, T.; Yamashita, H. Design and architecture of metal organic frameworks for visible light enhanced hydrogen production. *Appl. Catal. B Environ.* **2017**, *218*, 555–569. [CrossRef]

16. Zhang, T.; Lin, W. Metal–organic frameworks for artificial photosynthesis and photocatalysis. *Chem. Soc. Rev.* **2014**, *43*, 5982–5993. [CrossRef] [PubMed]

17. Deng, X.; Li, Z.; García, H. Visible Light Induced Organic Transformations Using Metal-Organic-Frameworks (MOFs). *Chem. Eur. J.* **2017**, *23*, 11189–11209. [CrossRef]

18. Wang, C.-C.; Li, J.-R.; Lv, X.-L.; Zhang, Y.-Q.; Guo, G. Photocatalytic organic pollutants degradation in metal–organic frameworks. *Energy Environ. Sci.* **2014**, *7*, 2831–2867. [CrossRef]

19. Dias, E.M.; Petit, C. Towards the use of metal–organic frameworks for water reuse: a review of the recent advances in the field of organic pollutants removal and degradation and the next steps in the field. *J. Mater. Chem. A* **2015**, *3*, 22484–22506. [CrossRef]

20. Fang, Y.; Ma, Y.; Zheng, M.; Yang, P.; Asiri, A.M.; Wang, X. Metal–organic frameworks for solar energy conversion by photoredox catalysis. *Coord. Chem. Rev.* **2018**, *373*, 83–115. [CrossRef]

21. Qiu, J.; Zhang, X.; Feng, Y.; Zhang, X.; Wang, H.; Yao, J. Modified metal-organic frameworks as photocatalysts. *Appl. Catal. B Environ.* **2018**, *231*, 317–342. [CrossRef]

22. Aguilera-Sigalat, J.; Bradshaw, D. Synthesis and applications of metal-organic framework–quantum dot (QD@MOF) composites. *Coord. Chem. Rev.* **2016**, *307*, 267–291. [CrossRef]

23. Li, R.; Hu, J.; Deng, M.; Wang, H.; Wang, X.; Hu, Y.; Jiang, H.-L.; Jiang, J.; Zhang, Q.; Xie, Y.; et al. Integration of an Inorganic Semiconductor with a Metal–Organic Framework: A Platform for Enhanced Gaseous Photocatalytic Reactions. *Adv. Mater.* **2014**, *26*, 4783–4788. [CrossRef]

24. Binh, N.T.; Thu, P.T.; Le, N.T.H.; Tien, D.M.; Khuyen, H.T.; Giang, L.T.K.; Huong, N.T.; Lam, T.D. Study on preparation and properties of a novel photo–catalytic material based on copper–centred metal–organic frameworks (Cu–MOF) and titanium dioxide. *Int. J. Nanotechnol.* **2015**, *12*, 447–455. [CrossRef]

25. Abedi, S.; Morsali, A. Ordered Mesoporous Metal–Organic Frameworks Incorporated with Amorphous TiO_2 As Photocatalyst for Selective Aerobic Oxidation in Sunlight Irradiation. *ACS Catal.* **2014**, *4*, 1398–1403. [CrossRef]

26. Li, R.; Wu, S.; Wan, X.; Xu, H.; Xiong, Y. Cu/TiO_2 octahedral-shell photocatalysts derived from metal–organic framework@semiconductor hybrid structures. *Inorg. Chem. Front.* **2016**, *3*, 104–110. [CrossRef]

27. Canivet, J.; Fateeva, A.; Guo, Y.; Coasne, B.; Farrusseng, D. Water adsorption in MOFs: Fundamentals and applications. *Chem. Soc. Rev.* **2014**, *43*, 5594–5617. [CrossRef]

28. Chen, C.; Wu, T.; Yang, D.; Zhang, P.; Liu, H.; Yang, Y.; Yang, G.; Han, B. Catalysis of photooxidation reactions through transformation between Cu^{2+} and Cu^{+} in TiO_2–Cu–MOF composites. *Chem. Commun.* **2018**, *54*, 5984–5987. [CrossRef]

29. Ahmed, A.; Robertson, C.M.; Steiner, A.; Whittles, T.; Ho, A.; Dhanak, V.; Zhang, H. Cu(I)Cu(II)BTC, a microporous mixed-valence MOF via reduction of HKUST-1. *RSC Adv.* **2016**, *6*, 8902–8905. [CrossRef]

30. Bahruji, H.; Bowker, M.; Davies, P.R.; Al-Mazroai, L.S.; Dickinson, A.; Greaves, J.; James, D.; Millard, L.; Pedrono, F. Sustainable H_2 gas production by photocatalysis. *J. Photochem. Photobiol. Chem.* **2010**, *216*, 115–118. [CrossRef]

31. Szanyi, J.; Daturi, M.; Clet, G.; Baer, D.R.; Peden, C.H.F. Well-studied Cu–BTC still serves surprises: Evidence for facile Cu^{2+}/Cu^+ interchange. *Phys. Chem. Chem. Phys.* **2012**, *14*, 4383–4390. [CrossRef]

32. Fu, Q.; Xie, K.; Tan, S.; Ren, J.M.; Zhao, Q.; Webley, P.A.; Qiao, G.G. The use of reduced copper metal–organic frameworks to facilitate CuAAC click chemistry. *Chem. Commun.* **2016**, *52*, 12226–12229. [CrossRef]

33. Loera-Serna, S.; Oliver-Tolentino, M.A.; de Lourdes López-Núñez, M.; Santana-Cruz, A.; Guzmán-Vargas, A.; Cabrera-Sierra, R.; Beltrán, H.I.; Flores, J. Electrochemical behavior of [$Cu_3(BTC)_2$] metal-organic framework: The effect of the method of synthesis. *J. Alloys Compd.* **2012**, *540*, 113–120. [CrossRef]

34. Dhumal, N.R.; Singh, M.P.; Anderson, J.A.; Kiefer, J.; Kim, H.J. Molecular Interactions of a Cu-Based Metal–Organic Framework with a Confined Imidazolium-Based Ionic Liquid: A Combined Density Functional Theory and Experimental Vibrational Spectroscopy Study. *J. Phys. Chem. C* **2016**, *120*, 3295–3304. [CrossRef]

35. Borfecchia, E.; Maurelli, S.; Gianolio, D.; Groppo, E.; Chiesa, M.; Bonino, F.; Lamberti, C. Insights into Adsorption of NH_3 on HKUST-1 Metal–Organic Framework: A Multitechnique Approach. *J. Phys. Chem. C* **2012**, *116*, 19839–19850. [CrossRef]

36. Wen, L.-L.; Wang, F.; Feng, J.; Lv, K.-L.; Wang, C.-G.; Li, D.-F. Structures, Photoluminescence, and Photocatalytic Properties of Six New Metal–Organic Frameworks Based on Aromatic Polycarboxylate Acids and Rigid Imidazole-Based Synthons. *Cryst. Growth Des.* **2009**, *9*, 3581–3589. [CrossRef]

37. Wen, L.; Zhao, J.; Lv, K.; Wu, Y.; Deng, K.; Leng, X.; Li, D. Visible-Light-Driven Photocatalysts of Metal–Organic Frameworks Derived from Multi-Carboxylic Acid and Imidazole-Based Spacer. *Cryst. Growth Des.* **2012**, *12*, 1603–1612. [CrossRef]

38. Schlichte, K.; Kratzke, T.; Kaskel, S. Improved synthesis, thermal stability and catalytic properties of the metal-organic framework compound $Cu_3(BTC)_2$. *Microporous Mesoporous Mater.* **2004**, *73*, 81–88. [CrossRef]

39. DeCoste, J.B.; Peterson, G.W.; Schindler, B.J.; Killops, K.L.; Browe, M.A.; Mahle, J.J. The effect of water adsorption on the structure of the carboxylate containing metal–organic frameworks Cu-BTC, Mg-MOF-74, and UiO-66. *J. Mater. Chem. A* **2013**, *1*, 11922–11932. [CrossRef]

40. Chen, X.; Mao, S.S. Titanium Dioxide Nanomaterials: Synthesis, Properties, Modifications, and Applications. *Chem. Rev.* **2007**, *107*, 2891–2959. [CrossRef]

41. Tranchemontagne, D.J.; Hunt, J.R.; Yaghi, O.M. Room temperature synthesis of metal-organic frameworks: MOF-5, MOF-74, MOF-177, MOF-199, and IRMOF-0. *Tetrahedron* **2008**, *64*, 8553–8557. [CrossRef]

 *catalysts*

Review

Titanium-Dioxide-Based Visible-Light-Sensitive Photocatalysis: Mechanistic Insight and Applications

Shinya Higashimoto

Department of Applied Chemistry, Faculty of Engineering, Osaka Institute of Technology, 5-16-1 Omiya, Asahi-ku, Osaka 535-8585, Japan; shinya.higashimoto@oit.ac.jp; Tel.: +81-(0)6-6954-4283

Received: 15 January 2019; Accepted: 14 February 2019; Published: 22 February 2019

Abstract: Titanium dioxide (TiO_2) is one of the most practical and prevalent photo-functional materials. Many researchers have endeavored to design several types of visible-light-responsive photocatalysts. In particular, TiO_2-based photocatalysts operating under visible light should be urgently designed and developed, in order to take advantage of the unlimited solar light available. Herein, we review recent advances of TiO_2-based visible-light-sensitive photocatalysts, classified by the origins of charge separation photo-induced in (1) bulk impurity (N-doping), (2) hetero-junction of metal (Au NPs), and (3) interfacial surface complexes (ISC) and their related photocatalysts. These photocatalysts have demonstrated useful applications, such as photocatalytic mineralization of toxic agents in the polluted atmosphere and water, photocatalytic organic synthesis, and artificial photosynthesis. We wish to provide comprehension and enlightenment of modification strategies and mechanistic insight, and to inspire future work.

Keywords: Titanium dioxide (TiO_2); visible-light-sensitive photocatalyst; N-doped TiO_2; plasmonic Au NPs; interfacial surface complex (ISC); selective oxidation; decomposition of VOC; carbon nitride (C_3N_4); alkoxide; ligand to metal charge transfer (LMCT)

1. Introduction

Titanium dioxide (TiO_2) is one of the most practical and prevalent photo-functional materials, since it is chemically stable, abundant (Ti: 10th highest Clarke number), nontoxic, and cost-effective. In recent years, a great deal of attention has been directed towards TiO_2 photocatalysis for useful applications such as photocatalytic mineralization of toxic agents in the polluted atmosphere and water, photocatalytic organic synthesis, and artificial photosynthesis [1–20].

The TiO_2 involving Ti^{3+} sites that are oxygen-deficient at the impurity level exhibits n-type semiconductor. The photocatalytic activities of TiO_2 strongly depend on crystal structures (anatase, brookite, and rutile), crystallinity, crystalline plane, morphology, particle sizes, defective sites, and surface OH groups. The valence band (V.B.) and conduction band (C.B.) of TiO_2 consist of O 2p and Ti 3d orbitals, respectively, and their band gap (forbidden band) is circa ~3.0–3.2 eV (~410–380 nm). Photo-irradiation ($hv > 3.2$ eV) of the TiO_2 photocatalyst leads to band gap excitation, resulting in charge separation of electrons into the C.B. and the holes in the V.B. These photo-formed electrons and holes simultaneously work as electron donors and acceptors, respectively, on the photocatalyst surface, thus enabling the photocatalytic reactions. Details are given in other articles and reviews [21–25]. UV light reaching the earth surface represents only a very small fraction (4%) of the solar energy available. Therefore, many researchers have endeavored to design several types of visible-light-responsive photocatalyst. In particular, TiO_2-based photocatalysts operating under visible light should be urgently designed and developed, in order to take advantage of the unlimited solar light available.

In the late 1990s, Anpo et al. first reported that TiO_2 doped with Cr, V, and Fe cations by ion implantation operates under visible light irradiation. They exhibited red shift of the band-edge of the

TiO_2, resulting in decomposition of NO into N_2, O_2, and N_2O [26]. This work accelerated subsequent works for the design and development of visible-light-responsive photocatalysts. Recently, much attention has been paid to visible-light-responsive TiO_2 prepared by: doping with nitrogen (N), carbon (C), and sulfur (S) ions etc.; surface plasmonic effects with Au or Ag nanoparticles (NPs); the interfacial surface complex (ISC); coupling with visible-light-sensitive hetero-semiconductors (cadmium sulfide, carbon nitride etc.); and dye-sensitized photocatalysts. In fact, some photocatalysts are considered to work under similar principles.

Along these backgrounds, this review focuses on the recent advances of the visible-light-sensitive TiO_2 photocatalyst. These advances have been classified by the origin of charge separation photo-induced in (1) the bulk impurity (N-doping), (2) hetero-junction of metal (Au NPs), and (3) the interfacial surface complex (ISC) (See Figure 1). They have been well characterized by several spectroscopic techniques, and applied for mineralization of volatile organic compounds (VOC), water splitting to produce H_2, and fine organic synthesis.

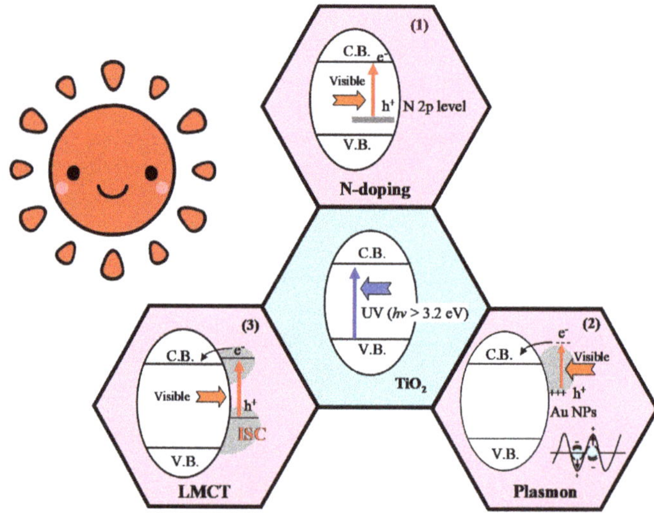

Figure 1. Visible-light-sensitive TiO_2 photocatalyst modified by (**1**) nitrogen-doping, (**2**) plasmonic Au nanoparticles (NPs), and (**3**) interfacial surface complex (ISC).

2. Nitrogen-doped TiO_2 Photocatalysts

In 1986, Sato and co-workers first explored the photocatalytic activity of nitrogen-doped TiO_2 (N-doped TiO_2) photocatalysts for the oxidation of gaseous ethane and carbon monoxide [27]. They found that N-doped TiO_2 photocatalyst exhibited a superior photocatalytic activity to pure TiO_2 under visible light irradiation. Later, in 2001, Asahi et al. demonstrated visible-light-induced complete photo-oxidation of gaseous CH_3CHO (one of VOCs) to CO_2 with an N-doped TiO_2 photocatalyst [28]. In this section, fundamental synthetic routes, characterizations, and application of photocatalytic reactions are highlighted.

2.1. Synthesis of N-doped TiO_2 Photocatalyst

N-doped TiO_2 was prepared by employing several procedures and materials. Details are given in Reference [13]. Preparation methods for N-doped TiO_2 photocatalysts can be classified into two categories: dry processes and wet processes.

2.1.1. Dry Processes

Typically, N-doped TiO_2 powder can be prepared by the nitrification of TiO_2 in an ammonia (NH_3) gas flow at high temperature [28,29]. The amount of N doping into the TiO_2 can be controlled by annealing temperatures in the range of 550–600 °C under an NH_3 flow. However, a large number of O vacancies are introduced into the N-doped TiO_2 with increasing annealing temperature, since the NH_3 decomposes into N_2 and H_2 at high temperature, and TiO_2 is simultaneously reduced by H_2 [30]. Figure 2 shows schematics of N-doping into TiO_2, accompanied by the formation of oxygen vacancies to exhibit the n-type semiconductor.

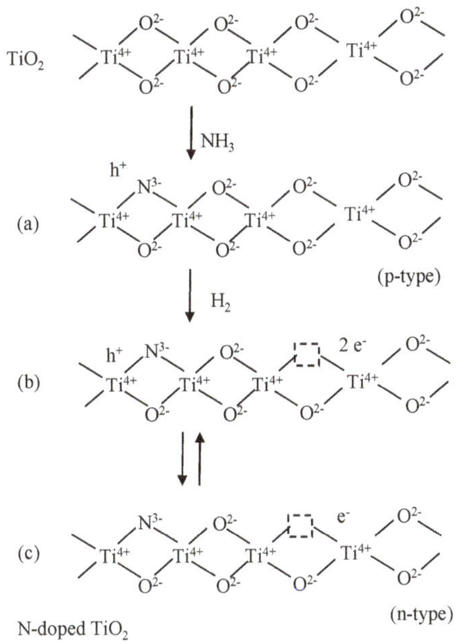

Figure 2. When the N^{3-} is replaced with lattice O^{2-} ions in the TiO_2 lattice, the hole (h^+) is formed in order to compensate for the charge balance (p-type semiconductor) (a). However, an oxygen vacancy is produced by the reduction with H_2, which is formed by the decomposition of NH_3 to produce an oxygen vacancy and excess electrons (b). As a consequence, N^{3-} doped into TiO_2 (N-doped TiO_2) involves electrons located at N 2p and Ti 3d sites at impurity levels (n-type semiconductor) (c).

2.1.2. Wet Processes

A sol-gel method can be employed for the preparation of N-doped TiO_2 powder. Typically, NH_3 aq. (NH_4OH) is added to a solution of titanium (IV) isopropoxide (TTIP) [31–33] to form titanium hydroxide involving N-species. The precipitate was dried, followed by calcination at ~400–450 °C in air to obtain a yellowish TiO_2 powder.

2.2. N-states in N-doped TiO_2

One of the major concerns is to understand the physico-chemical nature of the N species in N-doped TiO_2, which are responsible for the visible light sensitivity. They were characterized by density functional theory (DFT) calculations, X-rap photoelectron spectroscopy (XPS), Ultraviolet-visible (UV-vis) and electron paramagnetic resonance (EPR) spectroscopy.

2.2.1. DFT Calculations

DFT calculations demonstrated the electronic structures of the N-doped TiO$_2$ photocatalyst (see Figure 3). The substitution of N with lattice O of the N-doped TiO$_2$ exhibits band gap narrowing (circa 0.1 eV) caused by mixing orbitals of N 2p with O 2p, resulting in the negative shift of the valence band edge. On the other hand, the interstitial N is localized to impurity states (N 2p levels) above the V.B. (circa 0.7 eV) in the mid-band gap. Therefore, the oxidation power of photo-induced holes on the N 2p is lower than on the O 2p in the TiO$_2$ lattice.

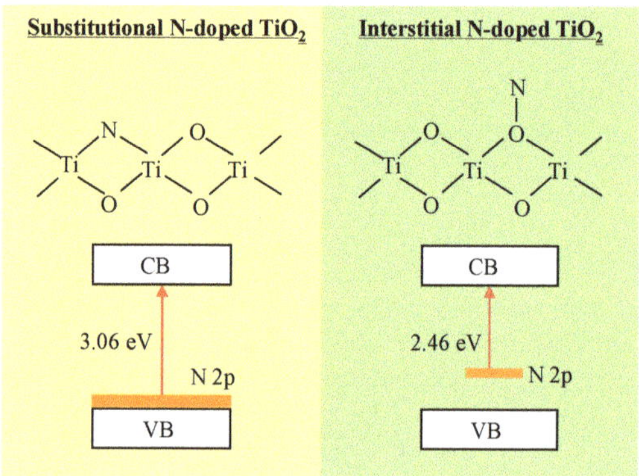

Figure 3. Schematic illustration of structures and their corresponding energy bands for substitutional and interstitial N species in the N-doped TiO$_2$, together with photo-induced electronic processes.

2.2.2. XPS Spectra

XPS analysis can confirm the oxidative states of the N species and bonding states in the N-doped TiO$_2$ (See Figure 4I). N 1s XPS peaks at a binding energy in the range of ~396–400 eV showed different oxidative states of the N species. By the combination of the DFT calculations [31], it was identified that the N 1s XPS peaks at ~396–397 eV are due to the substitution of N with the lattice O of TiO$_2$ [13,34], while those at ~399–400 eV are due to the interstitial N in the form of NO$_x$ or NH$_x$ [13,31,35,36].

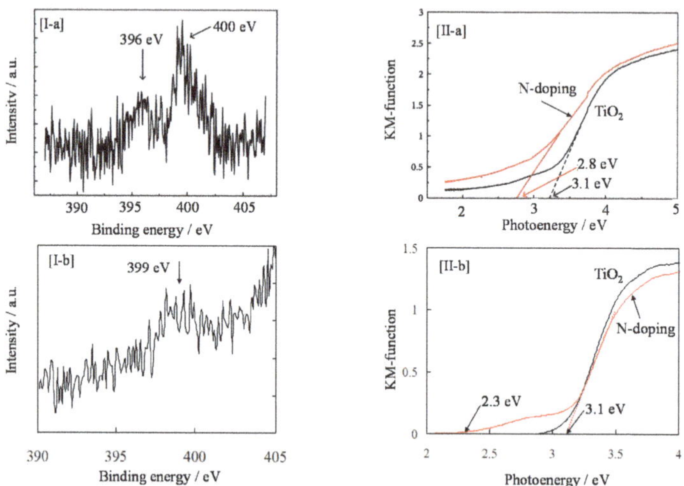

Figure 4. XPS [I] and UV-vis absorption spectra [II] of (a) N-doped TiO$_2$ nanoball film [34], and (b) N-doped TiO$_2$ prepared by the sol-gel method [36].

2.2.3. Optical Properties

The UV-vis absorption spectra of the N-doped TiO_2 are shown in Figure 4II. The N-doped TiO_2 with the substitution of N exhibited band gap narrowing from 3.1 to 2.8 eV. On the other hand, the N-doped TiO_2 prepared by the sol-gel method exhibited visible light absorption up to 540 nm (2.3 eV), due to the electronic transition from localized N doping level to the C.B. of the TiO_2, while band-narrowing was not observed. These results are in good agreement with the DFT calculations.

2.2.4. Electron Paramagnetic Resonance (EPR) Spectra

N species in the N-doped TiO_2 are present at either diamagnetic (N^-) or paramagnetic ($N^\bullet$) bulk centers, which are responsible for the visible light sensitivity [31,37]. The EPR measurements can detect the paramagnetic ($N^\bullet$) bulk centers (see Figure 5). One type, of three lines with a hyperfine tensor ($g = 2.006$ and $A = 32.0$ G) splitting by nuclear spin of nitrogen ($I = 1$), was observed. The signal intensity of $N^\bullet$ radicals increased when the light was turned on, while the signal intensity significantly decreased when the light was turned off. In general, the paramagnetic interaction between N species and O_2 makes EPR signals disappear. However, they were remarkably enhanced in the presence of O_2 under $\lambda > 420$ nm, while its signal intensity still remained to some extent even after the light was turned off. These results suggest that N-species are located in bulk inside the TiO_2, and visible light irradiation of the N-doped TiO_2 exhibits effective charge separation to form holes ($N^\bullet$ radicals) and electrons, which participate in the oxidation and reduction of reactant molecules, respectively.

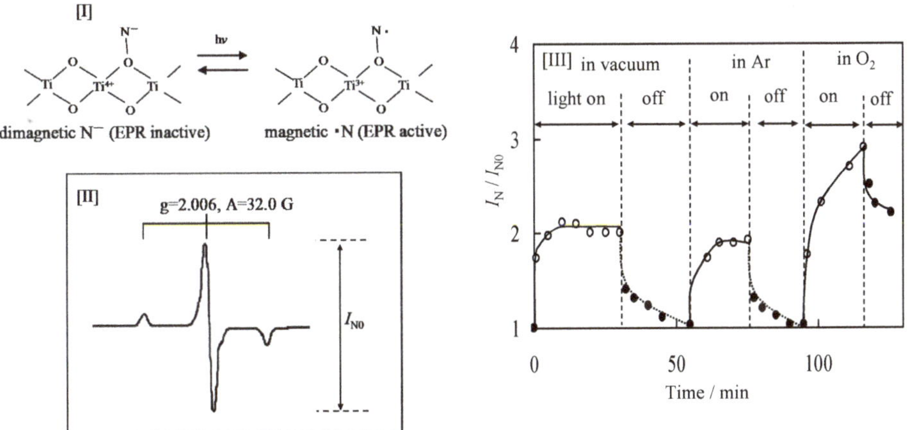

Figure 5. Schematic illustration of [I] formation of paramagnetic ·N by the excitation of diamagnetic N^- species. Electron paramagnetic resonance (EPR) signal [II] of ·N radicals on N–TiO_2, and the relative signal intensity of I_N/I_{N0} [III] under vacuum, in the presence of argon (Ar) or O_2 (400 Pa) [37]. I_{N0} and I_N show the intensity due to ·N radicals at the initial and measured time, respectively.

2.2.5. Photo-Electrochemical Properties

Nakamura et al. investigated the photo-electrochemical oxidation power of the N-doped TiO_2 by employing several electron donors [38]. Figure 6 shows that the photo-induced hole on the N 2p level can directly oxidize only I^- ions under visible light illumination, while I^-, SCN^-, Br^-, and H_2O are oxidized by the hole on the V.B. under UV light illumination. Therefore, the oxidation power of the holes induced on the N 2p level is lower than that of those on the O 2p on the V.B. Tang et al. studied the dynamics of photogenerated electrons and holes on the N-doped TiO_2 using transient absorption spectroscopy [39]. They concluded that the lack of activity of nanocrystalline N-doped TiO_2 film for photocatalytic water oxidation is due to rapid electron–hole recombination.

On the other hand, Higashimoto et al. investigated the photo-electrochemical reduction power of the N-doped TiO_2 (see Figure 7) [33]. When the N-doped TiO_2 was photo-excited under visible light irradiation, the photo-induced electrons were accumulated on the oxygen vacancies of TiO_2. Subsequently, when various kinds of redox species as electron acceptors were introduced into the photo-charged $N–TiO_2$, the accumulated electrons could reduce O_2 molecules, Pt^{4+}, Ag^+, and Au^{3+} ions, but not MV^{2+}, H^+, and Cu^{2+} ions. In principle, the N-doped TiO_2 has the potential to reduce H^+/H_2, but many oxygen vacancies involved in the bulk TiO_2 could influence the drastic charge recombination. In particular, photo-induced electrons trapped at the oxygen vacancies (mainly γ region) could reduce O_2 molecules to form such active oxygen species as hydrogen peroxide (H_2O_2), resulting in further oxidation of organic substrates.

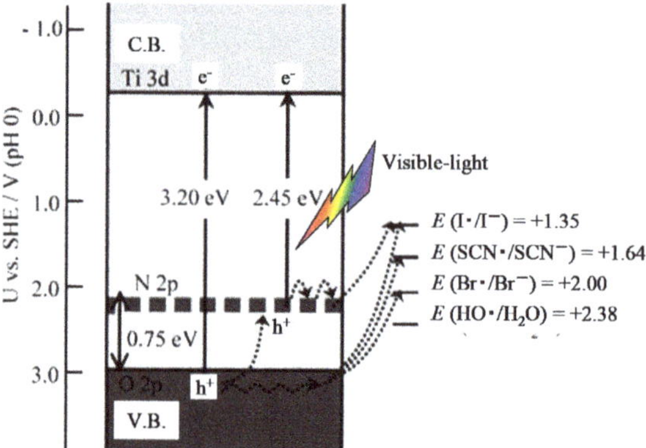

Figure 6. Schematic illustration of proposed energy bands for the N-doped TiO_2, together with some photo-induced electronic processes. *E*: equilibrium redox potentials for one electron transfer [38].

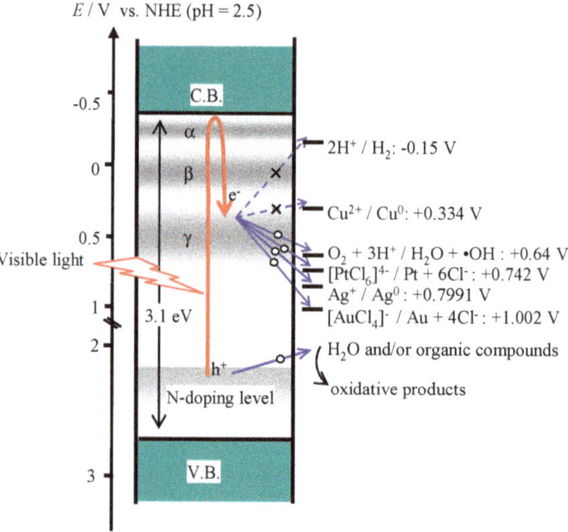

Figure 7. Energy levels for sub-band structures of N-doped TiO_2 and photo-induced charge transfer into various kinds of redox species under visible light irradiation. The energy levels of sub-bands at the α, β, and γ potential regions (oxygen vacancies) and N-doping levels are also shown. Oxygen vacancies were estimated from the photo-electrochemical measurements. Signs of circle and cross stand for energetically favorable and unfavorable electron transfers, respectively [33].

2.3. Application to Photocatalytic Decomposition of Volatile Organic Compounds (VOC)

Time profile for the photocatalytic decomposition of gaseous acetaldehyde on the N-doped TiO_2 is shown in Figure 8. The N-doped TiO_2 exhibited photocatalytic activity 5 times greater than TiO_2 under visible light irradiation, while they exhibited similar activities under UV light irradiation [28].

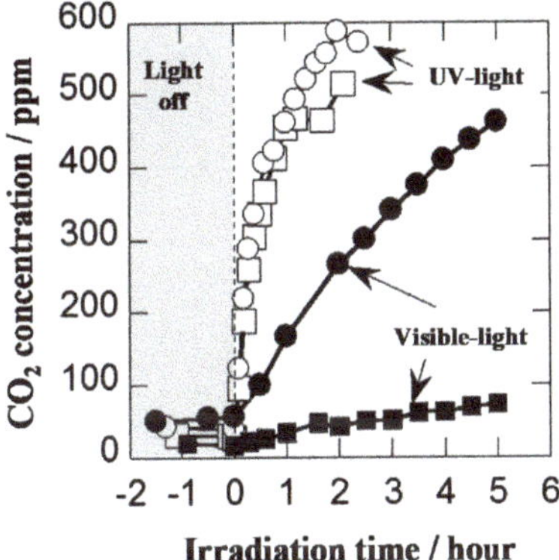

Figure 8. Photocatalytic decomposition of gaseous acetaldehyde on the N-doped TiO_2 photocatalyst. Evolved CO_2 concentration (○, ●, N-doped TiO_2; □, ■, TiO_2) [28].

Table 1 shows that the N-doped TiO_2 exhibited photocatalytic activity for the decomposition of several kinds of VOC into CO_2 under visible light irradiation ($\lambda > 420$ nm). It was observed that the N-doped TiO_2 exhibited photocatalytic activity for the decomposition of aldehydes, but little activity for alcohol, acid, ketone, and halogene compounds. The vanadium species was deposited on the N-doped TiO_2 (VCl_3/N-doped TiO_2) by impregnation method. As shown in Table 1, VCl_3/N-doped TiO_2 showed higher photocatalytic activity for the decomposition of all VOC, in particular, acetic acid or acetone by ~13–16 times more than N-doped TiO_2. Therefore, it was confirmed that vanadium species worked as the effective co-catalyst.

Table 1. Yields of CO_2 for the photocatalytic decomposition of various kinds of volatile organic compounds (VOC) in aqueous solutions with N–TiO_2 and VCl_3/N-doped TiO_2 under visible light irradiation ($\lambda > 420$ nm) for 3 h [40].

Entry	Reactant Molecules	Yields of CO_2/μmol	
		N-doped TiO_2	VCl_3/N-doped TiO_2
1	methanol [a]	0.2	1.1
2	ethanol [a]	0.3	0.5
3	formaldehyde [a]	4.6	21.6
4	acetaldehyde [a]	4.1	35.0
5	formic acid [a]	0.7	4.8
6	acetic acid [a]	1.2	17.0
7	acetone [a]	0.7	11.4
8	ethyl acetate [a]	1.3	10.6
9	dichloromethane [b]	2.4	4.1
10	trichloromethane [b]	1.5	4.1
11	1, 1-dichloroethane [b]	0.7	4.8
12	trans-1, 2-dichloroethylene [b]	1.0	5.7

Concentrations of VOC are (a) 0.5 M and (b) 50 mM.

Furthermore, effects of co-catalysts (48 metal ions using nitrate, sulfate, chloride, acetate, and oxide precursors) deposited on the N-doped TiO_2 for the photocatalytic activities were examined (See Figure 9) [40]. The bars marked in yellow exhibited higher photocatalytic activities than the N-doped TiO_2 by itself. In particular, N-doped-TiO_2-deposited Cu, Fe, V, and Pt oxides exhibited high photocatalytic activities. The local structures of the co-catalysts were characterized by XPS. It was observed that Cu loaded N-doped TiO_2 involves cuprous oxide (Cu_2O) or Cu hydroxides, Fe loaded N-doped TiO_2 involves clusters containing Fe–O bonds or Fe^{2+} hydroxide [41], and Pt loaded N-doped TiO_2 involves Pt^{4+}/Pt^{2+} species [36]. The redox potentials of co-catalysts such as V (+IV/+V), Fe (+II/+III), Cu (+I/+II), and Pt (+III/+IV) were in the range of circa +0.6 to +1.0 V vs. SHE, while the multi-electron reduction of O_2 leads to the formation of active oxygen species via $O_2 + 2H^+ + 2e^-$ / H_2O_2 (E_0 = +0.687 V vs. SHE). Therefore, the co-catalysts, such as Pt, Fe, Cu, and V species, enhance the photocatalytic activity due to the effective electron transfer to O_2 (O_2 reduction), resulting in the formation of active oxygen species.

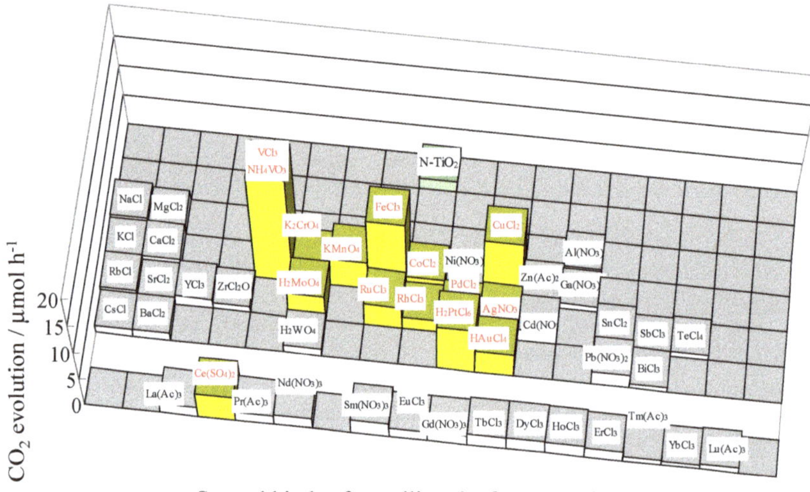

Figure 9. Photocatalytic activities for the decomposition of acetic acid under visible light irradiation ($\lambda > 420$ nm) on N-doped TiO_2, modified by various kinds of metal species as co-catalysts. Each metal salt used in this study is shown [40].

2.4. C_3N_4-Modified TiO_2 Compared with N-doped TiO_2

Several nitrogen sources such as urea, cyanamid, cyanuric acid, and melamine were employed for the preparation of N-containing TiO_2 photocatalyst, i.e., the TiO_2 surface is modified with polymerized carbon nitride (C_3N_4) [42–51]. The structures of the C, N-species strongly depend on their concentrations. If the C, N species are present in only a small amount, they act as a molecular photosensitizer. At higher amounts they form a C_3N_4 crystalline semiconductor, which chemically binds to TiO_2. The C_3N_4–TiO_2 was systematically synthesized by thermal condensation of cyanuric acid on the TiO_2 surface [51]. In fact, H_2 was evolved from TEA aq. on the C_3N_4–TiO_2 photocatalyst under visible light irradiation, while the N-doped TiO_2 did not exhibit H_2 production. From characterization of C_3N_4–TiO_2 by Fourier transformed-infrared (FT-IR), XPS, electrochemical measurements, and DFT calculations, the band structures and photo-induced charge separation mechanisms were demonstrated (Figure 10). The C_3N_4–TiO_2 was found to exhibit photo-induced charge separation through the hetero-coupling of semiconductors between C_3N_4 and TiO_2 on the surface. On the other hand, N-doped TiO_2 was photo-sensitized by bulk impurity of the N-doping. It can be assumed that many oxygen vacancies promoted the charge recombination, resulting in weak reduction power in the N-doped TiO_2.

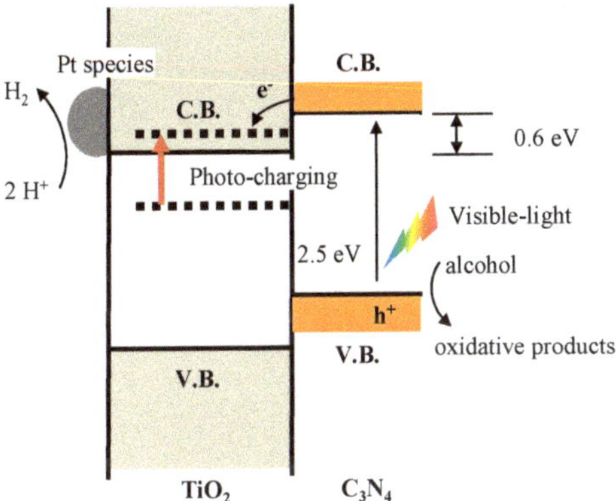

Figure 10. Photo-induced charge separation on the C_3N_4 deposited TiO_2 surface [51].

3. Plasmonic Au NPs Modified TiO_2

3.1. What Is Localized Surface Plasmon Resonance (LSPR)?

Localized surface plasmon resonance (LSPR) is an optical phenomenon generated by light when it interacts with conductive nanoparticles (NPs) that are smaller than the incident wavelength. The LSPR is induced by the collective oscillations of delocalized electrons in response to an external electric field. The resonance wavelength strongly depends on the size and shape of the NPs, the interparticle distance, and the dielectric property of the surrounding medium. The Au and Ag NPs exhibit unique plasmon absorption [52,53]. The plasmonic Ag NPs are considered to be unstable under illumination, and could be applicable to multi-colored rewritable devices. In this section, we focused on stable plasmonic Au NPs exploited for a visible-light-sensitive photovoltaic fuel cell or photocatalyst [54,55].

3.2. Preparation and Characterization of Au–TiO_2 Photocatalyst

3.2.1. Photodeposition (PD) Methods

By using the photocatalysis of TiO_2, metallic Au was deposited on the TiO_2 surface, accompanied by the oxidation of methanol [56,57] or ethanol [58]. Typically, TiO_2 powder was suspended in a 50 vol. % aqueous methanol in the presence of $HAuCl_4 \cdot 6H_2O$, purged of air with argon. The suspension was photoirradiated with UV light under magnetic stirring. The temperature of the suspension during photoirradiation was maintained at 298K. The Au/TiO_2 photocatalyst was centrifuged, washed with distilled water, dried at 393K, and ground in an agate mortar.

3.2.2. Colloid Photodeposition Operated in the Presence of a Hole Scavenger (CPH)

Colloidal Au NPs were prepared using the method reported by Frens [59]. In brief, mixtures of an aqueous tetrachloroauric acid ($HAuCl_4$) solution and sodium citrate were heated and boiled for 1 h. The color of the solution changed from deep blue to deep red. The citrate plays a role in the reduction of Au ions, and the capping agent in suppressing the aggregation of Au NPs. The suspension of TiO_2 in an aqueous solution of colloidal Au NPs and oxalic acid was then photo-irradiated at $\lambda > 300$ nm at 298 K under argon (Ar). The solids were recovered, washed, and dried to produce Au–TiO_2. Details are given in Reference [60].

3.2.3. Deposition Precipitation (DP) Method

Deposition–precipitation (DP) methods were employed for the deposition of a gold (III) species on the TiO_2 surface [61,62]. The $[AuCl(OH)_3]^-$, main species present at pH 8, adjusted by NaOH aq., reacts with hydroxyl groups of the TiO_2 surface to form a grafted hydroxyl–gold compound. The catalyst was then recovered, filtered, washed with deionized water, and dried. Finally, the powder was calcined at ~473–673 K in air.

3.2.4. Characterization of the Au–TiO_2 Photocatalyst

The Au–TiO_2 photocatalysts were typically characterized by the transmittance electron microscope (TEM) for the particle sizes, and UV-vis absorption for optical properties (See Table 2).

Table 2. Particle sizes of Au nanoparticles (NPs) and optical properties of the Au–TiO_2 prepared by several techniques.

Entry	Au Deposition Methods	Particle Sizes/nm	Top Peak/nm	Ref.
1	PD	~10–60	~530–610	[56–58]
2	CPH	~12–14	~550–560	[60,63–66]
		13	~550–620	[67]
3	DP	~2–6	~550–560	[61]
		< 5	550	[68–70]

Kowalska et al. [56,57] reported that Au–TiO_2 photocatalysts with different Au particle sizes (~10–60 nm) were prepared by photo-deposition (entry 1). The particle sizes of Au strongly depend on the particle sizes of the TiO_2 polycrystalline structure. The top peak of plasmonic absorption was in the range of ~530–610 nm, depending on the particle sizes of the Au NPs. Tanaka and Kominami et al. [60,63–66] reported unique CPH methods for the preparation of Au-TiO_2 (entry 2). The particle sizes were uniformed to be ~12–14 nm, which exhibits plasmonic absorption at ~550–560 nm. Thus, colloidal Au NPs were successfully loaded onto TiO_2 without change in the original particle size. Furthermore, the top peak of Au plasmon absorption was found to extend towards 620 nm by simple calcinations of the samples. This phenomenon is due to high contact area between TiO_2 and Au NPs without change of particle size [66]. Additionally, Naya et al. [67,68] and Shiraishi et al. [69] employed precipitation deposition methods to deposit small Au NPs (~2–6 nm) on TiO_2 (entry 3).

3.3. Application of LSPR of Au–TiO_2 to Several Photocatalytic Reactions

Au NPs deposited on TiO_2 have been used as visible-light-responsive photocatalysts for several chemical reactions: decomposition of VOCs, selective oxidation of an aromatic alcohol, direct water splitting, H_2 formation from sacrificial aqueous solutions, and reduction of organic compounds (see Table 3). Several research groups concluded that photocatalytic activities are induced by LSPR of the Au NPs. Some research indicates that small Au NPs (~5 nm) effectively work for the reactions [61,69]. Tanaka and Kominami et al. suggest that two types of Au particles of different sizes loaded onto TiO_2 exhibit different functionalities. That is, the larger Au particles contribute to strong light absorption, and the smaller Au particles act as a co-catalyst for H_2 evolution [63].

Table 3. Applications to several photocatalytic reactions on the Au–TiO$_2$ photocatalyst.

Entry	Photocatalytic Reactions	Au Deposition Methods	References
1	oxidations of 2-propanol and ethanol	PD	[56–58]
	oxidation of formic acid	CPH	[60]
2	oxidation of thiol to disulfide	DP	[67]
	oxidation of amine to imine	DP	[68]
	oxidation of aromatic alcohol to aldehyde	CPH	[66]
		DP	[69]
	oxidation of benzene to phenol	PD	[70]
3	H$_2$ formation from alcohols	CPH	[63,71]
	water splitting into H$_2$ and O$_2$	DP	[61]
		CPH	[64,72]
4	reduction of nitrobenzene to aniline	CPH	[65]

3.4. Application to a Photovoltaic Fuel Cell Operating under Visible Light Irradiation

The Au–TiO$_2$ films were found to exhibit the behavior of a photovoltaic fuel cell [54,55]. An anodic photocurrent was yielded on the Au–TiO$_2$ film as the visible light was irradiated, while the current was observed neither on a TiO$_2$ film under visible light irradiation, nor on the Au–TiO$_2$ film when the light was turned off. The short-circuit photocurrent density (J_{sc}) was strongly influenced by kinds of donors, and the photocurrent efficiency was maximized in the presence of Fe^{2+} ions. Furthermore, the photocurrent action spectra were closely fitted with the absorption spectrum of the Au NPs deposited on the TiO$_2$ film (See Figure 11).

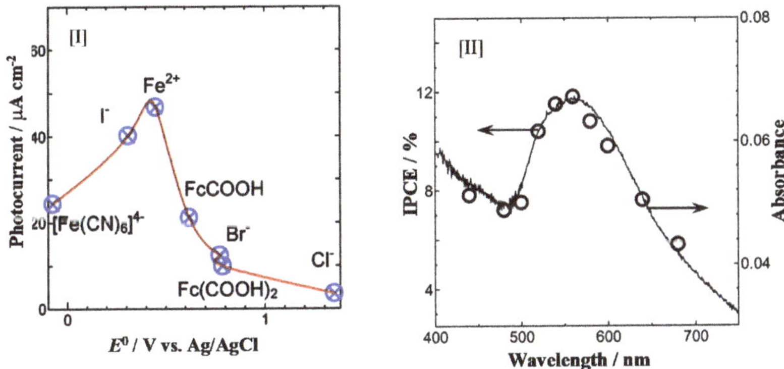

Figure 11. Short-circuit photocurrent densities [I] vs. apparent formal potential of different donors on the Au–TiO$_2$ photoanode in acetonitrile/ethylene glycol (v/v 60/40) containing 0.1 M LiNO$_3$ and 0.1 M donors; IPCE [II] of the Au–TiO$_2$ film in a N$_2$-saturated acetonitrile and ethylene glycol (v/v: 60/40) solution containing 0.1 M FeCl$_2$ and 0.05 M FeCl$_3$ [55].

3.5. Mechanisms of Charge Separation

The mechanism for the Au plasmon-induced charge separation is shown in Figure 12. Visible light irradiation generates the photo-excited state of the Au NPs by LSPR. The photo-excited electrons are injected into the C.B. of TiO$_2$, while the holes abstracted electrons from a donor in the solution. The Au NPs behave like an intrinsic semiconductor, and the Fermi levels of Au NPs and TiO$_2$ are leveled out, resulting in the formation of Schottky barrier at Au–TiO$_2$ junctions. This band model seems to be similar with dye-sensitized photo-anodic electrodes.

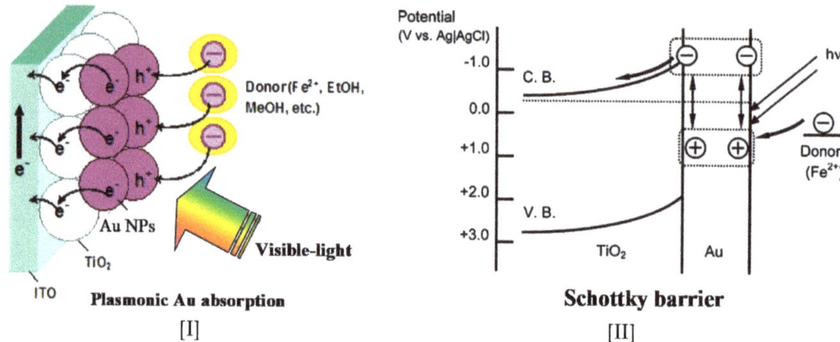

Figure 12. Schematic illustration [I] and its energy band levels [II] for the photo-induced charge separation on the Au–TiO$_2$ in the presence of donors [55].

Recently, Furube et al. studied the plasmon-induced charge transfer mechanisms between Au NPs and TiO$_2$ by means of femtosecond visible pump/infrared probe transient absorption spectroscopy [73]. The electron transfer from the Au NPs to the C.B. of TiO$_2$ was confirmed to occur within 50 fs, and that the electron injection yielded 20–50% upon 550 nm laser excitation.

4. Photo-Induced Interfacial Charge Transfer

4.1. Dye-Sensitized TiO$_2$ Photocatalysis

Dye sensitized TiO$_2$ photocatalysis was studied in the late 1990s. The Ru complex, [Ru(bipy)$_3$]$^{2+}$ grafted on the TiO$_2$ surface exhibits visible light absorption [74,75]. In this system, the excitation of the Ru complex induces electron transfer via metal–ligand charge transfer (MLCT). The photo-induced electrons are then transferred onto TiO$_2$, resulting in photocatalytic water splitting to produce H$_2$. The platinum-chloride-modified TiO$_2$ system was reported by Kisch et al. [76,77]. Photo-irradiation of Pt(IV) chloride exhibits visible-light absorption to generate the active center, (Pt^{4+}(Cl$^-$)$_4$ + $h\nu$ → Pt^{3+}Cl^0(Cl$^-$)$_3$. The photo-induced electrons are transferred from Pt^{3+} to C.B. of TiO$_2$ as reductive sites, while the Cl^0 work as the oxidative sites, resulting in the redox photocatalytic reactions. Important strategies to develop these types of photocatalysts are to design robust sensitizers adjusted with HOMO-LUMO levels.

4.2. Visible-Light-Responsive TiO$_2$ Photocatalyst Modified by Phenolic Organic Compounds

Strong interaction of phenolic groups in organic compounds with Ti–OH of the TiO$_2$ surface probably forms two types of interfacial surface complexes (ISC, Figure 13I), which exhibits visible light absorption via LMCT. The photocatalysis of the ISC is strongly influenced by the electronic structures of the ISC (Figure 13II): the ISC with EWG exhibits strong oxidizability under visible light irradiation, and it can favorably oxidize the TEA, together with H$_2$ evolution from deaerated TEA aqueous solutions [78]. The visible light response of the ISC is attributed to electronic excitation from the donor levels (0.7 V above V.B.) to the C.B. of TiO$_2$ (see Figure 14). Therefore, the electronic structures of sensitizers strongly influence the photocatalytic activities. Ikeda et al. [79] demonstrated that a TiO$_2$ photocatalyst modified with 1,1′-binaphthalene-2,2′-diol (bn(OH)$_2$) exhibited photocatalytic H$_2$ evolution from deaerated TEA aq. under visible light irradiation. Kamegawa et al. [80] designed a 2, 3-dihydroxynaphthalene (2,3-DN)-modified TiO$_2$ photocatalyst for the reduction of nitrobenzene to aminobenzene under visible light irradiation.

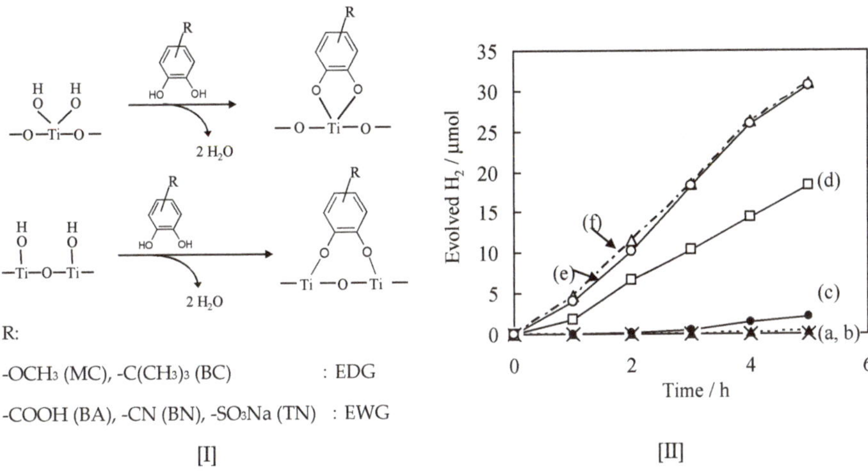

Figure 13. Schematic illustration for the formation of two types of ISCs [I], and photocatalytic H_2 evolution [II] from aq. TEA (10 vol. %) on (a) BC/TiO$_2$, (b) MC/TiO$_2$, (c) CA/TiO$_2$, (d) BA/TiO$_2$, (e) BN/TiO$_2$, and (f) TN/TiO$_2$ [78]. BC: 4-*t*-butyl catechol, MC: 3-methoxy catechol, CA: catecol, BA: 2,3-dihydroxy benzoic acid; BN: 3,4-dihydroxy benzonitrile; TN: tiron.

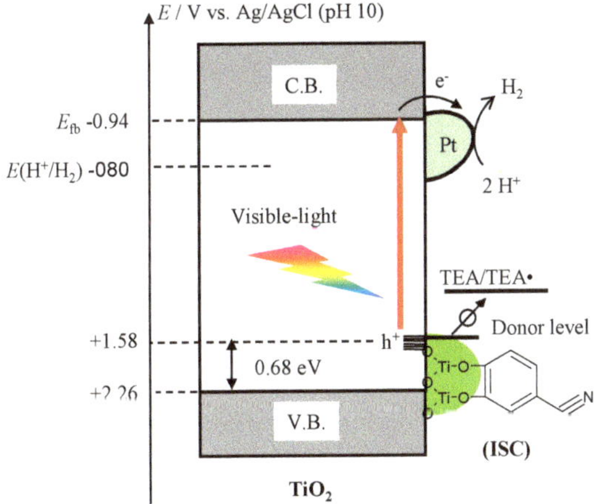

Figure 14. Schematic illustration of photo-induced charge separation on the BN/TiO$_2$ for H_2 evolution from TEA aq. in the presence of Pt as co-catalyst under visible light irradiation [78].

On the other hand, the phenolic compounds were degraded on the TiO$_2$ in the presence of O$_2$ under visible light ($\lambda > 420$ nm) illumination, producing Cl$^-$ and CO$_2$ [81]. The ISC formed by the interaction of phenolic compounds with TiO$_2$ exhibited self-degradation. It was proposed that an electronic transition occurs from the ISC to the C.B. of TiO$_2$ to form active oxygen species, which also participate in the oxidative degradation of phenolic compounds.

4.3. Interfacial-Surface-Complex-Mediated Visible-Light-Sensitive TiO$_2$ Photocatalysts

The interfacial surface complex (ISC)-mediated visible-light-sensitive TiO$_2$ photocatalyst was applied to selective oxidation of several aromatic alcohols [82–88]. Unlike to the ISC in Figure 14, reactant molecules adsorbed onto the TiO$_2$ surface (ISC) is activated under visible-light irradiation, and they are converted into products. Figure 15 shows reaction time profiles for the oxidation of benzyl alcohol in an acetonitrile solution suspended with TiO_2 photocatalyst in the presence of O_2 under visible light irradiation ($\lambda > 420$ nm). This reaction does not proceed without TiO_2 or irradiation. It was found that the amount of benzyl alcohol decreased with an increase in the irradiation time, while the amount of benzaldehyde increased. Neither benzoic acid nor CO_2 were formed as oxidative products. The yield of benzaldehyde reached circa 95%, and the carbon balance in the liquid phase was circa 95% after photo-irradiation for 4 h.

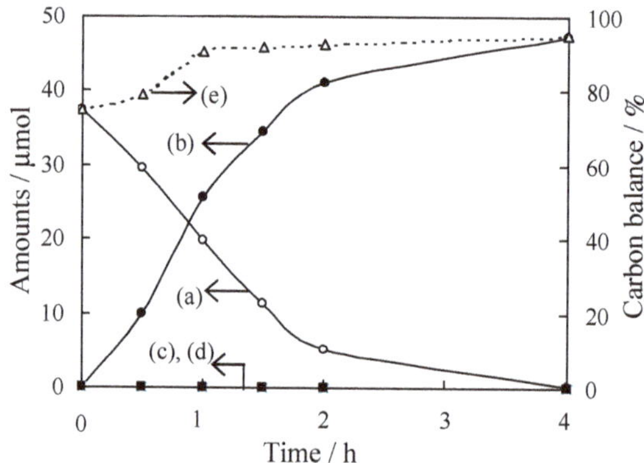

Figure 15. Selective oxidation of benzyl alcohol on TiO_2 (50 mg) under visible light irradiation [82]. The initial amount of benzyl alcohol was 50 µmol. Amounts of: benzyl alcohol (a); benzaldehyde (b); benzoic acid (c); CO_2 (d); and percentage of total organic compounds in solution (e).

Photocatalytic oxidation of benzyl alcohol and its derivatives into corresponding aldehydes was carried out with TiO_2 under visible light irradiation. Benzyl alcohol and its derivatives substituted by $-OCH_3$, $-Cl$, $-NO_2$, $-CH_3$, $-CF_3$, and $-C(CH_3)_3$ groups were successfully converted to corresponding aldehydes with a high conversion and high selectivity on TiO_2, while no other products were observed (See Table 4). However, the phenolic compound (entry 9) was deeply oxidized, since it strongly adsorbed on the TiO_2 surface [82].

Table 4. Chemoselective photocatalytic oxidation of different kinds of benzylic alcohols on TiO_2 [82].

Entry	R_1	R_2	Conversion (%)	Selectivity (%)
1	H	H	> 99	> 99
2	H	$C(CH_3)_3$	> 99	> 99
3	H	OCH_3	> 99	> 99
4	H	CH_3	> 99	> 99
5	H	Cl	> 99	> 99
6	H	NO_2	> 99	> 99
7	H	CF_3	> 99	> 99
8	CH_3	H	> 99	> 99
9	H	OH	> 85	23

4.3.1. What Is the Origin of the Visible Light Response?

The interaction of benzyl alcohol with TiO_2 was analyzed by FT-IR spectroscopy (See Figure 16). Characteristic features of the ISC are as follows: (i) a remarkable downward negative band at 3715 cm^{-1} attributed to the O–H stretching of the terminal OH group; (ii) a new band appeared at circa 1100 cm^{-1}, which is attributed to the C–O stretching of the alkoxide species formed by the interaction of benzyl alcohol with TiO_2, while that of benzyl alcohol by itself is 1020 cm^{-1}.

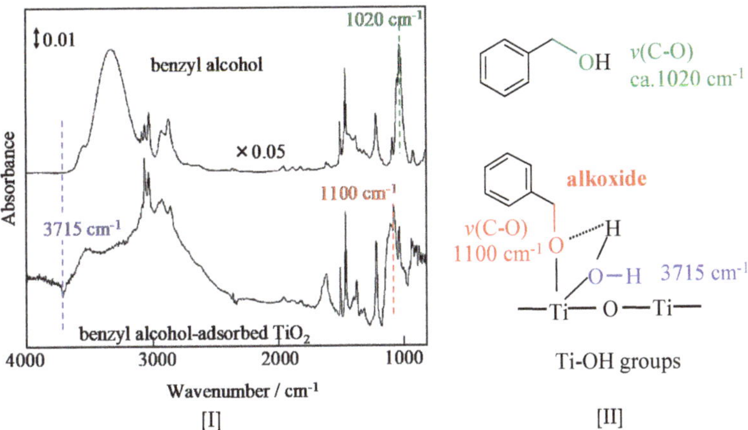

Figure 16. FT-IR spectra [I] of benzyl alcohol by itself and benzyl alcohol adsorbed on TiO_2; and [II] their peak identification [82].

When the TiO_2 was treated by diluted HF (aq), the IR band at 3715 cm^{-1} on the HF–TiO_2 drastically decreased, while the photocatalytic activity significantly decreased. The active sites were confirmed to be alkoxide by the interaction of benzyl alcohol with the terminal OH groups of TiO_2.

TiO_2 by itself exhibited absorption only in the UV region, which is attributed to the charge transition from V.B. to C.B. When the benzyl alcohol was adsorbed on TiO_2, absorption in the visible region could be observed. This absorption in the visible light region is assignable to the ISC through the LMCT (See Figure 17). The action spectra of apparent quantum yield (AQY) plots were fitted with the photo-absorption of TiO_2-adsorbed benzyl alcohol, suggesting that visible light absorption directly participated in the photocatalytic reactions.

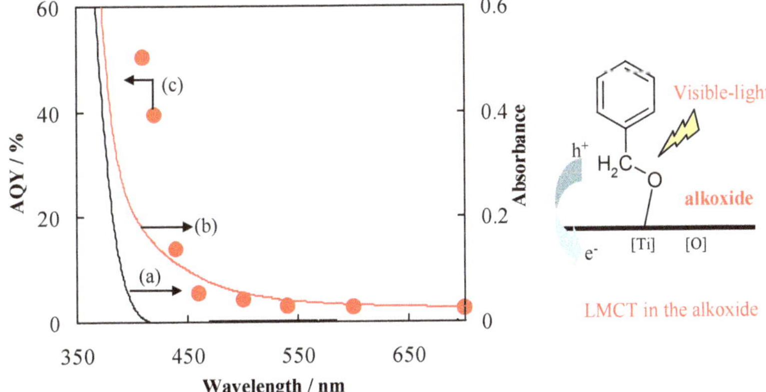

Figure 17. UV-vis absorption spectra of TiO_2 (a), TiO_2 adsorbed with benzyl alcohol (b), and apparent quantum yield (AQY) for the formation of benzaldehyde (c); and schematic illustration of photo-induced charge transfer through LMCT in the alkoxide [82].

DFT calculations [87] indicated the interaction of benzyl alcohol with surface hydroxyl groups on the TiO_2 surface, resulting in the formation of alkoxide species. The electron density contour maps for the alkoxide species are shown in Figure 18. The orbital #212 at -0.80 eV forms the V.B. of TiO_2, while #218 at $+2.25$ eV forms the C.B. One type of surface state consisting of the orbital (#215) originates with the alkoxide species ([Ti]–O–CH_2–ph) hybridized with the O2p AOs in the V.B. of the TiO_2. The energy gap between #215 and #218 (2.8 eV) was confirmed to be the origin for the visible light response.

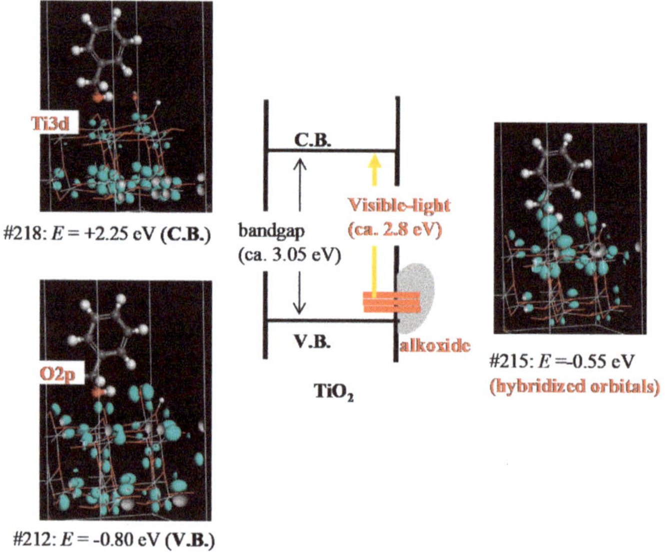

Figure 18. Photo-induced electron transfer from the hybridized orbital to the C.B. of TiO_2 under visible light irradiation [87]. Density maps of V.B., C.B., and hybridized orbital are shown here.

4.3.2. What Makes the High Selectivity for the Photocatalytic Reactions?

It was observed that benzyl alcohol is adsorbed on TiO_2 more favorably than benzaldehyde in a mixture of benzyl alcohol and benzaldehyde under dark conditions. This result indicates that the interaction between benzaldehyde and TiO_2 is fairly weak. According to DFT calculations [87,88], the interaction of benzyl alcohol with the TiO_2 surface formed a hybridized orbital, while benzaldehyde did not form orbital mixing. Therefore, once benzaldehyde was produced by the oxidation of benzyl alcohol, benzaldehyde was immediately released into the bulk solution, and was not oxidized further to benzoic acid or CO_2.

4.3.3. Reaction Mechanisms behind the Selective Photocatalytic Oxidation of Benzyl Alcohol

The photocatalytic activities for the oxidation of benzyl alcohol or α, α-d2 benzyl alcohol were investigated. The kinetic isotope effect (KIE) [$=k_{C\text{-}H}/k_{CD}$] was estimated to be 3.9 at 295 K. This result suggests that the process for the α-deprotonation is the rate determining step (RDS) for the overall reaction. From the experimental and theoretical studies by DFT calculations, one of the favorable reaction paths is depicted in Figure 19. When benzyl alcohol interacts with Ti–OH of the TiO_2, the alkoxide species (ISC) is formed on a Ti site (3). The ISC was photo-excited under visible light irradiation via LMCT of the ISC, which induces holes (h^+) and electrons (e^-). Subsequently, the electrons are transferred to O_2 to form superoxide anions (the bonding distance between O–O becomes longer), which induces α-deprotonation of the benzyl alcohol (4-5TS). Such hydro-peroxide species would further induce the de-protonation from another benzyl alcohol to form benzaldehyde (7-8TS), resulting in regeneration of the surface terminal OH groups. The consecutive generation of the terminal OH groups would, thus, be one of the key factors for the photocatalytic reactions.

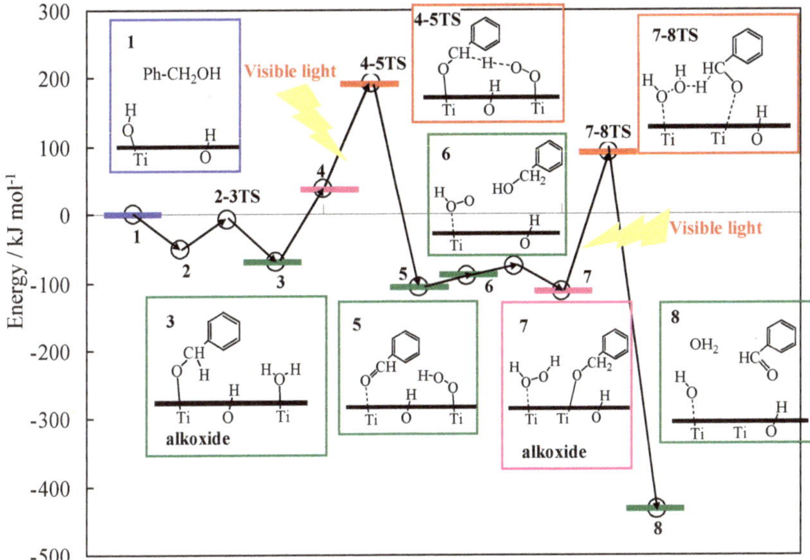

Figure 19. Possible reaction path for the selective oxidation of benzyl alcohol in the presence of O_2 on the TiO$_2$ under visible light irradiation [87].

4.4. Photocatalytic Oxidation of Benzyl Amine into Imine

Imines are important intermediates for the synthesis of pharmaceuticals and agricultural chemicals. Selective photocatalytic oxidation of benzyl amine into N-benzylidenebenzylamine takes place in the presence of O_2 on the TiO$_2$ at room temperature (Scheme 1) [89,90]. Several kinds of benzylic amines were examined, and they were converted into the corresponding imines, yielding circa 38–94% [89]. The origin of the visible light response is due to formation of amine oxide (ISC) through the interaction of benzylic amine onto the surface of TiO$_2$, and the ISC exhibits electronic transition from the localized N 2p orbitals of the amine oxide (ISC) to the C.B. of TiO$_2$. The photo-induced redox catalysis produces benzaldehyde in the presence of O_2. Subsequently, the condensation reaction of benzaldehyde with another benzyl amine forms N-benzylidenebenzylamine under dark conditions.

Scheme 1. Selective oxidation of benzyl amine into N-benzylidenebenzylamine on the TiO$_2$ photocatalyst under visible light irradiation.

5. Conclusions

This review focused on some fundamental issues behind the visible-light-sensitive TiO$_2$ photocatalysts, highlighting the bulk and/or surface electronic structures modified by doping with nitrogen anions; plasmonic Au NPs, and interfacial surface complexes (ISC) and their related photocatalysts. Tailoring the interface and bulk properties, including surface band bending, sub-band structure, surface state distribution, and charge separation, significantly reflects on the photocatalysis. We hope that this review has provided some useful contributions for the future design and development of novel photocatalytic systems employing TiO$_2$ as well as non-TiO$_2$ semiconductor materials with nanoscale levels. The applications of such photocatalytic systems could not only convert unlimited solar energy into chemical energy, but also protect our environment, leading to sustainable green chemistry.

Funding: This research received no external funding.

Conflicts of Interest: The authors declare no conflict of interest.

Abbreviations

NPs	nanoparticles
ISC	interfacial surface complex
VOCs	volatile organic compounds
V.B.	valence band
C.B.	conduction band
XPS	X-ray photoelectron spectroscopy
EPR	electron paramagnetic resonance
UV-vis	Ultraviolet-visible
LSPR	localized surface plasmon resonance
PD	photodeposition
CPH	colloid photodeposition by hole scavenger
DP	deposition precipitation
TEM	transmittance electron microscope
J_{SC}	short-circuit photocurrent
IPCE	incident photo to current efficiency
DFT	density functional theory
MLCT	metal to ligand charge transfer
FT-IR	Fourier transformed-infrared
KIE	kinetic isotope effect
LMCT	ligand to metal charge transfer

References

1. Honda, K.; Fijishima, A. Electrochemical Photolysis of Water at a Semiconductor Electrode. *Nature* **1972**, *238*, 37–38.
2. Inoue, T.; Fujishima, A.; Konishi, S.; Honda, K. Photoelectrocatalytic reduction of carbon dioxide in aqueous suspensions of semiconductor powders. *Nature* **1979**, *277*, 637–638. [CrossRef]
3. Kamat, P.V. Photochemistry on nonreactive and reactive (semiconductor) surfaces. *Chem. Rev.* **1993**, *93*, 267–300.
4. Fox, M.A.; Dulay, M.T. Heterogeneous photocatalysis. *Chem. Rev.* **1993**, *93*, 341–357.
5. Hoffman, M.R.; Martin, S.T.; Choi, W.; Bahnemann, D.W. Environmental applications of semiconductor photocatalysis. *Chem. Rev.* **1995**, *95*, 69–96. [CrossRef]
6. Fujishima, A.; Rao, T.N.; Tryk, A. Titanium dioxide photocatalysis. *J. Photochem. Photobio. C Photochem. Rev.* **2000**, *1*, 1–21. [CrossRef]
7. Anpo, M.; Takeuchi, M. The design and development of highly reactive titanium oxide photocatalysts operating under visible light irradiation. *J. Catal.* **2003**, *216*, 505–516. [CrossRef]
8. Chen, X.; Mao, S.S. Titanium Dioxide Nanomaterials: Synthesis, Properties, Modifications, and Applications. *Chem. Rev.* **2007**, *197*, 2891–2959. [CrossRef]
9. Shiraishi, Y.; Hirai, T. Selective organic transformations on titanium oxide-based photocatalysts. *J. Photochem. Photobiol. C Photochem. Rev.* **2008**, *9*, 157–170. [CrossRef]
10. Palmisano, G.; García-López, E.; Marcì, G.; Loddo, V.; Yurdakal, S.; Augugliaro, V.; Palmisano, L. Advances in selective conversions by heterogeneous photocatalysis. *Chem. Commun.* **2010**, *46*, 7074–7089. [CrossRef]
11. Schneider, J.; Matsuoka, M.; Takeuchi, M.; Zhang, J.; Horiuchi, Y.; Anpo, M.; Bahnemann, D.W. Understanding TiO$_2$ Photocatalysis: Mechanisms and Materials. *Chem. Rev.* **2014**, *114*, 9919–9986. [CrossRef]
12. Ma, Y.; Wang, X.; Jia, Y.; Chen, X.; Han, H.; Li, C. Titanium dioxide-based nanomaterials for: Photocatalytic fuel generations. *Chem. Rev.* **2014**, *114*, 9987–10043. [CrossRef]

13. Asahi, R.; Morikawa, T.; Irie, H.; Ohwaki, T. Nitrogen-doped titanium dioxide as visiblelight-sensitive photocatalyst: Designs, developments, and prospects. *Chem Rev.* **2014**, *114*, 9824–9852. [CrossRef]

14. Lang, X.; Chen, X.; Zhao, J. Heterogeneous visible light photocatalysis for selective organic transformations. *Chem. Soc. Rev.* **2014**, *43*, 473–486. [CrossRef]

15. Sang, L.; Zhao, Y.; Burda, C. TiO$_2$ Nanoparticles as Functional Building Blocks. *Chem. Rev.* **2014**, *114*, 9283–9318. [CrossRef]

16. Nosaka, Y.; Nosaka, A.Y. Generation and Detection of Reactive Oxygen Species in Photocatalysis. *Chem. Rev.* **2017**, *117*, 11302–11336. [CrossRef]

17. Kou, J.; Lu, C.; Wang, J.; Chen, Y.; Xu, Z.; Varma, R.S. Selectivity Enhancement in Heterogeneous Photocatalytic Transformations. *Chem. Rev.* **2017**, *117*, 1445–1514. [CrossRef]

18. Prakash, J.; Sun, S.; Swart, H.C.; Gupta, R.K. Noble metals-TiO$_2$ nanocomposites: From fundamental mechanisms to photocatalysis, surface enhanced Raman scattering and antibacterial applications. *Appl. Mater. Today* **2018**, *11*, 82–135. [CrossRef]

19. Wang, W.; Tadé, M.O.; Shao, Z. Nitrogen-doped simple and complex oxides for photocatalysis: A review. *Prog. Mater. Sci.* **2018**, *92*, 33–63. [CrossRef]

20. Yamashita, H.; Mori, K.; Kuwahara, Y.; Kamegawa, T.; Wen, M.; Verma, P.; Che, M. Single-site and nano-confined photocatalysts designed in porous materials for environmental uses and solar fuels. *Chem. Soc. Rev.* **2018**, *47*, 8072–8096. [CrossRef]

21. Ahmed, A.Y.; Kandiel, T.A.; Oekermann, T.; Bahnemann, D. Photocatalytic Activities of Different Well-defined Single Crystal TiO$_2$Surfaces: Anatase versus Rutile. *J. Phys. Chem. Lett.* **2011**, *2*, 2461–2465. [CrossRef]

22. Tanaka, K.; Capule, M.F.V.; Hisanaga, T. Effect of Crystallinity of TiO$_2$ on Its Photo-catalytic Action. *Chem. Phys. Lett.* **1991**, *187*, 73–76. [CrossRef]

23. Luttrell, T.; Halpegamage, S.; Tao, J.; Kramer, A.; Sutter, E.; Batzill, M. Why is anatase a better photocatalyst than rutile? - Model studies on epitaxial TiO$_2$ films. *Sci. Rep.* **2014**, *4*, 4043–4050. [CrossRef]

24. Gordon, T.R.; Cargnello, M.; Paik, T.; Mangolini, F.; Weber, R.T.; Fornasiero, P.; Murray, C.B. Nonaqueous Synthesis of TiO$_2$ Nanocrystals Using TiF$_4$ to Engineer Morphology, Oxygen Vacancy Concentration, and Photocatalytic Activity. *J. Am. Chem. Soc.* **2012**, *134*, 6751–6761. [CrossRef]

25. Zhang, Z.; Wang, C.-C.; Zakaria, R.J.; Ying, Y. Role of Particle Size in Nanocrystalline TiO$_2$-Based Photocatalysts. *J. Phys. Chem. B* **1998**, *102*, 10871–10878. [CrossRef]

26. Anpo, M.; Ichihashi, Y.; Takeuchi, M.; Yamashita, H. Design of unique titanium oxide photocatalysts by an advanced metal ion-implantation method and photocatalytic reactions under visible light irradiation. *Res. Chem. Intermed.* **1998**, *24*, 143–149. [CrossRef]

27. Sato, S. Photocatalytic activity of NO$_x$-doped TiO$_2$ in the visible light region. *Chem. Phys. Lett.* **1986**, *123*, 126–128. [CrossRef]

28. Asahi, R.; Morikawa, T.; Aoki, K.; Taga, Y. Visible-light photocatalysis in nitrogen-doped titanium oxides. *Science* **2001**, *293*, 269–271. [CrossRef]

29. Irie, H.; Watanabe, Y.; Hashimoto, K. Nitrogen-Concentration Dependence on Photocatalytic Activity of TiO$_{2-x}$N$_x$ Powders. *J. Phys. Chem. B* **2003**, *107*, 5483–5486. [CrossRef]

30. Shin, C.; Bugli, G.; Djega-Mariadassou, G. Preparation and characterization of titanium oxynitrides with high specific surface areas. *J. Solid State Chem.* **1991**, *95*, 145–155. [CrossRef]

31. Livraghi, S.; Paganini, M.C.; Giamello, E.; Selloni, A.; Di Valentin, C.; Pacchioni, G. Origin of Photoactivity of Nitrogen-Doped Titanium Dioxide under Visible Light. *J. Am. Chem. Soc.* **2006**, *128*, 15666–15671. [CrossRef]

32. Wang, J.; Zhu, W.; Zhang, Y.; Liu, S. An efficient two-step technique for nitrogen-doped titanium dioxide synthesizing: Visible-light-induced photodecomposition of methylene blue. *J. Phys. Chem. C* **2007**, *111*, 1010–1014. [CrossRef]

33. Higashimoto, S.; Azuma, M. Photo-induced charging effect and electron transfer to the redox species on nitrogen-doped TiO$_2$ under visible light irradiation. *Appl. Catal. B Environ.* **2009**, *89*, 557–562. [CrossRef]

34. Wang, H.; Hu, Y. The Photocatalytic Property of Nitrogen-Doped TiO$_2$ Nanoball Film. *Int. J. Photoenergy* **2013**. [CrossRef]

35. Diwald, O.; Thompson, T.L.; Zubkov, T.; Goralski, E.G.; Walck, S.D.; Yates, J.T., Jr. Photochemical Activity of Nitrogen-Doped Rutile TiO$_2$(110) in Visible Light. *J. Phys. Chem. B* **2004**, *108*, 6004–6008. [CrossRef]

36. Higashimoto, S.; Ushiroda, Y.; Azuma, M.; Ohue, H. Synthesis, characterization and photocatalytic activity of N-doped TiO_2 modified by platinum chloride. *Catal. Today* **2008**, *132*, 165–169. [CrossRef]

37. Higashimoto, S.; Ushiroda, Y.; Azuma, M. Mechanism for enhancement of visible light response on nitrogen-doped TiO_2 by modification with vanadium species. *J. Nanosci. Nanotechnol.* **2010**, *10*, 246–251. [CrossRef]

38. Nakamura, R.; Tanaka, T.; Nakato, Y. Mechanism for Visible Light Responses in Anodic Photocurrents at N-Doped TiO_2 Film Electrodes. *J. Phys. Chem. B* **2004**, *108*, 10617–10620. [CrossRef]

39. Tang, J.; Cowan, A.J.; Durrant, J.R.; Klug, D.R. Mechanism of O_2 production from water splitting: Nature of charge carriers in nitrogen doped nanocrystalline TiO_2 films and factors limiting O_2 production. *J. Phys. Chem. C* **2011**, *115*, 3143–3150. [CrossRef]

40. Higashimoto, S.; Tanihata, W.; Nakagawa, Y.; Azuma, M.; Ohue, H.; Sakata, Y. Effective photocatalytic decomposition of VOC under visible-light irradiation on N-doped TiO_2 modified by vanadium species. *Appl. Catal. A Gen.* **2008**, *340*, 98–104. [CrossRef]

41. Morikawa, T.; Ohwaki, T.; Suzuki, K.; Moribe, S.; Tero-Kubota, S. Visible-light-induced photocatalytic oxidation of carboxylic acids and aldehydes over N-doped TiO_2 loaded with Fe, Cu or Pt. *Appl. Catal. B Environ.* **2008**, *83*, 56–62. [CrossRef]

42. Sreethawong, T.; Laehsalee, S.; Chavadej, S. Use of Pt/N-doped mesoporous-assembled nanocrystalline TiO_2 for photocatalytic H_2 production under visible light irradiation. *Catal. Commun.* **2009**, *10*, 538–543. [CrossRef]

43. Dolat, D.; Quici, N.; Kusiak-Nejman, E.; Morawski, A.W.; Puma, G.L. One-step, hydrothermal synthesis of nitrogen, carbon co-doped titanium dioxide (N,C-TiO_2) photocatalysts. Effect of alcohol degree and chain length as carbon dopant precursors on photocatalytic activity and catalyst deactivation. *Appl. Catal. B Environ.* **2012**, *115*, 81–89. [CrossRef]

44. Virkutyte, J.; Varma, R.S. Visible light activity of Ag-loaded and guanidine nitrate-doped nano-TiO_2: Degradation of dichlorophenol and antibacterial properties. *RSC Adv.* **2012**, *2*, 1533–1539. [CrossRef]

45. Cong, Y.; Zhang, J.; Chen, F.; Anpo, M. Synthesis and Characterization of Nitrogen-Doped TiO_2 Nanophotocatalyst with High Visible Light Activity. *J. Phys. Chem. C* **2007**, *111*, 6976–6982. [CrossRef]

46. Yang, X.; Cao, C.; Erickson, L.; Hohn, K.; Maghirang, R.; Klabunde, K. Synthesis of visible-light-active TiO_2-based photocatalysts by carbon and nitrogen doping. *J. Catal.* **2008**, *260*, 128–133. [CrossRef]

47. Mitoraj, D.; Kisch, H. On the Mechanism of Urea–Induced Titania Modification. *Chem. Eur. J.* **2010**, *16*, 261–269. [CrossRef]

48. Chai, B.; Peng, T.; Mao, J.; Li, K.; Zan, L. Graphitic carbon nitride (g-C_3N_4)-Pt-TiO_2 nanocomposite as an efficient photocatalyst for hydrogen production under visible light irradiation. *Phys. Chem. Chem. Phys.* **2012**, *14*, 16745–16752. [CrossRef]

49. Han, C.; Wang, Y.; Lei, Y.; Wang, B.; Wu, N.; Shi, Q.; Li, Q. *In situ* synthesis of graphitic-C_3N_4 nanosheet hybridized N-doped TiO_2 nanofibers for efficient photocatalytic H_2 production and degradation. *Nano Res.* **2015**, *8*, 1199–1209. [CrossRef]

50. Yan, H.; Yang, H. TiO_2-g-C_3N_4 composite materials for photocatalytic H_2 evolution under visible light irradiation. *J. Alloy. Comp.* **2010**, *509*, L26–L29. [CrossRef]

51. Higashimoto, S.; Hikita, K.; Azuma, M.; Yamamoto, M.; Takahashi, M.; Sakata, Y.; Matsuoka, M.; Kobayashi, H. Visible Light-Induced Photocatalysis on Carbon Nitride Deposited Titanium Dioxide: Hydrogen Production from Sacrificial Aqueous Solutions. *Chin. J. Chem.* **2017**, *35*, 165–172. [CrossRef]

52. Zhang, Q.; Gangadharan, D.T.; Liu, Y.; Xu, Z.; Chaker, M.; Ma, D. Recent advancements in plasmon-enhanced visible light-driven water splitting. *J. Materiomics* **2017**, *3*, 33–50. [CrossRef]

53. Ohko, Y.; Tatsuma, T.; Fujii, T.; Naoi, K.; Niwa, C.; Kubota, Y.; Fujishima, A. Multicolour photochromism of TiO_2 films loaded with silver nanoparticles. *Nat. Mater.* **2003**, *2*, 29–31. [CrossRef] [PubMed]

54. Tian, Y.; Tatsuma, T. Plasmon-induced photoelectrochemistry at metal nanoparticles supported on nanoporous TiO_2. *Chem. Commun.* **2004**, *0*, 1810–1811. [CrossRef] [PubMed]

55. Tian, Y.; Tatsuma, T. Mechanisms and Applications of Plasmon-Induced Charge Separation at TiO_2 Films Loaded with Gold Nanoparticles. *J. Am. Chem. Soc.* **2005**, *127*, 7632–7637. [CrossRef] [PubMed]

56. Kowalska, E.; Abe, R.; Ohtani, B. Visible light-induced photocatalytic reaction of gold-modified titanium(IV) oxide particles: Action spectrum analysis. *Chem. Commun.* **2009**, *0*, 241–243. [CrossRef] [PubMed]

57. Kowalska, E.; Mahaney, O.O.P.; Abe, R.; Ohtani, B. Visible-light-induced photocatalysis through surface plasmon excitation of gold on titania surfaces. *Phys. Chem. Chem. Phys.* **2010**, *12*, 2344–2355. [CrossRef] [PubMed]

58. Kolinko, P.A.; Selishchev, D.S.; Kozlov, D.V. Visible Light Photocatalytic Oxidation of Ethanol Vapor on Titanium Dioxide Modified with Noble Metals. *Theor. Exp. Chem.* **2015**, *51*, 96–103. [CrossRef]

59. Frens, G. Controlled Nucleation for the Regulation of the Particle Size in Monodisperse Gold Suspensions. *Nat. Phys. Sci.* **1973**, *241*, 20–22. [CrossRef]

60. Tanaka, A.; Ogino, A.; Iwaki, M.; Hashimoto, K.; Ohnuma, A.; Amano, F.; Ohtani, B.; Kominami, H. Gold–Titanium(IV) Oxide Plasmonic Photocatalysts Prepared by a Colloid-Photodeposition Method: Correlation Between Physical Properties and Photocatalytic Activities. *Langmuir* **2012**, *28*, 13105–13111. [CrossRef]

61. Silva, C.G.; Juarez, R.; Marino, T.; Molinari, R.; Garcia, H. Influence of Excitation Wavelength (UV or Visible Light) on the Photocatalytic Activity of Titania Containing Gold Nanoparticles for the Generation of Hydrogen or Oxygen from Water. *J. Am. Chem. Soc.* **2011**, *133*, 595–602. [CrossRef] [PubMed]

62. Zanella, R.; Delannoy, L.; Louis, C. Mechanism of deposition of gold precursors onto TiO$_2$ during the preparation by cation adsorption and deposition–precipitation with NaOH and urea. *Appl. Catal. A Gen.* **2005**, *291*, 62–72. [CrossRef]

63. Tanaka, A.; Sakaguchi, S.; Hashimoto, K.; Kominami, H. Preparation of Au/TiO$_2$ with Metal Cocatalysts Exhibiting Strong Surface Plasmon Resonance Effective for Photoinduced Hydrogen Formation under Irradiation of Visible Light. *ACS Catal.* **2013**, *3*, 79–85. [CrossRef]

64. Tanaka, A.; Nakanishi, K.; Hamada, R.; Hashimoto, K.; Kominami, H. Simultaneous and Stoichiometric Water Oxidation and Cr(VI) Reduction in Aqueous Suspensions of Functionalized Plasmonic Photocatalyst Au/TiO$_2$–Pt under Irradiation of Green Light. *ACS Catal.* **2013**, *3*, 1886–1891. [CrossRef]

65. Tanaka, A.; Nishino, Y.; Sakaguchi, S.; Yoshikawa, T.; Imamura, K.; Hashimoto, K.; Kominami, H. Functionalization of a plasmonic Au/TiO$_2$ photocatalyst with an Ag co-catalyst for quantitative reduction of nitrobenzene to aniline in 2-propanol suspensions under irradiation of visible light. *Chem. Commun.* **2013**, *49*, 2551–2553. [CrossRef] [PubMed]

66. Tanaka, A.; Hashimoto, K.; Kominami, H. A very simple method for the preparation of Au/TiO$_2$ plasmonic photocatalysts working under irradiation of visible light in the range of 600–700 nm. *Chem. Commun.* **2017**, *53*, 4759–4762. [CrossRef]

67. Naya, S.; Teranishi, M.; Isobe, T.; Tada, H. Light wavelength-switchable photocatalytic reaction by gold nanoparticle-loaded titanium(IV) dioxide. *Chem. Commun.* **2010**, *46*, 815–817. [CrossRef]

68. Naya, S.; Kimura, K.; Tada, H. One-Step Selective Aerobic Oxidation of Amines to Imines by Gold Nanoparticle-Loaded Rutile Titanium(IV) Oxide Plasmon Photocatalyst. *ACS Catal.* **2013**, *3*, 10–13. [CrossRef]

69. Tsukamoto, D.; Shiraishi, Y.; Sugano, Y.; Ichikawa, S.; Tanaka, S.; Hirai, T. Gold Nanoparticles Located at the Interface of Anatase/Rutile TiO$_2$ Particles as Active Plasmonic Photocatalysts for Aerobic Oxidation. *J. Am. Chem. Soc.* **2012**, *134*, 6309–6315. [CrossRef]

70. Zheng, Z.; Huang, B.; Qin, X.; Zhang, X.; Dai, Y.; Whangbo, M.-H. Facile *in situ* synthesis of visible-light plasmonic photocatalysts M@TiO$_2$ (M = Au, Pt, Ag) and evaluation of their photocatalytic oxidation of benzene to phenol. *J. Mater. Chem.* **2011**, *21*, 9079–9087. [CrossRef]

71. Tanaka, A.; Sakaguchi, S.; Hashimoto, K.; Kominami, H. Preparation of Au/TiO$_2$ exhibiting strong surface plasmon resonance effective for photoinduced hydrogen formation from organic and inorganic compounds under irradiation of visible light. *Catal. Sci. Technol.* **2012**, *2*, 907–909. [CrossRef]

72. Tanaka, A.; Teramura, K.; Hosokawa, S.; Kominami, H.; Tanaka, T. Visible light-induced water splitting in an aqueous suspension of a plasmonic Au/TiO$_2$ photocatalyst with metal co-catalysts. *Chem. Sci.* **2017**, *8*, 2574–2580. [CrossRef] [PubMed]

73. Furube, A.; Du, L.; Hara, K.; Katoh, R.; Tachiya, M. Ultrafast Plasmon-Induced Electron Transfer from Gold Nanodots into TiO$_2$ Nanoparticles. *J. Am. Chem. Soc.* **2007**, *129*, 14852–14853. [CrossRef] [PubMed]

74. Borgarello, E.; Kiwi, J.; Pelizzetti, E.; Visca, M.; Grätzel, M. Photochemical cleavage of water by photocatalysis. *Nature* **1981**, *289*, 158–160. [CrossRef]

75. Vinodgopal, K.; Hua, X.; Dahlgren, R.L. Photochemistry of Ru(bpy)$_2$(dcbpy)$^{2+}$ on Al$_2$O$_3$ and TiO$_2$ surfaces. an insight into the mechanism of photosensitization. *J. Phys. Chem.* **1995**, *99*, 10883–10889. [CrossRef]

76. Sakthivel, S.; Kisch, H. Daylight photocatalysis by carbon-modified titanium dioxide. *Angew. Chem. Int. Ed.* **2003**, *42*, 4908–4911. [CrossRef] [PubMed]

77. Macyk, W.; Burgeth, G.; Kisch, H. Photoelectrochemical properties of platinum(IV) chloride surface modified TiO_2. *Photochem. Photobiol. Sci.* **2003**, *2*, 322–328. [CrossRef]

78. Higashimoto, S.; Nishi, T.; Yasukawa, M.; Azuma, M.; Sakata, Y.; Kobayashi, H. Photocatalysis of titanium dioxide modified by interfacial surface complexes (ISC) with different substituted groups. *J. Catal.* **2015**, *329*, 286–290. [CrossRef]

79. Ikeda, S.; Abe, C.; Torimoto, T.; Ohtani, B. Photochemical hydrogen evolution from aqueous triethanolamine solutions sensitized by binaphthol-modified titanium(IV) oxide under visible-light irradiation. *J. Photochem. Photobiol. A Chem.* **2003**, *160*, 61–67. [CrossRef]

80. Kamegawa, T.; Seto, H.; Matsuura, S.; Yamashita, H. Preparation of hydroxynaphthalene modified TiO_2 via formation of surface complexes and their applications in the photocatalytic reduction of nitrobenzene under visible-light irradiation. *ACS Appl. Mater. Interfaces* **2012**, *4*, 6635–6639. [CrossRef]

81. Kim, S.; Choi, W. Visible-light-induced photocatalytic degradation of 4-chlorophenol and phenolic compounds in aqueous suspension of pure titania: Demonstrating the existence of a surface-complex-mediated path. *J. Phys. Chem. B* **2005**, *109*, 5143–5149. [CrossRef] [PubMed]

82. Higashimoto, S.; Kitao, N.; Yoshida, N.; Sakura, T.; Azuma, M.; Ohue, H.; Sakata, Y. Selective photocatalytic oxidation of benzyl alcohol and its derivatives into corresponding aldehydes by molecular oxygen on titanium dioxide under visible light irradiation. *J. Catal.* **2009**, *266*, 279–285. [CrossRef]

83. Higashimoto, S.; Okada, K.; Morisugi, T.; Azuma, M.; Ohue, H.; Kim, T.-H.; Matsuoka, M.; Anpo, M. Effect of surface treatment on the selective photocatalytic oxidation of benzyl alcohol infrared study of hydroxy groups on coordinative defect sites. *Top. Catal.* **2010**, *53*, 578–583. [CrossRef]

84. Higashimoto, S.; Okada, K.; Azuma, M.; Ohue, H.; Terai, T.; Sakata, Y. Characteristics of the charge transfer surface complex on titanium(IV) dioxide for the visible light induced chemoselective oxidation of benzyl alcohol. *RSC Adv.* **2012**, *2*, 669–676. [CrossRef]

85. Higashimoto, S.; Suetsugu, N.; Azuma, M.; Ohue, H.; Sakata, Y. Efficient and selective oxidation of benzylic alcohol by O_2 into corresponding aldehydes on a TiO_2 photocatalyst under visible light irradiation: Effect of phenyl-ring substitution on the photocatalytic activity. *J. Catal.* **2010**, *274*, 76–83. [CrossRef]

86. Higashimoto, S.; Shirai, R.; Osano, Y.; Azuma, M.; Ohue, H.; Sakata, Y.; Kobayashi, H. Influence of metal ions on the photocatalytic activity: Selective oxidation of benzyl alcohol on iron (III) ion-modified TiO_2, using visible light. *J. Catal.* **2014**, *311*, 137–143. [CrossRef]

87. Kobayashi, H.; Higashimoto, S. DFT study on the reaction mechanisms behind the catalytic oxidation of benzyl alcohol into benzaldehyde by O_2 over anatase TiO_2 surfaces with hydroxyl groups: Role of visible-light irradiation. *Appl. Catal. B Environ.* **2015**, *170*, 135–143. [CrossRef]

88. Li, R.; Kobayashi, H.; Guo, J.; Fan, J. Visible-light induced high-yielding benzyl alcohol-to benzaldehyde transformation over mesoporous crystalline TiO_2: A self-adjustable photooxidation system with controllable hole-generation. *J. Phys. Chem. C* **2011**, *115*, 23408–23416. [CrossRef]

89. Lang, X.; Ma, W.; Zhao, Y.; Chen, C.; Ji, H.; Zhao, J. Visible-light-induced selective catalytic aerobic oxidation of amines into imines on TiO_2. *Chem. Eur. J.* **2012**, *18*, 2624–2631. [CrossRef]

90. Higashimoto, S.; Hatada, Y.; Ishikawa, R.; Azuma, M.; Sakata, Y.; Kobayashi, H. Selective Photocatalytic Oxidation of Benzyl Amine by O_2 into N-Benzylidenebenzylamine on TiO_2 Using Visible Light. *Curr. Org. Chem.* **2013**, *17*, 2374–2381. [CrossRef]

 *catalysts*

Review

Titanium Dioxide: From Engineering to Applications

Xiaolan Kang [†], Sihang Liu [†], Zideng Dai, Yunping He, Xuezhi Song and Zhenquan Tan *

School of Petroleum and Chemical Engineering, Dalian University of Technology, No. 2 Dagong Road,
New District of Liaodong Bay, Panjin, Liaoning 124221, China; kxl@mail.dlut.edu.cn (X.K.);
wdlsd@mail.dlut.edu.cn (S.L.); xiaodai@mail.dlut.edu.cn (Z.D.); yphe_04@mail.dlut.edu.cn (Y.H.);
songxz@dlut.edu.cn (X.S.)
* Correspondence: tanzq@dlut.edu.cn; Tel.: +86-427-263-1808
† These authors contributed equally to this work.

Received: 17 January 2019; Accepted: 10 February 2019; Published: 19 February 2019

Abstract: Titanium dioxide (TiO_2) nanomaterials have garnered extensive scientific interest since 1972 and have been widely used in many areas, such as sustainable energy generation and the removal of environmental pollutants. Although TiO_2 possesses the desired performance in utilizing ultraviolet light, its overall solar activity is still very limited because of a wide bandgap (3.0–3.2 eV) that cannot make use of visible light or light of longer wavelength. This phenomenon is a deficiency for TiO_2 with respect to its potential application in visible light photocatalysis and photoelectrochemical devices, as well as photovoltaics and sensors. The high overpotential, sluggish migration, and rapid recombination of photogenerated electron/hole pairs are crucial factors that restrict further application of TiO_2. Recently, a broad range of research efforts has been devoted to enhancing the optical and electrical properties of TiO_2, resulting in improved photocatalytic activity. This review mainly outlines state-of-the-art modification strategies in optimizing the photocatalytic performance of TiO_2, including the introduction of intrinsic defects and foreign species into the TiO_2 lattice, morphology and crystal facet control, and the development of unique mesocrystal structures. The band structures, electronic properties, and chemical features of the modified TiO_2 nanomaterials are clarified in detail along with details regarding their photocatalytic performance and various applications.

Keywords: TiO_2; energy band engineering; morphology modification; mesocrystals; applications

---

1. Introduction

Over the past several decades, the increasing severe energy shortages and environmental pollution have caused great concern worldwide. To achieve sustainable development of society, there is an urgent need to explore environmentally friendly technologies applicable to pollutant recovery and clean energy supplies. In the long-term, solar energy is an inexhaustible source of renewable energy; therefore, developing technologies and materials to enhance solar energy utilization is central to both energy security and environmental stewardship. In 1972, Fujishima and Honda first published a study for producing hydrogen on titanium dioxide (TiO_2) photoelectrodes under ultraviolet light illumination, which garnered worldwide attention [1,2]. From then on, semiconductor photocatalysis has been considered one of the most promising pathways to address both hydrogen production and pollution abatement. Photocatalysis can be widely used anywhere in the world, providing natural solar light or artificial indoor illumination is available [3].

Semiconductor materials are often used as photocatalysts [4]. According to band energy theory, the discontinuous band structure of semiconductors is composed of low energy valence bands filled with electrons, high-energy conduction bands, and band gaps. When the energy of the incident photons equals or exceeds the bandgap, the photoexcitation of electron–hole pairs and the consequential photocatalytic redox reaction take place [5]. The photocatalytic process mainly involves the steps of generation, separation, recombination, and surface capture of photogenerated electrons and hole pairs. Photochemical reactions occur on the surface of a solid catalyst, which includes two half-reaction oxidation reactions of photogenerated holes and reduction reactions of photogenerated electrons [6]. The specific process that occurs in semiconductors is described in Figure 1. During this process, a large proportion of charge carriers (e^-/h^+ pairs) recombine quickly at the surface and interior of the bulk material, leading to the dissipation of absorbed energy in the form of light (photon generation) or heat (lattice vibration). Therefore, these charge carriers cannot participate in the subsequent photocatalytic reactions, which is detrimental to the whole process [7].

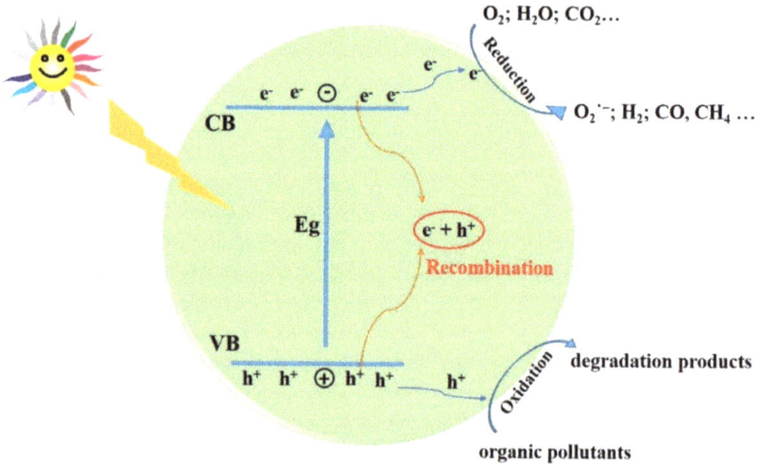

Figure 1. Photocatalytic process in semiconductor.

The electrons and holes that successfully migrate to the surface of the semiconductor without recombining can be involved in the reduction and oxidation reactions, respectively, which are the bases for photodegradation of organic pollutants and photocatalytic water splitting to produce H_2 [8]. As excellent oxidizers, the photogenerated holes can mineralize organic pollutants directly. In addition, the holes can also form hydroxyl radicals ($\bullet OH$) with strong oxidizing properties. Photoexcited electrons, on the other hand, can produce superoxide radicals ($O_2\bullet^-$) and $\bullet OH$. These free radicals and e^-/h^+ pairs are highly reactive and can induce a series of redox reactions. In addition, with respect to water splitting, photogenerated electrons can be captured by H+ in water to generate hydrogen, while holes will oxidize H_2O to form O_2 [9–11].

In general, to increase the activity of photocatalysts and utilize visible light more effectively, several requirements need to be satisfied. First, the light absorption process determines the amount of excited charges, which means that more charge carriers are likely to be accumulated on the surface if more light can be absorbed by the photocatalyst. Additionally, considering that ultraviolet (UV) light occupies less than 4% of sunlight's emission spectrum, while visible light accounts for approximately 40%, a smaller bandgap is necessary for a semiconductor to absorb solar energy across a broad range of spectra. Therefore, improving the optical absorption properties has become a common purpose for photocatalyst design to enhance their overall activity [12]. In addition, the position of conduction bands (CBs) and valence bands (VBs) is critical, which are responsible for the production of active species, such as $\bullet OH$, $HO_2\bullet$, H_2O_2, and $O_2\bullet^-$. Furthermore, the photogenerated electrons and holes should be transported and separated efficiently in the photocatalyst because the fast recombination of charge carriers will otherwise result in low reactivity. Finally, the as-prepared photocatalytic materials and their modification processes should be environmentally friendly and economical [13].

Since 1972, TiO$_2$ has been intensively investigated due to its thermal and chemical stability, superhydrophilicity, low toxicity, and natural geologic abundance. Compared with other semiconductor materials, TiO$_2$ is of ubiquitous interest across many research fields and for many applications [14], such as photodegradation of pollutants and hazardous materials, photolysis (splitting) of water to yield H$_2$, artificial photosynthesis, etc. Nevertheless, the poor visible light absorption and fast electron–hole recombination, as well as the sluggish transfer kinetics of the charge carriers to the surrounding media, considerably limit the photocatalytic activities of TiO$_2$. Hence, during the past few decades, much effort has been devoted to overcoming these problems by, for example, reducing e$^-$/h$^+$ pair recombination and improving the optical absorption properties by energy band regulation, morphology control, and the construction of heterogeneous junctions [15].

In this review, we mainly focus on the regulation of the electronic structure and modification of the micromorphology of TiO$_2$ nanomaterials to achieve property enhancements that could be applicable to a variety of potential applications.

2. Energy Band Engineering of TiO$_2$

The absorption of incident light and redox potential of TiO$_2$ mainly depend on its energy band configuration [16]. To utilize solar energy more effectively, it is necessary to explore and develop longwave-light-sensitive TiO$_2$ photocatalysts with excellent performance on the basis of energy band engineering [17]. A better understanding of the electronic structure of TiO$_2$ is important for band gap modification. The molecular orbital bonding energy diagram in Figure 2 clearly shows the fundamental features of anatase TiO$_2$ [18]. The chemical bonding of anatase TiO$_2$ can be deconstructed into Ti, e.g., Ti t_{2g} (d_{yz}, d_{xz}, and d_{xy}), O p_σ (in the Ti$_3$O cluster plane), and O p_π (out of the Ti$_3$O cluster plane). The upper valence bands include three main regions: the σ bonding, which is located at the bottom, is the most stable bond type, and arises from the hybridization of Ti, e.g., O p_σ; the hybridization of the O p_π and Ti d_{yz} (or d_{xz}) orbitals constitutes the middle energy region of π bonding; and the higher energy region in the top of the valence bands, which is dominated by the O p_π orbitals. The conduction band is composed of Ti 3d and 4s, and the bottom of the conduction bands is composed of the isolated Ti d_{xy} orbitals [19,20]. For the purpose of narrowing the bandgap of TiO$_2$, three basic approaches of adjusting the VBs or CBs or the continuous modification of the VBs and CBs of the anatase are shown in Figure 3.

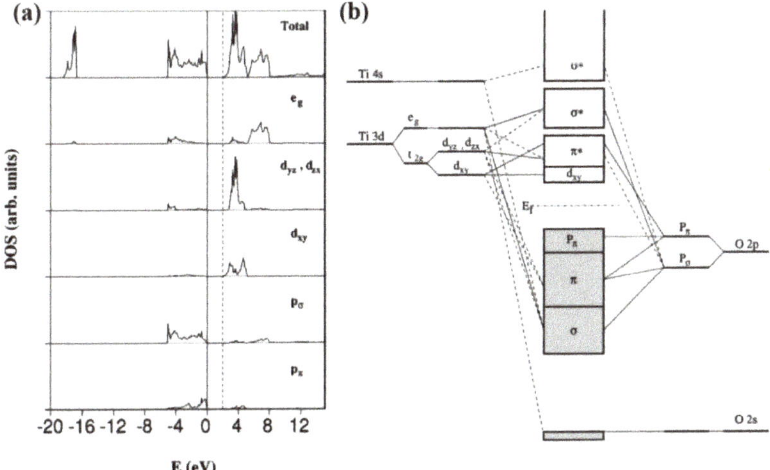

Figure 2. (a) Total and projected densities of states (DOS) of the anatase TiO$_2$ structure and (b) molecular orbital bonding structure for anatase TiO$_2$ [18]. Copyright 2004 The American Physical Society.

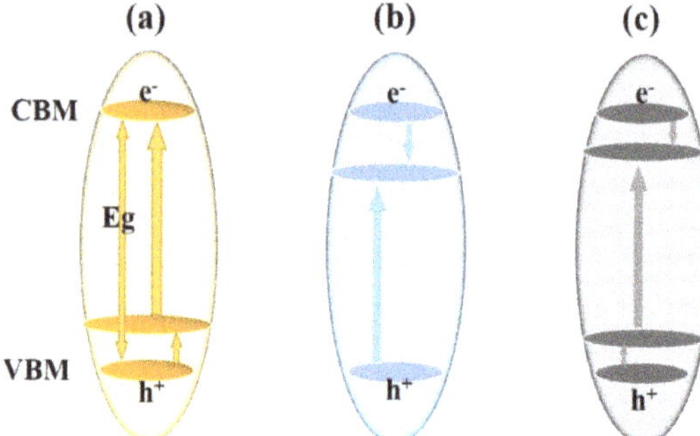

Figure 3. Three schemes of the band gap modifications of TiO_2 match the solar spectrum: (**a**) a higher shift in valence band maximum (VBM); (**b**) a lower shift in conduction band minimum (CBM); and (**c**) continuous modification of both VBM and CBM.

2.1. Doping of TiO_2

To extend the visible light response of TiO_2 and improve its photocatalytic activities, various modification strategies, such as dye sensitization, impurity or intrinsic doping or semiconductor coupling, have been developed [21–23]. Among them, introducing impurity ions into the TiO_2 crystal lattice to substitute the host anions and/or cations has earned much attention in the past decade.

By means of physical or chemical methods, researchers have been able to introduce a variety of ions into the TiO_2 matrix, where they change the band structure of TiO_2 by inducing impurity states within the bandgap [2], as shown in Figure 4. In general, ion doping contributes to the improved activities of TiO_2 in three ways: (1) by narrowing the bandgap and promoting the adsorption of the main region of the solar spectrum, such as doping with N, S, C, B, etc. [24,25]; (2) by improving the conductivity of TiO_2 and the mobility of charge carriers, the increased charge traps can reduce bulk recombination and separate photogenerated electrons and holes more efficiently (e.g., Zn, Fe, and Y) [26]; and (3) by altering the conduction band position of TiO_2 with certain metal ion dopants, such as Zr^{4+}, Nb^{5+}, and W^{6+}, which further affects the carrier transfer properties [27].

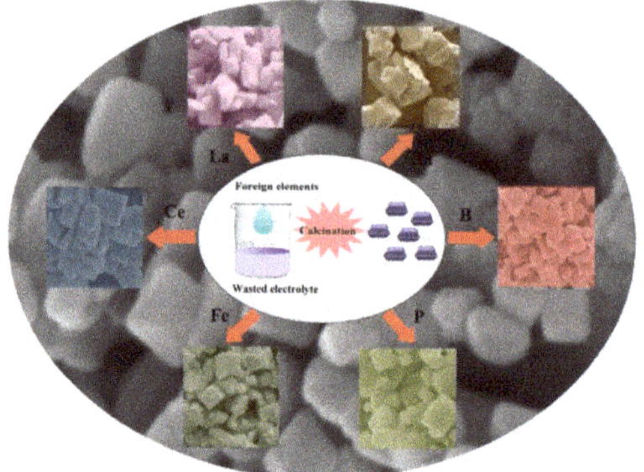

Figure 4. TiO_2 nanoparticles with different doping elements [2]. Copyright 2014 American Chemical Society.

TiO_2 doping can be doped with a variety of metal ions, including transition metal and rare earth metal ions. For transition metal dopants, such as Fe, Mn, V, Cu, and Cr, both delocalized and localized impurity states will be created within the band gap of TiO_2 along the crystal field splitting of metal 3d orbitals [28–30]. Mizushima et al. determined impurity levels of 1.9 to 3.0 eV below CBM by doping V, Cr, Mn, and Fe based on a large number of experimental results, and they suggested that cation vacancies may lead to these impurity states [31]. An early work by Borgarello et al. in 1982 reported that Cr^{3+}-doped TiO_2 nanoparticles (investigated for properties of photocatalytic hydrogen evolution) exhibit excellent absorption of visible light in the range of 400 to 550 nm. They believed that the 3d electrons of Cr^{3+} were excited into the conduction band of TiO_2, thus inducing a visible light response [32]. Doping TiO_2 with certain earth rare metal ions represents another promising method to prolong the recombination time of charge carriers and improve their separation efficiency. The 4f electrons in most rare earth elements can give rise to the formation of a multielectron configuration, which acts as a shallow trap for photogenerated electrons and holes [33]. Furthermore, the use of rare earth metal ion dopants in TiO_2 tends to facilitate the utilization of solar light from ultraviolet to infrared light regions. Li et al. prepared a series of Ce-doped TiO_2 nanoparticles by the sol–gel method. The characterization results showed that Ce ions entered the TiO_2 matrix at Ti sites, leading to the formation of impurity states, as shown in Figure 5. In addition, enhanced separation of the photogenerated charge carriers was also realized due to the coexistence of Ce^{3+} and Ce^{4+} dopant ions [34].

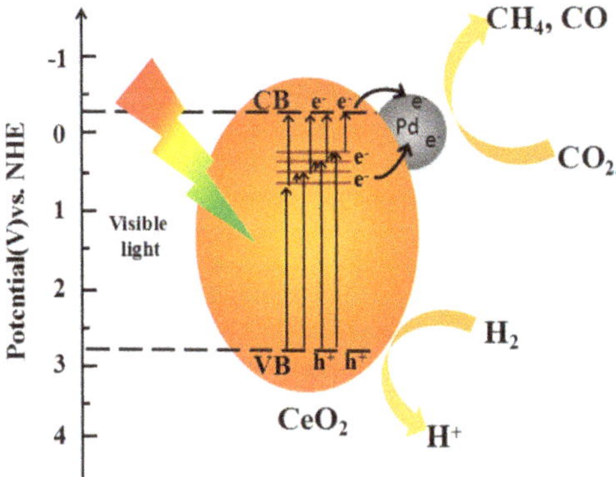

Figure 5. Band energy structure and charge transfer [34]. Copyright 2017 American Chemical Society.

Anandan et al. studied the photodegradation of monocrotophos under visible light irradiation with La-doped TiO_2. They associated rapid mineralization with the enhanced separation of electrons and holes by doping La^{3+} into the TiO_2 matrix, which subsequently generated a large number of •OH radicals along with the trapping of excess holes at the surface [35]. In contrast, based on the density functional theory calculation method, Sun et al. worked extensively on the changes of the electronic structure and the photocatalytic activity of TiO_2 after introducing substitutional La dopants. Their calculations demonstrate that the enhanced visible light absorption of La–TiO_2 mainly arises from adsorbed La on the TiO_2 surface rather than from substitutional La doping [36]. Notably, not all kinds of dopants give rise to positive consequences. Chio et al. systematically studied 21 kinds of metal ion-doped TiO_2 materials and their application with respect to various photocatalytic reactions [37]. The results associated with model reactions for the photocatalytic reduction of carbon tetrachloride and the photodegradation of chloroform indicated that only the doping of certain ions, such as Fe^{3+}, Ru^{3+}, Re^{5+}, V^{4+}, and Mo^{5+}, increased reactivity. In addition, the study demonstrated that optimizing the content and placement of the dopant ions content play a positive role in affecting photocatalytic activity. Despite the robust photoactivity of certain metal ion-doped TiO_2 catalysts, some inevitable problems remain and need to be considered. The metal-doped nanomaterials have been shown to suffer from unstable optical properties and thermal instability, in addition to the need to use expensive ion implantation equipment to produce these enhanced materials [38]. Furthermore, the localized d-electron state formed in the band gap of TiO_2 may become the recombination center of photogenerated electron–hole pairs, thereby leading to a decline in the photocatalytic activity.

Recently, the non-metal doping of nitrogen (N), sulfur (S), carbon (C), fluorine (F), iodine (I), and phosphorus (P) has been extensively studied due to their relatively high photostability and photoelectric properties [39]. However, in comparison to metal-doped TiO_2, the role of the non-metal dopants as recombination centers of charge carriers might be minimized. By replacing the oxygen atoms in the TiO_2 lattice, the non-metal elements can significantly narrow the bandgap and thereby improve the visible light response of TiO_2. In addition, impurity states can be formed near the valence band edge alone with non-metal doping, as displayed in Figure 6. Instead of acting as recombination centers, these occupied levels can be regarded as shallow traps that effectively separate photogenerated electron–hole pairs [40].

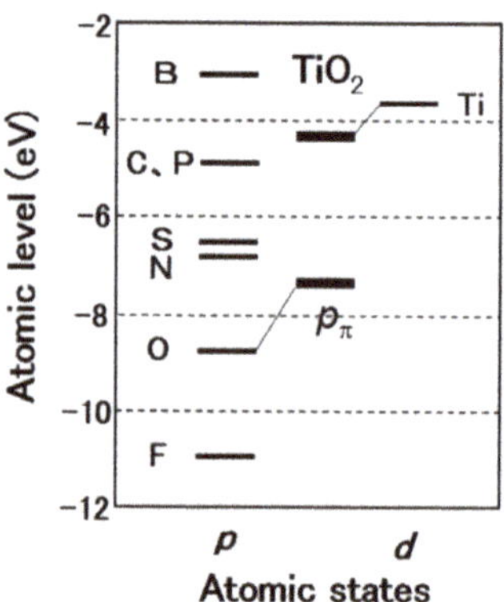

Figure 6. Comparison of atomic p levels among anions. The band gap of TiO_2 is formed between the O 2pπ and Ti 3d states [39]. Copyright 2014 American Chemical Society.

In 2001, Asahi et al. first published research on N-doped TiO_2 nanomaterials, which initiated a wave of studies related to non-metal-doped photocatalysts [41]. In a similar work, Zhao et al. reported highly active N-doped TiO_2 nanotubes for CO_2 reduction. Despite the tubular structure with a large surface area providing more surface active sites, the N dopants contributed more to the improved photocatalytic activity. It was found that a redshift of the light absorption and a color center were achieved with N-doped TiO_2 nanotubes because N atoms can substitute for the lattice O atoms of TiO_2, thereby reducing its bandgap and resulting in a ~4 times higher visible light photocatalytic CO_2 reduction activity in comparison to pure TiO_2 nanotubes [42]. Irie et al. prepared C-doped TiO_2 nanoparticles by oxidizing TiC powder, and the efficiency of decomposing gaseous isopropanol under visible light was significantly improved [43]. S-doped anatase TiO_2 with a high surface area was obtained by Li et al. They treated pure TiO_2 using a supercritical strategy and used the materials for methylene blue degradation under visible light irradiation. S atoms with large diameters are difficult to dope into the TiO_2 lattice, but X-ray photoelectron spectroscopy (XPS) detected the existence of S–Ti–O bonds, which introduced lattice defects, acting as shallow traps for electrons and reducing carrier recombination [44]. Li et al. mixed HIO_3 with tetrabutyl titanate and hydrolyzed the samples directly to obtain I-doped TiO_2, which significantly boosted its visible light performance [45].

Although various non-metal ions are used for doping modification of TiO_2, N doping is still one of the most widely used methods to modify the electronic structure and to extend light absorption to the visible range [46]. However, researchers have not yet come to a complete agreement regarding the mechanisms associated with the N doping enhancements. In the literature, it is not difficult to find studies stating that it is not only the dopant concentration but also the dopant location in the TiO_2 lattice (surface or bulk, substitutional, and interstitial) that ultimately determines the photocatalytic properties [17,47]. In the case of N-doped TiO_2 nanomaterials, some researchers believe that only the substitution of O^{2-} by N^{3-} with high dopant concentrations can elevate the valence band edge, bringing about the desired band gap narrowing [48,49]. However, others suggest that the doping of N will induce oxygen vacancies in TiO_2 and that the enhanced visible light adsorption is associated with the local state induced in the band gap, rather than the generally believed theory that the introduction of N into the TiO_2 lattice can reduce its band gap, as shown in Figure 7 [50].

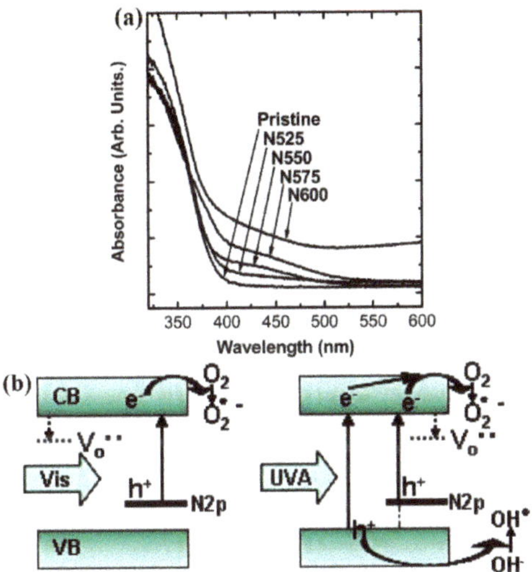

Figure 7. (a) Diffuse reflectance spectra of the anatase TiO_2 nanobelts before and after heat treatment in ammonia gas flow at different temperatures and (b) the band structure of N-doped-TiO_2 under visible and UV light irradiation [50]. Copyright © 2009 American Chemical Society.

As another widely studied non-metal-doped TiO_2, F-doped TiO_2 also shows promising potential for photocatalytic applications. Zhang et al. obtained F-doped TiO_2 mesocrystals through the topological transformation of $TiOF_2$ precursors. An in situ characterization technique was adopted to detect the doping process. The results showed that the doping of F was accompanied by the formation of oxygen defects, which ensured a higher visible light response [51]. Park et al. added sodium fluoride to aqueous TiO_2 suspensions to obtain surface fluorinated TiO_2, and a series of characterizations showed that neither an improvement in crystallinity nor a redshift of the band edge was achieved, but the photocatalytic oxidation of phenol and Acid Orange was considerably enhanced. They attributed such photocatalytic improvement to fluorine surface modification, which enhances free •OH radical-mediated oxidation pathways [19]. Similar to the doping of N, the reason for the observed high performance upon F doping is still undetermined. Some studies suggest that instead of entering the TiO_2 lattice, fluorine ions adsorbed on the surface of TiO_2 can increase the wettability and surface acidity, which is beneficial to the adsorptivity and e^-/h^+ separation of the oxide [20]. Other researchers hold the opinion that a tail state in the band gap of TiO_2 is formed by F doping, which favors the more efficient utilization of incident light. Recently, an increasing number of studies proposed that a charge compensation effect induced by F doping brings about the formation of a certain amount of oxygen vacancies and Ti^{3+} in TiO_2, resulting in the enhanced absorption of visible light [52,53]. Although the principle of F doping is not very clear, the proper doping level of F can effectively improve the activity of TiO_2.

2.2. Intrinsic Defect Formation

In 2011, a black TiO_2 with a narrowed bandgap (approximately 1.5 eV) and fabricated by hydrogenation reduction was reported to achieve absorption of full spectrum sunlight and improved photocatalytic activity [54]. Unsurprisingly, this discovery has aroused worldwide scientific interest and paved the way towards intrinsic defect modification. Creating intrinsic defects in the TiO_2 lattice is a kind of self-structural modification that includes surface disorder layers, Ti^{3+}/oxygen vacancy self-doping, formation of surface Ti–OH, and incorporation of doped-Consequentially, considerable changes in surface properties and electronic and crystal structures are often achieved in this process [55–57]. Furthermore, studies in terms of defect engineered TiO_2 have confirmed that these intrinsic defects are emerging as a promising attribute for improving the separation of electrons and holes, outperforming, in some cases, other kinds of modified TiO_2 nanomaterials [58].

Since the study by Chen et al., various methods have been developed to induce defects in TiO_2, including direct reduction of TiO_2; that is, the currently reported H_2, Al, Na, Mg, $NaBH_4$, hydrides, imidazoles, etc. can effectively transfer modify pure TiO_2 nanomaterials into their defect engineered counterparts under certain conditions [59,60]. In addition, electrochemical reduction and high-energy particle bombardment (such as photon beam and H_2 plasma or electron beam) are widely used to induce TiO_2 defects. Partial oxidation from low-valence-state Ti species such as TiH_2, TiO, $TiCl_3$, TiN, and even Ti foil represents another promising approach, fulfilling the needs for highly active TiO_{2-x} photocatalysts [61]. Liu et al. prepared rice-shaped Ti^{3+} self-doped TiO_{2-x} nanoparticles through mild hydrothermal treatment of TiH_2 in H_2O_2 aqueous solution, and proposed a unique "surface oxide-interface diffusion–redox mechanism" (as shown in Figure 8) to explain the formation process of TiO_{2-x} [62]. The defect types and their formation mechanism in TiO_{2-x} are closely related to the preparation methods. Generally, the Ti–H bond is present only in hydrogen-reduced TiO_{2-x}, while the surface disorder layer causes severe damage to the TiO_2 structure. Thus, relatively strong reduction conditions are required, such as high temperature/pressure hydrogen reduction, aluminothermic reduction, hydrogen plasma treatment, etc. Surface Ti–OH, Ti^{3+}, and oxygen vacancies commonly exist in most defective TiO_2 nanostructures [63].

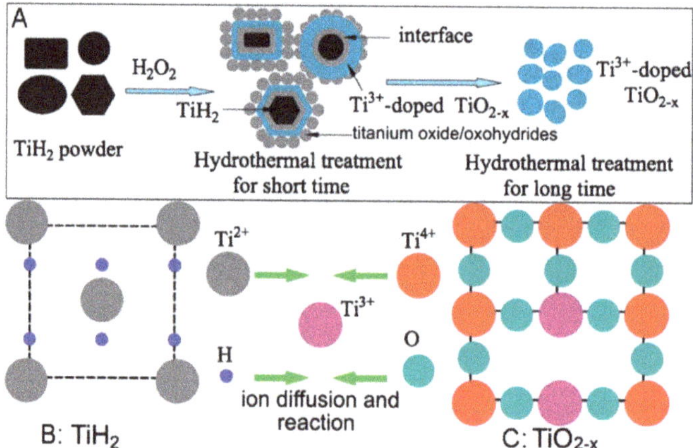

Figure 8. (**A**) Schematic of the formation mechanisms for the rice-shaped Ti^{3+} self-doped TiO_{2-x} nanoparticles. (**B,C**) The interface diffusion–redox diagram. The green arrows indicate ion diffusion [62]. Copyrighted 2014 The Royal Society of Chemistry.

The dominant mechanism involved in improving photocatalytic performance by inducing intrinsic defects into TiO_2 can be explained, both experimentally and theoretically, to be the regulation of the band structure of TiO_2 and boosted charge separation and transport. For black TiO_2, band tail states and shallow dopant states can be formed to reduce its band gap and further increase its optical absorption properties. Chen et al. observed a disordered surface layer in black TiO_2 nanocrystals after a hydrogenation treatment, as shown in Figure 9. From the high-resolution transmission electron microscopy (HRTEM) spectra, it can be readily observed that the straight lattice fringes are bent at the edge of the particles, and the lattice spacing is no longer uniform, indicating that the hydrotreated black TiO_2 nanoparticles possess a "crystal-disordered" core–shell structure. Such a disordered layer is believed to facilitate the introduction of the tail state at the top of the valence band and the bottom of the conduction band, consequently yielding a redshift of the light absorption [54]. Moreover, because the disorder layer exhibits a set of properties that are distinct from those of their crystalline counterparts, rapid charge separation could be realized when the amorphous layer closely contacts crystalline TiO_2. The lattice distortions tend to blueshift the VBM while having less impact on CBM. Therefore, the photogenerated holes accumulate in the thin disordered shell and participate in the photocatalytic reactions immediately; electrons are widely spread in both the shell and core regions. This result highlights the strong synergistic effect on charge transfer between the crystalline and disordered parts [64].

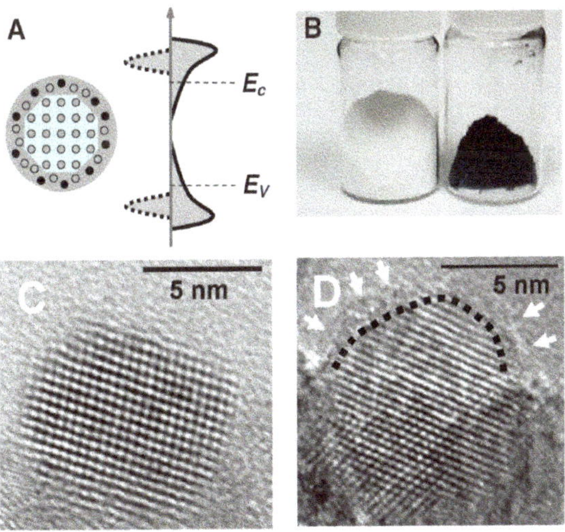

Figure 9. (**A**) Schematic illustration of the structure and electronic DOS of a semiconductor in the form of a disorder-engineered nanocrystal with dopant incorporation. (**B**) A photo comparing unmodified white and disorder-engineered black TiO_2 nanocrystals. (**C,D**) HRTEM images of TiO_2 nanocrystals before and after hydrogenation, respectively [54]. Copyright 2011 American Association for the Advancement of Science.

For Ti^{3+}/oxygen vacancy incorporation and H-doping in reduced TiO_{2-x}, the hybridization of Ti-3d, O-2p and H-1s orbitals results in the mid-gap states formation below the CBM and the Fermi level's upshift [65,66]. The extra electrons in either Ti^{3+} or oxygen vacancies are inclined to occupy the empty states of Ti ions, forming new Ti 3d bands below the CBM. With a further increase in defect concentration, the 3d band shifts deeper and finally results in multiple bands in the CBM. Moreover, the existence of multiple mid-gap states as well as the associated derivate (surface Ti–OH) can also function as extra carrier trap sites or carrier scavengers to prolong the lifetime of electrons and holes [67]. The high concentration of electron donors will greatly improve the conductivity of materials and promote the transfer of carriers [68]. Wang et al. treated pure white TiO_2 with hydrogen plasma to fabricate H-doped black TiO_2 for photodegradation of methyl orange under visible light irradiation. The as-prepared samples showed a degradation rate 2.5 times that of the white counterpart [69]. Sinhamahapatra et al. reported a novel controlled magnesiothermic reduction to synthesize reduced TiO_{2-x} under 5% H_2/Ar atmosphere [70]. During this process, the band position and band gap, surface defects and oxygen vacancies can be well regulated to maximize the optical adsorption in the visible and infrared regions and minimize the charge recombination centers. As shown in Figure 10, a new controlled magnesium thermal reduction method to synthesize and reduce black TiO_2 under 5% H_2/Ar atmosphere. The material has the best band gap and band position, oxygen vacancy, surface defect, and charge recombination center, and the optical absorption in visible and infrared regions is improved obviously. These synergistic effects enable the defective TiO_{2-x} with Pt as a co-catalyst to produce H_2 at a rate of 43 mmol h^{-1} g^{-1} under the full solar wavelength light illumination, superior to other reported photocatalysts for hydrogen production.

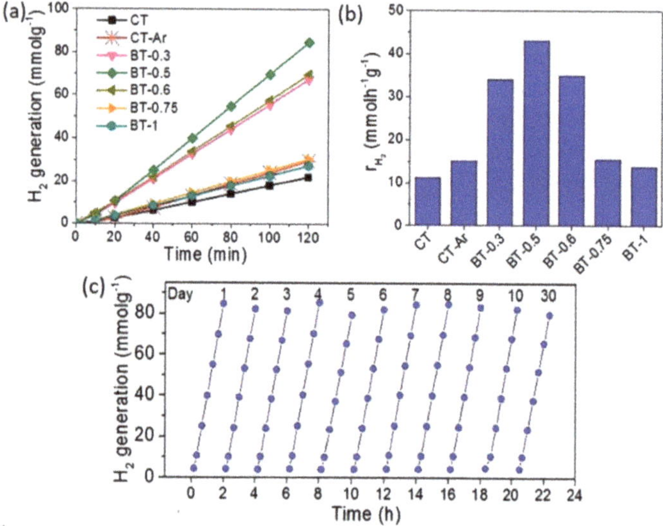

Figure 10. (a) H_2 generation profile, (b) rate (rH_2) of hydrogen generation for different samples, and (c) the stability study of the sample BT-0.5 under the full solar wavelength range of light [70]. Copyright 2015 The Royal Society of Chemistry.

To date, numerous strategies, either common or uncommon, have been developed to introduce various kinds of dopants or defects into the TiO_2 matrix. However, considering its highly stable nature, most methods are rigorous and energy-consuming, and are contrary to the sustainable and environmentally friendly development criteria. Therefore, an increasing number of studies are dedicated to seek convenient, economical, energy efficient, and environmentally friendly methods for the structural modification of TiO_2 [71]. In our recent studies, we developed a facile photoreduction strategy to induce intrinsic defects into anatase TiO_2 to modulate its band structure, thereby extending the absorption of incident light to the visible region. As shown in Figure 11, the band gap was narrowed to 2.7 eV, and the color changed to earth yellow after the photoreduction treatment. NH_4TiOF_3 mesocrystals were adopted as precursors, which can release fluorine and nitrogen ions during the topological transformation process. Thus, non-metal ion doping (i.e., F and N ions) was also achieved simultaneously, further improving the transport and separation of photogenerated charge carriers. The as-prepared $NF-TiO_{2-x}$ exhibited excellent photocatalytic degradation and photoelectrochemical efficiency under visible light irradiation compared to pristine TiO_2 [72,73].

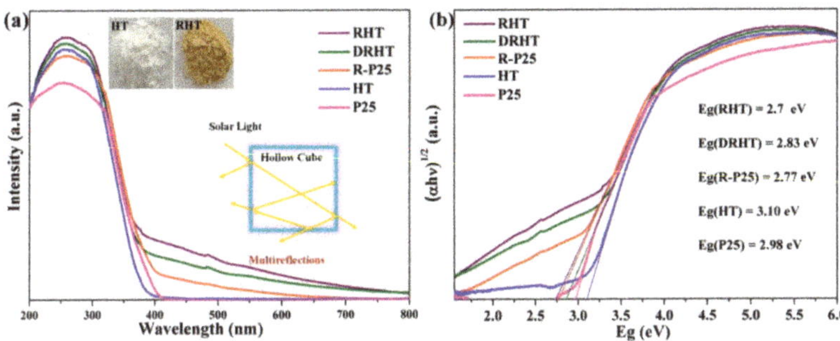

Figure 11. (a) UV–Vis diffuse reflectance spectra and (b) Tauc plot for band gap determination [73]. Copyright 2018 Springer Nature Publishing AG.

3. Morphology Modification

It is well known that the photocatalytic performance of semiconductors is closely related to their structural and morphological characteristics at the nanoscale, including their size, dimensionality, pore structure and volume, specific surface area, exposed surface facets, and crystalline phase content [74]. During the past few decades, numerous promising structure engineering strategies have been developed to fabricate highly active photocatalysts with the desired morphology and structure. Among them, particular emphasis has been placed on controlling and optimizing the structural dimensionality of a given semiconductor to improve its photocatalytic efficiency.

Zero-dimensional TiO_2 nanospheres are the most widely studied TiO_2-based materials because of their high specific surface area and attractive pore structures [75–77]. Figure 12 shows a classic ripening approach to synthesize hollow nanospheres [75]. As photocatalytic reactions take place on the surface of the photocatalyst, TiO_2 nanoparticles with smaller sizes are inclined to provide more reactive sites, resulting in better photocatalytic performance. Moreover, due to the quantum size effect, the photogenerated electrons and holes in the bulk regions are able to migrate to the surface of TiO_2 nanoparticles via shorter distances, thereby considerably reducing the carrier quench rate [78]. TiO_2 nanospheres are also good candidates as light captors, and their structural features enable as much light as possible to access the interior, resulting in amazing light harvesting capabilities. However, it should be mentioned that the diffusion length of photogenerated electrons and holes must be longer than the particle size to avoid the recombination of the dominant carriers on the surface of the photocatalyst, which is very important for achieving efficient charge carrier dynamics [79].

One-dimensional (1D) nanostructures, including nanotube (NT), nanorod (NR), nanobelt (NB), and nanowire (NW), have become a popular research topic in recent years. They have been extensively studied because of their distinct optical, electronic and chemical properties. Despite some similar features with nanoparticles, such as quantum confinement effects and large surface area, 1D nanomaterials possess many unique properties, which are hard for other categories of structured materials to achieve. For example, 1D nanostructures restrict the migration of electrons and protons by allowing the lateral confinement of electrons/protons and guide their transport in the axial direction [80,81]. Furthermore, excellent flexibility and mechanical properties enable them to be easily used and recycled. In this regard, 1D TiO_2-ordered nanostructures are promising not only for constructing highly active photocatalytic systems but also for building blocks for various (photo)electrochemical devices, such as batteries, fuel cells, solar cells, and photoelectrochemical cells. To further optimize the photocatalytic reactivity of 1D TiO_2 nanomaterials, one can precisely regulate the aspect ratio (the ratio of length to diameter) or modify these 1D nanostructures with novel strategies to accelerate electron transport and separation processes, as well as to enhance the capture of incident light; TiO_2 nanotubes are examples of these materials [82]. Through the electrochemical anodization process, it is possible to precisely control the tube crystal structure (anatase, rutile, or amorphous) and tube geometry (diameter and length), as shown in Figure 13a, or direct the tube arrangements to obtain a defined tube-to-tube interspace (Figure 13b). For the sake of extending the scope of application, constructing flow through membranes with TiO$_2$ nanotubes is a good choice (Figure 13c). Other modifications for minimizing charge carrier annihilation and boosting light harvesting are illustrated in Figure 13d–i, ranging from self-decoration to surface alterations to energy band engineering.

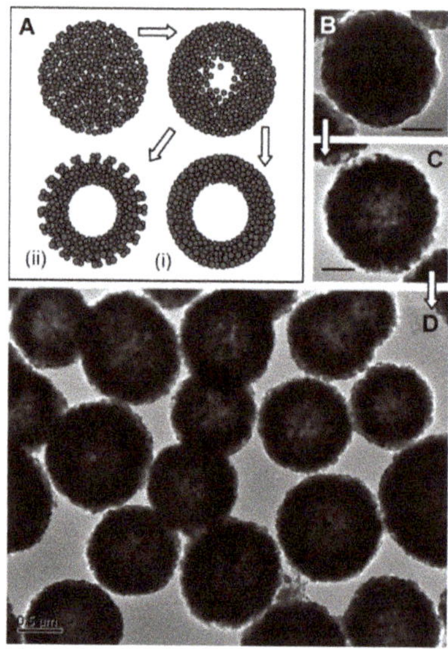

Figure 12. (**A**) Schematic illustration (cross-sectional views) of the ripening process and two types (i and ii) of hollow structures. Evolution (TEM images) of TiO$_2$ nanospheres synthesized with 30 mL of TiF$_4$ (1.33 mM) at 180 °C with different reaction times: (**B**) 2 h (scale bar = 200 nm), (**C**) 20 h (scale bar = 200 nm), and (**D**) 50 h (scale bar = 500 nm) [75]. Copyright 2004 American Chemical Society.

TiO$_2$ nanosheets, nanoflakes, and thin films consist of titania-based two-dimensional nanomaterials, which have flat surfaces and high aspect ratios. The lateral size of some nanomaterials is controllable, ranging from the sub-micrometer or even nanometer level to several tens of micrometers with thicknesses of 1–10 nm. Such structures provide TiO$_2$ nanomaterials with several unique characteristics, such as excellent adhesion to substrates, low turbidity and high smoothness [83]. Furthermore, when exposed to UV light irradiation, TiO$_2$ 2D nanomaterials exhibit superhydrophilicity, which leads to a variety of potential applications, such as self-cleaning coatings and electrodes in photoelectronic devices [84]. Notably, considering that photocatalytic reactions always occur on the surface of catalysts, the exposed crystal facets are of great importance in determining the photocatalytic performance. Accordingly, developing TiO$_2$ crystals with different active facets is highly desirable in many applications. In general, TiO$_2$ nanocrystals have three basic low-index exposed facets—{101}, {001}, and {010}—with surface energy relationships of {001}, 0.90 J m^{-2} > {100}, 0.53 J m^{-2} > {101}, 0.44 J m^{-2} [85,86]. Therefore, as the most thermodynamically stable facets, the {001} crystal facet is dominant among most anatase TiO$_2$ nanomaterials, reducing the overall surface energy of the material. In 2008, Yang et al. first reported TiO$_2$ single crystals with 47% highly active {001} facets exposed to HF as capping agents [87]. This work has attracted considerable global attention. Since then, TiO$_2$ with various ratios of exposed {001} facets have been successfully fabricated [88]. Meanwhile, other active planes, such as {010}, {111}, and {110}, have also been reported and widely used in water splitting, solar cells, artificial light synthesis and other fields, as shown in Figure 14 [89]. Zheng et al. obtained {001} facet-oriented anatase by facile heat treatment of a tetrabutyl titanate, absolute ethanol, and HF mixture.

Such a material with 85% {001} facets exhibited much higher photocatalytic activity in comparison to commercial P25 materials [90].

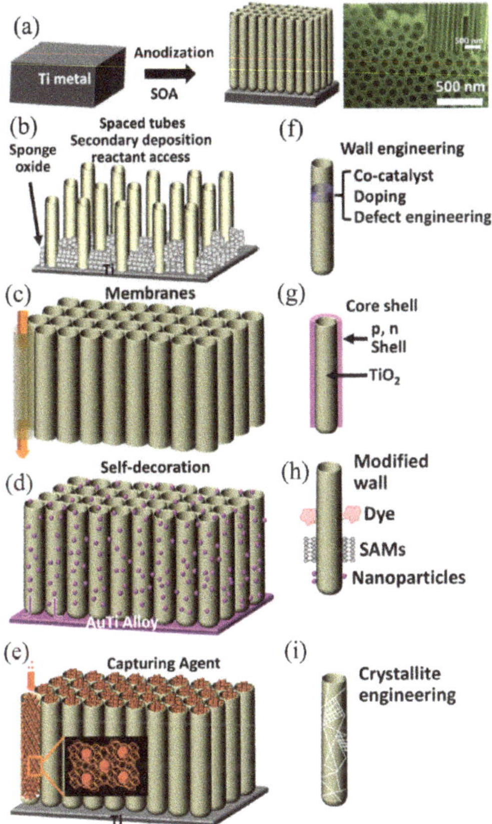

Figure 13. Schematic drawing of (**a**,**b**) formation and (**c**−**i**) modification of anodic nanotube arrays (as discussed in the text) [82]. Copyright 2017 American Chemical Society.

During the process of photocatalytic reactions, oxidation predominantly occurs in the {001} facets, while reduction occurs in the {101} crystal plane of TiO_2 because the {101} facet (with relatively low surface energy) tends to attract more electrons. Electron holes subsequently accumulate in the {001} plane, facilitating the space separation of electron–hole pairs [91]. In addition, Ti atoms of the {001} plane exist mainly in the form of 5-coordination, which can provide more active sites that more readily attract free reactant molecules than the {101} plane. Thus, when a certain proportion of {001} crystal facets are exposed, the photocatalytic activity increases rapidly. Nevertheless, it is not always the case that a higher {001} crystal face exposure ratio results in improved catalytic performance. Studies have reported that the photocatalytic activity is compromised when the proportion of {001} facets exceeds 71% [89]. In addition, faceted TiO_2 photocatalysts suffer from weak visible light utilization due to their large band gap. Hence, the modification of the electronic structure of faceted TiO_2 to fully utilize sunlight and promote the migration and separation of electron/hole pairs is highly desirable. Wang et al. prepared Ti^{3+} self-doped TiO_2 mesoporous nanosheets dominated by {001} facets with supercritical technology. They associated the extended region of incident light absorption with the introduction of Ti^{3+} [91]. Using an ionic liquid as a surface control agent, Biplab et al.

synthesized microporous TiO$_2$ nanocrystals with exposed {001} facets. After depositing Pt on the surface, the hydrogen production rate in visible irradiation was greatly improved [92].

Figure 14. Summary of main shapes and applications (i.e., lithium ion batteries, photocatalytic hydrogen evolution, photodegradation, and solar cells) of anatase, rutile, and brookite TiO$_2$ crystals with their surfaces consisting of different Facets [89]. Copyright 2014 American Chemical Society.

A three-dimensional TiO$_2$ hierarchical structure based on intrinsic shape-dependent properties has been the central focus of many recent studies. Designed and fabricated 3D TiO$_2$ nanomaterials commonly incorporate interconnected structures, hollow structures and hierarchical superstructures constructed from small dimensional building blocks [93]. Most of these novel structures include larger spatial dimensions and more varied morphologies. The high surface-to-volume ratio provides a more efficient diffusion path for reactant molecules, enabling the contaminant molecules to enter the framework of the photocatalyst for efficient purification, separation, and storage. In addition, the unique optical characteristic is of particular interest because many of these architectures have distinctive physicochemical properties favorable for incident light utilization. For example, when light is irradiated onto the surface of the TiO$_2$ hierarchical structure, photons are scattered multiple times, so the probability of the catalyst absorbing photons is increased; this phenomenon is known as the "trapping effect" and is illustrated in Figure 15 [94].

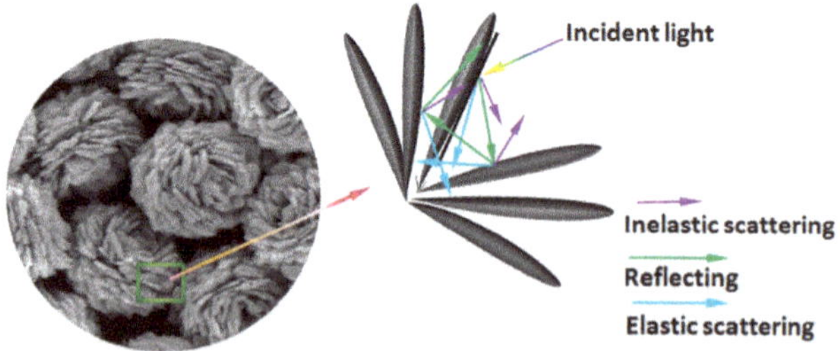

Figure 15. Schematic diagram of the reflecting and scattering effects in hierarchical microspheres [94]. Copyright 2014 The Royal Society of Chemistry.

The hollow structure TiO_2 nanomaterials have attracted considerable attention due to their amazing light harvesting ability, low density, and large specific surface area. The hollow structure, on the one hand, is capable of providing a large amount of space to accommodate more reactant molecules, thereby increasing the effective contact between the catalyst and the reactants. On the other hand, incident light inside the cavity can undergo multiple reflections to capture more light, as shown in Figure 16 [95]. Kondo et al. obtained TiO_2 hollow nanospheres through hydrothermal and calcination processes with polymer polyethylene cationic balls as templates. The as-prepared photocatalyst had more favorable activity than its commercial counterparts with respect to decomposing isopropanol [96]. In the following work, an ultrathin TiO_2 shell-like structure was prepared in a similar manner with a shell thickness of approximately 5 nm. The morphology of the TiO_2 hollow materials prepared by the hard template method is relatively uniform, and the composition and thickness of the shells are adjustable. However, the preparation process is complicated and requires multiple execution steps to be realized. Moreover, the hollow structure may be destroyed when the template is removed. Therefore, alternative strategies, including soft templates and non-template methods, have played an increasingly important role in the development of hollow structure TiO_2 nanomaterials in recent years.

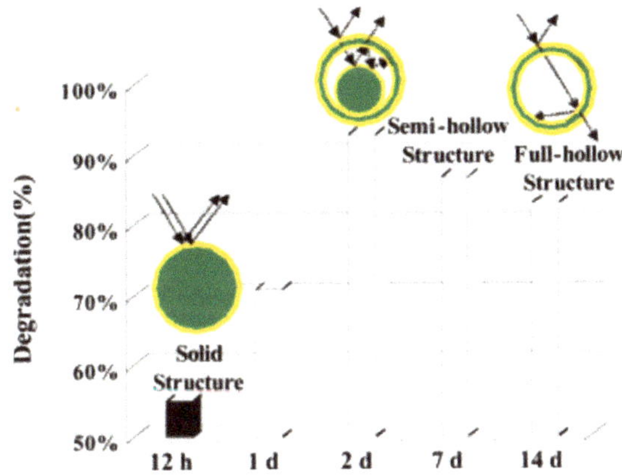

Figure 16. Comparison of photocatalytic activities of titania spheres with solid, sphere-in-sphere, and hollow structures [95]. Copyright 2007 American Chemical Society.

Li et al. prepared hollow TiO_2 nanospheres with high photocatalytic activity by a template-free process. The increased catalytic activity is mainly due to the multiple reflections of incident light inside the TiO_2 sphere, which extends the optical path [97]. Multichannel TiO_2 hollow nanofibers were constructed by Zhao et al. for degrading gaseous acetaldehyde, and the specific surface area of this material increased rapidly as the number of channels increased. They proposed that the multichannel hollow structures induced both an inner trap effect on gaseous molecules and a multiple-reflection effect on incident light, which were the main reasons for the improved photocatalytic activity of TiO_2 hollow fibers [98]. Shang et al. synthesized submicron-sized TiO_2 hollow spheres from a mixture of $TiCl_4$, alcohols, and acetone by a template-free solvothermal method. Control of the sphere size was achieved by adjusting the ratio of ethanol to acetone. Based on a series of characterizations, they suggested a possible formation mechanism for the hollow structure: the tiny anatase phase TiO_2 nanoparticles with poor crystallinity form through a hydrolysis reaction, due to the very high surface energy, and then quickly aggregate to form spheres. The increased water promotes the crystallinity of particles in the spherical shell, while the internal particles dissolve and migrate to the spherical shell, leading to the formation of highly crystalline TiO_2 hollow spheres [99]. An intriguing work carried out by Kang et al. to establish hierarchical anatase TiO_2 nanocubes with hollow structures has been reported recently. Instead of seeking complicated templates or surfactants, they directly converted NH_4TiOF_3 mesocrystals to hollow spiny TiO_2 with a high specific area and photodegradation activity [73].

4. TiO$_2$ Mesocrystals

It is widely accepted that for TiO_2-based photocatalytic materials, large crystallites result in high structural coherence, which benefits the transfer and separation of electron–hole pair, while the availability of plentiful reaction sites is dependent on obtaining large specific surface areas. However, producing a structure that simultaneously satisfies the requirements of large crystallites and high surface area is extremely challenging. Fortunately, the advent of mesocrystals is a promising material that may meet the challenge [100]. Mesocrystals were first proposed by Cölfen and Antonietti in 2005, and since then have received increased attention [101]. Different from the classical single crystals in which the crystal lattice of the entire sample is continuous with no grain boundaries and polycrystals whose units do not have the same orientation, mesocrystals are a new kind of superstructure material that follow a nonclassical crystallization process involving crystallographically ordered assemblies of nanocrystal building blocks. The relevant formation mechanisms of TiO_2 mesocrystals reported thus far mainly include topotactic transformation, mineral bridges, nanoparticle alignment with organic matrices, physical ordering, space constraints, and self-similar growth [100]. Different methods may give rise to different structures and morphologies, but the as-prepared TiO_2 mesocrystals are usually single-crystal-like structures with high porosity, surface area, and crystallinity; they are considered periodically hierarchical structures that are similar to sophisticated biominerals. All of these features pave the way for a wide range of applications, such as catalysis and energy storage and conversion [102].

Fabrication and modification strategies for TiO_2 mesocrystals have developed rapidly in recent years. Due to the similar structure between NH_4TiOF_3 and TiO_2, preparing TiO_2 mesocrystals through topotactic transformation from NH_4TiOF_3 represents an innovative process. As illustrated in Figure 17, the critical parameters in the {001} facets of both NH_4TiOF_3 and TiO_2 are quite similar, with an average lattice mismatch of 0.02%. The position of titanium atoms in the {001} plane of TiO_2 is similar to NH_4TiOF_3, but in NH_4TiOF_3, these are separated by ammonium ions in a lamellar structure. Hence, it is reasonable to use NH_4TiOF_3 as a starting material, transforming it into TiO_2 mesocrystals by thermal decomposition or aqueous hydrolysis with H_3BO_3 [71].

Based on this mechanism, Majima et al. performed extensive studies on tailoring TiO_2 mesocrystals with versatile structures and morphologies, as well as postmodifications to further improve their photocatalytic efficiency. For example, to investigate the anisotropic electron flow in different facets and to maximize their separation during the photocatalytic reaction, Zhang et al. controllably synthesized a specific facet-dominated TiO_2 superstructure with NH_4F as an orientation-directing agent [103]. Under UV light irradiation, mesocrystals with different facet ratios showed different reactivity orders in the photooxidation of 4-chlorophenol, i.e., {001} > {101} (by 1.7 times), and photoreduction, i.e., {101} > {001} (by 2–3 times).

Moreover, constructing the composite of MoS_2 and TiO_2 mesocrystals, as well as the co-catalyst selective modification on TiO_2, also showed the desired separation of photogenerated charge carriers during the hydrogen evolution reaction [104]. In terms of extending the absorption of incident light to the visible region, Zhang et al. tried doping or codoping non-mental elements into the TiO_2 matrix to examine the effects on its electronic structure and band gap. An in situ fluorine-doped TiO_2 superstructure was recently realized. F doping into TiO_2 mesocrystals for the incorporation of active color centers facilitates visible light harvesting and accelerates charge separation for hydrogen generation [51]. They further introduced nitrogen and fluorine codopants into {001} facet-oriented TiO_2 mesocrystals during topochemical transformation for photoreduction of Cr(VI) under visible light illumination. The extended optical light absorption could be attributed to doped nitrogen, which introduces the isolated mid-gap state. The high yield of hydroxyl radicals and preferential adsorption are correlated with fluorine doping, as confirmed by the comparison between untreated TiO_2 with

TiO$_2$ washed in NaOH aqueous solution. The synergistic effect on charge separation and trapping was suggested through a femtosecond time-resolved diffused reflectance (TDR) measurement [105]. As shown in Figure 18, the g-C$_3$N$_4$ nanosheet/TiO$_2$ mesocrystal metal-free composite was successfully constructed by Elbanna et al. [106]. The as-prepared sample exhibited an excellent hydrogen evolution rate under visible light irradiation without any noble metal co-catalyst. Then, they further broadened the light capture of the TiO$_2$ mesocrystals to include near-infrared regions. Au nanorods (NRs) with various aspect ratios were loaded onto the surface of TiO$_2$ by the ligand exchange method. Different aspect ratios resulted in different incident light absorption and photogenerated electron transfer. The highest photocatalytic activity of Au NRs and TMC composites reached 924 μmol h^{-1} g^{-1} under visible-near-infrared (NIR) light irradiation [107].

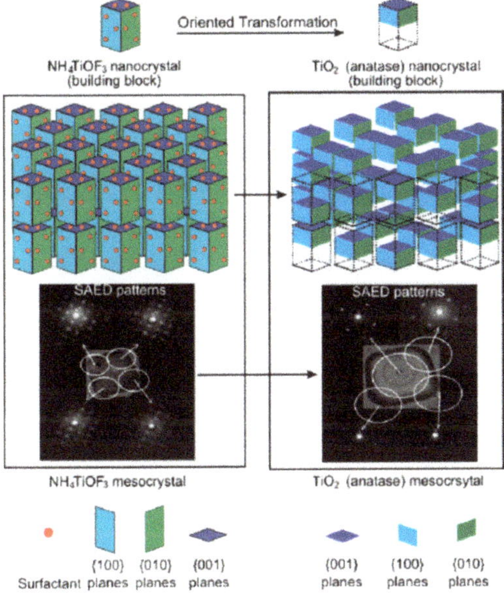

Figure 17. Illustration of the oriented transformation of NH$_4$TiOF$_3$ mesophyte to TiO$_2$ (anatase) mesocrystal [71]. Copyright 2008 American Chemical Society.

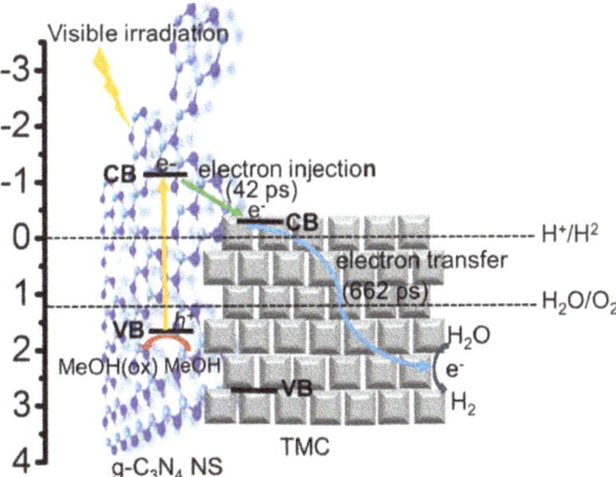

Figure 18. Representative scheme of electron injection and movement in g-C3N4 NS (31 wt %)/TMC during visible-light irradiation [106]. Copyright 2017 American Chemical Society.

Considering the aforementioned merits of mesocrystal nanomaterials, we recently tried different approaches to further improve the optical absorption properties of TiO_2 mesocrystals, in addition to their enhanced transfer and separation properties. Oxygen vacancies and N dopants were successfully introduced into the TiO_2 lattice with a facile low temperature calcination process [108], as shown in Figure 19. NH_4TiOF_3 mesocrystal nanocubes were used as precursors in our system, and topological transformation from NH_4TiOF_3 to TiO_2 facilitated the release and doping of nitrogen. Oxygen vacancies were also readily produced in the inert heating atmosphere. The significantly improved photodegradation and photoelectrochemical performance under visible light irradiation may be associated with the unique structure of mesocrystals as well as the introduction of foreign and intrinsic defects.

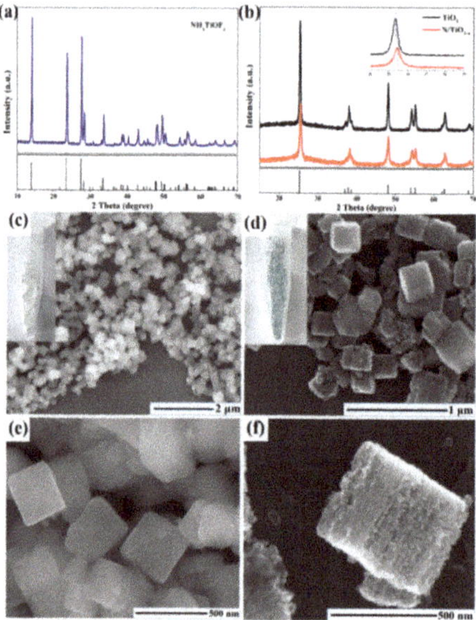

Figure 19. Schematic representation of the synthesis of TiO_x nanosheets. X-ray powder diffraction (XRD) pattern of (**a**) NH_4TiOF_3 and (**b**) N/TiO_{2-x}. SEM images of (**c,e**) NH_4TiOF_3 and (**d,f**) N/TiO_{2-x} [108]. Copyright 2019 The Royal Society of Chemistry.

5. Separation of Charges

Since metals and metal oxides have different working functions, resulting in the formation of a Schottky potential barrier, an effective modification method is to deposit precious metals (Ag, Au, or Pt) on the surface of metal oxides.

Choi et al. presented Ag/TiO_2 by a photodeposition method [109]. Due to the different transfer rates of interface charges between electrons and holes to redox species in water, excessive charges can accumulate on photocatalysts [110,111]. By depositing Ag, which can provide a temporary home for excessive electrons, the composite utilized the electron storage capacity to promote the separation of electrons and holes to reduce Cr(VI) in the following dark period. Li et al. prepared a sandwich structure with $CdS-Au-TiO_2$ on a fluorine-doped tin oxide (FTO) substrate [112]. In this composite structure, Au nanoparticles not only acted as an electronic relay between CdS quantum dots (QDs) and TiO_2 to increase charge separation occurring on a long-time scale but also served as a plasma photosensitizer that prolonged the photoconversion to improve the absorption range of light. The rate of charge transfer and reverse transfer depends on the relative energy of the hot plasma electrons to the Schottky barrier [112]. The PEC performance is represented in Figure 20.

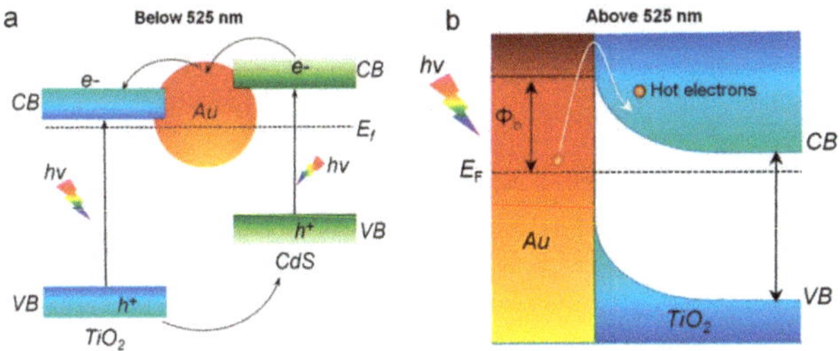

Figure 20. (**a**) Electron relay effect of Au nanoparticles, facilitating the charge transfer from CdS QDs to TiO_2 nanorods under the irradiation of incident solar light with a wavelength <525 nm. (**b**) Plasmonic energy transfer from the excited Au nanoparticles to TiO_2 through hot electron transfer under the irradiation of incident solar light with a wavelength >525 nm. CB = conduction band, VB = valence band, E_F = Fermi energy level, and Φ_b = Schottky barrier [112]. Copyright 2014 American Chemical Society.

Precious metal deposition can greatly improve the performance of catalysts, but the scarcity of precious metals dramatically limits this modification method and makes it difficult to achieve industrial-scale production. In this case, the search for an inexpensive and efficient doped composite has also attracted much attention. Carbon, abundant on earth, has good electrical conductivity, and its combination with TiO_2 can result in excellent photocatalytic performance. Wang et al. demonstrated TiO_2–carbon nanoparticles by the sol–gel method and then synthesized core–shell-structured TiO_2 and amorphous carbon [113]. This unique morphology and structure result in the modified TiO_2 sample exhibiting enhanced responsiveness and excellent photocatalytic activity. Due to the rapid charge transfer in the carbon shell, both the carrier separation efficiency and the photodegradation of pollutants in water is improved. The reduced TiO_2 is also more efficient in the production of H_2 due to its correct edge position.

6. Application of TiO_2 Nanomaterials

Over the past several years, semiconductors, especially titanium dioxide, have been widely used as photocatalysts. It is well known that there are three main steps associated with the photocatalysis process: (1) generation of electrons and holes after the absorption of photons; (2) separation and migration of the charge; and (3) transition of the charge and reaction between the carriers and the reagent. To date, TiO_2 has been mainly applied in the areas of environmental conservation, new energy resources, and so on. In this section, we will focus on recent progress in these photocatalytic applications of TiO_2.

6.1. Applications in the Environment

6.1.1. Degradation of Aqueous Pollutants

Industrial development is often accompanied by pollution of the environment, especially water. Photocatalytic water treatment using heterogeneous semiconductors under visible light is considered an eco-friendly technology. Photocatalysis involves the generation of large numbers of electrons and holes on the surface of TiO_2 after the absorption of photons; the photogenerated holes have considerable oxidizing capacity and can degrade almost all organic contaminants including carbon dioxide (CO_2). However, due to its own deficiencies, such as a wide bandgap and fast recombination of electrons and holes, TiO_2 cannot make full use of sunlight to remove the pollutants in water. Wang et al. reported hydrogenation by TiO_2 nanosheets with exposed {001} facets maintained by the formation of Ti–H bonds [114]. By annealing the fine-sized pristine hydrothermal product under a high-pressure hydrogen atmosphere, the hydrogenation of F-modified anatase TiO_2 nanosheets (with exposed high percentages of {001} facets) was achieved. Under UV–Vis and visible light irradiation, this material decomposed methylene blue (MB) faster than P25 and pristine TiO_2, as shown in Figure 21.

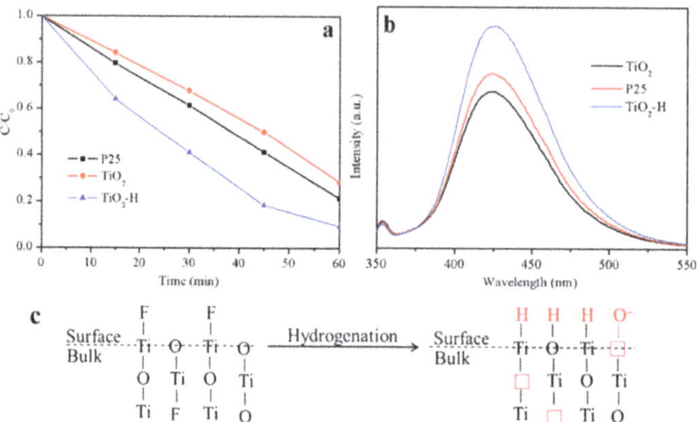

Figure 21. Photocatalytic decomposition of MB (**a**) and •OH generation measurement (**b**) of TiO_2 and TiO_2–H under UV–Vis light irradiation. Schematic illustration (**c**) of the hydrogenation effect on the structural change in TiO_2 and TiO_2–H [114]. Copyright 2012 The Royal Society of Chemistry.

Plodinec et al. applied black TiO_2 nanotube arrays with Ag nanoparticles, which promoted hydrogenation for the degradation of salicylic acid [115]. The photocatalyst can degrade salicylic acid effectively, and its photocatalytic performance far exceeds that of TiO_2 nanotubes and commercial TiO_2 P25 (the reference material used for the modeling of photocatalytic processes). Ling et al. prepared TiO_2 nanoparticles (with diameters of 10–23 nm) that exhibited photocatalytic activity [116]. The initial degradation rate of phenol by a TiO_2 nanocatalyst was 6 times higher than that achieved with H_2O_2 alone, and the addition of H_2O_2 to TiO_2 can increase the initial concentration of hydroxyl radicals and accelerate the degradation rate. Hao et al. developed a $TiO_2/WO_3/GO$ nanocomposite (via a hydrothermal synthesis), which presented excellent optical absorbance and displayed excellent photocatalytic activity for the degradation of bisphenol A [117].

In addition to the oxidizing capacity, the photogenerated electrons on TiO_2 have strong reducing capacity to remove pollutants, such as Cd(II), Hg(II), As(V), and Cr(VI), from water; these cations can be reduced into less toxic metallic or ion states. Dusadee et al. fabricated a titania-decorated reduced graphene oxide (TiO_2·rGO) nanocomposite via a hydrothermal process [110]. Studies on reducing the toxic Cr^{6+} (hexavalent chromium) ion toxicity using the titanium dioxide x/rGO numerical control have found that photocatalytic reduction of toxic Cr^{6+} generally increases with the increase in x. In addition, since rGO accelerates electron transport, the combination of photoexcited electrons and holes decreases leads to an increased duration of photocatalytic activity [118]. TiO_2 has facilitated many pollutant degradation processes such as the reduction of nitrate, the degradation of acid fuchsin, the decomposition of acetaldehyde, and the dechlorination of CCl_4 [119–122]. Due to the continued proliferation of environment pollutants, TiO_2 and other nanostructured materials should be vigorously developed in the future to improve the degradation of pollutants by photocatalysis.

6.1.2. Degradation of Air Pollutants

Just as industrial and technological developments can result in water pollution, so too can the atmosphere be adversely impacted by toxic pollutants that are emitted from chemical manufacturing plants, power plants, industrial facilities, transportation technologies, etc. Air pollution impacts the health of the global environment and the array of species that live within it, and new techniques are sought to reduce harmful airborne emissions. Highly efficient oxidation and reduction during photocatalysis are considered to be an effective method to degrade inorganic and organic air pollutants to improve air quality [123–125]. Similarly, TiO_2 is considered the most promising photocatalyst. Kakeru et al. prepared TiO_2 nanoparticles with palladium sub-nanoclusters (<1 nm) using the flame aerosol technique [126]. Under sunlight, these materials can remove NOx at approximately 3 to 7 times the rate of commercial TiO_2 (P25, Evonik) (without Pd). Natércia et al. prepared new composite materials of TiO_2 (P25) and N-doped carbon quantum dots (P25/NCQD) by a hydrothermal method, which was first used as the photooxidation catalyst of NO under the irradiation of ultraviolet and visible light [127]. The experiment showed that the conversion rate of the P25/NCQD composite material (27.0%) was more than twice that of P25 (10%) without modification, and the selectivity in visible light increased from 37.4% to 49.3%. The photocatalytic performance of the composite material in the UV region was also better than that of P25. Zeng et al. reported a H_2 reduction strategy to produce $H–TiO_2$ materials (with enhanced oxygen vacancy concentrations and distributions) that can promote formaldehyde decomposition in the dark [128]. Research of TiO_2-based photocatalysts has also been conducted to facilitate removal of tetrachloroethylene [129], acetone [130], benzene [131], phenol [73], etc. from the atmosphere.

6.2. Applications in Energy

6.2.1. Photocatalytic Hydrogen Generation

With the extensive use of nonrenewable fossil fuels, mankind is facing an unprecedented energy crisis. The photogenerated electrons on TiO_2 have strong reducing capacity, enabling hydrogen production from the photocatalytic splitting of water. Moreover, hydrogen combustion produces only water and no harmful emissions, and therefore its potential as a truly clean energy source has received considerable attention since it was discovered. Zou et al. reported a self-modified TiO_2 material with paramagnetic oxygen vacancies [132]. For the synthesis of V_o-TiO_2 (V_o: denotes a paramagnetic oxygen vacancy), they chose a porous amorphous TiO_2 material as a precursor that possessed a high surface area of 543 $m^2\ g^{-1}$. The precursor was calcined in the presence of imidazole and hydrochloric acid at an elevated temperature in air to obtain the V_o-TiO_2 material [132]. The V_o-TiO_2 sample (for H_2 evolution from water) used methanol as a sacrificial reagent under visible light ($\geq$400 nm) at room temperature, and the H_2 production rate was approximately 115 $\mu mol\ h^{-1}\ g^{-1}$, which is substantially higher than that achieved with V_o-Ti^{3+}-TiO_2 (32 $\mu mol\ h^{-1}\ g^{-1}$). Zhou et al. introduced an ordered mesoporous black TiO_2 material that utilized a thermally stable and high surface area mesoporous TiO_2 as the hydrogenation precursor for treatment at 500 °C [133]. The samples possessed a relatively high surface area of 124 $m^2\ g^{-1}$ and exhibited a photo response that extended from ultraviolet to visible light. As shown in Figure 22, the ordered mesoporous black TiO_2 material exhibits a high solar-driven hydrogen production rate (136.2 $\mu mol\ h^{-1}$), which is almost twice as high as that of pristine mesoporous TiO_2 (76.6 $\mu mol\ h^{-1}$). Zhong et al. constructed a covalently bonded oxidized graphitic C_3N_4/TiO_2 heterostructure that markedly increased the visible light photocatalytic activity for H_2 evolution by nearly a factor of approximately 6.1 compared to a simple physical mixture of TiO_2 nanosheets and O-g-C_3N_4 [134].

6.2.2. Photocatalytic CO_2 Reduction into Energy Fuels

In addition to reducing water to hydrogen, the photogenerated electrons on TiO_2 are capable of generating valuable solar energy fuels, such as CH_4, HCO_2H, CH_2O, CH_3OH, and CO_2, which are considered highly viable energy sources that can alleviate the problems associated with the production of greenhouse gases from the combustion of fossil fuels. Slamet et al. prepared Cu-doped TiO_2 through an improved impregnation method for photocatalytic CO_2 reduction [135]. Both the distribution of copper on the catalyst surface and the grain size of copper–titania catalysts (crystallite size of approximately 23 nm) were uniform, and it was determined that Cu doping can greatly enhance the photocatalytic performance of TiO_2 with respect to CO_2 reduction. Liu et al. found that copper-loaded titania photocatalysts, prepared via a one-pot, sol–gel synthesis method, comprised highly dispersed copper and that CO_2 photoreduction exhibited a strong volcano dependence on Cu loading, which reflected the transition from 2-dimensional CuO_x nanostructures to 3-dimensional crystallites; optimum CH_4 production was observed for 0.03 wt.% Cu/TiO_2 [136].

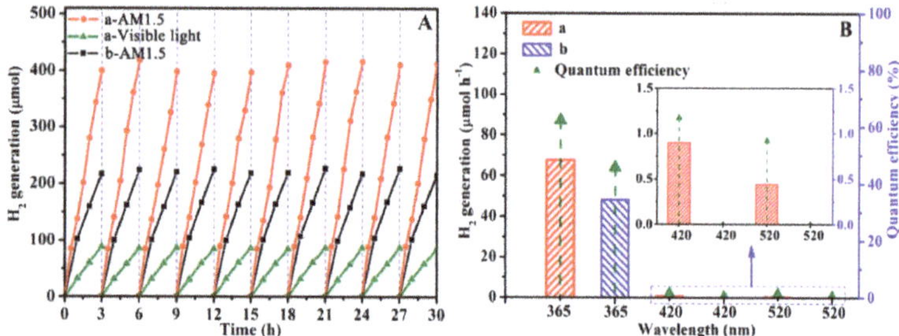

Figure 22. Photocatalytic hydrogen evolution of ordered mesoporous black TiO_2 (a) and pristine ordered mesoporous TiO_2 materials (b). (**A**) Cycling tests of photocatalytic hydrogen generation under AM 1.5 and visible light irradiation. (**B**) The photocatalytic hydrogen evolution rates under single-wavelength light and the corresponding QE. The inset enlarges the QE of single-wavelength light at 420 and 520 nm [133]. Copyright 2014 American Chemical Society.

6.2.3. Solar Batteries

Since semiconductors absorb photons to produce photonic carriers and the photonic carriers move and separate at the same time, electric energy can be obtained through charge transport. TiO_2 can also be applied to dye-sensitized solar cells, Li-ion batteries, Na-ion batteries, and supercapacitors. Liu et al. synthesized a spring-like $Ti@TiO_2$ nanowire array wire that could be used as a photoanode in dye-sensitized solar cells; this configuration exhibited a conversion efficiency maintenance rate of more than 95.95% [137]. Another study reported the use of anatase TiO_2 nanotubes on rutile TiO_2 nanorod arrays as photoanodes in quantum dot-sensitized solar cells, which have a small thickness of 1 μm and an excellent solar energy conversion efficiency of approximately 1.04%; this is almost 2.7 times higher than the conversion efficiencies measured for solar cells using the original TiO_2 nanorod array photoanodes, as shown in Figure 23 [138]. Chen et al. implemented a $C@TiO_2$ nanocomposite as the anode material for lithium-ion batteries, which utilize the esterification of ethylene glycol with acetic acid in the presence of potassium chloride. Li-ion batteries utilizing the $C@TiO_2$ nanocomposite anode exhibited excellent rate performance and specific capacity (237 mA h^{-1} g^{-1}), and a coulomb efficiency (CE) of approximately 100% after 100 cycles [139]. Su et al. synthesized anatase TiO_2 via a template approach for use as the anode in Na-ion batteries; use of the template-synthesized TiO_2 resulted in better battery performance in comparison to that achieved when amorphous and rutile TiO_2 was used as the anode material. Compared to other crystalline phases of titanium dioxide, anatase titanium dioxide produced the highest capacity, 295 mA h^{-1} g^{-1}, in the second cycle, tested at a current density of 20 mA g^{-1} [140]. Kim et al. developed a black-colored TiO_2 nanotube array synthesized by electrochemical self-doping of an amorphous TiO_2 nanotube array and N_2 annealing; the material exhibited good stability, high capacitance, and electrocatalytic performance, and is an excellent material for supercapacitors and oxide anodes [141].

6.2.4. Supercapacitors

Yang et al. developed a hybrid material, covalently coupled ultrafine $H–TiO_2$ nanocrystals/ nitrogen-doped graphene, via the hydrothermal route [142]. Due to the strong interaction between $H–TiO_2$ nanocrystals and NG plates, the high structural stability of the $H–TiO_2$ nanocrystal aggregation is inhibited. At the same time, the NG matrix plays the role of electron conductor and mechanical skeleton, imparting good stability and electrochemical activity on most of the well-dispersed ultrafine

H-TiO$_2$ nanocrystals [142]. The material exhibited a high reversible specific capacity of 385.2 F g^{-1} at 1 A g^{-1} and excellent cycling stability with 98.8% capacity retention. Parthiban et al. reported a blue titanium oxide (B-TiO2) nanostructure that was applied via a one-pot hydrothermal route and hydrothermal oxidation [143]. The B–TiO$_2$ nanostructure indicated excellent cycling stability with approximately 90.2% capacitance retention after 10,000 charge–discharge cycles.

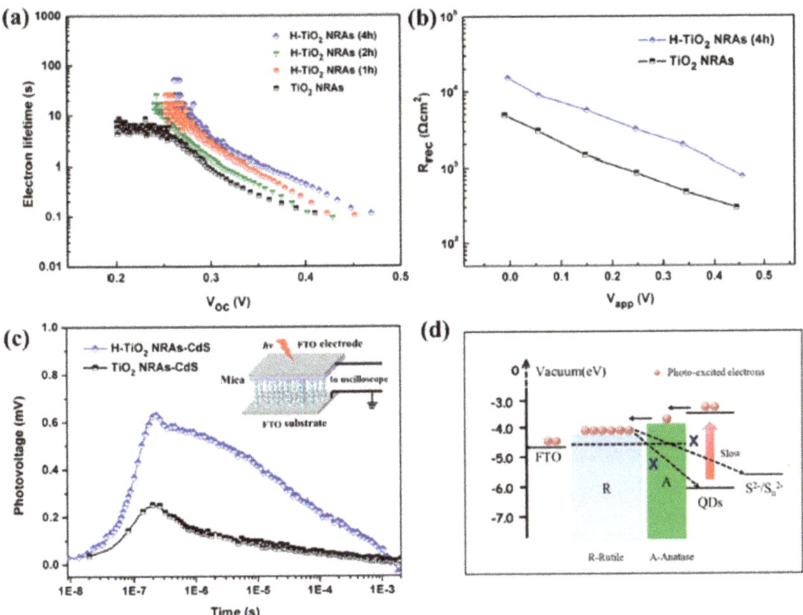

Figure 23. (**a**) Electron lifetime as a function of Voc for TiO2 NRA and H–TiO$_2$ NRA electrodes with various reaction times. (**b**) Recombination resistance (Rrec) of the QDSCs made from TiO$_2$ NRAs and H-TiO$_2$ NRAs at various forward biases in the dark. (**c**) Transient photovoltage responses of CdS–TiO$_2$ NRAs and CdS–H-TiO2 NRAs. The wavelength of the laser pulse was 532 nm. Inset: schematic setup of TPV measurements. (**d**) Schematic configuration for our device showing the interfacial charge transfer and recombination processes [138]. Copyright 2015 The Royal Society of Chemistry.

6.3. Other Applications

6.3.1. Antibacterial and Wound Healing

It is generally believed that electron–hole pairs formed under light illumination, such as •O^{2-} and •OH, not only destroy all chemical contaminants but also kill microorganisms. Liu et al. proposed a TiO$_2$/Ag$_2$O heterostructure (produced by a facile in situ precipitation route) to enhance antibacterial activities [144]. Yu et al. synthesized a TiO$_2$/BTO/Au heterostructured nanorod arrays (exhibiting piezophototronic and plasmonic effects) by using a simple process that combined hydrothermal and PVD methods. This material can be used as an antibacterial coating for efficient light driven in vitro/in vivo sterilization and wound healing [145].

6.3.2. Drug Delivery Carriers

TiO$_2$ has the advantages of nontoxicity, stability, biocompatibility, and natural abundance. The preparation of TiO$_2$ with a high specific surface area can be advantageous in drug delivery carrier applications. Johan et al. controlled the kinetics of drug delivery from mesoporous titania thin films via surface energy and pore size control [146]. Different pore sizes ranging from 3.4 nm to 7.2 nm were achieved by the use of different structural guiding templates and expansive agents. In addition, by attaching dimethyl silane to the pore wall, the surface energy of the pore wall could be altered. The results indicated that the pore size and surface energy had significant effects on the adsorption and release kinetics of alendronate [146]. Biki et al. designed silica-supported mesoporous titania nanoparticles (MTN) coated with hyaluronic acid to cure breast cancer by effectively delivering doxorubicin (DOX) to the cancer cells [147]. Guo et al. deposited (onto the surface of MTN) hyaluronic acid and cyclic pentapeptide (ADH-1), which target CD44-overexpressing tumor cells and selectively inhibit the function of N-cadherin, respectively, to overcome the drug resistance of tumors [148].

Recently, Nakayama et al. found that H_2O_2-treated TiO_2 can enhance the ability to produce reactive oxygen species (ROS) in response to X-ray irradiation [149]. As shown in Figure 24, the atomic packing factor (APF) intensity indicated that hydroxyl radical production in the TiOx (H_2O_2-treated TiO_2) nanoparticles increased in a radiation dose-dependent manner in comparison to that of the non-H_2O_2-treated TiO_2 nanoparticles. This behavior allows H_2O_2-treated TiO_2 nanoparticles to act as potential agents for enhancing the effects of radiation in the treatment of pancreatic cancer. Dai et al. designed and synthesized a novel nanodrug delivery system for the synergistic treatment of lung cancer [150]. They loaded DOX onto H_2O_2-treated TiO_2 nanosheets. In this way, chemotherapy and radiotherapy were combined effectively for the synergistic therapy of cancers.

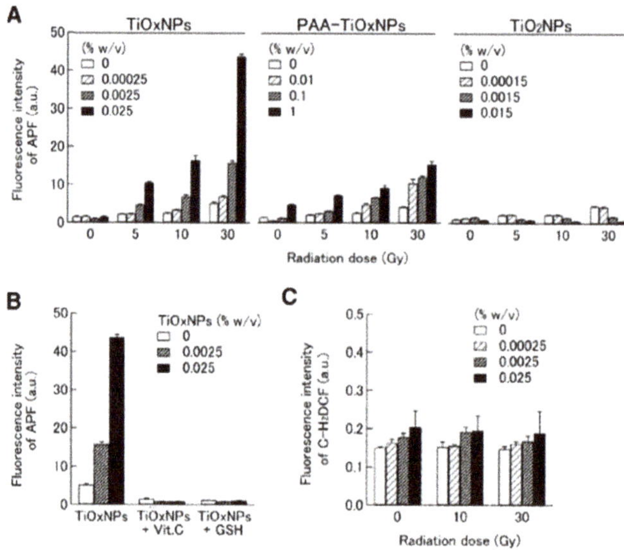

Figure 24. ROS production by the TiOxNPs, PAA-TiOxNPs, and TiO_2 NPs under X-ray irradiation. (**A**) Atomic packing factor (APF) intensity indicating that hydroxyl radical production in the TiOxNPs and the PAA-TiOxNPs increased in a radiation dose-dependent manner, but that of the TiO_2 NPs did not. Irradiated radiation doses were 0, 5, 10, and 30 Gy. Data are shown as the mean $\pm$ SD from 5 independent experiments. (**B**) Production and scavenging of ROS by 1 mM vitamin C (Vit. C) or 1 mM glutathione (GSH). Histograms show the mean $\pm$ SD calculated from 5 independent experiments. (**C**) Hydrogen peroxide production from the TiOxNPs under X-ray irradiation [149]. Copyright 2016 Springer Nature Switzerland AG.

7. Conclusions

As discussed in this review article, TiO_2-based nanomaterials with wide band gaps have advantages associated with natural geologic abundance, nontoxicity and stability but they also exhibit inherent deficiencies and limitations related to ineffective visible light responses and other photocatalytic properties. The present review aimed to summarize key studies related to the marked enhancement of the photocatalytic performance of TiO_2 by analyzing its electrical structure and photocatalytic reaction process. We have highlighted TiO_2 photocatalysts with well-defined electrical and structure design, as well as tailored facets, dimensions, and remarkable morphologies, which are promising with respect to enhancing the photocatalytic properties of TiO_2. All works presented in this review has enabled the authors to obtain an in-depth understanding of the TiO_2 photocatalytic process, and the critical design of TiO_2 nanostructures with enhanced light absorption, high surface area, desired photostability, and charge carrier dynamics. We hope that this review will guide the future development of more robust TiO_2-based photocatalysts for large-scale applications.

Finally, photocatalysis technology is one of the most active research fields in the world in recent years. However, photocatalysis technologies based on TiO_2 semiconductor still suffer from several key scientific and technological problems, such as low solar energy utilization rate, inferior quantum yield, and difficult recovery, which greatly restricts its wide application in industry. The fundamental solution to improve solar energy absorption is energy band engineering, designing and regulating the bandgap to optimize the harvesting of incident photons. Narrow bandgap and direct semiconductor are more likely to make use of low energy light, but they are restricted by very high electron and hole recombination rate and the incompatible band-edge position. High quantum yield is inevitable for an idea photocatalysis in practical solar engineering, but it cannot be achieved simply doping or inducing intrinsic defects. More works are needed to do to search high quantum yield. All of the above problems depend on the deepening of basic research. Although at present, photocatalysis technology is still a long way from large-scale production and application, its huge potential excellent performance provides a good way for our development. In the near future, with the breakthrough of these key issues, the practical application of nano-photocatalytic materials will certainly be realized to improve our environment, provide cleaner energy, and bring more convenience to our daily life.

Author Contributions: X.K. and S.L. collected references, prepared figures, and wrote the original draft of the manuscript, they contributed equally to this work; Z.D. and Y.H. collected references and analyzed the data; X.S. gave valuable advice; Z.T. acted as a project director and contributed to subsequent revisions. All authors agreed to the final version of the paper.

Funding: This research was funded by the National Natural Science Foundation of China grant number 21571028, 21601027, the Fundamental Research Funds for the Central Universities grant number DUT16TD19, DUT17LK33, DUT18LK28 and the Education Department of the Liaoning Province of China grant number LT2015007.

Conflicts of Interest: The authors declare no conflicts of interest.

References

1. Fujishima, A.; Honda, K. Electrochemical photolysis of water at a semiconductor electrode. *Nature* **1972**, *238*, 37. [CrossRef] [PubMed]
2. Long, L.; Zhang, A.; Yang, J.; Zhang, X.; Yu, H. A green approach for preparing doped TiO_2 single crystals. *ACS Appl. Mater. Interfaces* **2014**, *6*, 16712–16720. [CrossRef]
3. Nakata, K.; Fujishima, A. TiO_2 photocatalysis: Design and applications. *J. Photochem. Photobio. C* **2012**, *13*, 169–189. [CrossRef]
4. Tong, H.; Ouyang, S.; Bi, Y.; Umezawa, N.; Oshikiri, M.; Ye, J. Nano-photocatalytic materials: Possibilities and challenges. *Adv. Mater.* **2012**, *24*, 229–251. [CrossRef]
5. Grätzel, M. Photoelectrochemical cells. *Nature* **2001**, *414*, 338. [CrossRef] [PubMed]
6. Ma, Y.; Wang, X.; Jia, Y.; Chen, X.; Han, H.; Li, C. Titanium dioxide-based nanomaterials for photocatalytic fuel generations. *Chem. Rev.* **2014**, *114*, 9987–10043. [CrossRef] [PubMed]
7. Schneider, J.; Matsuoka, M.; Takeuchi, M.; Zhang, J.; Horiuchi, Y.; Anpo, M.; Bahnemann, D.W. Understanding TiO_2 photocatalysis: Mechanisms and materials. *Chem. Rev.* **2014**, *114*, 9919–9986. [CrossRef]
8. Wang, M.; Iocozzia, J.; Sun, L.; Lin, C.; Lin, Z. Inorganic-modified semiconductor TiO_2 nanotube arrays for photocatalysis. *Energ. Environ. Sci.* **2014**, *7*, 2182–2202. [CrossRef]
9. Kalyanasundaram, K. Photochemical applications of solar energy: Photocatalysis and photodecomposition of water. *Photochemistry* **2013**, *41*, 182–265.

10. Liu, B.; Yang, J.; Zhao, X.; Yu, J. The role of electron interfacial transfer in mesoporous nano-TiO_2 photocatalysis: A combined study of in situ photoconductivity and numerical kinetic simulation. *Phys. Chem. Chem. Phys.* **2017**, *19*, 8866–8873. [CrossRef]

11. Liu, B.; Zhao, X.; Terashima, C.; Fujishima, A.; Nakata, K. Thermodynamic and kinetic analysis of heterogeneous photocatalysis for semiconductor systems. *Phys. Chem. Chem. Phys.* **2014**, *16*, 8751–8760. [CrossRef] [PubMed]

12. Ravelli, D.; Dondi, D.; Fagnoni, M.; Albini, A. Photocatalysis. A multi-faceted concept for green chemistry. *Chem. Soc. Rev.* **2009**, *38*, 1999–2011. [CrossRef] [PubMed]

13. Bai, S.; Jiang, J.; Zhang, Q.; Xiong, Y. Steering charge kinetics in photocatalysis: Intersection of materials syntheses, characterization techniques and theoretical simulations. *Chem. Soc. Rev.* **2015**, *44*, 2893–2939. [CrossRef] [PubMed]

14. Gao, M.; Zhu, L.; Ong, W.L.; Wang, J.; Ho, G.W. Structural design of TiO_2-based photocatalyst for H_2 production and degradation applications. *Catal. Sci. Technol.* **2015**, *5*, 4703–4726. [CrossRef]

15. Šuligoj, A.; Arčon, I.; Mazaj, M.; Dražić, G.; Arčon, D.; Cool, P.; Štangar, U.L.; Tušar, N.N. Surface modified titanium dioxide using transition metals: Nickel as a winning transition metal for solar light photocatalysis. *J. Mater. Chem. A* **2018**, *6*, 9882. [CrossRef]

16. Hou, Y.; Liu, S.; Zhang, J.; Cheng, X.; Wang, Y. Facile hydrothermal synthesis of TiO_2-Bi_2WO_6 hollow superstructures with excellent photocatalysis and recycle properties. *Dalton Trans.* **2014**, *43*, 1025–1031. [CrossRef] [PubMed]

17. Ansari, S.A.; Khan, M.M.; Ansari, M.O.; Cho, M.H. Nitrogen-doped titanium dioxide (N-doped TiO_2) for visible light photocatalysis. *New J. Chem.* **2016**, *40*, 3000–3009. [CrossRef]

18. Asahi, R.; Taga, Y.; Mannstadt, W.; Freeman, A.J. Electronic and optical properties of anatase TiO_2. *Phys. Rev. B* **2000**, *61*, 7459–7465. [CrossRef]

19. Park, H.; Choi, W. Effects of TiO_2 surface fluorination on photocatalytic reactions and photoelectrochemical behaviors. *J. Phys. Chem. B* **2004**, *108*, 4086–4093. [CrossRef]

20. Xu, J.; Ao, Y.; Fu, D.; Yuan, C. Low-temperature preparation of F-doped TiO_2 film and its photocatalytic activity under solar light. *Appl. Surf. Sci.* **2008**, *254*, 3033–3038. [CrossRef]

21. Niu, M.; Cui, R.; Wu, H.; Cheng, D.; Cao, D. Enhancement mechanism of the conversion effficiency of dye-sensitized solar cells based on nitrogen-, fluorine-, and iodine-doped TiO_2 photoanodes. *J. Phys. Chem. C* **2015**, *119*, 13425–13432. [CrossRef]

22. Liu, H.; Li, Y.; Yang, Y.; Mao, M.; Zeng, M.; Lan, L.; Yun, L.; Zhao, X. Highly efficient UV–Vis-infrared catalytic purification of benzene on $CeMn_xO_y$/TiO_2 nanocomposite, caused by its high thermocatalytic activity and strong absorption in the full solar spectrum region. *J. Mater. Chem. A* **2016**, *4*, 9890–9899. [CrossRef]

23. Narayan, H.; Alemu, H.; Macheli, L.; Thakurdesai, M.; Rao, T.K. Synthesis and characterization of Y^{3+}-doped TiO_2 nanocomposites for photocatalytic applications. *Nanotechnology* **2009**, *20*, 255601. [CrossRef] [PubMed]

24. Wang, X.; Yuan, X.; Wang, D.; Dong, W.; Dong, C.; Zhang, Y.; Lin, T.; Huang, F. Tunable synthesis of colorful nitrogen-doped titanium oxide and its application in energy storage. *ACS Appl. Energy Mater.* **2018**, *1*, 876–882. [CrossRef]

25. Niu, P.; Wu, T.; Wen, L.; Tan, J.; Yang, Y.; Zheng, S.; Liang, Y.; Li, F.; Irvine, J.T.S.; Liu, G.; et al. Substitutional carbon-modified anatase TiO_2 decahedral plates directly derived from titanium oxalate crystals via topotactic transition. *Adv. Mater.* **2018**, *30*, e1705999. [CrossRef] [PubMed]

26. Zhao, X.; Liu, X.; Yu, M.; Wang, C.; Li, J. The highly efficient and stable Cu, Co, Zn-porphyrineTiO$_2$ photocatalysts with heterojunction by using fashioned one-step method. *Dyes Pigments* **2017**, *136*, 648–656. [CrossRef]

27. Hachiya, A.; Takata, S.; Komuro, Y.; Matsumoto, Y. Effects of V-ion doping on the photoelectrochemical properties of epitaxial TiO_2(110) thin films on Nb-doped TiO_2 (110) single crystals. *J. Phys. Chem. C* **2012**, *116*, 16951–16956. [CrossRef]

28. Klosek, S.; Raftery, D. Visible light driven V-doped TiO_2 photocatalyst and its photooxidation of ethanol. *J. Phys. Chem. B* **2001**, *105*, 2815–2819. [CrossRef]

29. Zahid, M.; Papadopoulou, E.L.; Suarato, G.; Binas, V.D.; Kiriakidis, G.; Gounaki, I.; Moira, O.; Venieri, D.; Bayer, I.S.; Athanassiou, A. Fabrication of visible light-induced antibacterial and self-cleaning cotton fabrics using manganese doped TiO_2 nanoparticles. *ACS Appl. Bio Mater.* **2018**, *1*, 1154–1164. [CrossRef]

30. Taguchi, T.; Ni, L.; Irie, H. Alkaline-resistant titanium dioxide thin film displaying visible-light-induced superhydrophilicity initiated by interfacial electron transfer. *Langmuir* **2013**, *29*, 4908–4914. [CrossRef]

31. Mizushima, K.; Tanaka, M.; Asai, A.; Iida, S.; Goodenough, B. Impurity levels of iron-group ions in TiO_2(II). *J. Phys. Chem. Solids* **1979**, *40*, 1129–1140. [CrossRef]

32. Borgarello, E.; Kiwi, J.; Grätzel, M.; Pelizzetti, E.; Viscald, M. Visible light induced water cleavage in colloidal solutions of chromium-doped titanium dioxide particles. *J. Am. Chem. Soc.* **1982**, *104*, 2996–3002. [CrossRef]

33. Xu, D.; Feng, L.; Lei, A. Characterizations of lanthanum trivalent ions/TiO_2 nanopowders catalysis prepared by plasma spray. *J. Colloid Interface Sci.* **2009**, *329*, 395–403. [CrossRef] [PubMed]

34. Li, N.; Zhou, X.; Liu, M.; Wei, L.; Shen, Q.; Bibi, R.; Xu, C.; Ma, Q.; Zhou, J. Enhanced visible light photocatalytic hydrogenation of CO_2 into methane over a Pd/Ce-TiO_2 nanocomposition. *J. Phys. Chem. C* **2017**, *121*, 25795–25804. [CrossRef]

35. Anandan, S.; Ikuma, Y.; Murugesan, V. Highly active rare-earth-metal La-doped photocatalysts: Fabrication, characterization, and their photocatalytic activity. *Int. J. Photoenergy* **2012**, *10*, 921412. [CrossRef]

36. Sun, L.; Zhao, X.; Cheng, X.; Sun, H.; Li, Y.; Li, P.; Fan, W. Synergistic effects in La/N codoped TiO_2 anatase (101) surface correlated with enhanced visible-light photocatalytic activity. *Langmuir* **2012**, *28*, 5882–5891. [CrossRef] [PubMed]

37. Choi, J.; Park, H.; Hoffman, M.R. Effects of single metal-ion doping on the visible-light photoreactivity of TiO_2. *J. Phys. Chem. C* **2010**, *114*, 783–792. [CrossRef]

38. Di Paola, A.; Ikeda, S.; Marcì, G.; Ohtani, B.; Palmisano, L. Transition metal doped TiO_2: Physical properties and photocatalytic behaviour. *Int. J. Photoenergy* **2001**, *3*, 171. [CrossRef]

39. Asahi, R.; Morikawa, T.; Irie, H.; Ohwaki, T. Nitrogen-doped titanium dioxide as visible-light-sensitive photocatalyst: Designs, developments, and prospects. *Chem. Rev.* **2014**, *114*, 9824–9852. [CrossRef]

40. Muhich, C.L.; Westcott, J.Y.; Fuerst, T.; Weimer, A.W.; Musgrave, C.B. Increasing the photocatalytic activity of anatase TiO_2 through B, C, and N doping. *J. Phys. Chem. C* **2014**, *118*, 27415–27427. [CrossRef]

41. Asahi, R.; Morikawa, T.; Ohwaki, T.; Aoki, K.; Taga, Y. Visible-light photocatalysis in nitrogen-doped titanium oxides. *Science* **2001**, *13*, 269–271. [CrossRef]

42. Zhao, Z.; Fan, J.; Wang, J.; Li, R. Effect of heating temperature on photocatalytic reduction of CO_2 by N-TiO_2 nanotube catalyst. *Catal. Commun.* **2012**, *21*, 32–37. [CrossRef]

43. Irie, H.; Watanabe, Y.; Hashimoto, K. Nitrogen-concentration dependence on photocatalytic activity of $TiO_{2-x}N_x$ powders. *J. Phys. Chem. B* **2003**, *107*, 5483–5486. [CrossRef]

44. Li, H.; Zhang, X.; Huo, Y.; Zhu, J. Supercritical preparation of a highly active S-doped TiO_2 photocatalyst for methylene blue mineralization. *Environ. Sci. Technol.* **2007**, *41*, 4410–4414. [CrossRef] [PubMed]

45. Liu, G.; Zhao, Y.; Sun, C.; Li, F.; Lu, G.Q.; Cheng, H.M. Synergistic effects of B/N doping on the visible-light photocatalytic activity of mesoporous TiO_2. *Angew. Chem. Int. Ed.* **2008**, *47*, 4516–4520. [CrossRef] [PubMed]

46. Peighambardoust, N.S.; Asl, S.K.; Mohammadpour, R.; Asl, S.K. Band-gap narrowing and electrochemical properties in N-doped and reduced anodic TiO_2 nanotube arrays. *Electrochim. Acta* **2018**, *270*, 245–255. [CrossRef]

47. Lynch, J.; Giannini, C.; Cooper, J.K.; Loiudice, A.; Sharp, I.D.; Buonsanti, R. Substitutional or interstitial site-selective nitrogen doping in TiO_2 nanostructures. *J. Phys. Chem. C* **2015**, *119*, 7443–7452. [CrossRef]

48. Di Valentin, C.; Finazzi, E.; Pacchioni, G.; Selloni, A.; Livraghi, S.; Paganini, M.C.; Giamello, E. N-doped TiO_2: Theory and experiment. *Chem. Phys.* **2007**, *339*, 44–56. [CrossRef]

49. Wu, Y.; Lazic, P.; Hautier, G.; Perssonb, K.; Ceder, G. First principles high throughput screening of oxynitrides for water-splitting photocatalysts. *Energy Environ. Sci.* **2013**, *6*, 157–168. [CrossRef]

50. Wang, J.; Tafen, D.N.; Lewis, J.P.; Hong, Z.; Manivannan, A.; Zhi, M.; Li, M.; Wu, N. Origin of photocatalytic activity of nitrogen-doped TiO_2 nanobelts. *J. Am. Chem. Soc.* **2009**, *131*, 12290–12297. [CrossRef]

51. Zhang, P.; Tachikawa, T.; Fujitsuka, M.; Majima, T. In situ fluorine doping of TiO_2 superstructures for efficient visible-light driven hydrogen generation. *ChemSusChem* **2016**, *9*, 617–623. [CrossRef] [PubMed]

52. Seo, H.; Baker, L.R.; Hervier, A.; Kim, J.; Whitten, J.L.; Somorjai, G.A. Generation of highly n-type titanium oxide using plasma fluorine insertion. *Nano Lett.* **2011**, *11*, 751–756. [CrossRef] [PubMed]

53. Yu, W.; Liu, X.; Pana, L.; Li, J.; Liu, J.; Zhang, J.; Li, P.; Chen, C.; Sun, Z. Enhanced visible light photocatalytic degradation of methylene blue by F-doped TiO_2. *Appl. Surf. Sci.* **2014**, *319*, 107–112. [CrossRef]

54. Chen, X.; Liu, L.; Yu, P.Y.; Mao, S.S. Increasing solar absorption for photocatalysis with black hydrogenated titanium dioxide nanocrystals. *Science* **2011**, *331*, 746–749. [CrossRef] [PubMed]

55. Song, H.; Li, C.; Lou, Z.; Ye, Z.; Zhu, L. Effective formation of oxygen vacancies in black TiO_2 nanostructures with efficient solar-driven water splitting. *ACS Sustain. Chem. Eng.* **2017**, *5*, 8982–8987. [CrossRef]

56. Hu, Y.H. A highly efficient photocatalyst–hydrogenated black TiO_2 for the photocatalytic splitting of water. *Angew. Chem. Int. Ed.* **2012**, *51*, 12410–12412. [CrossRef] [PubMed]

57. Lin, L.; Huang, J.; Li, X.; Abass, M.A.; Zhang, S. Effective surface disorder engineering of metal oxide nanocrystals for improved photocatalysis. *Appl. Catal. B Environ.* **2017**, *203*, 615–624. [CrossRef]

58. Liu, N.; Haublein, V.; Zhou, X.; Venkatesan, U.; Hartmann, M.; Mackovic, M.; Nakajima, T.; Spiecker, E.; Osvet, A.; Frey, L.; Schmuki, P. "Black" TiO_2 nanotubes formed by high-energy proton implantation show noble-metal-co-catalyst free photocatalytic H_2-evolution. *Nano Lett.* **2015**, *15*, 6815–6820. [CrossRef]

59. Wang, G.; Wang, H.; Ling, Y.; Tang, Y.; Yang, X.; Fitzmorris, R.C.; Wang, C.; Zhang, J.Z.; Li, Y. Hydrogen-treated TiO_2 nanowire arrays for photoelectrochemical water splitting. *Nano Lett.* **2011**, *11*, 3026–3033. [CrossRef]

60. Wang, X.; Fu, R.; Yin, Q.; Wu, H.; Guo, X.; Xu, R.; Zhong, Q. Black TiO_2 synthesized via magnesiothermic reduction for enhanced photocatalytic activity. *J. Nanopart. Res.* **2018**, *20*, 89. [CrossRef]

61. Dong, J.; Han, J.; Liu, Y.; Nakajima, A.; Matsushita, S.; Wei, S.; Gao, W. Defective black TiO_2 synthesized via anodization for visible-light photocatalysis. *ACS Appl. Mater. Interfaces* **2014**, *6*, 1385–1388. [CrossRef] [PubMed]

62. Liu, X.; Gao, S.; Xu, H.; Lou, Z.; Wang, W.; Huang, B.; Dai, Y. Green synthetic approach for Ti^{3+} self-doped TiO_{2-x} nanoparticles with efficient visible light photocatalytic activity. *Nanoscale* **2013**, *5*, 1870–1875. [CrossRef] [PubMed]

63. Chen, S.; Wang, Y.; Li, J.; Hu, Z.; Zhao, H.; Xie, W.; Wei, Z. Synthesis of black TiO_2 with efficient visible-light photocatalytic activity by ultraviolet light irradiation and low temperature annealing. *Mater. Res. Bull.* **2018**, *98*, 280–287. [CrossRef]

64. Liu, L.; Yu, P.Y.; Chen, X.; Mao, S.S.; Shen, D.Z. Hydrogenation and disorder in engineered black TiO_2. *Phys. Rev. Lett.* **2013**, *111*, 065505. [CrossRef] [PubMed]

65. Fujishima, A.; Zhang, X.; Tryk, D.A. TiO_2 photocatalysis and related surface phenomena. *Surf. Sci. Rep.* **2008**, *63*, 515–582. [CrossRef]

66. Pan, X.; Yang, M.Q.; Fu, X.; Zhang, N.; Xu, Y.J. Defective TiO_2 with oxygen vacancies: Synthesis, properties and photocatalytic applications. *Nanoscale* **2013**, *5*, 3601–3614. [CrossRef] [PubMed]

67. Tan, H.; Zhao, Z.; Niu, M.; Mao, C.; Cao, D.; Cheng, D.; Feng, P.; Sun, Z. A facile and versatile method for preparation of colored TiO_2 with enhanced solar-driven photocatalytic activity. *Nanoscale* **2014**, *6*, 10216–10223. [CrossRef]

68. Li, S.; Qiu, J.; Ling, M.; Peng, F.; Wood, B.; Zhang, S. Photoelectrochemical characterization of hydrogenated TiO_2 nanotubes as photoanodes for sensing applications. *ACS Appl. Mater. Interfaces* **2013**, *5*, 11129–11135. [CrossRef]

69. Wang, Z.; Yang, C.; Lin, T.; Yin, H.; Chen, P.; Wan, D.; Xu, F.; Huang, F.; Lin, J.; Xie, X.; Jiang, M. H-doped black titania with very high solar absorption and excellent photocatalysis enhanced by localized surface plasmon resonance. *Adv. Funct. Mater.* **2013**, *23*, 5444–5450. [CrossRef]

70. Sinhamahapatra, A.; Jeon, J.-P.; Yu, J.-S. A new approach to prepare highly active and stable black titania for visible light-assisted hydrogen production. *Energ. Environ. Sci.* **2015**, *8*, 3539–3544. [CrossRef]

71. Zhou, L.; Boyle, D.S.; O'Brien, P. A facile synthesis of uniform NH_4TiOF_3 mesocrystals and their conversion to TiO_2 mesocrystals. *J. Am. Chem. Soc.* **2008**, *130*, 1309–1320. [CrossRef]

72. Kang, X.; Han, Y.; Song, X.; Tan, Z. A facile photoassisted route to synthesis N, F-codoped oxygen-deficient TiO_2 with enhanced photocatalytic performance under visible light irradiation. *Appl. Surf. Sci.* **2018**, *434*, 725–734. [CrossRef]

73. Kang, X.; Song, X.-Z.; Han, Y.; Cao, J.; Tan, Z. Defect-engineered TiO_2 hollow spiny nanocubes for phenol degradation under visible light irradiation. *Sci. Rep.* **2018**, *8*, 5904. [CrossRef]

74. Li, X.; Yu, J.; Jaroniec, M. Hierarchical photocatalysts. *Chem. Soc. Rev.* **2016**, *45*, 2603–2636. [CrossRef] [PubMed]

75. Yang, H.G.; Zeng, H.C. Preparation of hollow anatase TiO_2 nanospheres via Ostwald ripening. *J. Phys. Chem. B* **2004**, *108*, 3492–3495. [CrossRef]

76. Pan, J.H.; Zhang, X.; Du, A.J.; Sun, D.D.; Leckie, J.O. Self-etching reconstruction of hierarchically mesoporous F-TiO$_2$ hollow microspherical photocatalyst for concurrent membrane water purifications. *J. Am. Chem. Soc.* **2008**, *130*, 11256–11257. [CrossRef] [PubMed]

77. Cao, J.; Song, X.-Z.; Kang, X.; Dai, Z.; Tan, Z. One-pot synthesis of oleic acid modified monodispersed mesoporous TiO$_2$ nanospheres with enhanced visible light photocatalytic performance. *Adv. Powder Technol.* **2018**, *29*, 1925–1932. [CrossRef]

78. Banerjee, A.N. The design, fabrication, and photocatalytic utility of nanostructured semiconductors: Focus on TiO$_2$-based nanostructures. *Nanotechnol. Sci. Appl.* **2011**, *4*, 35–65. [CrossRef]

79. Lee, S.-Y.; Park, S.-J. TiO$_2$ photocatalyst for water treatment applications. *J. Ind. Eng. Chem.* **2013**, *19*, 1761–1769. [CrossRef]

80. Pauzauskie, P.J.; Yang, P. Nanowire photonics. *Mater. Today* **2006**, *9*, 36–45. [CrossRef]

81. Yan, R.; Gargas, D.; Yang, P. Nanowire photonics. *Nature Photonics* **2009**, *3*, 569. [CrossRef]

82. Zhou, X.; Liu, N.; Schmuki, P. Photocatalysis with TiO$_2$ nanotubes: "Colorful" reactivity and designing site-specific photocatalytic centers into TiO$_2$ nanotubes. *ACS Catal.* **2017**, *7*, 3210–3235. [CrossRef]

83. Shibata, T.; Sakai, N.; Fukuda, K.; Ebina, Y.; Sasaki, T. Photocatalytic properties of titania nanostructured films fabricated from Titania nanosheets. *Phys. Chem. Chem. Phys.* **2007**, *9*, 2413–2420. [CrossRef] [PubMed]

84. Shichi, T.; Katsumata, K.-I. Development of photocatalytic self-cleaning glasses utilizing metal oxide nanosheets. *J. Surf. Finish. Soc. Jpn* **2010**, *61*, 30–35. [CrossRef]

85. Ong, W.-J.; Tan, L.-L.; Chai, S.-P.; Yong, S.-T.; Mohamed, A.R. Highly reactive {001} facets of TiO$_2$-based composites: Synthesis, formation mechanism and characterization. *Nanoscale* **2014**, *6*, 1946–2008. [CrossRef] [PubMed]

86. Sun, L.; Zhao, Z.; Zhou, Y.; Liu, L. Anatase TiO$_2$ nanocrystals with exposed {001} facets on graphene sheets via molecular grafting for enhanced photocatalytic activity. *Nanoscale* **2012**, *4*, 613–620. [CrossRef] [PubMed]

87. Yang, H.G.; Sun, C.H.; Qiao, S.Z.; Zou, J.; Liu, G.; Smith, S.C.; Cheng, H.M.; Lu, G.Q. Anatase TiO$_2$ single crystals with a large percentage of reactive facets. *Nature* **2008**, *453*, 638–641. [CrossRef]

88. Tan, Z.; Sato, K.; Takami, S.; Numako, C.; Umetsu, M.; Soga, K.; Nakayama, M.; Sasaki, R.; Tanaka, T.; Ogino, C.; et al. Particle size for photocatalytic activity of anatase TiO$_2$ nanosheets with highly exposed {001} facets. *RSC Adv.* **2013**, *3*, 19268–19271. [CrossRef]

89. Liu, G.; Yang, H.G.; Pan, J.; Yang, Y.Q.; Lu, G.Q.; Cheng, H.M. Titanium dioxide crystals with tailored facets. *Chem. Rev.* **2014**, *114*, 9559–9612. [CrossRef]

90. Zheng, Z.; Huang, B.; Qin, X.; Zhang, X.; Dai, Y.; Jiang, M.; Wang, P.; Whangbo, M.H. Highly efficient photocatalyst: TiO$_2$ microspheres produced from TiO$_2$ nanosheets with a high percentage of reactive {001} facets. *Chemistry* **2009**, *15*, 12576–12579. [CrossRef]

91. Wang, J.; Zhang, P.; Li, X.; Zhu, J.; Li, H. Synchronical pollutant degradation and H$_2$ production on a Ti^{3+}-doped TiO$_2$ visible photocatalyst with dominant (001) facets. *Appl. Catal. B Environ.* **2013**, *134–135*, 198–204. [CrossRef]

92. Banerjee, B.; Amoli, V.; Maurya, A.; Sinha, A.K.; Bhaumik, A. Green synthesis of Pt-doped TiO$_2$ nanocrystals with exposed (001) facets and mesoscopic void space for photo-splitting of water under solar irradiation. *Nanoscale* **2015**, *7*, 10504–10512. [CrossRef] [PubMed]

93. Fattakhova-Rohlfing, D.; Zaleska, A.; Bein, T. Three-dimensional titanium dioxide nanomaterials. *Chem. Rev.* **2014**, *114*, 9487–9558. [CrossRef] [PubMed]

94. Xiong, T.; Dong, F.; Wu, Z. Enhanced extrinsic absorption promotes the visible light photocatalytic activity of wide band-gap (BiO)2CO3 hierarchical structure. *RSC Adv.* **2014**, *4*, 56307–56312. [CrossRef]

95. Li, H.; Bian, Z.; Zhu, J.; Zhang, D.; Li, G.; Huo, Y.; Li, H.; Lu, Y. Mesoporous titania spheres with tunable chamber stucture and enhanced photocatalytic activity. *J. Am. Chem. Soc.* **2007**, *129*, 8406–8407. [CrossRef] [PubMed]

96. Kondo, Y.; Yoshikawa, H.; Awaga, K.; Murayama, M.; Mori, T.; Sunada, K.; Bandow, S.; Iijima, S. Preparation, photocatalytic activities, and dye-sensitized solar-cell performance of submicron-scale TiO$_2$ hollow spheres. *Langmuir* **2008**, *24*, 547–550. [CrossRef] [PubMed]

97. Wang, L.; Takayoshi, S.; Ebina, Y.; Kurashima, K.; Watanabe, M. Fabrication of controllable ultrathin hollow shells by layer-by-layer assembly of exfoliated ritania nanosheets on polymer templates. *Chem. Mater.* **2002**, *14*, 4827–4832. [CrossRef]

98. Zhao, T.; Liu, Z.; Nakata, K.; Nishimoto, S.; Murakami, T.; Zhao, Y.; Jiang, L.; Fujishima, A. Multichannel TiO_2 hollow fibers with enhanced photocatalytic activity. *J. Mater. Chem.* **2010**, *20*, 5095–5099. [CrossRef]

99. Shang, S.; Jiao, X.; Chen, D. Template-free fabrication of TiO_2 hollow spheres and their photocatalytic properties. *ACS Appl. Mater. Interfaces* **2012**, *4*, 860–865. [CrossRef]

100. Song, R.Q.; Colfen, H. Mesocrystals–ordered nanoparticle superstructures. *Adv. Mater.* **2010**, *22*, 1301–1330. [CrossRef]

101. Colfen, H.; Antonietti, M. Mesocrystals: Inorganic superstructures made by highly parallel crystallization and controlled alignment. *Angew. Chem. Int. Ed.* **2005**, *44*, 5576–5591. [CrossRef] [PubMed]

102. Cai, J.; Qi, L. TiO_2 mesocrystals: Synthesis, formation mechanisms and applications. *Sci. China Chem.* **2012**, *55*, 2318–2326. [CrossRef]

103. Zhang, P.; Tachikawa, T.; Bian, Z.; Majima, T. Selective photoredox activity on specific facet-dominated TiO_2 mesocrystal superstructures incubated with directed nanocrystals. *Appl. Catal. B Environ.* **2015**, *176–177*, 678–686. [CrossRef]

104. Zhang, P.; Tachikawa, T.; Fujitsuka, M.; Majima, T. Efficient charge separation on 3D architectures of TiO_2 mesocrystals packed with a chemically exfoliated MoS_2 shell in synergetic hydrogen evolution. *Chem. Commun.* **2015**, *51*, 7187–7190. [CrossRef] [PubMed]

105. Zhang, P.; Fujitsuka, M.; Majima, T. TiO_2 mesocrystal with nitrogen and fluorine codoping during topochemical transformation: Efficient visible light induced photocatalyst with the codopants. *Appl. Catal. B Environmental* **2016**, *185*, 181–188. [CrossRef]

106. Elbanna, O.; Fujitsuka, M.; Majima, T. g-C_3N_4/TiO_2 Mesocrystals composite for H_2 evolution under visible-light irradiation and its charge carrier dynamics. *ACS Appl. Mater. Interfaces* **2017**, *9*, 34844–34854. [CrossRef] [PubMed]

107. Ossama Elbanna, S.K. Mamoru Fujitsuka and Tetsuro Majima, TiO_2 mesocrystals composited with gold nanorods for highly efficient visible-NIR-photocatalytic hydrogen production. *Nano Energy* **2017**, *17*, 1842.

108. Kang, X.; Song, X.-Z.; Liu, S.; Pei, M.; Wen, W.; Tan, Z. In situ formation of defect-engineered N-doped TiO_2 porous mesocrystal for enhanced photo-degradation and PEC performance. *Nanoscale Adv.* **2019**. [CrossRef]

109. Choi, Y.; Ko, M.S.; Bokare, A.D.; Kim, D.-H.; Bahnemann, D.W.; Choi, W. Sequential process combination of photocatalytic oxidation and dark reduction for the removal of organic pollutants and Cr(VI) using Ag/TiO_2. *Environ. Sci. Technol.* **2017**, *51*, 3975–3981. [CrossRef]

110. Kim, S.; Park, H. Sunlight-harnessing and storing heterojunction TiO_2/Al_2O_3/WO_3 electrodes for night-time applications. *RSC Adv.* **2013**, *3*, 17551–17558. [CrossRef]

111. Tatsuma, T.; Saitoh, S.; Ohko, Y.; Fujishima, A. TiO_2-WO_3 photoelectrochemical anticorrosion system with an energy storage ability. *Chem. Mater.* **2001**, *13*, 2838–2842. [CrossRef]

112. Li, J.; Cushing, S.K.; Zheng, P.; Senty, T.; Meng, F.; Bristow, A.D.; Manivannan, A.; Wu, N. Solar hydrogen generation by a CdS-Au-TiO_2 sandwich nanorod array enhanced with Au nanoparticle as electron relay and plasmonic photosensitizer. *J. Am. Chem. Soc.* **2014**, *136*, 8438–8449. [CrossRef] [PubMed]

113. Hu, L.; Zhang, Y.; Zhang, S.; Li, B. A transparent TiO_2-C@TiO_2-graphene free-standing film with enhanced visible light photocatalysis. *RSC Adv.* **2016**, *6*, 43098–43103. [CrossRef]

114. Wang, W.; Ni, Y.; Lu, C.; Xu, Z. Hydrogenation of TiO_2 nanosheets with exposed {001} facets for enhanced photocatalytc activity. *RSC Adv.* **2012**, *2*, 8286–8288.

115. Plodinec, M.; Grcic, I.; Willinger, M.G.; Hammud, A.; Huang, X.; Panzic, I.; Gajovic, A. Black TiO_2 nanotube arrays decorated with Ag nanoparticles for enhanced wisible-light photocatalytic oxidation of salicylic acid. *J. Alloys Comp.* **2019**, *776*, 883–896. [CrossRef]

116. Ling, H.; Kim, K.; Liu, Z.; Shi, J.; Zhu, X.; Huang, J. Photocatalytic degradation of phenol in water on as-prepared and surface modified TiO_2 nanoparticles. *Catal. Today* **2015**, *258*, 96–102. [CrossRef]

117. Hao, X.; Li, M.; Zhang, L.; Wang, K.; Liu, C. Photocatalyst TiO_2/WO_3/GO nano-composite with high efficient photocatalytic performance for BPA degradation under visible light and solar light illumination. *J. Ind. Eng. Chem.* **2017**, *55*, 140–148. [CrossRef]

118. Khamboonrueang, D.; Srirattanapibul, S.; Tang, I.-M.; Thongmee, S. $TiO_2 \cdot$rGO nanocomposite as a photo catalyst for the reduction of Cr^{6+}. *Mater. Res. Bull.* **2018**, *107*, 236–241. [CrossRef]

119. Ren, H.-T.; Jia, S.-Y.; Zou, J.-J.; Wu, S.-H.; Han, X. A facile preparation of Ag_2O/P25 photocatalyst for selective reduction of nitrate. *Appl. Catal. B Environ.* **2015**, *176–177*, 53–61. [CrossRef]

120. Fan, C.; Chen, C.; Wang, J.; Fu, X.; Ren, Z.; Qian, G.; Wang, Z. Black hydroxylated titanium dioxide prepared via ultrasonication with enhanced photocatalytic activity. *Sci. Rep.* **2015**, *5*, 11712. [CrossRef]

121. Vequizo, J.J.M.; Matsunaga, H.; Ishiku, T.; Kamimura, S.; Ohno, T.; Yamakata, A. Trapping-induced enhancement of photocatalytic activity on brookite TiO_2 powders: Comparison with anatase and rutile TiO_2 powders. *ACS Catal.* **2017**, *7*, 2644–2651. [CrossRef]

122. Kim, S.; Moon, G.-h.; Kim, G.; Kang, U.; Park, H.; Choi, W. TiO_2 complexed with dopamine-derived polymers and the visible light photocatalytic activities for water pollutants. *J. Catal.* **2017**, *346*, 92–100. [CrossRef]

123. Huang, H.L.; Lee, W.G.; Wu, F.S. Emissions of air pollutants from indoor charcoal barbecue. *J. Hazard. Mater.* **2016**, *302*, 198–207. [CrossRef] [PubMed]

124. Rodrigues, S.; Ranjit, K.T.; Uma, S.; Martyanov, I.N.; Klabunde, K.J. Single-step synthesis of a highly active visible-light photocatalyst for oxidation of a common indoor air pollutant: Acetaldehyde. *Adv. Mater.* **2005**, *17*, 2467–2471. [CrossRef]

125. Lyu, J.; Zhu, L.; Burda, C. Considerations to improve adsorption and photocatalysis of low concentration air pollutants on TiO_2. *Catal. Today* **2014**, *225*, 24–33. [CrossRef]

126. Fujiwara, K.; Müller, U.; Pratsinis, S.E. Pd subnano-clusters on TiO_2 for solar-light removal of NO. *ACS Catal.* **2016**, *6*, 1887–1893. [CrossRef]

127. Martins, N.C.T.; Ângelo, J.; Girão, A.V.; Trindade, T.; Andrade, L.; Mendes, A. N-doped carbon quantum dots/TiO_2 composite with improved photocatalytic activity. *Appl. Catal. B Environ.* **2016**, *193*, 67–74. [CrossRef]

128. Zeng, L.; Song, W.; Li, M.; Zeng, D.; Xie, C. Catalytic oxidation of formaldehyde on surface of H-TiO_2/H-C-TiO_2 without light illumination at room temperature. *Appl. Catal. B Environ.* **2014**, *147*, 490–498. [CrossRef]

129. Yamazakia, S.; Tsukamoto, H.; Araki, K.; Tanimura, T.; Tejedor-Tejedor, I.; Anderson, M.A. Photocatalytic degradation of gaseous tetrachloroethylene on porous TiO_2 pellets. *Appl. Catal. B Environ.* **2001**, *33*, 109–117. [CrossRef]

130. Hernández-Alonso, M.D.; Tejedor-Tejedor, I.; Coronado, J.M.; Anderson, M.A.; Soria, J. Operando FTIR study of the photocatalytic oxidation of acetone in air over TiO_2-ZrO_2 thin film. *Catal. Today* **2009**, *143*, 364–373. [CrossRef]

131. Xu, Y.-J.; Zhuang, Y.; Fu, X. New insight for enhanced photocatalytic Activity of TiO_2 by doping carbon nanotubes: A case study on degradation of benzene and methyl orange. *J. Phys. Chem. C.* **2010**, *114*, 2669–2676. [CrossRef]

132. Zou, X.; Liu, J.; Su, J.; Zuo, F.; Chen, J.; Feng, P. Facile synthesis of thermal- and photostable titania with paramagnetic oxygen vacancies for visible-light photocatalysis. *Chemistry* **2013**, *1*, 2866–2873. [CrossRef]

133. Zhou, W.; Li, W.; Wang, J.Q.; Qu, Y.; Yang, Y.; Xie, Y.; Zhang, K.; Wang, L.; Fu, H.; Zhao, D. Ordered mesoporous black TiO(2) as highly efficient hydrogen evolution photocatalyst. *J. Am. Chem. Soc.* **2014**, *136*, 9280–9283. [CrossRef] [PubMed]

134. Zhong, R.; Zhang, Z.; Yi, H.; Zeng, L.; Tang, C.; Huang, L.; Gu, M. Covalently bonded 2D/2D O-g-C_3N_4/TiO_2 heterojunction for enhanced visible-light photocatalytic hydrogen evolution. *Appl. Catal. B Environ.* **2018**, *237*, 1130–1138. [CrossRef]

135. Slamet; Nasution, H.W.; Purnama, E.; Kosela, S.; Gunlazuardi, J. Photocatalytic reduction of CO_2 on copper-doped Titania catalysts prepared by improved-impregnation method. *Catal. Commun.* **2005**, *6*, 313–319. [CrossRef]

136. Liu, D.; Fernández, Y.; Ola, O.; Mackintosh, S.; Maroto-Valer, M.; Parlett, C.M.A.; Lee, A.F.; Wu, J.C.S. On the impact of Cu dispersion on CO_2 photoreduction over Cu/TiO_2. *Catal. Commun.* **2012**, *25*, 78–82. [CrossRef]

137. Liu, G.; Wang, H.; Wang, M.; Liu, W.; Ardhi, R.E.A.; Zou, D.; Lee, J.K. Study on a stretchable, fiber-shaped, and TiO_2 nanowire array-based dye-sensitized solar cell with electrochemical impedance spectroscopy method. *Electrochim. Acta* **2018**, *267*, 34–40. [CrossRef]

138. Liu, B.; Sun, Y.; Wang, X.; Zhang, L.; Wang, D.; Fu, Z.; Lin, Y.; Xie, T. Branched hierarchical photoanode of anatase TiO_2 nanotubes on rutile TiO_2 nanorod arrays for efficient quantum dot-sensitized solar cells. *J. Mater. Chem. A* **2015**, *3*, 4445–4452. [CrossRef]

139. Chen, J.; Li, Y.; Mu, J.; Zhang, Y.; Yu, Z.; Han, K.; Zhang, L. C@TiO_2 nanocomposites with impressive electrochemical performances as anode material for lithium-ion batteries. *J. Alloys Comp.* **2018**, *742*, 828–834. [CrossRef]

140. Su, D.; Dou, S.; Wang, G. Anatase TiO$_2$: Better anode material than amorphous and rutile phases of TiO$_2$ for Na-ion batteries. *Chem. Mater.* **2015**, *27*, 6022–6029. [CrossRef]

141. Kim, C.; Kim, S.; Lee, J.; Kim, J.; Yoon, J. Capacitive and oxidant generating properties of black-colored TiO$_2$ nanotube array fabricated by electrochemical self-doping. *ACS Appl. Mater. Interfaces* **2015**, *7*, 7486–7491. [CrossRef] [PubMed]

142. Yang, S.; Lin, Y.; Song, X.; Zhang, P.; Gao, L. Covalently coupled ultrafine H-TiO$_2$ nanocrystals/nitrogen-doped graphene hybrid materials for high-performance supercapacitor. *ACS Appl. Mater. Interfaces* **2015**, *7*, 17884–17892. [CrossRef] [PubMed]

143. Pazhamalai, P.; Krishnamoorthy, K.; Mariappan, V.K.; Kim, S.J. Blue TiO$_2$ nanosheets as a high-performance electrode material for supercapacitors. *J. Colloid Interface Sci.* **2019**, *536*, 62–70. [CrossRef] [PubMed]

144. Liu, B.; Mu, L.; Han, B.; Zhang, J.; Shi, H. Fabrication of TiO$_2$/Ag$_2$O heterostructure with enhanced photocatalytic and antibacterial activities under visible light irradiation. *Appl. Surf. Sci.* **2017**, *396*, 1596–1603. [CrossRef]

145. Yu, X.; Wang, S.; Zhang, X.; Qi, A.; Qiao, X.; Liu, Z.; Wu, M. Heterostructured nanorod array with piezophototronic and plasmonic effect for photodynamic bacteria killing and wound healing. *Nano Energy* **2018**, *46*, 29–38. [CrossRef]

146. Karlsson, J.; Atefyekta, S.; Andersson, M. Controlling drug delivery kinetics from mesoporous titania thin films by pore size and surface energy. *Int. J. Nanomed.* **2015**, *10*, 4425–4436. [CrossRef] [PubMed]

147. Gupta, B.; Poudel, B.K.; Ruttala, H.B.; Regmi, S.; Pathak, S.; Gautam, M.; Jin, S.G.; Jeong, J.H.; Choi, H.G.; Ku, S.K.; et al. Hyaluronic acid-capped compact silica-supported mesoporous titania nanoparticles for ligand-directed delivery of doxorubicin. *Acta Biomater.* **2018**, *80*, 364–377. [CrossRef]

148. Guo, Z.; Zheng, K.; Tan, Z.; Liu, Y.; Zhao, Z.; Zhu, G.; Ma, K.; Cui, C.; Wang, L.; Kang, T. Overcoming drug resistance with functional mesoporous titanium dioxide nanoparticles combining targeting, drug delivery and photodynamic therapy. *J. Mater. Chem. B* **2018**, *6*, 7750–7759. [CrossRef]

149. Nakayama, M.; Sasaki, R.; Ogino, C.; Tanaka, T.; Morita, K.; Umetsu, M.; Ohara, S.; Tan, Z.; Nishimura, Y.; Akasaka, H.; et al. Titanium peroxide nanoparticles enhanced cytotoxic effects of X-ray irradiation against pancreatic cancer model through reactive oxygen species generation in vitro and in vivo. *Radiat. Oncol.* **2016**, *11*, 91. [CrossRef]

150. Dai, Z.; Song, X.Z.; Cao, J.; He, Y.; Wen, W.; Xu, X.; Tan, Z. Dual-stimuli-responsive TiOx/DOX nanodrug system for lung cancer synergistic therapy. *RSC Adv.* **2018**, *8*, 21975. [CrossRef]

 *catalysts* 

Article

Synergistic Effect of Photocatalytic Degradation of Hexabromocyclododecane in Water by UV/TiO$_2$/persulfate

Qiang Li [1,2], Lifang Wang [1], Xuhui Fang [3], Li Zhang [2,*], Jingjiu Li [2] and Hongyong Xie [2,*]

[1] School of Management, Northwestern Polytechnical University, 127 West Youxi Road, Xian 710072, China; failureend@163.com (Q.L.); lifang@nwpu.edu.cn (L.W.)

[2] Research Center of Resource Recycling Science and Engineering, School of Environmental and Materials Engineering, Shanghai Polytechnic University, Shanghai 201209, China; lijingjiu123@163.com

[3] Centre Testing International Pinbiao (Shanghai) Co., Ltd., 1996 New Jinqiao Road, Shanghai 201206, China; fangxuwh@126.com

* Correspondence: zhangli@sspu.edu.cn (L.Z.); hyxie@sspu.edu.cn (H.X.); Tel.: +86-021-5021-1210 (L.Z.); +86-021-5021-1231 (H.X.)

Received: 4 January 2019; Accepted: 31 January 2019; Published: 18 February 2019

Abstract: In this work, the elimination of hexabromocyclododecane (HBCD) is explored by using photodegradation of the UV/TiO$_2$ system, the UV/potassium persulfate (KPS) system, and the homo/heterogeneous UV/TiO$_2$/KPS system. The experimental results show that the dosages of TiO$_2$ and potassium persulfate have optimum values to increase the degradation degree. HBCD can be almost completely degraded and 74.3% of the total bromine content is achieved in the UV/TiO$_2$/KPS homo/heterogeneous photocatalysis, much more than in the UV/persulfate system and the UV/TiO$_2$ system. Roles of radicals SO$_4{}^{\bullet-}$ and OH$^\bullet$ in the photocatalysis systems are discussed based on experimental measurements. The high yield of the concentration of bromide ions and decreased pH value indicates that synergistic effects exist in the UV/TiO$_2$/KPS homo/heterogeneous photocatalysis, which can mineralize HBCD into inorganic small molecules like carboxylic acids, CO$_2$ and H$_2$O, thus much less intermediates are formed. The possible pathways of degradation of HBCD in the UV/TiO$_2$/KPS system were also analyzed by GC/MS. This work will have practical application potential in the fields of pollution control and environmental management.

Keywords: hexabromocyclododecane; environmental management; photocatalysis; advanced oxidation processes

---

1. Introduction

Hexabromocyclododecane (HBCD) is a high bromine content additive flame retardant that is mainly used in polystyrene electrical equipment, insulation boards, resin, polyester fabric, synthetic rubber coating, and so on [1]. Studies have shown that HBCD is a potential endocrine disruptor, and it has immunotoxicity, neurotoxicity, and cytotoxicity [2]. The presence of HBCD was detected in environmental samples such as water, atmosphere, sediment, soil, food, and even in the human body in breast milk and plasma, as it can be enriched through the food chain, causing persistent pollution [3]. The hazard of HBCD and its pollution in the environment have caused widespread concern, and it is of great significance to develop a strategy for the elimination of HBCD pollution [4]. However, the molecular structure of HBCD with a ring structure is relatively stable, making it heat-resistant, UV-resistant, and difficult to be decomposed in the natural environment [5]. Methods for eliminating HBCD in the environment include microbial degradation, ultrasonic degradation, chemical reduction, phytoremediation, and mechanical ball milling [6–8]. However, these methods have not been practically used because of the harsh reaction conditions, high energy consumption, low efficiency, and secondary pollution [9,10]. Nowadays, advanced oxidation processes (AOPs) have been widely used for the elimination of organic pollutants of water or gas, using the highly reactive chemical species like hydroxyl radicals ($OH^\bullet$) to oxidize most of pollutants into small molecular substances that are harmless to the environment, such as CO_2, H_2O and so on [11]. Increasing the number of hydroxyl radicals could increase the efficiency of the AOPs reactions. Some types of AOPs based on UV, H_2O_2/UV, O_3/UV and H_2O_2/O_3/UV combinations use photolysis of H_2O_2 and ozone to produce $OH^\bullet$, while the heterogeneous UV/TiO_2 photocatalysis and homogeneous photo-Fenton are based on the use of a wide-band gap semiconductor and addition of H_2O_2 to dissolved iron salts that produce $OH^\bullet$ under UV irradiation, respectively [12]. Among AOPs, the UV/TiO_2 heterogeneous photocatalysis has gradually attracted the interest of scientists in elimination of toxic pollutions due to its efficiency, low-cost and broad applicability [13]. The photocatalytic technology can be briefly described as follows: under UV irradiation, the electron in TiO_2 was excited and transferred from the valence band (VB) to the conduction band (CB), resulting in the formation of high energy electron-hole pairs; the electrons may also react with O_2 and generate a superoxide ion ($O_2^{\bullet-}$), while holes were captured by surface hydroxyl groups (OH^-) on the photocatalyst surface to yield $OH^\bullet$ [14,15]. However, the photogenerated electron-hole pairs are easy to combine within a very short time of $10^{-9}{\sim}10^{-12}$ s, which results in a lower photocatalytic degradation efficiency [16]. In order to solve this problem, scientists have conducted lots of meaningful and in-depth research. For instance, Aronne et al. found that high Ti^{3+} self-doping TiO_{2-x} not only has a wide range of visible light responses, but also has a low recombination rate of electron–hole pairs [17]; Sannino et al. fabricated hybrid TiO_2–acetylacetonate amorphous gel-derived material with stably adsorbed superoxide radical ($O_2^{\bullet-}$) active in oxidative degradation of organic pollutants in the absence of any light irradiation [18,19].

It has been reported that using the strong oxidant of persulfate ion ($S_2O_8^{2-}$) (with redox potential of 2.05 V) is effective for degrading organic pollutions in water solution through direct chemical oxidation [20]. The $S_2O_8^{2-}$ can be activated via thermal, UV light, or redox decomposition to generate the stronger oxidant of sulfate radicals (E_0 = (2.5–3.1) V vs. NHE) [21–23]. It's worthwhile to note that both the persulfate ion and sulfate radicals ($SO_4^{\bullet-}$) can be dissolved in water, so the free radicals and contaminants in water can be contracted at the molecular level, leading to a higher reaction rate. For instance, Li et al. have found that addition of persulphate to UV/TiO_2 could improve the photocatalytic degradation of tetrabromobisphenol A and other pollutants [24–26]. Therefore, it is necessary to combine persulfate and UV/TiO_2 photocatalytic techniques to increase the mineralization of HBCD.

In this work, degradation of HBCD under UV/TiO_2, UV/potassium persulfate (KPS), and UV/TiO_2/KPS systems were investigated. Effects of TiO_2 and KPS dosage have been examined on degradation degree of HBCD. The photodegradation efficiency and the yield of bromide ion were tested to evaluate the mineralization of HBCD. The intermediates were analyzed by GC/MS to study the degradation mechanism. Based on experimental measurements, roles of radicals $SO_4^{\bullet-}$ and $OH^\bullet$ in the photocatalysis systems were also discussed.

2. Results and Discussions

2.1. Determination of TiO₂ Dosages

The amount of catalysts added in the solution needs to be matched to the number of contaminants in the photocatalytic process, so the dosing weight range of the catalyst were determined. Figure 1 shows the effects of different TiO_2 dosages on the photodegradation rate of HBCD. Under the condition of no addition of TiO_2, the degradation rate was only 21.5% at 180 min. Having increased the catalyst dosage of TiO_2 to 100 mg/L, the photocatalytic efficiency also increased to 82.93%. Further increasing the dosage of TiO_2 more than 100 mg/L, the photocatalytic efficiency decreased. The dosage of the addition increased the suspended particles in the solution, and greatly reduced the utilization of light, resulting in the partial catalyst not being fully activated during the photocatalysis, so the photocatalytic efficiency decreased [24,27]. In addition, the HBCD adsorbed on the catalysts in dark is less than 10%, so the free radical (OH$^\bullet$) reaction dominates the rate of degradation reaction in the UV/TiO$_2$ system.

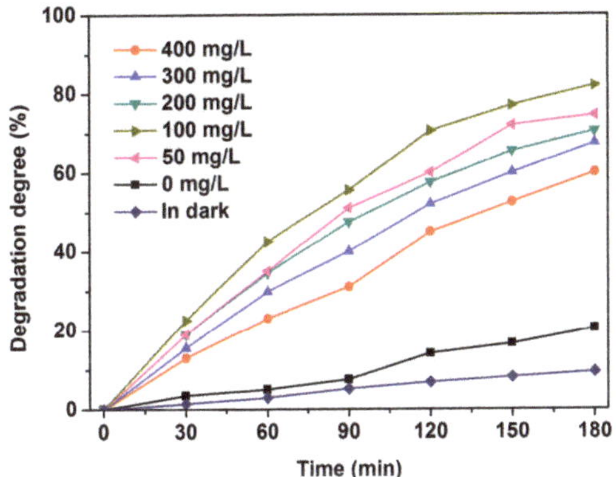

Figure 1. Effects of different TiO$_2$ dosages on the photodegradation rate of HBCD (the initial concentration of HBCD is 25 mg/L, and KPS dosage is 0 mg/L).

2.2. Effect of KPS Dosage

Figure 2 shows the effect of different K$_2$S$_2$O$_8$ dosages on the photodegradation rate of HBCD. The addition of K$_2$S$_2$O$_8$ can effectively improve the degradation efficiency of HBCD, but its degradation efficiency increases first and then decreases with further increasing K$_2$S$_2$O$_8$ concentration, and the highest degradation efficiency occurs at 4 mM. When the K$_2$S$_2$O$_8$ dosage was more than 4 mM, the degradation efficiency of HBCD decreased with the increase of persulfate dosage. When the K$_2$S$_2$O$_8$ concentrations in the solution are between 0 and 4 mM, the main reactions in the UV/TiO$_2$/KPS system are as follows [28]:

$$S_2O_8{}^{2-} + UV \rightarrow 2SO_4{}^{\bullet -} \tag{1}$$

$$TiO_2 + UV \rightarrow e_{CB}^- + h_{VB}^+ \tag{2}$$

$$S_2O_8{}^{2-} + e_{CB}^- \rightarrow 2SO_4{}^{\bullet -} \tag{3}$$

$$H_2O + h_{VB}^+ \rightarrow OH^\bullet + H^+ \tag{4}$$

The advanced oxidation process relies on the amount of free radicals and is reflected in the degradation rate of the contaminants. The strong oxidizing agents of sulfate radicals (SO$_4{}^{\bullet -}$) and hydroxyl radicals (OH$^\bullet$) generated by the above reactions increase with increasing KPS concentration in the solution. But increasing the KPS dosage further to 8 mM will lead to a surplus of reactants (S$_2$O$_8{}^{2-}$), which may deplete lots of OH$^\bullet$ and SO$_4{}^{\bullet -}$, and lead to the decrease of the degradation degree by the following two reactions [29]:

$$S_2O_8{}^{2-} + OH^\bullet \rightarrow S_2O_8{}^{\bullet -} + OH^- \tag{5}$$

$$S_2O_8{}^{2-} + SO_4{}^{\bullet -} \rightarrow SO_4{}^{2-} + S_2O_8{}^{\bullet -} \tag{6}$$

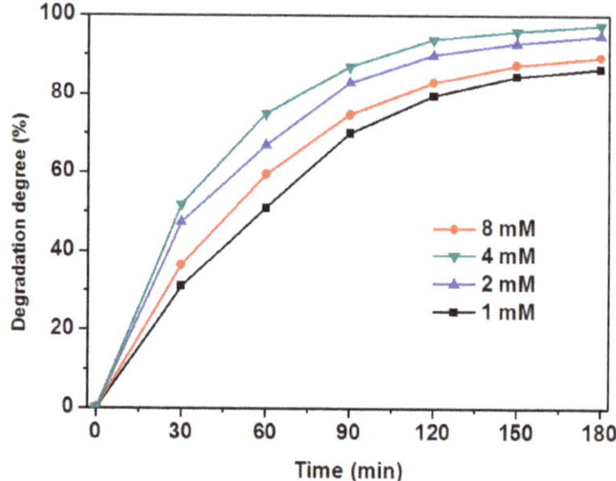

Figure 2. Effects of different KPS dosages on the photodegradation rate of HBCD (the initial concentration of HBCD is 25 mg/L, and TiO$_2$ dosage is 100 mg/L).

2.3. Kinetic Analysis of Different Reaction Systems

Figure 3 shows the degradation effect of HBCD in the three systems of "UV/TiO$_2$ (TiO$_2$: 100 mg/L)", "UV/K$_2$S$_2$O$_8$ (KPS: 4 mM)", and "TiO$_2$ (TiO$_2$: 100 mg/L) + K$_2$S$_2$O$_8$ (KPS: 4 mM)", respectively. The initial HBCD concentration and light source in the three systems were all the same (25 mg/L, 100 W mercury lamp). The degradation degree of HBCD over time in 180 min is shown in Figure 3a. The degradation degree for the UV/TiO$_2$/KPS photocatalytic system was 87.6% at 90 min, but the degradation rates for the UV/TiO$_2$ photocatalytic system and UV/KPS system were only 56.8% and 52.5% at the same time. The above experimental results show that the degradation effect of UV/TiO$_2$/KPS photocatalytic system on HBCD is far superior to that of the UV/TiO$_2$ system and the UV/KPS system.

The kinetic model was used to study the degradation dynamic behavior of the three different systems [30], $-\ln(C_0/C) = kt$, where k is the reaction apparent rate constant and t is the light irradiation time. Figure 3b shows the effect of different systems on the kinetics of HBCD under irradiation for 180 min. The three reaction systems are all fit to pseudo-first-order kinetics, and the k values for the UV/KPS system, the UV/TiO$_2$ system, and the UV/TiO$_2$/KPS system are 0.0065, 0.0080, and 0.0174 min^{-1}, respectively (Figure 3b). Obviously, the k value of the UV/TiO$_2$/KPS system is far higher than those of the UV/KPS system and the UV/TiO$_2$ system, indicating that the degradation efficiency of the UV/TiO$_2$/KPS photocatalytic system is much higher than that of the UV/KPS photocatalytic system and the UV/TiO$_2$ photocatalytic system. The photocatalytic process of the UV/TiO$_2$ system contains an adsorption and free radical (O$_2^{\bullet-}$, OH$^{\bullet}$, etc.) reaction [21]. The HBCD adsorbed on the catalysts in dark is less than 10%, so the free radical reaction dominates the rate of degradation reaction in the UV/TiO$_2$ system. The UV/KPS system also relies on sulfate radicals (SO$_4^{\bullet-}$, S$_2$O$_8^{\bullet-}$, etc.) excited by UV light to degrade pollutants [22,24]. In the UV/TiO$_2$/KPS system, more free radicals were present and the free radical reaction is more complicated. S$_2$O$_8^{2-}$ can be excited by photogenerated electrons on the surface of the catalyst to generate sulfate radicals (SO$_4^{\bullet-}$), while SO$_4^{\bullet-}$ can react with OH$^-$ to produce OH$^{\bullet}$ [21–24]. In the three systems, the degradation rates are all determined by the reactions between free radicals and contaminant molecules, while the intensity of the UV light (100 W) and the initial concentration (25.00 mg/L) of the contaminants in the three systems are all the same, so all reaction systems could be in line with pseudo-first-order kinetics. Figure S1 (Supporting Information) shows the degradation of HBCD over the UV/TiO$_2$/KPS system with three time cycling uses. The TiO$_2$ photocatalysts could be easily recovered by sedimentation and reused, which would greatly promote their industrial application in eliminating organic pollutants from wastewater.

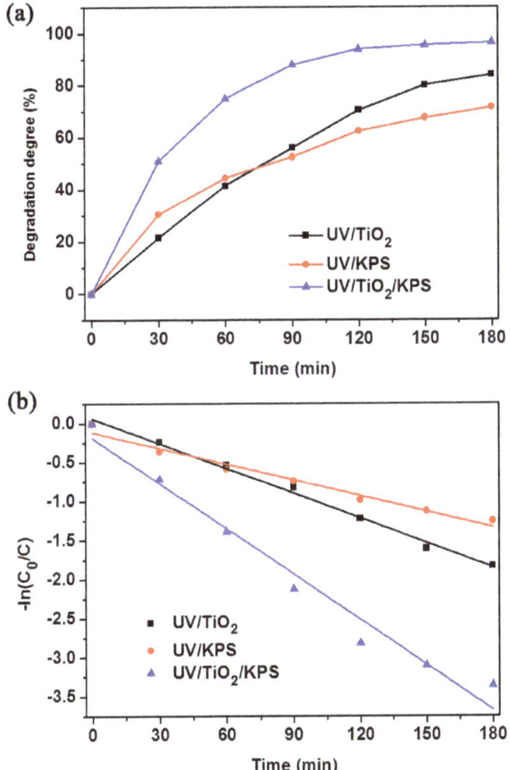

Figure 3. The degradation degree (**a**) and kinetic linear simulation curves of the removal of HBCD (**b**) in UV/TiO$_2$, UV/KPS, and UV/TiO$_2$/KPS systems. (The initial concentration of HBCD is 25.0 mg/L, KPS dosage is 4 mg/L, and TiO$_2$ dosage is 100 mg/L).

2.4. The Mineralization Degree of HBCD

Measuring the concentration of bromide ion is a practical strategy to evaluate the amount of intermediates and the mineralization degree of HBCD [24]. Figure 4 shows the change of bromide ion concentration during the degradation of HBCD in the UV/TiO$_2$/KPS system. The initial HBCD concentration, TiO$_2$ dosage, and K$_2$S$_2$O$_8$ dosage were 25.0 mg/L, 100 mg/L, and 4 mM, respectively.

As can be seen from Figure 4, with the prolongation of degradation time, the concentration of bromide ion in the solution increased continuously. When the reaction was carried out for 3.0 h, the concentration of bromide ions in the solution was 13.8 mg/L, which accounted for 74.3% of the total bromine content of HBCD in the solution. It can be seen that the yield of bromine ion by UV/TiO$_2$/KPS system is much better than that of UV/TiO$_2$ system (12.3 mg/L) and UV/KPS system (11.9 mg/L). The significantly increased bromide ion concentration yield indicates that there exist synergistic effects in the UV/TiO$_2$/KPS photocatalysis as described in the previous Formulas (1)–(4), which can mineralize HBCD into inorganic small molecules relatively thorough, thus much fewer intermediates are formed in UV/TiO$_2$/KPS homo/heterogeneous photocatalysis.

Figure 5 shows the change of pH in solution over time during HBCD degradation. It can be seen that the pH value of the solution gradually decreases from 6.53 to 3.72 with the increasing of the degradation time within 180 min. It may be due to the partial consumption of OH$^-$ in the solution, since OH$^-$ can easily react with SO$_4^{\bullet -}$ to produce OH$^\bullet$. The CO$_2$ gas generated during the mineralization of HBCD subsequently dissolved in the water, which also lead to a decrease in pH.

At the same time, there are some small molecules of carboxylic acids generated in the degradation of HBCD, which also cause the decrease in pH. As mentioned above, the degradation rate of HBCD reached 96.5% when the reaction proceeded to 180 min. It is indicated that the intermediate in the solution is rapidly decomposed into small molecular of carboxylic acids, and further mineralized to CO_2 and H_2O, so that the pH of the solution continued to decrease as the reaction time prolonged. The increasing concentration of bromide ions in the solution and the decreasing pH value indicate that HBCD is highly mineralized in $UV/TiO_2/KPS$ homo/heterogeneous photocatalysis.

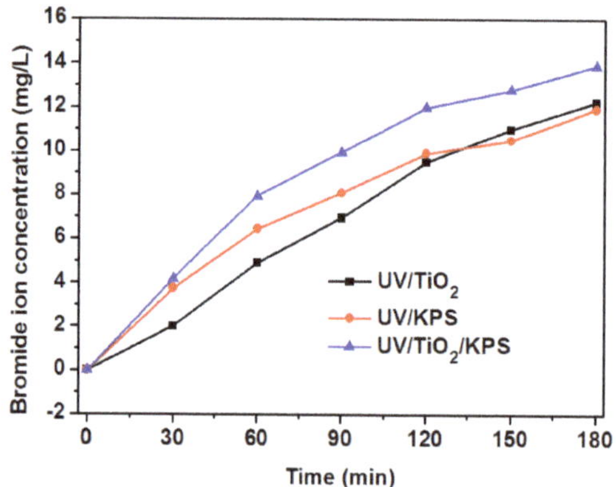

Figure 4. The change of bromide ion in water samples from different time points (the initial concentration of HBCD is 25.0 mg/L, KPS dosage is 4 mg/L, and TiO_2 dosage is 100 mg/L).

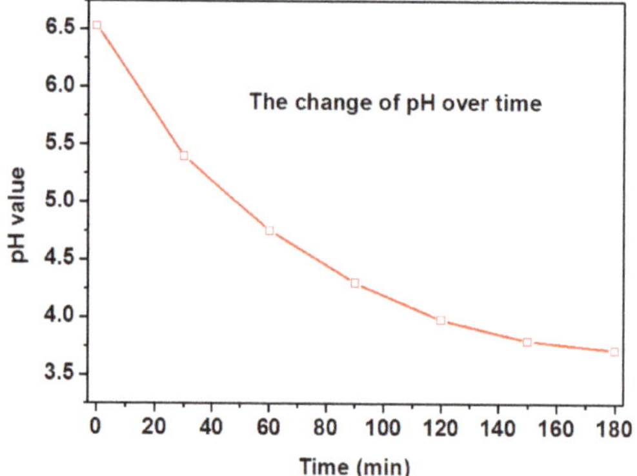

Figure 5. The change of pH over time (the initial concentration of HBCD is 25.0 mg/L, KPS dosage is 4 mg/L, and TiO_2 dosage is 100 mg/L).

2.5. The Mechanism of Photodegradation of HBCD

Figure 6 shows the mass spectrum of the intermediates obtained by GC-MS analysis. The solution was sampled during the degradation of HBCD in the $UV/TiO_2/KPS$ system at 90 min, with the reaction conditions the same as mentioned above. The mass spectrum of degradation products were tetrabromocyclododecene (A), dibromocyclododecadiene (B), 1,5,9-Cyclododecatriene (C), 1,2-Epoxy-5,9-cyclododecadiene (D), dibromo-epoxy-cyclododecene (E), 4,5-dibromooctanedioic acid (F), and succinic acid (G), respectively [31–33].

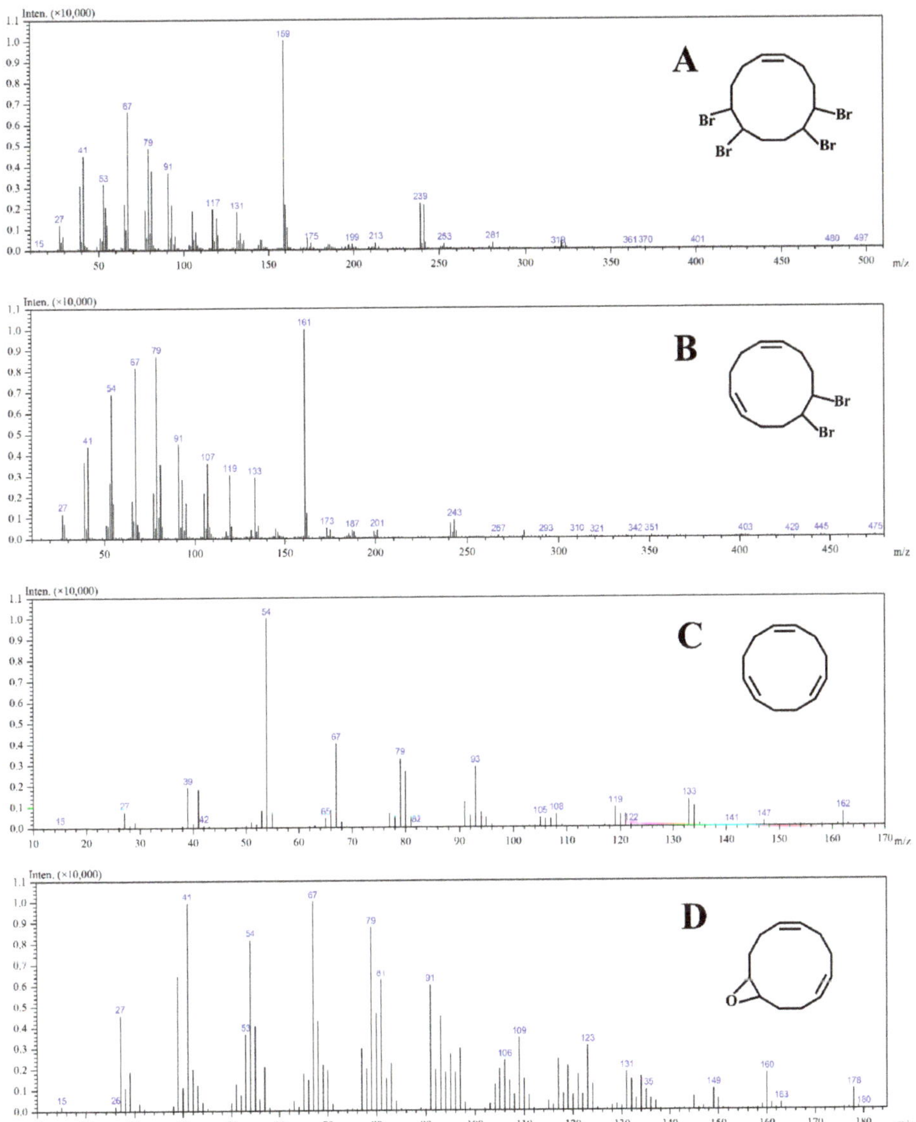

Figure 6. *Cont.*

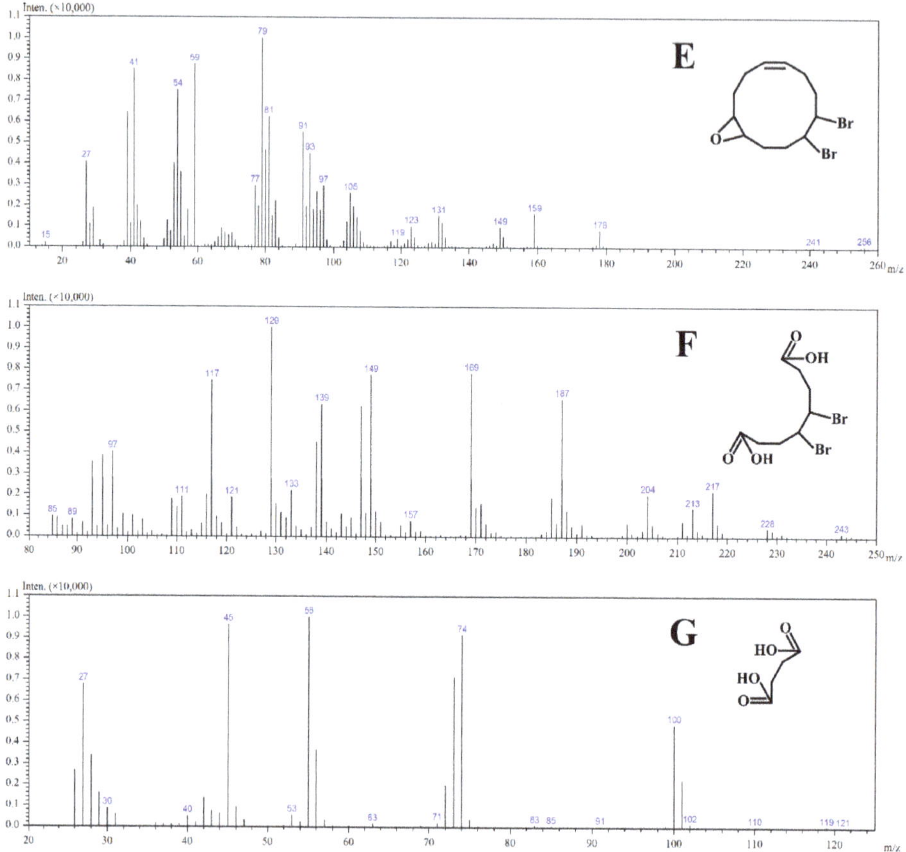

Figure 6. The mass spectrum of the intermediates in the UV/TiO$_2$/KPS system at 90 min ((**A**) Tetrabromocyclododecene; (**B**) Dibromocyclododecadiene; (**C**) 1,5,9-Cyclododecatriene; (**D**) 1,2-Epoxy-5,9-cyclododecadiene; (**E**) Dibromo-epoxy-cyclododecene; (**F**) 4,5-dibromooctanedioic acid; (**G**) succinic acid).

By analyzing the degradation products of GC/MS, the possible degradation pathway of HBCD in UV/TiO$_2$/KPS system is determined, as shown in Figure 7. Under the action of active free radicals, two adjacent C–Br bonds in the molecular structure of HBCD undergo cleavage and debromination to form carbon-carbon double bonds, thus the compounds A, B, and C were obtained successively [34]. Compound C can be directly oxidized to D or oxidized to G by double bond cleavage. In addition, the compound B can also be oxidized to form the compound E, or oxidized to F and G by double bond cleavage [35]. The compound F can also be further debrominated and oxidized to form G. Succinic acid (G) is a small molecule, and it can be easily degraded by free radicals (like SO$_4$$^{\bullet-}$ and OH$^{\bullet}$) into carboxylic acids, CO$_2$, and H$_2$O in the following time. The adsorption of intermediate species on the surface of TiO$_2$ may cover the active sites, which may result in a decrease in catalytic efficiency. But in this work, the initial concentration of HBCD is very low (25 mg/L), and under the irradiation of UV irradiation, the surface of the titanium dioxide is hydrophilic, so the organic intermediates are more easily dispersed into the water-methanol mixed solution under strong stirring.

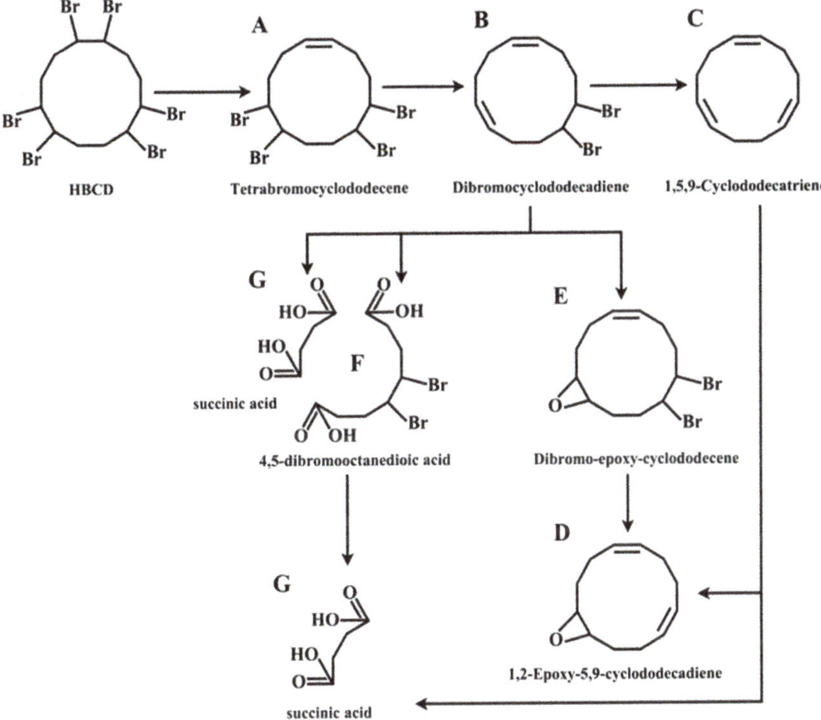

Figure 7. Possible pathways of degradation of HBCD in UV/TiO$_2$/KPS systems.

3. Materials and Methods

3.1. Reagents

Ethanol (HPLC grade) and acetonitrile (HPLC grade) were supplied by LABSCIENCE (Reno, NV, USA) and TEDIA (Nashville, TN, USA), respectively. HBCD (99.0%), sodium carbonate (99.8%), sodium bicarbonate (99.5%), sodium nitrite (99.0%), potassium persulfate (KPS, 99%), dichloromethane (HPLC grade), anhydrous sodium sulfate (99.0%) and methanol (HPLC grade) were supplied by Sinopharm Chemical Reagent Co., Ltd (Shanghai, China). All reagents were used as received without further purification. TiO$_2$ nanoparticles were laboratory-made, as described in the previous literature [26].

The preparation of HBCD stock solution was as follows: accurately weigh 0.05 g of HBCD powder into 100 mL volumetric flasks, and add chromatographically pure methanol to the 100 mL mark. After dissolving, the HBCD stock solution with the concentrations of 500.00 mg/L was obtained, and then it was stored in a refrigerator at 4 °C for later use. The HBCD stock solution was diluted by ultrapure water to different concentrations for drawing the peak area-concentration standard curve, and it also be used as pollutants in the photodegradation experiments.

3.2. Photodegradation of HBCD

The photoreactor was supplied by Xujiang Electromechanical Plant (XPA-7, Nanjing, China). For determination of TiO$_2$ dosages, a 50 mL HBCD water-methanol mixture solution with the concentration of 25.00 mg/L was added into a quartz tube, and then TiO$_2$ powder with different dosages (0~400 mg/L) were also added into the tube. For determination of the effect of KPS dosage, a 50 mL HBCD water-methanol mixture solution with the concentration of 25.00 mg/L was added into a quartz tube, and then TiO_2 powder with the dosage of 100 mg/L and KPS with different dosages (1~8 mM) were also added into the tube. For the kinetic analysis, the photodegradation of HBCD in the three systems of "UV/TiO_2 (TiO_2: 100 mg/L)", "UV/$K_2S_2O_8$ (KPS: 4 mM)", and "TiO_2 (TiO_2: 100 mg/L) + $K_2S_2O_8$ (KPS: 4 mM)" were performed, respectively. In all of the above experimental systems, after all the reagents were completely added, the mixed solution was placed in dark and stirred for 60 min to allow all the reagents to be uniformly mixed and to achieve adsorption equilibrium between TiO_2 particles (if any) and HBCD in the solution system. Then, turn on the cooling water and the 100 W mercury light source to start the photocatalytic reaction. Quickly take 2 mL of the sample at intervals of 30 min, place it in a tube containing 2 mL of methanol, mix well by shaking, and filter through a 0.22 μm filter. The filtrate was loaded into the sample vial for analysis in a liquid chromatograph. The average of 3 parallel determinations was taken as the concentration of each sample.

For the measurement of bromide ion concentration, a sample solution was quenched right after the sample was taken out by using a same volume of 0.2 M sodium nitrite solution. Then supernatant and the TiO_2 nanoparticles were separated in the same method. The supernatant was used to measure the concentration of bromide ion.

3.3. Analysis Method

Concentrations of HBCD were measured by a high-performance liquid chromatography (HPLC, LC-20AD, Shimadzu, Kyoto, Japan) instrument equipped with UV–vis detector set at 210 nm. The mobile phase was acetonitrile/water (85/15 (v/v)) and the flow rate was maintained at 1.0 mL/min. The HPLC chromatogram of HBCD was shown in Figure S2 (Supporting Information). According to the change of concentration of HBCD before and after degradation of the reaction system, the degradation rate of HBCD was calculated. The calculation was as follows:

$$\eta_{HBCD} = \frac{C_0 - C_t}{C_0} \times 100\% \tag{7}$$

where C_0 represents the initial concentration of HBCD in the reaction system and C_t represents the concentration of HBCD in the system at time t.

Concentration of bromide ion was measured by a Diane Ion chromatograph (ICS1100, Dionex, Sunnyvale, CA, USA) with an IonPac AS23 anion analytical column (250 mm × 4.0 mm × 5 μm, Dionex, Sunnyvale, CA, USA) and a Dionex IonPac AG22 anion protective column (50 mm × 4 mm, Dionex, Sunnyvale, CA, USA). The peak area-concentration standard curve of Br ion was plotted using potassium bromide powder as the bromine source. Leaching solution was 4.5 mM Na_2CO_3 and 1.4 mM $NaHCO_3$ with a flow rate of 1.2 mL/min.

The intermediates were qualitatively analyzed by a gas chromatography/mass spectrometry (GC/MS, Shimadzu QP2010 plus). The inlet temperature of GC is 200 degrees with the column type of DB-5MS capillary column (30 m × 0.25 mm × 0.25 μm, Agilent, Santa Clara, CA, USA). The injection volume is 1 μL, and the carrier gas is high purity nitrogen (99.999%). The ion source temperature, the electron bombardment energy, and the scanning mode of the mass spectrometer is 240 °C, 70 eV, and full scan mode (15~500 m/z), respectively. The sample solution was pretreated by extraction by dichloromethane and passed through anhydrous sodium sulfate, then it was concentrated by evaporation to about 1 mL under nitrogen, and passed through a 0.45 μm filter before the GC/MS analysis.

4. Conclusions

Degradation of HBCD is investigated in the UV/TiO_2, UV/KPS, and UV/TiO_2/KPS system by measurement of the concentrations of HBCD and bromide ion. HBCD can be almost completely degraded and 74.3% of the total bromine content is achieved in the UV/TiO_2/KPS homo/heterogeneous photocatalysis, much more than in the UV/KPS system and the UV/TiO_2

system. The $SO_4^{\bullet-}$ produced in persulphate and $OH^{\bullet}$ radicals produced in TiO_2 photocatalysis have synergistic effects in the degradation of HBCD in the $UV/TiO_2/KPS$ homo/heterogeneous photocatalysis. The high yield of the concentration of bromide ions in the solutions indicates that fewer intermediates are formed in the $UV/TiO_2/KPS$ homo/heterogeneous photocatalysis of HBCD. The efficient $UV/TiO_2/KPS$ homo/heterogeneous system would provide great impetus to pollution control and environmental management.

Supplementary Materials: The following are available online at http://www.mdpi.com/2073-4344/9/2/189/s1, Figure S1: Photocatalysis for HBCD degradation in the $UV/TiO_2/KPS$ system with three time cycling uses, Figure S2: The HPLC chromatogram of HBCD.

Author Contributions: Conceptualization, Q.L.; methodology, L.W.; formal analysis, L.Z.; investigation, X.F.; resources and data curation, J.L.; writing—original draft preparation, Q.L.; writing—review and editing, L.Z.; and H.X.

Funding: This research was funded by Natural Science Foundation of China (No. 21806101), Natural Science Foundation of Shanghai (Nos. 16ZR1412600, 15ZR1401200), Gaoyuan Discipline of Shanghai-Environmental Science and Engineering (Resource Recycling Science and Engineering), Innovation Research Grant (13YZ130) and Leading Academic Discipline Project (J51803) from the Shanghai Education Committee.

Conflicts of Interest: The authors declare no conflicts of interest.

References

1. Almughamsi, H.; Whalen, M.M. Hexabromocyclododecane and tetrabromobisphenol A alter secretion of interferon gamma (IFN-gamma) from human immune cells. *Arch. Toxicol.* **2016**, *90*, 1695–1707. [CrossRef] [PubMed]

2. Alaee, M.; Arias, P.; Sjodin, A.; Bergman, A. An overview of commercially used brominated flame retardants, their applications, their use patterns in different countries/regions and possible modes of release. *Environ. Int.* **2003**, *29*, 683–689. [CrossRef]

3. Jeannerat, D.; Pupier, M.; Schweizer, S.; Mitrev, Y.N.; Favreau, P.; Kohler, M. Discrimination of hexabromocyclododecane from new polymeric brominated flame retardant in polystyrene foam by nuclear magnetic resonance. *Chemosphere* **2016**, *144*, 1391–1397. [CrossRef] [PubMed]

4. Li, L.; Weber, R.; Liu, J.G.; Hu, J.X. Long-term emissions of hexabromocyclododecane as a chemical of concern in products in China. *Environ. Int.* **2016**, *91*, 291–300. [CrossRef]

5. Hunziker, R.W.; Gonsior, S.; Macgregor, J.A.; Desjardins, D. Fate and effect of hexabromocyclododecane in the environment. *Organohalogen Compd.* **2004**, *66*, 2300–2305.

6. Stiborova, H.; Vrkoslavova, J.; Pulkrabova, J.; Poustka, J.; Hajslova, J.; Demnerova, K. Dynamics of brominated flame retardants removal in contaminated wastewater sewage sludge under anaerobic conditions. *Sci. Total Environ.* **2015**, *533*, 439–445. [CrossRef] [PubMed]

7. Wagoner, E.R.; Baumberger, C.P.; Peverly, A.A.; Peters, D.G. Electrochemical reduction of 1, 2, 5, 6, 9, 10-hexabromocyclododecane at carbon and silver cathodes in dimethylformamide. *J. Electroanal. Chem.* **2014**, *713*, 136–142. [CrossRef]

8. Zhang, K.; Huang, J.; Wang, H.; Liu, K.; Yu, G.; Deng, S.B.; Wang, B. Mechanochemical degradation of hexabromocyclododecane and approaches for the remediation of its contaminated soil. *Chemosphere* **2014**, *116*, 40–45. [CrossRef] [PubMed]

9. Takigami, H.; Watanabe, M.; Kajiwara, N. Destruction behavior of hexabromocyclododecanes during incineration of solid waste containing expanded and extruded polystyrene insulation foams. *Chemosphere* **2014**, *116*, 24–33. [CrossRef] [PubMed]

10. Jondreville, C.; Cariou, R.; Meda, B.; Dominguez-Romero, E.; Omer, E.; Dervilly-Pinel, G.; Le Bizec, B.; Travel, A.; Baeza, E. Accumulation of a-hexabromocyclododecane (alpha-HBCDD) in tissues of fast- and slow-growing broilers (*Gallus domesticus*). *Chemosphere* **2017**, *178*, 424–431. [CrossRef]

11. Guo, Y.G.; Lou, X.Y.; Xiao, D.X.; Xu, L.; Wang, Z.H.; Liu, J.S. Sequential reduction-oxidation for photocatalytic degradation of tetrabromobisphenol A: Kinetics and intermediates. *J. Hazard. Mater.* **2014**, *241–242*, 301–306. [CrossRef] [PubMed]

12. Saien, J.; Ojaghloo, Z.; Soleymani, A.R.; Rasoulifard, M.H. Homogeneous and heterogeneous AOPs for rapid degradation of Triton X-100 in aqueous media via UV light, nano titania hydrogen peroxide and potassium persulfate. *Chem. Eng. J.* **2011**, *167*, 172–182. [CrossRef]

13. Salari, D.; Niaei, A.; Aber, S.; Rasoulifard, M.H. The photooxidative destruction of CI basic yellow 2 using UV/$S_2O_8^{2-}$ process in a rectangular continuous photoreactor. *J. Hazard. Mater.* **2009**, *166*, 61–66. [CrossRef] [PubMed]

14. Zhang, L.; Zhang, Q.H.; Xie, H.Y.; Guo, J.; Lyu, H.L.; Li, Y.G.; Sun, Z.G.; Wang, H.Z.; Guo, Z.H. Electrospun titania nanofibers segregated by graphene oxide for improved visible light photocatalysis. *Appl. Catal. B Environ.* **2017**, *201*, 470–478. [CrossRef]

15. Zhang, L.; Li, Y.G.; Xie, H.Y.; Wang, H.Z.; Zhang, Q.H. Efficient mineralization of toluene by W-doped TiO_2 nanofibers under visible light irradiation. *J. Nanosci. Nanotechnol.* **2015**, *15*, 2944–2951. [CrossRef]

16. Zhang, L.; Li, Y.G.; Zhang, Q.H.; Shi, G.Y.; Wang, H.Z. Fast synthesis of highly dispersed anatase TiO_2 nanocrystals in a microfluidic reactor. *Chem. Lett.* **2011**, *40*, 1371–1373. [CrossRef]

17. Aronne, A.; Fantauzzi, M.; Imparato, C.; Atzei, D.; De Stefano, L.; D'Errico, G.; Sannino, F.; Rea, I.; Pirozzi, D.; Elsener, B.; et al. Electronic properties of TiO_2-based materials characterized by high Ti^{3+} self-doping and low recombination rate of electron–hole pairs. *RSC Adv.* **2017**, *7*, 2373–2381. [CrossRef]

18. Sannino, F.; Pernice, P.; Imparato, C.; Aronne, A.; D'Errico, G.; Minieri, L.; Perfetti, M.; Pirozzi, D. Hybrid TiO_2–acetylacetonate amorphous gel-derived material with stably adsorbed superoxide radical active in oxidative degradation of organic pollutants. *RSC Adv.* **2015**, *5*, 93831–93839. [CrossRef]

19. Sannino, F.; Pernice, P.; Minieri, L.; Gamandona, G.A.; Aronne, A.; Pirozzi, D. Oxidative Degradation of Different Chlorinated Phenoxyalkanoic Acid Herbicides by a Hybrid ZrO_2 Gel-Derived Catalyst without Light Irradiation. *ACS Appl. Mater. Interfaces* **2015**, *7*, 256–263. [CrossRef]

20. Guo, Y.G.; Zhou, J.; Lou, X.Y.; Liu, R.L.; Xiao, D.X.; Fang, C.L.; Wang, Z.H.; Liu, J.S. Enhanced degradation of Tetrabromobisphenol A in water by a UV/base/persulfate system: Kinetics and intermediates. *Chem. Eng. J.* **2014**, *254*, 538–544. [CrossRef]

21. Neta, P.; Huie, R.E.; Ross, A.B. Rate constants of inorganic radicals in aqueous-solution. *J. Phys. Chem. Ref. Data* **1988**, *17*, 1027–1284. [CrossRef]

22. Xu, J.; Meng, W.; Zhang, Y.; Lei, L.; Guo, C.S. Photocatalytic degradation of tetrabromobisphenol A by mesoporous BiOBr: Efficacy, products and pathway. *Appl. Catal. B Environ.* **2011**, *107*, 355–362. [CrossRef]

23. Zhao, J.Y.; Zhang, Y.B.; Quan, X.; Chen, S. Enhanced oxidation of 4-chlorophenol using sulfate radicals generated from zero-valent iron and peroxydisulfate at ambient temperature. *Sep. Purif. Technol.* **2010**, *71*, 302–307. [CrossRef]

24. Li, Q.; Wang, L.F.; Zhang, L.; Xie, H.Y. Rapid degradation of tetrabromobisphenol A under the UV/TiO_2/KPS systems in alkaline aqueous solutions. *Res. Chem. Intermed.* **2018**. [CrossRef]

25. Ahmadi, M.; Ghanbari, F.; Moradi, M. Photocatalysis assisted by peroxymonosulfate and persulfate for benzotriazole degradation: Effect of pH on sulfate and hydroxyl radicals. *Water Sci. Technol.* **2015**, *72*, 2095–2102. [CrossRef] [PubMed]

26. Xie, H.Y.; Zhu, L.P.; Wang, L.L.; Chen, S.W.; Yang, D.D.; Yang, L.J.; Gao, G.L.; Yuan, H. Photodegradation of benzene by TiO_2 nanoparticles prepared by flame CVD process. *Particuology* **2011**, *9*, 75–79. [CrossRef]

27. Yang, H.; Zhou, S.L.; Yin, M.L.; Pi, L.L.; Zeng, J.; Yi, B. Parameters effect on photocatalytic kinetics of carbofuran in TiO_2 aqueous solution. *China Environ. Sci.* **2013**, *33*, 82–87.

28. Varanasi, L.; Coscarelli, E.; Khaksari, M.; Mazzoleni, L.R.; Minakata, D. Transformations of dissolved organic matter induced by UV photolysis, hydroxyl radicals, chlorine radicals, and sulfate radicals in aqueous-phase UV-Based advanced oxidation processes. *Water Res.* **2018**, *135*, 22–30. [CrossRef] [PubMed]

29. Li, W.; Jain, T.; Ishida, K.; Liu, H.Z. A mechanistic understanding of the degradation of trace organic contaminants by UV/hydrogen peroxide, UV/persulfate and UV/free chlorine for water reuse. *Environ. Sci. Water Res. Technol.* **2017**, *3*, 128–138. [CrossRef]

30. Elmolla, E.S.; Chaudhuri, M. Degradation of amoxicillin, ampicillin and cloxacillin antibiotics in aqueous solution by the UV/ZnO photocatalytic process. *J. Hazard. Mater.* **2010**, *173*, 445–449. [CrossRef] [PubMed]

31. Barontini, F.; Cozzani, V.; Cuzzola, A.; Petarca, L. Investigation of hexabromocyclododecane thermal degradation pathways by gas chromatography/mass spectrometry. *Rapid Commun. Mass Spectrom.* **2001**, *15*, 690–698. [CrossRef] [PubMed]

32. Zhou, D.N.; Chen, L.; Wu, F.; Wang, J.; Yang, F. Debromination of hexabromocyclododecane in aqueous solutions by UV-C irradiation. *Fresenius Environ. Bull.* **2012**, *21*, 107–111.
33. Yu, Y.; Zhou, D.; Wu, F. Mechanism and products of the photolysis of hexabromocyclododecane in acetonitrile–water solutions under a UV-C lamp. *Chem. Eng. J.* **2015**, *281*, 892–899. [CrossRef]
34. Zhao, Y.Y.; Zhang, X.H.; Sojinu, O.S. Thermodynamics and photochemical properties of alpha, beta, and gamma-hexabromocyclododecanes: A theoretical study. *Chemosphere* **2010**, *80*, 150–156.
35. Tso, C.P.; Shih, Y.H. The transformation of hexabromocyclododecane using zerovalent iron nanoparticle aggregates. *J. Hazard. Mater.* **2014**, *277*, 76–83. [CrossRef] [PubMed]

Article

Photocatalytic Degradation of Microcystins by TiO₂ Using UV-LED Controlled Periodic Illumination

Olivia M. Schneider [1,†], Robert Liang [1,2,*,†], Leslie Bragg [3], Ivana Jaciw-Zurakowsky [1], Azar Fattahi [1], Shasvat Rathod [1], Peng Peng [4], Mark R. Servos [3] and Y. Norman Zhou [1,2]

[1] Centre for Advanced Materials Joining, Department of Mechanical and Mechatronics Engineering, University of Waterloo, Waterloo, ON N2L 3G1, Canada; omschnei@edu.uwaterloo.ca (O.M.S.); ivana.jaciwzurakowsky@edu.uwaterloo.ca (I.J.-Z.); azar.fattahi@uwaterloo.ca (A.F.); s2rathod@edu.uwaterloo.ca (S.R.); nzhou@uwaterloo.ca (Y.N.Z.)

[2] Waterloo Institute of Nanotechnology, University of Waterloo, Waterloo, ON N2L 3G1, Canada

[3] Department of Biology, University of Waterloo, Waterloo, ON N2L 3G1, Canada; leslie.bragg@uwaterloo.ca (L.B.); mservos@uwaterloo.ca (M.R.S.)

[4] School of Mechanical Engineering and Automation, International Research Institute for Multidisciplinary Science, Beihang University, 37 Xueyuan Rd, Beijing 100191, China; peng.peng@uwaterloo.ca

* Correspondence: rliang@uwaterloo.ca; Tel.: +1-519-888-4567 (ext. 33326)

† Denotes equal contribution.

Received: 18 January 2019; Accepted: 11 February 2019; Published: 14 February 2019

Abstract: Toxic microcystins (MCs) produced by freshwater cyanobacteria such as *Microcystis aeruginosa* are of concern because of their negative health and economic impacts globally. An advanced oxidation process using UV/TiO₂ offers a promising treatment option for hazardous organic pollutants such as microcystins. The following work details the successful degradation of MC-LA, MC-LR, and MC-RR using a porous titanium–titanium dioxide (PTT) membrane under UV-LED light. Microcystin quantitation was achieved by sample concentration and subsequent LC–MS/MS analysis. The PTT membrane offers a treatment option that eliminates the need for the additional filtration or separation steps required for traditional catalysts. Controlled periodic illumination was successfully used to decrease the total light exposure time and improve the photonic efficiency for a more cost-effective treatment system. Individual degradation rates were influenced by electrostatic forces between the catalyst and differently charged microcystins, which can potentially be adjusted by modifying the solution pH and the catalyst's isoelectric point.

Keywords: *Microcystis aeruginosa*; microcystin; controlled periodic illumination; titanium dioxide; advanced oxidation process

---

1. Introduction

Cyanobacteria are a phylum of phototrophic bacteria capable of producing toxic blooms. *Microcystis aeruginosa* is a common freshwater cyanobacteria which produces microcystins (MC), a group of cyanotoxins with strong hepatotoxic effects. With the increased eutrophication of freshwater resources worldwide, the prevalence of these toxic blooms is a growing concern [1–3]. Over the last several years they have been linked to a variety of both animal and human poisonings globally, including in Canada, Australia, the United Kingdom, China, and Africa [4]. The World Health Organization deems the acceptable level of microcystin-LR, the most common microcystin, in drinking water to be 1 µg L⁻¹ [4]. For example, a review document produced by Health Canada in 2016 concluded that toxic algal blooms impact drinking water safety in the majority of Canadian provinces [5]. These toxic blooms also cause significant economic losses in affected areas by impeding tourism and fishing, lowering property values, and requiring expensive preventative strategies and monitoring [2,6,7].

Recent studies show that advanced oxidation processes (AOPs) such as ultraviolet (UV) light/TiO_2 photocatalysis can break down microcystins [8–12]. When TiO_2 particles are irradiated by UV light, electrons in the valence band are excited to the conduction band, creating electron–hole pairs. These electron–hole pairs can either undergo redox reactions directly with small organic molecules or with water to form reactive oxygen species (ROS) such as hydroxyl radicals, which then participate in redox reactions with small organic molecules [13–20]. UV/TiO_2 photocatalysis is a promising AOP for treating microcystins in water because the TiO_2 is catalytic, providing a constant oxidant source without having to be replenished, as is the case in UV/H_2O_2 AOPs.

Traditional slurry TiO_2 reactors are impractical for water treatment because they require an additional filtration step to remove the catalyst. In order to address this issue, a porous titanium–titanium dioxide membrane was used in this study. The membrane is made of a porous titanium sheet, oxidized, and calcinated to produce TiO_2 structures on its surface. In order to show the hydroxyl radical production of the catalyst under UV illumination, the conversion of terephthalic acid (TPA) to 2-hydroxyterephthalic acid (HTPA) was quantified by fluorescence [16]. The trade-off of using the membrane is that it has less surface area than the equivalent mass of powder TiO_2, decreasing adsorption and lowering the degradation rate [21,22].

Previous work also suggests that the photonic efficiency of the process can be improved using controlled periodic illumination (CPI) [19,23,24]. The improved efficiency under CPI conditions can be compared to a phenomenon called Parrondo's paradox, where alternating two less favorable conditions results in a more favorable outcome. The improved efficiency, when using a catalyst such as the porous titanium–titanium dioxide (PTT) membranes, can be explained by mass-transfer limitations. Because the membrane has a limited surface area for adsorption, the rate of the reaction may be limited by this surface area at high LED duty cycles [25]. In this case, periodically illuminated lighting conditions (within the mass-transfer limit) will be equally effective. In the case of photon-limited reactions (for example, slurry reactors), mass-transfer limitations would not apply because the reaction rate is faster than the adsorption rate.

Typical light sources such as mercury and xenon lamps require mechanical shutters to produce CPI, take time to warm up, and lack efficiency and reliability. UV-LEDs are ideal light sources in this case because high-frequency CPI can be generated using pulse-width modulation (PWM) through a microcontroller. The microcontroller also allows for easy optimization of the light frequency, including the implementation of dual-frequency profiles that may exhibit a synergistic effect. By decreasing the cumulative light exposure time in UV/TiO_2 photocatalysis, the efficiency of the system and the life span of the light source can be increased without sacrificing performance. A more energy-efficient system would be particularly advantageous for the practical application of larger-scale water treatment.

The following study details the removal of MC-LA, MC-LR, and MC-RR from water using UV/TiO_2 photocatalytic degradation with porous titanium–titanium dioxide membranes. Degradation under each set of conditions, for individual and cumulative microcystin concentrations, was monitored using LC–MS/MS. Controlled periodic illumination at frequencies of 50, 5, 0.5, and dual 0.05 and 25 Hz were all considered, with the goal of improving the photonic efficiency of the AOP.

2. Results and Discussion

2.1. PTT Membrane Characterization

The PTT membrane characterization results have been described in previous works [16,17]. In summary, Raman spectra and XRD indicated mainly anatase TiO_2, with some rutile TiO_2 and titanium. The experimental band-gap energy of 3.0 eV also indicated that the PTT membranes were primarily composed of crystalline TiO_2. The isoelectric point of the PTT membrane was also determined to be 6.0 using a SurPASS™ electrokinetic analyzer.

2.2. TPA Conversion

A summary of the TPA conversion under continuous UV, 0.5, 25, and dual 0.05 and 25 Hz can be seen in Figure 1. As expected, k_1 was highest for continuous UV light because the cumulative UV-LED exposure was twice that of the other lighting conditions. In order to properly compare degradation relative to the electrical energy consumed, electrical energy per order (E_{EO}) was calculated. Although the dual lighting conditions had the lowest degradation rates, they also had the lowest E_{EO}, making dual lighting the most efficient set of conditions for TPA conversion. The other two frequencies tested also had lower E_{EO} values than that for continuous light, making continuous UV-LED exposure the least efficient of the four conditions tested for TPA conversion. These results are in agreement with Parrondo's paradox and show that CPI is a viable method for improving the efficiency of the photocatalytic AOP.

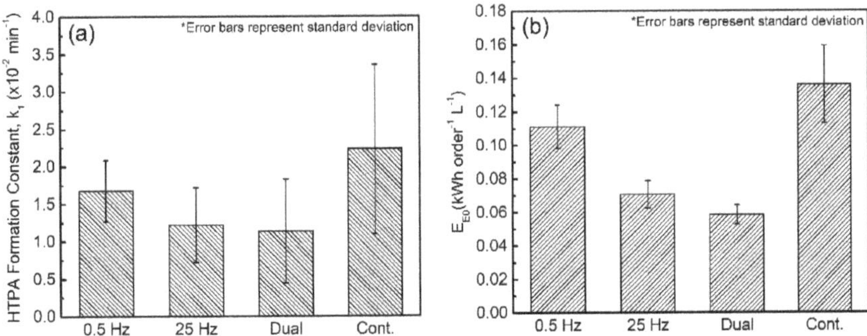

Figure 1. (**a**) Rate constants and (**b**) electrical energy per order for terephthalic acid (TPA) conversion under various UV-LED conditions.

2.3. Degradation of Microcystins under Continuous Light

Experiments testing the membrane under dark conditions and testing UV exposure without the membrane showed no degradation of the microcystins (see the Supplementary Information). This verifies that the photocatalytic AOP was responsible for the microcystin degradation. All three microcystins showed successful degradation when treated with UV light and PTT membranes, with a cumulative rate of -0.00453 min^{-1}. MC-LA degraded the fastest, with a rate of -0.00841 min^{-1}. Both MC-LR and MC-RR degraded at less than half that rate (-0.00350 min^{-1} and -0.00332 min^{-1}, respectively). This difference in the degradation rate can be explained by the difference in adsorption, which is influenced by relative charge.

The unmodified experimental pH was determined to be 5 and remained consistent throughout the course of the experiment. At this experimental pH, each microcystin has a different charge depending on the variable amino acids in the ring structure. A summary of these charges is included in Table 1. At experimental pH the PTT membrane is positively charged. These relative charges can explain the significant difference seen between degradation rates for MC-LA and MC-LR or MC-RR. In solution, MC-LA is the most oppositely charged to the PTT membrane and experiences the greatest electrostatic attraction. MC-LR is also negatively charged (though not as strongly) and will experience less significant electrostatic attraction. Inversely, MC-RR is positively charged and will experience electrostatic repulsion from the PTT membrane. The increased electrostatic attraction experienced by MC-LA will increase its adsorption onto the PTT membrane and result in a faster degradation rate. Since MC-LR and MC-RR compete with MC-LA for limited adsorption sites on the PTT membrane, the two microcystins that experience less electrostatic attraction will not adsorb as well and will have slower degradation rates. The influence of these interactions is reflected in the relative degradation rates under continuous UV illumination, where MC-LR and MC-RR have degradations rates less than half that of MC-LA.

Table 1. Microcystin charge at experimental pH.

Compound	Charge at pH 5[a]
MC-LA	−1.9332
MC-LR	−0.9329
MC-RR	0.0567

[a] Charge was calculated by chemicalize.org.

These results are consistent with previous studies by Arlos et al. and Liang et al., which showed that electrostatic forces between the pollutant and catalyst have a significant influence on the degradation rate [16,25]. The influence of electrostatic forces demonstrates the importance of considering the pH and the charge of target pollutants when treating water [26]. Degradation rates are highly pH-dependent, so the pH of the water being treated must be considered, especially in practical applications. In future water treatment designs, this information can be used to tune the isoelectric point of the catalyst to improve the degradation of desired pollutants.

2.4. Degradation of Microcystins under CPI

Using UV-LED PWM, the following CPI conditions were examined: 0.5, 5, 50, and dual 0.05 and 25 Hz. The calculated E_{EO} for these conditions, as well as those for continuous UV light for comparison, can be seen in Figure 2. Because MC-LA degraded preferentially to MC-LR and MC-RR, the change in concentration of MC-LR and MC-RR was subtle and lacked linearity (see Table S1 and Figure S1). For this reason, there were significantly larger errors associated with their calculated E_{EO}. As a result, the cumulative microcystin E_{EO} was considered when comparing different lighting conditions. As predicted by the TPA results, several of the controlled periodic illumination conditions presented more energy efficient options. Both the 5 and 0.5 Hz UV lighting had significantly lower E_{EO} than the continuous UV lighting. Dual frequency lighting, which is an equal combination of 0.05 and 25 Hz, had a comparable E_{EO} to continuous UV.

Among the lighting conditions tested, the least degradation overall was observed at 50 Hz (cumulative $k_{app} = -7.83 \times 10^{-4}\,min^{-1}$, see Table S1). Very little degradation of MC-LA and MC-RR occurred, and no degradation of MC-LR occurred (see Figures S1 and S2). Given the low magnitude of degradation and the associated margin of error, it is difficult to discern any trend in degradation between different microcystins under 50 Hz UV illumination, as was done for continuous UV. This significant decrease in overall degradation resulted in a correspondingly high E_{EO} for microcystin degradation under 50 Hz UV. In general, the E_{EO} increased as the frequency increased, meaning lower-frequency lighting conditions were more efficient. In dual lighting, the combination of both high and low frequencies balanced each other and resulted in an insignificant net change relative to continuous UV.

Interestingly, the improvement of reaction efficiency under CPI agrees with Parrondo's paradox. In Parrondo's paradox, alternating between two less favorable conditions yields a more favorable result. In this case, the two less favorable UV-LED conditions were off (which did not contribute to the AOP) and on (which was inefficient). By alternating these two conditions at different frequencies, the time the UV-LED was on decreased by 50%. The favorable outcome was that the E_{EO} under CPI decreased, demonstrating improved efficiency.

Although the PTT membrane provided a more practical option for water treatment, the reduced surface area relative to a slurry reactor imposed mass-transfer limitations. Because the rates of adsorption and desorption to the surface of the catalyst were significantly slower than the rate of electron–hole pair formation and recombination, the process of adsorption and desorption was rate-limiting [24,27]. This had a significant impact on the photonic efficiency of the system. In using CPI, the dark period allowed for the equilibration of the untreated pollutant molecules on adsorption sites, without wasting energy [18,23,28]. This improved the photonic efficiency of the photocatalytic system. The results indicated that CPI is a viable method for improving the efficiency of photocatalytic

AOPs used to treat organic pollutants and toxins, though further optimization of conditions is required. Improving the photonic efficiency of the process makes it more energy efficient and prolongs the life of the light source. These characteristics are particularly attractive in the water treatment industry because they reduce costs.

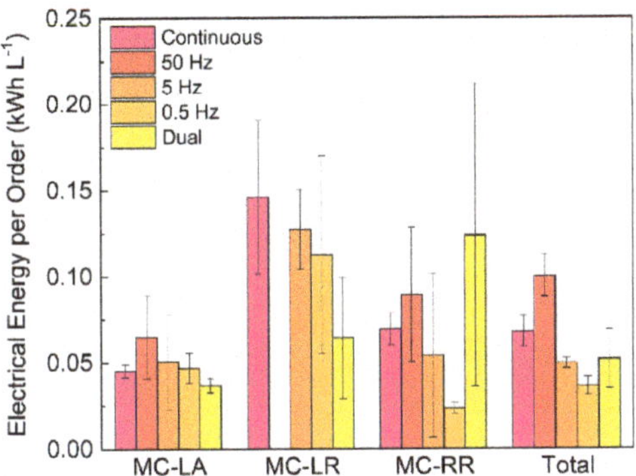

Figure 2. Electrical energy per order for microcystin degradation under continuous and controlled periodic illumination (CPI) conditions.

3. Materials and Methods

3.1. Reagents and Chemicals

Microcystins and nodularin (Cayman Chemicals, Ann Arbor, MI, USA) were dissolved in UHPLC-grade methanol (VWR International, Mississauga, ON, Canada) and stored at $-20\ °C$. PTT membrane synthesis required hydrogen peroxide (Sigma-Aldrich, St. Louis, MO, USA) and 0.254-mm-thick porous titanium (PTi) sheets (Accumet Materials, Ossining, NY, USA). For LC–MS analysis, HPLC-grade ammonium fluoride (Sigma-Aldrich, St. Louis, MO, USA) and HPLC-grade acetonitrile (Thermo Fisher Scientific, Waltham, MA, USA) were used. For measuring hydroxyl radical formation, TPA and sodium hydroxide were purchased from Sigma-Aldrich, St. Louis, MO, USA. Water was purified using a Milli-Q®Integral water purification system (EMD Millipore, Burlington, MA, USA) (18.2 mΩ·cm resistivity at 25 $°C$).

3.2. PTT Membrane Synthesis and Characterization

PTT membrane synthesis and characterization methods are described in previous works [16,17]. In short, PTi membranes were cut into 5-cm diameter discs, cleaned and oxidized in a hydrogen peroxide solution at 80 $°C$, and then calcined at 600 $°C$. Material characterization methods included micro-Raman spectroscopy (He–Ne laser λ = 632.8 nm, Renishaw, Wotton-under-Edge, UK), scanning electron microscopy (FE-SEM LEO 1550, Carl Zeiss Microscopy, Jena, Germany), and X-ray diffraction (XPERT-PRO, Malvern Panalytical, Malvern, UK).

3.3. Experimental Setup for Microcystin Degradation

A volume of MC-LA, MC-LR, and MC-RR stock solution was evaporated to dryness under nitrogen gas and then reconstituted to 2 µg/L in MilliQ water for a reaction solution. The PTT membranes were suspended in 0.4-L beakers on metal stands, 1.5 cm under the solution surface with a volume of 0.3 L.

Many studies of UV/TiO$_2$ photocatalysis use methanol as a carrier solvent when preparing aqueous pollutant solutions [10,16,17,29–35]. More recent studies show that methanol has a significant effect on photocatalytic degradation because it acts as a hydroxyl radical scavenger, even at low concentrations [36,37]. In order to replicate the effects of methanol under typical experimental conditions, all reactions were conducted in 0.02% methanol.

Reactions took place using a UV-LED source with an average irradiance of 2.18 mW cm^{-2} under continuous illumination and 1.08 mW cm^{-2} under a 50% duty cycle (measured 18 cm from the light source using Thorlabs PM100-USB power meter, S120VC 200–1100 nm, 50 mW). Reaction solutions were stirred at 600 rpm. The solution surface was initially 10.5 cm below the light source. A diagram of the reaction setup is shown in Figure 3a with the UV-LED (LED-Engin LZ1, 1000 mA) spectral power distribution peaking at 365 nm and the total radiation included angle of 105° (90% of the total radiant flux).

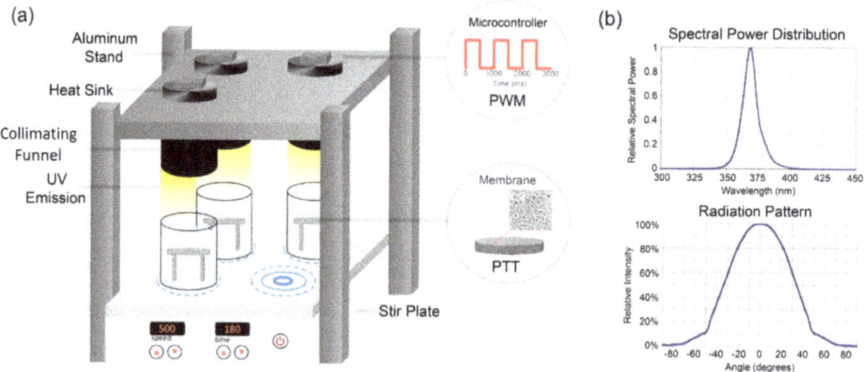

Figure 3. (a) Experimental setup for UV/TiO$_2$ reactors and (b) UV-LED spectral power distribution and radiation pattern. Abbreviations: PWM—pulse-width modulation, PTT—porous titanium–titanium dioxide.

Reactions were equilibrated in the dark for 1 h before UV-LED irradiation, with a total reaction time of 6 h. Samples of 4 mL were taken every hour. Each set of conditions was repeated in triplicate. Arduino microcontrollers and LEDSEEDUINO LED current drivers were used to program the UV-LEDs for 0.5, 5, and 50 Hz as well as a dual frequency (0.05 and 25 Hz alternating for equal periods). The pulsed-width modulation script used to program the different conditions can be found in the Supplementary Information. All UV-LED flashing sequences had a duty cycle of 50%. The duty cycle is the ratio of the time on to the time off, as described by the following equation:

$$D = \frac{PW}{T} \times 100\%$$

where D is the duty cycle expressed as a percentage, PW is the pulse width duration, and T is the period of the wave.

At low adsorbate concentrations, the following equation can be used to approximate Langmuir–Hinshelwood kinetics [38]:

$$-r = \frac{dC}{dt} = -k_{app}C.$$

The equation can then be rearranged and integrated to give the following [25]:

$$\ln\left(\frac{C}{C_o}\right) = k_{app}t$$

where C (g L^{-1}) is the analyte concentration at time t (min), C_o (g L^{-1}) is the initial analyte concentration at $t = 0$, and k_{app} (min^{-1}) is the apparent kinetic rate. The slope of a plot of the equation gives the

k_{app}. OriginLabPro (version 8.0, OriginLab, Northhampton, MA, USA, 2018) was used to complete the linear regression analysis to determine the rates for cumulative and individual compounds.

3.4. Experimental Setup for TPA Conversion

The experimental method is derived from previous work [16]. In brief, a solution of 5 mM TPA was made in 6 mM NaOH. Under the same conditions as described in the previous section, 300 mL of this solution was placed in a beaker with a PTT membrane. UV-LED irradiation began after a 1-h dark equilibration, and samples were taken at various time points over 4 h. The following lighting conditions were tested: Continuous, 0.05, 25, and dual 0.05 and 25 Hz. All frequencies were programmed with a duty cycle of 50%. HTPA concentrations were quantified by fluorescence using a plate reader (SpectraMax M3, Molecular Devices, San Jose, CA, USA) with an excitation wavelength of 315 nm and emission from 350 nm to 550 nm. The intensity value was taken from the peak of the spectrum.

HTPA was the first degradation product of TPA, so its concentration increased sharply at the beginning of the reaction. As more HTPA was produced, it also degraded into more oxidized products. These rates can be described by the following kinetic model [39]:

$$C_{HTPA} = \frac{k_1}{k_2}\left(1 - e^{-k_2 t}\right)$$

where k_1 is the zeroth-order rate of HTPA formation, k_2 is the pseudo first-order degradation rate of HTPA, C_{HTPA} is the concentration of HTPA in mol L^{-1}, and t is time in minutes.

3.5. Electrical Energy per Order

To more accurately compare the efficiency of lighting conditions with different duty cycles, electrical energy per order (E_{EO}) was calculated. E_{EO} is the energy in kWh needed to decrease the microcystin or TPA concentration by one order of magnitude in a liter of water. E_{EO} was calculated by the following equations [40,41]:

$$E_{EO}(MC) = \frac{1000 \cdot P \cdot t}{V \cdot log\left(\frac{C_i}{C_f}\right)}$$

$$E_{EO}(HTPA) = \frac{1000 \cdot P \cdot t}{V \cdot k_2}$$

where P is the power dissipated over the treatment process in kW, t is the reaction time in min, V is the reaction volume in L, k_2 is the degradation rate of HTPA, and C_i and C_f are the initial and final microcystin concentrations, respectively.

3.6. Sample Preparation and Analysis

Each 4-mL sample was spiked with nodularin (NOD), a toxin similar in structure to microcystins, to 0.8 µg L^{-1} for an internal standard [1]. The spiked samples were then evaporated to dryness in a Rocket Evaporator (Thermo Scientific) and reconstituted in 160 µL of UHPLC-grade methanol. Prepared samples were stored at $-20\ °C$ until analysis.

Microcystin quantitation was achieved with LC–MS/MS using an Agilent 1200 HPLC and 3200 quadrupole ion trap (QTRAP) mass spectrometer with electrospray ionization (ABSciex). Specific mass spectrometry parameters are summarized in Table 2. In order to achieve separation, a Poroshell 120 SB-C18 column (4.6 × 150 mm, 2.7 µm, Agilent Technologies) was used with 0.5 mM ammonium fluoride and acetonitrile (ACN) at 1 mL min^{-1} and 40 °C, with 20-µL sample injections. For the mobile phase gradient, 10% ACN was held for 0.5 min, which was increased to 100% ACN over 4.5 min and held for 1 min. The mobile phase composition was then returned to 10% ACN over 0.5 min and equilibrated for 3.5 min before the next injection. The calibration curves for each microcystin were linear from 0.5 to 500 µg L^{-1}.

Table 2. Mass spectrometry parameters for the detection of microcystins.

	Q1[a] (Da)	Q3[b] (Da)	Time (ms)	DP[c] (volts)	EP[d] (volts)	CE[e] (volts)	CXP[f] (volts)	CEP[g] (volts)	Retention Time (min)
NOD	825.563	135.3	150	96	12	75	4	40	4.55
MC-LA	911.395	135.2	150	51	12	81	4	36	4.62
MC-LR	995.699	135.1	150	116	12	99	4	36	4.67
MC-RR	519.960	135.2	150	131	7	41	4	26	4.88

[a] First quadrupole, [b] third quadrupole, [c] declustering potential, [d] entrance potential, [e] collision energy, [f] collision cell exit potential, and [g] collision cell entrance potential.

4. Conclusions

In this study three common microcystins, MC-LA, MC-LR, and MC-RR, were successfully degraded in water using a UV/TiO$_2$ photocatalytic AOP. E$_{EO}$ values calculated for TPA conversion determined continuous UV-LED illumination to be less efficient than all the CPI conditions tested, demonstrating the potential of CPI to improve the efficiency of the photocatalytic AOP. Under continuous illumination, the negatively charged MC-LA degraded at more than twice the rate of MC-LR or MC-RR because it preferentially adsorbed onto the positive PTT membrane catalyst. The pH dependence of the degradation rates suggests that the isoelectric point of the catalyst can be tuned to improve the degradation of target pollutants in water of a known pH, given the compound charges. CPI conditions of 0.5, 5, and 50 Hz as well as dual 0.5 and 25 Hz with a 50% duty cycle were also examined for treating microcystins. When considering the cumulative microcystin solution, the 0.5 and 5 Hz CPI conditions were determined to be more efficient than continuous UV light based on the calculated E$_{EO}$. These results can be explained by mass-transfer limitations, where the rate of adsorption and desorption onto the surface of the catalyst limits photonic efficiency. The results of this work indicate that the use of CPI has the potential to improve the energy efficiency and light source life span in photocatalytic AOPs and is worth further investigation. Improving these parameters makes photocatalytic AOPs more attractive as a large-scale water treatment solution because they have the potential to greatly decrease costs.

Supplementary Materials: The following are available online at http://www.mdpi.com/2073-4344/9/2/181/s1, Table S1: Calculated degradation rates. Poor fit with the linear regression model due to insignificant degradation is seen in membrane-only and UV-only conditions, as well as in 50 Hz MC-LR; Figure S1: Change in microcystin concentration over the course of various UV/TiO$_2$ treatments; Figure S2: Linear regression plots for the calculation of degradation rate; Figure S3: Chromatogram demonstrating the separation of MC-LA, MC-LR, MC-RR, and NOD.

Author Contributions: Conceptualization, R.L.; methodology, O.M.S. and L.B.; investigation, O.M.S., A.F., and I.J.-Z.; data curation, O.M.S. and S.R.; writing—original draft preparation, O.M.S.; writing—review and editing, O.M.S., A.F., S.R., and R.L.; visualization, R.L.; supervision, Y.N.Z., P.P. and M.R.S.; funding acquisition, Y.N.Z., M.R.S., and R.L.

Funding: This research was funded by the Natural Sciences and Engineering Research Council of Canada (grant number: STPG-494554-2016) through a strategic project grant and the Schwartz–Reisman Foundation through the Waterloo Institute of Nanotechnology—Technion University grant.

Acknowledgments: The authors would like to thank the Natural Sciences and Engineering Research Council of Canada, the Schwartz–Reisman Foundation, and Waterloo Institute of Nanotechnology for their financial support.

Conflicts of Interest: The authors have no conflict of interest to declare.

References

1. Dawson, R.M. The Toxicology of Microcystins. *Toxicon* **1998**, *36*, 953–962. [CrossRef]

2. Merel, S.; Walker, D.; Chicana, R.; Snyder, S.; Baurès, E.; Thomas, O. State of knowledge and concerns on cyanobacterial blooms and cyanotoxins. *Environ. Int.* **2013**, *59*, 303–327. [CrossRef] [PubMed]

3. Umehara, A.; Takahashi, T.; Komorita, T.; Orita, R.; Chio, J.-W.; Takenaka, R.; Mabuchi, R.; Park, H.-D.; Tsutsumi, H. Widespread dispersal and bio-accumulation of toxic microcystins in benthic marine ecosystems. *Chemoshpere* **2017**, *167*, 492–500. [CrossRef] [PubMed]

4. Chambon, P.; Lund, U.; Galal-Gorchev, H.; Ohanian, E. *Guidelines for Drinking-Water Quality Volume 2— Health Criteria and Other Supporting Information*, 2nd ed.; Kenny, J., Galal-Gorchev, H., Eds.; World Health Organisation: Geneva, Switzerland, 1998; Volume 2.

5. Committee on Drinking Water. *Cyanobacterial Toxins in Drinking Water*; Committee on Drinking Water: Ottawa, ON, Canada, 2016.

6. Wolf, D.; Klaiber, H.A. Bloom and bust: Toxic algae's impact on nearby property values. *Ecol. Econ.* **2017**, *135*, 209–221. [CrossRef]

7. Dyson, K.; Huppert, D.D. Regional economic impacts of razor clam beach closures due to harmful algal blooms (HABs) on the Pacific coast of Washington. *Harmful Algae* **2010**, *9*, 264–271. [CrossRef]

8. Sharma, V.K.; Triantis, T.M.; Antoniou, M.G.; He, X.; Pelaez, M.; Han, C.; Song, W.; O'shea, K.E.; De La Cruz, A.A.; Kaloudis, T.; et al. Destruction of microcystins by conventional and advanced oxidation processes: A review. *Sep. Purif. Technol.* **2012**, *91*, 3–17. [CrossRef]

9. Liu, I.; Lawton, L.A.; Bahnemann, D.W.; Liu, L.; Proft, B.; Robertson, P.K.J. The photocatalytic decomposition of microcystin-LR using selected titanium dioxide materials. *Chemosphere* **2009**, *76*, 549–553. [CrossRef] [PubMed]

10. Shephard, G.S.; Om, S.S.O.; De Villiers, D.; Engelbrecht, W.J.; E El, G.; Wessels, F.S. Degradation of microcystin toxins in a falling film photocatalytic reactor with immobilized titanium dioxide catalyst. *Water Res.* **2002**, *36*, 140–146. [CrossRef]

11. Cornish, B.J.P.A.; Lawton, L.A.; Robertson, P.K.J. Hydrogen peroxide enhanced photocatalytic oxidation of microcystin-LR using titanium dioxide. *Appl. Catal. B Environ.* **2000**, *25*, 59–67. [CrossRef]

12. Lawton, L.A.; Robertson, P.K.J.; Cornish, B.J.P.A.; Marr, I.L.; Jaspars, M. Processes influencing surface interaction and photocatalytic destruction of microcystins on titanium dioxide photocatalysts. *J. Catal.* **2003**, *213*, 109–113. [CrossRef]

13. Rizzo, L.; Meric, S.; Guida, M.; Kassinos, D.; Belgiorno, V. Heterogenous photocatalytic degradation kinetics and detoxification of an urban wastewater treatment plant effluent contaminated with pharmaceuticals. *Water Res.* **2009**, *43*, 4070–4078. [CrossRef] [PubMed]

14. Liang, R.; Hu, A.; Li, W.; Zhou, Y.N. Enhanced degradation of persistent pharmaceuticals found in wastewater treatment effluents using TiO2 nanobelt photocatalysts. *J. Nanopart. Res.* **2013**, *15*, 1990. [CrossRef]

15. Martínez, C.; Canle L., M.; Fernández, M.I.; Santaballa, J.A.; Faria, J. Aqueous degradation of diclofenac by heterogeneous photocatalysis using nanostructured materials. *Appl. Catal. B Environ.* **2011**, *107*, 110–118. [CrossRef]

16. Arlos, M.J.; Liang, R.; Hatat-Fraile, M.M.; Bragg, L.M.; Zhou, N.Y.; Servos, M.R.; Andrews, S.A. Photocatalytic decomposition of selected estrogens and their estrogenic activity by UV-LED irradiated TiO2 immobilized on porous titanium sheets via thermal-chemical oxidation. *J. Hazard. Mater.* **2016**, *318*, 541–550. [CrossRef] [PubMed]

17. Arlos, M.J.; Hatat-Fraile, M.M.; Liang, R.; Bragg, L.M.; Zhou, N.Y.; Andrews, S.A.; Servos, M.R. Photocatalytic decomposition of organic micropollutants using immobilized TiO$_2$ having different isoelectric points. *Water Res.* **2016**, *101*, 351–361. [CrossRef] [PubMed]

18. Ku, Y.; Shiu, S.-J.; Wu, H.-C. Decomposition of dimethyl phthalate in aqueous solution by UV–LED/TiO$_2$ process under periodic illumination. *J. Photochem. Photobiol. A Chem.* **2017**, *332*, 299–305. [CrossRef]

19. Buechler, K.J.; Nam, C.H.; Zawistowski, T.M.; Noble, R.D.; Koval, C.A. Design and Evaluation of a Novel-Controlled Periodic Illumination Reactor To Study Photocatalysis. *Ind. Eng. Chem. Res.* **1999**, *38*, 1258–1263. [CrossRef]

20. Schneider, J.; Matsuoka, M.; Takeuchi, M.; Zhang, J.; Horiuchi, Y.; Anpo, M.; Bahnemann, D.W. Understanding TiO$_2$ Photocatalysis: Mechanisms and Materials. *Chem. Rev.* **2014**, *114*, 9919–9986. [CrossRef]

21. Manassero, A.; Satuf, M.L.; Alfano, O.M. Photocatalytic reactors with suspended and immobilized TiO$_2$: Comparative efficiency evaluation. *Chem. Eng. J.* **2017**, *326*, 29–36. [CrossRef]

22. Hegedűs, P.; Szabó-Bárdos, E.; Horváth, O.; Szabó, P.; Horváth, K. Investigation of a TiO$_2$ photocatalyst immobilized with poly(vinyl alcohol). *Catal. Today* **2017**, *284*, 179–186. [CrossRef]

23. Sczechowski, J.G.; Koval, C.A.; Noble, R.D. Evidence of critical illumination and dark recovery times for increasing the photoefficiency of aqueous heterogeneous photocatalysis. *J. Photochem. Photobiol. A Chem.* **1993**, *74*, 273–278. [CrossRef]

24. Tokode, O.; Prabhu, R.; Lawton, L.A.; Robertson, P.K.J. Controlled periodic illumination in semiconductor photocatalysis. *J. Photochem. Photobiol. A Chem.* **2016**, *319–320*, 96–106. [CrossRef]

25. Liang, R.; Van Leuwen, J.C.; Bragg, L.M.; Arlos, M.J.; Li Chun Fong, L.C.M.; Schneider, O.M.; Peng, P.; Servos, M.R.; Zhou, Y.N. Utilizing UV-LED pulse width modulation on TiO$_2$ advanced oxidation processes to enhance the decomposition efficiency of pharmaceutical micropollutants. *Chem. Eng. J.* **2019**, *361*, 439–449. [CrossRef]

26. Friedmann, D.; Mendive, C.; Bahnemann, D. Environmental TiO$_2$ for water treatment: Parameters affecting the kinetics and mechanisms of photocatalysis. *Appl. Catal. B Environ.* **2010**, *99*, 398–406. [CrossRef]

27. Memming, R. Photoinduced charge transfer processes at semiconductor electrodes and particles. In *Topics in Current Chemistry*; Springer: Berlin/Heidelberg, Germany, 1994; Volume 169, pp. 105–181, ISBN 978-3-540-57565-8.

28. Sczechowski, J.G.; Koval, C.A.; Noble, R.D. A Taylor vortex reactor for heterogeneous photocatalysis. *Chem. Eng. Sci.* **1995**, *50*, 3163–3173. [CrossRef]

29. Miranda-García, N.; Maldonado, M.I.; Coronado, J.M.; Malato, S. Degradation study of 15 emerging contaminants at low concentration by immobilized TiO$_2$ in a pilot plant. *Catal. Today* **2010**, *151*, 107–113. [CrossRef]

30. Miranda-García, N.; Suárez, S.; Sánchez, B.; Coronado, J.M.; Malato, S.; Maldonado, M.I. Photocatalytic degradation of emerging contaminants in municipal wastewater treatment plant effluents using immobilized TiO2 in a solar pilot plant. *Appl. Catal. B Environ.* **2011**, *103*, 294–301. [CrossRef]

31. Sun, W.; Li, S.; Mai, J.; Ni, J. Initial photocatalytic degradation intermediates/pathways of 17α-ethynylestradiol: Effect of pH and methanol. *Chemosphere* **2010**, *81*, 92–99. [CrossRef] [PubMed]

32. Nasuhoglu, D.; Berk, D.; Yargeau, V. Photocatalytic removal of 17α-ethinylestradiol (EE2) and levonorgestrel (LNG) from contraceptive pill manufacturing plant wastewater under UVC radiation. *Chem. Eng. J.* **2012**, *185–186*, 52–60. [CrossRef]

33. Kralchevska, R.; Milanova, M.; Bistan, M.; Pintar, A.; Todorovsky, D. The photocatalytic degradation of 17α-ethynylestradiol by pure and carbon nanotubes modified TiO$_2$ under UVC illumination. *Open Chem.* **2012**, *10*, 1137–1148. [CrossRef]

34. Marinho, B.A.; de Liz, M.V.; Lopes Tiburtius, E.R.; Nagata, N.; Peralta-Zamora, P.; Iguchi, T.; Kubota, Y.; Fujishima, A. TiO$_2$ and ZnO mediated photocatalytic degradation of E2 and EE2 estrogens. *Photochem. Photobiol. Sci.* **2013**, *12*, 678–683. [CrossRef] [PubMed]

35. Fernández, R.L.; McDonald, J.A.; Khan, S.J.; Le-Clech, P. Removal of pharmaceuticals and endocrine disrupting chemicals by a submerged membrane photocatalysis reactor (MPR). *Sep. Purif. Technol.* **2014**, *127*, 131–139. [CrossRef]

36. Nosaka, Y.; Nosaka, A.Y. *Photocatalysis and Water Purification: From Fundamentals to Recent Applications*; John Wiley & Sons: Hoboken, NJ, USA, 2013; pp. 3–23.

37. Paul, T.; Miller, P.L.; Strathmann, T.J. Visible-light-mediated TiO$_2$ photocatalysis of fluoroquinolone antibacterial agents. *Environ. Sci. Technol.* **2007**, *41*, 4720–4727. [CrossRef] [PubMed]

38. Turchi, C.S.; Ollis, D.F. Photocatalytic degradation of organic water contaminants: Mechanisms involving hydroxyl radical attack. *J. Catal.* **1990**, *122*, 178–192. [CrossRef]

39. Cernigoj, U.; Kete, M.; Stangar, U.L. Development of a fluorescence-based method for evaluation of self-cleaning properties of photocatalytic layers. *Catal. Today* **2010**, *151*, 46. [CrossRef]

40. Gora, S.; Liang, R.; Zhou, Y.N.; Andrews, S. Settleable engineered titanium dioxide nanomaterials for the removal of natural organic matter from drinking water. *Chem. Eng. J.* **2018**, *334*, 638–649. [CrossRef]

41. Liang, R.; Li Chun Fong, L.C.M.; Arlos, M.J.; Van Leeuwen, J.; Shahnam, E.; Peng, P.; Servos, M.R.; Zhou, Y.N. Photocatalytic degradation using one-dimensional TiO$_2$ and Ag-TiO$_2$ nanobelts under UV-LED controlled periodic illumination. *J. Environ. Chem. Eng.* **2017**, *5*, 4365–4373. [CrossRef]

 *catalysts*

Article

Development of TiO$_2$-Carbon Composite Acid Catalyst for Dehydration of Fructose to 5-Hydroxymethylfurfural

Morongwa Martha Songo [1,2], **Richard Moutloali** [2] **and Suprakas Sinha Ray** [1,2,*]

[1] DST-CSIR National Centre for Nanostructured Materials, Council for Scientific and Industrial Research, Pretoria 0001, South Africa; msongo@csir.co.za

[2] Department of Applied Chemistry, University of Johannesburg, Doornfontein, Johannesburg 2028, South Africa; rmoutloali@uj.ac.za

* Correspondence: rsuprakas@csir.co.za; Tel.: +27-12-841-2388

Received: 28 December 2018; Accepted: 29 January 2019; Published: 31 January 2019

Abstract: A TiO$_2$-Carbon (TiO$_2$C) composite was prepared using the microwave-assisted method and sulfonated using fuming sulfuric acid to produce a TiO$_2$C solid acid catalyst. The prepared solid acid catalyst was characterised using scanning electron microscopy, Brunauer-Emmett-Teller analysis, Fourier transform infrared spectroscopy, and X-ray diffraction. Crystallinity analysis confirmed that TiO$_2$C has an anatase structure, while analysis of its morphology showed a combination of spheres and particles with a diameter of 50 nm. The TiO$_2$C solid acid catalyst was tested for use in the catalytic dehydration of fructose to 5-hydroxymethylfurfural (5-HMF). The effect of reaction time, reaction temperature, catalyst dosage, and solvent were investigated against the 5-HMF yield. The 5-HMF yield was found to be 90% under optimum conditions. The solid acid catalyst is very stable and can be reused for four catalytic cycles. Hence, the material has great potential for use in industrial applications and can be used for the direct conversion of fructose to 5-HMF because of its high activity and high reusability.

Keywords: TiO$_2$C composite; acid catalyst; dehydration; fructose; 5-Hydroxymethylfurfural

1. Introduction

Increase in the usage of fossil fuels for the production of chemicals and energy has caused not only a rise in greenhouse gas emissions into the atmosphere, but also water pollution, thereby contributing to the growing number of serious health issues. These side effects have spurred researchers to look for alternative ways of producing valuable chemicals and energy using sustainable and renewable resources. Biomass has been identified as a promising resource for the sustainable production of valuable chemical feedstocks and fuels [1]. 5-hydroxymethylfurfural (5-HMF) has been recognised as a key intermediate in the production of biorenewable chemicals [2] and has been classified as a "platform molecule" because it is an important chemical building block used to derive a variety of chemicals, pharmaceuticals, and furane-based polymers [3].

5-HMF is usually synthesised from the dehydration of fructose or glucose using homogeneous organic acids or heterogeneous solid acid catalysts [4]. Although the use of homogeneous catalysts produces high fructose conversion and high dehydrations of 5-HMF, there are drawbacks with regard to separation, recycling, and equipment corrosion [5,6]. Heterogeneous acid catalysts such as mesoporous zirconium phosphate [7,8], sulfonated metal oxides [9], magnetite mixed-metal oxides [10], and functionalised silica nanoparticles [11,12] have been used for the dehydration of fructose to 5-HMF. Although the use of some of these materials has promising results, the studies cited pointed out the remaining challenge of finding catalysts that are highly active, selective, chemically and structurally stable during repeated use, and do not lead to any side reactions (by-products). Biomass-derived carbonaceous materials were found to be promising candidates for catalytic support application because of their low cost of production and thermal stability [13]. For example, Wang et al. [14] used a sulfonated carbonaceous solid acid catalyst for the dehydration of fructose into 5-HMF in dimethyl sulfoxide (DMSO). Their catalyst was very efficient and effective in that it converted 96.1% of the fructose for a high 5-HMF yield of 93.4%. Zhao et al. [15] further explored a sulfonated carbon sphere solid acid catalyst that converted 100% of the fructose in DMSO solvent at 160 °C for 1.5 h to produce 90% 5-HMF. Wang et al. [16] used C-based solid acid catalysts to catalyse dehydration of fructose in DMSO for 1.5 h at 130 °C to achieve 91.2% 5-HMF. Guo et al. [17] used a lignin-derived carbonaceous catalyst to convert fructose into 5-HMF under microwave irradiation in the mixture of DMSO and ionic liquid at 110 °C for 10 min for a fructose conversion of 98% and a 5-HMF yieldof 84%. Hu et al. [18] explored the use of a magnetic lignin-derived carbonaceous acid catalyst for the catalytic conversion of fructose into 5-HMF in DMSO solvent. Use of the magnetic C-based catalyst achieved a 100% fructose conversion and a 5-HMF yieldof 81.1% under optimum reaction conditions.

Studies have found that the anatase phase of TiO_2 can catalyse the dehydration of carbohydrates such as fructose and glucose into 5-HMF and that the performance of the TiO_2 catalyst is strongly controlled by its morphological and structural properties [3,19–21]. For example, the use of nanostructured TiO_2 remarkably favoured the production of 5-HMF from glucose and fructose [19,20], with its high activity being ascribed to the morphology of the nanoparticles [20].

In this paper, we report the synthesis, characterisation, and application of TiO_2-Carbon sphere (TiO_2C) composite for the catalysis of the dehydration of fructose into 5-HMF in DMSO solvent. To the best of our knowledge, there have been no reports on the use of TiO_2C as a catalyst for the dehydration of fructose into 5-HMF. We synthesised the TiO_2C composites using a microwave-assisted method. Because TiO_2 has redox acidic sites, Bronsted acid sites were introduced by sulfonating the TiO_2C composites with concentrated sulfuric acid. Scanning electron microscopy (SEM), Brunauer-Emmett-Teller (BET) analysis, X-ray diffraction, and Fourier transform infrared (FTIR) spectroscopy were used to characterise the TiO_2C acid catalyst. This solid acid catalyst was tested in DMSO solvent as a catalyst for the dehydration of fructose into 5-HMF. To achieve a higher dehydration of 5-HMF, reaction parameters such as reaction temperature, reaction time, and catalyst amount were optimised. Moreover, the ability to recycle the catalyst was studied to evaluate its catalytic stability.

2. Experimental and Methods

2.1. Materials

All chemicals used for synthesising the TiO_2 nanoparticles (NPs) and TiO_2C composites were purchased from Sigma-Aldrich Corporation (St. Louis, MO, USA). Titanium (IV) isopropoxide (TTIP) and fructose were used as precursors in the preparation of TiO_2 and TiO_2C composites. Polyvinylpyrrolidone (PVP) was used as a surfactant and ammonium hydroxide was used as a reducing agent for the preparation of TiO_2 NPs. Sulfuric acid was used for sulfonating the TiO_2C solid acid catalyst.

2.2. Preparation of TiO_2 Nanoparticles Using the Sol-Gel Method

Ethanol (50 mL) was mixed with 17.79 g of TTIP and stirred for 30 min. PVP (4 g) was added to the mixture, which then was stirred at 70 °C until it was completely dissolved. Next, 4 mL of $NH_3 \cdot 2H_2O$ was added. The solution was heated at 90 °C under reflux conditions for 24 h. The resultant sample was washed with water and dried at 100 °C for 24 h.

2.3. Preparation of Carbon Spheres Using the Microwave-Assisted Method

The carbon precursor (fructose) was mixed with deionised (DI) water and stirred at room temperature (25 to 27 °C) for 2 h. The obtained mixture was transferred into microwavable plastic vessels that were then placed into the ultraclave high-performance microwave reactor and allowed to react at 180 °C for 2 h.

2.4. Preparation of TiO_2C Composites Using the Microwave-Assisted Method

TiO_2 nanoparticles were added to 50 mL of DI water. The solution was sonicated at 30 °C for at least 1 h, after which the fructose was added. The pH of the solution was adjusted to 2. The solution was stirred at room temperature (25 to 27 °C) for 2 h and then microwaved in the ultraclave high-performance microwave reactor for another 2 h at 180 °C. The resultant sample was washed three times with DI water, dried at 150 °C for 24 h, and then calcined at 500 °C for 2 h. Different amounts of TiO_2 nanoparticles (1, 5, and 10%) were loaded onto the carbon spheres. For example, to prepare the 1%TiO_2C nanocomposite, 0.1 g of TiO_2 nanoparticles was added to 9.9 g of fructose, whereas to prepare the 5%TiO_2C and 10%TiO_2C, the mass ratio of TiO_2 nanoparticles to fructose was 0.5:9.5 and 1:9, respectively. The TiO_2C composites loaded with 1, 5, and 10% TiO_2 were denoted as 1%TiO_2C, 5%TiO_2C, and 10%TiO_2C, respectively.

2.5. Preparation of the Sulfonated Carbon and TiO_2C Acid Catalysts

TiO_2C was sulfonated by adding 1 g of TiO_2C composite to 25 ml of concentrated sulfuric acid and stirring the solution at 120 °C for 24 h under reflux conditions. The sulfonated samples, denoted as TiO_2C_S, were washed with DI water and dried at 150 °C for 24 h. The carbon spheres were sulfonated using the same method and were denoted as C_S.

2.6. Surface Acid Measurement Tests

A total of 100 mg of C_S and TiO_2C_S composite acid catalysts was mixed with 50 ml of DI water and 40 ml of 0.01 M NaOH and stirred at room temperature (25 to 27 °C) overnight [22]. The solution was filtered and titrated with 0.1 N of oxalic acid using phenolphthalein as the indicator.

2.7. Characterisation

Elemental composition and surface morphology analyses were conducted using the AURIGA® scanning electron microscope (Zeiss, Oberkochen, Germany). The crystallinity and phase composition were investigated using the PanAlytical X'Pert Pro (Panalytical Ltd., Eindhoven, The Netherlands). The functional groups of the catalysts were measured in the wavelength range of 550–4000 cm^{-1} using a Spectrum 100 FTIR spectrometer (PerkinElmer, Waltham, MA, USA). The specific surface area, pore volume, and pore size were measured using the ASAP 2020 BET analysis system (Micromeritics Instruments Corp., Norcross, GA, USA).

2.8. Catalytic Testing

Fructose (0.5 g), DMSO (6 ml), and catalyst (0.1 g) were placed in a vessel and allowed to react for 1 h at 120 °C using the ultraclave high-performance microwave reactor. After the reaction, the solution was centrifuged to separate the solid catalyst particles from the liquid, and the supernatant liquid was analysed using a Lambda UV-Vis spectrometer Model 750s (PerkinElmer, Shelton, CT, USA) at the wavelength of 284 nm.

2.9. Reusability Studies

Reusability of the materials was studied using the dehydration of fructose into 5-HMF in DMSO as an illustrative reaction. The C_S and TiO_2C_S composite acid catalysts were allowed to react for 1 h in separate vessels and then were separated from the solution by centrifugation. The recovered catalysts were then washed thoroughly with DI water and ethanol and dried for 4 h at 90 °C. This process was repeated four times, with the 5-HMF yield calculated each time.

2.10. Hot Filtration Tests

The 1%TiO$_2$C_S, 5%TiO$_2$C_S, and 10%TiO$_2$C_S acid catalysts were used to perform hot filtration tests to determine the heterogeneous nature of the acid catalyst. In a typical procedure, 0.5 g of fructose and 0.1 g of TiO$_2$C_S were added to 6 ml of DMSO and allowed to react for 30 min using the ultraclave high-performance microwave reactor. After the reaction, the solution was allowed to cool down and then was centrifuged to separate the catalyst from the reaction mixture. The solution was analysed using a UV-Vis spectrometer at a wavelength of 284 nm. The same reaction was repeated for 1 h using the solution filtrate without adding the substrate.

3. Results and Discussion

3.1. BET Surface Area and Pore Volume of the Catalysts

The BET surface area and pore volume of the carbon spheres and the various compositions of the TiO$_2$C composites are presented in Table 1. The carbon spheres had a surface area of 517 m^2/g before sulfonation. However, a reduction in surface area was observed after adding TiO$_2$ NPs to the carbon due to TiO$_2$ NPs blocking some of the micropores of the carbon. The TiO$_2$C composite loaded with the lowest dosage of TiO$_2$ NPs (i.e., 1%TiO$_2$C) had a larger surface area and pore volume of 413 m^2/g and 0.26 cm^3/g, respectively, than the composites loaded with higher amounts of TiO$_2$ (i.e., 5%TiO$_2$C and 10%TiO$_2$C). The surface area of the TiO$_2$C composite decreased when a high loading of TiO$_2$ NPs was introduced onto the carbonaceous support. The 1%TiO$_2$C composite had the highest pore volume of 0.26 cm^3/g compared to that of 5%TiO$_2$C and 10%TiO$_2$C composites. Both the surface area and the pore volume were affected by functionalising the carbon spheres and TiO$_2$C with concentrated sulfuric acid. After sulfonation, the surface area of the carbon spheres decreased from 517 to 167 m^2/g and that of the 1%TiO$_2$C, 5%TiO$_2$C, and 10%TiO$_2$C composites decreased from 413 to 83, 273 to 61, and 202 to 59 m^2/g, respectively. Pore volume also decreased. The reduction in surface area and pore volume might be attributed to the SO$_3$H groups that were attached to the pores of the carbon spheres and TiO$_2$C composites during sulfonation. Other researchers reported similar results [21,22]. For example, Tamborini et al. [21] synthesised sulfonated porous carbon materials and used them for the production of biodiesel. The surface areas of the different synthesised carbons (PC100S and PC200S) were 630 and 695 m^2/g, respectively. The sulfonation process decreased the surface area of PC100S to 470 m^2/g and that of PC200S to 140 m^2/g. The pore volume of the carbon materials also decreased from 0.92 to 0.77 cm^3/g and from 1 to 0.34 cm^3/g, respectively. Sulfonation corroded the microporosity and mesoporosity of these carbon materials. Liu et al. [22] prepared a carbon-based acid catalyst and used it for the esterification of acetic acid with ethanol. After sulfonation, the surface area and pore volume of the activated carbon decreased from 751 to 602 m^2/g and from 0.47 to 0.38 cm^3/g, respectively. Their findings suggested that SO$_3$H groups were grafted onto the pore spaces of the activated carbons.

Table 1. Textural properties and hot filtration results of carbon spheres and TiO$_2$C nanocomposites.

Catalyst before Sulfonation				Catalyst after Sulfonation		
	S_{BET} (m^2/g)	V_p (cm^3/g)		S_{BET} (m^2/g)	V_p (cm^3/g)	Sulfonic Groups (m·molg^{-1}) [a]
C	517	-	C_S	167	-	1.34
1%TiO$_2$C	413	0.26	1%TiO$_2$C_S	83	0.07	1.46
5%TiO$_2$C	273	0.13	5%TiO$_2$C_S	61	0.04	1.55
10%TiO$_2$C	202	0.10	10%TiO$_2$C_S	59	0.03	1.49

C: Carbon; S_{BET}: Surface Area; V_p: Pore Volume; S_: Sulfonated; [a] Attained by titration with NaOH.

The acid strength of C_S, 1%TiO$_2$C_S, 5%TiO$_2$C_S, and 10%TiO$_2$C_S was calculated to be 1.46, 1.55, 1.49, and 1.36 mmol g^{-1}, respectively, as seen in Table 1. The acid distribution results confirmed that SO$_3$H acid sites were introduced onto the surface of the prepared catalysts, verifying the strong adsorption bands of the S=O group at 1180 and 1008 cm^{-1}, which were associated with the SO$_3^-$ groups in the FTIR spectra discussed in Section 3.3.

The heterogeneous nature of the 1%TiO$_2$C_S, 5%TiO$_2$C_S, and 10%TiO$_2$C_S acid catalysts was evaluated using hot filtration tests conducted at optimum conditions. The results showed that leaching of SO$_3$H functional groups did not occur during the dehydration of fructose to 5-HMF. The 5-HMF dehydration was about 80% after 30 min of reaction. The reaction was repeated again for 1 h and showed that the HMF dehydration did not improve beyond 80%. These findings confirmed that the TiO$_2$C_S acid catalysts are heterogeneous and that the functional groups were strongly attached to the surface of the TiO$_2$C_S catalysts, so leaching did not occur.

3.2. Surface Morphology

The SEM images of the neat carbon and neat TiO$_2$C composites are shown in Figure 1a–d, and those of the C_S and TiO$_2$C_S acid catalysts are shown in Figure 1e–h. The neat carbon was spherical and had a smooth surface, with the spheres agglomerated and interconnected (Figure 1a). The neat 1%TiO$_2$C_S, 5%TiO$_2$C_S, and 10%TiO$_2$C_S composites consisted of a combination of interconnected spheres and small TiO$_2$ particles which did not have a definite shape, and were aggregated (Figure 1b–d). Moreover, a high amount (5 and 10 wt.%) of TiO$_2$ NPs loading promoted the formation of irregular TiO$_2$ particles in the TiO$_2$C composites. After sulfonation, the particles were found to have shrunk, as seen in Figure 1f–h, whereas in a case of C_S the spheres were found to have enlarged as shown in Figure 2e.

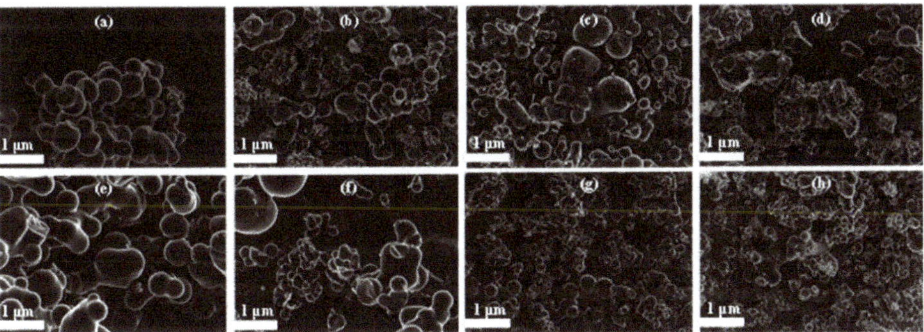

Figure 1. SEM low magnification images of (**a**) neat C, (**b**) neat 1%TiO$_2$C, (**c**) neat 5%TiO$_2$C, (**d**) neat 10%TiO$_2$C, (**e**) C_S solid acid catalyst, (**f**) 1%TiO$_2$C_S solid acid catalyst, (**g**) 5%TiO$_2$C_S solid acid catalyst, and (**h**) 10%TiO$_2$C_S solid acid catalyst.

Figure 2 shows the high magnification SEM images of (a) C_S solid acid catalyst, (b) 1%TiO$_2$C_S, (c) 5%TiO$_2$C_S, (d) 10%TiO$_2$C_S and (e) sulfonated TiO$_2$. From Figure 2a, we can see that the C_S solid acid catalyst had a smooth surface, whereas the 1%TiO$_2$C_S and 5%TiO$_2$_S showed the presence of small traces of TiO$_2$ particles which were deposited onto the carbon surface. In the 10%TiO$_2$C_S sample (Figure 2d) TiO$_2$ particles were deposited on the surface of the carbon. However, these TiO$_2$ particles were not homogeneously distributed onto the surface of the carbon. In the sulfonated TiO$_2$ sample (Figure 2e); SEM revealed that TiO$_2$ particles had irregular shaped and were agglomerated.

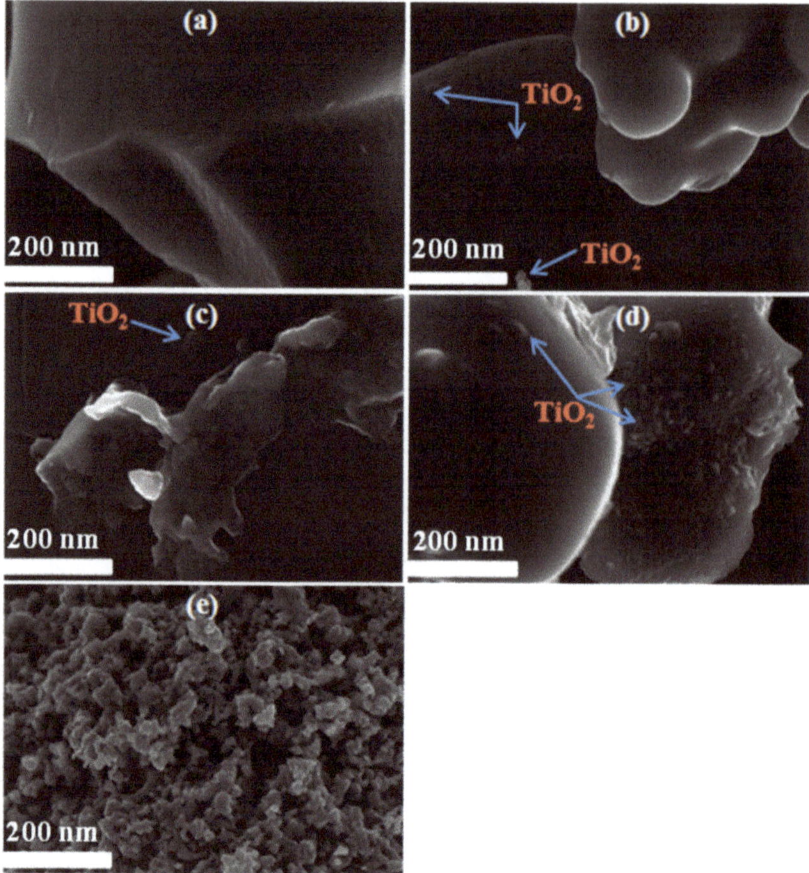

Figure 2. SEM high magnification images of (**a**) C_S solid acid catalyst, (**b**) 1%TiO$_2$C_S solid acid catalyst, (**c**) 5%TiO$_2$C_S solid acid catalyst, (**d**) 10%TiO$_2$C_S solid acid catalyst and (**e**) sulfonated TiO$_2$.

The energy-dispersive spectroscopy (EDS) spectra of the neat carbon and neat TiO$_2$C composites are shown in parts (a) to (d) of Figure 3, and those of the carbon spheres and TiO$_2$C after sulfonation are shown in parts (e) to (h) of Figure 3. The EDS spectrum of the carbon spheres contains only C and O peaks, whereas that of the composite samples contains the Ti peak, confirming that TiO$_2$ was present in the samples. The elemental weight percent of C decreased with an increase in TiO$_2$ loading. After sulfonating the carbon, the C, O, and S peaks are present in the spectrum (Figure 3e), indicating that the carbon spheres were neat and no impurities were detected in the sample. The spectra of the TiO$_2$C_S acid catalysts [parts (f) to (h) of Figure 3] showed the presence of C, O, Ti, and S, and the elemental weight percentage of Ti increased with an increase in the loading percentage of TiO$_2$. However, the weight percentage of C decreased with an increase in the loading percentage of TiO$_2$. No other impurity elements were found in the prepared solid acid catalysts. The presence of S in all the solid acid catalysts indicated that the samples were successfully functionalised with sulfuric acid. Elemental composition analysis revealed that all the prepared acid catalysts contained sulfonic acid in the form of SO$_3$H groups [23], further proving the presence of sulfonic groups, as indicated in Table 1. Wang et al. [14] also confirmed the presence of sulfonic groups on the surface of carbon materials after sulfonation. The TiO$_2$C_S composites had C, O, Ti, and S, and their elemental weight percent of Ti increased with an increase in the loading percentage of Ti. However, the weight percent of C decreased with an increase in the loading percentage of Ti.

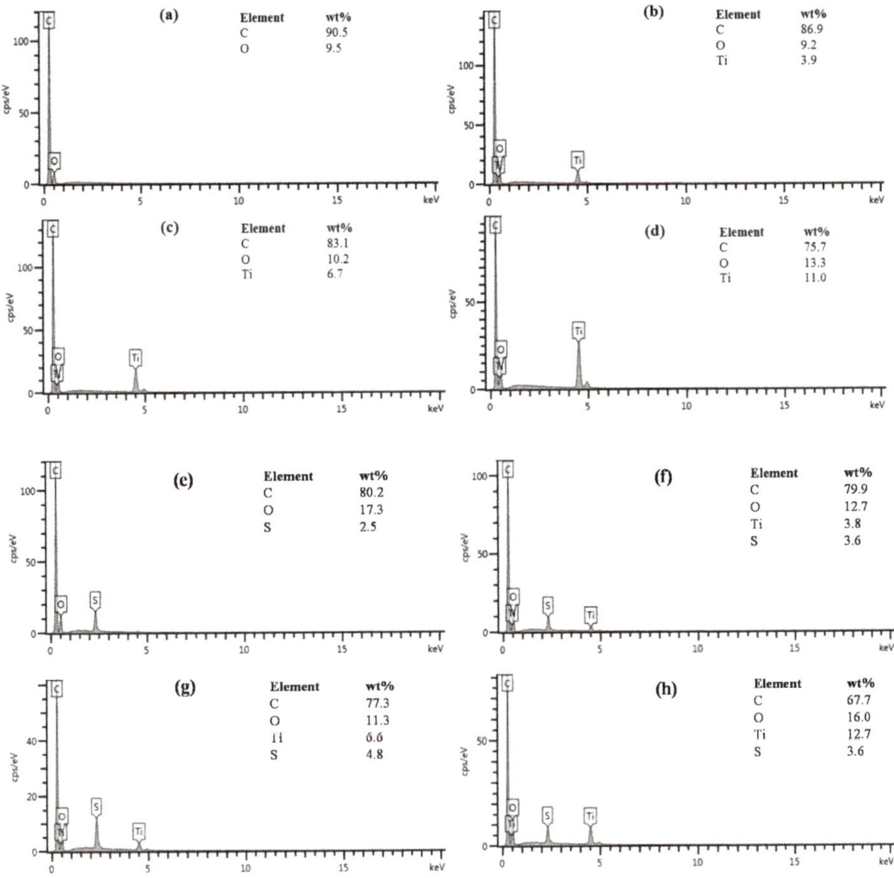

Figure 3. EDS spectra and elemental composition of (**a**) neat C, (**b**) neat 1%TiO$_2$C, (**c**) neat 5% TiO$_2$C, (**d**) neat 10% TiO$_2$C, (**e**) C_S solid acid catalyst, (**f**) 1%TiO$_2$C_S solid acid catalyst, (**g**) 5% TiO$_2$C_S solid acid catalyst, and (**h**) 10% TiO$_2$C_S solid acid catalyst.

3.3. Chemical Analysis Using FTIR

The FTIR spectra of the neat, calcined, and sulfonated carbon spheres are shown in Figure 4a. The presence of oxygen groups is demonstrated by the bands at 3000–3600 and 1710 cm^{-1}, which were attributed to –OH stretching and C=O vibrations, respectively [3,24]. The 875–750 cm^{-1} band was assigned to the aromatic C–H group, and the presence of aromatic rings was confirmed by the band at 1620 cm^{-1}, which was assigned to C=C vibrations [24]. Additional bands at 1180 and 1008 and 1106−1168 cm^{-1} were observed in the spectrum of the sulfonated carbon spheres. These bands were assigned to the symmetric stretching vibration of S=O groups, which are associated with the SO$_3^-$ groups [25] and C=S stretching [26], respectively as shown in Figure 4a. These sulfonic peaks in the spectrum of the sulfonated carbon spheres indicate that -SO$_3$H functional groups were attached to the carbon spheres. The FTIR spectra of the neat, calcined, and sulfonated 1%TiO$_2$C, 5%TiO$_2$C, and 10%TiO$_2$C (Figure 4b–d) show bands at 1700 and 1200 cm^{-1}, which were attributed to the C=O and

C–O stretching of the carboxyl group [27]. A strong adsorption band of the S=O group at 1008 cm^{-1}, which was associated with the SO$_3{}^-$ groups, was observed.

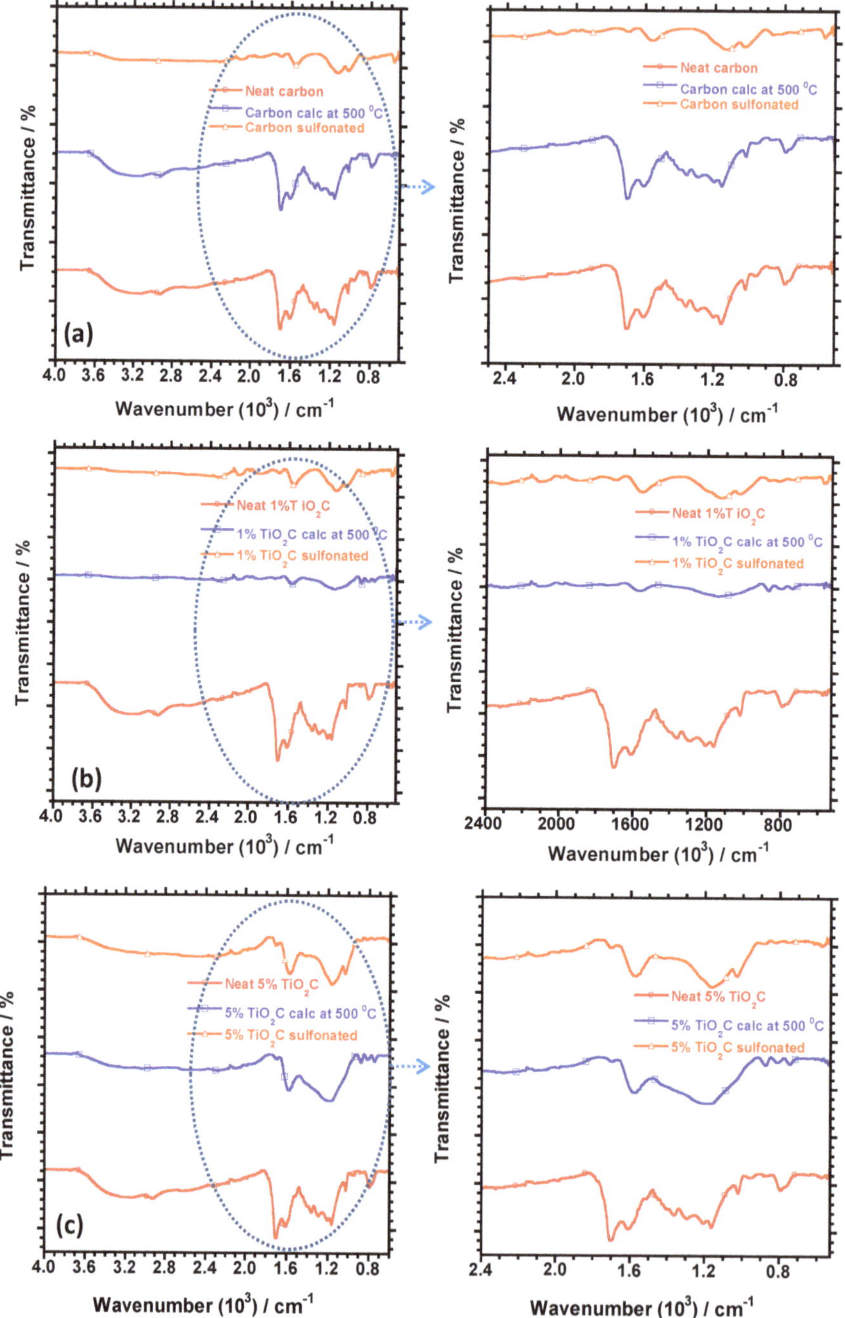

Figure 4. *Cont.*

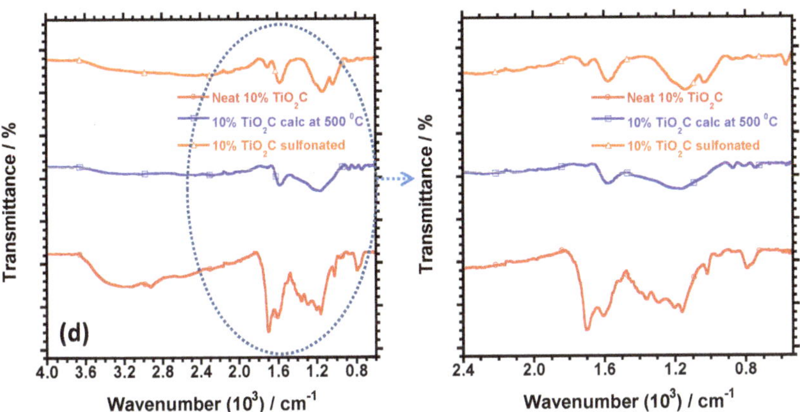

Figure 4. FTIR spectra of neat, calcined and sulfonated (**a**) carbon, (**b**) 1%TiO$_2$C, (**c**) 5%TiO$_2$C, and (**d**) 10%TiO$_2$C.

3.4. Structural Characterisation Using XRD

The XRD patterns of the calcined carbon spheres and TiO$_2$C composites are shown in Figure 5a. The XRD pattern of the neat carbon spheres showed broad peaks at 25° and 43°, which were indexed to the [002] and [100] characteristic phases of amorphous carbonaceous material, whereas the TiO$_2$C sample which was loaded with the least dosage of TiO$_2$ (1%TiO$_2$C), was found to be amorphous, due to the effect of high concentration of the carbon which was present in the sample. The XRD pattern of the crystalline structure of the 5%TiO$_2$C and 10%TiO$_2$C composites had sharp peaks at 25.2°, 37.83°, 48.2°, 54.8°, 62.2°, and 70.2°, which were indexed to the [101], [004], [200], [105], [204], and [116] phases of anatase TiO$_2$ [28]. After sulfonation, the peaks for the carbon spheres and 1%TiO$_2$C shifted, as shown in Figure 5b. Moreover, the acid treatment (sulfonation) caused the [100] plane peak to disappear. The 5%TiO$_2$C and 10%TiO$_2$C composites were not affected by the sulfonation.

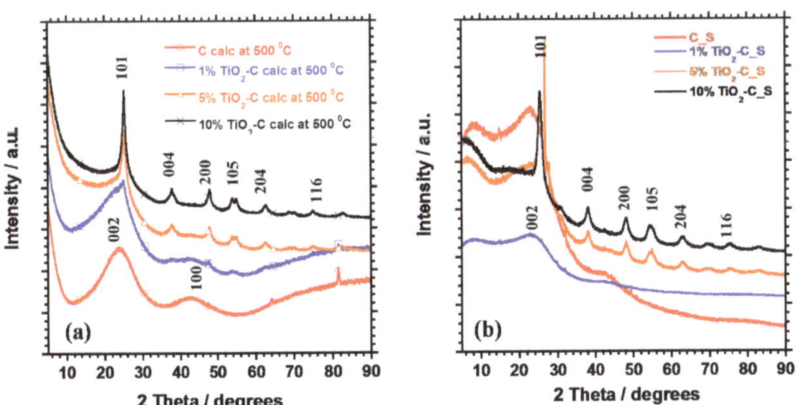

Figure 5. XRD results for carbon spheres and TiO$_2$C composites (**a**) that were calcined at 500 °C and (**b**) after sulfonation.

3.5. Catalytic Testing

3.5.1. Effect of Different Solvents on the Dehydration of Fructose into 5-HMF

Dehydration of fructose into 5-HMF was tested using different alcoholic solvents (such as isopropanol, ethanol, and methanol), water, and DMSO at 120 °C for 60 min (Table 2). All the alcoholic solvents produced a low 5-HMF yield of <12% on all the solid catalysts tested for this reaction. Water was also found to be an ineffective solvent. DMSO was the most effective solvent. The HMF yield was 84% for the carbon solid acid catalyst, 91% for 1%TiO_2C_S, 92% for 5%TiO_2C_S, and 95% for 10%TiO_2C_S. Thus, the HMF dehydration increased as the TiO_2 content increased. These results suggest that DMSO is a suitable solvent for use in the dehydration of fructose into 5-HMF when using TiO_2C solid acid catalysts. DMSO acted both as a solvent and as a reaction mediator; hence a high dehydration of 5-HMF was achieved.

Table 2. Dehydration of fructose in different solvents.

Catalyst	HMF Dehydration (%) [UV-Vis at the Measured Absorbance of 284 nm]				
	Methanol	DI Water	Ethanol	Isopropanol	DMSO
C	3	5	11	1	84
1%TiO_2C	3	6	4	9	91
5%TiO_2C	3	1	6	0	92
10%TiO_2C	0	4	12	11	95

Reaction conditions: Substrate, fructose; catalyst amount, 0.1 g; temperature, 120 °C; time, 60 min.

Table 3 shows a comparison of the TiO_2C composite acid catalyst synthesised in this work with other catalysts reported in published works on dehydration of fructose into 5-HMF in DMSO using the microwave-assisted method. We found that our designed TiO_2C_S acid catalysts produced the highest 5-HMF yield. The highest 5-HMF yield that we achieved at 120 °C, a temperature lower than those used in other studies, was 91% for 1%TiO_2C_S and 92 and 95% for 5%TiO_2C_S and 10%TiO_2C_S, respectively. To our knowledge, this is the first report of the use of TiO_2C_S composites as effective solid acid catalysts for the conversion of fructose into 5-HMF using the microwave-assisted method. De et al. [19] used mesoporous TiO_2 nanomaterial to catalyse the dehydration of D-fructose into 5-HMF in DMSO solvent under microwave-assisted heating. This reaction was conducted at 130 °C for 2 min for a 49.2% 5-HMF yield was achieved. Dutta et al. [15] also used mesoporous TiO_2 nanoparticles for the same reaction at 140 °C for 5 min and reported a yield of53.4%. Use of carbonaceous acid catalysts for this reaction achieved a 100% conversion of fructose and 90% 5-HMF yield at 160 °C for 1.5 h. Wang et al. [16] used carbon-based solid acid catalysts to catalyse the dehydration of fructose into 5-HMF at 130 °C for 1.5 h for a 5-HMF yield of about 91.2%. Hu et al. [18] explored a magnetic lignin-derived carbonaceous acid catalyst for the catalysed conversion of fructose into 5-HMF and achieved a 5-HMF yield of 81.1% with 100% fructose conversion.

Table 3. Comparison of results of the dehydration of fructose (substrate) into 5-HMF in DMSO using carbonaceous and TiO_2-based solid acid catalysts.

Catalyst	Catalyst Mass (g)	Substrate Mass (g)	T (°C)	Time (min)	5-HMF Yield (%)		Ref.
					Uv-Vis [a]	HPLC [b]	
C_S	0.1	0.5	120	60	84	-	This work
1%TiO_2C_S	0.1	0.5	120	60	91	-	This work
5%TiO_2C-S	0.1	0.5	120	60	92	-	This work
10%TiO_2C_S	0.1	0.5	120	60	95	-	This work
TiO_2	0.05	0.1	130	2	49.5	47.8	[19]
TiO_2	0.1	0.05	140	5	53.4	-	[20]
CS	0.1	0.5	160	90	-	90	[15]
C	0.4	0.5	130	90	-	91.2	[16]
Magnetic lignin-derived carbon (MLC)-SO_3H	0.05	0.1	130	40	-	81.1	[18]

[a] 5-HMF dehydration was calculated by UV-vis at the measured absorbance of 284 nm; [b] 5-HMF dehydration measured by HPLC.

3.5.2. Effect of Reaction Temperature on HMF Dehydration

The effect of reaction temperature on the catalytic transformation of fructose to HMF was carried out at 25, 60, 80, 100 and 120 °C. The reaction was conducted using 1%TiO_2C_S, 5%TiO_2C_S,

10%TiO$_2$C_S, C_S solid acid catalysts and without the catalyst (non-catalytic reaction). The results are shown in Figure 6, which suggests that no 5-HMF was formed when the reaction was carried out in the absence of a catalyst on all the reaction temperatures that were studied. For the carbon solid acid catalyst and reaction temperature of 25 °C, the 5-HMF yield was <8%. When the temperature was raised to 60 °C, the 5-HMF yield slightly increased to 35%. At 80 °C, the 5-HMF yield improved to 71%. The highest 5-HMF yield of 85% was achieved at 120 °C. The effect of reaction temperature was also tested using the 1%TiO$_2$C_S, 5%TiO$_2$C_S, and 10%TiO$_2$C_S solid acid catalysts. The 5-HMF yield was low when the reaction was conducted at 25 °C, then increased as the reaction temperature increased. The highest HMF yields of 91, 92, and 95% were achieved at 120 °C with the use of 1%TiO$_2$C_S, 5%TiO$_2$C_S, and 10%TiO$_2$C_S solid acid catalysts, respectively. Thus, the best temperature for 5-HMF production using these composite catalysts was 120 °C. Compared to the performance of the carbon solid acid catalyst, an improved HMF yield was achieved with the use of 1, 5, and 10%TiO$_2$C_S solid acid catalysts.

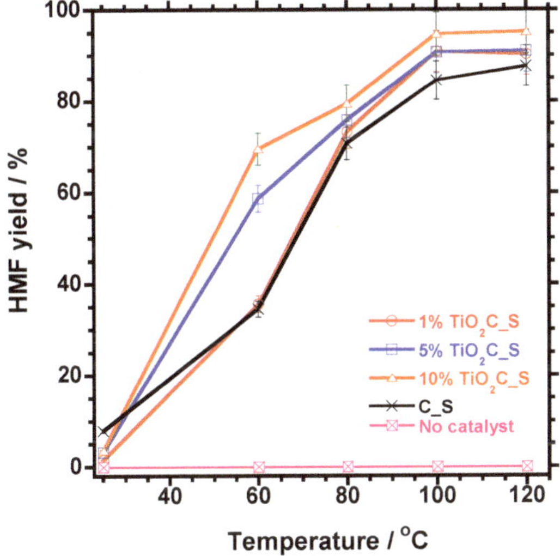

Figure 6. Effect of reaction temperature on fructose dehydration into HMF using sulfonated carbon and TiO$_2$C composite catalysts in DMSO solvent.

3.5.3. Effect of Reaction Time on 5-HMF Dehydration

The effect of reaction time on the catalysed dehydration of fructose into 5-HMF was studied using different times of 15, 30, 60, and 120 min. The reaction was conducted using the prepared solid acid catalysts and no catalyst (non-catalytic reaction). The results are shown in Figure 7. No 5-HMF was formed when the reaction was carried out without a catalyst for all four reaction times. When the reaction was performed with the C_S solid acid catalyst, increasing the reaction time from 15 to 30 min improved the 5-HMF yield from 25% to 46%. The HMF dehydration increased rapidly to 85% after conducting the reaction for 60 min. However, performing the reaction for 120 min caused a slight drop in the 5-HMF yield to 83%. When conducting the dehydration reaction using 1%TiO$_2$C_S, 5%TiO$_2$C_S, and 10%TiO$_2$C_S solid acid catalysts, increasing the reaction time from 15 to 30 min increased the 5-HMF yield from 27 to 49%, 28 to 58%, and 29 to 68%, respectively. Increasing the reaction time to 60 min significantly improved the 5-HMF yield to >90% for all the TiO$_2$C composite acid catalysts. Finally, for a reaction time of 120 min, the HMF dehydration for 1%TiO$_2$C_S, 5%TiO$_2$C_S, and 10%TiO$_2$C_S solid acid catalysts was calculated to be 88, 92, and 93%, respectively.

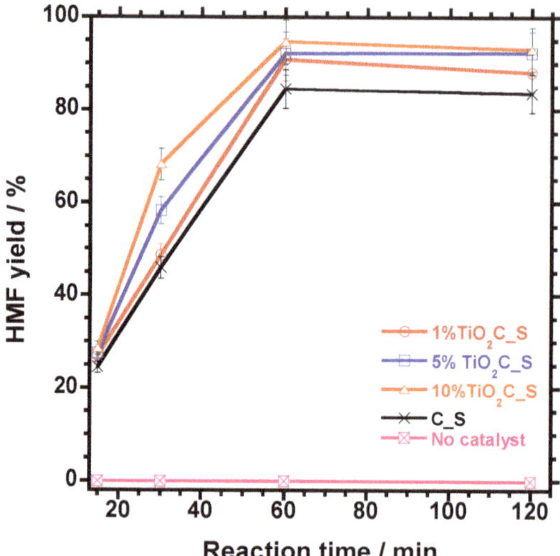

Figure 7. Effect of reaction time on the dehydration of fructose into HMF in DMSO solvent.

3.5.4. Effect of Catalyst Amount on HMF Dehydration

The effect of the amount of catalyst on the 5-HMF yield was studied, with the results presented in Figure 8. The reaction was carried out using catalyst dosages of 0.02, 0.05, 0.1, and 0.2 g. The results indicated that no 5-HMF was formed when this reaction was conducted in the absence of a catalyst on all the catalyst dosages that were studied. The 5-HMF yield increased from 34 to 59% when the amount of the C_S solid acid catalyst increased from 0.02 to 0.05 g under the same reaction conditions. An increase in the catalyst dosage to 0.1 g increased the 5-HMF yield to 85%. However, increasing the catalyst dosage to 0.2 g resulted in a reduction of the 5-HMF yield to 82%. The effect of catalyst concentration was also tested using the 1%TiO$_2$C_S, 5%TiO$_2$C_S, and 10%TiO$_2$C_S solid acid catalysts. An increase in the catalyst dosage from 0.02 to 0.1 g drastically improved the 5-HMF yield of the 1%TiO$_2$C_S, 5%TiO$_2$C_S, and 10%TiO$_2$C_S solid acid catalysts to 91, 92, and 95%, respectively. However, when the catalyst dosage increased to 0.2 g, the 5-HMF yield of the 5%TiO$_2$C_S, and 10%TiO$_2$C_S solid acid catalysts slightly decreased to 91 and 92% respectively, whereas in a case of the 1%TiO$_2$C_S the 5-HMF yield increased to 93%. This decrease in 5-HMF dehydration at a high catalyst dosage for the 5%TiO$_2$C_S and 10%TiO$_2$C_S, could be attributed to the excess acid active sites that promote both the dehydration reaction and the formation of by-products such as humins [29]. The 5-HMF yields did not increase with an increase in catalyst dosage for the C_S, 5%TiO$_2$C_S and 10%TiO$_2$C_S solid acid catalysts.

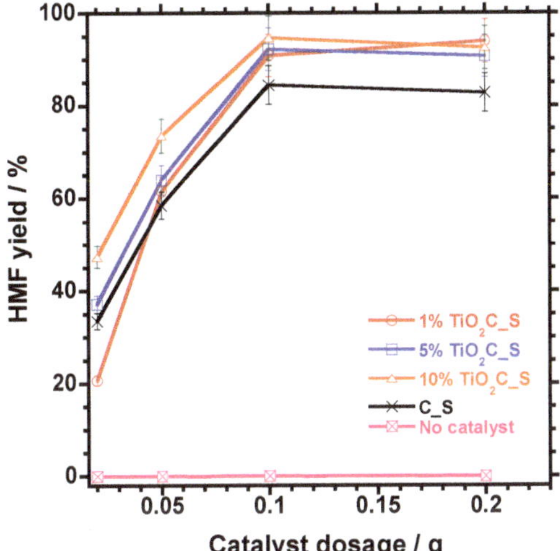

Figure 8. Effect of catalyst dosage on the dehydration of fructose into HMF in DMSO solvent.

3.5.5. Reusability of the Catalyst

The reusability of the catalysts was studied, with the results presented in Figure 9. After the first run, the 5-HMF yields were 94, 97, and 93% for 1%TiO$_2$C_S, 5%TiO$_2$C_S, and 10%TiO$_2$C_S, respectively. The 5-HMF yield slightly decreased after the fourth run. The 5-HMF yield of the 1%TiO$_2$C_S and 10%TiO$_2$C_S composite acid catalysts decreased by 3% and that of 5%TiO$_2$C_S decreased by 5%. These results indicate that the TiO$_2$C composite acid catalysts are highly stable and can be reused for the dehydration of fructose.

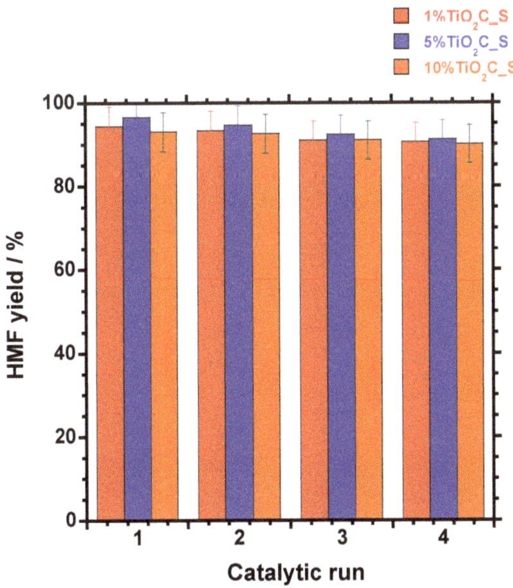

Figure 9. Recycling of the 1%TiO$_2$C_S, 5%TiO$_2$C_S, and 10%TiO$_2$C_S acid catalysts used in the dehydration of fructose into HMF. Conditions: reaction time = 60 min, reaction temperature = 120 °C, catalyst concentration = 0.1 g, amount of fructose = 0.5 g, volume of DMSO = 6 mL).

4. Conclusions

TiO$_2$C solid acid catalysts were successfully prepared using the microwave-assisted method and then sulfonated with concentrated sulfuric acid. Microwave heating increases the rate of dehydration of fructose. The dehydration of fructose is closely related to the acidity of the catalyst. Among the solvents tested for use in the dehydration of fructose into 5-HMF, DMSO performed the best in terms of 5-HMF dehydration. The reaction temperature, reaction time, and catalyst dosage were found to have an effect on the 5-HMF dehydration. The solid acid catalysts synthesised in this work were highly stable and heterogeneous. Moreover, this study was the first time such TiO$_2$C solid acid catalysts were used for dehydrating fructose into 5-HMF.

Author Contributions: M.M.S. design the concept and wrote the first draft of manuscript. R.M. went through manuscript and provided comments. S.S.R. critically reviewed and corrected the manuscript.

Funding: The authors are grateful to the Department of Science and Technology (DST, project no. HGERA8X) and the Council for Scientific and Industrial Research (CSIR, project no. HGER74p) of South Africa for financial support.

References

1. Démolis, A.; Essayem, N.; Rataboul, F. Synthesis and applications of alkyl levulinates. *ACS Sustain. Chem. Eng.* **2014**, *2*, 1338–1352. [CrossRef]
2. McNeff, C.V.; Nowlan, D.T.; McNeff, L.C.; Yan, B.; Fedie, R.L. Continuous production of 5-hydroxymethylfurfural from simple and complex carbohydrates. *Appl. Catal. A Gen.* **2010**, *384*, 65–69. [CrossRef]
3. Qi, X.; Watanabe, M.; Aida, T.M.; Smith, R.L., Jr. Catalytical conversion of fructose and glucose into 5-hydroxymethylfurfural in hot compressed water by microwave heating. *Catal. Commun.* **2008**, *9*, 2244–2249. [CrossRef]
4. Wang, J.; Ren, J.; Liu, X.; Xi, J.; Xia, Q.; Zu, Y.; Lu, G.; Wang, Y. Direct conversion of carbohydrates to 5-hydroxymethylfurfural using Sn-Mont catalyst. *Green Chem.* **2012**, *14*, 2506–2512. [CrossRef]
5. Moreau, C.; Finiels, A.; Vanoye, L. Dehydration of fructose and sucrose into 5 hydroxymethylfurfural in the presence of 1-H-3-methyl imidazolium chloride acting both as solvent and catalyst. *J. Mol. Catal. A Chem.* **2006**, *253*, 165–169. [CrossRef]
6. Asghari, F.S.; Yoshida, H. Acid catalysed production of 5-hydroxymethyl furfural from d-Fructose in Subcritical Water. *Ind. Eng. Chem. Res.* **2006**, *45*, 2163–2173. [CrossRef]
7. Xu, H.; Miao, Z.; Zhao, H.; Yang, J.; Zhao, J.; Song, H.; Liang, N.; Chou, L. Dehydration of fructose into 5-hydroxymethylfurfural by high stable ordered mesoporous zirconium phosphate. *Fuel* **2015**, *145*, 234–240. [CrossRef]
8. Jain, A.; Shore, A.M.; Jonnalagadda, S.C.; Ramanujachary, K.V.; Mugweru, A. Conversion Of fructose, glucose and sucrose to 5-hydroxymethyl-2-furfural over mesoporous zirconium phosphate catalyst. *Appl. Catal. A Gen.* **2015**, *489*, 72–76. [CrossRef]
9. Kılıc, E.; Yılmaz, S. Fructose dehydration to 5-hydroxymethylfurfural over sulfated TiO$_2$–SiO$_2$, Ti-SBA-15, ZrO$_2$, SiO$_2$, and activated carbon catalysts. *Eng. Chem. Res.* **2015**, *54*, 5220–5225. [CrossRef]
10. Wang, S.; Zhang, Z.; Liu, B. Catalytic conversion of fructose and 5-hydroxymethylfurfural into 2,5-Furandicarboxylic acid over a recyclable Fe$_3$O$_4$−CoO$_x$ magnetite nanocatalyst. *ACS Sustain. Chem. Eng.* **2015**, *3*, 406–412. [CrossRef]
11. Yang, Z.; Qi, W.; Huang, R.; Fang, J.; Su, R.; He, Z. Functionalized silica nanoparticles for conversion of fructose to 5-hydroxymethylfurfural. *Chem. Eng. J.* **2016**, *296*, 209–216. [CrossRef]
12. Morales, G.; Paniagua, M.; Melero, J.A.; Iglesias, J. Efficient production of 5-ethoxymethylfurfural from fructose by sulfonic mesostructured silica using DMSO as co-solvent. *Catal. Today* **2017**, *279*, 305–316. [CrossRef]
13. Hu, B.; Wang, K.; Wu, L.; Yu, S.H.; Antonietti, M.; Titirici, M.M. Engineering carbon materials from the hydrothermal carbonization process of biomass. *Adv. Mater.* **2010**, *22*, 813–828. [CrossRef] [PubMed]

14. Wang, J.; Zhang, Y.; Wang, Y.; Zhu, L.; Cui, H.; Yi, W. Catalytic fructose dehydration to 5-hydroxymethylfurfural over sulfonated carbons with hierarchically ordered pores. *J. Fuel Chem. Technol.* **2016**, *44*, 1341–1348. [CrossRef]

15. Zhao, J.; Zhou, C.; He, C.; Dai, Y.; Jia, X.; Yang, Y. Efficient dehydration of fructose to 5 hydroxymethylfurfural over sulfonated carbon sphere solid acid catalysts. *Catal. Today* **2016**, *264*, 123–130. [CrossRef]

16. Wang, J.; Xu, W.; Ren, J.; Liu, X.; Lu, G.; Wang, Y. Efficient catalytic conversion of fructose into hydroxymethylfurfural by a novel carbon-based solid acid. *Green Chem.* **2011**, *13*, 2678–2681. [CrossRef]

17. Guo, F.; Fang, Z.; Zhou, T.J. Conversion of fructose and glucose into 5-hydroxymethylfurfural with lignin-derived carbonaceous catalyst under microwave irradiation in dimethyl sulfoxide–ionic liquid mixtures. *Bioresour. Technol.* **2012**, *112*, 313–318. [CrossRef] [PubMed]

18. Hu, L.; Tang, X.; Wu, Z.; Linc, L.; Xu, J.; Xu, N.; Dai, B. Magnetic lignin-derived carbonaceous catalyst for the dehydration of fructose into 5-hydroxymethylfurfural in dimethyl sulfoxide. *Chem. Eng. J.* **2015**, *263*, 299–308. [CrossRef]

19. De, S.; Dutta, S.; Patra, A.K.; Bhaumik, A.; Saha, B. Self-assembly of mesoporous TiO_2 nanospheres via aspartic acid templating pathway and its catalytic application for 5-hydroxymethyl-furfural synthesis. *J. Mater. Chem.* **2011**, *21*, 17505–17510. [CrossRef]

20. Dutta, S.; De, S.; Patra, A.K.; Sasidharan, M.; Bhaumik, A.; Saha, B. Microwave assisted rapid conversion of carbohydrates into 5-hydroxymethylfurfural catalysed by mesoporous TiO_2 nanoparticles. *Appl. Catal. A Gen.* **2011**, *409–410*, 133–139. [CrossRef]

21. Tamborini, L.H.; Casco, M.E.; Militello, M.P.; Silvestre-Albero, J.; Barbero, C.A.; Acevedo, D.F. Sulfonated porous carbon catalysts for biodiesel production: Clear effect of the carbon particle size on the catalyst synthesis and properties. *Fuel Process. Technol.* **2016**, *149*, 209–217. [CrossRef]

22. Liu, X.Y.; Huang, M.; Ma, H.L.; Zhang, Z.Q.; Gao, J.M.; Zhu, Y.L.; Han, X.J.; Guo, X.Y. Preparation of a carbon-based solid acid catalyst by sulfonating activated carbon in a chemical reduction process. *Molecules* **2010**, *15*, 7188–7196. [CrossRef] [PubMed]

23. Hou, Q.; Li, W.; Ju, M.; Liu, L.; Chen, Y.; Yang, Q. One-pot synthesis of sulfonated graphene oxide for efficient conversion of fructose into HMF. *RCS Adv.* **2016**, *16*, 104016–104024. [CrossRef]

24. Qi, X.; Guo, H.; Li, L.; Smith, R.L., Jr. Acid-catalyzed dehydration of fructose into 5-hydroxymethylfurfural by cellulose derived amorphous carbon. *ChemSusChem* **2012**, *5*, 2215–2220. [CrossRef] [PubMed]

25. Sevilla, M.; Fuertes, A.B. The production of carbon materials by hydrothermal carbonization of cellulose. *Carbon* **2009**, *47*, 2281–2289. [CrossRef]

26. Sun, Y.; Zhao, J.; Wang, J.; Tang, N.; Zhao, R.; Zhang, D.; Guan, G.; Li, K. Sulfur doped millimetre sized microporous activated carbon spheres derived from sulfonated poly(styrene-devinylbenzene) for CO_2 capture. *J. Phys. Chem. C* **2017**, *121*, 10000–10009. [CrossRef]

27. Rao, C.N.R. Contribution to the infrared spectra of organosulphur compounds. *Can. J. Chem.* **1964**, *42*, 36–42. [CrossRef]

28. Yu, D.; Bo, B.; Yunhua, H. Fabrication of TiO_2@yeast-carbon hybrid composites with the raspberry like structure and their synergistic adsorption photocatalysis performance. *J. Nanomater.* **2013**, *2013*, 851417. [CrossRef]

29. Kansal, S.K.; Sood, S.; Umar, A.; Mehta, S.K. Photocatalytic degradation of Eriochrome Black T dye using well-crystalline anatase TiO_2 nanoparticles. *J. Alloys Compd.* **2013**, *581*, 392–397. [CrossRef]

 *catalysts*

Article

Bleached Wood Supports for Floatable, Recyclable, and Efficient Three Dimensional Photocatalyst

Yuming He [1], Huayang Li [1], Xuelian Guo [2] and Rongbo Zheng [1,*]

[1] Yunnan Province Key Laboratory of Wood Adhesives and Glued Products, College of Chemical Engineering, Southwest Forestry University, Kunming 650224, China; hymingin@163.com (Y.H.); Li876770996@outlook.com (H.L.)
[2] Wetland college, Southwest Forestry University, Kunming 650224, China; guoxuelian2009@hotmail.com
* Correspondence: zhengrbzy@hotmail.com

Received: 21 December 2018; Accepted: 18 January 2019; Published: 26 January 2019

Abstract: To suppress the agglomeration of a photocatalyst, facilitate its recovery, and avoid photolysis of dyes, various support materials such as ceramic, carbon, and polymer have been investigated. However, these support materials pose the following additional challenges: ceramic supports will settle down at the bottom of their container due to their high density, while the carbon support will absorb the UV-vis light for its black color. Herein, we propose a floatable, UV transmitting, mesoporous bleached wood with most lignin removal to support P25 nanoparticles (BP-wood) that can effectively, recyclable, three dimensional (3D) photocatalytic degrade dyes such as methylene blue (MB) under ambient sunlight. The BP-wood has the following advantages: (1) The delignification makes the BP-wood more porous to not only quickly transport MB solutions upstream to the top surface, but is also decorated with P25 nanoparticles on the cell wall to form a 3D photocatalyst. (2) The delignification endows the BP-wood with good UV transmittance to undergo 3D photocatalytic degradation under sunlight. (3) It can float on the surface of the MB solution to capture more sunlight to enhance the photodegradation efficiency by suppressing the photolysis of MB. (4) It has comparable or even better photocatalytic degradation of 40 mg/L and 60 mg/L MB than that of P25 nanoparticles suspension. (5) It is green, recyclable, and scalable.

Keywords: bleached wood support materials; 3D photocatalyst; UV transmittance; floatable; recyclable

---

1. Introduction

In order to rapidly, efficiently, and cost-effectively remove dyes from industrial waste water [1], various technologies, such as physical adsorption [2], photocatalytic degradation, chemical oxidation, and membrane filtration, have been implemented, among which photocatalytic degradation has been demonstrated to be of high efficiency [3–5]. Heterogeneous photocatalysis is based on the use of UV light with a wavelength shorter than 380 nm to stimulate a semiconductor material (i.e., TiO_2 with band gap of ca. 3.2 eV, corresponding to radiation of UV light with a wavelength of about 380 nm) to excite the electrons from the valence band to the conduction band to generate electron–hole pairs, which serve as the oxidizing and reducing agents to photocatalytic degrade dyes [6]. The efficiency of TiO_2 was reported to be influenced by many factors, such as crystalline structure [7–10], particle size [10–13], and doping with the other ions [14–17]. However, there are some disadvantages in the use of TiO_2 nanoparticle suspension during photocatalytic processes: it tends to agglomerate at high concentrations, and is difficult to separate and recycle from the solution [18,19]. To overcome these disadvantages, TiO_2 can be supported on a material that suppresses the agglomeration and facilitates its further recovery. In this context, various support materials, such as ceramic (i.e., molecular sieves, silica, zeolite, and clay) [19–22], carbon (i.e., activated carbon, carbon nanotube, graphene, and graphite) [18,23–26], and polymer (i.e., chitosan, polyamide, polyester) [27–29] have been investigated. However, these support materials pose additional challenges. For instance, ceramic supports will settle down at the bottom of their container due to their high density, while the dyes in the upper solution will absorb the UV light to have photolysis instead of photocatalytic degradation by the photocatalyst. The carbon support will absorb the UV-vis light due to its black color [30–32]. Thus, it is still a challenge to provide an excellent support material with low density to float on the dye solutions' surface to efficiently exploit UV light, 3D porous structure to support photocatalysts nanoparticles, transmit UV light and transport dye upstream.

Wood, an earth-abundant, natural, low density, hierarchical, mesoporous material, has been widely used as the template to prepare TiO_2 nanomaterials [33,34], the substrate to coat with TiO_2 nanoparticle to enhance weathering performance [35], and the support to decorate with palladium nanoparticles for efficient wastewater treatment [36]. With its mesoporous structure, wood is comprised of numerous long, partially aligned lumens as well as nanochannels along its growth direction, facilitating its floatation on the solutions' surface, decoration nanoparticles on the cell wall, and bulk treatment as water flows through the entire mesoporous wood [36]. However, it is difficult to exploit wood as a photocatalyst support due to the 20–30% lignin, whose absorption ranges from 300 nm to 600 nm [31].

The objective of the current study is to test the hypothesis that the bleached wood (i.e., cellulose-based hierarchical porous structure of wood obtained via delignification) can also be exploited as an alternative photocatalyst support material. Herein, P25, a commercial TiO_2 photocatalyst was coated on three kinds of wood-based supports, namely bleached wood with P25 (BP-wood), half-bleached wood with P25 (HBP-wood), and natural wood (N-wood) with P25 (NP-wood). The first two are obtained by removing about 50% lignin and 95% lignin, respectively. Notably, the mesoporous structure of N-wood is well maintained even after removing 95% lignin, and possess transmittance with UV light. The photocatalytic activity of above three wood-based catalysts are investigated in aqueous solution by using methylene blue (MB) dye as a model contaminant under ambient sunlight illumination. The mesoporous structure of the bleached wood is demonstrated to play important roles in the photocatalytic degradation since the wood is composed of 50% vessel channels and 20% fiber channels, which provide a pathway to quickly transport MB solution onto the top surface to be photodegraded with the P25 nanoparticles under sunlight illumination. Moreover, the P25 nanoparticles can be penetrated into the wood cell wall to form 3D photocatalitic composites to further enhance the photodegradation with the illumination of transmitted UV light. The experimental result shows that it has better photocatalytic degradation of 60 mg/L MB than that of P25 nanoparticles suspension.

2. Results and Discussion

Scheme 1a illustrates the preparation process of BP-wood and the photocatalytic process. In order to remove lignin, the natural basswood is delignified by H_2O_2 steam. Scheme 1b demonstrates the approximate photodegradation mechanism of BP-wood-supported catalysts. The degraded materials are continuously transported from the bottom of the BP-wood to the top and inside, forming a 3D catalytic mechanism under the permeation of the light source. As for the control group (P25 nanoparticles are directly added to the MB solution), shown in Scheme 1c. P25 is easily wrapped with the light-absorbing dye in the solution, so that the dye undergoes weak photolysis under illumination, which has a certain degree of influence on the photocatalytic degradation of P25.

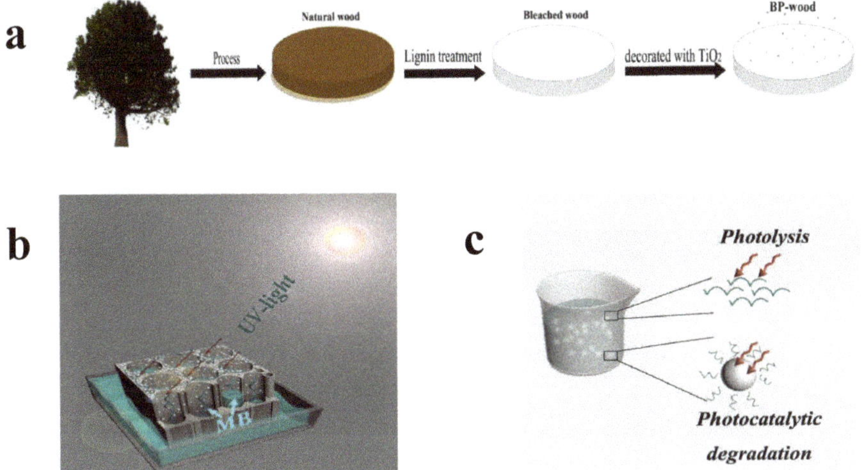

Scheme 1. Material preparation and usage. (**a**) Sketch of BP-wood preparation. (**b**) Wood absorbs methylene blue (MB) from the bottom of the contact surface and sends MB to the location of lower concentration by capillary action and transpiration of the pipeline. With the provision of ultraviolet light from sunlight, TiO$_2$, once exposed to MB, immediately produces an effective photocatalytic degradation. There is also a degradation process of MB inside the timber pipe and inside the pipe wall. Due to the higher transmission of UV light in BP-wood, the P25 penetrating into the interior of the wood also plays a role of catalyzer. (**c**) Schematic diagram of a control group-added P25 particles directly.

The lignin content can be decreased from 22.5% (N-wood) to 12.3% (Half B-wood), to 1.01% (B-wood) when the H$_2$O$_2$ steam time prolong to 1 h and 4 h, respectively (shown in Figure 1a). The mechanical strength of BP-wood is shown in Figure 1b. For the wet BP-wood with a thickness of 5 mm, the fracture strength is about 0.4 MPa. It is lower than that of N-wood and dry BP-wood (2.4 MPa), which was strong enough to be carried out in the photocatalytic process. From Figure 1c,d, we can see that massive microscale pores were generated in the cell wall and cell wall corners after delignification compared with N-wood, which will provide sites to P25 nanoparticles, and a pathway to quickly transport dye solution upstream to the B-wood's top surface. Furthermore, P25 nanoparticles dispersed in aqueous solutions were coated on the surface of N-wood, Half B-wood, and B-wood to obtain P25 nanoparticles supported on the N-wood (NP-wood), Half B-wood (HBP-wood), and B-wood (BP-wood), respectively. As shown in Figure 1e, Raman spectra revealed that peak intensity of B-wood at 1300, 1602, and 1730 cm^{-1} decreased compared with that of N-wood, which further demonstrated the removal of most of the lignin in B-wood [37]. Moreover, the degradation of the cellulose was negligible, while both lignin and hemicellulose were dramatically removed, as shown in our previous work [38]. The color changed from yellow to white during delignification process, shown in Figure 1f.

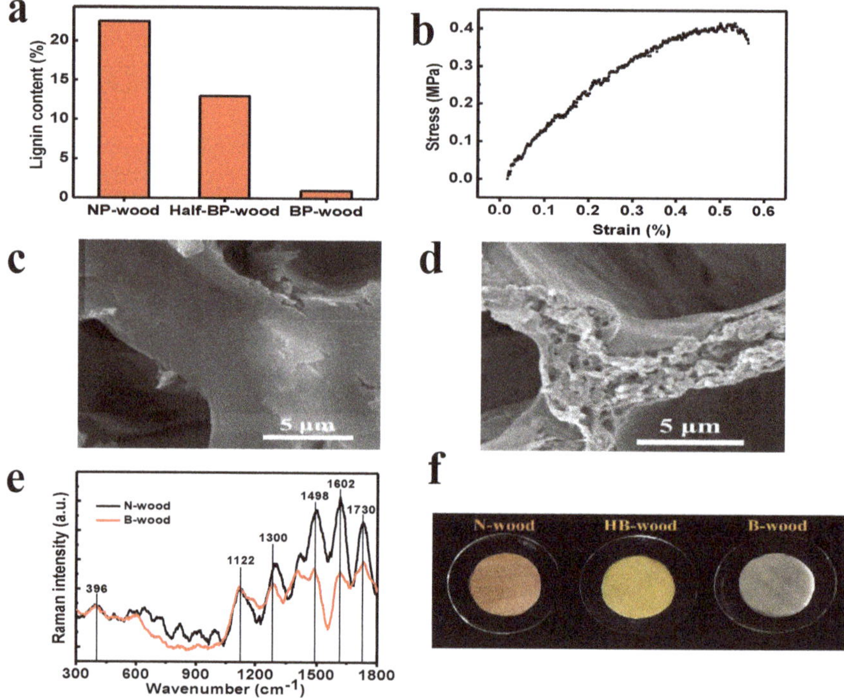

Figure 1. (**a**) Lignin content and (**b**) Mechanical strength of BP-wood in wet state. SEM images of cell wall corners and the middle lamella of (**c**) Natural basswood (N-wood) and (**d**) B-wood. (**e**) Raman characterization of N-wood, B-wood. The images were obtained through baseline corrected and normalized. (**f**) Photos of wood after removal of different content of lignin.

In order to investigate the water transportation capacity, we designed a dye transportation experiment. N-wood and B-wood are put into a dye solution to observe the distance the dye arrives after a certain time. As shown in Figure 2a, the dye in the B-wood reaches to a larger distance than that of the N-wood, which indicates that the B-wood exhibits better material transportation capabilities than that of N-wood. The speed of dye transportation in the B-wood and the N-wood are determined to be 2.2 and 6.3 mm/min, respectively (Figure 2b).

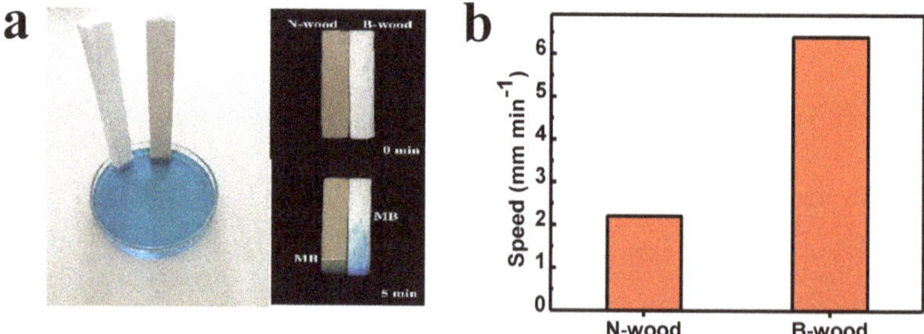

Figure 2. (**a**) The MB transport distance through B-wood and N-wood after 0, 5 min. (**b**) The MB transport speed through B-wood and N-wood during 5 min.

As shown in Figure 3a–d, P25 nanoparticles are not only decorated on the top surface of B-wood, but also penetrate into the cell wall of B-wood due to its mesoporous structure, which results in three-dimensional (3D) P25-wood composites. The light transmittance of wet BP-wood and NP-wood in the range of 200–800 nm are shown in Figure 3e. It is worth noting that, in the ultraviolet range of 300–400 nm, BP-wood still has a light transmittance of 0.5–20%, while NP-wood does not have light transmittance before about 550 nm due to the existence of 22.5% lignin.

To demonstrate the 3D photocatalytic features of BP-wood, the P25 nanoparticles coated on the B-wood surface were purposely removed to preserve the P25 nanoparticles which had penetrated into the interior of the B-wood (BI-wood). 10 mg/L methylene blue (MB) aqueous solution was photocatalytically degraded with BI-wood, and B-wood under ambient sunlight. The re-plotted linear graph of $\ln(c_0/c) \sim t$ shown in Figure 3f indicates that the photocatalytic degradation of MB with P25 decorated inside the BP-wood follows roughly the pseudo-first-order reaction [7]. The rate constants were determined to be 0.35, 0.21, and 0.18 h^{-1} for BI-wood, B-wood, and methylene blue (MB) aqueous solution, respectively. That is, the photocatalytic degradation of BI-wood is better than absorption of B-wood and the photolysis of MB, which indicates that the bleached, delignified wood can be used as 3D photocatalyst support due to its mesoporous structure, and UV transmittance.

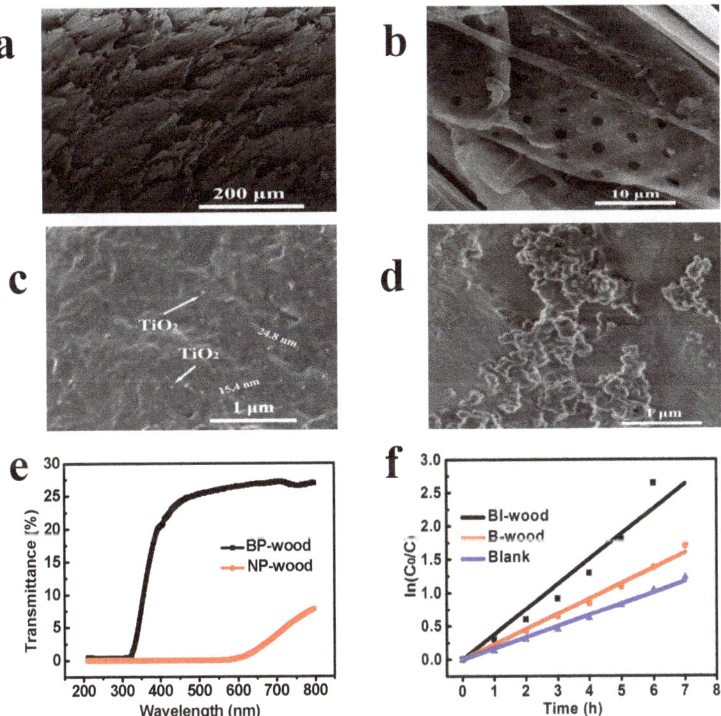

Figure 3. SEM images of B-wood's (**a**) top surface and (**b–d**) cross section decorated with P25 nanoparticles. (**e**) Optical transmittance of BP-wood and NP-wood. (**f**) Comparison of photocatalytic degradation of BI-wood with TiO_2 removed at the top and B-wood and blank control groups (where C_0 is the initial concentration of the dye solution and C is the concentration of dye at corresponding time) [6].

The photocatalytic properties of BP-wood, HBP-wood, and NP-wood were examined by measuring the photodegradation of 20 mg/L MB under ambient sunlight. As shown in Figure 4a, all wood-based photocatalysts, including BP-wood, HBP-wood, and NP-wood can float on the surface of the MB solution, and the P25 nanoparticles coating the top surface of the BP-wood are directly exposed to sunlight. Figure 4b shows the MB photolysis and photodegradation kinetic curves for reactions in which P25, NP-wood, HBP-wood, and BP-wood are used as photocatalysts. Overall, the photocatalytic activity increased when the lignin content decreasing. The re-plotted linear graph of $\ln(c_0/c) \sim t$ shown in Figure 4b indicates that the rate constants were determined to be 0.52, 0.41, 0.21, and 0.08 h^{-1} for P25, BP-wood, HBP-wood, and NP-wood, respectively. Combining with the corresponding lignin content, we can conclude that the photocatalytic activity of wood supported P25 increases with the decreasing of lignin content. Notably, the photolysis of MB with 20 mg/L is very small. After photodegradation, both BP-wood and NP-wood were taken out from the solutions. It is clear that, after the photodegradation, the P25 nanoparticles' suspension leads to a turbidity inside the entire beaker, which indicates the difficulty to be separated and recycled (Figure 4c). As for NP-wood, the solution after photodegradation exhibits yellow color due to the leaching of N-wood [32]. However, it is clean and pollution-free for BP-wood, which reveals the clean and environmental benign. We further exam the BP-wood and NP-wood after photodegradation, shown in Figure 4d,e. Compared with the blue color of the interior of NP-wood, BP-wood appears pure white without MB molecules. That is, the MB molecules inside BP-wood are also photocatalytical degraded, which further demonstrates the 3D photocatalyst feature of BP-wood. As for NP-wood, although there was photocatalytic degradation occurring on the surface, the MB molecules absorbed in the porous wood still remained.

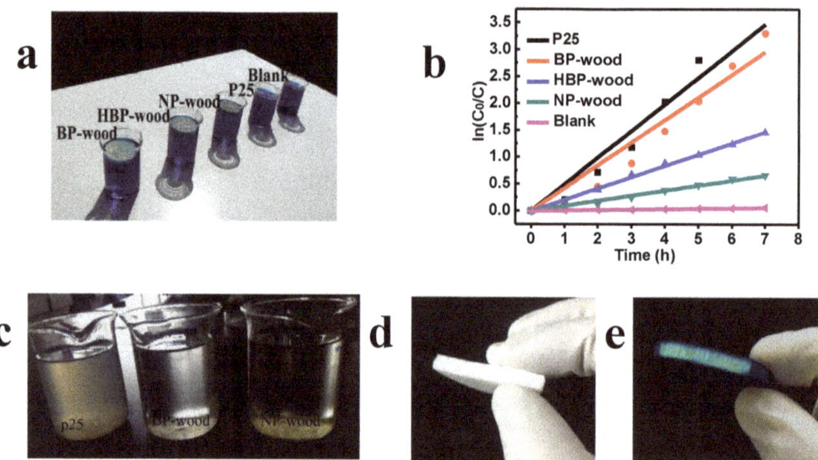

Figure 4. (**a**) Photo of photocatalytic degradation devices under sunlight. (**b**) Photodegradation of MB monitored as the normalized concentration change versus irradiation time under sunlight. Photo of (**c**) MB solutions, (**d**) BP-wood, and (**e**) NP-wood after photodegradation of MB solution under sunlight.

We also characterize the photocatalytic degradation of the high-concentration MB solution with BP-wood under ambient sunlight. From Figure 5a,b, we can see that the photodegradation performance of the BP-wood is comparable to or even better than those of P25 suspension when the concentration of MB increases to 40 mg/L and 60 mg/L, respectively. As we know, MB molecules can be degraded by either photolysis or photocatalytic degradation. With the MB concentration increased, MB molecules will absorb more UV light to be degraded by photolysis, which decreases the photocatalytic degradation efficiency of P25 suspension, while the effect on BP-wood is negligible due to its floatability. Figure 5c further demonstrates that the enhancement factor of the BP-wood versus P25 suspension increased with the increasing of MB concentration.

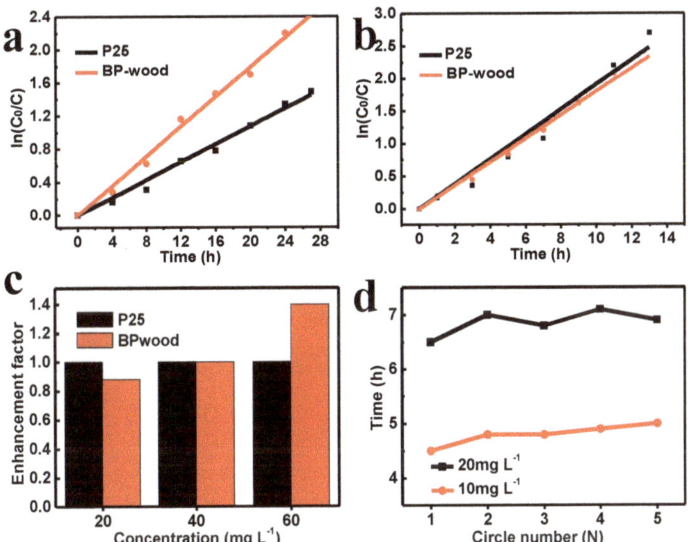

Figure 5. Photodegradation of MB solutions with (**a**) 40 mg/L, (**b**) 60 mg/L monitored as the normalized concentration change versus irradiation time in the presence of P25 and BP-wood under ambient sunlight. (**c**) Enhancement factor of the use of BP-wood compared with P25. (**d**) Recycling performance of BP-wood.

The photodegradation process of BP-wood consists of the following steps. Firstly, the BP-wood floats on the surface of MB aqueous solutions due to its low density. Secondly, the MB solutions will transport to the top surface of the BP-wood via aligned channels to make contact with P25 nanoparticles. Thirdly, MB molecules will be photocatalytical degraded via P25 nanoparticles under UV light with a wavelength shorter than 380 nm in sunlight. Fourth, MB molecules will continuously accumulate in both top surface and the interior of BP-wood via a concentration gradient to continuous photodegradation. It should be noted that the P25 nanoparticles decorated into the cell wall of BP-wood also exhibit photocatalytic activity since UV light can be transmitted into the interior of BP-wood.

Thanks to its large size and 0.4 MPa mechanical strength, BP-wood can be easily recycled to photodegrade the MB solution under ambient sunlight. After the degradation, the BP wood was taken out and kept under ambient conditions. As shown in Figure 5d, our BP-wood exhibited excellent recyclable performance: during the 5 circles, it takes 6.5, 7.0, 6.8, 7.1, and 6.9 h respectively to achieve photodegradation of a 20 mg/L MB solution. There is no significant decline in efficiency during the photodegradation process.

3. Materials and Methods

3.1. Materials and Chemicals

Basswood was used in this study. P25 was bought from Degussa AG. H_2O_2, MB, anhydrous ethanol ethanol were bought from Sigma Chemicals (Shanghai, China).

3.2. Preparation of N-Wood and B-Wood

Natural basswood (N-wood) slices with size of $\pi \times 20 \times 20 \times 5$ mm^3 were obtained by cutting along the direction perpendicular to the growth of wood. Half-B-wood and B-wood were obtained by H_2O_2 steam delignification of above-mentioned N-wood at 100 °C for 1 and 4 h, respectively [38]. After rinsed with water, and ethanol for three times, they were dried at 50 °C for 4 h.

3.3. Preparation of NP-Wood and BP-Wood

The P25 nanoparticles were dispersed in deionized water and ultrasonically dispersed for 10 min. After overnight, the upper 5 mL * 3 g/L P25 suspension was coated on the N-wood, Half-B-wood, and B-wood to form NP-wood, HBP-wood, and BP-wood.

3.4. Photocatalytic Activity Measurement

The photocatalytic activity of the aforementioned samples was investigated by placing NP-wood, HBP-wood, and BP-wood on the surface of MB aqueous solution to measure the P25-assisted photodegradation of 100 mL MB aqueous solutions. At the same time, 0 mL and 5 mL * 3 g/L P25 suspension was added into 100 mL MB aqueous solutions as the control group. Then, they were irradiated under ambient sunlight. Finally, the concentration of MB after illuminating for a certain time was monitored by measuring the absorbance of the solutions (which were centrifuged at 2000 rpm to remove P25) at 664 nm.

3.5. Characterization

Scanning electron microscopy images were determined with a Nova NanoSEM 450, Lincoln, Ne, USA. The accelerating voltage was 15 kV. The UV–vis absorption was measured on Cary 500 Scan UV–vis–NIR spectrophotometer (Harbor, CA, USA). The transmittance of the material comes from the ultraviolet visible spectrophotometer. The UV-visible spectrophotometer model is U-4100 Spectrophotometer, Hitachi (Tokyo, Japan). A universal mechanical test machine was used to measure the mechanical properties with the SUNS UTM-5000 electronic universal testing machine (Shenzhen, China). The size of the test sample was $10 \times 1 \times 0.5$ cm^3. And Raman spectra were obtained from LabRam HR Evolution, Horiba, France.

4. Conclusions

To summarize, a floatable, recyclable, efficient, UV light permeable, environmentally friendly, and 3D photocatalyst was easily synthesized through decoration with P25 nanoparticles on both the surface and in the interior of bleached wood. The bleached wood was obtained by removing most lignin from N-wood through H_2O_2 steam delignification. The delignification not only endows bleached wood with UV light transmittance, but also provides a highway to transport the MB solution up to the top surface. The as-made BP-wood photocatalyst shows a high photocatalytic degradation of 60 mg/L MB solution under ambient sunlight, better than that of P25 nanoparticles suspension. It also exhibits excellent recyclability due to its large size, floatability, and 0.4 MPa mechanical strength. The present work opens up an efficacious avenue for designing bleached wood-based recyclable, floatable, UV permeable, and efficient 3D photocatalyst for environmental pollution.

Author Contributions: Conceptualization, R.Z. and Y.H.; methodology, Y.H.; software, R.Z.; validation, R.Z. and X.G.; formal analysis, H.L.; investigation, X.G.; resources, R.Z.; data curation, R.Z.; writing—original draft preparation, Y.H.; writing—review and editing, R.Z.; visualization, R.Z.; supervision, R.Z.; project administration, R.Z. and X.G.; funding acquisition, R.Z. and X.G.

Funding: This research was funded by the Joint Special Project of Agricultural Basic Research in Yunnan (2017FG001036). And the APC was funded by the National Natural Science Foundation of China, grant number 41563008, 31100420.

References

1. Ren, Z.J.; Umble, A.K. Water treatment: Recover wastewater resources locally. *Nature* **2016**, *529*, 25. [CrossRef] [PubMed]

2. Maity, S.K.; Rana, M.S.; Bej, S.K.; Ancheyta-Juárez, J.; Murali Dhar, G.; Prasada Rao, T.S.R. TiO_2–ZrO_2 mixed oxide as a support for hydrotreating catalyst. *Catal. Lett.* **2001**, *72*, 115–119. [CrossRef]

3. Ali, I. New generation adsorbents for water treatment. *Chem. Rev.* **2012**, *112*, 5073–5091. [CrossRef] [PubMed]
4. Panizza, M.; Cerisola, G. Direct and mediated anodic oxidation of organic pollutants. *Chem. Rev.* **2009**, *109*, 6541–6569. [CrossRef] [PubMed]
5. Huang, L.; Chen, J.; Gao, T.; Zhang, M.; Li, Y.; Dai, L.; Qu, L.; Shi, G. Reduced graphene oxide membranes for ultrafast organic solvent nanofiltration. *Adv. Mater.* **2016**, *28*, 8669–8674. [CrossRef] [PubMed]
6. Gratzel, M. *Photocatalysis: Fundamentals and Applications*; Serpone, N., Pelizzetti, E., Eds.; Wiley: New York, NY, USA, 1989; p. 123.
7. Zheng, R.B.; Meng, X.W.; Tang, F.Q. Synthesis, characterization and photodegradation study of mixed-phase titania hollow submicrospheres with rough surface. *Appl. Surf. Sci.* **2009**, *255*, 5989–5994. [CrossRef]
8. Nishimoto, S.I.; Ohtani, B.; Kajiwara, H.; Kagiya, T. ChemInform abstract: Correlation of the crystal structure of titanium dioxide prepared from titanium tetra-2-propoxide with the photocatalytic activity for redox reactions in aqueous propan-2-ol and silver salt solutions. *J. Chem. Soc. Faraday Trans.* **1985**, *16*, 61–68. [CrossRef]
9. Fox, M.A.; Dulay, M.T. Heterogeneous photocatalysis. *Chem. Rev.* **1993**, *93*, 341–357. [CrossRef]
10. Tanaka, K.; Hisanaga, T.; Rivera, A.P. *Photocatalytic Purification and Treatment of Water and Air*; Ollis, D.F., Al-Ekabi, H., Eds.; Elsevier: Amsterdam, The Netherlands, 1993; p. 169.
11. Yoneyama, H.; Yamanaka, S.; Haga, S. Photocatalytic activities of microcrystalline titania incorporated in sheet silicates of clay. *J. Phys. Chem.* **1989**, *93*, 4833–4837. [CrossRef]
12. Zhang, Z.; Wang, C.C.; Zakaria, R.; Ying, J. Role of particle size in nanocrystalline TiO_2 -based photocatalyst. *J. Phys. Chem. B* **1998**, *102*, 10871–10878. [CrossRef]
13. Tsai, S.-J.; Cheng, S. Effect of TiO_2 crystalline structure in photocatalytic degradation of phenolic contaminants. *Catal. Today* **1997**, *33*, 227–237. [CrossRef]
14. Paola, A.D.; Ikeda, S.; Marcì, G.; Ohtani, B.; Palmisano, L. Photocatalytic degradation of organic compounds in aqueous systems by transition metal doped polycrystalline TiO_2. *Catal. Today* **2002**, *75*, 171–176. [CrossRef]
15. Hu, C.C.; Hsu, T.C.; Kao, L.H. One-step cohydrothermal synthesis of nitrogen-doped titanium oxide nanotubes with enhanced visible light photocatalytic activity. *Int. J. Photoenergy* **2012**, 391958. [CrossRef]
16. Chen, X.; Mao, S.S. Titanium dioxide nanomaterials synthesis, properties, modifications, and applications. *Chem. Rev.* **2007**, *107*, 2891–2959. [CrossRef] [PubMed]
17. Al Dahoudi, N.; Zhang, Q.; Cao, G. Alumina and hafnia ALD Layers for a niobium-doped titanium oxide photoanode. *Int. J. Photoenergy* **2012**, 401393. [CrossRef]
18. Zheng, R.B.; Meng, X.W.; Tang, F.Q. A general protocol to coat titania shell on carbon-based composite cores using carbon as coupling agent. *J. Solid State Chem.* **2009**, *182*, 1235–1240. [CrossRef]
19. Hsien, Y.H.; Chang, C.F.; Chen, Y.H.; Cheng, S. Photodegradation of aromatic pollutants in water over TiO_2 supported on molecular sieves. *Appl. Catal. B Envion.* **2001**, *31*, 241–249. [CrossRef]
20. Sampath, S.; Uchida, H.; Yoneyama, H. Photocatalytic degradation of gaseous pyridine over zeolite-supported titanium dioxide. *J. Catal.* **1994**, *149*, 189–194. [CrossRef]
21. Van Grieken, R.; Aguado, J.; López-Muñoz, M.J.; Marugán, J. Synthesis of size-controlled silica-supported TiO_2 photocatalysts. *J. Photochem. Photobiol. A Chem.* **2012**, *148*, 315–322. [CrossRef]
22. Paul, B.; Martens, W.N.; Frost, R.L. Immobilised anatase on clay mineral particles as a photocatalyst for herbicides degradation. *Appl. Clay Sci.* **2012**, *57*, 49–54. [CrossRef]
23. Baek, M.H.; Yoon, J.W.; Hong, J.S.; Suh, J.K. Application of TiO_2-containing mesoporous spherical activated carbon in a fluidized bed photoreactor—Adsorption and photocatalytic activity. *Appl. Catal. A Gen.* **2013**, *45*, 222–229. [CrossRef]
24. Chaturvedi, S.; Dave, P.N.; Shah, N.K. Applications of nano-catalyst in new era. *J. Saudi Chem. Soc.* **2016**, *16*, 307–325. [CrossRef]
25. Hsieh, S.H.; Chen, W.J.; Wu, C.T. Pt-TiO_2/Graphene photocatalysts for degradation of AO_7 dye under visible light. *Appl. Surf. Sci.* **2015**, *340*, 9–17. [CrossRef]
26. Li, D.; Jia, J.; Zhang, Y.; Wang, N.; Guo, X.; Yu, X. Preparation and characterization of Nano-graphite/TiO_2 composite photoelectrode for photoelectrocatalytic degradation of hazardous pollutant. *J. Hazard. Mater.* **2016**, *315*, 1–10. [CrossRef] [PubMed]
27. Hamdi, A.; Boufi, S.; Bouattour, S. Phthalocyanine/chitosan-TiO_2 photocatalysts: Characterization and photocatalytic activity. *Appl. Surf. Sci.* **2015**, *339*, 128–136. [CrossRef]

28. Lei, Y.; Zhang, C.; Lei, H.; Huo, J. Visible light photocatalytic activity of aromatic polyamide dendrimer/TiO_2 composites functionalized with spirolactam-based molecular switch. *J. Colloid Interface Sci.* **2013**, *406*, 178–185. [CrossRef]

29. Mejía, MI.; Marín, JM.; Restrepo, G.; Rios, LA.; Pulgarín, C.; Kiwi, J. Preparation, testing and performance of a TiO_2/polyester photocatalyst for the degradation of gaseous methanol. *Appl. Catal. B* **2010**, *94*, 166–172. [CrossRef]

30. Zhu, M.W.; Li, Y.J.; Chen, G.; Jiang, F.; Yang, Z.; Luo, X.G.; Wang, Y.B.; Lacey, S.D.; Dai, J.Q.; Wang, C.W.; et al. Tree-inspired design for high-efficiency water extraction. *Adv. Mater.* **2017**, *29*, 1704107. [CrossRef]

31. Liu, H.; Chen, C.J.; Chen, G.; Kuang, Y.D.; Zhao, X.P.; Song, J.W.; Jia, C.; Xu, X.; Hitz, E.; Xie, H.; et al. High-performance solar steam device with layered channels: Artificial tree with a reversed design. *Adv. Energy Mater.* **2017**, *8*, 1701616. [CrossRef]

32. Chen, C.; Li, Y.; Song, J.; Yang, Z.; Kuang, Y.; Hitz, E.; Jia, C.; Gong, A.; Jiang, F.; Zhu, J.Y.; et al. Highly flexible and efficient solar steam generation device. *Adv. Mater.* **2017**, *29*, 1701756. [CrossRef]

33. Sun, Q.F.; Lu, Y.; Tu, J.C.; Li, J. Bulky macroporous TiO_2 photocatalyst with cellular structure via facile wood-template method. *Int. J. Photoenergy* **2013**, *649540*. [CrossRef]

34. Gao, L.K.; Gan, W.T.; Li, J. Preparation of heterostructured WO_3/TiO_2 catalysts from wood fibers and its versatile photodegradation abilities. *Sci. Rep.* **2017**, *7*, 1102. [CrossRef] [PubMed]

35. Zheng, R.B.; Tshabalala, M.A.; Li, Q.Y.; Wang, H.Y. Construction of hydrophobic wood surfaces by room temperature deposition of rutile (TiO_2) nanostructures. *Appl. Surf. Sci.* **2015**, *28*, 453–458. [CrossRef]

36. Chen, F.J.; Gong, A.S.; Zhu, M.W.; Hu, L.B. Mesoporous, three-dimensional wood membrane decorated with nanoparticles for highly efficient water treatment. *ACS Nano* **2017**, *11*, 4275–4282. [CrossRef] [PubMed]

37. Jana, S.; Vanessa, S.; Tobias, K.; Ingo, B. Characterization of wood derived hierarchical cellulose scaffolds for multifunctional applications. *Materials* **2018**, *11*, 517. [CrossRef]

38. Li, H.Y.; Guo, X.L.; He, Y.M.; Zheng, R.B. A green, steam-modified delignification method to low lignin delignified wood for thick, large, highly transparent wood composites. *J. Mater. Res.* **2018**. [CrossRef]

 *catalysts*

Article

Ru-Ti Oxide Based Catalysts for HCl Oxidation: The Favorable Oxygen Species and Influence of Ce Additive

Jian Shi [1,2], Feng Hui [1,2], Jun Yuan [1,2], Qinwei Yu [1,2], Suning Mei [1,2], Qian Zhang [1,2], Jialin Li [1,2], Weiqiang Wang [1,2], Jianming Yang [1,2,*] and Jian Lu [1,2,*]

[1] State Key Laboratory of Fluorine & Nitrogen Chemicals, Xi'an, Shaanxi 710065, China; a38860075@126.com (J.S.); huifeng.hyff@126.com (F.H.); luckyyjjun@163.com (J.Y.); qinweiyu204@163.com (Q.Y.); meisuning@aliyun.com (S.M.); qz450945428@163.com (Q.Z.); lijialin95126@gmail.com (J.L.); wqwang07611@163.com (W.W.)

[2] Xi'an Modern Chemistry Research Institute, Xi'an, Shaanxi 710065, China

* Correspondence: yangjm204@163.com (J.Y.); lujian204@gmail.com (J.L.); Tel.: +86-29-88291367 (J.Y.)

Received: 16 December 2018; Accepted: 17 January 2019; Published: 22 January 2019

Abstract: Several Ru-Ti oxide-based catalysts were investigated for the catalytic oxidation of HCl to Cl_2 in this work. The active component RuO_2 was loaded on different titanium-containing supports by a facile wetness impregnation method. The Ru-Ti oxide based catalysts were characterized by XRD, N_2 sorption, SEM, TEM, H_2-TPR, XPS, and Raman, which is correlated with the catalytic tests. Rutile TiO_2 was confirmed as the optimal support even though it has a low specific surface area. In addition to the interfacial epitaxial lattice matching and epitaxy, the extraordinary performance of Ru-Ti rutile oxide could also be attributed to the favorable oxygen species on Ru sites and specific active phase-support interactions. On the other hand, the influence of additive Ce on the RuO_2/TiO_2-rutile was studied. The incorporation of Ce by varied methods resulted in further oxidation of RuO_2 into $RuO_2^{\delta+}$ and a modification of the support structure. The amount of favorable oxygen species on the surface was decreased. As a result, the Deacon activity was lowered. It was demonstrated that the surface oxygen species and specific interactions of the Ru-Ti rutile oxide were critical to HCl oxidation.

Keywords: Ru-Ti oxide catalysts; HCl oxidation; oxygen species; Ce incorporation; active phase-support interactions

---

1. Introduction

The treatment of the huge amount of excess hydrogen chloride byproduct has become a challenging and demanding problem in the chlorine-based chemical industry, as a byproduct HCl is environmentally undesirable and has a very restricted market [1,2]. The method using heterogeneously catalyzed HCl oxidation (Deacon process) to recycle chlorine is regarded as a low energy-consuming and sustainable route for the more efficient Cl_2 industry [1]. The reaction is exothermic and reversible, which is shown as follows.

$$4HCl + O_2 \overset{cat.}{\leftrightarrow} 2Cl_2 + 2H_2O \ \Delta H_{r,298} = -28.5 \text{ kJ} \cdot \text{mol}^{-1}_{HCl}$$

Ru-based catalysts are commonly considered the most active for this process. RuO_2 supported on rutile TiO_2 and SnO_2 (cassiterite) with an excellent activity and outstanding lifetime have been successively reported by Sumitomo [2] and Bayer [3], respectively. Great attention has been paid to Ru-Ti oxide catalysts in various catalytic reactions besides the Deacon process, including oxidation of propane [4], N_2O decomposition [5], selective methanation of CO [6], CO oxidation [7], and aqueous-phase ketonization of acetic acid [8,9]. The Ce-based catalysts also exhibit Deacon activity and outstanding stability in the process [10,11]. Related studies have demonstrated that CeO_2 can accelerate the catalyst reoxidation step by supplying oxygen donor/storage sites [10]. Although the single Ce-based or Cu-based catalyst shows a limited activity, the combination of Ce-Cu can boost the overall HCl oxidation performance [12,13]. The build-up of Ce-Ti oxide can also enhance the oxygen storage capacity [14,15]. The CeO_2/TiO_2 catalyst showed excellent activity for selective catalytic reduction of NO with NH_3, where the Ce-O-Ti species were confirmed to be the active sites [16–19]. The concentration of surface adsorbed oxygen presented a positive correlation with the catalytic activity in the NH_3-SCR reaction [16,20]. However, the combination of Ru-Ce-Ti for HCl catalytic oxidation has not been reported to the best of our knowledge.

Despite a number of studies concerning supported Ru-based catalysts for one-step HCl oxidation [21,22], there are only a few research studies about the details of active phase-support interactions in the Ru-Ti oxide-based catalysts [2,23]. Moreover, little attention was paid to the influence of oxygen species in this catalytic process. In addition to the lattice matching and epitaxial growth of RuO_2 on the substrate [2,21], the extraordinary performance of Ru-Ti rutile oxide for the Deacon process still requires a more sufficient interpretation.

In this article, the performance of different shaped Ru-Ti oxide based catalysts, resembling the forms of the industrial reality, are compared in the Deacon process. Detailed characterizations are performed to investigate the special oxygen species and interactions of the catalysts. The favorable oxygen species and specific active phase-support interactions ensure high Deacon activity of RuO_2/TiO_2-rutile. The role of Ce in the Ru-Ti rutile oxide system was also investigated. The addition of Ce decreased favorable oxygen species and affected the active phase-support interactions between RuO_2 and TiO_2, by evolving the Ru-O-Ce structure and enhancing the positive charge density of Ru sites. The Deacon activity was lowered as a result. It can be deduced that the electronic interactions between RuO_2 and rutile TiO_2 are critical for the gas-phase oxidation of HCl to Cl_2. The findings in this work may be a reference value for the design and tailor of Ru-Ti oxide based catalysts toward better Deacon activity.

2. Results and Discussion

2.1. Morphology and Phase Structure

The crystal structures of the catalysts were analyzed by X-ray diffraction (XRD). The XRD patterns of the supported catalysts only exhibit characteristic diffraction peaks of the TiO_2 supports (Figure 1). Due to the low loading and high dispersion of the Ru species, the RuO_2 phase were logically not detected. In the XRD pattern of $RuO_2/TiO(OH)_2$, much broader peaks can be observed than those of RuO_2/TiO_2-a, which implies that the support that originated from the $TiO(OH)_2$ precursor has a much lower crystallinity as well as average crystallite size.

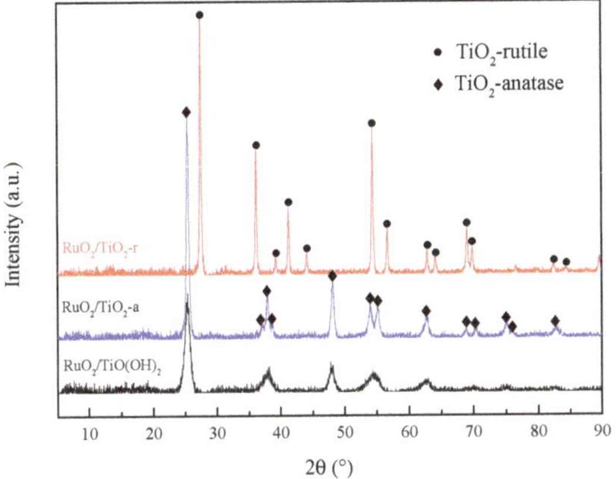

Figure 1. XRD patterns of the RuO_2/TiO_2 catalysts with the supports of: TiO_2-rutile, TiO_2-anatase, and $TiO(OH)_2$ (as the precursor).

The SEM micrographs in Figure 2 show that the anatase TiO_2 and the corresponding catalysts (35 to 50 nm) have a smaller particle size than the rutile of 45 to 60 nm. The particle size of rutile TiO_2 has not changed much after loading RuO_2 (Figures 2a and 2c), while some coagulation can be observed on anatase TiO_2 and the average size has increased from 40.3 nm to 48.5 nm (Figures 2b and 2d). The supported catalysts were also scrutinized by TEM. As displayed in Figure 3a, some dark edges and layers were observed on the substrate of RuO_2/TiO_2-r. They were presumed to be the dispersed RuO_2 phase, which was consistent with the literature [2,4]. On the other hand, some aggregation can be observed in RuO_2/TiO_2-a (Figure 3b). The EDX elemental mapping proves the existence of highly dispersed Ru species on the TiO_2-r support (Figure 4).

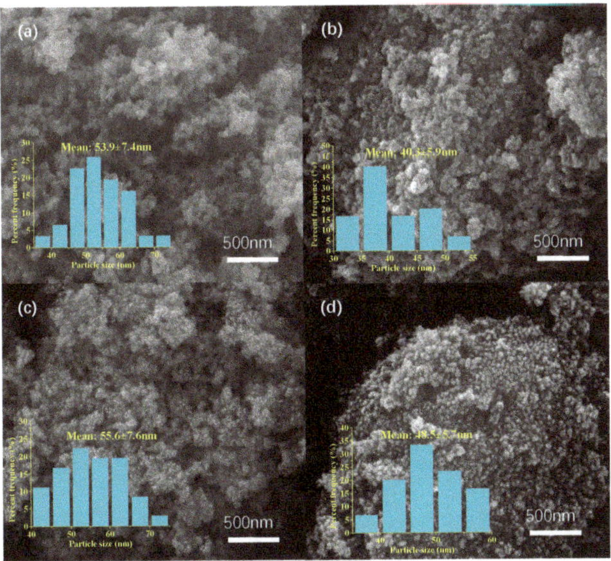

Figure 2. SEM micrographs of the rutile (**a**) and anatase TiO_2 supports (**b**) used in this study, RuO_2/TiO_2-r (**c**), RuO_2/TiO_2-a (**d**).

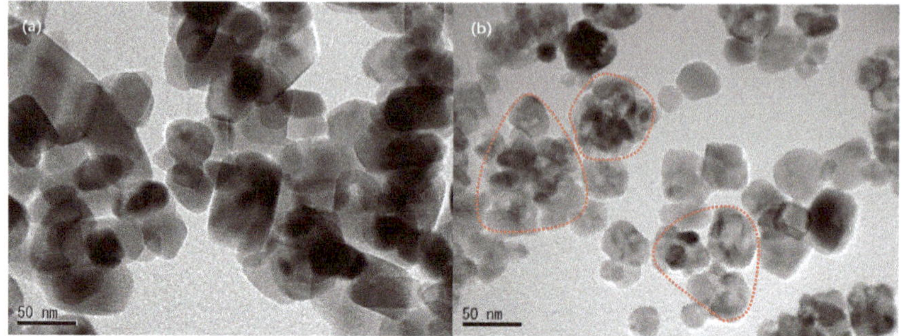

Figure 3. TEM images of RuO_2/TiO_2-r (**a**) and RuO_2/TiO_2-a (**b**).

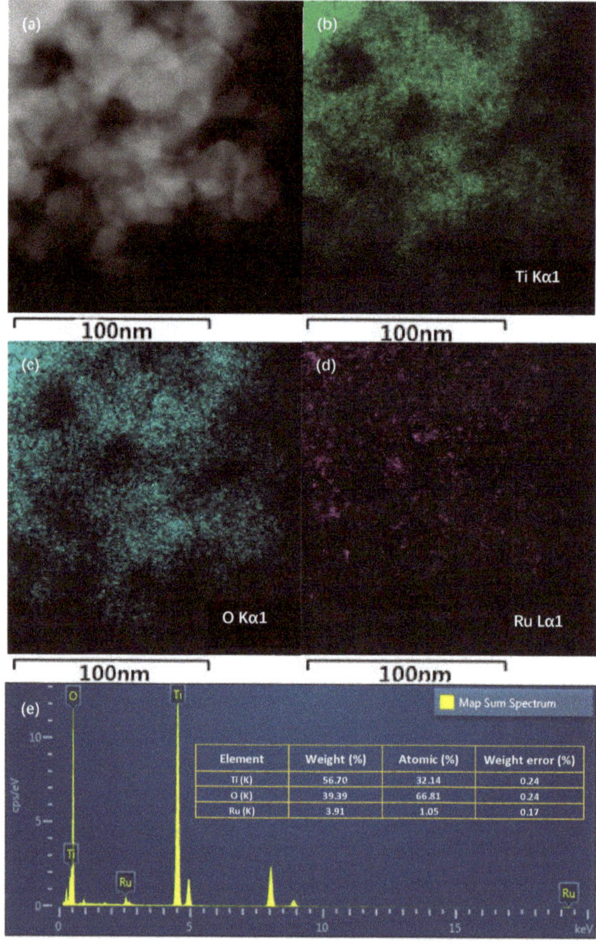

Element	Weight (%)	Atomic (%)	Weight error (%)
Ti (K)	56.70	32.14	0.24
O (K)	39.39	66.81	0.24
Ru (K)	3.91	1.05	0.17

Figure 4. TEM-EDX elemental mapping of RuO_2/TiO_2-r: (**a**) representative TEM image, (**b**) Ti (Kα1) green color, (**c**) O (Kα1) blue color, (**d**) Ru (Lα1) pink color, and (**e**) EDX (Energy-dispersive X-ray) result of the selected area.

2.2. Characterization of Oxygen Species and Interfacial Interactions

2.2.1. H$_2$-TPR Analysis

Temperature programmed reduction (TPR) by H$_2$ was employed in this study to distinguish specific oxygen species and estimate the oxygen storage capacity [11,14]. The H$_2$-TPR profiles of the bulk RuO$_2$ and supported catalysts are shown in Figure 5. In this scenario, all the supported catalysts were loaded with an Ru content of 0.5 wt%. The bulk RuO$_2$ sample was reduced in the range of 130 to 230 °C. It should be noted that the RuO$_2$ sample used in characterization was obtained from RuCl$_3$·3H$_2$O calcined at 350 °C for 8 h, the phase of which was verified by XRD (see Figure S1). The RuO$_2$ phase prepared by this method has a preferential (1 0 1) plane rather than (1 1 0) (PDF #65-2824). It may be the origin of different reduction temperatures as discussed in the previous study [21]. The H$_2$-TPR profile of RuO$_2$/TiO$_2$-r contains three peaks in the range of 90 to 200 °C and one peak from 330 to 450 °C. The former three peaks are assigned to the reduction of RuO$_2$, while the last one is attributed to the partial reduction of the TiO$_2$ surface [24]. Ru-Ce/TiO$_2$-r and RuO$_2$/TiO$_2$-a exhibit a similar reduction peak of titania, except that the peak shifts to a higher temperature were similar to RuO$_2$/TiO$_2$-r. Note the peak in the range 370 to 500 °C of Ru-Ce/TiO$_2$-r includes the reduction of ceria. The H$_2$-TPR profiles of pure TiO$_2$ in rutile and anatase phase are shown in Figure S2, which confirms the partial reduction of TiO$_2$ support.

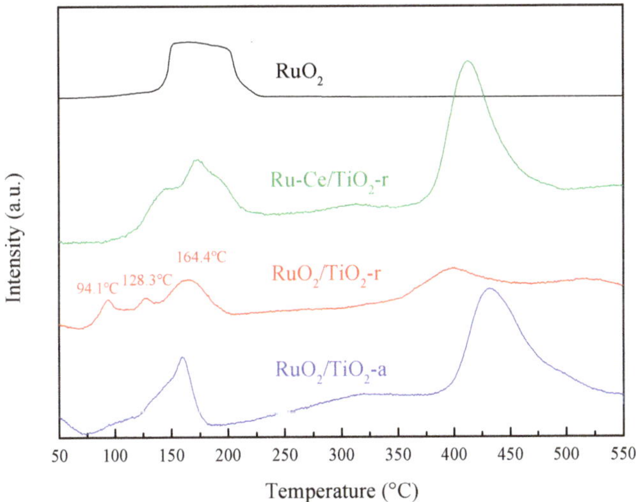

Figure 5. H$_2$-TPR profiles of the bulk and supported RuO$_2$ catalysts.

In the range of 50 to 200 °C, the oxygen species of RuO$_2$ were subjected to reduction. Generally, the reduction temperature of the same phase is related with the particle size. In fact, oxygen species with different reducibility can be a more essential perspective. We know that smaller particles expose more surface species. For our catalysts, RuO$_2$ phase is mainly distributed on the surface of TiO$_2$ support and more surface species mean more surface oxygen species. Among these oxygen species, it is quite probable that species with more coordination numbers are more difficult to reduce.

For sample RuO$_2$/TiO$_2$-r, the peaks (from 90 to 200 °C) are evidently distinguished from the other two catalyst samples. We deduce that the three peaks are assigned to the reduction of top oxygen, bridge oxygen, and bulk oxygen of RuO$_2$ with H$_2$, from a low to a high reduction temperature. These three types of oxygen are coordinated to 1, 2, 3 Ru atoms, respectively. The former two oxygen species are located at the surface and are significant to the Deacon process [21,25,26]. For sample RuO$_2$/TiO$_2$-a, only one reduction peak was detected in the range from 90 to 200 °C, which was attributed to the elimination of bulk oxygen. It could be rationalized in the following way. RuO_2 could not grow epitaxially on TiO_2-a due to a huge difference of lattice matching. Thus, the RuO_2 active phase mainly exists as bigger particles on TiO_2-a rather than films in RuO_2/TiO_2-r [2]. In this case, the bulk oxygen species of RuO_2 prevailed, which was coordinated to 3 Ru atoms and was the most difficult to reduce. From Figure 5, it can be observed that Ru-Ce/TiO_2-r and RuO_2/TiO_2-a have a better oxygen storage capacity than RuO_2/TiO_2-r, especially in the high temperature range.

2.2.2. XPS Analysis

XPS analysis was performed to study the surface species and electronic structure of the catalyst samples. The survey spectra verified the complete removal of chlorine in all the catalysts (not shown). The XPS peaks of O 1s were deconvoluted to analyze the different types of O species in the supported catalysts (Figure 6). The O 1s peaks were mainly composed of signals corresponding to the chemisorbed oxygen (O_{β_1}) and the lattice oxygen (O_{β_1}, O_{β_2}) [27–29]. XPS data of chemisorbed oxygen are listed in Table 1. The proportion of chemisorbed oxygen (O_α) on the surface exhibited a dependence on Ru loading (Figure 6a–c). RuO_2/TiO_2-r exhibited a higher amount of chemisorbed oxygen with the increase of Ru loading, while a much lower content of chemisorbed oxygen was detected in RuO_2/TiO_2-a (Figure 6d, Table 1). The Ti 2p core-level spectra of RuO_2/TiO_2-r also show a relevance with the Ru content (Figure S3), where the XPS peaks are broadened and shifted to a lower binding energy with the increase of ruthenium. The chemical environment change of the Ti sites can be ascribed to the interactions and electronic effects among Ti, Ru, and O atoms. A slight interfacial charge transfer from RuO_2 to TiO_2-r may lead to the binding energy shift of the Ti 2p peaks [30,31].

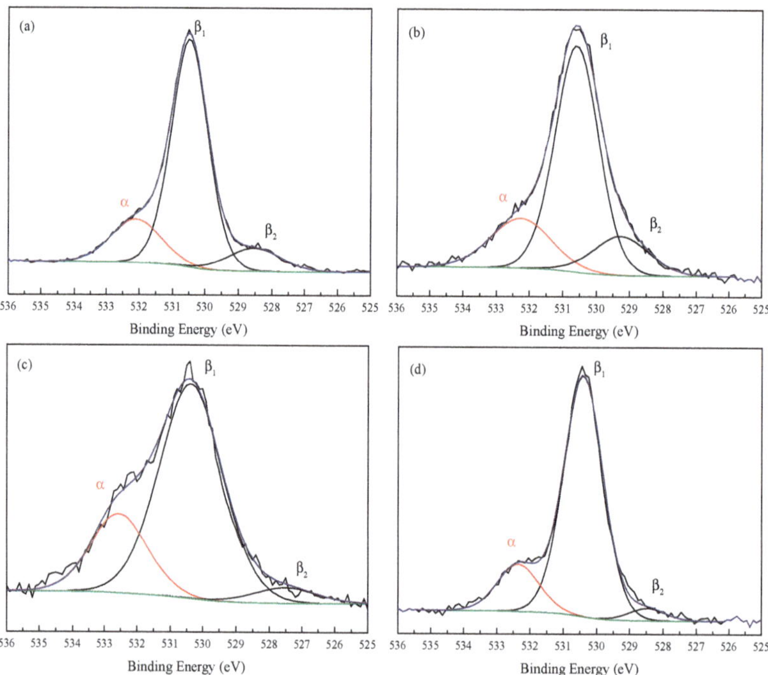

Figure 6. XPS profiles and fitting curves of O 1s peaks: RuO_2/TiO_2-r of (**a**) 0.3, (**b**) 0.5, (**c**) 1.0 wt% Ru, RuO_2/TiO_2-a of (**d**) 0.5 wt% Ru.

Table 1. Chemisorbed oxygen (O_α) in the RuO_2/TiO_2 catalysts.

Sample	E_b of O_α (eV)	O_α/O_T (%)
0.3 wt%-RuO_2/TiO_2-r	532.11	18.29
0.5 wt%-RuO_2/TiO_2-r	532.27	20.12
1.0 wt%-RuO_2/TiO_2-r	532.57	23.70
0.5 wt%-RuO_2/TiO_2-a	532.36	16.51

The Ru 3d spectra for RuO_2-CeO_2/TiO_2-r catalysts are presented in Figure 7. The signal of Ru $3d_{5/2}$ core-level, attributed to RuO_2 or $RuO_2^{\delta+}$, was detected in the region of 281.2–282.9 eV. Meanwhile, the peaks around 280.5 eV appeared for Ru-2Ce/TiO_2-r and Ru-2Ce-C/TiO_2-r, which were assigned to Ru^0 [3]. The peaks of C 1s and Ru $3d_{3/2}$ appeared to overlap [32,33]. The Ru $3d_{5/2}$ peaks shifted toward a higher binding energy when the preparation methods were altered. In the spectra of Ce 3d, an overall shift towards lower binding energies was observed in the similar sequence of Ru-2Ce-R/TiO_2-r, Ru-2Ce/TiO_2-r, and Ru-2Ce-C/TiO_2-r (Figure S4). It can be inferred that electrons are transferred from RuO_2 to CeO_2, which results in the further oxidation of RuO_2 into $RuO_2^{\delta+}$. As shown in Figure 5, the reduction peaks of Ru-Ce/TiO_2-r in the low temperature range (90 to 200 °C) are one fewer than those of RuO_2/TiO_2-r, which is ascribed to the elimination of top oxygen on the RuO_2 surface (vide supra). The existence of top oxygen is critical to the Deacon reaction with Ru-based catalysts [25,26]. In RuO_2-CeO_2/TiO_2-r, Ru-O-Ce linkage was likely formed, which induced the decrease of the active sites in the Ru-Ti rutile oxide system. The generation of Ru^0 in Ru-2Ce/TiO_2-r and Ru-2Ce-C/TiO_2-r also indicated the decrease of the active sites for HCl oxidation. The decline of chemisorbed oxygen from XPS data is also consistent with the deduction above (see Figure S5, Table S1).

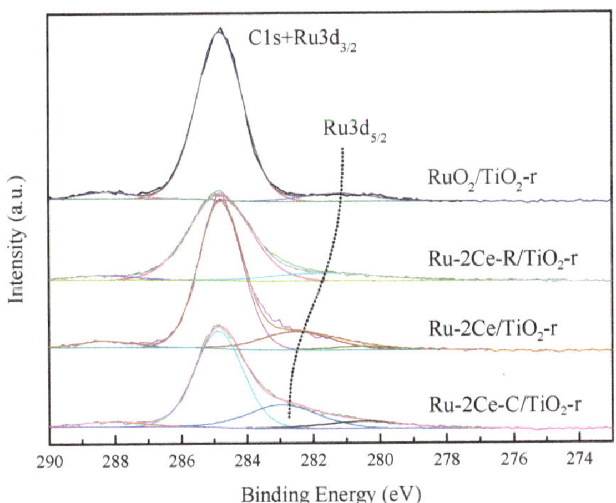

Figure 7. XPS spectra and peak fitting curves of Ru 3d of the supported RuO_2 and Ru-Ce/Ti oxide catalysts.

2.2.3. Raman Analysis

The active phase-support interactions were further confirmed by Raman characterization. As shown in Figure 8a, the characteristic bands of rutile TiO_2 are observed at 234, 441, and 606 cm^{-1}, which can be assigned to the multiple photon scattering process, the E_g (planar O-O vibration), and A_{1g} (Ti-O stretch) Raman-active modes, respectively [32,34,35]. Since the RuO_2 Raman bands overlapped with those of TiO_2-r, only the E_g mode of RuO_2 at 515 cm^{-1} could be distinguished [5], which indicates the existence of the RuO_2 phase in the catalyst. The declination of Raman signals of TiO_2-r was ascribed to the decreased amount of Ti-O-Ti structure, which resulted from the formation of Ru-O-Ti structure with the increase of Ru loading. The relative intensity of the RuO_2 Raman signal at 515 cm^{-1} was enhanced at the same time, which coincided with the change of the active component loading. Moreover, the RuO_2 Raman bands shifted from 515 to 507 cm^{-1} with the increase of Ru loading, while a clear blue-shift towards higher wavenumbers (from 234 cm^{-1} to 258 cm^{-1}) was observed in the spectrum of the support (Figures 8b and 8c). The Raman shifts suggest that the rutile structure of the TiO_2 support was modified according to the changed RuO_2 crystal size and interfacial interactions, substantially as a result of the mechanical strains generated from the differences between the rutile phase of RuO_2 and TiO_2 support [6]. The formation of Ru-O-Ti linkage corresponded with the results from XPS analysis. The interactions between Ce and Ti were also confirmed by Raman characterization on the set of Ce-containing RuO_2/TiO_2-r catalysts. The characteristic bands of planar Ti-O vibration and O-O stretch for TiO_2-r exhibited a slight blue shift (see Figure S6) in accordance with the binding energy shift of the Ru $3d_{5/2}$ core-level peak. It indicated that the chemical environment of the rutile support was also affected by Ce addition.

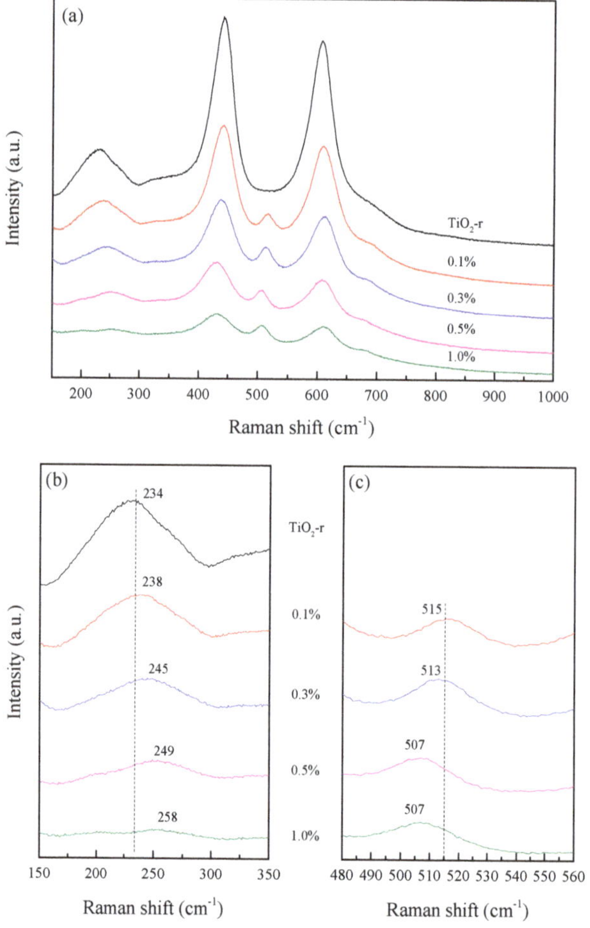

Figure 8. Raman spectra of TiO_2-r and RuO_2/TiO_2-r with 0.1, 0.3, 0.5, and 1.0 wt% Ru loading (**a**) and partial enlarged views of TiO_2-r (**b**) and RuO_2 (**c**).

2.3. Catalytic Performance of Ru-Ti Oxide Based Catalysts

2.3.1. Catalytic Activity of RuO$_2$/TiO$_2$-r Catalyst

The influence of Ru loading on HCl conversion is depicted in Figure 9a. It can be noted that the increase of Ru loading contributes to the reaction conversion to a certain extent. The catalysts with 0.5 wt% and 1.0 wt% Ru loadings achieved similar conversion. As presented in Figure 9b, the ratio of O$_2$/HCl also plays a critical role in the reaction, especially when the value is less than 1.0 vol./vol. Since oxygen re-adsorption is recognized as the rate-determining step under lean oxygen condition, a higher O$_2$ partial pressure has been proven to be beneficial for Cl$_2$ production [26].

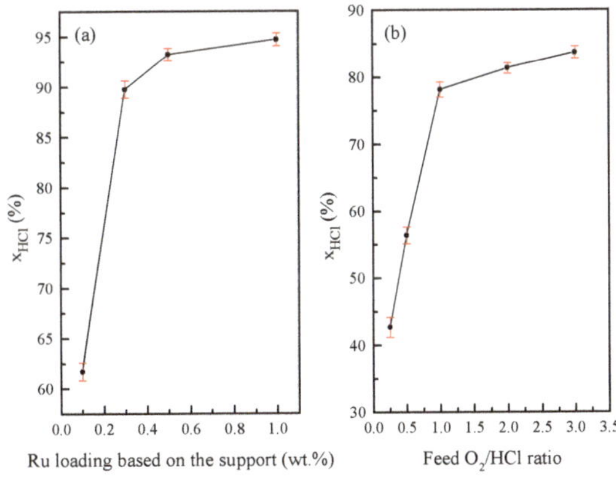

Figure 9. The influence of Ru loading at 350 °C (**a**) and feed O$_2$/HCl ratio at 320 °C (**b**) on HCl conversion. Note volumetric flowrate HCl/O$_2$ = 1:2 for (**a**).

Figure 10 shows HCl conversion at different reaction temperatures on catalysts with 0.5 wt% and 1.0 wt% Ru loading. For both catalysts, the conversion was improved with the elevation of the reaction temperature. When comparing the two catalysts with different Ru loadings over 300 °C, we found that the conversion was not proportional to the loading amount of the active component. It implied that the Ru-specific activity declined with the increasing loading. As indicated by Figure S3, a slight interfacial charge transfer from RuO$_2$ to TiO$_2$-r shows up with the increase of Ru content. This charge transfer was deduced not to be beneficial to Ru-specific activity.

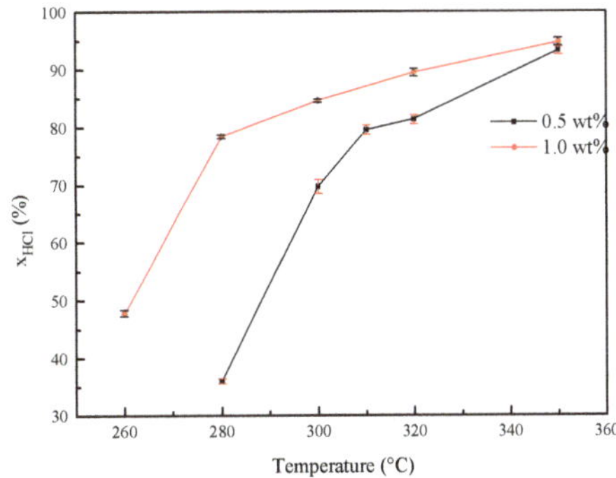

Figure 10. HCl conversion on different reaction temperatures and Ru loading.

2.3.2. Comparison of Ru-Ti Oxide Based Catalysts

The catalytic activities for HCl oxidation over Ru-Ti oxide based catalysts using different supports are compared in Table 2. Although RuO_2/TiO_2-r has the smallest surface area ($28\ m^2 \cdot g^{-1}$) among all the Ru-Ti oxide based catalysts, its catalytic activity turned out to be the best. Notably, the specific surface area of the supports seemed to be less critical in the RuO_2/TiO_2-based catalytic system. The performances of catalysts differed significantly when the support was changed, even though the supports were all based on the Ti-O structure.

Table 2. Characterization and catalytic activity data.

Reaction Temperature	No.	Catalyst [a]	BET Surface Area [c] $(m^2 \cdot g^{-1})$	HCl Conversion (%)	STY [d] $(g_{Cl_2} \cdot g_{Ru}^{-1} \cdot h^{-1})$	TOF (h^{-1}) [g]
	1	RuO_2/TiO_2-r	28 (28)	93.5	57.3	163.2
350 °C	2	RuO_2/TiO_2-a	44 (79)	53.6	32.8	93.4
	3	$RuO_2/TiO(OH)_2$ [b]	193 (309)	21.1	12.9	36.7
	4	RuO_2/TiO_2-r	28	81.0	49.6	141.3
	5	Ru-Ce/TiO_2-r [e]	27	74.3	45.5	129.6
320 °C	6	Ru-2Ce-R/TiO_2-r [f]	26	73.2	44.8	127.6
	7	Ru-2Ce/TiO_2-r	27	66.3	40.6	115.6
	8	Ru-2Ce-C/TiO_2-r	27	61.8	37.9	107.9

[a] 0.5 wt% Ru loading based on the support or support precursor. [b] The precursor of support is $TiO(OH)_2$. [c] Determined by N_2 adsorption, surface area of the support in brackets. [d] The space time yield defined as the gram of Cl_2 produced per gram of Ru per hour. [e] The molar ratio of Ru/Ce is 1 and 0.5, denoted as Ru-Ce and Ru-2Ce respectively. [f] The catalysts prepared by different methods are distinguished by suffixes –R and -C, which refer to the impregnation of TiO_2-r support with Ru first and Ce first, respectively. The catalysts without suffixes above are prepared by co-impregnation with Ru and Ce. [g] Calculated based on the mole of HCl reacted per hour per mole of Ru.

Herein, we attempt to explicate the influence of favorable oxygen species and interfacial interactions of the Ru-Ti oxide system on the Deacon process, by correlating the characterization results with catalytic performances. With the increase of Ru loading, the linkage of Ru-O-Ti appeared to be more abundant, which was confirmed by the results from XPS and Raman spectra. Subsequently, the amount of chemisorbed oxygen species increased. The chemisorbed oxygen is intimately related with coordinatively unsaturated ruthenium atoms [26], which provide critical active sites and promote oxygen activation. It can partially explain the higher activity of RuO_2 when loaded on TiO_2-r. Since RuO_2 can grow epitaxially on the rutile titania, the exposure of more active sites is favored.

A series of Ru-Ce-Ti oxide catalysts were prepared by different methods, as described in the experiment. From Table 2, it can be affirmed that the incorporation of Ce reduced the activity of Ru-Ti rutile oxide catalysts. The catalytic activity declined with the increase of Ce loading by comparing the results of No. 4, 5, and 7 in Table 2. When exchanging the impregnation sequence of Ru and Ce, it was found that performing Ru impregnation in the first place was beneficial for improving the catalytic performance to some extent (see No. 6, 7, and 8).

2.3.3. The Influence of Ce on the Ru-Ti Rutile Oxide System

The characterizations and catalytic performances of Ce-containing RuO_2/TiO_2-r catalysts further confirmed the significance of the active phase-support interactions for the RuO_2/TiO_2-r system. When Ce was introduced to the TiO_2-r support prior to Ru, a greater change of the RuO_2 and TiO_2-r structure was incurred. The formation of Ti-O-Ce and Ru-O-Ce linkages remarkably affected the interfacial interactions and electronic structure of the RuO_2/TiO_2-r system, which was corroborated by XPS and Raman characterizations. The new linkages restricted the active sites of coordinatively unsaturated ruthenium and the transport of oxygen species. The amount of chemisorbed oxygen evidently decreased when improving the priority of the introduction of Ce (Figure S4, Table S1). The correlated catalytic activity declined, as presented in Table 2 (No. 4, 6, 7, and 8).

The introduction of Ce triggered further oxidation of RuO_2 into $RuO_2{}^{\delta+}$ and could be unfavorable for H_2O desorption and Cl recombination. Higher positive charge density of Ru sites induced easier adsorption of the reactants. Therefore, the active-phase surface was more likely to be poisoned by adsorbates. Since HCl oxidation proceeds on RuO_2 via a Langmuir-Hinshelwood reaction mechanism, adsorbed HCl dehydrogenates through a hydrogen transfer to produce Cl and OH species in on-top positions [26]. The recombination of neighboring on-top Cl atoms to form the desired Cl_2 product is regarded as the rate-determining step. Nevertheless, the existence of Ce strengthened the dissociative adsorption of HCl so that the liberation of Cl_2 restricted the activity, which was also observed on the $IrO_2(110)$ surface [36,37]. Although Ce-based catalysts showed Deacon activity themselves, the active temperature for the Deacon process was generally reaching 430 °C or more, considering the higher energy requirements for Cl activation and recombination [10,11]. Therefore, ceria itself contributed to little Deacon activity for RuO_2/TiO_2-r below 350 °C in this study.

On the other hand, although Ce provided more reducible oxygen species (Figure 5), it seemed that the oxygen species with a high reduction temperature were not crucial for the Deacon reaction. The oxygen species of support with a lower reduction temperature (330 to 350 °C) in RuO_2/TiO_2-r, as a result of interfacial interactions between the phases, was speculated to be beneficial for facilitating activation and transport of oxygen species for the active phase. Moreover, compared to RuO_2/TiO_2-r, the most readily reducible oxygen species (90 to 100 °C) in Ru-Ce/TiO_2-r disappeared. The on-top oxygen occupying the coordinatively unsaturated ruthenium sites are mentioned in the discussion of the H_2-TPR results. The introduction of Ce enhanced the positive charge density of Ru sites, which might cause the easier formation of the bridge and bulk oxygen of RuO_2 other than the on-top oxygen. Because of the higher positive charge density of Ru sites, O was more inclined to bond to Ru atoms and the coordination number of O with Ru on the RuO_2 surface likely increased. More evidence may be provided by further characterizations on fine structure and corresponding computational studies.

3. Materials and Methods

3.1. Preparation of Catalysts

All the reagents were of an analytical grade, supplied by Aladdin (Shanghai, China), and used as received without further purification, except for $TiO(OH)_2$ from Tuoboda Titanium Dioxide Products Co. (Wuxi, China). Ru-Ti oxide based catalysts were prepared by a facile wetness impregnation method as follows. First, $RuCl_3 \cdot 3H_2O$ was dissolved in a mixed solution with an equal volume of water and ethanol. After 3 min of ultrasonic mixing, support or the support precursor was added and the suspension was stirred for 16 h at room temperature. Then the mixture was evaporated under vacuum in a rotary evaporator. The obtained powder was dried at 120 °C for 12 h and was then tableted into cylinders with a diameter of 5 mm. Lastly, the sample were calcined at 350 °C in static air for 8 h. The nominal loading of Ru was 0.1, 0.3, 0.5, and 1.0 wt% on the support basis. The supports and support precursors included rutile and anatase TiO_2 and $TiO(OH)_2$. Henceforth, the rutile and anatase TiO_2 polymorphs are abbreviated as TiO_2-r and TiO_2-a, respectively. The RuO_2-CeO_2/TiO_2-r catalysts were all loaded with 0.5 wt% Ru on the rutile TiO_2 basis. They were prepared by almost the same method as RuO_2/TiO_2 except that $Ce(NO_3)_3 \cdot 6H_2O$ was introduced to the solution when dissolving $RuCl_3 \cdot 3H_2O$. The molar ratios of Ru/Ce were 1.0 and 0.5, which were denoted as Ru-Ce/TiO_2-r and Ru-2Ce/TiO_2-r, respectively.

In order to investigate the effect of Ce on the Ru-Ti oxide system, the RuO_2-CeO_2/TiO_2-r catalysts were also prepared by changing the sequence of Ce introduction, where the molar ratios of Ru/Ce were 0.5. The powder from the rotary evaporator (vide supra) was added to the solution of $Ce(NO_3)_3 \cdot 6H_2O$. Then the suspension was stirred, evaporated, dried, tableted, and calcined as described before. The impregnation sequence of Ru and Ce was also exchanged for obtaining another catalyst. These two catalysts were distinguished by suffixes -R and -C, namely Ru-2Ce-R/TiO_2-r and Ru-2Ce-C/TiO_2-r. The catalysts without the suffixes were prepared by co-impregnation with Ru and Ce.

3.2. Characterization of Catalysts

Powder X-ray diffraction (XRD) patterns were recorded with an Empyrean, PANalytical X-ray diffractometer (Almelo, The Netherlands), with Cu Kα radiation ($\lambda = 0.154056$ nm) at 40 kV and 40 mA. The diffraction patterns were taken in the 2θ range of 5 to 90° with a step size of 0.02°. Specific surface areas of the samples were measured by N_2 physisorption at 77 K using a Micromeritics ASAP 2020 instrument (Norcross, GA, USA). The surface area was determined by the Brunauer-Emmett-Teller (BET) method. The morphology and particle size of prepared catalysts and supports were studied by a scanning electron microscope (SEM, FEI Quanta 600FEG, operated at 20 kV, Hillsboro, OR, USA) and a transmission electron microscope (TEM, FEI Tecnai G2 F20, operated on 200 kV, Hillsboro, OR, USA). The energy-dispersive X-ray spectroscopy (EDX, Oxford INCA Energy IE350, Oxford, UK) mapping method was applied to determine the elemental distributions of different components in the catalysts.

Temperature programmed reduction of hydrogen (H_2-TPR) was performed at an AutoChem II 2950 instrument (Micromeritics, Norcross, GA, USA) equipped with a thermal conductivity detector (TCD). Furthermore, 100 mg catalyst was heated and programmed from 50 °C to 550 °C (or 800 °C for the pure support) at a rate of 10 °C·min^{-1} in a gas flow of 5 vol.% H_2/Ar of 50 cm^3 STP min^{-1}. Raman spectra were collected on a confocal Raman microscope (inVia Raman Microscope, Renishaw plc, Wotton-under-Edge, UK) with a 785 nm laser diode (Renishaw plc, Wotton-under-Edge, UK). X-ray photoelectron spectra (XPS) were recorded on a Thermo ESCALAB 250 spectrometer (Waltham, MA, USA) with a monochromatized Al Kα (1486.6 eV) radiation and a passing energy of 50 eV. The binding energies were calibrated by the C 1s signal of adventitious carbon at 284.8 eV.

All the characterizations were performed on catalyst samples with a 0.5% Ru loading unless otherwise specified.

3.3. Catalytic Tests

The catalytic oxidation of HCl to Cl_2 was investigated in a Hastelloy alloy (HC-276®) fixed-bed reactor with a diameter of 30 mm at an ambient pressure. The upstream lines of the set-up were also made from Hastelloy alloy (HC-276®) in order to prevent the corrosion of the reactor, while the downstream lines were made from Teflon® to improve corrosion resistance. In addition, a 25-g shaped cylinder catalyst was loaded into the reactor. Thereafter, the reaction feed (HCl flow = 80 cm^3 STP min^{-1} and volumetric flowrate HCl/O_2 = 1:2, unless otherwise specified) was continuously introduced. The reaction temperature was controlled in the range of 330 to 430 °C (± 1.0 °C). The data obtained by each test were the average of at least three steady-state measurements. The total chlorine balance was confirmed with an accuracy of $\pm 2\%$. Blank support without a loading active component showed negligible activity under the corresponding reaction condition. The reaction effluent was absorbed by excessive potassium iodide solution and analyzed by iodometry and acid-base titration to measure the generated Cl_2 and unreacted HCl. The conversion of HCl was calculated based on the detected results.

4. Conclusions

A series of Ru-Ti oxide based catalysts have been investigated for HCl oxidation in this research. It was clarified that the special oxygen species and active phase-support interactions of RuO_2/TiO_2 were significant for the Deacon process. RuO_2 film grows epitaxially on rutile TiO_2. This produces more active sites and oxygen species on the catalyst surface. More importantly, the assembly of RuO_2 and rutile TiO_2 generates coordinatively unsaturated ruthenium sites and bridge oxygen, which are efficient for the Deacon reaction. The interactions of Ru-Ti were confirmed by characterizations and correlated to the amount of the active component. The reducible oxygen species of the rutile TiO_2 may facilitate activation and transport of oxygen species during the active phase. It is inferred that the rutile support is likely involved in the catalytic reaction rather than merely acting as an inert support. On the other hand, the incorporation of Ce altered the electronic structure of the RuO_2/TiO_2-r system. The formation of Ru-O-Ce linkage decreased the amount of favorable oxygen species and increased the positive charge of Ru sites, which restricted the recombination of Cl atoms and Cl_2 elimination. To achieve a better catalytic performance, a more delicate tuning of the RuO_2/TiO_2-r system by considering moderate positive charge density of Ru sites is required.

Supplementary Materials: The following are available online at http://www.mdpi.com/2073-4344/9/2/108/s1, Figure S1: XRD patterns of the self-made RuO_2 and the corresponding intensity line in red from PDF 65-2824, Figure S2: H_2-TPR profiles of the pure TiO_2 in rutile and anatase, Figure S3: XPS profiles of Ti 2p for RuO_2/TiO_2-r with 0.3, 0.5, and 1.0 wt% Ru, Figure S4: XPS spectra of Ce 3d for Ru-Ce/Ti oxide catalysts, Figure S5: XPS spectra of O 1s for the supported RuO_2 and Ru-Ce/Ti oxide catalysts, Table S1: Chemisorbed oxygen (O_α) in the supported RuO_2 and Ru-Ce/Ti oxide catalysts, Figure S6: Raman spectra of the supported RuO_2 and Ru-Ce/Ti oxide catalysts.

Author Contributions: Conceptualization, J.S., J.Y. (Jianming Yang) and J.L. (Jian Lu); Data curation, J.S., J.Y. (Jun Yuan), Q.Y., S.M., Q.Z. and W.W.; Formal analysis, J.Y. (Jianming Yang) and Q.Y.; Funding acquisition, J.Y. (Jianming Yang) and J.L. (Jian Lu); Investigation, J.S., F.H. and J.L. (Jialin Li); Methodology, J.S., F.H., J.Y. (Jun Yuan) and S.M.; Resources, J.S., Q.Z., J.L. (Jialin Li) and W.W.; Software, F.H.; Supervision, J.Y. (Jianming Yang) and J.L. (Jian Lu); Validation, J.Y. (Jianming Yang) and J.L. (Jian Lu); Writing – original draft, J.S.; Writing – review & editing, J.S., J.Y. (Jianming Yang) and J.L. (Jian Lu).

Funding: The research was funded by Key Research & Development Plan Projects of Shanxi Province (Nos. 2017ZDXM-GY-073, 2017ZDXM-GY-042, 2017ZDXM-GY-070).

Conflicts of Interest: The authors declare no conflicts of interest.

References

1. Hammes, M.; Valtchev, M.; Roth, M.B.; Stöwe, K.; Maier, W.F. A search for alternative Deacon catalysts. *Appl. Catal. B Environ.* **2013**, *132–133*, 389–400. [CrossRef]

2. Seki, K. Development of RuO_2/Rutile-TiO_2 Catalyst for Industrial HCl Oxidation Process. *Catal. Surv. Asia* **2010**, *14*, 168–175. [CrossRef]

3. Amrute, A.P.; Mondelli, C.; Schmidt, T.; Hauert, R.; Pérez-Ramírez, J. Industrial RuO_2-Based Deacon Catalysts: Carrier Stabilization and Active Phase Content Optimization. *ChemCatChem* **2013**, *5*, 748–756. [CrossRef]

4. Debecker, D.P.; Farin, B.; Gaigneaux, E.M.; Sanchez, C.; Sassoye, C. Total oxidation of propane with a nano-RuO_2/TiO_2 catalyst. *Appl. Catal. A Gen.* **2014**, *481*, 11–18. [CrossRef]

5. Lin, Q.; Huang, Y.; Wang, Y.; Li, L.; Liu, X.Y.; Lv, F.; Wang, A.; Li, W.-C.; Zhang, T. RuO_2/rutile-TiO_2: A superior catalyst for N_2O decomposition. *J. Mater. Chem. A* **2014**, *2*, 5178–5181. [CrossRef]

6. Martínez Tejada, L.M.; Muñoz, A.; Centeno, M.A.; Odriozola, J.A. In-situ Raman spectroscopy study of Ru/TiO_2 catalyst in the selective methanation of CO. *J. Raman Spectrosc.* **2016**, *47*, 189–197. [CrossRef]

7. Jiao, Y.; Jiang, H.; Chen, F. RuO_2/TiO_2/Pt Ternary Photocatalysts with Epitaxial Heterojunction and Their Application in CO Oxidation. *ACS Catal.* **2014**, *4*, 2249–2257. [CrossRef]

8. Pham, T.N.; Shi, D.; Sooknoi, T.; Resasco, D.E. Aqueous-phase ketonization of acetic acid over Ru/TiO_2/carbon catalysts. *J. Catal.* **2012**, *295*, 169–178. [CrossRef]

9. Aranda-Pérez, N.; Ruiz, M.P.; Echave, J.; Faria, J. Enhanced activity and stability of Ru-TiO_2 rutile for liquid phase ketonization. *Appl. Catal. A Gen.* **2017**, *531*, 106–118. [CrossRef]

10. Farra, R.; García-Melchor, M.; Eichelbaum, M.; Hashagen, M.; Frandsen, W.; Allan, J.; Girgsdies, F.; Szentmiklósi, L.; López, N.; Teschner, D. Promoted Ceria: A Structural, Catalytic, and Computational Study. *ACS Catal.* **2013**, *3*, 2256–2268. [CrossRef]

11. Amrute, A.P.; Mondelli, C.; Moser, M.; Novell-Leruth, G.; López, N.; Rosenthal, D.; Farra, R.; Schuster, M.E.; Teschner, D.; Schmidt, T.; et al. Performance, structure, and mechanism of CeO_2 in HCl oxidation to Cl_2. *J. Catal.* **2012**, *286*, 287–297. [CrossRef]

12. Amrute, A.P.; Larrazábal, G.O.; Mondelli, C.; Pérez-Ramírez, J. $CuCrO_2$ Delafossite: A Stable Copper Catalyst for Chlorine Production. *Angew. Chem. Int. Ed.* **2013**, *52*, 9772–9775. [CrossRef] [PubMed]

13. Fei, Z.; Liu, H.; Dai, Y.; Ji, W.; Chen, X.; Tang, J.; Cui, M.; Qiao, X. Efficient catalytic oxidation of HCl to recycle Cl_2 over the CuO–CeO_2 composite oxide supported on Y type zeolite. *Chem. Eng. J.* **2014**, *257*, 273–280. [CrossRef]

14. Dutta, G.; Waghmare, U.V.; Baidya, T.; Hegde, M.S.; Priolkar, K.R.; Sarode, P.R. Origin of Enhanced Reducibility/Oxygen Storage Capacity of $Ce_{1-x}Ti_xO_2$ Compared to CeO_2 or TiO_2. *Chem. Mater.* **2006**, *18*, 3249–3256. [CrossRef]

15. Sun, P.; Guo, R.; Liu, S.; Wang, S.; Pan, W. Enhancement of the low-temperature activity of Ce / TiO_2 catalyst by Sm modification for selective catalytic reduction of NO_x with NH_3. *Mol. Catal.* **2017**, *433*, 224–234. [CrossRef]

16. Fei, Z.; Yang, Y.; Wang, M.; Tao, Z.; Liu, Q.; Chen, X.; Cui, M.; Zhang, Z.; Tang, J.; Qiao, X. Precisely fabricating Ce-O-Ti structure to enhance performance of Ce-Ti based catalysts for selective catalytic reduction of NO with NH3. *Chem. Eng. J.* **2018**, *353*, 930–939. [CrossRef]

17. Zhang, Z.; Chen, L.; Li, Z.; Li, P.; Yuan, F.; Niu, X.; Zhu, Y. Activity and SO_2 resistance of amorphous Ce_aTiO_x catalysts for the selective catalytic reduction of NO with NH_3: In situ DRIFT studies. *Catal. Sci. Technol.* **2016**, *6*, 7151–7162. [CrossRef]

18. Ding, J.; Zhong, Q.; Zhang, S. A New Insight into Catalytic Ozonation with Nanosized Ce–Ti Oxides for NO_x Removal: Confirmation of Ce–O–Ti for Active Sites. *Ind. Eng. Chem. Res.* **2015**, *54*, 2012–2022. [CrossRef]

19. Li, P.; Xin, Y.; Li, Q.; Wang, Z.; Zhang, Z.; Zheng, L. Ce–Ti Amorphous Oxides for Selective Catalytic Reduction of NO with NH_3: Confirmation of Ce–O–Ti Active Sites. *Environ. Sci. Technol.* **2012**, *46*, 9600–9605. [CrossRef] [PubMed]

20. Shan, W.; Geng, Y.; Zhang, Y.; Lian, Z.; He, H. A CeO_2/ZrO_2-TiO_2 Catalyst for the Selective Catalytic Reduction of NO_x with NH_3. *Catalysts* **2018**, *8*, 592. [CrossRef]

21. Hevia, M.A.G.; Amrute, A.P.; Schmidt, T.; Pérez-Ramírez, J. Transient mechanistic study of the gas-phase HCl oxidation to Cl_2 on bulk and supported RuO_2 catalysts. *J. Catal.* **2010**, *276*, 141–151. [CrossRef]

22. Teschner, D.; Farra, R.; Yao, L.; Schlögl, R.; Soerijanto, H.; Schomäcker, R.; Schmidt, T.; Szentmiklósi, L.; Amrute, A.P.; Mondelli, C.; et al. An integrated approach to Deacon chemistry on RuO_2-based catalysts. *J. Catal.* **2012**, *285*, 273–284. [CrossRef]

23. Kondratenko, E.V.; Amrute, A.P.; Pohl, M.-M.; Steinfeldt, N.; Mondelli, C.; Pérez-Ramírez, J. Superior activity of rutile-supported ruthenium nanoparticles for HCl oxidation. *Catal. Sci. Technol.* **2013**, *3*, 2555. [CrossRef]

24. Perkas, N.; Zhong, Z.Y.; Chen, L.W.; Besson, M.; Gedanken, A. Sonochemically prepared high dispersed Ru/TiO_2 mesoporous catalyst for partial oxidation of methane to syngas. *Catal. Lett.* **2005**, *103*, 9–14. [CrossRef]

25. Zweidinger, S.; Crihan, D.; Knapp, M.; Hofmann, J.P.; Seitsonen, A.P.; Weststrate, C.J.; Lundgren, E.; Andersen, J.N.; Over, H. Reaction Mechanism of the Oxidation of HCl over RuO_2 (110). *J. Phys. Chem. C* **2008**, *112*, 9966–9969. [CrossRef]

26. López, N.; Gómez-Segura, J.; Marín, R.P.; Pérez-Ramírez, J. Mechanism of HCl oxidation (Deacon process) over RuO_2. *J. Catal.* **2008**, *255*, 29–39. [CrossRef]

27. Yang, S.; Zhu, W.; Wang, X. Influence of the structure of TiO_2, CeO_2, and CeO_2-TiO_2 supports on the activity of Ru catalysts in the catalytic wet air oxidation of acetic acid. *RARE Met.* **2011**, *30*, 488–495. [CrossRef]

28. Yang, S.; Feng, Y.; Wan, J.; Zhu, W.; Jiang, Z. Effect of CeO_2 addition on the structure and activity of RuO_2/γ-Al_2O_3 catalyst. *Appl. Surf. Sci.* **2005**, *246*, 222–228. [CrossRef]

29. Zeng, Y.; Zhang, S.; Wang, Y.; Liu, G.; Zhong, Q. The effects of calcination atmosphere on the catalytic performance of Ce-doped TiO_2 catalysts for selective catalytic reduction of NO with NH_3. *RSC Adv.* **2017**, *7*, 23348–23354. [CrossRef]

30. Kim, J.; Kim, J.Y.; Park, B.G.; Oh, S.J. Photoemission and x-ray absorption study of the electronic structure of $SrRu_{1-x}Ti_xO_3$. *Phys. Rev. B* **2006**, *73*. [CrossRef]

31. Wang, S.; Liu, B.; Zhu, Y.; Ma, Z.; Liu, B.; Miao, X.; Ma, R.; Wang, C. Enhanced performance of TiO_2-based perovskite solar cells with Ru-doped TiO_2 electron transport layer. *Sol. Energy* **2018**, *169*, 335–342. [CrossRef]

32. Nguyen-Phan, T.-D.; Luo, S.; Vovchok, D.; Llorca, J.; Sallis, S.; Kattel, S.; Xu, W.; Piper, L.F.J.; Polyansky, D.E.; Senanayake, S.D.; et al. Three-dimensional ruthenium-doped TiO_2 sea urchins for enhanced visible-light-responsive H_2 production. *Phys. Chem. Chem. Phys.* **2016**, *18*, 15972–15979. [CrossRef] [PubMed]

33. Sui, C.; Niu, X.; Wang, Z.; Yuan, F.; Zhu, Y. Activity and deactivation of Ru supported on $La_{1.6}Sr_{0.4}NiO_4$ perovskite-like catalysts prepared by different methods for decomposition of N_2O. *Catal. Sci. Technol.* **2016**, *6*, 8505–8515. [CrossRef]

34. Balachandran, U.; Eror, N.G. Raman spectra of titanium dioxide. *J. Solid State Chem.* **1982**, *42*, 276–282. [CrossRef]

35. Narksitipan, S.; Thongtem, S. Preparation and characterization of rutile TiO_2 films. *J. Ceram. Process. Res.* **2012**, *13*, 35–37.

36. Over, H.; Schomäcker, R. What Makes a Good Catalyst for the Deacon Process? *ACS Catal.* **2013**, *3*, 1034–1046. [CrossRef]

37. Moser, M.; Mondelli, C.; Amrute, A.P.; Tazawa, A.; Teschner, D.; Schuster, M.E.; Klein-Hoffman, A.; López, N.; Schmidt, T.; Pérez-Ramírez, J. HCl Oxidation on IrO_2-Based Catalysts: From Fundamentals to Scale-Up. *ACS Catal.* **2013**, *3*, 2813–2822. [CrossRef]

 *catalysts*

Review

Titanium Dioxide (TiO$_2$) Mesocrystals: Synthesis, Growth Mechanisms and Photocatalytic Properties

Boxue Zhang, Shengxin Cao, Meiqi Du, Xiaozhou Ye *, Yun Wang and Jianfeng Ye *

Department of Chemistry, College of Science, Huazhong Agricultural University, Wuhan 430070, China; bx1058779150@hotmail.com (B.Z.); shengxincao@hotmail.com (S.C.); dumeiqi@webmail.hzau.edu.cn (M.D.); wangyun@mail.hzau.edu.cn (Y.W.)

* Correspondence: xzye@mail.hzau.edu.cn (X.Y); jianfengye@mail.hzau.edu.cn (J.Y.); Tel.: +86-27-8728 4018 (J.Y.)

Received: 10 December 2018; Accepted: 11 January 2019; Published: 16 January 2019

Abstract: Hierarchical TiO$_2$ superstructures with desired architectures and intriguing physico-chemical properties are considered to be one of the most promising candidates for solving the serious issues related to global energy exhaustion as well as environmental deterioration via the well-known photocatalytic process. In particular, TiO$_2$ mesocrystals, which are built from TiO$_2$ nanocrystal building blocks in the same crystallographical orientation, have attracted intensive research interest in the area of photocatalysis owing to their distinctive structural properties such as high crystallinity, high specific surface area, and single-crystal-like nature. The deeper understanding of TiO$_2$ mesocrystals-based photocatalysis is beneficial for developing new types of photocatalytic materials with multiple functionalities. In this paper, a comprehensive review of the recent advances toward fabricating and modifying TiO$_2$ mesocrystals is provided, with special focus on the underlying mesocrystallization mechanism and controlling rules. The potential applications of as-synthesized TiO$_2$ mesocrystals in photocatalysis are then discussed to shed light on the structure–performance relationships, thus guiding the development of highly efficient TiO$_2$ mesocrystal-based photocatalysts for certain applications. Finally, the prospects of future research on TiO$_2$ mesocrystals in photocatalysis are briefly highlighted.

Keywords: TiO$_2$; photocatalysis; mesocrystals; synthesis; modification

---

1. Introduction

Semiconductor-based photocatalysis is well known to be one of the most effective approaches to alleviate the serious conundrums of global energy exhaustion, as well as environmental deterioration, by utilizing the inexhaustible solar energy [1–7]. Among various kinds of semiconductors, Titanium dioxide (TiO$_2$) is the most attractive one as a photocatalyst owing to its high photoreactivity, outstanding chemical stability, easy availability, and cheap price [8–15]. Despite tremendous efforts having been made toward the fabrication of TiO$_2$ materials, as well as the investigation of their photocatalytic properties, real applications of TiO$_2$ in photocatalysis are still largely hampered by the wide band gap of TiO$_2$ (e.g., 3.2 eV for anatase and brookite, 3.0 eV for rutile), which can merely absorb ultraviolet radiation (accounting for < 5% of solar light), and the fast recombination of photoinduced charge carriers, which leads to low quantum efficiency [16–21]. It is always a hot topic in the research area of materials chemistry and photocatalysis to manipulate the morphology and architecture of TiO$_2$ to achieve extended light response and facilitate photogenerated electron-hole separation, thus realizing remarkably enhanced photocatalytic activity in various applications [22–26].

Recently, it has been well demonstrated that building highly ordered superstructures from nanocrystal building blocks is very important for fabricating new materials and devices, as this kind of nanoparticle assembly can not only display properties and functions associated with individual nanoparticles, but can also exhibit new collective properties and advanced tunable functions [27–32]. In particular, mesocrystals, a new type of ordered superstructure built from crystallographically oriented nanocrystal subunits, have drawn significant research interest since the concept of "mesocrystal" was first introduced in 2005 [33,34]. These unique ordered superstructures were initially identified from the studies of the structural characteristics and growth mechanisms of biominerals, and were proposed to be formed through a non-classical, particle-mediated growth process, namely, mesoscale transformation, rather than the conventional classical, atom/ion-mediated crystallization route (Figure 1). Subsequently, the mesocrystal concept evolved from the classical mesocrystals, which were generated via the aforementioned mesoscale transformation process, to all the hierarchical materials built from crystallographically oriented nanocrystal subunits regardless of the mechanism of formation. Despite the flourishing emergence of reports on the fabrication of mesocrystals, the history of mesocrystal synthesis is closely related to the continuous exploitation of mesocrystals with new compositions and the persistent development of synthetic procedures having advantages in terms of low cost, convenience in handling, and easiness in compositional and structural control [35–41].

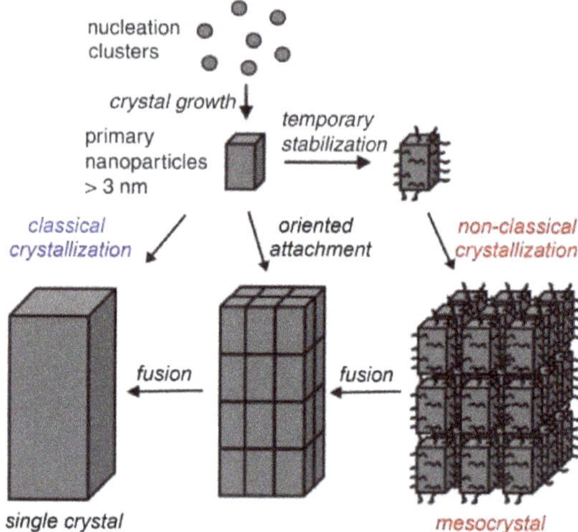

Figure 1. Schematic illustration of the single-crystal formation from classical crystallization, oriented attachment and non-classical crystallization. Reprinted with permission from [33]. Copyright John Wiley & Sons Inc., 2005.

To date, mesocrystals with a broad range of compositions involving metal oxides (e.g., TiO_2 [42–68], ZnO [69–85], Fe_2O_3 [86–95], CuO [96–101], SnOx [102,103], Co_3O_4 [104–108], Ag_2O [109]), metal chalcogenides (e.g., ZnS [110], PbS [111–113], Ag_2S [114], PbSe [115]), metals (e.g., Au [116–118], Ag [119], Cu [120], Pt [121,122], Pd [123]) have been produced, as introduced in some previous reviews [124–126]. Among these mesocrystals, TiO_2 mesocrystals are widely accepted to be particularly promising in photocatalytic applications [127–152]. It is noted that the high internal porosity and high surface areas of TiO_2 mesocrystals can be beneficial for the adsorption of reagents and provide more active sites for the subsequent photocatalytic reactions, while the well-oriented nanocrystal alignment provides effective conduction pathways and significantly enhances charge transport and separation with TiO_2 particles [135,153]. Although significant attention has been directed to fabricating TiO_2 mesocrystals with controlled morphologies, the realization of TiO_2 mesocrystals is always a challenging task, probably because the titanium precursors used are highly reactive, and it is rather difficult to precisely control the growth dynamic of TiO_2 crystals. Additionally, considering the wide band gap of the pristine TiO_2 materials, it is also demanding to modify the mesostructure of TiO_2 mesocrystals to realize broadened light absorption, thus achieving highly efficient photocatalysis in various applications.

In this review article, we first summarize numerous attempts toward the fabrication of TiO_2 mesocrystals. Four representative synthetic routes, namely, oriented topotactic transformation, growth on substrates, organic-additive-assisted growth in solution, and direct additive-free synthesis in solution, are presented one by one, with a special focus being channeled towards the underlying mesocrystallization mechanism and its controlling rules. The construction of doped TiO_2 mesocrystals, as well as TiO_2 mesocrystal-based heterostructures, is also covered in this review. The potential applications of the resultant TiO_2 mesocrystal-based materials in photocatalysis are then introduced to gain a deep understanding of the structure–performance relationships, thus providing useful guidelines for rationally designing and fabricating highly efficient TiO_2 mesocrystal-based photocatalysts for certain applications. Finally, some future research directions in the research area are briefly discussed and summarized.

2. Synthesis TiO₂ Mesocrystals

2.1. Oriented Topotactic Transformation

Early reports on the fabrication of TiO_2 mesocrystals were based on topotactic transformation from pre-synthesized NH_4TiOF_3 mesocrystals, as the titanium precursors used (e.g., $TiCl_4$, titanium tetrabutoxide (TBOT), titanium tetraisopropanolate (TTIP)) are normally highly reactive, making it rather challenging to manipulate the growth process of TiO_2 crystals upon direct syntheses. In 2007, O'Brien's group disclosed the first preparation of TiO_2 mesocrystals. In a synthetic procedure, NH_4TiOF_3 mesocrystals were first prepared in the $(NH_4)_2TiF_6$ and H_3BO_3 aqueous solution with the assistance of a nonionic surfactant (e.g., Brij 56, Brij 58, or Brij 700). After being washed with H_3BO_3 solution or sintered in air at 450 °C, the as-formed NH_4TiOF_3 mesocrystals were successfully transformed into anatase TiO_2 mesocrystals, with the original platelet-like shapes well preserved [42,43]. Such a topotactic transformation could proceed mainly because of the crystal structure similarity between NH_4TiOF_3 and anatase TiO_2 crystals (less than 0.02% in an average lattice mismatch), and the as-synthesized NH_4TiOF_3 mesocrystals could thus serve as a crystallographically matched template for the subsequent formation of TiO_2 mesocrystals (Figure 2). Owing to the great effectiveness of the methodology, NH_4TiOF_3 mesocrystals with a variety of morphologies were obtained by simply adjusting the reaction parameters, giving rise to a series of morphology-preserved anatase TiO_2 mesocrystals [44,45,137,141,143]. In addition, single-crystalline NH_4TiOF_3 crystals could also be utilized as a template for the oriented topotactic formation of anatase TiO_2 mesocrystals. For instance, by annealing a thin layer of aqueous solution containing TiF_4, NH_4F, and NH_4NO_3 on a Si wafer, nanosheet-shaped anatase TiO_2 mesocrystals enclosed by a high percentage of (001) facets were produced (Figure 3) [135]. Despite the one-step characteristic of the synthetic process, single-crystalline NH_4TiOF_3 nanosheets were actually first generated in the precursor solution at low annealing temperatures, which could then be easily transformed into anatase TiO_2 upon further increase in annealing temperature. With large quantities of N and F elements removed, the volume of the crystals decreased. Pores would form within the particles, resulting in anatase TiO_2 mesocrystals consisting of anatase nanocrystals predominantly enclosed by (001) facets.

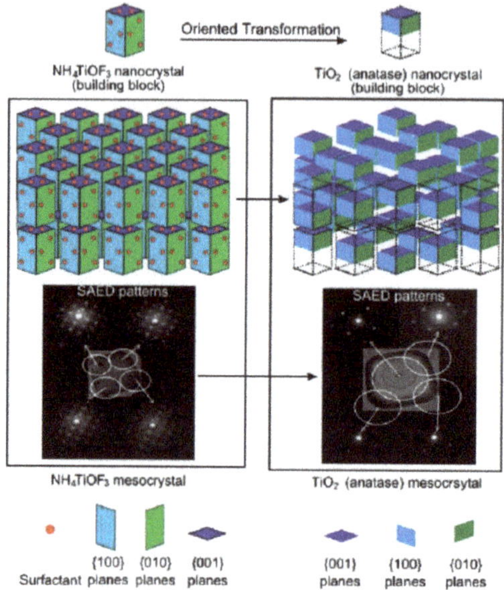

Figure 2. Schematic illustration of oriented topotactic transformation of NH_4TiOF_3 mesocrystal to anatase TiO_2 mesocrystal. The electron diffraction (SAED) patterns of the selected area illustrate single-crystal-like diffraction behavior for both samples. Reprinted with permission from [43]. Copyright American Chemical Society, 2008.

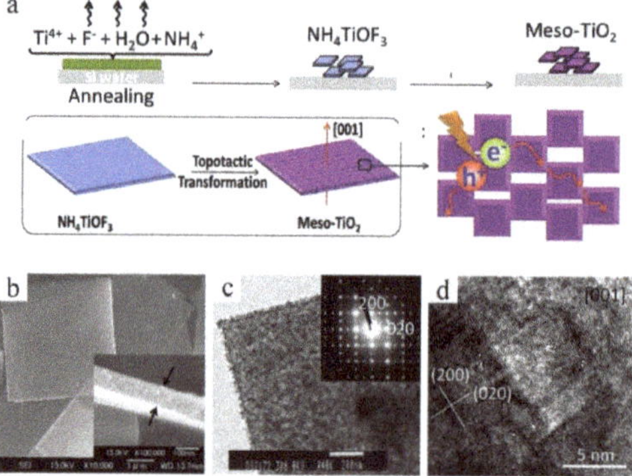

Figure 3. (**a**) Schematic presentation of oriented topotactic formation of anatase TiO_2 mesocrystals with dominant (001) facets; (**b**) SEM; (**c**) TEM; and (**d**) HRTEM images of anatase mesocrystals. The inset displays the related SAED pattern. Reprinted with permission from [135]. Copyright American Chemical Society, 2012.

Most recently, Qi's group proposed a new topotactic transformation method for fabricating anatase TiO_2 mesocrystals [154]. In their synthetic procedure, (010)-faceted orthorhombic titanium-containing precursor nanosheet arrays were firstly synthesized on conducting FTO glass substrate through solvothermally treating 0.1 M $K_2TiO(C_2O_4)_2$ in mixed solvents of deionized water and diethylene glycol. After a further hydrothermal treatment, the as-formed precursor nanosheet arrays could be readily converted to (001)-faceted anatase TiO_2 nanosheet arrays. It was revealed that the lattice match between the orthorhombic precursor crystal and the tetragonal anatase crystal accounted for the topotactic transformation from (010)-faceted precursor nanosheets to (001)-faceted anatase TiO_2 nanosheets (Figure 4).

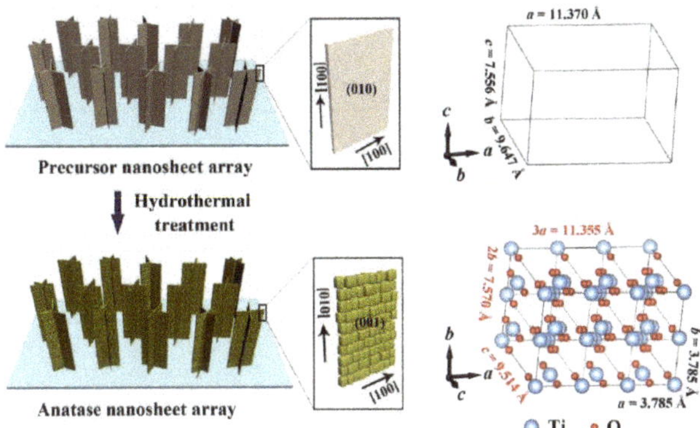

Figure 4. Schematic presentation of topotactic transformation from (010)-faceted precursor nanosheet arrays to (001)-faceted anatase TiO_2 nanosheet arrays on the basis of crystal lattice matchment between orthorhombic precursor crystal and tetragonal anatase crystal. Reprinted with permission from [154]. Copyright Springer, 2017.

2.2. Growth on Substrates

As presented above, topotactic transformation has been well demonstrated to be a very useful method to construct TiO_2 mesocrystals. However, precursors suitable for such a topotactic transformation are mainly limited to NH_4TiOF_3, and it is rather difficult to realize the morphological manipulation of the resultant TiO_2 mesocrystals at will. Therefore, it is highly desirable to explore facile solution-phase routes toward the direct fabrication of TiO_2 mesocrystals, since these kinds of syntheses are normally advantageous in light of their low cost, easy modulation of morphology, and great potential for environmentally benign production of inorganic materials. In 2008, Zeng's group first utilized multiwalled carbon nanotubes (CNTs) as substrate to grow anatase TiO_2 mesocrystals with controllable surface coverage [155]. It was revealed that the as-formed [001]-oriented petal-like anatase mesocrystals were uniformly distributed on CNTs, with TiO_2 nanocrystal building blocks having diameters in the range of 2–4 nm and mesopores having a very uniform size distribution centered at 2.5 nm. Additionally, by employing graphene nanosheets as a template to control the growth dynamic of TiO_2, uniform mesoporous anatase TiO_2 nanospheres were successfully generated and anchored on the graphene nanosheets (Figure 5) [156]. It is noteworthy that in comparison to the conventionally generated porous particles constructed by randomly aggregated anatase nanocrystals, the thus-formed mesoporous nanospheres were single-crystal-like. Detailed investigation on the growth process of the mesoporous anatase nanospheres revealed that such a graphene-nanosheet-assisted mesocrystallization route actually involved the nucleation of anatase TiO_2 on graphene nanosheets and subsequent oriented aggregation of tiny nanocrystals onto pre-anchored nuclei to reduce the total surface energy of anatase crystals. As a result, mesoporous mesocrystals of anatase TiO_2 would finally form. Moreover, Qi's group reported the fabrication of two-dimensional (2D) nanoarray structures constructed from mesocrystalline rutile TiO_2 nanorods on Ti substrate via a simple solution-phase synthesis [66]. These nanorod arrays were obtained by hydrothermally treating the aqueous solution of TBOT and HCl. It was revealed that during the growth process of the mesocrystalline rutile TiO_2 nanorod arrays, stem nanorods were first grown onto Ti substrate due to the high concentration of titanium-containing precursors, and with the consumption of the precursors, the resulting low concentration of reactant was responsible for the growth of the tiny nanotips with continuous crystal lattices, resulting in the final mesocrystalline rutile TiO_2 nanorods with a hierarchical architecture.

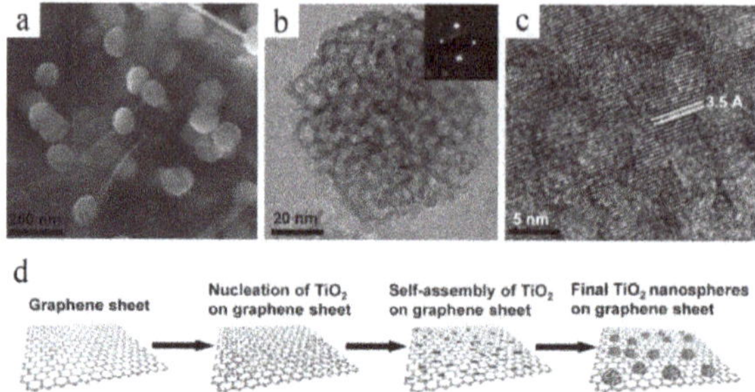

Figure 5. (**a**) SEM, (**b**) TEM, and (**c**) HRTEM images of mesoporous anatase TiO$_2$ nanospheres on graphene nanosheets. The inset is the SAED pattern related to a single nanosphere; (**d**) Schematic illustration of the growth mechanism of mesoporous anatase nanospheres. Reprinted with permission from [156]. Copyright John Wiley & Sons Inc., 2011.

2.3. Organic-Additive-Assisted Growth in Solution

Apart from the aforementioned solid templates or substrates, various organic additives could also be utilized to guide the formation of TiO$_2$ mesocrystals. In 2009, Yu's group first prepared hollow-sphere-shaped rutile TiO$_2$ mesocrystals assembled by nanorod subunits via a facile hydrothermal synthesis by using TiCl$_4$ as the titanium source and N, N'-dicyclohexylcarbodiimide (DCC) and L-serine as biological additives (Figure 6) [46]. It was proposed that such hollow-sphere-shaped mesocrystals were actually formed through a distinctive crystallization and transformation process, which involved the appearance of polycrystalline aggregates at the initial stage of reaction, mesoscale transformation to sector-shaped mesocrystals, further transformation of mesocrystals to nanorod bundles upon end-to-end and side-by-side oriented attachment accompanied by assembly of sectors to solid spheres, and final generation of hollow spheres via Ostwald ripening. Later on, with the assistance of organic small molecules of glacial acetic acid (HAc) and benzoic acid, rod-like anatase TiO$_2$ mesocrystals were successfully fabricated via a simple solvothermal route [127]. These mesocrystals were proposed to be formed through the well-known oriented attachment, and the mesocrystallization process was found to be carried out under the synergism of hydrophobic bonds, p-p interactions and "mixed-esters-templates". Furthermore, Gao's group synthesized spindle-shaped mesoporous anatase TiO$_2$ mesocrystals by utilizing peroxotitanium as the titanium source and polyacrylamide (PAM) as the polymer additive to adjust the growth process of TiO$_2$ [129]. They proposed that these anatase mesocrystals were formed via TiO$_2$-PAM co-assembly, accompanied by an amorphous-to-crystalline transformation.

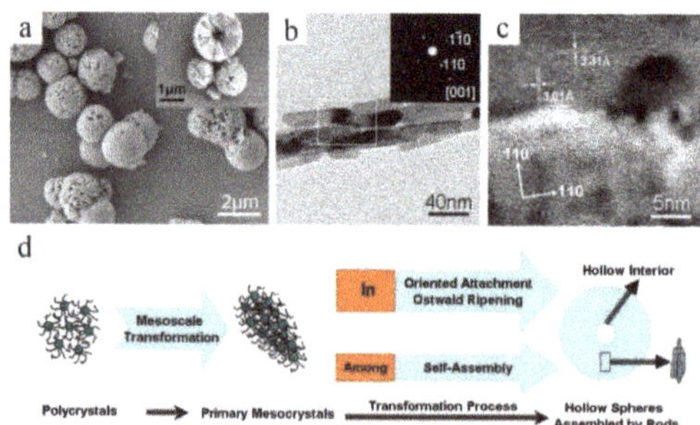

Figure 6. (**a**) SEM, (**b**) TEM, and (**c**) HRTEM images of hollow spheres of rutile TiO$_2$ mesocrystals. The inset in (**a**) is a magnified SEM image and the inset in (**b**) shows the related SAED pattern. (**d**) Schematic illustration of the formation mechanism of the rutile TiO$_2$ mesocrystals. Reprinted with permission from [46]. Copyright American Chemical Society, 2009.

In 2011, Tartaj's group developed a method based on inverse microemulsions to produce sub-100 nm sphere-like mesocrystalline nanostructures, which involved a two-stage temperature program [132]. In the first stage, the reaction at a low temperature (60 °C) triggered inverse microemulsions, resulting in thermal destabilization via forming nanomicellar structures smaller than 100 nm. The subsequent partial hydrolysis of TiOSO$_4$ produced sub-100-nm sphere-shaped TiO$_2$ frameworks through replicating those nanomicellar structures. In the second stage, increasing the reaction temperature to 80 °C or higher generated mesocrystalline TiO$_2$ architectures with interstitial porosity partially filled with surfactants. After the removal of the interstitial surfactants, mesoporosity was generated and uniform spherical-shaped mesocrystalline architectures of anatase TiO$_2$ with particle sizes ranging from 50 to 70 nm were produced finally. Later on, this method was extended to fabricate spherical-shaped mesoporous anatase TiO$_2$ mesocrystals with a much smaller size of 25 nm [133].

Recently, Zhao's group reported a facile evaporation-driven oriented assembly method to fabricate mesoporous anatase TiO$_2$ microspheres (~800 nm in diameter) with radially oriented hexagonal mesochannels and single-crystal-like pore walls (Figure 7) [64]. The synthesis started with the liquid-liquid phase separation, which was induced by the preferential evaporation of the solvent of tetrahydrofuran (THF) at a relatively low temperature (40 °C), and spherical-shaped PEO-PPO-PEO/TiO$_2$ oligomer composite micelles with PPO segments as the core and titania-associated PEO segments as the shell formed at the liquid-liquid phase interface. Upon further evaporation of THF at 40 °C, the concentration of the spherical micelles increased, leading to the formation of uniform mesoporous TiO$_2$ microspheres assembled by composite micelles (step 1 and 2). As the evaporation temperature increased to 80 °C, the continuous evaporation of the residual THF and hydrolyzed solvents from TBOT precursor drove the oriented growth of both mesochannels and nanocrystal building blocks from the initially formed spherical composite micelles along the free radial and restricted tangential direction within the TiO$_2$ microspheres (step 3). Radially oriented mesoporous anatase TiO$_2$ microspheres with single-crystal-like pore walls were produced after removal of the triblock copolymer templates finally (step 4). It is noteworthy that by simply adjusting the reaction parameters, mesoporous, single-crystal-like, olive-shaped, anatase TiO$_2$ mesocrystals constructed by ultrathin nanosheet subunits could also be synthesized [65].

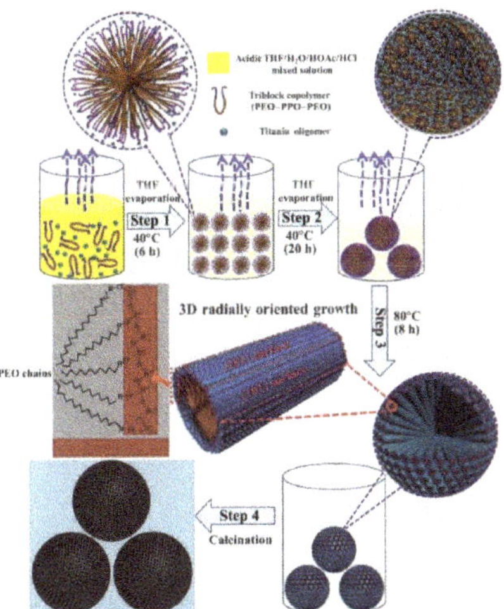

Figure 7. Schematic presentation of the formation process of mesoporous anatase TiO$_2$ microspheres with radially oriented hexagonal mesochannels and single-crystal-like pore walls through evaporation-driven oriented assembly. Reprinted with permission from [64]. Copyright American Chemical Society, 2015.

2.4. Direct Additive-Free Growth in Solution

Considering that the introduction of solid substrates or organic additives into the reaction system is unfavorable for the large-scale production of mesocrystals, it is, therefore, highly desirable to explore facile additive-free synthetic approaches toward functional mesocrystals with controllable crystallinity, porosity, morphology, and architecture. In 2011, Qi's group reported the first additive-free synthesis of nanoporous anatase TiO_2 mesocrystals with a spindle-shaped morphology, single-crystal-like structure, and tunable sizes via solvothermal treatment of the solution of TBOT in HAc, followed by calcination in air to remove the residual organics (Figure 8) [47]. These mesocrystals were illustrated to be elongated along the [001] direction, having lengths mainly in the range of 300–450 nm and diameters of 200–350 nm. It was revealed that under the solvothermal conditions, the reaction between TBOT and HAc firstly generated unstable titanium acetate complexes through ligand exchange/substitution, accompanied by the release of C_4H_9OH. The subsequent esterification reaction between thus-formed C_4H_9OH and the solvent HAc produced H_2O molecules slowly. Then, Ti-O-Ti bonds were formed via both nonhydrolytic-condensation and hydrolysis-condensation processes, resulting in transient amorphous fiber-like precursor. As the reaction continued, crystallized flower-like precursor was generated at the expense of the fiber-like precursor. This crystallized flower-like precursor acted as a reservoir to continuously release soluble titanium-containing species to generate tiny anatase nanocrystals. These tiny anatase nanocrystals underwent oriented aggregation along the [001] direction, together with some lateral attachment along some side facets of (101) facets, accompanied by the entrapment of in situ produced butyl acetate. As a result, [001]-elongated, spindle-shaped, anatase mesocrystals were produced when the reaction time was long enough. Further calcination in air would remove the butyl acetate residuals, consequently yielding nanoporous anatase TiO_2 mesocrystals.

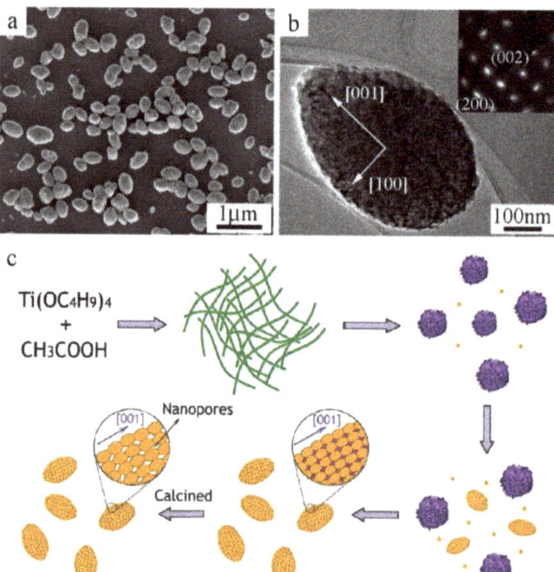

Figure 8. (**a**) SEM and (**b**) TEM images of nanoporous anatase TiO_2 mesocrystals obtained via solvothermal treatment of the solution of TBOT in HAc, followed by thermal treatment in air. The inset is the related SAED pattern of a single mesocrystal. (**c**) Proposed formation mechanism of nanoporous anatase TiO_2 mesocrystals. Reprinted with permission from [47]. Copyright American Chemical Society, 2011.

After half a month of Qi's pioneering work, Lu's group disclosed the fabrication of anatase TiO_2 mesocrystals with a single-crystal-like structure, high specific surface area, preferential exposure of highly reactive (001) crystal facets, and controllable mesoporous network [130]. As shown in Figure 9, by hydrothermal treating the solution of $TiOSO_4$ in *tert*-butyl alcohol, anatase TiO_2 nanocrystals were firstly generated, the (001) facets of which were preferably adsorbed by SO_4^{2-} anions. Subsequent oriented attachment of the anatase nanocrystal building blocks created anatase clusters with the (001) facets well protected (step 1). Upon further attachment of the building blocks, anatase TiO_2 mesocrystals preferentially exposed by (001) facets and having a disordered mesoporous network were finally produced (step 2). It is noteworthy that when the growth was confined in a scaffold with ordered pore channels, such as mesoporous silica containing 2D (SBA-15, *P6mm* space group) and three-dimensional (3D) (KIT-6, *Ia3d* space group) ordered mesopores, the subsequent scaffold removal would lead to TiO_2 crystals with replicated 2D hexagonal (step 3) or 3D (step 4) ordered network structure, respectively. More interestingly, such a novel methodology could be extended to fabricating mesoporous single-crystal-like structures with other compositions (e.g., ZrO_2, CeO_2, etc.), thus providing promising materials for various applications.

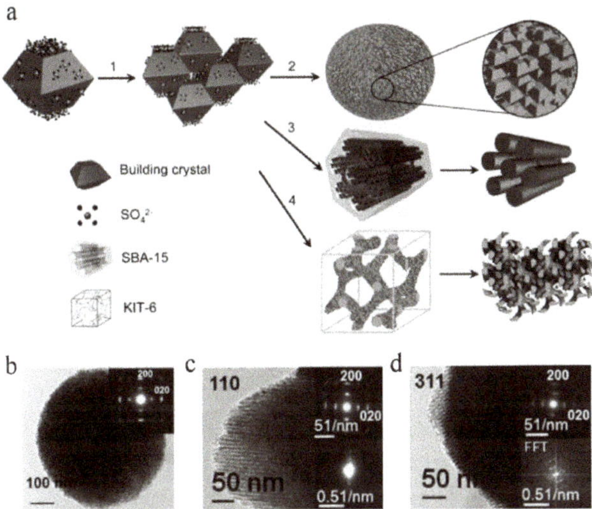

Figure 9. (**a**) Synthesis of mesoporous single-crystal-like anatase TiO_2 mesocrystals. (1) Formation of anatase clusters through oriented attachment of anatase nanocrystal building blocks with (001) facets preferably adsorbed by SO_4^{2-} ions. (2) Further attachment of the building blocks resulting in mesocrystals with preferential exposed (001) facets and disordered mesoporous structure. Mesocrystals with ordered mesoporous structure were prepared by a confined growth of the anatase crystals in (3) SBA-15 (mesoporous silica with 2D ordered pore channels) and (4) KIT-6 (mesoporous silica with 3D ordered pore channels) followed by scaffold removal. TEM images of anatase mesocrystals with disordered mesopores (**b**), mesoporous mesocrystals grown within SBA-15 (**c**) and KIT-6 (**d**) followed by removal of the scaffold. The insets in (**b**–**d**) show the related SAED and FFT patterns. Reprinted with permission from [130]. Copyright John Wiley & Sons Inc., 2011.

The above two groups' fascinating work opened a promising avenue for the facile synthesis of porous anatase mesocrystals. An increasing number of reports of the direct fabrication of TiO_2 mesocrystals in solutions without any additives have been disclosed in recent years. For example, Leite's group proposed a kinetically controlled crystallization process to produce anatase TiO_2 mesocrystals with a truncated bipyramidal morphology, which was realized through a nonaqueous sol-gel reaction between $TiCl_4$ and *n*-octanol [131]. By adopting a similar method to adjust the hydrolysis dynamic of

TTIP in an oxalic acid aqueous solution, hierarchical rutile TiO_2 mesocrystals were produced [48]. Zhao's group developed a facile synthetic approach to fabricate regular shaped anatase TiO_2 mesocrystals with controllable proportion of (001) and (101) facets [136]. These anatase TiO_2 mesocrystals were prepared by solvothermally treating the solution of TTIP in formic acid (FA), and the exposed (101)/(001) ratio could be adjusted via simply varying the duration of solvothermal treatment. Most recently, our group proposed a novel synthetic procedure for producing spindle-shaped, single-crystal-like, anatase TiO_2 mesocrystals, which was realized by controlling the hydrolysis rate of $TiCl_3$ in the green solvent PEG-400 (Figure 10) [150]. These mesocrystals constructed by ultrafine nanocrystals (~1.5–4.5 nm in size) were revealed to be spindle-shaped and elongated along the [001] direction, having lengths predominantly of 50–85 nm and diameters of 20–40 nm. It was proposed that at the initial stage of the reaction, the chelation of PEG-400 to titanium centers firstly resulted in the formation of a titanium precursor. This chelated titanium precursor then underwent hydrolysis-condensation reaction in the presence of water to form Ti-O-Ti bonds, accompanied by the gradual oxidation of Ti^{3+} to Ti^{4+} by the dissolved oxygen, yielding numerous tiny anatase nanocrystals. These tiny anatase nanocrystals were temporarily stabilized by the solvent PEG-400 molecules and underwent oriented attachment along the [001] direction, together with some lateral attachment along some side facets of (101) facets, resulting in the formation of mesocrystalline anatase aggregates elongated along the [001] direction. It is worth noting that continuous oriented attachment of tiny anatase nanocrystals on the preformed elongated mesocrystalline aggregates occurred when reaction time was prolonged, and well-defined spindle-shaped anatase TiO_2 mesocrystals were produced when the reaction time was extended to 5 h.

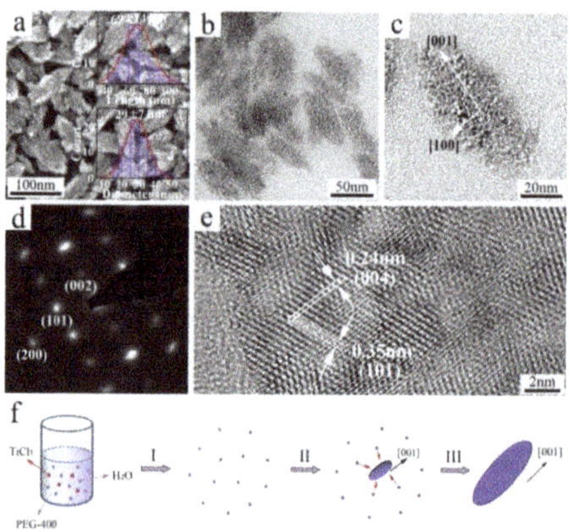

Figure 10. (**a**) SEM and (**b,c**) TEM images of anatase TiO_2 mesocrystals obtained via hydrolysis reaction of $TiCl_3$ in PEG-400. The insets in (**a**) are the related particle size distributions of the mesocrystals. (**d**) SAED pattern recorded on the anatase mesocrystal shown in (**c**); (**e**) HRTEM image of anatase mesocrystal; (**f**) A tentative mechanism for the formation of anatase mesocrystals. Reprinted with permission from [150]. Copyright American Chemical Society, 2017.

In addition to the widely employed titanium sources of TBOT, TTIP, $TiOSO_4$, and $TiCl_3$, it has been well proved that titanate precursors could also be utilized for the fabrication of TiO_2 mesocrystals. In 2012, Wei's group reported the synthesis of unique ultrathin-nanowire-constructed rutile TiO_2 mesocrystals through direct transformation from hydrogen titanate nanowire precursors (Figure 11) [61]. These hydrogen titanate nanowire precursors were prepared by hydrothermally treating the anatase TiO_2 in KOH solution, followed by acid washing. Then the precipitated hydrogen titanate nanowires were dispersed in HNO_3 aqueous solution and kept at 50 °C for 7 days, generating single-crystal-like rutile TiO_2 mesocrystals having lengths of about 300 nm and diameters 60–80 nm. It was proposed that such rutile mesocrystals were actually formed via face-to-face oriented attachment of ultrathin hydrogen titanate nanowire building blocks, accompanied by the conversion from hydrogen titanate precursor into rutile

TiO$_2$. To further modify the morphology of the rutile TiO$_2$ mesocrystals, Wei's group introduced the surfactant of sodium dodecyl benzene sulfonate (SDBS) into the reaction solution [62]. They found that SDBS played a vital role in the oriented self-assembly process, and rutile mesocrystals with controllable morphologies were successfully fabricated by varying the adding amount of SDBS. Specifically, uniform octahedral rutile TiO$_2$ mesocrystals 100–300 nm in size were obtained when the titanate/SDBS ratio was set at 0.09, while nanorod-shaped rutile TiO$_2$ mesocrystals were fabricated when the titanate/SDBS ratio increased to 0.15. Interestingly, the morphology and crystalline phase of the TiO$_2$ mesocrystals were demonstrated to be adjustable upon using different counterions to manipulate the growth dynamic of TiO$_2$ [63]. If the conversion of titanate nanowire precursors was carried out in HCl aqueous solution instead of HNO$_3$, dumbbell-shaped rutile TiO$_2$ superstructures composed of loose nanowire subunits were prepared, whereas anatase TiO$_2$ mesocrystals with a quasi-octahedral or truncated-octahedral morphology were obtained from H$_2$SO$_4$ aqueous solution. Such a novel synthetic procedure could also be extendable for the preparation of TiO$_2$ mesocrystals with other crystal phases. For example, by using amorphous titanates as titanium precursor and oxalic acid as structure-directing agent, novel brookite TiO$_2$ mesocrystals were successfully fabricated, as well [157].

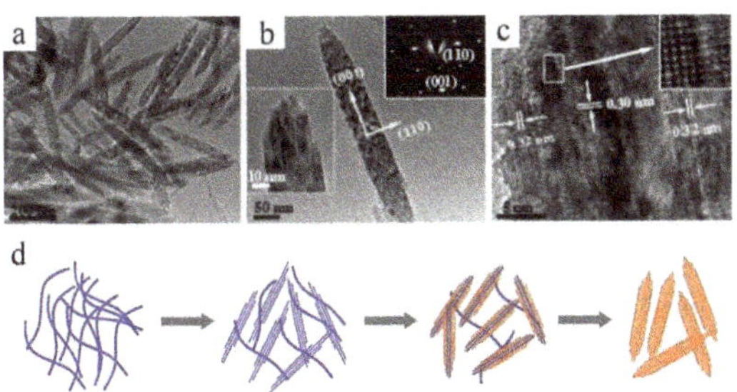

Figure 11. (**a**,**b**) TEM and (**c**) HRTEM images of rutile TiO$_2$ mesocrystals formed by conversion of titanate nanowire precursors in HNO$_3$ aqueous solution without any additives. The lower left inset in (**b**) is an enlarged TEM image, and the upper right inset is the SAED pattern related to the whole particle. (**d**) Schematic illustration of a tentative mechanism for the formation of rutile TiO$_2$ mesocrystals. Reprinted with permission from [61]. Copyright Royal Society of Chemistry, 2012.

3. Modification of TiO$_2$ Mesocrystals

3.1. Fabrication of Doped TiO$_2$ Mesocrystals

As mentioned above, the pristine TiO$_2$ can merely absorb ultra-violet irradiation owing to its wide band gap; continuous efforts have thus been channeled towards developing visible-light-responsive TiO$_2$ photocatalysts for various applications [8–13,16–21]. In addition to the well-known dye sensitization, the modification of TiO$_2$ with impurity doping was demonstrated to exhibit visible-light-responsive photocatalytic reactivity and showed improved stability upon light irradiation [11,16,19]. Considering the novel structural characteristics of TiO$_2$ mesocrystals, the fabrication of metal- or nonmetal-doped TiO$_2$ mesocrystals may give rise to ideal photocatalysts for particle applications, and thus has drawn considerable research interest [158–161]. For example, Majima's group successfully prepared N-doped anatase TiO$_2$ mesocrystals by solvothermal treatment of the pre-synthesized TiO$_2$ mesocrystals with triethanolamine [158]. Owing to the high internal porosity and high specific surface area of TiO$_2$ mesocrystals, the element of N could diffuse into the pores easily and was adsorbed on the surface. In addition, by stirring TiO$_2$ mesocrystals in NaF aqueous solution at room temperature, F-doped anatase TiO$_2$ mesocrystals could also be fabricated. It was proposed that surface fluorination via ligand exchange between F$^-$ and surface OH groups on TiO$_2$ occurred during the stirring process, resulting in the incorporation of F into TiO$_2$ mesocrystals. Combining these two doping strategies together would lead to the formation of N, F-codoped anatase TiO$_2$ mesocrystals without changing the morphology, crystallinestructure, and surface area of TiO$_2$ mesocrystals (Figure 12). Apart from the nonmetal-doped TiO$_2$ mesocrystals, it was demonstrated that metal-doped TiO$_2$ mesocrystals could also be synthesized. Wei's group prepared pure rutile TiO$_2$ mesocrystals first, and then hydrothermally treated them in aqueous niobium oxalate solution. After a certain period of hydrothermal treatment, homogeneous Nb-doped rutile TiO$_2$ mesocrystals could finally be produced [161].

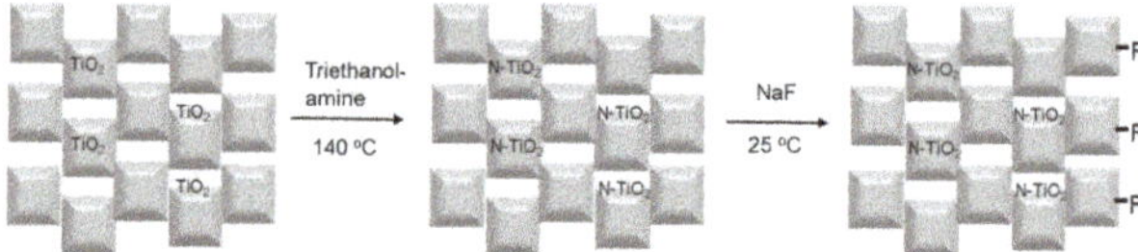

Figure 12. Proposed synthetic route toward N, F-codoped anatase TiO$_2$ mesocrystals. Reprinted with permission from [158]. Copyright Elsevier, 2016.

Recently, the introduction of oxygen vacancies or Ti^{3+} ions into TiO$_2$ to produce oxygen-deficient/Ti^{3+} self-doped TiO$_2$ mesostructures has been well accepted to be one of the most efficient ways to extend the light absorption region of TiO$_2$ to visible light [162–166]. Different from traditional doping strategies, introducing oxygen vacancies or Ti^{3+} ions is a unique doping method that can maintain the characteristic nature of TiO$_2$. At the same time, this kind of doping also improves the electroconductivity of TiO$_2$, thereby facilitating charge transportation within TiO$_2$ particles [162,164,167]. In this regard, great efforts have been made toward preparing oxygen-deficient/Ti^{3+} self-doped TiO$_2$ mesocrystals [65,136,150,168]. A good example in this area is that Zhao's group reported a facile evaporation-driven oriented assembly route combined with post thermal treatment in N$_2$ atmosphere to fabricate ultrathin-nanosheet-assembled olive-shaped mesoporous anatase TiO$_2$ mesocrystals (Figure 13) [65]. These mesoporous mesocrystals were illustrated to have high surface area (~189 m^2/g), large pore volume (0.56 cm^3/g), and abundant oxygen vacancies or unsaturated Ti^{3+} sites. Additionally, by thermally treating the anatase TiO$_2$ mesocrystals precipitated from the PEG-400/TiCl$_3$ mixed solution in vacuum, our group successfully synthesized Ti^{3+} self-doped, single-crystal-like, spindle-shaped, anatase TiO$_2$ mesocrystals [150]. Moreover, by reducing the pre-synthesized TiO$_2$ mesocrystals with NaBH$_4$, oxygen-deficient sheet-like anatase TiO$_2$ mesocrystals were also synthesized [168].

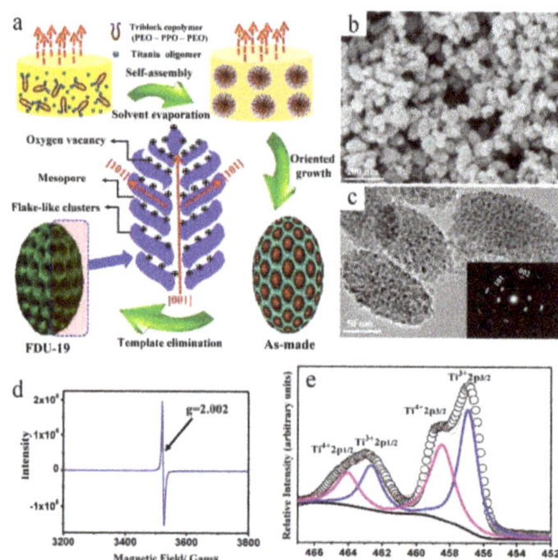

Figure 13. (a) Schematic illustration of the growth process of Ti^{3+} self-doped olive-shaped mesoporous anatase TiO$_2$ mesocrystals through evaporation-driven oriented assembly process; (b) SEM image, (c) TEM image, (d) EPR spectra, and (e) Ti2p XPS core-level spectra of Ti^{3+} self-doped olive-shaped mesoporous anatase TiO$_2$ mesocrystals. The inset in (c) is the SAED pattern of an individual mesocrystal. Reprinted with permission from [65]. Copyright American Chemical Society, 2015.

3.2. Construction of TiO$_2$ Mesocrystal-Based Heterostructures

Apart from the above-mentioned doping strategies, the coupling of TiO$_2$ mesocrystals with appropriate foreign elements to construct TiO$_2$ mesocrystal-based heterostructures is considered to be another effective way to enhance the light absorbance capability as well as inhibit the photoinduced charge carrier recombination [17,18,21]. Hitherto, various kinds of foreign elements have been successfully utilized to modify anatase TiO$_2$ mesocrystals [59,60,169–183]. For example, Sun's group successfully fabricated spindle-like TiO$_2$/CdS composites by uniformly distributing CdS nanoparticles onto nanoporous anatase mesocrystals via the simple hydrothermal and hot-injection methods [170]. Bian's group produced CdS quantum dot (QD)-decorated anatase TiO$_2$ mesocrystals preferably enclosed by (001) facets via the facile solvothermal treatment of TiOSO$_4$ in *tert*-butyl alcohol, followed by modification with CdS QDs via a simple ion-exchange treatment [175]. Majima's group applied a simple photodeposition method to deposit noble metal (Au, Pt) nanoparticles onto the pre-synthesized sheet-like anatase TiO$_2$ mesocrystals and realized the fabrication of novel metal-semiconductor superstructure nanocomposites [169]. Similarly, by adopting by a facile impregnation method, they were also able to deposite Au nanoparticles onto TiO$_2$ mesocrystals and fabricate promising plasmonic photocatalysts [172]. Moreover, to broaden the light-responsive region of TiO$_2$ mesocrystals to near-infrared (NIR) light, they also loaded Au nanorods with controllable size and tunable surface plasmon resonance (SPR) band onto anatase TiO$_2$ mesocrystals through the well-known ligand exchange method [179]. It is noteworthy that in addition to the deposition of guest elements onto the pre-synthesized anatase TiO$_2$ mesocrystals, anatase TiO$_2$ mesocrystals with desired morphologies could also be grown on various kinds of substrates. Tang's group introduced graphene oxide (GO) nanosheets into the reaction solution of TBOT in HAc. They found that after a solvothermal treatment at elevated temperatures, spindle-shaped anatase TiO$_2$ mesocrystals were successfully grown on the reduced graphene nanosheets [171]. Later on, Lu's group dispersed a certain amount of graphene into the reaction system of TiOSO$_4$ in *tert*-butyl alcohol. Upon microwave treatment of the obtained suspension, anatase TiO$_2$ mesocages with a single-crystal-like structure were found to be evenly anchored on graphene nanosheets [59]. Most recently, our group demonstrated that through in situ growth of nanosized defective anatase TiO$_{2-x}$ mesocrystals (DTMCs) on g-C$_3$N$_4$ nanosheets (NSs), a novel 3D/2D DTMC/g-C$_3$N$_4$ NS heterostructure with tight interfaces could be formed (Figure 14) [183].

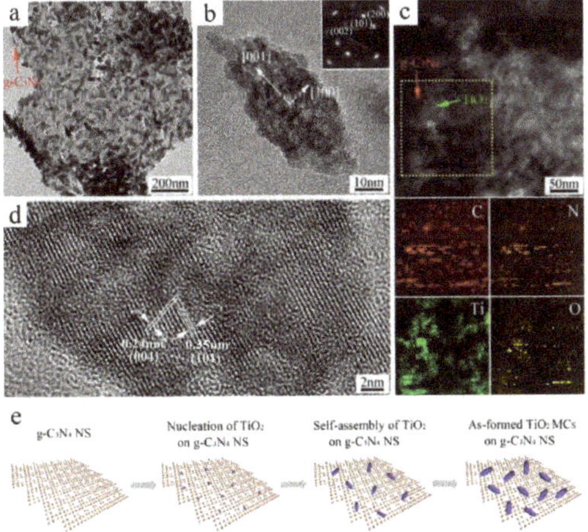

Figure 14. (**a,b**) TEM and (**c**) HRTEM images of 33.3% g-C$_3$N$_4$/DTMCs. The inset is the SAED pattern related to the whole particle. (**d**) HAADF-TEM image with elemental mapping of 33.3% g-C$_3$N$_4$/DTMCs. (**e**) Schematic presentation of the in situ growth of TiO$_2$ mesocrystals on a g-C$_3$N$_4$ nanosheet. Reprinted with permission from [183]. Copyright John Wiley & Sons Inc., 2018.

4. TiO$_2$ Mesocrystals for Photocatalytic Applications

4.1. Bare TiO$_2$ Mesocrystals for Photocatalytic Applications

Owing to the novel structural characteristics of mesocrystals, it is speculated the as-synthesized TiO$_2$ mesocrystals can be a promising candidate for photocatalytic applications. Liu's group first reported that the precipitated rod-like anatase TiO$_2$ mesocrystals delivered relatively higher photoreactivity toward the removal of methyl orange (MO) than the corresponding commercial P25 counterpart [127]. They ascribed the remarkably improved photocatalytic activity of the sample to its relatively high surface area, which could provide abundant sites for adsorption capability of MO. Yu's group proposed that the TiO$_2$ mesocrystals obtained in their additive-free reaction system possessed a well-crystallized rutile phase, low band gap energy and fast electron transfer property, and could exhibit high and stable photocatalytic activity for the removal of NO [128]. Lu's group evaluated the photoreactivity of the obtained single-crystal-like anatase TiO$_2$ mesocages and found that those unique TiO$_2$ mesocages with 3D ordered mesoporous channels exhibited superior photocatalytic activity toward oxidizing toluene to benzaldehyde and cinnamyl alcohol to cinnamaldehyde relative to that of TiO$_2$ mesocages with 2D ordered mesoporous channels, TiO$_2$ mesocages with disordered mesoporous channels, polycrystalline TiO$_2$, and P25 [130]. Leite's group claimed that the combination of high surface area and high crystallinity of the recrystallized mesocrystals can be more advantageous in photocatalytic applications than the corresponding disordered aggregate of nanocrystals [131].

Despite of the great efforts mentioned above toward the investigation of the photoreactivity of TiO$_2$ mesocrystals, it wasn't until 2012 that Majima's group first illustrated the photoelectronic properties of TiO$_2$ superstructures, in order to shed light on the intrinsic relationships between structural ordering and photoreactivity [135]. In their study, plate-like anatase TiO$_2$ mesocrystals synthesized via a topotactic transformation were selected as the target objects. These TiO$_2$ mesocrystals were built from crystallographically ordered anatase TiO$_2$ nanocrystal subunits and had a high surface area and high percentage of exposed highly reactive (001) facets. The photoconductive atomic force microscopy and time-resolved diffuse reflectance spectroscopy (DRS) were adopted to measure the charge transportation within the anatase mesocrystals, and the obtained results were compared with the reference anatase nanocrystals having similar surface area. It was consequently demonstrated that such a novel structure of anatase mesocrystals could exhibit largely enhanced charge separation and have remarkably long-lived charges, and thus could deliver greatly enhanced photoconductivity and photoreactivity (Figure 15). In 2015, Bian's group carefully evaluated the influence of intercrystal misorientation within anatase TiO$_2$ mesocrystals on the photoreactivity of the sample. They concluded that the misorientation of nanocrystal building blocks within anatase mesocrystals was harmful for the effective separation of photogenerated charge carriers and thus largely suppressed the photocatalytic efficiencies (Figure 16) [184]. Recently, Hu's group reported that the photocatalytic properties of anatase TiO$_2$ mesocrystals were actually largely dependent on the interfacial defects of intergrains within the particles [152]. They found that anatase TiO$_2$ mesocrystal photocatalysts exhibited much higher photocatalytic activity toward organic degradation and hydrogen evolution in comparison to single-crystalline crystals and poly crystalline crystals, which can be attributed to the presence of an appropriate number of interfacial defects at the intergrains and the facilitated charge carrier transport across the highly oriented interfaces. Moreover, it is inferred that the photoreactivity of the resultant anatase TiO$_2$ mesocrystal could be further optimized by regulation of defects, which could be simply achieved through annealing in redox atmospheres.

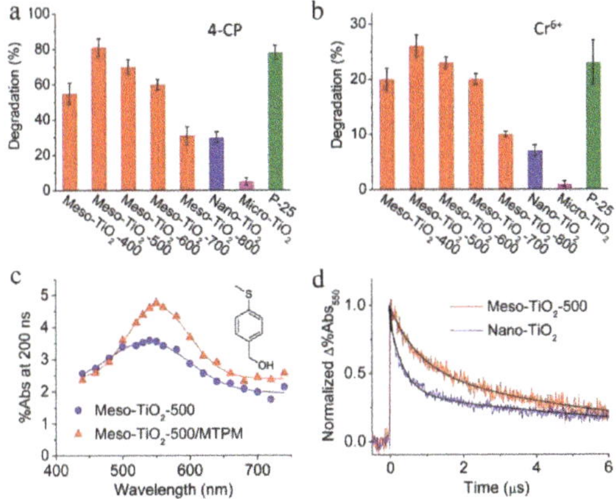

Figure 15. Photodegradation of (**a**) 4-CP and (**b**) Cr(VI) using various kinds of TiO_2 as catalysts. (**c**) Time-resolved diffuse reflectance spectra observed at 200 ns after the laser flash (355-nm) during the photolysis of Meso-TiO_2-500 in the absence and presence of 10 mM 4-(methylthio) phenyl methanol (MTPM) as the probe molecule to estimate the lifetime of the charge-separated state in acetonitrile. (**d**) Differential time traces of %Abs at 550 nm obtained in the presence of 10 mM MTPM for different TiO_2 samples in acetonitrile. Reprinted with permission from [135]. Copyright American Chemical Society, 2012.

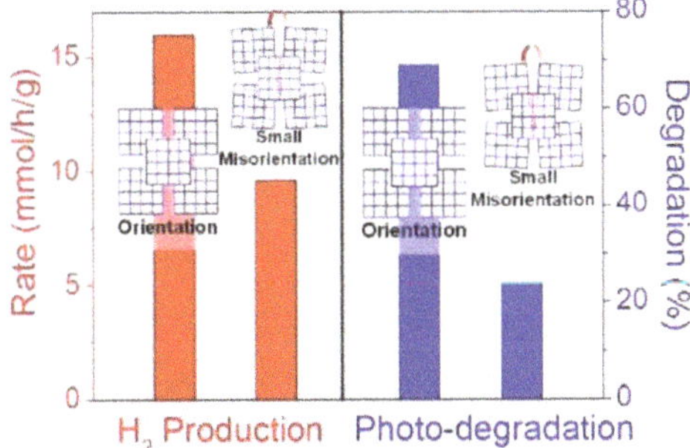

Figure 16. Rates comparison of phenol photodegradation and H_2 production upon TiO_2 mesocrystals built from well-ordered (red column) and less-ordered (blue column) orientation of nanocrystal subunits. Reprinted with permission from [184]. Copyright American Chemical Society, 2015.

4.2. Doped TiO$_2$ Mesocrystals for Photocatalytic Applications

Although a number of reports have demonstrated that TiO_2 mesocrystals can exhibit obviously enhanced photocatalytic performance in various applications, their real application is still hampered by the limited light absorbance of the pristine TiO_2 with a wide band gap. By utilizing the commonly used doping strategy, the thus-prepared doped TiO_2 mesocrystals can therefore become visible-light responsive, thus displaying enhanced visible-light-driven photoreactivity [136,150,158,159,168]. In 2016, Majima's group investigated the photoreactivity of N, F-codoped anatase TiO_2 mesocrystals. They found that, owing to the synergetic effect of N and F doping, the as-prepared product exhibited high visible-light-driven photoreactivity for degradating RhB and 4-nitrophenol (4-NP) [158]. Our group demonstrated that the obtained Ti^{3+} self-doped anatase TiO_2 mesocrystals showed much higher visible-light-driven photoreactivity toward removing NO and Cr (VI) compared with that of Ti^{3+} self-doped anatase nanocrystal counterparts. Such a photoreactivity enhancement was mainly due to the intrinsic self-doping nature, high crystallinity, as well as high porosity of the anatase mesocrystals (Figure 17) [150]. Most recently, Majima's group applied femtosecond time-resolved DRS and single-particle photoluminescence (PL) measurements to characterize reduced TiO_2 mesocrystals to get deep understanding of the correlation between oxygen deficiency, photogenerated charge transfer, and photoreactivity of the material [168]. They confirmed the enhanced light absorption through forming oxygen vacancies did not always result in higher photoreactivity, and an appropriate amount of oxygen vacancies was required to improve the photogenerated charge carrier separation, thus giving rise to optimized photoreactivity.

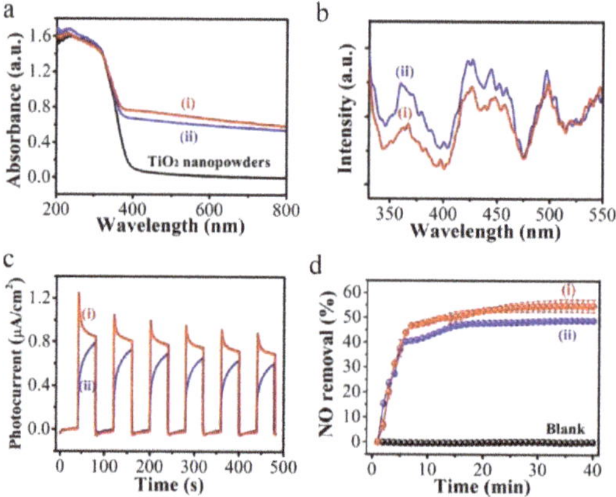

Figure 17. (**a**) UV-Vis DRS, (**b**) PL emission spectra, and (**c**) photocurrent intensity of (i) anatase mesocrystals and (ii) anatase nanocrystals of TiO_2 self-doped with Ti^{3+}. (**d**) Visible-light-driven photodegradation of NO upon (i) anatase mesocrystals and (ii) anatase nanocrystals self-doped with Ti^{3+}. Reprinted with permission from [150]. Copyright John Wiley & Sons Inc., 2017.

4.3. Composited TiO_2 Mesocrystals for Photocatalytic Applications

In addition to the aforementioned doping strategy, the coupling of TiO_2 mesocrystals with appropriate foreign materials to construct TiO_2-mesocrystal-based heterostructures is considered to be another useful methodology to broaden the light absorbance region of the material to visible light or even near-infrared (NIR) light, as well as to facilitate the mobility of photogenerated charge carriers within the particle [169–183]. For example, by utilizing CdS nanocrystals to modify spindle-shaped nanaporous anatase TiO_2 mesocrystals, Sun's group combined the advantages of the individual material, including (1) augmented specific surface area to provide more absorption and reactive sites; (2) TiO_2 mesocrystal substrate with high crystallinity and porosity to facilitate charge transport; (3) uniform distribution of CdS nanocrystals on mesocrystal surface and pores to facilitate charge transfer, and isolate photoinduced electrons and holes in two distinct materials; (4) tight contact between anatase mesocrystals and CdS nanocrystals to minimize the photo-corrosion and leaching off of CdS nanocrystals; and (5) extension of the photo-response of the material [170]. As expected, this unique spindle-shaped TiO_2/CdS photocatalyst exhibited relatively high visible-light-driven activity toward photodegradation of RhB. Bian's group reported that by decorating CdS QDs onto TiO_2 mesocrystals with a high percentage of exposed (001) facets, considerably high visible-light-driven photoreactivity could be achieved when selectively oxidizing various kinds of alcohols to their corresponding aldehydes [175]. Such an enhancement of the photoreactivity could be attributed to CdS QDs with improved photosensitization, porous mesostructure with high surface area, and exposed (001) facets with high surface energy and large quantities of oxygen vacancies, which could promote light absorbance in the visible light region, reactant molecule adsorption and activation, as well as photogenerated charge carrier separation. Majima's group claimed that superior electron transport and enhanced photoreactivity could be realized upon fabricating noble metal (Au, Pt) nanoparticle-loaded nanoplate-shaped anatase TiO_2 mesocrystals [169]. They proposed that most of the photogenerated electrons could migrate from the dominant surface to the edge of the TiO_2 mesocrystal with the reduction reactions mainly occurring at its lateral surfaces containing (101) facets, as illustrated by single-molecule fluorescence spectroscopy. The as-fabricated metal-semiconductor nanocomposites were found to display significant enhancement of the photocatalytic reaction rate in organic degradation and hydrogen production. More interestingly, by utilizing Au nanorods to modify anatase TiO_2 mesocrystal superstructures, highly efficient photocatalytic hydrogen production under visible-NIR-light irradiation could be obtained [179]. This efficient hydrogen production could be attributed to the SPR of Au nanorods which injected electrons into anatase TiO_2 mesocrystals and the facilitated charge transport within mesocrystal particles. Apart from the adjustment of deposited guest particles, it was also demonstrated that efficient defect-state-induced hot electron transfer could be found in the as-prepared Au nanoparticles/reduced TiO_2 mesocrystal photocatalysts, which lead to the enhanced photoreactivity of the photocatalyst in removing methylene blue (MB) [182]. Most recently, our group evaluated the photoreactivity of the 3D/2D DTMC/g-C_3N_4 NS heterostructure with chemically bonded tight interfaces and found that the as-fabricated composite photocatalyst displayed much higher visible-light-driven photoreactivity toward removing the pollutants of MO and Cr(VI) than the corresponding DTMCs and g-C_3N_4 NSs counterparts (Figure 18) [183]. Systematic characterization results indicated that such an enhancement in the photoredox ability of the composite photocatalyst was based on the direct Z-scheme charge separation, as verified by the ·OH-trapping experiment.

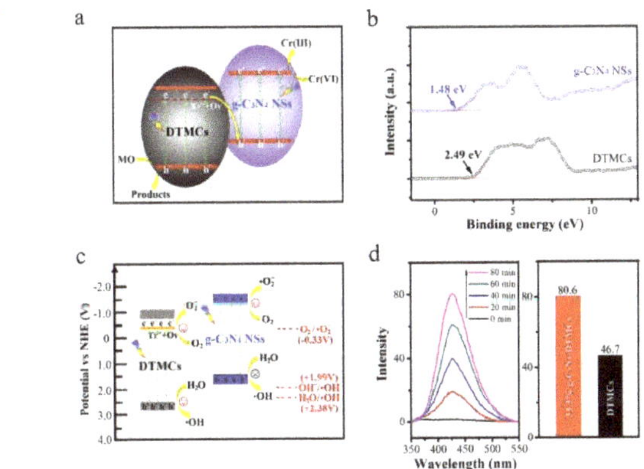

Figure 18. (a) Proposed Z-scheme charge-carrier transfer within DTMC/g-C_3N_4 composite. (b) XPS valence band spectra and (c) schematic electronic band structures of DTMCs and g-C_3N_4 NSs. (d) ·OH-trapping PL spectra of DTMCs/g-C_3N_4 and the corresponding fluorescence intensity upon DTMCs/g-C_3N_4 in comparison to DTMCs. Reprinted with permission from [183]. Copyright John Wiley & Sons Inc., 2018.

5. Summary and Outlook

In this paper, we have summarized some recent progress in fabricating TiO_2 mesocrystals, with special efforts being directed toward illustrating the underlying mesocrystallization process and its controlling rules. Four representative routes toward the fabrication of TiO_2 mesocrystals have been illustrated: oriented topotactic transformation, growth on substrates, organic-additive-assisted growth in solution, and direct additive-free synthesis in solution. In line with the flourishing emergence of reports on the fabrication of TiO_2 mesocrystals, the trends of TiO_2 mesocrystal synthesis are always related to the continuous exploitation of synthetic procedures having advantages like low cost, convenience in handling, and easiness of compositional and structural control. Apart from the fabrication of bare TiO_2 mesocrystals, the construction of doped TiO_2 mesocrystals, as well as TiO_2 mesocrystal-based heterostructures, are both considered to be promising strategies to further enhance the performance of TiO_2 mesocrystals in various applications, and thus have also been covered in this review. Taking into account the novel structural characteristics of TiO_2 meoscrystals, such as high crystallinity, high porosity, and oriented nanocrystal assembly, the potential applications of the resultant TiO_2 mesocrystal-based materials in photocatalysis have been discussed to gain a deep understanding of the structure-performance relationships, which can provide useful guidelines for designing and fabricating highly efficient TiO_2 mesocrystal-based photocatalysts for certain applications.

Despite great success having been achieved in the fabrication of TiO_2 mesocrystals, the related mesocrystallization process of TiO_2 mesocrystals is still not fully understood, and deserves further investigation. It remains an ongoing task to figure out the specific reason for the well-ordered alignment of TiO_2 nanocrystal building blocks in certain circumstances and develop facile, reproducible, and environmentally benign synthetic approaches toward TiO_2 mesocrystals with desired morphologies and architectures. In addition, it should be pointed out that compared with the synthesis of TiO_2 mesocrystals, the application of thus-produced TiO_2 mesocrystals in photocatalysis is much less explored, suggesting the high demand of a deep investigation into TiO_2 mesocrystal-based photocatalysts in various applications. For example, although overall enhancement of photoctalytic activity of TiO_2 mesocrystals has been demonstrated in recent years, the real mechanism for the photoreactivity enhancement in certain applications has not yet been fully understood. It is a necessity to thoroughly examine the relationship between the structure and photocatalytic properties of TiO_2 mesocrystals, which can guide the rational design and fabrication of TiO_2 mesocrystals with desired morphologies and architectures to fully satisfy the needs of specific applications in the future. In addition, the exploration of TiO_2 mesocrystal-based photocatalysts in some more challenging application areas, such as selective CO_2 reduction, ammonia synthesis, and methanol activation, deserves significant research attention to fully excavate their potential in photocatalytic applications.

Author Contributions: J.Y. and X.Y. chose the topic; J.Y., X.Y., B.Z., S.C., M.D., and Y.W. wrote and revised the article.

Funding: Financial support from National Natural Science Foundation of China (21603079, 21503085), Natural Science Foundation of Hubei Province (2015CFB175, 2015CFB233), Da Bei Nong Group Promoted Project for Young Scholar of HZAU (2017DBN010), and Fundamental Research Funds for the Central Universities (2662015QC042) is gratefully acknowledged.

Conflicts of Interest: The authors declare no conflict of interest.

References

1. Keane, D.A.; McGuigan, K.G.; Ibáñez, P.F.; Polo-López, M.I.; Byrne, J.A.; Dunlop, P.S.M.; O'Shea, K.; Dionysiou, D.D.; Pillai, S.C. Solar photocatalysis for water disinfection: Materials and reactor design. *Catal. Sci. Technol.* **2014**, *4*, 1211–1226. [CrossRef]

2. Spasiano, D.; Marotta, R.; Malato, S.; Fernandez-Ibanez, P.; Somma, I.D. Solar photocatalysis: Materials, reactors, some commercial and pre-industrialized applications. A comprehensive approach. *Appl. Catal. B Environ.* **2015**, *170–171*, 90–123. [CrossRef]

3. Chen, D.; Zhang, X.; Lee, A.F. Synthetic strategies to nanostructured photocatalysts for CO_2 reduction to solar fuels and chemicals. *J. Mater. Chem. A* **2015**, *3*, 14487–14516. [CrossRef]

4. Marszewski, M.; Cao, S.; Yu, J.; Jaroniec, M. Semiconductor-based photocatalytic CO_2 conversion. *Mater. Horiz.* **2015**, *2*, 261–278. [CrossRef]

5. Chen, S.; Takata, T.; Domen, K. Particulate photocatalysts for overall water splitting. *Nat. Rev. Mater.* **2017**, *2*, 17050. [CrossRef]

6. Zhu, S.; Wang, D. Photocatalysis: Basic principles, diverse forms of implementations and emerging scientific opportunities. *Adv. Energy Mater.* **2017**, *7*, 1700841. [CrossRef]

7. Christoforidis, K.C.; Fornasiero, P. Photocatalytic hydrogen production: A rift into the future energy supply. *ChemCatChem* **2017**, *9*, 1523–1544. [CrossRef]

8. Nakata, K.; Fujishima, A. TiO_2 photocatalysis: Design and applications. *J. Photochem. Photobiol. C Photochem. Rev.* **2012**, *13*, 169–189. [CrossRef]

9. Lan, Y.; Lu, Y.; Ren, Z. Mini review on photocatalysis of titanium dioxide nanoparticles and their solar applications. *Nano Energy* **2013**, *2*, 1031–1045. [CrossRef]

10. Schneider, J.; Matsuoka, M.; Takeuchi, M.; Zhang, J.; Horiuchi, Y.; Anpo, M.; Bahnemann, D.W. Understanding TiO$_2$ photocatalysis: Mechanisms and materials. *Chem. Rev.* **2014**, *114*, 9919–9986. [CrossRef]

11. Asahi, R.; Morikawa, T.; Irie, H.; Ohwaki, T. Nitrogen-doped titanium dioxide as visible-light-sensitive photocatalyst: Designs, developments, and prospects. *Chem. Rev.* **2014**, *114*, 9824–9852. [CrossRef] [PubMed]

12. Ma, Y.; Wang, X.; Jia, Y.; Chen, X.; Han, H.; Li, C. Titanium dioxide-based nanomaterials for photocatalytic fuel generations. *Chem. Rev.* **2014**, *114*, 9987–10043. [CrossRef] [PubMed]

13. Kapilashrami, M.; Zhang, Y.; Liu, Y.-S.; Hagfeldt, A.; Guo, J. Probing the optical property and electronic structure of TiO$_2$ nanomaterials for renewable energy applications. *Chem. Rev.* **2014**, *114*, 9662–9707. [CrossRef] [PubMed]

14. Wang, X.; Li, Z.; Shi, J.; Yu, Y. One-dimensional titanium dioxide nanomaterials: Nanowires, nanorods, and nanobelts. *Chem. Rev.* **2014**, *114*, 9346–9384. [CrossRef]

15. Li, W.; Wu, Z.; Wang, J.; Elzatahry, A.A.; Zhao, D. A perspective on mesoporous TiO$_2$ materials. *Chem. Mater.* **2014**, *26*, 287–298. [CrossRef]

16. Pelaez, M.; Nolan, N.T.; Pillai, S.C.; Seery, M.K.; Falaras, P.; Kontos, A.G.; Dunlop, P.S.M.; Hamilton, J.W.J.; Byrne, J.A.; O'Shea, K.; et al. A review on the visible light active titanium dioxide photocatalysts for environmental applications. *Appl. Catal. B Environ.* **2012**, *125*, 331–349. [CrossRef]

17. Park, H.; Park, Y.; Kim, W.; Choi, W. Surface modification of TiO$_2$ photocatalyst for environmental applications. *J. Photochem. Photobiol. C Photochem. Rev.* **2013**, *15*, 1–20. [CrossRef]

18. Zhang, G.; Kim, G.; Choi, W. Visible light driven photocatalysis mediated via ligand-to-metal charge transfer (LMCT): An alternative approach to solar activation of titania. *Energy Environ. Sci.* **2014**, *7*, 954–966. [CrossRef]

19. Etacheri, V.; Valentin, C.D.; Schneider, J.; Bahnemann, D.; Pillai, S.C. Visible-light activation of TiO$_2$ photocatalysts: Advances in theory and experiments. *J. Photochem. Photobiol. C Photochem. Rev.* **2015**, *25*, 1–29. [CrossRef]

20. Gao, M.; Zhu, L.; Ong, W.; Wang, J.; Ho, G.W. Structural design of TiO$_2$-based photocatalyst for H$_2$ production and degradation applications. *Catal. Sci. Technol.* **2015**, *5*, 4703–4726. [CrossRef]

21. Colmenares, J.C.; Varma, R.S.; Lisowski, P. Sustainable hybrid photocatalysts: Titania immobilized on carbon materials derived from renewable and biodegradable resources. *Green Chem.* **2016**, *18*, 5736–5750. [CrossRef]

22. Zhou, W.; Fu, H. Mesoporous TiO$_2$: Preparation, doping, and as a composite for photocatalysis. *ChemCatChem* **2013**, *5*, 885–894. [CrossRef]

23. Wang, M.; Ioccozia, J.; Sun, L.; Lin, C.; Li, Z. Inorganic-modified semiconductor TiO$_2$ nanotube arrays for photocatalysis. *Energy Environ. Sci.* **2014**, *7*, 2182–2202. [CrossRef]

24. Ola, O.; Maroto-Valer, M.M. Review of material design and reactor engineering on TiO$_2$ photocatalysis for CO$_2$ reduction. *J. Photochem. Photobiol. C Photochem. Rev.* **2015**, *24*, 16–42. [CrossRef]

25. Ge, M.; Li, Q.; Cao, C.; Huang, J.; Li, S.; Zhang, S.; Chen, Z.; Zhang, K.; Al-Deyab, S.S.; Lai, Y. One-dimensional TiO$_2$ nanotube photocatalysts for solar water splitting. *Adv. Sci.* **2017**, *4*, 1600152. [CrossRef]

26. Zhang, X.; Wang, Y.; Liu, B.; Sang, Y.; Liu, H. Heterostructures construction on TiO$_2$ nanobelts: A powerful tool for building high-performance photocatalysts. *Appl. Catal. B Environ.* **2017**, *202*, 620–641. [CrossRef]

27. Mann, S. Self-assembly and transformation of hybrid nano-objects and nanostructures under equilibrium and non-equilibrium conditions. *Nat. Mater.* **2009**, *8*, 781–792. [CrossRef] [PubMed]

28. Nie, Z.; Petukhova, A.; Kumacheva, E. Properties and emerging applications of self-assembled structures made from inorganic nanoparticles. *Nat. Nanotechnol.* **2010**, *5*, 15–25. [CrossRef]

29. Talapin, D.V.; Lee, J.-S.; Kovalenko, M.V.; Shevchenko, E.V. Prospects of colloidal nanocrystals for electronic and optoelectronic applications. *Chem. Rev.* **2010**, *110*, 389–458. [CrossRef]

30. Liu, J.-W.; Liang, H.-W.; Yu, S.-H. Macroscopic-scale assembled nanowire thin films and their functionalities. *Chem. Rev.* **2012**, *112*, 4770–4799. [CrossRef]

31. Klinkova, A.; Choueiri, R.M.; Kumacheva, E. Self-assembled plasmonic nanostructures. *Chem. Soc. Rev.* **2014**, *43*, 3976–3991. [CrossRef]

32. Cargnello, M.; Johnston-Peck, A.C.; Diroll, B.T.; Wong, E.; Datta, B.; Damodhar, D.; Doan-Nguyen, V.V.T.; Herzing, A.A.; Kagan, C.R.; Murray, C.B. Substitutional doping in nanocrystal superlattices. *Nature* **2015**, *524*, 450–455. [CrossRef] [PubMed]

33. Cölfen, H.; Antonietti, M. Mesocrystals: Inorganic superstructures made by highly parallel crystallization and controlled alignment. *Angew. Chem. Int. Ed.* **2005**, *44*, 5576–5591. [CrossRef] [PubMed]

34. Cölfen, H.; Antonietti, M. *Mesocrystals and Nonclassical Crystallization*; John Wiley & Sons: Chichester, UK, 2008.

35. Zhou, L.; O'Brien, P. Mesocrystals: A new class of solid materials. *Small* **2008**, *4*, 1566–1574. [CrossRef] [PubMed]

36. Song, R.-Q.; Cölfen, H. Mesocrystals-ordered nanoparticle superstructures. *Adv. Mater.* **2010**, *22*, 1301–1330. [CrossRef]

37. Fang, J.; Ding, B.; Gleiter, H. Mesocrystals: Syntheses in metals and applications. *Chem. Soc. Rev.* **2011**, *40*, 5347–5360. [CrossRef] [PubMed]

38. Zhou, L.; O'Brien, P. Mesocrystals-properties and applications. *J. Phys. Chem. Lett.* **2012**, *3*, 620–628. [CrossRef]

39. Uchaker, E.; Cao, G. Mesocrystals as electrode materials for lithium-ion batteries. *Nano Today* **2014**, *9*, 499–524. [CrossRef]

40. Tachikawa, T.; Majima, T. Metal oxide mesocrystals with tailored structures and properties for energy conversion and storage applications. *NPG Asia Mater.* **2014**, *6*, e100. [CrossRef]

41. Bergström, L.; Sturm (née Rosseeva), E.V.; Salazar-Alvarez, G.; Cölfen, H. Mesocrystals in biominerals and colloidal arrays. *Acc. Chem. Res.* **2015**, *48*, 1391–1402. [CrossRef]

42. Zhou, L.; Boyle, D.S.; O'Brien, P. Uniform NH_4TiOF_3 mesocrystals prepared by an ambient temperature self-assembly process and their topotaxial conversion to anatase. *Chem. Commun.* **2007**, 144–146. [CrossRef]

43. Zhou, L.; Smyth-Boyle, D.; O'Brien, P. A facile synthesis of uniform NH_4TiOF_3 mesocrystals and their conversion to TiO_2 mesocrystals. *J. Am. Chem. Soc.* **2008**, *130*, 1309–1320. [CrossRef] [PubMed]

44. Feng, J.; Yin, M.; Wang, Z.; Yan, S.; Wan, L.; Li, Z.; Zou, Z. Facile synthesis of anatase TiO_2 mesocrystal sheets with dominant {001} facets based on topochemical conversion. *CrystEngComm* **2010**, *12*, 3425–3429. [CrossRef]

45. Inoguchi, M.; Afzaal, M.; Tanaka, N.; O'Brien, P. The poly(ethylene glycol) assisted preparation of NH_4TiOF_3 mesocrystals and their topotactic conversion to TiO_2. *J. Mater. Chem.* **2012**, *22*, 25123–25129. [CrossRef]

46. Liu, S.-J.; Gong, J.-Y.; Hu, B.; Yu, S.-H. Mesocrystals of rutile TiO_2: Mesoscale transformation, crystallization, and growth by a biologic molecules-assisted hydrothermal process. *Cryst. Growth Des.* **2009**, *9*, 203–209. [CrossRef]

47. Ye, J.; Liu, W.; Cai, J.; Chen, S.; Zhao, X.; Zhou, H.; Qi, L. Nanoporous anatase TiO_2 mesocrystals: Additive-free synthesis, remarkable crystalline-phase stability, and improved lithium insertion behavior. *J. Am. Chem. Soc.* **2011**, *133*, 933–940. [CrossRef] [PubMed]

48. Wang, H.; Liu, Y.; Liu, Z.; Xu, H.; Deng, Y.; Shen, H. Hierarchical rutile TiO_2 mesocrystals assembled by nanocrystals-oriented attachment mechanism. *CrystEngComm* **2012**, *14*, 2278–2282. [CrossRef]

49. Zhen, M.; Guo, X.; Gao, G.; Zhou, Z.; Liu, L. Rutile TiO_2 nanobundles on reduced graphene oxides as anode materials for Li ion batteries. *Chem. Commun.* **2014**, *50*, 11915–11918. [CrossRef]

50. Wang, H.; Sun, L.; Wang, H.; Xin, L.; Wang, Q.; Liu, Y.; Wang, L. Rutile TiO_2 mesocrystallines with aggregated nanorod clusters: Extremely rapid self-reaction of the single source and enhanced dye-sensitized solar cell performance. *RSC Adv.* **2014**, *4*, 58615–58623. [CrossRef]

51. Fu, X.; Wang, B.; Chen, C.; Ren, Z.; Fan, C.; Wang, Z. Controllable synthesis of spherical anatase mesocrystals for lithium ion batteries. *New J. Chem.* **2014**, *38*, 4754–4759. [CrossRef]

52. Zhou, Y.; Wang, X.; Wang, H.; Song, Y.; Fang, L.; Ye, N.; Wang, L. Enhanced dye-sensitized solar cells performance using anatase TiO_2 mesocrystals with the Wulff construction of nearly 100% exposed {101} facets as effective light scattering layer. *Dalton Trans.* **2014**, *43*, 4711–4719. [CrossRef] [PubMed]

53. Hong, Z.; Zhou, K.; Zhang, J.; Huang, Z.; Wei, M. Facile synthesis of rutile TiO_2 mesocrystals with enhanced sodium storage properties. *J. Mater. Chem. A* **2015**, *3*, 17412–17416. [CrossRef]

54. Amarilla, J.M.; Morales, E.; Sanz, J.; Sobrados, I.; Tartaj, P. Electrochemical response in aprotic ionic liquid electrolytes of TiO_2 anatase anodes based on mesoporous mesocrystals with uniform colloidal size. *J. Power Sources* **2015**, *273*, 368–374. [CrossRef]

55. Hong, Z.; Zhou, K.; Huang, Z.; Wei, M. Iso-oriented anatase TiO_2 mesocages as a high performance anode material for sodium-ion storage. *Sci. Rep.* **2015**, *5*, 11960. [CrossRef]

56. Wu, D.; Cao, K.; Wang, H.; Wang, F.; Gao, Z.; Xu, F.; Guo, Y.; Jiang, K. Tunable synthesis of single-crystalline-like TiO$_2$ mesocrystals and their application as effective scattering layer in dye-sensitized solar cells. *J. Colloid Interface Sci.* **2015**, *456*, 125–131. [CrossRef]

57. Wu, Q.; Yang, X.; Zhou, W.; Gao, Q.; Lu, F.; Zhuang, J.; Xu, X.; Wu, M.; Fan, H.J. "Isofacet" anatase TiO$_2$ microcages: Topotactic synthesis and ultrastable Li-ion storage. *Adv. Mater. Interfaces* **2015**, *2*, 1500210. [CrossRef]

58. Hong, Z.; Hong, J.; Xie, C.; Huang, Z.; Wei, M. Hierarchical rutile TiO$_2$ with mesocrystalline structure for Li-ion and Na-ion storage. *Electrochim. Acta* **2016**, *202*, 203–208. [CrossRef]

59. Le, Z.; Liu, F.; Nie, P.; Li, X.; Liu, X.; Bian, Z.; Chen, G.; Wu, H.B.; Lu, Y. Pseudocapacitive sodium storage in mesoporous single-crystal-like TiO$_2$-graphene nanocomposite enables high-performance sodium-ion capacitors. *ACS Nano* **2017**, *11*, 2952–2960. [CrossRef] [PubMed]

60. Peng, Y.; Le, Z.; Wen, M.; Zhang, D.; Chen, Z.; Wu, H.B.; Li, H.; Lu, Y. Mesoporous single-crystal-like TiO$_2$ mesocages threaded with carbon nanotubes for high-performance electrochemical energy storage. *Nano Energy* **2017**, *35*, 44–51. [CrossRef]

61. Hong, Z.; Wei, M.; Lan, T.; Jiang, L.; Cao, G. Additive-free synthesis of unique TiO$_2$ mesocrystals with enhanced lithium-ion intercalation properties. *Energy Environ. Sci.* **2012**, *5*, 5408–5413. [CrossRef]

62. Hong, Z.; Wei, M.; Lan, T.; Cao, G. Self-assembled nanoporous rutile TiO$_2$ mesocrystals with tunable morphologies for high rate lithium-ion batteries. *Nano Energy* **2012**, *1*, 466–471. [CrossRef]

63. Hong, Z.; Xu, Y.; Liu, Y.; Wei, M. Unique ordered TiO$_2$ superstructures with tunable morphology and crystalline phase for improved lithium storage properties. *Chem. Eur. J.* **2012**, *18*, 10753–10760. [CrossRef]

64. Liu, Y.; Che, R.; Chen, G.; Fan, J.; Sun, Z.; Wu, Z.; Wang, M.; Li, B.; Wei, J.; Wei, Y.; et al. Radially oriented mesoporous TiO$_2$ microspheres with single-crystal–like anatase walls for high-efficiency optoelectronic devices. *Sci. Adv.* **2015**, *1*, e1500166. [CrossRef]

65. Liu, Y.; Luo, Y.; Elzatahry, A.A.; Luo, W.; Che, R.; Fan, J.; Lan, K.; Al-Enizi, A.M.; Sun, Z.; Li, B.; et al. Mesoporous TiO$_2$ mesocrystals: Remarkable defects-induced crystallite-interface reactivity and their in situ conversion to single crystals. *ACS Cent. Sci.* **2015**, *1*, 400–408. [CrossRef] [PubMed]

66. Cai, J.; Ye, J.; Chen, S.; Zhao, X.; Zhang, D.; Chen, S.; Ma, Y.; Jin, S.; Qi, L. Self-cleaning, broadband and quasi-omnidirectional antireflective structures based on mesocrystalline rutile TiO$_2$ nanorod arrays. *Energy Environ. Sci.* **2012**, *5*, 7575–7581. [CrossRef]

67. Dai, H.; Zhang, S.; Hong, Z.; Li, X.; Xu, G.; Lin, Y.; Chen, G. Enhanced photoelectrochemical activity of a hierarchical-ordered TiO$_2$ mesocrystal and its sensing application on a carbon nanohorn support scaffold. *Anal. Chem.* **2014**, *86*, 6418–6424. [CrossRef]

68. Dai, H.; Zhang, S.; Gong, L.; Li, Y.; Xu, G.; Lin, Y.; Hong, Z. The photoelectrochemical exploration of multifunctional TiO$_2$ mesocrystals and its enzyme-assisted biosensing application. *Biosens. Bioelectron.* **2015**, *72*, 18–24. [CrossRef]

69. Li, Z.; Gessner, A.; Richters, J.-P.; Kalden, J.; Voss, T.; Kuebel, C.; Taubert, A. Hollow zinc oxide mesocrystals from an ionic liquid precursor (ILP). *Adv. Mater.* **2008**, *20*, 1279–1285. [CrossRef]

70. Liu, Z.; Wen, X.D.; Wu, X.L.; Gao, Y.J.; Chen, H.T.; Zhu, J.; Chu, P.K. Intrinsic dipole-field-driven mesoscale crystallization of core-shell ZnO mesocrystal microspheres. *J. Am. Chem. Soc.* **2009**, *131*, 9405–9412. [CrossRef]

71. Wu, X.L.; Xiong, S.J.; Liu, Z.; Chen, J.; Shen, J.C.; Li, T.H.; Wu, P.H.; Chu, P.K. Green light stimulates terahertz emission from mesocrystal microspheres. *Nat. Nanotechnol.* **2011**, *6*, 103–106. [CrossRef]

72. Distaso, M.; Klupp Taylor, R.N.; Taccardi, N.; Wasserscheid, P.; Peukert, W. Influence of the counterion on the synthesis of ZnO mesocrystals under solvothermal conditions. *Chem. Eur. J.* **2011**, *17*, 2923–2930. [CrossRef] [PubMed]

73. Distaso, M.; Segets, D.; Wernet, R.; Taylor, R.K.; Peukert, W. Tuning the size and the optical properties of ZnO mesocrystals synthesized under solvothermal conditions. *Nanoscale* **2012**, *4*, 864–873. [CrossRef]

74. Hosono, E.; Tokunaga, T.; Ueno, S.; Oaki, Y.; Imai, H.; Zhou, H.; Fujihara, S. Crystal growth process of single-crystal-like mesoporous ZnO through a competitive reaction in solution. *Cryst. Growth Des.* **2012**, *12*, 2923–2931. [CrossRef]

75. Liu, M.-H.; Tseng, Y.-H.; Greer, H.F.; Zhou, W.; Mou, C.-Y. Dipole field guided orientated attachment of nanocrystals to twin-brush ZnO mesocrystals. *Chem. Eur. J.* **2012**, *18*, 16104–16113. [CrossRef] [PubMed]

76. Sun, S.; Zhang, X.; Zhang, J.; Song, X.; Yang, Z. Unusual designated-tailoring on zone-axis preferential growth of surfactant-free ZnO mesocrystals. *Cryst. Growth Des.* **2012**, *12*, 2411–2418. [CrossRef]

77. Waltz, F.; Wissmann, G.; Lippke, J.; Schneider, A.M.; Schwarz, H.-C.; Feldhoff, A.; Eiden, S.; Behrens, P. Evolution of the morphologies of zinc oxide mesocrystals under the influence of natural polysaccharides. *Cryst. Growth Des.* **2012**, *12*, 3066–3075. [CrossRef]

78. Wang, H.; Xin, L.; Wang, H.; Yu, X.; Liu, Y.; Zhou, X.; Li, B. Aggregation-induced growth of hexagonal ZnO hierarchical mesocrystals with interior space: Nonaqueous synthesis, growth mechanism, and optical properties. *RSC Adv.* **2013**, *3*, 6538–6544. [CrossRef]

79. Wang, S.-S.; Xu, A.-W. Template-free facile solution synthesis and optical properties of ZnO mesocrystals. *Cryst. Eng. Commun.* **2013**, *15*, 376–381. [CrossRef]

80. Peng, Y.; Wang, Y.; Chen, Q.-G.; Zhu, Q.; Xu, A.W. Stable yellow ZnO mesocrystals with efficient visible-light photocatalytic activity. *CrystEngComm* **2014**, *16*, 7906–7913. [CrossRef]

81. Liu, J.; Hu, Z.-Y.; Peng, Y.; Huang, H.-W.; Li, Y.; Wu, M.; Ke, X.-X.; Tendeloo, G.V.; Su, B.-L. 2D ZnO mesoporous single-crystal nanosheets with exposed {0001} polar facets for the depollution of cationic dye molecules by highly selective adsorption and photocatalytic decomposition. *Appl. Catal. B Environ.* **2016**, *181*, 138–145. [CrossRef]

82. Liu, M.-H.; Chen, Y.-W.; Liu, X.; Kuo, J.-L.; Chu, M.-W.; Mou, C.-Y. Defect-mediated gold substitution doping in ZnO mesocrystals and catalysis in CO oxidation. *ACS Catal.* **2016**, *6*, 115–122. [CrossRef]

83. Wang, H.; Wang, C.; Chen, Q.; Ren, B.; Guan, R.; Cao, X.; Yang, X.; Duan, R. Interface-defect-mediated photocatalysis of mesocrystalline ZnO assembly synthesized in-situ via a template-free hydrothermal approach. *Appl. Surf. Sci.* **2017**, *412*, 517–528. [CrossRef]

84. Liu, M.-H.; Chen, Y.-W.; Lin, T.-S.; Mou, C.-Y. Defective mesocrystal ZnO-supported gold catalysts: Facilitating CO oxidation via vacancy defects in ZnO. *ACS Catal.* **2018**, *8*, 6862–6869. [CrossRef]

85. Liang, S.; Gou, X.; Cui, J.; Luo, Y.; Qu, H.; Zhang, T.; Yang, Z.; Yang, Q.; Sun, S. Novel cone-like ZnO mesocrystals with coexposed (10-11) and (000-1) facets and enhanced photocatalytic activity. *Inorg. Chem. Front.* **2018**, *5*, 2257–2267. [CrossRef]

86. Park, G.-S.; Shindo, D.; Waseda, Y.; Sugimoto, T. Internal structure analysis of monodispersed pseudocubic hematite particles by electron microscopy. *J. Colloid Interface Sci.* **1996**, *177*, 198–207. [CrossRef]

87. Ahniyaz, A.; Sakamoto, Y.; Bergström, L. Magnetic field-induced assembly of oriented superlattices from maghemite nanocubes. *Proc. Natl. Acad. Sci. USA* **2007**, *104*, 17570–17574. [CrossRef]

88. Fang, X.-L.; Chen, C.; Jin, M.-S.; Kuang, Q.; Xie, Z.-X.; Xie, S.-Y.; Huang, R.-B.; Zheng, L.-S. Single-crystal-like hematite colloidal nanocrystal clusters: Synthesis and applications in gas sensors, photocatalysis and water treatment. *J. Mater. Chem.* **2009**, *19*, 6154–6160. [CrossRef]

89. An, Z.; Zhang, J.; Pan, S.; Yu, F. Facile template-free synthesis and characterization of elliptic a-Fe$_2$O$_3$ superstructures. *J. Phys. Chem. C* **2009**, *113*, 8092–8096. [CrossRef]

90. Chen, J.S.; Zhu, T.; Li, C.M.; Lou, X.W. Building hematite nanostructures by oriented attachment. *Angew. Chem. Int. Ed.* **2011**, *50*, 650–653. [CrossRef]

91. Ma, J.; Teo, J.; Mei, L.; Zhong, Z.; Li, Q.; Wang, T.; Duan, X.; Lian, J.; Zheng, W. Porous platelike hematite mesocrystals: Synthesis, catalytic and gas-sensing applications. *J. Mater. Chem.* **2012**, *22*, 11694–11700. [CrossRef]

92. Duan, X.; Mei, L.; Ma, J.; Li, Q.; Wang, T.; Zheng, W. Facet-induced formation of hematite mesocrystals with improved lithium storage properties. *Chem. Commun.* **2012**, *48*, 12204–12206. [CrossRef]

93. Fei, X.; Li, W.; Shao, Z.; Seeger, S.; Zhao, D.; Chen, X. Protein biomineralized nanoporous inorganic mesocrystals with tunable hierarchical nanostructures. *J. Am. Chem. Soc.* **2014**, *136*, 15781–15786. [CrossRef]

94. Cai, J.; Chen, S.; Ji, M.; Hu, J.; Ma, Y.; Qi, L. Organic additive-free synthesis of mesocrystalline hematite nanoplates via two-dimensional oriented attachment. *CrystEngComm* **2014**, *16*, 1553–1559. [CrossRef]

95. Agthe, M.; Plivelic, T.S.; Labrador, A.; Bergström, L.; Salazar-Alvarez, G. Following in real time the two-step assembly of nanoparticles into mesocrystals in levitating drops. *Nano Lett.* **2016**, *16*, 6838–6843. [CrossRef]

96. Liu, B.; Zeng, H.C. Mesoscale organization of CuO nanoribbons: Formation of "dandelions". *J. Am. Chem. Soc.* **2004**, *126*, 8124–8125. [CrossRef]

97. Yao, W.-T.; Yu, S.-H.; Zhou, Y.; Jiang, J.; Wu, Q.-S.; Zhang, L.; Jiang, J. Formation of uniform CuO nanorods by spontaneous aggregation: Selective synthesis of CuO, Cu$_2$O, and Cu nanoparticles by a solid-liquid phase arc discharge process. *J. Phys. Chem. B* **2005**, *109*, 14011–14016. [CrossRef]

98. Xu, M.; Wang, F.; Ding, B.; Song, X.; Fang, J. Electrochemical synthesis of leaf-like CuO mesocrystals and their lithium storage properties. *RSC Adv.* **2012**, *2*, 2240–2243. [CrossRef]

99. Jia, B.; Qin, M.; Zhang, Z.; Cao, Z.; Wu, H.; Chen, P.; Zhang, L.; Lu, X.; Qu, X. The formation of CuO porous mesocrystal ellipsoids via tuning the oriented attachment mechanism. *CrystEngComm* **2016**, *18*, 1376–1383. [CrossRef]

100. Zhang, J.; Cui, Y.; Qin, Q.; Zhang, G.; Luo, W.; Zheng, W. Nanoporous CuO mesocrystals: Low-temperature synthesis and improved structure-performance relationship for energy storage system. *Chem. Eng. J.* **2018**, *331*, 326–334. [CrossRef]

101. Hu, J.; Zou, C.; Su, Y.; Li, M.; Han, Y.; Kong, E.S.-W.; Yang, Z.; Zhang, Y. Ultrasensitive NO_2 gas sensor based on hierarchical Cu_2O/CuO mesocrystals nanoflower. *J. Mater. Chem. A* **2018**, *6*, 17120–17131. [CrossRef]

102. Zhao, J.; Tan, R.; Guo, Y.; Lu, Y.; Xu, W.; Song, W. SnO mesocrystals: Additive-free synthesis, oxidation, and top-down fabrication of quantum dots. *CrystEngComm* **2012**, *14*, 4575–4577. [CrossRef]

103. Chen, S.; Wang, M.; Ye, J.; Cai, J.; Ma, Y.; Zhou, H.; Qi, L. Kinetics-controlled growth of aligned mesocrystalline SnO_2 nanorod arrays for lithium-ion batteries with superior rate performance. *Nano Research* **2013**, *6*, 243–252. [CrossRef]

104. Liu, Y.; Zhu, G.; Ge, B.; Zhou, H.; Yuan, A.; Shen, X. Concave Co_3O_4 octahedral mesocrystal: Polymer-mediated synthesis and sensing properties. *CrystEngComm* **2012**, *14*, 6264–6270. [CrossRef]

105. Wang, F.; Lu, C.; Qin, Y.; Liang, C.; Zhao, M.; Yang, S.; Sun, Z.; Song, X. Solid state coalescence growth and electrochemical performance of plate-like Co_3O_4 mesocrystals as anode materials for lithium-ion batteries. *J. Power Sources* **2013**, *235*, 67–73. [CrossRef]

106. Su, D.; Dou, S.; Wang, G. Mesocrystal Co_3O_4 nanoplatelets as high capacity anode materials for Li-ion batteries. *Nano Res.* **2014**, *7*, 794–803. [CrossRef]

107. Hassen, D.; El-Safty, S.A.; Tsuchiya, K.; Chatterjee, A.; Elmarakbi, A.; Shenashen, M.A.; Sakai, M. Longitudinal hierarchy Co_3O_4 mesocrystals with high-dense exposure facets and anisotropic interfaces for direct-ethanol fuel cells. *Sci. Rep.* **2016**, *6*, 24330. [CrossRef]

108. Cao, W.; Wang, W.; Shi, H.; Wang, J.; Cao, M.; Liang, Y.; Zhu, M. Hierarchical three-dimensional flower-like Co_3O_4 architectures with a mesocrystal structure as high capacity anode materials for long-lived lithium-ion batteries. *Nano Res.* **2018**, *11*, 1437–1446. [CrossRef]

109. Fang, J.; Leufke, P.M.; Kruk, R.; Wang, D.; Scherer, T.; Hahn, H. External electric field driven 3D ordering architecture of silver (I) oxide meso-superstructures. *Nano Today* **2010**, *5*, 175–182. [CrossRef]

110. Belman, N.; Israelachvili, J.N.; Li, Y.; Safinya, C.R.; Ezersky, V.; Rabkin, A.; Sima, O.; Golan, Y. Hierarchical superstructure of alkylamine-coated ZnS nanoparticle assemblies. *Phys. Chem. Chem. Phys.* **2011**, *13*, 4974–4979. [CrossRef]

111. Querejeta-Fernandez, A.; Hernandez-Garrido, J.C.; Yang, H.; Zhou, Y.; Varela, A.; Parras, M.; Calvino-Gamez, J.J.; Gonzalez-Calbet, J.M.; Green, P.F.; Kotov, N.A. Unknown aspects of self-assembly of PbS microscale superstructures. *ACS Nano* **2012**, *6*, 3800–3812. [CrossRef]

112. Simon, P.; Rosseeva, E.; Baburin, I.A.; Liebscher, L.; Hickey, S.G.; Cardoso-Gil, R.; Eychmüller, A.; Kniep, R.; Carrillo-Cabrera, W. PbS-organic mesocrystals: The relationship between nanocrystal orientation and superlattice array. *Angew. Chem. Int. Ed.* **2012**, *51*, 10776–10781. [CrossRef]

113. Simon, P.; Bahrig, L.; Baburin, I.A.; Formanek, P.; Röder, F.; Sickmann, J.; Hickey, S.G.; Eychmüller, A.; Lichte, H.; Kniep, R.; et al. Interconnection of nanoparticles within 2D superlattices of PbS/oleic acid thin films. *Adv. Mater.* **2014**, *26*, 3042–3049. [CrossRef]

114. De la Rica, R.; Velders, A.H. Biomimetic crystallization of Ag_2S nanoclusters in nanopore assemblies. *J. Am. Chem. Soc.* **2011**, *133*, 2875–2877. [CrossRef] [PubMed]

115. Nagaoka, Y.; Chen, O.; Wang, Z.; Cao, Y.C. Structural control of nanocrystal superlattices using organic guest molecules. *J. Am. Chem. Soc.* **2012**, *134*, 2868–2871. [CrossRef] [PubMed]

116. Soejima, T.; Kimizuka, N. One-pot room-temperature synthesis of single-crystalline gold nanocorolla in water. *J. Am. Chem. Soc.* **2009**, *131*, 14407–14412. [CrossRef]

117. Fang, J.; Du, S.; Lebedkin, S.; Li, Z.; Kruk, R.; Kappes, M.; Hahn, H. Gold mesostructures with tailored surface topography and their self-assembly arrays for surface-enhanced raman spectroscopy. *Nano Lett.* **2010**, *10*, 5006–5013. [CrossRef]

118. You, H.; Ji, Y.; Wang, L.; Yang, S.; Yang, Z.; Fang, J.; Song, X.; Ding, B. Interface synthesis of gold mesocrystals with highly roughened surfaces for surface-enhanced Raman spectroscopy. *J. Mater. Chem.* **2012**, *22*, 1998–2006. [CrossRef]

119. Fang, J.; Ding, B.; Song, X. Self-assembly mechanism of platelike silver mesocrystal. *Appl. Phys. Lett.* **2007**, *91*, 083108. [CrossRef]

120. Cao, Y.; Fan, J.; Bai, L.; Hu, P.; Yang, G.; Yuan, F.; Chen, Y. Formation of cubic Cu mesocrystals by a solvothermal reaction. *CrystEngComm* **2010**, *12*, 3894–3899. [CrossRef]

121. Li, T.; You, H.; Xu, M.; Song, X.; Fang, J. Electrocatalytic properties of hollow coral-like platinum mesocrystals. *ACS Appl. Mater. Interfaces* **2012**, *4*, 6942–6948. [CrossRef]

122. Zhong, P.; Liu, H.; Zhang, J.; Yin, Y.; Gao, C. Controlled Synthesis of octahedral platinum-based mesocrystals by oriented aggregation. *Chem. Eur. J.* **2017**, *23*, 6803–6810. [CrossRef] [PubMed]

123. Huang, X.; Tang, S.; Yang, J.; Tan, Y.; Zheng, N. Etching growth under surface confinement: An effective strategy to prepare mesocrystalline Pd nanocorolla. *J. Am. Chem. Soc.* **2011**, *133*, 15946–15949. [CrossRef] [PubMed]

124. Cai, J.; Qi, L. TiO_2 mesocrystals: Synthesis, formation mechanisms and applications. *Sci. China Chem.* **2012**, *55*, 2318–2326. [CrossRef]

125. Hong, Z.; Wei, M. Recent progress in preparation and lithium-ion storage properties of TiO_2 mesocrystals. *J. Chin. Chem. Soc.* **2015**, *62*, 209–216. [CrossRef]

126. Zhang, P.; Tachikawa, T.; Fujitsuka, M.; Majima, T. The development of functional mesocrystals for energy harvesting, storage, and conversion. *Chem. Eur. J.* **2018**, *24*, 6295–6307. [CrossRef]

127. Li, L.; Liu, C.-Y. Organic small molecule-assisted synthesis of high active TiO_2 rod-like mesocrystals. *CrystEngComm* **2010**, *12*, 2073–2078. [CrossRef]

128. Zhang, D.; Li, G.; Wang, F.; Yu, J.C. Green synthesis of a self-assembled rutile mesocrystalline photocatalyst. *CrystEngComm* **2010**, *12*, 1759–1763. [CrossRef]

129. Liu, X.; Gao, Y.; Cao, C.; Luo, H.; Wang, W. Highly crystalline spindle-shaped mesoporous anatase titania particles: Solution-phase synthesis, characterization, and photocatalytic properties. *Langmuir* **2010**, *26*, 7671–7674. [CrossRef]

130. Bian, Z.; Zhu, J.; Wen, J.; Cao, F.; Huo, Y.; Qian, X.; Cao, Y.; Shen, M.; Li, H.; Lu, Y. Single-crystal-like titania mesocages. *Angew. Chem. Int. Ed.* **2011**, *123*, 1137–1140. [CrossRef]

131. Da Silva, R.O.; Gonçalves, R.H.; Stroppa, D.G.; Ramirez, A.J.; Leite, E.R. Synthesis of recrystallized anatase TiO_2 mesocrystals with Wulff shape assisted by oriented attachment. *Nanoscale* **2011**, *3*, 1910–1916. [CrossRef]

132. Tartaj, P. Sub-100 nm TiO_2 mesocrystalline assemblies with mesopores: Preparation, characterization, enzyme immobilization and photocatalytic properties. *Chem. Commun.* **2011**, *47*, 256–258. [CrossRef]

133. Tartaj, P.; Amarilla, J.M. Multifunctional response of anatase nanostructures based on 25 nm mesocrystal-like porous assemblies. *Adv. Mater.* **2011**, *23*, 4904–4907. [CrossRef]

134. Jiao, W.; Wang, L.; Liu, G.; Lu, G.Q.; Cheng, H.-M. Hollow anatase TiO_2 single crystals and mesocrystals with dominant {101} facets for improved photocatalysis activity and tuned reaction preference. *ACS Catal.* **2012**, *2*, 1854–1859. [CrossRef]

135. Bian, Z.; Tachikawa, T.; Majima, T. Superstructure of TiO_2 crystalline nanoparticles yields effective conduction pathways for photogenerated charges. *J. Phys. Chem. Lett.* **2012**, *3*, 1422–1427. [CrossRef]

136. Chen, Q.; Ma, W.; Chen, C.; Ji, H.; Zhao, J. Anatase TiO_2 mesocrystals enclosed by (001) and (101) facets: Synergistic effects between Ti^{3+} and facets for their photocatalytic performance. *Chem. Eur. J.* **2012**, *18*, 12584–12589. [CrossRef]

137. Liu, Y.; Zhang, Y.; Li, H.; Wang, J. Manipulating the formation of NH_4TiOF_3 mesocrystals: Effects of temperature, surfactant, and pH. *Cryst. Growth Des.* **2012**, *12*, 2625–2633. [CrossRef]

138. Aoyama, Y.; Oaki, Y.; Ise, R.; Imai, H. Mesocrystal nanosheet of rutile TiO_2 and its reaction selectivity as a photocatalyst. *CrystEngComm* **2012**, *14*, 1405–1411. [CrossRef]

139. Zhou, L.; Chen, J.; Ji, C.; Zhou, L.; O'Brien, P. A facile solid phase reaction to prepare TiO_2 mesocrystals with exposed {001} facets and high photocatalytic activity. *CrystEngComm* **2013**, *15*, 5012–5015. [CrossRef]

140. Yao, X.; Liu, X.; Liu, T.; Wang, K.; Lu, L. One-step and large-scale synthesis of anatase TiO_2 mesocrystals along [001] orientation with enhanced photocatalytic performance. *CrystEngComm* **2013**, *15*, 10246–10254. [CrossRef]

141. Guo, Y.; Li, H.; Chen, J.; Wu, X.; Zhou, L. TiO$_2$ mesocrystals built of nanocrystals with exposed {001} facets: Facile synthesis and superior photocatalytic ability. *J. Mater. Chem. A* **2014**, *2*, 19589–19593. [CrossRef]

142. Chen, J.; Li, G.; Zhang, H.; Liu, P.; Zhao, H.; An, T. Anatase TiO$_2$ mesocrystals with exposed (001) surface for enhanced photocatalytic decomposition capability toward gaseous styrene. *Catal. Today* **2014**, *224*, 216–224. [CrossRef]

143. Fang, Z.; Long, L.; Hao, S.; Song, Y.; Qiang, T.; Geng, B. Mesocrystal precursor transformation strategy for synthesizing ordered hierarchical hollow TiO$_2$ nanobricks with enhanced photocatalytic property. *CrystEngComm* **2014**, *16*, 2061–2069. [CrossRef]

144. Lai, L.-L.; Huang, L.-L.; Wu, J.-M. K$_2$TiO(C$_2$O$_4$)$_2$-mediated synthesis of rutile TiO$_2$ mesocrystals and their ability to assist photodegradation of sulfosalicylic acid in water. *RSC Adv.* **2014**, *4*, 49280–49286. [CrossRef]

145. Zhang, P.; Tachikawa, T.; Bian, Z.; Majima, T. Selective photoredox activity on specific facet-dominated TiO$_2$ mesocrystal superstructures incubated with directed nanocrystals. *Appl. Catal. B* **2015**, *176–177*, 678–686. [CrossRef]

146. Hu, D.; Zhang, W.; Tanaka, Y.; Kusunose, N.; Peng, Y.; Feng, Q. Mesocrystalline nanocomposites of TiO$_2$ polymorphs: Topochemical mesocrystal conversion, characterization, and photocatalytic response. *Cryst. Growth Des.* **2015**, *15*, 1214–1225. [CrossRef]

147. Fu, X.X.; Ren, Z.M.; Fan, C.Y.; Sun, C.X.; Shi, L.; Yu, S.Q.; Qian, G.D.; Wang, Z.Y. Designed fabrication of anatase mesocrystals constructed from crystallographically oriented nanocrystals for improved photocatalytic activity. *RSC Adv.* **2015**, *5*, 41218–41223. [CrossRef]

148. Lai, L.-L.; Wu, J.-M. Hollow TiO$_2$ microspheres assembled with rutile mesocrystals: Low-temperature one-pot synthesis and the photocatalytic performance. *Ceram. Int.* **2015**, *41*, 12317–12322. [CrossRef]

149. Fu, X.; Fan, C.; Yu, S.; Shi, L.; Wang, Z. TiO$_2$ mesocrystals with exposed {001} facets as efficient photocatalysts. *J. Alloys Compd.* **2016**, *680*, 80–86. [CrossRef]

150. Tan, B.; Zhang, X.; Li, Y.; Chen, H.; Ye, X.; Wang, Y.; Ye, J. Anatase TiO$_2$ mesocrystals: Green synthesis, in situ conversion to porous single crystals, and self-doping Ti^{3+} for enhanced visible light driven photocatalytic removal of NO. *Chem. Eur. J.* **2017**, *23*, 5478–5487. [CrossRef]

151. Tang, C.; Liu, L.; Li, Y.; Bian, Z. Aerosol spray assisted assembly of TiO$_2$ mesocrystals into hierarchical hollow microspheres with enhanced photocatalytic performance. *Appl. Catal. B* **2017**, *201*, 41–47. [CrossRef]

152. Wang, H.; Chen, Q.; Luan, Q.; Duan, R.; Guan, R.; Cao, X.; Hu, X. Photocatalytic properties dependent on the interfacial defects of intergrains within TiO$_2$ mesocrystals. *Chem. Eur. J.* **2018**, *24*, 17105–17116. [CrossRef] [PubMed]

153. Hong, Z.; Dai, H.; Huang, Z.; Wei, M. Understanding the growth and photoelectrochemical properties of mesocrystals and single crystals: A case of anatase TiO$_2$. *Phys. Chem. Chem. Phys.* **2014**, *16*, 7441–7447. [CrossRef] [PubMed]

154. Zhang, Y.; Cai, J.; Ma, Y.; Qi, L. Mesocrystalline TiO$_2$ nanosheet arrays with exposed {001} facets: Synthesis via topotactic transformation and applications in dye-sensitized solar cells. *Nano Res.* **2017**, *10*, 2610–2625. [CrossRef]

155. Liu, B.; Zeng, H.C. Carbon nanotubes supported mesoporous mesocrystals of anatase TiO$_2$. *Chem. Mater.* **2008**, *20*, 2711–2718. [CrossRef]

156. Li, N.; Liu, G.; Zhen, C.; Li, F.; Zhang, L.; Cheng, H.-M. Battery performance and photocatalytic activity of mesoporous anatase TiO$_2$ nanospheres/graphene composites by template-free self-assembly. *Adv. Funct. Mater.* **2011**, *21*, 1717–1722. [CrossRef]

157. Zhang, W.; Shen, D.; Liu, Z.; Wu, N.-L.; Wei, M. Brookite TiO$_2$ mesocrystals with enhanced lithium-ion intercalation properties. *Chem. Commun.* **2018**, *54*, 11491–11494. [CrossRef]

158. Elbanna, O.; Zhang, P.; Fujitsuka, M.; Majima, T. Facile preparation of nitrogen and fluorine codoped TiO$_2$ mesocrystal with visible light photocatalytic activity. *Appl. Catal. B* **2016**, *192*, 80–87. [CrossRef]

159. Zhang, P.; Fujitsuka, M.; Majima, T. TiO$_2$ mesocrystal with nitrogen and fluorine codoping during topochemical transformation: Efficient visible light induced photocatalyst with the codopants. *Appl. Catal. B* **2016**, *185*, 181–188. [CrossRef]

160. Primc, D.; Niederberger, M. Synthesis and formation mechanism of multicomponent Sb-Nb:TiO$_2$ mesocrystals. *Chem. Mater.* **2017**, *29*, 10113–10121. [CrossRef]

161. Lan, T.; Zhang, W.; Wu, N.-L.; Wei, M. Nb-doped rutile TiO$_2$ mesocrystals with enhanced lithium storage properties for lithium ion battery. *Chem. Eur. J.* **2017**, *23*, 5059–5065. [CrossRef]

162. Naldoni, A.; Allieta, M.; Santangelo, S.; Marelli, M.; Fabbri, F.; Cappelli, S.; Bianchi, C.L.; Psaro, R.; Santo, V.D. Effect of nature and location of defects on bandgap narrowing in black TiO_2 nanoparticles. *J. Am. Chem. Soc.* **2012**, *134*, 7600–7603. [CrossRef] [PubMed]

163. Pan, X.; Yang, M.-Q.; Fu, X.; Zhang, N.; Xu, Y.-J. Defective TiO_2 with oxygen vacancies: Synthesis, properties and photocatalytic applications. *Nanoscale* **2013**, *5*, 3601–3614. [CrossRef] [PubMed]

164. Chen, X.; Liu, L.; Huang, F. Black titanium dioxide (TiO_2) nanomaterials. *Chem. Soc. Rev.* **2015**, *44*, 1861–1885. [CrossRef]

165. Ullattil, S.G.; Narendranath, S.B.; Pillai, S.C.; Periyat, P. Black TiO_2 nanomaterials: A review of recent advances. *Chem. Eng. J.* **2018**, *343*, 708–736. [CrossRef]

166. Zhou, W.; Fu, H. Defect-mediated electron-hole separation in semiconductor photocatalysis. *Inorg. Chem. Front.* **2018**, *5*, 1240–1254. [CrossRef]

167. Chen, X.; Liu, L.; Yu, P.Y.; Mao, S.S. Increasing solar absorption for photocatalysis with black hydrogenated titanium dioxide nanocrystals. *Science* **2011**, *331*, 746–750. [CrossRef] [PubMed]

168. Elbanna, O.; Fujitsuka, M.; Kim, S.; Majima, T. Charge carrier dynamics in TiO_2 mesocrystals with oxygen vacancies for photocatalytic hydrogen generation under solar light irradiation. *J. Phys. Chem. C* **2018**, *122*, 15163–15170. [CrossRef]

169. Bian, Z.; Tachikawa, T.; Kim, W.; Choi, W.; Majima, T. Superior electron transport and photocatalytic abilities of metal-nanoparticle-loaded TiO_2 superstructures. *J. Phys. Chem. C* **2012**, *116*, 25444–25453. [CrossRef]

170. Gao, P.; Liu, J.; Zhang, T.; Sun, D.D.; Ng, W. Hierarchical TiO_2/CdS "spindle-like" composite with high photodegradation and antibacterial capability under visible light irradiation. *J. Hazard. Mater.* **2012**, *229–230*, 209–216. [CrossRef]

171. Yang, X.; Qin, J.; Li, Y.; Zhang, R.; Tang, H. Graphene-spindle shaped TiO_2 mesocrystal composites: Facile synthesis and enhanced visible light photocatalytic performance. *J. Hazard. Mater.* **2013**, *261*, 342–350. [CrossRef]

172. Bian, Z.; Tachikawa, T.; Zhang, P.; Fujitsuka, M.; Majima, T. Au/TiO_2 superstructure-based plasmonic photocatalysts exhibiting efficient charge separation and unprecedented activity. *J. Am. Chem. Soc.* **2014**, *136*, 458–465. [CrossRef] [PubMed]

173. Tachikawa, T.; Zhang, P.; Bian, Z.; Majima, T. Efficient charge separation and photooxidation on cobalt phosphate-loaded TiO_2 mesocrystal superstructures. *J. Mater. Chem. A* **2014**, *2*, 3381–3388. [CrossRef]

174. Zhang, P.; Tachikawa, T.; Fujitsuka, M.; Majima, T. Efficient charge separation on 3D architectures of TiO_2 mesocrystals packed with a chemically exfoliated MoS_2 shell in synergetic hydrogen evolution. *Chem. Commun.* **2015**, *51*, 7187–7190. [CrossRef]

175. Li, X.; Wang, J.; Men, Y.; Bian, Z. TiO_2 mesocrystal with exposed (001) facets and CdS quantum dots as an active visible photocatalyst for selective oxidation reactions. *Appl. Catal. B* **2016**, *187*, 115–121. [CrossRef]

176. Han, T.; Wang, H.; Zheng, X. Gold nanoparticle incorporation into nanoporous anatase TiO_2 mesocrystal using a simple deposition-precipitation method for photocatalytic applications. *RSC Adv.* **2016**, *6*, 7829–7837. [CrossRef]

177. Yan, D.; Liu, Y.; Liu, C.-Y.; Zhang, Z.-Y.; Niea, S.-D. Multi-component in situ and in-step formation of visible-light response C-dots composite TiO_2 mesocrystals. *RSC Adv.* **2016**, *6*, 14306–14313. [CrossRef]

178. Tang, H.; Chang, S.; Jiang, L.; Tang, G.; Liang, W. Novel spindle-shaped nanoporous TiO_2 coupled graphitic g-C_3N_4 nanosheets with enhanced visible-light photocatalytic activity. *Ceram. Int.* **2016**, *42*, 18443–18452. [CrossRef]

179. Elbanna, O.; Kim, S.; Fujitsuka, M.; Majima, T. TiO_2 mesocrystals composited with gold nanorods for highly efficient visible-NIR-photocatalytic hydrogen production. *Nano Energy* **2017**, *35*, 1–8. [CrossRef]

180. Elbanna, O.; Fujitsuka, M.; Majima, T. g-C_3N_4/TiO_2 mesocrystals composite for H_2 evolution under visible-light irradiation and its charge carrier dynamics. *ACS Appl. Mater. Interfaces* **2017**, *9*, 34844–34854. [CrossRef]

181. Yu, X.; Fan, X.; An, L.; Liu, G.; Li, Z.; Liu, J.; Hu, P.A. Mesocrystalline Ti^{3+}-TiO_2 hybridized g-C_3N_4 for efficient visible-light photocatalysis. *Carbon* **2018**, *128*, 21–30. [CrossRef]

182. Xue, J.; Elbanna, O.; Kim, S.; Fujitsuka, M.; Majima, T. Defect state-induced efficient hot electron transfer in Au nanoparticles/reduced TiO_2 mesocrystal photocatalysts. *Chem. Commun.* **2018**, *54*, 6052–6055. [CrossRef]

183. Tan, B.; Ye, X.; Li, Y.; Ma, X.; Wang, Y.; Ye, J. Defective anatase TiO_{2-x} mesocrystal growth in situ on g-C_3N_4 nanosheets: Construction of 3D/2D Z-scheme heterostructures for highly efficient visible-light photocatalysis. *Chem. Eur. J.* **2018**, *24*, 13311–13321. [CrossRef]

184. Chen, F.; Cao, F.; Li, H.; Bian, Z. Exploring the important role of nanocrystals orientation in TiO_2 superstructure on photocatalytic performances. *Langmuir* **2015**, *31*, 3494–3499. [CrossRef]

Article

Adsorption and Photocatalytic Decomposition of Gaseous 2-Propanol Using TiO$_2$-Coated Porous Glass Fiber Cloth

Sayaka Yanagida [1,*], Kentaro Hirayama [2], Kenichiro Iwasaki [2,3] and Atsuo Yasumori [2,3]

[1]   Center for Crystal Science and Technology, University of Yamanashi, 7-32 Miyamae, Kofu 400-8511, Japan
[2]   Department of Material Science and Technology, Tokyo University of Science, 6-3-1 Niijuku, Katsushika-ku, Tokyo 125-8585, Japan; kentaro1218hr@gmail.com (K.H.); iwasaki@rs.tus.ac.jp (K.I.); yasumori@rs.noda.tus.ac.jp (A.Y.)
[3]   Photocatalysis International Research Center, Research Institute for Science and Technology, Tokyo University of Science, 2641 Yamazaki, Noda-shi, Chiba 278-8510, Japan
*   Correspondence: syanagida@yamanashi.ac.jp; Tel.: +81-55-220-8723

Received: 5 December 2018; Accepted: 4 January 2019; Published: 14 January 2019

Abstract: Combinations of TiO$_2$ photocatalysts and various adsorbents have been extensively investigated for eliminating volatile organic compounds (VOCs) at low concentrations. Herein, TiO$_2$ and porous glass cloth composites were prepared by acid leaching and subsequent TiO$_2$ dip-coating of the electrically applied glass (E-glass) cloth, and its adsorption and photocatalytic ability were investigated. Acid leaching increased the specific surface area of the E-glass cloth from 1 to 430 m^2/g while maintaining sufficient mechanical strength for supporting TiO$_2$. Further, the specific surface area remained large (290 m^2/g) after TiO$_2$ coating. In the photocatalytic decomposition of gaseous 2-propanol, the TiO$_2$-coated porous glass cloth exhibited higher adsorption and photocatalytic decomposition ability than those exhibited by the TiO$_2$-coated, non-porous glass cloth. The porous composite limited desorption of acetone, which is a decomposition intermediate of 2-propanol, until 2-propanol was completely decomposed to CO$_2$. The CO$_2$ generation rate was affected by the temperature condition (15 or 35 °C) and the water content (2 or 18 mg/L); the latter also influenced 2-propanol adsorption in photocatalytic decomposition. Both the conditions may change the diffusion and adsorption behavior of 2-propanol in the porous composite. As demonstrated by its high adsorption and photocatalytic ability, the composite (TiO$_2$ and porous glass cloth) effectively eliminates VOCs, while decreasing the emission of harmful intermediates.

Keywords: photocatalyst; microporous material; composite; adsorption; air purification; TiO$_2$; porous glass

1. Introduction

Indoor air pollution by volatile organic compounds (VOCs) is considered to be a serious health problem. Construction materials emit various VOCs, including formaldehyde, acetaldehyde, toluene, xylene, hexane, acetone, and 2-propanol, which can cause the sick building syndrome even at VOC concentrations lower than 1 ppm [1]. Further, the VOC concentrations in the work spaces are strictly limited to prevent health damage from prolonged exposure. However, the imposed concentration limits of the major solvents are in the range of 10–10^3 ppm [2]. Titanium dioxide (TiO$_2$) photocatalysts have been extensively investigated in VOC decomposition because of their strong oxidation abilities under ultraviolet (UV) irradiation and because of their ability to decompose various organic compounds [3–6]. However, the VOC decomposition rates are observed to be insufficient at considerably low concentrations because the photocatalytic decomposition reaction tends to be diffusion controlled under such conditions [7].

A promising solution to this problem is to combine TiO_2 with an appropriate adsorbate material. Further, the proposed mechanism of VOC elimination by composites can be given as follows. The concentration of VOC molecules can be obtained based on the adherence of VOC molecules in bulk air to the adsorbate surface. The VOC molecules in bulk air are concentrated on the adsorbate surface by adsorption. The adsorbed molecules are diffused to the vicinity of the TiO_2 surface and are photo-catalytically decomposed by TiO_2 [8,9]. The adsorbate is a porous material, including activated carbon [9,10], zeolite [9,11,12], silica gel [13], or mesoporous silica [14,15]. The composites require a free-standing structure for maintaining sufficient mechanical strength in practical applications. Further, because it is generally difficult to prepare the composites from porous adsorbates, they are used in powder form or are supported on another substrate [12]. Porous glass is an adsorbate candidate exhibiting sufficient mechanical strength for TiO_2 support. Thus, a composite photocatalyst of TiO_2 and porous glass exhibits superior photocatalytic activity to powder TiO_2, or non-porous composite photocatalyst in case of gaseous and aqueous systems [16–19].

The electrically applied glass (E-glass) fiber is a mass-produced glass fiber with low alkaline metal content. E-glass fiber and its cloth are primarily used to form the fiber-reinforced plastics. The typical composition of E-glass fiber is 52–62 wt % SiO_2, 16–25 wt % CaO, 12–16 wt % Al_2O_3, 0–10 wt % B_2O_3, and 0–5 wt % MgO, along with small amounts of alkaline metal oxides (TiO_2 and Fe_2O_3) and fluorides [20]. When the E-glass fiber is corroded in an acid solution [21,22], a porous structure is formed. The pores are formed by the dissolution of Al_2O_3, B_2O_3, and the ions of alkaline metals and alkali earth metals, leaving the SiO_2 structure. The acid-leached E-glass fiber and cloth are the porous materials that had been investigated in previously conducted studies [23–25]. The properties of porous glass cloth, such as the high adsorption capacity for VOCs, high optical transparency, low weight, and sufficient mechanical strength, are considered to be advantageous for photocatalyst support. Very low alkali metal content is another advantage because the diffusion of alkaline metal ions from the glass substrate to the TiO_2 photocatalyst deteriorates the photocatalytic activity of TiO_2 [26,27]. Kitamura et al. prepared air filters by combining the TiO_2 photocatalyst and the porous glass fiber cloth that was prepared by acid leaching [25]. They further examined the filter's ability to eliminate gaseous formaldehyde; however, they did not measure the concentration of the decomposition products. Therefore, they could not discuss the contributions of adsorption and photocatalytic decomposition in the elimination process.

In this study, a composite material was fabricated from the TiO_2 photocatalyst and porous glass fiber cloth; further, its adsorption and photocatalytic decomposition ability has been examined with respect to gaseous air-diluted 2-propanol. The TiO_2–porous glass cloth composite was prepared by acid leaching the glass cloth and by subsequently dip-coating it using TiO_2. The pore formation process was investigated by observing the microstructure, analyzing the chemical composition, and measuring the N_2 adsorption of the acid-leached glass cloth. Further, the effects of TiO_2 loading on the composite properties were also investigated. Subsequently, the adsorption and photocatalytic properties of the TiO_2-coated porous glass cloth were investigated using 300 ppm of gaseous 2-propanol. The concentration changes in the 2-propanol, acetone as an intermediate oxidative product of 2-propanol, and CO_2 as a final product of 2-propanol decomposition were measured during the adsorption and photocatalytic decomposition. Further, the effects of temperature and water content on the adsorption and photocatalytic decomposition abilities of the composite were also investigated because these factors affect the photocatalytic decomposition rates of 2-propanol [28–31] and acetone [32,33], and because they may affect the adsorption and diffusion of organic molecules in the porous glass. Finally, based on the experimental results, we discuss the adsorption and photocatalytic decomposition mechanisms of the TiO_2-coated porous glass cloth.

2. Results and Discussion

2.1. Microstructure Analysis and Mechanical Strength Test

The micro-structural change of the E-glass cloth that was prepared by acid leaching and TiO_2 coating was investigated using field emission scanning electron microscopy (FE–SEM). Hereafter, the unleached and leached samples are referred to as the non-porous and porous glass samples, respectively. Panels (a–c) of Figure 1 denote the cross-sectional images of the non-porous and porous glass fibers comprising cloth samples. After leaching for 1 h, the structure of the glass fibers changed from uniform and dense with no pores (Figure 1a), to a dense core surrounded by a porous surface layer with a thickness of 2–3 μm (Figure 1b). After 3 h of leaching, the glass fiber was observed to become completely porous. The SEM images indicate that the porous structure was formed from the surface to the center as the leaching time increased, as previously reported by Tanaka et al. [24]. Panels (d–f) of Figure 1 denote the side views of the non-porous and porous glass fibers. A larger number of cracks developed in the porous fibers when compared to that in the non-porous fibers. The cracks that were observed on the 1-h leached fibers were mainly observed to be perpendicular to the fiber elongation direction and propagated only in the porous layer. This indicated that the cracks were generated by the tensile stress in the porous layer. During the drying process, the generation of strong capillary forces in the porous structure could shrink the porous surface layer. The different shrinkage ratios between the core and the porous layer could further result in crack generation. In contrast, the cracks on the 3-h leached fibers progressed along the fiber elongation direction and near the fiber center. Further, the mechanism of crack generation differed between the samples that were prepared after 1 and 3 h of leaching. The residual stress that was derived from the fiber spinning process may explain this type of crack generation, as will be discussed in our subsequent report. Despite the generation of clacking by acid leaching, the flexibility of the porous glass cloth was maintained at a high level. When the 3-h leached glass cloth was curved with a curvature radius of 3 mm, the glass cloth did not break; further, its form was recovered without any deformation. Panels (g–i) of Figure 1 denote the side views of the TiO_2-coated fibers. Agglomerates of the TiO_2 nanoparticles are observed on the surfaces, and the coating is observed to be in-homogeneous. Because the glass cloth samples have high fiber density, homogeneous coating of each fiber by dip-coating the cloth sample would be difficult. Further, the sampling process before FE–SEM observation can cause some damage to the TiO_2 coating. Figure S1 depicts the surface and fracture cross-section of TiO_2 coating on the porous glass cloth at a position where plenty of TiO_2 is loaded and where only little sample damage is observed. The TiO_2 coating exhibited a porous structure, and VOCs can penetrate the coating.

2.2. Crystalline Phase of TiO₂

The TiO_2-coated glass cloth samples, and the TiO_2 powder sample that was prepared from TiO_2 suspension used in the dip-coating process, were ground with a pestle in an agate mortar, and their X-ray diffraction (XRD) patterns were measured. Figure 2 depicts the XRD patterns of the aforementioned samples. The TiO_2 suspension comprised anatase (TiO_2) nanoparticles and solvent; therefore, the TiO_2 powder sample prepared from the TiO_2 suspension exhibited anatase peaks (Figure 2a). The crystalline size that was estimated from Scherrer's equation using 101 reflection is 6.9 nm. The TiO_2-coated non-porous glass cloth exhibited no anatase peak (Figure 2b). This can be attributed to the lesser amounts of TiO_2, and will be explained in the following elemental analysis in Section 2.3. The broad peaks that can be observed at around 26° denote the glass halo, whereas the peaks that can be observed at around 42° denote the graphite oxide generated by incomplete organic molecule combustion. The TiO_2-coated 3-h leached porous glass cloth did not exhibit the strongest 101 reflection of anatase because it was overwrapped with glass halo at around 24° (Figure 2c). However, the 220, 105, 211, and 204 reflections of anatase can be clearly observed. Based on these results, the TiO_2 on the porous sample was observed to be in the anatase form. Further, the TiO_2 on the non-porous sample was also in the anatase form because it was prepared in the same manner as the porous sample.

It indicates that the use of black light lump (center wavelength: 365 nm), in the following photocatalytic decomposition experiments, is suitable for the photo-excitation of this composite because anatase exhibits a bandgap of 3.2 eV (388 nm) [34].

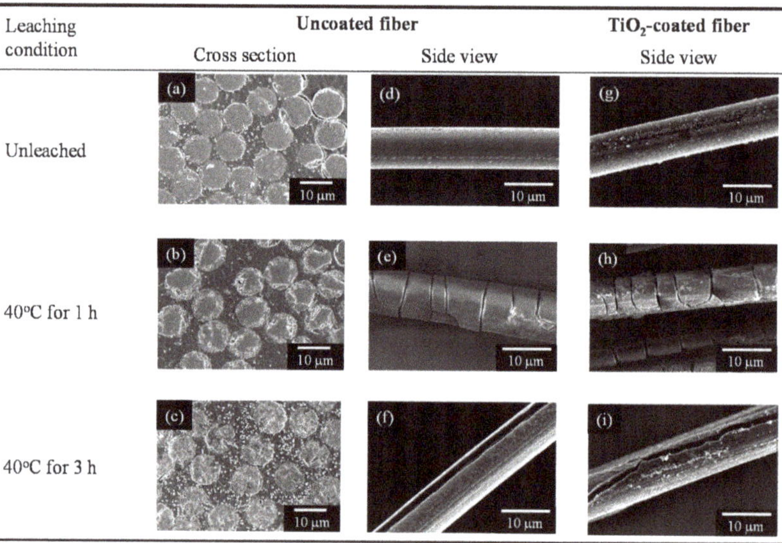

Figure 1. SEM images of the non-porous and porous glass fibers before (left and center panels) and after (right panel) TiO$_2$ coating.

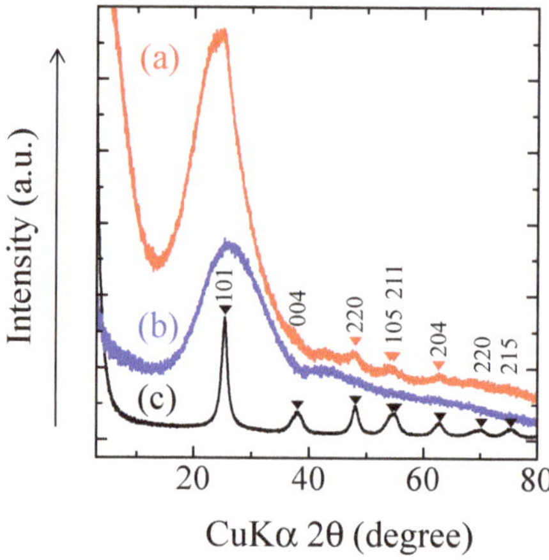

Figure 2. Powder X-Ray diffraction (XRD) pattern of the ground samples: (**a**) dried and heated TiO$_2$ suspension; (**b**) TiO$_2$-coated non-porous glass cloth; and (**c**) TiO$_2$-coated porous glass cloth. *hkl* are reflection indices of anatase.

2.3. Chemical Composition Analysis

Figure 3 depicts the chemical compositions of the non-porous and porous glass cloths that have been estimated using the X-ray fluorescence (XRF) measurements. In this figure, the porous cloths were obtained after leaching for 1, 3, and 12 h. While estimating the compositions of the 3- and 12-h leached glass cloths, sodium oxide was assumed to be zero because the sample pellets were prepared using Na$_2$CO$_3$. With an increase in leaching time, the composition ratios of the alkali earth metal oxides, boron oxide, alkali metal oxides, and alumina decreased, whereas the silica content was maintained to be almost constant. The leaching of the non-silica components from the glass fiber resulted in the formation of the observed microporous structures in the glass cloths Panels (b,c,e,f) of Figure 1. The composition almost ceased to change after 3 h of leaching (the time of core disappearance in the acid-leached fibers).

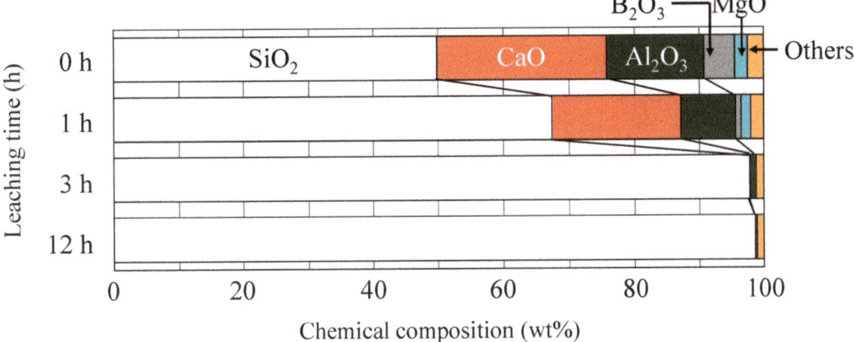

Figure 3. Chemical compositions of the non-porous (0-h leaching time) and porous glass cloths estimated from the X-ray fluorescence (XRF) measurements. The "Others" category includes TiO$_2$, Na$_2$O, K$_2$O, Fe$_2$O$_3$, and SrO.

Further, the TiO$_2$ contents in the TiO$_2$-coated non-porous and porous glass cloths were evaluated using XRF. Figure 4 plots the TiO$_2$ content and weight loss of the glass cloth as functions of the acid leaching time. The weight loss of the glass cloth was calculated as follows:

$$\text{Weight loss } (\%) = \frac{Weight\ change\ of\ glass\ cloth\ after\ acid\ leaching}{Weight\ of\ glass\ cloth\ before\ leaching} \times 100$$

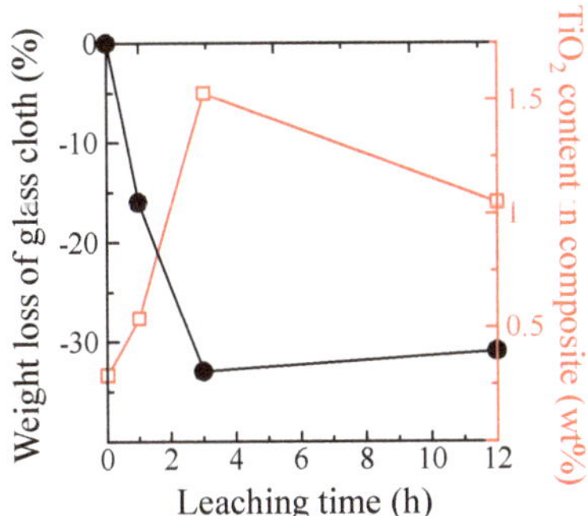

Figure 4. TiO$_2$ content in the TiO$_2$-coated porous glass cloth (open squares) and weight loss of the porous glass cloth (filled circles) versus the acid leaching time.

Both the TiO$_2$ loading amount and weight loss of the glass cloth increased at the maximum leaching time of 3 h. Even if the TiO$_2$ loading amount was unaffected by the leaching time, the TiO$_2$ composition ratio was observed to increase because of the weight loss of the glass cloth. However, the increase in the TiO$_2$ composition ratio exceeded the expected value corresponding to this weight loss, demonstrating the remaining factors that were responsible for this increase in TiO$_2$ composition ratio. As will be discussed subsequently, the TiO$_2$ particles cannot penetrate the pores in the glass fiber structure; therefore, the increased volume of pores cannot be explained by the increased loading amount of TiO_2. However, the large cracks that are observed in the SEM images (Figure 1e,f) are considered to be the likely support sites of the TiO_2 particles in the dip-coating process. Further, the pore formation can also increase the TiO_2 content by increasing the surface roughness or the wettability of the glass surface. In contrast, the increase in leaching time from 3 to 12 h decreases the TiO_2 amount. However, the factors that can affect the TiO_2 loading, weight loss, cracking, and surface conditions were not significantly altered in this time range. Therefore, the loading amount of TiO_2 was observed to be saturated rather than decreased.

2.4. N_2 Adsorption

Figure 5 depicts the N_2 adsorption and desorption isotherms of the non-porous and the 3-h leached porous glass cloths before and after TiO_2 coating. The non-porous samples (Figure 5a) yielded a type III adsorption isotherm, confirming the absence of micrometer- or nanometer-sized pores [35]. However, the porous samples (Figure 5b) yielded a type I adsorption isotherm, indicating the existence of nanometer-sized pores (<2 nm) [35]. The sample that was prepared in an extended leaching time (12 h) yielded a type I adsorption isotherm, indicating that the pore size will not be significantly changed by prolonging the leaching time. Figure 6 plots the specific surface areas (Sg) of the TiO_2-coated and uncoated glass cloth samples versus the acid leaching time. The Sg values of both the samples increased up to approximately 4 h of leaching time and remained almost constant thereafter, indicating that the microporous structure was completely formed after 4 h of leaching. However, in the SEM images, the fiber core disappeared after 3 h of leaching (Figure 1c), implying that the microporous structure continued to evolve after the fiber center was subjected to acid leaching. After the disappearance of the core, Sg was probably increased by the leaching of the non-silica components that remained in the porous glass cloth and by the precipitation of silica gel in the pores. The latter process is typically observed during the leaching of the phase-separated borosilicate glasses [36,37].

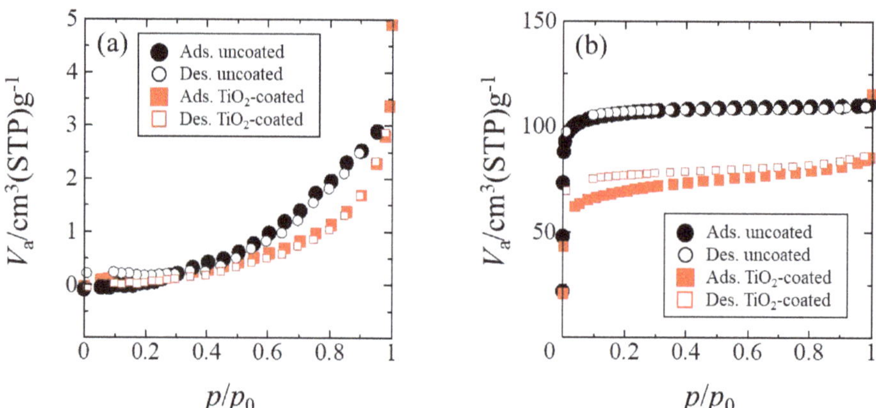

Figure 5. Adsorption (Ads.) and desorption (Des.) isotherms of (**a**) a non-porous glass cloth and (**b**) a porous glass cloth before and after TiO_2 coating.

As depicted in Figure 5b, the TiO_2 coating reduced the adsorbed/desorbed N_2 in the low pressure region or the isotherm. This decrease corresponds to the lower Sg value of the TiO_2-coated samples when compared to that of the uncoated samples depicted in Figure 5. This decrease cannot be ascribed to the filling of nanometer-sized pores by TiO_2 particles because even the primary particle size of TiO_2 (6 nm) is considerably larger than the estimated pore diameter (<2 nm). To elucidate the reason that the Sg decreased during the TiO_2 coating process, we investigated the manner in which heat treatment after dip-coating affected the Sg of the samples. The porous glass cloth that was prepared by 3 h of leaching was subjected to the same heat treatment as that used in the dip-coated samples

(300 °C for 2 h), and its measured Sg was compared with that of the samples obtained before heat treatment. The heat treatment decreased the Sg from 370 to 200 m^2/g, indicating that heating was mainly responsible for the loss of Sg in the TiO_2-coated sample. To verify the effects of coating and heating on the porous properties, we investigated the pore-size distribution in the samples. Figure S2 depicts the adsorption isotherms of the uncoated and TiO_2-coated porous glass cloths and their pore-size distributions estimated by Saito–Foley fitting [38,39]. The majority of the pores were observed to be less than 1 nm in diameter, and the TiO_2 coating and heating process reduced the volume of the large pores. The reduced Sg and enlarged pores in the heat-treated sample may have arisen from dehydration condensation of the silanol groups on the surface of the pores [40]. However, because the subsequent heat treatment ensured that a high specific surface area of the TiO_2-coated glass cloth was retained, the organic-molecule adsorption ability of the TiO_2-coated sample should not have been significantly degraded.

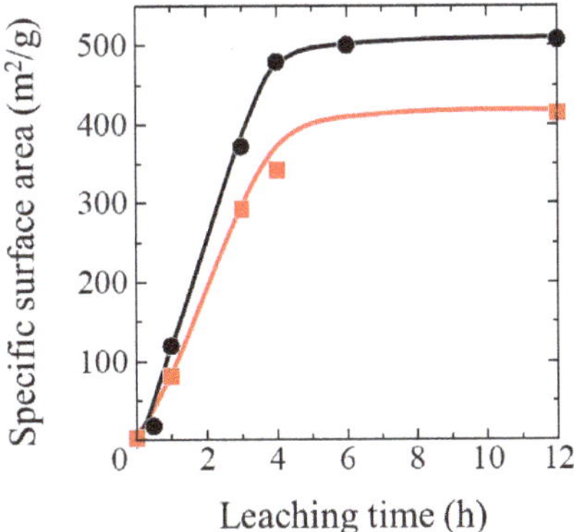

Figure 6. Specific surface area versus leaching time of the TiO_2-coated porous glass cloths (squares) and the uncoated porous glass cloth (circles).

2.5. Adsorption and Photocatalytic Ability of 2-Propanol

First, the adsorption and photocatalytic decomposition abilities of the TiO_2-coated porous glass cloth, that was prepared by 3 h of leaching, and the TiO_2-coated non-porous glass cloth were compared at 15 °C and a low water content (2 mg/L). Figure 7a plots the 2-propanol concentration versus time under dark conditions. The non-porous sample gradually reduced the concentration of gaseous 2-propanol during the initial 60 min, whereas the porous sample rapidly reduced the 2-propanol concentration within the initial 10 min, reducing it to lower than the detection limit in 20 min. These results indicate the strong absorbency of the porous sample for gaseous 2-propanol. The panels (b,c) of Figure 7 depict the temporal changes of 2-propanol concentration under UV light irradiation for the TiO_2-coated non-porous and porous glass cloths, respectively, after the adsorption experiment is conducted in dark conditions. Further, the photocatalytic oxidation decomposition of gaseous 2-propanol tends to desorb acetone from the TiO_2 surface; consequently, acetone is formed as a typical decomposition intermediate of this process [33]. The acetone concentration that was desorbed from the TiO_2-coated non-porous glass cloth increased to its maximum at 2.5 h of irradiation time and gradually decreased. Meanwhile, the CO_2 concentration monotonically increased with the UV light irradiation time. However, acetone was not detected from the porous glass sample, and the CO_2 concentration rapidly increased until approximately 1.5 h of the irradiation time; further, it gradually increased up to 4 h of irradiation time. These results indicate that the acetone that was generated on the porous sample was not desorbed from the surface but was decomposed to CO_2. Further, a similar reaction process has been reported for the TiO_2–zeolite composite [12]. The rapidly increasing amount of CO_2 indicates the effectiveness of photocatalytic decomposition during the early stages because of the 2-propanol concentration in the composite.

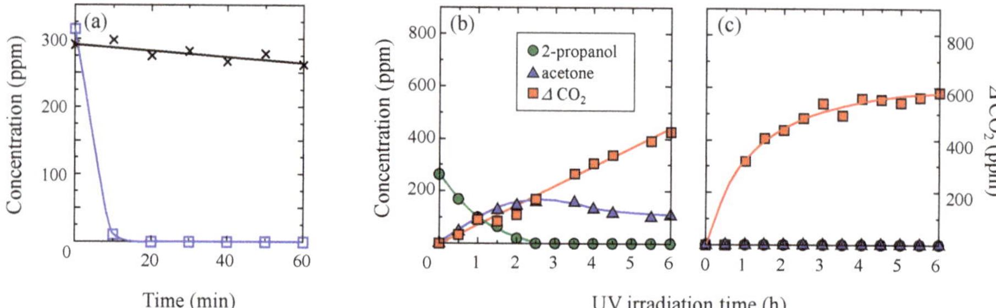

Figure 7. (**a**) Temporal concentration changes of gaseous 2-propanol under dark conditions for the TiO_2-coated non-porous (crosses) and porous (open squares) glass cloths. (**b,c**) Concentrations of gaseous 2-propanol, acetone, and CO_2 under subsequent UV light irradiation for the TiO_2-coated non-porous and porous glass cloths, respectively. The porous sample was prepared by 3 h of leaching. The dark adsorption and photocatalytic deposition experiments were conducted at 15 °C and in the presence of low water content (2 mg/L) in a 2-propanol atmosphere.

Further, the effects of temperature and water content on the adsorption and decomposition abilities of the TiO_2-coated porous glass cloth were investigated. Figure 8a depicts the trends of the 2-propanol adsorption under dark conditions in 2-propanol atmospheres with both high (18 mg/L) and low (2 mg/L) water contents. Both the experiments were conducted at 35 °C. A higher amount of 2-propanol was adsorbed from dry air when compared to that adsorbed from moist air. In the presence of high water contents, the alcohol and water molecules compete to be adsorbed on the silanol sites of the silica surface [41], thereby reducing the amount of adsorbed 2-propanol. The concentration changes of 2-propanol, acetone, and CO_2 in dry and moist atmospheres during UV irradiation are depicted in Panels (b,c) of Figure 8. Acetone was not detected under either condition, which was similar to that observed in the experiment that was conducted at 15 °C in the presence of a low water content (Figure 7c). Meanwhile, more CO_2 was generated at 35 °C than at 15 °C with low water content. In addition, at 35 °C, slightly more CO_2 was generated under the high-water atmosphere when compared to that generated under the low-water atmosphere.

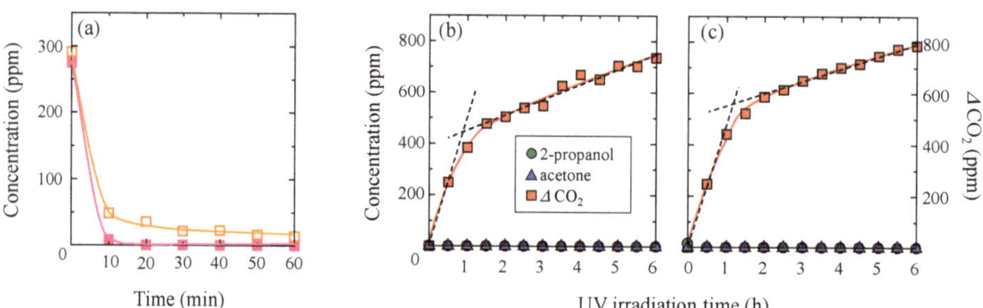

Figure 8. (**a**) Temporal concentration changes of 2-propanol during the dark storage of TiO_2-coated porous glass cloth in gaseous 2-propanol with 2-mg/L water content (closed squares) and 18-mg/L water content (open squares). (**b,c**) The concentration changes of 2-propanol, acetone, and CO_2 during UV irradiation of the TiO_2-coated non-porous glass cloth in gaseous 2-propanol with 2-mg/L and 18-mg/L water contents, respectively. All the experiments were conducted at 35 °C.

The rate of CO_2 generation by the decomposition of 2-propanol in the TiO_2-coated porous glass cloth (Figures 6c and 7b,c) changed after approximately 1.5 h of UV irradiation. No rate change could be observed in the non-porous sample (Figure 6b), confirming that the rate change phenomenon originated from the porous structure. The proposed decomposition model is depicted in Figure 9. In the TiO_2-coated porous glass cloth, the TiO_2 particles are mainly supported on the external surface of the porous glass fibers and do not penetrate the pores. Therefore, the photocatalytic reaction occurs on the external fiber surface, whereas the porous interior provides the adsorption sites of 2-propanol, acetone, and water. When the dried sample is inserted into a reactor filled with gaseous 2-propanol, the 2-propanol and water molecules are competitively adsorbed on the SiO_2 walls of the pores. Wu et al. clarified that the water and alcohol molecules emit comparatively large adsorption heat when they are adsorbed onto a bare SiO_2 surface than that emitted when they are adsorbed onto or when they form cluster with the molecules that have already being adsorbed onto the SiO_2 surface [42]. This observation indicates that the weakly and strongly adsorbed molecules coexist on the SiO_2 walls of the pores (Figure 9a). Under UV illumination, the 2-propanol molecules near the TiO_2 surface are decomposed to CO_2 via the acetone intermediate. In the early phase of photocatalytic decomposition, the weakly adsorbed 2-propanol molecules on the SiO_2 walls are preferentially desorbed and diffuse from the interior of the fiber to near the TiO_2 surface. Further, 2-propanol diffusion provides a sufficient supply of 2-propanol for sustaining CO_2 generation at a high rate (Figure 9b). The reaction rate gradually decreases as the supply of weakly adsorbed 2-propanol reduces. In the later phase, the CO_2 generation rate is limited by the desorption and diffusion of the strongly adsorbed 2-propanol molecules (Figure 9c). This limiting rate corresponds to the decelerated CO_2 generation rate after 1.5 h of light illumination in the photocatalytic decomposition of 2-propanol on the TiO_2-coated porous glass cloth (Figures 7c and 8b,c).

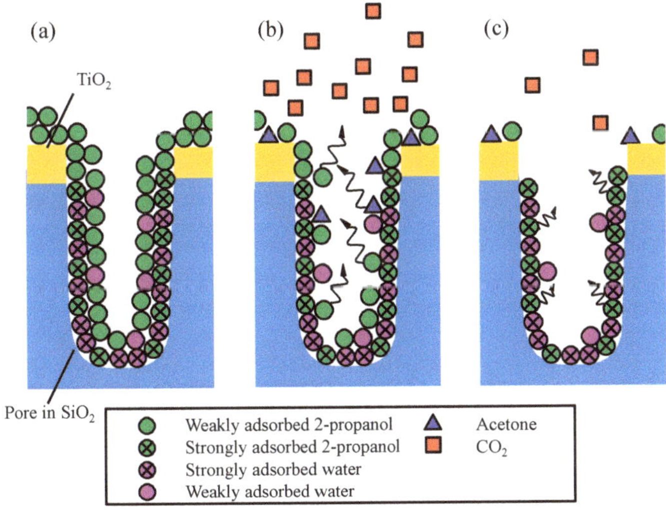

Figure 9. Schematics of (**a**) adsorption, (**b**) early-phase photocatalytic oxidation, and (**c**) late-phase photocatalytic oxidation on the TiO_2-coated porous glass cloth.

Based on this decomposition model, we compared the results of photocatalytic decomposition at 15 °C and 35 °C under a 2-propanol atmosphere with a low (2 mg/L) water content (Figures 7c and 8b). During the early phase of decomposition, the CO_2 generation rate was observed to be only slightly higher at 35 °C than that at 15 °C; however, during the later phase of decomposition, the CO_2 decomposition rates of the two conditions were observed to be clearly different. More specifically, the CO_2 generation rates after 2–6 h of UV light irradiation were 18 and 54 ppm/h at 15 °C and 35 °C, respectively. The higher CO_2 generation rate observed at 35 °C can be explained by the accelerated diffusion of 2-propanol from inside the porous glass fibers to the TiO_2 surface by a higher temperature.

The effect of water content was further evaluated from the photocatalytic decomposition results at 35 °C in 2-propanol atmospheres with low (2 mg/L) and high (18 mg/L) water contents (Figure 8b,c). After 30 min of light illumination, the CO_2 concentrations in the dry and moist atmospheres were observed to be identical; however, the CO_2 generation rate changed at CO_2 concentrations of 440 and 560 ppm in dry and moist atmospheres, respectively. This result indicates that the CO_2 generation rate changed at a later stage of the decomposition process in the moist condition. This difference may be obtained from the condition of 2-propanol adsorption in porous glass fibers. As shown in the decomposition model (Figure 9), the weakly and strongly adsorbed 2-propanol molecules coexisted in the pores. In the moist condition, the abundant water molecules are expected to occupy a large portion of the strong adsorption sites on the bare SiO_2 surface, thereby decreasing the ratio of the amount of strongly adsorbed 2-propanol to that of the weakly adsorbed 2-propanol. Because the weakly adsorbed 2-propanol molecules are easily desorbed from the SiO_2 walls, a sufficient supply of 2-propanol can continue the generation of CO_2 at a constant rate for a certain time. In contrast, in the dry condition, the proportion of strongly adsorbed 2-propanol should be relatively high. Once the weakly adsorbed 2-propanol molecules have been consumed in the early phase, the CO_2 generation rate will gradually decrease.

We further discuss the reason why the CO_2 generation rates were identical during the earliest phase of 2-propanol decomposition. In the surface-reaction limited situation, the rate of mass transportation is generally greater than the rate of surface reaction; therefore, the decomposition rate is considered to be independent of the reactant concentration. In the mass-transportation limited situation, the rate of surface reaction is greater than the rate of mass transportation, and the decomposition rate is related to the reactant concentration near the reaction site [7]. In our experiments, the concentration of weakly adsorbed 2-propanol in the porous glass fiber was expected to be dependent on the water content; however, the initial CO_2 generation rate remained constant. This indicated that the photocatalytic decomposition was limited by the surface reaction on TiO_2 in this phase.

Finally, the effect of water on the radical reaction at the photo-illuminated TiO_2 surface is discussed. In the photocatalytic reaction on TiO_2, the adsorbed water molecules react with the photogenerated carriers at the TiO_2 surface and change to hydroxyl radicals [43]. These hydroxyl radicals further diffuse and decompose the organic compounds near the TiO_2 surface; therefore, a certain amount of water vapor can enhance the photocatalytic decomposition rate of organic compounds such as 2-propanol [29,44,45]. Further, excess water vapor prevents the adsorption of organic molecules on the TiO_2 surface, thereby decreasing the photocatalytic decomposition rate [29,32,44]. However, as noted above, the initial CO_2 generation rates estimated from the CO_2 concentration after 30 min of UV irradiation were observed to remain the same under both dry and moist conditions. It indicated that the CO_2 generation rates under both conditions were very similar because the amount of water in the gaseous phase was not significantly different under both conditions. The majority of the water in the present reactor was expected to be adsorbed on the large Sg of the porous glass fiber cloth. Large water adsorption will decrease the water concentration in the gaseous phase, which can be used to plausibly explain the similar CO_2 generation rates at high and low water contents in the earliest phase of 2-propanol decomposition.

In the later phase of the decomposition, the CO_2 generation rates were estimated in the UV light irradiation time ranging from 2 to 6 h. The rates were 54 and 46 ppm/h under low and high water content conditions, respectively. Within this range, the water content did not largely influence the CO_2 generation rates because the rates were limited by the diffusion of the strongly adsorbed 2-propanol molecules.

3. Materials and Methods

3.1. Materials

The glass fiber cloth was provided by Arisawa Manufacturing Co. Ltd. (Joetsu, Japan). Figure 10 is a photomicrograph of the glass fiber cloth. It has a plain weave structure, and approximately 400 E-glass fibers of 9.1-μm diameters form the warp and weft thread; further, the thread density, the number of warp and the weft thread per certain area of woven fabric was 44×32 in 25×25 mm^2. The thickness of the cloth was 180 μm, and the weight per unit area was 203.5 g/m^2. The TiO$_2$ (anatase) nanoparticles were dispersed in 2-propanol (TKD-701, 17.0 wt %, d = 6 nm, TAYCA Co., Osaka, Japan). Further, reagent-grade aqueous hydrochloric acid (HCl; 5 mol/L), ethanol (99.5 wt %), 2-propanol (99.7 wt %), and sodium carbonate (99.5 wt %) were supplied by Wako Pure Chemical Industries Ltd. (Osaka, Japan) and were used without any purification.

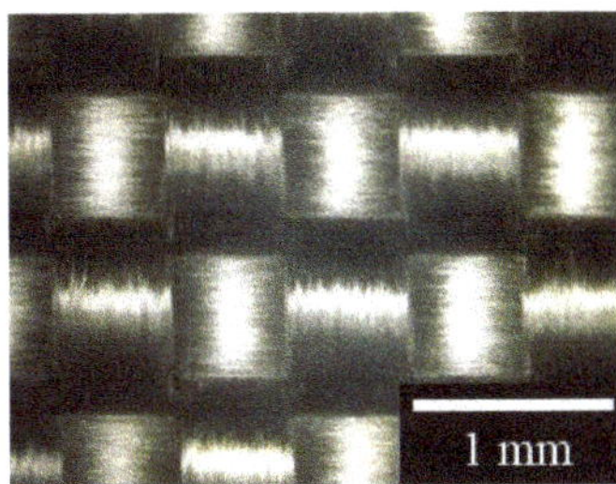

Figure 10. Photograph of the magnified surface of the E-glass cloth.

3.2. Preparation of the TiO$_2$-Coated Porous Glass Cloth

The organic compounds on the fibers of two pieces of glass cloth (ca. (30×35) mm$^2 \times 2$, total weight 0.19 g) were eliminated by heating at 500 °C for 1 h. The heat-treated cloth samples were immersed in 13.3 ml of 2.5 mol/L HCl aqueous solution at 40 °C for 0.5, 1, 3, 4, 6, or 12 h in a screw-capped perfluoroalkoxy alkane (PFA) container without stirring. After leaching by HCl, the samples were washed several times in distilled water and were further immersed in 80 ml of distilled water for 10 min. This immersion process was repeated, replacing the water between each immersion, until the pH of the immersing water reached ca. 7 (in this case, after two washes). The samples were washed with ethanol and immersed in sufficient ethanol to perform solvent exchange in the sample pores. After 10 min in ethanol, the samples were dried at 120 °C for 10 min to form the porous glass cloth. A reference sample with a non-porous structure was prepared by heating the glass fiber cloth at 500 °C for 1 h without subsequent acid leaching. This sample was used as the non-porous glass cloth.

TiO$_2$ was coated on both the porous and non-porous glass cloths using the conventional dip-coating method. The TiO$_2$ coating solution was prepared by diluting the TiO$_2$ suspension to 1 wt % in 2-propanol. The glass cloth samples were dipped in the diluted TiO$_2$ suspension and were further pulled up at 1.0 mm/s. The dip-coated samples were subsequently dried at room temperature and at 120 °C for 10 min; further, they were finally heated at 300 °C for 2 h. The obtained samples were referred to as the TiO$_2$-coated porous and non-porous glass cloths. On the other hand, as a reference in XRD measurement, the TiO$_2$ powder sample was also prepared from the TiO$_2$ suspension by drying at 120 °C and subsequently heating at 300 °C for 2 h.

3.3. Characterization

The surface morphologies of the samples were observed by field emission scanning electron microscopy (FE–SEM, Hitachi S-2400 and S-5200, Hitachi High-Technologyies, Tokyo, Japan). Before the FE–SEM observation, the samples were platinum-coated using a sputtering method. The XRD patterns were collected using CuKα radiation (λ = 0.15406 nm, monochromatized by Ni filter) by an X-ray diffractometer (MiniFlex 600, Rigaku Co., Tokyo, Japan) that was operated at 40 kV and 15 mA. The chemical compositions of the samples were analyzed using an X-ray fluorescence (XRF) spectrometer (ZSX Primus μ, Rigaku Co., Tokyo, Japan). For the XRF measurements, glass disks were prepared from the TiO_2-coated and uncoated samples as follows. First, 150 mg of the sample was ground with a pestle in an alumina mortar and was formed into a pellet by uniaxial pressing. For acid leaching (by varying the leaching time from 2 to 12 h), the porous samples were mixed with 25 mg of Na_2CO_3 as the flux. The pressed pellets were sintered by a 3-step heating process (700 °C for 2 h, 800 °C for 2 h, and 900 °C for 4 h), yielding the glass disks that were required to perform XRF analysis. Further, the porous properties of the samples were determined from the N_2 gas adsorption isotherms measured at 77 K (BELLSOAP mini II, BEL Japan Inc., Osaka, Japan). Prior to performing the N_2 gas adsorption measurements, all the samples were dried in vacuo at 120 °C for 2 h. Their specific surface areas were calculated using the Brunauer–Emmet–Teller (BET) multi-plot method. To analyze their pore-size distributions, the samples were dried in vacuo at 140 °C for 10 h, and the N_2 gas adsorption was measured at a very low pressure range (from 10^{-3} Pa, BELSORP-max-N-VP-CM, BEL Japan Inc., Osaka, Japan). The pore-size distribution was estimated by Saito–Foley fitting using the zeolite Y standard.

The adsorption and photocatalytic decomposition properties of 2-propanol were examined using the TiO_2-coated porous glass cloth and the TiO_2-coated non-porous glass cloth as reference. The samples were cut into divisions of 0.042 ± 0.001 g for performing the experiments. The cut samples were pre-treated by UV–vis light irradiation under a black light (BL) lamp (FL15BLB, Toshiba Lighting & Technology Co., Yokosuka, Japan, peak wavelength: 365 nm, light intensity at 365 nm: 3.5 mW/cm^2 at the sample surface) for 24 h and were subsequently heated at 120 °C in vacuo to eliminate the adsorbed organic molecules and water. Further, the concentrations of 2-propanol and water vapor were adjusted using the apparatus depicted in Figure 11. The air-diluted 2-propanol vapor that was produced by a calibration gas generator (Permeater PD-1B, GASTECH Co., Ayase, Japan) was mixed with the humid air, that was produced through two steps of water bubbling. The resulting mixed gas contained 300 ppm of 2-propanol and 2 or 18 mg/L of water. The 2-propanol concentration in the mixed gas was confirmed by a gas chromatograph (GC-8A, Shimadzu Co., Kyoto, Japan) using a thermal conductivity detector (TCD), a porous polymer beads column (Sunpak-A, 2 m, 160 °C, Shinwa Chemical Industries Ltd., Kyoto, Japan), and He carrier gas (20 mL/min). The water contents in the mixed gas were also confirmed in the gas detector tube (No.6, GASTEC Co., Ayase, Japan). The mixed gas flowed into a gas-tight bag containing two glass vial reactors (diameter: 4 cm; height: 6 cm, volume: approximately 65 mL). After 50 min of gas flow, the pre-treated cloth sample was transferred into one of the vials, and both the vial reactors were immediately sealed with a gas-tight septum. The sealed vials containing the sample, as well as the mixed gas or mixed gas only were transferred to an incubator (maintained at 15 °C or 35 °C) and left for 1 h in the dark. During this time, 2-propanol was adsorbed without conducting a photocatalytic reaction. The glass vials were illuminated by UV–vis light under a BL lamp (with 1.0 mW/cm^2 of light intensity at the sample position with no shielding of the glass vial) for 6 h in the incubator. The concentrations of 2-propanol, acetone, and CO_2 in the glass vial were determined at 10- and 30-min intervals during the 2-propanol adsorption and photocatalytic decomposition, respectively, by gas chromatography.

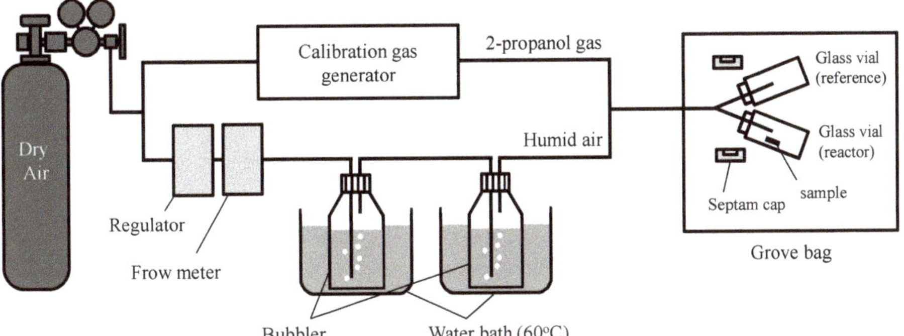

Figure 11. Apparatus settings for controlling the water content in gaseous 2-propanol.

4. Conclusions

A composite material of TiO_2 and porous glass cloth was prepared by acid leaching of an E-glass cloth and subsequent dip-coating of a TiO_2 photocatalyst. Working inward from the fiber surface, acid leaching resulted in the formation of a porous shell structure during an early stage. The fibers were observed to become completely porous after 3 h of leaching; however, the specific surface area of the acid-leached E-glass cloth continued to increase for another hour. Consequently, the specific surface area was maximized after 4 h of leaching. The compositional change was observed in the acid-leached E-glass by performing XRF analysis; Al_2O_3 and B_2O_3 were eliminated along with the ions of alkaline metals and alkali earth metals, leaving mainly the SiO_2 structure. During the photo-catalytic decomposition of 2-propanol, the TiO_2-coated porous glass cloth exhibited considerable adsorption ability with respect to 2-propanol and generated CO_2 at a higher rate than that of the TiO_2-coated non-porous glass cloth. The TiO_2-coated porous glass cloth also adsorbed acetone until the decomposition to CO_2 was completed. Increasing the temperature from 15 to 35 °C clearly increased the CO_2 generation rate of the TiO_2-coated porous glass cloth because the diffusion rate of the reactant molecules was accelerated. Further, increasing the moisture content from 2 to 18 mg/L at 35 °C slightly decreased the amount of adsorbed 2-propanol and delayed the change in the rate-controlling step from surface reaction to mass transportation. This indicated that the competitive adsorption of water and 2-propanol in the porous glass fiber decreased the amount of strongly adsorbed 2-propanol in the moist atmosphere. Finally, the porous glass cloth that was prepared from a commercial E-glass cloth provided a sufficiently strong TiO_2 support with a high specific surface area. The TiO_2-coated porous glass cloth can adsorb and photo-catalytically degrade VOCs such as 2-propanol and acetone. Therefore, it is considered to be a strong candidate for ensuring the practical elimination of gaseous organic pollutants.

Supplementary Materials: The following are available online at http://www.mdpi.com/2073-4344/9/1/82/s1. Figure S1: (a) Surface and (b) fracture cross section of the TiO_2 coating on the porous glass cloth, Figure S2: (a) N_2 adsorption isotherms of the non-coated and TiO_2-coated porous glass cloths and (b) pore-size distributions estimated by Saito–Foley fitting using the adsorption potential for N_2 on zeolite Y. V_a: adsorbed volume, d_p: pore diameter, V_p: pore volume.

Author Contributions: Conceptualization, S.Y.; methodology, S.Y. and A.Y.; investigation, K.H., K.I., and S.Y.; writing–original draft preparation, S.Y.; writing–review and editing A.Y. and K.I.; supervision, A.Y.

Funding: This work was supported in part by Nippon Sheet Glass Foundation for Materials Science and Engineering. A part of the publication cost was supported by the Gender Equality Office, University of Yamanashi.

Conflicts of Interest: The authors declare no conflict of interest.

References

1. Kostiainen, R. Volatile organic compounds in the indoor air of normal and sick houses. *Atmos. Environ.* **1995**, *29*, 693–702. [CrossRef]
2. Osha Annotated Table Z-1. Available online: http://www.webcitation.org/71UI8WUO9 (accessed on 7 August 2018).
3. Hoffmann, M.R.; Martin, S.T.; Choi, W.; Bahnemann, D.W. Environmental applications of semiconductor photocatalysis. *Chem. Rev.* **1995**, *95*, 69–96. [CrossRef]
4. Pichat, P.; Disdier, J.; Hoang-Van, C.; Mas, D.; Goutailler, G.; Gaysse, C. Purification/deodorization of indoor air and gaseous effluents by TiO_2 photocatalysis. *Catal. Today* **2000**, *63*, 363–369. [CrossRef]
5. Wang, S.; Ang, H.M.; Tade, M.O. Volatile organic compounds in indoor environment and photocatalytic oxidation: State of the art. *Environ. Int.* **2007**, *33*, 694–705. [CrossRef] [PubMed]
6. Mamaghani, A.H.; Haghighat, F.; Lee, C.-S. Photocatalytic oxidation technology for indoor environment air purification: The state-of-the-art. *Appl. Catal. B Environ.* **2017**, *203*, 247–269. [CrossRef]
7. Ohko, Y.; Fujishima, A.; Hashimoto, K. Kinetic analysis of the photocatalytic degradation of gas-phase 2-propanol under mass transport-limited conditions with a TiO_2 film photocatalyst. *J. Phys. Chem. B* **1998**, *102*, 1724–1729. [CrossRef]
8. Takeuchi, M.; Hidaka, M.; Anpo, M. Efficient removal of toluene and benzene in gas phase by the TiO_2/y-zeolite hybrid photocatalyst. *J. Hazard. Mater.* **2012**, *237*, 133–139. [CrossRef]
9. Yoneyama, H.; Torimoto, T. Titanium dioxide/adsorbent hybrid photocatalysts for photodestruction of organic substances of dilute concentrations. *Catal. Today* **2000**, *58*, 133–140. [CrossRef]
10. Ao, C.H.; Lee, S.C. Combination effect of activated carbon with TiO_2 for the photodegradation of binary pollutants at typical indoor air level. *J. Photochem. Photobiol. A Chem.* **2004**, *161*, 131–140. [CrossRef]
11. Mo, J.; Zhang, Y.; Xu, Q.; Yang, R. Effect of TiO_2/adsorbent hybrid photocatalysts for toluene decomposition in gas phase. *J. Hazard. Mater.* **2009**, *168*, 276–281. [CrossRef]
12. Yasumori, A.; Yanagida, S.; Sawada, J. Preparation of a titania/x-zeolite/porous glass composite photocatalyst using hydrothermal and drop coating processes. *Molecules* **2015**, *20*, 2349–2363. [CrossRef] [PubMed]
13. Zhang, M.; An, T.; Fu, J.; Sheng, G.; Wang, X.; Hu, X.; Ding, X. Photocatalytic degradation of mixed gaseous carbonyl compounds at low level on adsorptive TiO_2/SiO_2 photocatalyst using a fluidized bed reactor. *Chemosphere* **2006**, *64*, 423–431. [CrossRef] [PubMed]
14. Kang, M.; Hong, W.-J.; Park, M.-S. Synthesis of high concentration titanium-incorporated nanoporous silicates (ti-nps) and their photocatalytic performance for toluene oxidation. *Appl. Catal. B Environ.* **2004**, *53*, 195–205. [CrossRef]
15. Tasbihi, M.; Štangar, U.L.; Škapin, A.S.; Ristić, A.; Kaučič, V.; Tušar, N.N. Titania-containing mesoporous silica powders: Structural properties and photocatalytic activity towards isopropanol degradation. *J. Photochem. Photobiol. A Chem.* **2010**, *216*, 167–178. [CrossRef]
16. Anpo, M.; Aikawa, N.; Kubokawa, Y.; Che, M.; Louis, C.; Giamello, E. Photoluminescence and photocatalytic activity of highly dispersed titanium oxide anchored onto porous vycor glass. *J. Phys. Chem.* **1985**, *89*, 5017–5021. [CrossRef]
17. Yamashita, H.; Ichihashi, Y.; Harada, M.; Stewart, G.; Fox, M.A.; Anpo, M. Photocatalytic degradation of 1-octanol on anchored titanium oxide and on TiO_2 powder catalysts. *J. Catal.* **1996**, *158*, 97–101. [CrossRef]
18. Yamashita, H.; Honda, M.; Harada, M.; Ichihashi, Y.; Anpo, M.; Hirao, T.; Itoh, N.; Iwamoto, N. Preparation of titanium oxide photocatalysts anchored on porous silica glass by a metal ion-implantation method and their photocatalytic reactivities for the degradation of 2-propanol diluted in water. *J. Phys. Chem. B* **1998**, *102*, 10707–10711. [CrossRef]
19. Yazawa, T.; Machida, F.; Kubo, N.; Jin, T. Photocatalytic activity of transparent porous glass supported TiO_2. *Ceram. Int.* **2009**, *35*, 3321–3325. [CrossRef]
20. Wallenberger, F.T. Commercial and experimental glass fibers. In *Fiberglass and Glass Technology: Energy-Friendly Compositions and Applications*; Wallenberger, F.T., Bingham, P.A., Eds.; Springer: Boston, MA, USA, 2010; pp. 3–90. ISBN 978-1-4419-0736-3.
21. Caddock, B.D.; Evans, K.E.; Masters, I.G. Diffusion behaviour of the core-sheath structure in e-glass fibres exposed to aqueous HCl. *J. Mater. Sci.* **1989**, *24*, 4100–4105. [CrossRef]

22. Li, H.; Gu, P.; Watson, J.; Meng, J. Acid corrosion resistance and mechanism of E-glass fibers: Boron factor. *J. Mater. Sci.* **2013**, *48*, 3075–3087. [CrossRef]

23. Kiwi-Minsker, L.; Yuranov, I.; Siebenhaar, B.; Renken, A. Glass fiber catalysts for total oxidation of co and hydrocarbons in waste gases. *Catal. Today* **1999**, *54*, 39–46. [CrossRef]

24. Tanaka, H.; Kuraoka, K.; Yamanaka, H.; Yazawa, T. Development and disappearance of microporous structure in acid treated e-glass fiber. *J. Noncryst. Solids* **1997**, *215*, 262–270. [CrossRef]

25. Kitamura, T.; Ino, J.; Masuda, R.; Fukuchi, H.; Tougeda, H.; Nippon Sheet Glass, Co. Ltd.; Nippon Muki, Co. Ltd. Photocatalyst Supporting Glass Fiber Textile, Manufacturing Method of the Same and Air Filter Apparatus Using the Same. Jpn. Kokai Tokkyo Koho (unexamined patent publication) 2004-002176, 8 January 2004.

26. Aubry, E.; Ghazzal, M.N.; Demange, V.; Chaoui, N.; Robert, D.; Billard, A. Poisoning prevention of TiO_2 photocatalyst coatings sputtered on soda-lime glass by intercalation of sinx diffusion barriers. *Surf. Coat. Technol.* **2007**, *201*, 7706–7712. [CrossRef]

27. Yu, J.; Zhao, X. Effect of substrates on the photocatalytic activity of nanometer TiO_2 thin films. *Mater. Res. Bull.* **2000**, *35*, 1293–1301. [CrossRef]

28. Rekoske, J.E.; Barteau, M.A. Kinetics and selectivity of 2-propanol conversion on oxidized anatase TiO_2. *J. Catal.* **1997**, *165*, 57–72. [CrossRef]

29. Hager, S.; Bauer, R. Heterogeneous photocatalytic oxidation of organics for air purification by near uv irradiated titanium dioxide. *Chemosphere* **1999**, *38*, 1549–1559. [CrossRef]

30. Chang, C.-P.; Chen, J.-N.; Lu, M.-C. Characteristics of photocatalytic oxidation of gaseous 2-propanol using thin-film TiO_2 photocatalyst. *J. Chem. Technol. Biotechnol.* **2004**, *79*, 1293–1300. [CrossRef]

31. Vildozo, D.; Ferronato, C.; Sleiman, M.; Chovelon, J.-M. Photocatalytic treatment of indoor air: Optimization of 2-propanol removal using a response surface methodology (RSM). *Appl. Catal. B Environ.* **2010**, *94*, 303–310. [CrossRef]

32. Kim, S.B.; Hong, S.C. Kinetic study for photocatalytic degradation of volatile organic compounds in air using thin film TiO_2 photocatalyst. *Appl. Catal. B Environ.* **2002**, *35*, 305–315. [CrossRef]

33. Coronado, J.M.; Zorn, M.E.; Tejedor-Tejedor, I.; Anderson, M.A. Photocatalytic oxidation of ketones in the gas phase over TiO_2 thin films: A kinetic study on the influence of water vapor. *Appl. Catal. B Environ.* **2003**, *43*, 329–344. [CrossRef]

34. Reddy, K.M.; Manorama, S.V.; Reddy, A.R. Bandgap studies on anatase titanium dioxide nanoparticles. *Mater. Chem. Phys.* **2002**, *78*, 239–245. [CrossRef]

35. Kaneko, K. Determination of pore size and pore size distribution. *J. Membrane Sci.* **1994**, *96*, 59–89. [CrossRef]

36. Elmer, T.H.; Nordberg, M.E.; Carrier, G.B.; Korda, E.J. Phase separation in borosilicate glasses as seen by electron microscopy and scanning electron microscopy. *J. Am. Ceram. Soc.* **1970**, *53*, 171–175. [CrossRef]

37. Tanaka, H.; Yazawa, T.; Eguchi, K.; Nagasawa, H.; Matsuda, N.; Einishi, T. Precipitation of colloidal silica and pore size distribution in high silica porous glass. *J. Noncryst. Solids* **1984**, *65*, 301–309. [CrossRef]

38. Saito, A.; Poley, H.C. Argon porosimetry of selected molesular sieves: Experiments and examination of the adapted Horvath-Kawazoe model. *Microporous Mater.* **1995**, *3*, 531–542. [CrossRef]

39. Saito, A.; Poley, H.C. Curvature and parametric sensitivity in models for adsorption in micropores. *AIChE J.* **1991**, *37*, 429–436. [CrossRef]

40. Zhuravlev, L.T. The surface chemistry of amorphous silica. Zhuravlev model. *Colloids Surf. A Physicochem. Eng. Asp.* **2000**, *173*, 1–38. [CrossRef]

41. Chuiko, A.A.; Lobanov, V.V.; Grebenyuk, A.G. Structure of disperse silica surface and electrostatic aspects of adsorption. In *Colloidal Silica: Fundamentals and Applications*; Bergna, H.E., Roberts, W.O., Eds.; CRC Press: Boca Raton, FL, USA, 2005; pp. 331–360. ISBN 9780824709679.

42. Wu, D.; Guo, X.; Sun, H.; Navrotsky, A. Energy landscape of water and ethanol on silica surfaces. *J. Phys. Chem. C* **2015**, *119*, 15428–15433. [CrossRef]

43. Murakami, Y.; Kenji, E.; Nosaka, A.Y.; Nosaka, Y. Direct detection of oh radicals diffused to the gas phase from the uv-irradiated photocatalytic TiO_2 surfaces by means of laser-induced fluorescence spectroscopy. *J. Phys. Chem. B* **2006**, *110*, 16808–16811. [CrossRef]

44. Luo, Y.; Ollis, D.F. Heterogeneous photocatalytic oxidation of trichloroethylene and toluene mixtures in air: Kinetic promotion and inhibition, time-dependent catalyst activity. *J. Catal.* **1996**, *163*, 1–11. [CrossRef]
45. Bouazza, N.; Lillo-Ródenas, M.; Linares-Solano, A. Photocatalytic activity of TiO_2-based materials for the oxidation of propene and benzene at low concentration in presence of humidity. *Appl. Catal. B Environ.* **2008**, *84*, 691–698. [CrossRef]

Article

Enhanced Photocatalytic Reduction of Cr(VI) by Combined Magnetic TiO$_2$-Based NFs and Ammonium Oxalate Hole Scavengers

Yin-Hsuan Chang [1] and Ming-Chung Wu [1,2,3,*]

[1] Department of Chemical and Materials Engineering, Chang Gung University, Taoyuan 33302, Taiwan; cgu.yinhsuanchang@gmail.com
[2] Green Technology Research Center, Chang Gung University, Taoyuan 33302, Taiwan
[3] Division of Neonatology, Department of Pediatrics, Chang Gung Memorial Hospital, Linkou, Taoyuan 33305, Taiwan
* Correspondence: mingchungwu@cgu.edu.tw; Tel.: +886-3211-8800 (ext. 3834)

Received: 19 December 2018; Accepted: 5 January 2019; Published: 10 January 2019

Abstract: Heavy metal pollution of wastewater with coexisting organic contaminants has become a serious threat to human survival and development. In particular, hexavalent chromium, which is released into industrial wastewater, is both toxic and carcinogenic. TiO$_2$ photocatalysts have attracted much attention due to their potential photodegradation and photoreduction abilities. Though TiO$_2$ demonstrates high photocatalytic performance, it is a difficult material to recycle after the photocatalytic reaction. Considering the secondary pollution caused by the photocatalysts, in this study we prepared Ag/Fe$_3$O$_4$/TiO$_2$ nanofibers (NFs) that could be magnetically separated using hydrothermal synthesis, which was considered a benign and effective resolution. For the photocatalytic test, the removal of Cr(VI) was carried out by Ag/Fe$_3$O$_4$/TiO$_2$ nanofibers combined with ammonium oxalate (AO). AO acted as a hole scavenger to enhance the electron-hole separation ability, thereby dramatically enhancing the photoreduction efficiency of Cr(VI). The reaction rate constant for Ag/Fe$_3$O$_4$/TiO$_2$ NFs in the binary system reached 0.260 min^{-1}, 6.95 times of that of Ag/Fe$_3$O$_4$/TiO$_2$ NFs in a single system (0.038 min^{-1}). The optimized Ag/Fe$_3$O$_4$/TiO$_2$ NFs exhibited high efficiency and maintained their photoreduction efficiency at 90% with a recyclability of 87% after five cycles. Hence, taking into account the high magnetic separation behavior, Ag/Fe$_3$O$_4$/TiO$_2$ NFs with a high recycling capability are a potential photocatalyst for wastewater treatment.

Keywords: TiO$_2$; magnetic property; photocatalyst; reusable; photoreduction

1. Introduction

With the advancement of various industries comes serious industrial water pollution. Such wastewaters usually contain a complicated mixture of constituents, often involving the co-existence of multiple contaminants such as heavy metals and organic pollutants. With the development of electroplating, metallurgy, leathermaking and more, heavy metal pollution has become a serious threat to human survival and development. One such heavy metals released into industrial wastewater is Cr(VI), which is both toxic and carcinogenic. It has been the first type of carcinogen listed by the World Health Organization's International Cancer Research Institute since 2012. Cr(VI) is easily accumulated in living organisms and can result in vomiting, liver damage, and severe diarrhea. Compared to Cr(VI), trivalent chromium (Cr(III)) is less toxic and more vital for animals and humans [1,2]. The conventional approach for the reduction or removal of Cr(VI) includes electrochemical precipitation [3,4], adsorption [5,6], bacterial reduction [7,8], ion exchange [9,10], photoreduction [11–16], etc. Compared to the above methods, photocatalytic reactions are considered a clean and promising technology owing to its highly efficient photoreduction of Cr(VI) to the less harmful Cr(III).

TiO_2 is a well-known photocatalyst widely applied for environmental purification due to its advantages, such as its highly active photocatalytic properties, chemical inertness, environmental-friendliness, non-toxicity, and cost-effectiveness [17–24]. It shows great potential in solving the difficult problem of reducing Cr(VI) to Cr(III) in industrial wastewaters. Though TiO_2 demonstrates a high photocatalytic performance, it is difficult to recycle following the photocatalytic reaction. Traditional separation approaches such as filtration and centrifugation have been widely adopted. However, the recycling efficiency is hindered by the loss of photocatalysts. Considering the secondary pollution caused by the photocatalysts, combining TiO_2 with Fe_3O_4 to form magnetic composite materials for the magnetic separation under modest magnetic fields has been seen as a benign and effective resolution [25–28]. To date, there have been many facile methods used to synthesize magnetic iron oxides/TiO_2 hybrid nanomaterial such as sol–gel, metal–organic chemical vapor deposition, the seed-mediated method, and hydrothermal treatment. In spite of introducing magnetic separation by doping Fe_3O_4, the photocatalytic performance could be further enhanced by modifying the shape of Fe_3O_4 to increase the active surface area [29]. In addition, modifying the structure of TiO_2 is also a common method used to enhance photocatalytic performance. Furthermore, combing the ultrafine Fe_3O_4 with one-dimension TiO_2 nanofibers can provide a superior charge transport in a one-dimensional direction, and show high activity.

A great deal of literature has indicated that incorporating Fe_3O_4 into TiO_2 does not improve the photocatalytic properties of TiO_2 as expected [30–32]. The crystallinity of TiO_2 depends on the calcination process, which plays a crucial role in the photocatalytic performance. At the same time, calcination also decreases the saturation magnetization of Fe_3O_4. With the increasing calcination temperature, Fe_3O_4, which has a superparamagnetic phase, would undergo a phase transition to γ-Fe_2O_3 and finally become α-Fe_2O_3, which has a soft ferromagnetic phase [33]. Another problem is the small bandgap of Fe_3O_4, which leads to the fast electron-hole pair recombination in Fe_3O_4/TiO_2 composite material [31]. Therefore, in order to enhance the photocatalytic activity and to maintain the magnetic properties, a lot of research has focused on doping metals to obtain the desired effect [34–38]. In particular, doping Ag into TiO_2 not only enhances the separation of electron-hole pairs, but also maintains the magnetic performance of the Fe_3O_4/TiO_2 composite material. For the Ag-doped TiO_2, the Ag dopants act as the photo-generated electron trapper that enhances the separation of the electron-hole pair and even creates a local electrical field to facilitate electron excitation [39–43].

In this study, in order to achieve both a high photocatalytic activity and a high magnetic property, we prepared Ag and Fe_3O_4 co-doped TiO_2 nanofibers (Ag/Fe_3O_4/TiO_2 NFs) via hydrothermal synthesis followed by a calcination treatment. The Ag/Fe_3O_4/TiO_2 NFs were studied systematically through synchrotron X-ray diffractometer, UV-Vis spectroscopy, field emission scanning electron microscopy (FESEM), and transmission electron microscopy (TEM). For the photocatalytic test, the removal of Cr(VI) was carried out by Ag/Fe_3O_4/TiO_2 NFs combined with ammonium oxalate (AO). Hence, taking into account the high magnetic separation behavior, Ag/Fe_3O_4/TiO_2 NFs with a high recycling capability are a potential photocatalyst for wastewater treatment.

2. Results

Prior to combining magnetic NPs into TiO_2, a basic characterization of Fe_3O_4 was investigated and summarized in Figure 1. The synchrotron X-ray diffractometer was applied to characterize the crystal structure of the Fe_3O_4 NPs as shown in Figure 1a. The characteristic peaks could be indexed to standard Fe_3O_4 (JCPDS No. 019-0629). The method used to determine the bandgap of Fe_3O_4 NPs from the diffusion reflectance is shown schematically in Figure 1b. It was calculated according to $[F(R)h\nu]^{1/2}$ versus the energy of incident light based on Kubelka–Munk function spectra, $F(R)$. According to Figure 1b, the band gap of Fe_3O_4 NPs was ~0.8 eV. The magnetic property of the Fe_3O_4 NPs were investigated using a (Superconducting quantum interference device magnetometer) SQUID at 10 K. The magnetic hysteresis loop shown in Figure 1c indicates the ferromagnetic property that exists in Fe_3O_4 NPs. The inset of Figure 1d shows the magnetic separation of Fe_3O_4 NPs from the aqueous dispersion attracted by the Nd-Fe-B magnets. The collected Fe_3O_4 NPs indicated that it could be controlled by an applied magnetic field.

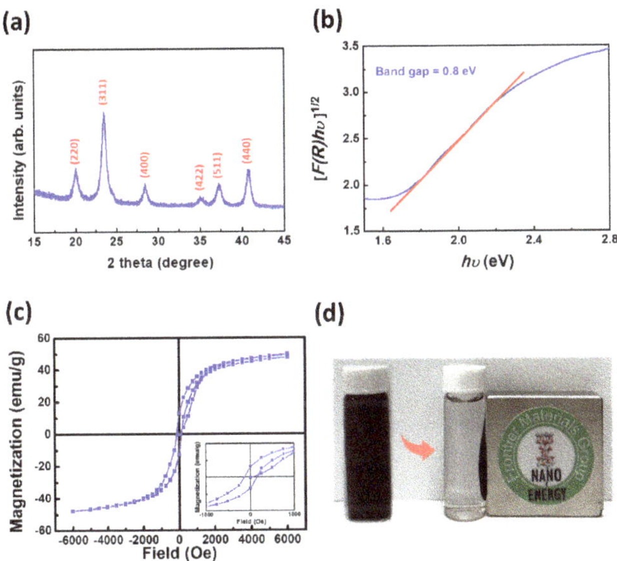

Figure 1. (**a**) Synchrotron X-ray pattern; (**b**) Tauc plot for the indirect band gap; (**c**) magnetic hysteresis loop measured at 10 K of as-synthesized Fe_3O_4 NPs; and (**d**) the Fe_3O_4 suspensions before and after magnetic attraction.

The calcination temperature for the magnetic material is seen as an important factor. For example, as the calcination temperature exceeds 600 °C, it results in a phase transformation from magnetite (Fe_3O_4) to maghemite (γ-Fe_2O_3) to hematite (α-Fe_2O_3) conversions. This phase transformation behavior would cause magnetic material to lose its magnetic properties. In order to maintain the ability to magnetically separate the synthesized TiO_2, the calcination temperature was fixed at 550 °C for the study. Figure 2a shows the synchrotron X-ray patterns Fe_3O_4/TiO_2 with various doping concentrations that depend on the Fe_3O_4/TiO_2 ratio (wt %). The characteristic peaks, which centered at 2θ around 16.71°, 24.27°, 24.81°, 25.31°, 31.33°, 34.99° and 35.71°, could be indexed to anatase phase TiO_2. Pristine TiO_2 exhibited characteristic peaks at 2θ around 18.94°, 19.75° and 22.01° that could be identified as TiO_2 low-temperature phase, β-TiO_2 monoclinic. With the incorporation of Fe_3O_4, the anatase phase TiO_2 became the only phase in the crystal structure, and no characteristic peaks from other phases could be detected. In addition, the radius of an Fe ion (Fe^{2+} ~ 0.76 Å, Fe^{3+} ~ 0.64 Å) is slightly smaller than that of a Ti ion (Ti^{4+} ~ 0.68 Å), indicating that some of doped Fe ion might enter to interstitial voids of TiO_2 lattice [30,44]. The Fe ions in the TiO_2 lattice would act as carrier traps, leading to the electron-hole recombination. Taking into account the recombination phenomenon, the photocatalytic performance could be affected when Fe ions were incorporated into the catalyst. When the doping concentration reached 25.0 wt %, both anatase TiO_2 and Fe_3O_4 peaks were detected. The excessive Fe_3O_4 NPs in the Fe_3O_4-TiO_2 lead to non-uniform doping and to the decrease of the crystallinity of TiO_2. From the magnetic hysteresis loop shown in Figure 2b, as the amount of Fe_3O_4 increased, the magnetization increased as well. In addition, the magnetization was proportional to the doping amount. To optimize the doping concentration, the photocatalytic activity was measured by photoreduction of Cr(VI) in $K_2Cr_2O_7$ aqueous solution under UV-B irradiation.

The photoreduction of Cr(VI) using TiO_2-based catalyst usually follows Langmuir–Hinshelwood kinetics. It can be mathematically simplified to first-order kinetics in the early stage described as $\ln(C_0/C) = kt$, where C_0 is the initial concentration of Cr(VI) in $K_2Cr_2O_7$, C is the remaining Cr(VI) concentration at various times, k is the apparent reaction rate constant, and t is the photodegradation time. The blank experiment was performed under the same conditions but without the existence of the photocatalyst. For the dark experiment, 10.0 wt %-Fe_3O_4/TiO_2 was also tested in dark conditions to observe the adsorption–desorption behavior. From Figure 2c, 10.0 wt %-Fe_3O_4/TiO_2 calcined at 550 °C showed the highest photoreduction performance among other Fe_3O_4/TiO_2 photocatalysts due to the highest crystallinity among the Fe_3O_4/TiO_2 series. High crystallinity can hinder the recombination of photoexcited electrons and holes and thus result in high photocatalytic activity. With further increasing the Fe_3O_4 doping concentration to 15.0 wt % and 25.0 wt %, the excessive dopant might destroy the lattice of TiO_2, thus decreasing the crystallinity of TiO_2 dramatically and form the impurity phases composed of Fe_3O_4, γ-Fe_2O_3 and α-Fe_2O_3. In addition, all of the Fe_3O_4/TiO_2 showed poorer performance compared to the pristine TiO_2, which is in accordance with the XRD spectra. The Fe^{3+} as carrier traps leading to recombination phenomenon and decreased the photocatalytic performance compared with pristine TiO_2.

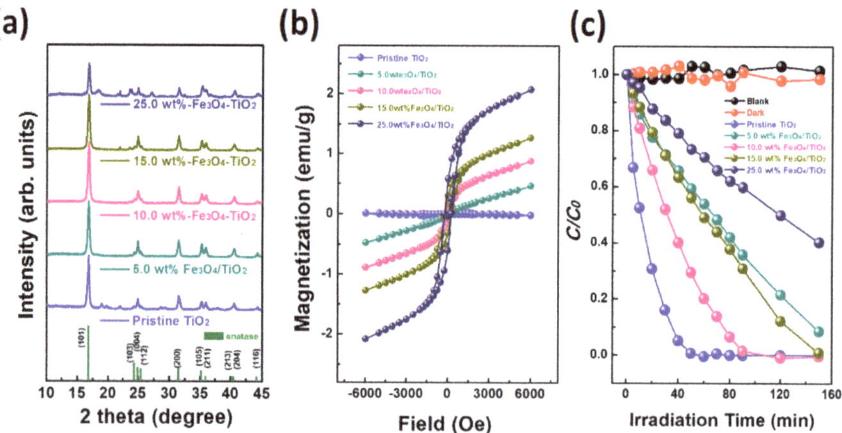

Figure 2. Dependence on the Fe_3O_4/TiO_2 weight ratio (**a**) Synchrotron X-ray patterns; (**b**) magnetic hysteresis loop measured at 10 K; and (**c**) the C/C_0 curves for the photoreduction of Cr(VI) in $K_2Cr_2O_7$ aqueous solution under UV-B irradiation using pristine TiO_2 and Fe_3O_4-TiO_2 with various doping concentrations calcined at 550 °C.

Ag was co-doped with 10.0 wt % of Fe_3O_4 into TiO_2 to improve the electron-hole separation further. Figure 3a shows the synchrotron X-ray patterns of the Ag/Fe_3O_4/TiO_2 series with various Ag doping concentrations that depended on the amount of Ag (mol %) co-doped with 10 wt % Fe_3O_4/TiO_2. The characteristic peaks of Ag/Fe_3O_4/TiO_2 could all be assigned to anatase phase TiO_2 without any Ag signal. The results indicated that the incorporation of Fe_3O_4 and Ag did not destroy the crystal structure of TiO_2. The magnetic hysteresis loop (Figure 3b) illustrates that as the amount of Ag increased, the magnetization decreased. When the excessive Ag dopant was 10.0 mol %, it resulted in a decay of saturation magnetization compared to Fe_3O_4/TiO_2, due to the contribution of the volume of non-magnetic material to the total sample volume. Therefore, the magnetism of the 10.0 mol % Ag/Fe_3O_4/TiO_2 was too low for magnetic separation by adding a magnetic field. Figure 3c demonstrates the C/C_0 curves for photoreduction of Cr(VI) under UV-B irradiation over the Ag/Fe_3O_4/TiO_2 series with different Ag doping concentrations. The blank experiment was also performed under the same conditions but without the presence of the photocatalyst. For the dark experiment, 5.0 mol % $Ag/Fe_3O_4/TiO_2$ was also tested in dark conditions to eliminate the adsorption–desorption behavior. The 10.0 mol % $Ag/Fe_3O_4/TiO_2$ showed the highest photoreduction performance, even higher than that of pristine TiO_2. Although 10.0 mol % $Ag/Fe_3O_4/TiO_2$ possessed the highest reduction performance, after considering the ability to be magnetically separated, we selected the 5.0 mol % doping level as the optimal photocatalyst. We could also observe that the photoreduction for Ag co-doped with 10 wt % Fe_3O_4/TiO_2 showed the higher performance after the incorporation of Ag compared to 10 wt % Fe_3O_4/TiO_2. This enhancement could be interpreted by the energy level theory, namely that the conduction band of Fe_3O_4 is lower than the conduction band of TiO_2, so the conduction band of TiO_2 becomes an electron capture position. With the further introduction of Ag into Fe_3O_4/TiO_2, Ag could act as another electron trap to enhance the electron-hole separation ability [45].

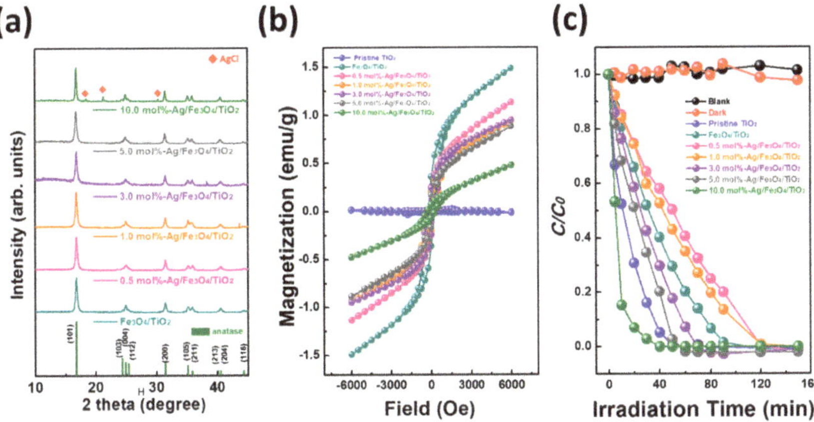

Figure 3. Dependence on the amount of Ag (mol %) co-doped with 10.0 wt % Fe_3O_4/TiO_2. (**a**) Synchrotron X-ray patterns; (**b**) magnetic hysteresis loop measured at 10 K; and (**c**) the C/C_0 curves for the photoreduction of Cr(VI) in $K_2Cr_2O_7$ aqueous solution under UV-B irradiation using pristine TiO_2 and $Ag/Fe_3O_4/TiO_2$ with various Ag doping concentration calcined at 550 °C.

After the optimization process, pristine TiO_2, 10.0 wt % Fe_3O_4/TiO_2 (Fe_3O_4/TiO_2) and 5.0 mol % $Ag/Fe_3O_4/TiO_2$ ($Ag/Fe_3O_4/TiO_2$) were compared. The FESEM images of TiO_2-based NFs before and after combining magnetic NPs and Ag are shown in Figure 4. The image shows that the surface of the pristine TiO_2 was very clean and smooth (Figure 4a). When incorporated with Fe_3O_4 NPs, there was no significant morphological change for the Fe_3O_4/TiO_2 (Figure 4b). For $Ag/Fe_3O_4/TiO_2$, the surface became relatively rough and some particles aggregated on it (Figure 4c). On increasing the silver content, the surface charge of TiO_2-based material would gradually decrease. With small amounts of Ag dopant, Ag_2O and AgO might disperse on the surface of TiO_2-based material. When increasing Ag doping concentration, the decrease in surface charge was attributed to an agglomeration of the silver species and a reduction to Ag^0 on the TiO_2 surface [46]. The EDS-characterized elemental compositions and the corresponding results are listed in Table 1. For Fe_3O_4/TiO_2, the ratio of Fe/Ti and Ag/Ti were ~2.9% and ~0.0%, respectively. After incorporating Ag, the ratio of Fe/Ti was ~3.1%, which was approximately the same as Fe_3O_4/TiO_2, and the ratio of Ag/Ti increased to 0.4%. The corresponding ratios of Fe/Ti and Ag/Ti illustrated the existence of Ag in the $Ag/Fe_3O_4/TiO_2$, together with the leading component Ti and Fe. The distinct signals of these elements present in the spectrum confirmed the successful inclusion of Ag ions into the host TiO_2 lattice.

Table 1. The corresponding ratios of Fe/Ti and Ag/Ti for pristine TiO_2, Fe_3O_4/TiO_2 and $Ag/Fe_3O_4/TiO_2$.

Sample	Fe/Ti (%)	Ag/Ti (%)
Pristine TIO_2	0.0	0.0
Fe_3O_4/TiO_2	2.9	0.0
$Ag/Fe_3O_4/TiO_2$	3.1	0.4

Figure 4. SEM images of (**a**) pristine TiO_2; (**b**) Fe_3O_4/TiO_2 and (**c**) $Ag/Fe_3O_4/TiO_2$.

The Kubelka–Munk function spectra of TiO_2-based materials are shown in Figure 5a. Pristine TiO_2 only showed absorption behavior in the UV range. However, compared to pristine TiO_2, the $F(R)$ spectra of Fe_3O_4/TiO_2 and $Ag/Fe_3O_4/TiO_2$ showed an obvious extension to the visible light region, and the band gap energy also decreased from 3.1 eV to 2.1 eV and 2.0 eV, respectively (Figure 5b). This could be ascribed to the introduction of Fe_3O_4. During the calcination process, the introduced Fe^{3+} could exchange with the lattice position of Ti^{4+} and therefore form an impurity band. Fe_3O_4/TiO_2 and $Ag/Fe_3O_4/TiO_2$ with a decreased forbidden bandwidth could successfully narrow the band gap for the higher absorption behavior in the visible region. This enhanced absorption behavior could generate a lot of photo-excited electrons and holes for photocatalytic reactions.

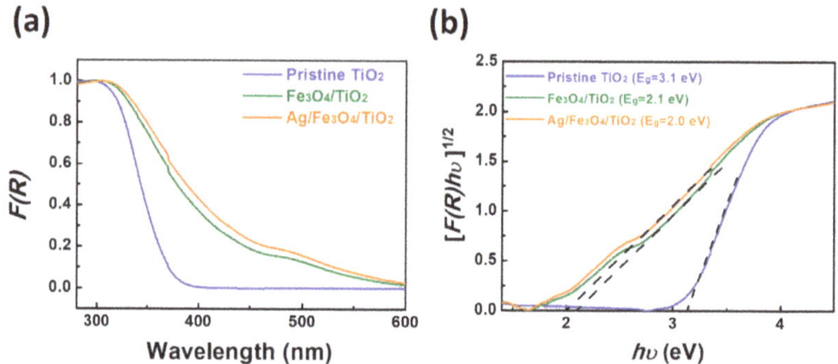

Figure 5. (**a**) Kubelka–Munk function spectra and (**b**) Tauc plot for the indirect band gap of pristine TiO_2, Fe_3O_4/TiO_2 and $Ag/Fe_3O_4/TiO_2$.

The photocatalytic activity test was examined by photoreduction of Cr(VI) to Cr(III). The photoreduction pathways of Cr(VI) on the surface of TiO_2 through UV irradiation can be described by the following reaction sequence (Equations (1)–(6)). After UV light irradiation, photo-excited electron-hole pairs are generated. During the photoreduction reaction of Cr(VI), electrons dominate the entire reaction. Meanwhile, the hole will oxidize H_2O to form the reactive oxygen species OH, which will further react with Cr(III) to generate Cr(VI).

$$TiO_2 + h\upsilon \rightarrow h^+ + e^- \tag{1}$$

$$Cr_2O_7{}^{2-} + 14H^+ + 6e^- \rightarrow 2Cr^{3+} + 7H_2O \tag{2}$$

$$e^- + h^+ \rightarrow recombination \tag{3}$$

$$h^+ + H_2O \rightarrow \cdot OH + H^+ \tag{4}$$

$$h^+ + OH^- \rightarrow \cdot OH \tag{5}$$

$$3 \cdot OH + Cr^{3+} \rightarrow 3OH^- + Cr^{6+} \tag{6}$$

It is unfavorable to reduce Cr(VI) to Cr(III) while Cr(VI) participates in the reaction alone, due to the electron-hole recombination and the oxidation of Cr(III). Figure 6a shows the photoreduction of Cr(VI) over pristine TiO_2, Fe_3O_4/TiO_2 and $Ag/Fe_3O_4/TiO_2$ in a single system for which only Cr(VI) existed in the initial condition. The reduction of Cr(VI) was greatly promoted by the coexistence of ammonium oxalate (AO), and the corresponding results for single systems are also plotted for comparison (Figure 6b). AO is a type of hole scavenger that is widely used for detecting reactive oxygen species during the photocatalytic reaction in order to better understand the reaction mechanism. Therefore, AO would capture the photogenerated holes during the photocatalysis reaction, leaving the photogenerated electrons on the surface of the TiO_2-based NFs. With the help of AO, the separation of the electron-hole was greatly facilitated and thus the reduction performance of Cr(VI) was enhanced. The poor enhancement of pristine TiO_2 compared with Fe_3O_4/TiO_2 and $Ag/Fe_3O_4/TiO_2$ could be due to the bandgap of each sample. A decrease in the bandgap for Fe_3O_4/TiO_2 and $Ag/Fe_3O_4/TiO_2$ resulted in a greater absorption of photons, which was beneficial for the production of electrons and holes required for the photocatalytic reactions. However, the photoexcited electron-hole pair in the Fe_3O_4/TiO_2 and $Ag/Fe_3O_4/TiO_2$ favored a transfer to Fe_3O_4. Holes can provide a faster reaction route with AO, rather than recombining with the electron. Further, the residual electron on the surface of Fe_3O_4/TiO_2 and $Ag/Fe_3O_4/TiO_2$ can reduce Cr(VI) to Cr(III). Therefore, the photoreduction performance for Fe_3O_4/TiO_2 and $Ag/Fe_3O_4/TiO_2$ showed a dramatic enhancement. The reaction rate constant for $Ag/Fe_3O_4/TiO_2$ in binary system achieved 0.260 min^{-1}, which was 6.95 times that of $Ag/Fe_3O_4/TiO_2$ in a single system at 0.038 min^{-1}. These results confirmed the synergetic promotion effect of ammonium oxalate.

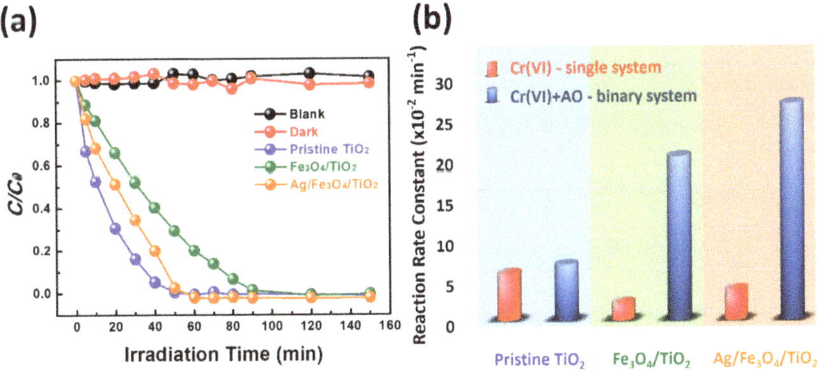

Figure 6. Photocatalytic reaction in (**a**) Cr(VI) single system and (**b**) Cr(VI) + AO binary system with pristine TiO_2, Fe_3O_4/TiO_2, and $Ag/Fe_3O_4/TiO_2$.

The stability and recyclability of the photocatalyst is an important index for practical application. In order to examine the stability and recyclability of $Ag/Fe_3O_4/TiO_2$, the photoreduction of Cr(VI) was repeated five times. Each time, the photocatalysts were recycled by adding a magnetic field. This exhibited a slight decay of reduction efficiency after each cycle, which accounted for the weight loss during every recycle process. After five cycles, the photoreduction efficiency was maintained at 90% (Figure 7a), and the amount of the remaining photocatalyst was 87% (Figure 7b). The stability and recyclability tests proved that the Cr(VI) photoreduction efficiency over $Ag/Fe_3O_4/TiO_2$ has consistently high stability and recyclability. Therefore, $Ag/Fe_3O_4/TiO_2$ is a potential photocatalyst for wastewater treatment.

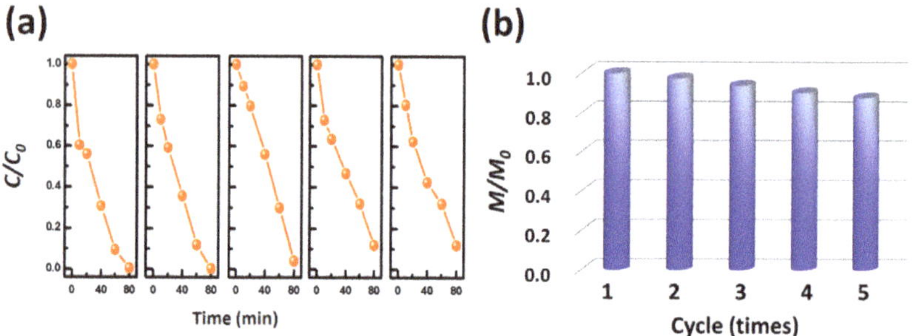

Figure 7. (**a**) Stability and (**b**) recyclability test of $Ag/Fe_3O_4/TiO_2$ for the photocatalytic reduction of Cr(VI) over five cycles.

3. Materials and Methods

3.1. Synthesis of Fe_3O_4 Magnetic NPs

The synthesis of Fe_3O_4 NPs was carried out by the co-precipitation method, in which the iron(II) chloride ($FeCl_2·4H_2O$, Acros, 99+%) and iron(III) chloride (($FeCl_3·6H_2O$, Acros, 99+%) were used as the raw materials with a molar proportion of 1:2. First, they were dissolved in deionized water and preheated to 60 °C. After that, a 10 M sodium hydroxide aqueous solution (NaOH) acting as a precipitation reagent was added into the mixture solution under continuous stirring for 1 h. The Fe_3O_4 suspension was magnetically separated and washed with deionized water repeatedly until the pH was 7. Finally, the product was air dried at 60 °C.

3.2. Synthesis of $Ag/Fe_3O_4/TiO_2$ NFs

The TiO_2-based NFs were synthesized by hydrothermal method and crystallized by heat treatment. First, 2.5 g anatase phase TiO_2 powder (98%, Sigma-Aldrich, St. Louis, MO, USA), as-synthesized Fe_3O_4 NPs, and silver nitrate ($AgNO_3$, extra pure, Choneye, Taipei, Taiwan) with various stoichiometric ratios were suspended into separate 62.5 mL of 10 M NaOH. The suspension was dispersed uniformly into an ultrasonic bath. After that, the reactants were transferred into a polytetrafluoroethylene-lined autoclave for thermal treatment at 150 °C for 24 h to obtain sodium titanate ($Na_2Ti_3O_7$). Then, various forms of $Na_2Ti_3O_7$ were washed with 0.10 M hydrochloric acid (HCl, 37%, Sigma-Aldrich, St. Louis, MO, USA) to exchange the sodium ion for protons. Finally, the sodium hydrogen titanate ($Na_xH_{2-x}Ti_3O_7$) was filtered and air dried at 80 °C. The dried $Na_xH_{2-x}Ti_3O_7$ was calcined at 550 °C for 12 h at a 5 °C/min heating rate to obtain magnetic TiO_2-based NFs.

3.3. Characterization

To observe the crystal structure, the synchrotron X-ray spectra were collected from 5° to 45° of 2θ with a scan rate of 0.02°/s and a wavelength of ~ 1.025 Å. The Kubelka–Munk function, *F(R)*, spectra were measured and recorded by UV/Vis spectrophotometer (Jacso, V-650, Tokyo, Japan) from 200 to 900 nm wavelength. The magnetic properties of Fe_3O_4 NPs and magnetic TiO_2-based NFs were measured at 10 K temperature using a SQUID magnetometer (MPMS3, Quantum Design, San Diego, CA, USA). The microstructure was characterized by transmission electron microscopy (TEM, spherical-aberration corrected ULTRA-HRTEM, JEM-ARM200FTH, JEOL Ltd., Tokyo, Japan). The morphology and atomic ratio of TiO_2-based NFs were measured by FE-SEM (SU8010, Hitachi, Tokyo, Japan) equipped with EDS (XFlash Detector 5030, Bruker AXS, Karlsruhe, Germany).

3.4. Photocatalytic Measurement

For the measurement of the photoreduction of Cr(VI), 20.0 mg of magnetic TiO_2-based photocatalyst was dispersed into 150.0 mL of potassium dichromate ($K_2Cr_2O_7$, 0.0167 M, Fisher Scientific, CA, USA) with an initial concentration of 1.0 ppm at ambient conditions. As the control group, 20.0 mg of pristine TiO_2 was also dispersed into 150.0 mL $K_2Cr_2O_7$ with an initial concentration of 1.0 ppm at ambient conditions. The two UV-B light lamps (G15T8E, λ_{max} ~312 nm, 8.0 W, Sankyo Denki, Osaka, Japan) were placed ~10.0 cm above the reaction system. Before exposure to light irradiation, the suspensions were put in the dark for 30 min in order to achieve the adsorption equilibrium and thus minimize the surface adsorption behavior. The concentration of retained Cr(VI) was measured by the diphenylcarbazide method. By comparing the intensity of the Cr(VI) characteristic peak located at λ = 540 nm with the calibration curve examined previously, we can obtain its corresponding concentration. In order to examine the mechanism of Cr(VI) photoreduction, 142.2 µL tert-butanol ((CH_3)$_3$COH, $\geq$99.0%, J.T.Baker, Phillipsburg, NJ, USA) and 24.0 mg ammonium oxalate ($C_2H_8N_2O_4$, 98%, Vetec, trademark of Sigma-Aldrich, St. Louis, MO, USA) were added into $K_2Cr_2O_7$ in the beginning, respectively. The stability and recyclability of the photocatalysts were measured by cycling experiments. After the Cr(VI) photoreduction for each cycle, the magnetic TiO_2 was collected by Nd-Fe-B magnet wrapped with PVC film. After removing the magnetic field, the magnetic TiO_2 was washed three times with ethanol to remove residual ions and molecules and then dried at 80 °C. The fresh 1.0 ppm $K_2Cr_2O_7$ aqueous solution was mixed with the used photocatalyst to perform the second run of photoactivity testing. Similarly, the photocatalyst was recycled to perform the third, fourth, and fifth tests.

4. Conclusions

In this study, we successfully synthesized Ag and Fe_3O_4 co-doped TiO_2 NFs using hydrothermal synthesis followed by thermal treatment in order to achieve high photocatalytic performance and a feasible recycle process. The synthesized $Ag/Fe_3O_4/TiO_2$ exhibited a relatively narrower band gap (2.0 eV) than that of pristine TiO_2 (3.1 eV). For the photoreduction of Cr(VI), electrons dominated the photoreduction efficiency. The photocatalytic process paired with ammonium oxalate could greatly facilitate the separation of electron-hole pairs and thus enhance the reduction rate of Cr(VI). After five cycles of the stability and recyclability test, the photoreduction efficiency was maintained at 90%, and the amount of remaining photocatalyst was maintained at 87%. Consequently, taking into account the high magnetic separation behavior and the high stability, $Ag/Fe_3O_4/TiO_2$ showed great potential to be used for practical wastewater treatment.

Author Contributions: Y.-H.C. performed the research and analyzed the data; Y.-H.C. and M.-C.W. wrote the paper; M.-C.W. was the supervisor and revised the paper. All authors read and approved the final manuscript.

Funding: This research was funded by the Ministry of Science and Technology, Taiwan (MOST 106-2221-E-182-057-MY3 and MSOT 107-2119-M-002-012), Green Technology Research Center, Chang Gung University (QZRPD181) and Chang Gung Memorial Hospital, Linkou (BMRPC74 and CMRPD2H0171).

Acknowledgments: The authors appreciate Wei-Fang Su at National Taiwan University and the Ming-Tao Lee group (TLS BL13A1) at National Synchrotron Radiation Research Center for useful discussion.

Conflicts of Interest: The authors declare no conflict of interest.

References

1. Costa, M. Toxicity and Carcinogenicity of Cr(VI) in Animal Models and Humans. *Crit. Rev. Toxicol.* **1997**, *27*, 431–442. [CrossRef] [PubMed]
2. Rowbotham, A.L.; Levy, L.S.; Shuker, L.K. Chromium in the Environment: An Evaluation of Exposure of the UK General Population and Possible Adverse Health Effects. *J. Toxicol. Environ. Health* **2000**, *3*, 145–178. [CrossRef]

3. Kongsricharoern, N.; Polprasert, C. Electrochemical Precipitation of Chromium (Cr^{6+}) from an Electroplating Wastewater. *Water Res.* **1995**, *31*, 109–117. [CrossRef]

4. Hunsom, M.; Pruksathorn, K.; Damronglerd, S.; Vergnes, H.; Duverneuil, P. Electrochemical Treatment of Heavy Metals (Cu^{2+}, Cr^{6+}, Ni^{2+}) from Industrial Effluent and Modeling of Copper Reduction. *Water Res.* **2005**, *39*, 610–616. [CrossRef] [PubMed]

5. Etemadi, M.; Samadi, S.; Yazd, S.S.; Jafari, P.; Yousefi, N.; Aliabadi, M. Selective Adsorption of Cr(VI) Ions from Aqueous Solutions Using Cr^{6+}-Imprinted Pebax/Chitosan/GO/Aptes Nanofibrous Adsorbent. *Int. J. Biol. Macromol.* **2017**, *95*, 725–733. [CrossRef] [PubMed]

6. Yogeshwaran, V.; Priya, A.K. Removal of Hexavalent Chromium (Cr^{6+}) Using Different Natural Adsorbents—A Review. *J. Chromatogr. Sep. Tech.* **2017**, *8*, 1000392. [CrossRef]

7. Ohtake, H. New Biological Method for Detoxification and Removal of Hexavalent Chromium. *Water Res.* **1992**, *25*, 395–402. [CrossRef]

8. Bennett, R.M.; Cordero, P.R.F.; Bautista, G.S.; Dedeles, G.R. Reduction of Hexavalent Chromium Using Fungi and Bacteria Isolated from Contaminated Soil and Water Samples. *Chem. Ecol.* **2013**, *29*, 320–328. [CrossRef]

9. Xing, Y.; Chen, X.; Wang, D. Electrically Regenerated Ion Exchange for Removal and Recovery of Cr(VI) from Wastewater. *Environ. Sci. Technol.* **2007**, *41*, 1439–1443. [CrossRef]

10. Mekatel, H.; Amokrane, S.; Benturki, A.; Nibou, D. Treatment of Polluted Aqueous Solutions by Ni^{2+}, Pb^{2+}, Zn^{2+}, Cr^{+6}, Cd^{+2} and Co^{+2} Ions by Ion Exchange Process Using Faujasite Zeolite. *Procedia Eng.* **2012**, *33*, 52–57. [CrossRef]

11. Cai, J.; Wu, X.; Zheng, F.; Li, S.; Wu, Y.; Lin, Y.; Lin, L.; Liu, B.; Chen, Q.; Lin, L. Influence of TiO_2 Hollow Sphere Size on Its Photo-Reduction Activity for Toxic Cr(VI) Removal. *J. Colloid Interface Sci.* **2017**, *490*, 37–45. [CrossRef] [PubMed]

12. Lu, D.; Yang, M.; Fang, P.; Li, C.; Jiang, L. Enhanced Photocatalytic Degradation of Aqueous Phenol and Cr(VI) over Visible-Light-Driven Tbxoy Loaded TiO_2-Oriented Nanosheets. *Appl. Surf. Sci.* **2017**, *399*, 167–184. [CrossRef]

13. Wang, L.; Zhang, C.; Gao, F.; Mailhot, G.; Pan, G. Algae Decorated TiO_2/Ag Hybrid Nanofiber Membrane with Enhanced Photocatalytic Activity for Cr(VI) Removal under Visible Light. *Chem. Eng. J.* **2017**, *314*, 622–630. [CrossRef]

14. Li, Y.; Liu, Z.; Wu, Y.; Chen, J.; Zhao, J.; Jin, F.; Na, P. Carbon Dots-TiO_2 Nanosheets Composites for Photoreduction of Cr(VI) under Sunlight Illumination: Favorable Role of Carbon Dots. *Appl. Catal. B* **2018**, *224*, 508–517. [CrossRef]

15. Wang, W.; Lai, M.; Fang, J.; Lu, C. Au and Pt Selectively Deposited on {001}-Faceted TiO_2 toward SPR Enhanced Photocatalytic Cr(VI) Reduction: The Influence of Excitation Wavelength. *Appl. Surf. Sci.* **2018**, *439*, 430–438. [CrossRef]

16. Ngo, A.; Nguyen, H.; Hollmann, D. Criticial Assessment of the Photocatalytic Reduction of Cr(VI) over Au/TiO_2. *Catalysts* **2018**, *8*, 606. [CrossRef]

17. Tan, L.-L.; Ong, W.-J.; Chai, S.-P.; Mohamed, A.R. Visible-Light-Activated Oxygen-Rich TiO_2 as Next Generation Photocatalyst: Importance of Annealing Temperature on the Photoactivity toward Reduction of Carbon Dioxide. *Chem. Eng. J.* **2016**, *283*, 1254–1263. [CrossRef]

18. Busiakiewicz, A.; Kisielewska, A.; Piwoński, I.; Batory, D. The Effect of Fe Segregation on the Photocatalytic Growth of Ag Nanoparticles on Rutile TiO_2 (001). *Appl. Surf. Sci.* **2017**, *401*, 378–384. [CrossRef]

19. Low, J.; Cheng, B.; Yu, J. Surface Modification and Enhanced Photocatalytic CO_2 Reduction Performance of TiO_2: A Review. *Appl. Surf. Sci.* **2017**, *392*, 658–686. [CrossRef]

20. Tan, L.-L.; Ong, W.-J.; Chai, S.-P.; Mohamed, A.R. Photocatalytic Reduction of CO_2 with H_2O over Graphene Oxide-Supported Oxygen-Rich TiO_2 Hybrid Photocatalyst under Visible Light Irradiation: Process and Kinetic Studies. *Chem. Eng. J.* **2017**, *308*, 248–255. [CrossRef]

21. Zhang, Y.; Gu, D.; Zhu, L.; Wang, B. Highly Ordered Fe^{3+}/TiO_2 Nanotube Arrays for Efficient Photocataltyic Degradation of Nitrobenzene. *Appl. Surf. Sci.* **2017**, *420*, 896–904. [CrossRef]

22. Duan, Y.; Liang, L.; Lv, K.; Li, Q.; Li, M. TiO_2 Faceted Nanocrystals on the Nanofibers: Homojunction TiO_2 Based Z-Scheme Photocatalyst for Air Purification. *Appl. Surf. Sci.* **2018**, *456*, 817–826. [CrossRef]

23. Makal, P.; Das, D. Self-Doped TiO_2 Nanowires in TiO_2-B Single Phase, TiO_2-B/Anatase and TiO_2-Anatase/Rutile Heterojunctions Demonstrating Individual Superiority in Photocatalytic Activity under Visible and UV Light. *Appl. Surf. Sci.* **2018**, *455*, 1106–1115. [CrossRef]

24. Wu, M.-C.; Hsiao, K.-C.; Chang, Y.-H.; Chan, S.-H. Photocatalytic Hydrogen Evolution of Palladium Nanoparticles Decorated Black TiO_2 Calcined in Argon Atmosphere. *Appl. Surf. Sci.* **2018**, *430*, 407–414. [CrossRef]

25. Jing, J.; Li, J.; Feng, J.; Li, W.; Yu, W.W. Photodegradation of Quinoline in Water over Magnetically Separable Fe_3O_4/TiO_2 Composite Photocatalysts. *Chem. Eng. J.* **2013**, *219*, 355–360. [CrossRef]

26. Li, Z.-J.; Huang, Z.-W.; Guo, W.-L.; Wang, L.; Zheng, L.-R.; Chai, Z.-F.; Shi, W.-Q. Enhanced Photocatalytic Removal of Uranium(VI) from Aqueous Solution by Magnetic TiO_2/Fe_3O_4 and Its Graphene Composite. *Environ. Sci. Technol.* **2017**, *51*, 5666–5674. [CrossRef] [PubMed]

27. Yu, Y.; Yan, L.; Cheng, J.; Jing, C. Mechanistic Insights into TiO_2 Thickness in $Fe_3O_4@TiO_2$-GO Composites for Enrofloxacin Photodegradation. *Chem. Eng. J.* **2017**, *325*, 647–654. [CrossRef]

28. Liu, M.-C.; Liu, B.; Sun, X.-Y.; Lin, H.-C.; Lu, J.-Z.; Jin, S.-F.; Yan, S.-Q.; Li, Y.-Y.; Zhao, P. Core/Shell Structured $Fe_3O_4@TiO_2$-DNM Nanospheres as Multifunctional Anticancer Platform: Chemotherapy and Photodynamic Therapy Research. *J. Nanosci. Nanotechnol.* **2018**, *18*, 4445–4456. [CrossRef] [PubMed]

29. Wu, W.; Xiao, X.; Zhang, S.; Ren, F.; Jiang, C. Facile Method to Synthesize Magnetic Iron Oxides/TiO_2 Hybrid Nanoparticles and Their Photodegradation Application of Methylene Blue. *Nanoscale Res. Lett.* **2011**, *6*, 533. [CrossRef]

30. Shojaei, A.F.; Shams-Nateri, A.; Ghomashpasand, M. Magnetically Recyclable $Fe^{3+}/TiO_2@Fe_3O_4$ Nanocomposites Towards Degradation of Direct Blue 71 under Visible-Light Irradiation. *IET Micro Nano Lett.* **2017**, *12*, 161–165. [CrossRef]

31. Jia, X.; Dai, R.; Lian, D.; Han, S.; Wu, X.; Song, H. Facile Synthesis and Enhanced Magnetic, Photocatalytic Properties of One-Dimensional Ag@Fe_3O_4-TiO_2. *Appl. Surf. Sci.* **2017**, *392*, 268–276. [CrossRef]

32. Wang, H.; Fei, X.; Wang, L.; Li, Y.; Xu, S.; Sun, M.; Sun, L.; Zhang, C.; Li, Y.; Yang, Q.; et al. Magnetically Separable Iron Oxide Nanostructures-TiO_2 Nanofibers Hierarchical Heterostructures: Controlled Fabrication and Photocatalytic Activity. *New J. Chem.* **2011**, *35*, 1795–1802. [CrossRef]

33. Jafari, A.; Shayesteh, S.F.; Salouti, M.; Boustani, K. Effect of Annealing Temperature on Magnetic Phase Transition in Fe_3O_4 Nanoparticles. *J. Magn. Magn. Mater.* **2015**, *379*, 305–312. [CrossRef]

34. Wang, W.-K.; Chen, J.-J.; Gao, M.; Huang, Y.-X.; Zhang, X.; Yu, H.-Q. Photocatalytic Degradation of Atrazine by Boron-Doped TiO_2 with a Tunable Rutile/Anatase Ratio. *Appl. Catal. B* **2016**, *195*, 69–76. [CrossRef]

35. Rossi, G.; Pasquini, L.; Catone, D.; Piccioni, A.; Patelli, N.; Paladini, A.; Molinari, A.; Caramori, S.; O'Keeffe, P.; Boscherini, F. Charge Carrier Dynamics and Visible Light Photocatalysis in Vanadium-Doped TiO_2 Nanoparticles. *Appl. Catal. B* **2018**, *237*, 603–612. [CrossRef]

36. Wu, M.-C.; Wu, P.-Y.; Lin, T.-H.; Lin, T.-F. Photocatalytic Performance of Cu-Doped TiO_2 Nanofibers Treated by the Hydrothermal Synthesis and Air Thermal Treatment. *Appl. Surf. Sci.* **2018**, *430*, 390–398. [CrossRef]

37. Avilés-García, O.; Espino-Valencia, J.; Romero-Romero, R.; Rico-Cerda, J.; Arroyo-Albiter, M.; Solís-Casados, D.; Natividad-Rangel, R. Enhanced Photocatalytic Activity of Titania by Co-Doping with Mo and W. *Catalysts* **2018**, *8*, 631. [CrossRef]

38. Ramírez-Sánchez, I.; Bandala, E. Photocatalytic Degradation of Estriol Using Iron-Doped TiO_2 under High and Low UV Irradiation. *Catalysts* **2018**, *8*, 625. [CrossRef]

39. Ali, T.; Ahmed, A.; Alam, U.; Uddin, I.; Tripathi, P.; Muneer, M. Enhanced Photocatalytic and Antibacterial Activities of Ag-Doped TiO_2 Nanoparticles under Visible Light. *Mater. Chem. Phys* **2018**, *212*, 325–335. [CrossRef]

40. Mandari, K.K.; Do, J.Y.; Police, A.K.R.; Kang, M. Natural Solar Light-Driven Preparation of Plasmonic Resonance-Based Alloy and Core-Shell Catalyst for Sustainable Enhanced Hydrogen Production: Green Approach and Characterization. *Appl. Catal. B* **2018**, *231*, 137–150. [CrossRef]

41. Chen, S.-H.; Chan, S.-H.; Lin, Y.-T.; Wu, M.-C. Enhanced Power Conversion Efficiency of Perovskite Solar Cells Based on Mesoscopic Ag-Doped TiO_2 Electron Transport Layer. *Appl. Surf. Sci.* **2019**, *469*, 18–26. [CrossRef]

42. Wu, M.-C.; Liao, Y.-H.; Chan, S.-H.; Lu, C.-F.; Su, W.-F. Enhancing Organolead Halide Perovskite Solar Cells Performance through Interfacial Engineering Using Ag-Doped TiO_2 Hole Blocking Layer. *Sol. RRL* **2018**, *2*, 1800072. [CrossRef]

43. Akel, S.; Dillert, R.; Balayeva, N.; Boughaled, R.; Koch, J.; El Azzouzi, M.; Bahnemann, D. Ag/Ag_2O as a Co-Catalyst in TiO_2 Photocatalysis: Effect of the Co-Catalyst/Photocatalyst Mass Ratio. *Catalysts* **2018**, *8*, 647. [CrossRef]

44. Wen, L.; Liu, B.; Zhao, X.; Nakata, K.; Murakami, T.; Fujishima, A. Synthesis, Characterization, and Photocatalysis of Fe-Doped TiO_2: A Combined Experimental and Theoretical Study. *Int. J. Photoenergy* **2012**, *2012*, 368750. [CrossRef]

45. Zhan, J.; Zhang, H.; Zhu, G. Magnetic Photocatalysts of Cenospheres Coated with Fe_3O_4/TiO_2 Core/Shell Nanoparticles Decorated with Ag Nanopartilces. *Ceram. Int.* **2014**, *40*, 8547–8559. [CrossRef]
46. Mogal, S.; Gandhi, V.G.; Mishra, M.; Tripathi, S.; Shripathi, T.; Joshi, P.; Shah, D. Single-Step Synthesis of Silver-Doped Titanium Dioxide: Influence of Silver on Structural, Textural, and Photocatalytic Properties. *Ind. Eng. Chem. Res.* **2014**, *53*, 5749–5758. [CrossRef]

 *catalysts*

Article

Photocatalytic Hydrogen Production Under Near-UV Using Pd-Doped Mesoporous TiO$_2$ and Ethanol as Organic Scavenger

Bianca Rusinque, Salvador Escobedo and Hugo de Lasa *

Chemical Reactor Engineering Centre (CREC), Faculty of Engineering, Western University, London, ON N6A 5B9, Canada; brusinqu@uwo.ca (B.R.); selfa.iq@gmail.com (S.E.)
* Correspondence: hdelasa@uwo.ca; Tel.: +1-519-661-2149

Received: 20 November 2018; Accepted: 21 December 2018; Published: 2 January 2019

Abstract: Photocatalysis can be used advantageously for hydrogen production using a light source (near-UV light), a noble metal-doped semiconductor and an organic scavenger (2.0 v/v% ethanol). With this end, palladium was doped on TiO$_2$ photocatalysts at different metal loadings (0.25 to 5.00 wt%). Photocatalysts were synthetized using a sol-gel method enhancing morphological properties with a soft template precursor. Experiments were carried out in the Photo-CREC Water II reactor system developed at CREC-UWO (Chemical Reactor Engineering Centre- The University of Western Ontario) Canada. This novel unit offers hydrogen storage and symmetrical irradiation allowing precise irradiation measurements for macroscopic energy balances. Hydrogen production rates followed in all cases a zero-order reaction, with quantum yields as high as 30.8%.

Keywords: Photo-CREC Water II reactor; Palladium; TiO$_2$; Hydrogen production; Quantum Yield

---

1. Introduction

Hydrogen is a key energy carrier that will likely play an important role in the transportation sector by 2050 [1]. It is considered an environmentally friendly energy vector due to its zero CO$_2$ and zero noxious gas emissions when combusted [2]. Through a photocatalytic water-splitting process, hydrogen can be produced using water and a light source as primary resources [3]. Furthermore, a sacrificial organic agent is required to allow the photocatalytic reaction to occur, forming the desired products [4].

Common sacrificial agents include methanol, triethanolamine, ethanol, acids and inorganic compounds [5]. Ethanol, as one of the most-investigated sacrificial agents, provides high quantum efficiencies and will be used in this work as scavenger. The use of ethanol as a sacrificial agent is advantageous given that it can be easily produced from renewable biomass (fermentation processes), making it available and inexpensive [6].

Photocatalytic hydrogen production with sacrificial organic agents proceeds as follows: (a) absorbed photons surpass the energy band gap and generate excited electron-hole pairs [7], and (b) photoexcited electron-hole pairs can be separated due to the sacrificial agent presence. This allows the formation of hydrogen with minimum electron-hole pair recombination [8] and (c) hydroxyl groups from dissociated water lead OH· radical formation and contribute to the conversion of the scavenger [9]. The "in-series-parallel" reaction network was described in detailed by our research team in [10].

Titanium dioxide (TiO$_2$) is a well-known photocatalyst capable of absorbing light and producing electron-hole pairs to accelerate the rate of a water-splitting reaction [11]. TiO$_2$ has been the most used material due to its stability, resistance to corrosion, cleanliness (no pollutant), availability in nature and inexpensiveness compared to other semiconductors [12]. It can be found in three allotropic phases—anatase, rutile, and brookite—where the anatase phase is the most photoactive phase reported [13]. Doping noble metals on TiO_2 allows (a) increasing the efficiency of the hydrogen evolution reaction, (b) narrowing the band gaps, and (c) improving the optoelectronic semiconductor properties [14]. Furthermore, doping TiO_2 with noble metals such as Pd helps to promote energy levels near the band edges or mid-gap states, which may reduce the effective bandgap energy [15]. Therefore, the effect of Pd on TiO_2 can be attributed to the shift of Fermi levels in the composite material [16–20].

One of the most important parameters in photocatalytic reactors and photocatalysts is the quantum yield (QY). This parameter establishes the process efficiency relating the photogenerated radical rate over the absorbed photons rate [21]. Using the QY, the doped photocatalysts of the present study were evaluated.

Previous studies by our research team considered Pt doped on TiO_2 obtaining modest quantum yield efficiencies in the 0.7–8.0% range when using undoped DP-25 and 1.0% Pt-impregnated DP-25, respectively [22]. However, the present work emphasizes the photocatalytic hydrogen production using doped Pd, as a much less expensive dopant on mesoporous TiO_2. This semiconductor material, $Pd–TiO_2$, is employed under near-UV light, in the Photo-CREC Water II reactor unit developed at CREC (Chemical Reactor Engineering Centre). Results obtained are of significant value due to the high quantum yields obtained in the system.

2. Results and Discussion

2.1. Photocatalyst Characterization

2.1.1. Brunauer–Emmett–Teller (BET) Surface Area

Using a BET surface area unit (Micrometrics, ASAP 2010), the photocatalysts were analyzed with nitrogen. Furthermore, the adsorption–desorption isotherms of type IV were generated and the V_p total photocatalyst pore volume was also calculated with the liquid nitrogen adsorbed at the P/P_o relative pressure of 0.99 [23]. Table 1 reports the specific surface area, the average pore diameter and the specific pore volume for mesoporous TiO_2 prepared using F-127 template. It is shown that when using this template, the morphological properties of TiO_2 are improved as follows: (a) specific surface areas are increased, (b) average pore diameters are augmented, and (c) specific pore volumes are increased. Furthermore, one can also notice that the $F-127–TiO_2$, displays both higher porosity and specific surface area than the TiO_2 itself.

Table 1. Surface area and pore diameter using template Pluronic F-127.

Photocatalyst	S^{BET} $(m^2\,g^{-1})$	Dp^{BJH} $(4Vp^{BJH}/S^{BET})$ (nm)	Vp^{BJH} (cm^3g^{-1})
Degussa P-25	59	7.5	0.25
F-127–TiO_2 500 °C	140	17.5	0.61

According to Table 2, the best results in specific surface area were obtained with the mesoporous TiO_2 photocatalysts. These photocatalysts display a clear increment of the specific surface area and specific pore volume (D_p) when compared to Degussa P-25 (commercial titania photocatalyst that is commonly used in photocatalytic reactions). However, when using doped Pd on TiO_2, there was a modest reduction in specific surface area and a mild change in pore diameter attributed to a possible and moderate blocking of the TiO_2 pores with Pd [24].

Table 2. Surface area and pore diameter using palladium.

Photocatalyst	S^{BET} $(m^2 \, g^{-1})$	$Dp^{\,BJH}$ $(4Vp^{BJH}/S_{BET})$ (nm)	Vp^{BJH} $(cm^3 g^{-1})$
Anatase	11	7.3	0.05
Rutile	5	4.7	0.05
Degussa P-25	59	7.5	0.25
F-127–TiO$_2$-500 °C	140	17.5	0.61
F-127–0.25 wt% Pd–TiO$_2$ 500 °C	131	16.5	0.53
F-127–0.50 wt% Pd–TiO$_2$ 500 °C	124	16.8	0.52
F-127–1.0 wt% Pd–TiO$_2$ 500 °C	123	21.2	0.65
F-127–2.5 wt% Pd–TiO$_2$ 500 °C	122	19.9	0.60
F-127–5.0 wt% Pd–TiO$_2$ 500 °C	119	18.9	0.56

The Barrett–Joyner–Halenda (BJH) method was also used to determine the pore size distribution, by utilizing N_2 as an adsorbate and as a desorption isotherm. In all cases, a bimodal pore volume distribution was observed. The largest pore sizes in the 16–35 nm range were achieved with F-127–1.0 wt% Pd–TiO$_2$-500 °C.

2.1.2. Pulse Hydrogen Chemisorption

Table 3 reports hydrogen chemisorption showing the effect of metal loading on metal dispersion. When Pd is used as a dopant, it is shown that higher metal loadings lead to reduced metal dispersion.

Table 3. Chemisorption analysis: metal dispersion.

Photocatalyst	Metal Dispersion (%)
F-127–0.25 wt% Pd–TiO$_2$ 500 °C	75
F-127–0.50 wt% Pd–TiO$_2$ 500 °C	27
F-127–1.0 wt% Pd–TiO$_2$ 500 °C	26
F-127–2.5 wt% Pd–TiO$_2$ 500 °C	12
F-127–5.0 wt% Pd–TiO$_2$ 500 °C	8

2.1.3. X-Ray Diffraction (XRD)

Figure 1 reports XRD diffractograms for TiO$_2$ synthesized using the F-127 template and with varying palladium loadings. Anatase and rutile XRDs are reported as references. In this respect, one can observe that anatase peaks are at the 2θ diffraction angles of 25°, 38°, 48°, 54°, 63°, 69°, 70.5° and 75° corresponding to the planes (101), (004), (200), (105), (204), (116), (220) and (215) [JCPDS No. 73-1764]. For rutile, there are a couple of noticeable peaks at 54° and 67° corresponding to the planes (201) and (301) [JCPDS No. 34-0180].

Furthermore, Figure 1 also shows that the XRD diffractograms for Pd-doped TiO$_2$ were consistent where a significant anatase XRD peak was observed. The nature of the desirable anatase peaks in this semiconductor was confirmed with a 99.7% anatase from Aldrich reference sample [25]. Anatase for all photocatalyst was the dominant TiO$_2$ crystalline phase assumed as 100% with no rutile being present.

Pd peaks were also identified and recorded at 40.12° (111) and 46.66° (200) at the 2θ angles [JCPDS No. 87-0638]. One should observe that, in principle, a third peak at 2θ = 68.1° (220) might be recorded when using Pd as a dopant. However, this peak may overlap with anatase and as a result cannot be used for Pd identification [26].

On the other hand, the crystallite sizes for each photocatalyst were determined using the Scherrer equation. On this basis, the mesoporous photocatalysts displayed crystallite sizes between 9 and 14 nm. Lattice parameters of the tetragonal anatase unit cell were also calculated and are reported in Appendix B.

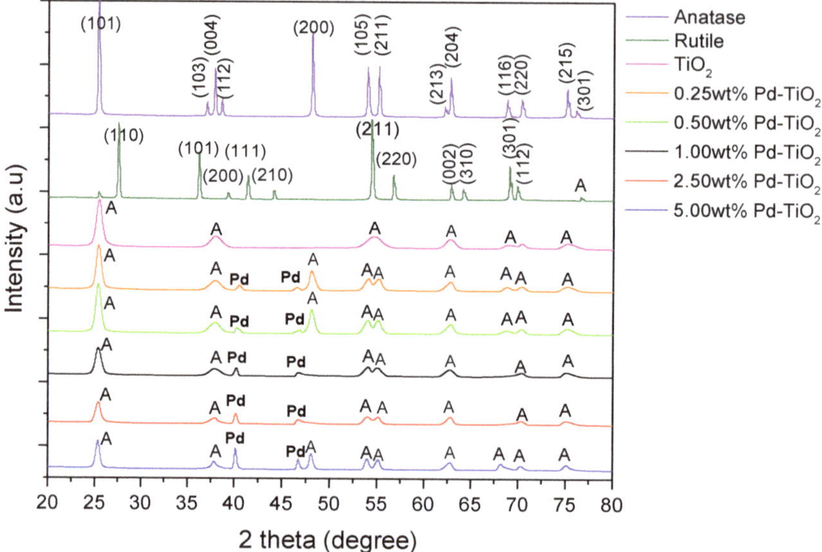

Figure 1. X-ray diffraction (XRD) diffractograms of photocatalyst doped with Pd. XRDs overlapped for comparison. A = anatase, Pd = palladium.

2.1.4. Band Gap

Figure 2 reports that increasing the Pd content slightly augments the band gap. However, and regarding the observed results, one can see in all cases a significant reduction in the band gaps for TiO_2 doped with lower Pd loadings versus the band gaps for undoped TiO_2. The best band gaps achieved were 2.51 for 0.25 wt% Pd and 2.55 eV for 0.50–1.00 wt% Pd–TiO_2.

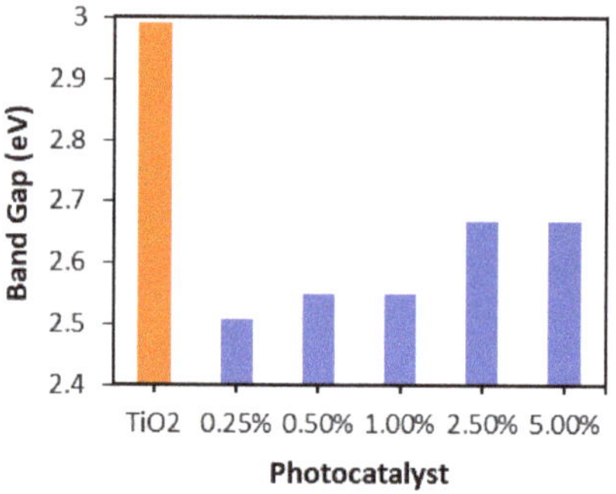

Figure 2. Effect of Pd loading on the optical band gap.

By applying the Kubelka–Munk (K–M) model and following the Tauc plot methodology, the band gaps were determined. Figure 3 reports the changes of the "$(\alpha h v)^{1/2}$" function versus the photon energy "hv", with α representing the absorption coefficient, h being the Planck constant (6.34×10^{34} J s/photon) and v denoting the radiation frequency. It should also be noted that $v = c/\lambda$, where c is the speed of light under vacuum (3.00×10^8 m/s^2). If the straight-line methodology is applied for the band gap calculation as shown with the red line, one can see that the intersection of this line with the abscissa provides the wavelength corresponding to the semiconductor band gap. Furthermore, the Tauc plots (Figure 3) were developed for Pd-doped TiO_2 photocatalysts using the F-127 template and at a 500 °C calcination temperature.

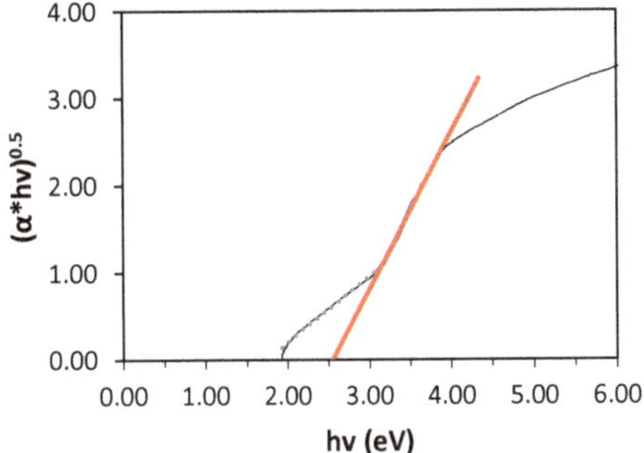

Figure 3. Band gap calculation using the Tauc plot methodology and the straight-line extrapolation for 1.0 wt% Pd–TiO$_2$.

2.1.5. X-Ray Photoelectron Spectroscopy (XPS)

The 1.0 wt% Pd–TiO$_2$ photocatalyst was analyzed using the XPS technique. Figure 4 shows the XPS for Pd(0) and PdO. In each case, one can see double peaks as follows: (a) Pd (0) 3d5/2, with 334.70 eV and 3d3/2 at 339.96 eV binding energies, (b) PdO 3d5/2 with 336.46 eV and 3d3/2 at 341.72 eV. This yields a consistent 48.6% of Pd (0) and 51.4% of PdO, at the two binding energy ranges considered.

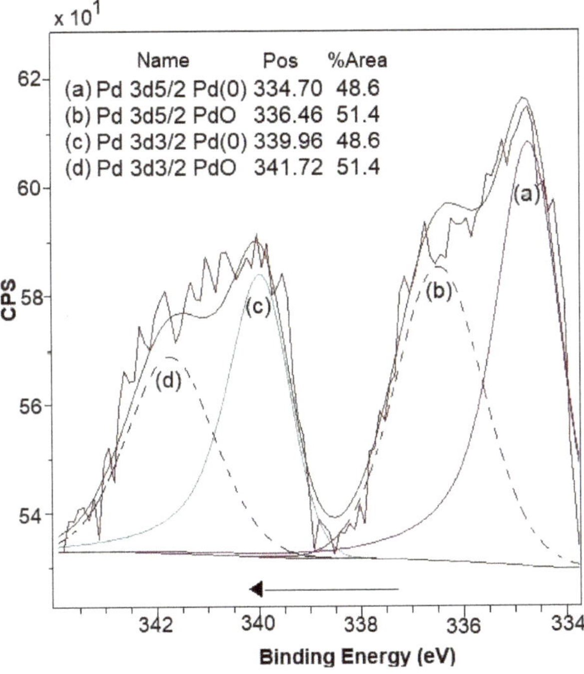

Figure 4. High-resolution X-ray photoelectron spectroscopy (XPS) spectra for 1.00 wt% Pd–TiO$_2$. Note: Full lines represent Pd (0) at (a) 3d5/2 and (c) 3d3/2. Broken lines represent PdO at (b) 3d5/2 and (d) 3d3/2.

It was also observed that titanium and oxygen were present as major components in the photocatalyst mesoporous support as TiO$_2$ species. Titanium was detected at a binding energy position of 454.45 eV, while oxygen was identified at 525.85 eV. These bands fell outside the Pd and PdO binding energies as shown in Figure 4, avoiding any possible inadequate band assignment.

In conclusion, the XPS data of Figure 4 shows the significant Pd (0) availability, and points towards possible future improvements of the synthesized photocatalyst via enhanced Pd reduction.

2.2. Macroscopic Radiation Energy Balance (MREB)

Photocatalytic reactors operate based on emitted photons. These photons are absorbed by a circulating semiconductor slurry suspension. To be able to establish the absorbed radiation in the Photo-CREC Water II Reactor, one must develop a macroscopic radiation balance for accurate energy efficiency calculations [27].

The macroscopic balance estimates the photons absorbed as the difference between the incident photons and the combined scattered and transmitted photons [28].

$$P_a = P_i - P_{bs} - P_t \tag{1}$$

where Pa is the rate of absorbed photons, for which it is desired to be as high as possible; P_i is the rate of photons reaching the reactor at the inner Pyrex glass surface and is calculated according to Equation (1) in Einstein s^{-1}; P_{bs} is the rate of backscattered photons; and P_t is the rate of transmitted photons. All these variables can be expressed using the Einstein s^{-1} units.

Furthermore:

$$P_i = P_0 - P_{a-wall} \tag{2}$$

with P_0 in Einstein s^{-1} being the rate of photons emitted by the lamps as per $P_{a\text{-}wall}$ in Einstein s^{-1}, which accounts for the rate of backscattered photons absorbed by the Pyrex glass walls.

In addition, P_0 can be calculated as:

$$P_0 = \int_{\lambda_1}^{\lambda_2} \lambda \int_0^L \int_0^{2\pi} q(\theta, Z, \lambda) r \, d\theta \, dz \, d\lambda \tag{3}$$

where $q\,(\theta, z, \lambda)$ is the radiative flux (J s^{-1} m^{-3}), λ represents the wavelength (nm), r stands for the radial coordinate (m), z denotes the axial coordinate (m), h is the Planck's constant (J s), and c represents the speed of light (m s^{-1}). The term $q\,(\theta, z, \lambda)$ is determined using the spectrometer.

Furthermore, when photocatalytic experiments are performed in the Photo-CREC Water II (PCW-II) reactor, photons are absorbed and scattered in the reacting medium. As a result, a backscattering has to be accounted for. A possible approach to calculate backscattering is to establish the difference between P_i and the rate of photons transmitted when the catalyst concentration approaches zero ($P_t \,|\, _{C \to 0+}$):

$$P_{bs} = P_i - P_t |_{c \to 0^+} \tag{4}$$

Equation (4) assumes that photons are backscattered on the TiO$_2$ particle layer close to the inner surface of the transparent Pyrex walls surface. Equation (4) also assumes that no other backscattered photons contribute to P_{bs}.

Additionally, for P_t determination, Equation (5) considers that transmitted radiation can be defined as the addition of normal scattered photons and forward scattered photons:

$$P_t = P_{ns} - P_{fs} \tag{5}$$

One should note that ($P_{fs} + P_{ns}$) can be measured by employing aluminum polished collimators, which capture radiation reaching the measuring point, with large view angles [28].

Thus, to assess P_a as in Equation (1), macroscopic balances using near-UV light were established at the central axial position using a 0.15 g/L photocatalyst concentration. Figure 5 reports measurements for various TiO$_2$ photocatalysts with different metal loadings.

According to Table 4 and Figure 5, one can observe that additions of Pd on TiO_2 show that (a) lower Pd levels (0.25 to 1.00 wt%) lead to an increased P_a and high absorption efficiencies compared to undoped TiO_2, and (b) higher Pd levels (2.50 and 5.00 wt% Pd) give smaller P_a and reduced absorption efficiency. These findings are in line with an increased rate of transmitted photons when using low Pd loadings as well as incremental photon backscattering when using high Pd loadings.

Table 4. Absorbed photon rates on TiO_2 photocatalysts at 0.15 g/L of photocatalyst concentration.

Near-UV Light	P_a (Einstein/s)
TiO_2	3.11×10^6
0.25 wt% Pd	3.18×10^{-6}
0.50 wt% Pd	3.52×10^{-6}
1.00 wt% Pd	5.11×10^{-6}
2.50 wt% Pd	3.77×10^{-6}
5.00 wt% Pd	3.76×10^{-6}

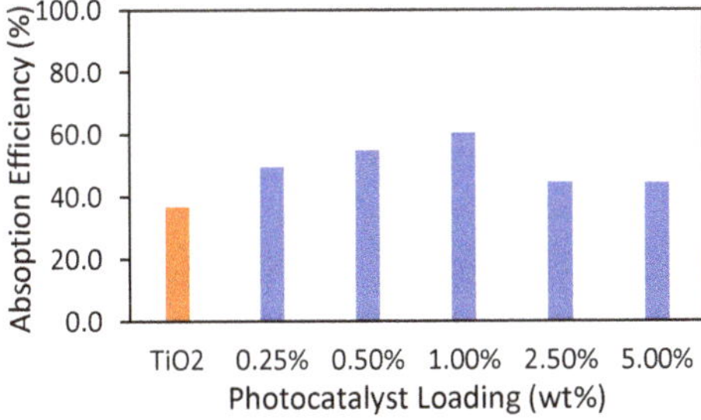

Figure 5. Absorption efficiency on TiO_2 photocatalysts at different metal loadings under near-UV light.

2.3. Hydrogen Production

2.3.1. Effect of Palladium Loadings

Palladium was used as co-catalyst to dope the structure of the TiO_2 photocatalyst. This metal enhances the hydrogen production, as compared to the undoped mesoporous TiO_2. Nobel metal crystallites reduce the band gap and facilitate electron capture [29]. As a result, Pd reduces the recombination between holes and electrons, promoting better photocatalytic water-splitting performances [30].

Figure 6 reports the influence of Pd on TiO_2 in terms of cumulative hydrogen volume.

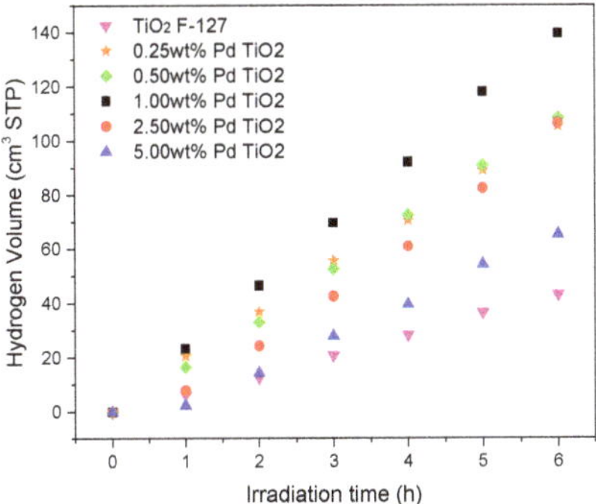

Figure 6. Cumulative hydrogen volume using Pd at different metal loadings (0.25, 1.50, 1.00, 2.50 and 5.00 wt%). Conditions: photocatalyst concentration 0.15 g/L, 2.0 v/v% ethanol, pH = 4 ± 0.05 and near-UV light. Standard deviation: ±3.0%.

Figure 6 shows there is a maximum volume of 140 cm^3 STP (standard temperature and pressure) of hydrogen produced in six hours when using 1.00 wt% Pd on TiO_2. This volume is slightly higher than the maximum volume of hydrogen produced when using 0.25 wt% and 0.50 wt% Pd–TiO_2, and three times the volume of hydrogen obtained for undoped TiO_2. One should also note as well that this volume is close to the 128 cm^3 STP of hydrogen produced when platinum is used as a dopant under the same reaction conditions but with a much larger metal loading (5.00 wt% Pt) on TiO_2 [31].

Furthermore, one should note that the 140 cm^3 STP of hydrogen produced in six hours with 1.00 wt% Pd on TiO_2 decreased up to 60 cm^3 STP when using higher Pd loadings (2.50 wt% Pd and 5.00 wt% Pd). The macroscopic radiation energy balance provides an explanation showing that at the higher Pd loadings, there is increased irradiation backscattering, with greater irradiation being reflected and, as a consequence, light absorption being reduced. This is in contrast with the lower than 1.00 wt% Pd loadings evaluated, where the absorption efficiency, as well as the rate of transmitted photons, increases. Thus, a diminished irradiation absorption given by 2.50 wt% Pd and 5.00 wt% Pd negatively affects the photocatalyst performance [32].

In agreement with this, at the lower palladium loadings studied (0.25, 0.50 and 1.00 wt%) good metal dispersion, mildly affected specific surface area and pore structure were achieved [33]. On the other hand, for 2.50 and 5.00 wt% Pd–TiO_2, poorer metal dispersion with larger metal crystallite sizes were observed, with this being in line with the lower photocatalytic activity [34].

In all cases, palladium-doped TiO_2 showed a consistent steady linear trend. The hydrogen production rate displayed consistent zero-order kinetics, with no noticeable photocatalytic decay. This material is stable for extended irradiation periods and no apparent deactivation for 24 h following an "in series-parallel" reaction mechanism shown in detail in [22].

These results show that palladium at 1.00 wt% loading can produce valuable hydrogen yields, with this being an excellent replacement for platinum. As well, Pd can be considered more advantageous than Pt, given that Pd is less expensive (only 20–25% of the cost of platinum). Furthermore, and given the premise of nominal 1.00 wt% Pd–TiO_2, photocatalyst X-ray fluorescence spectrometry (XRF) was used to confirm the nominal loading. The observed XRF value was 1.17 wt% Pd on mesoporous TiO_2.

2.3.2. Effect of Catalyst Concentration on Hydrogen Production

Considering that 1.00 wt% Pd–TiO_2 showed the best performance in terms of hydrogen production, additional experiments were carried out to determine the influence of the catalyst concentration during photoreaction.

Figure 7 displays four different slurry concentrations of the 1.00 wt% Pd–TiO_2 photocatalyst: 0.15, 0.30, 0.50 and 1.00 g/L. These experiments were studied during 6 h of irradiation. One can observe that the runs with 1.00 g/L showed the highest hydrogen production. Thus, given these results, it can be considered that when higher photocatalyst slurry concentrations are used, more photocatalyst electron-holes are provided, with this promoting better hydrogen production.

As a result, it was observed that the hydrogen production rate increased 54% when photocatalyst concentration was augmented seven times from 0.15 to 1.00 g/L. However, despite this hydrogen production increase, this could be considered a modest improvement only, given that the photocatalyst needed and the related cost was significantly augmented. Therefore, a photocatalyst concentration of 0.15 g/L was considered as a best choice and was selected for further experimentation.

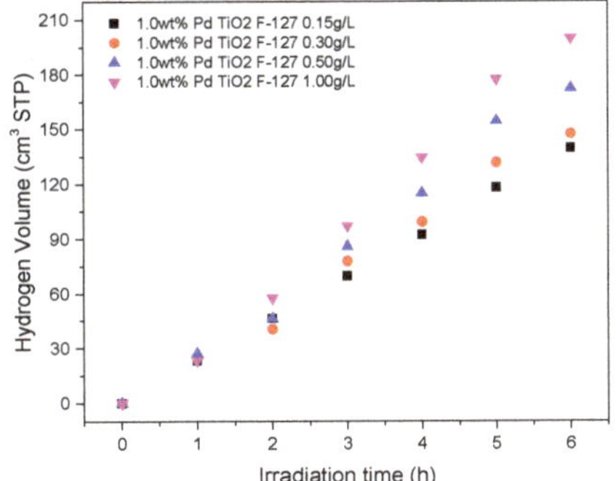

Figure 7. Cumulative hydrogen production using 1.0 wt% Pd–TiO$_2$ at different catalyst concentrations (0.15, 0.30, 0.50 and 1.0 g/L). Conditions: 2.0 v/v% ethanol, pH = 4 ± 0.05 and near-UV light. Standard deviation: ±4.0%.

2.3.3. Effect of Photo-CREC Water II Atmosphere using Argon and CO$_2$

Before starting water-splitting runs, the reactor gas chamber was purged with an inert gas to remove the oxygen from the air, avoiding combustion reactions. Argon was used initially as the inert gas given this is heavier than oxygen facilitating its displacement [35]. On the other hand, CO$_2$ was also used in separate runs in the reactor gas chamber to determine its possible influence on water dissociation reactions.

According to Figure 8, using argon as an inert gas and utilizing 1.00 wt% Pd–TiO$_2$, yielded 140 cm^3 STP of hydrogen after six hours of irradiation. On the other hand, when a CO$_2$ atmosphere was employed, only 80 cm^3 STP of hydrogen was obtained. One should note that under a CO$_2$ atmosphere and due to the competition of the CO$_2$ photoreduction with the hydrogen production, a lower net hydrogen formation can be explained.

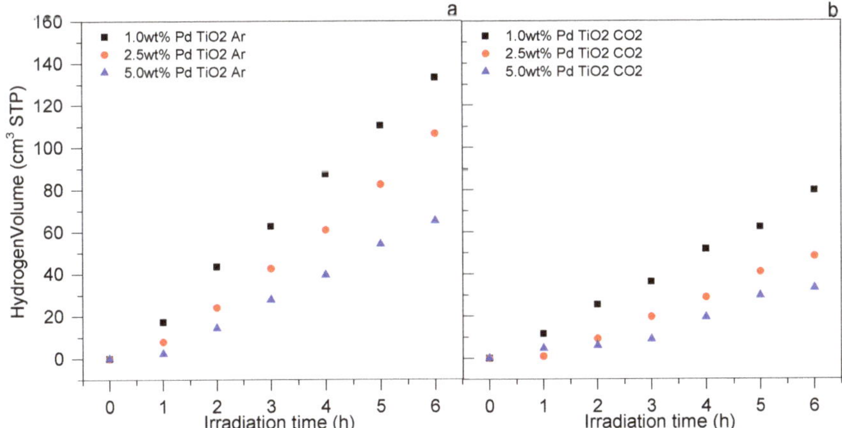

Figure 8. Cumulative hydrogen production using xPd–TiO$_2$ (x = 1.00, 2.50 and 5.00 wt%) and under two atmospheres: (**a**) argon and (**b**) CO$_2$. Conditions: photocatalyst concentration 0.15 g/L, 2.0 v/v% ethanol, pH = 4 ± 0.05 and near-UV light. Standard deviation: (**a**) ±3.0%, (**b**) ±3.4%.

Regarding CO$_2$ during the six h of irradiation, it was observed that it steadily augmented under an argon atmosphere reaching 0.4 cm^3 STP. On the other hand, when the runs were performed under a CO$_2$ atmosphere, the CO$_2$ increment was limited to 0.01 cm^3 STP. These findings support the view that there is competition between CO$_2$ photoreduction and CO$_2$ formation via ethanol OH· radical scavenging. It is assumed that these gas phase CO_2 findings could be also be influenced by the enhanced CO_2 solubility in water–ethanol [36].

2.3.4. Effect of Sacrificial Agent Concentration

As a scavenger, ethanol offers important advantages, such as the photogeneration of electron-holes, limiting electron-site recombination and improving photocatalytic activity. Ethanol can donate electrons to scavenge the valence holes and suppresses the reverse reaction [37].

Experiments were performed at 1.00 wt% Pd and three ethanol concentrations (1.0, 2.0, 4.0 v/v%) under an argon atmosphere and with 0.15 g/L of photocatalyst concentration. This was done to evaluate the effect of the ethanol concentration on hydrogen production. Figure 9 reports the influence of increasing ethanol from 2.0–4.0% on hydrogen production rates.

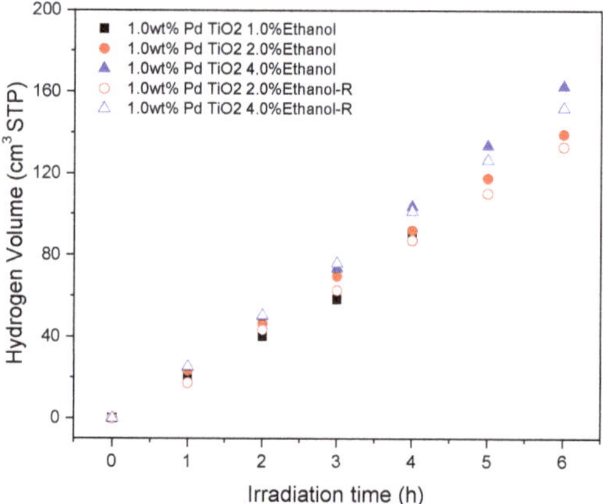

Figure 9. Hydrogen volume using 1.00 wt% Pd at 1.0, 2.0 and 4.0 v/v% ethanol. Conditions: photocatalyst concentration 0.15 g/L, argon atmosphere, pH = 4 ± 0.05 and near-UV light, R = repeat. Standard deviation: ±6.5%.

As shown in Figure 9, the highest hydrogen formation rate was obtained at the highest ethanol concentration. However, these important ethanol concentrations changes did not influence hydrogen production significantly. This was particularly true between 2.0 and 4.0 v/v% ethanol concentration. Therefore, 2.0 v/v% was considered fully adequate and was the selected concentration of the ethanol scavenger used for further studies.

2.3.5. By-Products Formation

There are several by-products generated from the water-splitting reaction in the gas phase. Detected by-products include methane, ethane, acetaldehyde and CO_2. To quantify these by-products, gas samples were taken hourly from the gas port located in the storage tank. They were analyzed using a Shimadzu gas chromatograph (GC) unit. All the experiments were repeated at least three times to secure reproducibility.

One can thus see that as soon as the photo-redox reaction starts, all these by-products, together with hydrogen, increase progressively as is shown in Figure 10. In the liquid phase, ethanol was also measured using a Shimadzu HPLC. One can observe in Figure 11 a balanced consumption-formation of ethanol, with a net stable ethanol concentration. This occurs when hydrogen is being produced using the 1.00 wt% Pd–TiO$_2$ photocatalyst.

The observed trends could be considered a promising result, showing that none or little additional scavenger is required in subsequent runs once the initial ethanol is fed to the Photo-CREC Water II reactor unit.

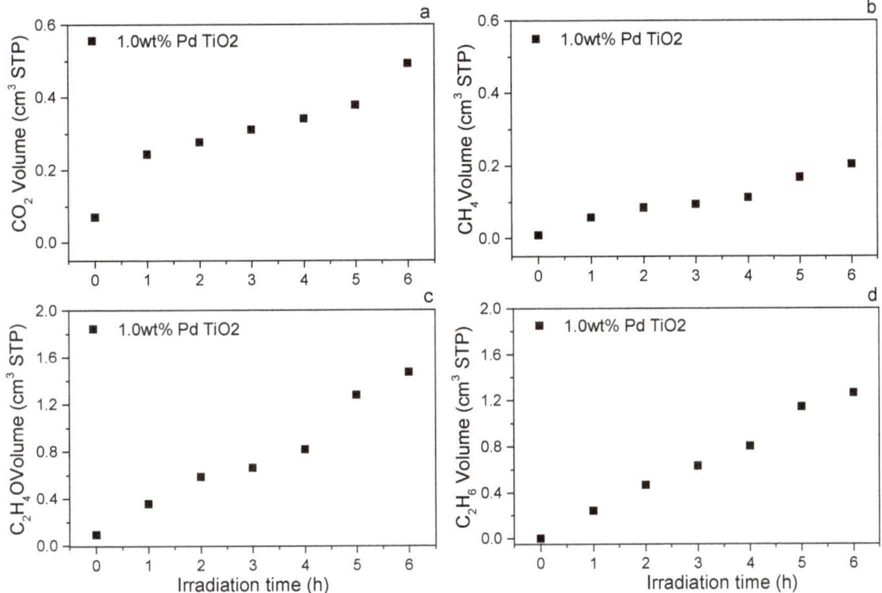

Figure 10. Hydrocarbon profiles of (**a**) carbon dioxide (CO_2), (**b**) methane (CH_4), (**c**) acetaldehyde (C_2H_4O) and (**d**) ethane (C_2H_6) at 1.00 wt% Pd. Conditions: Photocatalyst concentration 0.15 g/L, 2.0 v/v% ethanol, argon atmosphere, pH = 4 ± 0.05 and near-UV light. Standard deviation: (a) ±4.1%, (b) ±4.7%, (c) ±5.1%, (d) ±6.3%.

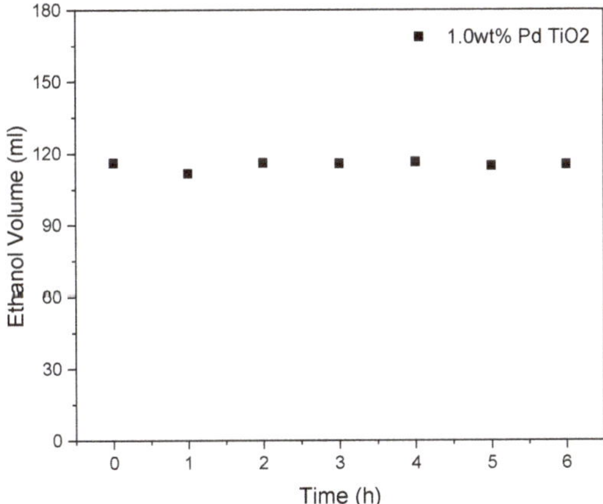

Figure 11. Ethanol changes with irradiation time. Conditions: photocatalyst concentration 0.15 g/L, argon atmosphere, 2.0 v/v% ethanol, pH = 4 ± 0.05 and near-UV light. Standard deviation: ±3.0%.

2.4. Quantum Yield (QY) evaluation

The quantum yield (QY) is the most important parameter to establish the energy utilization efficiency in photocatalytic reactors [38]. In terms of hydrogen production, quantum yield can be defined as the hydrogen radical production rate over the absorbed photon rate on the photocatalyst surface. According to this definition, QY can be determined as follows:

$$QY_{H\bullet} = \frac{moles\ of\ H\bullet/s}{moles\ of\ photons\ absorbed\ by\ the\ photocatalyst/s} \tag{6}$$

Equation (6) is equivalent to:

$$\%QY = \frac{\left[\frac{dN_H}{dt}\right]}{P_a} \times 100 \tag{7}$$

where $\frac{dN_H}{dt}$ represents the rate of moles of hydrogen radicals formed at any time during the photocatalyst irradiation.

To use Equation (7) the assessment of P_a or the moles of absorbed photons is required. This can be accomplished by using the macroscopic radiation energy balance (MREB) in the Photo-CREC Water II reactor as proposed by Escobedo et al. [39]. Appendix C provides a calculation sample to assess the QY.

2.4.1. Effect of Pd Addition on Quantum Yields

The quantum yield evaluation for different TiO_2 photocatalysts involves rigorous macroscopic radiation energy balances. These calculations require the assessment of the P_t transmitted, the P_i incident, and the P_{bs} backscattered photons using the macroscopic radiation energy balance as described in Section 2.2. With this information and using Equation (1), the P_a was calculated.

Furthermore, for every experiment and once the lamp is turned on, the rate of moles of hydrogen can be established. On this basis, QY% can be calculated using Equation (7).

Table 5 and Figure 12 report QY% for the mesoporous photocatalysts doped with palladium at different metal loadings (0.25, 0.50, 1.00, 2.50 and 5.00 wt%) under the following conditions: (a) photocatalyst slurry concentrations of 0.15 g/L, (b) 2.0 v/v% ethanol, (c) pH = 4 ± 0.05 and (d) near-UV light.

Table 5. Quantum yield (QY) for the Pd–TiO₂ photocatalyst when using 0.15 g/L. All reported data are average values of three repeats.

Semiconductor	QY (%)
F–127 TiO_2	5.0
F-127–0.25 wt% Pd–TiO_2	13.7
F-127–0.50 wt% Pd–TiO_2	12.8
F-127–1.00 wt% Pd–TiO_2	10.9
F-127–2.50 wt% Pd–TiO_2	9.6
F-127–5.00 wt% Pd–TiO_2	8.5

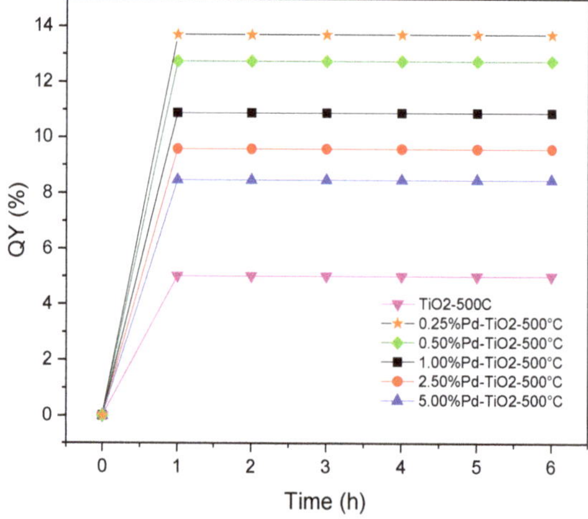

Figure 12. QY% at various irradiation times under near-UV light and 0.15 g/L of photocatalyst concentration, and using Pd at different loadings (0.25, 0.50, 1.00, 2.50 and 5.00 wt%).

There is a significant increase of QY% with 0.25, 0.50 and 1.00 wt% Pd–TiO$_2$, whereas higher Pd loadings led to a decrease of QY%. These results are in line with the QY% of 8% reported by Escobedo when Pt addition proceeds [39].

Figure 12 reports that QY% displays consistent trends for Pd-doped TiO$_2$ photocatalysts: (a) during the first hour of irradiation, QY% increased progressively until it reached a stable value; and (b) during the following six hours of irradiation, QY% remained unchanged, with this showing a steady performance of the photocatalysts under study.

2.4.2. Effect of Catalyst Concentration on Quantum Yields

Considering the QY% observed for the 1.00 wt% Pd–TiO$_2$ during hydrogen production, further QY% evaluations were developed by changing the photocatalyst concentration in the slurry. Table 6 and Figure 13 report the QY% obtained, by augmenting the photocatalyst concentration, under the following conditions: (a) 2.0 v/v% ethanol as scavenger organic compound, (b) pH = 4 ± 0.05 and (c) near-UV light irradiation.

Table 6. Quantum yield for 1.00 wt% Pd–TiO$_2$ photocatalyst at different photocatalyst concentrations in the slurry.

Catalyst Concentration (g/L)	QY (%)
0.15	10.9
0.30	14.5
0.50	22.4
1.00	30.8

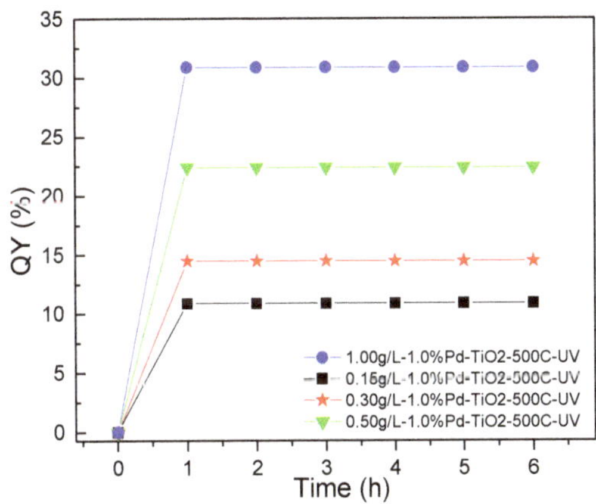

Figure 13. QY% at various irradiation times using near-UV irradiation and 0.15, 0.30, 0.50 and 1.00 g/L photocatalyst concentrations. Note: Loading was1.00 wt% Pd on TiO$_2$.

Figure 13 provides QY% for different photocatalyst concentrations. Here, it was again observed that there was a noticeable increase of the QY% in the first hour of irradiation, followed by a stable QY% in the next 5 h of irradiation. Constant QY% during the 1 to 6-h irradiation period was assigned to the steady hydrogen formation rate, linked to consistent zero-order reaction kinetics with no photocatalyst activity decay observed in all cases.

3. Experimental Methods

The photocatalysts of the present study were synthesized using the sol–gel methodology and doped with palladium. Different techniques were utilized to characterize the doped semiconductors as follows: (a) BET for specific surface area, (b) chemisorption for crystallite size, (c) x-ray diffraction for crystallographic structure and (d) UV-vis absorption for band gaps. The prepared semiconductors were evaluated in a Photo-CREC Water II reactor unit.

3.1. Photocatalyst Synthesis

The sol–gel method can be used for photocatalyst synthesis by converting monomers into colloids (sol phase), and thus promoting a gel structure formation [40]. The sol–gel method for TiO_2 synthesis can be modified, leading to improvements in photocatalyst structural properties such as particle diameter and surface area. Therefore, this also leads to improved photocatalytic activity [41].

Some copolymers, such as Pluronic® F-127 and Pluronic® P-123, formed by chains of ethylene oxide and propylene oxide, can be used for TiO_2 synthesis as soft templates. These templates optimize the pore structure network during semiconductor preparation, enhancing pore size distribution, enlarging the surface area, controlling the purity, homogeneity, and morphology of mesoporous materials [42].

Rusinque shows that the Pluronic F-127 template has a greater impact than the Pluronic P-123 template on TiO_2 photoactivity, increasing the hydrogen production up to 86% [31]. Thus, considering the Pluronic F-127 advantage over Pluronic P-123 for hydrogen production, further experiments were carried out using only copolymer Pluronic F-127.

The sol–gel method adopted used the following reagents: (a) ethanol USP (C_2H_5OH) from commercial alcohols, (b) hydrochloric acid (HCl, 37% purity), (c) Pluronic F-127, (d) anhydrous citric acid, (e) titanium (IV) isopropoxide, and (f) palladium (II) chloride ($PdCl_2$, 99.9% purity). All the reagents were obtained from Sigma Aldrich, with photocatalyst preparation effected according to the methodology proposed by Guayaquil et al. [43].

Figure 14 describes the sol–gel synthesis as follows: (a) Step 1: In 400 mL of ethanol, 33 g of hydrochloric acid and 20 g of Pluronic F-127 were added until dissolution, under continuous stirring for 1 h. (b) Step 2: 6.30 g of citric acid were dissolved in 20 mL of water for posterior addition to the initial suspension to mix them together for 1 h in order to set the pH at 0.75. (c) Step 3: 28.5 g of titanium (IV) isopropoxide was dissolved in ethanol and added dropwise to the mixture. Finally, palladium (II) chloride was incorporated at different loadings (0.25 to 5.00 wt% Pd). (d) Step 4: The resulting sol–gel suspension was stirred for 24 h and then calcined at 500 °C for 6 hours under an air atmosphere. The copolymer was evaporated during the thermal treatment and an ordered mesoporous titanium framework was formed [23].

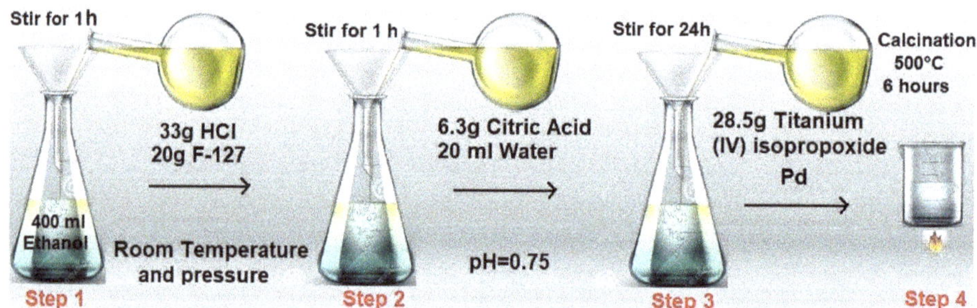

Figure 14. Photocatalyst preparation process describing the four steps considered for Pd-doped mesoporous.

3.2. Equipment

The Photo-CREC Water II (PCW-II) reactor is a novel unit used for water splitting reactions and therefore, hydrogen production. It is a 5.7 L slurry batch reactor configured with two concentric tubes: (a) an inner tube made from transparent borosilicate (Pyrex) and (b) an outer tube made from opaque polyethylene. The fluorescent lamp is placed inside this inner Pyrex tube. Furthermore, the suspended photocatalyst flows in the annular space between the outer polyethylene tube and the inner Pyrex transparent tube which only absorbs 5%) of the near-UV light emitted by the lamp [44]. See Appendix A for a detailed lamp characterization.

The PCW-II unit is equipped with a storage feed tank where the photocatalyst suspension is always kept sealed under agitation. This tank has 2 ports for periodic liquid and gas phase sampling. Figure 15 describes the main components of PCW-II: (a) the Photo-CREC Water II Reactor, (b) the centrifugal pump, (c) the sealed storage tank, and (d) the electrical circuit powering the near-UV light lamp.

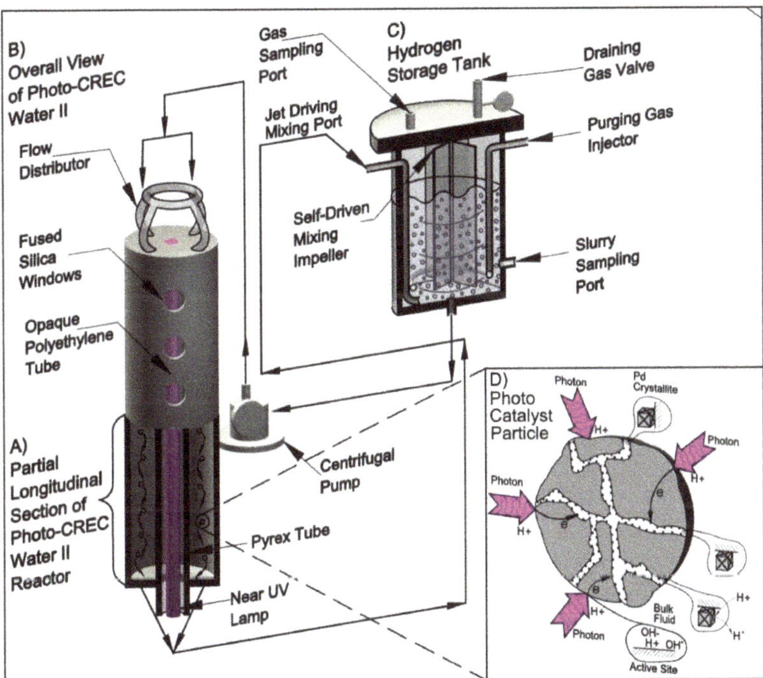

Figure 15. Schematic representation of the Photo-CREC Water II Reactor with a H_2 Mixing/Storage Tank: (**A**) partial longitudinal cross-section of the PCW- II unit showing the down flow circulation of the slurry in the annular channel, (**B**) overall view of PCW-II showing windows, near UV lamp and recirculation pump (**C**) hydrogen storage tank with its components, (**D**) detail of a photocatalyst particle.

The emitted radiation spectra of the lamp used inside the Photo-CREC Water II was established using a Stellar Net EPP2000-25 spectrometer (StellarNet Inc.). The light source is a polychromatic black light blue (BLB) Ushio UV lamp (15 W, 0.305 A, 55 V) with a spectral peak at 368 nm in the 300–420 nm emission range [45].

3.3. Photocatalyst Characterization

Photocatalyst specific surfaces areas were determined using a BET surface area analyzer (Micrometrics, ASAP 2010) at −195 °C. Each photocatalyst was degassed at 300 °C during a period of 3 h. The BET analysis was developed using nitrogen to generate the adsorption–desorption equilibrium isotherms and to establish the isotherm inflection point. The BJH (Barrett–Joyner–Halenda) method was used to determine the pore size distribution, by utilizing the desorption isotherm with N_2 as an adsorbate.

By using the Micromeritics AutoChem II Analyzer for pulse chemisorption, one can calculate the fraction of dispersed metal and average active metal crystallite size [46]. Furthermore, to identify the phases of a crystalline material, X-ray diffraction (XRD) was used [47]. The XRD spectra were analyzed in a Rigaku Rotating Anode X-Ray Diffractometer (Rigaku) perated at 45 kV and 160 mA. The scans were taken between 20–80°, with a step size of 0.02° and a dwell time of 2 s/step.

In order to determine the characteristic band gap associated to each photocatalyst an UV-VIS-NIR spectrophotometer (Shimadzu UV-3600) was used [48]. $BaSO_4$ was utilized as a reference sample. Kubelka–Munk (K–M) developed a Tauc plot methodology that was followed to establish the corresponding band gaps [49]. X-ray photoelectron spectroscopy (XPS) analysis was also used to identify the elemental composition and the chemical state of each element in the synthesized photocatalyst [50].

3.4. Hydrogen Production

Pd-doped TiO_2 photocatalysts were evaluated using the Photo-CREC Water II reactor equipped with the BLB near-UV lamp for 6 hours of continuous irradiation. This lamp was turned on 30 min before initiating the photoreaction. The hydrogen storage/mixing tank was loaded with 6000 mL of water. Ethanol was used as an organic scavenger and the pH was adjusted to 4 ± 0.05 with H_2SO_4 [2M] keeping the photoreaction under acidic conditions, which favours available H^+ for water splitting process [39].

Following this step, the photocatalyst was loaded at a specific weight concentration ensuring that most of the radiation was absorbed in the slurry medium. The photocatalyst was subjected to sonication, which reduces the formation of particle agglomerates and promotes homogeneous mixing. Argon gas was circulated to guarantee an inert atmosphere at the beginning of the reaction.

3.5. Analytical Techniques

The gas phase was analyzed with a Shimadzu GC2010 gas chromatograph using argon (Praxair 99.999%) as gas carrier. It has 2 detectors, a flame ionization detector (FID) and a thermal conductivity detector (TCD). This unit was equipped with a HayeSepD 100/120 mesh packed column (9.1 m × 2 mm × 2 μm nominal SS) used for the separation of hydrogen from air. This equipment detects hydrogen (H_2), carbon monoxide (CO), carbon dioxide (CO_2), methane (CH_4) and other hydrocarbon organic species.

A Shimadzu HPLC model UFLC (ultra-fast liquid chromatography) system was utilized to characterize the liquid phase. This analytical technique allows the liquid mobile phase (0.1% H_3PO_4) to transport the sample through a column (Supelcogel C-610H 30cm × 7.8mm ID) containing a stationary phase. It selectively separates individual compounds (i.e., ethanol) from water for further detection. This quantitative analysis is performed by employing the RID (refractive index detector) 10A due to polar nature of ethanol.

Both the GC and the HPLC analytical techniques were used simultaneously. Samples were taken at different irradiation times.

4. Conclusions

(a) The TiO_2 mesoporous photocatalysts of the present study were prepared using a F-127 template and following a sol–gel methodology. It was found that the mesoporous prepared using a F-127 template displayed a good photocatalytic performance.

(b) The prepared Pd–TiO_2 photocatalysts were characterized using BET, XRD, UV-VIS and XPS. On this basis it was proven that energy band gaps were significantly affected with Pd addition, and that binding energies showed significant contribution of the Pd (0) on the doped-palladium TiO_2.

(c) Macroscopic radiation energy balances were successfully employed to establish photon absorption rates and radiation absorption efficiencies in the PCW-II unit. For the Pd–TiO_2 semiconductors, photon absorption efficiencies were in the 45 and 60% range under near-UV light.

(d)   The formation of hydrogen using Pd–TiO_2 photocatalysts followed, in all cases, steady zero-order kinetics with no apparent photocatalyst activity decay.

(e)   The prepared Pd–TiO_2 photocatalysts under near UV-light were shown to be adequate for hydrogen production reaching up to 210 cm^3 STP when using the 1.00 wt%-Pd on TiO_2. This photocatalyst showed a best QY% of 30.8%.

Author Contributions: Conceptualization, investigation and supervision, H.d.L.; proposed methodology and supervision, S.E.; validation, formal analysis and writing, B.R.

Funding: This research was funded by Natural Sciences and Engineering Research Council of Canada (NSERC) and the University of Western Ontario, grant given to Hugo de Lasa.

Acknowledgments: We would like to gratefully thank Florencia de Lasa who assisted with the editing and the drafting of the graphical abstract of the present article.

Conflicts of Interest: The authors declare no conflict of interest.

Nomenclature

CO_2	Carbon dioxide
CH_4	Methane
C_2H_6	Ethane
C_2H_4O	Acetaldehyde
c	Speed of light (3.0×10^8 m/s)
D_p	Pore diameter (cm)
e-	Electron
h+	Hole
h	Planck's constant (6.63×10^{34} J/s)
E_{bg}	Energy band gap (eV)
E_{av}	Average energy of a photon (kJ/mol photon)
F-127	Poly (ethylene oxide)/poly (propylene oxide)/poly (ethylene oxide)
H•	Hydrogen radical
H_2O	Water
$I(\lambda)$	Intensity of light (W/cm^2)
OH-	Hydroxide ions
OH•	Hydroxide radicals
P-123	Poly (ethylene glycol)-block-poly (propylene glycol)-block-poly (ethylene glycol)
P_0	Rate of photons emitted by the BLB lamp (einstein/s)
P_a	Rate of absorbed photons (einstein/s)
$P_{a\text{-wall}}$	Rate of photons absorbed by the inner pyrex glass (einstein/s)
P_{bs}	Rate of backscattered photons exiting the system (einstein/s)
Pd	Palladium
$PdCl_2$	Palladium II chloride
PEO	Poly (ethylene oxide)
P_{fs}	Rate of forward-scattered radiation (einstein/s)
P_i	Rate of photons reaching the reactor inner surface (einstein/s)
P_{ns}	Rate of transmitted non-scattered radiation (einstein/s)
PPO	Poly (propylene oxide)
Pt	Rate of transmitted photons (einstein/s)
Pt	Platinum
$q\,(\theta, z, \lambda, t)$	Net radiative flux over the lamp emission spectrum ($\mu W/cm^2$)
t	Time (h)
TiO_2	Titanium dioxide
V	Total volume of the gas chamber (5716 cm^3)
W	Weight (g)
Wt%	Weight percent (% m/m)

Greek symbols

θ	Diffraction angle, also scattering angular angle (o)
λ	Wave length (nm)
φ	Quantum Yield Efficiency (%)

Acronyms

BJH	Barrett–Joyner–Halenda model
BLB	Black light blue lamp
BET	Brunauer–Emmett–Teller Surface Area Method
CB	Conduction band
DP25	Degussa P25 (TiO_2)
JCPDS	International Centre for Diffraction Data
MIEB	Macroscopic Irradiation Energy Balance
PCW-II	Photo CREC Water II reactor
PC	Photocatalyst concentration
STP	Standard temperature and pressure (273 K and 1 atm)
UV	Ultraviolet
VB	Valence band
B_g	Band gap

Appendix A. Lamp Characterization

Figure A1 reports the spectrum of the polychromatic BLB Ushio near-UV lamp, with an observed output power of 1.61 W and an average of 325.1 kJ/photon mole of emitted photon energy.

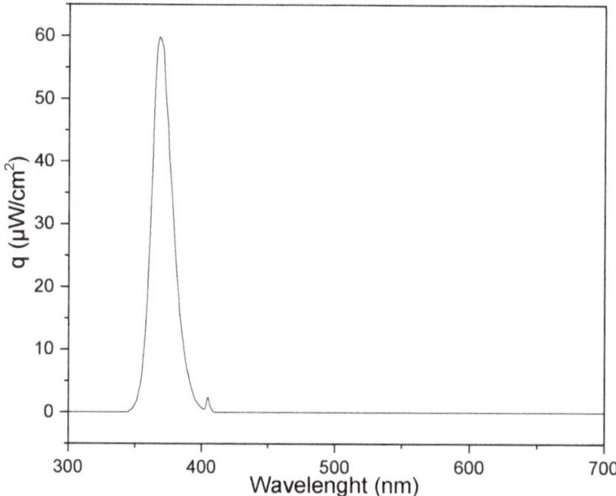

Figure A1. Near-UV Lamp Irradiation Spectrum.

The average emitted photon energy was calculated using the recorded irradiation spectra as follows [51]:

$$E_{av} = \frac{\int_{\lambda_{min}}^{\lambda_{max}} I(\lambda)\, E(\lambda)\, d\lambda}{\int_{\lambda_{min}}^{\lambda_{max}} I(\lambda)\, d\lambda} \tag{A1}$$

where,

$$E(\lambda) = \frac{hc}{\lambda} \tag{A2}$$

With h being the Planck constant (6.34×10^{-34} J s/photon), c representing the speed of light in a vacuum (3.00×10^8 m/s^2) and λ denoting the wavelength expressed in nanometers (nm). I is the emitted photons intensity (W/cm^2), assessed as $I(\lambda) \approx q(\theta, z, \lambda, t)\, d\lambda$ and measured with a spectrophotoradiometer. The irradiance is represented by $q\ (\theta, z, \lambda, t)\ d\lambda$ and given by the lamps spectra as shown in Figure A2.

The average emitted photon Energy was calculated as shown in Equation (A1)

$$E_{av} = \frac{\int_{\lambda_{min}}^{\lambda_{max}} I\ (\lambda)\ E(\lambda)\ d\lambda}{\int_{\lambda_{min}}^{\lambda_{max}} I\ (\lambda)\ d\lambda} = \frac{\int_{\lambda_{min}}^{\lambda_{max}} \frac{hc}{\lambda} * q\ (\theta, z, \lambda, t)\ d\lambda}{\int_{\lambda_{min}}^{\lambda_{max}} q\ (\theta, z, \lambda, t)\ d\lambda} \tag{A3}$$

$$E_{av} = 5.36 \times 10^{-19} J/mol\ photon = 325.1\ KJ/mol\ photo \tag{A4}$$

Regarding the PCW-II, the axial distribution of the radiative flux was determined. Figure A2 reports the near-UV lamp axial radiation distribution. One can observe that the radiation profile shows no significant changes in radiation levels in the central section of the PCW-II. On the other hand, significant radiation decay can be seen approaching the endpoints of the lamp [52].

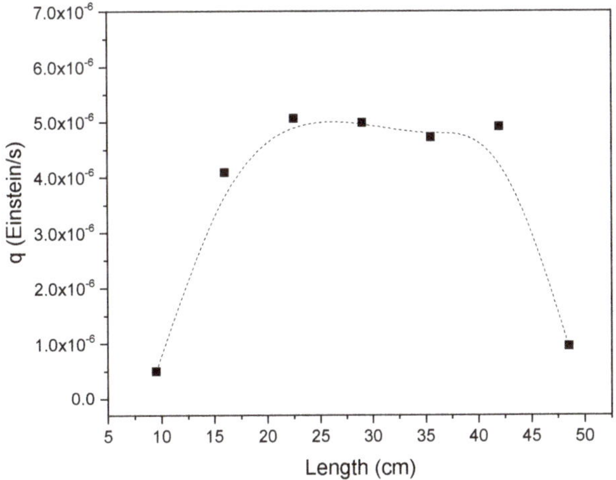

Figure A2. Near-UV Lamp Axial Distribution.

Appendix B. Semiconductor Crystallite Sizes and Lattice Parameters

The crystallite sizes were determined using the Scherrer equation as reported in the enclosed Table A1. On this basis the mesoporous photocatalysts displayed crystallite sizes between 9 and 14 nm.

Table A1. Photocatalyst Crystallite Sizes.

Photocatalyst	Crystallite Size (nm)
TiO2	9
TiO2 0.25 wt% Pd 500 °C	11
TiO2 0.50 wt% Pd 500 °C	11
TiO2 1.00 wt% Pd 500 °C	11
TiO2 2.50 wt% Pd 500 °C	13
TiO2 5.00 wt% Pd 500 °C	14

Furthermore, the calculated a, b and c lattice constants of the tetragonal anatase unit cell are shown in Table A2 indicating that pure anatase was successfully obtained with the phase structures maintained at $\alpha = \beta = \gamma = 90°$ angles. These resulting a, b, and c parameters are in closed agreement with those reported in the literature [53]. Note that lattice parameters a = b ≠ c and these were calculated for Anatase phase (h k l) = (1 0 1).

Table A2. Lattice Parameters for TiO_2 and Pd doped TiO_2.

Photocatalyst	a = b	c	2θ (deg)	d (Å)
TiO_2 [53]	3.7821	9.5022	25.33	3.5139
TiO_2 500 °C (our study)	3.7679	9.5002	25.41	3.5025
TiO_2 0.25 wt% Pd 500 °C	3.7832	9.4833	25.33	3.5139
TiO_2 0.50 wt% Pd 500 °C	3.7858	9.4737	25.31	3.5155
TiO_2 1.00 wt% Pd 500 °C	3.7825	9.5099	25.32	3.5147
TiO_2 2.50 wt% Pd 500 °C	3.7748	9.4713	25.38	3.5065
TiO_2 5.00 wt% Pd 500 °C	3.7691	9.4809	25.41	3.5025

Appendix C. Quantum Yield Calculation

As stated in Section 2.4, QY% can be defined as the number of moles of hydrogen radical produced per absorbed photons on the photocatalyst surface:

$$\%QY = \frac{\left[\frac{dN_H}{dt}\right]}{Pa} \times 100 \tag{A5}$$

where $\frac{\left[\frac{dN_H}{dt}\right]}{Pa}$ represents the rate of moles of hydrogen radicals formed and P_a stands for the moles of photons absorbed.

As well, and according to the Macroscopic Irradiation Energy Balances (MIEB) in the Photo-CREC Water Reactor II, P_a was calculated as follows:

$$P_a = P_i - P_{bs} - P_t \tag{A6}$$

where, P_i is the rate of photons reaching the reactor at the inner reactor surface, P_{bs} represents the rate of backscattered photons, and P_t is the rate of transmitted photons (Einstein s^{-1}).

A sample calculation is given below considering a hydrogen production rate of 0.2494 μmol/cm^3 h using: (a) 1.0 wt.% Pd-TiO_2, (b) a photocatalyst concentration of 1.0 g/L, (c) ethanol at 2.0 v/v%, (d) pH = 4 ± 0.05, (e) near-UV Light, (f) gas phase volume in the reactor of 5716 cm^3 and (g) Pa = 2.57 × 10^{-6} Einstein/s.

$$QY_{H\bullet} = \frac{2 * (0.2494 \times 10^{-6}\ mol/cm^3h) * (5716\ cm^3) * (6.022 \times 10^{23}\ photon/mol\ H_2) * (1h/3600s)}{2.57 \times 10^{17}\ photon/s} \tag{A7}$$

$$\%QY_{H\bullet} = 30.8\%$$

References

1. Ramesohl, S.; Merten, F. Energy system aspects of hydrogen as an alternative fuel in transport. *Energy Policy* **2006**, *34*, 1251–1259. [CrossRef]
2. Barreto, L.; Makihira, A.; Riahi, K. The hydrogen economy in the 21st century: A sustainable development scenario. *Int. J. Hydrogen Energy* **2003**, *28*, 267–284. [CrossRef]
3. Maeda, K.; Teramura, K.; Lu, D.; Takata, T.; Saito, N.; Inoue, Y.; Domen, K. Photocatalyst releasing hydrogen from water. *Nature* **2006**, *440*, 295. [CrossRef] [PubMed]
4. Galińska, A. Photocatalytic Water Splitting over Pt−TiO_2 in the Presence of Sacrificial Reagents. *Energy Fuels* **2005**, *19*, 1143–1147. [CrossRef]
5. Wang, M.; Shen, S.; Li, L.; Tang, Z.; Yang, J. Effects of sacrificial reagents on photocatalytic hydrogen evolution over different photocatalysts. *J. Mater. Sci.* **2017**, *52*, 5155–5164. [CrossRef]
6. López, C.R.; Melián, E.P.; Méndez, J.A.O.; Santiago, D.E.; Rodríguez, J.M.D.; Díaz, O.G. Comparative study of alcohols as sacrificial agents in H_2production by heterogeneous photocatalysis using Pt/TiO_2 catalysts. *J. Photochem. Photobiol. A Chem.* **2015**, *312*, 45–54. [CrossRef]
7. Mills, A. An overview of semiconductor photocatalysis. *J. Photochem. Photobiol. A Chem.* **1997**, *108*, 1–35. [CrossRef]

8. Abe, R. Significant effect of iodide addition on water splitting into H_2 and O_2 over Pt-loaded TiO_2 photocatalyst: Suppression of backward reaction. *Chem. Phys. Lett.* **2003**, *371*, 360–364. [CrossRef]

9. Mills, A. Photosensitised dissociation of water using dispersed suspensions of n-type semiconductors. *J. Chem. Soc. Faraday Trans. 1 Phys. Chem. Condens. Phases* **1982**, *12*, 3659–3669. [CrossRef]

10. Escobedo Salas, S. Photocatalytic Water Splitting Using a Modified Pt-TiO_2. Kinetic Modeling and Hydrogen Production Efficiency. Ph.D. Thesis, The University of Western Ontario, London, ON, Canada, August 2013.

11. Khan, M.M.; Adil, S.F.; Al-Mayouf, A. Metal oxides as photocatalysts. *J. Saudi Chem. Soc.* **2015**, *19*, 462–464. [CrossRef]

12. Haider, A.J. Exploring potential Environmental applications of TiO_2 Nanoparticles. *Energy Procedia* **2017**, *119*, 332–345. [CrossRef]

13. Chin, W.L.; Low, F.W.; Chong, S.W.; Hamid, S.B.A. An Overview: Recent Development of Titanium Dioxide Loaded Graphene Nanocomposite Film for Solar Application. *Curr. Org. Chem.* **2015**, *19*, 1882–1895.

14. Yang, J. Roles of Cocatalysts in Photocatalysis and Photoelectrocatalysis. *Acc. Chem. Res.* **2013**, *46*, 1900–1909. [CrossRef]

15. Moslah, C.; Kandyla, M.; Mousdis, G.A.; Petropoulou, G.; Ksibi, M. Photocatalytic Properties of Titanium Dioxide Thin Films Doped with Noble Metals (Ag, Au, Pd, and Pt). *Phys. Status Solidi Appl. Mater. Sci.* **2018**, *215*, 1–7. [CrossRef]

16. García-Zaleta, D.S.; Torres-Huerta, A.M.; Domínguez-Crespo, M.A.; García-Murillo, A.; Silva-Rodrigo, R.; González, R.L. Influence of Phases Content on Pt/TiO_2, Pd/TiO_2 Catalysts for Degradation of 4-Chlorophenol at Room Temperature. *J. Nanomater.* **2016**, *2016*, 1805169. [CrossRef]

17. Subramanian, V.; Wolf, E.E.; Kamat, P.V. Catalysis with TiO_2/Gold Nanocomposites. Effect of Metal Particle Size on the Fermi Level Equilibration. *J. Am. Chem. Soc.* **2004**, *126*, 4943–4950. [CrossRef]

18. Santara, B.; Pal, B.; Giri, P.K. Signature of strong ferromagnetism and optical properties of Co doped TiO_2 nanoparticles. *J. Appl. Phys.* **2011**, *110*, 114322. [CrossRef]

19. Khairy, W.; Zakaria, M. Effect of metal-doping of TiO_2 nanoparticles on their photocatalytic activities toward removal of organic dyes. *Egypt. J. Pet.* **2014**, *23*, 419–426. [CrossRef]

20. Sobana, N.; Muruganadham, M.; Swaminathan, M. Nano-Ag particles doped TiO_2 for efficient photodegradation of Direct azo dyes. *J. Mol. Catal. A Chem.* **2006**, *258*, 124–132. [CrossRef]

21. Cassano, A.E.; Martin, C.A.; Brandi, R.J.; Alfano, O.M. Photoreactor Analysis and Design: Fundamentals and Applications. *Ind. Eng. Chem. Res.* **1995**, *34*, 2155–2201. [CrossRef]

22. Escobedo, S.; Serrano, B.; Calzada, A.; Moreira, J.; de Lasa, H. Hydrogen production using a platinum modified TiO_2 photocatalyst and an organic scavenger. Kinetic modeling. *Fuel* **2016**, *181*, 438–449. [CrossRef]

23. Yu, J.C.; Wang, X.; Fu, X. Pore-Wall Chemistry and Photocatalytic Activity of Mesoporous Titania Molecular Sieve Films. *Chem. Mater.* **2004**, *16*, 1523–1530. [CrossRef]

24. Pan, X.; Xu, Y.J. Defect-mediated growth of noble-metal (Ag, Pt, and Pd) nanoparticles on TiO_2 with oxygen vacancies for photocatalytic redox reactions under visible light. *J. Phys. Chem. C* **2013**, *117*, 17996–18005. [CrossRef]

25. Zhang, J.; Zhou, P.; Liu, J.; Yu, J. New understanding of the difference of photocatalytic activity among anatase, rutile and brookite TiO_2. *Phys. Chem. Chem. Phys.* **2014**, *16*, 20382–20386. [CrossRef] [PubMed]

26. Rodriguez-Vindas, D. Synthesis of palladium with different nanoscale structures by sputtering deposition onto fiber templates. *J. Nanophotonics* **2008**, *2*, 021925. [CrossRef]

27. Moreira, J.; Serrano, B.; Ortiz, A.; de Lasa, H.; de Lasa, H. Evaluation of Photon Absorption in an Aqueous TiO_2 Slurry Reactor Using Monte Carlo Simulations and Macroscopic Balance. *Ind. Eng. Chem. Res.* **2010**, *49*, 10524–10534. [CrossRef]

28. Salaices, M.; Serrano, B.; de Lasa, H.I. Experimental evaluation of photon absorption in an aqueous TiO_2 slurry reactor. *Chem. Eng. J.* **2002**, *90*, 219–229. [CrossRef]

29. Thornton, J.M.; Raftery, D. Efficient photocatalytic hydrogen production by platinum-loaded carbon-doped cadmium indate nanoparticles. *ACS Appl. Mater. Interfaces* **2012**, *4*, 2426–2431. [CrossRef]

30. Yoshida, H.; Hirao, K.; Nishimoto, J.; Shimura, K.; Kato, S.; Itoh, H.; Hattori, T. Hydrogen production from methane and water on platinum loaded titanium oxide photocatalysts. *J. Phys. Chem. C* **2008**, *112*, 5542–5551. [CrossRef]

31. Rusinque, B. Hydrogen Production by Photocatalytic Water Splitting Under Near-UV and Visible Light Using Doped Pt and Pd TiO$_2$. Master Thesis, The University of Western Ontario, London, ON, Canada, September 2018.

32. Zhang, N.; Liu, S.; Fu, X.; Xu, Y.J. Synthesis of M@TiO2 (M = Au, Pd, Pt) core-shell nanocomposites with tunable photoreactivity. *J. Phys. Chem. C* **2011**, *115*, 9136–9145. [CrossRef]

33. Riyapan, S.; Boonyongmaneerat, Y.; Mekasuwandumrong, O.; Yoshida, H.; Fujita, S.; Arai, M.; Panpranot, J. Improved catalytic performance of Pd/TiO$_2$in the selective hydrogenation of acetylene by using H2-treated sol-gel TiO$_2$. *J. Mol. Catal. A Chem.* **2014**, *383–384*, 182–187. [CrossRef]

34. Akbayrak, S.; Tonbul, Y.; Özkar, S. Nanoceria supported palladium(0) nanoparticles: Superb catalyst in dehydrogenation of formic acid at room temperature. *Appl. Catal. B Environ.* **2017**, *206*, 384–392. [CrossRef]

35. Borodin, V.B.; Tsygankov, A.A.; Rao, K.K.; Hall, D.O. Hydrogen production byAnabaena variabilisPK84 under simulated outdoor conditions. *Biotechnol. Bioeng.* **2000**, *69*, 478–485. [CrossRef]

36. Dalmolin, I.; Skovroinski, E.; Biasi, A.; Corazza, M.L.; Dariva, C.; Oliveira, J.V. Solubility of carbon dioxide in binary and ternary mixtures with ethanol and water. *Fluid Phase Equilib.* **2006**, *245*, 193–200. [CrossRef]

37. Puangpetch, T.; Sreethawong, T.; Yoshikawa, S.; Chavadej, S. Hydrogen production from photocatalytic water splitting over mesoporous-assembled SrTiO$_3$nanocrystal-based photocatalysts. *J. Mol. Catal. A Chem.* **2009**, *312*, 97–106. [CrossRef]

38. Ibrahim, H.; de Lasa, H. Novel photocatalytic reactor for the destruction of airborne pollutants reaction kinetics and quantum yields. *Ind. Eng. Chem. Res.* **1999**, *38*, 3211–3217. [CrossRef]

39. Escobedo, S.; Serrano, B.; de Lasa, H. Serrano and H. de Lasa. Quantum Yield with Platinum Modified TiO$_2$ Photocatalysts for Hydrogen Prodcution. *Appl. Catal. B. Environ.* **2013**, *140*, 523–536. [CrossRef]

40. Guo, S.P.; Li, J.C.; Xu, Q.T.; Ma, Z.; Xue, H.G. Recent achievements on polyanion-type compounds for sodium-ion batteries: Syntheses, crystal chemistry and electrochemical performance. *J. Power Sources* **2017**, *361*, 285–299. [CrossRef]

41. Brinker, J.; Schere, G.W. *Sol-Gel Science: The Physics and Chemistry of Sol-Gel Processing*; Academic Press Inc.: San Diego, CA, USA, 1990.

42. He, X. *Recent Progress in Fabrication of Nanostructured Carbon Monolithic Materials*; Elsevier: Oxford, UK, 2017.

43. Guayaquil-Sosa, J.F.; Serrano-Rosales, B.; Valadés-Pelayo, P.J.; de Lasa, H. Photocatalytic hydrogen production using mesoporous TiO$_2$ doped with Pt. *Appl. Catal. B Environ.* **2017**, *211*, 337–348. [CrossRef]

44. De Lasa, H.; Serrano, B.; Salaices, M. *Photocatalytic Reaction Engineering*; Springer Scicence: New York, NY, USA, 2005.

45. Ushio. *UV-B Blacklight & Blacklight Blue*; Catalogue: Cypress, CA, USA.

46. *AutoChem 2920 Automated Catalyst Characterization System Operator's Manual*; AutoChem: Norcross, GA, USA, 2014.

47. Warren, B.E. *X-Ray Diffraction*; Dover Publications: New York, NY, USA, 1990.

48. *UV-VIS-NIR Spectrophotometer*; Shimadzu: Tokyo, Japan, 2018.

49. Slav, A. Optical characterization of TiO$_2$ -Ge nanocomposite films obtained by reactive magnetron sputtering. *Dig. J. Nanomater. Biostructures* **2011**, *6*, 915–920.

50. Briggs, D. X: X-Ray Photoelectron Spectroscopy. In *Handbook of Adhesion*, 2nd ed.; 2005; pp. 621–622.

51. Serrano, B.; Ortíz, A.; Moreira, J.; de Lasa, H.I. Energy efficiency in photocatalytic reactors for the full span of reaction times. *Ind. Eng. Chem. Res.* **2009**, *48*, 9864–9876. [CrossRef]

52. Salaices, M.; Serrano, B.; De Lasa, H. Photocatalytic conversion of Organic pollutants extinction coefficients and quantum effiencies. *Ind. Eng. Chem. Res.* **2001**, *40*, 5455–5464. [CrossRef]

53. Treacy, J.P.W.; Hussain, H.; Torrelles, X.; Grinter, D.C.; Cabailh, G.; Bikondoa, O.; Nicklin, C.; Selcuk, S.; Selloni, A.; Lindsay, R.; et al. Geometric structure of anatase Ti O$_2$(101). *Phys. Rev. B* **2017**, *95*, 1–7. [CrossRef]

 *catalysts*

Article

Comparing the Efficiency of N-Doped TiO$_2$ and N-Doped Bi$_2$MoO$_6$ Photo Catalysts for MB and Lignin Photodegradation

Ricardo Rangel [1,*], Verónica Janneth Cedeño [1], Jaime Espino [1], Pascual Bartolo-Pérez [2], Geonel Rodríguez-Gattorno [2] and Juan José Alvarado-Gil [2]

[1] División de estudios de Posgrado, Facultad de Ingeniería Química, Universidad Michoacana de S.N.H. Morelia Z.P. 58060, Michoacán, Mexico; cedegarci@gmail.com (V.J.C.); jespinova@yahoo.com.mx (J.E.)
[2] CINVESTAV-IPN, Unidad Mérida. Mérida Z.P.97310, Yucatán, Mexico; pascual@cinvestav.mx (P.B.-P.); geonelr@gmail.com (G.R.-G.); jjag09g@gmail.com (J.J.A.-G.)
* Correspondence: rrangel@umich.mx; Tel.: +52-443-327-3584

Received: 27 November 2018; Accepted: 13 December 2018; Published: 19 December 2018

Abstract: In this study, we tested the efficiency of nitrogen-doped titanium dioxide (N-TiO$_2$) and nitrogen-doped bismuth molybdate (N-Bi$_2$MoO$_6$) compounds as photocatalysts capable of degrading methylene blue and lignin molecules under irradiation with ultraviolet (UV) and visible light (VIS). Moreover, we compared TiO$_2$ and Bi$_2$MoO$_6$ catalysts with N-TiO$_2$ and N-Bi$_2$MoO$_6$ compounds using chemical coprecipitation. The catalysts were prepared starting from Ti(OCH$_2$CH$_2$CH$_3$)$_4$, Bi(NO$_3$)$_3$·5H$_2$O, and (NH$_4$)$_6$Mo$_7$O$_{24}$ reagents. N-doping was achieved in a continuous reflux system, using ethylene diamine as a nitrogen source. The resulting materials were characterized using Scanning Electron Microscopy (SEM), X-Ray diffraction (XRD), Fourier transform infrared spectroscopy (FTIR), and X-ray photoelectron spectroscopy (XPS). Additionally, we observed the decrease in particle size after processing the compounds in the reflux system. The results regarding photocatalytic degradation tests show a remarkable effect for nitrogen doped samples, achieving 90% of lignin degradation.

Keywords: photocatalysis; Titanium dioxide; bismuth molybdate; lignin; UV light; visible light

---

1. Introduction

Advanced oxidation processes (AOP) are commonly seen as an alternative to degraded environmental water pollutants, based on their effectiveness, high reactivity, non-selectivity, and their extensive variety of applications [1]. Since its inception [2], titanium oxide remains one of the most effective and versatile compounds for photocatalytic applications, even if traditional compounds like ZnO [3], CdS [4], SnO$_2$ [5], ZnS [6], and BiVO$_4$ [7] have been tested to improve their efficiency or performance in comparison to TiO$_2$. Another aspect commonly used to enhance photocatalytic yield is the process manipulation that produces different structural features of TiO$_2$ particles, such as specific surface area or shape. Some of these manipulations include chemical coprecipitation [8], sol-gel combustion method [9], spray drying [10], or microwave heating [11]. Most of those studies were used to reduce or control the crystal size and had the purpose of increasing the volume/length ratio to maximize the exposed surface, thus increasing the active sites to achieve the catalytic process more effectively, in terms of reactivity [12]. In the majority of reported studies, the results are marginal in comparison to TiO$_2$. Only few compounds, including ZnO or Bi$_2$MoO$_6$, have exhibited promising results in enhancing the catalytic activity in at least one order of magnitude [13,14]. ZnO is as a robust candidate for wastewater treatment, due to its similarity with TiO$_2$, in aspects such as charge carrier dynamics upon bandgap excitation and the generation of reactive oxygen species in aqueous suspensions. However, inherent details like its wide band gap and massive charge carrier recombination has limited their massive usage [3].

Bismuth molybdate (Bi_2MoO_6) has been profusely studied in the past as a catalyst in oxidation reactions [15]. At the present time, Bi_2MoO_6 has attracted a lot of attention in photocatalysis research, due to their effectiveness to degrade organic pollutants under UV irradiation. The Bi_2MoO_6 morphology [16,17] and production methods include sol-gel [18], solvothermal synthesis [19], and co-precipitation [20], among others.

An alternative method to increase the catalytic activity of some photocatalytic materials is through the inclusion of some relevant dopant elements, either by using transition metal cations to replace metal sites [21] or by inserting nonmetal anions like F [22], C [23], or N [24].

The purpose is to create new electronic states between the valence and conduction bands, facilitating the electron conduction to enhance the photocatalytic efficiency and also favoring the electron-hole formation. In addition, those new states promote the shifting towards the visible region of the electromagnetic spectrum. In this regard, nitrogen-doping has been considered one of most effective approaches to improve photocatalytic activity of TiO_2 in the visible region [25]. The methods currently reported to achieve it superficially or by replacing atoms include ion implantation [26], physical vapor deposition [27], and spray pyrolysis [28], as well as variants of the known process of sol-gel. An alternative method to include nitrogen is the refluxing methodology, which consists of inducing intimate contact between the nitrogen source precursor (hydrazine, urea, ethylene diamine, etc.) and the catalyst inside a closed system of reflux at constant temperature. As a result of that methodology, it is possible to obtain catalysts containing nitrogen [29]. The amount of nitrogen that can be introduced depends on factors including the type of precursor, the reflux temperature, and the time involved.

The purpose of the present research was to study the efficiency of titanium oxide and bismuth molybdate, both nitrogen-doped, to be tested as photocatalysts capable of degrading the methylene blue (MB) dye, which is considered as a model molecule for degradation studies.

Additionally, we studied the photodegradation of lignin using both kinds of compounds. Lignin is a compound commonly found in wastewater from the wood process manufacturing. Usually, lignin is partially solubilized with acid and their product is concentrated and burned. Even so, some residual amount of lignin remains bonded to cellulose, which in many cases, is discarded to body waters affecting the environment and altering the photosynthetic processes. Our hypothesis is that nitrogen doping on TiO_2 and Bi_2MoO_6 will make the photocatalytic oxidation process more efficient, this achieving the lignin degradation.

2. Results

2.1. Scanning Electron Microscopy (SEM)

The morphology of samples was analyzed using Scanning Electron Microscopy (SEM). The Figure 1a displays a SEM image of TiO_2, which shows spherical and defined morphology. The N-doped TiO_2 compound, presented in Figure 1b, shows agglomerates of nano particles in a range of 0.1 to 0.5 μm. The image 1c of Bi_2MoO_6 displays laminar particles of an N-doped Bi_2MoO_6 compound. Figure 1d exhibits defined, elongated bar-like shaped particles with smooth edges.

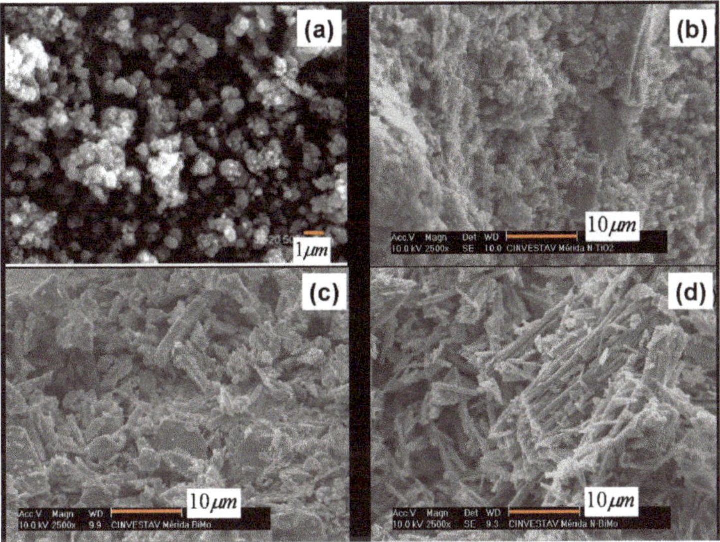

Figure 1. Scanning Electron Microscopy (SEM) photo-micrograph obtained at 5000X of (**a**) titanium dioxide (TiO_2), (**b**) nitrogen-doped titanium dioxide (N-TiO_2), (**c**) bismuth molybdate (Bi_2MoO_6), and (**d**) nitrogen-doped bismuth molybdate (N-Bi_2MoO_6) Bi_2MoO_6 catalysts.

2.2. X-ray Diffraction Analysis

X-ray analyses were carried out in order to capture pristine TiO_2 and Bi_2MoO_6 catalysts. I addition, they were carried out for their corresponding nitrogen doped compounds. For TiO_2 compounds, X-ray diffraction (XRD) patterns are displayed in Figure 2. It can be observed that this sample matches well with the anatase phase reported for TiO_2. For the N-doped TiO_2 sample, wider peaks are observed, as are changes in the intensity of the (101) diffraction plane, revealing that after N-doping through the reflux system, TiO_2 particles become smaller in size, providing the characteristic spectrum of an amorphous material. Figure 3 shows the diffraction pattern for Bi_2MoO_6 compounds, which reveals how the orthorhombic phase, also known as γ-Bi_2MoO_6 phase, was stabilized in these compounds. The crystallite size was calculated for every system using the Scherrer equation, which is as follows:

$$L = \frac{k\lambda}{W cos\Theta} \tag{1}$$

where k is a constant that depends on particle morphology (usually $k = 1.0$ for cubic or nearly-cubic systems); λ is the Cu, K_α radiation (nm); W is the full width at half-maximum (rad); and Θ is the diffraction angle (deg). The (101) reflection was used to perform calculations through Equation (1) for TiO_2 compounds, while the (131) reflection was used to obtain the crystallite size for Bi_2MoO_6 compounds. The results were 8.6 nm for TiO_2, 4.01 nm for N-doped TiO_2, 16.14 nm for γ-Bi_2MoO_6. and 7.92 nm for N-doped Bi_2MoO_6. Briefly, it appeared that nitrogen doping inhibited the crystal growth.

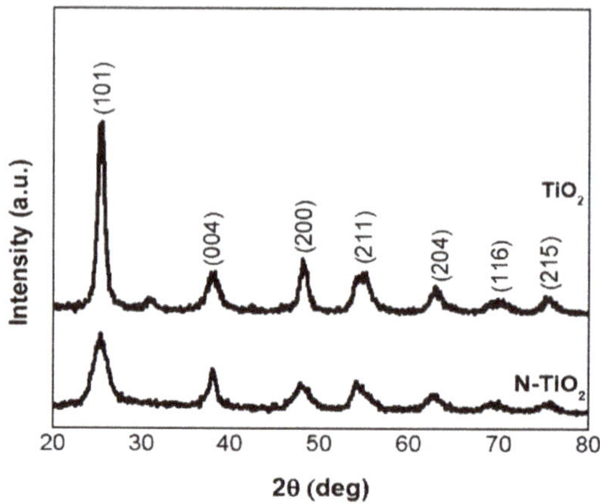

Figure 2. X-ray diffraction (XRD) patterns of TiO_2 and N-TiO_2.

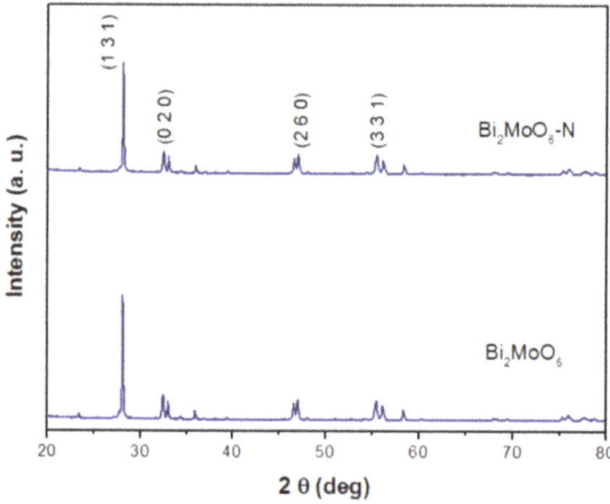

Figure 3. XRD patterns for Bi_2MoO_6 and Bi_2MoO_6-N catalysts.

2.3. Specific Surface Area Determination

The textural properties of TiO_2, Bi_2MoO_6, and N-doped catalysts are summarized in Table 1. It was appreciated that after doping TiO_2 and Bi_2MoO_6, their surface area changed in comparison to the pristine compounds. A dramatic change was observed for the TiO_2 compound which, after doping, decreased their BET area by 37%, while specific surface for Bi_2MoO_6 area was reduced to 47%. Apparently, the nitration process in both systems entails amorphization and pore reduction.

Table 1. Specific surface area for TiO_2, Bi_2MoO_6, TiO_2-N, and Bi_2MoO_6-N catalysts.

Compound	BET Surface Area (m^2/g)
TiO_2	117.0
N-TiO_2	73.7
Bi_2MoO_6	3.8
N-Bi_2MoO_6	2.0

Through the reflux procedure, carried out for at least 2 h, the constant movement of the particles in intimate contact with the ethylene-diamine and 1-hexanol resulted in a separation of the initially obtained particles, modifying the size of aggregates towards smaller values, which impacted the specific surface area. The resulting particles were smaller on average, especially in comparison to pristine undoped TiO_2 or Bi_2MoO_6 compounds, in the order of nanometers. While this process occurred, nitrogen molecules were fixed on the surface blocking pores and contributed to an apparent decrease in the specific surface area.

2.4. Diffuse Reflectance Measurements

Because the use of visible energy is necessary to test these systems, it is important to determine if there are any energy gap reductions after conducting the doping process. Thus, the diffuse reflectance spectra were obtained (Figure 4) and transformed into F(R), with a magnitude proportional to the extinction coefficient through the Kubelka-Munk function. In this case, R is the absolute reflectance of the sampled layer. Therefore, it is inferred that by using the following equation, the energy gap could be obtained graphically.

$$[F(R)hv]^2 = C(hv - Eg) \tag{2}$$

where Eg is the energy gap for every sample.

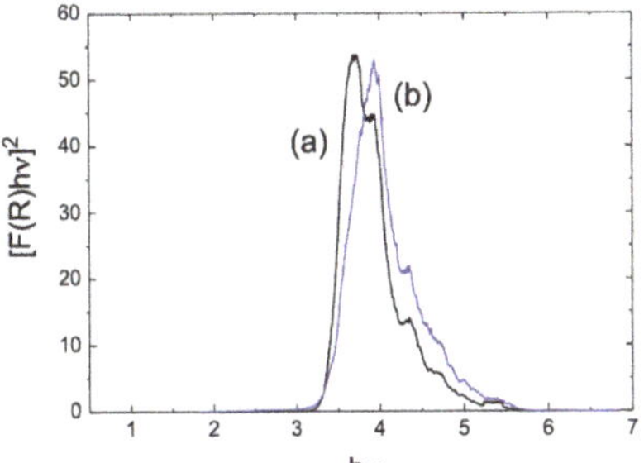
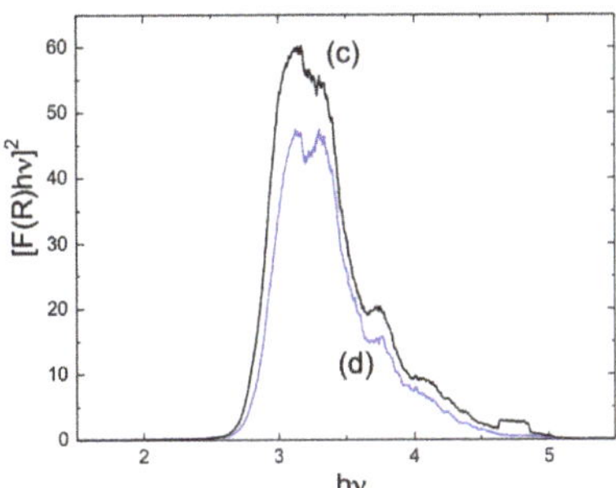

Figure 4. Reflectance diffuse measurements for (**a**) N-TiO$_2$, (**b**) TiO$_2$, (**c**) N-Bi$_2$MoO$_6$, and (**d**) Bi$_2$MoO$_6$, compounds.

The values obtained for the band gap are summarized in Table 2, where it is appreciated that, in both cases, a decreasing occurs after the nitrogen doping. These results indicate the possibility of using the developed compounds as a photocatalyst in the visible region of the electromagnetic spectrum, which will be demonstrated later in this work.

Table 2. Gap values for TiO_2, Bi_2MoO_6, N-TiO_2, and N-Bi_2MoO_6 catalysts.

Compound	Experimental Gap, eV	Reported Gap, eV
TiO_2	3.17	3.20
N-TiO_2	2.96	
Bi_2MoO_6	2.84	2.90
N-Bi_2MoO_6	2.73	

2.5. X-ray Photoelectron Spectroscopy (XPS)

XPS analyses were performed in samples with the purpose of establishing present elements and especially to determine if nitrogen doping is detected in TiO_2 or Bi_2MoO_6 doped compounds. Figure 5 shows the XPS spectra for TiO_2 and TiO_2-N compounds. The lower image, in this figure, corresponds to the pristine TiO_2, where the peaks O1s and Ti 2p are pointed out, as well as the C 1s. This last peak corresponds to small traces of carbon. In the upper part of Figure 5, it can be seen the corresponding image to N-TiO_2. On it, O1s, Ti 2p which were detected, and the N 1s signal was also found. In addition, a high-resolution scanning analysis was from 390 to 410 eV was performed, aimed at demonstrating the presence of nitrogen, which is shown as inset in the same figure. The XPS results for the Bi_2MoO_6 compounds are included in Figure 6, where the XPS spectra was acquired from 0–700 eV. The image shows the $3p_{3/2}$, 3d, and 4p states for Mo and Bi 4f, but also the O 1s and N 1 transitions are pointed out. In the upper left, an inset corresponding to the high-resolution energy window for N has been included, in order to emphasize their presence. In this way, for both cases, it is demonstrated that the procedure to introduce or impregnate nitrogen has been carried out successfully. Our results are in agreement with those results reported by other authors regarding the position of the N peak [26–28].

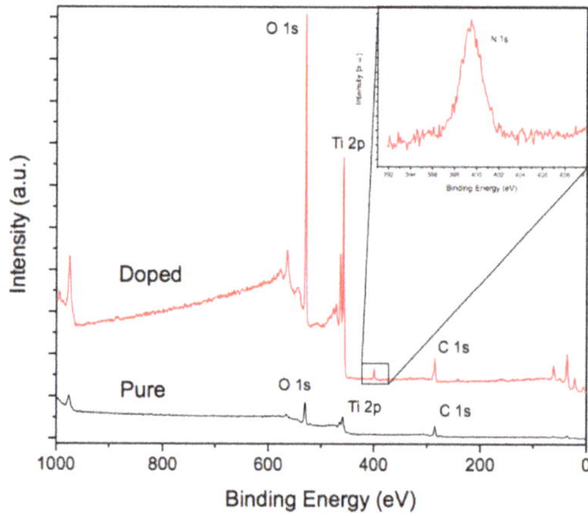

Figure 5. X-ray Photoelectron Spectroscopy (XPS) analysis for TiO_2 and N-TiO_2 catalysts.

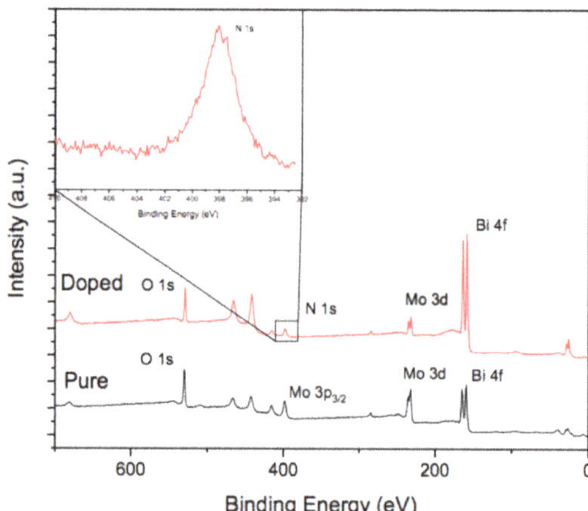

Figure 6. XPS analysis for Bi_2MoO_6 and N-Bi_2MoO_6 catalysts.

The quantification of the elements through the integration under the curve of each of their corresponding reflections gave us the following values summarized in Table 3.

It is observed that titanium oxide compounds are closer to the TiO_2 stoichiometry in comparison to Bi_2MoO_6 compounds. The reason is that usually bismuth tends to segregate close to the surface in the calcination stage, altering in some proportion the desired stoichiometry, as has been discussed in previous works [19,20]. It is important to note that the nitrogen content for N-Bi_2MoO_6 is larger than the value obtained for N-TiO_2. However, it will be shown in the next section that N-TiO_2 exhibits better performance in terms of catalytic activity. This means that the proportion of nitrogen has specific effects for every compound, as will be demonstrated latter.

Table 3. Atomic percent values from XPS analyses for TiO_2, Bi_2MoO_6, $N-TiO_2$, and $N-Bi_2MoO_6$ catalysts.

Compound	Ti2p	O1s	N1s	Bi	Mo
TiO_2	22.73	67.77	-	-	-
$N-TiO_2$	27.01	67.05	5.94	-	-
Bi_2MoO_6	-	61.86	-	24.12	14.02
$N-Bi_2MoO_6$	-	52.3	13.38	25.29	9.04

3. Photocatalytic Activity Tests

Figures 7–9 show photocatalytic activity measurements and solutions containing TiO_2 or Bi_2MoO_6 compounds. The reaction under ultraviolet (UV) and visible light (VIS) was evaluated separately for both, specifically methylene blue and lignin samples, which were measured to have an optical absorption at 660 nm and 289 nm, respectively. Pristine TiO_2 or Bi_2MoO_6 compounds were tested for MB degradation for about 120 min in the UV range (Figure 7). The degradation reached for TiO_2 was 80%, while for Bi_2MoO_6 it was nearly 70%. On the other hand, both compounds showed an improved degradation behavior when doped with nitrogen, reaching 90% of MB degradation using $N-Bi_2MoO_6$ and 93% for $N-TiO_2$. After the MB tests, both Bi_2MoO_6 and TiO_2 compounds were studied for lignin degradation, which is a more complex organic structure, when compared to MB. The degradation attained for Bi_2MoO_6 was about 62%; the degradation attained for TiO_2 was about 70% (Figure 8). In the case of nitrogen-doped compounds, the degradation was 82% for Bi_2MoO_6–N and 93% for $N-TiO_2$. Figure 8 summarizes the results obtained for lignin degradation that used TiO_2, $N-TiO_2$, Bi_2MoO_6, and $N-Bi_2MoO_6$ catalysts under visible light. Regarding the TiO_2 compound, it was less sensitive to VIS irradiation in such a way that only 12% degradation for TiO_2 and 25% for Bi_2MoO_6 were found. These results reveal the complex nature of the lignin molecule, which is hard to degrade and frequently results in compounds derived from this degradation, such as formic acid, ketones, and aldehydes, among others [30–33]. An interesting result was found for lignin degradation when using $N-TiO_2$ and $N-Bi_2MoO_6$ catalysts (Figure 9), where a 32% degradation for $N-TiO_2$ catalyst and 38% for $N-Bi_2MoO_6$ compound can be observed. Interestingly, nitrogen doping on both compounds, has been beneficial. However, the nitrogen doping in TiO_2 makes this material more efficient, which can be related to the reduction of the energy gap, which would be able to provide additional electronic states among conduction and valence bands.

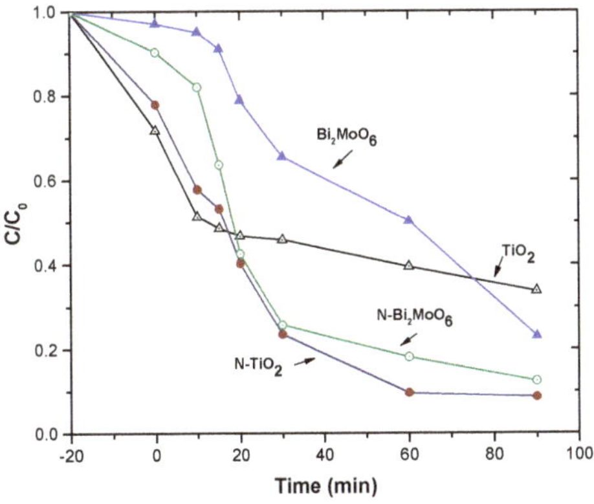

Figure 7. Methylene blue degradation in presence of TiO_2, $N-TiO_2$, Bi_2MoO_6, and $N-Bi_2MoO_6$ catalysts under visible light (VIS) energy irradiation.

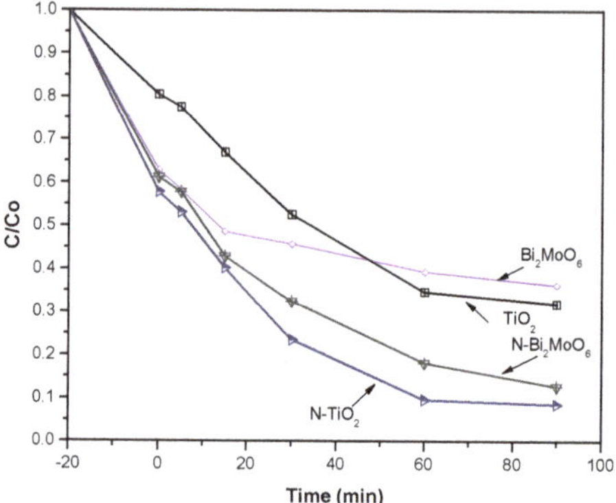

Figure 8. Lignin degradation in presence of TiO_2, $N\text{-}TiO_2$, Bi_2MoO_6, and $N\text{-}Bi_2MoO_6$ catalysts under ultraviolet (UV) energy irradiation.

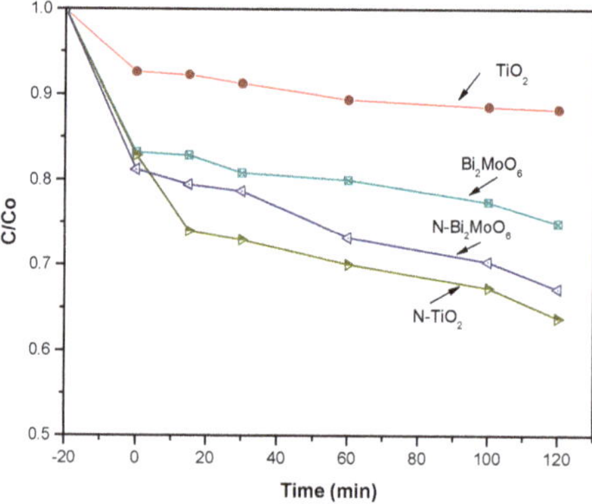

Figure 9. Lignin degradation in presence of TiO_2, $N\text{-}TiO_2$, Bi_2MoO_6, and $N\text{-}Bi_2MoO_6$ catalysts under VIS energy irradiation.

4. Methods and Materials

4.1. Synthesis of TiO_2 and Bi_2MoO_6 Catalysts

A TiO_2 material was based on colloidal TiO_2 obtained from $Ti[OCH(CH_3)_2]_4$ hydrolysis and stirring the suspension to obtain the gel. The resulting material was filtered and oven dried at $100\ °C$ and finally calcined at $450\ °C$. The Bi_2MoO_6 compound was obtained using chemical coprecipitation starting with high purity $Bi(NO_3)_3\cdot5H_2O$ and $(NH_4)_6Mo_7O_{24}$, diluted in distilled water and adjusting the pH 7.0 using either NH_4OH or HNO_3. Afterward, the precipitated Bi_2MoO_6 was washed and oven-dried at $100\ °C$ and calcined at $400\ °C$.

4.2. Synthesis of N-doped TiO_2 and N-doped Bi_2MoO_6

In order to obtain N-doped Bi_2MoO_6 and N-doped TiO_2 catalysts, both compounds separately, were added with 25 mL of a 2 M ethylene diamine solution and 150 mL of 1-hexanol, in a continuous refluxing system, stirring it for 2 h. Afterward, N-doped Bi_2MoO_6 and N-doped TiO_2 materials were filtered, washed several times with water, and dried in an oven at $200\ °C$. Upon drying, both compounds yielded an intense yellowish powder.

4.3. Characterization Details

The obtained compounds were characterized using X-ray diffraction (Siemens, D-5000 model), operating at 30 keV and 20 mA, with a step size $0.02°/\min$ from 10 to $70°$ (2θ). The images were obtained in a SEM JSM-6400 JEOL Noran Instruments, at 20 keV and 10^{-6} Torr. The diffuse reflectance spectra (R) data were obtained using a UV-visible spectrophotometer (AvaSpec-2048), equipped with an integrating sphere (Ocean Optics, Mod. ISP-50-8-R-GT), equipped with a deuterium halogen light source (Mod. AvaLight DH-S-BAL). Specific surface area was measured in a Micrometrics Gemini 2060 RIG-100, model at 77 K using the BET method. For the XPS analyses, samples were excited with Al and Ka X-rays with an energy of 1486.6 eV. The spectrometer was calibrated using the Cu 2p3/2 (932.4 eV) and Cu 3p3/2 (74.9 eV) lines. Binding energy calibration was based on C 1s at 284.6 eV.

4.4. Photocatalytic Evaluation

Catalytic activity was tested as previously described [30,31]. Briefly, the reaction was carried out in a batch micro reactor provided with an oxygen flow, to generate superoxide radicals and prevent electron recombination. The solution was previously stirred for 20 min without the presence of light in order to reach a stable MB or lignin absorbance on the photocatalyst surface. Then, the solution was irradiated with the source of light, magnetically stirred, and air was introduced once the reaction system was started. For MB studies, 0.1 g of every catalyst were placed into a beaker containing an aqueous solution of 0.2 g/L of MB. Regarding the lignin degradation, 0.2 g of this compound (Sigma-Aldrich, PM = 28,000 g/mol) was mixed with 15 mL of NaOH (Sigma-Aldrich, México city, México); 0.05 M was used to dissolve the samples, where then the samples were placed into a beaker containing 500 mL of deionized water. The pH = 8 was adjusted to enhance the photo catalytic reaction, according to reference [32]. For every catalyst, the reaction was achieved for 80 min. Samples were taken from the reactor system at 20 min intervals to follow the course of the reaction. Samples were centrifuged for 5 min at 220 rpm, in order to separate the catalyst from the solution to determine the progress of the reaction. A UV light lamp was used with a wavelength of 365 nm for the photocatalytic reaction. In the case of the visible light experiments, the irradiation was performed using a UV-VIS light source of 200 W, provided with a Xe arc lamp (Oriel). The concentration was monitored through a UV-VIS Hach Dr/4000u spectrophotometer at a wavelength of 289 nm for lignin and 660 nm for MB.

5. Discussion

After the refluxing process, the dissolution and recrystallization processes of a dispersed solid under reflux (nitrogen doping) was expected. It was substantiated by the well-known Ostwald's ripening. Often, the Ostwald's ripening includes the large crystals growing at the cost of smaller ones (i.e., coarsening). In the present case, the presence of the amine might have changed the expected growth by favoring the nucleation of new parties as part of the recrystallizing processes. According to Classical Nucleation Theory (CNT), a phase transition (i.e., the crystallization of a new phase within another) can be rationalized as result of two main opposed contributions. On one hand, the driving force for the process is universally identified in chemical reactions with the chemical potential difference ($\Delta\mu$); on the other hand, the work spent to form the new surface was related to the new phase (associated with the interface energy, σ, and the area created, A) [34]. Hence, the Gibbs free energy for homogeneous nucleation was, $\Delta G = \Delta\mu + A\sigma$, where spontaneous nucleation will depend on the balance between both energies' contribution. In a heterogeneous nucleation, a surface area already exists and acts as nucleation site with lower contribution from the second term. Therefore, $\Delta G_{\text{heterogeneous}}$ becomes a fraction of $\Delta G_{\text{homogeneous}}$ [35]. This explains the well-known "coarsening" phenomena. However, the coarsening should compete with nucleation of new particles, as the amine acts as a surfactant and also decreases the second term by lowering the interfacial energy. Therefore, diminishing the particle size is intuitively expected under present conditions.

Furthermore, we cannot discard a possible amorphization process as result of the nitrogen inclusions within the matrix of the solids. Recrystallization accompanied nitrogen impurification might cause displacements of Wyckoff positions expected for the spatial groups of both solids; this could occur with the consequent strain increase that would change the Full wide half maximum (FWHM) of the reflections.

The XPS analyses have demonstrated the incorporation of nitrogen in both N-TiO$_2$ and N-Bi$_2$MoO$_6$ compounds. The nitrogen signal is located close to 400 eV. However, it is worth mentioning that some differences arise when this peak is closely analyzed for every compound. In the case of nitrogen doping for the TiO$_2$ compound, one peak is located at 397.3 eV and another in 400.7 eV. The first is attributed to substitutional or interstitial impurities (corresponding to Ti-N bonds that substitute O by N in the lattice); the other transition, located in 400.7 eV, can be attributed to molecularly chemisorbed (superficial) nitrogen. Regarding the Bi$_2$MoO$_6$ catalyst, something similar occurs when the high-resolution peak that corresponds to nitrogen is analyzed, as the peak found at 396 eV corresponds to shallow surface nitrogen, while the nitrogen signal at 398 eV can be assigned to interstitial nitrogen. In general, nitrogen doping has been beneficial in most cases, because the nitrogen doped samples showed the best degradation performance in comparison to the pristine Bi$_2$MoO$_6$ or TiO$_2$ catalysts. As expected, lignin degradation was more difficult to carry out in comparison to MB, due to the complexity of the lignin molecule. However, the N-TiO$_2$ sample showed a 90% degradation for lignin when using UV radiation. In the case of experiments carried out using visible radiation, it was found that by using an N-Bi$_2$MoO$_6$ compound, 30% degradation was attained; in the case of N-TiO$_2$, however, a 35% of lignin degradation was reached.

6. Conclusions

TiO$_2$ and Bi$_2$MoO$_6$ N-doped photocatalysts were successfully synthesized and our XPS analyses demonstrate that nitrogen doping was carried out efficiently. It was also found that the N-TiO$_2$ catalyst exhibited a better performance in terms of MB or lignin degradation. Even if N-TiO$_2$ and N-Bi$_2$MoO$_6$ catalysts provide good efficiency for MB and lignin degradation, nitrogen doped TiO$_2$ is the best catalyst to degrade lignin. It is demonstrated that nitrogen doping in both compounds, is an effective way to improve their degradation performance. It was also shown that nitrogen doping provides the possibility of using both catalysts under visible light.

Author Contributions: Conceptualization, R.R.; methodology, V.C.; formal analysis, P.B.; investigation, V.C.; resources, J.E.; data curation, J.E.; writing—original draft preparation, R.R.; writing—review and editing, G.R.; and J.A.

Funding: This research received no external funding.

Acknowledgments: R. Rangel acknowledges financial support from CIC-UMSNH under project 2018. Also thanks to W. Cahuich from Cinvestav-IPN for SEM images. Also to LANBIO of Cinvestav, Merida for allowing acces to their facilities.

Conflicts of Interest: The authors declare not having conflict of interest.

References

1. Fernández, C.; Larrechi, M.S.; Callao, M.P. An analytical over-view of processes for removing organic dyes from wastewater effluents. *Trends Anal. Chem.* **2010**, *29*, 1202–1211. [CrossRef]
2. Fujishima, A.; Honda, K. Electrochemical photolysis of water at a semiconductor electrode. *Nature* **1972**, *238*, 37–38. [CrossRef] [PubMed]
3. Kumar, G.; Rao, K.S.R.K. Zinc oxide based photocatalysis: Tailoring Surface–Bulk structure and related interfacial charge carrier dynamics for better environmental applications. *RSC Adv.* **2015**, *5*, 3306–3351. [CrossRef]
4. Huang, Y.; Sun, F.; Wu, T.; Wu, Q.; Huang, Z.; Su, H.; Zhang, Z. Photochemical preparation of CdS hollow microspheres at room temperature and their use in visible-light photocatalysis. *J. Solid State Chem.* **2011**, *184*, 644–648. [CrossRef]

5. Abdelkadera, E.; Nadjia, L.; Naceur, B.; Noureddine, B. SnO_2 foam grain-shaped nanoparticles: Synthesis, characterization and UVA light induced photocatalysis. *J. Alloys Compd.* **2016**, *679*, 408–419. [CrossRef]

6. Ye, Z.; Kong, L.; Chen, F.; Chen, Z.; Lin, Y.; Liu, C. A comparative study of photocatalytic activity of ZnS photocatalyst for degradation of various dyes. *Optik* **2018**, *164*, 345–354. [CrossRef]

7. Ullaha, S.; Ferreira-Neto, E.P.; Hazra, C.; Parveen, R.; Rojas-Mantilla, H.D.; Calegaro, M.L.; Serge-Correales, Y.E.; Rodrigues-Filho, U.P.; Ribeiro, S.J.L. Broad spectrum photocatalytic system based on $BiVO_4$ and $NaYbF_4:Tm^{3+}$ upconversion particles for environmental remediation under UV-vis-NIR illumination. *Appl. Catal. B Environ.* **2019**, *243*, 121–135. [CrossRef]

8. Sanchez-Martinez, A.; Ceballos-Sanchez, O.; Koop-Santa, C.; López-Mena, E.R.; Orozco-Guareño, E.; García-Guaderrama, M. N-doped TiO_2 nanoparticles obtained by a facile coprecipitation method at low temperature. *Ceram. Int.* **2018**, *44*, 5273–5283. [CrossRef]

9. Moustakas, N.G.; Kontos, A.G.; Likodimos, V.; Katsaros, F.; Boukos, N.; Tsoutsou, D.; Dimoulas, A.; Romanos, G.E.; Dionysiou, D.D.; Falaras, P. Inorganic-organic core-shell titania nanoparticles for efficient visible light activated photocatalysis. *Appl. Catal. B Environ.* **2013**, *130–131*, 14–24. [CrossRef]

10. Khan, H.; Rigamonti, M.G.; Patience, G.S.; Boffito, D.C. Spray dried TiO_2/WO_3 heterostructure for photocatalytic applications with residual activity in the dark. *Appl. Catal. B Environ.* **2018**, *226*, 311–323. [CrossRef]

11. Nunes, D.; Pimentel, A.; Pinto, J.V.; Calmeiro, T.R.; Nandy, S.; Barquinha, P.; Pereira, L.; Carvalho, P.A.; Fortunato, E.; Martins, R. Photocatalytic behavior of TiO_2 films synthesized by microwave irradiation. *Cat. Today* **2016**, *278*, 262–270. [CrossRef]

12. Ramakrishnan, V.M.; Natarajan, M.; Santhanam, A.; Asokan, V.; Velauthapillai, D. Size controlled synthesis of TiO_2 nanoparticles by modified solvothermal method towards effective photo catalytic and photovoltaic applications. *Mater. Res. Bull.* **2018**, *97*, 351–360. [CrossRef]

13. Rangel, R.; Cedeño, V.; Ramos-Corona, A.; Gutierrez, R.; Alvarado-Gil, J.J.; Ares, O.; Bartolo-Perez, P.; Quintana, P. Tailoring surface and photocatalytic properties of ZnO and nitrogen-doped ZnO nanostructures using microwave-assisted facile hydrothermal synthesis. *Appl. Phys.* **2017**, *123*, 552. [CrossRef]

14. Phuruangrat, A.; Dumrongrojthanath, P.; Thongtem, S.; Thongtem, T. Synthesis and characterization of visible light-driven W-doped Bi_2MoO_6 photocatalyst and its photocatalytic activities. *Mater. Lett.* **2017**, *194*, 114–117. [CrossRef]

15. Rangel, R.; Maya, R.; García, R. Novel $[Ce_{1-x}La_xO_2, La_{2-y}Ce_yO_3]/Bi_2Mo_{0.9}W_{0.1}O_6$ Catalysts for CO Oxidation at low temperature. *Catal. Sci. Technol.* **2012**, *2*, 639–642. [CrossRef]

16. Geng, B.; Wei, B.; Gao, H.; Xu, L. Ag_2O nanoparticles decorated hierarchical Bi_2MoO_6 microspheres for efficient visible light photocatalysts. *J. Alloys Compd.* **2017**, *699*, 783–787. [CrossRef]

17. Guo, J.; Shi, L.; Zhao, J.; Wang, Y.; Yuan, X. Enhanced visible-light photocatalytic activity of Bi2MoO6 nanoplates with heterogeneous Bi_2MoO_{6-x} and Bi_2MoO_6 core-shell structure. *App. Catal. B Environ.* **2018**, *224*, 692–704. [CrossRef]

18. Umapathy, V.; Manikandan, A.; Antony, S.A.; Ramu, P.; Neeraja, P. Structure, morphology and opto-magnetic properties of Bi_2MoO_6 nano-photocatalyst synthesized by sol-gel method. *Trans. Nonferrous Metals Soc. China* **2015**, *25*, 3271–3278. [CrossRef]

19. Bi, J.; Wu, L.; Li, J.; Li, Z.; Wang, X.; Fu, X. Simple solvothermal routes to synthesize nanocrystalline Bi_2MoO_6 photocatalysts with different morphologies. *Acta Mater.* **2007**, *55*, 4699–4705. [CrossRef]

20. Martínez-de la Cruz, A.; Obregón Alfaro, S. Synthesis and characterization of γ-Bi_2MoO_6 prepared by co-precipitation: Photoassisted degradation of organic dyes under vis-irradiation. *J. Mol. Catal. A Chem.* **2010**, *320*, 85–91. [CrossRef]

21. Jin, S.; Hao, H.; Gan, Y.; Guo, W.; Li, H.; Hu, X.; Hou, H.; Zhang, G.; Yan, S.; Gao, W.; et al. Preparation and improved photocatalytic activities of Ho^{3+}/Yb^{3+} co-doped Bi_2MoO_6. *Mater. Chem. Phys.* **2017**, *199*, 107–112. [CrossRef]

22. Yu, Ch.; Wu, Z.; Liu, R.; Dionysiou, D.D.; Yang, K.; Wang, Ch.; Liu, H. Novel fluorinated Bi_2MoO_6 nanocrystals for efficient photocatalytic removal of water organic pollutants under different light source illumination. *App. Catal. B Environ.* **2017**, *209*, 1–11. [CrossRef]

23. Xing, Y.; Gao, X.; Ji, G.; Liu, Z.; Du, C. Synthesis of carbon doped Bi_2MoO_6 for enhanced photocatalytic performance and tumor photodynamic therapy efficiency. *Appl. Surf. Sci.* **2019**, *465*, 369–382. [CrossRef]

24. Asashi, R.; Morikawa, T.; Ohwaki, T.; Aoki, K.; Taga, Y. Visible light photocatalysis in nitrogen-doped titanium oxide. *Science* **2001**, *293*, 269–271. [CrossRef] [PubMed]

25. Wang, J.; Tafen, D.; Lewis, J.; Hong, Z.; Manivannan, A.; Zhi, M.; Wu, N. Origin of photocatalytic activity of nitrogen-doped TiO_2 nanobelts. *J. Am. Chem. Soc.* **2009**, *131*, 12290–12297. [CrossRef] [PubMed]

26. Yoshida, T.; Niimi, S.; Yamamoto, M.; Ogawa, S.; Nomoto, T.; Yagi, S. Characterization of nitrogen ion implanted TiO_2 photocatalysts by XAFS and XPS. *Nucl. Instrum. Methods Phys. Res. Sect. B: Beam Interact. Mater. At.* **2015**, *365*, 79–81. [CrossRef]

27. Manova, D.; Franco-Arias, L.; Hofele, A.; Alani, I.; Kleiman, A.; Asenova, I.; Decker, U.; Marquez, A.; Mändl, S. Nitrogen incorporation during PVD deposition of TiO_2:N thin films. *Surf. Coat. Technol.* **2017**, *312*, 61–65. [CrossRef]

28. Boningaria, T.; Reddy-Inturia, S.N.; Suidan, D.; Smirniotisa, P.G. Novel one-step synthesis of nitrogen-doped TiO_2 by flame aerosol technique T for visible-light photocatalysis: Effect of synthesis parameters and secondary nitrogen (N) source. *Chem. Eng. J.* **2018**, *350*, 324–334. [CrossRef]

29. Tran, V.A.; Truong, T.T.; Pham-Phan, T.A.; Nguyen, T.N.; Huynh, T.V.; Agresti, A.; Pescetelli, S.; Le, T.K.; Carlo, A.; Lund, T.; et al. Application of nitrogen-doped TiO_2 nano-tubes in dye-sensitized solar cells. *Appl. Surf. Sci.* **2017**, *399*, 515–522. [CrossRef]

30. Rangel, R.; López-Mercado, G.J.; Bartolo-Pérez, P.; García, R. Nanostructured-[CeO_2, La_2O_3, C]/TiO_2 catalysts for lignin photodegradation. *Sci. Adv. Mater.* **2012**, *4*, 573–578. [CrossRef]

31. Rangel, R.; García-Espinoza, J.D.; Espitia-Cabrera, I.; Alvarado-Gil, J.J.; Quintana, P.; Bartolo-Pérez, P.; Trejo-Tzab, R. Synthesis of Mesoporous of $N_yTi_{1-x}Ce_xO_{2-y}$ Structures and its Visible Light Induced Photocatalytic Performance. *Nano* **2013**, *8*, 1350051–1350061. [CrossRef]

32. Dahm, A.; Lucian, A. Titanium dioxide catalyzed photodegradation of lignin in industrial effluents. *Ind. Eng. Chem. Res.* **2004**, *43*, 7996–8000. [CrossRef]

33. Gazi, S.; Hung Ng, W.K.; Ganguly, R.; Putra, A.M.; Hirao, H.; Soo, H.S. Selective photocatalytic C–C bond cleavage under ambient conditions with earth abundant vanadium complexes. *Chem. Sci.* **2015**, *6*, 7130–7142. [CrossRef] [PubMed]

34. La Mer, V.K; Dinegar, R.H. Theory, Production and Mechanism of Formation of Monodispersed Hydrosols. *Ind. Eng. Chem.* **1950**, *72*, 4847–4854.

35. Wnek, W.J. The simulation of precipitation kinetics. *Powder Technol.* **1978**, *20*, 289–293. [CrossRef]

Article

Highly Selective Photocatalytic Reduction of o-Dinitrobenzene to o-Phenylenediamine over Non-Metal-Doped TiO$_2$ under Simulated Solar Light Irradiation

Hamza M. El-Hosainy [3,4], Said M. El-Sheikh [1], Adel A. Ismail [1,2,*], Amer Hakki [3], Ralf Dillert [3], Hamada M. Killa [5], Ibrahim A. Ibrahim [1] and Detelf W. Bahnemann [3,6,*]

1   Department of Nanomaterials and Nanotechnology, Central Metallurgical R & D Institute, Cairo 11421, Egypt; selsheikh2001@gmail.com (S.M.E.-S.); ibrahimahmedcmrdi01@gmail.com (I.A.I.)

2   Nanotechnologyand and Advanced Materials Program, Energy & Building Research Center, Kuwait Institute for Scientific Research (KISR), P.O. Box 24885, Safat 13109, Kuwait

3   Institut für Technische Chemie, Leibniz Universität Hannover, Callinstr. 3, D-30167 Hannover, Germany; hamzaelhosainy@gmail.com (H.M.E.-H.); a.hakki@abdn.ac.uk (A.H.); dillert@iftc.uni-hannover.de (R.D.)

4   Institute of Nanoscience & Nanotechnology, Kafrelsheikh University, Kafrelsheikh 33516, Egypt

5   Faculty of Science, Zagazig University, Zagazig 44519, Egypt; hamadakilla48@gmail.com

6   Laboratory "Photoactive Nanocomposite Materials", Saint-Petersburg State University, Ulyanovskaya str. 1, Peterhof, Saint-Petersburg 198504, Russia

*   Correspondence: aaismail@kisr.edu.kw (A.A.I.); bahnemann@iftc.uni-hannover.de (D.W.B.)

Received: 21 November 2018; Accepted: 6 December 2018; Published: 9 December 2018

Abstract: Photocatalytic reduction and hydrogenation reaction of o-dinitrobenzene in the presence of oxalic acid over anatase-brookite biphasic TiO$_2$ and non-metal-doped anatase-brookite biphasic TiO$_2$ photocatalysts under solar simulated light was investigated. Compared with commercial P25 TiO$_2$, the prepared un-doped and doped anatase-brookite biphasic TiO$_2$ exhibited a high selectivity towards the formation of o-nitroaniline (85.5%) and o-phenylenediamine ~97%, respectively. The doped anatase-brookite biphasic TiO$_2$ has promoted photocatalytic reduction of the two-nitro groups of o-dinitrobenzene to the corresponding o-phenylenediamine with very high yield ~97%. Electron paramagnetic resonance analysis, Transient Absorption Spectroscopy (TAS) and Photoluminescence analysis (PL) were performed to determine the distribution of defects and the fluorescence lifetime of the charge carriers for un-doped and doped photocatalysts. The superiority of the doped TiO$_2$ photocatalysts is accredited to the creation of new dopants (C, N, and S) as hole traps, the formation of long-lived Ti^{3+} defects which leads to an increase in the fluorescence lifetime of the formed charge carriers. The schematic diagram of the photocatalytic reduction of o-dinitrobenzene using the doped TiO$_2$ under solar light was also illustrated in detail.

Keywords: photocatalysis; non-metal- doped TiO$_2$; nitroaromatic compounds; reduction; selectivity

---

1. Introduction

The great challenge for the modern chemical industry is to drive chemical reactions employing a sustainable, green, and eco-friendly process using renewable energy sources. Therefore, the development of new strategies to obtain fine chemicals in a fast, clean, and efficient approach is of great significance and still requires considerable efforts. Heterogeneous photocatalysis is one of the promising and eco-friendly approaches that satisfies these requirements. This method is one of the important processes that encourages the use of sunlight as the source for chemical conversion processes [1–4]. Moreover, the photo-induced organic transformations by solar-driven photocatalysis can produce specific products with high selectivity and lower cost compared to conventional methods [5–7]. The reduction of nitroaromatic compounds to the corresponding amino compounds is one of these photo-induced organic transformations with TiO_2 is [3–6] which has attracted significant attention because of the importance of these amino compounds as intermediates of numerous valuable compounds such as dyes and medicines [8]. Most of the reported studies in this field have used alcohols as reaction media, hole scavenger, and hydrogen source for photocatalytic hydrogenation of nitroaromatics using TiO_2 photocatalyst under inert gas atmosphere [9,10]. However, the oxidation products of alcohols may react with the reduction product of the nitro compounds which affect the selectivity and the yield of the desired amino compound. Thus, of other hole scavengers, oxalic acid is preferred since it is easily dissolved in water, which can be used as reaction media, and CO_2 is the only byproduct of its oxidation [4,11–13]. TiO_2 (P-25) was employed as a photocatalyst for the photoreduction of dinitro compound o-dinitrobenzene to corresponding mono and diamino compounds (o-nitroaniline and o-phenylenediamine) under solar light irradiation [14]. The results showed formation of low yield and selectivity of 55% o-nitroaniline and 30% o-phenylenediamine.

On the other hand, the reduction of nitroaromatic compounds by non-metal-doped TiO_2 under solar light is rarely investigated. One example is the use of N-doped TiO_2 together with KI for the reduction of o-nitrophenol in the existence of methanol under solar light [3]. The results revealed that N-TiO_2 has a low efficiency for reduction of nitroaromatic compounds under solar light irradiation. The phase type of TiO_2 also has a great effect on the photocatalytic selectivity. Rutile TiO_2 displays higher activity and selectivity than anatase TiO_2 and P25 TiO_2 for selective hydrogenation of nitroaromatic compounds [15]. Furthermore, compared with rutile TiO_2, P25 TiO_2 has the ability to complete photocatalytic reduction of m-dinitrobenzene to the corresponding m-phenylenediamine in the deaerated aqueous iso-propanol under 4 h of UV light irradiation [16]. Herein, we report for the first time the use of anatase-brookite biphasic TiO_2 and non-metal (C, N and S)-doped anatase-brookite biphasic TiO_2 for the selective hydrogenation of o-dinitrobenzene to the corresponding o-nitroaniline and o-phenylenediamine under solar simulator light, respectively. To the best of our knowledge, there are no reports showing the high selectivity and formation of o-phenylenediamine by using non-metal-doped anatase-brookite biphasic TiO_2 under solar light irradiation. The expected schematic diagram and mechanism for the photocatalytic reduction of o-dinitrobenzene was also interpreted.

2. Results and Discussion

Similar to our previous work [17], XRD analysis proved the formation of anatase and brookite biphase TiO_2 with compositions ~75% and 25%, respectively. The surface area and pore size for un-doped and doped samples amounted to 226.2 and 85.1 $m^2\ g^{-1}$ and 2.2 nm and 3.6 nm, respectively. Therefore, these results show the prepared TiO_2 samples have a mesoporous structure. The particle size for un-doped and doped samples were 5–10 nm and 10–15 nm, respectively. The XPS analysis has proved the existence of C, N, S in the doped sample. UV-Vis. spectroscopy displayed a red absorption shift for the doped sample, reflecting that the band gap value of doped TiO_2 sample decreased from 3.1 to 2.9 eV. Herein, the feasibility of using these materials for photocatalytic reduction of o-dinitrobenzene in aqueous oxygen-free solutions under solar light irradiation was also conducted.

2.1. Reduction of O-Dinitrobenzene to O-Nitroaniline and O-Phenylenediamine

Firstly, initial experiments for reduction of o-dinitrobenzene were carried out with an aqueous solution containing TiO_2 samples in the existence of oxalic acid as a hole scavenger.

Figure 1 represents the time-dependent change in the concentration of o-dinitrobenzene and its photocatalytic products in its aqueous solution containing either un-doped (a) or doped TiO_2 (b) in presence of oxalic acid as hole scavenger during the irradiation with solar simulated light. It is clearly observed that the concentration of o-dinitrobenzene dramatically decreases with increasing of the photoirradiation time for both T and DT samples. 9 h were needed to achieve the complete conversion of o-dinitrobenzene when employing T as the photocatalyst, whereas only 7 h were enough in the case of DT sample. Interestingly, the concentration of corresponding monoamino compound (o-nitroaniline) increases gradually with prolonged photoirradiation time when employing the un-doped photocatalyst (T). However, the photo-catalytically produced (o-nitroaniline) undergoes further reduction and hydrogenation to produce the corresponding diamino product (o-phenylenediamine) when DT was employed as the photocatalyst. Yield and selectivity of o-nitroaniline and o-phenylenediamine employing either T or DT samples are displayed in Figure 2a,b, respectively. In the case of T sample, o-nitroaniline is only selective as a result of hydrogenation of o-dinitrobenzene (Figure 2a). The yield and selectivity boost with the increase of photoirradiation time reaching ~88.5% within 13 h (see Figure 2a). On the other hand, DT photocatalyst shows a higher yield and selectivity ~97% of o-phenylenediamine as a result of reduction and hydrogenation of the two-nitro groups of the o-dinitrobenzene after only 9 h irradiation as displayed in Figure 2b. By comparison, the commercial P25 TiO_2 was tested for photocatalytic reduction of o-dinitrobenzene under solar simulator light. The results showed the formation of non-selective reduction products from ≈13% o-nitroaniline and ≈86.5% o-phenylenediamine within 13 h under solar simulator light as displayed in Figure 2c. Irradiation of aqueous solution containing o-dinitrobenzene with TiO_2 samples and oxalic acid under solar light produced o-nitroaniline and o-phenylenediamine as a reduction product. On the other hand, no reduction products were obtained from the aqueous solution containing o-dinitrobenzene without using photocatalyst or light and/or oxalic acid, respectively. This means that these parameters are essential for reduction of o-dinitrobenzene to the corresponding o-phenylenediamine.

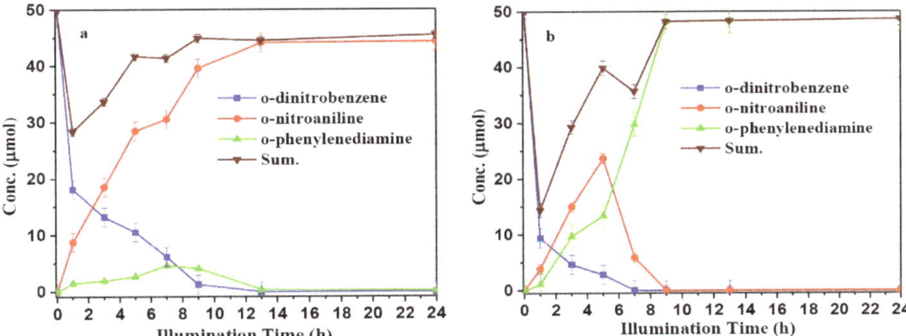

Figure 1. Time-dependent change in the concentration of substrate and products in aqueous solution of (**a**) un-doped TiO_2 (T sample) Note Conc.: Concentration and (**b**) (C, N, S)-doped TiO_2 (DT sample) in the presence of oxalic acid as hole scavenger during photoirradiation under simulated solar light, Note Conc.: Concentration and reaction conditions: 25 mg TiO_2 samples, 50 µmol o-dinitrobenzene, 250 µmol oxalic acid, 5 cm^3 deionized water, Ar.

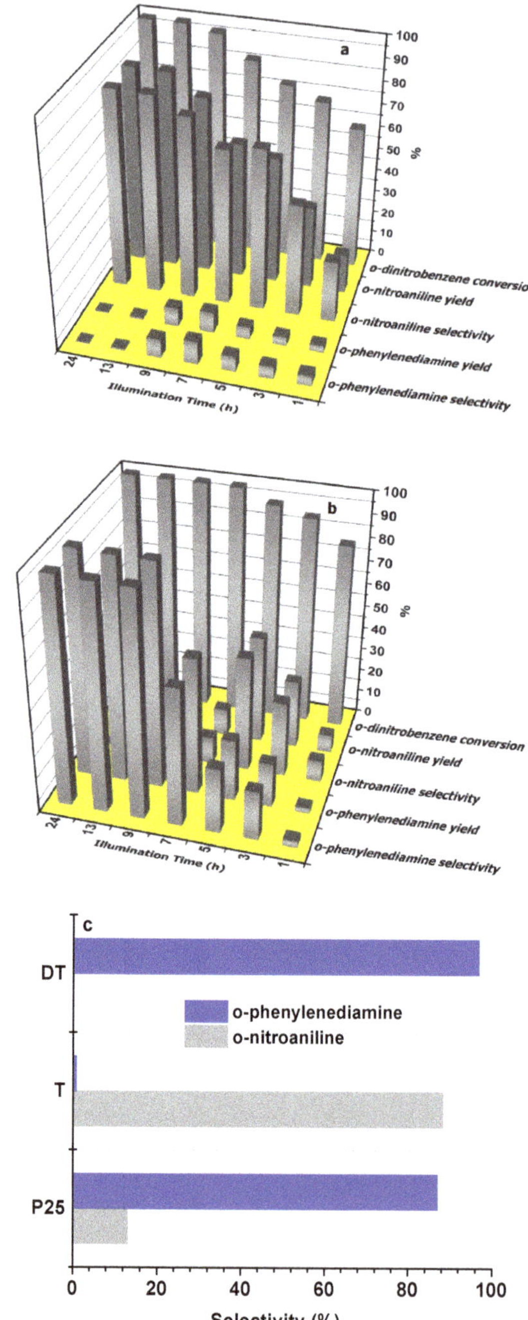

Figure 2. Yield and Selectivity for (**a**) T, (**b**) DT samples and (**c**) Selectivity for P25 compared with other samples, reaction conditions: 25 mg TiO_2 samples, 50 µmol o-dinitrobenzene, 250 µmol oxalic acid, 5 cm^3 deionized water, Ar.

It is well known that the light-induced six-electron reduction of a one-nitro group of the nitroaromatic compound to the corresponding amino compound in the presence of TiO_2 occurs via a sequence of electron transfer, protonation, and dehydrogenation reactions [18]. Thus, the complete reduction of two-nitro groups to two-amino groups requires twelve electrons and twelve protons. This usually occurs via the formation of hydroxylamine and/or nitrosobenzene as intermediates. However, neither nitrosobenzene nor N-phenyl hydroxylamine were detected in our cases. This might be explained by the fact that DT photocatalyst expedites the conversion of the nitro-to-amine through hydrogenation reactions (i.e., via hydrogen species derived from oxalic acid). This inhibits side reactions and facilitates selective o-phenylenediamine production. Therefore, with DT sample, photoirradiation leads to complete transformation of o-dinitrobenzene to the corresponding o-phenylenediamine with high yield and selectivity. It is also important to mention that the reduction of the second nitro group is usually more difficult that the first one and therefore it requires stronger reducing agent. The doped anatase/brookite biphasic TiO_2 (DT sample) showed the high ability to complete the reduction of the two-nitro group of the dinitro compound to diamino compound (o-phenylenediamine). Therefore, compared with commercial P25 TiO_2, un-doped and doped samples formed a selective reduction product from o-nitroaniline and o-phenylenediamine, respectively.

The observed difference in the selectivity of the photocatalytic conversion of o-dinitrobenzene employing the un-doped and doped materials can be attributed to the following different factors:

Firstly, this can be accredited to decrease in the band gap for the doped sample (2.9 eV) compared to the un-doped one (3.1 eV) 1 to enhancement its absorption capacity under solar simulator light (see UV-Vis. analysis, Figure S1) [17]. By non-metal doping, the O2P orbitals of TiO_2 mixes with the dopants 2P orbitals of C, N and S forming a new mid-gap above the valence band of TiO_2 (see Scheme 1, see XPS analysis, Figure S2) which leads to decrease its band gap. Briefly, as illustrated in our previous work [17], XPS analysis revealed C, N and S are doped with TiO_2 and carbon is also located on the surface (see Figure S2). Figure 2a illustrates the presence peaks of S2p with binding energy located at 168.5 eV for S^{6+}. Besides, Figure S2b displays the N1s peaks for the doped sample. It is clear that there are two constituent peaks at around 399.7 and 401.8 eV, without the peak at 396–397 eV definitely assigned to substitutional nitrogen. In the meantime, a peak observed at around 401 eV was credited to interstitial N-doping. Moreover, non-metal dopants lead to formation of Ti^{3+} defects. This is due to the charge difference between N (-3) and O (-2) when N atoms bonded to Ti atoms [19]. The different electronic interactions of Ti with N anions may result in partial electron transfer from the N to Ti which may form Ti^{3+} defects. The formation of these Ti^{3+} species was verified using XPS and electron paramagnetic resonance (EPR) analyses (see Figures 3 and 4). From XPS analysis of the doped sample, the $Ti2p$ spectrum revealed a slight negative shift of the two peaks at 457.7 eV ($Ti2p_{3/2}$) and 463.4 eV ($Ti2p_{1/2}$) with respect to Ti^{4+} (458 eV, $Ti2p_{3/2}$ and 463.7 eV, $Ti2p_{1/2}$) in un-doped sample (see Figure 3). This shift revealed the formation of Ti^{3+} species [20]. This new Ti^{3+} species/ defects also enhance the electronic states for TiO_2 by the formation of isolated defect energy level from Ti^{3+} below the bottom of conduction band for TiO_2 as displayed in Scheme 1 [21]. From EPR analysis, for the doped DT sample, the resonances at g values of less than 2.0 (1.96–1.92) are attributed to photogenerated electrons stabilized in Ti cations located at crystallization defects as shown in Figure 4. These trapped electrons could reduce Ti^{4+}, cause the formation of Ti^{3+} paramagnetic species [22]. In general, the surface Ti^{3+} has considerably lesser g factors value than those found in bulk. Additionally, the signal shapes for surface Ti^{3+} is commonly broad, but in the inner (bulk) Ti^{3+} has a narrow axially symmetric signal [23]. Thus, the g value of 1.92 was credited to the surface Ti^{3+} species. Moreover, the g value of 1.943 and 1.961 was associated with the formation of bulk Ti^{3+} [24–26]. Therefore, from XPS and EPR analyses, we can assume that the band gap for the doped sample decreased not only by non-metal dopants (C, N, and S) but also via formation of Ti^{3+} defects. Subsequently, all these new-formed bands lead to enhancement of the absorption capacity of the doped sample under solar simulator light. Consequently, this leads to enhancing the photocatalytic activity of the doped sample for complete reduction of o-dinitrobenzene to the corresponding o-phenylenediamine.

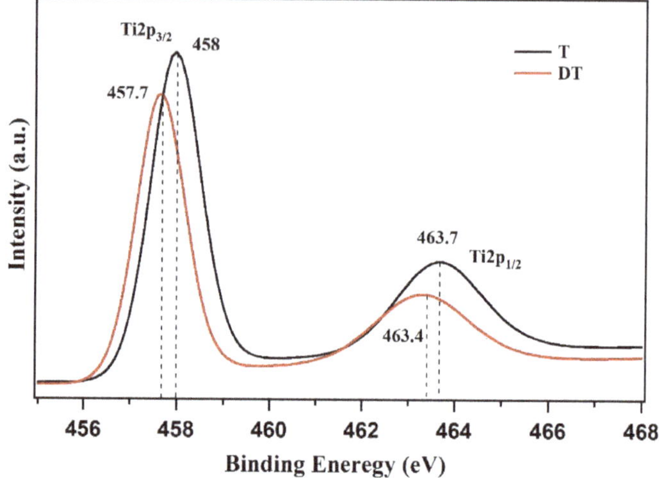

Figure 3. XPS detailed scans in the energy regions of Ti2p for T and DT samples.

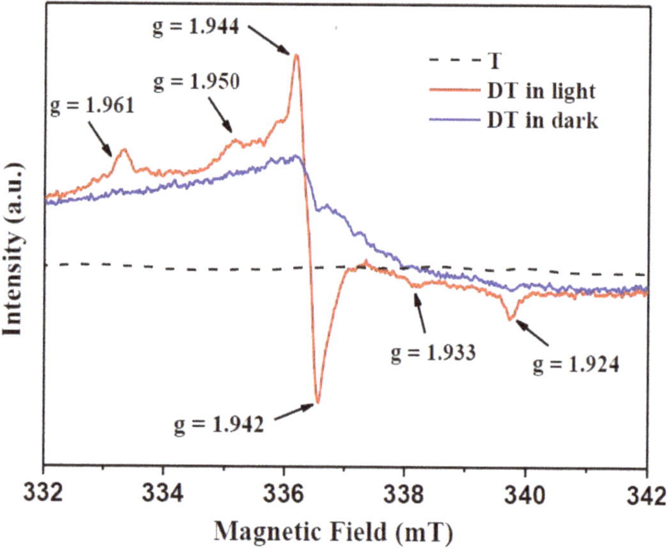

Figure 4. EPR spectra of T and DT samples, the DT sample recorded at dark and under UV irradiation (after 5 min) at room temperature. Instrument setting: operating at 9.41 GHz field modulation. modulation amplitude: 0.2 mT, power: 10 mW, gain: 5.

Secondly, these new electronic states act as electron-hole traps which leads to an increase in their lifetime by reducing the electron-hole recombination, resulting in an enhancement of the photocatalytic activity for the doped sample. The lifetimes and charge carrier trapping can by determined using laser flash photolysis [27]. The absorption time profile noticed at the selected wavelength (600 nm) for the un-doped and doped samples is shown in Figure 5. It can be clearly noticed that the initial decay for the un-doped T sample is faster than that of the doped DT sample. This can be attributed to presence of the long-lived Ti^{3+} species. Moreover, the amount of the generated charge carriers upon irradiating the doped sample is higher than that formed in the un-doped one. Overall, the doped sample demonstrates the utmost significant charge generation and the maximum number of hole–electron pairs available to participate in surface redox reactions with adsorbed species. This finding is in very good agreement with the photoreactivity results, as the above doped sample seemed to be the most photoactive for the studied photoreduction reaction. Combining the results of XPS, EPR and TAS analyses, it can be deduced that the non-metal dopants caused the formation of surface/ bulk Ti^{3+} in DT sample. Therefore, this new defect results in an enhancement in the absorption capacity of the material for complete the reduction of o-dinitrobenzene to the corresponding o-phenylenediamine under solar simulated light. Another evidence on the effect of doping on the charge carrier's lifetime can be gained from the PL analysis. Figure 6 shows the PL spectra of T and DT samples at excitation wavelength (259 nm) using Xe lamp at room temperature for further evidence out finding results. The PL spectra for both samples are similar with different intensities. The PL spectra of T and DT samples revealed several emission peaks, the maximum and centered one at about 470 nm, which were referred to as the shallow energy level excitonic PL phenomenon [28]. Moreover, we can notice that the PL intensity of the doped sample decreased compared to the un-doped one. The lower PL signals for the doped sample may indicate the lower electron-hole recombination rate and the higher separation efficiency and this result agrees with TAS measurements. From the above, it can be concluded that the enhancement of the photocatalytic activity of the doped sample using solar light irradiation is not only due to the formation of a mid-gap level via non-metal dopants (C, N, S) above the valence band, but also due to the formation of isolated defect energy level (Ti^{3+}) below the bottom of the conduction band of doped sample. This finding leads to a decrease of the band gap, decrease of the charge recombination, and increase of the life time of the charge carriers for the doped sample and consequently leads to enhancement of the complete reduction of o-dinitrobenzene to the corresponding o-phenylenediamine as shown in Scheme 1. Thirdly, hydrogen species (maybe associated with Ti^{3+}) derived from oxalic acid facilitates the complete reduction of o-dinitrobenzene (see Scheme 1). Moreover, the formed Ti^{3+} atoms act as active sites for the reduction of o-dinitrobenzene and o-nitroaniline to the corresponding o-phenylenediamine. These surface Ti^{3+} atoms behave as an adsorption site for o-dinitrobenzene and o-nitroaniline via an electron donation and as a trapping site for photogenerated electron formed in conduction band (see Scheme 1) [15,16,29]. Therefore, these avenues facilitate the achievement of the reduction process of the two-nitro group of o-dinitrobenzene to the corresponding o-phenylenediamine. Finally, the high crystallinity and the mesoporosity leads to improvement of the photocatalytic activity of the doped sample (see Figure S3) [17]. This attribute to the doped sample was calcined at 450 °C. Up to calcination, the organic remains in TiO_2 matrix disintegrated and formed a highly mesoporous material with pore-size diameter 3.6 nm compared with the un-doped T sample with pore-size diameter about 2.2 nm [17]. This mesoporous nature for the doped sample facilitated the adsorption capacity of the nitro aromatic compounds. Consequently, this enhanced the photocatalytic activity of the doped sample for complete reduction of o-dinitrobenzene under solar stimulator light. For all the above-mentioned reasons, it can be shown that the doped sample has versatile properties and great ability for highly selective photocatalytic reduction of o-dinitrobenzene to the corresponding o-phenylenediamine under solar simulator light.

On the other hand, one of the essential parameters in photocatalytic applications in an aqueous medium is the stability and reusability of the prepared samples at the end of the reaction. The stability of the doped sample can be investigated by monitoring the UV-Vis analysis at the end of the reaction. The results show that there is no change in the reflectance behavior of the doped sample as shown in Figure S1. This clearly revealed DT sample has good stability. Furthermore, the reusability for the doped sample was investigated after four cycles (Figure 7). The o-phenylenediamine yield is slightly decreased owing to a little amount of photocatalyst loss during product separation.

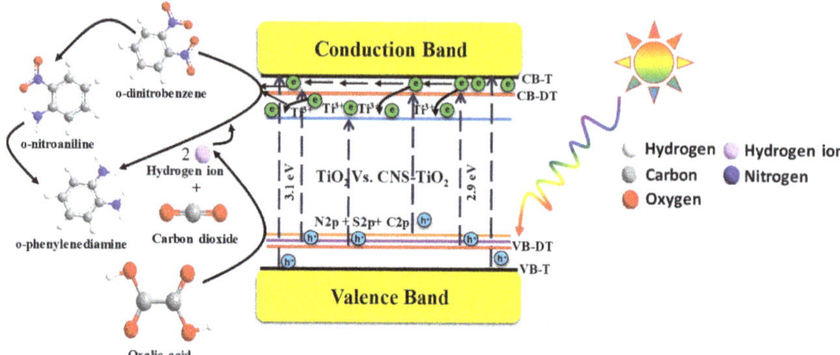

Scheme 1. Suggested mechanism for the effect of non-metal dopants (C, N, S) and Ti^{3+} surface defects in the photocatalytic conversion of o-dinitrobenzene to the corresponding o-phenylenediamine under solar light.

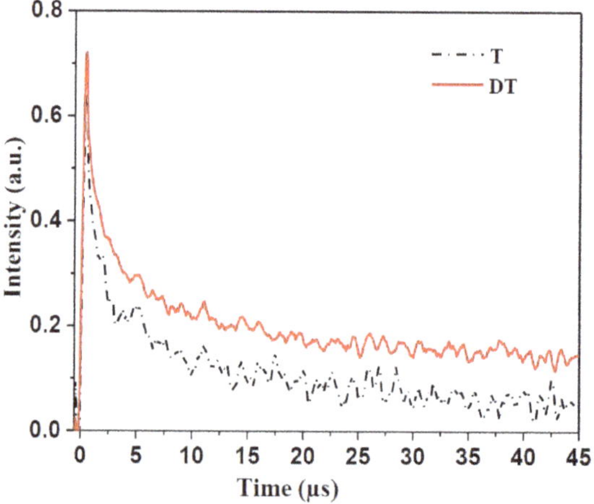

Figure 5. Absorption time profile of T and DT samples at 600 nm.

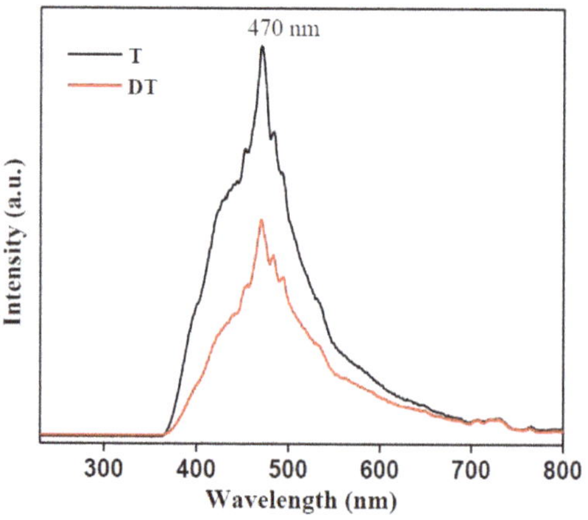

Figure 6. PL spectra of T and DT samples.

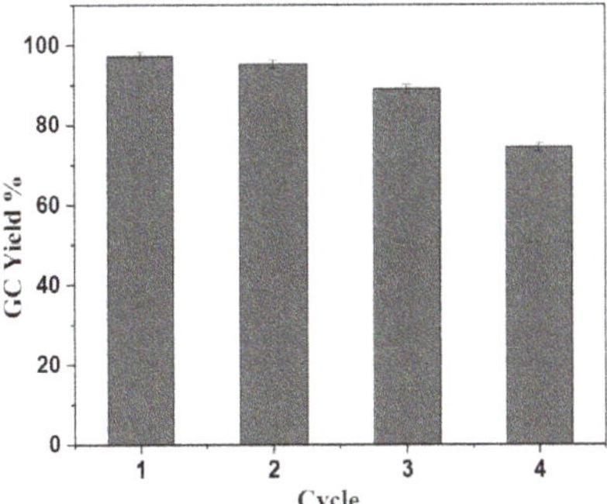

Figure 7. Reusability and photocatalytic efficiency of the DT aqueous solution for reduction of o-dinitrobenzene to corresponding o-phenylenediamine under solar simulated light irradiation after 24 h, reaction conditions: 25 mg DT sample, 50 µmol o-dinitrobenzene, 250 µmol oxalic acid, 5 cm^3 deionized water, Ar Experimental section.

2.2. Materials and Chemicals

The triblock copolymer surfactant poly (ethylene glycol)-poly (propylene glycol)- poly (ethylene glycol) (P-123, M wt. ~5800), titanium tert-butoxide $Ti(OC(CH_3)_3)_4$ (TBOT), thiourea ($\geq$99%), Triton-X 100, polyethylene glycol (10,000 MW), sodium sulfate (>99%), oxalic acid dihydrate ($\geq$99%), dichloromethane (High-performance liquid chromatography (HPLC) grade, >99.9%), ethanol (99.8%), o-dinitrobenzene ($\geq$99%), o-nitroaniline (98%) and o-phenylene diamine (99%) were purchased from Sigma-Aldrich, Darmstadt, Germany and were used as received.

2.3. Photocatalysts Preparation

The preparation procedure of un-doped anatase/brookite biphase TiO_2 and (C, N, S)-doped anatase/brookite biphase TiO_2 was published [17]. Un-doped anatase/brookite biphase TiO_2 was produced via sol-gel method using TBOT as a TiO_2 source and P123 as a directing agent. Then, the prepared TiO_2 powder was mixed with thiourea in a weight ratio of 1:1 and calcined in a covered vessel at 450 °C for 1 h to get (C, N, S) -doped anatase/ brookite biphase TiO_2. The obtained samples were donated as T and DT for un-doped anatase/ brookite biphase TiO_2 and (C, N, S)-doped anatase/ brookite biphase TiO_2.

2.4. Sample Characterization

EPR spectra were recorded at room temperature on a MiniScope X-band EPR spectrometer (MS400 Magnettech GmbH, Berlin, Germany) operating at 9.41 GHz field modulation. modulation amplitude: 0.2 mT, power: 10 mW, gain: 5. The experimental EPR spectra acquisition and simulation were carried out. The surface chemical composition of the samples was determined using X-ray Photoelectron Spectroscopy, Thermo Fisher Scientific K-Alpha XPS system (Waltham, MA, USA) with X-ray source −Al Ka micro-focused mono-chromator. The binding energies of surface adventitious carbon calibrated to the C1s peak at 284.4 $\pm$ 0.1 eV. Spectrofluorophotometer (RF-5301 PC, Shmidzu, Tokyo, Japan) was used to determine the photoluminescence (PL) spectra of the samples at room temperature with excitation wavelength 259 nm. Nanosecond diffuse reflectance transient absorption spectroscopy measurements were performed using an experimental set-up as reported previously [30]. For measurements, all powders were purged for $\frac{1}{2}$ h with N_2 prior to the measurements.

2.5. Photocatalytic Reaction Procedure

The photocatalytic reactions were carried out in a sealed glass snap-cap bottle (23 mm in diameter and 75 mm in length) with contentious stirring. 25 mg TiO_2 (un-doped or doped or P25) were suspended in 5 cm^3 of deionized water containing 50 μmol of the o-dinitrobenzene and ~250 μmol oxalic acid. The mixture was stirred in the glass snap-cap bottle in the dark with Ar being purged for 15 min to remove molecular oxygen. Then the mixture was irradiated for 24 h using solar simulator (SOL1200 lamp, UV (A) was measured by *Dr.* K *Hönle* UV (A)-detector (Munich, Germany) to be *20 mW/cm^2*). Afterward, the excess amount of oxalic acid was neutralized by adding desired amount of NH$_4$OH followed by extraction of the reactant and products from the aqueous phase by dichloromethane to be quantitatively and qualitatively analyzed by Gas Chromatograph-Mass Spectrometry (GC/MS) and GC with Flame Ionization Detector (GC-FID), respectively, after filtration through 0.2 μm filter. Shimadzu GC/MS-QP 5000 (Tokyo, Japan) equipped with a 30 m Rxi-5ms ($d = 0.32$ mm) capillary column with operating temperatures programmed: injection temperature 310 °C, oven temperature 120 °C (hold 2 min) from 120 to 280 °C at a rate of 10 °C min^{-1}, 280 °C (hold 15 min) in splitless mode, injection volume was 3.0 μL with helium as a carrier gas was used to qualitative analysis. Shimadzu GC 2010 (Tokyo, Japan) equipped with a Rtx-5 ($d = 0.25$ mm) capillary column and an FID detector was used to define the concentration of the reactant and of the products. Operating temperatures programmed: injection temperature 250 °C, oven temperature 70 °C (hold 2 min) from 70 to 280 °C at a rate of 10 °C min^{-1}, in splitless mode. Injection volume was 2.0 μL with nitrogen as the carrier gas. The concentrations of the reactant, besides the products, were evaluated according to the calibration curves prepared with authentic standards.

3. Conclusions

Mesostructured anatase-brookite biphase un-doped TiO$_2$ and (C, N, S) doped anatase-brookite biphase TiO$_2$ photocatalysts have various selectivities towards the reduction of o-dinitrobenzene in aqueous solution in the presence of oxalic acid as a hole scavenger under solar simulator light. Compared with commercial P25 TiO$_2$, the un-doped material showed a good selectivity (85.5%) towards the reduction of just the one-nitro group, i.e., towards the production of o-nitroaniline. On the other hand, (C, N, S) the doped sample displayed a high selectivity (97%) towards the complete reduction of the two-nitro group in o-dinitrobenzene to the corresponding o-phenylenediamine. The superiority of the doped TiO$_2$ photocatalysts is attributed to the formation of new dopants (C, N, S) as hole traps, the formation of Ti^{3+} defects and increase in the lifetime of the charge carriers, which leads to enhancement of the absorption capacity under solar simulator light. Furthermore, the surface Ti^{3+} atoms of doped TiO$_2$ act as the adsorption site for nitroaromatics and the trapping site for photogenerated electrons formed on the conduction band. This finding leads to the acceleration of rapid nitro-to-amine reduction/hydrogenation and the complete formation of o-phenylenediamine, while suppressing side reactions.

Supplementary Materials: The following are available online at http://www.mdpi.com/2073-4344/8/12/641/s1, Figure S1: (a) UV-Vis absorption spectra (b) Tauc plots of modified Kubelka-Munk function of samples T, and DT before and after reusing., Figure S2: XPS detailed scans in the energy regions of (a) S2p, (b) N1s and (c) C1s for DT sample, Figure S3: (a) Low angle XRD patterns and (b) Wide angle XRD patterns for the T and DT samples.

Author Contributions: Conceptualization, H.M.E.-H., S.M.E.-S., A.H. and A.A.I.; methodology, H.M.E.-H.; formal analysis and data curation, H.M.E.-H., A.H. and S.M.E.-S.; writing—original draft preparation, H.M.E.-H., S.M.E.-S. and A.A.I.; writing—review and editing, H.M.E.-H., R.D., S.M.E.-S., A.H. and A.A.I.; resources, D.W.B. and S.M.E.-S.; supervision, H.M.K., I.A.I. and D.W.B.

Funding: This work was supported by short cycle 5, Science & Technology Development Fund in Egypt (STDF fellowship) under Grant no. ID 12282.

Acknowledgments: This work was supported by short cycle 5, Science & Technology Development Fund in Egypt (STDF fellowship) under Grant no. ID 12282. H. El-Hosainy acknowledges Institut für Technische Chemie, Leibniz Universität Hannover, Germany for hosting him during the current research work. The publication of this article was funded by the Open Access Fund of the Leibniz Universität Hannover.

Conflicts of Interest: The authors declare no conflicts of interest.

References

1. Ferry, J.L.; Glaze, W.H. Photocatalytic reduction of nitro organics over illuminated titanium dioxide: Role of the TiO_2 surface. *Langmuir* **1998**, *14*, 3551–3557. [CrossRef]
2. Shiraishi, Y.; Hirai, T. Selective organic transformations on titanium oxide-based photocatalysts. *J. Photochem. Photobiol. C Photochem. Rev.* **2008**, *9*, 157–170. [CrossRef]
3. Wang, H.; Yan, J.; Chang, W.; Zhang, Z.; Wang, H. Practical synthesis of aromatic amines by photocatalytic reduction of aromatic nitro compounds on nanoparticles N-doped TiO_2. *Catal. Commun.* **2009**, *10*, 989–994. [CrossRef]
4. Imamura, K.; Iwasaki, S.; Maeda, T.; Hashimoto, K.; Ohtani, B.; Kominami, H. Photocatalytic reduction of nitrobenzenes to aminobenzenes in aqueous suspensions of titanium (IV) oxide in the presence of hole scavengers under deaerated and aerated conditions. *Phys. Chem. Chem. Phys.* **2011**, *13*, 5114–5119. [CrossRef] [PubMed]
5. Hakki, A.; Dillert, R.; Bahnemann, D.W. Photocatalysis as an Auspicious Synthetic Route towards Nitrogen Containing Organic Compounds. *Curr. Org. Chem.* **2013**, *17*, 2482–2502. [CrossRef]
6. Ragaini, F.; Cenini, S.; Gasperini, M. Reduction of nitrobenzene to aniline by CO/H_2O, catalysed by $Ru_3(CO)_{12}$/chelating diimines. *J. Mol. Catal. A Chem.* **2001**, *174*, 51–57. [CrossRef]
7. Longo, C.; Alvarez, J.; Fernández, M.; Pardey, A.J.; Moya, S.A.; Baricelli, P.; Mdleleni, M.M. Water as hydride source in the reduction of nitrobenzene to aniline catalyzed by cis-$[Rh(CO)_2(2\text{-picoline})_2](PF_6)$ in aqueous 2-picoline under CO atmosphere: Kinetics study. *Polyhedron* **2000**, *19*, 487–493. [CrossRef]
8. Blaser, H.-U.; Steiner, H.; Studer, M. Selective Catalytic Hydrogenation of Functionalized Nitroarenes: An Update. *ChemCatChem.* **2009**, *1*, 210–221. [CrossRef]
9. Mahdavi, F.; Bruton, T.C.; Li, Y. Photoinduced reduction of nitro compounds on semiconductor particles. *J. Org. Chem.* **1993**, *58*, 744–746. [CrossRef]
10. Tada, H.; Ishida, T.; Takao, A.; Ito, S. Drastic enhancement of TiO_2-photocatalyzed reduction of nitrobenzene by loading Ag clusters. *Langmuir* **2004**, *20*, 7898–7900. [CrossRef]
11. Li, Y.; Wasgestian, F. Photocatalytic reduction of nitrate ions on TiO_2 by oxalic acid. *J. Photochem. Photobiol. A* **1998**, *112*, 255–259. [CrossRef]
12. Kominami, H.; Nakaseko, T.; Shimada, Y.; Furusho, A.; Inoue, H.; Murakami, S.; Kera, Y.; Ohtani, B. Selective photocatalytic reduction of nitrate to nitrogen molecules in an aqueous suspension of metal-loaded titanium (IV) oxide particles. *Chem. Commun.* **2005**, *23*, 2933–2935. [CrossRef] [PubMed]
13. Kominami, H.; Furusho, A.; Murakami, S.; Inoue, H.; Kera, Y.; Ohtani, B. Effective Photocatalytic Reduction of Nitrate to Ammonia in an Aqueous Suspension of Metal Loaded Titanium (IV) Oxide Particles in the Presence of Oxalic Acid. *Catal. Lett.* **2001**, *76*, 31–34. [CrossRef]
14. Zand, Z.; Kazemi, F.; Hosseini, S. Development of chemoselective photoreduction of nitro compounds under solar light and blue LED irradiation. *Tetrahedron Lett.* **2014**, *55*, 338–341 [CrossRef]
15. Shiraishi, Y.; Togawa, Y.; Tsukamoto, D.; Tanaka, S.; Hirai, T. Highly efficient and selective hydrogenation of nitroaromatics on photoactivated rutile titanium dioxide. *ACS Catal.* **2012**, *2*, 2475–2481. [CrossRef]
16. Kaur, J.; Pal, B. 100% selective yield of m-nitroaniline by rutile TiO_2 and m-phenylenediamine by P25-TiO_2 during m-dinitrobenzene photoreduction. *Catal. Commun.* **2014**, *53*, 25–28. [CrossRef]
17. El-Sheikh, S.M.; Zhang, G.; El-Hosainy, H.M.; Ismail, A.A.; Dionysiou, D.D. High performance, sulfur, nitrogen and carbon doped mesoporous anatase-brookite TiO_2 photocatalyst for the removal of microcystin-LR under visible light irradiation. *J. Hazard. Mater.* **2014**, *280*, 723–733. [CrossRef]
18. Hakki, A.; Dillert, R.; Bahnemann, D. Photocatalytic conversion of nitroaromatic compounds in the presence of TiO_2. *Catal. Today* **2009**, *144*, 154–159. [CrossRef]
19. Wang, J.; Li, H.; Li, H.; Zou, C. Mesoporous $TiO_{2-x}A_y$ (A = N, S) as a visible-light-response photocatalyst. *Solid State Sci.* **2010**, *12*, 490–497. [CrossRef]
20. Etacheri, V.; Seery, M.K.; Hinder, S.J.; Pillai, S.C. Highly Visible Light Active $TiO_{2-x}N_x$ Heterojunction Photocatalysts. *Chem. Mater.* **2010**, *22*, 3843–3853. [CrossRef]
21. Justicia, I.; Ordejon, P.; Canto, G.; Mozos, J.L.; Fraxedes, J.; Battiston, G.A. Designed Self-Doped Titanium Oxide Thin Films for Efficient Visible-Light Photocatalysis. *Adv. Mater.* **2002**, *14*, 1399–1402. [CrossRef]
22. Yang, G.; Jiang, Z.; Shi, H.; Xiao, T.; Yan, Z. Preparation of highly visible-light active N-doped TiO_2 photocatalyst. *J. Mater. Chem.* **2010**, *20*, 5301–5309. [CrossRef]

23. Xiong, L.-B.; Li, J.-L.; Yang, B.; Yu, Y. Ti^{3+} in the surface of titanium dioxide: Generation, properties and photocatalytic application. *J. Nanometer.* **2012**, *2012*. [CrossRef]

24. Tan, H.; Zhao, Z.; Niu, M.; Mao, C.; Cao, D.; Cheng, D.; Feng, P.; Sun, Z. A facile and versatile method for preparation of colored TiO_2 with enhanced solar-driven photocatalytic activity. *Nanoscale* **2014**, *6*, 10216–10223. [CrossRef] [PubMed]

25. Yan, Y.; Han, M.; Konkin, A.; Koppe, T.; Wang, D.; Andreu, T.; Chen, G.; Vetter, U.; Ramon Morante, J.; Schaaf, P. Slightly hydrogenated TiO_2 with enhanced photocatalytic performance. *J. Mater. Chem. A* **2014**, *2*, 12708–12716. [CrossRef]

26. Grabstanowicz, L.R.; Gao, S.; Li, T.; Rickard, R.M.; Rajh, T.; Liu, D.-J.; Xu, T. Facile oxidative conversion of TiH_2 to high-concentration Ti^{3+}-self-doped rutile TiO_2 with visible-light photoactivity. *Inorg. Chem.* **2013**, *52*, 3884–3892. [CrossRef] [PubMed]

27. Bahnemann, D.; Henglein, A.; Lilie, J.; Spanhel, L. Flash photolysis observation of the absorption spectra of trapped positive holes and electrons in colloidal titanium dioxide. *J. Phys. Chem.* **1984**, *88*, 709–711. [CrossRef]

28. Li, D.; Jia, J.; Zheng, T.; Cheng, X.; Yu, X. Construction and characterization of visible light active Pd nano-crystallite decorated and C-N-S-co-doped TiO_2 nanosheet array photoelectrode for enhanced photocatalytic degradation of acetylsalicylic acid. *Appl. Catal. B* **2016**, *188*, 259–271. [CrossRef]

29. Kaur, J.; Singh, R.; Pal, B. Influence of coinage and platinum group metal co-catalysis for the photocatalytic reduction of m-dinitrobenzene by P25 and rutile TiO_2. *J. Mol. Catal. A Chem.* **2015**, *397*, 99–105. [CrossRef]

30. Schneider, J.; Nikitin, K.; Wark, M.; Bahnemann, D.W.; Marschall, R. Improved charge carrier separation in barium tantalate composites investigated by laser flash photolysis. *Phys. Chem. Chem. Phys.* **2016**, *18*, 10719–10726. [CrossRef]

Article

Enhanced Photocatalytic Activity of Titania by Co-Doping with Mo and W

Osmín Avilés-García [1], Jaime Espino-Valencia [1,*], Rubí Romero-Romero [2], José Luis Rico-Cerda [1], Manuel Arroyo-Albiter [1], Dora Alicia Solís-Casados [2] and Reyna Natividad-Rangel [2,*]

[1] Facultad de Ingeniería Química, Universidad Michoacana de San Nicolás de Hidalgo, Edif. V1, Ciudad Universitaria, Morelia 58060, Michoacán, Mexico; agosmin@gmail.com (O.A.-G.); jlceri@yahoo.com.mx (J.L.R.-C.); albitmanuel@gmail.com (M.A.-A)

[2] Centro Conjunto de Investigación en Química Sustentable, UAEMéx-UNAM, Universidad Autónoma del Estado de México, Km 14.5 Carretera Toluca-Atlacomulco, San Cayetano, Piedras Blancas, 50200 Toluca, MEX, Mexico; rubiromero99@gmail.com (R.R.R.); solis_casados@yahoo.com.mx (D.A.S.-C.)

* Correspondence: rnatividadr@uaemex.mx (R.N.-R.); jespinova@yahoo.com.mx (J.E.-V.); Tel.: +52-722-2766610 (ext. 7723) (R.N.-R.); +52-443-3223500 (ext. 2002) (J.E.-V.)

Received: 22 October 2018; Accepted: 19 November 2018; Published: 6 December 2018

Abstract: Various W and Mo co-doped titanium dioxide (TiO_2) materials were obtained through the EISA (Evaporation-Induced Self-Assembly) method and then tested as photocatalysts in the degradation of 4-chlorophenol. The synthesized materials were characterized by thermogravimetric analysis (TGA), Fourier transform infrared (FTIR) spectroscopy, X-ray diffraction (XRD), Raman spectroscopy (RS), N_2 physisorption, UV-vis diffuse reflectance spectroscopy (DRS), X-ray photoelectron spectroscopy (XPS), and transmission electron microscopy (TEM). The results showed that the W-Mo-TiO_2 catalysts have a high surface area of about 191 m^2/g, and the presence of an anatase crystalline phase. The co-doped materials exhibited smaller crystallite sizes than those with one dopant, since the crystallinity is inhibited by the presence of both species. In addition, tungsten and molybdenum dopants are distributed and are incorporated into the anatase structure of TiO_2, due to changes in red parameters and lattice expansion. Under our experimental conditions, the co-doped TiO_2 catalyst presented 46% more 4-chlorophenol degradation than Degussa P25. The incorporation of two dopant cations in titania improved its photocatalytic performance, which was attributed to a cooperative effect by decreasing the recombination of photogenerated charges, high radiation absorption capacity, high surface areas, and low crystallinity. When TiO_2 is co-doped with the same amount of both cations (1 wt.%), the highest degradation and mineralization (97% and 74%, respectively) is achieved. Quinones were the main intermediates in the 4-chlorophenol oxidation by W-Mo-TiO_2 and 1,2,4-benzenetriol was incompletely degraded.

Keywords: W-Mo dopants; titanium dioxide; nanoparticles; photocatalytic activity

---

1. Introduction

Among the advanced oxidation processes (AOPs), heterogeneous photocatalysis is considered as an efficient method for the degradation of organic pollutants in water and air [1,2]. Numerous semiconductors have been investigated as photocatalysts. Among the semiconductors used, titanium dioxide (TiO_2) is the most promising and widely studied material for photocatalytic applications due to its chemical stability, high efficiency, photostability, high oxidizing power, abundance, nontoxicity, and low cost [3,4]. The main quality of TiO_2 is attributed to oxidative power of hydroxyl radicals generated when the electrons are photoexcited by UV light absorption [5]. However, the recombination of the charge pair (holes and electrons) should not be ignored because it decreases the photoactivity. In order to avoid this and improve the photocatalytic activity, the photogenerated charges must be trapped on TiO_2 surface, thus reducing recombination [6]. In this context, several techniques such as doping with metals and non-metals [7], dye sensitization [8], deposition with noble metals [9], and coupled semiconductor [10] have been assessed. Among these studies, doping has shown positive effects on titania because it gives unique electronic and structural properties that translate into better activity. Doping with non-metallic ions usually introduces energy levels above the valence band of the semiconductor for photon absorption in the visible-light region [11]. Alternatively, the use of dopant metals promotes charge transfer and separation of photogenerated charges [12]. In addition, concentration and distribution of dopant ions in TiO_2 are factors that must be considered for a good photocatalytic performance [13]. The incorporation of two types of cations into TiO2 lattice and its photocatalytic performance has been reported in several studies [14,15]. Estrellan et al. [14] reported that when the TiO2 is co-doped with iron and niobium, the photocatalytic efficiency is improved due to synergistic actions between the doping species, which favor the e−/h+ generation and reduce the recombination rate. Shi et al. [15] reported enhanced photoactivity of titania with iron and cerium by co-doping, which is due to cooperative effects of both dopants, by broadening the absorption spectrum and retarding the recombination of the photogenerated charges.

The synthesis of the catalysts was carried out by Evaporation-Induced Self-Assembly (EISA) method, which allows obtaining mesoporous structures with high surface areas [16]. On the other hand, 4-chlorophenol oxidation was chosen as the reaction to evaluate the synthesized materials, because it is considered as a model molecule for photocatalytic evaluations [17]. However, it is important to mention that the removal of chlorophenols can be carried out by other techniques, such as hydrodechlorination, which has been the subject of many investigations [18,19]. 4-chlorophenol is a pollutant commonly found in the effluents from industries related to insecticides, dyes, plastics, herbicides, detergents, wood preservatives, and petroleum reforming [20]. In addition, it is classified by the U.S. Environmental Protection Agency (USEPA) as a very toxic pollutant in these effluents, since it causes damage to human health and aquatic environments, so its effective removal is of great interest [21].

In this work, samples of TiO_2 co-doped with transition metals tungsten (W) and molybdenum (Mo) were synthesized by the EISA method. In addition, these materials have not been reported previously. The resulted solids were characterized and their photocatalytic activity evaluated in the degradation and mineralization of 4-chlorophenol. This was conducted with the main objective of demonstrating a synergistic effect in terms of improved photoactivity by co-doping compared to W-doped and Mo-doped TiO_2 catalysts. Results were compared to those obtained with commercial Degussa P25 TiO_2.

2. Results and Discussion

2.1. Photocatalysts Characterization

Thermal analysis (TGA) of the synthesized titania without dopants is shown in Figure 1. A weight loss (9%) from room temperature to 200 °C can be observed, which is assigned to desorption of water and residual organic solvents [22]. Subsequently, a significant weight loss (34%) between 200 °C and 300 °C is attributed to the elimination of P123 organic surfactant. Finally, the dehydroxylation process is observed above 300 °C. The thermogravimetric graph of the TiO_2 sample co-doped with 1 wt.% tungsten and 1 wt.% molybdenum is shown in Figure 1. For comparison purposes, it can be observed that the incorporation of dopant cations into titania reduces the percentage of total weight loss, which resulted in a better thermal stability, as previously observed by Hussain et al. [23]. The applied heat treatment (400 °C as maximum temperature) ensured the complete elimination of the organic surfactant in all synthesized samples. On the other hand, infrared analysis (FTIR) before and after heat treatment were carried out to corroborate the P123 elimination (see Figure 2). Before thermal treatment, the characteristic signals of the surfactant by IR at 1090, 1640, 2850, and 3250 cm^{-1} are attributed to the C-H, H-O-H, O-H, and C-C bonds vibrations, respectively [24]. After heat treatment, these characteristic bands disappear, which confirms the complete elimination of the organic compound incorporated during the synthesis.

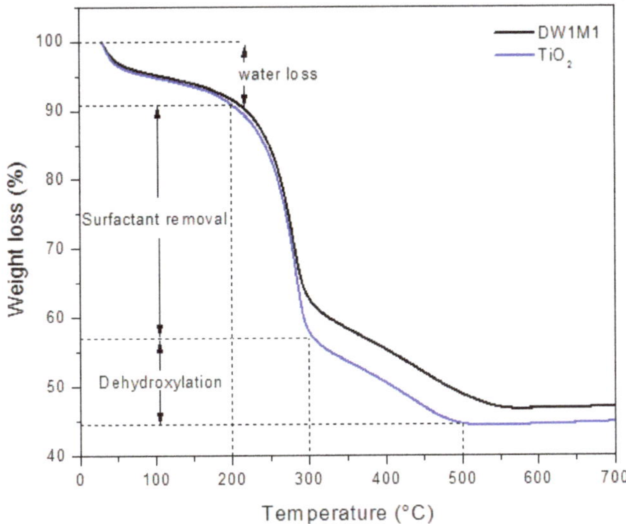

Figure 1. Thermogravimetric analysis of pure TiO_2 and TiO_2 co-doped with 1 wt.% tungsten and molybdenum (DW1M1).

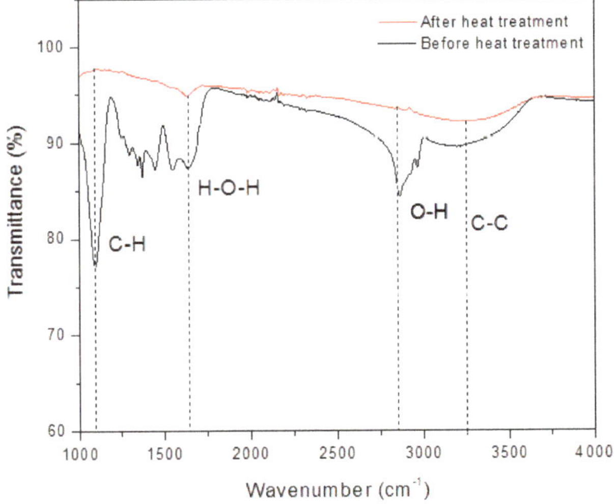

Figure 2. Infrared spectra of pure TiO_2 sample before and after thermal treatment.

Figure 3 shows the X-ray diffraction (XRD) patterns of all synthesized samples. The materials exhibited reflections at $2\theta \approx 25.4°$, $37.8°$, $48.1°$, $54.2°$, $55.2°$, and $62.6°$, which are representative of the crystalline planes (101), (004), (200), (105), (211), and (204) of the anatase TiO_2 phase, respectively [25]. The average crystallite sizes according to the Scherrer equation, as well as the lattice distortion of the samples are shown in Table 1. The crystal size of TiO_2 (8.6 nm) and DM2 (8.6 nm) is higher compared to the co-doped titania samples, whose values are in the range between 6.6 nm and 8.3 nm. This suggests that the inclusion of tungsten and molybdenum into the TiO_2 matrix suppresses crystalline growth. No relationship was observed due to the increase in weight percentage of a second cation by co-doping with respect to the average crystallite size. The lattice parameters and the unit cell volume of the materials are shown in Table 1. As can be seen, the incorporation of W and Mo in the titania by co-doping further increases the c-axis in comparison with the mono-doping, which indicates a greater expansion of the crystal lattice. Since the ion radii of Mo^{6+} (0.062 nm) and W^{6+} (0.060 nm) are very similar to that of Ti^{4+} (0.0605 nm), they can replace titanium within the anatase structure [26]. These changes in the lattice parameters are associated with lattice distortion due to the inclusion of dopants in the TiO_2 matrix.

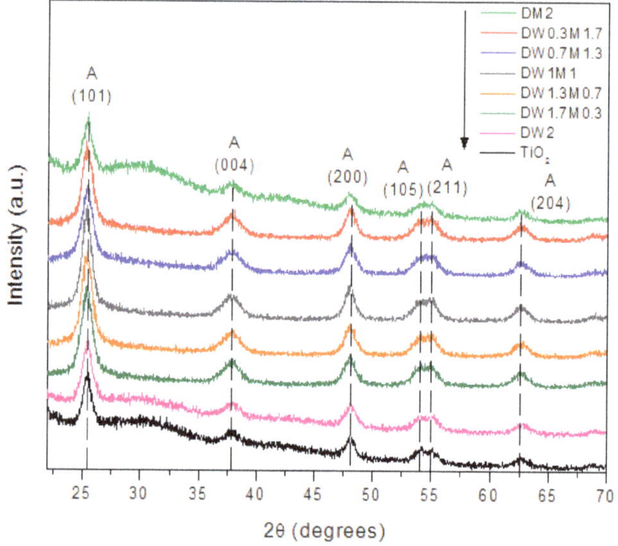

Figure 3. X-ray diffraction patterns of all prepared catalysts.

Table 1. Crystal size, lattice distortion and lattice parameters of the synthesized materials.

Sample	Average Crystallite Size (nm)	Lattice Distortion (ε)	Lattice Parameters		
			a (nm)	c (nm)	V (nm³)
TiO2	8.6	0.0183	0.377	0.931	0.132
DM2	8.6	0.0183	0.378	0.933	0.133
DW0.3M1.7	7.7	0.0207	0.378	0.982	0.140
DW0.7M1.3	6.6	0.0238	0.378	0.943	0.135
DW1M1	8.3	0.0191	0.379	0.937	0.134
DW1.3M0.7	7.3	0.0216	0.378	0.947	0.136
DW1.7M0.3	7.1	0.0224	0.379	0.959	0.138
DW2	8.1	0.0195	0.379	0.933	0.134

Figure 4 shows Raman spectra (RS) of TiO_2, W-TiO_2, Mo-TiO_2, and W-Mo-TiO_2 samples. As observed in Figure 4a, some peaks centered at 144, 197, 397, 517, and 640 cm^{-1} are assigned to the $E_{g(1)}$, $E_{g(2)}$, $B_{1g(1)}$, (A_{1g} + $B_{1g(2)}$), and $E_{g(3)}$ vibration modes, respectively, corresponding to anatase crystalline phase [27]. This suggests that, after doping with W and/or Mo cations, the anatase phase is maintained on the TiO_2 surface, which is in accordance with the XRD results. The position and intensity of the most intense Raman peak at 144 cm^{-1} change when titania is mono-doped and co-doped with these kinds of cations (see Figure 4b), which is attributed to changes in lattice parameters and unit cell volume by XRD analysis (see Table 1) [28]. This shift Raman is related to the increase in the c lattice parameter and the lattice expansion, which is due to the incorporation of W and/or Mo dopant ions into anatase, generating structural distortion [29]. In addition, the absence of new Raman bands in the spectra implies that there are no additional phases, which means that the dopants have been incorporated and they are well distributed in the crystal structure of TiO$_2$ [30].

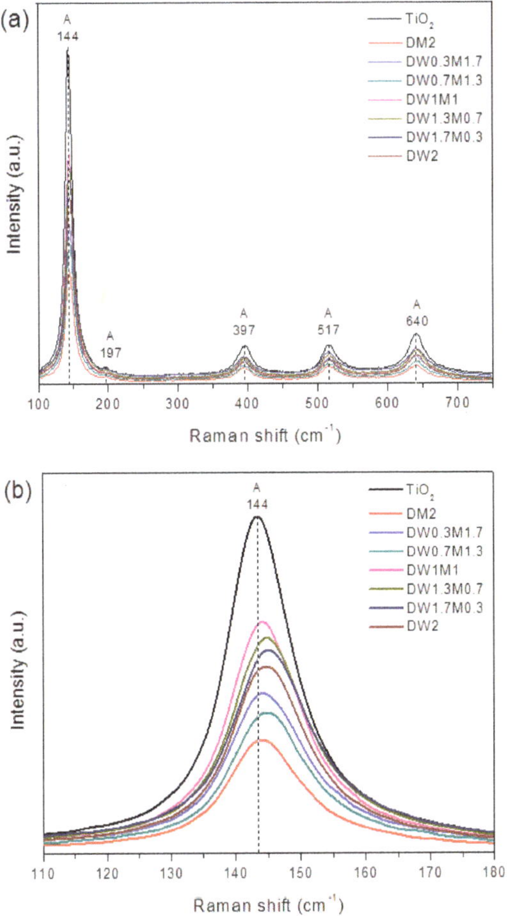

Figure 4. (**a**) Raman spectra and (**b**) Raman peak at 144 cm^{-1} of all synthesized materials.

The nitrogen adsorption-desorption isotherms and pore size distributions of all prepared materials are shown in Figures 5 and 6, respectively. The samples exhibited type IV isotherm, which is characteristic of mesoporous materials [31] according to the IUPAC (International Union of Pure and Applied Chemistry) classification. Type H2 hysteresis loop with monomodal pore size distribution is attributed to porous structures formed by interconnected networks. Table 2 presents the specific surface area, average pore diameter, and pore volume of all samples. The specific surface area of the co-doped samples is higher than the doped ones (DW2 and DM2), which is probably due to the incorporation of W and Mo in the crystal structure as well as the decrease in crystallinity. The maximum specific surface area of 191 m^2/g is obtained with sample DW0.7M1.3. No relationship was observed between the surface area and the concentration of a second dopant cation, however, when W or Mo is incorporated in a small amount (0.33 wt.%) by co-doping (DW1.7M0.3 and DW0.3M1.7 samples), the specific surface area increases by approximately 15% with respect to mono-doped TiO$_2$. High surface areas may offer more adsorption sites for photocatalytic reactions, which could result in better photoactivity [32]. Additionally, the average pore diameter of the co-doped catalysts remained constant (5.6 nm) and with a value between that of the mono-doped catalysts (4.6 nm and 6.1 nm), with the exception of sample DW1M1 that exhibited the highest average pore diameter of 6.6 nm and the maximum pore volume of 0.391 cm^3/g.

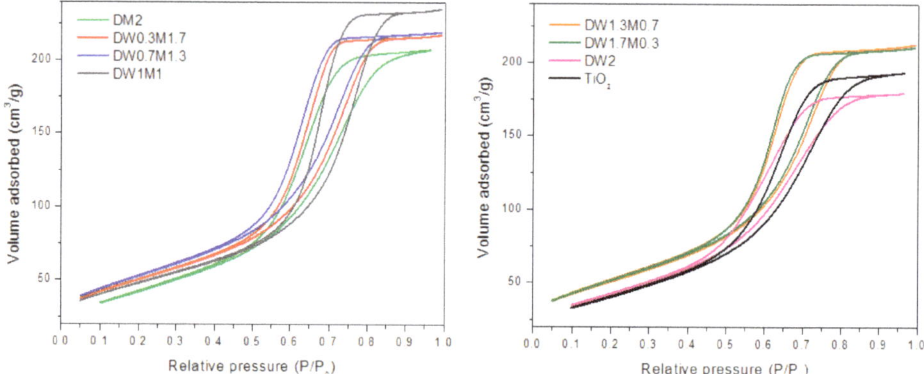

Figure 5. N_2 adsorption-desorption isotherms of undoped, doped, and co-doped TiO_2 with different weight percentages of W and Mo ions.

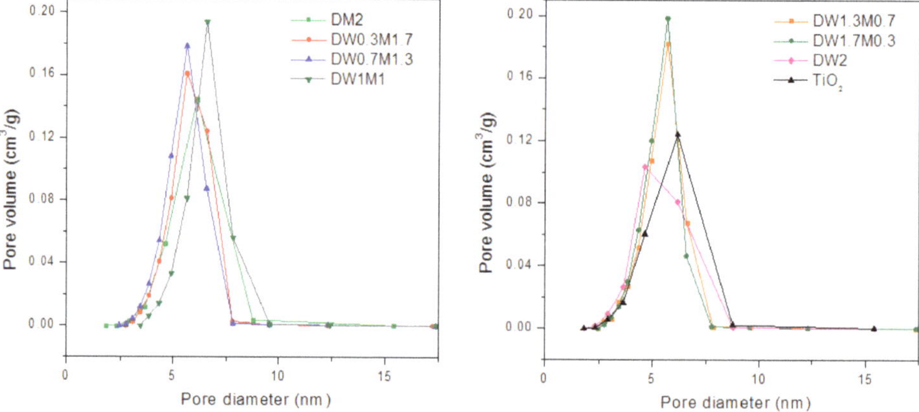

Figure 6. Pore size distributions of undoped, doped, and co-doped TiO_2 with different weight percentages of W and Mo ions.

Table 2. Textural properties of TiO_2, W-TiO_2, Mo-TiO_2, and W-Mo-TiO_2 catalysts.

Catalyst	Specific Surface Area (m^2/g)	Average Pore Diameter (nm)	Pore Volume (cm^3/g)
TiO_2	144	6.1	0.328
DM2	161	6.1	0.354
DW0.3M1.7	181	5.6	0.362
DW0.7M1.3	191	5.6	0.365
DW1M1	172	6.6	0.391
DW1.3M0.7	185	5.6	0.352
DW1.7M0.3	188	5.6	0.352
DW2	160	4.6	0.307

The catalysts were analyzed by UV-vis diffuse reflectance spectroscopy (DRS) to estimate their band gap energy. Figure 7a shows the UV-vis absorption spectra for all materials. As can be seen, the concentration of molybdenum cations in the titania gradually shifts the absorption edge towards the long wavelength region. Figure 7b shows the Tauc plots from the Kubelka-Munk function [33]. The band gap and its corresponding wavelength are presented in Table 3 for all synthesized samples. TiO$_2$ doped with 2 wt.% tungsten did not show a significant reduction in band gap energy (3.08 eV). However, titania doped with 2 wt.% molybdenum showed a slight shift towards visible radiation absorption at 452 nm and a band gap energy of 2.74 eV. The incorporation and increase of tungsten by co-doping does not favor the band gap reduction, since the "d" orbitals of W are located into the conduction band of TiO$_2$, making it difficult to generate energy levels under this band [34]. On the other hand, the increase of molybdenum by co-doping favors the band gap reduction due to charge transfer transitions between the "d" orbitals of molybdenum and TiO$_2$ located under the conduction band. The incorporation of this type of dopant cations (W and Mo) does not exhibit cooperative effects by co-doping towards an effective reduction of the band gap in anatase, so that its absorption of energy for electronic excitation is between the limits of mono-doped materials (DW2 and DM2 samples) at 403–452 nm.

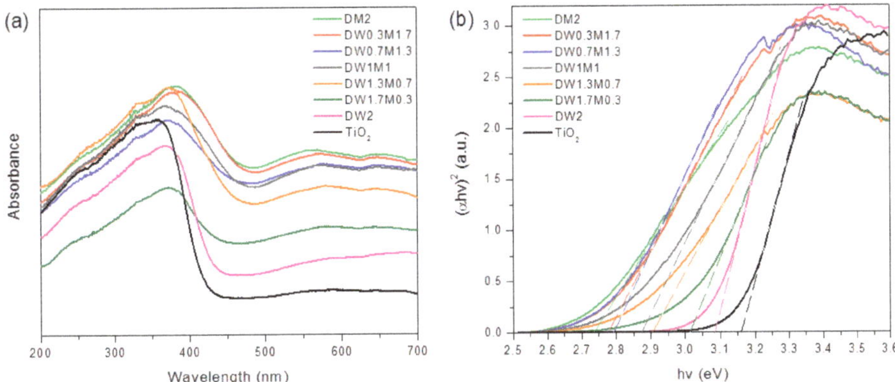

Figure 7. (**a**) UV-vis absorbance spectra and (**b**) band gap energies (Tauc plots) of synthesized catalysts.

Table 3. Band gap energy and wavelength of synthesized materials.

Sample	Band Gap (eV)	Wavelength (nm)
TiO$_2$	3.16	392
DM2	2.74	452
DW0.3M1.7	2.78	446
DW0.7M1.3	2.80	443
DW1M1	2.87	432
DW1.3M0.7	2.90	428
DW1.7M0.3	3.01	412
DW2	3.08	403

X-ray photoelectron spectroscopy (XPS) was used to determine the chemical state of the elements on the surface of the synthesized materials. Figure 8 shows the XPS spectra of Ti 2p in samples TiO$_2$, DM2, DW2, and DW1M1. Two peaks located with binding energies at 458.1 eV and 463.8 eV are assigned to Ti 2p$_{3/2}$ and Ti 2p$_{1/2}$ states, respectively, and these correspond to Ti^{4+} into TiO$_2$ lattice. A slight shift towards higher binding energies at 0.3 eV after doping and co-doping with W and Mo cations is evidence that these dopant metals are part of the anatase crystalline structure [35]. The O 1s XPS spectra for the sample co-doped with 1 wt.% W and 1 wt.% Mo is shown in Figure 9a. The peak at 529.8 eV is attributed to the Ti-O bond in TiO$_2$, whereas that at 532.2 eV is assigned to surface hydroxyl groups. In photocatalysis, these hydroxyl groups play an important role, since they react with the holes generated during photoexcitation to produce hydroxyl radicals, which degrade organic compounds [36]. Figure 9b presents the XPS spectrum in the Mo 3d region for sample DW1M1. In this figure, two main peaks, attributed to Mo $3d_{3/2}$ and Mo $3d_{5/2}$, show contributions of Mo^{6+} at 233.2 eV and 234.7 eV, and Mo^{5+} at 232.8 eV and 234.3 eV. The percentage of these two species in the catalyst surface is 77.5% and 22.5% for Mo^{6+} and Mo^{5+}, respectively. The XPS spectrum shown in Figure 9c corresponds to the W 4f region. As it is observed, the surface of sample DW1M1 is composed of 91.7% W^{6+} and 8.3% W^{5+} at 36.0 eV and 34.3 eV, respectively. These signals confirm the presence of both W and Mo dopants on the surface layers of TiO_2, which are involved in the photocatalytic reactions.

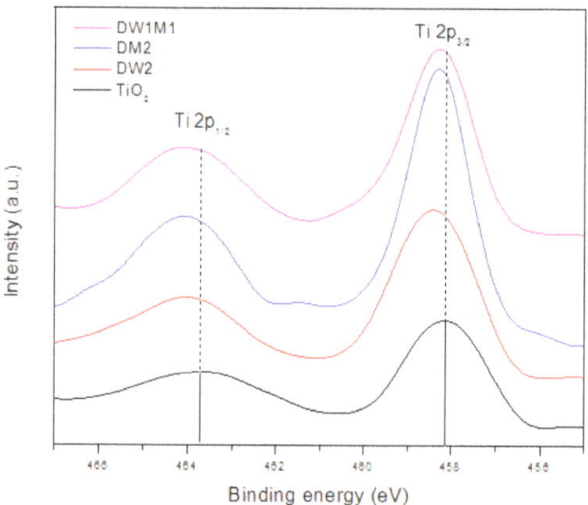

Figure 8. XPS spectra of Ti 2p for undoped, doped, and co-doped TiO_2 samples with W and Mo.

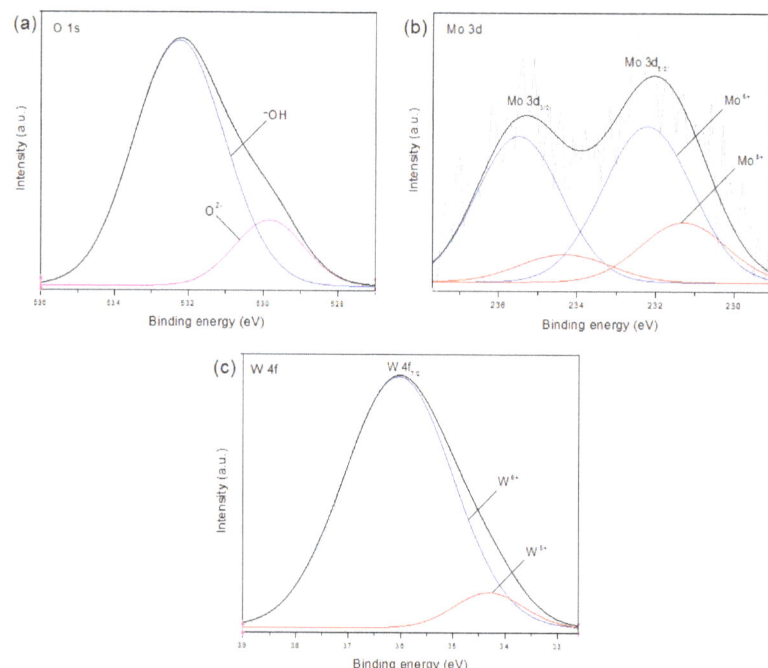

Figure 9. XPS spectra of (**a**) O 1s, (**b**) Mo 3d, and (**c**) W 4f for co-doped TiO_2 sample with W and Mo cations.

Figure 10 shows the transmission electron micrographs, the high resolution transmission micrographs, as well as the electron diffraction patterns of mono-doped samples DM2 and DW2, and co-doped sample DW1M1. The TEM images in Figure 10a,d,g for mono-doped and co-doped titania with Mo and W cations show individual nanoparticles from five to 12 nm in size. The HRTEM micrographs for samples DM2 and DW2 in Figure 10b,e, exhibit interlayer spacing (0.35 nm) that corresponds to the (101) plane of anatase crystalline structure. In the HRTEM image for sample co-doped with 1 wt.% of both cations (Figure 10h); the (101) and (004) planes of anatase phase can be observed [37]. Furthermore, the SAED patterns shown in Figure 10c,f,i indicate a sequence of rings, which are characteristic of the crystalline planes (101), (004), (200), and (105). All these results are consistent with those obtained by the XRD analysis.

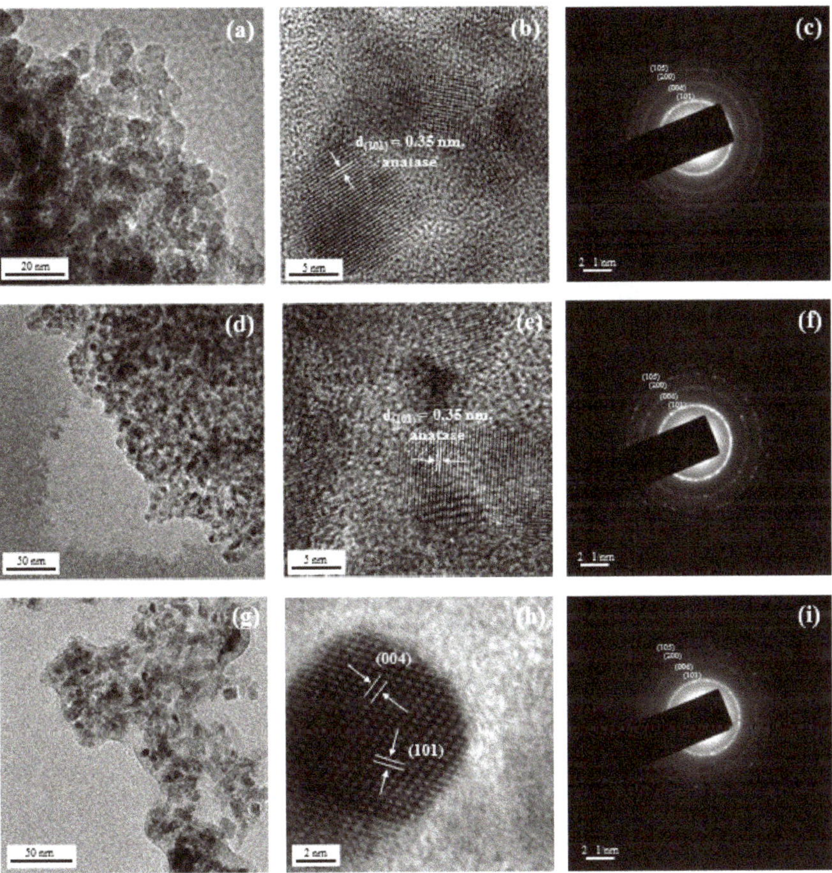

Figure 10. TEM images, HRTEM images and SAED patterns of samples (**a–c**) DM2, (**d–f**) DW2, and (**g–i**) DW1M1.

In order to visualize the dopant atoms in the anatase crystalline structure, Figure 11 shows STEM-HAADF micrographs with corrected aberration of synthesized catalysts DW1, DM1, and DW1M1. It can be observed in Figure 11a (W-mono-doped TiO_2) that the tungsten atoms are well distributed and they are part of well-defined atomic columns within the anatase phase. However, when titania is mono-doped with molybdenum atoms (Figure 11b), these can have two locations: (i) As being part of the crystalline structure with a good distribution (yellow circles), and (ii) forming some dispersed agglomerates that are not part of the atomic structure (orange circles). On the other hand, when TiO_2 is co-doped with W and Mo, the dispersion of both atoms inside the crystalline lattice is favored (atomic agglomerations disappear) (see Figure 11c).

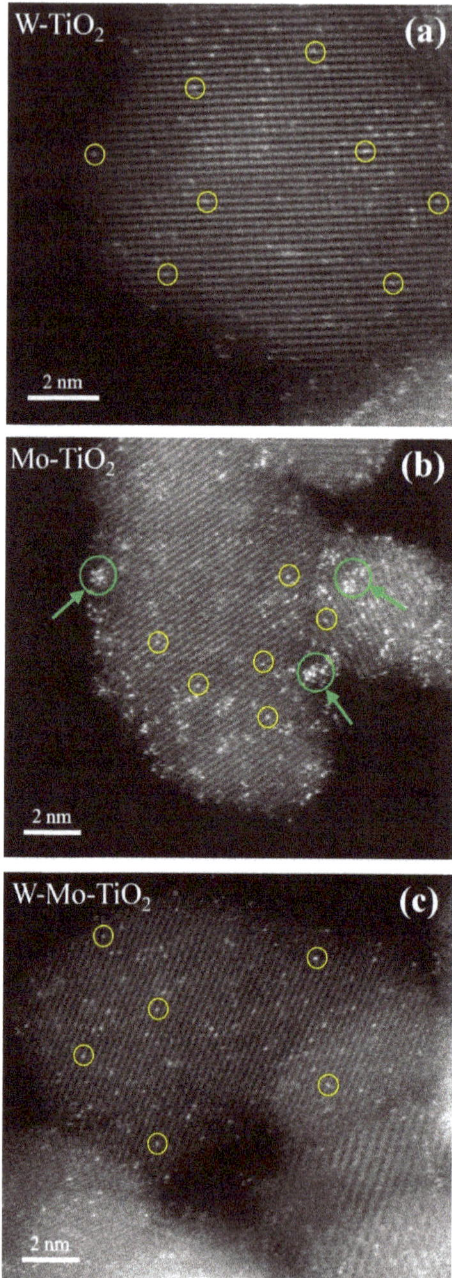

Figure 11. Aberration-corrected STEM-HAADF images of TiO$_2$ mono-doped with (**a**) W or (**b**) Mo, and (**c**) co-doped with both cations.

2.2. Photocatalytic Tests

All synthesized samples of pure titania, mono-doped, and co-doped with molybdenum and tungsten cations, as well as the commercial catalyst Degussa P25 were evaluated in the photocatalytic degradation of 4-chlorophenol (4CP). Figure 12 shows the photodegradation profiles with 100 min of reaction as total time. It can be seen that pure synthesized TiO_2 and Degussa P25 exhibit the lowest percentages of 4CP degradation. With respect to the co-doped TiO_2 samples, these show better degradation than the mono-doped materials (DM2 and DW2), which might be because of synergistic effects by the two doping species, thus improving their photoactivity. In addition, the co-doped samples exhibited higher surface areas and smaller crystallite sizes than the mono-doped catalysts, which increase the sites density for substrates adsorption during photocatalytic activity [38]. The degradation is gradually improved when the amount of tungsten increases from 0.3 wt.% to 1 wt.%, achieving more than 97% for the DW1M1 sample. Concentrations above 1 wt.% W in co-doped TiO_2 do not favor photocatalytic activity, which results in less degradation. While the samples co-doped with high molybdenum concentration (DW0.3M1.7 and DW0.7M1.3) exhibited higher radiation absorption, the significant improvement in photoactivity is very likely due to suitable concentration of both dopants. The co-doped TiO_2 with W:Mo = 1:1 ratio exhibits a synergistic effect between the species towards the best photocatalytic activity, since it presents a good radiation absorption capacity (see Figure 7) that favors the generation of e^-/h^+ pairs, as well as the ability to reduce the recombination processes of the photogenerated charges [39]. The process to inhibit the recombination (Equation (2)) after the e^-/h^+ pair generation (Equation (1)) is shown in Equations. (3) and (4). The electrons can be trapped by the most stable states of W and Mo, the W^{6+} and Mo^{6+} species, respectively, reaching a local charge compensation that leads to a longer lifetime of the generated holes [40]. In this way, the holes can be trapped by the adsorbed H_2O or by the -OH groups on the TiO_2 surface (Equations (5) and (6)) (see Figure 9a), allowing the generation of more •OH radicals, which are responsible for the oxidation of organic compounds (Equation (7)). On the other hand, the redox potentials of the doping species involved should not be left aside. In this case, the potentials of the Mo^{6+}/Mo^{5+} and W^{6+}/W^{5+} pairs are +0.4 V and −0.03 V, respectively [41,42], and they are more positive than the potential of the TiO_2 conduction band (−0.51 V) [43], so their electronic capture is favored when both are present in titania. The prepared catalysts showed no 4CP removal due to adsorption, and likewise, the photolysis effect was negligible.

$$W\text{-}Mo\text{-}TiO_2 + h\nu \rightarrow e^- + h^+ \tag{1}$$

$$e^- + h^+ \rightarrow TiO_2 \tag{2}$$

$$W^{6+} + e^- \rightarrow W^{5+} \tag{3}$$

$$Mo^{6+} + e^- \rightarrow Mo^{5+} \tag{4}$$

$$H_2O + h^+ \rightarrow \bullet OH + H^+ \tag{5}$$

$$-\!\!-OH + h^+ \rightarrow \bullet OH \tag{6}$$

$$\bullet OH + 4CP \rightarrow \rightarrow CO_2 + H_2O + Cl^- \tag{7}$$

The photocatalytic degradation profiles of 4CP were adjusted to a pseudo-first-order kinetics [44]. Apparent kinetic constants as well as half-life for all catalysts are summarized in Table 4. It can be seen that the maximum degradation is achieved by using the titania co-doped at 1 wt.% of both cations with a 4CP half-life of approximately 20 min, where about 75% of the initial concentration of 4CP is degraded in 40 min.

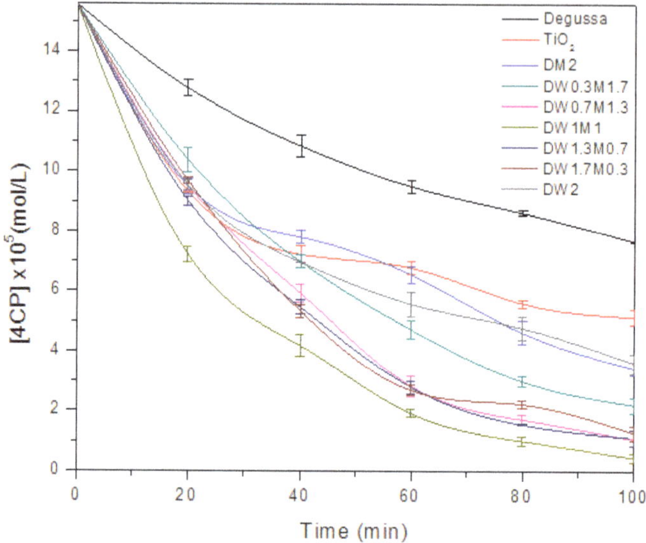

Figure 12. 4-Chlorophenol photodegradation profiles over TiO_2, Mo-TiO_2, W-TiO_2, W-Mo-TiO_2, and Degussa P25 catalysts. $[4CP]_0 = 15.56 \times 10^{-5}$ M, $C_{cat} = 2 \times 10^{-4}$ kg/L, $pH_0 = 2$, T = 298 K.

Table 4. Removal percentages, apparent kinetic constant, half-time and regression coefficient for 4CP degradation.

Catalyst	4CP Degradation (%)	$k \times 10^2$ (min^{-1})	$t_{1/2}$ (min)	r^2
DM2	76	1.31	52.8	0.97
DW0.3M1.7	86	1.98	34.9	0.99
DW0.7M1.3	93	2.75	25.2	0.99
DW1M1	97	3.49	19.9	0.99
DW1.3M0.7	93	2.73	25.4	0.99
DW1.7M0.3	92	2.51	27.6	0.99
DW2	75	1.26	54.9	0.95
TiO_2	67	1.04	66.8	0.90
Degussa P25	51	0.67	102.9	0.97

The percentage of mineralization, which is estimated by the total organic carbon content (TOC), is an important parameter of the degree of deep or complete oxidation of 4-chlorophenol to carbon dioxide and water (Equation (8)) [45].

$$C_6H_5ClO + \text{W-Mo-}TiO_2 + h\nu \rightarrow \text{Intermediates} \rightarrow CO_2 + H_2O + Cl^- \tag{8}$$

Figure 13 shows the 4CP mineralization percentages using the TiO_2 catalysts mono and co-doped with W and/or Mo. W-Mo-co-doped TiO_2 with a tungsten weight ratio of W/(W + Mo) = 0.5 showed the highest percentage of mineralization achieving more than 74%. On the other hand, the mono-doped samples (DM2 and DW2) exhibited average mineralization percentages of 50% and 53%, respectively.

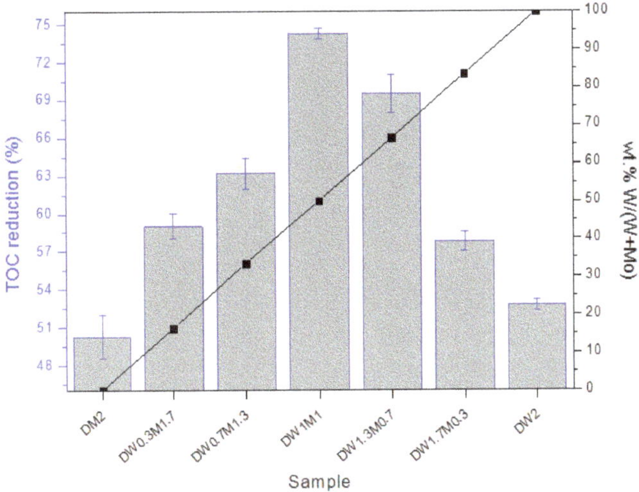

Figure 13. TOC reduction percentages and amount of tungsten dopant on mono-doped and co-doped TiO$_2$. [4CP]$_0$ = 15.56 × 10^{-5} M, C$_{cat}$ = 2 × 10^{-4} kg/L, pH$_0$ = 2, T = 298 K.

Table 5 shows the photocatalytic performance of the synthesized sample DW1M1 in comparison with other reported co-doped TiO$_2$ materials. It can be seen that, although the reaction conditions are a little different, the synthesized material of W-Mo-TiO$_2$ exhibits a higher degradation/mineralization percentage of 4-chlorophenol, as well as enhanced kinetic parameters.

Table 5. Photocatalytic performance of W-Mo-co-doped TiO$_2$ compared to other reported co-doped TiO2 materials.

Photo-Catalyst	[4CP]$_0$ × 10^5 (mol/L)	Radiation	4CP Degradation (%)	4CP Mineralization (%)	k × 10^2 (min^{-1})	Ref.
W-Mo-TiO$_2$	15.56	UV	97 (100 min)	74 (100 min)	3.49	-
B-N-TiO$_2$	5.00	UV	98 (120 min)	-	3.41	[46]
N-F-TiO$_2$	7.78	VL	72 (300 min)	-	-	[47]
C-W-TiO$_2$	50.00	VL	-	57 (300 min)	-	[48]

The intermediate compounds formed during the photocatalytic degradation of 4-chlorophenol were determined by HPLC. Figure 14 shows the concentration profiles of the main intermediaries when the catalyst DW1M1 is used. The name and structure of these identified intermediaries are summarized in Table 6. The aromatic ring of 4-chlorophenol can be attacked by hydroxyl radicals in several positions, and for this reason some mechanistic pathways are possible towards the formation of these species (see Figure 15). Furthermore, the generation of hydroxylated compounds is clear evidence of the involvement of hydroxyl radicals in the degradation of 4CP [49].

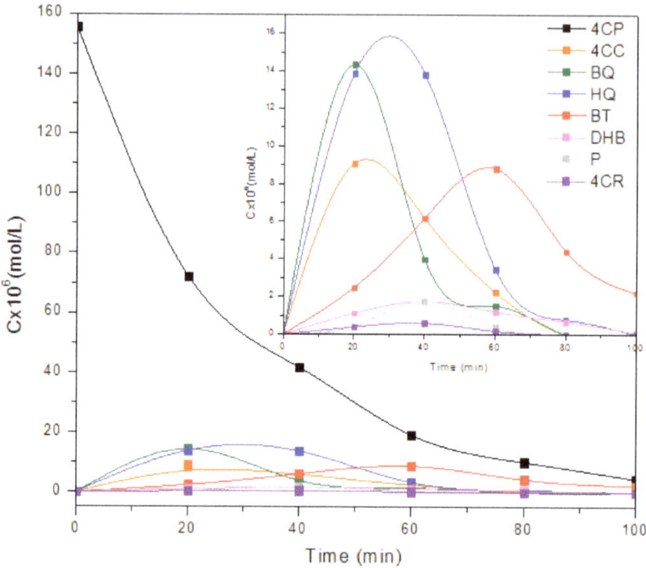

Figure 14. Concentration profiles of intermediates formed during 4CP photodegradation over W-Mo-TiO$_2$. [4CP]$_0$ = 15.56 × 10^{-5} M, C$_{cat}$ = 2 × 10^{-4} kg/L, pH$_0$ = 2, T = 298 K.

Table 6. Intermediate compounds formed in the photocatalytic degradation of 4-chlorophenol over TiO$_2$ co-doped with W and Mo.

Intermediate Compound ID	Name	Molecular Structure
HQ	Hydroquinone 1,4-benzenediol 1,4-dihydroxybenzene	
BQ	Benzoquinone 1,4-benzoquinone p-benzoquinone Quinone	
4CC	4-chlorocatechol 4-chloro-1,2-benzenediol	
BT	1,2,4-benzenetriol Hydroxyhydroquinone	
DHB	1,2-dihydroxybenzene 1,2-benzenediol Catechol Pyrocatechol	
P	Phenol Hydroxybenzene	
4CR	4-chlororesorcinol 1,3-dihydroxy-4-chlorobenzene	

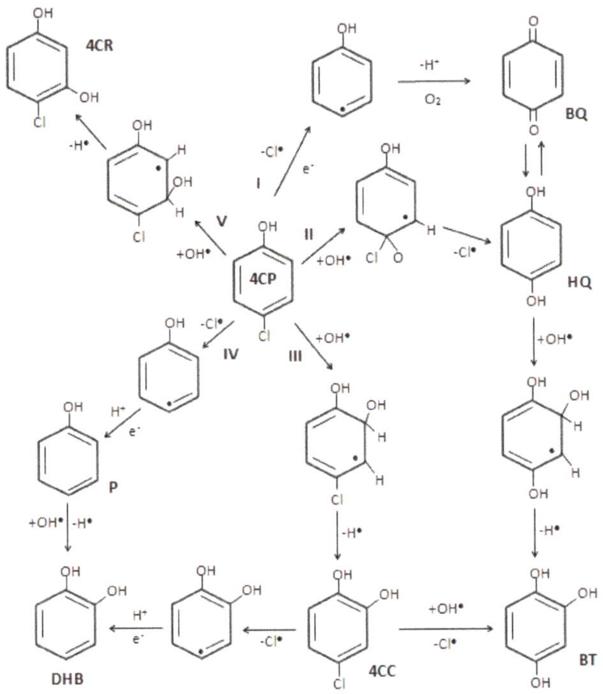

Figure 15. Proposed photocatalytic degradation pathway of 4-chlorophenol by W-Mo-TiO$_2$.

According to Figure 15, the oxidation process of 4CP is favored towards the quinone route (routes I and II) with maximum concentrations in the first minutes of reaction. Hydroquinone is generated by hydroxylation and dechlorination of 4CP in para position of the aromatic ring, which can be oxidized to benzoquinone. Additionally, benzoquinone can be formed from the hydroxyphenyl radical via route I [50].

4CP hydroxylation in ortho position generates 4-chlorocatechol (route III). The *para*-dechlorination of 4CC forms the 1,2-dihydroxyphenyl radical [51], which, when combined with a proton and an electron, produces 1,2-dihydroxybenzene. The peak with the highest concentration of 1,2,4-benzenetriol is found at 60 min, because its formation is after hydroxylation of hydroquinone (route II) or 4-chlorocatechol (route III). The oxidation process of 4CP can be initiated by the attack of the C-Cl bond (route IV) to form the hydroxyphenyl radical [52], which subsequently combined with a proton generates phenol. On the other hand, DHB (1,2-dihydroxybenzene) can also be formed by *ortho*-hydroxylation of phenol. However, this oxidation route is not favored due to low concentrations of phenol and 1,2-dihydroxybenzene observed during photocatalytic tests (see Figure 14). Similarly, *meta*-hydroxylation of 4CP generates very little 4-chlororesorcinol concentration (route V). Finally, the aromatic ring of these intermediate compounds is opened by subsequent oxidations to form aliphatic species (organic acids), which are later mineralized to carbon dioxide and water according to Equation (8) [53].

3. Materials and Methods

3.1. Chemicals

Titanium (IV) butoxide [Ti(OC$_4$H$_9$)$_4$, 97%], ammonium metatungstate [(NH$_4$)$_6$H$_2$W$_{12}$O$_{40}$·xH$_2$O] and ammonium heptamolybdate [(NH$_4$)$_6$Mo$_7$O$_{24}$·4H$_2$O] served as metal precursors Ti, W, and Mo, respectively. Pluronic P123 [EO$_{20}$PO$_{70}$EO$_{20}$] was used as a surfactant. Ethanol (C$_2$H$_6$O, 99.5%) and nitric acid (HNO$_3$, 70%) were employed as solvents and catalysts, respectively. 4-chlorophenol (ClC$_6$H$_4$OH, 99%) and deionized water were used during the photocatalytic tests. All reagents were of analytical grade and supplied by Sigma-Aldrich (St. Louis, MO, USA). Methanol and acetonitrile (HPLC grade) were used to prepare the mobile phases during the chromatographic analysis.

3.2. Preparation of Undoped, Doped and Co-Doped TiO$_2$

Pure TiO$_2$ was prepared by mixing ethyl alcohol and Ti(OC$_4$H$_9$)$_4$ under stirring for 15 min. This mixture was added to the P123 surfactant with constant stirring for 30 min and finally nitric acid was dropwise incorporated to the reaction media. The homogenous mixture was kept under vigorous stirring for 3 h. The molar ratio of reagents Ti(OBu)$_4$:C$_2$H$_6$O:P123: HNO$_3$ was maintained constant at 1:18.71:0.018:3.55. To prepare co-doped TiO$_2$, tungsten and/or molybdenum precursors were added before incorporating the HNO$_3$. The resulting solution was dried in a rotary evaporator until a solid was formed. This was subsequently calcined at 300 °C for 1 h (surfactant removal) and then at 400 °C for 4 h with a heating rate of 1 °C/min. Various samples with different weight percentages of dopant were synthetized. However, the total amount of dopant (s) was kept constant at 2 wt.% with respect to titania. The catalysts were labeled as DWAMB. W stands for tungsten and M for molybdenum. In addition, A and B are the weight percentages in samples. The DM2 and DW2 catalysts contain 2 wt.% molybdenum and 2 wt.% tungsten, respectively. Commercial TiO$_2$ (Degussa P25) was used for comparison.

3.3. Characterization of Photocatalysts

The temperature for the removal of the surfactant during heat treatment was determined by thermogravimetry (TGA 4000, Pyris, Perkin Elmer, Waltham, MA, USA) with N$_2$ gas flow at 20 mL/min in a range of 30 to 700 °C (5 °C/min). Before and after heat treatment, Infrared spectroscopy (FT-IR, Perkin Elmer Spectrum Two, Waltham, MA, USA) was used. The FT-IR spectra were acquired in the range of 4000–1000 cm^{-1} and using a resolution of 1 cm^{-1}.

All synthesized samples were analyzed with an X-ray diffractometer (Bruker Advance D8, Billerica, MA, USA) to determine the crystalline properties. Cu Kα radiation at 1.5406 Å was used as the X-ray source. The voltage and current applied were 30 kV and 25 mA, respectively. The diffraction patterns were determined over 2θ range of 20°–70° with a resolution of 0.02° 2θ/16 s. The average crystallite size (D) was estimated by the Scherrer equation D = 0.9λ/β cosθ, the lattice distortion (d) by d = β/4 tgθ, and the lattice parameters according to the Bragg equation 2dhkl sinθ = λ, where λ is the applied wavelength, θ is the Bragg angle, and β is the FWHM value. Raman spectroscopy was also used to determine the structural properties. For this purpose, a Renishaw micro-Raman spectrometer, provided with a laser of 514 nm, was used.

Average pore diameters and specific surface areas of the synthesized materials were obtained through nitrogen adsorption–desorption isotherms on a Quantachrome (Boynton Beach, FL, USA) Autosorb-1 at 77 K by BJH and BET methods, respectively. Before measurements, the materials were degassed out for 2 h at 200 °C.

The band gaps of all samples were determined through UV-vis Diffuse Reflectance Spectroscopy (DRS) using a Perkin Elmer spectrophotometer (Lambda 35, Waltham, MA, USA) equipped with an integration sphere (Labsphere rsape-20). The Kubelka-Munk function $\alpha = (1 - R)^2/2R$ was used to transform the reflectance spectra into absorption spectra, which were then used to estimate the band gap energies by constructing Tauc plots of hv vs. $(\alpha hv)^2$ and extrapolating the linear part to $(\alpha hv)^2$ equal to zero.

The chemical states of the surface elements present in the materials were examined by X-ray photoelectron spectroscopy (JEOL JPS-9200 spectrometer, Akishima, TYO, Japan) with an Al Kα X-ray source. The energy of C 1s at 284.6 eV was used as a reference for charge correction during the estimation of the binding energies.

High resolution-transmission electron microscopy (HR-TEM, JEOL-2100 at 200 kV, Akishima, TYO, Japan) and selected area electron diffraction (SAED) were used to determine the particle size and crystalline structure. High-angle annular dark-field scanning microscopy with a spherical aberration corrector (ac-HAADF-STEM, JEOL 2200FS+CS, Akishima, TYO, Japan) was used to visualize the doping atoms in the crystalline structures.

3.4. Photocatalytic activity and analysis

The photoactivity of all catalysts was evaluated in the degradation of 4-chlorophenol. The constant reaction conditions used were 2×10^{-4} kg/L of catalyst loading, 20×10^{-6} kg/L of 4CP and temperature of 298 K. A UV lamp was used as a radiation source with a wavelength and intensity of 254 nm and 4500 μW/cm^2, respectively. The 4-chlorophenol solution was adjusted to an initial pH of 2 and stirred at 1000 rpm to keep the catalyst in suspension. Aliquots were withdrawn from the reactor every 20 min and subsequently centrifuged to remove the catalyst. The solutions were then analyzed on a UV-vis spectrophotometer (Perkin Elmer Lambda 25, Waltham, MA, USA) to determine the amount of 4-chlorophenol at 280 nm, according to its absorbance peak. The total organic carbon concentration in the solution was measured by using a Shimadzu TOC-L analyzer to evaluate the 4CP mineralization. The identification of intermediate compounds during 4CP degradation was carried out by HPLC (Varian 230, isocratic mode, Santa Clara, CA, USA). Ascentis Express C18 column (2.7 μm, 3 cm $\times$ 4.6 mm, Sigma-Aldrich) with a mobile phase methanol/water (20/80 v/v, 1.0 mL min^{-1}) and Eclipse XDB-C18 column (5 μm, 15 cm $\times$ 4.6 mm, Agilent, Santa Clara, CA, USA) with a mobile phase acetonitrile/water (10/90 v/v, 0.6 mL/min) were used at 25 °C.

4. Conclusions

TiO$_2$ nanoparticles were prepared by mono and co-doping with molybdenum and tungsten cations by the Evaporation-Induced Self-Assembly (EISA) method. All synthesized materials presented anatase crystalline phase and larger specific surface areas than the commercial photocatalyst Degussa P25. The presence of a second cation by co-doping increased the surface area and decreased the crystallinity, as well as changes in lattice parameters due to structural distortion, since the doping species were successfully incorporated into TiO$_2$ lattice. The dopant cations exhibited oxidation states with valence 5+ and 6+ in the titania. Low tungsten concentrations by co-doping favored the reduction of the anatase band gap. Raman analysis showed that no additional phases by co-doping were created, and both dopants are distributed inside the anatase. All synthesized materials exhibited photocatalytic activity in the oxidation of 4-chlorophenol and all of them outperformed commercial TiO2 P25. Synergistic effect with an appropriate concentration of both dopant cations improved photocatalytic activity by reducing the recombination of photogenerated charges and by increasing the absorption of radiation to promote their generation. The experiments performed at ambient temperature and during 100 min showed that titania, co-doped with 1 wt.% W and 1 wt.%, Mo exhibited the best photoactivity among the catalysts studied, with 97% and 74% of degradation and mineralization of 4-chlorophenol, respectively. The oxidation route of 4-chlorophenol was favored towards quinones as main intermediates via hydroxyl radicals. Remaining TOC (26%) is attributed to 1,2,4-benzenetriol as secondary intermediate and aliphatic compounds.

Author Contributions: J.E.-V. and R.N.-R. designed the experiments and supervised the project; O.A.-G. performed the experiments and analyzed the data; D.A.S.-C. contributed with characterization of materials; O.A.-G. wrote the manuscript; All authors contributed to a review of the manuscript before submission.

Funding: This research was funded by PROMEP-Mexico (financial support through project 103.5/13/S257) and CONACYT-Mexico (project 269093). Scholarship 378292 by CONACYT.

Acknowledgments: Authors are grateful to CCIQS from UAEM-UNAM and CIMAV-Mexico for the granted support. The technical support of Gustavo López Téllez, Alfredo Rafael Vílchis Néstor, Uvaldo Hernández, Citlalit Martínez and Carlos Elías Ornelas Gutiérrez is also acknowledged.

References

1. Finčur, N.L.; Krstić, J.B.; Šibul, F.S.; Šojić, D.V.; Despotović, V.N.; Banić, N.D.; Agbaba, J.R.; Abramović, B.F. Removal of alprazolam from aqueous solutions by heterogeneous photocatalysis: Influencing factors, intermediates, and products. *Chem. Eng. J.* **2017**, *307*, 1105–1115. [CrossRef]

2. Verbruggen, S.W. TiO$_2$ photocatalysis for the degradation of pollutants in gas phase: From morphological design to plasmonic enhancement. *J. Photochem. Photobiol. C Photochem. Rev.* **2015**, *24*, 64–82. [CrossRef]

3. Linsebigler, A.L.; Lu, G.; Yates, J.T. Photocatalysis on TiO$_2$ surfaces: Principles, mechanisms, and selected results. *Chem. Rev.* **1995**, *95*, 735–758. [CrossRef]

4. Fujishima, A.; Zhang, X.; Tryk, D.A. TiO$_2$ photocatalysis and related surface phenomena. *Surf. Sci. Rep.* **2008**, *63*, 515–582. [CrossRef]

5. Kong, M.; Li, Y.; Chen, X.; Tian, T.; Fang, P.; Zheng, F.; Zhao, X. Tuning the relative concentration ratio of bulk defects to surface defects in TiO$_2$ nanocrystals leads to high photocatalytic efficiency. *J. Am. Chem. Soc.* **2011**, *133*, 16414–16417. [CrossRef]

6. Liqiang, J.; Honggang, F.; Baiqi, W.; Dejun, W.; Baifu, X.; Shudan, L.; Jiazhong, S. Effects of Sn dopant on the photoinduced charge property and photocatalytic activity of TiO$_2$ nanoparticles. *Appl. Catal. B Environ.* **2006**, *62*, 282–291. [CrossRef]

7. Yang, G.; Yan, Z.; Xiao, T. Low-temperature solvothermal synthesis of visible-light-responsive S-doped TiO$_2$ nanocrystal. *Appl. Surf. Sci.* **2012**, *258*, 4016–4022. [CrossRef]

8. Ding, H.; Sun, H.; Shan, Y. Preparation and characterization of mesoporous SBA-15 supported dye-sensitized TiO$_2$ photocatalyst. *J. Photochem. Photobiol. A Chem.* **2005**, *169*, 101–107. [CrossRef]

9. Wu, Y.; Liu, H.; Zhang, J.; Chen, F. Enhanced photocatalytic activity of nitrogen-doped titania by deposited with gold. *J. Phys. Chem. C* **2009**, *113*, 14689–14695. [CrossRef]

10. Bayati, M.R.; Golestani-Fard, F.; Moshfegh, A.Z. Photo-degradation of methelyne blue over V$_2$O$_5$–TiO$_2$ nano-porous layers synthesized by micro arc oxidation. *Catal. Lett.* **2010**, *134*, 162–168. [CrossRef]

11. Yang, G.; Jiang, Z.; Shi, H.; Xiao, T.; Yan, Z. Preparation of highly visible-light active N-doped TiO$_2$ photocatalyst. *J. Mater. Chem.* **2010**, *20*, 5301–5309. [CrossRef]

12. Wilke, K.; Breuer, H.D. The influence of transition metal doping on the physical and photocatalytic properties of titania. *J. Photochem. Photobiol. A Chem.* **1999**, *121*, 49–53. [CrossRef]

13. Choi, W.; Termin, A.; Hoffmann, M.R. The role of metal ion dopants in quantum-sized TiO$_2$: Correlation between photoreactivity and charge carrier recombination dynamics. *J. Phys. Chem.* **1994**, *98*, 13669–13679. [CrossRef]

14. Estrellan, C.R.; Salim, C.; Hinode, H. Photocatalytic activity of sol–gel derived TiO$_2$ co-doped with iron and niobium. *React. Kinet. Catal. Lett.* **2009**, *98*, 187–192. [CrossRef]

15. Shi, Z.; Lai, H.; Yao, S.; Wang, S. Photocatalytic activity of Fe and Ce co-doped mesoporous TiO$_2$ catalyst under UV and visible light. *J. Chin. Chem. Soc.* **2012**, *59*, 614–620. [CrossRef]

16. Soler-Illia, G.D.A.; Louis, A.; Sanchez, C. Synthesis and characterization of mesostructured titania-based materials through evaporation-induced self-assembly. *Chem. Mater.* **2002**, *14*, 750–759. [CrossRef]

17. Al-Ekabi, H.; Serpone, N. Kinetics studies in heterogeneous photocatalysis. I. Photocatalytic degradation of chlorinated phenols in aerated aqueous solutions over titania supported on a glass matrix. *J. Phys. Chem.* **1988**, *92*, 5726–5731. [CrossRef]

18. Chang, W.; Kim, H.; Lee, G.Y.; Ahn, B.J. Catalytic hydrodechlorination reaction of chlorophenols by Pd nanoparticles supported on graphene. *Res. Chem. Int.* **2016**, *42*, 71–82. [CrossRef]

19. Ruiz-García, C.; Heras, F.; Calvo, L.; Alonso-Morales, N.; Rodriguez, J.J.; Gilarranz, M.A. Platinum and N-doped carbon nanostructures as catalysts in hydrodechlorination reactions. *Appl. Catal. B Environ.* **2018**, *238*, 609–617. [CrossRef]

20. Hernandez, S.R.; Kergaravat, S.V.; Pividori, M.I. Enzymatic electrochemical detection coupled to multivariate calibration for the determination of phenolic compounds in environmental samples. *Talanta* **2013**, *106*, 399–407. [CrossRef]

21. Janda, V.; Svecova, M. By-products in drinking water disinfection. *Chem. List* **2000**, *94*, 905–908.

22. Li, B.; Zhao, Z.; Gao, F.; Wang, X.; Qiu, J. Mesoporous microspheres composed of carbon-coated TiO$_2$ nanocrystals with exposed {001} facets for improved visible light photocatalytic activity. *Appl. Catal. B Environ.* **2014**, *147*, 958–964. [CrossRef]

23. Hussain, S.T.; Siddiqa, A.; Siddiq, M.; Ali, S. Iron-doped titanium dioxide nanotubes: A study of electrical, optical, and magnetic properties. *J. Nanopart. Res.* **2011**, *13*, 6517–6525. [CrossRef]
24. Lan, X.; Wang, L.; Zhang, B.; Tian, B.; Zhang, J. Preparation of lanthanum and boron co-doped TiO_2 by modified sol–gel method and study their photocatalytic activity. *Catal. Today* **2014**, *224*, 163–170. [CrossRef]
25. Hsieh, C.-T.; Fan, W.-S.; Chen, W.-Y.; Lin, J.-Y. Adsorption and visible-light-derived photocatalytic kinetics of organic dye on Co-doped titania nanotubes prepared by hydrothermal synthesis. *Sep. Purif. Technol.* **2009**, *67*, 312–318. [CrossRef]
26. Nešić, J.; Manojlović, D.D.; Anđelković, I.; Dojčinović, B.P.; Vulić, P.J.; Krstić, J.; Roglić, G.M. Preparation, characterization and photocatalytic activity of lanthanum and vanadium co-doped mesoporous TiO_2 for azo-dye degradation. *J. Mol. Catal. A Chem.* **2013**, *378*, 67–75. [CrossRef]
27. Ohsaka, T. Temperature dependence of the Raman spectrum in anatase TiO_2. *J. Phys. Soc. Jpn.* **1980**, *48*, 1661–1668. [CrossRef]
28. Štengl, V.; Velická, J.; Maříková, M.; Grygar, T.M. New generation photocatalysts: how tungsten influences the nanostructure and photocatalytic activity of TiO_2 in the UV and visible regions. *ACS Appl. Mater. Int.* **2011**, *3*, 4014–4023. [CrossRef]
29. Wu, Y.; Zhang, J.; Xiao, L.; Chen, F. Properties of carbon and iron modified TiO_2 photocatalyst synthesized at low temperature and photodegradation of acid orange 7 under visible light. *Appl. Surf. Sci.* **2010**, *256*, 4260–4268. [CrossRef]
30. Shojaie, A.F.; Loghmani, M.H. La^{3+} and Zr^{4+} co-doped anatase nano TiO_2 by sol-microwave method. *Chem. Eng. J.* **2010**, *157*, 263–269. [CrossRef]
31. Liu, D.; Wu, Z.; Tian, F.; Ye, B.-C.; Tong, Y. Synthesis of N and La co-doped TiO_2/AC photocatalyst by microwave irradiation for the photocatalytic degradation of naphthalene. *J. Alloys Compd.* **2016**, *676*, 489–498. [CrossRef]
32. Wang, Q.; Jiang, H.; Zang, S.; Li, J.; Wang, Q. Gd, C., N and P quaternary doped anatase-TiO_2 nano-photocatalyst for enhanced photocatalytic degradation of 4-chlorophenol under simulated sunlight irradiation. *J. Alloys Compd.* **2014**, *586*, 411–419. [CrossRef]
33. Kubelka, P.; Munk, F. Ein beitrag zur optik der farbanstriche. *Z. Tech. Phys.* **1931**, *12*, 593–601.
34. Gutiérrez-Alejandre, A.; Ramírez, J.; Busca, G. The electronic structure of oxide-supported tungsten oxide catalysts as studied by UV spectroscopy. *Catal. Lett.* **1998**, *56*, 29–33. [CrossRef]
35. Li, J.; Li, B.; Li, J.; Liu, J.; Wang, L.; Zhang, H.; Zhang, Z.; Zhao, B. Visible-light-driven photocatalyst of La–N-codoped TiO_2 nano-photocatalyst: Fabrication and its enhanced photocatalytic performance and mechanism. *J. Ind. Eng. Chem.* **2015**, *25*, 16–21. [CrossRef]
36. Zhang, J.; Xu, L.J.; Zhu, Z.Q.; Liu, Q.J. Synthesis and properties of (Yb, N)-TiO_2 photocatalyst for degradation of methylene blue (MB) under visible light irradiation. *Mater. Res. Bull.* **2015**, *70*, 358–364. [CrossRef]
37. Khan, H.; Berk, D. Characterization and mechanistic study of Mo^{+6} and V^{+5} codoped TiO_2 as a photocatalyst. *J. Photochem. Photobiol. A Chem.* **2014**, *294*, 96–109. [CrossRef]
38. Wang, Q.; Xu, S.; Shen, F. Preparation and characterization of TiO_2 photocatalysts co-doped with iron (III) and lanthanum for the degradation of organic pollutants. *Appl. Surf. Sci.* **2011**, *257*, 7671–7677. [CrossRef]
39. Wang, Z.; Chen, C.; Wu, F.; Zou, B.; Zhao, M.; Wang, J.; Feng, C. Photodegradation of rhodamine B under visible light by bimetal codoped TiO_2 nanocrystals. *J. Hazard. Mater.* **2009**, *164*, 615–620. [CrossRef]
40. Wang, S.; Bai, L.N.; Sun, H.M.; Jiang, Q.; Lian, J.S. Structure and photocatalytic property of Mo-doped TiO_2 nanoparticles. *Powder Technol.* **2013**, *244*, 9–15. [CrossRef]
41. Luo, S.-Y.; Yan, B.-X.; Shen, J. Enhancement of photoelectric and photocatalytic activities: Mo doped TiO_2 thin films deposited by sputtering. *Thin Solid Films* **2012**, *522*, 361–365. [CrossRef]
42. Tae Kwon, Y.; Yong Song, K.; In Lee, W.; Jin Choi, G.; Rag Do, Y. Photocatalytic behavior of WO_3-loaded TiO_2 in an oxidation reaction. *J. Catal.* **2000**, *191*, 192–199. [CrossRef]
43. Park, H.; Park, Y.; Kim, W.; Choi, W. Surface modification of TiO_2 photocatalyst for environmental applications. *J. Photochem. Photobiol. C Photochem. Rev.* **2013**, *15*, 1–20. [CrossRef]
44. Wang, Z.; Liu, X.; Li, W.; Wang, H.; Li, H. Enhancing the photocatalytic degradation of salicylic acid by using molecular imprinted S-doped TiO_2 under simulated solar light. *Ceram. Int.* **2014**, *40*, 8863–8867. [CrossRef]
45. Sharma, S.; Mukhopadhyay, M.; Murthy, Z.V.P. Rate parameter estimation for 4-chlorophenol degradation by UV and organic oxidants. *J. Ind. Eng. Chem.* **2012**, *18*, 249–254. [CrossRef]

46. Yuan, J.; Wang, E.; Chen, Y.; Yang, W.; Yao, J.; Cao, Y. Doping mode, band structure and photocatalytic mechanism of B–N-codoped TiO_2. *Appl. Surf. Sci.* **2011**, *257*, 7335–7342. [CrossRef]

47. Li, X.; Zhang, H.; Zheng, X.; Yin, Z.; Wei, L. Visible light responsive N-F-codoped TiO_2 photocatalysts for the degradation of 4-chlorophenol. *J. Environ. Sci.* **2011**, *23*, 1919–1924. [CrossRef]

48. Neville, E.M.; Mattle, M.J.; Loughrey, D.; Rajesh, B.; Rahman, M.; MacElroy, J.M.D.; Sullivan, J.A.; Thampi, K.R. Carbon-Doped TiO_2 and Carbon, Tungsten-Codoped TiO_2 through Sol–Gel Processes in the Presence of Melamine Borate: Reflections through Photocatalysis. *J. Phys. Chem. C* **2012**, *116*, 16511–16521. [CrossRef]

49. Shamaila, S.; Sajjad, A.K.L.; Chen, F.; Zhang, J. Synthesis and characterization of mesoporous-TiO_2 with enhanced photocatalytic activity for the degradation of chloro-phenol. *Mater. Res. Bull.* **2010**, *45*, 1375–1382. [CrossRef]

50. Theurich, J.; Lindner, M.; Bahnemann, D.W. Photocatalytic degradation of 4-chlorophenol in aerated aqueous titanium dioxide suspensions: A kinetic and mechanistic study. *Langmuir* **1996**, *12*, 6368–6376. [CrossRef]

51. Gaya, U.I.; Abdullah, A.H.; Zainal, Z.; Hussein, M.Z. Photocatalytic treatment of 4-chlorophenol in aqueous ZnO suspensions: Intermediates, influence of dosage and inorganic anions. *J. Hazard. Mater.* **2009**, *168*, 57–63. [CrossRef] [PubMed]

52. Lipczynska-Kochany, E.; Kochany, J.; Bolton, J.R. Electron paramagnetic resonance spin trapping detection of short-lived radical intermediates in the direct photolysis of 4-chlorophenol in aerated aqueous solution. *J. Photochem. Photobiol. A Chem.* **1991**, *62*, 229–240. [CrossRef]

53. Pozan, G.S.; Kambur, A. Removal of 4-chlorophenol from wastewater: Preparation, characterization and photocatalytic activity of alkaline earth oxide doped TiO2. *Appl. Catal. B Environ.* **2013**, *129*, 409–415. [CrossRef]

Article

Photocatalytic Degradation of Estriol Using Iron-Doped TiO$_2$ under High and Low UV Irradiation

Irwing M. Ramírez-Sánchez [1] **and Erick R. Bandala** [2,3,*]

[1] Department of Civil, Architectural and Environmental Engineering, The University of Texas at Austin, Austin, TX 78712, USA; irwingmoises@gmail.com

[2] Desert Research Institute (DRI), 755 E. Flamingo Road, Las Vegas, NV 89119-7363, USA

[3] Graduate Program Hydrologic Sciences, University of Nevada, Reno, NV 89557, USA

* Correspondence: erick.bandala@dri.edu; Tel.: + 1-(702)-862-5395

Received: 29 September 2018; Accepted: 28 November 2018; Published: 5 December 2018

Abstract: Iron-doped TiO$_2$ nanoparticles (Fe-TiO$_2$) were synthesized and photocatalitically investigated under high and low fluence values of UV radiation. The Fe-TiO$_2$ physical characterization was performed using X-ray Powder Diffraction (XRD), Brunauer–Emmett–Teller (BET) surface area analysis, Transmission Electron Microscopy (TEM), Scanning Electron Microscopy (SEM), Diffuse Reflectance Spectroscopy (DRS), and X-ray Photoelectron Spectroscopy (XPS). The XPS evidenced that the ferric ion (Fe^{3+}) was in the TiO$_2$ lattice and unintentionally added co-dopants were also present because of the precursors of the synthetic method. The Fe^{3+} concentration played a key role in the photocatalytic generation of hydroxyl radicals ($^{\bullet}$OH) and estriol (E3) degradation. Fe-TiO$_2$ accomplished E3 degradation, and it was found that the catalyst with 0.3 at.% content of Fe (0.3 Fe-TiO$_2$) enhanced the photocatalytic activity under low UV irradiation compared with TiO$_2$ without intentionally added Fe (zero-iron TiO$_2$) and Aeroxide$^{\circledR}$ TiO$_2$ P25. Furthermore, the enhanced photocatalytic activity of 0.3 Fe-TiO$_2$ under low UV irradiation may have applications when radiation intensity must be controlled, as in medical applications, or when strong UV absorbing species are present in water.

Keywords: iron-doped TiO$_2$; photocatalytic activity; low UV irradiation; hydroxyl radical; estriol

---

1. Introduction

In recent years, society and the scientific community have concerned of Emerging Contaminants (ECs, also called Contaminants of Emerging Concern), which are chemicals that threaten the environment, human health, and water safety and are not currently covered by existing local or international water quality regulations [1]. ECs include chemical species such as algae toxins, illegal drugs, industrial compounds, flame retardants, food additives, nanoparticles, pharmaceuticals (human and veterinary), personal care products, pesticides, biocides, steroids, synthetic and natural hormones, and surfactants [2].

Natural hormones (e.g., estrone (E1), 17β-estradiol (E2), and estriol (E3)) as ECs are susceptible of persisting and bioaccumulating in the environment, and could induce endocrine disruption in humans and wildlife (vertebrates [3–5] and invertebrates [6,7]). Natural attenuation, drinking water purification, and conventional municipal wastewater treatment processes are either incapable or only partially capable of removing estrogens from water [8]. As result, water treatment techniques are being developed to manage, reduce, degrade, and mineralize low-concentrated ECs (including natural estrogen) in drinking and wastewater [9]. Advanced Oxidation Processes (AOPs) are promising techniques to treat ECs in aqueous phase, which include well-known processes such as Fenton and Fenton-like processes, UV/H$_2$O$_2$, ozonation, and photocatalysis using semiconductors, peroxone processes (H$_2$O$_2$/O$_3$), and cavitation [10,11]. Although there are many known AOPs, since Coleman's work [12], photocatalysis using titanium dioxide (TiO_2) has been identified as one of the most effective methods to degrade estrogens in water [13]. Several reports recognized that TiO_2 can degrade estrogens, which prevents increases in estrogenic activity in water [14,15] and partially or completely mineralizing estrogens [14,16].

Titanium dioxide is the most commonly used photocatalyst because of its reasonable optical and electronic properties, good photocatalytic activity, insolubility in water, chemical and photochemical stability, nontoxicity, low cost, and high efficiency in pollutant mineralization [17–20]. However, the band gap energy (E_g) of TiO_2, frequently reported as 3.2 eV [21], restrains the photocatalytic activation to energy sources with a portion of spectrum emission below 387.5 nm [22].

In general the photocatalytic mechanism is as shown in Figure 1. According to Density Functional Theory (DFT) computations, the valence band (VB) and conduction band (CB) of pure TiO_2 are mainly composed of O2p orbitals and Ti3d orbitals, respectively. Hence, the Fermi level (EF) is located in the middle of the band gap (BG), indicating that VB is full filled while CB is empty [23]. When using photons with energy higher than 3.2 eV, photoexcitation of the semiconductor promotes electrons from VB to CB creating a charge vacancy or hole (h^+) in the VB. The h^+ in the VB can react with hydroxide ion to form hydroxyl radical ($^\bullet$OH) or can also be filled by donor absorbed organic molecule (OM_{ads}). Photogenerated electrons in the CB can be transferred to acceptor of electrons and bring about $^\bullet$OH.

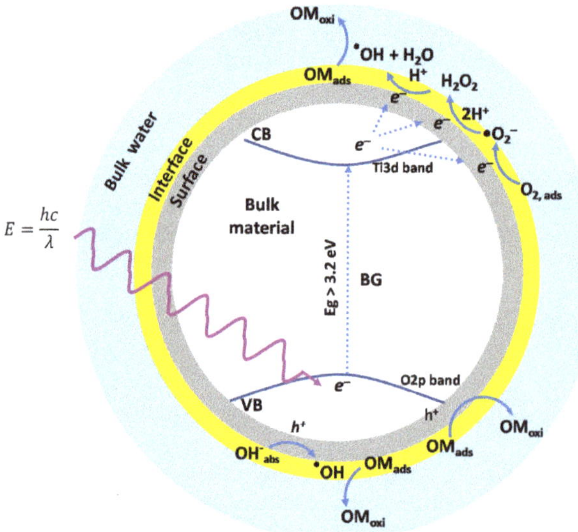

Figure 1. Photocatalytic mechanism of TiO_2 for $^\bullet$OH generation. Where E_g: Band gap energy; E: photon energy; OM_{ads}: adsorbed organic molecule; and OM_{oxi}: oxidized organic molecule.

Consequently, reducing the photon energy needed for TiO_2 photoactivation has been the focus of the scientific community until now. Doping is one of the techniques that has been tested to control or modify the surface properties or internal structure of TiO_2. Doping introduces a foreign element into TiO_2 to cause an impurity state in the band gap. The most frequently used doping materials are transition-metal cations (e.g., Cr, V, Fe, and Ni) at Ti sites, and anions (e.g., N, S, and C) at O sites [24]. Among anion- and cation-dopants, the ferric ion (Fe^{3+}) is one of the most often used because the ionic radius of Fe^{3+} (0.69 A) is similar to Ti^{4+} (0.745 A) [25]. Therefore, Fe^{3+} can be easily incorporated into the TiO_2 crystal lattice.

The main reported effects of iron-doped TiO_2 is a rapid increase in photocatalytic activity that increases with increased Fe doping, which then reaches a maximum value, and finally decreases with further increased Fe content [23,26–37]. However, detrimental effects have been also reported because of high Fe content [38,39] or agglomerated Fe-TiO$_2$ nanoparticles [40,41].

Although several theoretical and experimental Fe-TiO$_2$ studies have been developed, the trade-off between doping ratio and radiation intensity is scarcely mentioned. Furthermore, Fe-TiO$_2$ photocatalyst has rarely been considered to be a useful technique for the degradation of E3 [42].

In this work, Fe-TiO$_2$ nanoparticles were synthesized to increase the understanding of the relationship between doping ratio and radiation intensity for hydroxyl radical ($^\bullet$OH) generation and E3 degradation. Therefore, we investigated the photocatalytic degradation of E3 using Fe-TiO$_2$ under high and low UV irradiation. We highlight the term low UV irradiation to avoid confusion with the term "photocatalytic processes under visible light" because we did not intentionally use UV cutoff filters for the experiments.

2. Results and Discussion

2.1. Characterization of Iron-Doped TiO$_2$

Figure 2 shows X-ray Photoelectron Spectroscopy (XPS) general spectra of TiO$_2$ without added Fe (zero-iron TiO$_2$) and Fe-TiO$_2$ materials (b, c, and d). For the experimental condition used, Fe did not affect the bonding structure between titanium and oxygen because the main peaks for all samples were Ti2p and O1s with the proportion 1:2.2, which is in agreement with the atomic formula of TiO$_2$.

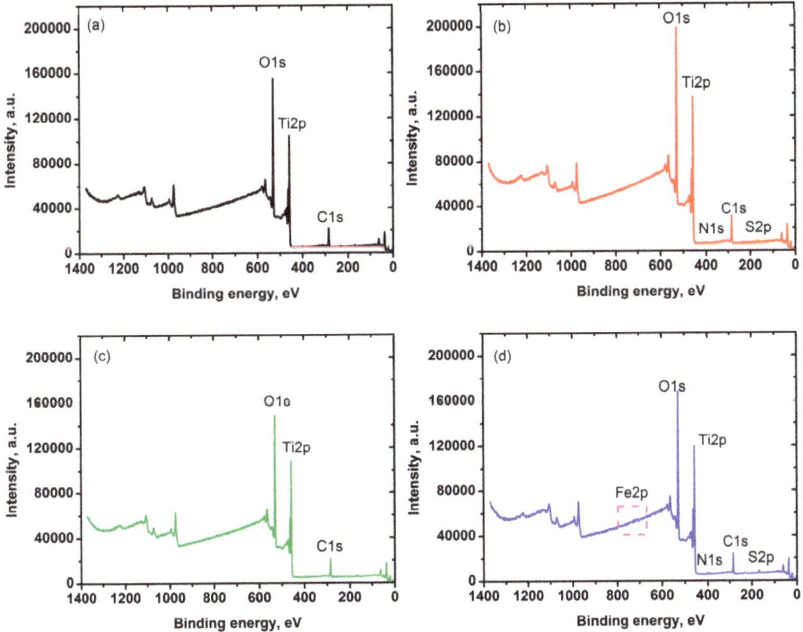

Figure 2. X-ray Photoelectron Spectroscopy (XPS) general spectra for zero-iron TiO$_2$ (**a**), 0.3 Fe-TiO$_2$ (**b**), 0.6 Fe-TiO$_2$ (**c**), and 1.0 Fe-TiO$_2$ (**d**).

XPS detected unintentionally added elements such as carbon, sulfur, and nitrogen (Table 1) as co-dopants of zero-iron TiO$_2$ and Fe-TiO$_2$, which were introduced into TiO$_2$ via precursors of the synthesis. Carbon and sulfur could come from sodium dodecyl sulfate (SDS), and nitrogen could come from iron (III) nitrate (Fe(NO$_3$)$_3$·9H$_2$O) and HNO$_3$, all of them used in the synthesis process.

Table 1. Surface elemental composition determined by XPS.

Material	Atomic % of Elements (at.%)					
	Ti2p	O1s	C1s	Fe2p	S2p	N1s
Zero-iron TiO_2	24.4	52.9	21.3	0	1.4	-
0.3 Fe-TiO_2	23.8	51.1	22.9	0.3	1.1	0.8
0.6 Fe-TiO_2	23.9	53.1	22.5	0.6	-	-
1.0 Fe-TiO_2	23.5	52.5	20.6	1	1.5	0.9

High-resolution XPS spectra for the iron region (Figure 3) was studied only for 1.0 Fe-TiO_2 because no Fe2p signals were detected for zero-iron TiO_2, 0.3 Fe-TiO_2, or 0.6 Fe-TiO_2. The deconvolution of high-resolution XPS spectra (Figure 3) was developed for previously reported peaks of Fe^{2+} and Fe^{3+} [43]. Shirley baseline was subtracted before peak fitting. The Gaussian–Lorentzian mix function was used with a 40% factor. Charge compensation was set by the O1s peak charge with -0.58 eV. As a result, the correlation between the experimental signal and the theoretic model ($\Sigma\chi^2$) was 8.43×10^{-2}.

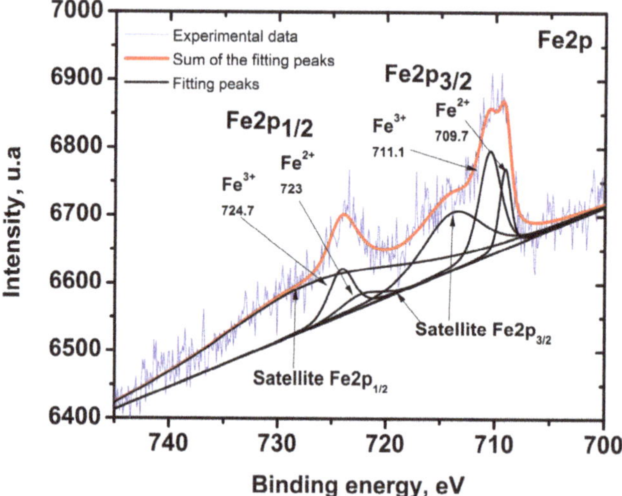

Figure 3. High-resolution XPS spectra for the iron region for 1.0 Fe-TiO_2.

According to the theoretical model (sum of fitting peaks), both Fe^{3+} and Fe^{2+} were present in the lattice of 1.0 Fe-TiO_2. We suggest that Fe^{3+} was incorporated into the lattice of TiO_2 to form Ti–O–Fe bonds, because the ionic radius of Fe^{3+} (0.69 A) is similar to the ionic radius of Ti^{4+} (0.745 A) [25]. The XPS technique detected Fe^{2+} because Fe^{3+} underwent reduction to Fe^{2+} during XPS measurement in vacuum [44].

The band gap energy (E_g) obtained with the Kubelka–Monk method (Figure 4) for Aeroxide® TiO_2 P25 was 3.2 eV, which is consistent with the value reported previously [45]. For Aeroxide® TiO_2 P25 E_g, red-shifts were detected as 0.22, 0.24, 0.25, and 0.3 eV for zero-iron TiO_2, 0.3 Fe-TiO_2, 0.6 Fe-TiO_2, and 1.0 Fe-TiO_2, respectively, which is consistent with values reported by Shi et al. of 0.25 eV [46] and with density functional theory calculations that suggested the hybridized band of Ti3d and Fe3d reduces E_g approximately 0.3–0.5 eV [44], or 0.2–0.34 eV [47].

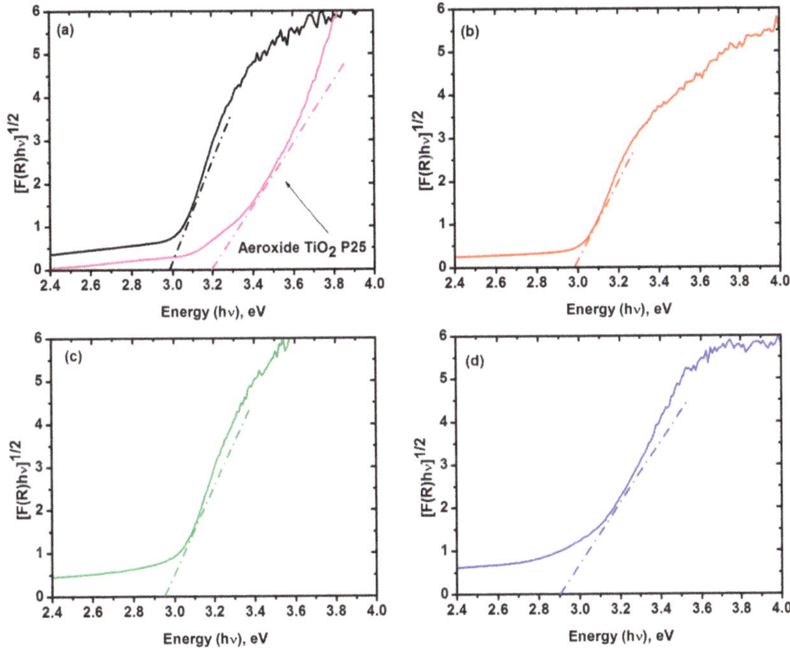

Figure 4. Band gap energy (E_g) by the Kubelka–Monk method. Zero-iron TiO_2 (**a**), 0.3 Fe-TiO_2 (**b**), 0.6 Fe-TiO_2 (**c**), and 1.0 Fe-TiO_2 (**d**).

For zero-iron TiO_2, E_g for Fe-TiO_2 materials (Table 2) decreased as long as the Fe content increased, so the Fe content generated red-shift. For Aeroxide® TiO_2 P25 E_g, the red-shift of Fe-TiO_2 agreed with previously reported values, but it agreed less for zero-iron TiO_2. Therefore, red-shift was not only related to Fe content, but also to the synthesis method and unintentionally co-doped TiO_2.

Table 2. Structural and optical properties of zero-iron TiO_2, and Fe-TiO_2.

Material	E_g		Anatase: Rutile	Particle Size	Surface Area	Pore Size	High UV	Low UV
	eV	nm	%	nm	$m^2\,g^{-1}$	nm	%	%
Aeroxide® TiO_2 P25	3.2 *	387.5 *	80:20 *	21 *	50 ± 15 *	17.5 *	36.4	0.8
Zero-iron TiO_2	2.98	416.1	73.1:26.9	6.6	66.5	8.4	99.26	7.64
0.3 Fe-TiO_2	2.96	418.9	77.9:21.1	6.9	77.6	1.2	99.40	8.21
0.6 Fe-TiO_2	2.95	420.3	78.8:21.2	7.1	73.0	1.4	99.42	8.77
1.0 Fe-TiO_2	2.90	427.6	76.3:23.7	6.9	83.1	9.4	99.43	10.63

* According to the manufacturer.

XRD patterns in Figure 5 revealed zero-iron TiO_2 and Fe-TiO_2 materials had both anatase and rutile phases. No XRD Fe_2O_3 peaks (2θ equal to 33.0°, 35.4°, 40.7°, 43.4°, and 49.2°) were observed, concluding that Fe^{3+} replaced Ti^{4+} in the TiO_2 crystal framework [48,49]. The synthesis method allowed uniform distribution of Fe within TiO_2. The anatase:rutile phase ratio calculated by Spurr and Myers' method showed that zero-iron TiO_2 and Fe-TiO_2 materials were a mixture of anatase and rutile phases (Table 2). The amount of anatase was less in Fe-TiO_2 materials than in Aeroxide® TiO_2 P25. The smaller proportion of anatase could lead to a reduction of photocatalytic activity because the anatase phase has higher photocatalytic activity than rutile TiO_2 [50,51]. However, it is accepted that the optimal photocatalytic activity of TiO_2 is reached with an optimal mixture of anatase and rutile phases [52]. Moreover, the increased anatase proportion in 0.3 Fe-TiO_2 and 0.6 Fe-TiO_2 compared with zero-iron TiO_2 could improve photocatalytic activity. The increased anatase proportion was attributable to Fe doping disturbing the arrangements of TiO_2 phases [53]. This trend has also been observed when Fe-doped TiO_2 was synthesized using sol-gel [54] or co-precipitation methods [32].

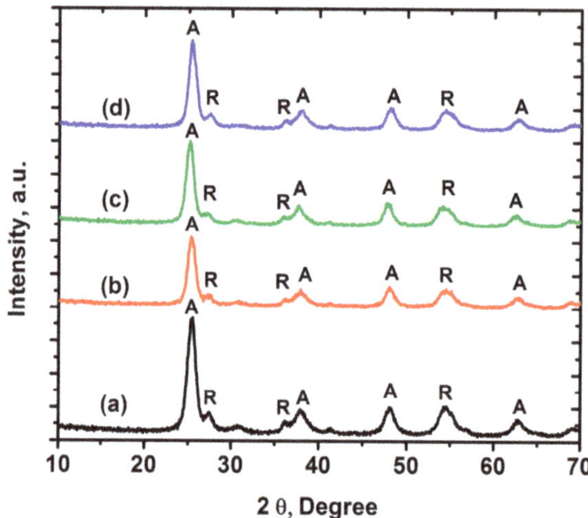

Figure 5. XRD patterns for zero-iron TiO_2 (**a**), 0.3 Fe-TiO_2 (**b**), 0.6 Fe-TiO_2 (**c**), and 1.0 Fe-TiO_2 (**d**), where A is Anatase and R is Rutile phases.

The average particle size of Fe-TiO_2 materials obtained by Scherrer's formula was 6.9 nm, which is less than the particle size of Aeroxide® TiO_2 P25 (Table 2). Fe-TiO_2 materials should increase photocatalytic activity because of their higher surface area and the short migration distance of the photogenerated charge carriers (electron/hole (e^-/h^+)) from the bulk material to the surface.

Further BET analysis (Figure 6) confirmed that average surface area of Fe-TiO_2 materials was 77.9 m^2 g^{-1}, higher than zero-iron TiO_2 and Aeroxide® TiO_2 P25. BET isotherms followed a type IV shape according to the Langmuir classification, which is associated with the characteristics of mesoporous material [55]. The observed hysteresis is probably due to gas cooperative adsorption or condensation inside the pores of material [56]. BET analysis showed pore sizes (Table 2) were in the mesoporous range (2–50 nm, according to IUPAC classification) for zero-iron TiO_2 and 1.0 Fe-TiO_2, and the microporous range (0.2–2 nm, according to IUPAC classification) for 0.3 Fe-TiO_2 and 0.6 Fe-TiO_2. Mesoporous pore size should facilitate the mass transfer of reactants and products in the reaction system, so photocatalytic improvement based on this property could improve zero-iron TiO_2 and Fe-TiO_2 materials with respect to Aeroxide® TiO_2 P25 [31].

Patra et al. [49] developed a similar nanoparticle synthesis procedure, which generated surface area values ranging from 126 to 385 m^2 g^{-1} and mesoporous size distribution values ranging from 3.1 to 3.4 nm. Particles obtained in our work were different, probably because of the application of a mild thermal treatment and the use of SDS at critical micelle concentration as a template.

Figure 7 shows SEM images of agglomerated and assembled nanoparticles of zero-iron TiO_2. The different amounts of Fe in the TiO_2 lattice changed neither the particle size nor the morphology of the zero-iron TiO_2. Although the average pore size allowed an increase of the superficial area, agglomeration could lead to lower photocatalytic activity.

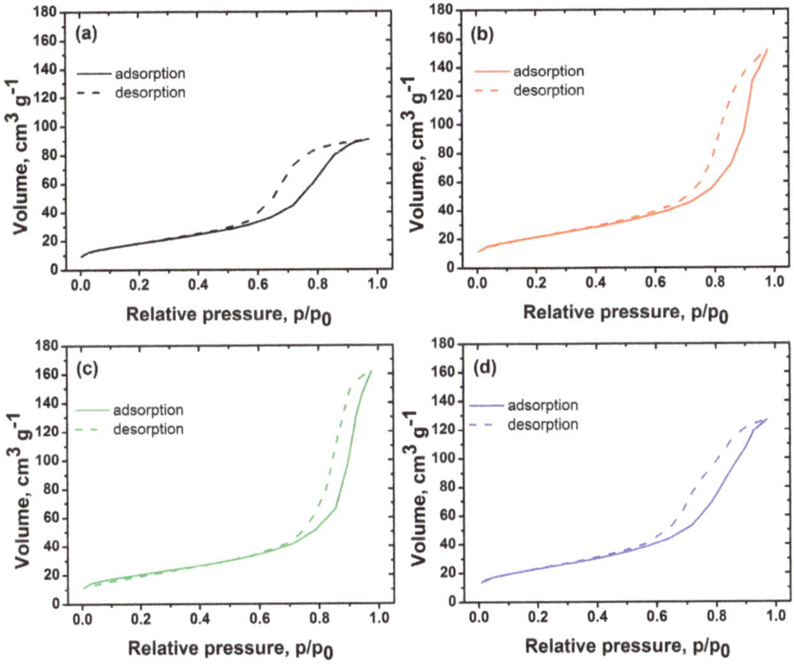

Figure 6. Brunauer–Emmett–Teller (BET) isotherms for zero-iron TiO_2 (**a**), 0.3 Fe-TiO_2 (**b**), 0.6 Fe-TiO_2 (**c**), and 1.0 Fe-TiO_2 (**d**).

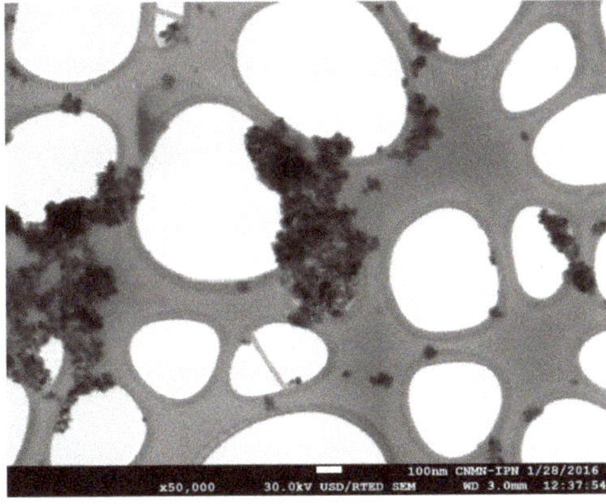

Figure 7. SEM image of zero-iron TiO_2 after mechanical grinding and sonication.

Transmission electron microscopy (TEM) images confirmed nanoparticle clusters and particle sizes of zero-iron TiO_2 (Figure 8b) and 0.3 Fe-TiO_2 (Figure 8a) between 5 and 10 nm (between 1.2 and 9.4 nm according to Scherrer's formula). The lattice fringe spacing was 0.35 nm, as shown in Figure 8b, which was consistent with the d-spacing (101) of anatase [25]. The lattice fingers of the nanoparticles showed that Fe-TiO_2 materials were highly crystallized.

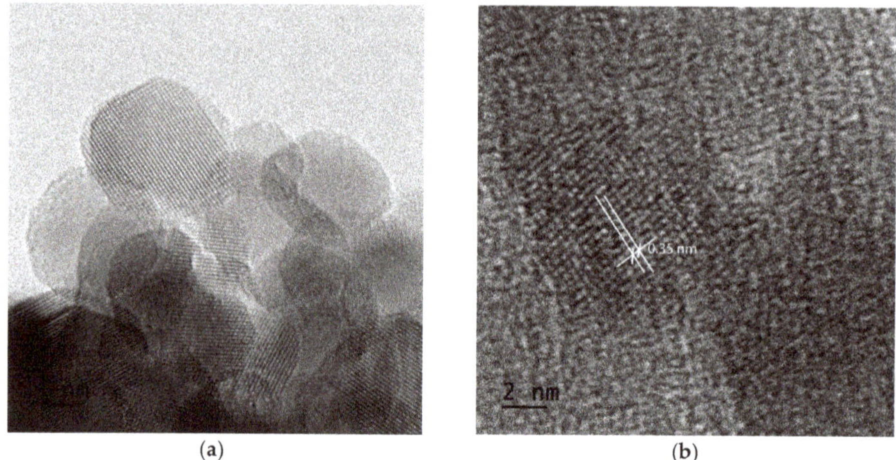

Figure 8. Transmission electron microscopy (TEM) image of 0.3 Fe-TiO$_2$ (**a**) and zero-iron TiO$_2$ (**b**).

2.2. Characterization of Irradiation Source

Figure 9 shows the emission spectra of irradiation sources used in this study. Using the main peaks reported for a fluorescent lamp (Figure 9a), the calibration of the spectrometer generated an R^2 value equal to 0.999. The emission spectrum of the GE F15T8 BLB lamp (Figure 9b) was in the 356–410 nm range. However, the emission spectrum of the GE F15T8 D lamp (Figure 9c) was continuous broadband between 380 and 750 nm. The light intensity of the GE F15T8 lamp was reported to be between 3440 µW cm^{-2} [57] and 4000 µW cm^{-2} [58], from which 6% was UV radiation [59]. The intensity of the GE F15T8 lamp was 1500 µW cm^{-2}. This lamp has an internal coating that absorbs 78% of visible light (as specified by the manufacturer) in the spectrum below 400 nm, as shown in Figure 9b. Therefore, the GE F15T8 BLB and GE F15T8 D lamps were designated as high and low UV irradiation sources, respectively.

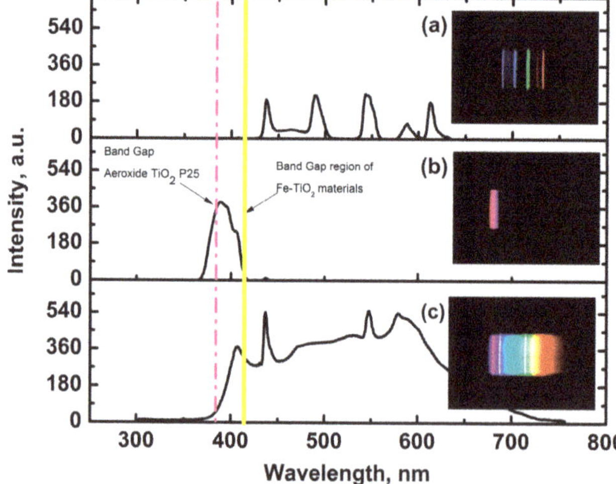

Figure 9. Emission spectrum and intensity graph of the irradiation source of Tecnolite fluorescent lamp (**a**), GE F15T8 BLB lamp (**b**), and GE F15T8 D lamp (**c**).

Because E_g of Aeroxide$^{®}$ TiO$_2$ P25 is 3.2 eV (387.5 nm), see Figure 9, both the GE F15T8 BLB and GE F15T8 D lamps emitted photons that could photoactivate Aeroxide$^{®}$ TiO$_2$ P25. However, the proportion of the emission spectrum that Aeroxide$^{®}$ TiO$_2$ P25 could use for photocatalytic activity was different. An approximation of the amount of radiative intensity used for photocatalytic activity was obtained with the area under the curve-spectrum below the E_g value. Consequently, Aeroxide$^{®}$ TiO$_2$ P25 could take advantage of 36.4% of the emission spectrum of the GE F15T8 BLB lamp and 0.8% of the emission spectrum of the GE F15T8 D lamp. Table 2 lists amount of radiative spectrum used by zero-iron TiO$_2$ and Fe-TiO$_2$ materials according to each E_g.

Based on morphological and crystalline structure analysis, the favorable characteristics to enhance photocatalytic activity of Fe-TiO$_2$ material are effective insertion of the Fe^{3+} ion into the TiO$_2$ lattice, red-shift (2.90–2.96 eV), nanoparticle size (6.9–7.1 nm), specific surface area (73.0–83.1 nm), pore size (1.2–9.4 nm), and radiation absorbance below the equivalent E_g wavelength (8.21–10.63% of daylight lamp spectrum). Its main disadvantageous characteristics are expected to be high particle agglomeration and lower anatase phase compared with zero-iron TiO$_2$. Further, photocatalytic activity is very sensitive to crystalline array and particle size and shape; differences in the density of hydroxyl groups on the particle surface and the number of water molecules hydrating the surface; the surface area and surface charge; differences in the number and nature of trap sites; the dopant concentration, localization, and chemical state of the dopant ions; radiation intensity; particle aggregation and superficial charge; and scavenger species in media [39,60]. Consequently, material characterization alone could not predict photocatalytic activity [28]. Therefore, in this research, we used the *N,N*-dimethyl-p-nitrosoaniline (pNDA) probe and E3 to evaluate the photocatalytic activity by following $^{•}$OH production, which is one of the most significant reactive oxygen species (ROS), and E3, which is an EC.

2.3. Hydroxyl Radical Generation under High and Low UV Irradiation

The generation of $^{•}$OH was measured using pNDA, which is a well-characterized $^{•}$OH scavenger as mentioned in Section 3.5. In brief, pNDA undergoes bleaching when reacting with $^{•}$OH according to Muff et al. mechanism of the oxidation of pNDA by $^{•}$OH [61].

In this work, pNDA bleaching followed a pseudo-first-order equation, so the apparent rate constant was calculated by $\ln(C/C_0) = k_1 t$, where C_0 is the initial concentration, C is the reaction concentration at a given time, and k_1 is the pseudo-first-order reaction rate constant. The slope of the plot after applying a linear fit represents the rate constant, k_1.

Because the relationship between pNDA bleaching and $^{•}$OH production follows a 1:1 stoichiometry [61], the steady state of $^{•}$OH generation ($[^{•}OH]_{ss}$) can be considered equal to the initial velocity (r_0) according to Equation (1) and reported in Table 3:

$$\left.\frac{[pNDA]}{dt}\right|_{t=0} = r_0 = [^{•}OH]_{ss} \tag{1}$$

Fe-TiO$_2$ materials showed a similar anatase:rutile phase ratio, particle size, and specific surface area, and therefore the variation in r_0 values was due to the difference of Fe content inside TiO$_2$. The generation of $^{•}$OH radicals (r_0) was feasible using zero-iron TiO$_2$, Fe-TiO$_2$ materials, and Aeroxide$^{®}$ TiO$_2$ P25 under both high (Figure 10a) and low UV irradiation (Figure 10b).

Table 3. $^{\bullet}$OH generation rate of zero-iron TiO_2 and Fe-TiO_2.

Catalyst	at.%	Load	High UV Irradiation			Low UV Irradiation		
			k_1	R^2	r_0	k_1	R^2	r_0
		$mg\ L^{-1}$	min^{-1}		$\mu M_{\bullet OH}\ min^{-1}$	min^{-1}		$\mu M_{\bullet OH}\ min^{-1}$
TiO_2 Aeroxide$^{\circledR}$ P25	-	20	0.06	0.988	0.49	0.012	0.989	0.105
Zero-iron TiO_2	0	320	0.056	0.993	0.49	0.005	0.973	0.045
0.3 Fe-TiO_2	0.3	320	0.067	0.998	0.58	0.004	0.990	0.042
0.6 Fe-TiO_2	0.6	320	0.031	0.998	0.28	0.002	0.999	0.025
1.0 Fe-TiO_2	1	320	0.004	0.987	0.04	0.00002	0.891	0.0002

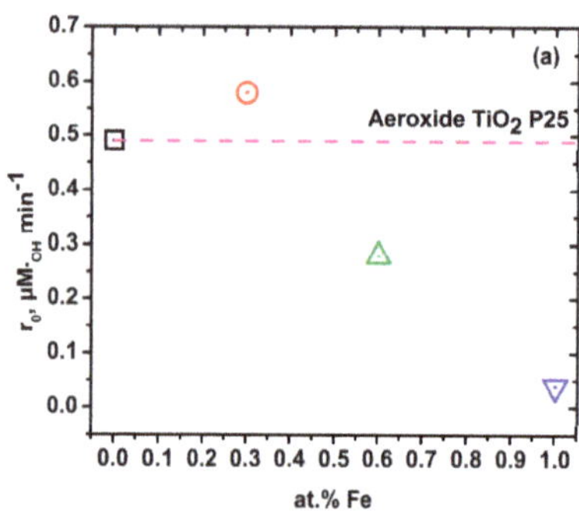

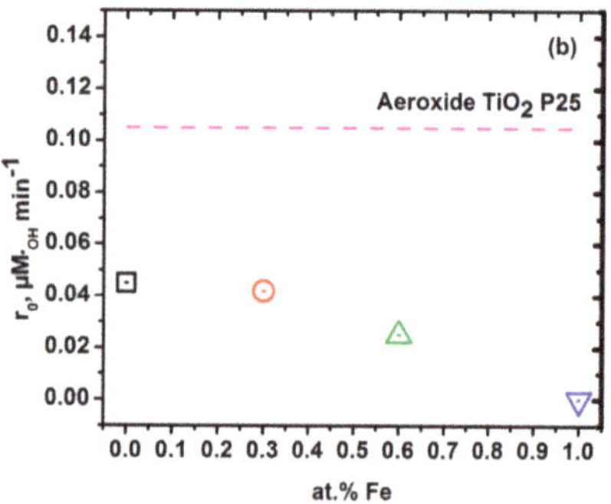

Figure 10. $^{\bullet}$OH generation (initial velocity) of zero-iron TiO_2 and Fe-TiO_2 under high UV irradiation (**a**) and low UV Irradiation (**b**); where □ zero-iron TiO_2, ☉ 0.3 Fe-TiO_2, △ 0.6 Fe-TiO_2, and 1.0 Fe-TiO_2 ▽ at pH 6 ± 0.1, and 20 °C.

When high UV irradiation was used, the maximum r_0 was 0.58 $\mu M_{\bullet OH}\ min^{-1}$ for 0.3 Fe-TiO_2. The enhancement in photocatalytic activity of 0.3 at.% Fe-TiO_2, compared with zero-iron TiO_2 was by the extended lifetime values of the photogenerated charge carriers (e^- and h^+) produced by Fe^{3+} ions, which played a role as charge carriers trapped at or near the particle surface. The trapping mechanisms are shown in Equations (2)–(5) [62].

$$Fe^{3+} + e_{cd}^{-} \rightarrow Fe^{2+} \qquad \text{electron trap} \tag{2}$$

$$Fe^{2+} + Ti^{4+} \rightarrow Fe^{3+} + Ti^{3+} \qquad \text{migration} \tag{3}$$

$$Fe^{3+} + h_{vb}^{+} \rightarrow Fe^{4+} \qquad \text{hole trap} \tag{4}$$

$$Fe^{4+} + OH^{-} \rightarrow Fe^{3+} + {}^{\bullet}OH \qquad \text{migration} \tag{5}$$

The mechanism suggested for $^{\bullet}$OH generation is shown in Figure 11. When TiO_2 contains a Fe^{3+} ion, the Fe3d orbitals split into two bands, one is a hybrid band (A2g) and one is midgap band (T2g), which induce a new localized BG state [23]. Therefore, when TiO_2 absorbs photons with energy less than 3.2 eV, photoexcitation of the semiconductor promotes an electron from the VB to the midgap band (T2g), also called a shallow trap, creating an electron-hole pair. The hole in the valence band (VB) can react with hydroxide ions to form $^{\bullet}$OH, absorbed organic molecules, or trap Fe^{3+} following Equations (4) and (5). Additionally, photogenerated electrons in the midgap band (T2g) can be transferred to Fe^{3+} following a dark redox reaction at the interface, as suggested by Neubert et al. [63] and consequently bring about $^{\bullet}$OH.

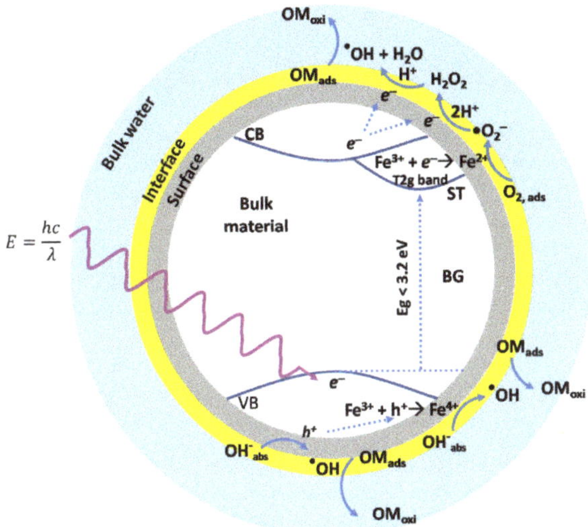

Figure 11. Photocatalytic mechanism of Fe-TiO$_2$ and $^\bullet$OH generation. E_g is band gap energy, E is photon energy, OM$_{ads}$ is adsorbed organic molecule, OM$_{oxi}$ is oxidized organic molecule.

Increasing the Fe^{3+} doping content of Fe-TiO$_2$ to 0.6 and 1.0 at.%, Fe-TiO$_2$ was unfavorable to the photocatalytic activity because the additional Fe^{3+} doping in the TiO$_2$ sample inhibited the extended lifetime of charge carriers, acted as recombination sites and consequently decreased the photocatalytic efficiency [29], as proposed in Equations (6)–(9) [39].

$$Fe^{2+} + h_{vb}^{+} \rightarrow Fe^{3+} \qquad \text{recombination} \tag{6}$$

$$Fe^{4+} + e_{cd}^{-} \rightarrow Fe^{3+} \qquad \text{recombination} \tag{7}$$

$$Fe^{4+} + Fe^{2+} \rightarrow 2Fe^{3+} \qquad \text{recombination} \tag{8}$$

$$Fe^{4+} + Ti^{3+} \rightarrow Fe^{3+} + Ti^{4+} \qquad \text{recombination} \tag{9}$$

When low UV irradiation conditions were used, the r_0 values for zero-iron TiO$_2$ and Fe-TiO$_2$ materials were lower than the value estimated for Aeroxide$^®$ TiO$_2$ P25. Compared with the effects of high UV irradiation, the reduction in r_0 value observed was related both to pNDA adsorption of UV-visible radiation (lowered the number of photons available to activate the photocatalyst), and the augmented Fe content, which increased the recombination rate.

2.4. Photocatalytic Degradation of Estriol under High and Low UV Irradiation

E3 photocatalytic degradation curves are shown in Figure 12a,b using both high and low UV irradiation, respectively. In both cases, E3 photocatalytic degradation followed a pseudo-first-order model and the rate constant, k_1 (Table 4), was obtained by fitting experimental data to ln ([E3]/[E3$_0$]) = k_1t. Fe content influenced k_1 for both high and low UV irradiation.

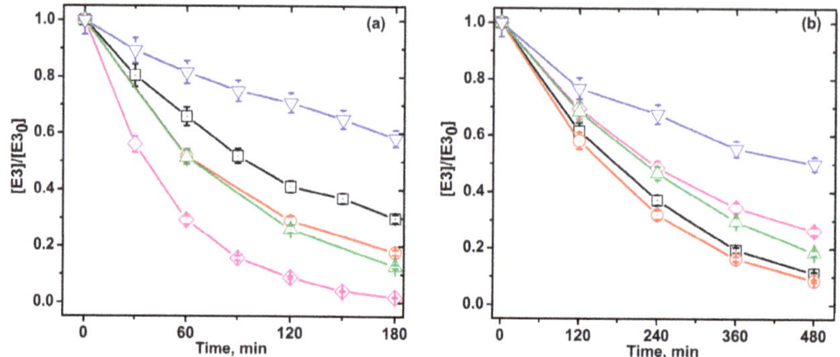

Figure 12. Photocatalytic degradation of E3 under high UV irradiation (**a**), and low UV irradiation (**b**); where ▣ zero-iron TiO_2, ⊙ 0.3 Fe-TiO_2, △ 0.6 Fe-TiO_2, ▽ 1.0 Fe-TiO_2, and ◇ Aeroxide® TiO_2 P25; at pH 6 ± 0.1, and 20 °C.

Table 4. Kinetic values of E3 degradation using zero-iron TiO_2 and Fe-TiO_2.

Catalyst	Load	High UV Irradiation			Low UV Irradiation		
		k_1	R^2	$r_{0,E3}$	k_1	R^2	$r_{0,E3}$
	mg L^{-1}	min^{-1}		μM_{E3} min^{-1}	min^{-1}		μM_{E3} min^{-1}
TiO_2 Aeroxide® P25	20	0.021	0.996	0.21	0.0029	0.992	0.030
Zero-iron TiO_2	320	0.007	0.997	0.069	0.0045	0.991	0.040
0.3 Fe-TiO_2	320	0.009	0.994	0.090	0.0050	0.992	0.042
0.6 Fe-TiO_2	320	0.011	0.997	0.099	0.0034	0.999	0.030
1.0 Fe-TiO_2	320	0.003	0.979	0.027	0.0016	0.987	0.012

Figure 13 shows the pseudo-first-order rate constant (k_1) of E3 photocatalytic degradation. In general, the photocatalytic activity first increased and then decreased as the Fe concentration increased, which is similar to the behavior found with the •OH probe in Section 2.3 and has been previously reported using other organic molecules [23,29,64].

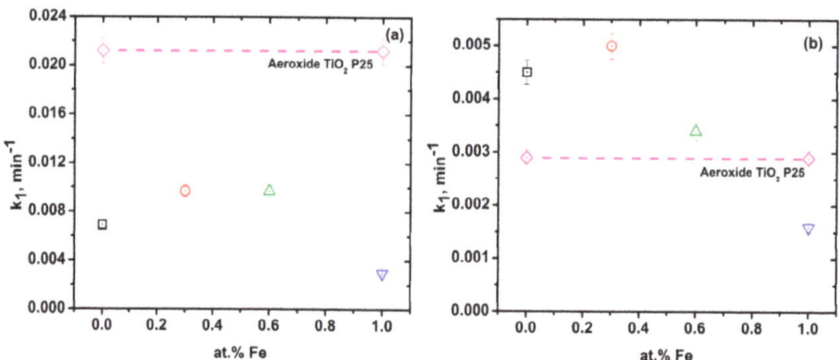

Figure 13. Photocatalytic reaction rate (k_1) for degradation of E3 under high UV irradiation (**a**), and low UV irradiation (**b**); where ▣ zero-iron TiO_2, ⊙ 0.3 Fe-TiO_2, △ 0.6 Fe-TiO_2, and ▽ 1.0 Fe-TiO_2; at pH 6 ± 0.1, and 20 °C.

Under high UV irradiation (Figure 13a), 0.6 Fe-TiO_2 k_1 was higher than for zero-iron TiO_2, 0.3 Fe-TiO_2, and 1.0 Fe-TiO_2. The increase in photocatalytic performance of 0.6 Fe-TiO_2 was related with the increase in the lifetime of electron-hole pairs because Fe created additional energy levels near the conduction band of TiO_2, as the mechanism suggests in Figure 11.

Under low UV irradiation (Figure 13b), zero-iron TiO_2, 0.3 Fe-TiO_2, and 0.6 Fe-TiO_2 showed more photocatalytic activity than Aeroxide® TiO_2 P25 because those materials had enhanced superficial properties, such as particle size, and superficial area, as mentioned in Section 2.1. Furthermore, 0.3 Fe-TiO_2 enhanced photocatalytic activities with k_1 values as high as 0.005 min^{-1}. The high photocatalytic activity of 0.3 Fe-TiO_2 was due to the synergistic effect of unintentionally added co-dopants, superficial properties, and Fe content that increased the lifetime of photogenerated charge carriers and the efficiency of electron transfer.

The photocatalytic degradation rate of E3 using Aeroxide® TiO_2 P25 was reported to be 0.25 min^{-1} [65], 0.134 min^{-1} [66], and 0.12 min^{-1} [67]. However, the experimental setups and catalyst loads were different. Besides these few studies, E3 degradation using Fe-TiO_2 nanoparticles is scarcely reported. Only comparing magnitudes of k_1, the first-order rates to degrade pharmaceuticals using Fe-TiO_2 nanoparticles were 0.001 min^{-1} for ibuprofen, 0.0015 min^{-1} for carbamazepine, and 0.0014 min^{-1} for sulfamethoxazole [68], which are in the order of magnitude obtained in this work (see Table 4).

Regarding unintentionally added co-dopants, Fe-TiO_2 co-doping demonstrated a synergistic effect to increase photocatalytic activity under visible light for sulfur [69], nitrogen [44], and $Fe_xTi_{1-x}O_{2-y}N_y$ co-doping [70]. Surface properties of the material, such as a particle size (6.9 nm) and surface area (77.6 $m^2\ g^{-1}$), also facilitated the mass transfer between interface, E3, and sub-products.

The relationship between the •OH radical system and E3 kinetic degradation was determined via linear fit between •OH initial rate generation ($r_{0,OH}$) and initial E3 degradation ($r_{0,E3}$). In general, the procedure to correlate $r_{0,OH}$ and $r_{0,E3}$ was first to sort pair values ($r_{0,OH}$, $r_{0,E3}$), and then fit the data to linear regression, as shown Figure 14a,b.

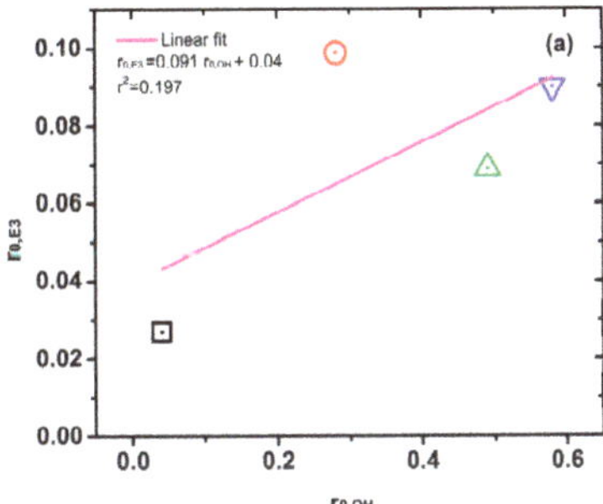
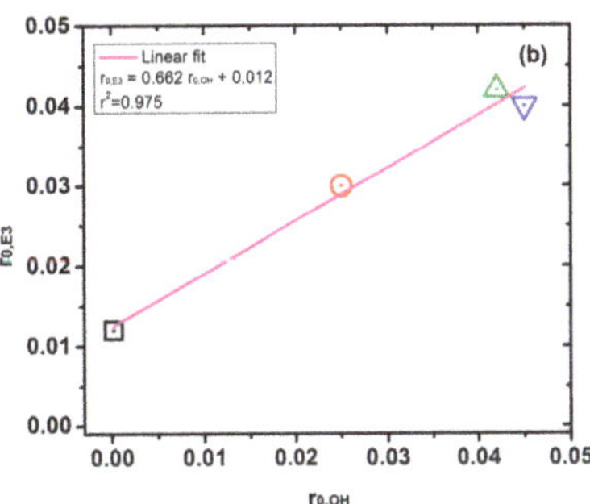

Figure 14. Correlation between •OH initial rate generation ($r_{0,OH}$) and initial E3 degradation ($r_{0,E3}$) under high UV irradiation (**a**), and low UV irradiation (**b**); where ▣ zero-iron TiO_2, ⊙ 0.3 Fe-TiO_2, △ 0.6 Fe-TiO_2, and ▽ 1.0 Fe-TiO_2; at pH 6 ± 0.1, and 20 °C.

Under high UV irradiation, the linear fit correlation was $r_{0,E3} = 0.091\ r_{0,OH} + 0.040$ with $R^2 = 0.197$. Under low UV irradiation, the linear fit correlation was $r_{0,E3} = 0.066\ r_{0,OH} + 0.012$ with $R^2 = 0.975$. The correlation between the pair ($r_{0,OH}$, $r_{0,E3}$) under high UV irradiation was too low to be considered a linear relationship. We suggest the low correlation was because not only •OH caused E3 degradation, but holes (h^+) or other reactive oxygen species also caused E3 degradation.

However, a linear relationship under low UV irradiation was attributable to •OH being the main reactive oxygen species responsible for photocatalytic activity. Therefore, the contribution of h^+ to photocatalytic activity was lower because oxidation power was lower due to reduced E_g. This suggestion supports the mechanisms proposed in Figure 11, in which adding Fe into the lattice of TiO_2 reduced the E_g with a consistent reduction of redox potential, as mentioned by others [28].

The main mechanism of E3 degradation under low UV irradiation was via electron (e^-) transfer to give rise •OH. Additionally, the enhanced photocatalytic activity of 0.3 Fe-TiO_2 under low UV irradiation provides evidence that the trapping-recombination mechanism of Fe-TiO$_2$ can be controlled by irradiation intensity. Therefore, we suggest that there is a trade-off between irradiation intensity, the trapping-recombination rate, and $^{\bullet}$OH production that is worthy of further research.

The efficiency resource of the Fe-TiO$_2$/Low UV system was obtained through dimensional analysis of the slope of the linear fit of data shown in Figure 14b. The units of slope are E3 moles degraded per $^{\bullet}$OH mol generated at initial time, so 0.662 E3 molecules underwent degradation when one $^{\bullet}$OH was generated for the photocatalytic system independent of Fe doping content in TiO$_2$. A sustainable process was also achieved, for which 0.3 Fe-TiO$_2$ since absorbed 8.21% of emission spectra of the lamp below the equivalent E_g wavelength over 0.8% or 7.64% of Aeroxide$^{®}$ TiO$_2$ P25 and zero-iron TiO$_2$, respectively.

2.5. Relationship between Fe Content and Kinetic Constant

Photonic efficiency has been suggested to increase linearly with the doping ratio due to the formation of the charge carrier trapping centers, while it concurrently decreases quadratically with the doping ratio because to the creation of recombination centers [71]. Alternatively, we suggest an empirical relationship between the E3 degradation pseudo-first-order rate constant (k_1) and Fe content (at.%) in TiO$_2$, as described in Equation (10):

$$k_1(\delta) = c\left[e^{-k_e(\delta+\alpha)} - e^{-k_a(\delta+\alpha)}\right] \tag{10}$$

where k_1 is the pseudo-first-order constant, k_e is the electron trap constant, k_a is the electron recombination constant, δ at.% is the Fe doping amount in TiO$_2$, and c and α are system constants. To solve the model described in Equation (10), a numerical approximation by root-mean-square error minimization method was used according to Equation (11):

$$\varepsilon = \sqrt{\frac{1}{n}\sum_i \left|\overline{[k_{1.i}]} - [k_{1.i}]\right|} \tag{11}$$

where $\overline{[k_{1.i}]}$ is the theoretical k_1 value, $[k_{1.i}]$ is the experimental k_1 value, n is the number of data, and ε is the root-mean-square error. The solution of Equation (10) was performed by simultaneously solving k_e, k_a, c, and α using Excel Solver$^{®}$ (Frontline Systems, NV, US). As an example, photocatalytic degradation of E3 under low UV irradiation was fitted to Equation (10), as shown in Figure 15.

The empirical model solved in Equation (12) shows that electron trap constant (k_e) overcome electron recombination (k_a) before optimal catalyst load. This model could lead to experimental work using iron-doped TiO$_2$ in which the optimal content of Fe gives rise to the maximum E3 degradation.

$$k_1(\delta) = -1.99\left[e^{-2.81(\delta+0.197)} - e^{-2.78(\delta+0.197)}\right] \tag{12}$$

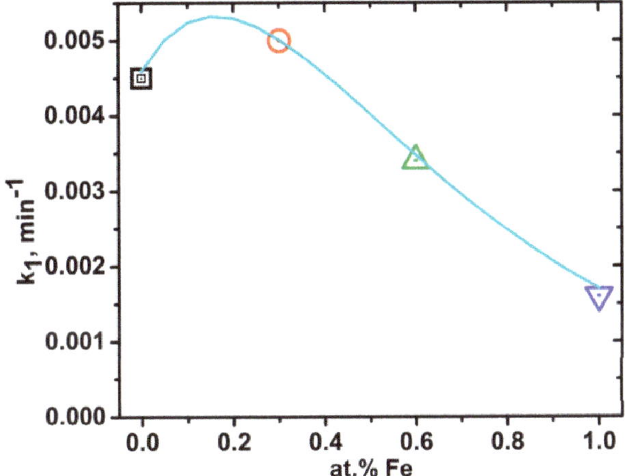

Figure 15. Experimental relationship between pseudo first order constant and at.% content; where ▣ zero-iron TiO_2, ⊙ 0.3 Fe-TiO_2, △ 0.6 Fe-TiO_2, and ▽ 1.0 Fe-TiO_2; at pH 6 ± 0.1; and 20 °C.

3. Materials and Methods

3.1. Reagents

Sigma-Aldrich (St. Louis, MO, USA) supplied estriol (E3, $C_{18}H_{24}O_3$, $\geq$97%), titanium isopropoxide (TTIP, $Ti[OCH(CH_3)_2]_4$, 97%), N,N-Dimethyl-4-nitrosoaniline (pNDA, also called RNO, $C_8H_{10}N_{2O}$, 97%), sodium dodecyl sulfate (SDS), and iron (III) nitrate (Fe(NO$_3$)$_3$·9H$_2$O, >99.99%). Aeroxide® TiO_2 P25 (formerly Degussa P25 with 50 ± 15 m^2 g^{-1} of the specific surface area, 21 nm of average particle size, 80:20 of anatase:rutile ratio according to the manufacturer) granted by Evonik Industries (Essen, Germany) was the photocatalytic standard. Fremont (CA, USA) supplied HNO_3, H_2SO_4, absolute ethanol, HPLC-grade methanol, and HPLC-grade water. All chemicals were used as received.

3.2. Photoreactor Setup

Figure 16 depicts the photoreactor, which was a cylindrical water-jacketed glass vessel (318 mL) with 102 mm and 63 mm of interior height and diameter, respectively. The horizontal and vertical position of the photoreactor was constant for all experiments. Lamps were set horizontally and centered above the photoreactor. Two 15 W GE F15T8 BLB lamps (also called black-light lamps, Boston, MA, USA) supplied high UV irradiation, and two 15 W GE F15T8 D lamps (also called daylight lamps) provided low UV irradiation. The overall system was in a closed box to avoid the effects of sunlight or any artificial radiation sources. Lamp emission spectra were measured using a lab-made spectrophotometer using a CMOS webcam with a diffraction grating of 1000 lines mm^{-1} [72,73]. Emission spectra calibration of the spectrophotometer was developed using a 9 W fluorescent lamp (Tecnolite, Jalisco, Mexico). The temperature of all experiments was set at 20 °C using a thermostatic bath with recirculation (Polystat, Cole-Palmer, Vernon Hills, IL, USA). An optical filter was not used in the experiments, so visible light condition was not simulated.

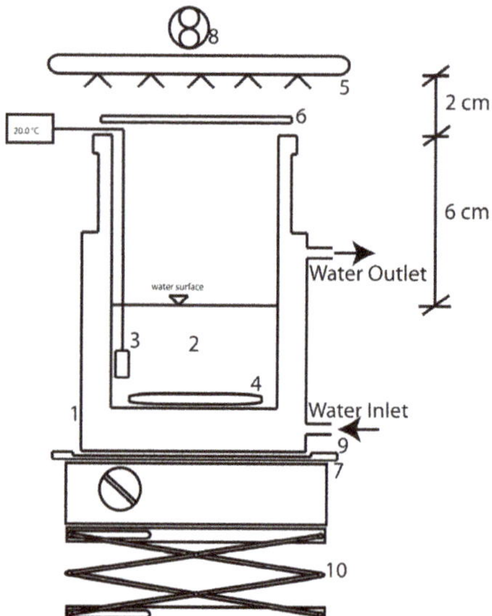

Figure 16. Scheme of photoreactor used for experiments: glass reactor (**1**), testing solution (**2**), temperature probe (**3**), spin bar (**4**), lamps (**5**), an optical filter (if needed) (**6**), stirring plate (**7**), cooling fan (**8**), horizontal position template (**9**), and lab jack lifting platform (**10**).

3.3. Synthesis of Materials

The synthesis method of iron-doped TiO$_2$ (Fe-TiO$_2$) materials followed the hydrothermal sol-gel synthetic approach proposed by Patra et al. with some differences in precursor and thermal treatment [49]. Our synthesis method used iron (III) nitrate instead of FeCl$_3$ and absolute ethanol instead of isopropyl alcohol. The thermal treatment was a programmed cycle of 31 h (increasing ramp-drying-increasing ramp-calcination-decreasing ramp) instead of direct calcination for 6 h. First, solution A was prepared by dissolving 1.44 g of SDS in 10 mL of deionized water. Then, four different solutions B were prepared to dissolve iron (III) nitrate in 2 mL of absolute ethanol ($\geq$99.8 %) and 3 mL of TTIP was added slowly. The amounts of iron (III) nitrate were 0, 0.4, 4.3, and 42.6 mg of Fe(NO$_3$)$_3$·9H$_2$O identified as zero-iron TiO$_2$, 0.3 Fe-TiO$_2$, 0.6 Fe-TiO$_2$, and 1.0 Fe-TiO$_2$, respectively. Once ready, solution A was continuously stirred and solution B was slowly dropped into solution A. The pH of the resulting mixture was adjusted to 1 using concentrated HNO$_3$ and stirred for 3 h. The mixture was kept at 3 °C for 36 h. The precipitated solid was collected by filtration using Whatman Quantitative Filter Paper Grade 42. The materials were simultaneously dried and calcinated with a programmed thermal treatment (Isotemp® Programmable Muffle Furnace, Fisher Scientific, Dubuque, IA, USA) following first the temperature increase from ambient temperature to 353 K, with a temperature ramp of 1 K min^{-1} that was held for 720 min. The temperature was then increased from 353 K to 773 K with a temperature ramp of 1 K min^{-1} that was held for 360 min. Finally, the temperature was decreased from 773 K to 353 K with a temperature ramp of -1 K min^{-1}, and then the furnace was turned off. The materials were washed with 50:50 methanol-water and dried to 377 K overnight.

3.4. Materials Characterization

X-ray photoelectron spectroscopy (XPS) was performed using a Thermo Fisher Scientific K-Alpha X-ray photoelectron spectrometer (Waltham, MA, USA) with a monochromatized Al K$_\alpha$ X-ray source (1487 V). The deconvolution of high-resolution XPS spectra was developed using the software XPSpeak 4.1. (Raymund W.M. Kwok, Shatin, Hong Kong).

UV-visible reflectance spectroscopy was obtained with Video–Barrelino integrating sphere coupled to Cary 50 spectrophotometer (Varian Inc, Palo Alto, CA, USA). Diffuse reflectance spectra were transformed using the Kubelka–Munk method to obtain E_g of zero-iron TiO_2 and Fe-TiO_2 materials. Kubelka–Munk method plots $(F(R)hv)^{1/2}$ versus hv, draws a tangent at the inflection point on the curve and estimates E_g with the hv value at the intersection with abscissa. In this case, $F(R)$ is a reflectance function equal to $(1 - R)^2/2R$, R is the reflectance percentage, h is the Planck's constant, and v is frequency.

XRD patterns were recorded in a Siemens D-5000 diffractometer (Munich, Germany) using Cu K_α radiation (λ = 1.54060 Å) from 10° to 85°. The procedure for phase identification used the QualX2.0 software with database developed by Altomare et al. [74]. The cards used for identification were 00-901-5929, 00-900-1681, and 00-900-4140 for anatase, rutile, and brookite, respectively. The quantification phases followed the method proposed by Spurr and Myers according to Equation (13):

$$f = \frac{1}{1 + 1.26\frac{I_R}{I_A}} \tag{13}$$

where f is the anatase percentage, I_A is intensity at a diffraction angle 2θ of 25.36°, and I_R is intensity at a diffraction angle 2θ of 27.46° [75].

The particle size was estimated by Scherrer's formula described in Equation (14), where β is the full width at half of the maximum of the diffraction peaks (radians), k is the shape constant, λ is the wavelength of the incident Cu K_α radiation (λ = 1.54060 Å), θ is the Bragg's angle (radians), and D is the particle size (Å).

$$D = \frac{k\,\lambda}{\beta\,\cos\theta} \tag{14}$$

Brunauer–Emmett–Teller (BET) isotherms were obtained in Nova Station A equipment (Quantachrome Instruments, Boynton Beach, FL, USA). The surface morphology was observed by SEM in a JEOL ultrahigh resolution field emission electron microscope JSM-7800 F (JEOL, Tokyo, Japan) with 20 kV accelerating voltage, and 3 mm WD. Transmission electron microscopy (TEM) images were obtained in a JEM-2100 LaB6 electron microscope (JEOL, Tokyo, Japan).

3.5. Hydroxyl Radical Generation

In this study, pNDA bleaching was selected as an $^\bullet$OH probe because pNDA was useful for measuring the photocatalytic performance of TiO_2 [51,76,77] because of the following advantages: (1) it is selective of the reaction of pNDA with $^\bullet$OH [78]; (2) its high reaction rate with $^\bullet$OH on the order of 10^{10} M^{-1} s^{-1} [51,79]; (3) its easy application through observable bleaching at 440 nm following Beer's Law, in which pNDA bleaching a yellowish solution to transparent; and (4) its 1:1 stoichiometry, meaning that one $^\bullet$OH can bleach one pNDA molecule [51,80–82].

The pNDA absorption (Figure 17) measurements were obtained using a UV-visible spectrophotometer (Hatch DR/4000U, Loveland, CO, USA) at 440 nm following Beer-Lambert law. The pNDA test solution was 10 µM initial concentration and pH 6.0 ± 0.1 adjusted using NaOH or HCl when needed. No buffer solutions were used because they can compete for $^\bullet$OH. Final pH was verified at the end of tests to discharge pH-pNDA bleaching.

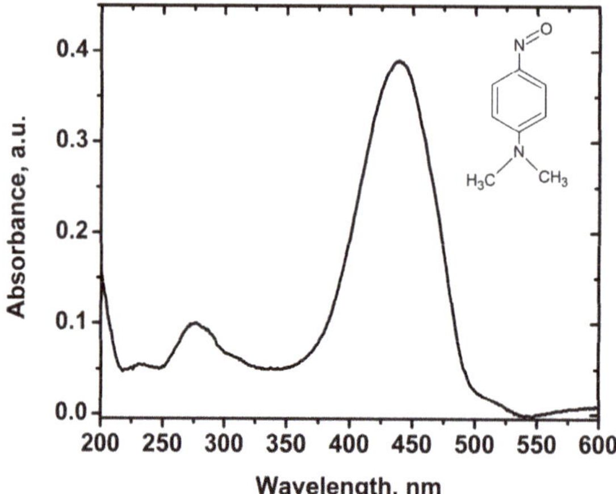

Figure 17. Structural formula and absorbance spectrum of *N*,*N*-dimethyl-p-nitrosoaniline (pNDA).

The photocatalytic standard was Aeroxide® TiO$_2$ P25, and the load was 20 mg L^{-1}. The choice of catalyst load was based on our previous work on $^\bullet$OH generation of Aeroxide® TiO$_2$ P25 [16]. For zero-iron TiO$_2$ and Fe-TiO$_2$ materials, the catalyst load used was 320 mg L^{-1}, which produced a $^\bullet$OH generation rate under high UV irradiation to set a baseline. Catalyst load differences were attributable to the aggregation of lab-made TiO$_2$, superficial properties, and optical properties of suspensions, as shown in Figure 18.

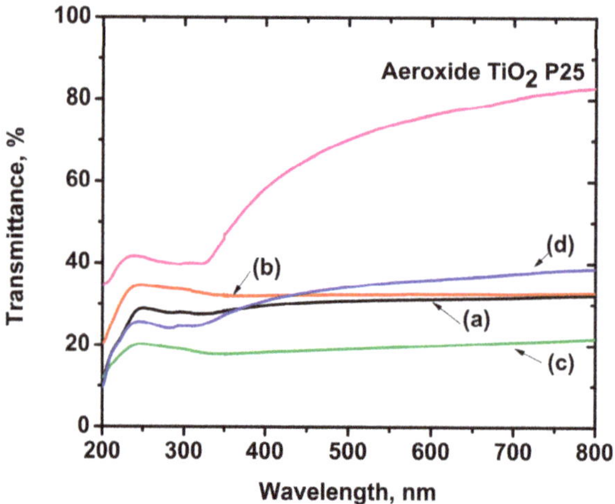

Figure 18. Suspension transmittance of Fe-TiO$_2$ material and Aeroxide TiO$_2$ P25; where zero-iron TiO$_2$ (**a**), 0.3 Fe-TiO$_2$ (**b**), 0.6 Fe-TiO$_2$ (**c**), and 1.0 Fe-TiO$_2$ (**d**).

The photocatalytic experiments were conducted as follows. First, a pNDA test solution was set at 20 °C, the catalyst was added, and the suspension was mixed for 20 min without radiation. To evaluate the adsorption of pNDA on TiO$_2$, an aliquot was withdrawn and centrifuged. Then, the system was fully illuminated, and aliquots were withdrawn after specific periods. Each sample was centrifuged at 6000 rpm for 15 min (Biofuge Primo, Sorvall, Hanau, Germany) and measured in the UV-visible spectrophotometer. Once the catalyst load was used and after the dark phase, no adsorption of pNDA was detected near the detection limit of UV-visible spectrophotometer.

3.6. Photolysis and Photocatalytic Degradation of E3

The initial E3 concentration was 10 µM because (1) this research was part of a project focused on the removal of E3 in water using sequentially coupled membrane filtration; (2) the solubility limit of E3 in water was previously reported to be 11.1 µM [83], and 45.1 µM [8,84], and (3) the sensitivity of the analytical techniques used in this work. The E3 solution was prepared to dissolve 2.88 mg of E3 in 1 L of deionized water by stirring at room conditions in the dark for six hours. Working solutions were stored in an amber flask.

Each photocatalytic experiment used 100 mL of E3 working solution. Initial pH was adjusted to obtain a similar surface charge of TiO_2 [85]. Depending on the initial water conditions, the initial pH value was adjusted to 6.0 ± 0.1 using NaOH or HCl when needed. A dark period (no radiation) was allowed for 20 min. Then, similar experimental conditions were carried out as described in Section 3.5. Additionally, the aliquots withdrawn from suspension were filtered using a 0.1 µm syringe filter (MillexVV, Millipore, Billerica, MA, USA). A blank experiment without irradiation and TiO_2 photocatalyst was conducted for comparison. The blank experiment showed that E3 cannot be degraded in absences of either TiO_2 or UV light. Once the catalyst was loaded and after the dark phase, no adsorption of E3 was detected near the detection limit of HPLC.

3.7. Analytical Methods

The E3 concentration was monitored using an HPLC system (Waters 1515; Milford, MA, USA) equipped with a UV detector (Waters 2787) that has an injection volume of 20 µL. The analytical method was performed in isocratic analytical mode using an Inertsil® ODS-3 column (GL Science, Tokyo, Japan; 150 mm × 4.6 mm, 5 µm) thermostated at 25 °C. The wavelength was at 280 nm according to E3 maximum absorbance. The mobile phase was methanol (49%) and deionized water (51%) at a flow rate of 1 mL min^{-1}. The retention time of E3 was 10 min, and the limit of E3 detection was 0.1 µM (0.029 mg L^{-1}). The detection limit was obtained by developing two calibration curves: the first between 10 and 0.1 and second between 1 and 0.01. Both calibration curves followed area = 2928[E3] with R^2 = 0.9899, but areas below 0.1 were not detected.

4. Conclusions

This study provided an understanding of the relationship between the Fe doping ratio and radiation intensity for •OH generation and estriol (E3) degradation. The main results were that:

- E3 degradation using 0.3 $Fe\text{-}TiO_2$ was feasible and can be improved by controlling irradiation intensity which was found closely related with light absorption and the catalytic reaction rate;
- the synthesis method and thermal treatment allowed nanoparticles with large superficial areas and the incorporation of iron ions into the TiO_2 lattice.; and
- changes in trapping recombination centers could be controlled with irradiation intensity to enhance the photocatalytic activity.

Therefore, our findings provide the opportunity to reconsider studies in which iron-doped TiO_2 impaired photocatalytic activity and to improve an application in which irradiation should be controlled. For example, $Fe\text{-}TiO_2$ can potentially be applied to medical uses in which low irradiation intensity should be used to avoid adverse effects in humans or wildlife, which has also been suggested by others [86]. In the field of water treatment, we propose that $Fe\text{-}TiO_2$ is an efficient material that could harvest low-energy photons to degrade and mineralize dyes [87], biocides [88], pharmaceuticals [89], industrial chemicals [90], and estrogens—as shown in this study—to create an energetically green water treatment process.

Author Contributions: Funding acquisition, E.R.B.; Investigation, I.M.R.-S.; Project administration, E.R.B.; Supervision, E.R.B.; Writing—original draft, I.M.R.-S.; Writing—review & editing, E.R.B.

Funding: This manuscript is based on work supported in part by ConTex postdoctoral program, which is an initiative of the University of Texas System and Mexico's National Council of Science and Technology (CONACYT). The research was partially funded by CONACYT under Project CB-2011/168285. The APC was funded by the Institutional Open Access Program (IAOP) between The University of Texas at Austin and Desert Research Institute (DRI) at Nevada.

Acknowledgments: The Aeroxide® P25 Evonik catalyst used for this work was provided by Intertrade S.A. de C.V., the supplier of Evonik Industries in Mexico. The authors thank L. Lartundo-Rojas, Raul Borja Urbi, Hugo Martinez Gutiérrez, and Joao Jairzinho Salinas Camargo for assistance in XPS spectroscopy, TEM images, SEM images, and absorption isotherms, respectively, all of whom are from Centro de Nanociencias y Micro y Nanotecnología (CNMN) of IPN, Mexico. The authors thank M.A. Quiroz Alfaro for his excellent technical help and for his permission to use materials and equipment at the UDLAP's electrochemical lab. The authors also thank Nicole Damon (DRI) for her editorial review.

Conflicts of Interest: The authors declare no conflict of interest.

References

1. Dulio, V.; van Bavel, B.; Brorström-Lundén, E.; Harmsen, J.; Hollender, J.; Schlabach, M.; Slobodnik, J.; Thomas, K.; Koschorreck, J. Emerging pollutants in the EU: 10 years of NORMAN in support of environmental policies and regulations. *Environ. Sci. Eur.* **2018**, *30*, 5. [CrossRef]

2. Mandaric, L.; Celic, M.; Marcé, R.; Petrovic, M. Introduction on Emerging Contaminants in Rivers and Their Environmental Risk. In *Emerging Contaminants in River Ecosystems: Occurrence and Effects under Multiple Stress Conditions*; Petrovic, M., Sabater, S., Elosegi, A., Barceló, D., Eds.; Springer International Publishing: Cham, Switzerland, 2016; pp. 3–25. ISBN 978-3-319-29376-9.

3. Houtman, C.J.; Legler, J.; Thomas, K. *Effect-Directed Analysis of Complex Environmental Contamination*; Brack, W., Ed.; Springer: Berlin/Heidelberg, Germany, 2011; pp. 237–265. ISBN 978-3-642-18384-3.

4. Dimogerontas, G.; Liapi, C. Endocrine Disruptors (Xenoestrogens): An Overview. In *Plastics in Dentistry and Estrogenicity: A Guide to Safe Practice*; Eliades, T., Eliades, G., Eds.; Springer: Berlin/Heidelberg, Germany, 2014; pp. 3–48. ISBN 978-3-642-29687-1.

5. Hileman, B. Environmental Estrogens linked to Reproductive Abnormalities, Cancer. *Chem. Eng. News Arch.* **1994**, *72*, 19–23. [CrossRef]

6. Prat, N.; Rieradevall, M.; Barata, C.; Munné, A. The combined use of metrics of biological quality and biomarkers to detect the effects of reclaimed water on macroinvertebrate assemblages in the lower part of a polluted Mediterranean river (Llobregat River, NE Spain). *Ecol. Indic.* **2013**, *24*, 167–176. [CrossRef]

7. Souza, M.S.; Hallgren, P.; Balseiro, E.; Hansson, L.A. Low concentrations, potential ecological consequences: Synthetic estrogens alter life-history and demographic structures of aquatic invertebrates. *Environ. Pollut.* **2013**, *178*, 237–243. [CrossRef]

8. Silva, C.P.; Otero, M.; Esteves, V. Processes for the elimination of estrogenic steroid hormones from water: A review. *Environ. Pollut.* **2012**, *165*, 38–58. [CrossRef]

9. Rodriguez-Narvaez, O.M.; Peralta-Hernandez, J.M.; Goonetilleke, A.; Bandala, E.R. Treatment technologies for emerging contaminants in water: A review. *Chem. Eng. J.* **2017**, *323*, 361–380. [CrossRef]

10. Gągol, M.; Przyjazny, A.; Boczkaj, G. Wastewater treatment by means of advanced oxidation processes based on cavitation—A review. *Chem. Eng. J.* **2018**, *338*, 599–627. [CrossRef]

11. Boczkaj, G.; Fernandes, A. Wastewater treatment by means of advanced oxidation processes at basic pH conditions: A review. *Chem. Eng. J.* **2017**, *320*, 608–633. [CrossRef]

12. Coleman; Eggins, B.; Byrne, J.A.; Palmer, F.L.; King, E. Photocatalytic degradation of 17-β-oestradiol on immobilised TiO_2. *Appl. Catal. B Environ.* **2000**, *24*, L1–L5. [CrossRef]

13. Ramirez-Sanchez, I.M.; Mendez-Rojas, M.A.; Bandala, E.R. CHAPTER 25 Photocatalytic Degradation of Natural and Synthetic Estrogens with Semiconducting Nanoparticles. In *Advanced Environmental Analysis: Applications of Nanomaterials*; The Royal Society of Chemistry: London, UK, 2017; Volume 2, pp. 153–177. ISBN 978-1-78262-906-1.

14. Ohko, Y.; Iuchi, K.; Niwa, C.; Tatsuma, T.; Nakashima, T.; Iguchi, T.; Kubota, Y.; Fujishima, A. 17β-Estradiol Degradation by TiO_2 Photocatalysis as a Means of Reducing Estrogenic Activity. *Environ. Sci. Technol.* **2002**, *36*, 4175–4181. [CrossRef]

15. Coleman, H.M.; Routledge, E.J.; Sumpter, J.P.; Eggins, B.R.; Byrne, J.A. Rapid loss of estrogenicity of steroid estrogens by UVA photolysis and photocatalysis over an immobilised titanium dioxide catalyst. *Water Res.* **2004**, *38*, 3233–3240. [CrossRef]

16. Ramírez-Sánchez, I.M.; Tuberty, S.; Hambourger, M.; Bandala, E.R. Resource efficiency analysis for photocatalytic degradation and mineralization of estriol using TiO_2 nanoparticles. *Chemosphere* **2017**, *184*, 1270–1285. [CrossRef]

17. Hashimoto, K.; Irie, H.; Fujishima, A. Photocatalysis: A Historical Overview and Future Prospects. *Jpn. J. Appl. Phys.* **2005**, *44*, 8269–8285. [CrossRef]

18. Fujishima, A.; Zhang, X.; Tryk, D.A. TiO_2 photocatalysis and related surface phenomena. *Surf. Sci. Rep.* **2008**, *63*, 515–582. [CrossRef]

19. Tong, A.Y.C.; Braund, R.; Warren, D.S.; Peake, B.M. TiO_2-assisted photodegradation of pharmaceuticals—A review. *Cent. Eur. J. Chem.* **2012**, *10*, 989–1027. [CrossRef]

20. Cassaignon, S.; Colbeau-Justin, C.; Durupthy, O. Titanium dioxide in photocatalysis. In *Nanomaterials: A Danger or a Promise?: A Chemical and Biological Perspective*; Springer: London, UK, 2013; pp. 153–188. ISBN 9781447142133.

21. Augugliaro, V.; Loddo, V.; Pagliaro, M.; Palmisano, G.; Palmisano, L. *Clean by Light Irradiation: Practical Applications of Supported TiO2*; RSC Publishing: Cambridge, UK, 2010; ISBN 1847558704.

22. Etacheri, V.; Di Valentin, C.; Schneider, J.; Bahnemann, D.; Pillai, S.C. Visible-light activation of TiO_2 photocatalysts: Advances in theory and experiments. *J. Photochem. Photobiol. C Photochem. Rev.* **2015**, *25*, 1–29. [CrossRef]

23. Wen, L.; Liu, B.; Zhao, X.; Nakata, K.; Murakami, T.; Fujishima, A. Synthesis, Characterization, and Photocatalysis of Fe-Doped TiO_2: A Combined Experimental and Theoretical Study. *Int. J. Photoenergy* **2012**, *2012*, 1–10. [CrossRef]

24. Yu, H.; Irie, H.; Hashimoto, K. Conduction band energy level control of titanium dioxide: Toward an efficient visible-light-sensitive photocatalyst. *J. Am. Chem. Soc.* **2010**, *132*, 6898–6899. [CrossRef]

25. Choi, W.; Termin, A.; Hoffmann, M.R. The role of metal ion dopants in quantum-sized TiO_2: Correlation between photoreactivity and charge carrier recombination dynamics. *J. Phys. Chem.* **1994**, *98*, 13669–13679. [CrossRef]

26. Kaur, T.; Sraw, A.; Wanchoo, R.K.; Toor, A.P. Visible–Light Induced Photocatalytic Degradation of Fungicide with Fe and Si Doped TiO_2 Nanoparticles. *Mater. Today Proc.* **2016**, *3*, 354–361. [CrossRef]

27. Zhao, B.; Mele, G.; Pio, I.; Li, J.; Palmisano, L.; Vasapollo, G. Degradation of 4-nitrophenol (4-NP) using Fe-TiO_2 as a heterogeneous photo-Fenton catalyst. *J. Hazard. Mater.* **2010**, *176*, 569–574. [CrossRef]

28. Yalçın, Y.; Kılıç, M.; Çınar, Z. Fe^{+3}-doped TiO_2: A combined experimental and computational approach to the evaluation of visible light activity. *Appl. Catal. B Environ.* **2010**, *99*, 469–477. [CrossRef]

29. Cai, L.; Liao, X.; Shi, B. Using Collagen Fiber as a Template to Synthesize TiO_2 and Fe_x/TiO_2 Nanofibers and Their Catalytic Behaviors on the Visible Light-Assisted Degradation of Orange II. *Ind. Eng. Chem. Res.* **2010**, *49*, 3194–3199. [CrossRef]

30. Li, J.; Xu, J.; Dai, W.L.; Li, H.; Fan, K. Direct hydro-alcohol thermal synthesis of special core-shell structured Fe-doped titania microspheres with extended visible light response and enhanced photoactivity. *Appl. Catal. B Environ.* **2009**, *85*, 162–170. [CrossRef]

31. Tong, T.; Zhang, J.; Tian, B.; Chen, F.; He, D. Preparation of Fe^{3+}-doped TiO_2 catalysts by controlled hydrolysis of titanium alkoxide and study on their photocatalytic activity for methyl orange degradation. *J. Hazard. Mater.* **2008**, *155*, 572–579. [CrossRef]

32. Ambrus, Z.; Balázs, N.; Alapi, T.; Wittmann, G.; Sipos, P.; Dombi, A.; Mogyorósi, K. Synthesis, structure and photocatalytic properties of Fe(III)-doped TiO_2 prepared from $TiCl_3$. *Appl. Catal. B Environ.* **2008**, *81*, 27–37. [CrossRef]

33. Cong, Y.; Zhang, J.; Chen, F.; Anpo, M.; He, D. Preparation, photocatalytic activity, and mechanism of nano-TiO_2 Co-doped with nitrogen and iron (III). *J. Phys. Chem. C* **2007**, *111*, 10618–10623. [CrossRef]

34. Adán, C.; Bahamonde, A.; Fernández-García, M.; Martínez-Arias, A. Structure and activity of nanosized iron-doped anatase TiO_2 catalysts for phenol photocatalytic degradation. *Appl. Catal. B Environ.* **2007**, *72*, 11–17. [CrossRef]

35. Yamashita, H.; Harada, M.; Misaka, J.; Takeuchi, M.; Neppolian, B.; Anpo, M. Photocatalytic degradation of organic compounds diluted in water using visible light-responsive metal ion-implanted TiO_2 catalysts: Fe ion-implanted TiO_2. *Catal. Today* **2003**, *84*, 191–196. [CrossRef]

36. Li, X.; Yue, P.-L.; Kutal, C. Synthesis and photocatalytic oxidation properties of iron doped titanium dioxide nanosemiconductor particles. *New J. Chem.* **2003**, *27*, 1264. [CrossRef]

37. Zhang, Z.; Wang, C.-C.; Zakaria, R.; Ying, J.Y. Role of Particle Size in Nanocrystalline TiO_2-Based Photocatalysts. *J. Phys. Chem. B* **1998**, *102*, 10871–10878. [CrossRef]

38. Litter, M.I.; Navío, J.A. Photocatalytic properties of iron-doped titania semiconductors. *J. Photochem. Photobiol. A Chem.* **1996**, *98*, 171–181. [CrossRef]

39. Fàbrega, C.; Andreu, T.; Cabot, A.; Morante, J.R. Location and catalytic role of iron species in TiO_2:Fe photocatalysts: An EPR study. *J. Photochem. Photobiol. A Chem.* **2010**, *211*, 170–175. [CrossRef]

40. Seabra, M.P.; Salvado, I.M.M.; Labrincha, J.A. Pure and (zinc or iron) doped titania powders prepared by sol-gel and used as photocatalyst. *Ceram. Int.* **2011**, *37*, 3317–3322. [CrossRef]

41. Abazović, N.D.; Mirenghi, L.; Janković, I.A.; Bibić, N.; Šojić, D.V.; Abramović, B.F.; Čomor, M.I. Synthesis and characterization of rutile TiO_2 nanopowders doped with iron ions. *Nanoscale Res. Lett.* **2009**, *4*, 518–525. [CrossRef]

42. Geissen, V.; Mol, H.; Klumpp, E.; Umlauf, G.; Nadal, M.; van der Ploeg, M.; van de Zee, S.E.A.T.M.; Ritsema, C.J. Emerging pollutants in the environment: A challenge for water resource management. *Int. Soil Water Conserv. Res.* **2015**, *3*, 57–65. [CrossRef]

43. Lin, T.C.; Seshadri, G.; Kelber, J.A. A consistent method for quantitative XPS peak analysis of thin oxide films on clean polycrystalline iron surfaces. *Appl. Surf. Sci.* **1997**, *119*, 83–92. [CrossRef]

44. Xing, M.; Wu, Y.; Zhang, J.; Chen, F. Effect of synergy on the visible light activity of B, N and Fe co-doped TiO_2 for the degradation of MO. *Nanoscale* **2010**, *2*, 1233. [CrossRef] [PubMed]

45. Lopez, R.; Gomez, R. Band-gap energy estimation from diffuse reflectance measurements on sol-gel and commercial TiO_2: A comparative study. *J. Sol-Gel Sci. Technol.* **2012**, *61*, 1–7. [CrossRef]

46. Shi, J.; Chen, G.; Zeng, G.; Chen, A.; He, K.; Huang, Z.; Hu, L.; Zeng, J.; Wu, J.; Liu, W. Hydrothermal synthesis of graphene wrapped Fe-doped TiO_2 nanospheres with high photocatalysis performance. *Ceram. Int.* **2018**, *44*, 7473–7480. [CrossRef]

47. Yu, J.; Xiang, Q.; Zhou, M. Preparation, characterization and visible-light-driven photocatalytic activity of Fe-doped titania nanorods and first-principles study for electronic structures. *Appl. Catal. B Environ.* **2009**, *90*, 595–602. [CrossRef]

48. Goswami, P.; Ganguli, J.N. Evaluating the potential of a new titania precursor for the synthesis of mesoporous Fe-doped titania with enhanced photocatalytic activity. *Mater. Res. Bull.* **2012**, *47*, 2077–2084. [CrossRef]

49. Patra, A.K.; Dutta, A.; Bhaumik, A. Highly ordered mesoporous TiO_2-Fe_2O_3 mixed oxide synthesized by sol-gel pathway: An efficient and reusable heterogeneous catalyst for dehalogenation reaction. *ACS Appl. Mater. Interfaces* **2012**, *4*, 5022–5028. [CrossRef] [PubMed]

50. Luttrell, T.; Halpegamage, S.; Tao, J.; Kramer, A.; Sutter, E.; Batzill, M. Why is anatase a better photocatalyst than rutile?—Model studies on epitaxial TiO_2 films. *Sci. Rep.* **2014**, *4*, 4043. [CrossRef] [PubMed]

51. Zang, L.; Qu, P.; Zhao, J.; Shen, T.; Hidaka, H. Photocatalytic bleaching of p-nitrosodimethylaniline in TiO_2 aqueous suspensions: A kinetic treatment involving some primary events photoinduced on the particle surface. *J. Mol. Catal. A Chem.* **1997**, *120*, 235–245. [CrossRef]

52. Othman, S.H.; Abdul Rashid, S.; Mohd Ghazi, T.I.; Abdullah, N. Fe-Doped TiO_2 Nanoparticles Produced via MOCVD: Synthesis, Characterization, and Photocatalytic Activity. *J. Nanomater.* **2011**, *2011*, 1–8. [CrossRef]

53. Teoh, W.Y.; Amal, R.; Mädler, L.; Pratsinis, S.E. Flame sprayed visible light-active Fe-TiO_2 for photomineralisation of oxalic acid. *Catal. Today* **2007**, *120*, 203–213. [CrossRef]

54. Pongwan, P.; Inceesungvorn, B.; Wetchakun, K.; Phanichphant, S.; Wetchakun, N. Highly efficient visible-light-induced photocatalytic activity of Fe-doped TiO_2 nanoparticles. *Eng. J.* **2012**, *16*, 143–151. [CrossRef]

55. Kruk, M.; Jaroniec, M. Gas adsorption characterization of ordered organic-inorganic nanocomposite materials. *Chem. Mater.* **2001**, *13*, 3169–3183. [CrossRef]

56. Limousin, G.; Gaudet, J.P.; Charlet, L.; Szenknect, S.; Barthès, V.; Krimissa, M. Sorption isotherms: A review on physical bases, modeling and measurement. *Appl. Geochem.* **2007**, *22*, 249–275. [CrossRef]

57. Carvalho, T.C.; La Cruz, T.E.; Tábora, J.E. A photochemical kinetic model for solid dosage forms. *Eur. J. Pharm. Biopharm.* **2017**, *120*, 63–72. [CrossRef] [PubMed]

58. Daugherty, J.P.; Hixon, S.C.; Yielding, K.L. Direct in vitro photoaffinity labeling of DNA with daunorubicin, adriamycin, and rubidazone. *BBA Sect. Nucleic Acids Protein Synth.* **1979**, *565*, 13–21. [CrossRef]

59. Hartman, P.E.; Biggley, W.H. Breakthrough of ultraviolet light from various brands of fluorescent lamps: Lethal effects on DNA repair-defective bacteria. *Environ. Mol. Mutagen.* **1996**, *27*, 306–313. [CrossRef]

60. Serpone, N. Relative photonic efficiencies and quantum yields in heterogeneous photocatalysis. *J. Photochem. Photobiol. A Chem.* **1997**, *104*, 1–12. [CrossRef]

61. Muff, J.; Bennedsen, L.R.; Søgaard, E.G. Study of electrochemical bleaching of p-nitrosodimethylaniline and its role as hydroxyl radical probe compound. *J. Appl. Electrochem.* **2011**, *41*, 599–607. [CrossRef]

62. Zhu, J.; Zheng, W.; He, B.; Zhang, J.; Anpo, M. Characterization of Fe-TiO$_2$ photocatalysts synthesized by hydrothermal method and their photocatalytic reactivity for photodegradation of XRG dye diluted in water. *J. Mol. Catal. A Chem.* **2004**, *216*, 35–43. [CrossRef]

63. Neubert, S.; Mitoraj, D.; Shevlin, S.A.; Pulisova, P.; Heimann, M.; Du, Y.; Goh, G.K.L.; Pacia, M.; Kruczała, K.; Turner, S.; et al. Highly efficient rutile TiO$_2$ photocatalysts with single Cu(II) and Fe(III) surface catalytic sites. *J. Mater. Chem. A* **2016**. [CrossRef]

64. Zhou, M.; Yu, J.; Cheng, B. Effects of Fe-doping on the photocatalytic activity of mesoporous TiO$_2$ powders prepared by an ultrasonic method. *J. Hazard. Mater.* **2006**, *137*, 1838–1847. [CrossRef]

65. Coleman, H.M.; Vimonses, V.; Leslie, G.; Amal, R. Removal of contaminants of concern in water using advance oxidation techniques. *Water Sci. Technol.* **2007**, *55*, 301–306. [CrossRef]

66. Coleman, H.M.; Abdullah, M.I.; Eggins, B.R.; Palmer, F.L. Photocatalytic degradation of 17[beta]-oestradiol, oestriol and 17[alfa]-ethynyloestradiol in water monitored using fluorescence spectroscopy. *Appl. Catal. B Environ.* **2005**, *55*, 23–30. [CrossRef]

67. Coleman, H.M.; Chiang, K.; Amal, R. Effects of Ag and Pt on photocatalytic degradation of endocrine disrupting chemicals in water. *Chem. Eng. J.* **2005**, *113*, 65–72. [CrossRef]

68. Lin, L.; Wang, H.; Jiang, W.; Mkaouar, A.R.; Xu, P. Comparison study on photocatalytic oxidation of pharmaceuticals by TiO$_2$-Fe and TiO$_2$-reduced graphene oxide nanocomposites immobilized on optical fibers. *J. Hazard. Mater.* **2017**, *333*, 162–168. [CrossRef] [PubMed]

69. Hamadanian, M.; Reisi-Vanani, A.; Behpour, M.; Esmaeily, A.S. Synthesis and characterization of Fe,S-codoped TiO$_2$ nanoparticles: Application in degradation of organic water pollutants. *Desalination* **2011**, *281*, 319–324. [CrossRef]

70. Naik, B.; Parida, K.M. Solar Light Active Photodegradation of Phenol over a Fe$_x$Ti$_{1-x}$O$_{2-y}$N$_y$ Nanophotocatalyst. *Ind. Eng. Chem. Res.* **2010**, *49*, 8339–8346. [CrossRef]

71. Bloh, J.Z.; Dillert, R.; Bahnemann, D.W. Zinc Oxide Photocatalysis: Influence of Iron and Titanium Doping and Origin of the Optimal Doping Ratio. *ChemCatChem* **2013**, *5*, 774–778. [CrossRef]

72. Lorenz, R.D. A simple webcam spectrograph. *Am. J. Phys.* **2014**, *82*, 169–173. [CrossRef]

73. Widiatmoko, E.; Widayani; Budiman, M.; Abdullah, M.; Khairurrijal. A simple spectrophotometer using common materials and a digital camera. *Phys. Educ.* **2011**, *46*, 332–339. [CrossRef]

74. Altomare, A.; Corriero, N.; Cuocci, C.; Falcicchio, A.; Moliterni, A.; Rizzi, R. QUALX2.0: A qualitative phase analysis software using the freely available database POW_COD. *J. Appl. Crystallogr.* **2015**, *48*, 598–603. [CrossRef]

75. Spurr, R.A.; Myers, H. Quantitative Analysis of Anatase-Rutile Mixtures with an X-ray Diffractometer. *Anal. Chem.* **1957**, *29*, 760–762. [CrossRef]

76. Kim, C.; Park, H.J.; Cha, S.; Yoon, J. Facile detection of photogenerated reactive oxygen species in TiO$_2$ nanoparticles suspension using colorimetric probe-assisted spectrometric method. *Chemosphere* **2013**, *93*, 2011–2015. [CrossRef]

77. Simonsen, M.E.; Muff, J.; Bennedsen, L.R.; Kowalski, K.P.; Søgaard, E.G. Photocatalytic bleaching of p-nitrosodimethylaniline and a comparison to the performance of other AOP technologies. *J. Photochem. Photobiol. A Chem.* **2010**, *216*, 244–249. [CrossRef]

78. Kraljic, I.; Trumbore, C.N. p-Nitrosodimethylaniline as an OH radical scavenger in radiation chemistry. *J. Am. Chem. Soc.* **1965**, *87*, 2547–2550. [CrossRef]

79. Farhataziz, A.B.R. *Selected Specific Rates of Reactions of Transients from Water in Aqueous Solutions III: Hydroxyl Radical and Perhydroxyl Radical and Their Radical Ions*; U.S. Department of Commerce: Washington, DC, USA, 1977.

80. Martínez-Huitle, C.A.; Quiroz, M.A.; Comninellis, C.; Ferro, S.; De Battisti, A. Electrochemical incineration of chloranilic acid using Ti/IrO_2, Pb/PbO_2 and Si/BDD electrodes. *Electrochim. Acta* **2004**, *50*, 949–956. [CrossRef]

81. Bors, W.; Michel, C.; Saran, M. On the nature of biochemically generated hydroxyl radicals. Studies using the bleaching of p-nitrosodimethylaniline as a direct assay method. *Eur. J. Biochem.* **1979**, *95*, 621–627. [CrossRef] [PubMed]

82. Barashkov, N.N.; Eisenberg, D.; Eisenberg, S.; Shegebaeva, G.S.; Irgibaeva, I.S.; Barashkova, I.I. Electrochemical chlorine-free AC disinfection of water contaminated with Salmonella typhimurium bacteria. *Russ. J. Electrochem.* **2010**, *46*, 306–311. [CrossRef]

83. Hurwitz, A.R.; Liu, S.T. Determination of aqueous solubility and pKa values of estrogens. *J. Pharm. Sci.* **1977**, *66*, 624–627. [CrossRef]

84. Ying, G.G.; Kookana, R.S.; Ru, Y.J. Occurrence and fate of hormone steroids in the environment. *Environ. Int.* **2002**, *28*, 545–551. [CrossRef]

85. Fernández-Ibáñez, P.; De Las Nieves, F.J.; Malato, S. Titanium Dioxide/Electrolyte Solution Interface: Electron Transfer Phenomena. *J. Colloid Interface Sci.* **2000**, *227*, 510–516. [CrossRef]

86. George, S.; Pokhrel, S.; Ji, Z.; Henderson, B.L.; Xia, T.; Li, L.; Zink, J.I.; Nel, A.E.; Mädler, L. Role of Fe doping in tuning the band gap of TiO_2 for the photo-oxidation-induced cytotoxicity paradigm. *J. Am. Chem. Soc.* **2011**, *133*, 11270–11278. [CrossRef]

87. Bhatu, M.N.; Lavand, A.B.; Malghe, Y.S. Visible light photocatalytic degradation of malachite green using modified titania. *J. Mater. Res. Technol.* **2018**. [CrossRef]

88. Tabasideh, S.; Maleki, A.; Shahmoradi, B.; Ghahremani, E.; McKay, G. Sonophotocatalytic degradation of diazinon in aqueous solution using iron-doped TiO_2 nanoparticles. *Sep. Purif. Technol.* **2017**, *189*, 186–192. [CrossRef]

89. Aba-Guevara, C.G.; Medina-Ramírez, I.E.; Hernández-Ramírez, A.; Jáuregui-Rincón, J.; Lozano-Álvarez, J.A.; Rodríguez-López, J.L. Comparison of two synthesis methods on the preparation of Fe, N-Co-doped TiO_2 materials for degradation of pharmaceutical compounds under visible light. *Ceram. Int.* **2017**, *43*, 5068–5079. [CrossRef]

90. Hemmati Borji, S.; Nasseri, S.; Mahvi, A.; Nabizadeh, R.; Javadi, A. Investigation of photocatalytic degradation of phenol by Fe(III)-doped TiO_2 and TiO_2 nanoparticles. *J. Environ. Health Sci. Eng.* **2014**, *12*, 101. [CrossRef] [PubMed]

Review

Compositing Two-Dimensional Materials with TiO$_2$ for Photocatalysis

Yu Ren [1,2], Yuze Dong [1,2], Yaqing Feng [1,2,*] and Jialiang Xu [1,3,*]

1   School of Chemical Engineering and Technology, Tianjin University, Yaguan Road 135, Tianjin 300350, China; renyu9505@163.com (Y.R.); yuze441295@tju.edu.cn (Y.D.)
2   Collaborative Innovation Center of Chemical Science and Engineering (Tianjin), Tianjin 300072, China
3   School of Materials Science and Engineering, Nankai University, Tongyan Road 38, Tianjin 300350, China
*   Correspondence: yqfeng@tju.edu.cn (Y.F.); jialiang.xu@nankai.edu.cn (J.X.)

Received: 12 November 2018; Accepted: 23 November 2018; Published: 28 November 2018

Abstract: Energy shortage and environmental pollution problems boost in recent years. Photocatalytic technology is one of the most effective ways to produce clean energy—hydrogen and degrade pollutants under moderate conditions and thus attracts considerable attentions. TiO$_2$ is considered one of the best photocatalysts because of its well-behaved photo-corrosion resistance and catalytic activity. However, the traditional TiO$_2$ photocatalyst suffers from limitations of ineffective use of sunlight and rapid carrier recombination rate, which severely suppress its applications in photocatalysis. Surface modification and hybridization of TiO$_2$ has been developed as an effective method to improve its photocatalysis activity. Due to superior physical and chemical properties such as high surface area, suitable bandgap, structural stability and high charge mobility, two-dimensional (2D) material is an ideal modifier composited with TiO$_2$ to achieve enhanced photocatalysis process. In this review, we summarized the preparation methods of 2D material/TiO$_2$ hybrid and drilled down into the role of 2D materials in photocatalysis activities.

Keywords: photocatalysis; 2D materials; TiO$_2$; composite

---

1. Introduction

With the massive consumption of fossil energy and serious environmental pollution problems, there is an urgent need for clean energy and more efficient ways to decompose pollutants. Photocatalysis is an advanced technology that uses photon energy to convert chemical reactions occurring under harsh conditions into reactions under mild conditions by appropriate photocatalyst, and thus emerged as recognizable fields such as hydrogen generation [1–4], sewage treatment [5–7], harmful gas removal [8,9], organic pollutant degradation [10–13] and carbon dioxide reduction [14–16].

Since the first report that TiO$_2$ electrode was applied for hydrogen production by Fujishima and Honda in 1972 [17], TiO$_2$ has attracted numerous attention in photocatalysis as a typical n-type semiconductor [18–21]. Being non-toxic, inexpensive, highly stable [22–24], TiO$_2$ is widely investigated in photocatalytic fields. Hoffman proposed the following general mechanism (Table 1) for heterogeneous photocatalysis on TiO$_2$ [25].

Table 1. Mechanism for heterogeneous photocatalysis on TiO_2.

Primary Process		Characteristic Times
charge-carrier generation	$TiO_2 + h\nu \rightarrow h_{VB}^+ + e_{cb}^-$	(fs)
charge-carrier trapping	$h_{VB}^+ + >Ti^{IV}OH \rightarrow \{>Ti^{IV}OH\}\bullet^+$ $e_{cb}^- + >Ti^{IV}OH \rightarrow \{>Ti^{III}OH\}$ $e_{cb}^- + >Ti^{IV} \rightarrow >Ti^{III}$	fast (10 ns) shallow trap (100 ps) (dynamic equilibrium) deep trap (10 ns) (irreversible)
charge-carrier recombination	$e_{cb}^- + \{>Ti^{IV}OH\}\bullet^+ \rightarrow >Ti^{IV}OH$ $h_{VB}^+ + \{>Ti^{III}OH\} \rightarrow Ti^{IV}OH$	slow (100 ns) fast (10 ns)
interfacial charge transfer	$\{>Ti^{IV}OH\}\bullet^+ + Red \rightarrow >Ti^{IV}OH + Red\bullet^+$ $e_{tr}^- + Ox \rightarrow Ti^{IV}OH + Ox\bullet^-$	slow (100 ns) very slow (ms)

Where $>TiOH$ represents the primary hydrated surface functionality of TiO_2, e_{cb}^- is a conduction band (CB) electron, etr^- is a trapped conduction band electron, h_{VB}^+ is a valence band (VB) hole, Red is an electron donor, Ox is an electron acceptor, $\{>Ti^{IV}OH\}\bullet^+$ is the surface-trapped VB hole (i.e., surface-bound hydroxyl radical), and $\{>Ti^{III}OH\}$ is the surface-trapped CB electron. Upon light irradiation, electrons transfer from VB to CB of TiO_2, while both electrons and holes can be trapped by primary hydrated surface functionality of TiO_2, achieving the separation of photo induced electrons and holes. At the same time, the recombination between electrons and holes exits, which competes with charge-carrier trapping process. The competition has thus a negative effect on later interfacial charge transfer. Deliberating on TiO_2 photocatalysis process, some drawbacks exit as following: (1) The wide bandgap of TiO_2 (3.2 eV) means that photons with adequate energy can only excite electrons in the VB to the CB of TiO_2, which limits its effective use of sunlight (UV region, $\lambda \leq 387$ nm); (2) The recombination of excited electrons and holes is inevitable while time for carrier recombination is much shorter than that for charge transfer. Therefore, the effective function of photoexcitation is suppressed greatly.

Considering the above two factors, the improvement of the photocatalytic efficiency of TiO_2 can be obtained through two aspects: the improvement of solar light utilization efficiency and the suppression of recombination of electron and hole pairs. In this text, surface modification and hybridization of TiO_2 such as noble metal loading [26–29] and semiconductor heterojunction [30–32] are effective methods to enhance the photocatalytic performance. The Schottky barrier formed at the interface between the noble metal material and TiO_2 can effectively promote the separation of photogenerated carriers. Similarly, the heterojunction structure can form a matching energy level at the semiconductor interface to suppress the recombination of photogenerated carriers. However, the opportunities of improvements in photocatalysis performances offered by these attempts are narrow, and thus limited their commercial and efficient application. In the past decade, two-dimensional (2D) materials have attracted more and more attention because of the flexible preparation methods, low price and superior physical and chemical properties. In particular, their high surface area, suitable bandgap, structural stability and high charge mobility [33–36] endow these 2D materials with remarkable performances for applications in photocatalysis [37–41]. When combined with TiO_2, not only the utilization of sunlight is improved, but also the matching between energy levels is formed to inhibit the recombination, and the large specific surface area provides support and active sites for the reaction. In this review, we summarize the recent advances of 2D material-TiO_2 composites, including synthesis methods, properties, and catalytic behaviors. Furthermore, the photocatalytic mechanism is deliberated in detail to elaborate the role of 2D materials in the photocatalytic processes.

2. 2D-Material Modified TiO_2

Based on the mobile dimension of electronics, it can be divided into zero-dimensional (0D) materials, one-dimensional (1D) materials, two-dimensional (2D) material and three-dimensional (3D) materials [36], while 2D materials represent an emerging class of materials that possess sheet-like structures with the thickness of only single or a few atom layers [42]. Compared with the bulk structures, the ultrathin 2D structure exhibits superior properties such as modification of energy level and larger adjustable surface area. The excellent properties of 2D materials make them widely used in many aspects [43–45]. When composited with TiO_2, the synergistic effect of the two can significantly improve the photocatalytic activity and thus 2D materials is ideal for TiO_2 photocatalysis.

2.1. Graphene Modified TiO$_2$

Since the first isolation by Geim and Novoselov in 2004, graphene has attracted significant attention [46–49]. Graphene is a 2D honeycomb construction consisting of carbon atoms. The thickness of graphene is only 0.335 nm, which is the thickness of a carbon atom layer. In the sp^2 hybrid distribution form, each carbon atom contributes an unbonded π electron, which can delocalize freely throughout the carbon atom 'net' to form an extended π bond. This construction endows graphene excellent properties such as high charge mobility (200,000 cm^2 V^{-1} s^{-1}), high thermal conductivity (5000 W m^{-1} K^{-1}), and large surface area [35], which is ideal for applications in sensors [50], energy conversion and storage [37], polymer composites [51], drug delivery systems [52], and environmental science [53]. When composited with TiO_2, graphene can accept photoinduced electrons from TiO_2 and thus greatly enhances the efficiency of carriers' separation [54–58].

2.1.1. The Synthesis of Graphene/TiO$_2$ Composites

Graphite oxide and graphene oxide (GO) intermediates are widely used in the process of combining graphene with other materials [59]. The most widely used technique is chemical reduction of GO as shown in Figure 1, which is usually conducted by Hummers' method [60]. Graphite is added to a strongly oxidizing solution such as HNO_3, $KMnO_4$, and H_2SO_4 to prepare graphite oxide and the oxygen-containing groups are introduced into the surface or edge of the graphite during the process. The sheets of graphite oxide were exfoliated to obtain GO. The presence of oxygen-containing groups allows GO to provide more surface modification active sites and larger specific surface areas for synthetic graphene-based composites. GO can be converted to reduced graphene oxide (RGO) by chemical reduction to remove these oxygen-containing group. During this process, the number of oxygen-containing groups on the GO decreases drastically, and the conjugated structure of the graphene base will be effectively restored. The presence of oxygen functionalities in GO allows interactions with the cations and provides reactive sites for the nucleation and growth of nanoparticles, which results in the rapid growth of various graphene-based composites. The preparation methods for graphene/TiO_2 composites are divided into ex-situ hybridization and in-situ growth, the difference between which is the process of TiO_2 formation.

- *Ex-situ hybridization.* The common procedure for ex-situ hybridization is to mix GO and modified TiO_2 with physical process such as ultrasound sonication and heat treatments. Rahmatollah et al. [62] reported a facile one-step solvothermal method to synthesize the TiO_2-graphene composite sheets by dissolving different mass ratios of GO and TiO_2 nanoparticles in anhydrous ethanol solution. Ultrasound irradiation was used to disperse the GO. Finally, a six-fold enhancement was observed in the photocurrent response compared to the improved photoelectrochemical performance (3%) with the pure TiO_2. Florina et al. [63] prepared graphene/TiO_2-Ag based composites as electrode materials. Similarly, GO suspensions were mixed with prepared TiO_2-Ag nanoparticles in NaOH solution. The suspensions were sonicated, dried and subjected to thermal treatment. However, the control of modification between the TiO_2 and graphene may lead to a decreased interaction between these two parts [64].

- *In-situ growth.* The in-situ growth method is widely used to prepare graphene-based composite materials, and the method can effectively avoid clustering of nanoparticles on the surface of graphene. According to different preparation process, it might be divided into reduction method, electrochemical deposition method, hydrothermal method and sol-gel method.

  - *Reduction method.* Usually, in a reduction method, GO and TiO_2 metal salts are mixed as precursors. By controlling the hydrolysis of the precursor, TiO_2 crystal nucleus grows on GO, while GO is reduced to obtain graphene-based TiO_2 composite materials [65]. In addition to the chemical reduction method, other commonly used reduction methods are photocatalytic reduction [66] and microwave-assisted chemical reduction [67].

- *Electrochemical deposition method.* In an electrochemical deposition method, graphene or reduced graphene is used as a working electrode in a dielectric solution containing a metal precursor or its compound [68].
- *Hydrothermal/solvothermal method.* A hydrothermal/solvothermal method is commonly used for preparing inorganic nanomaterials. It is generally carried out in a dispersion of GO. Under high temperature and high pressure, GO and titanium salt precursor are reduced simultaneously [69,70].
- *Sol-gel method.* The sol-gel method takes titanium alkoxide or titanium chloride as precursors, and it can be uniformly bonded with oxygen group on graphene, polycondensed to form a gel. Then TiO_2 nanoparticles are formed through calcining [71,72]. The sol-gel method can obtain loaded nanoparticles with higher uniformity of dispersion.

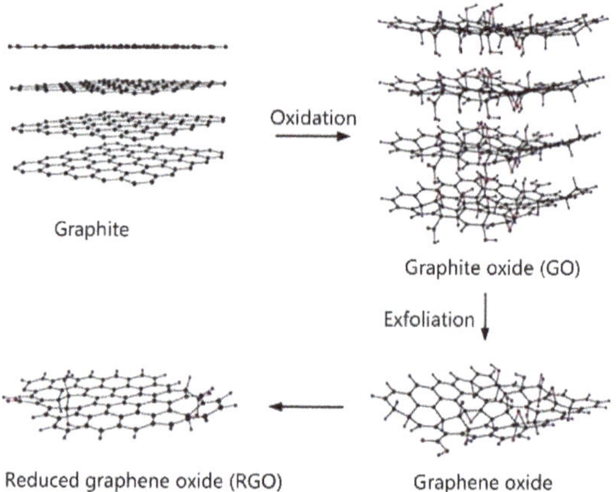

Figure 1. Preparation of graphene by chemical reduction of graphene oxide synthesized by Hummers' method. Reprinted with permission from [61]. Copyright 2011, Wiley-VCH.

2.1.2. The Role of Graphene in TiO_2 Photocatalysis

Due to the large bandgap, the photocatalysis process of pure TiO_2 can only be activated under UV light. Thus, the hybridization of graphene and TiO_2 is essential to ensure a broad light stimulation process. In graphene/TiO_2 system, electrons flow from TiO_2 to graphene through interface because of the higher Fermi level of TiO_2. Then graphene gains excess negative charges while TiO_2 has positive charges, leading to a space charge layer at the interface which is regarded as Schottky junction. The Schottky junction can serve as an electron trap to efficiently capture the photoinduced electrons [73] and thus enhance the photocatalysis activity. Meanwhile, the Schottky barrier also acts as the main obstruction for the electron transport from the graphene to TiO_2. Under visible light, electrons on Fermi level of graphene are irradiated and the Schottky barrier has to overcome to ensure the injection of electrons to conduct band of TiO_2. In the UV light irradiation process, graphene plays a role as electron acceptor and thus promotes the separation of electron-hole pairs [54] (Figure 2).

Different interface interactions have been extensively studied [55,56]. Compared with 0D-2D Degussa P25 (TiO_2)/graphene and 1D-2D TiO_2 nanotube/graphene composites, the 2D-2D TiO_2 nanosheet/graphene hybrid demonstrates higher photocatalytic activity toward the degradation of rhodamine B and 2,4-dichlorophenol under the UV irradiation [56]. The intimate and uniform contact between the two sheets-like nanomaterials allowed for the rapid injection of photogenerated electrons from the excited TiO_2 into graphene across the 2D-2D interface while achieving effective electron-hole pair separation and promoted radical's generation. In another example of RGO–TiO$_2$ hybrid, by having a narrower bandgap, the photo-response range of RGO–TiO$_2$ nanocomposites clearly extended from UV (~390 nm) to visible light (~480 nm), which offered a better utilization of visible light [55]. Raman spectra and other characterization revealed that the narrow bandgap was attributed to the Ti–O–C bond between the two components, and thus caught an intimate interaction between TiO$_2$ nanoparticles and RGO sheets. What's more, the up-conversion photoluminescence (UCPL) effect of RGO assists the light absorption, and enabled the efficient utilization of both UV light and visible light (Figure 3). It is worth to note that the surface area of RGO–TiO$_2$ was smaller than that of pure TiO$_2$ (P25), which revealed that the enhanced photocatalytic activity of RGO-TiO$_2$ was relevant to the improved conductivity and bandgap structure other than their surface area. RGO nanosheet can play a role in both charge transfer and active sites after doping with heteroatoms. TiO$_2$/nitrogen (N) doped reduced graphene oxide (TiO$_2$/NRGO) nanocomposites was applied to photoreduction of CO$_2$ with H$_2$O vapor in the gas-phase under the irradiation of a Xe lamp (the wavelength range of 250–400 nm) [57]. Compared with TiO$_2$, TiO$_2$/NRGO composites exhibited a narrower bandgap due to chemical bonding between TiO$_2$ and the specific sites of N-doped graphene. In the photoreduction of carbon dioxide, the function of nitrogen atoms varied in different chemical environments. The pyridinic-N and pyrrolic-N worked as active sites for CO$_2$ capture and activation while quaternary-N worked as an electron-mobility activation region for the effective transfer of photogenerated electrons from the CB of the TiO$_2$ [57] (Figure 4). The results reveal that the doped atoms can act as basic sites for anchoring target molecular, adjusting the electronic properties and local surface reactivity of graphene.

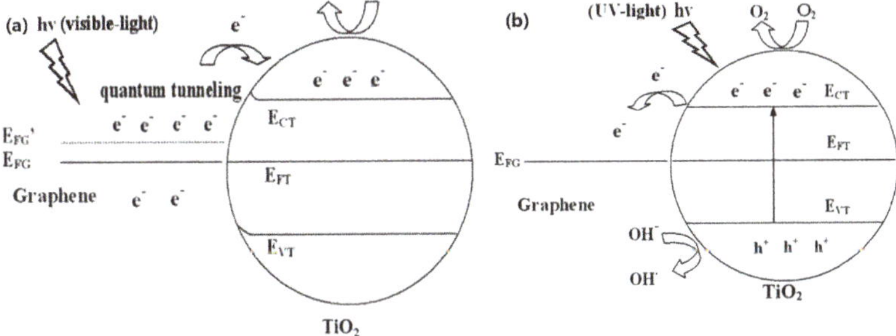

Figure 2. Photocatalytic mechanisms of graphene-TiO$_2$ composite under (**a**) visible light (**b**) UV light. Reprinted with permission from [54]. Copyright 2013, Elsevier.

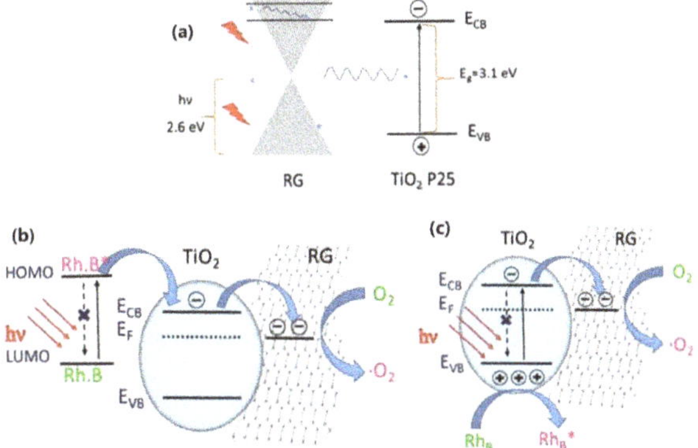

Figure 3. (**a**) Schematic of up-conversion photoluminescence (UCPL) mechanism for reduced graphene oxide (RGO)–TiO$_2$ nanocomposite under visible light (hv~2.6 eV) irradiation; (**b**,**c**) Schematics of proposed mechanism of Rh. B photodegradation. Reprinted with permission from [55]. Copyright 2017, Springer.

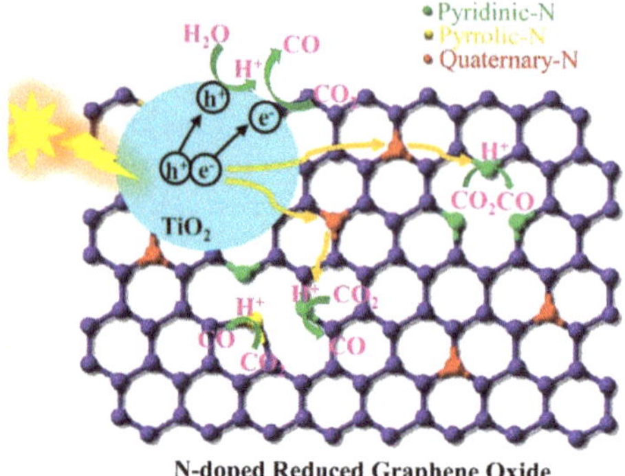

Figure 4. Reaction mechanisms for photoreduction of CO_2 with H_2O over TiO_2/NRGO-300 samples. Reprinted with permission from [57]. Copyright 2017, Elsevier. NRGO: nitrogen doped reduced graphene oxide.

Except for dimension factor and bonding interaction between graphene and TiO_2, a linkage is introduced to graphene/TiO_2 system to achieve better interfacial contact as well. A N-doping Graphene-TiO_2 composite nano-capsule for gaseous HCHO degradation was reported [58]. It indicated that wrapping with dopamine on the surface of TiO_2 enhanced interfacial contact between TiO_2 and melamine-doped graphene (MG) sheets, thus promoting the separation and mobility of photoinduced electrons and holes in TiO_2@MG-D. The dopamine acted as bridge between TiO_2 and MG, creating numerous migration channels for charges and restraining the recombination of electrons and holes (Figure 5). The introduction of linkage can effectively improve the weak interfacial contact and overcome the long distance of electron transport between the graphene and TiO_2, leading to raised separation and mobility of photoinduced electrons and holes and thus higher photocatalytic activity.

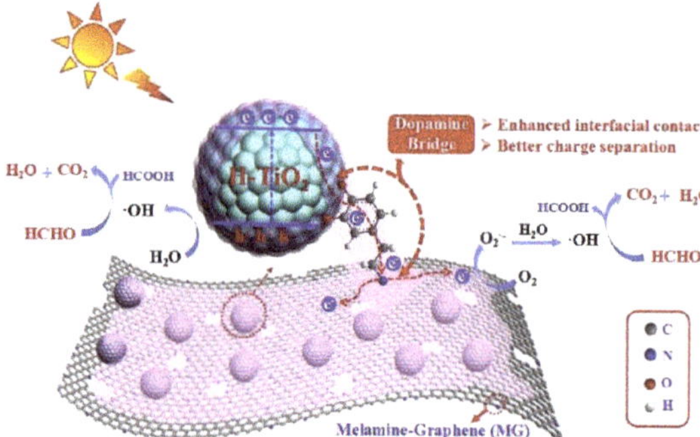

Figure 5. Schematic illustrations for dopamine bridged Melamine-Graphene/TiO_2 nanocapsule and photocatalytic degradation process of HCHO. Reprinted with permission from [58]. Copyright 2018, Elsevier.

Despite of electron accepter and electron storage, graphene can also act as a transport bridge between photocatalysts. For example, in the 2D ternary $BiVO_4$/graphene oxide (GO)/TiO_2 system, both the $BiVO_4$ and the TiO_2 were connected to GO forming a p-n heterogeneous structure. The CB of $BiVO_4$ was more negative than that of GO and the CB of GO was more negative than that of TiO_2; thus, the electrons generated from the CB of BiVO4 can transfer to the GO and then the electron further moved to the conduction band of TiO_2 (Figure 6). Therefore, the GO can enhance the effective separation of the photo-generated electron-hole pairs due to its superior electrical conductivity. Meanwhile, the large surface area of the GO is also beneficial for dye attachment [74].

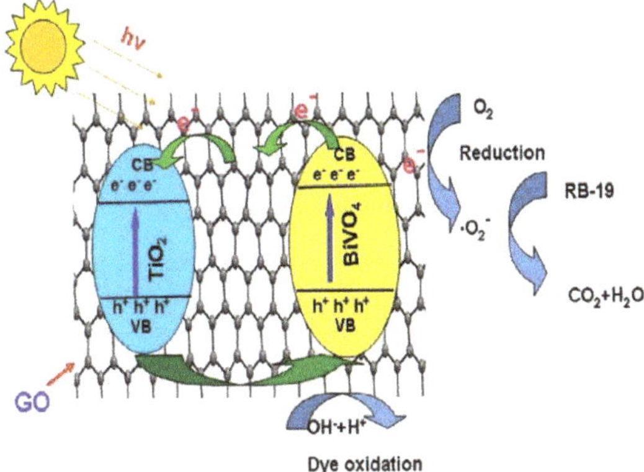

Figure 6. Photodegradation mechanism of BiVO$_4$/TiO$_2$/GO photocatalyst. Reprinted with permission from [74]. Copyright 2017, Elsevier.

2.2. Graphdiyne Modified TiO$_2$

Graphdiyne (GD) is a new carbon allotrope in which the benzene rings are conjugated by 1,3-diyne bonds to form a 2D planar network structure and features both sp and sp^2 carbon atoms. Since the successful synthesis by Li et al. [75], GD has evoked significant interest in various scientific fields because of unique mechanical, chemical and electrical properties [38,42,76–80]. GD shows potential for photocatalysis with its large surface area as well as high charge mobility. GD features an intrinsic bandgap and exhibits semiconducting property with a measured conductivity of 2.516 × 10^{-4} S·m^{-1} and was predicted to be the most stable structure among various diacetylenic non-natural carbon allotropes [81]. It also provides highly active sites for catalysis. Furthermore, GD with diacetylene linkage can be chemically bonded with TiO$_2$ [82–85]. Therefore, the TiO$_2$-graphdiyne composites can greatly improve the photocatalytic activity, and thus their application in photocatalysis has been explored recently [83,84,86].

2.2.1. The Synthesis of GD/TiO$_2$ Composites

The general preparation of GD film is through a coupling reaction in which hexaethynylbenzene (HEB) acts as precursor and copper foil serves as catalysis. Meanwhile, the copper foil provides a large planar substrate for the directional polymerization growth of the GD film (Figure 7). Despite of film, GD with different morphologies such as nanotube arrays, nanowires, nanowalls and nanosheets have been also prepared for diverse applications [87,88].

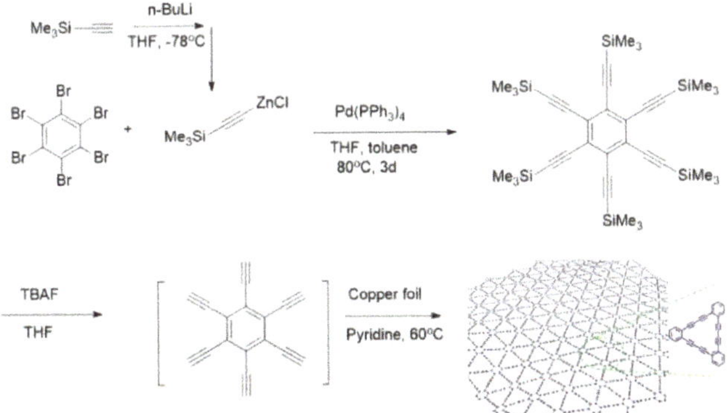

Figure 7. Preparation of graphdiyne (GD) film.

Ex-situ hydrothermal method is commonly used in preparation of GD/TiO$_2$ composites [83,84,86]. In general, the GD and TiO$_2$ are prepared separately. Then the pre-prepared GD and TiO$_2$ are mixed in H$_2$O/CH$_3$OH solvent. After stirring to obtain a homogeneous suspension, the suspension is placed in Teflon sealed autoclave and heated to combine the TiO$_2$ and GD. Being rinsed and dried, the GD/TiO$_2$ composites are obtained.

2.2.2. The Role of GD in TiO$_2$ Photocatalysis

Wang et al. [84] were the first to combine GD with TiO$_2$ for the enhancement of TiO$_2$ photocatalysis. The resultant GD-P25 composites exhibited higher visible light photocatalytic activity than those of the bare P25, P25-CNT (titania-carbon nanotube), and P25-GR (graphene) materials. By changing the weight percent of GD in the hybrid, the photocatalytic activity of P25-GD can be adjusted. It was speculated that the formation of chemical bonds between P25 and GD can effectively decrease the bandgap of P25 and extended its absorbable light range [84]. Namely, electrons in VB of TiO$_2$ can easily migrate to impurity band which is attributed to the insertion of carbon *p*-orbitals into the TiO$_2$ bandgap, and then transfer to CB of TiO$_2$ thus enhancing the photo-response activity. In order to further explore the role of GD, Yang et al. [83] investigated the chemical structures and electronic properties of TiO$_2$-GD and TiO$_2$-GR composites employing first-principles density functional theory (DFT) calculations. The results revealed that for the TiO$_2$ (001)-GR composite, O and atop C atoms could form C–O σ bond, which acted as a charge transfer bridge at the interface between TiO$_2$ and GR. Besides the C–O σ bond, another Ti-C π bond is also formed in TiO$_2$ (001)-GD composite, which makes GD combine with TiO$_2$ tightly and therefore enhances the charge transfer. In addition, calculated Mulliken charge for the surface of TiO$_2$ (001)-GD and TiO$_2$ (001)-GR suggested a stronger electrons' capture ability of former (Figure 8). The calculated results were in accordance with theoretical prediction that TiO$_2$ (001)-GD composites showed the highest photocatalysis performance among 2D carbon-based TiO$_2$ composites, confirming that GD could become a promising competitor in the field of photocatalysis. After that, Dong et al. prepared GD-hybridized nitrogen-doped TiO$_2$ nanosheets with exposed (001) facets (GD-NTNS) [86]. The doped N and incorporated GD efficiently narrowed the bandgap compared with pure TiO$_2$ and widened response range towards light from UV light to 420 nm visible light. The activity of the GD-NTNS photocatalyst presented the most superior performance compared with bare TiO$_2$ nanosheets (TNS) and nitrogen-doped TiO$_2$ nanosheets (NTNS) and GR-NTNS.

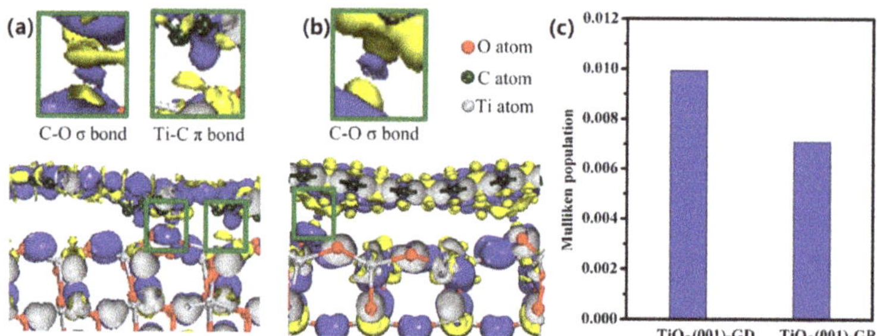

Figure 8. Plots of electron density difference at the composites interfaces: (**a**) TiO$_2$ (001)-GD; (**b**) TiO$_2$ (001)-GR; (**c**) Mulliken charge of GD or GR (graphene) surface in the composites. Reprinted with permission from [83]. Copyright 2013, American Chemical Society.

The mechanisms of photocatalysis enhancement by introducing GD remain to be understood. In general, with a lower Fermi level than the conduction band minimum of TiO$_2$, GD can be regarded as an electron pool which accept electrons excited from TiO$_2$ [84,89,90] (Figure 9). As a result, it prompts the charge carriers' separation and prevents electron-hole recombination. Moreover, GD can generate an impurity band and thus broaden the visible light absorption in TiO$_2$-GD composites [91–93].

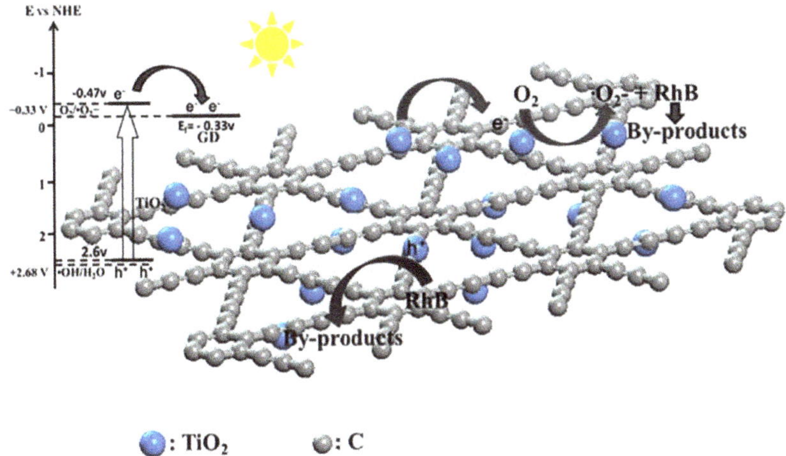

Figure 9. Schematic illustration for the possible mechanism of the visible light-driven photocatalytic degradation for the GD-NTNS composites. Reprinted with permission from [86]. Copyright 2018, Springer.

2.3. C_3N_4 Modified TiO_2

Graphitic carbon nitride (g-C_3N_4) is a 2D polymer material which shows broad application prospects in many fields, given the simple synthesis, rich source, along with unique electronic structure, good thermal stability and chemical stability. Its graphene-like structure is composed of triazine (C_3N_3) or tri-s-striazine (C_6N_7) allotropes units (Figure 10). The tri-s-striazine unit structure is more stable and thus draws in extensive studies [34]. Since the first report of g-C_3N_4 for water decomposition, g-C_3N_4 has attracted wide attention in photocatalyst [40]. The bandgap of g-C_3N_4 (2.6–2.7 eV) is moderate and the substantial nitrogen sites and ordered units structure endue g-C_3N_4 an ideal material to composite with TiO_2.

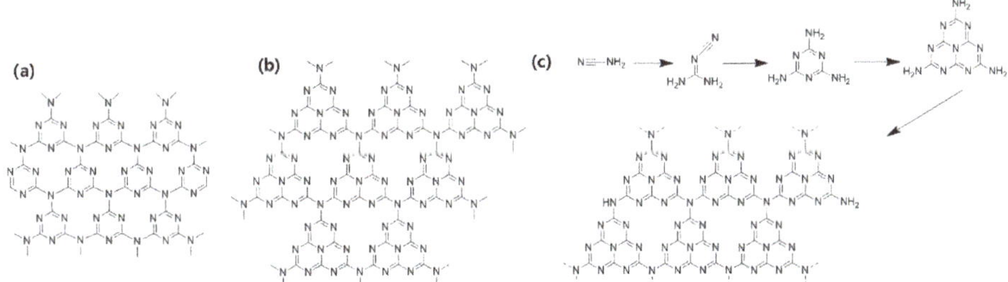

Figure 10. Triazine (**a**) and tri-s-striazine (**b**) allotropes units of g-C_3N_4; (**c**) The synthesis of g-C_3N_4.

2.3.1. The Synthesis of g-C_3N_4/TiO_2 Composites

In general, the synthesis of g-C_3N_4/TiO_2 composites can be also divided into ex-situ method and in-situ method.

- In the ex-situ way, both g-C_3N_4 and TiO_2 materials are pre-prepared, which can be integrated through physical process such as ball milling [94], solvent evaporation [95,96], etc. Though physical process is easy to operate under moderate conditions, some flaws also exist such as ununiformly dispersing and unstable structure.

- The in-situ method uses one of the materials as a substrate and then the other material grows on the surface of the substrate. For g-C_3N_4/TiO_2 composites, both materials can be regarded as substrates.

- When used as substrates, g-C$_3$N$_4$ is pre-prepared by calcinations of precursors. Solvothermal/ hydrothermal method is most common for the next step. After mixing g-C$_3$N$_4$ and titanates in a certain solvent, the solution is well dispersed and sealed in the Teflon-lined autoclave, followed by a solvothermal/hydrothermal treatment [97–99]. Furthermore, Atomic Layer Deposition (ALD) was applied to form thin TiO$_2$ films on g-C$_3$N$_4$ substrates. ALD involves the surface of a substrate exposed alternately to alternating precursor flow. Then the precursor molecule reacts with the surface in a self-limiting way, which guarantees that the reaction stops as all the reactive sites on the substrate reacted with the precursors. It is an effective way to control the thickness and homogeneity of deposited layer [100].

- When TiO$_2$ was used as substrates, calcination is widely used for the convenience and easy operation. In this process, the solid mixture of TiO$_2$ and pure urea or melamine or dicyandiamide powder are calcinated under fixed temperature to obtain g-C$_3$N$_4$/TiO$_2$ composites. Before calcination, the two components should be evenly dispersed by sonication [101], stirring [102], or grounding [103]. Recently, Tan et al. [104] reported another facile one-step way to prepare nanostructured g-C$_3$N$_4$/TiO$_2$ composite. As seen in Figure 11, melamine was at the bottom of the crucible while P25 was on the top of a cylinder put in the crucible. After a 4-h vapor deposition process, nanostructured g-C$_3$N$_4$/TiO$_2$ composite was obtained.

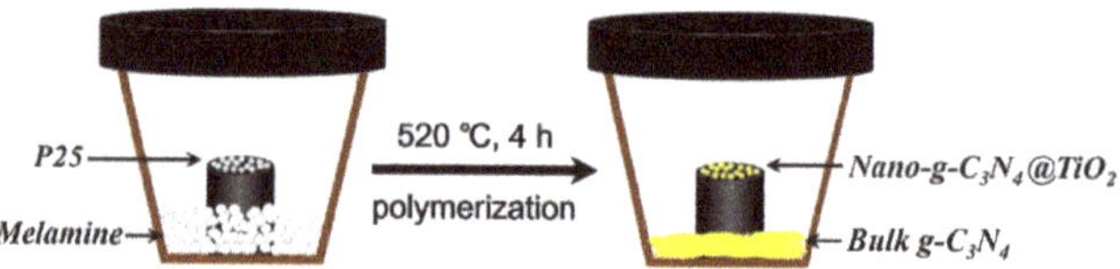

Figure 11. Vapor deposition process in the preparation of g-C3N4/TiO$_2$ composite. Reprinted with permission from [104]. Copyright 2018, Elsevier.

2.3.2. The Role of g-C$_3$N$_4$ in Photocatalysis

With a moderate bandgap of ~2.7 eV, g-C$_3$N$_4$ shows ability of photocatalyst under visible light, in contrast to TiO$_2$, which owns a large bandgap of 3.2 eV (Figure 12). However, because of the rapid recombination of photogenerated electron-hole pairs, the synergistic effect between g-C$_3$N$_4$ and TiO$_2$ plays important roles. In a photocatalyst system of g-C$_3$N$_4$/TiO$_2$ composites, the CB electrons of g-C$_3$N$_4$ transfer to the CB of TiO$_2$ and the VB holes of TiO$_2$ transfer to the VB of g-C$_3$N$_4$, which is a typical Type II system [41]. The electron/hole conduction mechanism can effectively separate electrons and holes, and thus enhances the separation efficiency and inhibit the recombination.

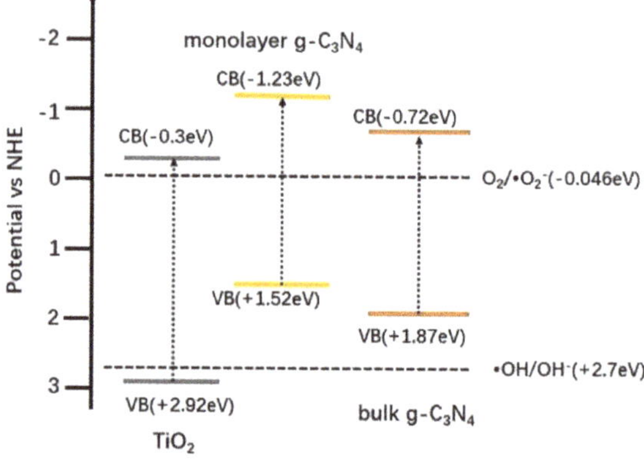

Figure 12. Bandgaps of TiO$_2$, monolayer g-C$_3$N$_4$ and bulk g-C$_3$N$_4$.

The structure plays a vital role in enhancing photocatalysis efficiency. g-C_3N_4 nanosheets (NS)-TiO_2 mesocrystals (TMC) composites was prepared by in-situ process [105]. Compared with bulk g-C_3N_4/TMC composites, the H_2 evolution rate of g-C_3N_4 (NS)/TMC was about six times higher, which was possibly due to a larger surface area of g-C_3N_4 (NS)/TMC ($57.4\ m^2g^{-1}$) than that of bulk g-C_3N_4/TMC ($34.3\ m^2g^{-1}$). What's more, the g-C_3N_4 nanosheets owned a lower surface defect density, given the surface defects normally is seen as recombination centers for photoinduced electrons and holes. However, surface area is not the unparalleled factor of promoted efficiency of photocatalyst, taking the fact that the surface area of g-C_3N_4 NS (31 wt%)/TMC ($57.4\ m^2g^{-1}$) and g-C_3N_4 NS (31 wt%)/P25 ($52.3\ m^2g^{-1}$) was nearly the same, as the H_2 evolution rate of g-C_3N_4 (NS)/TMC was about 7 times higher. Further research indicated that the tight interface between g-C_3N_4 NS and TMC facilitated the charge transfer, which is a flexible way to promote solar energy utilization of g-C_3N_4/TiO_2 photocatalyst.

Other structures like core-shell was lucubrated to create high photocatalytic activity towards many dyes [106]. After in-situ calcination and growth of cyanamide on the surface of TiO_2, a multiple direction contact structure of TiO_2@g-C_3N_4 hollow core@shell heterojunction photocatalyst (HTCN-1) was synthesized. The g-C_3N_4 nanosheets grew on the surface of TiO_2 caused closer contact between TiO_2 and g-C_3N_4 and a larger interfacial area, as confirmed by XPS analysis [106]. Compared with another core-shell type TiO_2@g-C_3N_4 (C-T) with unidirectional contact structures [107], HTCN-1 possessed higher efficiency in the charge separation and enhanced charge transfer. It demonstrated that multiple direction contact resulted in a large interfacial area, which would provide sufficient channels for efficient and rapid charge transfer (Figure 13) [106]. In another core-shell structure of g-C_3N_4/TiO_2 hybrid, Ag was introduced as interlayers to participate in electrical conduction and bridge the gap between g-C_3N_4 and TiO_2, facilitating the separation of photoexcited charge and reducing the recombination of the photogenerated electron hole (Figure 14) [108]. The surface area of the samples didn't change much upon the introduction of Ag ($228.4\ m^2g^{-1}$ and $210.3\ m^2g^{-1}$ for Ag/TiO_2 microspheres and nonsilver containing TiO_2, respectively). It was worth noting that low content of g-C_3N_4 (2%) in g-C_3N_4/Ag/TiO_2 microspheres had a larger surface area but lower photocatalytic activity than the g-C_3N_4 (4%)/Ag/TiO_2 microsphere sample [108]. The possible reason was that high content of g-C_3N_4 can generate more electron-hole pairs, leading to a higher photocatalytic activity. However, the g-C_3N_4 (6%)/Ag/TiO_2 microsphere sample showed decreased photocatalytic activity due to reduced surface area, which limited the contact between the catalyst and pollutant and thus lowered the photocatalytic reaction. It reflects that proper surface area is needed to provide both active sites and reaction sites.

The doping of g-C_3N_4 is another viable way to realize structure modification process. Sulfur was introduced to g-C_3N_4 nanostructures, and their photocatalytic performance was studied for decomposition of MO dye under visible light. The degradation efficiency over g-C_3N_4-TiO_2 composites (CNT) reached 61% within 90 min, while S-C_3N_4-TiO_2 composites (SCNT) reached nearly 100% within the same period [109]. SEM image showed a more transparent and thinner layer of S-C_3N_4 compared with g-C_3N_4 when composited with TiO_2, leading to an enhanced visible light absorption capability. On the other hand, unique bar-like structure of SCNT provided a pathway for carriers and isolate photon absorption with carriers' collection in perpendicular directions. Meanwhile, TiO_2 nanoparticles were more evenly dispersed on and inside S-C_3N_4 substrate in SCNT sample, which is beneficial for the interfacial carriers' transportation between S-C_3N_4 layer and TiO_2 particle [109]. Calculations revealed that the modified electronic structure with elevation of CB and VB values owing to doped sulfur, contributed to a higher driving force from CB of S-C_3N_4 to CB of TiO_2 and thus promoted the separation efficiency of electron-hole pairs (Figure 15). The doping of sulfur alternated both the structure and level distribution of C_3N_4, causing excellent separation efficiency of electron-hole pair when contacted with TiO_2.

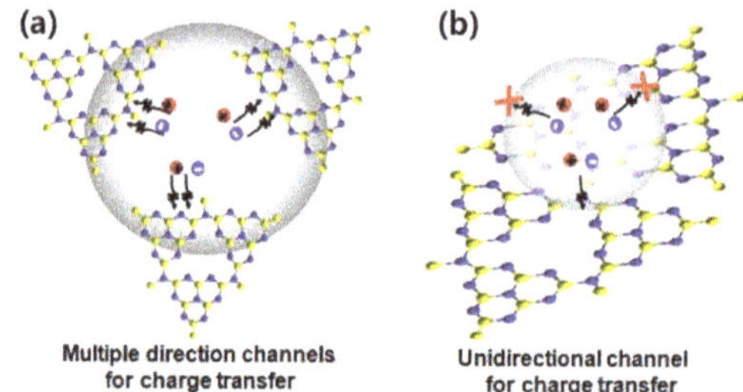

Figure 13. Structure of HTCN-1 (**a**) and C-T (**b**). Reprinted with permission from [106]. Copyright 2018, Elsevier.

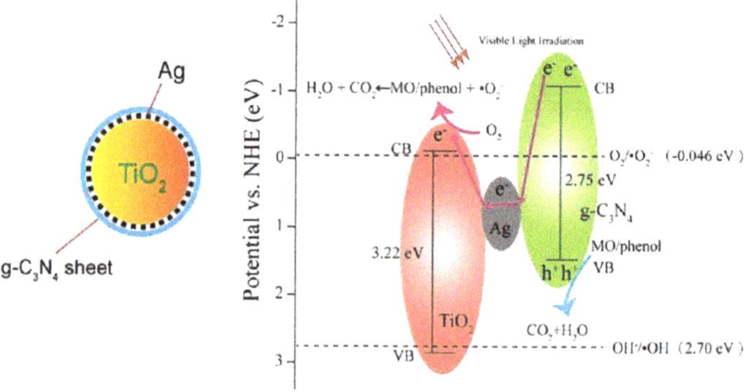

Figure 14. Photocatalytic mechanism scheme of g-C$_3$N$_4$/Ag/TiO$_2$ microspheres under visible light irradiation (>420 nm). Reprinted with permission from [108]. Copyright 2014, American Chemical Society.

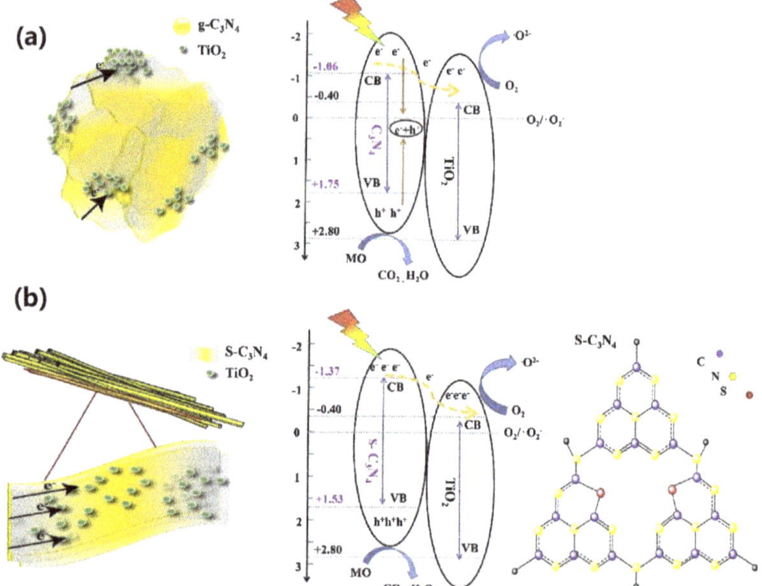

Figure 15. Mechanism of fast charge transfer at the interface between (**a**) C$_3$N$_4$-TiO$_2$ and (**b**) S-C$_3$N$_4$/TiO$_2$. Reprinted with permission from [109]. Copyright 2017, Elsevier.

2.4. MoS$_2$ Modified TiO$_2$

2D layered transition metal chalcogenides (TMCs) nanostructures spark a research boom due to its unique physical and chemical properties compared with other 2D materials. The usual formula of TMCs is MX$_2$, while M is transition metal and X is chalcogenide element, namely, S, Se, or Te. Because of the typical 2D structure with high surface-to-volume ratio and missing coordination at edge (Figure 16), TMCs exhibits high chemical sensitivity [36]. Considering its versatile physicochemical properties, TMCs can be applied in catalyst [41], energy storage [39], and biology [110]. Some TMCs such as WS$_2$ [111], TiS$_2$ [112] are also used in TiO$_2$ photocatalysis. Among TMCs, MoS$_2$ show extraordinary potential as semiconductors owing to its thickness dependent bandgap and natural abundance. When bulk MoS$_2$ are stripped into a single layer or several layers of nanosheets, the indirect bandgap (1.3 eV) can be converted to a direct bandgap (1.8 eV) [113] and show excellent performance in photocatalysis after compositing with TiO$_2$ [114]. Besides, its high surface-to-volume ratio makes up for the limitation of the low theoretical specific capacity of TiO$_2$. The synergy between MoS$_2$ and TiO$_2$ endows the TiO$_2$/MoS$_2$ composite superior performance compared to their single material.

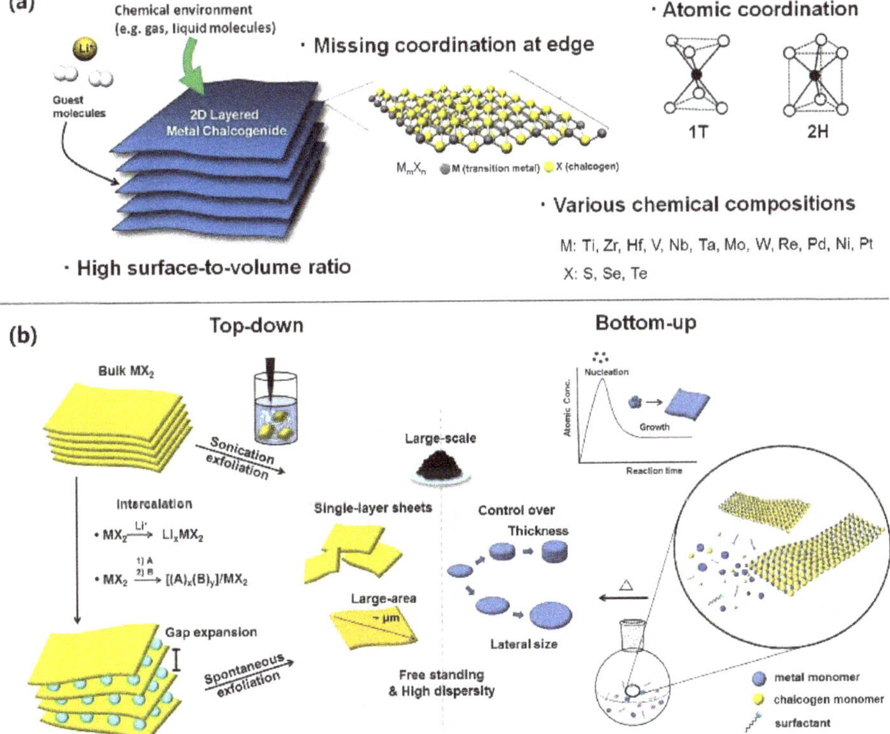

Figure 16. Structure (**a**) and solution-based preparation (**b**) of 2D layered transition metal chalcogenides (TMCs) nanosheets based on top-down and bottom-up approaches. Reprinted with permission from [36]. Copyright 2018, American Chemical Society.

2.4.1. The Synthesis of MoS$_2$/TiO$_2$ Composites

Similar to the synthesis methods of graphene/TiO$_2$ composite, the synthesis of MoS$_2$/TiO$_2$ composites is also divided into ex-situ methods and in-situ methods. For the in-situ method, TiO$_2$ and MoS$_2$ are synthesized separately, then the two are combined by various methods, such as hydrothermal/solvothermal assembly [115,116], mechanical method [117], drop-casting [118], or sol–gel [119], which can be also applied for in-situ methods [120,121]. The ex-situ method is simple and inexpensive, but the two compounds have poor dispersion and show weak interactions. Despite the same process as ex-situ method, there are chemical vapor deposition [122] and co-reduction precipitation [123] in in-situ process. Among them, the hydrothermal method is simple, easy to operate, and has good controllability, and thus is most commonly used in the preparation of MoS_2/TiO_2 composite materials. The in-situ reduction method uses one of the materials as a substrate and then coats or loads the other material. This involves the molybdenum disulfide as substrate or TiO_2 as a substrate. The following paragraphs will discuss the two kinds of composites.

- *MoS_2 as substrate.* In this process, MoS_2 are pre-prepared as substrate for the in-situ growth of TiO_2. Hydrothermal method is widely used in which tetrabutyl titanate serves as titanate source [124,125]. Recently, another approach has been developed to synthesize $MoS_2@TiO_2$ composites. Ren et al. [126] reported TiO_2-modified MoS_2 nanosheet arrays by the ALD process, coating a thin layer of TiO_2 on both the edge and basal planes of TiO_2 (Figure 17). It provides a new insight for the combination of sites at the basal planes of TiO_2.

- *TiO_2 composite as substrate.* For coated MoS_2/TiO_2 composites, TiO_2 are usually substrates. Liu et al. [127] reported a N-$TiO_{2-x}@MoS_2$ core-shell heterostructure composite. TBT and urea were used to prepare N-doped TiO_2 microspheres (N-TiO_2) with a smooth surface by hydrothermal method. Considering the growth of molybdenum sulfide on the TiO_2 substrate, specific morphology and growth sites of TiO_2 is needed. Sun et al. [128] took a targeted etching route to control the morphology of TiO_2/MoS_2 nanocomposites. Hollow microspheres structured TiO_2/MoS_2 showed a higher dye degradation activity due to a larger proportion of interface, compared to TiO_2/MoS_2 nanocomposites of yolk-shell structures. Other structures such as nanobelts and nanotubes have also been developed [129,130]. In addition to the morphology, the formation of a specific crystal structure of TiO_2 as a substrate has also got attention to prepare high performance MoS_2/TiO_2 composites [130,131]. He et al. [130] reported a few-layered 1T-MoS_2 coating on Si doped TiO_2 nanotubes (MoS_2/TiO_2 NTs hybrids) through hydrothermal process. Because of the higher catalytic activity of 1T phase of MoS_2 and Si doped TiO_2, MoS_2/TiO_2 NTs hybrids nanocomposites exhibited excellent photocatalytic activity.

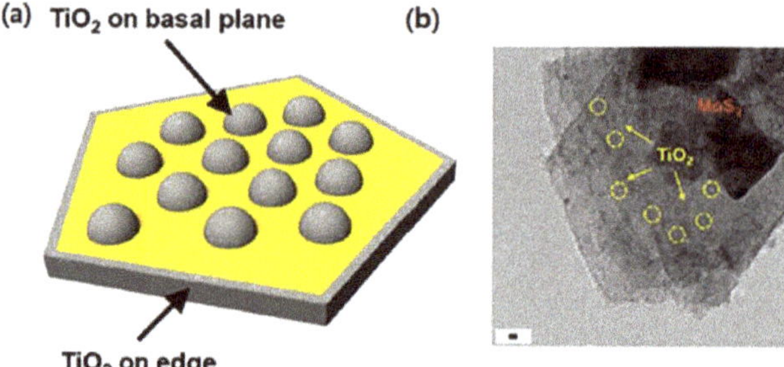

Figure 17. (**a**) Schematic illustration and (**b**) TEM image of the ALD TiO_2 coating on pristine MoS_2. Reprinted with permission from [126]. Copyright 2017, Wiley-VCH.

2.4.2. The Role of MoS_2 in TiO_2 Photocatalysis

During the photocatalysis process, electrons transfer through the interface between TiO_2 and MoS_2, and therefore the contact between the two is vital for photocatalytic activity. A strategy for construction of 3D semiconductor heterojunction structure by TiO_2 and 2D-structured MoS_2 is proposed to achieve increase of active sites and decrease of electron-hole pair combination [127,132]. For example, a 3D flower-like N-$TiO_{2-x}@MoS_2$ was obtained by hydrothermal method. Considering that the smooth TiO_2 nanosphere shows poor affinity when coated with MoS_2 nanosheets, TiO_2 was doped with N and

Ti^{3+}. X-ray photoelectron spectroscopy (XPS) shows the existence of electronic interactions between MoS_2 and $N-TiO_{2-x}$ and the strong heterostructure effect between the MoS_2 nanoflower and $N-TiO_{2-x}$ nanosphere [127]. Another study of 3D $TiO_2@MoS_2$ revealed that the formation of Ti-S bonds made TiO_2 nanoarrays firmly grasp MoS_2, thus affording a marvelous mechanical stability for the integrated architectures [133].

Different phase of MoS_2 exhibits various chemical and physical properties when combined with TiO_2. MoS_2 has two main phases, namely the metallic 1T phase and semiconducting 2H phase. As for 2H phase, the active site with catalytic activity is located at the edge of the MoS_2 layers and the basal surface of MoS_2 is catalytically inactive [134]. Therefore, the 1T phase of MoS_2 with active sites on both edge and basal planes attracts researchers' attention in recent years [118,125,131]. A typical schematic of MoS_2/TiO_2 composites for photocatalytic hydrogen production is shown in Figure 18. The $1T-MoS_2$ nanosheets not only provide extra reaction sites on the basal plane, but also play a role in electron delivery. Because of the active site distributing on the edge of $2H-MoS_2$ nanosheets, the photogenerated electron from TiO_2 needs a long-distance move before reacted with H_2O. This leaded to a lower diffusion rate compared with $1T-MoS_2/TiO_2$ composites and thus enhanced the separation efficiency of electron-hole pairs. Therefore, the $1T-MoS_2/TiO_2$ composites exhibited excellent photocatalytic activity as the hydrogen production rate of $1T-MoS_2/TiO_2$ was 5 and 8 times higher than those of bare TiO_2 and $1T-MoS_2/TiO_2$ [125]. In another research, $1T-MoS_2$ coated onto TiO_2 (001) composite (MST) was synthesized. DFT calculations suggested a closer distance between the interface electrons and MoS_2 surface than that of TiO_2 [131] (Figure 19). Therefore the photo-induced electrons can easily transfer to the conducting channel of MoS_2. Furthermore, the introduction of $1T-MoS_2$ prolonged the carrier lifetime remarkedly. All the factors led to an enhanced photocatalytic activity.

To further inhibit the recombination of electron-hole pairs, cocatalyst such as graphene is applied to MoS_2/TiO_2 system [115,135,136]. Xiang et al. employed $TiO_2/MoS_2/graphene$ composite as photocatalyst [135]. In this system, photo-inducted electrons transfer from VB to CB of TiO_2. Then the electrons are further injected into the graphene sheets or MoS_2 nanoparticles. What is more, graphene sheets can be seen as electrons transport 'highway' through which electrons move from VB of TiO_2 to MoS_2 (Figure 20). The cocatalyst of MoS_2 and graphene enhances the interfacial charge transfer rate, inhibits the recombination of electron-hole pairs and offers a host of active site for adsorption and reaction. Han et al. constructed 3D $MoS_2/P25/graphene$-aerogel networks. In addition to the above-mentioned advantages, 3D graphene porous architecture has a highly porous ultrafine nanoassembly network structure, excellent electric conductivity, and the maximization of accessible sites [115]. Recently, a 3D double-heterostructured photocatalyst was constructed by connecting a TiO_2-MoS_2 core-shell nanosheets (NSs) on a graphite fiber ($GF@MoS_2-TiO_2$) [136]. Mechanism of photocatalytic decomposition of dyes under both visible light and UV light was discussed (Figure 21). Anatase TiO_2 has a wide band gap (2.96 eV), while the band gap of MoS_2 is 1.8 eV. Because of the moderate bandgap of MoS_2, the electrons can be irradiated from VB to CB of MoS_2 and then inject into CB of TiO_2 or transfer to graphene through intimate double-heterojunction contact under visible light. Graphene acts as electrons accepter under both circumstance, leading to a high rate of charge separation and thus depress the charge recombination. The contact interfaces and synergy among graphene, TiO_2 and MoS_2 play an important role in the superior photocatalytic activities.

While the transfer of electrons are paid special attention, the role of capturing the holes are often ignored. To solve this problem, a $TiO_2/WO_3@MoS_2$ (TWM) hybrid Z-scheme photocatalytic system was structured. TiO_2 and WO_3 have the appropriate energy level matching to form the Z-scheme, while the position of VB in WO_3 is lower than the VB of TiO_2, and the CB of WO_3 is between the CB and VB of TiO_2 [137]. Under UV light irradiation, the VB electrons of all three parts are excited to corresponding CB level. The excited electrons on CB of TiO_2 then transfer to CB of MoS_2 for H_2 evolution, meanwhile the excited electrons on CB of WO_3 were inject to the VB of TiO_2 (Figure 22). This procedure suppressed the recombination of photoinduced electrons and holes in TiO_2, and therefore the photogenerated electrons and holes can be efficiently separated, which further leads to effective photocatalytic activity [137].

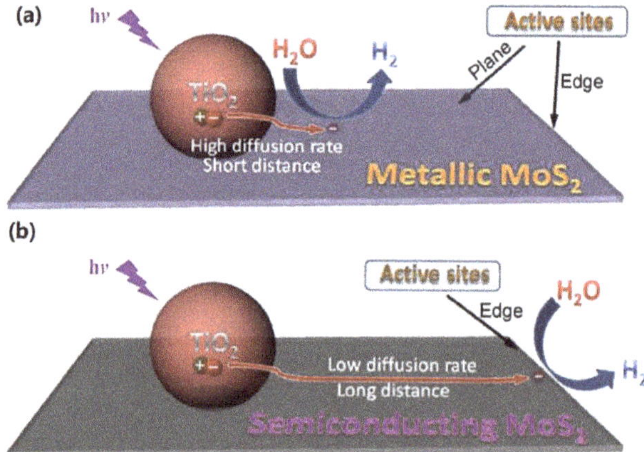

Figure 18. Schematic illustrating charge-transfer behavior and H_2 evolution active sites for (**a**) 1T-MoS$_2$/TiO$_2$ and (**b**) 2H-MoS$_2$/TiO$_2$. Reprinted with permission from [125]. Copyright 2014, Springer.

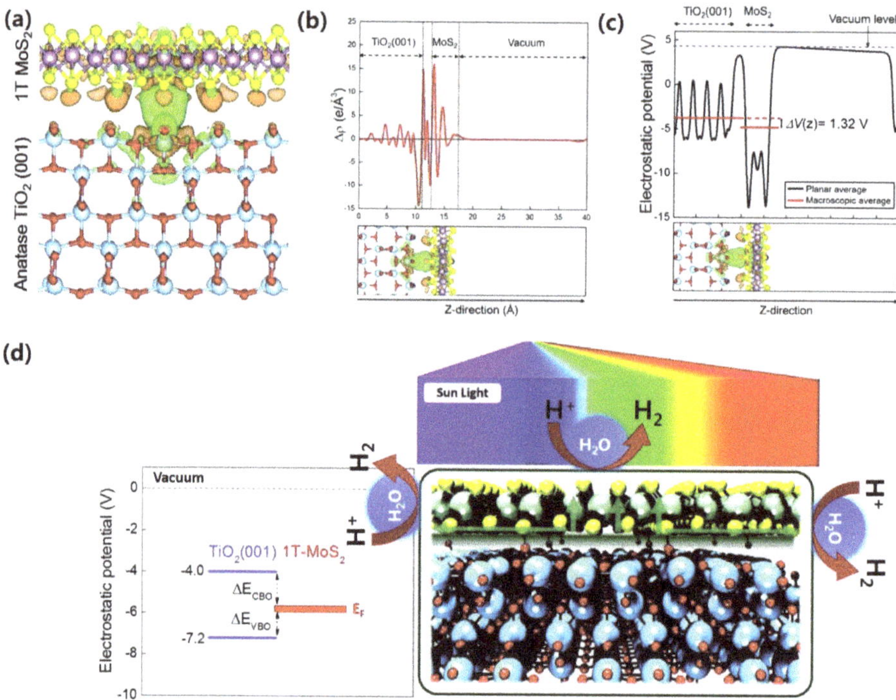

Figure 19. (**a**) The charge density difference, (**b**) electrostatic potential and differential charge density of the MoS$_2$/TiO$_2$(001) junction; (**c**) Planar-averaged differential electron density Dr(z) for MoS$_2$/TiO$_2$(001); (**d**) Photocatalytic mechanism for 1T-MoS$_2$/TiO$_2$. Reprinted with permission from [131]. Copyright 2017, the Owner Societies.

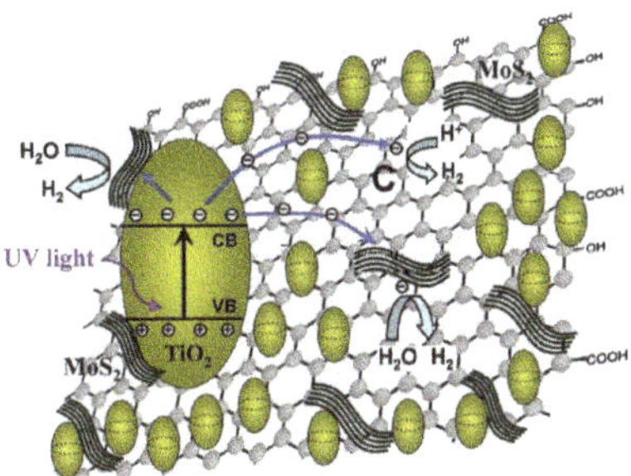

Figure 20. Schematic illustration of the charge transfer in $TiO_2/MoS_2/graphene$ composites. Reprinted with permission from [135]. Copyright 2012, American Chemical Society.

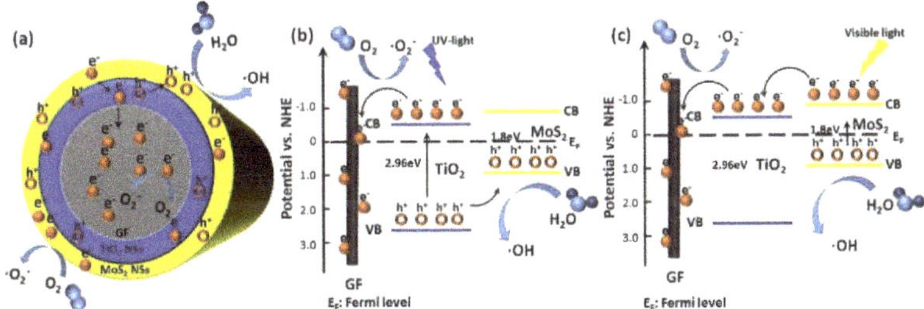

Figure 21. Structure (**a**) and schematic diagram of electron-hole separation mechanism upon UV (**b**) and visible light (**c**) excitation for 3D graphene@MoS$_2$-TiO$_2$ composites. Reprinted with permission from [136]. Copyright 2017, Elsevier.

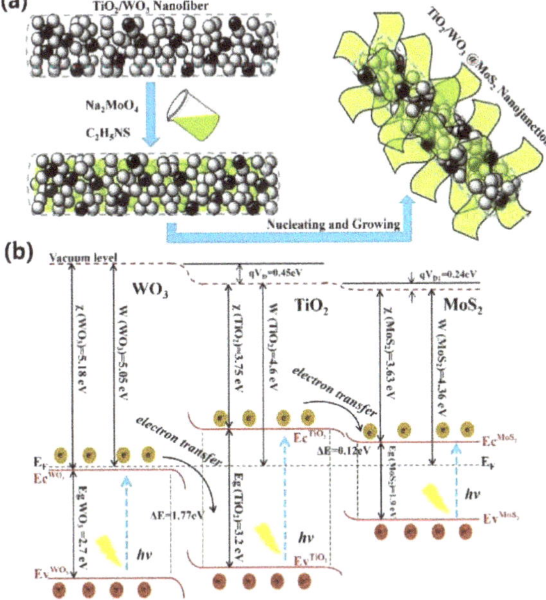

Figure 22. (**a**) Schematic illustration for the growth of MoS$_2$ nanosheets (**b**) Schematic diagram of the photocatalytic H$_2$ generation over the ternary TiO_2/WO_3@MoS$_2$ heterostructure composite. Reprinted with permission from [137]. Copyright 2017, Elsevier.

3. Conclusions

The coupling between TiO_2 and 2D material has proven to be an efficient approach to enhanced photocatalytic activity. Different methods vary the structures and surface contact of the hybrid and thus can modify the carrier separation process. The synergistic effects show that 2D material plays a vital role in photocatalysis when composited with TiO_2. First, 2D material can act as electrons accepter or bridge to conduct photoinduced electrons, and therefore represses the recombination of carriers efficiently. Second, the gigantic surface of 2D material provides substantial active sites for substrate capture and reaction, not to mention rapid electrons transfer rate. Third, the 2D material can be decorated to obtain expected properties, for example, non-metal doping to adjust the energy level, specific crystal structure to short the pathway for interfacial charge transfer, and defects or introduced functional group for substrate trapping. What's more, the interfacial heterojunction can adjust energy level to broaden light response range and improve solar utilization. To further enhance the separation efficiency of electron-hole pairs, other photocatalysts are introduced to construct co-catalyst systems among which Z-scheme system can raise the hole trapping rate to some extent, and thus offers a new point to improve the separation of carriers. All factors mentioned above highlight the critical role of 2D material in photocatalyst and the 2D material/TiO_2 hybrid is worth to get further insight for a wider range of applications.

Funding: This research was funded by the National Natural Science Foundation of China (21761132007, 21773168, and 51503143), the National Key R&D Program of China (2016YFE0114900), Tianjin Natural Science Foundation (16JCQNJC05000), Innovation Foundation of Tianjin University (2016XRX-0017), and Tianjin Science and Technology Innovation Platform Program (No. 14TXGCCX00017).

Conflicts of Interest: The authors declare no conflict of interest.

References

1. Lei, J.M.; Peng, Q.X.; Luo, S.P.; Liu, Y.; Zhan, S.Z.; Ni, C.L. A nickel complex, an efficient cocatalyst for both electrochemical and photochemical driven hydrogen production from water. *Mol. Catal.* **2018**, *448*, 10–17. [CrossRef]
2. Wang, Z.; Jin, Z.; Yuan, H.; Wang, G.; Ma, B. Orderly-designed Ni_2P nanoparticles on g-C_3N_4 and UiO-66 for efficient solar water splitting. *J. Colloid Interface Sci.* **2018**, *532*, 287–299. [CrossRef] [PubMed]
3. Vaiano, V.; Lara, M.A.; Iervolino, G.; Matarangolo, M.; Navio, J.A.; Hidalgo, M.C. Photocatalytic H_2 production from glycerol aqueous solutions over fluorinated Pt-TiO_2 with high {001} facet exposure. *J. Photochem. Photobiol. A* **2018**, *365*, 52–59. [CrossRef]
4. Liu, Y.L.; Yang, C.L.; Wang, M.S.; Ma, X.G.; Yi, Y.G. Te-doped perovskite $NaTaO_3$ as a promising photocatalytic material for hydrogen production from water splitting driven by visible light. *Mater. Res. Bull.* **2018**, *107*, 125–131. [CrossRef]
5. Cai, Q.; Hu, J. Effect of UVA/LED/TiO_2 photocatalysis treated sulfamethoxazole and trimethoprim containing wastewater on antibiotic resistance development in sequencing batch reactors. *Water Res.* **2018**, *140*, 251–260. [CrossRef] [PubMed]
6. Lee, C.G.; Javed, H.; Zhang, D.; Kim, J.H.; Westerhoff, P.; Li, Q.; Alvarez, P.J.J. Porous electrospun fibers embedding TiO_2 for adsorption and photocatalytic degradation of water pollutants. *Environ. Sci. Technol.* **2018**, *52*, 4285–4293. [CrossRef] [PubMed]
7. Zhou, Y.; Li, W.; Wan, W.; Zhang, R.; Lin, Y. W/Mo co-doped $BiVO_4$ for photocatalytic treatment of polymer-containing wastewater in oilfield. *Superlattice Microstruct.* **2015**, *82*, 67–74. [CrossRef]
8. Yu, J.; Wang, S.; Low, J.; Xiao, W. Enhanced photocatalytic performance of direct Z-scheme g-C_3N_4-TiO_2 photocatalysts for the decomposition of formaldehyde in air. *Phys. Chem. Chem. Phys.* **2013**, *15*, 16883–16890. [CrossRef] [PubMed]
9. Abou Saoud, W.; Assadi, A.A.; Guiza, M.; Bouzaza, A.; Aboussaoud, W.; Ouederni, A.; Soutrel, I.; Wolbert, D.; Rtimi, S. Study of synergetic effect, catalytic poisoning and regeneration using dielectric barrier discharge and photocatalysis in a continuous reactor: Abatement of pollutants in air mixture system. *Appl. Catal. B Environ.* **2017**, *213*, 53–61. [CrossRef]

10. Zhang, J.; Ma, Z. Ag_3VO_4/AgI composites for photocatalytic degradation of dyes and tetracycline hydrochloride under visible light. *Mater. Lett.* **2018**, *216*, 216–219. [CrossRef]

11. Huang, C.; Chen, H.; Zhao, L.; He, X.; Li, W.; Fang, W. Biogenic hierarchical MIL-125/TiO_2@SiO_2 derived from rice husk and enhanced photocatalytic properties for dye degradation. *Photochem. Photobiol.* **2018**, *94*, 512–520. [CrossRef]

12. Kumar, M.; Mehta, A.; Mishra, A.; Singh, J.; Rawat, M.; Basu, S. Biosynthesis of tin oxide nanoparticles using psidium guajava leave extract for photocatalytic dye degradation under sunlight. *Mater. Lett.* **2018**, *215*, 121–124. [CrossRef]

13. Dashairya, L.; Sharma, M.; Basu, S.; Saha, P. Enhanced dye degradation using hydrothermally synthesized nanostructured Sb_2S_3/rGO under visible light irradiation. *J. Alloy Compd.* **2018**, *735*, 234–245. [CrossRef]

14. Xiao, L.; Lin, R.; Wang, J.; Cui, C.; Wang, J.; Li, Z. A novel hollow-hierarchical structured Bi_2WO_6 with enhanced photocatalytic activity for CO_2 photoreduction. *J. Colloid Interface Sci.* **2018**, *523*, 151–158. [CrossRef] [PubMed]

15. Nie, N.; Zhang, L.; Fu, J.; Cheng, B.; Yu, J. Self-assembled hierarchical direct Z-scheme g-C_3N_4/ZnO microspheres with enhanced photocatalytic CO_2, reduction performance. *Appl. Surf. Sci.* **2018**, *441*, 12–22. [CrossRef]

16. Xia, P.; Zhu, B.; Yu, J.; Cao, S.; Jaroniec, M. Ultrathin nanosheet assemblies of graphitic carbon nitride for enhanced photocatalytic CO_2 reduction. *J. Mater. Chem. A* **2017**, *5*, 3230–3238. [CrossRef]

17. Fujishima, A.; Honda, K. Electrochemical photolysis of water at a semiconductor electrode. *Nature* **1972**, *238*, 37–38. [CrossRef] [PubMed]

18. Verbruggen, S.W. TiO_2 photocatalysis for the degradation of pollutants in gas phase: From morphological design to plasmonic enhancement. *J. Photochem. Photobiol. C Photochem. Rev.* **2015**, *24*, 64–82. [CrossRef]

19. Zhang, H.; Wang, Z.; Li, R.; Guo, J.; Li, Y.; Zhu, J.; Xie, X. TiO_2 supported on reed straw biochar as an adsorptive and photocatalytic composite for the efficient degradation of sulfamethoxazole in aqueous matrices. *Chemosphere* **2017**, *185*, 351–360. [CrossRef] [PubMed]

20. Geltmeyer, J.; Teixido, H.; Meire, M.; Van Acker, T.; Deventer, K.; Vanhaecke, F.; Van Hulle, S.; De Buysser, K.; De Clerck, K. TiO_2 functionalized nanofibrous membranes for removal of organic (micro)pollutants from water. *Sep. Purif. Technol.* **2017**, *179*, 533–541. [CrossRef]

21. Shayegan, Z.; Lee, C.-S.; Haghighat, F. TiO_2 photocatalyst for removal of volatile organic compounds in gas phase—A review. *Chem. Eng. J.* **2018**, *334*, 2408–2439. [CrossRef]

22. Drunka, R.; Grabis, J.; Jankovica, D.; Krumina, A.; Rasmane, D. Microwave-assisted synthesis and photocatalytic properties of sulphur and platinum modified TiO_2 nanofibers. *IOP Conf. Sér. Mater. Sci. Eng.* **2015**, *77*, 012010. [CrossRef]

23. Dong, C.; Song, H.; Zhou, Y.; Dong, C.; Shen, B.; Yang, H.; Matsuoka, M.; Xing, M.; Zhang, J. Sulfur nanoparticles in situ growth on TiO_2 mesoporous single crystals with enhanced solar light photocatalytic performance. *RSC Adv.* **2016**, *6*, 77863–77869. [CrossRef]

24. Wang, S.; Pan, L.; Song, J.J.; Mi, W.; Zou, J.J.; Wang, L.; Zhang, X. Titanium-defected undoped anatase TiO_2 with p-type conductivity, room-temperature ferromagnetism, and remarkable photocatalytic performance. *J. Am. Chem. Soc.* **2015**, *137*, 2975–2983. [CrossRef] [PubMed]

25. Hoffmann, M.R.; Choi, W.; Bahnemann, D.W. Environmental applications of semiconductor photocatalysis. *Chem. Rev.* **1995**, *95*, 69–96. [CrossRef]

26. Ma, Y.; Li, Z. Coupling plasmonic noble metal with TiO_2, for efficient photocatalytic transfer hydrogenation: M/TiO_2, (M = Au and Pt) for chemoselective transformation of cinnamaldehyde to cinnamyl alcohol under visible and 365 nm UV Light. *Appl. Surf. Sci.* **2018**, *452*, 279–285. [CrossRef]

27. Matos, J.; Llano, B.; Montana, R.; Poon, P.S.; Hidalgo, M.C. Design of Ag/ and Pt/TiO_2-SiO_2 nanomaterials for the photocatalytic degradation of phenol under solar irradiation. *Environ. Sci. Pollut. Res. Int.* **2018**, *25*, 18894–18913. [CrossRef] [PubMed]

28. Roberto, F.; Orc, M.B.; Luisa, D.; Giuseppe, C.; Leonardo, P.; Salvatore, S. Au/TiO_2-CeO_2 catalysts for photocatalytic water splitting and VOCs oxidation reactions. *Catalysts* **2016**, *6*, 121. [CrossRef]

29. Md Saad, S.K.; Ali Umar, A.; Ali Umar, M.I.; Tomitori, M.; Abd Rahman, M.Y.; Mat Salleh, M.; Oyama, M. Two-dimensional, hierarchical Ag-doped TiO_2 nanocatalysts: Effect of the metal oxidation state on the photocatalytic properties. *ACS Omega* **2018**, *3*, 2579–2587. [CrossRef]

30. Meryam, Z.; Benoit, S.; Mounira, M.; Ramzi, B.; Yu, W.; Wu, M.; Olivier, D.; Li, Y.; Su, B. ZnO quantum dots decorated 3DOM TiO$_2$ nanocomposites: Symbiose of quantum size effects and photonic structure for highly enhanced photocatalytic degradation of organic pollutants. *Appl. Catal. B Environ.* **2016**, *199*, 187–198. [CrossRef]

31. Luisa, M.P.; Sergio, M.-T.; Sónia, A.C.; Josephus, G.B.; José, L.F.; Adrián, M.T.; Silvab, J.L. Photocatalytic activity of functionalized nanodiamond-TiO$_2$ composites towards water pollutants degradation under UV/Vis irradiation. *Appl. Surf. Sci.* **2018**, *458*, 839–848. [CrossRef]

32. Prabhu, S.; Cindrella, L.; Kwon, O.J.; Mohanraju, K. Photoelectrochemical and photocatalytic activity of TiO$_2$-WO$_3$ heterostructures boosted by mutual interaction. *Mater. Sci. Semicond. Process.* **2018**, *88*, 10–19. [CrossRef]

33. Zhang, X.; Lai, Z.; Tan, C.; Zhang, H. Solution-processed two-dimensional MoS$_2$ nanosheets: Preparation, hybridization, and applications. *Angew. Chem. Int. Ed. Engl.* **2016**, *55*, 8816–8838. [CrossRef] [PubMed]

34. Thomas, A.; Fischer, A.; Goettmann, F.; Antonietti, M.; Müller, J.O.; Schlögl, R.; Carlsson, J.M. Graphitic carbon nitride materials: Variation of structure and morphology and their use as metal-free catalysts. *J. Mater. Chem.* **2008**, *18*, 4893–4908. [CrossRef]

35. Hisatomi, T.; Kubota, J.; Domen, K. Recent advances in semiconductors for photocatalytic and photoelectrochemical water splitting. *Chem. Soc. Rev.* **2014**, *43*, 7520–7535. [CrossRef] [PubMed]

36. Han, J.H.; Kwak, M.; Kim, Y.; Cheon, J. Recent advances in the solution-based preparation of two-dimensional layered transition metal chalcogenide nanostructures. *Chem. Rev.* **2018**, *118*, 6151–6188. [CrossRef] [PubMed]

37. Bonaccorso, F.; Colombo, L.; Yu, G.; Stoller, M.; Tozzini, V.; Ferrari, A.C.; Ruoff, R.S.; Pellegrini, V. 2D materials. Graphene, related two-dimensional crystals, and hybrid systems for energy conversion and storage. *Science* **2015**, *347*, 1246501. [CrossRef] [PubMed]

38. Jin, Z.; Yuan, M.; Li, H.; Yang, H.; Zhou, Q.; Liu, H.; Lan, X.; Liu, M.; Wang, J.; Sargent, E.H.; et al. Graphdiyne: An efficient hole transporter for stable high-performance colloidal quantum dot solar cells. *Adv. Funct. Mater.* **2016**, *26*, 5284–5289. [CrossRef]

39. Muller, G.A.; Cook, J.B.; Kim, H.S.; Tolbert, S.H.; Dunn, B. High performance pseudocapacitor based on 2D layered metal chalcogenide nanocrystals. *Nano Lett.* **2015**, *15*, 1911–1917. [CrossRef] [PubMed]

40. Wang, X.; Maeda, K.; Thomas, A.; Takanabe, K.; Xin, G.; Carlsson, J.M.; Domen, K.; Antonietti, M. A metal-free polymeric photocatalyst for hydrogen production from water under visible light. *Nat. Mater.* **2009**, *8*, 76–80. [CrossRef] [PubMed]

41. Su, T.; Shao, Q.; Qin, Z.; Guo, Z.; Wu, Z. Role of interfaces in two-dimensional photocatalyst for water splitting. *ACS Catal.* **2018**, *8*, 2253–2276. [CrossRef]

42. Huang, C.; Li, Y.; Wang, N.; Xue, Y.; Zuo, Z.; Liu, H.; Li, Y. Progress in research into 2D graphdiyne-based materials. *Chem. Rev.* **2018**, *118*, 7744–7803. [CrossRef]

43. Fiori, G.; Francesco, B.; Iannaccone, G.; Palacios, T.; Neumaier, D.; Seabaugh, A.; Banerjee, S.K.; Colombo, L. Electronics based on two-dimensional materials. *Nat. Nanotechnol.* **2014**, *9*, 768–779. [CrossRef] [PubMed]

44. Cao, X.; Tao, C.; Zhang, X.; Zhao, W.; Zhang, H. Solution-processed two-dimensional metal dichalcogenide-based nanomaterials for energy storage and conversion. *Adv. Mater.* **2016**, *28*, 6167–6196. [CrossRef] [PubMed]

45. Deng, D.; Novoselov, K.S.; Fu, Q.; Zheng, N.; Tian, Z.; Bao, X. Catalysis with two-dimensional materials and their heterostructures. *Nat. Nanotechnol.* **2016**, *11*, 218–230. [CrossRef] [PubMed]

46. Novoselov, K.S.; Geim, A.K.; Morozov, S.V.; Jiang, D.; Zhang, Y.; Dubonos, S.V.; Grigorieva, I.V.; Firsov, A.A. Electric field effect in atomically thin carbon films. *Science* **2004**, *306*, 666–669. [CrossRef] [PubMed]

47. Lee, C.; Wei, X.; Kysar, J.W.; Hone, J. Measurement of the elastic properties and intrinsic strength of monolayer graphene. *Science* **2008**, *321*, 385–388. [CrossRef] [PubMed]

48. Tang, B.; Hu, G.; Gao, H.; Hai, L. Application of graphene as filler to improve thermal transport property of epoxy resin for thermal interface materials. *Int. J. Heat Mass Transf.* **2015**, *85*, 420–429. [CrossRef]

49. Tang, B.; Guoxin, H.; Gao, H. Raman spectroscopic characterization of graphene. *Appl. Spectrosc. Rev.* **2010**, *45*, 369–407. [CrossRef]

50. Wang, Q.; Arash, B. A review on applications of carbon nanotubes and graphenes as nano-resonator sensors. *Comput. Mater. Sci.* **2014**, *82*, 350–360. [CrossRef]

51. Mittal, G.; Dhand, V.; Rhee, K.Y.; Park, S.-J.; Lee, W.R. A review on carbon nanotubes and graphene as fillers in reinforced polymer nanocomposites. *J. Ind. Eng. Chem.* **2015**, *21*, 11–25. [CrossRef]

52. Zhang, Q.; Wu, Z.; Li, N.; Pu, Y.; Wang, B.; Zhang, T.; Tao, J. Advanced review of graphene-based nanomaterials in drug delivery systems: Synthesis, modification, toxicity and application. *Mater. Sci. Eng. C Mater. Biol. Appl.* **2017**, *77*, 1363–1375. [CrossRef] [PubMed]

53. Chabot, V.; Higgins, D.; Yu, A.; Xiao, X.; Chen, Z.; Zhang, J. A review of graphene and graphene oxide sponge: Material synthesis and applications to energy and the environment. *Energy Environ. Sci.* **2014**, *7*, 1564–1596. [CrossRef]

54. Hu, G.; Tang, B. Photocatalytic mechanism of graphene/titanate nanotubes photocatalyst under visible-light irradiation. *Mater. Chem. Phys.* **2013**, *138*, 608–614. [CrossRef]

55. Chen, Y.; Dong, X.; Cao, Y.; Xiang, J.; Gao, H. Enhanced photocatalytic activities of low-bandgap TiO_2-reduced graphene oxide nanocomposites. *J. Nanopart. Res.* **2017**, *19*. [CrossRef]

56. Sun, J.; Zhang, H.; Guo, L.H.; Zhao, L. Two-dimensional interface engineering of a Titania-graphene nanosheet composite for improved photocatalytic activity. *ACS Appl. Mater. Interfaces* **2013**, *5*, 13035–13041. [CrossRef] [PubMed]

57. Lin, L.Y.; Nie, Y.; Kavadiya, S.; Soundappan, T.; Biswas, P. N-doped reduced graphene oxide promoted nano TiO_2 as a bifunctional adsorbent/photocatalyst for CO_2 photoreduction: Effect of N species. *Chem. Eng. J.* **2017**, *316*, 449–460. [CrossRef]

58. Zhu, M.; Muhammad, Y.; Hu, P.; Wang, B.; Wu, Y.; Sun, X.; Tong, Z.; Zhao, Z. Enhanced interfacial contact of dopamine bridged melamine-graphene/TiO_2 nano-capsules for efficient photocatalytic degradation of gaseous formaldehyde. *Appl. Catal. B Environ.* **2018**, *232*, 182–193. [CrossRef]

59. Xiang, Q.; Yu, J.; Jaroniec, M. Graphene-based semiconductor photocatalysts. *Chem. Soc. Rev.* **2012**, *41*, 782–796. [CrossRef] [PubMed]

60. Hummers, J.; Offeman, R.E. Preparation of graphitic oxide. *J. Am. Chem. Soc.* **1958**, *80*, 1339. [CrossRef]

61. Bai, H.; Li, C.; Shi, G. Functional composite materials based on chemically converted graphene. *Adv. Mater.* **2011**, *23*, 1089–1115. [CrossRef] [PubMed]

62. Rahimi, R.; Zargari, S.; Sadat Shojaei, Z. Photoelectrochemical investigation of TiO_2-graphene nanocomposites. In Proceedings of the International Electronic Conference on Synthetic Organic Chemistry, Tehran, Iran, 18 November 2014; pp. 1–10.

63. Pogacean, F.; Rosu, M.C.; Coros, M. Graphene/TiO_2-Ag based composites used as sensitive electrode materials for amaranth electrochemical detection and degradation. *J. Electrochem. Soc.* **2018**, *165*, B3054–B3059. [CrossRef]

64. Kuila, T.; Bose, S.; Mishra, A.K.; Khanra, P.; Kim, N.H.; Lee, J.H. Chemical functionalization of graphene and its applications. *Prog. Mater Sci.* **2012**, *57*, 1061–1105. [CrossRef]

65. Zhang, J.; Xiong, Z.; Zhao, X.S. Graphene–metal–oxide composites for the degradation of dyes under visible light irradiation. *J. Mater. Chem.* **2011**, *21*, 3634–3640. [CrossRef]

66. Fan, W.; Lai, Q.; Zhang, Q.; Wang, Y. Nanocomposites of TiO_2 and reduced graphene oxide as efficient photocatalysts for hydrogen evolution. *J. Phys. Chem. C* **2011**, *115*, 10694–10701. [CrossRef]

67. Fu, C.; Chen, T.; Qin, W.; Lu, T.; Sun, Z.; Xie, X.; Pan, L. Scalable synthesis and superior performance of TiO_2-reduced graphene oxide composite anode for sodium-ion batteries. *Ionics* **2015**, *22*, 555–562. [CrossRef]

68. Yan, Y.; Zhang, X.; Mao, H.; Huang, Y.; Ding, Q.; Pang, X. Hydroxyapatite/gelatin functionalized graphene oxide composite coatings deposited on TiO_2 nanotube by electrochemical deposition for biomedical applications. *Appl. Surf. Sci.* **2015**, *329*, 76–82. [CrossRef]

69. Anjusree, G.S.; Nair, A.S.; Nair, S.V.; Vadukumpully, S. One-pot hydrothermal synthesis of TiO_2/graphene nanocomposites for enhanced visible light photocatalysis and photovoltaics. *RSC Adv.* **2013**, *3*, 12933–12938. [CrossRef]

70. Shi, M.; Shen, J.; Ma, H.; Li, Z.; Lu, X.; Li, N.; Ye, M. Preparation of graphene–TiO_2 composite by hydrothermal method from peroxotitanium acid and its photocatalytic properties. *Colloids Surf. Physicochem. Eng. Aspects* **2012**, *405*, 30–37. [CrossRef]

71. Zhang, X.Y.; Li, H.P.; Cui, X.L.; Lin, Y. Graphene/TiO_2 nanocomposites: Synthesis, characterization and application in hydrogen evolution from water photocatalytic splitting. *J. Mater. Chem.* **2010**, *20*, 2801–2806. [CrossRef]

72. Farhangi, N.; Chowdhury, R.R.; Medina-Gonzalez, Y.; Ray, M.B.; Charpentier, P.A. Visible light active Fe doped TiO_2 nanowires grown on graphene using supercritical CO_2. *Appl. Catal. B Environ.* **2011**, *110*, 25–32. [CrossRef]

73. Peng, R.; Liang, L.; Hood, Z.D.; Boulesbaa, A.; Puretzky, A.; Ievlev, A.V.; Come, J.; Ovchinnikova, O.S.; Wang, H.; Ma, C.; et al. In-plane heterojunctions enable multiphasic two-dimensional (2D) MoS$_2$ nanosheets as efficient photocatalysts for hydrogen evolution from water reduction. *ACS Catal.* **2016**, *6*, 6723–6729. [CrossRef]

74. Zhu, Z.; Han, Q.; Yu, D.; Sun, J.; Liu, B. A novel p-n heterojunction of BiVO$_4$/TiO$_2$/GO composite for enhanced visible-light-driven photocatalytic activity. *Mater. Lett.* **2017**, *209*, 379–383. [CrossRef]

75. Li, G.; Li, Y.; Liu, H.; Guo, Y.; Li, Y.; Zhu, D. Architecture of graphdiyne nanoscale films. *Chem. Commun.* **2010**, *46*, 3256–3258. [CrossRef] [PubMed]

76. Matsuoka, R.; Sakamoto, R.; Hoshiko, K.; Sasaki, S.; Masunaga, H.; Nagashio, K.; Nishihara, H. Crystalline graphdiyne nanosheets produced at a gas/liquid or liquid/liquid interface. *J. Am. Chem. Soc.* **2017**, *139*, 3145–3152. [CrossRef] [PubMed]

77. Yin, X.P.; Wang, H.J.; Tang, S.F.; Lu, X.L.; Shu, M.; Si, R.; Lu, T.B. Engineering the coordination environment of single-atom platinum anchored on graphdiyne for optimizing electrocatalytic hydrogen evolution. *Angew. Chem. Int. Ed. Engl.* **2018**, *57*, 9382–9386. [CrossRef] [PubMed]

78. Parvin, N.; Jin, Q.; Wei, Y.; Yu, R.; Zheng, B.; Huang, L.; Zhang, Y.; Wang, L.; Zhang, H.; Gao, M.; et al. Few-layer graphdiyne nanosheets applied for multiplexed real-time DNA detection. *Adv. Mater.* **2017**, *29*. [CrossRef] [PubMed]

79. He, J.; Wang, N.; Cui, Z.; Du, H.; Fu, L.; Huang, C.; Yang, Z.; Shen, X.; Yi, Y.; Tu, Z.; et al. Hydrogen substituted graphdiyne as carbon-rich flexible electrode for lithium and sodium ion batteries. *Nat. Commun.* **2017**, *8*, 1172. [CrossRef] [PubMed]

80. Wang, K.; Wang, N.; He, J.; Yang, Z.; Shen, X.; Huang, C. Preparation of 3D architecture graphdiyne nanosheets for high-performance sodium-ion batteries and capacitors. *ACS Appl. Mater. Interfaces* **2017**, *9*, 40604–40613. [CrossRef] [PubMed]

81. Diederich, F. Carbon scaffolding-building acetylenic all-carbon and carbon-rich compounds. *Nature* **1994**, *369*, 199–207. [CrossRef]

82. Xue, Y.; Li, Y.; Zhang, J.; Liu, Z.; Zhao, Y. 2D graphdiyne materials: Challenges and opportunities in energy field. *Sci. China Chem.* **2018**, *61*, 765–786. [CrossRef]

83. Yang, N.; Liu, Y.; Wen, H.; Tang, Z.; Zhao, H.; Li, Y.; Wang, D. Photocatalytic properties of graphdiyne and graphene modified TiO$_2$: From theory to experiment. *ACS Nano* **2013**, *7*, 1504–1512. [CrossRef] [PubMed]

84. Wang, S.; Yi, L.; Halpert, J.E.; Lai, X.; Liu, Y.; Cao, H.; Yu, R.; Wang, D.; Li, Y. A novel and highly efficient photocatalyst based on P25-graphdiyne nanocomposite. *Small* **2012**, *8*, 265–271. [CrossRef] [PubMed]

85. Jia, Z.; Li, Y.; Zuo, Z.; Liu, H.; Huang, C.; Li, Y. Synthesis and properties of 2D carbon-graphdiyne. *Acc. Chem. Res.* **2017**, *50*, 2470–2478. [CrossRef] [PubMed]

86. Dong, Y.; Zhao, Y.; Chen, Y.; Feng, Y.; Zhu, M.; Ju, C.; Zhang, B.; Liu, H.; Xu, J. Graphdiyne-hybridized N-doped TiO$_2$ nanosheets for enhanced visible light photocatalytic activity. *J. Mater. Sci.* **2018**, *53*, 8921–8932. [CrossRef]

87. Li, G.; Li, Y.; Qian, X.; Liu, H.; Lin, H.; Chen, N.; Li, Y. Construction of tubular molecule aggregations of graphdiyne for highly efficient field emission. *J. Phys. Chem. C* **2011**, *115*, 2611–2615. [CrossRef]

88. Zhou, J.; Gao, X.; Liu, R.; Xie, Z.; Yang, J.; Zhang, S.; Zhang, G.; Liu, H.; Li, Y.; Zhang, J.; et al. Synthesis of graphdiyne nanowalls using acetylenic coupling reaction. *J. Am. Chem. Soc.* **2015**, *137*, 7596–7599. [CrossRef] [PubMed]

89. Li, Y.; Xu, L.; Liu, H.; Li, Y. Graphdiyne and graphyne: From theoretical predictions to practical construction. *Chem. Soc. Rev.* **2014**, *43*, 2572–2586. [CrossRef] [PubMed]

90. Thangavel, S.; Krishnamoorthy, K.; Krishnaswamy, V.; Raju, N.; Kim, S.J.; Venugopal, G. Graphdiyne–ZnO nanohybrids as an advanced photocatalytic material. *J. Phys. Chem. C* **2015**, *119*, 22057–22065. [CrossRef]

91. Long, M.; Tang, L.; Wang, D.; Li, Y.; Shuai, Z. Electronic structure and carrier mobility in graphdiyne sheet and nanoribbons: Theoretical predictions. *ACS Nano* **2011**, *5*, 2593–2600. [CrossRef] [PubMed]

92. Qi, H.; Yu, P.; Wang, Y.; Han, G.; Liu, H.; Yi, Y.; Li, Y.; Mao, L. Graphdiyne oxides as excellent substrate for electroless deposition of pd clusters with high catalytic activity. *J. Am. Chem. Soc.* **2015**, *137*, 5260–5263. [CrossRef] [PubMed]

93. Zhang, X.; Zhu, M.; Chen, P.; Li, Y.; Liu, H.; Li, Y.; Liu, M. Pristine graphdiyne-hybridized photocatalysts using graphene oxide as a dual-functional coupling reagent. *Phys. Chem. Chem. Phys.* **2015**, *17*, 1217–1225. [CrossRef] [PubMed]

94. Zhou, J.; Zhang, M.; Zhu, Y. Photocatalytic enhancement of hybrid C_3N_4/TiO_2 prepared via ball milling method. *Phys. Chem. Chem. Phys.* **2015**, *17*, 3647–3652. [CrossRef] [PubMed]
95. Gu, L.; Wang, J.; Zou, Z.; Han, X. Graphitic-C_3N_4-hybridized TiO_2 nanosheets with reactive {001} facets to enhance the UV- and visible-light photocatalytic activity. *J. Hazard. Mater.* **2014**, *268*, 216–223. [CrossRef] [PubMed]
96. Li, H.; Gao, Y.; Wu, X.; Lee, P.H.; Shih, K. Fabrication of heterostructured g-C_3N_4/Ag-TiO_2 hybrid photocatalyst with enhanced performance in photocatalytic conversion of CO_2 under simulated sunlight irradiation. *Appl. Surf. Sci.* **2017**, *402*, 198–207. [CrossRef]
97. Li, K.; Huang, Z.; Zeng, X.; Huang, B.; Gao, S.; Lu, J. Synergetic effect of Ti^{3+} and oxygen doping on enhancing photoelectrochemical and photocatalytic properties of TiO_2/g-C_3N_4 heterojunctions. *ACS Appl. Mater. Interfaces* **2017**, *9*, 11577–11586. [CrossRef] [PubMed]
98. Zhang, C.; Zhou, Y.; Bao, J.; Fang, J.; Zhao, S.; Zhang, Y.; Sheng, X.; Chen, W. Structure regulation of ZnS@g-C_3N_4/TiO_2 nanospheres for efficient photocatalytic H_2 production under visible-light irradiation. *Chem. Eng. J.* **2018**, *346*, 226–237. [CrossRef]
99. Song, J.; Wang, X.; Ma, J.; Wang, X.; Wang, J.; Xia, S.; Zhao, J. Removal of microcystis aeruginosa and microcystin-LR using a graphitic-C_3N_4/TiO_2 floating photocatalyst under visible light irradiation. *Chem. Eng. J.* **2018**, *348*, 380–388. [CrossRef]
100. Ricci, P.C.; Laidani, N.; Chiriu, D.; Salis, M.; Carbonaro, C.M.; Corpino, R. ALD growth of metal oxide on carbon nitride polymorphs. *Appl. Surf. Sci.* **2018**, *456*, 83–94. [CrossRef]
101. Ren, B.; Wang, T.; Qu, G.; Deng, F.; Liang, D.; Yang, W.; Liu, M. In situ synthesis of g-C_3N_4/TiO_2 heterojunction nanocomposites as a highly active photocatalyst for the degradation of orange ii under visible light irradiation. *Environ. Sci. Pollut. Res.* **2018**, *25*, 19122–19133. [CrossRef] [PubMed]
102. Wang, X.; Wang, F.; Bo, C.; Cheng, K.; Wang, J.; Zhang, J.; Song, H. Promotion of phenol photodecomposition and the corresponding decomposition mechanism over g-C_3N_4/TiO_2 nanocomposites. *Appl. Surf. Sci.* **2018**, *453*, 320–329. [CrossRef]
103. Wu, D.; Li, J.; Guan, J.; Liu, C.; Zhao, X.; Zhu, Z.; Ma, C.; Huo, P.; Li, C.; Yan, Y. Improved photoelectric performance via fabricated heterojunction g-C_3N_4/TiO_2/HNTS loaded photocatalysts for photodegradation of ciprofloxacin. *J. Ind. Eng. Chem.* **2018**, *64*, 206–218. [CrossRef]
104. Tan, Y.; Shu, Z.; Zhou, J.; Li, T.; Wang, W.; Zhao, Z. One-step synthesis of nanostructured g-C_3N_4/TiO_2 composite for highly enhanced visible-light photocatalytic H_2 evolution. *Appl. Catal. B Environ.* **2018**, *230*, 260–268. [CrossRef]
105. Elbanna, O.; Fujitsuka, M.; Majima, T. g-C_3N_4/TiO_2 mesocrystals composite for H_2 evolution under visible-light irradiation and its charge carrier dynamics. *ACS Appl. Mater. Interfaces* **2017**, *9*, 34844–34854. [CrossRef] [PubMed]
106. Guo, N.; Zeng, Y.; Li, H.; Xu, X.; Yu, H.; Han, X. Novel mesoporous TiO_2@g-C_3N_4 hollow core@shell heterojunction with enhanced photocatalytic activity for water treatment and H_2 production under simulated sunlight. *J. Hazard. Mater.* **2018**, *353*, 80–88. [CrossRef] [PubMed]
107. Jiang, Y.; Li, F.; Liu, Y.; Hong, Y.; Liu, P.; Ni, L. Construction of TiO_2 hollow nanosphere/g-C_3N_4 composites with superior visible-light photocatalytic activity and mechanism insight. *J. Ind. Eng. Chem.* **2016**, *41*, 130–140. [CrossRef]
108. Chen, Y.; Huang, W.; He, D.; Situ, Y.; Huang, H. Construction of heterostructured g-C_3N_4/Ag-TiO_2 microspheres with enhanced photocatalysis performance under visible-light irradiation. *ACS Appl. Mater. Interfaces* **2014**, *6*, 14405–14414. [CrossRef] [PubMed]
109. Zhao, Y.; Xu, S.; Sun, X.; Xu, X.; Gao, B. Unique bar-like sulfur-doped g-C_3N_4/TiO_2 nanocomposite: Excellent visible light driven photocatalytic activity and mechanism study. *Appl. Surf. Sci.* **2018**, *436*, 873–881. [CrossRef]
110. Cheng, L.; Liu, J.; Gu, X.; Gong, H.; Shi, X.; Liu, T.; Wang, C.; Wang, X.; Liu, G.; Xing, H.; et al. Pegylated WS_2 nanosheets as a multifunctional theranostic agent for in vivo dual-modal CT/photoacoustic imaging guided photothermal therapy. *Adv. Mater.* **2014**, *26*, 1886–1893. [CrossRef] [PubMed]
111. Zheng, L.; Xiao, X.; Li, Y.; Zhang, W. Enhanced photocatalytic activity of TiO_2 nanoparticles using WS_2/g-C_3N_4 hybrid as co-catalyst. *Trans. Nonferrous Met. Soc. China* **2017**, *27*, 1117–1126. [CrossRef]
112. He, H.Y. Solvothermal synthesis and photocatalytic activity of s-doped TiO_2 and TiS_2 powders. *Res. Chem. Intermed.* **2010**, *36*, 155–161. [CrossRef]

113. Shen, M.; Yan, Z.; Yang, L.; Du, P.; Zhang, J.; Xiang, B. MoS_2 nanosheet/TiO_2 nanowire hybrid nanostructures for enhanced visible-light photocatalytic activities. *Chem. Commun.* **2014**, *50*, 15447–15449. [CrossRef] [PubMed]

114. Hai, X.; Chang, K.; Pang, H.; Li, M.; Li, P.; Liu, H.; Shi, L.; Ye, J. Engineering the edges of MoS_2 (WS_2) crystals for direct exfoliation into monolayers in polar micromolecular solvents. *J. Am. Chem. Soc.* **2016**, *138*, 14962–14969. [CrossRef] [PubMed]

115. Han, W.; Zang, C.; Huang, Z.; Zhang, H.; Ren, L.; Qi, X.; Zhong, J. Enhanced photocatalytic activities of three-dimensional graphene-based aerogel embedding TiO_2 nanoparticles and loading MoS_2 nanosheets as co-catalyst. *Int. J. Hydrogen Energy* **2014**, *39*, 19502–19512. [CrossRef]

116. Nimbalkar, D.B.; Lo, H.-H.; Ramacharyulu, P.V.R.K.; Ke, S.C. Improved photocatalytic activity of rGO/MoS_2 nanosheets decorated on TiO_2 nanoparticles. *RSC Adv.* **2016**, *6*, 31661–31667. [CrossRef]

117. Wang, D.; Xu, Y.; Sun, F.; Zhang, Q.; Wang, P.; Wang, X. Enhanced photocatalytic activity of TiO_2 under sunlight by MoS_2 nanodots modification. *Appl. Surf. Sci.* **2016**, *377*, 221–227. [CrossRef]

118. Pi, Y.; Li, Z.; Xu, D.; Liu, J.; Li, Y.; Zhang, F.; Zhang, G.; Peng, W.; Fan, X. 1T-phase MoS_2 nanosheets on TiO_2 nanorod arrays: 3D photoanode with extraordinary catalytic performance. *ACS Sustain. Chem. Eng.* **2017**, *5*, 5175–5182. [CrossRef]

119. Yu, Y.; Wan, J.; Yang, Z.; Hu, Z. Preparation of the MoS_2/TiO_2/HMFS ternary composite hollow microfibres with enhanced photocatalytic performance under visible light. *J. Colloid Interface Sci.* **2017**, *502*, 100–111. [CrossRef] [PubMed]

120. Yuan, Y.-J.; Ye, Z.J.; Lu, H.W.; Hu, B.; Li, Y.-H.; Chen, D.Q.; Zhong, J.S.; Yu, Z.T.; Zou, Z.G. Constructing anatase TiO_2 nanosheets with exposed (001) facets/layered MoS_2 two-dimensional nanojunctions for enhanced solar hydrogen generation. *ACS Catal.* **2015**, *6*, 532–541. [CrossRef]

121. Liu, X.; Xing, Z.; Zhang, H.; Wang, W.; Zhang, Y.; Li, Z.; Wu, X.; Yu, X.; Zhou, W. Fabrication of 3D mesoporous black TiO_2/MoS_2/TiO_2 nanosheets for visible-light-driven photocatalysis. *ChemSusChem* **2016**, *9*, 1118–1124. [CrossRef] [PubMed]

122. He, H.; Lin, J.; Fu, W.; Wang, X.; Wang, H.; Zeng, Q.; Gu, Q.; Li, Y.; Yan, C.; Tay, B.K.; et al. MoS_2/TiO_2 edge-on heterostructure for efficient photocatalytic hydrogen evolution. *Adv. Funct. Mater.* **2016**, *6*, 1600464. [CrossRef]

123. Van Haandel, L.; Geus, J.W.; Weber, T. Direct synthesis of TiO_2-supported MoS_2 nanoparticles by reductive coprecipitation. *ChemCatChem* **2016**, *8*, 1367–1372. [CrossRef]

124. Ren, X.; Qi, X.; Shen, Y.; Xiao, S.; Xu, G.; Zhang, Z.; Huang, Z.; Zhong, J. 2D co-catalytic MoS_2 nanosheets embedded with 1D TiO_2 nanoparticles for enhancing photocatalytic activity. *J. Phys. D Appl. Phys.* **2016**, *49*, 315304. [CrossRef]

125. Bai, S.; Wang, L.; Chen, X.; Du, J.; Xiong, Y. Chemically exfoliated metallic MoS_2 nanosheets: A promising supporting co-catalyst for enhancing the photocatalytic performance of TiO_2 nanocrystals. *Nano Res.* **2014**, *8*, 175–183. [CrossRef]

126. Kim, Y.; Jackson, D.H.K.; Lee, D.; Choi, M.; Kim, T.W.; Jeong, S.-Y.; Chae, H.J.; Kim, H.W.; Park, N.; Chang, H.; et al. In situ electrochemical activation of atomic layer deposition coated MoS_2 basal planes for efficient hydrogen evolution reaction. *Adv. Funct. Mater.* **2017**, *27*, 1701825. [CrossRef]

127. Liu, X.; Xing, Z.; Zhang, Y.; Li, Z.; Wu, X.; Tan, S.; Yu, X.; Zhu, Q.; Zhou, W. Fabrication of 3D flower-like black n- TiO_{2-x}@MoS_2 for unprecedented-high visible-light-driven photocatalytic performance. *Appl. Catal. B Environ.* **2017**, *201*, 119–127. [CrossRef]

128. Sun, Y.; Lin, H.; Wang, C.; Wu, Q.; Wang, X.; Yang, M. Morphology-controlled synthesis of TiO_2/MoS_2 nanocomposites with enhanced visible-light photocatalytic activity. *Inorg. Chem. Front.* **2018**, *5*, 145–152. [CrossRef]

129. Pu, S.; Long, D.; Wang, M.Q.; Bao, S.J.; Liu, Z.; Yang, F.; Wang, H.; Zeng, Y. Design, synthesis and photodegradation ammonia properties of MoS_2@TiO_2 encapsulated carbon coaxial nanobelts. *Mater. Lett.* **2017**, *209*, 56–59. [CrossRef]

130. He, H.Y. Efficient hydrogen evolution activity of 1T- MoS_2/Si-doped TiO_2 nanotube hybrids. *Int. J. Hydrogen Energy* **2017**, *42*, 20739–20748. [CrossRef]

131. Han, H.; Kim, K.M.; Lee, C.W.; Lee, C.S.; Pawar, R.C.; Jones, J.L.; Hong, Y.R.; Ryu, J.H.; Song, T.; Kang, S.H.; et al. Few-layered metallic 1T-MoS$_2$/TiO$_2$ with exposed (001) facets: Two-dimensional nanocomposites for enhanced photocatalytic activities. *Phys. Chem. Chem. Phys.* **2017**, *19*, 28207–28215. [CrossRef] [PubMed]
132. Lin, T.; Kang, B.; Jeon, M.; Huffman, C.; Jeon, J.; Lee, S.; Han, W.; Lee, J.; Lee, S.; Yeom, G.; et al. Controlled layer-by-layer etching of MoS$_2$. *ACS Appl. Mater. Interfaces* **2015**, *7*, 15892–15897. [CrossRef] [PubMed]
133. Yu, L.; Xie, Y.; Zhou, J.; Li, Y.; Yu, Y.; Ren, Z. Robust and selective electrochemical reduction of CO$_2$: The case of integrated 3D TiO$_2$@MoS$_2$ architectures and Ti–S bonding effects. *J. Mater. Chem. A* **2018**, *6*, 4706–4713. [CrossRef]
134. Jaramillo, T.F.; Jorgensen, K.P.; Bonde, J.; Nielsen, J.H.; Horch, S.; Chorkendorff, I. Identification of active edge sites for electrochemical H$_2$ evolution from MoS$_2$ nanocatalysts. *Science* **2007**, *317*, 100–102. [CrossRef] [PubMed]
135. Xiang, Q.; Yu, J.; Jaroniec, M. Synergetic effect of MoS$_2$ and graphene as cocatalysts for enhanced photocatalytic H$_2$ production activity of TiO$_2$ nanoparticles. *J. Am. Chem. Soc.* **2012**, *134*, 6575–6578. [CrossRef] [PubMed]
136. Hu, X.; Zhao, H.; Tian, J.; Gao, J.; Li, Y.; Cui, H. Synthesis of few-layer MoS$_2$ nanosheets-coated TiO$_2$ nanosheets on graphite fibers for enhanced photocatalytic properties. *Sol. Energy Mater. Sol. Cells* **2017**, *172*, 108–116. [CrossRef]
137. Zhao, J.; Zhang, P.; Fan, J.; Hu, J.; Shao, G. Constructing 2D layered MoS$_2$ nanosheets-modified z-scheme TiO$_2$/WO$_3$ nanofibers ternary nanojunction with enhanced photocatalytic activity. *Appl. Surf. Sci.* **2018**, *430*, 466–474. [CrossRef]

Review

Heterogeneous Photocatalysis and Prospects of TiO$_2$-Based Photocatalytic DeNOxing the Atmospheric Environment

Nick Serpone

PhotoGreen Laboratory, Dipartimento di Chimica, Università di Pavia, via Taramelli 12, 27011 Pavia, Italy; nick.serpone@unipv.it; Tel.: +1-514-489-9551

Received: 18 October 2018; Accepted: 5 November 2018; Published: 16 November 2018

Abstract: This article reviews the efforts of the last two decades to deNOxify the atmospheric environment with TiO$_2$-based photocatalytic materials supported on various cementitious-like substrates. Prior to undertaking this important aspect of applied photocatalysis with metal-oxide emiconductor photocatalysts, however, it is pertinent to describe and understand the fundamentals of Heterogeneous Photocatalysis. The many attempts done in a laboratory setting to degrade (deNOxify) the major components that make up the NO$_x$, namely nitric oxide (NO) and nitrogen dioxide (NO$_2$), but most importantly the efforts expended in deNOxifying the real environment upon depositing titania-based coatings on various model and authentic infrastructures, such as urban roads, highway noise barriers, tunnels, and building external walls among others, are examined. Both laboratory and outdoor experimentations have been performed toward NO$_x$ being oxidized to form nitrates (NO$_3{}^-$) that remain adsorbed on the TiO$_2$-based photocatalytic surfaces (except in tunnels—*indoor* walls) but get subsequently dislodged by rain or by periodic washings of the infrastructures. However, no serious considerations have been given to the possible conversion of NO$_x$ via photocatalytic reduction back to N$_2$ and O$_2$ gases that would restore the atmospheric environment, as the adsorbed nitrates block the surface-active sites of the photocatalyst and when washed-off ultimately cause unduly damages to the environment.

Keywords: photocatalysis; deNOxing; Titania; photophysics; metal oxides; environment

---

1. Introduction

The atmosphere is a very complex matrix that, in addition to nitrogen (N$_2$) and oxygen (O$_2$) gases, so important to human survival, consists of various other pollutant gases albeit at much lower concentrations: (i) carbonaceous oxides (CO$_x$); (ii) nitrogen oxides (NO$_x$); (iii) sulfur oxides (SO$_x$); (iv) various hydrocarbons (HCs); and (v) particulate matter. These pollutants originate from both natural sources (e.g., volcanic eruptions, wildfires, lightning, and natural degradation of forests among others) and anthropogenic areas (fertilizers and livestock, farms, and urban areas), stationary sources (e.g., industries, power plants, and sewage treatment plants), and mobile sources (e.g., automobiles, trucks, buses, motorcycles, ships, and airplanes) (see Figure 1) [1]. The natural sources of chemical pollutants, however, are of lesser concerns as they are part of the natural environment equilibrium, contrary to the anthropogenic sources that keep increasing in number and concentration with the ever increasing global human population and society's continuous increasing demand for energy and associated technological advances.

Undoubtedly, the most important sources of air pollutants implicate the combustion of fossil fuels to produce energy (residential heating and electricity-generating power plants), together with major metallurgical industries, cement/construction industries, and the transportation sector. Figure 1 also identifies the primary pollutants from various sources: carbon monoxide, sulfur dioxide, ammonia, volatile organic compounds (VOCs), particulates and, relevant to the present article, the two major NO_x agents (NO and NO_2). Subsequently, through various interacting events that involve the Sun's radiation, secondary pollutants are generated, among which are sulfur trioxide, ozone, hydrogen peroxide, and sulfuric and nitric acids (the causes of acid rain). Another class of air pollutants generated from internal combustion engines and industrial fumes that react in the atmosphere with sunlight produce secondary pollutants that, in combination with the primary emissions, create photochemical smog [2].

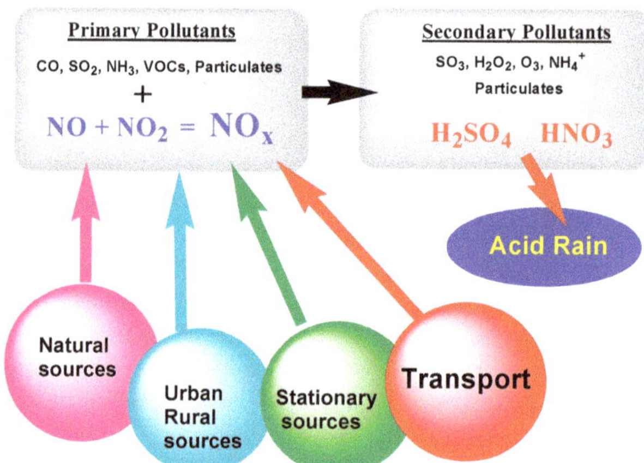

Figure 1. Graphic illustrating the natural and anthropogenic sources of atmospheric pollution together with primary and secondary atmospheric pollutants from various sources that ultimately lead to the formation of acid rain.

Pre-industrial concentrations of atmospheric nitrogen oxides have increased steadily from about 280 ppbv to ca. 320 ppbv until a decade ago (2010), with estimated annual emissions of 13.8 Tg of N per year (teragrams; 10^{12} g), of which ca. 70% is produced by nitrification and denitrification processes in undisturbed terrestrial environments and world's oceans, and ca. 3 Tg of N per year (~ 8%) from agricultural tillage, fertilizer use, and animal wastes [3].

The NO_x gases are formed in large measure in gasoline/diesel combustion engines and in power plants that use fossil fuels to produce electricity via high-temperature combustion/oxidation of the fuel's nitrogen with air oxygen. Initially, only NO is formed followed by formation of NO_2 after combustion in the exhaust and in the atmosphere in the presence of more O_2. Figure 2a reports the 2011 levels of NO_x emissions in the European Union [4], while Figure 2b reports the 2005 NO_x emission levels in the United States [5]. The major anthropogenic sources of nitrogen oxides are combustion engines (transportation sector) and the electricity/heating sectors.

Most of the tropospheric ozone is formed when NO_x, CO and VOCs react in the atmosphere in the presence of sunlight, and, although they might originate in urban areas, airstreams can carry the NO_x far from its sources causing ozone formation in less populated regions. Globally, a VOC whose atmospheric concentration has increased greatly during the last century (viz., methane) contributes to the formation of ozone [6]. A series of complex reactions that involve a VOC (e.g., CO) in the formation of ozone implicates oxidation of this VOC by a hydroxyl radical ($^\bullet$OH) [7] first to yield the radical species HO$^\bullet$CO (Equation (1)), which subsequently reacts with oxygen to produce the hydroperoxy radical HOO$^\bullet$ (Equation (2)) that later reacts with NO to give NO_2 (Equation (3)); the latter photolyzes in sunlight to NO and atomic O(^{3}P) (Equation (4)), which by reaction with oxygen yields ozone (O_3; Equation (5)).

$$OH + CO \rightarrow HO^\bullet CO \tag{1}$$

$$HO^\bullet CO + O_2 \rightarrow HOO^\bullet + CO_2 \tag{2}$$

$$HOO^\bullet + NO \rightarrow {}^\bullet OH + NO_2 \tag{3}$$

$$NO_2 + h\nu \rightarrow NO + O(^3P) \tag{4}$$

$$O(^3P) + O_2 \rightarrow O_3 \tag{5}$$

While the chemistry involving other VOCs might be more complex, the critical step that leads to ozone formation remains nonetheless the oxidation of NO to NO_2 by $HOO^\bullet$ radicals. Nitrogen dioxide also reacts with hydrocarbon molecules present in VOCs to produce yet another pollutant (peroxyacetyl nitrates; PAN), a component of photochemical smog that is mostly responsible for eye irritation and is more damaging to plants than ozone [8].

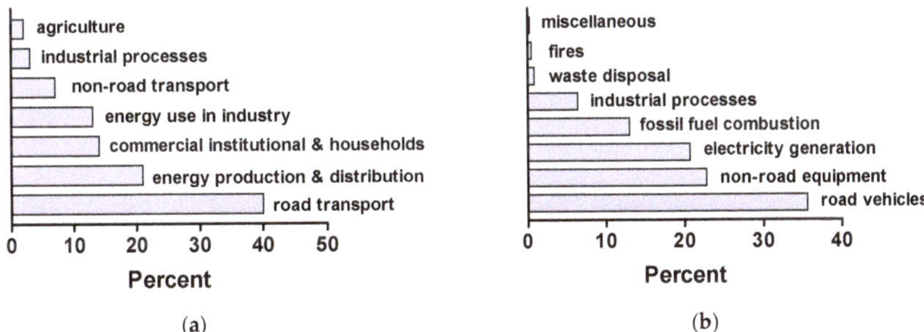

Figure 2. (**a**) Percent emission of NO_x agents from various sources for 2011 in the European Union (Source: *European Union emission inventory report 1990–2011 under the UNECE Convention on Long-range Trans-boundary Air Pollution* (LRTAP)). Reproduced from Ref. [4]. (**b**) Percent emission of NO_x agents for 2005 in the United States (Source: U.S. Environmental Protection Agency, *Air Emission Sources*, 4 November 2009). Reproduced from Ref. [5].

The NO_x family of pollutants (NO, NO_2, N_2O, and their derivatives) causes a wide range of health issues. Nitric oxide (NO) spreads to all parts of the respiratory system because of its low solubility in water, while the health effects of NO_2 are related to its ability to dissolve in moisture to produce HNO_3 acid—a strong mineral acid. Some of the acute health effects include eye irritation (stinging and watering), throat irritation (pungent smell, stinging nose, and coughing), lung irritation (coughing, wheezing and tight chest—difficulty in breathing), and asthma triggered in asthmatics [9,10]. The most serious acute effects occur after significant exposure to NO_2 causing: (a) *acute pulmonary edema*—fluid from damaged lung tissue pours out into air spaces preventing air from getting to deeper lung thereby causing choking (asphyxia); and (b) other chronic health effects such as *asthma* and *obliterative bronchiolitis*, in which the smallest air passages (the bronchioles) are seriously scarred and become distorted and blocked. Consequently, no one questions the need for NO_x-free clean air as essential to maintain/enhance an individual's health, and to maintain the integrity of the surrounding environment. In this regard, transformation of the two major NO_x species (NO and NO_2), indeed their suppression, has become a necessity as they underwrite (with the VOCs) the formation of hazardous secondary air pollutants and the accompanying photochemical smog.

Two cities where photochemical smog is not insignificant are the Greater Los Angeles (LA) area in the United States and Beijing, China. Home to nearly 19 million people and located in a geological basin confined by the Pacific Ocean and mountains, LA is the basin of considerable pollution caused by its car-centric culture, its bustling industries and ports, its sprawling development, and its sunny climate with often stagnant winds. It was only in the 1950s that hydrocarbons and NO_x were recognized as the source of photochemical smog (Figure 3); however, with the implementation of mandatory catalytic converters in automobiles in the last two decades, smog has been attenuated somewhat.

Figure 3. Photochemical smog appearing in the Greater Los Angeles area, USA. (Source: United States Geological Survey); see, e.g., https://serc.carleton.edu/eet/aura/case_study.html (accessed 10 November 2018).

With its rapid growth and home to nearly 20 million people, and being an important industrial hub, Beijing is a city where poor air quality has been for decades a regrettable fact of everyday life owing to the presence of significant quantities of particulate matter and photochemical smog, as (Figure 4) experienced by the author in the early 1990s; in subsequent trips several years later, however, this author experienced significant improvements of air quality but by no means have the pollutants and smog been totally eliminated.

Figure 4. Photochemical smog appearing on January 2016 in Beijing, China. Reproduced from https://i1.wp.com/dnnsociety.org/wp-content/uploads/2016/01/beijing.jpg?resize=845%2C450&ssl=1 (accessed 10 November 2018).

There have been many attempts to remediate the occurrence of NO_x and VOC species in polluted urban environments with TiO_2-based photocatalytic cementitious-like materials and photocatalytic coatings (paints) on various supports [11]. Several studies report on the performance of titania deposited on, or otherwise incorporated into cementitious substrates toward the minimization, if not suppression, of air pollutants (see, for example, Refs. [12–18]). Laboratory studies have shown, rather conclusively, that NO_x can be oxidized to nitrate anions [19,20], while VOCs can be converted into CO_2 and H_2O [21]. Of some concern, however, are studies that demonstrate the formation of harmful intermediates (e.g., nitrous acid, HONO), which are far more harmful to human health than either NO or NO_2 during the disposal of NO_x [22,23]. Not least is the potential that nitrates (NO_3^-) produced and deposited on the TiO_2 particulate surface in the disposal of NO_x may be implicated in *reNOxification* reactions; that is, back to NO_x [24–26] and formation of ozone [26] that would forestall the application of TiO_2-based photocatalytic surfaces to improve the quality of urban air environments.

The objective of this review article is to examine the various attempts at eliminating NO_x species in the urban environment produced mostly by vehicular traffic through application of commercially available titania-based photocatalytic materials, coatings and paints in tunnels, highways, highway noise barriers, and urban roads. However, before tackling that discussion, we describe briefly some fundamentals from basic research that underpin this TiO_2-based photocatalytic technology. In its pristine or modified form, TiO_2 has been the most popular and most extensively investigated photocatalyst, and is the primary source of modern third generation composite photoactive materials [27,28].

Photocatalytic processes occurring in heterogeneous systems are complex and multifarious starting from the absorption of photons by the solid photocatalyst, and ending with the evolution of reaction products. This complexity is particularly reflected in the terminology used to describe various characteristics of heterogeneous photocatalysis, which, although it has come to some maturity in recent years, continues to undergo extensive developments through efforts of many researchers from the fields of catalysis, photochemistry and materials science, among others. Accordingly, prior to tackling the many deNOxing efforts in cleaning up the atmospheric environment, an important aspect of applied photocatalysis with metal-oxide semiconductor photocatalysts, it is imperative to appreciate and understand some of the fundamentals underlying Heterogeneous Photocatalysis (following Sections 2 and 3)—the primary approach in these efforts.

2. Some Fundamentals of Heterogeneous Photocatalysis

Historically, Heterogeneous Photocatalysis is an interdisciplinary field at the intersection of Chemistry and Physics. It rests on four basic pillars (Figure 5): (a) heterogeneous catalysis; (b) photochemistry; (c) molecular/solid-state spectroscopy; and (d) materials science of semiconductor photocatalysts of interest (in the present context: metal oxides). Accordingly, it is worth looking into each of these pillars to assess how they have shaped present-day knowhow, particularly Pillars (a)–(c) (Section 2) together with the photophysics of metal-oxide semiconductors (Section 3).

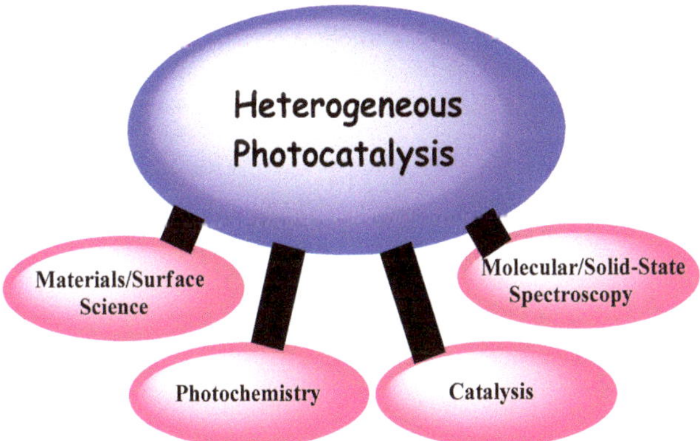

Figure 5. The four pillars that have had a great impact on the development of heterogeneous photocatalysis.

Currently, photocatalysis is best described as a change in the rate of a chemical reaction or its initiation under the action of ultraviolet, visible, or infrared radiation in the presence of a substance that absorbs light and is involved in the chemical transformation of the reaction partners; the photocatalyst is the substance that causes, by absorption of ultraviolet, visible, or infrared radiation, the chemical transformation of the reaction partners, repeatedly coming into intermediate chemical interactions with them and regenerating its chemical composition after each cycle of such interactions [29].

2.1. Influence of Catalysis on Photocatalysis—Comparisons and Contrasts

Some researchers consider photocatalysis a component of the field of Catalysis. However, some postulates that are typical of traditional catalysis are somewhat antagonistic to photocatalysis. For instance, by analogy with catalysis, one of the postulates would suggest that a photocatalytic process should favor only thermodynamically allowed chemical reactions. From this point of view, light should cause a decrease of the potential energy barrier relative to the dark catalytic reaction and thus accelerate the establishment of chemical equilibrium between reagents and products in the heterogeneous system. Photoreactions that take place in heterogeneous systems and are thermodynamically unfavorable (e.g., photolysis of water) are classified as non-photocatalytic but photosynthetic, despite similarities in all major steps of the photoprocesses. The sole reason for both types of photoreactions is the free energy of the actinic light. In fact, the action of the free energy of light turns the system into a thermodynamically open system so that, by definition, the concept of thermodynamic equilibrium is not applicable.

The definitions given to photocatalysis and photocatalysts are very similar to the definitions of catalysis and catalyst commonly used in conventional (thermal) catalysis. This similarity suggests a strong impact of catalysis to the field of photocatalysis. According to current thinking, photocatalysis is viewed as an alteration of the reaction rate in the presence of a substance that interacts repeatedly with reagents subsequent to which its original state is restored after each reaction cycle just like in catalytic processes. Thus, photocatalysis could, in principle, be considered as catalysis involving the action of light. Formally speaking then, the transformation of a reagent in a catalytic process may be exemplified by a simple chemical reaction (e.g., Equation (6)):

$$Cat + R \rightarrow Cat + P \tag{6}$$

where *Cat* is the catalyst, R is the reagent, and P is the product. For a photocatalytic process, we need only involve the interaction of the system with light (Equation (7)):

$$Cat + R + \text{h}\nu \rightarrow Cat + P \tag{7}$$

Certain similarities between catalysis and photocatalysis support such considerations, as both are characterized by the alteration of: (i) the reaction rate; and (ii) the reaction pathway; as well as (iii) by the essential role of adsorption of the reagent R and desorption of the product P in the (photo)catalytic cycle. In addition, the red-shift of the spectral limit of a photocatalytic process, in contrast to a catalyzed photochemical process, can be taken as an analog of the decrease of the activation energy required for the catalyzed reaction compared to the non-catalyzed reaction. Not surprisingly then, some terminology, major characteristics, and quantitative parameters typically used in catalysis have been adapted to the field of photocatalysis. For example, the terms activity and selectivity of a photocatalyst, as well as the terms *surface-active center, turnover number* and *turnover frequency* are examples transferred from catalysis to photocatalysis. Within the present context, the most important feature in photocatalysis borrowed from catalysis is the notion of a *surface-active center* that can best be described as a surface regular site or a surface defect site capable of initiating chemical transformations of the reactants [29].

In heterogeneous photocatalysis, surface-active centers (sites) initiate a chemical sequence subsequent to the centers being activated by absorption of (light) photons by the photocatalyst. That is, before the reaction cycle, the surface-active centers (S) on the photocatalyst are initially in their inactive ground state unable to react with other molecules. However, once the solid photocatalyst has been electronically excited by absorption of photons with the appearance of free charge carriers, electrons (e^-) and holes (h^+), or excitons (e°), this excitation energy may be localized on the surface-active centers to yield centers in their chemically active excited states (S^+ and S^-; Equation (8)), which can also form by interaction with the excitons (Equation (9)), or otherwise the photons may excite the surface-active sites directly to give S* (Equation (10)). The surface-active centers in their electronically activated excited state are then capable of initiating surface chemical reactions.

$$S + e^- \rightarrow S^- \tag{8a}$$

$$S + h^+ \rightarrow S^+ \tag{8b}$$

$$S + e^\circ \rightarrow S^- + h^+ \tag{9a}$$

$$S + e^\circ \rightarrow S^+ + e^- \tag{9b}$$

$$S + h\nu \rightarrow S^* \tag{10}$$

The photoactivated states of the surface-active centers can decay back to the ground state of the centers through different physical relaxation pathways: (a) by recombination of the charge carriers trapped at surface-active centers with free charge carriers of the opposite sign, or with free excitons (Equations (11) and (12)); (b) by thermal ionization when the activation energy of ionization is comparable to kT (Equation (13)); (c) photo-ionization in the spectral range of photexcitation corresponding to the absorption band of the surface-active centers (Equation (14)); and (d) by spontaneous deactivation (Equation (15)).

$$S^+ + e^- \rightarrow S \tag{11a}$$

$$S^- + h^+ \rightarrow S \tag{11b}$$

$$S^+ + e^\circ \rightarrow S + h^+ \tag{12a}$$

$$S^- + e^\circ \rightarrow S + e^- \tag{12b}$$

$$S^+ \rightarrow S + h^+ \tag{13a}$$

$$S^- \rightarrow S + e^- \tag{13b}$$

$$S^+ + h\nu \rightarrow S + h^+ \tag{14a}$$

$$S^- + h\nu \rightarrow S + e^- \tag{14b}$$

$$S^* \rightarrow S + h\nu' \tag{15}$$

Concurrently, chemical reactions of molecules with the photoactivated states of the surface-active centers (S*) also lead to the return of the centers back to their ground state (S; Equation (16)).

$$S^* + R \rightarrow S + P \tag{16}$$

In summary, there are two states of surface-active centers in heterogeneous photocatalysis: (i) the initial inactive ground state S; and (ii) the chemically active excited state(s) (S^+, S^- or S*). Both physical relaxation and chemical interactions of these excited state(s) with molecules return the centers back to their initial ground state.

2.1.1. Is the Process Photocatalytic or Stoichiometric?

The photocatalytic cycle begins by absorption of photons causing the activation of surface-active centers, followed by chemical reactions that terminate by desorption of product(s) from the active centers, and restoring the centers back to their initial ground state. This is essential in determining quantitatively the parameters borrowed from catalysis: namely, the *turnover frequency* (*TOF*) and the *turnover number* (*TON*) [29]. *TOF* refers to the number of photoinduced transformations (product formed or reactant consumed) per *catalytic site* per *unit time* as expressed by Equation (17); N_a is the number of catalyst active sites, and N is the number of photocatalytic transformations.

$$TOF = \frac{1}{N_a}\frac{dN}{dt} \tag{17}$$

It is important to recognize that *TOF* considers the number of *surface-active centers* in their *initial ground state* before photactivation. When the number of such active sites is unknown, the surface area is often used to normalize the number of turnovers—this is known as the *areal turnover frequency* [29].

In catalysis, *TOF* describes the activity of the catalytic centers. In photocatalysis, however, the rate of a chemical transformation of a molecule in a photocatalytic process depends on the light intensity, so that *TOF* is also light intensity-dependent and is generally taken as a characteristic feature of a photocatalyst. *TOF* can be used to compare the activities of various photocatalysts *only* if the photocatalysts were photoactivated under *identical conditions*. This is extremely important! Otherwise, any comparison of the activities of various photocatalysts will have no physical/chemical meaning.

Another turnover quantity taken from conventional catalysis is *TON* that describes the number of times an overall reaction (the photochemical transformation) goes through a photocatalytic cycle for a given period, t (Equation (18)). Again, it is important to recognize that it is *the number of photocatalytic centers* in their *ground state* that is relevant when assessing *TON* in photocatalysis.

$$TON = \frac{1}{N_a}\int_0^t \frac{dN}{dt}dt = \frac{N}{N_a} \tag{18}$$

Since the photocatalytic reaction rate, and thus the number of photoinduced chemical transformations, depends on light intensity, *TON* cannot be considered an intrinsic property of a photocatalyst. Nonetheless, *TON* is an important parameter in experimental studies of heterogeneous photochemical reactions as it shows whether the photochemical transformation is catalytic. Where *TON* > 1, the active center is restored back to its initial ground state after the chemical transformation cycle and is ready to initiate subsequent cycles: *the photochemical process is then said to be photocatalytic.* When *TON* $\leq$ 1 after the first cycle, the active center is completely deactivated and cannot initiate subsequent cycles of the chemical transformations; this means that the surface-active center only interacts with one molecule. In this case, *the process is a stoichiometric process*, as encountered in photostimulated adsorptions of molecules on solid surfaces.

If a stoichiometric surface reaction were to occur on the same surface-active center as a photocatalytic reaction, it would provide a method with which to account for the number of surface-active centers and thus aid in establishing the magnitude of *TON*. For instance, if a photostimulated adsorption of a molecule is the first step in a complex photocatalytic process that concludes the photocatalytic cycle at the photoadsorption step, then it is possible to determine the number of photoadsorbed molecules that corresponds to the number of surface-active centers involved in the photocatalytic reaction. This method was used by Emeline and coworkers [30] to assess quantitatively *TON* for the photooxidation of hydrogen by oxygen and the photoreduction of oxygen in the presence of hydrogen over solid particles of ZrO_2. The authors showed that the photostimulated adsorption of oxygen on zirconia, which by default is a *stoichiometric* reaction, is the first step in the photooxidation of hydrogen. Therefore, the maximal number of photoadsorbed oxygen molecules gives the number of surface-active centers for the photooxidation of hydrogen. They determined that *TON* > 1 for this process, and consequently the process was deemed to be a *photocatalytic process*.

2.1.2. Mechanistic Implications

Another common feature between catalytic and photocatalytic processes are the two general mechanisms of surface chemical reactions: (a) the Eley–Rideal mechanism (ER); and (b) the Langmuir–Hinshelwood (LH) mechanism. The Eley–Rideal mechanism assumes that molecules in the bulk solution phase interact with surface species (either pre-adsorbed molecules or surface-active centers), whereas the Langmuir–Hinshelwood mechanism presupposes that the reaction involves *pre-adsorbed* molecules and surface species (either pre-adsorbed molecules or surface-active centers). For decades, the LH mechanism was the favorite mechanism among experimentalists (although erroneously) because of the experimentally observed dependence of the rate of the photocatalytic process on reagent concentration (Equation (19)).

$$\frac{dC}{dt} = \frac{kKC}{1 + KC} \tag{19}$$

where k is the *apparent rate constant*, K is the *apparent equilibrium constant*, and C is the concentration of reactant molecules. This equation correlates with the Langmuir isotherm (Equation (20)):

$$\theta = \frac{K_L C}{1 + K_L C} \tag{20}$$

where K_L is the adsorption/desorption equilibrium constant ($K_L = k_{ads}/k_{des}$) provided the rate of the process is proportional to the surface coverage (θ) by pre-adsorbed molecules (Equation (21)). That is, the adsorption/desorption equilibrium is not disturbed during the photoprocess.

$$\frac{dC}{dt} = k\theta \tag{21}$$

If $K = K_L$, then by default the observation that the rate dependence followed Equation (19) was taken as an indicator that the process followed the LH mechanism. However, for many heterogeneous photoprocesses, it was established conclusively that $K \neq K_L$ and K depended on light intensity. Later, Emeline and coworkers [31,32] demonstrated that the dependence expressed by Equation (19) was also consistent with the ER mechanism, considering that the lifetime of the activated state of the surface-active centers is limited (see above). Moreover, if the physical relaxation pathway of the excited state of the active center involved recombination or photo-ionization, then K became dependent on light intensity in accord with experimental results (Figures 6 and 7).

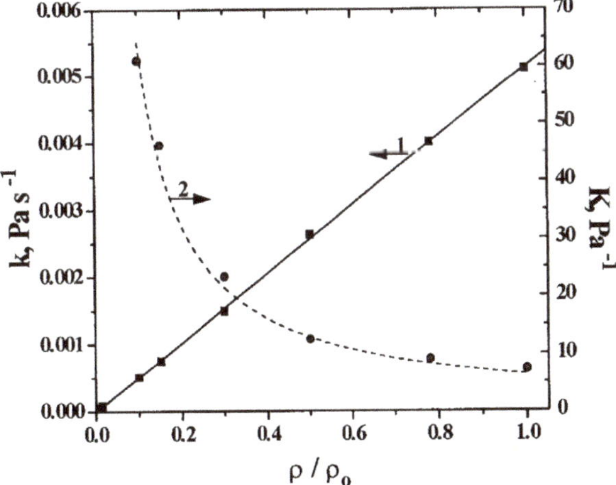

Figure 6. Dependence of the apparent constants k and K_L on light intensity for the photostimulated adsorption of oxygen on ZrO_2. Reproduced with permission from Emeline et al. [31]; Copyright 1998 by the American Chemical Society.

The approach based on a limited lifetime of the excited state of active centers is applicable to both ER and LH mechanisms, provided a *quasi*-steady-state is valid for the concentration of active centers in the activated state. The alternative view (i.e., LH mechanism) is also based on a *quasi*-steady-state approach for the concentration of adsorbed molecules and on a strong disruption of the adsorption equilibrium in accord with experimental data [33,34]. Kinetic measurements alone, as typically conducted in photocatalytic studies, cannot distinguish unambiguously which mechanism is operative in a photocatalytic process. Whatever the actual mechanism (ER or LH), however, the role of adsorption of a molecule on the surface is not insignificant and in fact is typical in heterogeneous photocatalysis. The only difference occurs in the ER mechanism, whereby photostimulated adsorption plays an important role rather than dark adsorption, with the latter so crucial to the LH mechanism.

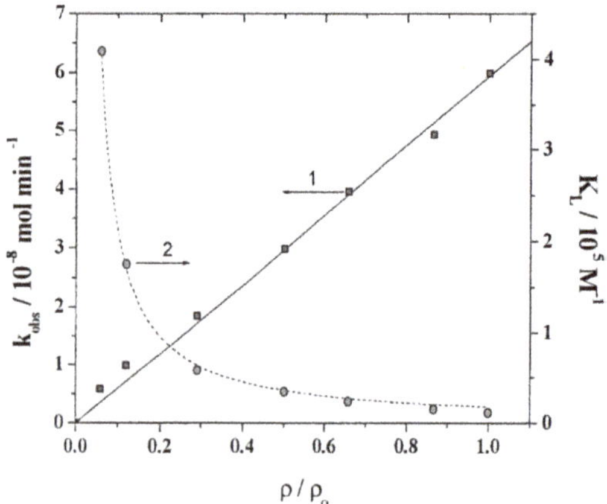

Figure 7. Dependence of the apparent constants k and K_L on light intensity for the photodegradation of phenol over TiO_2. Reproduced with permission from Emeline et al. [32]. Copyright 2000 by Elsevier B.V. (License No. 4452260016035).

In summary, heterogeneous catalysis has had a non-insignificant impact on the field of heterogeneous photocatalysis. Nonetheless, a serious discrepancy between these two phenomena remains. In catalysis, the catalyst is in its electronic ground state and is in thermodynamic equilibrium with the environment at a given temperature; it accelerates both direct and back reactions, and promotes a faster establishment of the reaction equilibrium. By contrast, in heterogeneous photocatalysis, the photocatalyst becomes active in an electronically excited state that is *not* in thermodynamic equilibrium with its surroundings, so that the sole relaxation pathway back to its ground state is acceleration of the direct reaction, because the photocatalytic heterogeneous system is an open system under a permanent flow of light energy as occurs in photochemistry.

2.2. Influence of Photochemistry on Photocatalysis—Comparisons and Contrasts

Photochemistry of heterogeneous systems is also an important pillar of photocatalysis, as it considers thermodynamically open systems. Thus, photocatalysis can also induce reactions that lead to light energy conversion and storage through formation of higher energy products, as occurs in natural photosynthesis [35,36]. Concepts taken from photochemistry suggest that the photocatalytic process occurs through electronically excited states of the photocatalyst prompted by light absorption. It should be noted that photocatalyzed reactions are often indistinguishable from photosensitized reactions examined in photochemistry [37]. Historically, many photocatalytic reactions were considered photosensitized reactions, so that what was once called a photosensitizer is now referred to as a photocatalyst. Two major characteristics of photosensitized reactions are also typical of photocatalytic processes: (i) a red-shift of the spectral limit of the photoreaction; and (ii) the alteration of the reaction pathway compared to that of the photochemical reaction. The red-shift of the spectral limit of a photoprocess and alteration of the reaction pathway are distinguishable fingerprints of heterogeneous photocatalytic reactions. Enhancing the ability of photocatalysts to sensitize photochemical reactions, especially toward visible light, is a major challenge in applied heterogeneous photocatalysis.

Photochemistry is concerned with the chemical effects of *ultraviolet, visible,* or *infrared* radiation while the *photocatalyst* is a substance that can produce on *absorption* of light chemical transformations of the reaction partners subsequent to formation of an *excited state* of the photocatalyst, which interacts repeatedly with the reaction partners forming reaction intermediates and regenerates itself after each cycle of such interactions [37]. Hence, a photocatalytic process (Equation (22)) is similar to a photochemical reaction (Equation (23)).

$$Cat + h\nu \rightarrow Cat^* + R \tag{22}$$

$$R + h\nu \rightarrow R^* \rightarrow P \tag{23}$$

where *Cat* and *Cat** denote the photocatalyst in the ground state and in its electronically excited states, respectively; R is the reagent in its ground electronic state; R^* is the reagent in the electronically excited state formed by absorption of photons; and P is the product of the photochemical/photocatalytic transformation.

Thus, a photocatalytic process could be viewed as a particular case of a photochemical reaction, whereby the photocatalyst plays the double role as one of the reagents and as one of the products subsequent to restoring its initial ground state.

2.2.1. Photocatalytic Versus Photochemical Processes

There are far more mechanistic similarities between photocatalytic and photochemical reactions than there are between photocatalyzed and catalyzed reactions. Both the photochemical reaction and the photocatalytic reaction require absorption of photons to form electronically excited states of one of the reaction partners, which cause distortions of the initial thermodynamic equilibrium and both reactions then occur through non-equilibrium states. This means that the subsequent sequence of molecular transformations can only proceed in one direction to restore the equilibrium state of the system and there is no path for a back reaction. Both photoreactions are characterized by the existence of physical relaxation pathways of the electronically excited states: radiative and non-radiative relaxation processes. For a solid photocatalyst, these relaxation processes are excitonic decay, free charge carrier recombination, and charge carrier trapping. In addition, all photocatalytic processes obey the general law of photochemistry: only absorbed photons initiate chemical transformations, and one absorbed photon results in only one elementary transformation. Accordingly, photocatalysis can be viewed as catalysis of a photochemical reaction. Consequently, three major characteristics used in photocatalysis have been adopted from photochemistry: (a) *spectral sensitivity of a photoprocess*; (b) *photoactivity of a photocatalyst*; and (c) *selectivity of a photocatalyst*.

Spectral Sensitivity of a Photoprocess

The spectral sensitivity of a photoprocess is described by the spectral range wherein photon absorption by the system initiates a chemical transformation; for a photocatalytic process this is the red spectral limit that corresponds to the lowest energy photons to initiate photocatalytic cycles. Note that the blue spectral limit of all photocatalysts expands into the vacuum UV region and so cannot be determined under typical conditions. We have often noted that the photocatalysis literature claims (albeit incorrectly) that photons with energy lower than bandgap of the photocatalyst cannot initiate chemical reactions, and so the red spectral limit of a photocatalytic process is taken as the bandgap energy. Theoretically, this is true only for ideal solids with no defect states, although even for an ideal solid structure there are always regular surface states (e.g., the Tamm and Shockley states) whose energies of photoexcitation could be lower than the bandgap energy. In practice, however, one deals with real solid photocatalysts whose structures possess various imperfections: for example, intrinsic point defects, impurity defects, dislocations, and defect surface states (e.g., add-atoms, corners, edges, steps etc.). Photoexcitation of such defect states results in a significant shift of the red limit of the photocatalytic effect toward lower energy photons compared to the bandgap energy—for the case of TiO_2, although the bandgap energies are 3.0 eV for the rutile polymorph and 3.2 eV for the anatase counterpart, the red limits for both oxidative and reductive pathways are 2.2 eV ($\lambda = 560$ nm), well below bandgap.

Accordingly, from a practical application of heterogeneous photocatalysis, the longer is the wavelength of the red spectral limit of the photocatalytic process, the larger is the fraction of sunlight that can be exploited for a higher overall activity of the photocatalytic system. Increasing the spectral sensitivity of a photocatalyst toward visible light is a major challenge in applied photocatalysis, currently being examined through modification of pristine photocatalysts by metal and non-metal doping and by physical modifications to form intrinsic defects, which give rise to extended *extrinsic absorption* of visible light. A detailed assessment of this issue and corresponding problems have been described elsewhere [38–40].

Photoactivity of a Photocatalyst: The Quantum Yield Φ

The photoactivity of photocatalysts describes the ability of a solid to transform the absorbed actinic light into a chemical sequence, for which the quantum yield (Φ) is its most relevant parameter. In heterogeneous photocatalysis, Φ is the ratio between the number of molecules that have reacted (N_m) to the number of photons actually absorbed ($N_{h\nu}$) at a given wavelength of the actinic light (Equation (24)), provided the photocatalytic process has reached a stationary state [29], otherwise Φ can be expressed by the differential form (Equation (25)):

$$\Phi = \frac{N_m}{N_{h\nu}} \tag{24}$$

$$\Phi = \frac{\frac{dN_m}{dt}}{\frac{dN_{h\nu}}{dt}} \tag{25}$$

Typically, Equation (25) is used since the reaction rate, and thus Φ, depends on such parameters as time, temperature, wavelength of the actinic light, concentration of reagent, light intensity, and the solution pH (among others). Regrettably, the rate of absorption of photons and the number of absorbed photons remain inaccessible quantities in heterogeneous photocatalysis because of experimental limitations. Consequently, the photonic yield (ξ) was introduced and defined in a manner similar to the quantum yield (Φ) with the main difference being that reference is made to the number of photons of a given wavelength of the actinic light ($N^\circ{}_{h\nu}$) *incident* on the photoreactor, and not on the number of photons absorbed by the photocatalyst under stationary conditions, as expressed by Equation (26) or in differential form by Equation (27) [28].

$$\xi = \frac{N_m}{N^o_{h\nu}} \tag{26}$$

$$\xi = \frac{\frac{dN_m}{dt}}{\frac{dN^o_{h\nu}}{dt}} \tag{27}$$

Both Φ and ξ are defined as in photochemistry [37]; the former is more appropriate, however, since only *absorbed photons* can initiate interfacial chemical reactions, thus Φ is an intrinsic characteristic of a photocatalyst. Nonetheless, ξ may be a more practical parameter to assess experimentally as it depends on the number of *incident photons* and not on the number of photons actually absorbed by the photocatalyst; ξ is also an intrinsic characteristic of a photocatalyst.

In the field of Chemical Engineering, the parameters often used are *quantum efficiency* and *photonic efficiency*; the former describes the rate of a given photophysical or photochemical process divided by the total *photon flux* absorbed [29]; it applies especially when using polychromatic radiation to activate the photocatalyst. The photonic efficiency describes the ratio of the rate of the photoreaction measured at $t = 0$ (initial rates) to the rate of incident photons within a given wavelength range [29]. Consequently, these parameters provide an estimate of the overall photoactivity of the photocatalysts, but provide no information regarding light absorption, photoexcitation mechanisms, and specific details as to the efficiencies of excitation transfer to the surface or to the initiation of surface chemical transformation as they denote engineering efficiencies.

We cannot overemphasize that the above parameters used in photocatalysis and borrowed from photochemistry characterize the efficiency of photocatalysts, unlike *TON* and *TOF* from catalysis which are light intensity-dependent. Photochemical parameters may also depend on the intensity of photoexcitation; however, experimental conditions can be controlled such as to obtain a linear dependence of the reaction rate on light intensity as established by Emeline and coworkers [30,31] to obtain the maximal photoefficiency of photocatalysts when all the quantum and photonic parameters become independent of light intensity. The quantum yield value (and other photochemical parameters) is essentially governed by the reaction rate of a heterogeneous photocatalytic reaction that depends on the stationary surface concentration of charge carriers (either electrons or holes) (Equations (28) and (29)).

$$R + e^- \rightarrow P \tag{28a}$$

$$R + h^+ \rightarrow P \tag{28b}$$

$$d[R]/dt = k\,[R]\,[e] \tag{29a}$$

$$d[R]/dt = k\,[R]\,[h] \tag{29b}$$

Selectivity of a Photocatalyst

The selectivity of a photocatalyst is the ability of a photocatalyst to drive a photoprocess toward a certain reaction product. Emeline and coworkers [41–44] established two possible causes for the selectivity of a photocatalyst when the rate of the surface photochemical reaction is dictated by the surface concentration of photocarriers (electrons for reduction, holes for oxidation).

The first originates from the ratio between the concentrations of electrons and holes at the surface of a photoactive material, so that alteration of this ratio caused by various factors leads to changes in the ratio between the efficiencies of surface redox reactions for molecules displaying both electron-acceptor and electron-donor behavior. For example, if a given reaction product (P_i) were formed by a reduction pathway during the photostimulated surface reaction with reagent (R) interacting with both electrons (e) and holes (h), the surface selectivity toward the formation of the given product (S_{Pi}) would then be expressed by Equation (30): $k_{e,i}$ and $k_{h,j}$ are the rate constants for a given reaction pathway. A similar expression can be formulated for the products formed by the oxidation reaction pathway.

$$S_{P_i} = \frac{\frac{d[P_i]}{dt}}{\frac{d[R]}{dt}} = \frac{k_{e,i}[e]}{\sum_i k_{e,i}[e] + \sum_j k_{h,j}[h]} \tag{30}$$

By default, it is assumed that the reaction rate does not depend on reagent concentration, which would correspond to saturation of the rate dependence on reagent concentration for the LH-like kinetics. Equation (30) clearly indicates that the selectivity of the photocatalyst is determined by the ratio between the surface concentrations of electrons and holes.

The second origin of selectivity is related to the formation of surface-active sites of a different type, which may display different activity and selectivity. Different types of surface-active sites can be formed through photoexcitation of localized surface and sub-surface electronic states (intrinsic defects and/or dopant states), or through formation on different types of surface structures (e.g., dominating surface plane). This type of selectivity is clearly observed for molecules that demonstrate only electron-donor or electron-acceptor behavior, although it is typical of any surface photochemical reaction. As such, the selectivity of the surface toward formation of a given reaction product can be expressed by Equation (31), which is determined by the ratio of a given rate constant characteristic of a given sort of active site to the sum of rate constants corresponding to all possible types of active sites on the surface of the photocatalyst (here rate constants are apparent constants and implicitly include the surface concentrations of active sites). Obviously, this factor (ratio of rate constants) is also significant for the selectivity in the earlier scenario (Equation (30)).

$$S_{P_i} = \frac{\frac{d[P_i]}{dt}}{\frac{d[R]}{dt}} = \frac{k_{e,i}}{\sum_i k_{e,i}} \tag{31}$$

Unlike macro crystals with their clearly manifested crystal faces, the dominant surface planes in nanoparticles of photocatalysts are strongly affected by the method and conditions of their synthesis. Concomitantly, in the course of photoprocesses, various planes of photocatalyst particles may display different selectivities, which may be due to different atomic structures, or magnitude and distribution of surface charge, or to a dominant type of defects (e.g., surface-active centers, among others) [45–49]. Hence, altering the nanoparticle surface structure via formation of various dominating faces with strong selectivity toward either oxidation or reduction can alter the overall selectivity of the photocatalyst.

2.2.2. Correlation between Activity and Selectivity

The correlation between activity and selectivity of a photocatalyst has been demonstrated by Emeline et al. [50] in the photodegradation of 4-chlorophenol taking place over irradiated TiO$_2$ with formation of three primary intermediates: (i) benzoquinone formed by a reductive pathway; (ii) 4-chlorocatechol formed by an oxidative pathway; and (iii) hydroquinone produced by both oxidative and reductive pathways. Figure 8 shows that the selectivity toward formation of hydroquinone and the activities of the photocatalyst are strongly correlated (r = 0.984; p = 0.0004).

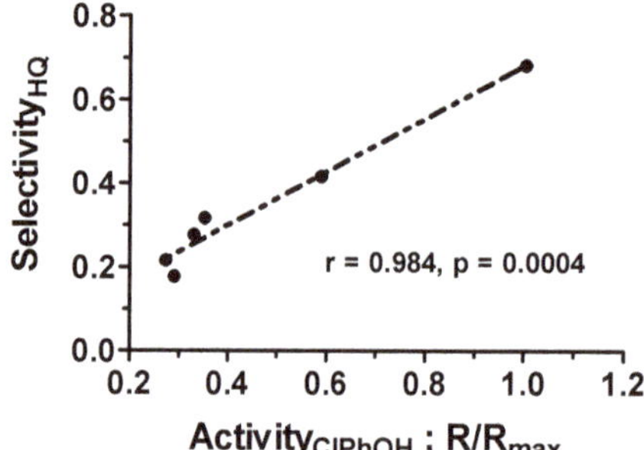

Figure 8. Correlation between activities of six tested TiO$_2$ photocatalysts in the photodegradation of 4-chlorophenol and their selectivity toward formation of hydroquinone. Reproduced with permission from Emeline et al. [50]. Copyright 2011 by Elsevier B.V. (License No.: 4452260940331).

According to the balance of charge, as expressed by Gerisher [51], a true (photo)catalytic process is characterized by the equality of the rates of consumption of electrons and consumption of holes in the overall reaction (Equation (32)). That is,

$$\frac{d[e]}{dt} = \frac{d[h]}{dt} \tag{32}$$

This charge balance is a required condition for the effective photocatalytic process, or else the deviation from catalytic equilibrium (Equation (32)) would result in the transformation of charge balance according to Equation (33) [38,52].

$$\frac{d[e]}{dt} + \frac{d[F]}{dt} = \frac{d[h]}{dt} + \frac{d[V]}{dt} \tag{33}$$

where F and V denote, respectively, electrons and holes trapped in bulk defects (color centers), which accelerate bulk charge carrier recombination and decrease the activity of the photocatalyst. Since formation of hydroquinone consumes both electrons and holes, these reaction pathways create a favorable condition for the photodegradation of the phenol being truly photocatalytic by suppressing bulk recombination. Therefore, the higher is the selectivity of the photocatalyst surface toward formation of hydroquinone, the higher is the activity of the photocatalyst during the photodegradation. In general, this rule can be formulated thus: a higher activity of photocatalysts can be expected provided both reductive and oxidative pathways occur with equally high efficiency.

2.3. Influence of Molecular Spectroscopy on Heterogeneous Photocatalysis

Another major pillar of heterogeneous photocatalysis is molecular spectroscopy of adsorbed molecules. Most pioneering studies focused on understanding the changes in molecular structure induced by adsorption of molecules on solid surfaces [53–55], with light being an active factor in transforming these adsorbates. The effects of photodissociation and photodesorption of adsorbed molecules have been established together with the photoadsorption of simple molecules on dispersed semiconductors [55,56].

The photocatalytic cycle begins with the absorption of light quanta by the solid photocatalyst and culminates with the chemical transformations of surface-adsorbed molecules, ultimately evolving reaction products into either the gaseous or liquid phase. The role of the photocatalyst and corresponding photophysical events taking place in solids are often treated in a simplistic manner. The ensemble of particles that absorb photons is the light harvesting system, whereby the photocatalyst particle is both a sensitizer and the source of intermediates—i.e., photoelectrons and photoholes. In addition, the intrinsic, that is, the fundamental absorption of light by the solids is of primordial importance in photocatalysis; this was a reasonable approach in most studies that were oriented on mechanistic investigations of chemical reactions, or else on practical applications of heterogeneous photocatalysis. The complexity and variety of photophysical processes in solid photocatalysts, together with the interdependence between physical and chemical events at the microparticles' and nanoparticles' surfaces, must always be kept in mind, even in applied heterogeneous photocatalysis. Accordingly, next we describe some relevant events that precede surface chemical reactions on the photocatalyst particle (Figure 9).

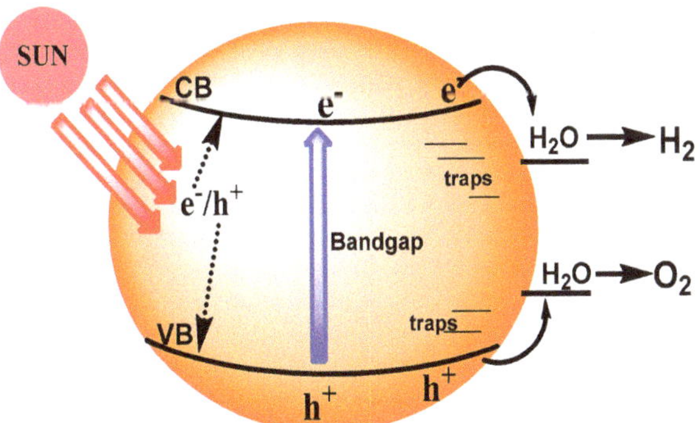

Figure 9. Schematic illustration of a photocatalytic semiconductor nanoparticulate system in converting light energy into a solar fuel (H$_2$) from the water splitting process.

2.3.1. Absorption of Light by Solid Photocatalysts

Absorption of electromagnetic radiation represents the transfer of energy from an electromagnetic field to a material (the photocatalyst) or to a molecular entity [29]. The transformation of light energy into the energy of electronic excitations (electrons, holes) of a solid photocatalyst is fundamental in heterogeneous photocatalysis, with these excitations being created as a result of absorption of photons in a solid during some time (lifetime of charge carriers) before relaxation (recombination of charge carriers), i.e., before transformation of the electronic energy into thermal energy. Electron–hole pairs in solids then transfer the energy within some sub-surface space, the depth of which depends on both the linear absorption coefficient of the photocatalyst and the diffusion length of the photocarriers and, in some particular cases, on particle size.

2.3.2. Quantities that Describe Light Absorption in Heterogeneous Photocatalysis

Dispersed (powdered) photocatalysts in contact with gaseous or liquid phases in irregular shaped reactors are typically the light absorbing media, and so require consideration of the absorption and scattering of light together with a few quantities that we now outline for the absorption of light used in experimental photocatalytic reactions.

Absorption of light is manifested by the decrease of the energy of the light beam when it passes through a medium because of the transformation of the photon's energy into another energy form. Experience has taught us that $I_R + I_T < I_0$ for light absorbing substances and so in accord with the energy conservation law, the intensities of the incident (I_0), transmitted (I_T) and specular reflected (I_R) beams are given by Equation (34),

$$I_0 = I_R + I_T + I_A \tag{34}$$

where I_A is the intensity of the absorbed light beam. For moderate intensities of the incident light beam and in the absence of additional input of energy to the substance, we have (Equation (35))

$$I_R = R \, I_0 \tag{35a}$$

$$I_T = T \, I_0 \tag{35b}$$

$$I_A = A \, I_0 \tag{35c}$$

The dimensionless coefficients R (here it denotes the reflectance), T (transmittance) and A (absorbance) in Equations (34) and (35a)–(35c) satisfy Equation (36),

$$R + T + A = 1 \tag{36}$$

Typically, absorbance can be determined optically using a dual-beam spectrophotometric technique such that the ratio I_T/I_0 is detected, while reflectance is compensated using a reference sample. From experience, we have that (Equation (37))

$$I_T = (1 - R)I_0 e^{-\alpha d} \tag{37}$$

And, from Equations (35a)–(35c) and (37), we obtain for A (Equation (38)) that

$$A = (1 - R)\left[1 - e^{-\alpha d}\right] \tag{38}$$

which describes the Lambert–Bugger law for absorbance by a solid parallel plate of thickness d; the quantity α is the linear (Naperian) absorption coefficient (in cm^{-1}). In general (Equation (39)),

$$I_{(x)} = (1 - R)I_0 e^{-\alpha x} \tag{39}$$

where $I_{(x)}$ is the intensity of the light beam a distance $\times$ from the illuminated surface of the optically uniform plate.

Equation (39) can also be obtained from Equation (40) which expresses the independence of the absorption coefficient α from the light intensity,

$$dI = -\alpha\, I_{(x)}\, dx \tag{40}$$

where dI is the fraction of light absorbed by a thin layer dx a distance $\times$ from the plate surface. The density of absorbed photons (Equation (41)) can be deduced from Equations (39) and (40). If the light beam of intensity I_0 were given in units of photons per unit area per second, then the units of dI/dx are $cm^{-3}\, s^{-1}$.

$$\frac{dI}{dx} = \alpha I_0 e^{-\alpha x} \tag{41}$$

The absorption coefficient α is a spectrally dependent parameter ($\alpha = \alpha_{(\lambda)}$), whose magnitude varies in a wide range for different solids and can reach values up to ca. 10^6 cm^{-1} for the fundamental absorption band but can be very small for extrinsic absorption bands of solids.

Where absorption of light is due to defects or to structural imperfections (i.e., in the *extrinsic* absorption region for a transparent solid), the linear absorption coefficient α is proportional to the concentration of defects n (Equation (42)); σ_a is the absorption cross section of a defect.

$$\alpha = \sigma_a\, n \tag{42}$$

Equation (42) represents Beer's law (for defects in solids); the law fails at very high defect concentrations (typically $> 10^{18}$ cm^{-3}) when the interaction between defects in the solid becomes significant.

Taking into account Equations (42) and (37)–(39), a comparison with corresponding formulas in UV-Vis absorption spectroscopy of solvents indicates that the wavelength-dependent quantity σ_a becomes the extinction coefficient that characterizes the optical properties of a given defect in the solid. The absorption cross-section, σ_a, for optical transitions in defects that are allowed by appropriate selection rules can reach values of $\sim 10^{-16}$ cm^2. Consequently, $\alpha \approx 10^2$ cm^{-1} for $n \approx 10^{17}$–10^{18} cm^{-3}; this is typical of imperfect photocatalyst particles. When the nominal concentration of regular entities (atoms, ions) in solids ($n \approx 10^{22}$ cm^{-3}) is multiplied by $\sigma_a \approx 10^{-16}$ cm^2 leads to a realistic estimation of the absorption coefficient ($\alpha \approx 10^6$ cm^{-1}) for fundamental absorption bands, despite breaching the conditions of Beer's law validity. Variation of α with wavelength of the actinic light is an important factor in the activity of solid photocatalysts and other spectrally dependent parameters.

2.3.3. Absorbance and Reflectance of Photocatalysts in Powdered Form

Powdered solids with sufficiently high BET surface areas ($S_{BET} = 10$–100 m^2 g^{-1}) are typically used in photocatalytic studies and correspond to characteristic mean particle sizes in the microscale to nanoscale comparable to the wavelength of the actinic light. Contrary to the optically uniform plate noted earlier, light scattering now plays a major role in the action of light on powdered photocatalysts. Because of light scattering, a powdered photocatalyst in pellet form is now opaque to light compared with the optically uniform plate of the same material, even in the spectral region where absorbance is very low (extrinsic absorption region). Consequently, most of the light is reflected back in various directions and gives rise to diffuse reflectance. The light intensity decreases with distance from the illuminated side of the pellet. In this case, the Lambert–Bugger law is not valid in its original form (Equation (6)) so that in practice the energy conservation law (Equation (34)) should be applied to powdered materials with some restraint since care must be taken in measuring the diffuse reflected light.

Diffuse reflectance spectroscopy has proven a useful method in experimental measurements of the optical absorption of powdered metal-oxide photocatalysts. For instance, the absorbance A of a sample and the diffuse reflectance coefficient, R, are related by Equation (43) for zero transmittance (i.e., for $T = 0$); in practice, this is valid when the thickness of the powdered sample is around 3–5 mm.

$$A = 1 - R \tag{43}$$

UV-Vis-NIR diffuse reflectance spectra, $R_{(\lambda)}$, are usually recorded with spectrophotometers equipped with an integrating sphere assembly and a standard reference sample of known diffuse reflectance spectrum. Accordingly, $R_{(\lambda)}$ or $A_{(\lambda)}$ spectra provide the spectral information about the initial state of a powdered sample.

Additional spectral information can be obtained from difference diffuse reflectance spectra, $\Delta R_{(\lambda)}$, when changes in reflection/absorption occur from controllable treatments of the photocatalyst (e.g., annealing or illumination in a vacuum or in the presence of gases) (Equation (44)).

$$\Delta R_{(\lambda)} = R_{1(\lambda)} - R_{2(\lambda)} \tag{44}$$

where $R_{1(\lambda)}$ and $R_{2(\lambda)}$ are the diffuse reflectance spectra measured, respectively, before and after the treatment. The case where $\Delta R_{(\lambda)} > 0$ corresponds to the treatment-induced absorption spectrum. Hence, from Equation (43) we obtain (Equation (45)):

$$\Delta R_{(\lambda)} = \Delta A_{(\lambda)} = A_2 - A_1 \tag{45}$$

Thus, alteration of the diffuse reflectance coefficient in photocatalytic experiments corresponds to changes in the absorbance of the photocatalyst sample, which for practical applications is more important than changes in the absorption coefficient (such as quantum yield measurements) as it provides the information about the spectral behavior of the whole heterogeneous system. Difference diffuse reflectance spectra ($\Delta R_{(\lambda)}$) not only increase significantly the accuracy of measurements but make information accessible with regard to absorption spectral shapes induced by whatever treatment the photocatalyst is subjected to. Hence, measurements of a set of induced absorption spectra of different spectral shapes allow for a numerical analysis of the spectral data that could be very important for absorption spectra consisting of several overlapping single absorption bands.

2.4. Intrinsic and Extrinsic Absorption of Light by Semiconductor Photocatalysts

Two different spectral regions of light absorption can be distinguished in the field of optics and in the photophysics of the solid state: (i) intrinsic or fundamental absorption; and (ii) extrinsic absorption of light. Intrinsic light absorption is due to photoinduced electronic transitions between occupied delocalized states in the valence band and unoccupied delocalized states in the conduction band of semiconductor photocatalysts [57,58].

The intensity of the inter-band photoexcited transitions, characterized by the coefficients α and σ_a, is determined by selection rules similar to those for atomic photoexcitation and by the additional requirement for a *quasi*-momentum conservation. As a case in point, photoinduced electronic transitions in TiO_2 from the top of the valence band (mostly oxygen p-states) to the bottom of the conduction band (mostly titanium d-states) are allowed transitions ($p \rightarrow d$ transitions) as are the $p \rightarrow s$ transitions in alkali halides. The intensities of photoinduced transitions from the VB to the CB in alkali halides are much greater than those observed in TiO_2. The reason for this difference in solids may be found in the momentum conservation law for electrons that requires the fulfillment of the condition expressed by Equation (46):

$$k_1 - k_2 = k_{h\nu} \tag{46}$$

where k_1 and k_2 are the wave vectors of electrons in the states between which the transition occurs and $k_{h\nu}$ is the wave vector of the photon which, when absorbed by the solid, causes the transition. To the extent that the momentum of photons resulting in the electronic transition in a typical spectral range of photoexcitation of a solid is much smaller than the *quasi*-momentum of electrons occupying the band states, the condition expressed by Equation (44) can be simplified to Equation (47) with the proviso that $k_{h\nu} \approx 0$.

$$\Delta k = k_1 - k_2 \approx 0 \tag{47}$$

The transition between states in the VB and CB bands that follows Equation (46) is referred to as a direct transition (i.e., the transition preserves the momentum of the electron) and is characterized by high intensity at the edge of the fundamental absorption. Where the positions of the maxima do not coincide in k-space (typical of TiO_2), direct transition between these states is then forbidden because momentum conservation is not preserved. Nevertheless, photoinduced electronic transitions between such states can occur via involvement of phonons that possess the required momentum. Accordingly,

$$\Delta k \pm k_{h\omega} \approx 0 \tag{48}$$

where $k_{h\omega}$ is the wave vector of either the absorbed or the emitted phonon. Such transition is referred to as an indirect transition characterized by lower intensity as it requires a three-body interaction of the electron, the photon and the phonon.

Extrinsic light absorption in solids may originate from the photoexcitation of such defect states as: (i) zero-dimensional (0-D) intrinsic point defects (e.g., vacancies, interstitials) and impurity atoms/ions; (ii) one-dimensional (1-D) linear defects (e.g., dislocations); and (iii) two-dimensional (2-D) states (e.g., intrinsic surface states).

Special interest in heterogeneous photocatalysis concerns the surface absorption of light associated with the electronic excitation of surface states, for which the energy of absorption corresponds to the extrinsic spectral region of the solids. In this regard, the greater the specific surface area is, the greater is the impact of surface absorption into extrinsic absorption. The important issue here is that the generation of charge carriers that may induce surface chemical reactions occurs only at the surface of the solid photocatalyst.

To recap, the various absorption bands corresponding to photoexcitation of different types of defects, impurities, and surface states can cover a wide spectral range of extrinsic light absorption. This photoexcitation generates free charge carriers or surface-localized excited states that can initiate interfacial chemical processes. Accordingly, the red spectral limit of a heterogeneous photochemical or photocatalytic reaction corresponds to photon energy much less than the bandgap energy of the solid photocatalysts. The photoactivity in the extrinsic absorption spectral region is typical of most metal-oxide photocatalysts, including TiO_2.

3. Photophysical Processes in Metal-Oxide Photocatalysts

3.1. Intrinsic Structural Point Defects in Metal Oxides

Point defects related to anion (oxygen) and cation (metal) vacancies in sub-lattices are the main types of defects in metal-oxide semiconductors/insulators. Structure, together with the optical and EPR properties of families of both oxygen vacancies (V_o) and cation vacancies (V_m) are now fairly understood for wide bandgap metal oxides such as MgO, Al_2O_3, and ZrO_2 (among others) [59–67].

3.1.1. Defects Related to Oxygen Vacancies (V_o)

There are three main types of defects in the oxygen sub-lattice of MgO related to oxygen vacancies, the so-called F-type centers: (i) doubly charged (with respect to the lattice) and optically silent anion oxygen vacancies (F^{2+} centers); (ii) anion vacancies with one trapped electron (F^+ centers); and (iii) two trapped electrons (F centers) [59–62]. A main feature of F^+ and F centers is a strong localization of the electron(s) within the vacancy. Semi-empirical quantum chemical calculations based on the INDO approach [61] have shown that the optimized geometry of oxygen vacancies with both bare and trapped electrons is characterized by a displacement of the nearest-neighbor cations in a direction away from the vacancy. Atomic relaxation is largest for F^{2+} centers (6.5% of regular Mg–O distance), whereas cation displacement for neutral F centers is <2% [61]. Calculations of ionic MgO crystals have shown that electrons trapped by V_o are indeed localized in the vacancy, and that the effective charge of F^+ and F centers in the ground state was −1.002 and −2.002, respectively, while the changes in the effective charges of Mg and O ions surrounding V_o were inconsequential with respect to regular ones [61].

In the wurtzite ZnO lattice, the value and direction of the relaxations of Zn atoms adjacent to an oxygen vacancy strongly depend on the charge state of the defect. For the F center (V_o^0), the Zn atoms are displaced inward by approximately 11–12%, while, for the F^{2+} center (V_o^{2+}), an outward relaxation of about 19–23% occurs (Figure 10) [64].

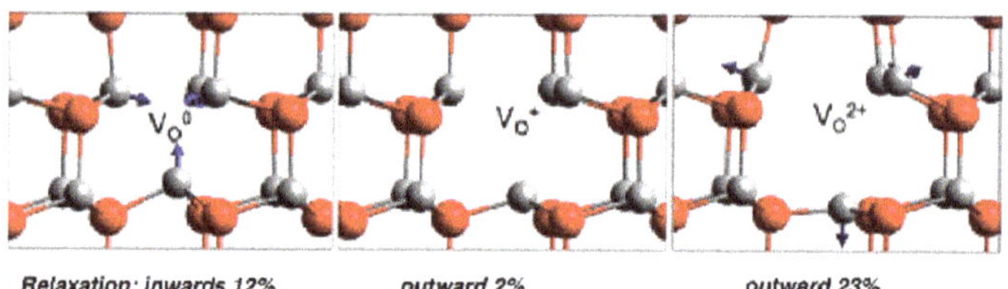

Figure 10. Ball and stick representation of the local atomic relaxations around the oxygen vacancy in ZnO in the (0), (1$^+$), and (2$^+$) charged states. Reproduced with permission from Seebauer and Kratzer [64]. Copyright 2007 by Elsevier B.V. (License No.: 4452270159193).

The optical absorption by F^+ and F centers and electron processes subsequent to absorption of light is of some importance in metal oxides, because the energy of a photostimulated electron transition from the ground state to the exited state and the position of the exited state within the bandgap relative to the bottom of the CB band determine the formation of free electrons in the CB band.

An explanation of the formation of free charge carriers by photoexcitation of F or F^+ centers had been proposed in two earlier studies by Kuznetsov and coworkers [68,69]; it is illustrated schematically in Figure 11. It was implied that the first photophysical event is the optically- activated electronic transition from the ground state F (or F^+) center to its excited state $(F)^*$ (or $(F^+)^*$). Free electrons in the conduction band then result from a thermally-activated electron transition from the F^* (or $(F^+)^*$) state to the conduction band, while the free holes in the valence band appear through a thermally-activated capture of an electron from the valence band by the lower level of the excited F^* (or $(F^+)^*$) center. After the photoformation of electron–hole pairs through such a scheme, the F^* (or $(F^+)^*$) center returns to its initial ground state that ensures the stability of these processes during the photoexcitation events. The thermal energy needed for the occurrence of such processes is about 0.1 eV for the 2.95 eV absorption band and ~ 0.5 eV for the 2.56 eV band that significantly exceeded the energy of phonons at the temperatures employed (bandgap energy E_g of VLA rutile was 3.05 eV at 90 K and 3.01 eV at 290 K). A recent study based on experimental results and literature data further led Kuznetsov and coworkers [70] to hypothesize that, following the prime optical excitation of defects in the solid (Ti^{3+} centers), the heat released during the non-radiative electron transitions dissipated into the nearest neighborhood of these centers with consequences (see Section 5). Localized non-equilibrated excitation of the phonon subsystem was equivalent to energies up to 1 eV.

Experimentally determined positions of the absorption band maxima of F and F^+ centers in several other metal oxides have been reported [60,61,71,72]. In accord with experimental results, theoretical calculations placed the ground state levels of both F and F^+ centers in MgO (bandgap, 7.8 eV) at 3 eV above the top of the VB band [61]; this means that the exited states of these centers are near the bottom of the CB band (experimentally, ca. 0.06 eV below CB; see Ref. [61] and references therein). In the case of TiO_2, the F centers have been located at 2.9 and 2.55 eV [61].

The ground state of the F center in α-Al_2O_3 is at 5.3 eV from the VB band, and thus the excited state F^* lies within the conduction band manifold [61]. The ground state of the F^+ center is distant from the VB band by 3.1 eV so that the exited state $(F^+)^*$, corresponding to the absorption band at 6.3 eV, lies within the CB band. Indeed, irradiation at 6.1 eV within the spectral range corresponding to the F^+ center's absorption band produces photoconductivity down to at least 10 K (see Ref. [63] and references therein). For other F^+-center's exited states corresponding to the 5.4 eV and 4.8 eV absorption bands are 1.0 eV and 1.7 eV distant, respectively, from the CB band. Thus, UV irradiation in the absorption bands of F and F^+ centers in MgO or F centers in α-Al$_2$O$_3$ result in the photo-ionization of the defects and ultimately initiation of surface reactions in the absence of any significant number of bulk electron traps.

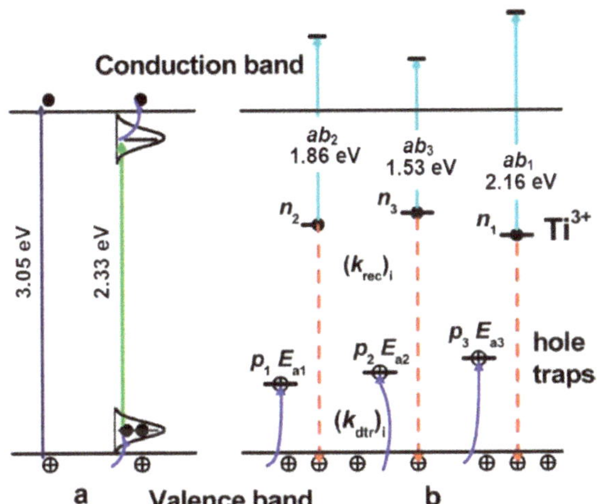

Figure 11. Illustration of the relative positions of the energy levels of: (**a**) $F°$ centers; and (**b**) hole traps and electron traps (e.g., Ti^{3+} centers) within the band gap of VLA TiO$_2$ rutile. The scheme represents a simplistic description of the formation of charge carriers via the intrinsic absorption at 3.05 eV and absorption in the band at 2.33 eV. For simplicity, only three electron and three-hole traps are shown in (**b**). The blue arrows in (**b**) indicate the thermostimulated release of photoholes, while the red arrows indicate the recombination of holes in the valence band with electron Ti^{3+} centers; Reproduced with permission from Kuznetsov et al. [68]. Copyright 2014 by the American Chemical Society.

Absorption spectra of strongly defective crystals of MgO and Al$_2$O$_3$ revealed additional absorption bands that were ascribed to aggregates of oxygen vacancies [61,73–76], the main types being F_2 centers (i.e., 2 V_Os with 4 trapped electrons), and F_2^+ and F_2^{2+} centers (two V_Os with three and two trapped electrons, respectively). Under irradiation with fast neutrons (E > 1.2 MeV) [75] or bombardment by Cu$^+$ or Ti$^+$ ions (E = 30 KeV) [74], F and F^+ centers were shown to be the dominating defects in the oxygen sub-lattice. It is evident that the family of intrinsic defects related to oxygen vacancies yields *in toto* broad absorption spectra covering the UV, visible and near-IR spectral regions.

A full understanding of the pathways of photoactivation of wide bandgap metal oxides requires a detailed examination of the photoconversion $F^+ \rightarrow F$. In his 1982 review article, Crawford called attention to a puzzling feature of the behavior of F^+ centers in Al$_2$O$_3$ when irradiated into the 4.8 eV and 5.4 eV absorption bands, which decreased the number of F^+ centers and increased the number of F centers (absorption band at 6.0 eV) [63]. Photoconversions $F^+ \rightarrow F$ were also found in MgO crystals and were accompanied by formation of hole centers [61–63], which excludes thermal ionization of the excited $(F^+)^*$ center followed by further electron trapping by another F^+ center. Hole formation was connected with the spontaneous electron capture from an O^{2-} adjacent to the $(F^+)^*$-center's empty ground state [63].

Semi-empirical calculations have also offered explanations for the $F^+ \rightarrow F$ photoconversions [61,62]. Theory predicted the existence of several *quasi*-local energy levels of the F^+ center in the upper part of the VB band, which consist mainly of atomic orbitals of O^{2-} ions surrounding the F^+ center. Under optical excitation, electron transition from these *quasi*-local levels to the ground state of the F^+ center yielded an F center and a hole in the VB band [61,62]; ultimately, the hole was trapped either by a cation vacancy V_m or by some impurity.

Predictable schemes of the photogeneration of electrons and holes on excitation of F-type centers in wide bandgap oxides can be summarized by Equations (49)–(51c).

$$F + h\nu \rightarrow F^* \rightarrow F^+ + e^-_{cb} \tag{49}$$

$$F^+ + h\nu \rightarrow F^{+*} \rightarrow F^{2+} + e^-_{cb} \tag{50}$$

$$F^+ + h\nu \rightarrow F^{+*} \tag{51a}$$

$$F^{+*} + (O^{2-})_{vb} \rightarrow F + (O^-)_{vb} \tag{51b}$$

$$(F^+)_{ql} + h\nu + (O^{2-})_{vb} \rightarrow F + (O^-)_{vb} \tag{51c}$$

Equations (49) and (50) represent cases of the photoionization of defect centers with formation of free electrons in the CB band. Equation (51b) corresponds to electron capture from an O^{2-} ion adjacent to the F^{+*}-center, whereas Equation (51c) represents electron excitation from *quasi*-local levels of an F^+-center (i.e., F^+_{ql}) in the VB band accompanied by formation of an F center.

Emeline and coworkers [77] reported a single maximum at 4.6 eV in the spectral dependence of the quantum yield of photoadsorption of O_2 on powdered MgO, whereas Kuznetsov et al. [78] and Lisachenko [79] reported the maximum to occur at 5.0 eV. Spectral dependences of Φ of photo-adsorption of H_2 and CH_4 on powdered MgO displayed maxima at 4.6–4.55 eV [77] and a well-resolved shoulder at 3.75 eV. The photogeneration of holes on excitation within the spectral range, corresponding to the absorption band of F^+ centers with maximum at 4.9 eV, was likely responsible for the photoadsorption; however, the spectral feature at 3.75 eV found no counterpart in the absorption spectra of F^+ and F centers. Only the absorption band of an F_2 center at 3.63 eV was near the feature at 3.75 eV, but nothing is known about the processes that followed the photon absorption by this center. Note that hole trap point defects related to cation vacancies in MgO absorbed light in the region 2.2–2.35 eV [67], far from the spectral feature just described.

Spectral efficiencies of the photoadsorption of O_2 and the photodissociation of N_2O on Al_2O_3 displayed a maximum at 5.6–5.4 eV and a shoulder at ~ 4.8 eV [79], in good agreement with the absorption bands of F^+ centers at 5.4 and 4.8 eV; as noted earlier, however, direct photoionization of these centers in Al_2O_3 in this spectral region appeared unlikely.

Although optical properties of F-type centers in wide bandgap metal oxides have been studied sufficiently both theoretically and experimentally, interpretation of the data on the spectral response subsequent to irradiation, even for simple photoreactions such as photoadsorption of O_2 and H_2, has encountered some problems.

Indeed, the spectral dependencies of the quantum yield of O_2 and H_2 photoadsorption in the visible and near-bandgap spectral regions for powdered TiO_2 reported by Cherkashin's group [80,81], by Emeline and coworkers [77,82], and by Komaguchi et al. [83] have been digitized; results are illustrated in Figures 12 and 13. The majority of the action spectra appear as a sum of two bands, the main band occurring at 3.0–2.9 eV with a half-bandwidth of 0.15–0.3 eV, independent of the type of gas used (O_2, H_2 or CH_4) and irrespective of the crystalline structure of TiO_2 (rutile or anatase). Spectra 2 and 3 in Figure 12 and 3 and 4 in Figure 13 display an additional band (seen as a shoulder) with maximum at ca. 2.5–2.7 eV. The action spectra of the photogeneration of paramagnetic $O_2^{-\bullet}$ (Figure 12, Curve 4) and $[O^-O_2]$ species (Figure 13, Curve 5) show broad spectra with a single maximum around 2.6–2.8 eV.

The action spectra of the photogeneration of electrons (photoadsorption of O_2 or photoformation of $O_2^{-\bullet}$) or of the photogeneration of holes (photoadsorption of H_2 and CH_4 or the photoformation of $[O^-O_2]$) originate from excitation of intrinsic defects, which have a biographical origin because all samples examined were nominally pure (undoped). Note that the rutile specimen displaying the broad action spectrum (Figure 13, Curve 5) had been strongly reduced by a H_2 heat treatment at 773 K;

it showed an absorption spectrum with a well-resolved shoulder at 2.7 eV and a broad absorption in the near-IR region (after bleaching the initial absorption in the presence of O_2) [83].

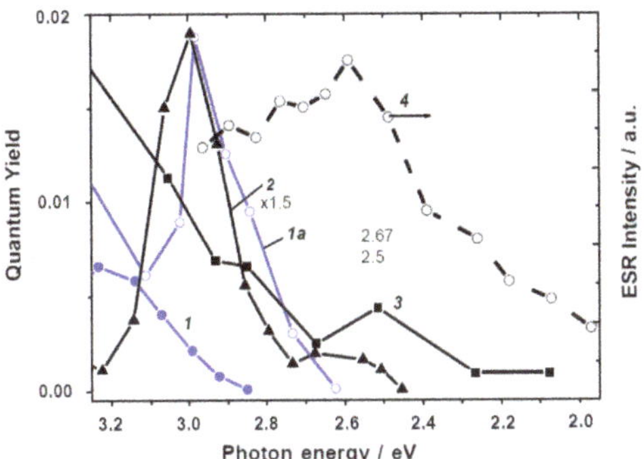

Figure 12. Spectral dependencies of the quantum yields of photostimulated adsorption of O_2 on the surface of TiO_2 specimens: Degussa P25 pre-treated at 600 K (Curve *1*) and at 850 K (Curve *1a*) [82], rutile (Aldrich) pretreated at 600 K (Curve *2*) [82], rutile pretreated at 820 K (Curve *3*) [80]. Spectral dependence of the photogeneration of $O_2^{-•}$ species at 77 K on the surface of rutile reduced by a H_2 heat treatment at 773 K (Curve *4*) [81].

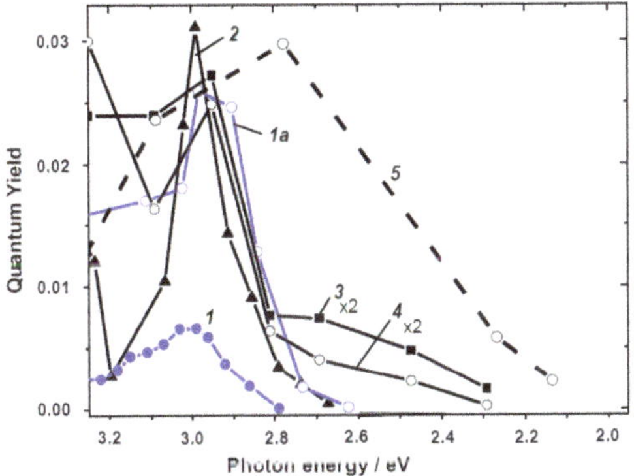

Figure 13. Spectral dependencies of the quantum yields of photostimulated adsorption of H_2 (Curves *1*, *1a*, *2*, and *3*) and CH_4 (Curve *4*) on the surface of TiO_2 specimens: Degussa P25 pre-treated at 600 K (Curve *1*) and at 850 K (Curve *1a*) [82], rutile (Aldrich) pre-treated at 600 K (Curve *2*) [82], rutile (Curves *3* and *4*) [77]. Spectral dependence of the efficiency of photogeneration of a hole species [O^-O_2] at 77 K on the surface of anatase pretreated at 800 K (Curve *5*) [81].

Four principal reasons led Kuznetsov and Serpone [84] to model the photoactivation of (modified) TiO_2 in the visible region: (1) the spectral position of the bands constituting the action spectra (i.e., bands at 3.0–2.9 and 2.5–2.7 eV); (2) the coincidence of the bands for the photo- generation of electrons and holes with the latter allowing the authors to propose that photoexcitation of intrinsic defects leads to the simultaneous generation of electrons and holes (electron–hole pairs); (3) the spectral position of the absorption bands at 2.95–2.75 eV and at 2.55–2.50 eV in the spectra of reduced TiO_2 attributed to *F* centers [84]; and (4) the known regularities of photoexcitation of *F* type centers and consequent physical processes. In this regard, the authors [84] proposed a simple reasonable mechanism to interpret the experimental results reported in Figures 12 and 13. In accord with their study of TiO_2 specimens, the first photophysical event of photoinduced absorption and photoreactions was light absorption by the F center that led to the formation of the corresponding excited state F^* (Equation (52)).

$$F + h\nu_{AB} \rightarrow F^* \tag{52}$$

$$F^* \rightarrow F^+ + e^-_{cb} \tag{53a}$$

$$F^* + \Delta \rightarrow F^+ + e^-_{cb} \tag{53b}$$

$$F^+ + (O^{2-})_{vb} \rightarrow F + (O^-)_{vb} \tag{54}$$

$$F^* + (O^{2-})_{vb} \rightarrow F + (O^-)_{vb} + e^-_{cb} \tag{55}$$

If the ground state of the F center lay within the bandgap near the VB band, then F^* should lie within or close to the CB band. The energy difference ($E_{bg} - h\nu_{AB}$) was only about 0.2–0.3 eV between the E_{bg} of the rutile/anatase TiO_2 and the AB1 band, and 0.55–0.60 eV between the E_{bg} and the AB2 band. The excited F^* center transforms spontaneously into the F^+ center and a conduction band electron (Equation (53a)), or else it does so through thermal stimulation (Equation (53b)). Formation of a hole supposes the capture of an electron by the F^+ center from the VB band, i.e., from the O^{2-} ion (Equation (54)). Equation (55) describes the synchronous formation of an electron–hole pair. Hence, Equations (52)–(55) infer the photogeneration of both CB band electrons and VB band holes under visible light irradiation, and recovery of the centers of light absorption during irradiation.

3.1.2. Defects Related to Cation Vacancies (V_m)

The main type of defects in the cation sub-lattice of metal oxides are defects related to cation vacancies, V_m, which have one or two holes trapped, i.e., V^- or V° centers, respectively. The atomic and electronic structure, the mechanism of optical excitation, and thus optical properties of hole V-type centers are defined by hole localization (trapping) at one of several equivalent oxygen ions surrounding the metal vacancy. The term bound small polaron also applies to emphasize the interconnection of carrier (hole) trapping at a single site and the stabilization of the hole at this site by a lattice distortion. INDO calculations [61] showed that, in the V^- center of MgO, the O^- ion is closer to the cation vacancy by 3% of the Mg–O distance in the perfect crystal. In the V° center, the two O^- ions that can capture two holes are located on the opposite side of the cation vacancy; they are shifted inward toward the vacancy by 2% of the regular Mg–O distance.

It is clear from the brief description of optical properties of photocatalytically active metal oxides that, in addition to the photoexcitation of solids in the fundamental absorption bands, there are many other pathways for the generation of electrons and holes. Concurrently, the same defects can play either negative or positive roles with respect to heterogeneous photocatalysis. For instance, they can act as recombination centers that reduce the concentration of photocarriers. However, being localized at the surface of photocatalyst particles, the defects are more likely to play the role of surface-active centers in photocatalytic processes.

3.1.3. Photogeneration, Recombination and Trapping of Charge Carriers in Photoactive Solids

Photoexcitation of solid photocatalysts in their fundamental absorption bands is the most appropriate type of photoexcitation in heterogeneous photocatalysis, as both electrons (reductants) and holes (oxidants) are generated. Concurrently, the generation of carriers also occurs as a result of light irradiation at wavelengths corresponding to extrinsic absorption bands, i.e., in the longer wavelength spectral region with respect to the fundamental absorption edge of the solids. The same is true of dopant-sensitized photocatalysts.

Recombination of electrons and holes can occur by two pathways: (i) direct recombination of free electrons from the conduction band with free holes from the valence band; and (ii) indirectly through the participation of defects (recombination centers). In the latter case, sequential trapping of free carriers of opposite charge by recombination centers takes place in each recombination cycle. The *quasi*-stationary concentrations of carriers established under stationary illumination of the photocatalyst depends on the concentration and trapping cross-section of the recombination centers. When photogeneration of carriers takes place in the bulk of the photocatalyst particle (a typical case), photoelectrons and photoholes reach the surface via diffusion. On the other hand, a fraction of the carriers is captured by various carrier traps, both in the bulk and at the surface of the solid particle in sufficiently high number, provided that trapping cross sections of one carrier takes place relatively slow (so-called *deep traps* of the *color center* type). Deep traps that capture the carriers may be centers of the *V*-type or of the *F*-type, depending on the charge of the captured carrier. Surface traps of this kind play the role of photocatalytic active centers. In relation to the chemical interaction with molecules, it is useful to distinguish two states of such centers—the inactive state (without trapped carriers) and the active state (trap is occupied by either an electron or a hole).

3.1.4. Trapping of Carriers by Defects

Carrier trapping by a defect is a good example of a perfect inelastic collision of carriers with the defect. The effectiveness of carrier trapping in solids is characterized by the phenomenological trapping cross-section, σ_{tr} (Equation (56)).

$$\frac{dN}{dt} = \sigma_{tr} v n \tag{56}$$

where dN/dt is the number of carriers trapped by the defect per unit time; v is the mean velocity of the carriers ($v \approx 10^7$ cm s^{-1}); and n is the concentration of thermal carriers (or stationary concentration of photocarriers). The quantity dN/dt has dimensions and notion of frequency; its inverse can be treated as the lifetime of a defect (unoccupied by a carrier) relative to a trapped carrier.

Trapping cross-sections are determined by the properties of the defects; they vary within a rather wide range from 10^{-21}–10^{-20} cm^2 (trapping is energetically favorable but an activation barrier exists) p to 10^{-11}–10^{-10} cm^2 (strongly attractive Coulomb center; scattering of excess energy is effective). Typical values of σ_{tr} lie in the range between 10^{-16} and 10^{-15} cm^2 for neutral defects.

3.1.5. Stationary Concentration of Photocarriers and Band-to-Band Recombination

Figure 14 outlines the generation and recombination of carriers of various kinds that determine the stationary concentration of photocarriers in wide bandgap solids [85].

In general, both electrons and holes generated in wide bandgap solids on absorption of photons with $h\nu > E_{bg}$ initially transit into states at some energy distance from the bottom of the CB and the top of the VB, respectively (Step 1 in Figure 14a). Thermal equilibrium between the crystal lattice and the photocarriers is established as a result of relaxation of the energy and momentum of carriers within the timescale of $\sim 10^{-10}$ s (Steps 2 and 2'). Subsequent lowering of the energy of carriers occurs via recombination and trapping. For ideal (non-defective) solids, only radiative and nonradiative band-to-band recombination is possible (Steps 3 and 3'). The nonradiative electron transitions CB $\rightarrow$ VB in wide bandgap solids ($E_{bg} \geq 3$ eV) have low probability since participation of many phonons with energy of 0.1 eV is required in the process. In general, the higher is E_{bg}, the higher is the probability of radiative transitions in solids with emission of photons manifested as an inter-band or edge luminescence. At the same time, the probability of radiative band-to-band transitions is rather low, since the momentum conservation law (see above) requires that the condition similar to that for band-to-band direct transitions be satisfied. Typically, band-to-band luminescence in wide bandgap solids with photon energy close to E_{bg} (Step 3, Figure 14a) is detected at moderate intensities of the exciting light for rather perfect crystals, for which competitive radiative recombination of carriers through defects (recombination centers, R) is suppressed, or in the case of intense photoexcitation of the crystal when a high concentration of photocarriers is achieved in the solid.

A more detailed treatment of band-to-band recombination was outside the scope of this review article, but suffices to note that band-to-band recombination in line with band-to-band optical excitation in TiO$_2$-based heterogeneous photocatalysis is often treated following Equations (57) and (58).

$$TiO_2 + h\nu \rightarrow (e^- + h^+) \tag{57}$$

$$(e^- + h^+) \rightarrow TiO_2 \text{ (band-to-band recombination)} \tag{58}$$

With regard to the effectiveness of band-to-band recombination, one should use caution in believing the bimolecular stage described by Equation (58) as being real (not symbolic) in kinetic studies of heterogeneous photocatalytic reactions, even though it can easily explain experimental non-linear dependencies of reaction rates on light intensities, which typically follow a square-root dependence.

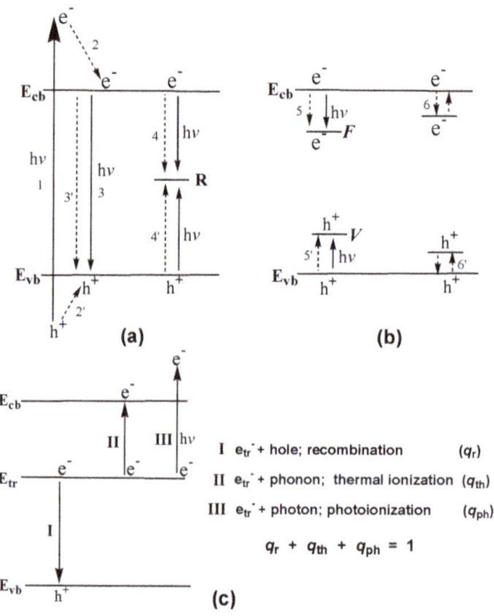

Figure 14. Schemes illustrating the processes of: recombination (**a**); and trapping (**b**) of photocarriers and events leading to discharge of the defect (for distinctness for defect with captured electron): Step 1: band-to-band optical transition with "hot" electron and "hot" hole generation, where the corresponding initial levels of excited electron and hole lay above the "bottom" of conduction band (E$_c$) and below the "selling" of the valence band (E$_v$); Steps 2 and 2': thermal relaxation of hot carriers; Steps 3 and 3': radiative and nonradiative band-to-band recombination; Steps 4 and 4': radiative and nonradiative recombination via defects; Steps 5 and 5': trapping of carriers with formation of color centers; and Steps 6 and 6': trapping and de-trapping of carriers by shallow traps (for further details, see text). (**c**) various stages I, II, and III regarding recombination, thermal ipomnization and photoionization, respectively. Adapted from Artemiev and Ryabchuk [85].

3.1.6. Recombination of Carriers via Defects

Recombination of photocarriers via recombination centers is the main pathway of carrier recombination in imperfect wide bandgap solids. In this case, a given recombination center R subsequently captures a free electron and a free hole, or vice versa, in a single recombination cycle (Steps 4 and 4' in Figure 14a). In an alternative approach (without the concept of holes), recombination via defects means a consequent two-step transition of electrons from the CB to an empty state of the defect and from the defect to an empty state of the VB, or from the defect to the VB and then from the CB to an empty state of the recombination center R. The excess energy of the electron is dissipated at both steps of the recombination. Typically, the dissipation of excess energy at one of the two stages occurs via a nonradiative pathway with the assistance of existing phonons. Normally, the probability of nonradiative transitions increases as the number of emitted phonons decreases so that the energy levels corresponding to efficient recombination centers typically lie near the middle of the energy gap in wide bandgap solids. Radiative transitions at any stage of the two-step recombination via a center R are responsible for recombination luminescence in solids.

3.1.7. Trapping of Carriers with Formation of Centers Similar to Color Centers

In addition to the recombination centers R, other types of centers can be distinguished in wide bandgap solids—e.g., the color centers of the *V*- and *F*-type and shallow (or thermal) traps (Figure 14b). Discrimination between centers is rather arbitrary. Following the work of Siline and Trukhin [86], we may assume that a center captures photocarriers with a probability p and loses them with probability q (units: number of events per unit time). The rate of formation of defects in a state with a trapped carrier dn/dt (units: $cm^{-3} s^{-1}$) is then given by Equation (59) with N_0 defining the initial concentration of empty trap centers.

$$\frac{dn}{dt} = p(N_0 - n) - qn \tag{59}$$

Equation (59) leads to the kinetics of formation of filled traps $n_{(t)}$ given by Equation (60), and to the stationary concentration of filled traps n_∞ by Equation (61),

$$n_{(t)} = N_0 \frac{p}{p+q}(1 - e^{-(p+q)t}) \tag{60}$$

$$n_\infty = N_0 \frac{p}{p+q} \tag{61}$$

Three possible events can occur for a carrier trapped at a trap center (Figure 14c):

1. Stage I, recombination via trapping of carriers of the opposite sign;
2. Stage II, thermal emission of carriers into the corresponding band; and
3. Stage III, photo-ionization of the trapped carrier by absorption of photons.

Accordingly, q in Equation (61) is given by,

$$q = q_r + q_{th} + q_{ph} \tag{62}$$

For simplicity, if the process defined by Stage III were neglected (i.e., $q_{ph} = 0$), and if $q = (q_r + q_{th})$ >> p, and q_r >> q_{th}, then n_∞ << N_0. In other words, filled traps do not accumulate in a solid in numbers compared with the initial number of empty traps N_0. Typically, these deep trap centers with high cross-section of carrier trapping that do not accumulate in illuminated solids are the recombination centers. In contrast to deep centers, carriers trapped in shallow traps with energy ~ kT from the CB (electrons) or from the VB (holes) can be detrapped via thermal emission with probability q_{th} such that $q_{th} \sim exp(-E/\kappa T)$ >> q_r.

Deep centers with a low probability of trapping a second carrier ($p \approx q$) accumulate in illuminated solids in sufficiently high number ($n_\infty \approx N_0$). Historically, they have been referred to as color centers. The ultimate concentration of color centers can reach values of $n_\infty = 10^{17}$–10^{18} cm^{-3} for some samples of wide bandgap solids. The term color center originated from the accumulation of such traps in crystals that display absorption bands in the visible spectral region. The result of these traps is that transparent (white when powder) solids become colored under UV illumination, with the color tending to be rather stable at ambient temperature. Such solids can be discolored by calcination, or by thermal bleaching (probability of this process given by $q_{th} \cong exp(-E/\kappa T)$ is high), or partly by photobleaching (q_{ph} is also high) via illumination of the crystalline specimen with light at the wavelengths corresponding to the absorption bands of the color centers.

3.1.8. Lifetime and Concentration of Free Charge Carriers

A major factor affecting processes in heterogeneous photocatalysis is the stationary (surface) concentration of charge carriers. As such, a consideration of some basic approaches that describe the processes responsible for the charge carrier concentration is worth noting.

The spatially uniform photogeneration of charge carriers occurs at a constant rate of generation, g, in some space of the solid bulk. The rate of the temporary alteration of the charge carrier concentration is then given by Equation (63),

$$\frac{dn}{dt} = g - \frac{n}{\tau} \tag{63}$$

where τ is the lifetime of the free charge carrier, independent of the charge carrier concentration. In the case of several types of trapping and recombination centers, the lifetime of the charge carriers can then be expressed by Equation (64), provided that the various centers do not interact with each other. Otherwise, the determination of the lifetime becomes rather more complex [87].

$$\frac{1}{\tau} = \sum_i \frac{1}{\tau_i} \tag{64}$$

where $\tau_i = 1/\sigma_{tr} v N_i$ is the lifetime of the charge carrier with respect to trapping by the defects of a given i-sort; σ_{tr} is the trapping cross-section; v is the velocity of the charge carrier; and N_i is the concentration of the defects of a given sort.

After an initial period of irradiation, recombination centers and trapping centers are filled with charge carriers to a stationary level; the lifetime of the charge carriers becomes pseudo-constant and its value determines the stationary concentration of the free charge carriers. Thus, $dn/dt = 0$, and the stationary concentration of charge carriers is then (Equation (65)),

$$n_\infty = g\tau \tag{65}$$

Obviously, the higher the concentration of recombination centers and the larger the trapping cross-section are, the shorter is the lifetime and the smaller is the concentration of free charge carriers. For semiconductors and isolators, the lifetime of photoinduced charge carriers can vary in a wide range from picoseconds (10^{-12} s) to milliseconds (10^{-3} s).

When stationary conditions are established for the charge carriers, one fraction of the carriers remains in the delocalized state, while the other fraction is trapped by the defects. The relation between charge carriers obeys the electroneutrality principle (Equation (66)),

$$n_e + n_{tr}^e = n_h + n_{tr}^h \tag{66}$$

where $n^e{}_{tr}$ and $n^h{}_{tr}$ are the concentrations of electrons and holes, respectively, trapped by the defects.

The lifetime of a trapped charge carrier is determined by the efficiency of either thermo- or photo-ionization of the center or by the efficiency of the recombination event. When thermo-ionization dominates, the lifetime τ_{th} (Equation (67)) varies from picoseconds for shallow traps with depth comparable to kT, to infinite time for color centers in wide bandgap insulators.

$$\tau_{th} = \frac{1}{q_{th}} \approx 10^{-12} e^{\left(\frac{E}{kT}\right)} \tag{67}$$

However, when the lifetime of the trapped charge carrier is determined by recombination (Equation (68)) then

$$\tau_r = \frac{1}{q_r} = \frac{1}{\sigma_r v n} \tag{68}$$

where σ_r is the recombination cross-section, v is the velocity of the charge carrier of the opposite sign, and n_i is the concentration of the charge carrier of the opposite sign. The higher is the concentration of the opposite charge carrier, the shorter is the lifetime of the trapped charge carrier.

Charge carrier trapping and recombination processes determine the stationary concentration of charge carriers at the surface. Moreover, considering that surface defects with trapped charge carriers can act as surface-active centers that initiate surface chemical sequences, the lifetime of trapped charge carriers on such surface defects corresponds to the lifetime of the chemically active states of the surface-active centers. Recombination processes then return the surface-active centers to their (initial) chemically inactive ground states.

4. Applied Photocatalysis: Laboratory-Scale deNOxing of NO$_x$ Agents (NO & NO$_2$)

As discussed earlier, nitrogen oxides (NO$_x$) are major atmospheric pollutants that play an important role in atmospheric chemistry, and have been the object of a significant number of investigations toward their minimization, if not complete removal from the environment. The concentration of NO$_x$ in polluted urban air is around 100 ppbv, whereas in the unpolluted troposphere, it ranges from 10 to 500 pptv [88]. Recall that NO$_x$ are emitted primarily from artificial sources (e.g., traffic, coal burning boilers, thermal power plants, and industries of various sorts) and from natural sources (e.g., biological degradation in soil and from lighting thunder). NO$_x$ participate in various environmental processes: for instance, in the formation of acid rain; in the greenhouse effect in synergy with sulfur oxides; in the formation of photochemical smog in the presence of CO and VOCs; in the depletion of stratospheric ozone; and in the formation of peroxyacetyl nitrates (PAN), all of which have negative effects on ecosystems and lead to non-insignificant human health issues. With regard to the latter, NO$_x$ pollutants cause problems in the respiratory tract that include lung edema and the reduction of the oxygen-caring capacity of blood—e.g., in the transformation of hemoglobin to methemoglobin.

No wonder then that significant efforts have been expended to reduce environmental NO$_x$ agents back to N$_2$ via a thermal technology using a variety of reductants (e.g., CO, hydrocarbons, H$_2$ and NH$_3$) in what is known as Selective Catalytic Reduction (SCR). While reduction of NO occurs around 100 °C in the presence of H$_2$ and a Pd-supported catalyst [89], other reactions require significantly greater temperatures. In fact, reduction of NO to N$_2$ through selective catalytic reduction with NH$_3$, and thus potentially treat NO$_x$ agents, the costs of the SCR technology for the construction and operation of a facility to treat NO$_x$ pollutants, together with the required consumption of energy, may prove prohibitively high. Nonetheless, despite the many efforts to eliminate the NO$_x$ emitted from the various sources noted earlier by SCR, the fact remains that the concentration of NO in air in Japan was nearly constant throughout the 1980s, and was often higher than the air quality standard set for NO$_2$, principally along heavily trafficked roads in densely populated areas [90]. This led to the development of a new technology for the disposal of NO$_x$ at sub-ppm level from air and from trafficked roads and tunnels, and other environmental sources that emit NO$_x$.

Recognition that plants and micro-organisms can easily consume nitrite (NO$_2^-$) and nitrate (NO$_3^-$) ions as raw materials for nitrogen assimilation provided a further impetus to examine alternative technologies to achieve a practical removal of dilute NO$_x$ agents from the environment using sunlight (UV-Visible) radiation at significantly lower costs. In this regard, Takeuchi and Isubuki [91] investigated the dry deposition of NO$_x$ onto the ground and found that the rate of adsorption of NO$_x$ on some soil particles was enhanced by photoillumination. Of the metal oxides constituting the soil particles, TiO$_2$ showed the highest activity for NO$_x$ adsorption under photoillumination with ca. 60% of NO$_x$ being captured as nitric acid (HNO$_3$) on the surface of TiO$_2$ particulates. Accordingly, the authors thought that the photocatalytic oxidation of NO$_x$ to HNO$_3$ by illuminated TiO$_2$ might be most advantageous to treat dilute environmental NO$_x$, as any extra reactants such as NH$_3$ were not required and HNO$_3$ could be trapped on the surface as nitrates.

One of the first studies to examine the fate of one of the NO_x agents, namely NO, in the presence of (Degussa) P-25 TiO_2 exposed to UV irradiation was reported in 1984 by Courbon and Pichat [92] who exposed isotopically labeled $N^{18}O$ at 295 K in the dark to pre-oxidized and pre-reduced TiO_2 powder; subsequent to UV illumination resulted in three phenomena: photoadsoprtion, photoexchange, and photodecomposition of NO to yield N_2O and, to a lesser extent, N_2. The formation of N_2O + N_2 corresponded to a photodecomposition of ca. 15% of the NO pressure (decrease) for a pre-reduced titania sample and ca. 20% for a pre-oxidized titania; N_2 formed only at the beginning and the percent $N_2{}^{16}O$ produced was initially greater for the pre-oxidized titania sample. This early study [92] confirmed that illumination of TiO_2 with UV light considerably increased the ease of detachment of surface oxygen atoms, as the isotopic hetero-exchange of $N^{18}O$ occurred at room temperature, while it required higher temperatures in the absence of bandgap (3.2 eV) illumination of the mostly anatase TiO_2. Adsorbed oxygen species were involved, as pre-oxidized titania exhibited higher initial efficiency; however, the instantaneous exchange with a pre-reduced titania sample in H_2 at 723 K showed that detachment also involved surface oxygen atoms that were replenished from NO. Another aspect of this study was the corroboration of the direct involvement of O^- species in photocatalytic oxidations over TiO_2 and other n-type semiconductors, since NO and O_2 played similar roles in yielding dissociated oxygen species active in both oxidation and oxygen isotopic exchange.

A later study (1985) by Hori and coworkers [93] demonstrated that $NO_2{}^-$ ions are oxidized to $NO_3{}^-$ with or without O_2 in aqueous suspensions of some semiconductor powders (Ag_2O, PbO, anatase TiO_2, Si, ZnO, SnO_2, CdS, and Bi_2O_3) under bandgap illumination; with TiO_2, 96% of nitrite was oxidized to nitrate in the presence of oxygen. Along similar lines, Anpo and coworkers [94] found that Cu^+ ions on SiO_2 (Cu^+/SiO_2 catalyst) could decompose NO molecules photocatalytically and stoichiometrically into N_2 and O_2 at 275 K, which they attributed to the significant role played by the excited state of the Cu^+ species; the photoreaction involved an electron transfer from the excited state of the Cu^+ ion into an anti-bonding π orbital of the NO molecule within the lifetime of its excited state. The relationship between the local structures of Cu^+ ions in zeolite and their photocatalytic reactivity in the decomposition of NO_x into N_2 and O_2 at 275 K was reviewed by Anpo and coworkers [95] after which Anpo's group [96] reported on the metal ion-implantation of TiO_2 with various transition metal ions that subsequent to calcination in oxygen at ca. 723 K resulted in a large shift of the absorption edge of TiO_2 toward visible light regions depending on the amount and type of metal ions implanted; the resulting metal ion-implanted TiO_2s proved active in the photocatalytic decomposition of NO to N_2, O_2 and N_2O at 275 K under irradiation with visible light at wavelengths longer than 450 nm.

Following their 1989 report [91], Ibusuki and Takeuchi [97] examined the photocatalytic destructive oxidation of NO to $NO_3{}^-$ using a mixture (200–250 mg) of TiO_2, activated carbon (AC) and Fe_2O_3 particles located in a flow-type photochemical reactor system (Figure 15) that was photo-illuminated by a cylindrical bank of 12 black lights (wavelength: 300–400 nm) [97]. The AC and Fe_2O_3 had a remarkable effect in increasing the catalytic activity for NO_x removal, likely due to their high adsorptive activity for NO and NO_2. The authors inferred that photo-illuminated TiO_2 generated reactive oxygen species that oxidized NO and NO_2, respectively, to NO_2 and $NO_3{}^-$, while activated carbon trapped NO_2 to allow enough time for TiO_2 to oxidize NO_2 to $NO_3{}^-$; it appears that Fe_2O_3 acted as a promoter for more NO/NO_2 molecules to be adsorbed on the surface of the titania photocatalyst [97].

In a further study, the Takeuchi group [98] examined the use of TiO_2 to eliminate NO_x in open air with the photocatalyst being activated by sunlight, but noted, however, that in so doing desorption of NO_2 occurred during the oxidative removal of NO; the NO_2 also needed to be suppressed as it is also a regulated pollutant. Although NO_x adsorb on activated carbon to be oxidized ultimately to $NO_3{}^-$, development of an activated carbonaceous photocatalytic material proved difficult. Accordingly, recognizing that thin films have many micropores they designed and prepared TiO_2 thin film photocatalysts by a dip-coating process using titanium alkoxide as the TiO_2 precursor and the polymer additive polyethylene glycol (PEG) of different molecular masses (PEG-300, PEG-600, PEG-1000) to give TiO$_2$-PX films with thicknesses of 1.0, 0.5 and 0.25 µm after calcination of the films at 450 °C for 1 h deposited on silica-coated glass plates.

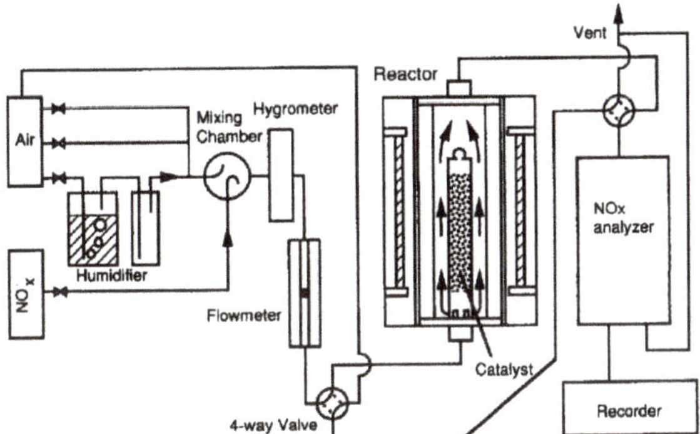

Figure 15. Schematic diagram of the flow-type reactor used for the heterogeneous photocatalytic reductions of NO$_x$ agents. Experimental conditions: 1–2 ppm NO/NO$_2$; flow rate, 500 mL min^{-1} for 5–10 h at different relative humidity (dry to 72%); reactor volume, 126 mL; pure O$_2$ or purified air; reaction temperature, ca. 310 K. Reproduced with permission from Ibusuki and Takeuchi [97]. Copyright 1994 by Elsevier B.V. (License No.: 4452271448487).

Table 1 summarizes the extent of NO removal [98]. Adsorbed NO was photooxidized to NO$_2$ by the thin films, while the produced NO$_2$ was re-photooxidized to NO$_3^-$ before it desorbed from the film surface.

Table 1. Extent of NO removal over 1.0 µm thick TiO$_2$-PX thin films irradiated at 365 nm (illuminated area, 100 cm^2; light irradiance, 0.38 mW cm^{-2}) for 12 h in a flow-type reactor (flow rate, 1.5 L min^{-1}; initial concentration of NO, 1.0 ppm; dry air).

Photocatalytic Thin Film	BET Surface Area (m^2 g^{-1})	Average NO Removal (%)
TiO$_2$-P0	112	71
TiO$_2$-P300	104	65
TiO$_2$-P600	118	70
TiO$_2$-P1000	141	81

Following reports that TiO$_2$ prepared by high-temperature hydrolysis of titanium tetra-alkoxides, Ti(OR)$_4$, in a hydrocarbon solvent was very active toward the photocatalytic dehydrogenation of *iso*-propanol in aqueous media under deaerated conditions [99] and mineralization of acetic acid under aerated conditions [100], Hashimoto et al. [101] prepared TiO$_2$ by the hydrolysis of titanium alkoxide in a hydrocarbon solvent, followed by calcination at various temperatures; the titania calcined at 300 °C proved most active for the photocatalytic oxidation of NO (Table 2), in comparison with P-25 titania. The photocatalytic oxidation was carried out in a fixed bed continuous flow Pyrex-glass reactor under atmospheric pressure with the TiO$_2$ (0.12 g) UV-irradiated with a 10-Wh black light; air contained 10 ppm of NO; flow rate, 110 mL min^{-1}. IR spectral results indicated that UV irradiation promoted the oxidation of NO in the presence of oxygen to yield nitrate species, while the data from ESR measurements for oxygen radicals showed that UV irradiation increased the number of O$_2^{-\bullet}$ adsorbed on the surface of titania in the presence of oxygen. These O$_2^{-\bullet}$ species vanished simultaneously with their exposure to NO, whereas the spectral intensity of the radical generated from secondary products of O$^-$ showed no change. The number of O$_2^{-\bullet}$ radical anions generated by UV irradiation reflected the photocatalytic oxidative activity of titania toward the oxidation of NO. The rate of formation of O$_2^{-\bullet}$ and the number of free electrons induced by UV irradiation decreased significantly with an increase in post-calcination temperature (Table 2) [101].

Table 2. Crystal size and surface area of the titania photocatalysts together with the rate of formation and the quantities of the $O_2^{-\bullet}$ radical anions **.

Photocatalyst	Calcination T (°C)	Crystal Type	Crystal Size (nm)	BET Surface Area (m^2 g^{-1})	Rate of $O_2^{-\bullet}$ Formation (μmol min^{-1})	$[O_2^{-\bullet}]$ (μmol g^{-1})
TiO$_2$	300	Anatase	10	133	2.9	7.2
TiO$_2$	550	Anatase	18	78	0.80	3.7
TiO$_2$	700	Anatase	26	34	0.06	0.25
TiO$_2$	800	Anatase + Rutile	47	8	0.01	0.07

** Adapted with permission from Hashimoto et al. [101]. Copyright 2000 by Elsevier Science S.A. (License No.: 4453260419845).

A photocatalytic reaction that takes place in a gas/solid reactor necessitates both the exposure of the catalysts to light irradiation and good contact between reactants and catalyst. In this regard, Lim and coworkers [102] noted that a two-dimensional fluidized-bed photoreactor not only increased the contact of catalyst and gas, but also enhanced UV light penetration compared with a packed bed reactor in which light could not easily penetrate the interior of the catalyst bed, so that it was important to design and fabricate a fluidized-bed photoreactor with higher light throughputs and lower pressure drops. Accordingly, they used: (i) an annular flow-type photoreactor; and (ii) a modified two-dimensional fluidized-bed photoreactor to examine the photocatalytic decomposition of NO. In the first case, two serial annular flow photoreactors were used to increase contact time between the gas and the photocatalyst (Figure 16).

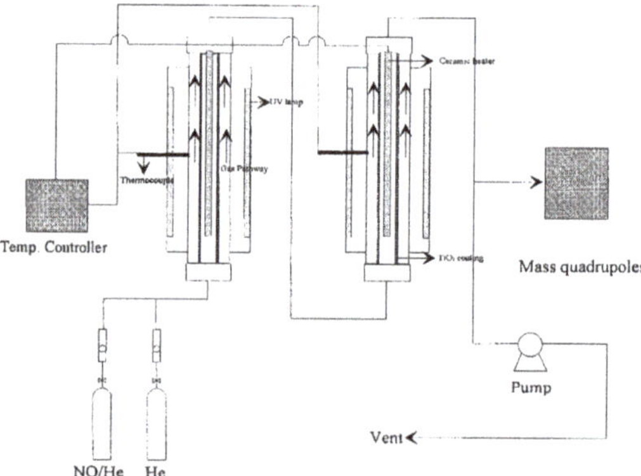

Figure 16. Schematic diagram of the annular flow-type photoreactor composed of two quartz glass tubes (height, 500 mm; diameters, 12 mm and 20 mm). Reproduced with permission from Lim and coworkers [102]. Copyright 2000 by Elsevier Science S.A. (License No.: 4452641174439).

By comparison, the modified two-dimensional fluidized-bed reactor (Figure 17) consisted of an annular-type reactor made of a larger quartz glass tube (internal diameter, 30 mm; height, 400 mm) in which a small diameter quartz tube (inner diameter, 20 mm; height, 375 mm) was located at the center of the larger tube such that the thickness of the annulus in the bed was 5 mm [102]. A quartz filter (100-mesh size) was used to distribute a uniform fluidization of the catalyst; a square mirror box surrounded the photoreactor to minimize loss of light irradiation and to improve utilization of reflected and deflected light.

In their study [102], the authors examined the effects of gas-residence time, initial NO concentration, reaction temperature and UV light source on the photocatalytic decomposition of NO carried out in the annular flow-type reactor. P-25 titania powder was used to cover a quartz tube (430 mm) by dipping it into a stirred 5% TiO$_2$ slurry solution and then air-dried for 24 h, after which the TiO$_2$-coated quartz tube was fired in a high-temperature furnace at 400 °C for 1 h; TiO$_2$ coating was repeated several times until the amount of TiO$_2$ deposited on the quartz tube reached 0.10 g. The quartz tube had been sandblasted previously to create a granular texture to anchor the fine TiO$_2$

powder. In addition, precursor solutions for coating TiO_2 on silica gel were prepared using titanium ethoxide as a precursor to prepare the TiO_2 sample.

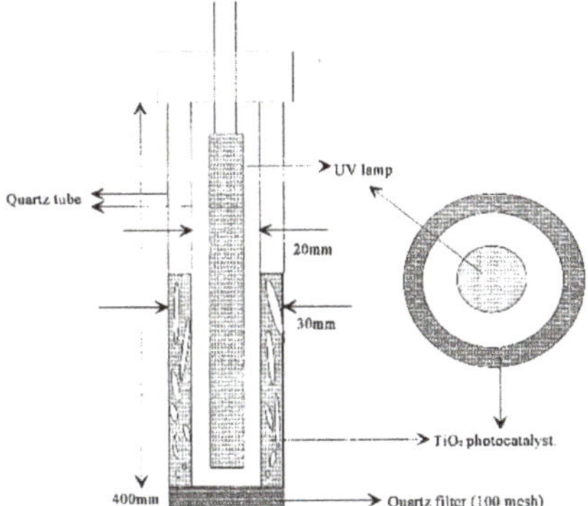

Figure 17. Modified two-dimensional fluidized-bed photoreactor. Reproduced with permission from Lim and coworkers [102]. Copyright 2000 by Elsevier Science S.A. (License No.: 4452641174439).

A gas stream (200 mL min^{-1}) of 138 ppmv NO in He in the annular flow-type reactor was irradiated by four UV lamps without TiO_2 photocatalyst at ambient temperature for 140 min with no variation in the NO concentration; in the presence of TiO_2, however, irradiation led to the decomposition of NO with formation of NO_2, N_2O and N_2 products [102]. The reaction rate followed the power law $R = R_o\,I^n$ with n = 0.48 and n = 0.87 depending on the UV intensity of the germicidal white lamp (254 nm) and the fluorescent black lamp (365 nm), respectively (gas flow rate, 100 mL min^{-1}; TiO_2 loading, 0.1 g; reaction temperature, 311 K; initial NO concentration, 50 ppm). Adsorption of nitrate on the surface of the photocatalyst increased with irradiation time leading to the deactivation of the photocatalyst. The decomposition of NO decreased linearly on increasing the initial NO concentration and on decreasing the residence time of gas in the photoreactor, so that it was necessary to increase the residence time of the gaseous reactant to provide effective contact of UV light, gaseous reactant and photocatalyst to obtain higher NO decomposition in the annular photoreactor.

In the modified two-dimensional fluidized-bed photoreactor, four reaction conditions (without TiO_2/SiO_2 and UV lamp on/off, with TiO_2/SiO_2 and UV lamp-on/off) were tested to confirm whether the decomposition of NO really took place by a photocatalytic process. Indeed, in the presence of TiO_2/SiO_2 and UV lamp-on, the NO concentration decreased indicating that it was in fact decomposed [101]. Decomposition of NO increased with decreasing initial NO concentration and increasing gas-residence time; the reaction rate increased with increasing UV light intensity. Clearly, the modified photoreactor displayed efficient contact between photocatalyst and reactant gas with good transmission of UV-light and, consequently, increased the NO decomposition efficiency (>70%) compared with the annular flow-type photoreactor. Hence, the former photoreactor was an effective tool with which to carry out significant NO decomposition with efficient utilization of photon energy [102].

Anpo and coworkers [96] had earlier prepared a TiO_2 photocatalyst that subsequent to the implantation of Cr ion and upon irradiation with visible light (>450 nm) decomposed NO into N_2, O_2, and N_2O under O_2-free conditions. Additionally, the Cr ion-implanted TiO_2 catalyst displayed the exact same photocatalytic efficiency as the original TiO_2 catalyst, albeit under UV irradiation. As a follow-up to this study, Nakamura et al. [103] examined the role of oxygen vacancies in the removal of NO under an oxidative atmosphere using a commercial TiO_2 (Ishihara ST-01; 100% anatase; crystallite size, 7 nm; nominal specific surface area, 300 m^2 g^{-1}) and hydrogen plasma-treated TiO_2 powders; the latter was photoactive up to 600 nm without a decrease in UV light activity. Reactions to remove 1.0 ppm of NO were carried out in a Pyrex glass flow reactor (500 cm^3) with irradiation from a

300-W Xe light source; the UVA (315–400 nm) irradiance at the photocatalyst surface was 0.03 mW cm^{-2} (flow rate, 1500 mL min^{-1}; TiO$_2$ loading, 0.20 g; total pressure, 760 Torr)—no removal of NO occurred without the metal oxide photocatalyst. NO was converted mainly to NO$_3^-$ (also less than 2% NO$_2^-$ formed) by oxidation over the TiO$_2$ powder; NO$_3^-$ ions accumulated on the catalyst surface. Electrons trapped in oxygen vacancies in the plasma-treated TiO$_2$ were detected under visible light irradiation (*F*-type color centers; ESR measurements displayed a signal at g = 2.004) with the number being proportional to the percent of NO$_x$ removed, which suggested that the number of trapped electrons determined the activity of the photocatalytic oxidation of NO to NO$_3^-$. The visible-light photocatalytic activity of the plasma-treated TiO$_2$ was due to photoexcitation of the *F*-type color centers with energy levels within the forbidden bandgap of the metal oxide (see Figure 18).

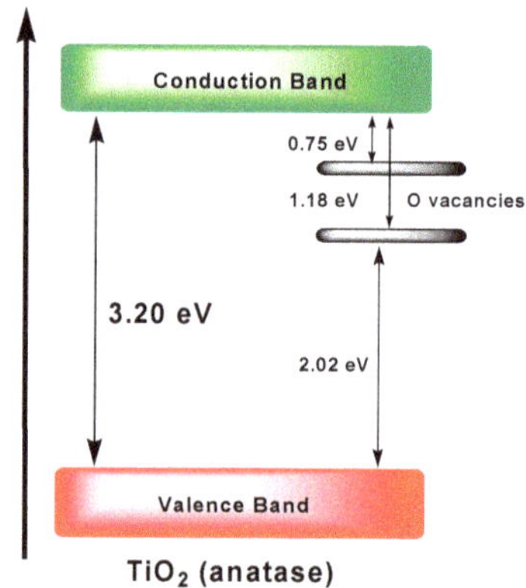

Figure 18. Proposed band structure model for the anatase TiO$_2$ with oxygen vacancies.

The Anpo group [104] investigated the photocatalytic decomposition of NO$_x$ (NO and NO$_2$) on five well-characterized standard reference ultrafine powdered TiO$_2$ photocatalysts (grain size, 0.02–1 μm) denoted TiO$_2$ (JRC-TIO-2, -3, -4, and -5) supplied by the Catalysis Society of Japan (properties summarized in Table 3) in a large-scale continuous flow reaction system (Figure 19a) with high efficiency. Special attention was expended on the effects of pretreatment and reaction conditions on the reaction and conversion rates of NO. The authors established that surface hydroxyl groups played a significant role as active sites in the decomposition of NO.

Table 3. Physicochemical properties of the standard reference TiO$_2$ photocatalysts (JRC-TiO-2, -3, -4, and -5) supplied by the catalysis Society of Japan **.

Catalysts (JRC-TiO-)	Surface Area (m^2 g^{-1})	Acid Concentration (μmol g^{-1})	Relative –OH Concentration	Bandgap (eV)
2 (anatase)	16	6	1.0	3.47
3 (anatase)	51	22	1.6	3.32
4 (anatase)	49	5	3.0	3.50
5 (rutile)	3	7	3.1	3.09

** Reproduced with permission from Zhang et al. [104]. Copyright 2001 by Academic Press (License No.: 4453251363124).

When used on a large scale for long periods in photoreactions, photocatalysts tended to lose, albeit gradually, their photocatalytic activity. In Anpo's study [104], after 2 h, conversion of NO for each photocatalyst leveled off and dropped to between 0.25 and 0.20 the photocatalytic activity observed initially, indicating a decline in photocatalytic activity of the TiO$_2$ in the decomposition of NO in the absence of O$_2$ and/or H$_2$O. Reaction products in the flow reaction system were N$_2$, O$_2$, and N$_2$O, just as occurred in a closed reaction system. The JRC-TIO-4 photocatalyst displayed the highest photocatalytic activity for the conversion of NO, while for the other three there were small differences:

JRC-TIO-4 (11%) >>> −3 (~ 2%) > −5 (1.8%) > −2 (1%) (see Figure 19b). The anatase TiO_2 catalyst with the larger surface area, wider bandgap, and numerous surface –OH groups (Table 3) exhibited the highest photocatalytic reactivity in the decomposition of NO, which the authors deduced that these were the principal factors that affected photocatalytic efficiency. The increased bandgap of JRC-TiO-4 was accompanied by a shift in the conduction band edge to higher negative energies, thus moving the redox potential to more negative values thereby enhancing photocatalytic reactivity. Moreover, surface –OH groups and/or physisorbed H_2O also played a significant role in the photocatalytic reactions through the facile formation of reactive •OH radicals. The intensity of the incident light is also an important factor that affects the kinetics of the photocatalytic decomposition. The quantum efficiency of the photocatalytic reaction was higher at the lower intensities of the incident light, and lower at higher intensities of the incident UV light; in addition, the efficiency of conversion of NO increased with increase in the O_2 flow rate.

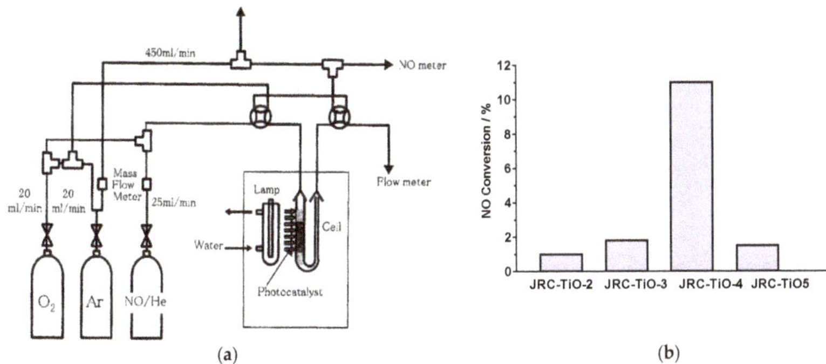

Figure 19. (**a**) Flow system for the photocatalytic reaction of NO_x. Conditions: 150 mg TiO_2; NO reactant gas (NO + He), 10 ppm; flow rate, 100 mL min^{-1}; irradiation time, 2 h; Toshiba SHL-100UV high-pressure Hg lamp; color filter, UV-27 (λ > 270 nm). (**b**) Conversion of the photocatalytic decomposition of NO on the standard reference TiO_2 photocatalysts at room temperature; adapted from Ref. [104]. Reproduced with permission from Zhang et al. [104]. Copyright 2001 by Academic Press (License No.: 4452650162755).

The activity of the JRC-TiO-4 photocatalyst was also tested by Tanaka and coworkers [105] in the photoassisted selective catalytic reduction of NO with ammonia (photo-SCR) at low temperature over irradiated TiO_2 in a flow reactor; the process was efficient and the adsorbed ammonia reacted with NO under irradiation of TiO_2 (Figure 20); note the nearly identical kinetics of formation of both N_2 and N_2O.

The total amount of N_2 formed was 0.23 mmol g_{cat}^{-1}, consistent with the amount of ammonia (0.24 mmol g_{cat}^{-1}) adsorbed over TiO_2 in equilibrium at 323 K. The kinetic experiment carried out under differential conditions in the pressure range 300 < p(NO), p(NH_3) < 2000 ppm, and the presence of excess O_2 affected the evolution rate of N_2 which depended only on partial pressure of NO; kinetics were first order on NO, and zeroth order on O_2 and NH_3, which strongly suggested that the rate-determining step was adsorption of NO to the irradiated TiO_2 adsorbing ammonia molecules [105]. To the extent that the selective catalytic reduction (SCR) with ammonia is a downhill reaction, it also proceeded in the dark at low temperature with a 20% conversion of NO. However, photoirradiation caused a remarkable enhancement of the activity: the evolution rate of N_2 gradually increased attaining a steady rate at ca. 80% conversion after 2 h of irradiation.

To achieve a further understanding of surface reactions involved in TiO_2-based photocatalysis, Dalton and coworkers [106] examined two titania samples (one of unknown source) using X-ray photoelectron spectroscopy and Raman spectroscopy to investigate the NO_x adsorbate reaction at the surface of these two TiO_2 substrates. The NO_x gas was composed of 109 ± 5 ppm of NO_x,

21.0 ± 0.4% O_2, the remaining ca. 79% being N_2; dry air was the mixer gas to dilute the NO_x (NO_x concentration, 10–100 ppm) during the reaction performed under UV exposure for 6 and 48 h in a glass vessel (ca. 3 mm thick; Figure 21) that allowed > 80% transmission of the radiation at λ = 320 nm. Formation of NO_3^- did not vary significantly with either exposure time or NO_x concentration. The authors [106] proposed a stepwise mechanism (Figure 22) in which the surface hydroxyls increased the efficacy of the process and participated by reacting with NO_x molecules to yield nitrate ions formed indirectly via initial reductive (formation of $O_2^{-\bullet}$ radical anions by conduction band electrons) and oxidative (formation of $^\bullet OH$ radicals by valence band holes) processes.

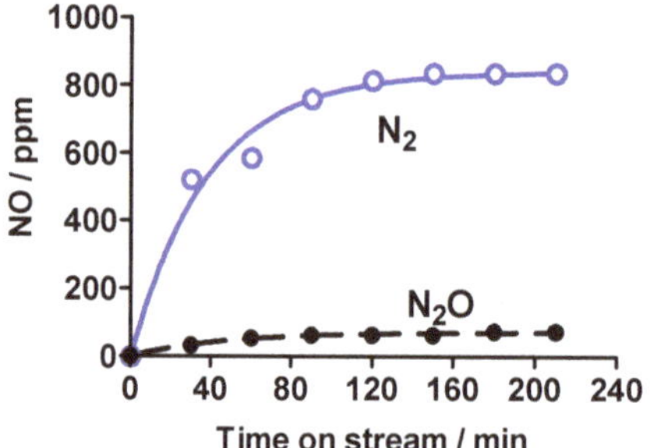

Figure 20. Outlet concentration of N_2 and N_2O in the SCR of NO with ammonia at 323 K under irradiation. Conditions: TiO_2 loading, 1.2 g; volume of catalyst bed, 1.5 mL; irradiation, 300-W ultra-high pressure Xe lamp reflection by a cold mirror; composition of reaction gas, 1000 ppm NO + 5% O_2, and balance was Ar gas; flow rate, 100 mL min^{-1}; k_{N_2} = 0.026 ± 0.003 min^{-1}, k_{N_2O} = 0.021 ± 0.003 min^{-1}. Adapted with permission from Tanaka et al. [105]. Copyright 2002 by the Royal Society of Chemistry (License No.: 4452660060880).

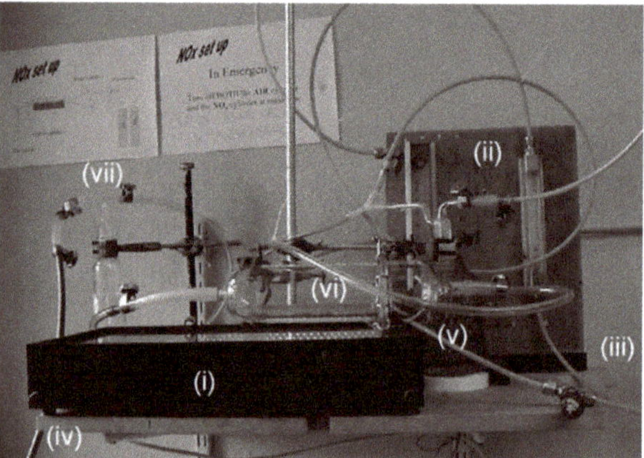

Figure 21. Apparatus for NO_x gas removal. Shown are: (i) UV exposure box; (ii) gas flow meters; (iii) NO_x gas and air mixer gas in; (iv) gas out; (v) water bubbler for air mixer gas; (vi) glass reaction vessel; and (vii) sodium hydroxide bubbler for excess NO_x removal. The system consisted of flow meters to allow an NO_x concentration of between 10 and 100 ppm when used in conjunction with the air mixer gas; a water bubbler to allow the reaction to be studied with wet or dry gas' a second bubbler containing aqueous NaOH was used after the reaction vessel to remove unreacted NO_x. Reproduced with permission from Dalton et al. [106]. Copyright 2002 by Elsevier Science Ltd. (License No.: 4452660699057).

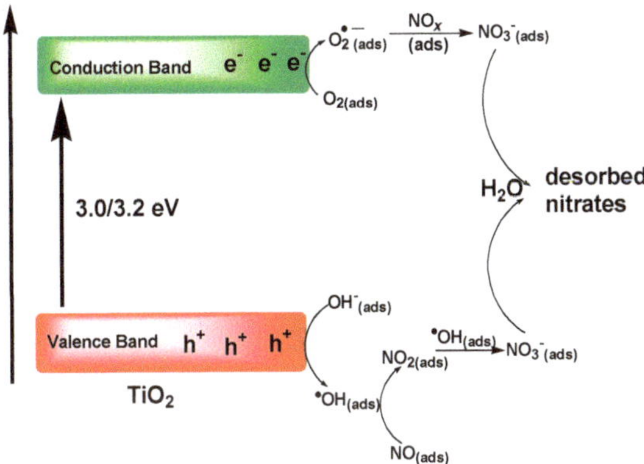

Figure 22. Suggested mechanism for the photocatalyzed oxidative removal of NO_x over the irradiated surface of the two TiO_2 photocatalytic substrates.

Dalton et al. [106] concluded that TiO_2 was effective at converting NO_x agents to NO_3^- and that XPS proved useful in quantifying the efficiency of the reaction, while Raman spectroscopy was a quick and simple way of ascertaining the surface crystal structure of the titania. XPS confirmed only one oxidation state of Ti on the untreated TiO_2 materials; however, the O_{1s} peak indicated the presence of two additional components of TiO_2: Ti–OH and Ti–OH_2. After exposure to UV radiation, XPS spectra revealed nitrogen peaks attributable to organic species (also present before reaction), to some unreacted NO adsorbed on the surface, and to nitrate anions.

Reactive nitrogen (NO_y) in the atmosphere consists of the sum of the two NO_x oxides (NO + NO_2) and all compounds produced by atmospheric oxidation of NO_x that include the minor species: HNO_3, HNO_2, the nitrate radical $NO_3{}^\bullet$, N_2O_5, peroxynitric acid HNO_4, peroxyacetyl nitrate (PAN) ($CH_3C(O)OONO_2$) and its homologs, and peroxyalkyl nitrates ($RC(O)OONO_2$) [107]. Such compounds can be regarded as reservoirs of NO_2 but apparently play no critical role in the formation of ozone O_3 that the precursors NO_2 and NO do. The oxidative removal of NO over irradiated TiO_2 catalyst was examined by Devahasdin and coworkers [107] at source levels (5–60 ppm) in a thin-film photoreactor systems (see Figure 23); the process involved a series of oxidation steps through the action of photoformed $^\bullet$OH radicals (NO → HNO_2 → NO_2 → HNO_3). Light intensity increased the capability to oxidize NO (from 0 to 0.8 mW cm^{-2}); the selectivity for NO_2 increased with light intensity for 5 ppm inlet NO but remained constant for 40 ppm inlet NO. The steady-state conversion of NO increased with relative humidity from 0 to 50% leveling off at higher relative humidity; the ratio of NO_2^- to NO_3^- from spent catalyst liquor decreased with irradiation time until steady state was reached.

Transient behavior of TiO_2 during the first 2 h of operation with the system setup of Figure 23 (conditions: space time: 12 s; inlet concentration, 40 ppm; light source, two 8-W black lamps; relative humidity, 50%; TiO_2 loading: 1.07 mg cm^{-2}) revealed that initially the conversion of NO was very high (ca. 95% after 0.5–3 min of irradiation depending on TiO_2 loading) and decreased approaching steady state after 6 h of operation; all the nitrogen was accounted for in the gas phase: NO out (26 ppm) + NO_2 out (14 ppm) = NO in (40 ppm). Conversion of NO was 35%; gas phase mass balance showed no N_2O formed in the reaction system under steady-state conditions [106]; NO_2 selectivity remained constant at 100% for 40 ppm inlet NO and increased with light intensity for 5 ppm inlet NO with a 50% relative humidity, which suggested that for 40 ppm inlet NO at steady state, all the NO should have been converted to NO_2. However, the authors [107] believed that, for the 5 ppm inlet NO, the true steady state had not yet been reached, so that increasing light intensity caused the HNO_3 to dissociate back to NO_2 and $^\bullet$OH and to promote NO_2 selectivity from 82% to 95%. The latter inference called attention for the first time to the possible reNOxification of the nitrates produced in the deNOxification of the environment (see Section 5).

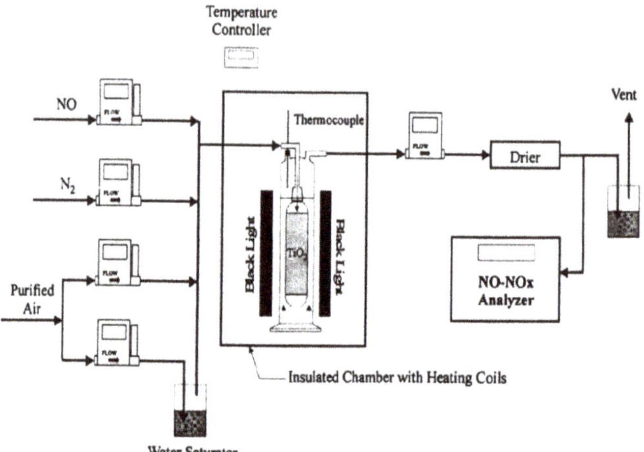

Figure 23. Experimental setup used in the disposal of NO_x. The system consisted of a thin-film photoreactor coated with Degussa P-25 TiO_2 (typical loading, 1.0 mg cm^2) irradiated with two 8 W or 25 W black lights from both sides; light intensity of 25 W bulbs varied with a dimming electronic ballast; reactor setup and light sources were contained in insulated chamber for control of temperature; light intensity measured with UVA radiometer (range, 320–390 nm) placed inside the reactor; NO and NO_2 measured with a chemiluminescent $NO–NO_x$ gas analyzer; initial NO concentrations were 28.5 and 472 ppm; NO_2 and $NO_3{}^-$ measured by chromatography. Reproduced with permission from Devahasdin and et al. [107]. Copyright 2003 by Elsevier Science B.V. (License No.: 4452661132706).

Toma and workers [108] reported using a test chamber built specifically for the TiO_2 (Degussa P-25; in powder or pellet form) photocatalytic decrease of NO_x. The experimental device consisted of three parts: (a) a chamber where gaseous NO_x were prepared in situ by chemical reaction of Cu powder with a dilute solution of HNO_3; (b) an environmental chamber; and (c) a NO_x analyzer. The pollutants were subsequently injected at ambient temperature into the environmental chamber (volume, ~0.4 m^3) until the concentration of NO_x reached 1–2 ppmv. A fan ensured homogenization of the gaseous pollutants in the environmental chamber. A polycarbonate photoreactor (100 mm × 100 mm × 50 mm box) equipped with a 70 × 70 mm Plexiglas window allowed light transmission from a 15-W daylight lamp (30% UVA, 4% UVB) placed inside the environmental chamber and crossed by the NO_x flow (flow rate, 0.6 L min^{-1}); NO_x concentrations were continuously monitored with a chemiluminescence NO_x analyzer.

For small TiO_2 powder quantities, conversion rates increased proportionally reaching maximal value at 0.2 g loading of TiO_2; at higher quantities of TiO_2 the decrease in NO_x remained constant and independent of TiO_2 powder amount. After 30 min of UV irradiation (surface, 54 cm^2), conversion rates were about 32–35% and 15–18%, respectively, when the mass of the catalyst varied from 0.2 to 1.2 g; maximal conversion was reached at 3.7–4 mg cm^{-2} of TiO_2 powder [108]. Exposing a TiO_2 pellet surface (mass, 0.4 to 1.2 g; 54 cm^2) to UV radiation from one side only led to a photocatalytic conversion of ca. 28–30% of NO and 10–12% of NO_x (Figure 24); the conversion efficiency increased with the surface area of the pellet. The amount of compressed TiO_2 powder and the thickness of the pellet had little influence on the extent of NO_x decomposition. Anatase TiO_2 showed better efficiency for the photocatalytic decrease of NO_x relative to rutile TiO_2, accounting for only 10% and 5%, respectively, for NO and NO_x removal.

We noted above that the photocatalytic decomposition of NO over TiO_2 reported in some of the literature led to the formation of N_2O as the main reaction product [92,94,98] with minor N_2, NO_2 and O_2 products. Only Anpo's group [96] reported the selectivity of NO photodecomposition over a TiO_2 photocatalyst to yield N_2O and N_2, and no other products. According to the views of Bowering and coworkers [109], use of only TiO_2 as the catalyst is not ideal for removing NO from the atmosphere as N_2O itself is also a regulated pollutant. As such, Tanaka et al. [105] reported that photoassisted selective catalytic reduction (photo-SCR) of NO over TiO_2 with NH_3 as a reductant was very selective towards N_2 formation, with relatively small amounts of N_2O. As NH_3 is also a pollutant, it would need to be eliminated from the exhaust gas, thereby causing an increase in overall costs of a system.

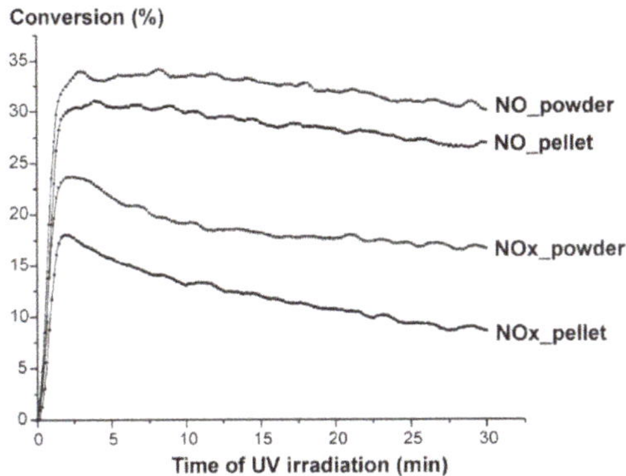

Figure 24. Comparative decrease of the NO_x removal performed on TiO_2 in the form of powder and pellets (plane surface, 54 cm^2). Reproduced with permission from Toma et al. [108]. Copyright 2004 by Springer-Verlag (License No.: 4453250060579).

Car exhaust and industrial emissions are mostly controlled using selective catalytic reduction (SCR) to convert NO_x to N_2. Accordingly, Bowering et al. [109] used CO as the reductant in eliminating NO photocatalytically with Degussa P-25 TiO_2 in a continuous flow reactor (Figure 25) with the objective to convert the NO_x preferentially into N_2 gas. The authors added TiO_2 powder (ca. 0.2 g) to acidified triply deionized water (TDW; 6 mL of 0.05 M HNO_3 in 500 mL of TDW) yielding a dispersion that was stirred for 12 h and then dried at 70 °C for 48 h, after which the resulting powders were calcined for 2 h at 120, 200, 450 or 600 °C. Subsequently, 25-mL fractions of the dispersion were evaporated at 70 °C onto degreased borosilicate glass slides; an amount of TiO_2 powder (~ 1 mg) was deposited on the slides and then calcined following the same methodology as for the powders.

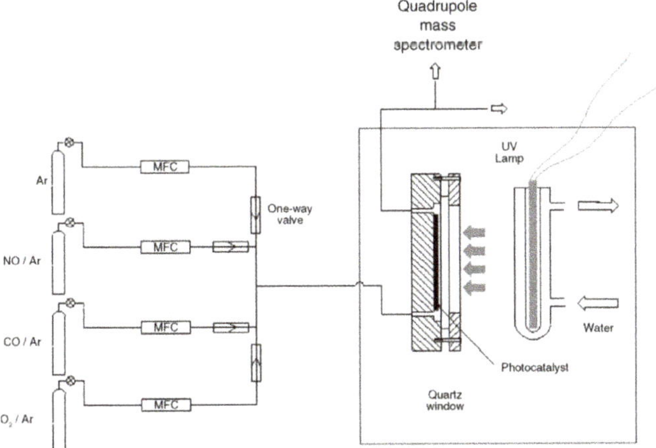

Figure 25. Schematic of the photoreactor used for testing the photocatalytic behavior of the various P-25 TiO_2 catalytic samples (see text). Reproduced with permission from Bowering et al. [109]. Copyright 2005 by Elsevier B.V. (License No.: 4452670211528).

The effect of calcination temperatures on the composition and crystallite sizes of P-25 photocatalysts, and the effect of pretreatment temperature on rate of NO conversion and selectivity for N_2 formation for NO decomposition and reduction reactions are presented in Table 4 [109]. Pretreatment (calcination) temperatures caused no appreciable change in phase composition; original composition (ca. 77 vol.% anatase, 23 vol.% rutile) was maintained even after treatment at 600 °C. The photocatalytic activity for both decomposition and reduction reactions decreased with increasing pretreatment temperature, which was attributed to removal of surface hydroxyl species that acted as active sites for reaction. The only products observed in the decomposition reactions were N_2 and N_2O; the selectivity for nitrogen formation remained constant (ca. 23%) regardless of pretreatment temperature. However, the presence of CO in the reaction gas had a dramatic effect on selectivity of the reactions with N_2 selectivity as high as 65%; in addition, an increase in the CO/NO ratio led to increased selectivity for N_2 formation.

Table 4. Effect of calcination temperatures on the composition and crystallite sizes of P-25 photocatalysts, and of the pretreatment temperature on the rate of NO conversion and selectivity for N_2 formation for NO decomposition and reduction reactions **.

Calcination T (°C)	Rutile (vol. %)	Crystallite Size (nm)		BET Surface Area ($m^2\,g^{-1}$)	Rate of NO Conversion ($\mu mol\,h^{-1}\,g_{cat}^{-1}$)		Selectivity for N_2 Formation (%)	
		Anatase	Rutile		Dec.	Red.	Dec.	Red.
70	23.0	28.0	66.3	51.92	1210	657	21	46
120	23.2	28.8	64.8	50.69	1107	560	25	49
200	23.0	28.2	62.7	49.87	983	467	21	48
450	23.0	28.3	59.6	49.54	550	243	26	26
600	28.0	30.5	69.6	48.24	430	240	30	25

** Adapted with permission from Bowering et al. [109]; Copyright 2005 by Elsevier B.V. (License No.: 4452670211528).

It is likely that under UV illumination electron transfer occurred from electron trapped centers into antibonding orbitals of adsorbed NO molecules, resulting in their decomposition and formation of $N_{(ads)}$ and $O_{(ads)}$ surface species, which can then scan the TiO_2 surface and react with other surface species (e.g., $NO_{(ads)}$, $N_{(ads)}$, $O_{(ads)}$) to form N_2O, NO_2, O_2 and N_2. To the extent that neither O_2 nor NO_2 was detected led the authors [109] to deduce that Reactions (69) and (70) did not occur on P-25 surfaces under decomposition conditions; the main surface reaction was Reaction (71), as N_2O was the major reaction product under these conditions.

$$NO_{(ads)} + O_{(ads)} \rightarrow NO_{2(ads}$$
$$\tag{69}$$

$$O_{(ads)} \rightarrow O_{2(ads)}$$
$$\tag{70}$$

$$NO_{(ads)} + N_{(ads)} \rightarrow N_2O_{(ads)}$$
$$\tag{71}$$

In the presence of CO on the photocatalyst surface, and under UV illumination, other reactions are possible between adsorbed CO and NO molecules together with reactions of CO with $N_{(ads)}$ and $O_{(ads)}$ atoms. No reaction occurred in the dark and under UV illumination without TiO_2 indicating that both TiO_2 and UV are required for adsorbed NO and CO species to react. Under decomposition conditions, the major reaction product was N_2O (~ 75%) with N_2 being the minor product (ca. 25%). On the other hand, under reduction conditions selectivity for N_2 formation increased (ca. 48%) at the pretreatment temperatures of 70 and 120 °C. However, at higher pretreatment temperatures, the selectivity was similar to that achieved in the absence of CO suggesting that the surface N_2 forming reaction was favored on a titanium surface rich in hydroxyl groups [109].

Germane to the previous study [109], Roy and coworkers [110] examined a photocatalytic route to destroy NO_x by developing a new Pd ion-substituted TiO_2 system ($Ti_{1-x}Pd_xO_{2-\delta}$) with which to reduce NO in the presence of CO via creation of redox adsorption sites and using anion oxygen vacancies on titania; the optimal Pd^{2+} ion concentration was 1 at.% in TiO_2 (anatase). Apparently, even though both NO and CO competed for the same Pd^{2+} adsorption sites, reduction of NO to N_2O was two orders of magnitude higher with the $Ti_{0.99}Pd_{0.01}O_{1.99}$ photocatalyst under ambient conditions than unsubstituted TiO_2; using UV irradiation with a 125-W high-pressure Hg lamp and an inlet 5000 ppm of NO in a flow-type reactor, the rate of reduction of NO was 0.53 μmol g^{-1} s^{-1}.

The photocatalytic deNOxing activities of TiO_2, N-doped TiO_2, Fe-loaded N-doped TiO_2, and Pt-loaded N-doped TiO_2 exposed to irradiation from monochrome LED lamps at various wavelengths have been investigated in some detail by Yin and coworkers [111], unlike many studies that have typically used 100–500 Watt high-pressure Hg or otherwise Xe light sources to activate titania-based photocatalysts. Bare $TiO_{2-x}N_x$ (denoted TiON) powders were prepared by treating 20-nm Ishihara ST-01 anatase TiO_2 in an NH_3 atmosphere at 600 °C for 3 h, followed by annealing at 300 °C for 2 h in humid air (N content, ca. 0.25 at.%); for comparison, the ST-01 anatase TiO_2 powder was heat-treated in air at 500 °C for 1 h (S-TiO_2)—BET specific surface areas were 57.7 m^2 g^{-1} and 100.7 m^2 g^{-1}, respectively, for TiON and S-TiO_2 powders. The Fe- and Pt-loaded N-doped TiO_2 systems were prepared by dispersing TiON powder in a HNO_3 aqueous solution containing either $Fe(NO_3)_3$ or $Pt(NH_3)_2(NO_3)_2$ at ambient temperature, followed by stirring for 1 h, heated at 150 °C to remove the water, and then calcined at 300 and 400 °C for 2 h, respectively (loading of Fe and Pt, 0.5 wt.%; systems denoted TiONFe and TiONPt; BET areas were 61.1 m^2 g^{-1} and 59.0 m^2 g^{-1}, respectively).

The specifics of irradiation from the four LED sources were (wavelength, irradiance): (i) red light LED (627 nm, 72.76 μmol m^{-2} s^{-1}); (ii) green light LED (530 nm, 125.12 μmol m^{-2} s^{-1}); (iii) blue light LED (445 nm, 76.22 μmol m^{-2} s^{-1}); and (iv) UV light LED (390 nm, 73.70 μmol m^{-2} s^{-1}). Different samples showed different wavelength dependencies; for instance, S-TiO_2 displayed excellent activity at 390 nm but very weak activity at 445 nm, while TiON showed excellent UV light (390 nm) and visible light-induced photocatalytic activity on exposure to 445 nm and 530 nm irradiation (Figure 26) [111]. By comparison, both TiONFe and TiONPt showed excellent deNOxing abilities even under red light irradiation at 627 nm. Specifically, TiONPt showed the highest $deNO_x$ abilities at all light wavelength ranges: about 37.8%, 36.8%, 28.2%, and 16.0% of NO_x was removed, respectively, under continuous irradiation by monochromatic light at 390 nm (UV LED), 445 nm (blue LED), 530 nm (green LED), and 627 nm (red LED).

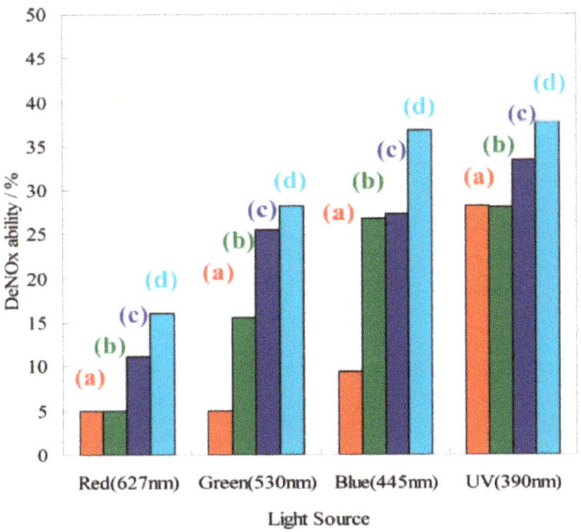

Figure 26. Photocatalytic $deNO_x$ abilities of samples under irradiation by various LED light sources: (a) S-TiO_2; (b) TiON; (c) TiONFe; and (d) TiONPt. Reproduced with permission from Yin et al. [111]. Copyright 2008 by the American Chemical Society.

Because of the newly formed N_{2p} level within the bandgap of titania above the O_{2p} valence band, N-doped titania displayed an extrinsic bandgap smaller than the intrinsic bandgap of titania such that TiON absorbed visible light. In addition, TiON displayed significant chemiluminescence emission attributed to formation of singlet oxygen $^1O_2$ relative to undoped S-TiO$_2$ which failed to display any light emission. For comparison, TiONFe and TiONPt also displayed relatively high chemiluminescence emission, albeit lower than TiON; the latter showed chemiluminescence intensity increases in the order: UV < blue < green < red.

For the undoped S-TiO$_2$ sample, the study of Yin et al. [111] showed a correlation between very weak chemiluminescence emission intensities and very low visible-light induced photocatalytic activity that they attributed to its relatively large intrinsic bandgap. By contrast, for TiON, TiONFe, and TiONPt, results demonstrated that the deNO$_x$ ability decreased with an increase in chemiluminescence emission intensity. Nonetheless, the photocatalytic deNO$_x$ activity of TiON was nearly the same as that of S-TiO$_2$ under 390 nm (UV LED) and 627 nm (red LED) irradiation. However, TiON exhibited greater activity than S-TiO$_2$ under 445 nm and 530 nm irradiation, but lower than TiONFe and TiONPt samples under every type of LED light irradiation. The authors ascribed this variation to different band structures and to the presence of Fe and Pt loaded onto the surface of the TiON.

Mechanistically, the sequence of events that led to deNOxing by these four titania samples was summarized [110] by the series of Reactions (72)–(77). Subsequent to irradiation of the titania that yields conduction band electrons (e$^-$) and valence band holes (h$^+$), formation of singlet oxygen $^1O_2$ (Equation (76)) competes with formation of superoxide radical anions (Equation (73)) and hydroxyl radicals (Equation (74)) in air (molecular oxygen; relative humidity, ca 25%). Fe and Pt loading on TiON increased charge transfer and charge separation on the surface of the TiONFe and TiONPt photocatalysts [111].

$$TiO_2 + h\nu \rightarrow e^- + h^+ \tag{72}$$

$$e^- + O_2 \rightarrow O_2^{-\bullet} \tag{73}$$

$$O_2^{-\bullet} + H^+ \rightarrow HOO^{\bullet} \tag{74}$$

$$H_2O + h^+ \rightarrow {}^{\bullet}OH + H^+ \tag{75}$$

$$O_2^{-\bullet} + h^+ \rightarrow {}^1O_2 \tag{76}$$

$$O_2^{-\bullet} + h^+ \rightarrow {}^3O_2 \tag{77}$$

In the present context, deNOxing reportedly occurred by oxidation of the NO$_x$ molecules via the active oxygen $^{\bullet}OH$ and $O_2^{-\bullet}$ species (Equations (78)–(80)), whereby NO is converted to NO$_2$ and subsequently to NO$_3^-$ ions.

$$NO + 2\,{}^{\bullet}OH \rightarrow NO_2 + H_2O \tag{78}$$

$$NO_2 + {}^{\bullet}OH \rightarrow NO_3^- + H^+ \tag{79}$$

$$NO + O_2^{-\bullet} \rightarrow NO_3^- \tag{80}$$

In a NO$_x$ atmosphere, the NO$_x$ molecules adsorb onto the photocatalyst's surface and then interact with the superoxide radical anions $O_2^{-\bullet}$ to form NO$_3^-$ (Equation (80)), as a result of which the NO$_x$ molecules consume $O_2^{-\bullet}$ and delay singlet oxygen formation (Equation (76)), thereby causing the chemiluminescence emission intensity in the NO$_x$ atmosphere to be much lower than in air. To recapitulate, Yin and coworkers [111] deduced that:

1. Nanosized titania exhibited very low deNO$_x$ ability under visible light irradiation, irrespective of their excellent UV light-induced (390 nm) deNO$_x$ ability.
2. N-doped titania displayed excellent photocatalytic activity under 445 nm and 530 nm light irradiation.

3. Fe and Pt loading improved the photocatalytic activity of N-doped TiO_2 under not only UV light but also long-wavelength visible-light irradiation (λ = 530 nm and λ = 627 nm).
4. Pt-loaded, N-doped titania possessed the best visible light- and UV-induced photocatalytic activity. In addition, Fe- and Pt-loaded N-doped titania exhibited relatively high quantum yields of deNOxing under long-wavelength LED light irradiation.

In their extensive 2009 review article on the catalytic abatement of NO_x in the environment, Roy and coworkers [112] focused mostly on thermal methods in the presence of suitable reducing agents, and briefly gave a short account of the alternative photocatalytic methodology at ambient conditions; summarized was also some of the earlier work reported by selected researchers noting that direct photocatalytic decomposition of NO would yield N_2 and O_2, which would indeed be the ideal outcome and sole products if that could be realized. Unfortunately, as noted above, different conditions and different titania-based photocatalysts lead to significantly different results that are worth recalling briefly. For instance,

1. Anpo and coworkers showed that metal ion-implanted TiO_2 decomposed NO photo- catalytically to N_2, O_2 and N_2O at 275 K under irradiation with visible light at wavelengths longer than 450 nm [96].
2. Lim et al. [102] found that the photocatalytic decomposition of NO over Degussa P-25 TiO_2 in an annular flow type reactor produces NO_2, N_2O and N_2, with the efficiency increasing with light intensity and residence time and decreasing with initial NO concentration.
3. Bowering et al. [109] showed that the photocatalytic activity of Degussa P-25 TiO_2 toward deNOxing decreased with increasing pretreatment temperature.
4. Roy and coworkers [110] reported that reduction of NO over the catalyst $Ti_{1-x}Pd_xO_{2-\delta}$ was two orders of magnitude greater than unsubstituted TiO_2. Direct NO decomposition into N_2 and N_2O occurred via dissociation of NO in the presence of UV radiation at room temperature yielding N_2, N_2O and O_2 with the O_2 evolved reacting with NO to give NO_2 that is adsorbed by the catalyst upon formation. Prolonged NO_2 adsorption makes the surface inactive for NO dissociation; NO dissociation resumed when CO was passed to scavenge the evolved dissociated O_2 [110].

On the other hand, the seminal review article by Skalska et al. [113] presented an extensive survey of NO_x emission control technologies for three major anthropogenic emission sources: power plants, vehicles and the chemical industry, and further described new and alternative methods such as a hybrid system of SCR (selective catalytic reduction) and O_3 injection, fast SCR, and electron beam gas treatment, among others. Also described was the influence of NO_x on the environment and human health. The main focus was put on NO_x control methods applied in the combustion of fossil fuels in power stations and mobile vehicles, together with methods used in the chemical industry; the authors emphasized the implementation of ozone and other oxidizing agents in NO_x oxidation.

Following these footsteps, Heo and coworkers [114] combined photocatalysis and SCR with hydrocarbons as reducing agents (HC/SCR) to improve the activity and durability of deNO$_x$ catalysts. The authors developed a photocatalytic HC/SCR system that exhibited high deNO$_x$ performance (54.0−98.6% NO_x conversion) at low temperatures (150−250 °C) using dodecane as the HC reductant over a hybrid SCR system that included a photocatalytic reactor (PCR) and a dual-bed HC/SCR reactor (Figure 27). The PCR generated the highly active oxidants O_3 and NO_2 from O_2 and NO in the feed stream, followed by subsequent formation of the highly efficient reducing oxygenated hydrocarbon (OHC), NH_3, and organo-nitrogen compounds. These reductants were key in enhancing the low-temperature deNO$_x$ performance of the dual-bed HC/SCR system containing Ag/Al_2O_3 and CuCoY in the front and rear bed of the reactor, respectively (Table 5). Moreover, the OHCs proved particularly effective for both NO_x reduction and NH_3 formation over the Ag/Al_2O_3 catalyst, while NH_3 and organo-nitrogen compounds were effective for the reduction of NO_x over CuCoY. The photocatalytic assisted hybrid HC/SCR system demonstrated an overall deNOxing performance comparable to that of the NH_3/SCR, thus its potential as a promising alternative to the current urea/SCR technology [114].

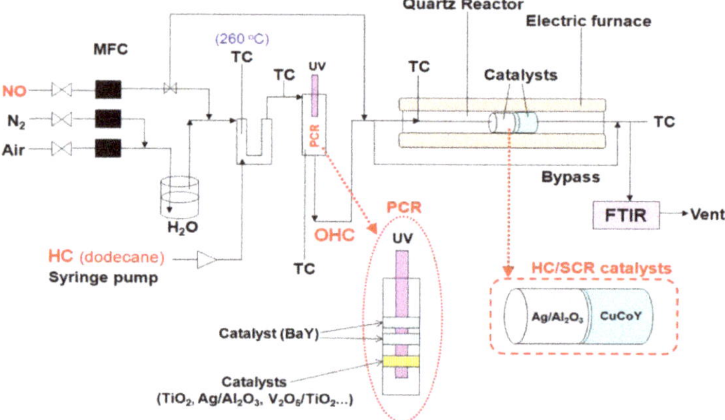

Figure 27. Schematic flow diagram of the hybrid PCR + HC/SCR reactor system. Reproduced with permission from Heo et al. [114]. Copyright 2013 by the American Chemical Society.

Table 5. Effect of catalysts on the performance of the PCR system at 150 °C. Feed gas composition: 200 ppm NO, 134 ppm $C_{12}H_{26}$, 6% O_2, 10% H_2O, and N_2 balance. Total flow rate = 500 mL min^{-1}; TiO_2 (anatase nanopowder) **.

	Catalyst	Yield of Reductants (%)		NO_x Conversion (%)
		Total OHCs	**NH₃**	
2 BaY +	TiO_2	35 (84 [a])	-	43
	V_2O_5/TiO_2	60	-	29
	Au/TiO_2	24	-	43
	Ag/TiO_2	18	-	55
	Pt/TiO_2	2	-	22
	CuCoY	19	-	37
	BaY	10	-	55
	Ag/Al_2O_3	16	6	54
Blank PCR (no catalyst)		45	-	-

[a] Yield of OHCs in the absence of NO. ** Reproduced with permission from Heo et al. [114]. Copyright 2013 by the American Chemical Society.

To the extent that the deNO$_x$ performance of the conventional HC/SCR catalyst was enhanced by both OHCs and NH$_x$-containing reductants, the two representative PCR catalysts V_2O_5/TiO_2 and Ag/Al_2O_3 were chosen by the authors [114] for further examination in the PCR + HC/SCR hybrid system: the V_2O_5/TiO_2 PCR for its superior OHC formation and the Ag/Al_2O_3 PCR for its formation of NH_3 and possibly organo-nitrogen compounds as precursors of NH_3.

Table 6 lists the conversions of NO and NO$_x$, the conversion of NO$_x$ to N_2, and the yields of NO_2, N_2O, and NH_3 during the reduction of NO with the PCR + HC/SCR system [114]. The NO$_x$ conversion to N_2 was estimated from the conversion of NO$_x$ and the yields of NO_2, N_2O, and NH_3 by the mass balance of nitrogen. The selectivity of the Ag/Al_2O_3 PCR + HC/SCR system for N_2 was 94%, 91%, and 82% at 200, 250, and 300 °C, respectively. The slight decrease N_2 in selectivity of the Ag/Al_2O_3 PCR + HC/SCR system at 300 °C was ascribed to increased formation of NH_3 by reaction of NO with OHCs over the HC/SCR reactor, since the PCR readily converted dodecane to OHCs. At 400 °C, the HC/SCR system alone completely reduced NO$_x$ with high N_2 selectivity up to 96% so that the PCR could be turned off and bypassed to save energy at temperatures above 400 °C.

Table 6. Conversion of NO and yields of NO_2, N_2O, and NH_3 during the reduction of NO over the PCR + HC/SCR system. Feed gas composition: 200 ppm NO, 134 ppm $C_{12}H_{26}$, 6% O_2, 10% H_2O, and N_2 balance. Gas hourly space velocity of the HC/SCR monolith reactor = 16,500 h^{-1} **.

System	Temperature (°C)	NO (%)	NO_2 (%)	N_2O (%)	NH_3 (%)	Total NO (%)	Estimated N_2 (%)
HC/SCR only	256	13	0	1	0	13	13
	303	62	0	1	1	62	60
	400	100	0	3	2	100	96
	500	53	7	1	1	46	44
Ag/Al_2O_3 PCR + HC/SCR	200	63	0	4	0	63	59
	250	99	0	5	4	99	91
	300	100	0	5	13	100	82
V_2O_5/TiO_2 PCR + HC/SCR	200	42	0	4	1	42	38
	250	95	0	5	0	95	91

** Reproduced with permission from Heo et al. [114]. Copyright 2013 by the American Chemical Society.

Key to the successful demonstration of this advanced $deNO_x$ process was the unique design and functionality of the PCR, which led to three major conclusions [114]: (1) PCR with catalysts was very efficient for both OHC formation and reduction of NO_x because of its dual function: in situ UV-induced formation of OHC and conversion of NO_x over the catalysts; (2) blank PCR (no catalyst) was very efficient for oxidation of NO to NO_2 and HC to OHC, but was inefficient for converting NO_x because of the absence of a catalyst; and (3) Ag/Al_2O_3 PCR (with BaY + Ag/Al_2O_3) produced OHC and NH_3 as intermediates that could be used subsequently to further convert NO_x in a downstream reactor containing a dual-bed catalyst such as Ag/Al_2O_3 (for OHC/SCR) and CuCoY (for NH_3/SCR).

5. Applied Photocatalysis: Prospective Attempts at DeNOxing the Atmospheric Environment

Energy-related emissions of nitrogen oxides continue to increase worldwide, standing close to 110 Mt in 2015 with the transportation sector accounting for 52%, followed by industry (26%) and power generation (14%). China (23 Mt) and the United States (13 Mt) accounted for ca. 33% of global NO_x emissions that year. According to the International Energy Agency [115], power generation in 2015 was a major source of worldwide emissions of nitrogen oxides (14% of total NO_x) with coal being the principal fuel responsible for 70% of those NO_x emissions; burning oil to generate electricity also produced significant quantities of NO_x. Natural gas-fired plants emitted fewer air pollutants than either coal-fired or oil-fired power plants; however, in 2015, gas-fired power generation emitted close to 20% of NO_x, while biomass played a negligible role in global power generation, although, in relative terms, it performed only slightly better than coal-fired plants for NO_x emissions. Manufacturing industries and other transformation sectors (e.g., refining and mining) accounted for ca. 30% of NO_x (28 Mt) in 2015. Process-related NO_x emissions were mostly released in cement making (1.5 kt of NO_x per Mt of cement that accounted for >50% of global process-related NO_x emissions) followed by pulp and paper production (1.2 kt NO_x Mt^{-1} of paper) (see Figure 28) [115]. Considering combustion and process emissions from a regional perspective, China was the largest emitter of NO_x, accounting for nearly 40% followed by the United States (11%). However, the United States witnessed considerable decreases in NO_x emissions in decades prior to 2015, while NO_x emissions from Chinese and Indian industries increased significantly. Together with ammonia, NO_x and SO_2 are the main precursors to formation of acid rain, which affects soil and water (with adverse impact on vegetation and animal life) and accelerates the deterioration of equipment and cultural heritage [115]. The presence of NO_x and volatile organic compounds (VOCs) in the environment leads to formation of ground-level ozone (O_3) under sunlight.

Thus, the considerable research interest witnessed over the last 2–3 decades to attenuate the extensive presence of NO_x in the environment is not surprising. In a 1999 Technical Bulletin, the United States Environmental Protection Agency (EPA) [116] described the various components that make up the NO_x pollutants, together with their properties, some of the health concerns, and how the environmental NO_x could be abated and controlled by external combustion—pollution prevention methods and add-on control technologies—that is, by non-photocatalytic technologies. In this regard, methods to reduce thermally the NO_x emissions at the origins include improved combustion techniques (e.g., fuel denitrogenation, modification to combustion methods, modification of operating conditions, and tail-end control processes) and installation of low-NO_x burners in process heaters and industrial heat and electricity generation plants. Another primary combustion technology is a fluidized-bed combustion technology for solid fuels, while end-of-pipe technologies focus on the removal of NO_x from flue gases by means of either physical separation or chemical reactions before their release to the atmosphere. Selective catalytic reduction (SCR; NH_3 as the reductant; presence of a combination of TiO_2 and V/W oxides as catalysts (Equations (81) and (82)) or selective non-catalytic reduction systems (SNR; urea or ammonia (Equations (83) and (84)) can significantly reduce NO_x in the flue gas of stationary sources into N_2 and H_2O.

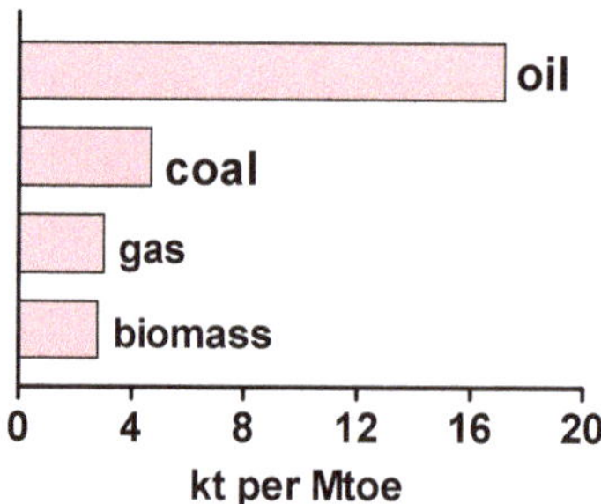

Figure 28. Global average NO_x emissions from various sources in 2015 in kilo tonnes (kt) per million tonnes oil equivalent (Mtoe). Plot made from selected data from the International Energy Agency [115].

Reduction of NO_x in the environment may not only involve SCR and SNR technologies, but also and of particular relevance herein is the TiO_2-based photocatalytic technology that is the subject of this section.

(a) Selective Catalytic Reduction (SCR)

$$4\,NO + 4\,NH_3 + O_2 + Cat. \rightarrow 4\,N_2 + 6\,H_2O \tag{81}$$

$$2\,NO_2 + 4\,NH_3 + O_2 + Cat. \rightarrow 3\,N_2 + 6\,H_2O \tag{82}$$

Conditions: Temperature: 300–400 °C; typical efficiencies: about 80%

(b) Selective Non-catalytic Reduction (SNR)

$$4\,NO + 4\,NH_3 + O_2 \rightarrow 4\,N_2 + 6\,H_2O \tag{83}$$

$$4\,NO + 2\,(NH_2)_2CO + O_2 \rightarrow 4\,N_2 + 4\,H_2O + 2\,CO_2 \tag{84}$$

Conditions: Temperature: 900–1000 °C; 40–60% reduction is obtained.

Several review articles have appeared in the last decade [117–119] that described, among others, some of the early attempts in the abatement of NO_x agents under indoor and outdoor experimental conditions. The 2008 article by Fujishima et al. [117] offered an overview of some highlights of TiO_2 photocatalysis, reviewed some of its origins, and indicated some useful applications: self-cleaning surfaces, water purification, air purification, self-sterilizing surfaces, anti-fogging surfaces, heat dissipation and heat transfer, anticorrosion applications, environmentally friendly surface treatment, photocatalytic lithography, photochromism of metal oxides, and microchemical systems. In the air purification application (i.e., deNOxing), the 2008 article briefly noted that some Japanese companies were considering covering roads with the TiO_2 photocatalyst, and removing the NO_x from automobile exhaust with sunlight using TiO_2-coated road bricks, prepared by mixing colloidal TiO_2 solutions with cement: the photo-road technology applied to no less than 14 different locations in Japan, one of which was the 7th belt highway in Tokyo (surface area covered, ca. 300 m^2); NO_x removed from this testing area was ca. 50–60 mg per day, equivalent to NO_x discharged by 1000 automobiles (Figure 29); however, no relative efficiency was provided with respect to total NO_x in the environment.

Figure 29. Usage of TiO_2-based photocatalytic material on roadway surfaces to convert nitrogen oxides NO_x to nitrate: finished roadway with the coated surface showing a lighter color—photo was courtesy of Fujita Road Construction Co., Ltd. to the authors of Ref. [117]. Reproduced with permission from Fujishima et al. [117]. Copyright 2008 by Elsevier B.V. (License No.: 4452671501245).

The 2013 article by Hanus and Harris [118] entertained some innovations of nanotechnology for the construction industry, most noteworthy being improvements in concrete strength, durability and sustainability being achieved with use of metal/metal-oxide nanoparticles and engineered nanoparticles (carbon nanotubes and carbon nanofibers), as well as environment-responsive anticorrosion coatings formed using nano-encapsulation techniques.

For their part, Fresno and coworkers [119] described achievements, near-future trends and critically assessed many photocatalytic materials on the basis of knowledge accumulated in pre-2014 years as to which materials or multicomponent systems, among the multitude of developments, could be taken as a ready consolidated technology or else as more likely to become a real alternative in the short term. Germane to this, they noted that a photocatalyst could be incorporated during material manufacturing either as an additive, most often the case of construction materials, or as a coating on an already conformed cementitious surface with the mechanical resistance of the coating, the optimal amount of photocatalyst (TiO_2) and binder, their impact on the properties of the materials and their long-term performance and aesthetic durability being factors to consider in their applications as carbonation can lead to deactivation after several months of use, a point we shall emphasize later with regard to a most celebrated example: the *Dives in Misericordia* church in Rome (see below) [120] built with Italcementi's TX-Active® photocatalytic concrete [121] for self-cleaning purposes and for reduction of NO_x pollution (among others). Incorporation of anatase titania into the wearing layer, and the use of a double-layered concrete with addition of the photocatalyst to the top layer were two of the possible strategies noted for this application to reduce NO_x. Most importantly, the authors [119] emphasized the need for further assessments of the durability of the photoactive coatings, and their capability of abating other air pollutants (e.g., the VOCs).

Along these lines, the discussion below will first emphasize the results from TiO_2-based photocatalytic deNOxing the environment with the photocatalyst deposited on cementitious substrates in an indoor laboratory setting using small photoreactors set in flow-through systems and UV light sources. This is then followed by deNOxing results from various outdoor settings in wide open air environments with sunlight as the light source to activate the photocatalytic surfaces.

5.1. Indoor DeNOxing Environment Tests with TiO_2 Photocatalytic Cementitious Surfaces

One of the first articles in the search for means to reduce NO_x from the various emission sources, which affected the air quality in Japan because of relatively high concentrations of NO_x, particularly along heavily trafficked roads in densely populated areas, was that of Ibusuki and Takeuchi in 1994 [94]. They proposed a new non-thermal technology for removing NO_x at sub-ppm level from the air of trafficked roads, tunnels and other environmental emission sources (see Section 4) following their earlier findings [91] that NO_x deposited onto the ground caused an enhancement of the rate of adsorption of NO_x on soil particles upon photoillumination, with TiO_2 exhibiting the highest activity for NO_x adsorption—about 60% of NO_x was captured as nitric acid on the TiO_2 surface. Their follow-up laboratory experiments indicated that up to 90% of NO could be removed using mixtures of TiO_2, activated carbon (AC), and iron oxide (Fe_2O_3) particulates in a flow-through reactor [96].

Along similar lines, at the 2000 JIPEA World Congress, Murata and coworkers [122] reported the development of interlocking cementitious bricks loaded with TiO_2 ($NOXER^©$) for the oxidative removal of NO_x under sunlight UV radiation, humidity, and NO_x concentrations that paralleled roadside environments. This novel technology at the time was implemented in a couple of Tokyo suburbs. The permeable interlocking bricks (Figure 30) were prepared by mixing aggregates, cement, TiO_2 powder and water in an appropriate template, and then cured at ambient temperature for one month, after which they were used for the photocatalytic indoor tests in a flow-through small PVC reactor (conditions in Figure 31).

After 12 h of UV irradiation, ca. 88% of NO was removed at a relative humidity (RH) of 10%, decreasing to 52% at 80% RH, whereas the quantity of NO removed increased with increase in UV intensity (from 10% at 0 W m^{-2} to 88% at 12 W m^{-2}). Varying the NO concentration from 0.05 to 5.0 ppm showed that in the 0.05–0.15 ppm range of NO—a range similar to roadside levels—a constant removal of ca 90% was observed decreasing to ca. 45% at 5 ppm of NO [122]. In an outdoor test in which NO_x from the roadside (ca. 0.5 ppm) was passed through the PVC reactor exposed to natural sunlight ($T = 17\,°C$ and RH = 47%) led to ca. 80% decrease in NO_x.

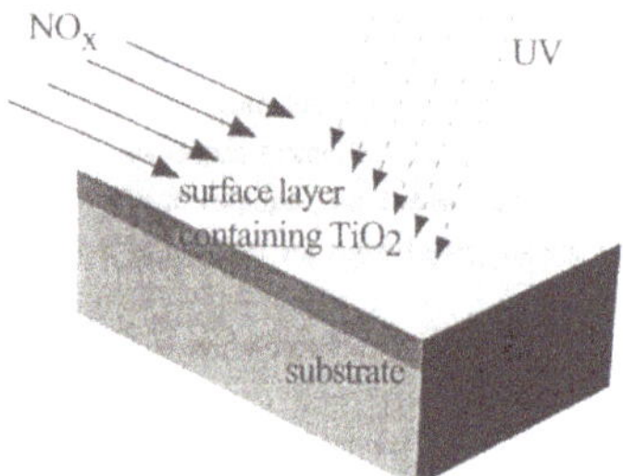

Figure 30. Paving brick for NO_x removal; dimensions of brick: 10 cm × 20 cm; thickness of the surface layer containing TiO_2: 5–7 mm. Reproduced from Murata et al. [122].

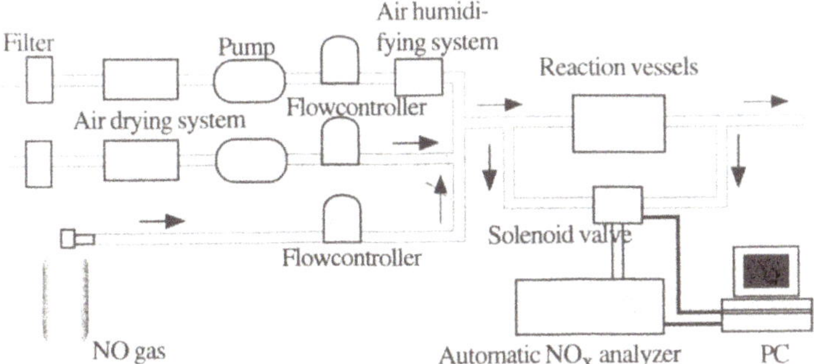

Figure 31. Indoor flow-through PVC reactor system for evaluating the performance of a brick in removing NO$_x$ from the environment. Simulated emission gas consisted of mixing dry air, dilute NO gas (1.0 ppm); flow rate was 3 L min^{-1}; air space between the brick and the Pyrex glass window was 5 mm; temperature, 25 °C; relative humidity RH, 50%; UV intensity from 2 black lights (300–400 nm), 6 W m^{-2}. Reproduced from Murata et al. [122].

Evidence of the effectiveness of NO$_x$ abatement was also carried out by the Italcementi Group using their TX-Active® photocatalytic TiO$_2$ deposited on a suitable substrate [121,123] and subsequently placed in a small reactor (top in Figure 32 [121]) of predetermined volume in which NO$_x$ gas was introduced, and then diluted with air to reach a certain pollutant concentration; the schematic of the flow-through reactor assembly is also displayed (bottom of Figure 32). The efficacy against NO$_x$ gases was also demonstrated during the project PICADA with tests conducted at the European Laboratory of Ispra (Italy), inside an Indoortron—an environmental chamber with people access and characterized by such controlled parameters as temperature, relative humidity, air quality and air exchange rate—to also study the fate of various other internal VOC contaminants. Figure 33 illustrates the time course of the removal of NO$_x$ with the reactor assembly of Figure 32. Evidently, complete NO$_x$ removal under the ISO conditions occurred within ca. 6 h of light (UV) irradiation in the presence of TX Active® in the absence of which no changes in NO$_x$ concentration occurred. Tests on the best formulations of a white and a grey photocatalytic paint with TX Active®, chosen for tunnel renovation (see below), showed a NO$_x$ abatement capacity of 88–90% after only 60 min under UV light irradiation in the reactor assembly of Figure 32; however, under similar conditions, the same paints necessitated nearly 26 h of UV irradiation to decolorize 70% of a rhodamine-B dye stain on the paint's surface [124].

An otherwise similar reactor assembly was used by Martinez and coworkers [125] to examine the degradation of NO present in the air by means of a photocatalytic oxidation process based on TiO$_2$ nanoparticles that had been incorporated in a polymer-matrix-based coating. The experimental setup consisted of a flow type reactor (Figure 34) adapted from the ISO 22197-1 standard; the final products detected were NO$_2$ in the gas phase and nitrate ions adsorbed on the photocatalytic surface. The photocatalyst (anatase-TiO$_2$) was a commercial slurry solution available from Evonik (Aerodisp® W740X). The coatings were primarily formulated for the surface treatment of building materials. To identify the possible influence of the nature of the substrate on the photocatalytic efficiency, various types of substrates were tested. The coatings were applied to: (i) mortars; (ii) glass plates; and (iii) non-absorbent cardboard materials. The wet thickness of coatings was 40 μm (other conditions are reported in Figure 34).

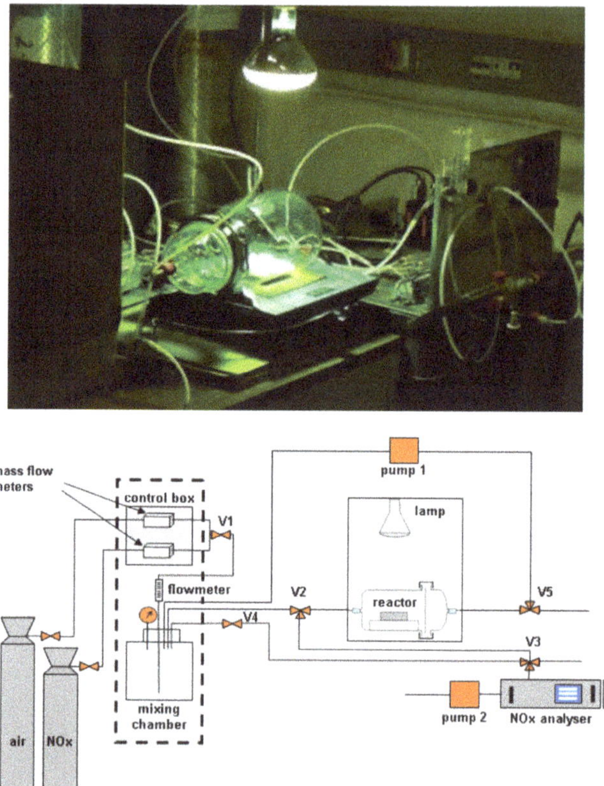

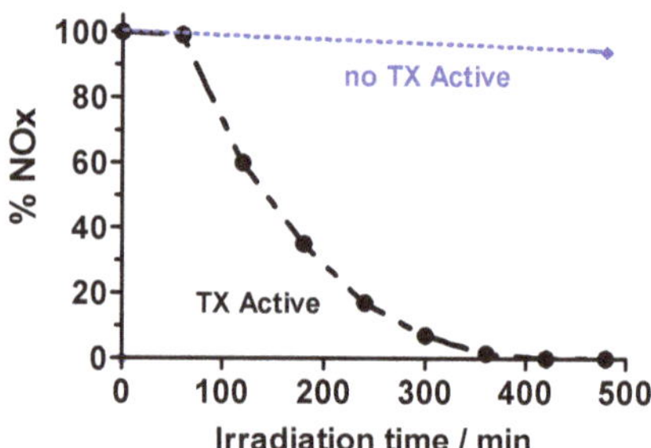

Figure 32. (**top**) Indoor reactor used to measure NO$_x$ abatement using the TX-Active photocatalytic TiO$_2$–based cementitious substrate Photograph reproduced from Borgarello Ref. [121]. (**bottom**) Schematic of the flow-through reactor assembly; reproduced from Guerrini and Peccati [123].

Figure 33. Plots illustrating the immediate destruction of NO$_x$ upon turning the light on and after 60 min of lamp stabilization of the chamber (recirculation tests). Plot made from data reported in Borgarello [121].

Results indicated that irradiation for 60 min and relative humidity of 60% led to ca. 25% of an initial concentration of 8.61 μmol of NO to be degraded at a flow rate of 1.5 L min^{-1} [125]. On mortar and glass substrates, the influence of increasing humidity on the degradation rates depended on the nature of the substrate and on initial NO concentrations; no significant influence of humidity was observed at initial NO concentrations of 400 and 1000 ppb, while a significant decrease in the kinetics was seen with a decrease of humidity at higher initial concentrations of NO (1500 and 2000 ppb). Generation of NO_2 on mortar was very low, because of good adsorption capacities of the supporting substrate. On glass, NO_x degradation rates decreased strongly on generation of NO_2 owing to competition between pollutant and humidity (water) for the adsorption sites [125].

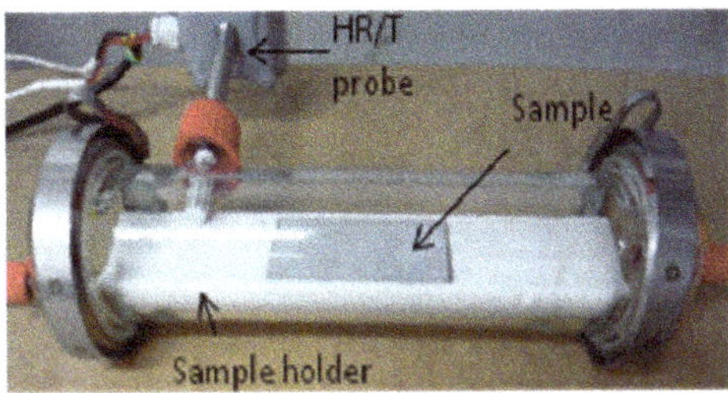

Figure 34. Cylindrical reactor (borosilicate-glass; dia. = 60 mm; length = 300 mm; high transparency to UV-A radiation; low adsorption capacity) used by Martinez and coworkers [125] to test the photocatalytic TiO_2 coatings toward the abatement of NO_x pollutants; coated and control samples (100×50 mm^2) were placed in the median plane of the reactor using a PTFE holder; gas circulated through the semi-cylindrical space between test piece and upper part of the reactor. Light source was a 300-W OSRAM Ultravitalux bulb with an emission spectrum close to that of daylight (light intensity = 5.8 W m^{-2}). Other experimental conditions: flow rate, 1.5 L min^{-1}; initial NO concentrations, 400–2000 ppb; relative humidity, 0–74%; temperature, 25 °C). Reproduced with permission from Martinez et al. [125]. Copyright 2011 by Elsevier Ltd. (License No.: 4452680979256).

Using a specifically-developed test apparatus (Figure 35), Staub de Melo and Triches [126] assessed the efficiency of a photocatalytic mortar under no less than 27 different environmental conditions: varying the relative air humidity (30%, 50% and 70%), the UVA radiation (10, 25 and 40 W m^{-2}), pollutant mass flow rate (1, 3 and 5 L min^{-1}), and initial concentration of NO (20 ppmv). Results showed that the higher were the levels of UVA radiation, the better was the performance of the mortar in degrading NO_x. By contrast, at higher relative humidity levels and flow rate caused a decrease in photocatalytic activity, which showed that environmental conditions have a significant impact on the efficiency of the photocatalytic mortar in the degradation of NO_x.

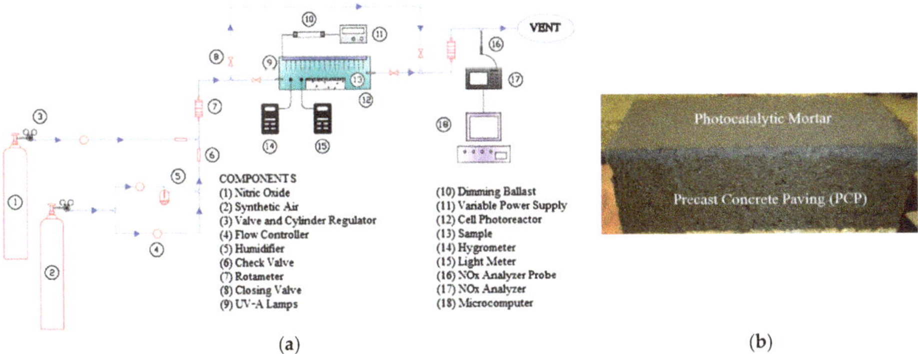

(**a**) (**b**)

Figure 35. (**a**) Photoreactor in a flow-through assembly for the measurement of photocatalytic activity of a photocatalytic mortar for the abatement of NO_x; and (**b**) photocatalytic mortar applied to a Precast Concrete Paving sample. Reproduced with permission from Staub de Melo and Triches [126]. Copyright 2012 by Elsevier Ltd. (License No.: 4452681429886).

The photocatalytic mortar was produced using Portland cement with Pozzolan (CP II Z 32); the catalyst was a nanometric rutile TiO_2 bar (dia. = 10 nm; length = 40 nm; 98% purity; specific surface area = 150 ± 10 m^2 g^{-1}; real density = 4.23 g cm^{-3}). Mortars with addition of 3%, 6% and 10% TiO$_2$ were investigated; layers of photocatalytic mortar with thicknesses of 3, 6 and 10 mm were applied to samples of precast concrete paving (PCP). The gas system consisted of dry air and 500 ppmv of NO stabilized with N$_2$ gas, which simulated a polluted atmosphere. Maximal removal of NO$_x$ was 50% for an initial 20 ppmv of NO, relative humidity of 50%, a flow rate of 1.0 L min^{-1} and a UVA irradiation (10 W m^{-2}) period of 25 min. A cost to photocatalytic efficiency evaluation led to a 3-mm coated mortar incorporating 3% of TiO$_2$. For the application of such materials, the authors [126] suggested that locations with lower relative humidity, high incidence solar radiation and little air mass movement should be sought in the field, as they would provide better conditions to achieve high efficiency of the TiO$_2$-coated precast concrete paving materials.

An examination of the past literature shows that different types of photoreactors have been used in standardization methods to quantify the activity of photocatalysts in air remediation with commercially available photocatalytic materials. Classically, the degradation of NO$_x$ has been a major subject of investigations because relatively simple and inexpensive chemiluminescence instruments are available to quantify NO$_x$ and because NO$_2$ is of crucial importance for urban air quality. Nonetheless, when investigating the photodegradation of NO$_x$ in laboratory settings, only NO was used because of its facile detection, its lower ability to adsorb on reactor surfaces, and because of slower dark reactions that might occur on photocatalytic surfaces [127]. To reduce the time for establishing adsorption equilibrium and to increase precision of the NO$_x$ data from low-sensitive instruments, unrealistically high NO concentrations (500–1000 ppbv) have been commonly used; and as analyses of reaction products require more sophisticated instrumentation, no reaction products (e.g., nitrite and nitrate) other than NO$_2$ are quantified. To the extent that NO$_2$ is an intermediate in the photocatalytic oxidation of NO by reaction with O$_2^{-\bullet}$/HOO$^\bullet$ radicals, the photocatalytic removal of NO$_x$, not just NO, is quantified by different standardization methods that are still under development (as we speak).

In this regard, Ifang and coworkers [127] demonstrated that transport limitations can lead to an underestimation of the activity, if fast heterogeneous reactions were investigated in bed photoreactors. When using stirred tank photoreactors, complex secondary chemistry can lead to an overestimation of the photocatalytic remediation of NO$_x$, if NO$_2$ were also present, not to mention that the quantities used for ranking the activity of photocatalysts in air remediation in the different methods currently used are not independent of experimental conditions, so that any inter- comparison between different methods or extrapolation to atmospheric conditions is a futile exercise. Consequently, the authors [127] proposed a modified method for quantifying air remediation activity of photocatalytic surfaces that would overcome such problems. The method is based on a bed flow reactor (Figure 36) that can easily be adapted to the ISO method. The extent of degradation of NO$_x$ on photocatalytic surfaces in continuous stirred tank reactors can be significantly influenced: (a) by the gas-phase Leigthon chemistry (Equations (85)–(87)); (b) by unwanted wall losses of reactive agents and products; and (c) by heterogeneous formation of products on reactor walls (e.g., HONO) [127]. In such a photoreactor, short reaction times of only a few seconds and more homogeneous inert surfaces (no fan, among others) should minimize the aforementioned issues.

$$NO_2 + h\nu \rightarrow NO + O(^3P) \tag{85}$$

$$O(^3P) + O_2 \rightarrow O_3 \tag{86}$$

$$NO + O_3 \rightarrow NO_2 + O_2 \tag{87}$$

The bed flow photoreactor was constructed of a single block of Teflon in which photocatalytic samples up to 40 cm $\times$ 5 cm $\times$ 1 cm could be investigated; recommended light sources were two 20-W UVA fluorescence lamps (300–500 nm, λ_{max} = 370 nm; length, 57 cm) mounted at variable distances to the reactor to adjust the irradiance level measured by a calibrated spectroradiometer [127]. Because of differences in photoactivity of commercial photocatalysts toward NO$_x$ (and VOCs), the authors suggested that at least one compound from each class be examined in standardization methods for air remediation. Moreover, to avoid saturation problems, laboratory experiments would have to be performed under relevant atmospheric conditions (i.e., for RH = 50%; reactant concentrations ≤ 100 ppb), and, as NO_2 is of much greater environmental importance compared to NO, the use of NO_2 as the test reactant was strongly recommended [127].

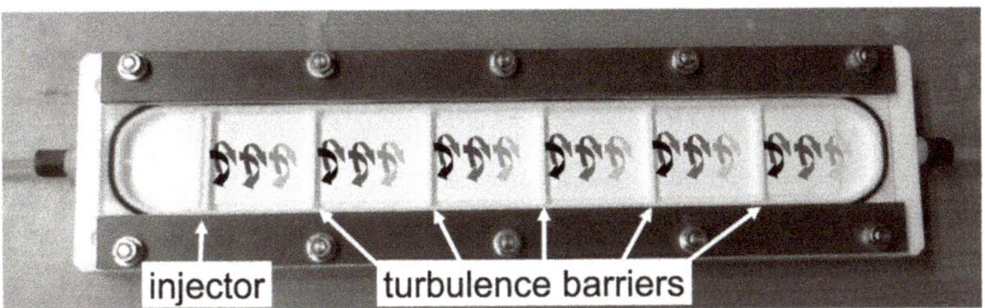

Figure 36. Modified bed flow photoreactor with movable injector and turbulence barriers. Reproduced with permission from Ifang et al. [127]. Copyright 2014 by Elsevier Ltd. (License No.: 4452690260303).

On their part, Zouzelka and Rathousky [128] investigated the photocatalytic activity of two commercial titania-based products: (1) Protectam *FN2* that consisted of ca. 74% of Evonik's Aeroxide P-25 TiO_2 powder and 26% of an inorganic binder; and (2) Aeroxide P-25 TiO_2 powder as photocatalytic coatings (10 μm thick) on concrete and plaster supports toward the abatement of NO and NO_2. Photocatalytic experiments on the coatings were performed in two types of flow reactors, one with laminar flow while the other with an ideally-mixed flow (Figure 37), under real world conditions in terms of temperature, relative humidity, irradiation intensity and pollutant concentrations. Results showed that the photocatalytic process reduced significantly the concentration of both NO_x agents in the air.

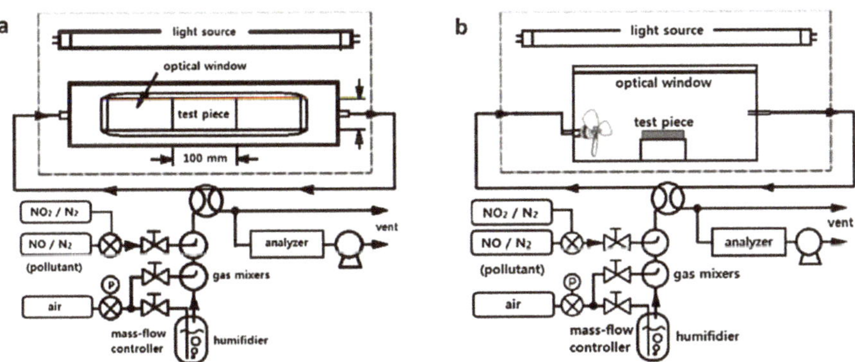

Figure 37. Experimental laminar flow (**a**); and ideally-mixed flow (**b**) reactors used to examine the photocatalytic oxidation of low concentrations of gas NO_x streams at some specified humidity. Reproduced with permission from Zouzelka and Rathousky [128]. Copyright 2017 by The Authors (open access license).

The decrease in the concentration of NO_x achieved in the steady-state for an inlet concentration of NO and NO_2 of 0.1 ppmv, corresponding to highly polluted urban air, was up to 75 μmol m^{-2} h^{-1} (for the Protectam *FN2*) and 50 μmol m^{-2} h^{-1} (for the Aeroxide P-25 TiO_2) at a flow rate of 3000 cm^3 min^{-1} and a relative humidity of 50%. Because of a conspicuous lack of data regarding the performance of photocatalytic coatings over long periods, the authors [128] also examined aged photocatalytic *FN2* coatings on a 300-m^2 concrete noise barrier that had been exposed to heavily-trafficked (ca. 30,000 vehicles a day) thoroughfare in Prague where the NO_x concentration reached 30–40 μg m^{-3}, often exceeding the permitted NO_2 limit of 40 μg m^{-3} (or 0.021 ppmv).

Experimentally, the area of irradiated photocatalytic surface was 50 cm^2 (5 cm × 10 cm); flow rate of air mixture was 3000 cm^3 min^{-1}; total volume of air treated in 24 h was 4.32 × 10^6 cm^3; volume of purified air and area of irradiated photocatalytic surface were the same in both photoreactors, although the reactors differed substantially in volume/irradiated area (65 times greater for the ideally-mixed flow reactor); volume of the ideally-mixing flow reactor was 5200 cm^3 (18 × 32 × 9 cm); the free volume of the laminar flow reactor was 80 cm^3 (5 × 32 × 0.5 cm); and linear streaming velocity of the gas was 0.2 m s^{-1}. The bandgaps of the TiO$_2$ in the two materials were 3.2 eV for the P-25 sample and 3.05 eV for the *FN2* sample, the red-shift in the latter being attributed to the effect of the binder on titania [128].

Comparison of the photocatalytic performance of P-25 and *FN2* materials reported in Figure 38 shows the reaction rate with the *FN2* coating to be greater than for the P-25, even though the quantity of TiO$_2$ in the *FN2* coating was lower. For instance, the reaction rate with the *FN2* coating on concrete in both laminar and ideally-mixed flow reactors at an inlet NO concentration of 1.0 ppmv was 40% and 49% higher, respectively, than the corresponding reaction rate of P-25 which, according to the authors [128], was likely due to the nearly twofold larger surface area of the *FN2* specimen (82 m^2 g^{-1}) relative to the P-25 sample (47 m^2 g^{-1}).

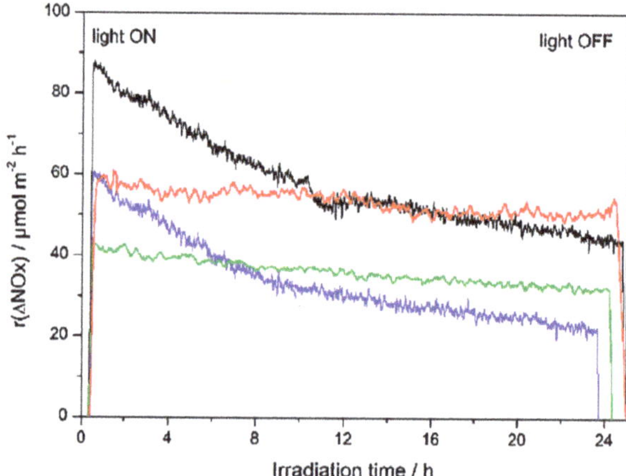

Figure 38. Comparison of the photocatalytic reaction rates in the degradation of NO$_x$ with the *FN2* and P-25 in both ideally-mixed and laminar flow reactors for an inlet NO$_2$ concentration of 0.1 ppmv and relative humidity (RH) of 50%. *FN2* (black) and P-25 (blue) coated on concrete in laminar flow reactor. *FN2* (red) and P25 (green) coated on plaster in ideally-mixed flow reactor. Reproduced with permission from Zouzelka and Rathousky [128]. Copyright 2017 by The Authors (open access license).

Evidently, the inlet concentration of NO had a substantial influence on the reaction rate in both reactors, as evidenced on comparing the data for the *FN2* coating in the laminar flow reactor in which the reaction rate was approximately proportional to the inlet NO concentration; in other words, for an inlet NO concentration of 1.0 ppmv, the rate was about ten times greater than for a concentration of 0.1 ppmv [128]. With regard to the effect of NO$_2$ concentration used (0.1 ppmv) the trends differed from those for NO. First, the difference in the reaction rate at the beginning of the reaction and in the steady-state was much smaller for both P-25 and *FN2* coatings when applied on plaster than on concrete. On plaster, the reaction rate was practically unchanged, whereas on concrete the decrease in rate was between 40% and 60%. Moreover, the steady-state reaction rate (after a 24 h period) on the coatings applied on plaster was consistently 1.5–1.8 times higher in comparison with the coatings on concrete.

Another important issue emphasized by Zouzelka and Rathousky [128] was the durability of the performance of the photocatalytic coatings under real-world conditions, as exemplified by the commercial photocatalytic coating Protectam *FN2* that maintained relatively high efficiency in removing NO_x from contaminated air even after two years under the harsh conditions noted above; this was likely due to good mechanical properties of the binder.

Moving from small benchtop laboratory photoreactors to actual field studies to investigate the efficiencies (activities) of photocatalytic concrete/mortar with embedded titania photocatalysts, simulation chambers have also proven suitable to test photocatalytic materials under controlled environmental conditions, as they represent a mid-way step between laboratory and actual environ-ment (a sort of pilot plant scale). Additionally, simulation chambers offer the opportunity to examine the impact of photocatalytic surfaces, also on secondary air chemistry, in contrast to the smaller flow-type reactors. In this regard, within the framework of the Life + project PhotoPAQ (2010–2014) that will be described later (Section 5.2), Mothes and coworkers [25] tested the behavior of O_3, NO_x, and selected VOCs (toluene and isoprene) on a photocatalytic cementitious coating material (with Italcementi's TX-Active$^®$) under UV irradiation and atmospherically relevant conditions (relative humidity, temperature, and realistic pollutant concentration) using the aerosol chamber LEipziger Aerosol Kammer (LEAK) at the Leibniz Institute for Tropospheric Research (TROPOS)—a 19 m^3 aerosol cylindrical chamber with a surface-to-volume ratio of 2.0 m^{-1} and illuminated with UV lamps (λ = 300–400 nm) with an average light intensity of 11 W m^{-2}.

The TX-Active$^®$ material was mixed with ultrapure water and subsequently applied manually on both sides of sand-blasted glass plates supported by a home-made device to obtain a rough but uniform surface (final thickness, ca. 3–4 mm), after which the material was cured for ca. 1 month. The cleaned coated glass plates were then inserted into a special aluminum rack installed inside the chamber—the area of the photocatalytically active surface was 6.65 m^2 (boosted specimen)—a blank specimen devoid of TX-Active$^®$ titania was also made for comparison. Both specimens were flushed with clean air (200 L min^{-1}) with the UV lamps turned ON for ca. 30 h to remove any potentially adsorbed pollutants; RH varied between dry air and 50%—the photocatalytic tests were performed following the 2007 ISO-22197-1 standard methodology. Results obtained for the NO and NO_x degradation experiments together with formation of O_3 at three RHs and at 293 K are displayed in Figure 39; dark period ca. 120 min, irradiation period also 120 min, initial concentrations of NO and NO_x = 40 ppb [25].

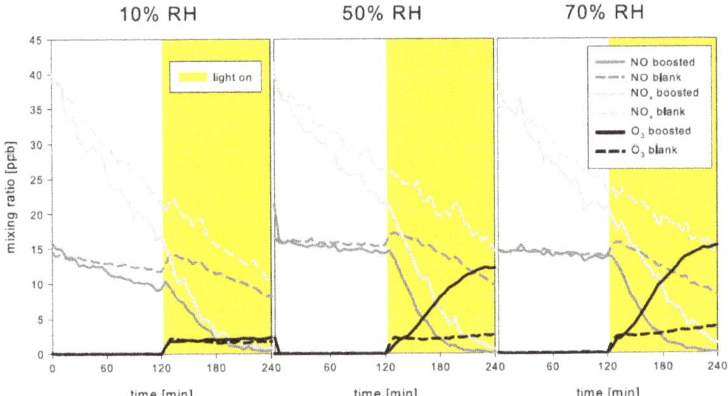

Figure 39. NO_x investigations under dark and light conditions in LEAK simulation chamber (RH = 10%, 50% and 70%; T = 293 K) to compare the behavior of the boosted (solid lines) and blank materials (dashed lines) for the time-dependent changes in the concentrations of NO, NO_x, and O_3 akin to environmental quantities. Reproduced with permission from Mothes et al. [25]. Copyright 2016 by Springer-Verlag Berlin Heidelberg (License No.: 4453250415918).

Contrary to the behavior of NO_x during the dark period, NO showed only a slightly different behavior at the different RHs; NO's behavior was nearly constant at 50% RH and 70% RH, whereas at 10% RH a slight decrease was seen that was attributed to slightly adsorbed NO on the specimens' surface [25]. Comparison of the blank and boosted materials revealed that UV irradiation had a significant influence on NO_x owing to the photocatalytic properties of TiO_2 embedded in the boosted material. The small increase of NO at the beginning of the light period was attributed by the authors [25] to photolysis of NO_2 according to the Leighton sequence described by Reactions (85)–(87).

Results from the chamber studies also showed a photocatalytic effect on O_3, evidenced by an increase of the geometric uptake coefficient from 5.2×10^{-6} for the inactive to 7.7×10^{-6} for the active material under irradiation. Measured first-order rate constants for NO_x under irradiation ranged from $2.6 \times 10^{-4}\ s^{-1}$ to $5.9 \times 10^{-4}\ s^{-1}$, significantly higher compared to the range for the inactive materials (7.3–$9.7 \times 10^{-5}\ s^{-1}$), thus demonstrating the photocatalytic effect. However, no significant photocatalytic degradation was observed for the VOCs (toluene, isoprene); the upper limit uptake coefficient for both was only 5.0×10^{-7}. Small carbonyl (C1–C5) gas-phase compounds were identified when using the photocatalytically active material, a result of the photocatalytic degradation of the organic additives [25]. In contrast to the uptake observed for pure O_3, a clear photocatalytic formation of O_3 was observed ($k_o(O_3) \approx 5 \times 10^7$ molecules $cm^{-3}\ s^{-1}$) during the experiments with NO_x (RH $\geq 50\%$). The authors further noted the necessity for detailed studies of heterogeneous reactions on such surfaces under more complex simulated atmospheric conditions as enabled by simulation chambers.

Although many studies have been done on TiO_2 photocatalytic cementitious substrates to deNOxify the environment—as we discuss below—nearly all studies neglected to consider the inter-relationship between the metal-oxide photocatalyst and the non-negligible chemistry that might occur on the cementitious support—in fact, this inter-relationship is particularly relevant for an appropriate understanding of deNOxification processes. In this regard, the investigations carried out by Macphee and Folli [129] on *Photocatalytic concretes—The interface between photocatalysis and cement chemistry* have opened up the proverbial Pandora's Box in which they addressed: (i) the photocatalytic mechanisms applicable to atmospheric depollution; (ii) the influence of doping the metal oxide; and (iii) the application of TiO_2-based photocatalysts to concrete. These authors further emphasized some of the points described earlier that the catalyst efficiency is influenced by several factors, none of which are negligible: for instance: (i) energy and intensity of the activating radiation incident on the photocatalyst; (ii) the number and relative energy positions of the electronic states in the photocatalyst, defined by the crystal structure and the redox potentials associated with the required redox processes; (iii) the charge carrier mobility within the semiconductor's conduction and valence bands; (iv) the kinetics of charge transfer processes; and (v) the accessible catalyst surface for adsorption of the pollutants, oxygen, and water [129].

Of greater importance, they noted that the chemistry of the cement environment is quite different from the ambient conditions normally prevalent in environmental photocatalysis, which will have a non-insignificant influence on the normal behavior of the catalysts in the concrete [129]. In freshly mixed cement, the high pH and the high ionic strength aqueous mix containing multiply charged ions dramatically modifies the surface chemistry and behavior of TiO_2 dispersions, and thus will have a strong impact on dispersion behavior and adsorption properties, as the mode of adsorption and strength of binding between an adsorbing molecule and a surface are conditioned by their relative charges. Additionally, pH changes during the early stages of mixing and ageing the cementitious substrate will also have an impact on the band edge positions in the semiconductor photocatalyst as well as the oxidation chemistry of NO_x. The alkalinity of wet cement/concrete is typically greater than pH 13, which can change rapidly at the surface as the concrete sets and the formwork is removed (Figure 40a) [129]. The highly alkaline aqueous phase can react with atmospheric CO_2 to produce CO_3^{2-} ions, causing the pH to decrease and induces surface deposition of calcite—a carbonate mineral and the most stable polymorph of calcium carbonate—that can obstruct deposited photocatalysts. Figure 40b–d displays the microscopic details of the surface calcite layer that obstructs TiO_2 clusters present in the hardened cementitious matrix.

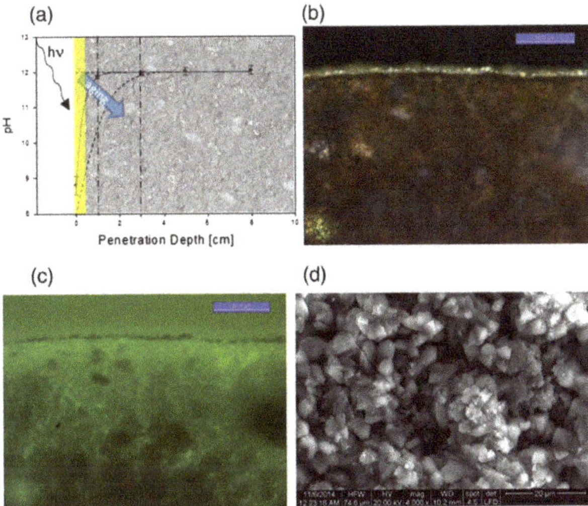

Figure 40. (a) Schematic illustration of pH changes and light penetration profile in the ageing concrete; (b) optical microscope image of the cross section of a concrete specimen containing TiO_2 under X-polarized light—note the surface exhibits a thin calcite layer that appears in a lighter colour; (c) optical microscope image of the cross section of a concrete specimen containing TiO_2 under fluorescent light—note that the calcite layer now appears dark green indicative of low porosity; and (d) SEM image of the top surface of a concrete specimen containing TiO_2 (same sample as in (b,c)) wherein the surface consists of a layer of small, closely spaced, euhedral calcite crystals—no TiO_2 photocatalytic clusters are exposed. Reproduced with permission from Macphee and Folli [129]. Copyright 2016 by Elsevier Ltd. (License No.: 4453091259877).

The alkali nature of the cementitious surface also has significant consequences on the chemistry taking place at the surface when NO_x molecules are adsorbed, even in the absence of any catalyst and under non-irradiating conditions, as both NO and NO_2 (NO_x molecules) can undergo an oxidative reaction with the alkali to yield nitrite and nitrate ions (Equations (88) and (89)) [130]. These two reactions must be considered whenever photocatalytic efficiencies are assessed in deNOxification processes.

$$2\,NO_2 + 2\,OH^- \rightarrow NO_2^- + NO_3^- + H_2O \tag{88}$$

$$NO_2 + NO + 2\,OH^- \rightarrow 2\,NO_2^- + H_2O \tag{89}$$

To address the geographical limitations of UV intensity and recognizing that sunlight is the light source to activate the photocatalyst in practical applications of photocatalytic concrete in deNOxification, the Macphee group [129,131] also investigated M-doped titania (M = W, Nb) photo-catalysts toward removal of NO_x because unmodified TiO_2 releases a significant quantity of toxic NO_2 in the deNOxification process, a problem rarely raised in several previous studies as NO_x abatement is commonly assessed only on the basis of NO removal rather than NO_2 or total NO_x removal. Accordingly, to highlight the problem, they performed a systematic study on a number of commercial titania powders and doped titania specimens by determining both the photocatalytic activity (i.e., the photonic efficiency ξ [29], as defined by Equation (90)) and the selectivity (S defined by Equation (91)) toward formation of NO_2 and NO_3^- in the oxidation of NO.

$$\xi = \frac{(C_{dark} - C_{irr})V p}{\rho_0 A R T} \tag{90}$$

$$S = \frac{\zeta_{NO_x}}{\zeta_{NO}} \qquad (91)$$

where C_{dark} and C_{irr} are the concentrations of the species of interest under dark and irradiated conditions, respectively; V is the volumetric flow rate; p is the pressure; ρ_0 is the photon flux impinging on the photocatalyst surface; A is the irradiated area; R the gas constant; and T is the Kelvin temperature.

In the case of W-doped TiO$_2$, Figure 41 demonstrates that while doping titania with tungsten significantly improved the selectivity toward formation of NO$_3^-$, it considerably reduced the photonic efficiency [131]. This raised the question as to how one could evaluate various photocatalysts toward NO$_x$ abatement when both activity and selectivity needed to be considered. To resolve this matter, the authors [129,131] defined a new figure of merit: the DeNO$_x$ index (Equation (92)), which was derived by assigning a toxicity value to both NO and NO$_2$ and then expressing the change in total toxicity rather than the concentration change of the individual NO$_x$ gases [129,131]:

$$\zeta_{deNO_x} = \zeta_{NO} - 3\zeta_{NO_2} \qquad (92)$$

The *deNO$_x$ index* is positive if the photocatalyst lowers the NO$_x$ toxicity.

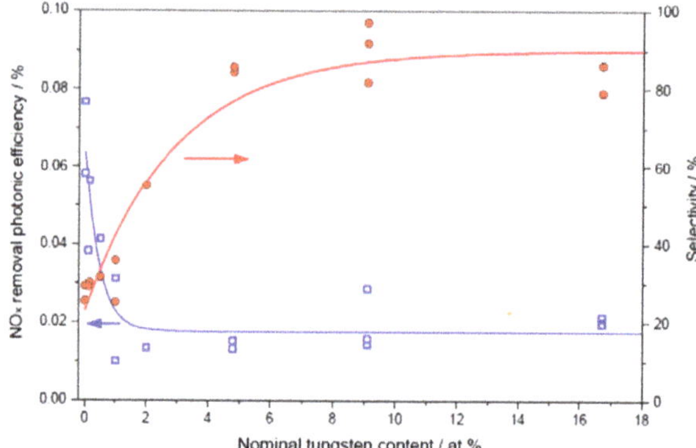

Figure 41. Plots illustrating the photonic efficiency of NO$_x$ removal and nitrate selectivity of W-modified TiO$_2$ illuminated under broadband radiation. Adapted with permission from Bloh et al. [131]. Copyright 2014 by the Royal Society of Chemistry (license ID: 4453100929571).

Figure 42a illustrates the results from several photocatalyst compositions and demonstrates a positive *deNO$_x$ index* at nominal W contents > 4.2 at.% for the W-doped TiO$_2$ [129]. A range of commercial TiO$_2$ photocatalysts were tested under comparable conditions giving a *DeNO$_x$ index* between 0 and −4000 representing poor nitrate selectivity and, most importantly, meaning that the more toxic NO$_2$ was generated [129]. These data were further supported by analyses on powder and mortar samples incorporating W-doped and Nb-doped TiO$_2$, as displayed in Figure 42b [129]. Interestingly, although undoped P-25 TiO$_2$ has been the workhorse in photocatalysis, it displayed a net negative effect owing to its high activity but low nitrate selectivity; that is, it efficiently converted NO to NO$_2$ rather than to NO$_3^-$ ions. By contrast, both W- and Nb-doped TiO$_2$ showed a higher tendency towards NO$_3^-$ and thus are better suited as remediation photocatalysts, whether in powdered form or embedded in mortars.

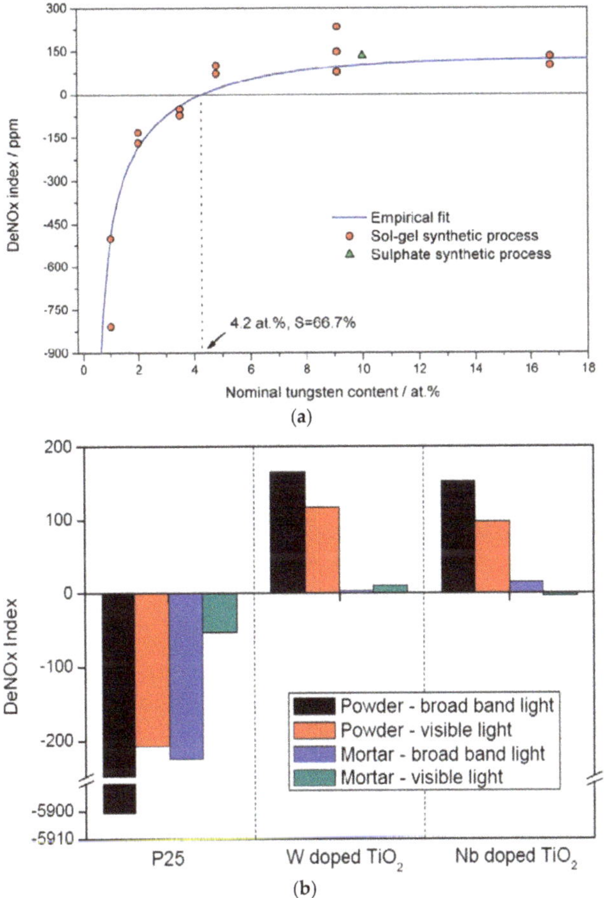

Figure 42. (a) deNO$_x$ index measured using W-doped TiO$_2$ under broad band radiation simulating sunlight; and (b) deNO$_x$ data on powder and mortar samples containing W-and Nb-doped TiO$_2$ (broad band and visible light exposure). Reproduced with permission from Macphee and Folli [129]. Copyright 2016 by Elsevier Ltd. (License No.: 4453091259877).

Continuing their interesting studies on photocatalytic concrete, in their most recent article, the Macphee group [132] examined the effect of photocatalyst placement as regards photocatalyst efficiencies in concrete technology, and pointed out that even though the technology represents a well-established concept and notwithstanding the significant opportunities for air quality improve-ments to be derived from the considerable concrete surfaces exposed to the atmosphere, especially in cities highly polluted by vehicle exhaust and industrial emissions, photocatalytic concretes have so far remained in the investigative sphere, rather than in the mainstream of applications. As with any commercial new technology, the likely barriers for widespread implementation may well be cost effectiveness that might emanate from the photocatalyst impact and the challenges in measuring directly what the impact of this new concrete technology may actually be on air quality. The challenges are indeed very complex. Accordingly, the authors [132] placed photocatalytic efficiencies into context by comparing performances of conventional photo-catalyst dispersions in surface mortar coatings vis-à-vis photocatalysts supported on surface-exposed aggregates as well as on the nature and impact of catalyst binding to aggregate supports. However, as we discuss below, the efficiencies in real-world environmental applications differ significantly from the efficiencies experienced in a laboratory setting where small photo-reactors have been used to measure the efficacies of various TiO_2 photocatalyst concrete specimens indoor, wherein experimental conditions can be controlled in contrast to outdoor environmental conditions, which are not only widely different, but are also uncontrollable conditions that one experiences in the real world. Nonetheless, their findings in photocatalyst placement as regards efficacies of photocatalytic concrete are worth noting [132].

In treating NO_x gases by TiO_2-based photocatalytic substrates, the currently accepted sequence is represented by Reaction (93); the greater is the degree of conversion of NO and NO_2 to nitrates, the greater is the catalytic activity of the substrates.

$$NO \rightarrow HNO_2 \rightarrow NO_2 \rightarrow NO_3^{-} \tag{93}$$

Although this oxidation sequence is thought to be the preferred reaction with respect to air quality, as claimed by the authors [132], a far better process to improve air quality would be the reduction of NO_x back to N_2 and O_2 as can be achieved by selective catalytic reduction (SCR) and selective non-catalytic reduction (SNCR) technologies (see above). Regardless, in TiO_2-based photocatalytic methods, both oxidative entities (valence band holes) and reductive entities (conduction band electrons) are generated upon sunlight UV activation of the TiO_2 photocatalyst, so that the oxidation sequence (Reaction (93)) may also lead to intermediates—for example, ozone and peroxyacyl nitrates—in addition to those in the sequence (Reactions (94)–(96)) produced in surface processes in the NO_2 production regime, and those produced in the sequence (Reactions (97)–(100)) in surface processes in the N_2O_5 production regime when ozone is present [133], all of which could be released into the atmospheric environment not forgetting that NO_2 is nearly three times more toxic than NO.

$$2\ NO_{2(ads)} \longleftrightarrow NO_3^{-}{}_{(ads)} + NO^{+}{}_{(ads)} \tag{94}$$

$$NO^{+}{}_{(ads)} + O^{2-}{}_{(ads)} \longleftrightarrow NO_2^{-}{}_{(ads)} \tag{95}$$

$$NO_2^{-}{}_{(ads)} + NO_{2(ads)} \longleftrightarrow NO_3^{-}{}_{(ads)} + NO_{(gas)} \tag{96}$$

$$TiO_2 + O_3 \rightarrow O\text{-}TiO_2 + O_2 \tag{97}$$

$$NO_{2(ads)} + O\text{-}TiO_2 \rightarrow NO_3 - TiO_2 \tag{98}$$

$$NO_{3(ads)} + NO_{2(g)} \rightarrow N_2O_{5(ads)} \tag{99}$$

$$N_2O_{5(ads)} \rightarrow N_2O_{5(g)} \tag{100}$$

Notwithstanding the above sequences, accumulation of NO_3^{-} on catalytic surfaces must be managed since adsorbed NO_3^{-} ions block catalytic sites from further NO/NO_2 adsorption—thus, affecting catalytic activity—as the nitrates back react to form NO_2 thereby causing nitrate selectivity to be diminished. This calls attention to the notion that *reNOxification* may be an important event in *deNOxification* processes. Accordingly, NO_3^{-} must be removed periodically by washings [132] at least on a weekly basis for certain infrastructures (tunnels, noise barriers, and buildings, among others) but especially for road tunnels where rain has no impact; this maintenance would carry increased costs. Moreover, where rain is sparse in the summer months as often occurs in Southern Italy and California (USA), such periodic washings may well be necessary. In addition, in conventional photo-catalytic concretes, nitrate washings via condensed atmospheric moisture represents but a mild risk from acidification of the cement (carbonated at the near surface) which may influence leaching-induced damage over time [132].

In their earlier articles [129,131], Macphee's group addressed various factors that limit photocatalytic efficiencies in cement-based systems in which the metal-oxide photocatalyst was incorporated in the mortar/concrete. They have now proposed to separate the chemistry taking place on the photocatalyst surface from the chemistry occurring on the cementitious support, so as to enhance photocatalytic efficiency by depositing the photocatalyst on the surface of the cementitious aggregate (Figure 43) [132]; that would require some sort of binder which unfortunately would also add to the complexity of the events as the binder may not only photodegrade but may also bring about changes to the electronic structure of the photocatalyst [128].

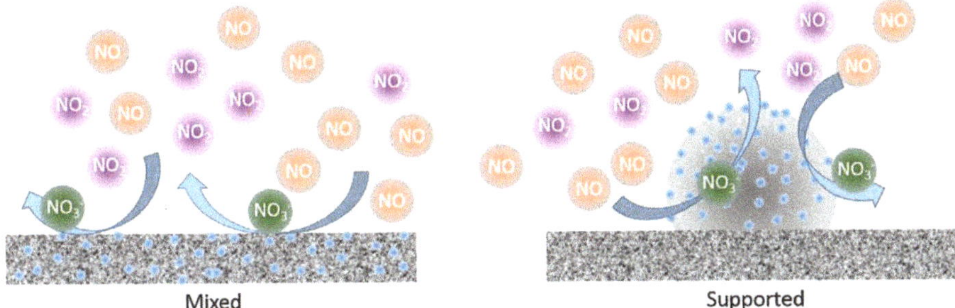

Figure 43. Conventional mortar/concrete with the photocatalyst embedded within the aggregate (mixed), and aggregate-supported configuration of the photocatalyst (blue) applications in proposed concrete technology whereby the photocatalyst is supported on sort of binder that bonds both the photocatalyst and the cementitious support. Reproduced with permission from Yang et al. [132]. Copyright 2018 by The Authors (open access license).

The aggregate used to support the prepared TiO_2 photocatalyst was quartz sand (1–2 mm diameter) that was subsequently treated in aqueous solution of $Ca(OH)_2$ followed by carbonation with CO_2 (aggregates denoted QST) [132]. The aggregates were placed in the flow-through reactor displayed in Figure 44 (monolayer; area, 5×10^{-3} m^2), after which they were irradiated with a 500-W Xe-lamp solar illuminator (photon flux at the aggregates, 3.05 µmol m^{-2} s^{-1}); other conditions were: 1 ppm NO gas in synthetic air; relative humidity, 40%; temperature, 25 °C; volumetric flow rate, 5×10^{-5} m^3 s^{-1} through the reactor.

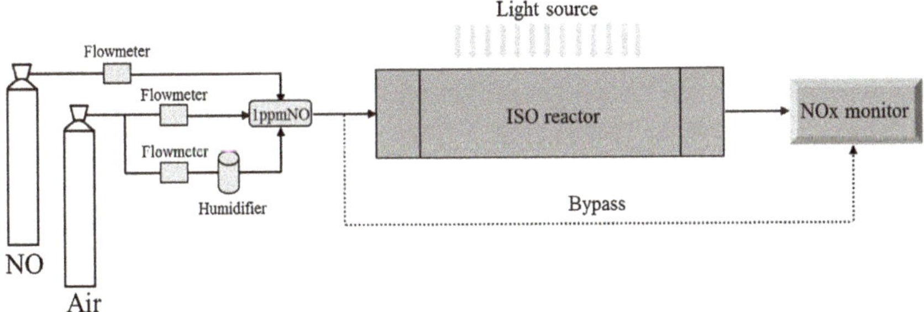

Figure 44. Schematic illustration of photoreactor used in testing the photocatalytic degradation of NO$_x$ gases. Reproduced with permission from Yang et al. [132]. Copyright 2018 by The Authors (open access license).

Three stages were identified by the authors in the process of converting NO to nitrate: (i) stabilization of NO concentrations on the QST specimens in the dark for achieving an adsorption/desorption equilibrium; (ii) the photocatalytic conversion of NO under illumination; and (iii) recovery of NO concentrations after illumination was terminated [132]. Results showed that the concentration of NO decreased rapidly on illumination, remaining nearly constant for a while, whereas the concentration of NO_2 increased significantly during this time; stopping the illumination caused the NO_2 concentration to go back to zero (Figure 45). The photonic efficiency ξ_{NO} increased initially with increasing TiO_2 mass fractions but became steady around 0.76%. Concomitantly, the ξ_{NO_x} and nitrate selectivity (S%) increased with increasing mass fraction of TiO_2, which led to a corresponding decrease of the rate of production of NO_2. Compared with the NO_x removal results for pure TiO_2, the estimated ξ_{NO} for the QST specimens was ca. 0.6%, while nitrate selectivity was ca. 40%; the supported TiO_2 particles presented higher photonic efficiency for NO removal, but lower nitrate selectivity owing to reduced NO_2 removal efficiency.

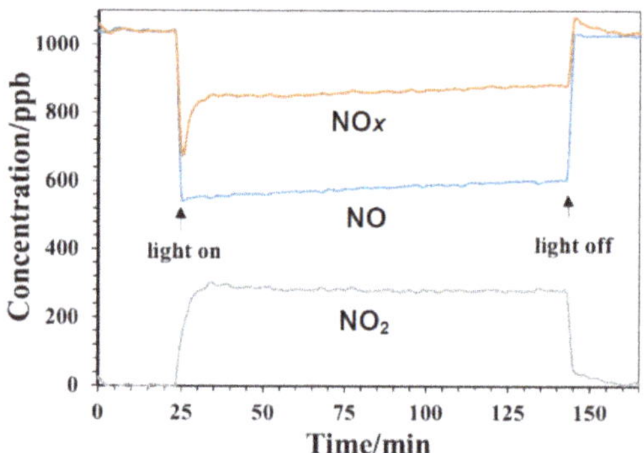

Figure 45. Concentration profiles for NO, NO_2 and NO_x (NO + NO_2) during the photocatalytic oxidation of NO on TiO_2 hydrosol (0.1 g) photocatalysts. Reproduced with permission from Yang et al. [132]. Copyright 2018 by the authors (open access license).

Clearly, the photocatalytic performance depended on the TiO_2 loading with the activity increasing with TiO_2 content up to ca. 0.76%, while nitrate selectivity increased across the range of TiO_2 loading used, and was strongly conditioned by the negative influence of Ti–O–Si bonds on selectivity and the effect of dilution as TiO_2 loading was increased. Washing to remove weakly bonded TiO_2 led to a decrease of the photocatalytic activity ξ_{NO} by about 30%, even though nearly half of the TiO_2 had been lost (only 0.34% remaining); however, selectivity was not significantly affected [132]. While such a low photocatalytic activity could be explained, in part, by fast charge carrier recombination and charge carrier trapping dynamics within intra-bandgap defect states in TiO_2 photocatalytic substrates, mechanical loss of titania particulates from the support, coupled with the observed activity under controlled indoor experimental conditions (small flow-through reactor), as reported by the authors [132], do not bode well to convince implementation of the photocatalytic concrete technology to outdoor real-world environments where, as noted earlier, the conditions are different and uncontrollable.

The results of the few studies reported herein with regard to the extent of removal of NO_x gases (NO and NO_2) are summarized in Table 7 [25,97,121,122,124–126,128,132,134] for indoor laboratory experiments carried out with small bed flow-through reactors employing a variety of different conditions from one study to the next. Various standardization protocols have been proposed in the last few years that have led to significant variation in the results—as should have been expected.

Table 7. Percent removal of NO_x gases carried out with TiO_2-based photocatalytic protocols in an indoor laboratory setting using small flow-through reactors under a variety of conditions.

Study	Gas	TiO_2-Based Photocatalytic Removal of NO_x Gases	Year Study Was Carried Out	Refs
1	NO_x	60–90%	1994	[97]
2	NO	52–88% 45–90%	2000	[122]
3	NO_x	88–90%	2009, 2011	[121,124]
4	NO	25%	2011	[125]
5	NO_x	50%	2012	[126]
6	NO_x	40–60%	2017	[128]
7	NO_x	35%; 66%	2016	[25]
8	NO NO_x	44% 16%	2018	[132]
9	NO	80%	2006	[134]

5.2. DeNOxing the OUTDOOR Environment with TiO$_2$ Photocatalytic Cementitious Materials

As shown above, many studies have been reported, albeit not all have been cited herein, on the photocatalytic abatement of NO_x in a laboratory setting wherein experimental conditions could be controlled. Relatively speaking, however, only a few field trials have been conducted outdoors in open air environment under real-world conditions; as we have noted, the conditions are not only different but, more importantly, they also cannot be controlled. As an example of the latter, some reduction of NO_x levels was achieved at the laboratory scale and in outdoor field experiments in Antwerp (Belgium) where a surface of ca. 10,000 m^2 was covered with photocatalytic pavement blocks—researchers that did the study were unable to conclude the extent to which NO_x was reduced because the measurement period was too short [135]. Additionally, an experiment to test the efficacy of photocatalytic TiO_2-mortar panels to degrade NO_x in a setting that involved artificial canyon streets showed NO_x concentrations to be reduced by as much as 37–82% depending on the conditions [136]. Likewise, a study carried out in 2006 in Bergamo (Italy) over a four-week period in a segment of a local street covered with photocatalytic paving stones showed a 30–40% reduction in the NO_x concentration compared to a similar section left untreated [137]; elsewhere, NO_x reduction levels of 20 to 50% have been quoted that depended on weather and traffic conditions [138]. A study done in the city of Segrate (Italy) [137] where a concrete road segment (7000 m^2) was constructed using a thin-layer of a photocatalytic mortar was said to display a 57% NO_x abatement level, while a similar study in an industrial section near the town of Calusco (Italy) reported a NO_x abatement level of 45% for an 8000 m^2 pavement built with photocatalytic concrete blocks. By contrast, field studies carried out in an artificial model street canyon in Petosino, Italy [15] and in the Leopold II tunnel in Brussels (Belgium) [14] reported photocatalytic reductions of NO_x well below measurement precision errors (1–2%). An "*indoor*" field study of the Umberto I Tunnel (Rome), for which the walls were coated with a photocatalytic paint and a new UV-Visible lighting system installed, showed more than 20% reduction of NO_x [138]. The latter two field studies [14,138] are discussed in some detail below.

Clearly, the many discrepancies in NO_x reduction levels when using photocatalytic paints, photocatalytic paving bricks, or photocatalytic mortar/concrete on streets, highways, or roadside noise barriers: (i) call attention to highly differing prevailing atmospheric conditions in field experiments; (ii) call into question how these field experiments were carried out; and (iii) call into question how the empirical results were interpreted. For that reason, we now describe the various attempts carried out using TiO_2-based photocatalytic substrates/aggregates in a few countries and discuss their findings with an emphasis on three major undertakings funded wholly or in part by the European Union: the PICADA Project (2002–2006); the Life$^+$-funded Project PhotoPAQ (2010–2014); and the LIFE MINOX-STREET Project (2014–2018).

One of the first applications of photocatalytic cement-based materials for self-cleaning purposes was Richard Meier's *Dives in Misericordia* Church project in Rome where Italcementi served as the principal technical sponsor, analyzed the concrete mix (Figure 46) [139], and provided continuous supervision during the construction period (1999–2001). The church was erected with three huge, totally white sails made of precast photocatalytic concrete blocks (Figure 47) that would ensure unparalleled and time-enduring white color to the built elements thanks to the self-cleaning properties of the final TX-Active® surfaces [140,141]. The photocatalytic cement-based product used was expected to maintain its *aesthetic characteristics unchanged over time, especially the color, even in the presence of aggressive urban environments.* Laboratory pilot-scale tests demonstrated that NO_x abatement with photocatalytic cement-based products was impacted by temperature, relative humidity, contact time of NO_x with the surface (flow velocity, height of the air flow above the sample, among others); it was also noted that reduction efficiency increased with longer contact times (larger surfaces, lower velocities, and higher turbulence/mixing), higher temperatures and lower relative humidity—however, in real situations, conditions such as high temperatures, no wind and no rain present the largest risk of ozone formation [142].

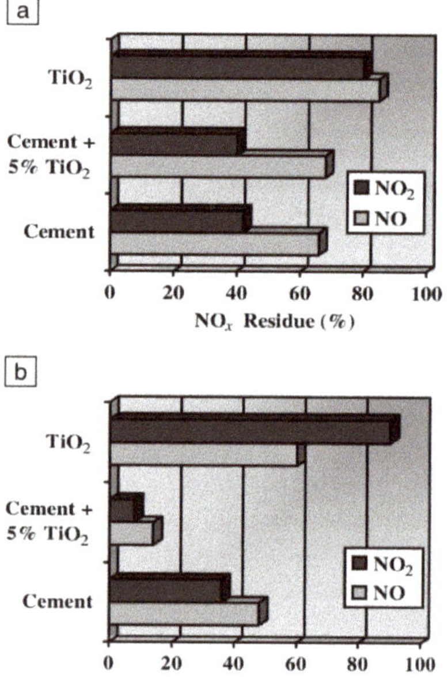

Figure 46. Laboratory tests of the anatase TiO_2 photocatalytic cement-based products used in the precast concrete panels in the sails of the *Dives in Misericordia* Church (Rome): (**a**) under dark conditions; and (**b**) under UV irradiation for 7 h. Note the increased reduction of NO and NO_2 under UV irradiation. Reproduced with permission from Cassar [139]. Copyright 2004 by the Materials Research Society (License No.: 4453250907572).

Guerrini and Corazza [142] claimed that a building element containing cement to which TiO_2 had been added was capable of maintaining its aesthetic appearance unaltered in time (see also [121]), thus contributing to reduce the dirtiness of surfaces exposed to specific polluted environments. A look at the photographs illustrated in Figure 48, however, shows that the long-term effectiveness of anatase TiO_2 photocatalytic cement-based external walls of the church failed the test of time.

Figure 47. *Dives in Misericordia* Church, Rome, constructed of TiO_2 self-cleaning and depolluting TX Active cement (inaugurated in 2003). Reproduced from https://es.i-nova.net/content?articleId=96804 or https://www.archdaily.com/20105/church-of-2000-richard-meier.

Another important sponsorship of the CTG-Italcementi Group was the renovation of the Umberto I Tunnel located in the center of Rome (dimensions: length, 347.7 m; width, 17 m; height: 8.5 m) for which renovation, carried out in the Summer of 2007, was deemed necessary because, among others, of the griminess of the *indoor vault* that was coated with an oily thickness of smog (Figure 49a) [123]. After a thorough clean-up, the tunnel walls were coated with a photocatalytic cement-based paint, with the gray paint applied on both sides of the tunnel (up to 1.80 m height from the road surface) and with a white paint for the remaining surface (total surface, 9000 m^2)—it is not clear, however, what the source of the paints was (Cimax Ecosystem Paint by Calci Idrate Marcellina srl, a photovoltaic product patented by Italcementi [123], or Airlite paint, as claimed by Ref. [143]). Pollution and weather conditions were expected to be less variable in the *indoor vault* of the tunnel than in a typical outdoor environment, thus the researchers thought would facilitate the evaluation of the photocatalytic depolluting action of the photocatalytic grey and white paints with a reduced number of parameters and making the interpretation of the data less complex. During the first campaign of July 2007, the extent of NO_x abatement was determined to be 20–25% in the center of the tunnel, whereas in the second campaign of September–October 2007, the NO_x abatement level was estimated at 51–64% based on a statistical approach [138]. Regardless, Figure 49b [123] displays a photograph of the renovated tunnel before inauguration day of 9 September 2007, while the photographs in Figure 50 show the characteristics of the Umberto I Tunnel in Rome nearly 11 years after the renovation. Evidently, the long-term effectiveness of the photocatalytic product is also an issue here, not to mention the degradation (blisters) of the painted cementitious walls.

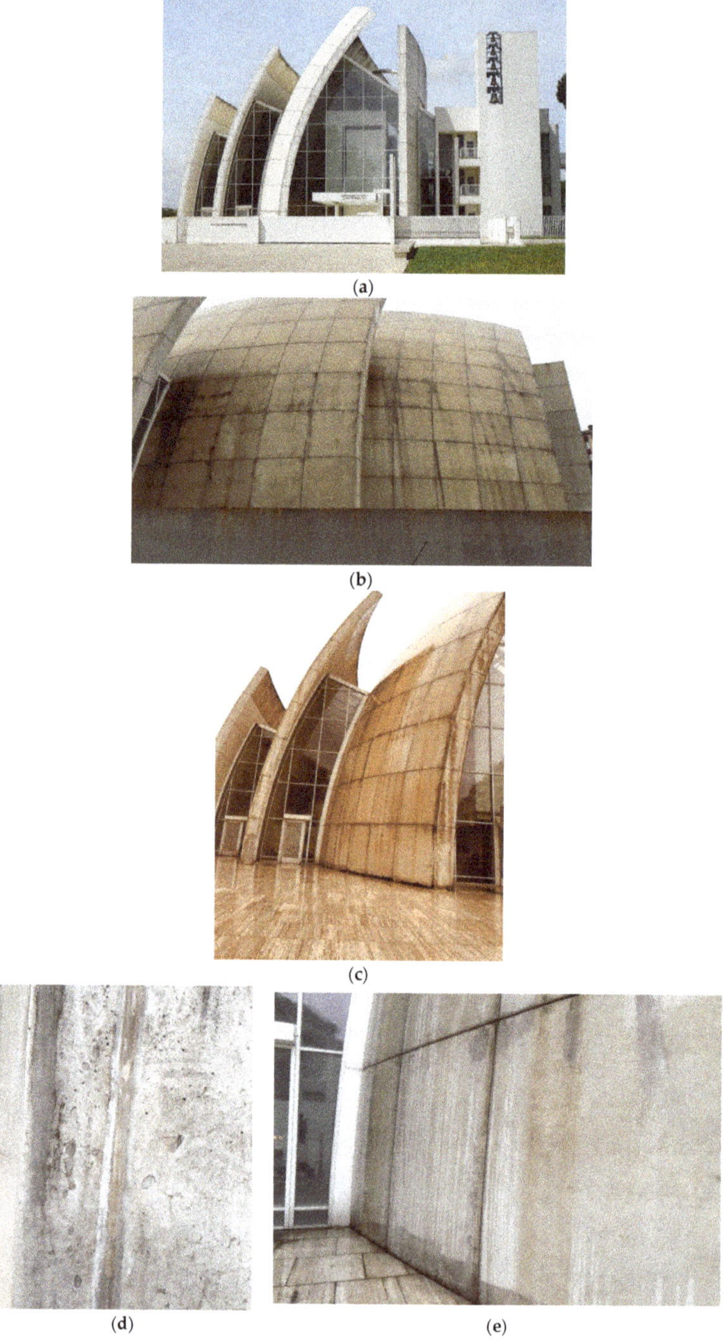

Figure 48. (**a**) Google Earth photograph of the *Dives in Misericordia* Church in Rome, Italy (unloaded Spring of 2017); and (**b–e**) photographs of the *Dives in Misericordia* Church taken on 24 February 2018 (Copyright by N. Serpone). Note the breaking-up of the cementitious layer on the outside sails.

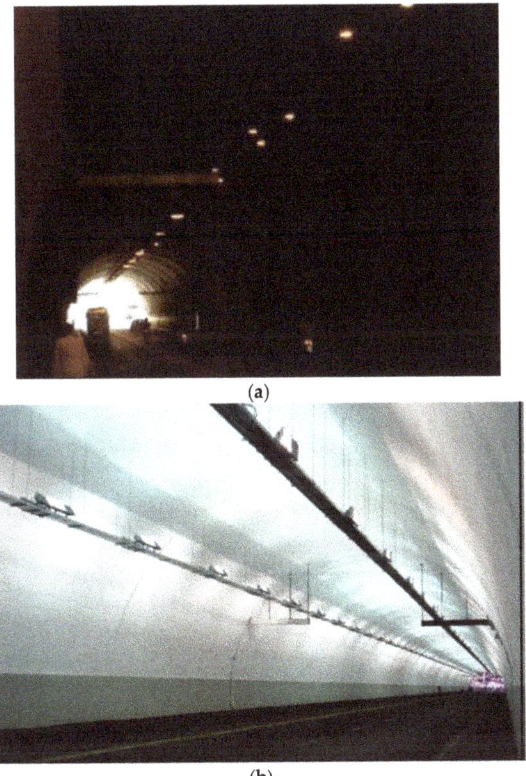

Figure 49. (a) Conditions of the Umberto I tunnel in Rome (Italy) before the 2007 renovation; and (b) tunnel after renovation but before inauguration day of 9 September 2007. Reproduced from Guerrini and Peccati [123].

As the adsorbed nitrates (NO_3^-) formed subsequent to the oxidative conversion of the NO_x gases on the surface of the photocatalytic cement-based paints cannot be washed off from the tunnel's *indoor walls*, a problem arises with regard to the chemistry that occurs on the walls as accumulation of adsorbed NO_3^- on the catalytic surfaces block the catalytic sites from further NO/NO_2 adsorption—thus, not only is the catalytic activity of the photocatalytic paint affected, but also the NO_3^- may be reduced catalytically back to NO_2. In other words, *reNOxification* is likely to occur subsequent to the *deNOxification* process, unless the adsorbed nitrates are removed regularly by washings [132], not possible by natural means in a tunnel configuration unless done manually and periodically by appropriate maintenance of the infrastructure (best done on a weekly basis, also because the lamps will be covered with dirt and dust from the traffic (see, e.g., Figure 57), thereby diminishing the UV irradiance of the lamps—compare Figures 49 and 50).

The principal goal of the European funded PICADA project (Photocatalytic Innovative Coverings Applications for Depollution Assessment) [134] was the development of a range of innovative materials that could easily be applied to various structures (buildings, tunnels, streets, roadside noise barriers, etc.) and that possessed despoiling and depolluting capabilities such as the TiO_2 semiconductor photocatalyst (in powder format) that had earlier shown excellent photocatalytic properties toward the oxidative/reductive destruction of organic contaminants in aqueous media and in air (VOCs), removal of toxic metals (Hg and Pb), recovery of precious metals (Ag, Au, Pt, Pd, and Rh) [144–148], and self-cleaning capabilities when deposited on glass and other supports (in colloidal format). The PICADA project involved a consortium (eight partners) that assembled both industry and research institutes from Greece, France, Italy, Denmark and Great Britain, and whose two main objectives were: (i) to develop and optimize industrial formulations of TiO$_2$ and application methodologies; and (ii) to establish local behavior models under different exposure conditions in realistic urban environments [134].

Figure 50. Photographs showing the status of the Umberto I Tunnel in Rome nearly 11 years after the renovation. Photographs taken 4 March 2018 at 17:00 (Copyright by N. Serpone).

Various cement-based and ready-to-use products within this project involved nano-sized TiO$_2$ and suitable additives/binders with products to be developed with regard to desoiling and depolluting efficiencies, ease-of-use, durability, and cost effectiveness; Table 8 gives a brief description of these products.

Table 8. Preparations and properties of the range of water-based products developed within the PICADA Project.

Products	Notation	Precursor Materials	Properties	Applications
Dry mix (water added on site)	B1	Sand; White ordinary Portland cement; lime; TiO_2; mineral binder	10 mm thick decorative mortar	Façade coatings; Interior applications
	B2	Filler; White ordinary Portland cement; TiO_2; mineral binder	1 mm mineral rendering; 100 µm thick paint	Thin decorative mineral façade coating; paint; Interior applications
Ready-to-use	C1	Siloxane binder; TiO_2; $CaCO_3$ filler	Translucent; resistant to photocatalytic effect of TiO_2	Coating
	C2	Siloxane binder, TiO_2	Opaque coating	Paints
	C3	Acrylic binder; TiO_2	Opaque coating	Paints; Indoor applications
	C4	Silicate binder; TiO_2	Opaque coating	Paints

Laboratory tests of these products toward NO_x destruction were carried out in a 0.45-m glass chamber equipped with a UV lamp 50 cm above the eight petri dishes and a fan to circulate the inlet air/NO_x mixture (NO_x, 200 ppb); the materials displayed significant performance toward destruction of NO_x (especially NO) that disappeared completely at the end of the experiments. By comparison, tests toward the destruction of NO_2 were carried out in a 20 L glass chamber with one petri dish per experiment; UV lamp was located 25 cm above the dish, air/NO_2 mix (NO_2, 200 ppb) was recirculated within a closed circuit; nitrite and nitrate ions were detected. Materials again performed well toward destruction of NO_2 with an efficiency ca. 10 times greater than a control material without TiO_2 [134].

Subsequently, three of the products (the 10-mm thick B1 mortar, the 1-mm thick B2 mineral coating, and the translucent C1 coating) were coated on a 4-m^2 glass surface and tested in a much larger chamber (23 °C; relative humidity, 50%) equipped with a fan to keep the air/NO_x (NO_x, 200 ppb) circulating; lamp was located 150 cm above the samples. Results of the latter tests showed that in all cases ca. 80% of NO was destroyed. The TiO_2-treated B1 mortar was also tested in a pilot-scale three-canyon streets (dimensions: 18 m long, 2 m wide and 5 m high) near Paris; NO_x gas was emitted from an engine into the test and reference canyons by a perforated pipe. The B1 mortar reduced NO_x by 40–80% relative to the reference canyon without the B1 test material [134]; another report with more details is also available [149].

Gurol [150] examined the results of the PICADA project and concluded (at that time) that there were several unanswered questions that needed to be addressed in a laboratory environment with regard to reaction rates, mechanisms, reaction products, fate of reaction products, and types of pollutants; the effect of various variables that include TiO_2 particle size, type of TiO_2, percent loading of TiO_2 in the mixture, thickness of the mixture required (penetration distance of reactions), temperature, humidity, and concentration of pollutants; identification and quantification of reaction products under various indoor and outdoor conditions to *establish that no hazardous chemicals were released from photocatalytic reactions*; determination of the useful life of the TiO_2-containing material; and evaluation of the effectiveness of TiO_2 in a colored matrix as all buildings are not necessarily painted white. In addition, laboratory experiments ought to be conducted in a fully-controlled system that can operate under steady-state and continuous-flow conditions (as opposed to batch lab experiments) to be able to simulate realistic conditions. Furthermore, Gurol [150] recommended that a conceptual process model be developed to describe mass transport and reaction of various pollutants under various realistic scenarios for outdoor atmospheric conditions, together with evaluating the sensitivity of the model to all possible variables and atmospheric conditions. Whether these issues were considered and examined is typically not disclosed by industries, as they too often maintain that the results are proprietary.

Nonetheless, additional concerns with regard to outdoor applications remain to this day: (i) the long-term effectiveness of TiO_2-containing materials; (ii) the possible desorption of pollutants at night time; (iii) the decrease of the effectiveness of TiO_2 over time if adsorbed reaction products and chemicals are not washed off from the surface; and (iv) formation of ozone during the degradation of NO_2 as reported by Maggos et al. [151].

Field studies undertaken in 2005–2009 in the Netherlands by an international panel of air quality experts sought innovative solutions to improve air quality on and around motorways in densely populated bottlenecks (hot spots) along some of the Dutch motorways. To this end, two series of practical trials were conducted using four panels with different TiO_2 coatings fitted to an existing noise barrier along the A1 highway at Terschuur, and on a later erected TiO_2-coated porous concrete noise barrier along the A28 motorway at Putten (Figure 51); measurements of NO_2 and NO_x performed under various weather conditions at both locations showed very low conversion rates of the NO_x gases [152] that were attributed to the short contact time between air and barrier, to the relatively unfavorable meteorological conditions (wind direction and light intensity), to the high relative humidity, and to frequent low temperatures in the Netherlands. The experts concluded that improved air quality with TiO_2 coatings on concrete had not been demonstrated!

Figure 51. Example of a TiO_2-coated concrete noise barrier erected along the A1 and A28 highways in Terschuur and Putten, the Netherlands. Reproduced from Ref. [152].

Road rehabilitation of Petersbergerstraße in Fulda (Hesse, Germany) was undertaken to investigate possible effects on the concentration of NO_2 emission by TiO_2-coated photocatalytic paving stones installed on sidewalks (ca. 800 m) on both sides of the street [153]. Measurements were carried out at two points (Points 1 and 2) across from each other and taken in June 2010 (Point 1) and in October 2010 (Point 2) at two different heights: 10 cm and 3 m above the road surface; for comparison, similar measurements were undertaken beyond the 800-m stretch as a reference. The average traffic volume on this street amounted to about 24,000–30,000 vehicles per day; proportion of light and heavy trucks and buses was 3.2% and 2.1% and 0.2%, whereas the car share was 88%, and motorcycles 6.5%. The whole TiO_2-coated surface was ca. 4500 m^2, a relatively small proportion compared to the remaining road surface of ca. 15,000 m^2. Relative to the reference section, the NO_2-reducing effectiveness appeared somewhat greater directly above the pavement surface (at 10 cm) than at the 3 m height; at Point 2 the effectiveness was, respectively, 17% and 9%, whereas at Point 1 the effectiveness was 3.5% at 10 cm above ground while the NO_2 level remained virtually unchanged at 3 m above. The large fluctuations in the individual values cast doubt as to whether the results were significant [153]. It must be noted that dirt, dust and oily layer on the sidewalks caused by the traffic and pedestrians had a non-insignificant impact on reducing the level of NO_x emitted. By contrast, Cristal Global (producer of CristalACTiV™; anatase TiO_2 nanoparticulate powder or colloidal sols) carried out a trial in 2006 in which TiO_2 was incorporated into paving slabs laid down at either side along the length of a street (covering ca. 1200 m^2) in the Borough of Camden (London, UK); a chemiluminescence analysis of NO_x showed a reduction of ca. 20% (no other details were given) [154].

A field trial conducted at a train station in Manila (Philippines) by coating a 4100-m^2 exterior wall with TiO_2-based photocatalytic paint (Cristal Global) showed that about 26 g of NO_x per 100 m^2 of painted surface was removed; it was also claimed that each painted square meter could remove 80 g of NO_x per year [154,155]. In another trial that ran over a four-year period, a 135 m^2 wall in London was treated with a Cristal photocatalytic paint; the company claimed a reduction of 60% of NO and 20% of NO_2 in the vicinity of the wall (again short in details regarding the exact physical steps taken in the trials). In addition, a depolluting 12-month trial in an indoor car park in 2007 in Paris, in which two Cristal specialist paints were used with a very active photocatalyst within the matrix of the coating, revealed (analysis of monthly nitrate accumulation) reduction levels of NO_x between 53% and 99% depending on paint type and lighting levels [154]. However, another trial carried out, this time in a courtyard behind the Central St. Martin's College of Art & Design (London, UK) by Cristal Global with one of its photocatalytic coatings, showed that, subsequent to chemiluminescence monitoring of NO, NO_2 and NO_x for nearly two years after the application of the photocatalytic coating, NO_x reductions amounted to 35–65% depending on time of year and local weather conditions [154].

With regard to processes that might take place on paint surfaces, Laufs and coworkers [156] examined the photocatalytic reactions of nitrogen oxides (NO_x = NO + NO_2) on commercial TiO_2-doped façade paints in a flow-tube photoreactor under simulated atmospheric conditions in a laboratory setting. Both NO and NO_2 were rapidly converted photocatalytically, albeit only on the photocatalytic paints but not on non-catalytic reference paints. Nitrous acid (HONO) was formed in the dark on all paints examined; however, HONO decomposed efficiently under irradiation only on photocatalytic samples, so that photocatalytic paint surfaces did not represent a daytime source of HONO, contrary to pure TiO_2 surfaces. Formation of adsorbed nitric acid/nitrate anion (HNO_3/NO_3^-) occurred with near unity yield. The mechanism proposed by Laufs et al. for the photocatalytic reactions of NO, NO_2, and HONO carried out on photocatalytic paint surfaces in a laboratory setting is displayed in Figure 52 [156].

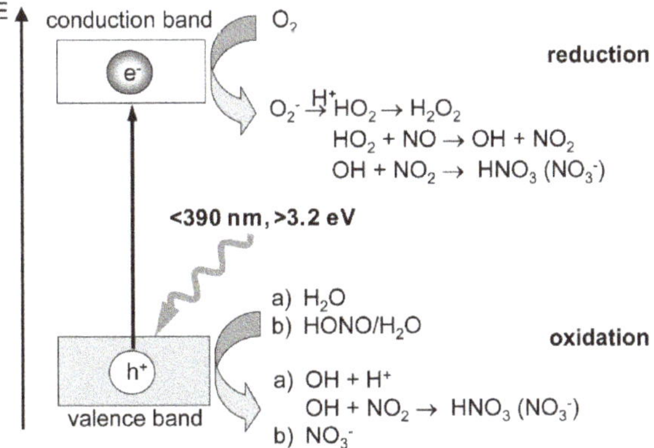

Figure 52. Postulated mechanism for the photocatalytic reactions of NO, NO_2 and HONO on photoactive paint surfaces. Reproduced with permission from Laufs et al. [156]. Copyright 2010 by Elsevier Ltd. (License No.: 4453150462958).

A full-scale outdoor field demonstration of air purifying pavement in Hengelo (The Netherlands) was carried out by Ballari and Brouwers [157] on the full width of a street surfaced with a concrete pavement containing C-doped TiO_2 (Kronos International; 4% w/w TiO_2 water suspension; 50 L of suspension covered a 750 m^2 surface; TiO_2 loading was 2.67 g m^{-2}) sprayed over a length of 150 m; for comparison, another part of the street (ca. 100 m) was paved with normal paving blocks; outdoor monitoring was done during 26 days over a period of more than a year. Prior and during the field measurements, the used blocks were examined simultaneously in a laboratory setting (small flow-through reactor) to assess performance. The first coating applied May 2010 gave good results in the laboratory setting (7.7% of NO and 6.9% of NO_x under visible light illumination) and in the field. NO_x levels were sampled at different heights: 5 cm (near the active surface), 30 cm (car exhaust height) and 1.5 m (the breathing zone) to assess the extent of deNOxing as a function of distance from the active surface. Unfortunately, the TiO_2 photocatalytic coating vanished after the blocks were exposed outdoors for 2.5 months due to normal wear, to vehicular traffic, to weather, and/or else due to solid dust and dirt deposits on the surface. A second coating was subsequently applied to the surfaced road in September 2010; after being exposed for 1.5 months to the street environment, the photocatalytic performance returned to values of the first coating (laboratory testing). On average, the extent of NO_x converted (to nitrates) outdoors determined by the chemiluminescence technique was 19.2 $\pm$ 17.8% (daily readings) and 28.3 $\pm$ 20.0% (afternoon readings)—note the high fluctuations in the samplings, not typically reported by many researchers; under ideal weather conditions (high radiation; low relative humidity) the decrease of NO_x was 45% [157].

We have seen thus far that although NO_x levels can be reduced effectively with TiO_2-based photocatalytic surfaces in an indoor laboratory setting, significant variable results have been more the rule than the exception with regard to NO_x reduction levels in an outdoor urban setting. For instance, NO_x reduction levels of 40–80% [134,136,149], 26–66% [137], and ca. 19% [157] have been reported in various outdoor field trials; however, as we have seen in the field trial in Putten (The Netherlands) [152], the measured NO_x reductions were at or well below detection limits. Such variations are likely the result: (i) of limited contact between the pollutant NO_x and the photo-catalytic surface; (ii) of variable features of field sites; (iii) of prevailing local atmospheric conditions (wind velocity, wind direction, relative humidity, light intensity, etc.); (iv) of the time of measurements; and (v) of the nature and source of TiO_2 photocatalytic products and their associated reactivity upon photo-activation.

Attempts to reduce NO_x levels in an external "indoor" environment such as a road tunnel, as was the case with the Umberto I tunnel in Rome [123,138], were expected to bring about certain benefits: reduced health issues of pedestrians that use the tunnel, no need for a ventilation system, and decrease of the contribution of pollutants to the surrounding areas. To ascertain the benefits of photocatalysis toward NO_x reductions in such indoor infrastructures and as part of field studies within the PhotoPAQ program, Gallus and coworkers [14] undertook an extensive field study of the Leopold II tunnel in Brussels (2011–2013) where photocatalytic cementitious coating materials (first campaign; Italcementi's TX-Active® Skim Coat) were applied on the side walls and ceiling of a test section (73 m long) of one tube of the Leopold II tunnel (Figure 53) [158], followed by a monitoring campaign and later by a third campaign in which the same section was extended to 160 m and covered by a novel photocatalytically more active mortar (Italcementi's TX-Active Skim Coat Boosted) [14]. Installed UV lighting (wavelength range, 315–420 nm) had an average irradiance on the active surfaces of 0.6 $\pm$ 0.3 W m^{-2} (TX-Active; second campaign) and 1.6 $\pm$ 0.8 W m^{-2} (TX-Active Boosted; third campaign).

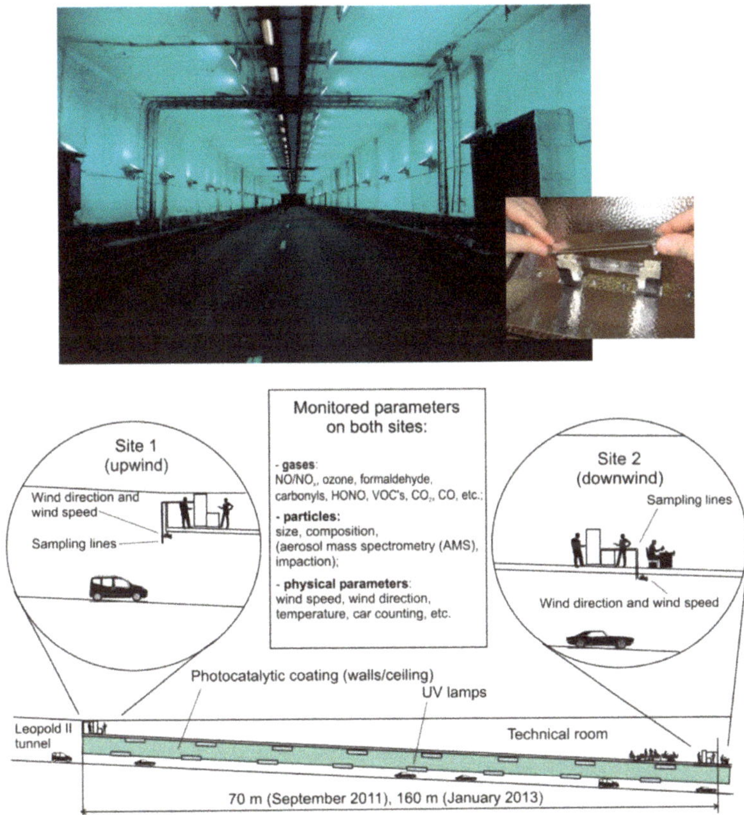

Figure 53. (**upper**) Condition of the Leopold II tunnel after coating the walls and also showing the lighting system; and (**lower**) schematic representation of the test sites in the Leopold II tunnel during the PhotoPAQ field trials. Reproduced with permission from Boonen et al. [158]. Copyright 2015 by Elsevier Ltd. (License No.: 4453110422490).

Most interesting in the field trial of Gallus et al. [14] was the highly recommendable experimental approach used to assess the level of photocatalytic reduction of NO_x gases inside the tunnel based on the realization that changes in atmospheric compositions, traffic flow, and dilution (i.e., different wind speeds) occur inside the road tunnel. Consequently, the authors monitored NO_x concentrations and normalized them to the photocatalytically inert tracer CO_2 gas that was emitted by the vehicular traffic at the same time as the NO_x (and others) to yield NO_x/CO_2 concentration ratios; the instruments used throughout the campaigns had typical detection limits of 1–2 ppb, precisions of ~ 1% and accuracies of ~ 7% and ~ 10% for NO and NO_2, respectively; CO_2 levels were measured by a nondispersive infrared absorption technique: detection limit, 2 ppm; precision, 1%; accuracy, 7%.

The NO_x/CO_2 ratios so obtained from the slopes of plots of NO_x against CO_2 were independent of the absolute pollution level and thus of the emissions from the vehicular source and variable dilution inside the tunnel. Comparison of such slopes obtained in a control setting (dark conditions or otherwise under light illumination but without the TiO_2 photocatalyst) relative to a setting in which the slopes of NO_x versus CO_2 plots obtained under light illumination should show significant variations if NO_x were photocatalytically converted (to nitrates and others, such as HONO). Moreover, changes in the NO_x/CO_2 ratio—e.g., between upwind and downwind of the active section—could then be attributed to a photocatalytic remediation of the pollutants. The demonstration of NO_x levels at different heights within the tunnel were also relevant in their approach; in this regard, Figure 54 shows NO_x levels at Site 2 (under normal traffic conditions, only air at this site was in contact with the photocatalytic surfaces) were sampled at 1.1 m and 3.2 m above the street level and then plotted against the corresponding data at the tunnel ceiling (4.4 m above the street level) [14]. No significant gradients were observed at Site 2 under normal driving conditions with similar concentrations at the different heights above the street (note the slopes of the lines are nearly unity for data at the two heights).

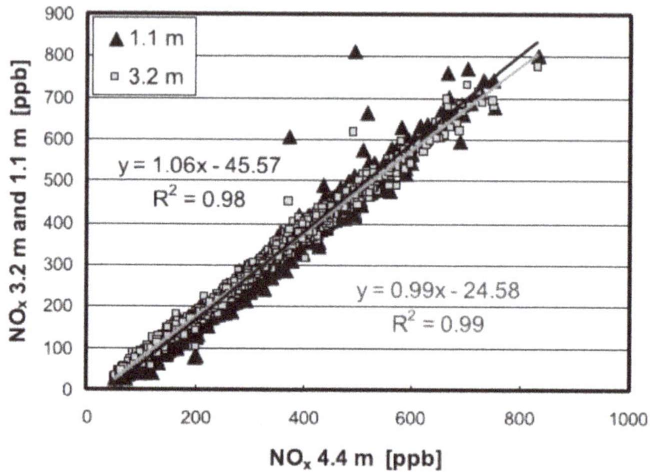

Figure 54. Plot of NO_x concentration (10-min averages) monitored at 1.1 m and 3.2 m above the ground surface against the corresponding data at the tunnel ceiling (4.4 m); data recorded at Site 2 (downwind; see Figure 53 (lower)) during the third campaign (160 m section). Reproduced with permission from Gallus et al. [14]. Copyright 2014 by Elsevier Ltd. (License No.: 4453111244188).

Plotting the NO_x data from the downwind Site 2 of the third campaign against CO_2 data yielded a slope that gave a NO_x/CO_2 ratio of $(3.09 \pm 0.04) \times 10^{-3}$ (Figure 55). More comprehensive results obtained under various conditions are summarized in Table 9 [14].

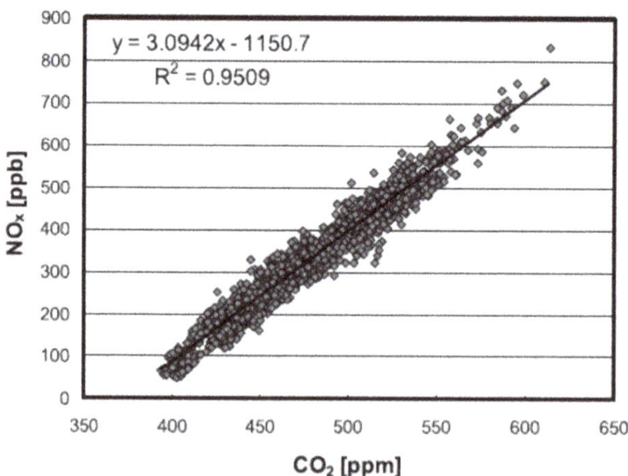

Figure 55. Plot of NO_x against CO_2 data (10-min averages) at Site 2 during the third campaign. Reproduced with permission from Gallus et al. [14]. Copyright 2014 by Elsevier Ltd. (License No.: 4453111244188).

Table 9. Average NO_x/CO_2 ratios obtained from experimental results in the photocatalytic remediation of NO_x in the Leopold II tunnel in Brussels; standard errors represent 2σ **.

Average NO_x/CO_2 Ratios from Site 1 and Site 2 in the Second and Third Campaigns with Photocatalytically-Active Surfaces and with Lights ON			
Campaign	Photocatalytic Material (TiO_2)	Site 1—upwind ($\times 10^{-3}$)	Site 2—downwind ($\times 10^{-3}$)
Second; 09/2011	TX-Active	3.03 ± 0.06	3.14 ± 0.07
Third; 01/2013	TX-Active Boosted	3.18 ± 0.08	3.10 ± 0.05

Average NO_x/CO_2 Ratios for High and Low Wind Speeds with Lights ON in the Third Campaign			
Wind speed		Site 1—upwind ($\times 10^{-3}$)	Site 2—downwind ($\times 10^{-3}$)
Low	<2 m s^{-1}	3.24 ± 0.13	3.23 ± 0.12
High	>2 m s^{-1}	2.93 ± 0.20	2.95 ± 0.10

Average NO_x/CO_2 Ratios at Downwind Site 2 with Lights ON or OFF in the Third Campaign			
Data used		Lights OFF	Lights ON
All		3.09 ± 0.06	3.10 ± 0.05
Low wind speed	<2 m s^{-1}	3.25 ± 0.10	3.23 ± 0.12
High wind speed	>2 m s^{-1}	2.99 ± 0.11	2.95 ± 0.10

** Adapted from Gallus et al. [14]. Copyright Elsevier Ltd. (License No.: 4453120115447).

Comparison of all upwind (Site 1) and downwind (Site 2) data from the second campaign showed a minor formation of NO_x (70 m test section; TX-Active used), whereas a small reduction was inferred at Site 2 from the third campaign (160-m section; TX-Active Boosted used) in qualitative accord with laboratory experiments on sample plates exposed to tunnel air. In the latter experiments, the authors [14] observed photocatalytic formation of NO_x on dirty tunnel samples when the TX-Active material was used (second campaign), whereas only a very small reduction in NO_x was measured on sample plates exposed to tunnel air when TX-Active Boosted was used. An examination of the combined errors of the NO_x/CO_2 ratios together with the precision of the duplicate instruments used ($\pm 2\%$ for NO_x) (see Table 9) led the authors to deduce that the observed differences in results from the second and third tunnel campaigns were insignificant.

To substantiate whether there was a potentially small photocatalytic reduction of NO_x from the upwind/downwind data from the third campaign, the authors [14] further examined the influence of reaction time on the photocatalytic NO_x remediation in the tunnel, in which a greater photocatalytic reduction was expected at lower wind speeds because of longer reaction times (residence times). However, no photocatalytic reduction of NO_x was observed on comparing upwind and downwind data for both sets of wind speeds (or reaction times), thereby precluding any (expected) increase in reduction of NO_x pollution at the lower wind speed. Comparing the NO_x/CO_2 ratios obtained with lights ON (active) and lights OFF (non-active) also showed no quantifiable remediation beyond experimental uncertainty, as also observed when only low wind speed data were used for which highest reduction was expected (Table 9). Accordingly, in view of the precision errors in the data analysis, the authors concluded that the extent of photocatalytic NO_x remediation in the 160 m more active tunnel section from the third campaign was at best $\leq 2\%$ [14].

A theoretical model showed that, in accord with the 20% NO_x reduction observed in the 350 m long tunnel in Rome [123,138], a reduction upper limit of $\leq 20\%$, would have been possible in the 160 m test section of the Leopold II tunnel conditions as relative humidity (RH, 50%), wind speed (1 m s^{-1}) and UVA light irradiance (10 W m^{-2}) were optimal, and deactivation of the tunnel's photocatalytic surfaces was disregarded. Unfortunately, deactivation of these surfaces was not insignificant under the heavily polluted tunnel conditions as demonstrated by laboratory experiments, not to mention that UVA irradiances of 0.6 and 1.6 W m^{-2} were far below the targeted value above 4 W m^{-2}, which was therefore a contributing factor in the deactivation phenomenon and further decreased the photocatalytic activity. Moreover, typical wind speed (ca. 3 m s^{-1} during daytime) and the cold and humid (RH, 70–90%) conditions during the third campaign of January 2013 also caused a decrease of the activity of the photocatalytic material. Interestingly, another simple model calculation [14] that used uptake kinetics determined from laboratory experiments under the polluted tunnel conditions indicated an upper limit of only if such 0.4% for the photocatalytic NO_x remediation, in fairly good accord with the experimental tunnel results of <2%. Not to be neglected, photocatalytic degradation of NO led to significant formation of NO_2, particularly under humid conditions, thereby lowering the expected NO_x reduction even further.

Thus far, the description of some field trials carried out under a variety of experimental approaches and conditions that could not be controlled, as the trials were performed in an open air environment, have led to a large variation in results associated with the extent of TiO_2-based photocatalytic removal of NO_x gases. Accordingly, it is worth summarizing the results, which are collected in Table 10 [14,15,134,136–139,149,152–157,159,160]. Clearly, the results are all over the place.

Table 10. Percent removal of NO_x gases in field trials carried out with TiO_2-based photocatalytic protocols in an open outdoor environment and under a variety of conditions.

Field Trial	Gas	Extent of Photocatalytic Removal	Structures	Year	Ref.
1	NO_x	37–82%	Model street canyon	2008	[136]
2	NO_x	30–40%	Local street	2007	[137]
3	NO_x	20–50%	Street	2012	[138]
4	NO_x	45–57%	Street	2007	[137]
5	**NO_x**	**≤2%**	**Model street canyon, Tunnel**	2015, 2015	[14,15]
6	NO_x	> 20%	Tunnel	2012	[138]
7 (a)	NO	55%	Church external wall	2004	[139]
	NO_2	32%	Church external wall		
8	NO_x	20–25%	Tunnel	2012	[138]
	NO_x	51–64%	Tunnel		
9	NO_x	40–80%	Model street canyon	2006, 2008	[134,136,149]
10	**NO_x**	**Not-measurable**	**Highways' noise barriers**	2010, 2009	[152,160]
11	NO_2	0–17%	Sidewalks/street	2012	[153]
12	NO_x	20%	Street	2006	[154]
13	NO	60%	Building external wall	2006, 2012	[154,155]
	NO_x	20%	Building external wall		
14	NO_x	53–99%	Car park	2006	[154]
15	NO_x	35–65%	Building external wall	2006	[154]
16	NO_x	1–37%	Street	2013	[157]
	NO_x	8–48%	Street		
17	NO_x	26–66%	Real urban street canyon	2007	[137]
18	NO_x	25–30%	Model street canyon	2010	[159]
19	NO_x	40–80%	Model street canyon	2006, 2008	[136,149]
20	NO_x	19%	Real urban street canyon	2013	[157]
21	**NO_x**	**Not-measurable**	**Street**	2012	[153]

(a) Photocatalytic material (TX-Active) tested in a laboratory setting.

It has been reported time and again that photocatalysis, especially with TiO_2-based commercial products, could degrade environmental pollutants as shown, within the present context, by the removal of NO_x gases from the environment not only in a laboratory setting (Table 7), but to some extent also outdoors under environmental conditions (Table 10). Taking artificial model street canyons as examples for the removal of NO_x, we have seen that the reported extent of NO_x removal varies from 25–30% to 40–80% and 37–82%, while in a real urban street canyon values of 19% and 26–66% have been reported (Table 10). By contrast, similar TiO_2-based products have shown no effect in removing environmental NO_x gases in urban streets or on the highways' noise barriers.

All the outdoors field trials expected the NO_x to be transformed into nitrates on the photocatalytic surfaces ultimately to be desorbed when raining. However, there are also reports that other intermediate species are likely to form also such as nitrous acid (HONO) [156,161–164], which is far more toxic than the NO_x pollutants, and not least is the potential for *reNOxification* and formation of ozone from the reaction of adsorbed nitrates with reducing agents (TiO_2 conduction band electrons) [164,165]. For these very reasons and as part of the PhotoPAQ investigative program, Gallus and coworkers [15] structured a two-step campaign to investigate the fate of NO_x gases outdoors in two artificial model street canyons in Petosino (Italy) (see Figure 56). In the first campaign, both model canyons had their side walls covered with a photocatalytically inactive fibrous cement, while in the second campaign one of the canyons had its side walls and ground surface covered with a photocatalytic cementitious coating material (Italcementi's TX-Active Skim Coat Boosted), while the other was used as the reference canyon; the experimental approach used was otherwise similar to the approach used to examine the fate of NO_x in the Leopold II tunnel [14]. The results of their investigations are reproduced herein to demonstrate, what we consider the best approach, their highly recommended protocol and their actual results, not simply the authors' conclusions [15].

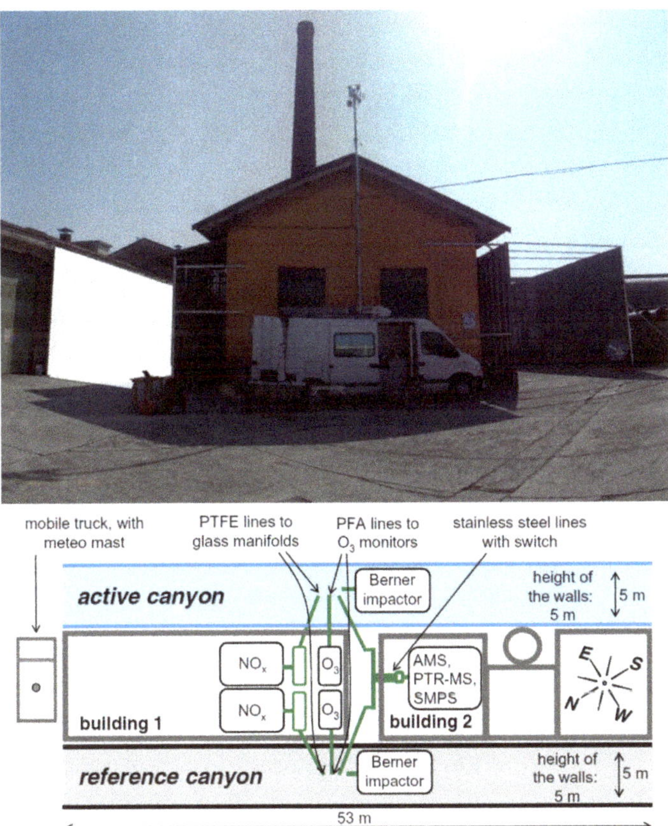

Figure 56. Image and schematic of the two artificial model street (reference and active) canyons with dimensions $5 \times 5 \times 53$ m (width $\times$ height $\times$ length) at an Italcementi industrial site in Petosino near Bergamo, Italy. Reproduced with permission from Gallus et al. [15]. Copyright 2015 by Springer-Verlag Berlin Heidelberg (License No.: 4453120887380).

Results from monitoring and analyzing the NO, NO_2 and NO_x gases (10-min averaged values) from the active canyon were plotted against the results obtained in the reference canyon (Figure 57), with the expectation that differences in the respective slopes (Table 11) between daytime data and nighttime data would reflect the photocatalytic effect independent of any artificial differences. Perusal of the slopes that are summarized in Table 11 indicates that only for the conversion of NO was there a hint of a photocatalytic effect ($-3.5 \pm 3.3\%$). In addition, comparison of the daytime and nighttime results led the authors [15] to infer that less than 2% (that is, $\leq 2\%$) of the NO_x was converted photocatalytically to nitrates (Table 12).

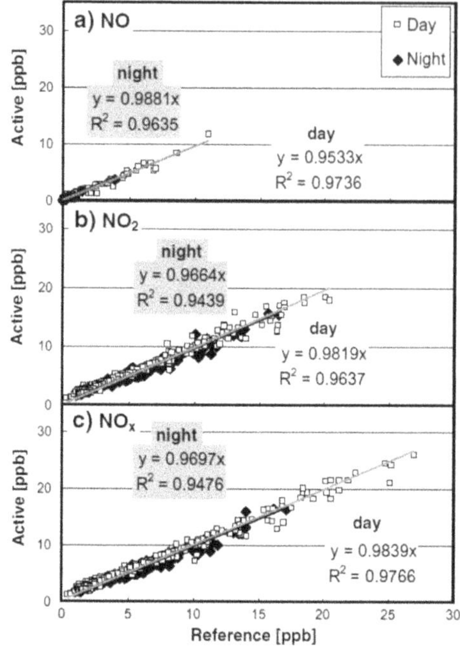

Figure 57. Plots of the 10 min averaged data from the active canyon against the reference canyon for NO, NO_2 and NO_x gases during nighttime and daytime periods. Reproduced with permission from Gallus et al. [15]. Copyright 2015 by Springer-Verlag Berlin Heidelberg (License No.: 4453120887380).

Table 11. Slopes from the plots of daytime and nighttime results for NO, NO_2 and NO_x gases from the active canyon versus the reference canyon; a negative value implies a photocatalytic remediation of that pollutant **.

Gaseous Pollutant	Daytime Results	Nighttime Results	(Day—Night) × 100 (% ± 2σ)
NO	0.953 ± 0.016	0.988 ± 0.029	−3.5 ± 3.3
NO_2	0.982 ± 0.013	0.966 ± 0.014	+1.6 ± 1.9
NO_x	0.984 ± 0.011	0.970 ± 0.014	+1.4 ± 1.8

** Reproduced with permission from Gallus et al. [15]. Copyright 2015 by Springer-Verlag Berlin Heidelberg (License No.: 4453121427505).

Table 12. Average concentrations of NO, NO_2, and NO_x in both canyons for daytime (06:00–20:30) and nighttime (20:30–06:00) and relative concentration differences; negative values imply a photocatalytic effect in the active canyon **.

		Concentrations (ppb ± 2σ)		
		NO	NO_2	NO_x
All data	(1 − (Ref./Act.) × 100 [%]	2 ± 17	−0.7 ± 6.1	−0.4 ± 6.0
Daytime	(1 − (Ref./Act.) × 100 [%]	0.6 ± 11.5	0.2 ± 6.1	0.3 ± 6.0
Nighttime	(1 − (Ref./Act.) × 100 [%]	23 ± 150	−2.4 ± 5.9	−2.0 ± 5.9

** Adapted with permission from Gallus et al. [15]. Copyright 2015 by Springer-Verlag Berlin Heidelberg (License No.: 4453121427505).

The field trial carried out in the Leopold II tunnel in Brussels found that the photocatalytic active surface was deactivated [14]. Accordingly, laboratory tests were undertaken to verify whether deactivation might also have occurred in the canyon trial campaign by taking samples of the active surface in the canyon (before and after) and using a flow-through reactor to examine their photocatalytic behavior toward degradation of the gas pollutants. Results showed that no deactivation of the canyon surfaces had occurred. Consequently, the authors [15] surmised that this was due to several evident differences in the model canyon campaign versus those from the Leopold II tunnel; that is,

1. Much lower pollution levels exist in the canyon than in the tunnel; in the latter case, the greater pollutant level (very dirty ambient conditions–dirt/dust/grime/air pollutants) blocked the active sites (see Figure 58).
2. Much higher UV irradiance was available in the canyon (sunlight UV in the canyon versus an artificial UV light system in the tunnel) causing the pollutants to degrade faster.
3. In the tunnel case, the cementitious surfaces were allowed to cure in the dark for several days, whereas photocatalytic and non-photocatalytic surfaces in the canyons were freshly exposed to sunlight UV radiation.

Figure 58. Sample of a photoactive TiO_2-based cementitious surface before and after one week in the Leopold II tunnel in Brussels. Courtesy of Dr. Falk Mothes of the Leibnitz Institute for Tropospheric Research (TROPOS), Germany.

Even though they expected to confirm NO_x removal levels of ca. 20%, if not greater (see Table 10), as reported in earlier field trials in open atmospheric environments, the findings that there was no significant photocatalytic remediation of NO_x (and VOCs, O_3, and particulate matter)—removal level of ca. 2% or less—was somewhat enigmatic to say the least [15], albeit consistent with non-measurable NO_x reduction levels found in highways (The Netherlands) and in a real urban street (Fulda, Germany). Such minute NO_x reductions were likely due to low surface-to-volume ratios (S_{active}/V) of the open structures; note that the S_{active}/V ratio is a very significant factor that reflects the available active surface versus the space volume occupied by gaseous pollutants, i.e., the limiting heterogeneous uptake of a species onto a solid surface [166,167], as is also the associated deposition velocity. Evidently, other factors must have played a role, and/or some differences in experimental approaches might aid in explaining such very low levels of NO_x reductions, namely:

1. Although the overall geometry of the PhotoPAQ canyon model was otherwise similar to earlier model street canyons, the dimensions were different with respect to smaller model sites (e.g., 5 m wide canyon versus the earlier 2.4-m wide canyons that led to unrealistically higher S_{active}/V ratios and as such to higher NO_x reduction levels of 20–80% (Table 10) versus ≤2%. Note that the large reduction levels of NO_x on photocatalytic TiO_2 surfaces experienced in laboratory settings (Table 7) are also explained by this surface-to-volume ratio factor which is considerably greater: typically, 167 m^{-1} for a small flow-through reactor in a laboratory setting versus 0.6 m^{-1} for the model street canyon in the open environment in the industrial zone of Petosino (Italy). If one were to scale down the earlier results [134,136,149,166,167] to real urban street canyon conditions, it would result in only approximately 5% NO_x reduction.
2. Previous field trials considered results of monitoring the fate of NO_x pollutants during daytime hours only, contrary to other field trials where the whole diurnal data were considered. NO_x levels during nighttime were often comparable to those during daytime (higher emissions during the

day were often compensated by stronger convective vertical dilution) so that the above estimate of ~ 5% based on daytime data was further reduced to ca. 3%, comparable to expectations typical of a main urban street [15].

3. The distance at which the NO_x gases were sampled from the photoactive surface is another important factor. Gallus et al. [15] recommended sampling NO_x at a distance of 3 m above the photoactive surface for urban network stations, contrary to earlier field trials where sampling of NO_x was done but a few centimeters from the active surface (5 cm to 1.5 m [148]; 30 cm to 1 m [137]; 50 cm [166,167])—indeed, since conversion of NO_x gases to nitrates requires the NO_x to be adsorbed onto the photocatalytically active surface, it is obvious that close to the surface is where the greatest change in NO_x concentration will be felt—this means that, as you stroll on a photocatalytic street surface, your feet will experience a healthier environment than your head.

4. High reductions of NO_x levels in earlier field trials [134,137,149,157,166,167] likely reflected some differences between active sites and reference sites in terms of the quantity of NO_x artificially injected into the active and reference canyons [134,136,149,166,167], whereas the field trial in the Petosino model canyon the NO_x gases were those present in the homogeneous industrial environment [15].

5. Where the active and reference model street canyons were distant from each other [137,157] may also have resulted in non-insignificant differences in the quantity of NO_x artificially injected and in pollution dispersion (e.g., different wind speeds and different wind direction), thus leading to non-insignificant uncertainties in the results.

With regard to the latter issue, model studies have re-evaluated the daytime results from the experimental street canyon in Bergamo and determined that the upper limit of photocatalytic NO_x remediation was more like 4–14% [160], rather than the claimed 26–66% [137], because of: (a) strong differences in the vehicle NO_x emissions at the active and reference sites with much higher vehicle fleet density at the reference site; and (b) different dispersion conditions (geometry of sites, micrometeorology). However, Gallus et al. [15] contended that, if in the model studies one were to assume a reasonable wind speed of $1 \ m \ s^{-1}$ and a realistic photocatalytic deposition velocity of NO_x on the active surface of $0.3 \ cm \ s^{-1}$, the upper limit of 4% [160] would be reduced further to <2% considering diurnal averages and transport limitations and more in line with real environmental results [152,153,168]. Clearly, the above factors and discussion call attention to the necessity of some care into how the data are treated and how the data are interpreted.

Several TiO_2-based photocatalytic materials were also tested in 2015 by Pujadas and coworkers [169] of CIEMAT (Madrid, Spain) within the framework of the 2013–2018 LIFE $MINO_x$-STREET European project. Tests were first carried out in a laboratory setting to choose the most photoactive TiO_2-based material (fresh sample showed 45% reduction of NO in a flow-type reactor, dropping to 20% for a used street sample caused by traffic and ageing over a month) so designed as to be used on bituminous mixtures with which to examine the effect of reducing NO_x levels in the two-way Boulevard Paseo de la Chopera, a real urban setting in the Municipality of Alcobendas (Madrid)—conditions were: ground level wind speed, $2 \ m \ s^{-1}$; relative humidity, <65%; solar radiation, >400 W m^{-2}; TiO_2-based coated active area, 1000 m^2; length of section, ca. 60 m. The experimental approach was similar to that used in the PhotoPAQ project in that the daytime and nighttime NO_x collected data in the median strip at two active sites (Site 2 and Site 3) and at the inactive site (Site 4) were plotted against NO_x data collected at the inactive Site 1; the plots yielded the slopes reported in Table 13 [169]. The similarities between the data collected in daytime and at nighttime during the most optimal measurement periods before and after application of the photocatalytic coating led the researchers to deduce that there was no photocatalytic reduction of NO levels, in line with observations made by others in a real urban setting (see above).

Table 13. Slopes from correlation plots of NO data (ppb $\pm$ 1σ) from three sampling sites (two active, one inactive) against inactive Site 1 for NO concentrations before and after implementation of the photocatalytic coating.

	Diurnal		Nocturnal	
Sampling Site	**Before Applying TiO$_2$-Based Coating**	**After Applying TiO$_2$-Based Coating**	**Before Applying TiO$_2$-Based Coating**	**After Applying TiO$_2$-Based Coating**
2 (active)	1.05 $\pm$ 0.08	1.07 $\pm$ 0.08	1.014 $\pm$ 0.005	0.983 $\pm$ 0.004
3 (active)	1.41 $\pm$ 0.17	1.03 $\pm$ 0.06	0.977 $\pm$ 0.008	0.915 $\pm$ 0.004
4 (inactive)	0.89 $\pm$ 0.06	0.90 $\pm$ 0.07	0.985 $\pm$ 0.005	0.946 $\pm$ 0.025

A real scale examination of NO$_x$ depollution by TiO$_2$-based photocatalytic sidewalk pavement and façade was performed by Pujadas and coworkers of CIEMAT [170] in an urban environment that resembled a model street canyon, albeit narrower, in which they also found no significant NO$_x$ reduction during their experiments in the sidewalk model in open air, even though in a laboratory setting the extent of NO$_x$ removal by the photocatalytic material was 65%. Nonetheless, a small ambient NO$_x$ reduction effect was observed on the photocatalytic façade (mimicking a building brick wall), although for only a very short time and very close to the active surface (e.g., <10 cm), and then only under very specific ambient and meteorological conditions.

At a presentation of their research at a conference in Barcelona (Spain) that was performed within the MINOx-STREET project, Palacios and coworkers [171] described results on urban street experiments, as well as experiments with model sidewalks and model building façades. Two principal conclusions are worth noting from that conference:

1. A great difficulty was experienced in urban scenarios to establish a possible cause-effect relationship between any observed ambient NO$_x$ reductions in the presence of photocatalytic surfaces. In road bituminous pavement (the Alcobendas experiment) and sidewalk scenarios, the NO$_x$ removal effect of photocatalytic materials had not been unequivocally demonstrated experimentally; however, in the case of the façade scenario that effect was documented, albeit the effect was seen only at distances very close to the wall surface.

2. Physicochemical characteristics of heterogeneous photocatalysis, such as low quantum yield for absorption of solar radiation by TiO$_2$, heterogeneous molecular processes, and the high dependence on ambient conditions helped to explain the weak macroscopic effects observed in open urban air. Consequently, a low NO$_x$ concentration reduction was only attained very close to photoactive surfaces with poor global incidence on ambient air.

3. The CIEMAT group [171] also investigated the NO$_x$ depolluting effect of TiO$_2$-based photocatalytic materials in a medium-scale tunnel reactor under semi-controlled conditions using 200 ppbv of NO and compared the results with those from a real-scale outdoor tunnel (street in Alcobendas) and from a laboratory-scale reactor—Table 14 summarizes the parameters and the extent of NO removed in all three cases [171].

Table 14. Parameters and results of NO removal in a medium-scale tunnel reactor (UVA irradiance, >40 W m^{-2}; relative humidity, <30%; dimensions, 0.4 $\times$ 0.4 $\times$ 10 m; photoactive surface, 0.4 $\times$ 10 m) compared to a real-scale tunnel and a small laboratory flow-through reactor.

Parameters	Outdoor Real-Scale Tunnel Reactor	Outdoor Medium-Scale Tunnel Reactor	Laboratory Flow-through Reactor
Deposition velocity (m s^{-1})	5.25 $\times$ 10^{-3}	2.05 $\times$ 10^{-3}	5.25 $\times$ 10^{-3}
Length (m)	60	100	0.1
Air velocity (m s^{-1})	1	0.33	0.20
Photoactive surface (m^2)	1000	4	0.005
S$_{active}$/V (m^{-1})	0.067	2.5	200
Residence time (s^{-1})	60	30.3	0.5
First order rate coefficient (s^{-1})	3.4 $\times$ 10^{-4}	5.07 $\times$ 10^{-3}	1.1
Average yield of NO removed (%)	~ 2	15 $\pm$ 4	41

6. Concluding Remarks

One of the objectives of this review article is to provide an introduction to the basic approaches and terminology inherently and commonly used in heterogeneous photocatalysis. Relevant characteristic features of solid semiconductor/insulator photocatalysts are examined, together with some initial and subsequent events that follow the absorption of photons by these solids. The relationship(s) between traditional heterogeneous catalysis, on the one hand, and photochemistry and molecular spectroscopy, on the other hand, with heterogeneous photocatalysis have been made. The closer interconnection of photocatalysis to photochemistry is emphasized. Where similar symbols are used to mean different things, they are nonetheless fully described in the context used. The other objective is to examine closely the various attempts made to apply the photocatalytic technology—albeit briefly with regard to the published literature on environmental remediation of contaminated ecosystems—through incorporation of the extensively investigated TiO_2 photocatalyst into various cementitious substrates (concrete, mortar, plaster, paints, etc.) applied to various infrastructures (highways, urban streets, building external walls and road tunnels, among others) and through the use of various coating materials.

With regard to the latter aspect, we have witnessed the external conditions of the *Dives in Misericordia Church* and the *Umberto I tunnel* in Rome (Italy) over a decade later from when the photoactive coatings were first applied to these structures. Evidently, deactivation of TiO_2-based photoactive surfaces may become an issue over time so that, as Gallus et al. [15] aptly noted, experiments should be carried out at the field site of interest in small scale experiments exposed to the prevailing atmospheric conditions of the site—and not in a laboratory setting miles away from the application—before giving any consideration to apply the photocatalytic technology to surfaces of large infrastructures. Only when these surfaces display high activity towards NO_x (NO and/or NO_2) removal, in particular, and degradation of VOC pollutants, in general, and no strong deactivation of the photoactive surface occurs under the prevailing specific field conditions, should photocatalysis be advocated as a possible technology to reduce NO_x and VOC levels in open urban environments.

We have also seen that the photocatalytic technology has demonstrated significant removal of NO_x pollutants in a laboratory setting, whereas the overall results from real-scale application in an open air environment have been, to say the least, highly variable, but quite disappointing at best even under appropriate approaches in carrying out the experiments. Such deviations from a laboratory setting to a real outdoor environmental setting is due to several factors that advantage the laboratory results: (1) a small flow-through photoreactor was used in the laboratory tests against a vast open environment reactor system; (2) the photocatalytic surface used in the laboratory was a clean surface against an outdoor surface that in relatively short time is ultimately covered with dust, filth etc., thereby blocking the surface active sites where the reaction between the NO_x and the oxidizing entities at these sites takes place—while rain might wash off the nitrates, the dust, and the filth from the open outdoor, that is not the case where the indoor walls of tunnels are concerned; (3) the rate of flow of the NO_x gases could be controlled in the laboratory tests, unlike the open environment where no control of wind speed and/or its direction are possible; (4) the UV/Visible radiance could be controlled in the laboratory versus lack of control of the sunlight impinging on the photocatalytic surfaces; (5) unlike the open environment, in laboratory tests, the flow rate of the NO_x gases in the photoreactor could be controlled which, therefore, affected the deposition velocity of the NO_x onto the photocatalytic surface (i.e., a competition between horizontal flow versus vertical flow) as well as the residence time spent by the NO_x gases within the reactor; (6) the S_{active}/volume ratio in the laboratory reactor was far more advantageous than it could be in the open environment; (7) the relative humidity within the laboratory reactor could also be controlled, unlike the open environmental reactor; and (8) laboratory tests were carried out for a relatively short period (minutes to a few hours) within which the photocatalytic surface retained its integrity versus the open environment where the wished-for expectations were that the photocatalytic surfaces should remain active for months if not years and beyond.

Nonetheless, despite the disadvantages experienced in the open environment, there is no reason to throw away the baby with the bath water. It suffices to carry out cost-efficiency analyses (as they may be geographically different) and compare them with other technologies available out there—for example, selective catalytic reduction (SCR) and selective non-catalytic reduction (SNCR) together with natural means (e.g., trees on road sides). Photocatalysis may yet be found more attractive on a cost-efficiency basis compared to other, perhaps more expensive methods if extra costs of photocatalytic surfaces associated with application to normal urban surfaces (roads, paints, roof tiles, etc.) were minimized by the industrial sector in the future, and if photocatalytic materials were applied when urban infrastructures were being renovated [15].

It cannot be overemphasized that the effective use of photocatalysts in cementitious substrates in the *deNOxification* of the environment requires serious considerations of some fundamental physical and chemical notions prior to any attempt at implementation of the TiO_2-based photo-catalytic methodology, and as appropriately emphasized by Macphee and Folli [129], the following concerns are worth noting and sharing:

1. Photocatalysis is a surface phenomenon that is impacted by the chemistry of the immediate environment, so that the cementitious photocatalytic surface must be engineered to maximize light absorption by the photocatalytic surface and for the reactants to access the surface.

2. Redox potentials of pollutants considered for elimination must match the semiconductor band edge positions (i.e., flatband potentials) of the conduction and valence bands for a successful deNOxification as the band edges are pH-dependent, especially for a metal-oxide photo-catalyst.

3. More fundamental research is needed toward optimizing photonic efficiencies of visible-light activation of photocatalysts if anion- and/or cation-doped photocatalysts were used, as geographical locations may be limiting the usage of photocatalytic concrete if conventional pristine TiO_2-based surfaces were used.

4. There is a need to maximize both the dispersion of photocatalytic particles in the cementitious substrate and the photoactive surface area toward target applications and agglomeration of particles in a porous structure should be minimized as it could block access of larger pollutant molecules to the internal photoactive surface.

5. Oxidative removal of NO_x pollutants by their conversion to nitrates occurs only if they are adsorbed on the photocatalytic surface; recall that the adsorption mode of molecules on the photoactive surface is highly pH-dependent (surface charge) and could thus affect the oxidative efficiency.

6. The $deNO_x$ index that combines photonic efficiency and selectivity into one environmental impact parameter for a given photocatalytic material should be considered whenever screening candidate photocatalytic materials; for instance, if this index for a photoactive surface were negative shows that the oxidative removal of NO favors formation of the more toxic product NO_2 rather than the NO_3^- species, as demonstrated for a conventional unselective TiO_2 (e.g., bare P-25 titania) that exhibited a large negative $deNO_x$ index [129], a sign that the catalytic process was ineffective. Consequently, both photocatalyst selectivity and photo-catalyst activity must be assessed to reduce emission of harmful by-products.

7. On the more fundamental side, the thermodynamics (energetics) are the most valuable tool in understanding and designing selective photocatalytic processes. Recall that, after the rapid e^-/h^+ pair recombination, the next fastest process is charge carrier trapping. This calls attentions that not only is the semiconductor band edge positions relevant, but also charge transfer from traps to adsorbate molecules need to be fully appreciated in interpreting photocatalytic processes as also the engineering and design of band structures of single semiconductors or semiconductor composites.

Additionally, using a tunnel as an example of an infrastructure in need of eliminating NO_x agents, Gallus et al. [14] also expressed the notion that a reasonable photocatalytic remediation of NO_x in a road tunnel can only be obtained under suitable/special experimental conditions, as photocatalytic remediation will not be effective where the photoactive materials are strongly deactivated under highly polluted conditions. Accordingly, photocatalytic materials should always be tested in small-scale bed-flow reactors on site using: (1) the UVA irradiance equivalent to that of the tunnel; and (2) the NO_x present in ambient air sucked from the tunnel as the reactant.

Only if NO_x were efficiently decomposed under these conditions, would larger scale applications in a road tunnel be worth considering. Regardless of the possible deactivation of photo-catalytic materials under high pollution conditions, model calculations have shown that applications of such materials are unlikely to result in significant remediation of NO_x to nitrates at low UVA irradiances and where high wind speeds and/or high relative humidity prevail. With regard to NO_x conversions in road tunnels, this calls further attention to the need for a careful characterization of tunnel conditions, for a quantification of possible deactivation of the photocatalytic materials, and for performing simple upper limit model calculations of expected NO_x reductions, together with a cost–benefit analysis well before considering application of photocatalytic materials and installation of costly UV lighting systems. These same considerations also apply to pollutant remediation in other infrastructures.

7. Recommendations

In practical applications of photocatalysis, conventional TiO_2-based photocatalytic surfaces have been used to oxidize NO_x to nitrate species; the latter species do not desorb spontaneously and consequently deactivate or block the surface-active centers of the photocatalyst from carrying out the next cycles. To avoid such deactivation, the nitrates (or nitric acid) should typically be washed away by rain [172]; however, the nitric acid is corrosive and could pollute the soil when its concentration at the site becomes too high. A promising way to resolve this problem, which the users of applied photocatalysis have failed to consider but known to occur in a laboratory setting since the first report by Courbon et al. [92] in 1984, would be to change the selectivity of the photocatalytic reaction so that the NO_x gases are converted back to N_2 and O_2 by some photoreduction pathway as reported some time ago by Anpo and coworkers [94,95], who used Cu^+ ions in SiO_2 or in zeolite to effect the photocatalytic reduction of NO_x. No deactivation of the active sites would occur for this photoreduction reaction since nitrogen and oxygen readily desorb from the surface [173]. The selectivity toward the photoreduction of NO could be improved greatly by reducing the hexacoordinated Ti^{4+} species (TiO_6 octahedra) to tetracoordinated Ti^{4+} species (TiO_4 tetrahedra) [174], as successfully achieved by depositing isolated TiO_4 clusters inside cavities of zeolite-Y using ion beam implantation [174,175].

In a more recent article, Wu and van de Krol [176] proposed a novel strategy to change the photocatalytic selectivity of TiO_2 by creating a large and stable concentration of oxygen vacancies in TiO_2 nanoparticles through thermal reduction in a reducing atmosphere; these oxygen vacancies were stabilized by doping the TiO_2 nanoparticles with an electron acceptor-type dopant such as Fe^{3+} which also greatly enhanced the activity of the photoreduction process. The authors [176] further showed that with this strategy NO was indeed photoreduced to N_2 and O_2 and that photooxidation of NO was largely suppressed. Moreover, photoreducing Fe^{3+} to Fe^{2+} provided a recombination pathway that suppressed nearly quantitatively the formation of NO_2 and consequently enhanced the selectivity of the reaction for N_2 formation [176]. The authors also alluded to formation of N_2 and O_2 via two different routes. One route would see a small amount of tetrahedrally coordinated Ti formed in the Fe-doped TiO_2 samples, which Anpo et al. [175,177] claimed as the active site for the catalytic decomposition of NO to N_2 and O_2 at Ti-modified zeolites. As most Ti^{IV} ions at the TiO_2 surface are fivefold-coordinated, a single oxygen vacancy created at or near the surface could lead to a fourfold-coordinated Ti^{4+} center; however, this would require a strong reduction of the Ti–O bond length that would be possible only at very high oxygen vacancy concentrations, which the authors [177] deemed an unlikely pathway and proposed the other route that implicated oxygen vacancies acting as the catalytic centers through the capture of the oxygen side of NO as illustrated in Figure 59 and summarized in Reactions (101)–(104); the associated experimental data are also displayed in Figure 59.

$$V_o^{++}{}_{(surf)} + 2\,e^- + NO_{(g)} \rightarrow O_{surf} - N \tag{101}$$

$$2\,O_{surf} - N \rightarrow 2\,O_{surf} + N_{2(g)} \tag{102}$$

$$2\,O_{surf} \rightarrow 2\,V_o^{x} + O_{2(g)} \tag{103}$$

$$V_o^{x} + 2h^+ \rightarrow V_o^{++} \tag{104}$$

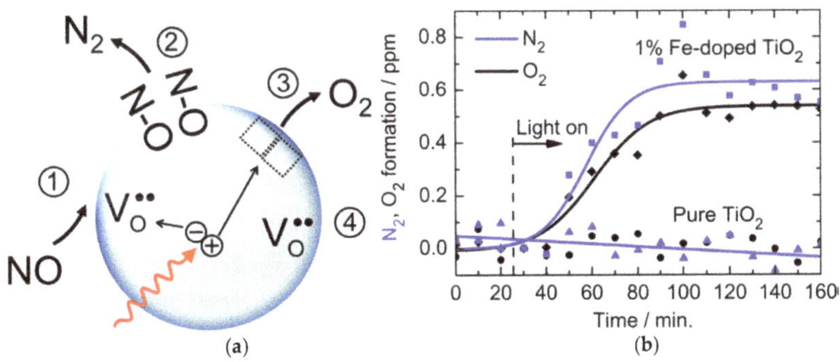

Figure 59. (**a**) Cartoon illustrating the possible pathway to reduce NO to N_2 and O_2 gases through the involvement of oxygen vacancies; and (**b**) photocatalytic conversion of NO to N_2 and O_2 over 1% Fe-doped TiO_2 under irradiation with UV light; concentration of NO: 100 ppm in He. Reproduced with permission from Ref. [176]. Copyright 2012 by the American Chemical Society.

Summing Reactions (101)–(104) yields the overall Reaction (105).

$$\begin{aligned} &\quad\quad 2\,V_o^{++} \\ 2\,NO_{(g)} + 4\,h\nu &\rightarrow N_{2(g)} + O_{2(g)} \end{aligned} \tag{105}$$

Although the conversion efficiency was somewhat modest (ca. 4.5% after 1050 min correspond-ing to a TON of ~2 NO molecules per O vacancy site), the Fe-doped TiO_2 photocatalyst showed no signs of deactivation as the NO conversion centers were not blocked by nitrate species [168], contrary to standard deNO$_x$ TiO_2-based photocatalysts that have to be washed away periodically.

In a most recent article, Cao and coworkers [178] investigated the adsorption of NO and the consequent reactions on differently treated rutile $TiO_2(110)$ surfaces using polarization/azimuth-resolved infrared reflection absorption spectroscopy. Apparently, surface defects (e.g., oxygen vacancies, V_o) and reconstructions on $TiO_2(110)$ had a strong effect on the reaction pathways of NO $\rightarrow$ N_2O conversion (N_2O is laughing gas). The pathway proposed involved a defect-free oxidized $TiO_2(110)$ surface in which two NO molecules are adsorbed on adjacent surface-pentacoordinated Ti (Ti_{5c}) sites first, which then couple to form a cis-$(NO)_2$/Ti&Ti dimer through the N$-$N bond of the dimer, and then are converted to N_2O species (or perhaps even to N_2 gas) [178].

Clearly, much fundamental research in TiO_2-based photocatalysis needs to be undertaken in the optic toward applications to environmental *deNOxification*, with special attention and efforts directed at titania doped with Fe, Cr, Co and Ni dopants that may yet prove interesting [176].

As a case in point, a recent article by Kuznetsov and coworkers [70] examined possible additional specific channels of photoactivation of solid semiconductors with regard to thermo-/photo-stimulated bleaching of photoinduced Ti^{3+} color centers in visible-light-active (VLA) photo-chromic rutile TiO_2, which an optical emission spectroscopic analysis had shown to contain 99.4 at.% Ti and 0.2 at.% Al as the principal impurity, together with 0.09 at.% Fe, 0.05 at.% Sn, 0.04 at.% Nb, and 0.03 at.% Cr as minor impurities. Considering that the prime photophysical process of photostimulated bleaching of Ti^{3+} color centers is absorption of light quanta by the Ti^{3+} centers, the authors [70] found that no selectivity of photostimulated bleaching of a certain type of Ti^{3+} centers could be ascertained, and that photogenerated holes captured at a set of traps were also participants in the photostimulated bleaching of these color centers. Based on current findings and earlier results, the authors hypothesized that the heat released during nonradiative electron transitions, following the prime photophysical processes of excitation and ionization of Ti^{3+} centers, dissipates in the nearest neighborhood of the Ti^{3+} centers and that localized nonequilibrated excitation of the phonon subsystem leads to thermal detrapping of the photoholes with different depths up to 1 eV. Subsequent recombination of free holes with trapped electrons from Ti^{3+} centers leads to the observable photostimulated bleaching of the color centers [70]. Based on experimental evidence, the authors further argued that following absorption of vis–NIR light by the color centers, the subsequent release of thermal energy accompanying nonradiative electron transitions provides an additional specific channel to photoactivate the VLA rutile TiO_2, in particular, and possibly other photocolorable metal-oxide semiconductors as well.

Following their interest of the photophysics of color centers in VLA rutile titania ceramics and titania powder resulting from the photoformation and separation of charge carriers, Kuznetsov et al. [179] noted that the action spectrum of the photoformation of Ti^{3+} centers at very low temperatures (90 K) accorded fully with the absorption spectra of intrinsic defects that consisted of a set of individual absorption bands that they attributed to several different Ti^{3+} centers. Analysis of the dependencies of the photoformation of separate centers on the wavelength of illumination and light exposure, which provided extraction of specific Ti^{3+} centers, led the authors to identify Ti^{3+}-based centers with excessive negative charge that formed at significantly high concentration upon maximal exposure of the titania specimens to Vis-light illumination: $(2Ti^{3+} + V_o^{2+}) \longleftrightarrow (Ti^{\delta+} + V_o^{2+})$ with $3 > \delta > 2$. They also showed from thermoprogrammed annealing (TPA) that the spectra of Ti^{3+} color centers in the range 90–500 K consisted of a set of first-order peaks corresponding to traps, whose depths ranged from ~ 0.2 eV (peak at 130 K in the powder specimen) to 1.06 eV (peak at 455 K in the ceramics specimen). The highest rate of recombination of holes released to the valence band with Ti^{3+} centers—an event attributed to $Ti^{\delta+}$ centers—provided TPA spectra that clearly manifested the existence of shallow traps. In addition, mass spectrometric experiments on the photoadsorption of molecular oxygen and photodesorption of photoadsorbed oxygen from the surface of powdered VLA titania specimens provided further evidence of the photoformation of electrons and holes in VLA TiO_2 under Vis-light illumination, and allowed the authors [179] to determine the kinetics of photo-desorption of O_2 under orange light illumination subsequent to photoadsorption of O_2 stimulated by blue light excitation. Those experiments provided further proof of the occurrence of another specific channel toward the photoactivation of VLA TiO_2 via photoexcitation of photoinduced Ti^{3+} color centers.

It is important to recognize that Ti^{3+}-based centers (i.e., $Ti^{\delta+}$ centers) appeared after many other Ti^{3+} centers had already been formed. In other words, such Ti^{3+}-based centers appeared at high density of Ti^{3+} centers (see below). Accordingly, specific properties of Ti^{3+}-related centers responsible for the (*extrinsic*) absorption bands at 1.56 eV and 1.26 eV were postulated to account for the excess negative charge characteristic of such Ti^{3+}-based centers. In line with the work of Déak and coworkers [180], the two adjacent Ti^{3+} centers located near a single oxygen vacancy forming a $(2Ti^{3+} + V_o^{2+})$ complex (Figure 60a) were taken by Kuznetsov et al. [179] as extra charged Ti^{3+} centers when compared to isolated Ti^{3+} centers (Figure 60b). The two Ti^{3+} species in the $(2Ti^{3+} + V_o^{2+})$ complex can, in principle, disproportionate to $(Ti^{2+} + Ti^{4+} + V_o^{2+})$ so that, in accord with the more generally accepted view, these extra-negatively charged Ti^{3+} centers are best referred to as $Ti^{\delta+}$ centers for which $3 > \delta > 2$. Clearly, the appearance of such $Ti^{\delta+}$ centers, whether photogenerated or resulting from the removal of the structural oxygen during a reduction event, is of lower probability because of the electrostatic repulsion of the two trapped electrons and the well-known instability of such centers to oxidation. Consequently, the formation and increase in the concentration of such photoinduced $Ti^{\delta+}$ centers appears to occur only at high density of photogenerated Ti^{3+} centers that ensue upon prolonged exposure to Vis-light illumination in the later stages of photocoloration (i.e., formation of Ti^{3+} color centers).

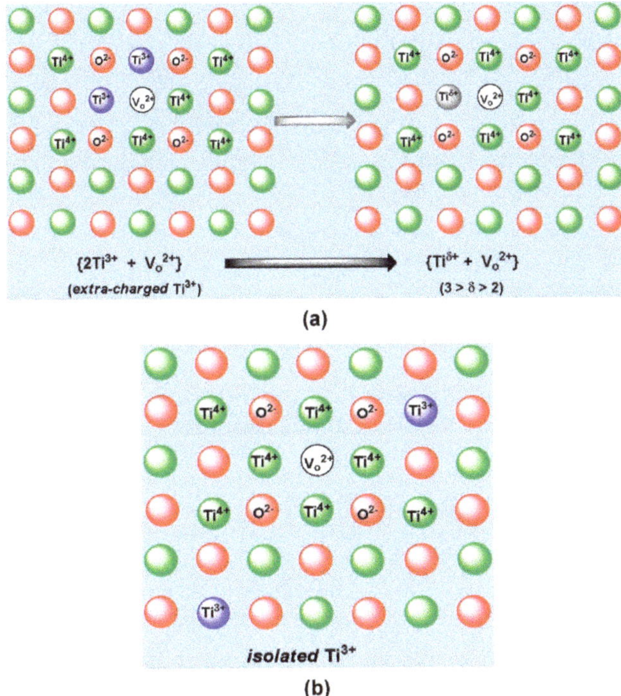

Figure 60. Cartoons representing the meaning of: (**a**) extra-negatively charged Ti^{3+} as being the complex formed between two adjacent Ti^{3+} species neighboring an oxygen vacancy (V_o) that can be viewed as a $Ti^{\delta+}$ center with $3 > \delta > 2$; and (**b**) isolated Ti^{3+} centers. Reproduced with permission from Kuznetsov et al. [70]. Copyright 2018 by Elsevier, B.V. (License No.: 4453641492289).

Germane to the above, the work of Déak and coworkers [180] showed that the first case scenario is that two self-trapped electrons in the (V_o^{2+} + $2e$) complexes are located at two equivalent first neighbors of the oxygen vacancy (extra-negatively charged Ti^{3+} → $Ti^{\delta+}$ centers; Figure 60a), while in the second scenario both electrons are more remote from the V_o^{2+} vacancy and are not in the same plane as the vacancy (isolated Ti^{3+} centers; Figure 60b); the energies of the vertical transitions of these self-trapped electrons to the conduction band are ca. 1.1 eV. Following this reasoning, a question arose as to why the growth of the number of such $Ti^{\delta+}$ centers was observed only under Vis-light illumination. This led Kuznetsov and coworkers [70] to focus attention on the differences in the spatial photoexcitation events that occur in the microparticle when illuminated in the UV and Vis spectral regions. Such differences had not heretofore been considered in the literature; their views of the events that occur under UV and Visible light illumination are summarized in Figure 61.

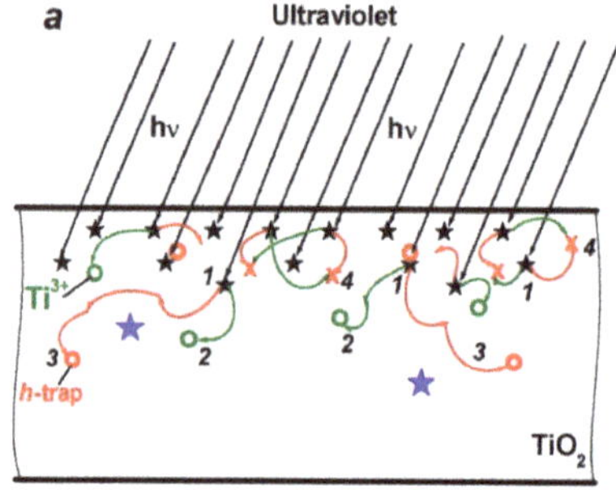

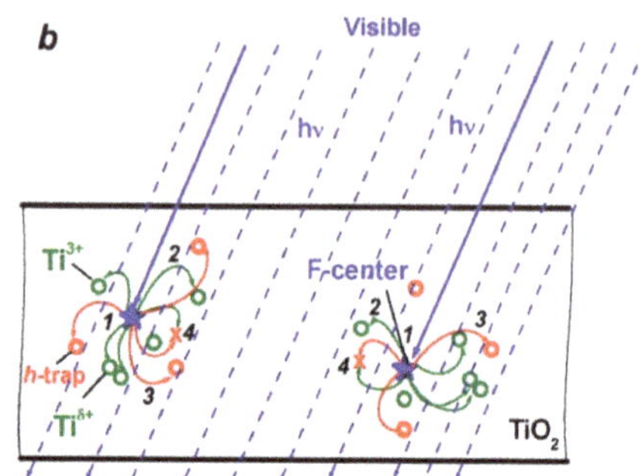

Figure 61. Illustration of the absorption of light quanta and formation of: (*1*) electron–hole pairs; (*2*) electron; and (*3*) hole transport and their localization at traps; together with (*4*) electron–hole recombination under: (**a**) UV irradiation; and (**b**) Vis-light irradiation of a microcrystalline particle of VLA TiO_2. In (**a**), the black stars denote the photogenerated electron–hole pairs, the blue stars the *F*-type centers (electron trapped in oxygen vacancy); the green circles the Ti^{3+} color centers, while the red circles refer to the trapped holes. In (**b**), the blue stars denote the *F*-type centers (electron trapped in oxygen vacancy), the green circles the Ti^{3+} color centers, while the red circles denote the trapped holes. Note the $Ti^{\delta+}$ centers in (**b**). Reproduced with permission from Kuznetsov et al. [70]. Copyright 2018 by Elsevier, B.V. (License No.: 4453641492289).

When $h\nu > E_g$, light quanta are absorbed spontaneously in solids in an arbitrary manner, each time producing e-h pairs at new spatial sites (black stars in Figure 61a) for which charge carrier transport and localization in the microparticle are determined by the distribution of charge carrier traps (Processes 2 and 3 in Figure 61a). At moderate UV-light irradiances, the authors [70] supposed that since every subsequent photoformation of charge carriers and trapping event occur in other spatial sites, a significant density of Ti^{3+} centers would not be reached. However, photoformation of electrons and holes can also be achieved on illumination in the Vis region at $h\nu \leq E_g$ when light quanta are absorbed by the native point defects (i.e., *F* (or F^+) centers); the latter are limited in number and are located at definite sites in the microparticle (Figure 61b). Moreover, because photoexcitation of *F* (or F^+) centers can produce charge carriers followed by their subsequent decay to their initial electronic states, as proposed in earlier studies [68,69], repetitive absorption of light quanta and photogeneration of electrons and holes occurs each time at the same spatial sites (*F* or F^+ centers) in the microparticle (blue stars in Figure 61b). Such considerations then lead to the reasonable inference that transport of carriers and occupation of traps (processes 2 and 3 in Figure 61b) start repetitively at the same sites in the microparticle. In that case, filling of the nearest neighbor *F* (or F^+) center traps facilitates the attainment of a high density of Ti^{3+} centers and the consequent formation of the $Ti^{\delta+}$ centers [70].

The above notwithstanding regarding the TiO_2-based technology, people intending to deNOxify the environment must first come to appreciate and understand the rich chemistry of nitrogen oxides, in general, and NO and NO_2, in particular, in a homogeneous phase and in hetero- geneous media.

For instance, NO dimerizes to N_2O_2 upon condensing to a liquid, although the association is weak and reversible [181]. In addition, to the extent that the enthalpy of formation of NO is endothermic, NO can easily undergo disproportionation back to its constituent elements N_2 and O_2 as might occur in catalytic converters—for example, Reaction (106) occurs over the zeolite Cu^{2+}-ZSM-5 [182].

$$2\,NO \rightarrow N_2 + O_2 \tag{106}$$

Nitric oxide is also thermodynamically unstable at 25 °C and 1 atm; under pressure, it decomposes readily in the temperature range 30–50 °C to yield NO_2 and N_2O (Reaction (107)) and may react either as NO_2 or as N_2O_3 [183].

$$3\,NO \rightarrow N_2O + NO_2 \tag{107}$$

When exposed to atmospheric oxygen, nitric oxide converts instantly to NO_2 (Reaction (108)) [181], which likely occurs via the intermediate ON–O–O–NO.

$$2\,NO + O_2 \rightarrow 2\,NO_2 \tag{108}$$

In water, NO reacts with oxygen and water to form nitrous acid HONO (Reaction (109)).

$$4\,NO + O_2 + 2\,H_2O \rightarrow 4\,HONO \tag{109}$$

Since both NO_2 and NO are radical species, they combine to form the intensely blue dinitrogen trioxide N_2O_3 (Reaction (110)) [184].

$$NO + NO_2 \rightleftarrows ON - NO_2 \tag{110}$$

Both the brown gas nitrogen dioxide, NO_2, and the colorless gas dinitrogen tetroxide, N_2O_4, exist in a strongly temperature-dependent equilibrium (Reaction (111)) for which $\Delta H = -57.23$ kJ mol^{-1}, with NO_2 being favored at higher temperatures, while N_2O_4 predominates at lower temperatures.

$$2\,NO_2 \rightleftarrows N_2O_4 \tag{111}$$

Because of the relatively weak N–O bond in NO_2, nitrogen dioxide is a relatively good oxidizing agent in aqueous media (Reaction (112); nearly comparable to Br_2 gas), which makes the mixed oxides NO_2 and N_2O_4—also known as nitrous fumes—react vigorously if not explosively with several compounds, particularly with hydrocarbons via hydrogen abstraction as a first step (Reaction (113)) [181].

$$N_2O_4 + 2\,H^+ + 2\,e^- \rightarrow 2\,HONO\ E^0 = 1.07\ V \tag{112}$$

$$NO_2 + RH \rightarrow R^\bullet + HONO \tag{113}$$

In aqueous media, NO_2 also hydrolyzes to form nitrous acid and nitric acid via Reaction (114), which is one of the steps in the industrial production of nitric acid from ammonia via the Ostwald process [185].

$$2\,NO_2 \rightleftarrows N_2O_4 \xrightarrow{H_2O} HONO + HNO_3 \tag{114}$$

Although Reaction (114) is negligibly slow at the low concentrations of NO_2 characteristically encountered in the ambient atmosphere, it does proceed upon uptake of NO_2 onto surfaces to produce gaseous HONO in outdoor and indoor environments [186].

Several studies have examined the interactions of NO_2 on the TiO_2 surface under various experimental conditions, such as different NO_2 partial pressures and various temperatures in the range 323–573 K [102,106,108,172]. All these studies reported production of NO in the gas phase, and formation of nitrates on the TiO_2 surface, albeit under photocatalytic conditions.

However, other aspects that seem to have been overlooked by many are the potential specific interactions between the two NO_x molecules and the TiO_2 surface under dark conditions, which need to be re-emphasized constantly.

In this regard, in their 2003 FTIR study carried out in the dark in borosilicate glass vessels, Finlayson-Pitts et al. [186] discovered that the loss of gaseous NO_2 was accompanied by formation of HONO, NO and N_2O; further FTIR studies also revealed the formation of HNO_3, N_2O_4 and NO_2^+ species, which led them to hypothesize that the symmetric form of the NO_2 dimer, N_2O_4, is taken up on the surface and isomerizes to the asymmetric form, $ONONO_2$, with the latter undergoing autoionization to $NO^+NO_3^-$. Apparently, it is the latter intermediate species that react with water to generate HONO and surface-adsorbed HNO_3. Subsequently, NO is generated by secondary reactions of HONO on the highly acidic surface. The authors further noted that a key aspect of this chemistry is that in the atmospheric boundary layer where human exposure occurs and many measurements of HONO and related atmospheric constituents (e.g., ozone) are made, a major component for this heterogeneous chemistry is the surface of buildings, roads, soils, vegetation and other materials [186].

A more recent investigation by Sivachandrian and coworkers [187] on the adsorption of NO and NO_2 molecules on the metal oxide TiO_2 at ambient temperature, specifically carried out under dark experimental conditions to avoid any photocatalytic interference, showed no significant adsorption of NO on TiO_2. By contrast, not only did NO_2 significantly influence the adsorption of VOCs and mineralization on the TiO_2 surface, but once the threshold surface coverage of NO_2 was reached at room temperature, the NO_2 adsorbed reactively on the TiO_2 surface by evolving NO in the gas phase. Quantitative measurements performed downstream of the reactor led the authors [187] to propose a new mechanism expressed by the Reactions (115)–(118) for the adsorption of NO_2 on TiO_2 at room temperature under dry air conditions:

$$2\,NO_{2(ads)} \rightleftharpoons NO_2^{-} + NO^{+}{}_{(ads)} \tag{115}$$

$$NO + O^{2-}\,(TiO_{2\,\text{Surface latice}}) \rightleftharpoons NO_2^{-}{}_{(ads)} \tag{116}$$

$$NO_2^{-}{}_{(ads)} + NO_{2(ads)} \rightleftharpoons NO_3^{-}{}_{(ads)} + NO_{(g)} \tag{117}$$

According to this sequence, the global reaction of NO_2 adsorption is then (Reaction (118)):

$$3\,NO_{2(ads)} \rightleftharpoons 2\,NO_3^{-}{}_{(ads)} + NO_{(g)} \tag{118}$$

Accordingly, the proposed NO_2 adsorption mechanism on TiO_2 at room temperature in the dark [187], together with the experimental observations, could be summarized thus: (i) three NO_2 molecules adsorb on TiO_2, produce two NO_3^{-} ions on the TiO_2 surface, and evolve one NO molecule in the gas phase; (ii) the ratio between consumed NO_2, TPD desorbed NO_2 subsequent to adsorption, and NO produced during NO_2 adsorption = 3:2:1; (iii) the NO_2 adsorption time (i.e., the TiO_2 surface coverage) significantly modified the nature of the adsorbed species at ambient temperature; (iv) the NO formation time was controlled principally by the surface coverage of NO_2^{-} and NO_3^{-} ions, rather than by the NO_2 inlet concentration; and (v) at higher NO_2 gas phase concentrations (greater than 35 ppm) at room temperature, the total amount of consumed NO_2 decreased as a result of self-poisoning of the sites by adsorbed NO_3^{-} species.

Similarly, Haubrich and coworkers [188] examined the interaction of NO_2 on rutile $TiO_2(110)$ and discovered that the presence of NO_2 and water led to the formation of multilayers under dark conditions of nitric acid, HNO_3, contrary to exposure of the surface to pure water after saturation of the surface with 200 mTorr of NO_2; no further growth of the AP-XPS (ambient pressure X-ray photoelectron spectroscopy) nitrate signals occurred under the latter conditions. Apparently, formation of HNO_3 requires weakly adsorbed NO_2 molecules, an important finding with important implications in environmental processes since their study [188] confirmed that metal oxides facilitate the formation of nitric acid under ambient humidity (in the dark) conditions typically encountered in atmospheric environments.

It is evident that there is much more that needs to be investigated to fully understand whatever events occur in and on the TiO_2 (and others) semiconductor photocatalyst. Laboratory experiments using models and solar simulators are just the beginning, after which what is needed is to bring the laboratory outdoors using actual environmental quantities of NO_x as the reagents and humid air as prevails in the environment being investigated for application of the photocatalytic substrates. A careful study of the levels of NO_x or its products at various distances from the photocatalytic surface, both vertically and horizontally, without precluding determination of all other environmental factors is also needed. The $deNO_x$ index has shown that metal-doped TiO_2 systems are also worth investigating further.

Funding: This research received no external funding.

Acknowledgments: We are grateful to Angelo Albini for his gracious hospitality in the PhotoGreen Laboratory at the University of Pavia, Italy.

Conflicts of Interest: The author declares no conflict of interest.

References

1. Engineering Alliance, Inc. Air Quality Services, Types_of_sources_02-2012. Available online: https://www.eaincglobal.com/air-quality/attachment/types_of_sources_02-2012/ (accessed on 12 October 2018).
2. Smog. Available online: https://en.wikipedia.org/wiki/Smog (accessed on 12 October 2018).
3. Popescu, F.; Ionel, I. Anthropogenic Air Pollution Sources. In *Air Quality*; Kumar, A., Ed.; InTech Europe: Rijeka, Croatia, August 2010; Chapter 1; pp. 1–22. ISBN 978-953-307-131-2. Available online: http://www.intechopen.com/books/airquality/anthropogenic-air-pollution-sources (accessed on 15 July 2018).
4. European Union Emission Inventory Report 1990–2011 under the UNECE Convention on Long-Range Trans-Boundary Air Pollution (LRTAP). Available online: http://www.icopal-noxite.co.uk/nox-problem/nox-pollution.aspx (accessed on 12 October 2018).
5. United States Environmental Protection Agency. Air Emission Sources. 4 November 2009. Available online: http://www.epa.gov/air/emissions/index.htm (accessed on 12 October 2018).
6. Tropospheric Ozone. Available online: https://en.wikipedia.org/wiki/Tropospheric_ozone#cite_note-5 (accessed on 12 October 2018).
7. Reeves, C.E.; Penkett, S.A.; Bauguitte, S.; Law, K.S.; Evans, M.J.; Bandy, B.J.; Monks, P.S.; Edwards, G.D.; Phillips, G.; Barjat, H.; et al. Potential for photochemical ozone formation in the troposphere over the North Atlantic as derived from aircraft observations during ACSOE. *J. Geophys. Res.* **2002**, *107*, 4707. [CrossRef]
8. Deziel, C. How Is Photochemical Smog Formed? Available online: https://sciencing.com/photochemical-smog-formed-6505511.html (accessed on 25 April 2017).
9. Health Effects of Nitrogen Oxides, Department of Employment, Economic Development and Innovation, Queensland Government, Australia. Available online: https://www.dnrm.qld.gov.au/data/assets/pdf_file/0020/212483/2-health-effects-of-nitrogen-dioxide.pdf (accessed on 12 October 2018).
10. McCarron, G. Air Pollution and human health hazards: A compilation of air toxins acknowledged by the gas industry in Queensland's Darling Downs. *Int. J. Environ. Stud.* **2018**, *75*, 171–185. [CrossRef]
11. Chen, J.; Poon, C.-S. Photocatalytic construction and building materials: From fundamentals to applications. *Build. Environ.* **2009**, *44*, 1899–1906. [CrossRef]
12. Mendoza, C.; Valle, A.; Castellote, M.; Bahamonde, A.; Faraldos, M. TiO_2 and TiO_2–SiO_2 coated cement: Comparison of mechanic and photocatalytic properties. *Appl. Catal. B* **2015**, *178*, 155–164. [CrossRef]
13. Pei, C.C.; Leung, W.W.F. Photocatalytic oxidation of nitrogen monoxide and o-xylene by TiO_2/ZnO/Bi_2O_3 nanofibers: Optimization, kinetic modeling and mechanisms. *Appl. Catal. B* **2015**, *174 175*, 515–525. [CrossRef]
14. Gallus, M.; Akylas, V.; Barmpas, F.; Beeldens, A.; Boonen, E.; Boréave, A.; Cazaunau, M.; Chen, H.; Daële, V.; Doussin, J.F.; et al. Photocatalytic depollution in the Leopold II tunnel in Brussels: NO_x abatement results. *Build. Environ.* **2015**, *84*, 125–133. [CrossRef]
15. Gallus, M.; Ciuraru, R.; Mothes, F.; Akylas, V.; Barmpas, F.; Beeldens, A.; Bernard, F.; Boonen, E.; Boréave, A.; Cazaunau, M.; et al. Photocatalytic abatement results from a model street canyon. *Environ. Sci. Pollut. Res.* **2015**, *22*, 18185–18196. [CrossRef] [PubMed]
16. Sikkema, J.K.; Ong, S.K.; Alleman, J.E. Photocatalytic concrete pavements: Laboratory investigation of NO oxidation rate under varied environmental conditions. *Constr. Build. Mater.* **2015**, *100*, 305–314. [CrossRef]
17. Gandolfo, A.; Bartolomei, V.; Gomez-Alvarez, E.; Tlili, S.; Gligorovski, S.; Kleffmann, J.; Wortham, H. The effectiveness of indoor photocatalytic paints on NO_x and HONO levels. *Appl. Catal. B* **2015**, *166–167*, 84–90. [CrossRef]
18. Martinez, T.; Bertron, A.; Escadeillas, G.; Ringot, E.; Simon, V. BTEX abatement by photocatalytic TiO_2-bearing coatings applied to cement mortars. *Build. Environ.* **2014**, *71*, 186–192. [CrossRef]
19. Folli, A.; Pade, C.; Hansen, T.B.; De Marco, T.; Macphee, D.E. TiO_2 photocatalysis in cementitious systems: Insights into self-cleaning and depollution chemistry. *Cem. Concr. Res.* **2012**, *42*, 539–548. [CrossRef]

20. Karapati, S.; Giannakopoulou, T.; Todorova, N.; Boukos, N.; Antiohos, S.; Papageorgiou, D.; Chaniotakis, E.; Dimotikali, D.; Trapalis, C. TiO$_2$ functionalization for efficient NO$_x$ removal in photoactive cement. *Appl. Surf. Sci.* **2014**, *319*, 29–36. [CrossRef]

21. Mo, J.; Zhang, Y.; Xu, Q.; Lamson, J.J.; Zhao, R. Photocatalytic purification of volatile organic compounds in indoor air: A literature review. *Atmos. Environ.* **2009**, *43*, 2229–2246. [CrossRef]

22. Bartolomei, V.; Sörgel, M.; Gligorovski, S.; Alvarez, E.G.; Gandolfo, A.; Strekowski, R.; Quivet, E.; Held, A.; Zetzsch, C.; Wortham, H. Formation of indoor nitrous acid (HONO) by light-induced NO$_2$ heterogeneous reactions with white wall paint. *Environ. Sci. Pollut. Res.* **2014**, *21*, 9259–9269. [CrossRef] [PubMed]

23. Langridge, J.M.; Gustafsson, R.J.; Griffiths, P.T.; Cox, R.A.; Lambert, R.M.; Jones, R.L. Solar driven nitrous acid formation on building material surfaces containing titanium dioxide: A concern for air quality in urban areas? *Atmos. Environ.* **2009**, *43*, 5128–5131. [CrossRef]

24. Ndour, M.; Conchon, P.; D'Anna, B.; Ka, O.; George, C. Photochemistry of mineral dust surface as a potential atmospheric renoxification process. *Geophys. Res. Lett.* **2009**, *36*, 1–4. [CrossRef]

25. Mothes, F.; Böge, O.; Herrmann, H. A chamber study on the reactions of O$_3$, NO, NO$_2$ and selected VOCs with a photocatalytically active cementitious coating material. *Environ. Sci. Pollut. Res.* **2016**, *23*, 15250–15261. [CrossRef] [PubMed]

26. Monge, M.E.; George, C.; D'Anna, B.; Doussin, J.-F.; Jammoul, A.; Wang, J.; Eyglunent, G.; Solignac, G.; Daële, V.; Mellouki, A. Ozone formation from illuminated titanium dioxide surfaces. *J. Am. Chem. Soc.* **2010**, *132*, 8234–8235. [CrossRef] [PubMed]

27. Emeline, A.V.; Kuznetsov, V.N.; Ryabchuk, V.K.; Serpone, N. On the way to the creation of next generation photoactive materials. *Environ. Sci. Pollut. Res.* **2012**, *19*, 3666–3675. [CrossRef] [PubMed]

28. Serpone, N.; Emeline, A.V. Semiconductor photocatalysis—Past, present, and future outlook. *J. Phys. Chem. Lett.* **2012**, *3*, 673–677. [CrossRef] [PubMed]

29. Braslavsky, S.E.; Braun, A.M.; Cassano, A.E.; Emeline, A.V.; Litter, M.I.; Palmisano, L.; Parmon, V.N.; Serpone, N. Glossary of terms used in photocatalysis and radiation catalysis (IUPAC Recommendations 2011). *Pure Appl. Chem.* **2011**, *83*, 931–1014. [CrossRef]

30. Emeline, A.V.; Panasuk, A.V.; Sheremetyeva, N.; Serpone, N. Mechanistic studies of the formation of different states of oxygen on irradiated ZrO$_2$ and the photocatalytic nature of photoprocesses from determination of turnover numbers. *J. Phys. Chem. B* **2005**, *109*, 2785–2792. [CrossRef] [PubMed]

31. Emeline, A.V.; Rudakova, A.V.; Ryabchuk, V.K.; Serpone, N. Photostimulated reactions at the surface of wide bandgap metal oxides {ZrO$_2$ and TiO$_2$}: Interdependence of rates of reactions on pressure- concentration and on light intensity. *J. Phys. Chem. B* **1998**, *102*, 10906–10916. [CrossRef]

32. Emeline, A.V.; Ryabchuk, V.K.; Serpone, N. Factors affecting the efficiency of a photocatalyzed process in aqueous metal-oxide dispersions. Prospect for distinguishing between the two kinetic models. *J. Photochem. Photobiol. A Chem.* **2000**, *133*, 89–97. [CrossRef]

33. Ollis, D.F. Kinetics of liquid phase photocatalyzed reactions: An illuminating approach. *J. Phys. Chem. B* **2005**, *109*, 2439–2444. [CrossRef] [PubMed]

34. Mills, A.; Wang, J.; Ollis, D.F. Kinetics of liquid phase semiconductor photoassisted reactions: Supporting observations for a pseudo-steady-state model. *J. Phys. Chem. B* **2006**, *110*, 14386–14390. [CrossRef] [PubMed]

35. Graetzel, M. (Ed.) *Energy Resources through Photochemistry and Catalysis*; Academic Press: New York, NY, USA, 1983; p. 573.

36. Clayton, R.K. Photosynthesis: Physical mechanisms and chemical patterns. *IUPAB Biophys. Ser.* **1980**, *4*, 281.

37. Braslavsky, S.E. Glossary of terms used in photochemistry. (IUPAC Recommendations 2006). *Pure Appl. Chem.* **2007**, *79*, 293–465. [CrossRef]

38. Emeline, A.V.; Sheremetyeva, N.V.; Khomchenko, N.V.; Kuzmin, G.N.; Ryabchuk, V.K.; Teoh, W.Y.; Amal, R. Spectroscopic studies of pristine and fluorinated nano-ZrO$_2$ in photostimulated heterogeneous processes. *J. Phys. Chem. C* **2009**, *113*, 4566–4583. [CrossRef]

39. Serpone, N. Is the band gap of pristine TiO$_2$ narrowed by anion- and cation-doping of titanium dioxide in second-generation photocatalysts? *J. Phys. Chem. B* **2006**, *110*, 24287–24293. [CrossRef] [PubMed]

40. Serpone, N.; Emeline, A.V.; Kuznetsov, V.N.; Ryabchuk, V.K. Visible-light-active titania photocatalysts. The case of N-doped TiO$_2$s—Properties and some fundamental issues. *Int. J. Photoenergy* **2008**, *1*, 1–19.

41. Serpone, N.; Emeline, A.V. Modeling heterogeneous photocatalysis by metal-oxide nanostructured semiconductor and insulator materials; Factors that affect the activity and selectivity of photocatalysts. *Res. Chem. Intermed.* **2005**, *31*, 391–432. [CrossRef]

42. Emeline, A.V.; Serpone, N. Spectral selectivity of photocatalyzed reactions on the surface of titanium dioxide nanoparticles. *J. Phys. Chem. B* **2002**, *106*, 12221–12226. [CrossRef]

43. Emeline, A.V.; Zhang, X.; Jin, M.; Murokami, T.; Fujishima, A. Spectral dependences of the activity and selectivity of N-doped TiO_2 in photodegradation of phenols. *J. Photochem. Photobiol. A Chem.* **2009**, *207*, 13–19. [CrossRef]

44. Emeline, A.V.; Kuzmin, G.N.; Serpone, N. Quantum yields and their wavelength-dependence in the photoreduction of O_2 and photooxidation of H_2 on a visible-light-active N-doped TiO_2 system. *Chem. Phys. Lett.* **2008**, *454*, 279–283. [CrossRef]

45. Murakami, N.; Chiyoya, T.; Tsubota, T.; Ohno, T. Switching redox site of photocatalytic reaction on titanium(IV) oxide particles modified with transition-metal ion controlled by irradiation wavelength. *Appl. Catal. A* **2008**, *348*, 148–152. [CrossRef]

46. Baye, E.; Murakami, N.; Ohno, T. Exposed crystal surface-controlled TiO_2 nanorods having rutile phase from $TiCl_3$ under hydrothermal conditions. *J. Mol. Catal. A Chem.* **2009**, *300*, 72–79. [CrossRef]

47. Murakami, N.; Kurihara, Y.; Tsubota, T.; Ohno, T. Shape-controlled anatase titanium(IV) oxide particles prepared by hydrothermal treatment of peroxo titanic acid in the presence of polyvinyl alcohol. *J. Phys. Chem. C* **2009**, *113*, 3062–3069. [CrossRef]

48. Baye, E.; Ohno, T. Exposed crystal surface-controlled rutile TiO_2 nanorods prepared by hydrothermal treatment in the presence of poly(vinyl)pyrrolidone. *Appl. Catal. B Environ.* **2009**, *91*, 634–639. [CrossRef]

49. Tachikawa, T.; Yamashita, S.; Majima, T. Evidence for crystal-face-dependent TiO_2 photocatalysis from single-molecule imaging and kinetic analysis. *J. Am. Chem. Soc.* **2011**, *133*, 7197–7204. [CrossRef] [PubMed]

50. Emeline, A.V.; Zhang, X.; Murakami, T.; Fujishima, A. Activity and selectivity of photocatalysts in photodegradation of phenols. *J. Hazard. Mater.* **2012**, *211–212*, 154–160. [CrossRef] [PubMed]

51. Gerisher, H.; Heller, A. The role of oxygen in photooxidation of organic molecules on semiconductor particles. *J. Phys. Chem.* **1991**, *95*, 5261–5267. [CrossRef]

52. Emeline, A.V.; Kataeva, G.V.; Panasuk, A.V.; Ryabchuk, V.K.; Sheremetyeva, N.; Serpone, N. Effect of surface photoreactions on the photocoloration of a wide band gap metal oxide: Probing whether surface reactions are photocatalytic. *J. Phys. Chem. B* **2005**, *109*, 5175–5185. [CrossRef] [PubMed]

53. Terenin, A.N. Optical investigations of the adsorption of gas molecules. *Uchenye Zapiski Leningrad. Gosudarst. Univ. Ser. Fiz. Nauk* **1939**, *5*, 26–40. (In Russian)

54. Terenin, A.N. Optical investigations of activated adsorption. *Z. Fiz. Khim.* **1940**, *14*, 1362–1369. (In Russian)

55. Kasparov, K.Ya.; Terenin, A. Optical investigations of activated adsorption. I. Photodecomposition of NH_3 adsorbed on catalysts. *Acta Physicochim. USSR* **1941**, *15*, 343–365.

56. Terenin, A.N.; Solonitzyn, Y. Action of light on the gas adsorption by solids. *Discuss. Faraday Soc.* **1959**, *28*, 28–35. [CrossRef]

57. Ashcroft, N.W.; Mermin, N.D. *Solid State Physics*; Holt, Rinehart and Winston: New York, NY, USA, 1976.

58. Pankove, J.I. *Optical Processes in Semiconductors*; Dover Publications: New York, NY, USA, 1971.

59. Stoneham, A.M. *Theory of Defects in Solids*; Clarendon Press: Oxford, UK, 1975.

60. Henderson, B.; Werts, J.E. *Defects in the Alkaline Earth Oxides*; Taylor & Francis Ltd.: London, UK, 1977; p. 152.

61. Kotomin, E.A.; Popov, A.I. Radiation-induced point defects in simple oxides. *Nucl. Instrum. Methods Phys. Res. B* **1998**, *141*, 1–15. [CrossRef]

62. Popov, A.I.; Kotomin, E.A.; Maier, J. Basic properties of the F-type centers in halides, oxides and perovskites. *Nucl. Instrum. Methods Phys. Res. B* **2010**, *268*, 3084–3089. [CrossRef]

63. Crawford, H.J. Recent developments in Al_2O_3 color-center research. *Semicond. Insul.* **1982**, *5*, 599–620.

64. Seebauer, E.G.; Kratzer, M.C. Charged point defects in semiconductors. *Mater. Sci. Eng. R* **2006**, *55*, 57–149. [CrossRef]

65. Schirmer, O.F. O^- bound small polarons in oxide materials. *J. Phys. Condens. Matter* **2006**, *18*, R667–R704. [CrossRef]

66. Schirmer, O.F. Holes bound as small polarons to acceptor defects in oxide materials: Why are their thermal ionization energies so high? *J. Phys. Condens. Matter* **2011**, *23*, 334218. [CrossRef] [PubMed]

67. Dolgov, S.A.; Kärner, T.; Lushchik, A.; Maaroos, A.; Nakonechnyi, S.; Shablonin, E. Trapped hole centers in MgO single crystals. *Phys. Solid State* **2011**, *53*, 1244–1252. [CrossRef]

68. Kuznetsov, V.N.; Emeline, A.V.; Glazkova, N.I.; Mikhaylov, R.V.; Serpone, N. Real-time in situ monitoring of optical absorption changes in visible-light-active TiO_2 under light irradiation and temperature-programmed annealing. *J. Phys. Chem. C* **2014**, *118*, 27583–27593. [CrossRef]

69. Kuznetsov, V.N.; Ryabchuk, V.K.; Emeline, A.V.; Mikhaylov, R.V.; Rudakova, A.V.; Serpone, N. Thermo- and photo-stimulated effects on the optical properties of rutile titania ceramic layers formed on titanium substrates. *Chem. Mater.* **2013**, *25*, 170–177. [CrossRef]

70. Kuznetsov, V.N.; Glazkova, N.I.; Mikhaylov, R.V.; Serpone, N. Additional specific channel of photo-activation of solid semiconductors. A revisit of the thermo-/photo-stimulated bleaching of photo-induced Ti^{3+} color centers in visible-light-active photochromic rutile titania. *J. Phys. Chem. C* **2018**, *122*, 13294–13303. [CrossRef]

71. Zecchina, A.; Lofthouse, M.G.; Stone, F.S. Reflectance spectra of surface states in magnesium oxide and calcium oxide. *J. Chem. Soc. Faraday Trans. 1 Phys. Chem. Condens. Phases* **1975**, *71*, 1476–1490. [CrossRef]

72. Zecchina, A.; Stone, F.S. Reflectance spectra of surface states in strontium oxide and barium oxide. *J. Chem. Soc. Faraday Trans. 1 Phys. Chem. Condens. Phases* **1976**, *72*, 2364–2374. [CrossRef]

73. Kristianpoller, N.; Rehavi, A.; Shmilevicha, A.; Weiss, D.; Chen, R. Radiation effects in pure and doped Al_2O_3 crystals. *Nucl. Instrum. Methods Phys. Res. B* **1998**, *141*, 343–346. [CrossRef]

74. Kortov, V.S.; Vainshtein, I.A.; Vokhmintsev, A.S.; Gavrilov, N.V. Spectroscopic characteristics of anionic centers in α-Al_2O_3 crystals bombarded by Cu^+ and Ti^+. *J. Appl. Spectrosc.* **2008**, *75*, 452–455. [CrossRef]

75. Izerrouken, M.; Benyahia, T. Absorption and photoluminescence study of Al_2O_3 single crystal irradiated with fast neutrons. *Nucl. Instrum. Methods Phys. Res. B* **2010**, *268*, 2987–2990. [CrossRef]

76. Itou, M.; Fujiwara, A.; Uchino, T. Reversible photoinduced interconversion of color centers in α-Al_2O_3 prepared under vacuum. *J. Phys. Chem. C* **2009**, *113*, 20949–20957. [CrossRef]

77. Emeline, A.V.; Kuzmin, G.N.; Purevdorj, D.; Ryabchuk, V.K.; Serpone, N. Spectral dependencies of the quantum yield of photochemical processes on the surface of wide band gap solids. 3. Gas/Solid systems. *J. Phys. Chem. B* **2000**, *104*, 2989–2999. [CrossRef]

78. Kuznetsov, V.N.; Lisachenko, A.A. Spectral manifestation of wide-band oxide own defects in photo-stimulated surface reactions. *Russ. J. Phys. Chem.* **1991**, *65*, 1328–1334.

79. Lisachenko, A. Photon-driven electron and atomic processes on solid-state surface in photoactivated spectroscopy and photocatalysis. *J. Photochem. Photobiol. A Chem.* **2008**, *196*, 127–137. [CrossRef]

80. Zakharenko, V.S.; Cherkashin, A.E.; Volodin, A.M.; Keier, N.P. Spectral dependence of oxygen and carbon monoxide photoadsorption on rutile. *React. Kinet. Catal. Lett.* **1979**, *10*, 325–332. [CrossRef]

81. Volodin, A.M.; Cherkashin, A.E.; Zakharenko, V.S. Influence of physically adsorbed oxygen on the separation of electron-hole pairs on anatase irradiated by visible light. *React. Kinet. Catal. Lett.* **1979**, *11*, 103–106. [CrossRef]

82. Emeline, A.V.; Smirnova, L.G.; Kuzmin, G.N.; Basov, L.L.; Serpone, N. Spectral dependence of quantum yields in gas/solid heterogeneous photosystems. Influence of anatase/rutile content on the photo- stimulated adsorption of dioxygen and dihydrogen on titania. *J. Photochem. Photobiol. A. Chem.* **2002**, *148*, 99–104.

83. Komaguchi, K.; Maruoka, T.; Nakano, H.; Imae, I.; Ooyama, Y.; Harima, Y. Electron-transfer reaction of oxygen species on TiO_2 nanoparticles induced by sub-band-gap illumination. *J. Phys. Chem. C* **2010**, *114*, 1240–1245. [CrossRef]

84. Kuznetsov, V.N.; Serpone, N. On the origin of the spectral bands in the visible absorption spectra of visible-light-active TiO_2 specimens: Analysis and assignments. *J. Phys. Chem. C* **2009**, *113*, 15110–15123. [CrossRef]

85. Artemiev, Y.M.; Ryabchuk, V.K. *Introduction to Heterogeneous Photocatalysis (A Textbook)*; Saint Petersburg State University: Saint Petersburg, Russia, 1999. (In Russian)

86. Siline, A.R.; Trukhin, A.N. *Point Defects and Elementary Excitations in Crystalline and Non-Crystalline SiO_2*; Zinatne: Riga, Latvia, 1985.

87. Ryvkin, S.M. *Photoelectric Effects in Semiconductors*; Consultants Bureau: New York, NY, USA, 1964.

88. Delany, A.C.; Dickerson, R.R.; Melchior, F.L.; Wartburg, A.F. Modification of a commercial NO_x detector for high sensitivity. *Rev. Sci. Instrum.* **1982**, *12*, 1899–1902. [CrossRef]

89. Wolf, C.A.; Nieuwenhuys, B.E. The NO–H_2 reaction over Pd(III). *Surf. Sci.* **2000**, *469*, 196–203. [CrossRef]

90. Environmental Agency of Japan. *Kankyo Hakusho*; State of Environment in Japan: Tokyo, Japan, 1991. (In Japanese)

91. Takeuchi, K.; Ibusuki, T. Heterogeneous Photochemical Reactions and Processes in the Troposphere. In *Encyclopedia of Environmental Control Technology*; Cheremisonoff, P.N., Ed.; Gulf Publishing: Houston, TX, USA, 1989; p. 279.

92. Courbon, H.; Pichat, P. Room-temperature interaction of $N^{18}O$ with ultraviolet-illuminated titanium dioxide. *J. Chem. Soc. Faraday Trans. 1* **1984**, *80*, 3175–3185. [CrossRef]

93. Hori, Y.; Nakatsu, A.; Suzuki, S. Heterogeneous photocatalytic oxidation of NO_2 in aqueous suspension of various semiconductor powders. *Chem. Lett.* **1985**, *14*, 1429–1432. [CrossRef]

94. Anpo, M.; Nomura, T.; Kitao, T.; Giamello, E.; Murphy, D.; Che, M.; Fox, M.A. Approach to de-NOx-ing photocatalysis. II. Excited state of copper ions supported on silica and photocatalytic activity for NO decomposition. *Res. Chem. Intermed.* **1991**, *15*, 225. [CrossRef]

95. Anpo, M.; Matsuoka, M.; Hanou, K.; Mishima, H.; Yamashita, H.; Patterson, H.H. The relationship between the local structure of copper(I) ions on Cu^{+}/zeolite catalysts and their photocatalytic reactivities for the decomposition of NO_x into N_2 and O_2 at 275 K. *Coord. Chem. Rev.* **1998**, *171*, 175–184. [CrossRef]

96. Anpo, M.; Ichihashi, Y.; Takeuchi, M.; Yamashita, H. Design of unique titanium dioxide photocatalysts by an advanced metal-ion implantation method and photocatalytic reactions under visible light irradiation. *Res. Chem. Intermed.* **1998**, *24*, 143–149. [CrossRef]

97. Ibusuki, T.; Takeuchi, K. Removal of low concentration nitrogen oxides through photoassisted heterogeneous catalysis. *J. Mol. Catal.* **1994**, *88*, 93–102. [CrossRef]

98. Negishi, N.; Takeuchi, K.; Ibusuki, T. Surface structure of the TiO_2 thin film photocatalyst. *J. Mater. Sci.* **1998**, *33*, 5789–5794. [CrossRef]

99. Kominami, H.; Matsuura, T.; Iwai, K.; Ohtani, B.; Nishimoto, S.; Kera, Y. Ultra-highly active titanium(IV) oxide photocatalyst prepared by hydrothermal crystallization from titanium(IV) alkoxide in organic solvents. *Chem. Lett.* **1995**, *24*, 693–694. [CrossRef]

100. Kominami, H.; Kato, J.; Kohno, M.; Kera, Y.; Ohtani, B. Photocatalytic mineralization of acetic acid in aerated aqueous suspension of ultra-highly active titanium(IV) oxide prepared by hydrothermal crystallization in toluene. *Chem. Lett.* **1996**, *25*, 1051–1052. [CrossRef]

101. Hashimoto, K.; Wasada, K.; Toukai, N.; Kominami, H.; Kera, Y. Photocatalytic oxidation of nitrogen monoxide over titanium(IV) oxide nanocrystals large size areas. *J. Photochem. Photobiol. A Chem.* **2000**, *136*, 103–109. [CrossRef]

102. Lim, T.H.; Jeong, S.M.; Kim, S.D.; Gyenis, J. Photocatalytic decomposition of NO by TiO_2 particles. *J. Photochem. Photobiol. A Chem.* **2000**, *134*, 209–217. [CrossRef]

103. Nakamura, I.; Negishi, N.; Kutsuna, S.; Ihara, T.; Sugihara, S.; Takeuchi, K. Role of oxygen vacancy in the plasma-treated TiO_2 photocatalyst with visible light activity for NO removal. *J. Mol. Catal. A Chem.* **2000**, *161*, 205–212. [CrossRef]

104. Zhang, J.; Ayusawa, T.; Minagawa, M.; Kinugawa, K.; Yamashita, H.; Matsuoka, M.; Anpo, M. Investigations of TiO_2 photocatalysts for the decomposition of NO in the flow system: The role of pretreatment and reaction conditions in the photocatalytic efficiency. *J. Catal.* **2001**, *198*, 1–8. [CrossRef]

105. Tanaka, T.; Teramura, K.; Arakaki, K.; Funabiki, T. Photoassisted NO reduction with NH_3 over TiO_2 photocatalyst. *Chem. Commun.* **2002**, *22*, 2742–2743. [CrossRef]

106. Dalton, J.S.; Janes, P.A.; Jones, N.G.; Nicholson, J.A.; Hallam, K.R.; Allen, G.C. Photocatalytic oxidation of NO_x gases using TiO_2: A surface spectroscopic approach. *Environ. Pollut.* **2002**, *120*, 415–422. [CrossRef]

107. Devahasdin, S.; Fan, C., Jr.; Li, K.; Chen, D.H. TiO_2 photocatalytic oxidation of nitric oxide: Transient behavior and reaction kinetics. *J. Photochem. Photobiol. A Chem.* **2003**, *156*, 161–170. [CrossRef]

108. Toma, F.L.; Bertrand, G.; Klein, D.; Coddet, C. Photocatalytic removal of nitrogen oxides via titanium dioxide. *Environ. Chem. Lett.* **2004**, *2*, 117–121. [CrossRef]

109. Bowering, N.; Walker, G.S.; Harrison, P.G. Photocatalytic decomposition and reduction reactions of nitric oxide over Degussa P25. *Appl. Catal. B Environ.* **2006**, *62*, 208–216. [CrossRef]

110. Roy, S.; Hegde, M.S.; Ravishankar, N.; Madras, G. Creation of redox adsorption sites by Pd^{2+} ion substitution in nano-TiO_2 for high photocatalytic activity of CO oxidation, NO reduction, and NO decomposition. *J. Phys. Chem. C* **2007**, *111*, 8153–8160. [CrossRef]

111. Yin, S.; Liu, B.; Zhang, P.; Morikawa, T.; Yamanaka, K.-I.; Sato, T. Photocatalytic oxidation of NO_x under visible LED light irradiation over nitrogen-doped titania particles with iron or platinum loading. *J. Phys. Chem. C* **2008**, *112*, 12425–12431. [CrossRef]
112. Roy, S.; Hegde, M.S.; Madras, G. Catalysis for NO_x abatement. *Appl. Energy* **2009**, *86*, 2283–2297. [CrossRef]
113. Skalska, K.; Miller, J.S.; Ledakowicz, S. Trends in NO_x abatement: A review. *Sci. Total Environ.* **2010**, *408*, 3976–3989. [CrossRef] [PubMed]
114. Heo, I.; Kim, M.K.; Sung, S.; Nam, I.-S.; Cho, B.K.; Olson, K.L.; Li, W. Combination of photocatalysis and HC/SCR for improved activity and durability of deNOx catalysts. *Environ. Sci. Technol.* **2013**, *47*, 3657–3664. [CrossRef] [PubMed]
115. International Energy Agency. *2016 Energy and Air Pollution*; World Energy Outlook Special Report: Paris, France, 2016.
116. *Nitrogen Oxides (NO_x), Why and How They Are Controlled*; Technical Bulletin; Report EPA-456/F-99-006R; United States Environmental Protection Agency: Research Triangle Park, NC, USA, 1999.
117. Fujishima, A.; Zhang, X.; Tryk, D.A. TiO_2 photocatalysis and related surface phenomena. *Surf. Sci. Rep.* **2008**, *63*, 515–582. [CrossRef]
118. Hanus, M.J.; Harris, A.T. Nanotechnology innovations for the construction industry. *Prog. Mater. Sci.* **2013**, *58*, 1056–1102. [CrossRef]
119. Fresno, F.; Portela, R.; Suarez, S.; Coronado, J.M. Photocatalytic materials: Recent achievements and near future trends. *J. Mater. Chem. A* **2014**, *2*, 2863–2884. [CrossRef]
120. Church of 2000/Richard Meier & Partners. Available online: https://www.archdaily.com/20105/church-of-2000-richard-meier (accessed on 12 October 2018).
121. Borgarello, E. *TX Active® Principio Attivo Fotocatalitico—APPROFONDIMENTO TECNICO*; Italcementi, Italcementi Group: Bergamo, Italy, October 2009; Available online: https://www.construction21.org/italia/data/sources/users/62/txactiveapprofondimentoottobre2009ita.pdf (accessed on 12 October 2018).
122. Murata, Y.; Kamitami, K.; Takeuchi, K. Air purifying blocks based on photocatalysis. In Proceedings of the JIPEA World Congress, Tokyo, Japan, 17–21 September 2000.
123. Guerrini, G.L.; Peccati, E. *TUNNEL "UMBERTO I" IN ROME Monitoring Program Results*; Report No. 24; CTG-Italcementi Group: Bergamo, Italy, 22 April 2008; Available online: http://www.tiocem.pl/files/references/TX_Active_Tunnel_Umberto_I_ENG.pdf (accessed on 28 September 2018).
124. Guerini, G.L. *Case Study: The Italcementi TX Active® Story*. Cristal Global Conference, London, UK, 17 November 2011. Available online: http://www.cristalactiv.com/uploads/speaker/Case_Study_The_Italcementi_TX_Active_Story_Gian_Luca_Guerrini.pdf (accessed on 12 October 2018).
125. Martinez, T.; Bertron, A.; Ringot, E.; Escadeillas, G. Degradation of NO using photocatalytic coatings applied to different substrates. *Build. Environ.* **2011**, *46*, 1808–1816. [CrossRef]
126. Staub de Melo, J.V.; Trichês, G. Evaluation of the influence of environmental conditions on the efficiency of photocatalytic coatings in the degradation of nitrogen oxides (NO_x). *Build. Environ.* **2012**, *49*, 117–123. [CrossRef]
127. Ifang, S.; Gallus, M.; Liedtke, S.; Kurtenbach, R.; Wiesen, P.; Kleffmann, J. Standardization methods for testing photocatalytic air remediation materials: Problems and solution. *Atmos. Environ.* **2014**, *91*, 154–161. [CrossRef]
128. Zouzelka, R.; Rathousky, J. Photocatalytic abatement of NO_x pollutants in the air using commercial functional coating with porous morphology. *Appl. Catal. B Environ.* **2017**, *217*, 466–476. [CrossRef]
129. Macphee, D.E.; Folli, A. Photocatalytic concretes—The interface between photocatalysis and cement chemistry. *Cem. Concr. Res.* **2016**, *85*, 48–54. [CrossRef]
130. Horgnies, M.; Dubois-Brugger, I.; Gartner, E.M. NO_x de-pollution by hardened concrete and the influence of activated charcoal additions. *Cem. Concr. Res.* **2012**, *42*, 1348–1355. [CrossRef]
131. Bloh, J.Z.; Folli, A.; Macphee, D.E. Photocatalytic NO_x abatement: Why the selectivity matters. *RSC Adv.* **2014**, *4*, 45726–45734. [CrossRef]
132. Yang, L.; Hakki, A.; Wang, F.; Macphee, D.E. Photocatalyst efficiencies in concrete technology: The effect of photocatalyst placement. *Appl. Catal. B Environ.* **2018**, *222*, 200–208. [CrossRef]
133. Erme, K.; Raud, J.; Jogi, I. Adsorption of nitrogen oxides on TiO_2 surface as a function of NO_2 and N_2O_5 fraction in the gas phase. *Langmuir* **2018**, *34*, 6338–6345. [CrossRef] [PubMed]

134. Official Presentation—Innovative Façade Coatings with De-soiling and De-polluting Properties. In *The PICADA Project—Photocatalytic Innovative Coverings Applications for Depollution Assessment*; EC Project No. GRD1-2001-40449; GTM Construction: Nanterre, France, 2006.

135. Boonen, E.; Beeldens, A. Recent photocatalytic applications for air purification in Belgium. *Coatings* **2005**, *4*, 553–573. [CrossRef]

136. Maggos, T.; Plassais, A.; Bartzis, J.G. Photocatalytic degradation of NO_x in a pilot street canyon configuration using TiO_2-mortar panels. *Environ. Monit. Assess.* **2008**, *136*, 35–44. [CrossRef] [PubMed]

137. Guerrini, G.L.; Peccati, E. Photocatalytic cementitious roads for de-pollution. In Proceedings of the RILEM International Symposium on Photocatalysis 'Environment and Construction Materials', Florence, Italy, 8–9 October 2007; pp. 179–186.

138. Guerrini, G.L. Photocatalytic performances in a city tunnel in Rome: NO_x monitoring results. *Constr. Build. Mater.* **2012**, *27*, 165–175. [CrossRef]

139. Cassar, L. Photocatalysis of cementitious materials: Clean buildings and clean air. *MRS Bull.* **2004**, *29*, 328–331. [CrossRef]

140. Cassar, L.; Pepe, C.; Tognon, G.; Guerrini, G.L.; Amadelli, R. White cement for architectural concrete possessing photocatalytic properties. In Proceedings of the 11th International Congress on the Chemistry of Cement, Durban, South Africa, 11–16 May 2003; Volume 4, p. 12.

141. Guerrini, G.L.; Plassais, A.; Pepe, C.; Cassar, L. Use of photocatalytic cementitious materials for self-cleaning applications. In Proceedings of the RILEM International Symposium on Photocatalysis 'Environment and Construction Materials', Florence, Italy, 8–9 October 2007; pp. 219–226.

142. Guerrini, G.L.; Corazza, F. White cement and photocatalysis Part 1: Fundamentals. In Proceedings of the First Arab International Conference and Exhibition on The Uses of White Cement, Cairo, Egypt, 28–30 April 2008; Available online: https://www.researchgate.net/publication/266358310_WHITE_CEMENT_AND_PHOTOCATALYSIS_PART_1_FUNDAMENTALS (accessed on 29 September 2018).

143. NO_x Gas. Available online: http://www.airlite.com/air-quality/nox-gas/ (accessed on 28 September 2018).

144. Borgarello, E.; Harris, R.; Serpone, N. Photochemical deposition and photorecovery of gold using semiconductor dispersions. A practical application. *Nouv. J. Chim.* **1985**, *9*, 743–747.

145. Borgarello, E.; Terzian, R.; Serpone, N.; Pelizzetti, E.; Barbeni, M. Photocatalyzed transformation of cyanide to thio-cyanate by rhodium-loaded cadmium sulfide in alkaline aqueous sulfide media. *Inorg. Chem.* **1986**, *25*, 2135–2137. [CrossRef]

146. Borgarello, E.; Serpone, N.; Emo, G.; Harris, R.; Pelizzetti, E.; Minero, C. Light-induced reduction of Rh(III) and Pd(II) on TiO_2 dispersions, and the selective photochemical separation and recovery of Au(III), Pt(IV), and Rh(III) from dilute solutions. *Inorg. Chem.* **1986**, *25*, 4499–4503. [CrossRef]

147. Serpone, N.; Borgarello, E.; Barbeni, M.; Pelizzetti, E.; Pichat, P.; Herrmann, J.-M.; Fox, M.A. Photo-chemical reduction of gold(III) on semiconductor dispersions of TiO_2 in the presence of cyanide ions: Disposal of CN^- with H_2O_2. *J. Photochem.* **1987**, *36*, 373–388. [CrossRef]

148. Serpone, N.; Ah-You, Y.K.; Tran, T.P.; Harris, R.; Pelizzetti, E.; Hidaka, H. AM1 simulated sunlight photoreduction and elimination of Hg(II) and CH_3Hg(II) chloride salts from aqueous suspensions of titanium dioxide. *Sol. Energy* **1987**, *39*, 491–498. [CrossRef]

149. PICADA PROJECT—Performance Process Protocol, Workpackage 7 January 2006. Available online: http://www.picada-project.com/domino/SitePicada/Picada.nsf/1f9d19927a32e752c12569ab002c7ff8/50905a4f28b6ae58c12571320033f015/$FILE/D20.pdf (accessed on 18 September 2018).

150. Gurol, M.D. Photocatalytic Construction Materials and Reduction in Air Pollutants, San Diego State University, San Diego, CA, USA, March 2006. Available online: https://www.csus.edu/calst/FRFP/PHOTO-CATALYTIC.pdf (accessed on 18 September 2018).

151. Maggos, Th.; Kotzias, D.; Bartzis, J.G.; Leva, P.; Bellintani, A.; Vasilakos, C. Investigations of TiO_2-containing construction materials for the decomposition of NO_x in environmental chambers. In Proceedings of the 5th International Conference on Urban Air Quality, Valencia, Spain, 29–31 March 2005.

152. The 2010 Report Dutch Air Quality Innovation Programme Concluded. Available online: https://laqm.defra.gov.uk/documents/Dutch_Air_Quality_Innovation_Programme.pdf (accessed on 29 September 2018).

153. Jacobi, S. NO_2-Reduzierung Durch Photokatalytisch Wirksame Oberflächen? Modellversuch Fulda, (Hesse, Germany) 2012. Available online: https://www.hlnug.de/fileadmin/dokumente/das_hlug/jahresbericht/2012/jb2012_059-066_I2_Jacobi_final.pdf (accessed on 30 September 2018).

154. Photocatalytic Titanium Dioxide—A Demonstrated and Proven Technology (Cristal ACTiv™), Cristal Global, London, UK. Available online: http://www.cristalactiv.com/uploads/case/casePhotocatalysis%20-%20English.pdf (accessed on 1 October 2018).

155. Burton, A. Titanium dioxide photocleans polluted air. *Environ. Health Perspect.* **2012**, *120*, A229. [CrossRef] [PubMed]

156. Laufs, S.; Burgeth, G.; Duttlinger, W.; Kurtenbach, R.; Maban, M.; Thomas, C.; Wiesen, P.; Kleffmann, J. Conversion of nitrogen oxides on commercial photocatalytic dispersion paints. *Atmos. Environ.* **2010**, *44*, 2341–2349. [CrossRef]

157. Ballari, M.M.; Brouwers, H.J.H. Full scale demonstration of air-purifying pavement. *J. Hazard. Mater.* **2013**, *254–255*, 406–414. [CrossRef] [PubMed]

158. Boonen, E.; Akylas, V.; Barmpas, F.; Boreave, A.; Bottalico, L.; Cazaunau, M.; Chen, H.; Daele, V.; De Marco, T.; Doussin, J.F.; et al. Construction of a photo- catalytic de-polluting field site in the Leopold II tunnel in Brussels. *J. Environ. Manag.* **2015**, *155*, 136–144. [CrossRef] [PubMed]

159. Fraunhofer. Clean Air by Airclean®. 2010. Available online: http://www.ime.fraunhofer.de/content/dam/ime/de/documents/AOe/2009_2010_Saubere%20Luft%20durch%20Pflastersteine_s.pdf (accessed on 11 May 2015).

160. Tera. In Situ Study of the Air Pollution Mitigating Properties of Photocatalytic Coating, Tera Environement, (Contract Number 0941C0978), Report for ADEME and Rhone-Alpe region, France. Available online: http://www.air-rhonealpes.fr/site/media/telecharger/651413 (accessed on 11 May 2015).

161. Gustafsson, R.J.; Orlov, A.; Griffiths, P.T.; Cox, R.A.; Lambert, R.M. Reduction of NO_2 to nitrous acid on illuminated titanium dioxide aerosol surfaces: Implications for photocatalysis and atmospheric chemistry. *Chem. Commun.* **2006**, 3936–3938. [CrossRef] [PubMed]

162. Ndour, M.; D'Anna, B.; George, C.; Ka, O.; Balkanski, Y.; Kleffmann, J.; Stemmler, K.; Ammann, M. Photoenhanced uptake of NO_2 on mineral dust: Laboratory experiments and model simulations. *Geophys. Res. Lett.* **2008**, *35*, L05812. [CrossRef]

163. Beaumont, S.K.; Gustafsson, R.J.; Lambert, R.M. Heterogeneous photochemistry relevant to the troposphere: H_2O_2 production during the photochemical reduction of NO_2 to HONO on UV-illuminated TiO_2 surfaces. *Chem. Phys. Chem.* **2009**, *10*, 331–333. [CrossRef] [PubMed]

164. Monge, M.E.; D'Anna, B.; George, C. Nitrogen dioxide removal and nitrous acid formation on titanium oxide surfaces—An air quality remediation process? *Phys. Chem. Chem. Phys.* **2010**, *12*, 8991–8998. [CrossRef] [PubMed]

165. Mothes, F.; Herrmann, H. Lab and field studies on photocatalysis to improve urban air quality—Results from the PhotoPAQ project. In Proceedings of the Life MINOx-STREET Project Ending Meeting: Results and Conclusions, CIEMAT, Madrid, Spain, 21 March 2018.

166. Flassak, T. Numerical simulation of the depollution effectiveness of photocatalytic coverings in street canyons. In Proceedings of the Photocatalysis: Science and Application for Urban Air Quality, The 2012 LIFE+ PhotoPaq Conference, Proticcio, Island of Corsica, France, 14–17 May 2012.

167. Bolte, G.; Flassak, T. Numerische simulation der wirksamkeit photokatalytisch aktiver betonoberflächen. In Proceedings of the Internationale Baustofftagung 18, Ibausil (Proceedings), Bauhaus-University Weimar, Weimar, Germany, 12–15 September 2012; Fischer, H.-B., Bode, K.-A., Beuthan, C., Eds.; Bauhaus-University Weimar: Weimar, Germany, 2012.

168. Pujadas, M.; Palacios, M.; Nunez, L.; German, M.; Fernandez-Pampillon, J.; Iglesias, J.D.; Santiago, J.L. Real scale demonstration of the depolluting capabilities of a photocatalytic pavement in a real urban area. In Proceedings of the 17th International Conference on Harmonization within Atmospheric Dispersion Modeling for Regulatory Purposes, Budapest, Hungary, 9–12 May 2016.

169. Pujadas, M.; Palacios, M.; Nunez, L.; German, M.; Fernandez-Pampillon, J.; Sanchez, B.; Santiago, J.L.; Sanchez, B.; Munos, R.; Moral, F.; et al. Real scale tests of the depolluting capabilities of a photocatalytic sidewalk pavement and a façade in an urban scenario. In Proceedings of the 18th International Conference on Harmonization within Atmospheric Dispersion Modelling for Regulatory Purposes, Bologna, Italy, 9–12 October 2017.

170. Palacios, M.; Pujadas, M.; Nunez, L.; Sanchez, B.S.; Santiago, J.L.; Martilli, A.; Suarez, S.; Cabrero, B.S. Monitoring and modeling NO_x removal efficiency of photocatalytic materials: A strategy for urban air quality management. In Proceedings of the Life-Platform Meeting on Air, Barcelona, Spain, 26–27 September 2017.

171. Pujadas, M.; Palacios, M.; Nunez, L.; Fernandez-Pampillon, J.; German, M. Characterization of the NO_x depolluting effect of photocatalytic materials in a medium-scale tunnel reactor. In Proceedings of the Air Quality Meeting, Barcelona, Spain, 12–16 March 2018.

172. Wang, H.; Wu, Z.; Zhao, W.; Guan, B. Photocatalytic oxidation of nitrogen oxides using TiO_2 loading on woven glass fabric. *Chemosphere* **2007**, *66*, 185–190. [CrossRef] [PubMed]

173. Anpo, M.; Zhang, S.G.; Mishima, H.; Matsuoka, M.; Yamashita, H. Design of photocatalyst encapsulated within the zeolite framework and cavities for the decomposition of NO into N_2 and O_2 at normal temperature. *Catal. Today* **1997**, *39*, 159–168. [CrossRef]

174. Anpo, M.; Takeuchi, M.; Ikeue, K.; Dohshi, S. Design and development of titanium oxide photocatalysts operating under visible and UV light irradiation: The applications of metal ion-implantation techniques to semiconducting TiO_2 and Ti/zeolite catalysts. *Curr. Opin. Solid State Mater. Sci.* **2002**, *6*, 381–388. [CrossRef]

175. Yamashita, H.; Ichihashi, Y.; Zhang, S.G.; Matsumura, Y.; Souma, Y.; Tatsumi, T.; Anpo, M. Photocatalytic decomposition of NO at 275 K on titanium oxide catalysts anchored within zeolite cavities and framework. *Appl. Surf. Sci.* **1997**, *121*, 305–309. [CrossRef]

176. Wu, Q.; van de Krol, R. Selective photoreduction of nitric oxide to nitrogen by nanostructured TiO_2 photocatalysts: Role of oxygen vacancies and iron dopant. *J. Am. Chem. Soc.* **2012**, *134*, 9369–9375. [CrossRef] [PubMed]

177. Hu, Y.; Martra, G.; Zhang, J.; Higashimoto, S.; Coluccia, S.; Anpo, M. Characterization of the local structures of Ti-MCM-41 and their photocatalytic reactivity for the decomposition of NO into N_2 and O_2. *J. Phys. Chem. B* **2006**, *110*, 1680–1685. [CrossRef] [PubMed]

178. Cao, Y.; Yu, M.; Qi, S.; Ren, Z.; Yan, S.; Hu, S.; Xu, M. Nitric oxide reaction pathways on rutile $TiO_2(110)$: The influence of surface defects and reconstructions. *J. Phys. Chem. C* **2018**. [CrossRef]

179. Kuznetsov, V.N.; Glazkova, N.I.; Mikhaylov, R.V.; Kozhevina, A.V.; Serpone, N. Photophysics of color centers in visible-light-active rutile titania. Evidence of the photoformation and trapping of charge carriers from advanced diffuse reflectance spectroscopy and mass spectrometry. *Catal. Today* **2018**. [CrossRef]

180. Déak, P.; Aradi, B.; Frauenheim, T. Quantitative theory of the oxygen vacancy and carrier self-trapping in bulk TiO_2. *Phys. Rev. B* **2012**, *86*, 195206. [CrossRef]

181. Cotton, F.A.; Wilkinson, G. *Advanced Inorganic Chemistry*, 5th ed.; John Wiley & Sons: New York, NY, USA, 1988.

182. Iwamoto, M.; Furukawa, H.; Mine, Y.; Uemura, F.; Mikuriya, S.-I.; Kagawa, S. Copper(II) ion-exchanged ZSM-5 zeolites as highly active catalysts for direct and continuous decomposition of nitrogen monoxide. *J. Chem. Soc. Chem. Commun.* **1986**, 1272–1273. [CrossRef]

183. Chiu, K.W.; Savage, P.D.; Wilkinson, G.; Williams, D.J. Nitrosation of alkenes by nitric oxide: Crystal structures of bis-(1-nitroso-2-nitro-cyclohexane) and bis-(1-nitroso-2-nitro-1-phenylethane). *Polyhedron* **1985**, *4*, 1941–1945. [CrossRef]

184. Greenwood, N.N.; Earnshaw, A. *Chemistry of the Elements*, 2nd ed.; Butterworth-Heinemann: Oxford, UK, 1997.

185. Thiemann, M.; Scheibler, E.; Wiegand, K.W. Nitric Acid, Nitrous Acid, and Nitrogen Oxides. In *Ullmann's Encyclopedia of Industrial Chemistry*; Wiley-VCH: Weinheim, Germany, 2005.

186. Finlayson-Pitts, B.J.; Wingen, L.M.; Sumner, A.L.; Syomin, D.; Ramazan, K.A. The heterogeneous hydrolysis of NO_2 in laboratory systems and in outdoor and indoor atmospheres: An integrated mechanism. *Phys. Chem. Chem. Phys.* **2003**, *5*, 223–242. [CrossRef]

187. Sivachandrian, L.; Thevenet, F.; Gravejat, P.; Rousseau, A. Investigation of NO and NO_2 adsorption mechanisms on TiO_2 at room temperature. *Appl. Catal. B Environ.* **2013**, *142–143*, 196–204. [CrossRef]

188. Haubrich, J.; Quiller, R.G.; Benz, L.; Liu, Z.; Friend, C.M. In Situ ambient pressure studies of the chemistry of NO_2 and water on rutile $TiO_2(110)$. *Langmuir* **2010**, *26*, 2445–2451. [CrossRef] [PubMed]

 *catalysts*

Article

Mechanistic Study on Facet-Dependent Deposition of Metal Nanoparticles on Decahedral-Shaped Anatase Titania Photocatalyst Particles

Kenta Kobayashi [1], Mai Takashima [1,2,*], Mai Takase [3] and Bunsho Ohtani [1,2]

[1]   Graduate School of Environmental Science, Hokkaido University, Sapporo 060-0810, Japan; kobayashi.k@cat.hokudai.ac.jp (K.K.); ohtani@cat.hokudai.ac.jp (B.O.)
[2]   Institute for Catalysis, Hokkaido University, Sapporo 001-0021, Japan
[3]   Graduate School of Engineering, Muroran Institute of Technology, Mizumoto-cho, Muroran 050-8585, Japan; mai@mmm.muroran-it.ac.jp
*   Correspondence: takashima.m@cat.hokudai.ac.jp; Tel.: +81-11-706-9130

Received: 11 October 2018; Accepted: 9 November 2018; Published: 13 November 2018

Abstract: Facet-selective gold or platinum-nanoparticle deposition on decahedral-shaped anatase titania particles (DAPs) exposing {001} and {101} facets via photodeposition (PD) from metal-complex sources was reexamined using DAPs prepared with gas-phase reaction of titanium (IV) chloride and oxygen by quantitatively evaluating the area deposition density on {001} and {101} and comparing with the results of deposition from colloidal metal particles in the dark (CDD) or under photoirradiation (CDL). The observed facet selectivity, more or less {101} preferable, depended mainly on pH of the reaction suspensions and was almost non-selective at low pH regardless of the deposition method, PD or CDL, and the metal-source materials. Based on the results, the present authors propose that facet selectivity is attributable to surface charges (zeta potential) depending on the kind of facets, {001} and {101}, and pH of the reaction mixture and that this concept can explain the observed facet selectivity and possibly the reported facet selectivity without taking into account facet-selective reaction of photoexcited electrons and positive holes on {101} and {001} facets, respectively.

Keywords: decahedral-shaped anatase titania particles; {001} and {101} facets; facet-selective metal photodeposition; pH dependence; zeta potential; facet-selective reaction

---

1. Introduction

The term "charge separation" is one of the most attractive and convenient terms and/or concepts for researchers in the field of heterogeneous photocatalysis [1,2]. Since charge separation, i.e., spatial separation of a photoexcited electron in the conduction band and a positive hole in the valence band, is a kind of physical process and since the separated charges recombine easily with each other within a very short time period if there is no subsequent chemical reaction consuming those charges, direct observation of the single physical process of charge separation seems practically impossible. Time-resolved pump-probe spectroscopy using a femtosecond laser system has been reported to show such charge separation, i.e., accumulation of trapped photoexcited electrons was completed within a laser pulse and only the decay (disappearance) of those separated charges could be followed. To the best of the authors' knowledge, however, there has been no reported evidence of the primary step of charge separation itself. Then, why has "charge separation" been believed to occur in photocatalyst materials under photoirradiation? One possible reason is speculation assuming a mechanism as an analogy of a photoelectrochemical reaction of semiconductor electrodes, in which there is a space charge layer, i.e., an electric field in a semiconductor being in contact with the electrolyte; for n-type semiconductors such as metal oxides, positive holes and photoexcited electrons are made to migrate to the surface and the bulk of an electrode, respectively [2,3]. It should be noted that the depth of the space charge layer depends on the donor density of the electrode material, and it is known that ordinary (non-doped) semiconductor particles such as "white" titanium (IV) oxide particles have a negligible donor density to make the depth larger than the particle size, i.e., there may not be a space charge layer in those particles. Another reason for believing the occurrence of "charge separation" is simple; this concept is very convenient for interpretation of results of photocatalytic reactions; for example, high and low quantum efficiencies or photocatalytic activities have been explained by high and low extents of charge separation without showing direct evidence as described above.

What we observe (or can observe) is the results of chemical reactions that follow physical processes, photoexcitation and possible charge separation. Since the fate of electron-positive hole pairs is limited to an alternative, chemical reaction or recombination, the pairs that are not used to liberate photocatalytic-reaction products must disappear by mutual recombination [4]. Based on a simple kinetic assumption, the overall efficiency of electron-hole utilization is regulated by the ratio of the rate of this alternative; the efficiency must be high and low when the ratio is high and low, respectively. A frequently found misconception in papers on photocatalysis is low efficiency (low activity) being attributed only to faster recombination; slower electron/positive hole transfer to a substrate(s) can reduce efficiency/activity even though recombination occurs at a constant rate. Another misconception involves the recognition of charge recombination as a counter backward process of charge separation, i.e., charge recombination occurs because charges are not spatially separated. Thus, charge separation, a possible physical process just after photoabsorption, has been conveniently used in the interpretation of results of chemical reactions detected in chemical analysis without any support or evidence.

Anyway, if electron-positive hole pairs are created in the bulk of photocatalyst particles, charge separation would be expected to occur only when there is an internal electric field (IEF) in each particle; it seems impossible to separate negative and positive charges without an IEF overcoming the attractive electrostatic force between them. However, as described above, such an IEF does not seem to exist in not heavily (or negligibly) doped semiconducting materials, and even if there is an IEF from the surface to the bulk of a particle, a charge, electron or positive hole, separated to the bulk cannot react with a surface-adsorbed substrate(s). Therefore, when charge separation induced by an IEF is expected, a photocatalyst particle must have (i) two kinds of surfaces with different potentials and (ii) a smoothly changing bulk structure from one surface to the other surface forming a potential slope, i.e., an IEF in the bulk, though such a fine structure, especially providing (ii), seems unrealistic.

Titanium (IV) oxide (titania) is one of the most promising photocatalyst materials and is well known to exhibit high level of photocatalytic activity in various kinds of heterogeneous photocatalytic reactions. Three kinds of crystalline polymorphs of titania (with negligible occurrence of $TiO_2(B)$ [5,6]) have been found: Anatase, rutile and brookite as natural minerals in characteristic octahedral, complexed faceted and hexagonal plate-like crystal shapes, respectively. For natural anatase crystals, though they are predominantly octahedral crystals exposing eight equivalent {101} facets, decahedral-shaped crystals exposing an additional two {001} facets have occasionally been found. On the other hand, when titania particles were prepared in various procedures, octahedral-shaped anatase particles have been rarely found except for reports from the authors' group [7,8], presumably because natural crystals might grow very slowly for satisfying the thermodynamic requirement to expose only the most stable (lowest energy) facets, {101}. However, detailed analysis revealed that decahedral-shaped anatase particles (DAPs) existed in titania particles, e.g., Evonik (previously Degussa) P25 [9], though selective preparation of DAPs had not been reported before the publications of independent works by Yang and coworkers [10] and the present authors' group [11] using a hydrothermal reaction with a structure-controlling agent (SCA) and using a gas-phase reaction of titanium (IV) chloride and oxygen, respectively.

It has been believed that DAPs exposing two {001} and eight {101} facets have a high level of photocatalytic activity because photoexcited electrons and positive holes migrate to the {101} and {001} facets, respectively, resulting in efficient charge separation [12–23]. However, since such "charge separation" cannot be observed directly as mentioned above, the reason seems to be just speculation or a hypothesis proposed on the basis of microscopic observations of metal- and metal oxide-deposited DAPs through photocatalytic reduction and oxidation from their precursors on {101} and {001} facets, respectively [12–20]. There are also reports about facet-selective metal/metal-oxide deposition on the surface of rutile titania and other metal-oxide particles with a polyhedral shape [24–27]. These results have been attributed to facet-selective reduction and oxidation on {101} and {001} facets, i.e., as far as the authors know, all reports except for one [28] have indicated that photoexcited electrons and positive holes migrate selectively to {101} and {001} facets, respectively. It seems that this "facet-selective redox (FSR) hypothesis" has become established, and studies are now focusing on the possible mechanism, e.g., facet-dependent band positions [12–14], though the basic assumptions that (i) electron-hole pairs are created in the bulk and then move to the surface and (ii) electrons and holes migrate to different facets depending on the band energies seem inconsistent with (a) observation by femtosecond pump-probe laser spectroscopy of electrons and holes being trapped quickly in the surface states [29] and (b) the fact that charges, electrons and holes, cannot undergo separation and/or directional migration to escape from their electrostatic attraction without an electric field in the bulk of each particle.

However, before discussing the mechanism of FSR, experimental results of photocatalytic metal/metal-oxide deposition leading to the FSR hypothesis may have several problems. For example, (a) the effect of possibly remaining SCAs, used in order to prepare facetted particles, is neglected, (b) the facet-selective deposition has been evaluated only qualitatively using a few scanning electron microscopic images, without showing the number (or volume) ratio of deposits depending on the type of facets and (c) there has been no discussion of the possible migration of metal/metal-oxide deposits after photodeposition.

On the basis of the above-mentioned background, FSR on facetted anatase titania particles was re-examined in this study using (1) DAPs prepared by gas-phase reaction of titanium (IV) chloride and oxygen without using SCAs through (2) quantitative analysis by counting the number of deposited particles in order to evaluate the surface-density ratio of facet-selective deposition via an ordinary photocatalytic reaction and (3) photoassisted deposition from colloidal metal particles; points (1), (2) and (3) were introduced/employed to solve (or suppress) problems (a), (b) and (c), respectively.

2. Results and Discussion

2.1. Decahedral-Shaped Anatase Titania Particles (DAPs) Used in This Study

The sample particles, DAPs, were prepared by a previously reported procedure, coaxial-flow gas-phase reaction of titanium (IV) chloride and oxygen as schematically shown in Figure 1. One of the features of this DAP sample is its high level of purity, possibly because only titanium, chlorine and oxygen are involved in the preparation system, i.e., no SCAs and even hydrogen sources such as water or organic compounds are included. Although the reason why DAPs exposing only {101} and {001} facets are selectively prepared has not been clarified yet, it is speculated that DAPs are liberated as a lower surface area/volume ratio only exposing two kinds of facets ({101} and {001}) as the lowest and appreciably low surface energy [30,31], based on the observation that the observed oblateness (ratio of the short side to the long side of the {101} trapezoid (b/a: see Materials and Methods 3.5.)) was ca. 0.7 and it was not changed even when the preparation conditions were modified (See Materials and Methods). The lowest surface area (nm^2)-volume (nm^3) ratio (SV ratio; 8.2 nm^{-1}) of ideal decahedral particles is expected to be obtained with oblateness of ca. 0.5 assuming the same surface energy of {101} and {001} facets. This oblateness of the lowest SV ratio seems smaller than the observed ratio, though a plot of SV ratio as a function of oblateness seems parabolic as shown in Section 3.5 and the SV ratio at 0.7 oblateness, 8.9 nm^{-1}, seems not so high compared with the SV ratio at 0.5. Although there is still a possibility that the sole by-product of the gas-phase reaction, chlorine, remained on the {001} facets even after thorough washing with water and this led to lowering of the {001} surface energy and high oblateness, it can be stated that the particle shape is not controlled by the surface energy (stability) of each particle, not surface modifiers, and this sample seems suitable for study on FSR behavior. In any case, even if the observed oblateness was 0.7, more than 70% of the entire exposed surface area of a DAP was {101}.

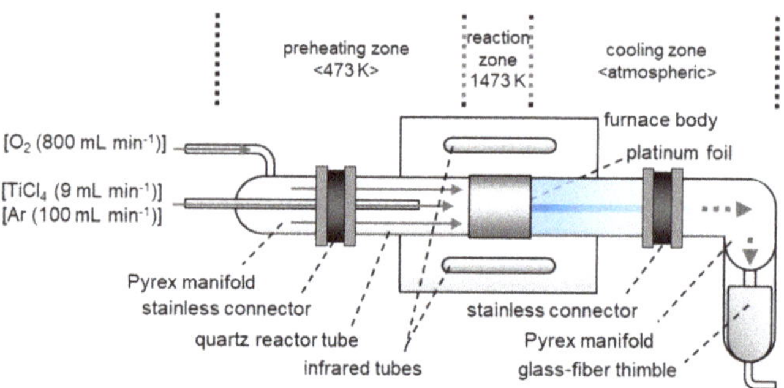

Figure 1. Schematic representation of the DAP synthesis process by coaxial-flow gas-phase reaction of titanium (IV) chloride and oxygen.

Figure 2 shows a representative electron-microscopic image of the sample. In this image, all of the particles have a sharp-edged decahedral shape with sizes of ca. 50–200 nm. Ridges in these observed DAPs looked white due to the "edge effect", i.e., secondary-electron emission occurs preferably at sharp edges. In other words, the DAP samples used in this study have sharp edges. Other images for the samples showed a similar trend except for a few particles with a non-decahedral shape, which might be rutile or non-crystalline titania particles as described below. The crystallite size of anatase evaluated from the XRD patterns with the Scherrer equation was ca. 75 nm (d_{101}: 71 nm, d_{004}: 68 nm and d_{200}: 87 nm), which was almost the same as the above-mentioned particle size evaluated from SEM images. This rough coincidence suggests that each DAP appearing in SEM images was a single crystal anatase particle.

Figure 2. FE-SEM image of DAPs synthesized by coaxial-flow gas-phase reaction.

The crystalline content of the DAP sample used was shown by Rietveld analysis of the diffraction pattern to be 89% anatase and 4% rutile with a 7% non-crystal component.

2.2. Deposition of Metal Particles on DAPs

SEM images of gold and platinum-deposited DAPs through photodeposition (PD; (a) and (d)) from precursors and colloid deposition in the dark (CDD; (b) and (e)) and that under photoirradiation (CDL; (c) and (f)) are shown in Figure 3. The roughly estimated particle sizes of both gold and platinum deposits, shown in Figure 3, were in the range of 4–12 nm (Table 1) and no distinct difference in the size depending on the kind of facets was observed. For metal particles photodeposited from their precursors, the size was 4–5 nm regardless of the kind of metals, though the possibility of small (<1 nm) particle formation could not be excluded due to the resolution of SEM analysis in this study. The average particle size of the original gold colloid was ca. 12 nm and this was not changed by loading under photoirradiation and in the dark, while the original size of platinum colloid particles was ca. 5 nm. The two-times larger size after deposition in the dark is attributable to possible aggregation of a few platinum particles, though each of them was observed as one particle.

Figure 3. Representative FE-SEM images of DAPs deposited with gold (**a**–**c**) and platinum (**d**–**f**) by photodeposition (PD) from their precursors (HAuCl$_4$ and H$_2$PtCl$_6$, respectively) (**a**,**d**), deposition from their colloids in the dark (CDD; **b**,**e**) and under photoirradiation (CDL;**c**,**f**).

Table 1. Summary of the results for deposition of metals on DAPs.

Entry	Source	Amount [1] (wt %)	Medium	Size [2]/nm	pH [3]	$D_{\{001\}}$ [4]/10^{-4} nm^{-2}	$D_{\{101\}}$ [5]/10^{-4} nm^{-2}	S [6]
PD [7](Au)	HAuCl$_4$	0.5	MeOH [8]	5	— [9]	3.9	6.4	0.61
CDD [10](Au)	Au colloid	2.0	water	12	— [9]	0.065	0.37	0.18
CDL [11](Au)	Au colloid	0.5	water	12	7.6	0.011	0.25	0.04
PD [7](Pt)	H$_2$PtCl$_6$	0.5	MeOH [8]	4	7.2	4.5	16.0	0.28
PD [7](Pt/CA)	H$_2$PtCl$_6$	0.5	MeOH [8] + CA [12]	5	2.5	5.1	8.2	0.62
CDD [10](Pt)	Pt colloid	2.0	water	10	— [9]	0.12	0.48	0.24
CDL [11](Pt)	Pt colloid	0.5	water	5	7.0	0.76	16.0	0.05
CDL [11](Pt)	Pt colloid	2.0	water	5	8.9	0.676	22.0	0.03
CDL [11](Pt/CA)	Pt colloid	0.5	water +CA [12]	4	2.4	6.6	9.0	0.73

[1] Amount as metal. [2] Roughly estimated average size of metal deposits assuming a spherical shape. [3] Measured after deposition. [4] Area deposition density of metal deposits for {001} facets. [5] Area deposition density of metal deposits for {101} facets. [6] Facet selectivity (=$D_{\{001\}}/D_{\{101\}}$). [7] Photodeposition. [8] 50 vol % aqueous methanol. [9] Not measured. [10] Colloid deposition in the dark. [11] Colloid deposition under UV irradiation. [12] Citric acid (0.1 mol L^{-1}).

As a general trend, gold and platinum particles were deposited preferably on {101} facets. In order to evaluate the facet selectivity quantitatively, facet selectivity (s) was defined as the ratio of number (not volume) density per unit area, i.e., area density, of metal deposits on {001} ($D_{\{001\}}$) and {101} ($D_{\{001\}}$) facets determined by counting more than 100 DAPs in several SEM images for each sample (See Materials and Methods). Since the exposed surface was predominantly {101} facets for DAPs used in this study, a comparison without consideration of the surface areas of the two kinds of facets does not seem to make sense.

For the photodeposited samples, the s values of gold and platinum (in the absence of citric acid) were 0.61 and 0.28, respectively (Table 1), both of which were below 1, i.e., photodeposition proceeded preferably on {101} facets. Although these selectivities were not 0%-or-100%, the tendency of preferential photodeposition on {101} facets seems consistent with previously reported results.

2.3. Mechanism of Metal Deposition on DAPs and FSR

Although it has not been discussed so far, the mechanism of photocatalytic metal-particle deposition does not seem to be straightforward since the formation of deposited metal particles composed of a large number of metal atoms requires a process to make metal-metal bonds. At least three steps may be included. (1) The first step is reduction of precursor metal-complex anions ($AuCl_4^-$ and $PtCl_6^{2-}$) adsorbed on the surface of photocatalyst particles by photoexcited electrons along with oxidation of a sacrificial hole scavenger, methanol in the present case, by positive holes. (2a) The next step is migration of low-valent or metal atoms and/or clusters to grow into metal particles (as detected in SEM analysis). It is expected that the larger the size of meal clusters or nanoparticles is, the lower is the possibility of surface migration. (2b) The next step is reduction of metal precursor anions on the surface of formed metal nanoparticles being in contact with titania. Both steps (2a) and (2b) are probable, and facet-selective deposition is regulated by (i) facet-selective migration of photoexcited electrons and positive holes in step (1), (ii) facet-selective adsorption of metal precursor anions in step (1/2b) and/or (iii) surface migration and facet-selective attachment (fixation) of metal clusters/nanoparticles in steps (2a/2b). At least one of them, if there are no other possible reasons, can lead to a value of s that is different from unity, though previous reports suggested reason (i) without showing evidence that excludes the possibility of (ii) and (iii).

In order to check the possibility of (ii) and (iii), deposition of metal particles on DAPs from gold and platinum colloid solutions was examined in the dark (CDD) and under photoirradiation (CDL). It is well known that colloidal particles prepared using citric acid, which was used in this study, are stabilized by coverage of their surface with citric acid and repulsion with negative charges with carboxylate groups, i.e., the metal-particle surface is negatively charged. For CDD, as shown in Table 1, the area deposition density (D) was lower than that with PD presumably because there seems to be no driving force for colloid particles to settle on the surface (other than interaction with possible protonated hydroxyl groups on the DAP surface) and a higher concentration (2 wt %) of the colloid solution was necessary to count the number of deposited metal-particles. On the other hand, CDL gave a high D compared to that with CDD. Based on the fact that hydrogen and carbon dioxide production was detected and the fact that the particle size of CDL-deposited platinum was almost half of that of CDD-deposited platinum, oxidative decomposition of surface-covering citric acid to expose the bare metal surface proceeded as a counter reaction of hydrogen evolution by photoexcited electrons.

One of the interesting features is that s was less than unity in all cases (PD, CDD and CDL), suggesting that the facet selectivity was governed by the metal-nanoparticle fixation (attachment) process, not the redox process. Another interesting feature is that CDL gave very low s, i.e., high {101} selectivity, which was induced by lowering $D_{\{001\}}$ and enhancement of $D_{\{101\}}$ in gold and platinum deposition results, respectively. It seems that these results cannot be interpreted by the conventional mechanism including facet-selective migration of electrons and positive holes if the above-mentioned photocatalytic oxidative decomposition of citric acid on the surface of metal particles induces deposition and no migration of metal particles occurs after the deposition.

2.4. Influence of pH on Metal Deposition on DAPs

Then, what governs the observed facet selectivity? One possible candidate is the surface charge depending on the kind of facets; in all of the deposition methods, PD, CDD and CDL, negatively charged substances, metal-precursor anions or metal colloid particles, are deposited on the surface and such deposition must be influenced by the surface charge of DAPs.

Figure 4 shows representative SEM images of platinum-deposited DAPs obtained by (a) PD and (b) CDL under acidic conditions. Citric acid was chosen for acidification of the reaction medium since colloidal metal particles contained this acid as a surface stabilizer. As easily seen in those images, platinum particles were deposited both on {001} and {101} facets and this tendency was evaluated by counting more than 100 DAPs as shown in Table 1 (PD(Pt/CA) and CDL(Pt/CA)); the values of *s* were greatly increased by the acidification to 0.62 and 0.73, respectively, which were almost non-selective. On the other hand, a markedly low *s* value, high {101} selectivity, was observed when pH of the colloidal suspension was basic (pH = 8.9) by using a higher concentration of platinum colloid (2 wt %) (Table 1). Thus, *s* depended strongly on pH of suspensions in both PD and CDL platinization.

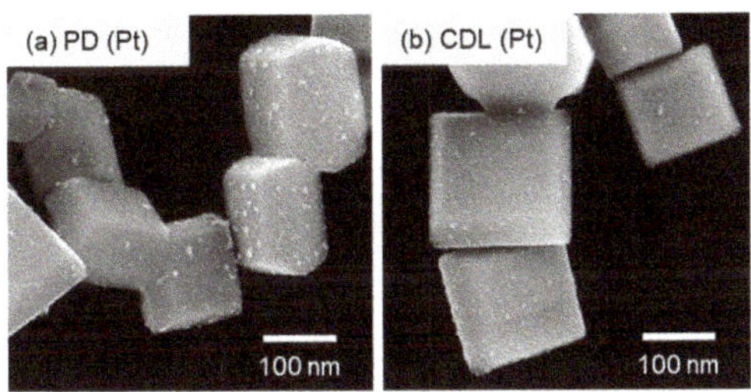

Figure 4. FE-SEM images of metal-deposited DAPs under acidic conditions by: (**a**) photodeposition; (**b**) colloid deposition under UV irradiation.

The effect of pH (acidification) is summarized in Figure 5a. The plots for PD and CDL platinization processes in Figure 5a seem to show resemblance, i.e., their pH dependence may originate from the same effect. The drastic decrease in the *s* value along with pH was caused by the drastic (Note that the *D* plots are shown in a logarithmic scale.) decrease in $D_{\{001\}}$, while $D_{\{101\}}$ was constant or slightly increased with an increase in pH raise regardless of the procedure, PD or CDL, and deposition amount, 0.5 or 2.0 wt %.

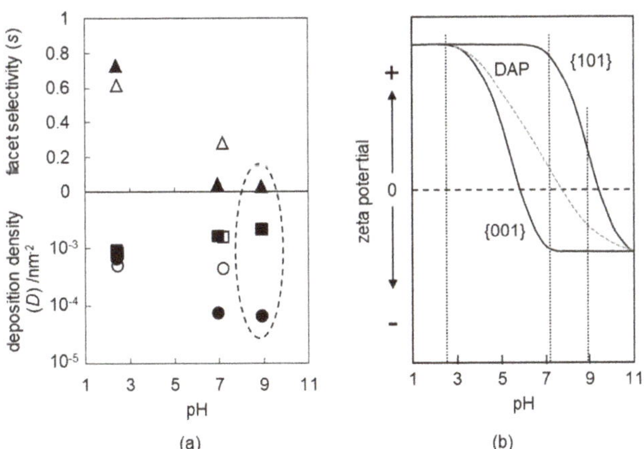

Figure 5. (**a**) pH dependence of facet selectivity (*s*) and deposition density (*D*) in PD (open symbols) and CDL (closed symbols) platinum deposition. For deposition-density plots, circles and squares correspond to the density of {001} and {101} facets, respectively. Plots in the dotted oval reflect samples deposited with a 2 wt % platinum-colloid solution. (**b**) Hypothetical pH-dependent zeta-potential curves for {101} and {001} facets. A dashed line is a rough sketch of actual zeta-potential measurement of a DAP sample giving point of zero charge at ca. 7. It should be noted that the DAP sample used in this zeta-potential change was prepared under slightly different reaction conditions.

One of the possible reasons for such pH dependence is pH-dependent and facet-dependent surface charges and, assuming negatively charged species $PtCl_6^-$ anions or negatively charged platinum colloidal particles are approaching the surface for platinum deposition, changes in surface charge by pH for {101} and {001} facets are different. It is well known that there are appreciable amounts of hydroxyl groups on the surfaces of metal-oxide particles, and protonation/deprotonation depending on pH of a surrounding medium gives protonated ($-OH_2^+$), neutral ($-OH$) and deprotonated ($-O^-$) forms, resulting in the observed zeta-potential curves; the surface of titania particles is positively charged at low pH, decreased by pH increase and then negatively charged at high pH.

A hypothesis is that a zeta-potential curve for {001} facets is shifted to the lower-pH side from that of {101} facets as depicted in Figure 5b, being consistent with the actual zeta-potential curve for a DAP sample shown in the figure as a rough sketch. As has been reported for the facet-selective surface charge for bismuth oxybromide [26], different facets with different surface energies may have different surface charges.

The hypothetical zeta-potential curves for {101} and {001} facets can reasonably interpret the observed pH-dependent s and deposition density (D) shown in Figure 5a as follows. At low pH, both {101} and {001} facets are positively charged with protonated surface hydroxyls ($-OH_2^+$) to induce attraction of negatively charged precursor anions/colloidal particles and thereby non facet-selective deposition occurs at pH = 2.4–2.5. At neutral pH, the average charge on {001} facets is decreased, while {101} facets are still positively charged to decrease s at pH ca. 7. A further increase in pH leads to negatively charged ($-O^-$) {001} facets to give negligible facet selectivity at pH = 8.9. At higher pH, it is expected that both {101} and {001} facets are negatively charged, resulting in negligible deposition densities. Under such high pH conditions, however, large platinum particles were formed and the number of deposits was very small (data not shown), presumably because precursor small metal particles or colloidal particles were aggregated by neutralization of surface negative charges by sodium cations.

Although there has been no experimental evidence for the above-mentioned hypothetical zeta-potential curves due to the lack of a technique for measuring zeta-potentials of each facet on a DAP, the observed facet selectivity in PD and CDL-induced deposition of platinum nanoparticles on DAPs can be consistently explained without taking "charge separation" (FSR) into account. Since there also seems to be no direct evidence for FSR, i.e., speculated only from the position of metal and/or metal oxide deposition, the above-mentioned effect of facet-dependent variation in surface charge may still be a possible reason for the facet-selective (or non-selective) deposition of metals as an alternative of FSR.

2.5. Influence of Stirring Operation on the Deposition Density of Metal Nanoparticles on DAPs

When the above-mentioned interaction between the charged surface (facets) and precursor anions or metal nanoparticles is assumed, the possible detachment of the photodeposited metal nanoparticles should be examined. Figure 6 shows the effect of post-irradiation stirring in the dark on Ds for (a) gold and (b) platinum deposition. Although the plots seemed rather scattered, the densities were decreased by post-irradiation stirring except for platinum deposition on {101} facets. For gold deposition, the densities on both facets were decreased, but it seemed that gold deposits on {001} facets tended to be detached faster than those on {101} facets to result in a lower facet-selectivity value. On the other hand, platinum deposits were more stable than gold deposits and the trend of decrease in D was not obvious for both {101} and {001} facets. The observation of a decrease in the number of deposited metal particles by post-irradiation stirring in the dark indicates that the deposited metal particles can be detached and thereby possibly aggregated to larger particles.

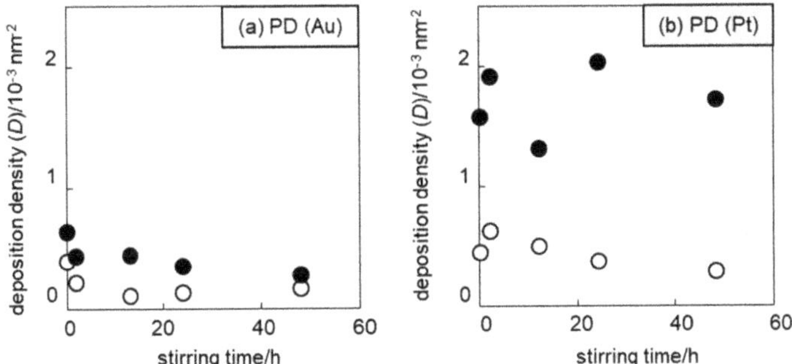

Figure 6. Change in deposition density (D) by post-irradiation stirring of PD processes for (**a**) gold and (**b**) platinum. Open and closed circles correspond to D of {001} and {101} facets, respectively.

Figure 7 shows the change in the particle size distribution of gold nanoparticles deposited on each facet of DAPs with the time of stirring in the dark after PD. Although only ca. 200 gold particles were counted in each distribution, the number of which seems too small for reliable statistical analysis, the shape of the distribution patterns seemed to change with stirring time. However, a change in the average particle size on each facet was not obvious and thereby the detachment of deposited gold particles proceeded almost homogeneously for both {101} and {001} facets. Being consistent with the results showing that {001} facets tend to release gold nanoparticles faster than do {101} facets as shown in Figure 6, the average gold-nanoparticle size on {001} facets was a slightly smaller than that on {101} facets, i.e., {101} facets might be able to keep larger particles than those can be kept on {001} facets.

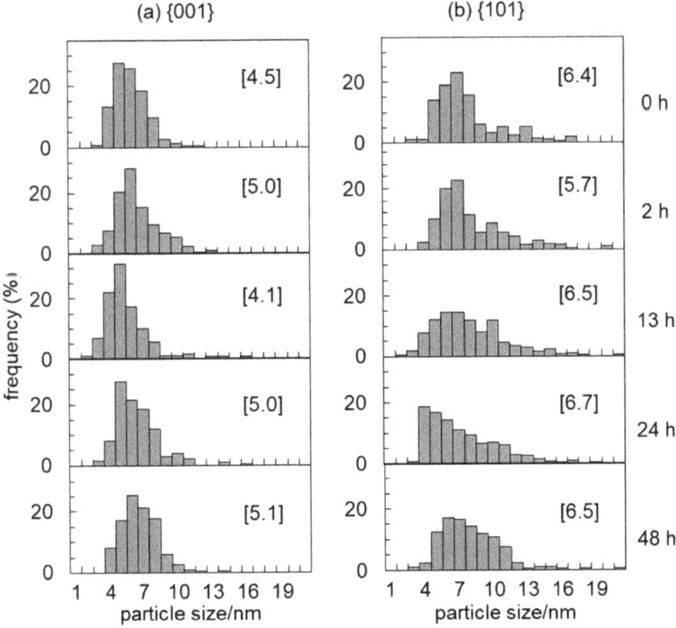

Figure 7. Change in distribution of gold-particle size on (**a**) {001} and (**b**) {101} facets by stirring in the dark for (top) 0 h, 2 h, 13 h, 24 h and (bottom) 48 h. Frequency was standardized to be 100% for the 0-h samples and summation of distribution is proportional to the number of remaining gold nanoparticles. Figures in square brackets show average particle size in the unit of nm.

Thus, the results suggested that the distribution of metal nanoparticles on {001} and {101} facets depends on the size and that metal particles that have been deposited can be detached during the process of PD.

2.6. Photodeposition of Metal Particles on DAPs under No Stirring Conditions

Since, as described in the preceding section, metal particles deposited in PD processes can be detached from the DAP surface, further experiments using PD without magnetic stirring were performed to reduce the detachment, and the results are summarized in Table 2. In these experiments, DAPs were fixed on a glass plate and irradiated in aqueous methanol containing metal complexes (2 wt % as metal) (See Materials and Methods). Ds were evaluated neglecting DAPs without any metal deposits because unlike PD under the condition of magnetic stirring, only the surface layer of the DAP film absorbs light for metal deposition. Even though such modification in the D evaluation scheme was adopted and a higher concentration of the metal source was used, the actual densities were comparable or even lower than those of PD with magnetic stirring (Table 1). In all of the cases shown in Table 2, s was in the middle range, 0.4–0.6, i.e., ambiguous facet selectivity. Although it is difficult to compare the facet selectivity with that obtained for deposition with magnetic stirring (Table 1) and although facet selectivity for gold deposition without stirring was even decreased, the above-mentioned ambiguous facet selectivity again suggested less probable FSR.

Table 2. Summary of results on PD deposition of metals on DAPs without magnetic stirring.

Entry	Source	Amount [1] (wt %)	Medium	Size [2]/nm	pH [3]	$D_{\{001\}}$ [4]$/10^{-4}$ nm^{-2}	$D_{\{101\}}$ [5]$/10^{-4}$ nm^{-2}	s [6]
PD [7](Au)	HAuCl$_4$	2.0	MeOH [8]	5	— [9]	0.28	0.75	0.38
PD [7](Pt)	H$_2$PtCl$_6$	2.0	MeOH [8]	4	7.2	1.3	3.3	0.39
PD [7](Pt)	[Pt(NH$_3$)$_4$]Cl$_2$	2.0	MeOH [8]	5	7.4	2.0	3.5	0.58

[1] Amount as metal. [2] Roughly estimated average size of metal deposits assuming spherical shape. [3] Measured after deposition. [4] Area deposition density of metal deposits for {001} facets. [5] Area deposition density of metal deposits for {101} facets. [6] Facet selectivity (= $D_{\{001\}}/D_{\{101\}}$). [7] Photodeposition. [8] 50 vol % aqueous methanol. [9] Not measured.

Another feature seen in Table 2 is that when a cationic precursor, $[Pt(NH_3)_4]^{2+}$, was used, the Ds for platinum were comparable or even higher than those for deposition with an anionic precursor, $PtCl_6^{2-}$. This fact suggests that the precursor for deposition which is affected by the surface charge, based on the assumption that D is governed by the surface charge, is small metal particles/clusters, not source metal-complex ions, e.g., $PtCl_6^{2-}$ or $[Pt(NH_3)_4]^{2+}$; in the initial stage of PD, small metal particles/clusters are created followed by migration of these precursors with possible particle growth to be fixed on the surface, and the position of deposition in the second step is regulated by surface charges depending on the kind of facets, {101} or {001}. It should be noted that this proposed mechanism does not exclude the possibility of FSR in the first step, and even though facet-selective metal deposition is observed, it seems that this does not prove the occurrence of FSR.

3. Materials and Methods

3.1. Preparation and Characterization of DAP Samples

DAP samples were prepared by a gas-phase reaction of titanium (IV) chloride (TiCl$_4$; Wako, Tokyo, Japan) and oxygen (O$_2$) using a coaxial-flow gas-phase reactor as reported previously (modified from the original procedure [32]). A brief description of the procedure is as follows. An argon (Ar; >99.99%; purified by a Shimadzu (Kyoto, Japan) GLC Click-on Triple (hydrocarbon, oxygen and moisture) trap) stream (100 mL min^{-1}, 453 K) containing 1 vol % TiCl$_4$ (Wako, Tokyo, Japan)(quantitatively introduced by a syringe feeder) vapor and an O$_2$ stream (>99.5%, dried and purified by a Shimadzu (Kyoto, Japan) GLC Click-on Combi (hydrocarbon and moisture) trap; 800 mL min^{-1}) are introduced into a quartz reactor tube as inside and outside, respectively, coaxial flow and heated from platinum foil (3.0 cm), wrapped around the quartz reactor tube, the temperature of which is kept at 1473 K by infrared lamps (Advance Riko VHT-E44, Yokohama., Japan; totally 2 kW maximum). The preheated outside O_2 flow is heated by the 1473-K wrapped platinum foil prior to the inside $TiCl_4$/Ar stream and expanded toward the center to react with $TiCl_4$ ($TiCl_4 + O_2 \rightarrow TiO_2 + 2Cl_2$). The resultant white titania smoke flows at the center of the reactor tube (see Figure 1) and is collected by a glass-fiber filter thimble (Whatman high-purity glass microfiber extraction thimble, 25 mm × 90 mm, Tokyo, Japan). The white product is washed with water five times to remove possibly adsorbed chlorine and then freeze-dried (EYELA FDU-2100, Tokyo, Japan) under vacuum (<10 Pa) for 24 h.

The DAP samples were characterized by X-ray diffractometry (XRD) and scanning electron microscopy (SEM) with a Rigaku SmartLab X-ray diffractometer with CuK_α radiation (40 kV, 30 mA, Rigaku, Akishima, Japan) and a JEOL JSM-7400F microscope (JEOL, Akishima, Japan), respectively. The details of XRD measurements are as follows: A DAP sample and 20 wt % nickel oxide (NiO; Wako, Tokyo, Japan) as an internal crystalline standard [33] were mixed thoroughly in an agate mortar, and the XRD pattern of the mixture was recorded with a scanning rate of $1.0°$ min^{-1} and steps of $0.008°$ in the 2θ range of 10–90°. Recorded diffractograms were analyzed using the software PDXL 2 (Version 2.6.1.2, Rigaku, Akishima, Japan) including a RIETAN-FP Rietveld analysis package [33]. Crystallite size, i.e., primary particle size, was estimated by the Scherrer equation with corrected average peak width of anatase 101, 004 and 200 peaks at 2θ of ca. 25.4°, 37.8° and 48.0°, respectively. The detailed conditions and procedure for SEM analysis of the DAP samples are described in the following sections.

3.2. Photodeposition of Metal Nanoparticles on DAP Samples from Metal Complexes

In the process of PD of platinum and gold, a 30-mL solution of 50 vol % aqueous methanol (Wako, Tokyo, Japan) containing hydrogen hexachloroplatinum(IV) (H_2PtCl_6, Wako, Tokyo, Japan) or aqueous hydrogen tetrachlorogold(III) ($HAuCl_4$, Wako, Tokyo, Japan) (0.5 wt % (or 2.0 wt %) as metal) was poured in a glass tube containing 0.015 g of DAP. In some experiments, tetraammineplatinum (II) chloride ([$Pt(NH_3)_4$]Cl_2, Wako, Tokyo, Japan) was used instead of H_2PtCl_6, and citric acid (Wako, Tokyo, Japan; 0.1 mol L^{-1} in a suspension) was added to acidify the suspension. The suspension was sonicated to be homogenized, deaerated by argon bubbling, and then irradiated by a 400-W mercury arc (>290 nm; Eiko-sha 400) with vigorous magnetic stirring at 1000 rpm. After 15-min irradiation and 2-h irradiation for platinum and gold, respectively, pH of the suspension was measured using a pH meter (Horiba pH meter LAQUA twin, Kyoto, Japan) and the powder was recovered by centrifugation, washed three times with Milli-Q water, and freeze-dried under vacuum (<10 Pa) for 24 h.

3.3. Photodeposition of Metal Nanoparticles on DAP Samples from Metal Colloids

Platinum and gold colloid solutions were prepared following the reported procedures for platinum [34] and gold [35], respectively. For platinum colloid, a 196-mL portion of an aqueous H_2PtCl_6 solution (0.30 mmol L^{-1}) was heated to be refluxed by a mantle heater (MS-ES-3, As one, Osaka, Japan) under magnetic stirring. Then 4.0 mL of aqueous sodium-citrate solution (0.84 mol L^{-1}) was added and the reaction mixture was kept boiling for 45 min. After being cooled down rapidly in an ice bath, excess citric acid and inorganic salts in the resultant colloidal solutions were removed by being passed through an ion exchange resin (Organo Amberlite MB-1, Tokyo, Japan)-packed column. For gold colloid, a 202-mL portion of an aqueous $HAuCl_4$ solution (1.0 mmol L^{-1}) was heated to be refluxed, and then 24 mL of sodium-citrate solution (0.039 mol L^{-1}) was added followed by maintenance of reflux for 30 min. The workup procedure was the same as that for the above-mentioned platinum colloid preparation.

Deposition of platinum nanoparticles and deposition of gold nanoparticles under CDL or CDD were performed using 0.015 g and 0.044 g, respectively, of DAP suspended in a 5.0-mL colloid solution containing the required amount of the metal. The reaction mixture was kept at 298 K with magnetic stirring under photoirradiation, with the same setup as that for PD, or in the dark. The resultant powder was recovered by centrifugation, washed three times with Milli-Q water, and freeze-drying for 24 h.

3.4. Photodeposition of Metal Nanoparticles on DAP Samples without Agitation

A 0.20-mL portion of a sonicated DAP suspension (10 mg mL^{-1}) was poured onto a glass plate and dried in vacuum at ambient temperature for 24 h. The DAP-coated glass plate was immersed in 50 vol % aqueous methanol containing a metal complex, H_2PtCl_6, $HAuCl_4$ or $[Pt(NH_3)_4]Cl_2$ (2 wt % as metal), and irradiated by a mercury arc at >290 nm for 2 h. After metal deposition, the DAP-coated glass plate was dried at room condition without washing procedure.

3.5. Evaluation of Deposition Densities and Facet Selectivity of Metal Deposition

The metal nanoparticle-deposited DAPs were analyzed by electron microscopy using a field emission-type scanning electron microscope (FE-SEM; JEOL JSM-7400M, Yokohama, Japan) in a mode of secondary electron image (SEI) with operating conditions of 5.0–10.0-kV electron-acceleration voltage, 10.0-μA current and 3–6-mm working distance. Evaluation of D and s was performed by counting the number, not volume, of metal deposits per unit area in FE-SEM images as follows. First, the number of deposited metal nanoparticles ($N_{\{001\}}$ and $N_{\{101\}}$) and total area of deposited facets ($S_{\{001\}}$ and $S_{\{101\}}$) were measured using several SEM images for {001} and {101} facets, respectively. In order to keep statistical reliability and reproducibility, more than 100 metal-deposited DAPs were counted. For the area measurement, the following equations were used to estimate the area of each facet, $S_{\{001\}}$ and $S_{\{101\}}$, with the measured lengths of two ridges of a DAP, long (a) and short (b) sides of a {101} trapezoid (Figure 8a). Then the Ds on each facet, $D_{\{001\}}$ and $D_{\{101\}}$, were obtained as N/S, and s was calculated as $D_{\{001\}}/D_{\{101\}}$. Values of s of more than 1, 1 and less than 1 mean {001} selective, non selective and {101} selective, respectively.

$$S_{\{001\}} = b^2 \tag{1}$$

$$S_{\{101\}} = (a + b) \times h/2 = (a + b) \times (a - b) \times \tan 69.7°/4 \tag{2}$$

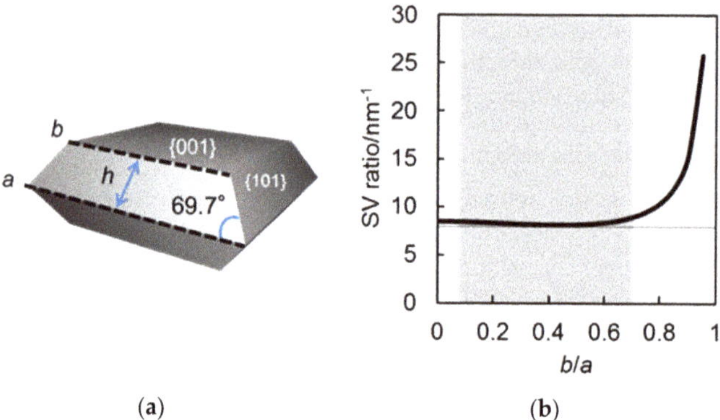

Figure 8. (**a**) Assumed dimension of a DAP and (**b**) surface area-volume (SV) ratio as a function of oblateness, b/a, of an ideal {101} trapezoid. The grey part corresponds to the oblateness range giving an SV ratio within excess 10% of the minimum value (8.2 nm^{-1}).

The surface area-volume (SV) ratio, S_{total}/V, of a DAP was calculated using the dimension shown in Figure 8b.

4. Conclusions

As described above, metal-nanoparticle deposition was reexamined using DAPs synthesized by gas-phase reaction of titanium (IV) chloride and oxygen in the absence of a so-called SCA to obtain the following three significant aspects.

One is that the frequently reported almost perfect (0%-or-100%) facet selectivity for photocatalytic deposition, PD, i.e., reductive metal deposition and oxidative metal-oxide deposition on {101} and {001} facets, respectively, due to FSR could not be observed, at least for the DAP samples used in this study, though there seemed to be a tendency of {101}-selective deposition. The difference from previously reported results may be due to (i) the conclusion of 0%-or-100% selectivity in previous studies by using only one or a few microscopic images matching the FSR concept or (ii) an appreciable difference in the DAP surface structures, e.g., our samples being influenced by the negligibly remaining chlorine or the surfaces of previous samples prepared through liquid-phase processes being covered by an SCA or the others.

The second aspect is that the change in the observed facet selectivity with different reaction conditions was similar to that in deposition of metals, gold and platinum, from CDD or CDL. The FSR concept cannot be applied to CDD since photoexcitation of titania is not induced in this process. Although the detailed mechanism, at least why metal deposition from colloids was enhanced by photoirradiation, has not yet been clarified, the colloid-stabilizing agent citric acid was decomposed along with hydrogen and carbon dioxide evolution and thereby oxidative decomposition/removal of citric acid covering colloidal metal particles may lead to deposition. In such a case, oxidative deposition of metal nanoparticles should be observed on {001} facets, not the actually observed {101} facets, according to the FSR concept.

The third aspect is pH-dependent change in facet selectivity of platinum-nanoparticle deposition in both PD and CDL processes; {101}-preferable facet selectivity in neutral and basic pH conditions became ambiguous at low pH. Based on the assumption that {001} facets are more acidic, i.e., easily releasing protons to bear negative surface charges, than are {101} facets and that metal nanoparticles are created by the assembly of small atomic or cluster-sized metal precursors, which migrate on the surface in detachment and re-attachment cycles, it is thought that the difference in surface charges depending on the kind of facets and deposition conditions, e.g., pH, accounts for the observed facet selectivity in all of the PD, CDD and CDL processes.

On the basis of these aspects, it can be concluded that the concept of FSR, facet-selective reaction of photoexcited electrons and positive holes, does not seem to be necessary to explain the change in facet selectivity observed in this study and that the results can be consistently interpreted by the possible surface charges depending on the kind of facets, {001} and {101}. Since the procedure and conditions for preparation of DAPs in this study were actually different from those used in previous studies, there might be another mechanism for the reported facet selectivity. Furthermore, the results of this study, non 0%-or-100% facet selectivity, do not suggest that only 0%-or-100% selectivity results were chosen in previous studies since possibly different surface structures of DAPs used in previous studies might have led to 0%-or-100% selectivity. However, the present authors propose here that the above-mentioned surface charge-dependent deposition of platinum and gold nanoparticles can explain the results shown in this article and may be expanded to the previously reported results.

Author Contributions: All authors: draft preparation, writing and editing. K.K.: performing all of the experiments and data analyses. M.T. (Mai Takase) and B.O.: design, maintenance and instruction of the gas-phase reaction for preparation of DAPs. M.T. (Mai Takashima): project management and maintenance of FE-SEM. B.O.: project leading and fund acquisition.

Funding: This research was partly supported by a Grant-in-Aid for Scientific Research (A) (Grant Numbers: 15H0220106 and 18H0392308) from Japan Society for the Promotion of Science (JSPS).

Acknowledgments: FE-SEM analyses of samples were carried out using a JEOL JSM-7400F electron microscope at Global Facility Center, Creative Research Institution, Hokkaido University Technical assistance and support for construction and maintenance of the gas-phase reactor for DAP synthesis and the other instrumental setups by the Technical Division of Institute for Catalysis, Hokkaido University are acknowledged.

Conflicts of Interest: The authors declare no conflict of interest.

References

1. Ohtani, B. Preparing Articles on Photocatalysis—Beyond the Illusions, Misconceptions, and Speculation. *Chem. Lett.* **2008**, *37*, 216–229. [CrossRef]
2. Ohtani, B. Revisiting the fundamental physical chemistry in heterogeneous photocatalysis: Its thermodynamics and kinetics. *Phys. Chem. Chem. Phys.* **2014**, *16*, 1788–1797. [CrossRef] [PubMed]
3. Ohtani, B. Photocatalysis A to Z—What we know and what we do not know in a scientific sense. *J. Photochem. Photobiol. C Photochem. Rev.* **2010**, *11*, 157–178. [CrossRef]
4. Ohtani, B. Titania Photocatalysis beyond Recombination: A Critical Review. *Catalysts* **2013**, *3*, 942–953. [CrossRef]
5. Banfield, J.F.; Veblen, D.R.; Smith, D.J. The identification of naturally occurring TiO_2(B) by structure determination using high-resolution electron microscopy, image simulation, and distance-least-squares refinement. *Am. Mineralogist* **1991**, *76*, 343–353.
6. Banfield, J.F.; Bischoff, B.L.; Anderson, M.A. TiO_2 accessory minerals: Coarsening, and transformation kinetics in pure and doped synthetic nanocrystalline materials. *Chem. Geol.* **1993**, *110*, 211–231. [CrossRef]
7. Amano, F.; Yasumoto, T.; Prieto-Mahaney, O.-O.; Uchida, S.; Shibayama, T.; Ohtani, B. Photocatalytic activity of octahedral single-crystalline mesoparticles of anatase titanium (IV) oxide. *Chem. Comm.* **2009**, 2311–2313. [CrossRef] [PubMed]
8. Amano, F.; Yasumoto, T.; Prieto-Mahaney, O.-O.; Uchida, S.; Shibayama, T.; Terada, Y.; Ohtani, B. Highly Active Titania Photocatalyst Particles of Controlled Crystal Phase, Size, and Polyhedral Shapes. *Top. Catal.* **2010**, *53*, 455–461. [CrossRef]
9. Ohtani, B.; Iwai, K.; Nishimoto, S.-I.; Sato, S. Role of Platinum Deposits on Titanium (IV) Oxide Particles: Structural and Kinetic Analyses of Photocatalytic Reaction in Aqueous Alcohol and Amino Acid Solutions. *J. Phys. Chem. B* **1997**, *101*, 3349–3359. [CrossRef]
10. Yang, H.G.; Sun, C.H.; Qiao, S.Z.; Zou, J.; Liu, G.; Smith, S.C.; Cheng, H.M.; Lu, G.Q. Anatase TiO_2 single crystals with a large percentage of reactive facets. *Nature* **2008**, *453*, 638. [CrossRef] [PubMed]
11. Amano, F.; Prieto-Mahaney, O.-O.; Terada, Y.; Yasumoto, T.; Shibayama, T.; Ohtani, B. Decahedral Single-Crystalline Particles of Anatase Titanium (IV) Oxide with High Photocatalytic Activity. *Chem. Mater.* **2009**, *21*, 2601–2603. [CrossRef]
12. Ye, L.Q.; Liu, J.Y.; Tian, L.H.; Peng, T.Y.; Zan, L. The replacement of {101} by {010} facets inhibits the photocatalytic activity of anatase TiO_2. *Appl. Catal. B. Environ.* **2013**, *134*, 60–65. [CrossRef]
13. Zheng, Z.; Huang, B.; Lu, J.; Qin, X.; Zhang, X.; Dai, Y. Hierarchical TiO_2 Microspheres: Synergetic Effect of {001} and {101} Facets for Enhanced Photocatalytic Activity. *Chem. Eur. J.* **2011**, *17*, 15032–15038. [CrossRef] [PubMed]
14. Meng, A.; Zhang, J.; Xu, D.; Cheng, B.; Yu, J. Enhanced photocatalytic H_2-production activity of anatase TiO_2 nanosheet by selectively depositing dual-cocatalysts on {101} and {001} facets. *Appl. Catal. B. Environ.* **2016**, *198*, 286–294. [CrossRef]
15. Sun, D.; Yang, W.; Zhou, L.; Sun, W.; Li, Q.; Shang, J.K. The selective deposition of silver nanoparticles onto {101} facets of TiO_2 nanocrystals with co-exposed {001}/{101} facets, and their enhanced photocatalytic reduction of aqueous nitrate under simulated solar illumination. *Appl. Catal. B. Environ.* **2016**, *182*, 85–93. [CrossRef]
16. Murakami, N.; Kurihara, Y.; Tsubota, T.; Ohno, T. Shape-Controlled Anatase Titanium (IV) Oxide Particles Prepared by Hydrothermal Treatment of Peroxo Titanic Acid in the Presence of Polyvinyl Alcohol. *J. Phys. Chem. C* **2009**, *113*, 3062–3069. [CrossRef]
17. Chamtouri, M.; Kenens, B.; Aubert, R.; Lu, G.; Inose, T.; Fujita, Y.; Masuhara, A.; Hofkens, J.; Uji-i, H. Facet-Dependent Diol-Induced Density of States of Anatase TiO_2 Crystal Surface. *ACS Omega* **2017**, *2*, 4032–4038. [CrossRef] [PubMed]
18. Liu, C.; Han, X.; Xie, S.; Kuang, Q.; Wang, X.; Jin, M.; Xie, Z.; Zheng, L. Enhancing the Photocatalytic Activity of Anatase TiO_2 by Improving the Specific Facet-Induced Spontaneous Separation of Photogenerated Electrons and Holes. *Chem. Asian. J.* **2012**, *8*, 282–289. [CrossRef] [PubMed]

19. Wang, W.; Lai, M.; Fang, J.; Lu, C. Au and Pt selectively deposited on {001}-faceted TiO_2 toward SPR enhanced photocatalytic Cr(VI) reduction: The influence of excitation wavelength. *Appl. Surf. Sci.* **2018**, *439*, 430–438. [CrossRef]

20. Murakami, N.; Kawakami, S.; Tsubota, T.; Ohno, T. Dependence of photocatalytic activity on particle size of a shape-controlled anatase titanium (IV) oxide nanocrystal. *J. Mol. Catal. A. Chem.* **2012**, *358*, 106–111. [CrossRef]

21. Xiong, Z.; Lei, Z.; Chen, X.; Gong, B.; Zhao, Y.; Zhang, J.; Zheng, C.; Wu, J.C.S. CO_2 photocatalytic reduction over Pt deposited TiO_2 nanocrystals with coexposed {101} and {001} facets: Effect of deposition method and Pt precursors. *Catal. Commun.* **2017**, *96*, 1–5. [CrossRef]

22. Jiang, Z.; Ding, D.; Wang, L.; Zhang, Y.; Zan, L. Interfacial effects of MnOx-loaded TiO_2 with exposed {001} facets and its catalytic activity for the photoreduction of CO_2. *Catal. Sci. Technol.* **2017**, *7*, 3065–3072. [CrossRef]

23. Ohno, T.; Sarukawa, K.; Matsumura, M. Crystal faces of rutile and anatase TiO_2 particles and their roles in photocatalytic reactions. *New. J. Chem.* **2002**, *26*, 1167–1170. [CrossRef]

24. Li, R.; Zhang, F.; Wang, D.; Yang, J.; Li, M.; Zhu, J.; Zhou, X.; Han, H.; Li, C. Spatial separation of photogenerated electrons and holes among {010} and {110} crystal facets of $BiVO_4$. *Nat. Commun.* **2013**, *4*, 1432. [CrossRef] [PubMed]

25. Zhou, C.; Wang, S.; Zhao, Z.; Shi, Z.; Yan, S.; Zou, Z. A Facet-Dependent Schottky-Junction Electron Shuttle in a $BiVO_4${010}–Au–Cu_2O Z-Scheme Photocatalyst for Efficient Charge Separation. *Adv. Funct. Mater.* **2018**, *28*, 1801214. [CrossRef]

26. Guo, Y.; Siretanu, I.; Zhang, Y.; Mei, B.; Li, X.; Mugele, F.; Huang, H.; Mul, G. pH-Dependence in facet-selective photo-deposition of metals and metal oxides on semiconductor particles. *J. Mater. Chem. A* **2018**, *6*, 7500–7508. [CrossRef]

27. Zhang, Q.; Li, R.; Li, Z.; Li, A.; Wang, S.; Liang, Z.; Liao, S.; Li, C. The dependence of photocatalytic activity on the selective and nonselective deposition of noble metal cocatalysts on the facets of rutile TiO_2. *J. Catal.* **2016**, *337*, 36–44. [CrossRef]

28. Wenderich, K.; Klaassen, A.; Siretanu, I.; Mugele, F.; Mul, G. Sorption-Determined Deposition of Platinum on Well-Defined Platelike WO_3. *Angew. Chem. Int. Edit.* **2014**, *53*, 12476–12479. [CrossRef]

29. Ohtani, B.; Bowman, R.M.; Colombo, D.P., Jr.; Kominami, H.; Noguchi, H.; Uosaki, K. Femtosecond Diffuse Reflectance Spectroscopy of Aqueous Titanium (IV) Oxide Suspension: Correlation of Electron-Hole Recombination Kinetics with Photocatalytic Activity. *Chem. Lett.* **1998**, *27*, 579–580. [CrossRef]

30. Lazzeri, M.; Vittadini, A.; Selloni, A. Structure and energetics of stoichiometric TiO_2 anatase surfaces. *Phys. Rev. B* **2001**, *63*, 155409. [CrossRef]

31. Lazzeri, M.; Vittadini, A.; Selloni, A. Erratum: Structure and energetics of stoichiometric TiO_2 anatase surfaces. *Phys. Rev. B* **2002**, *65*, 119901. [CrossRef]

32. Janczarek, M.; Kowalska, E.; Ohtani, B. Decahedral-shaped anatase titania photocatalyst particles: Synthesis in a newly developed coaxial-flow gas-phase reactor. *Chem. Eng. J.* **2016**, *289*, 502–512. [CrossRef]

33. Wei, Z.; Kowalska, E.; Ohtani, B. Influence of Post-Treatment Operations on Structural Properties and Photocatalytic Activity of Octahedral Anatase Titania Particles Prepared by an Ultrasonication-Hydrothermal Reaction. *Molecules* **2014**, *19*, 19573–19587. [CrossRef] [PubMed]

34. Kajita, M.; Kuwabara, T.; Hasegawa, D.; Yagi, M. Element-saving preparation of an efficient electrode catalyst based on self-assembly of Pt colloid nanoparticles onto an ITO electrode. *Green Chem.* **2010**, *12*, 2150–2152. [CrossRef]

35. Frens, G. Controlled Nucleation for Regulation of Particle-Size in Monodisperse Gold Suspensions. *Nature* **1973**, 20–22. [CrossRef]

Article

Solid-Phase Photocatalytic Degradation of Polyvinyl Borate

Hafize Nagehan Koysuren

Department of Environmental Engineering, Ahi Evran University, Kirsehir 40100, Turkey;
hnkoysuren@gmail.com; Tel.: +90-386-280-3855

Received: 28 September 2018; Accepted: 24 October 2018; Published: 26 October 2018

Abstract: In this study, polymer composites based on polyvinyl borate (PVB) with titanium dioxide (TiO_2) nanoparticles were prepared through the condensation reaction of polyvinyl alcohol and boric acid in the presence of TiO_2 nanoparticles. The solid-phase photocatalytic degradation of the polymer composites under UV light irradiation was investigated and compared with that of the pure PVB with the aid of weight loss measurements. The introduction of the photocatalyst nanoparticles in PVB enhanced the solid-phase photocatalytic degradation of the polymer matrix under UV light irradiation. The structural and morphological properties of PVB/TiO_2 composites were analyzed by transmission electron microscopy (TEM), scanning electron microscopy (SEM), Fourier transform infrared spectroscopy (FTIR), thermogravimetric analysis (TGA), and UV-Vis spectroscopy, respectively. FTIR analysis revealed that PVB synthesis was successfully carried out in the presence of the photocatalyst nanoparticles. According to the morphological analyses, TiO_2 nanoparticles were well dispersed in the PVB matrix.

Keywords: solid-phase photocatalytic degradation; polyvinyl borate; titanium dioxide

---

1. Introduction

With the speeding up of the industrial process based on the plastic technology, a large amount of plastic waste is directly discharged into the nature, inducing the accumulation of large amounts of toxic organic compounds in our daily lives and bringing with it an enormous threat to human health [1]. Most of the plastic based organic pollutants are burned for disposal, producing toxic gases. Recycling is applied to some plastics, but the recycling cost is high and requires high technology equipment [2]. The degradation of the plastic waste has attracted more attention in recent years and most of the studies have focused mainly on biodegradation. However, certain types of plastic materials are biodegradable under both aerobic and anaerobic conditions and the degradation rate of the most used plastics under natural environmental conditions is too low to apply in practice [3]. Most of the plastics are very difficult to degrade by the ordinary processing methods [1]. The solid-phase photocatalytic process is an ideal method to degrade plastic waste by using the inexhaustible solar light energy [1].

The solid-phase photocatalytic degradation process is based on solar energy corresponding to the band energy of the photocatalyst and the following photo-generated electron transfer. There are many kinds of semiconductor photocatalysts and titanium dioxide (TiO_2) is the most commonly used among them because of its high photostability, non-toxicity, low cost, and high activity [4]. TiO_2 as a photocatalyst has attracted great attention in various fields like air and waste water treatment, hydrogen fuel production, metal anti-corrosion, antibacterial activity, and self-purification. TiO_2 and other semiconductor atoms possess a valence band, which is occupied with stable energy electrons, and a conduction band, which is empty. The band gap energy, which is the energy difference between the top energy state of the valence band and the bottom energy state of the conduction band in semiconductors, is utilized to emit light inside TiO_2 to induce a redox reaction on its surface, which is known as the photocatalytic reaction [5]. When photons of sun light with energy greater than the band gap of TiO_2 are absorbed by TiO_2 semiconductor, one electron from the valance band rises into the conduction band, generating a photoinduced electron-hole pair. This electron-hole pair transfers to the photocatalyst surface, where it reacts with the surface-absorbed molecules like water and oxygen to form active radicals. These active radicals can degrade the plastic based organic pollutants through oxidation reactions [2]. Plastic waste is discharged into nature, where it is directly subjected to sunlight in open air. Therefore, it is noteworthy to study the solid-phase photocatalytic degradation of plastics under the sunlight.

The significance of immobilization of the TiO_2 photocatalyst in polymer matrix for the solid-phase photocatalytic degradation of plastics is highlighted by the recent studies. In literature, the photocatalytic degradation of polystyrene [6,7], polyaniline [8], polyvinyl chloride [9,10], low-density polyethylene [2], polyvinyl alcohol [3], and poly (methyl methacrylate) [11] was investigated in the presence of TiO_2 nanoparticles. The solid-phase photocatalytic degradation rate of plastics including TiO_2 nanoparticles was much faster than the simple photolysis of pure plastics. Polyvinyl borate is a synthetic polymer prepared through the condensation reaction of polyvinyl alcohol and boric acid. Polyvinyl alcohol is crosslinked with boric acid to improve the thermal and mechanical properties of the polymer. In addition, polyvinyl alcohol is a water-soluble polymer and crosslinking the polymer with boric acid enhances the moisture resistance of the matrix [12]. A few studies have been performed on crosslinking of polyvinyl alcohol with boric [4,12–14]. In contrast to polyvinyl alcohol, PVB is a non-environmental polymer and it degradation is also significant in terms of human health and the environment. There is no study in the literature on the solid-phase photocatalytic degradation of PVB. In this study, PVB composites were prepared through the condensation reaction of polyvinyl alcohol and boric acid in the presence of TiO_2 nanoparticles. The photocatalytic activity of PVB/TiO_2 composites was evaluated by investigating the solid-phase photocatalytic degradation of the polymer matrix in the ambient air under ultraviolet light irradiation.

2. Results and Discussion

Figure 1 illustrates the FTIR spectra of pure PVA and PVB/TiO_2 composite, including 10 wt.% of the photocatalyst nanoparticles. The broad absorption band around $3200 \ cm^{-1}$ was related to the O–H group of polyvinyl alcohol, which formed complexes with boron-containing oxyanions during the crosslinking reactions [13]. On the other hand, this broad band could also be related to the stretching band of O–H, which might be due to unreacted O–H groups of polyvinyl alcohol (Figure 1a) [14]. The absorption peaks at $2923 \ cm^{-1}$ and $1338 \ cm^{-1}$ are related to the stretching bond of C–H (Figure 1a) and the peak at $1724 \ cm^{-1}$ was assigned to the stretching bond of C=O (Figure 1a) [14]. The absorption peaks at $1299 \ cm^{-1}$ and $1133 \ cm^{-1}$ were assigned to the stretching vibrations of B–O–C bonds. This absorption peak provided strong evidence for the condensation reaction between polyvinyl alcohol and boric acid to synthesize PVB (Figure 1a) [14]. In addition, the peak at $1430 \ cm^{-1}$ was attributed to the stretching vibration of the B–O bond, which might be due to the unreacted boric acid (Figure 1a). Figure 1b illustrates FTIR spectrum of the composite. According to Figure 1a,b, the spectrum of the composite matched with that of the pure polymer. Characteristic absorption peaks of PVB, the stretching vibrations of B–O–C bonds, were also observed on the spectrum of the composite (Figure 1b). Different from the spectrum of pure PVB, there was a broad absorption band between $600 \ cm^{-1}$ and $900 \ cm^{-1}$, which was attributed to Ti–O stretching vibrations (Figure 1b) [15]. This absorption band proved the presence of TiO_2 nanoparticles in PVB matrix. The small absorption peak of pure PVB, present at around $1600 \ cm^{-1}$ and related to physically absorbed moisture, could also be seen on the spectrum of the composite (Figure 1a,b) [15]. This peak intensity became wider and increased. Hence, TiO_2 contribution might increase the moisture content of the composite.

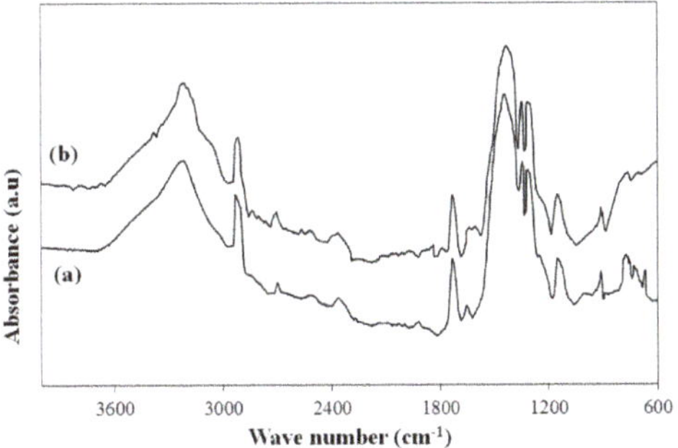

Figure 1. Fourier transform infrared spectroscopy (FTIR) spectra of (**a**) pure polyvinyl borate (PVB) and (**b**) PVB/TiO$_2$ (10 wt.%) composite.

Figure 2 shows TGA curves of pure polymer and the composite, containing 10 wt.% of the photocatalyst nanoparticles. The TGA curve of the pure polymer contains two main degradation steps between 300–400 °C. One of these degradation steps is the large step, corresponding to deacetylation reactions, and the other is the small step, corresponding to chain scission reactions [12]. TGA curves revealed that the photocatalyst nanoparticles slightly enhanced the thermal stability of PVB. The onset temperature of degradation is higher than that of pure PVB. Pure PVB and PVB/TiO$_2$ composite exhibited total weight losses of 42% and 39%, respectively, in the temperature range between 250 °C and 450 °C.

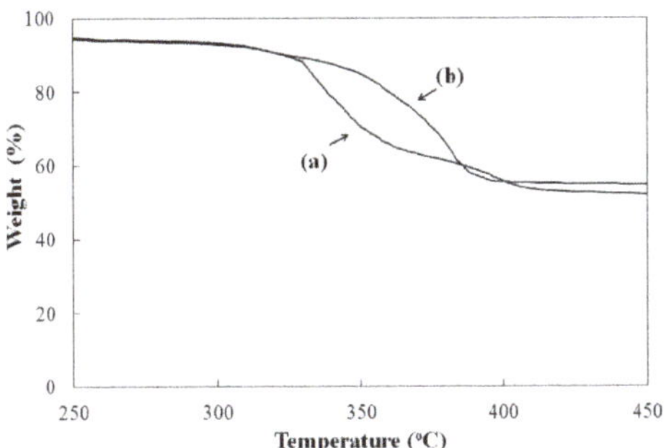

Figure 2. Thermogravimetric analysis (TGA) curves of (**a**) pure PVB and (**b**) PVB/TiO$_2$ (10 wt.%) composite.

SEM analyses of the composites, containing 5 wt.%, 10 wt.%, and 15 wt.% of TiO$_2$, respectively, were carried out to visualize the distribution of the photocatalyst nanoparticles in PVB matrix. Figure 3 illustrates SEM images of the composites. TiO$_2$ aggregates, which consisted of hundreds of nanoparticles, could be seen on the composite surfaces, which suggested that TiO$_2$ nanoparticles tended to agglomerate. During the synthesis of the composite, insufficient mixing might be performed to lead to a partial aggregated morphology. TEM images supported the stated thought (Figure 4). TiO$_2$ aggregates were well distributed in the polymer matrix and the size of most aggregates was less than 500 nm. According to SEM and TEM images, the polymer matrix held TiO$_2$ nanoparticles in intimate contact form, which is also important for enhanced photocatalytic activity.

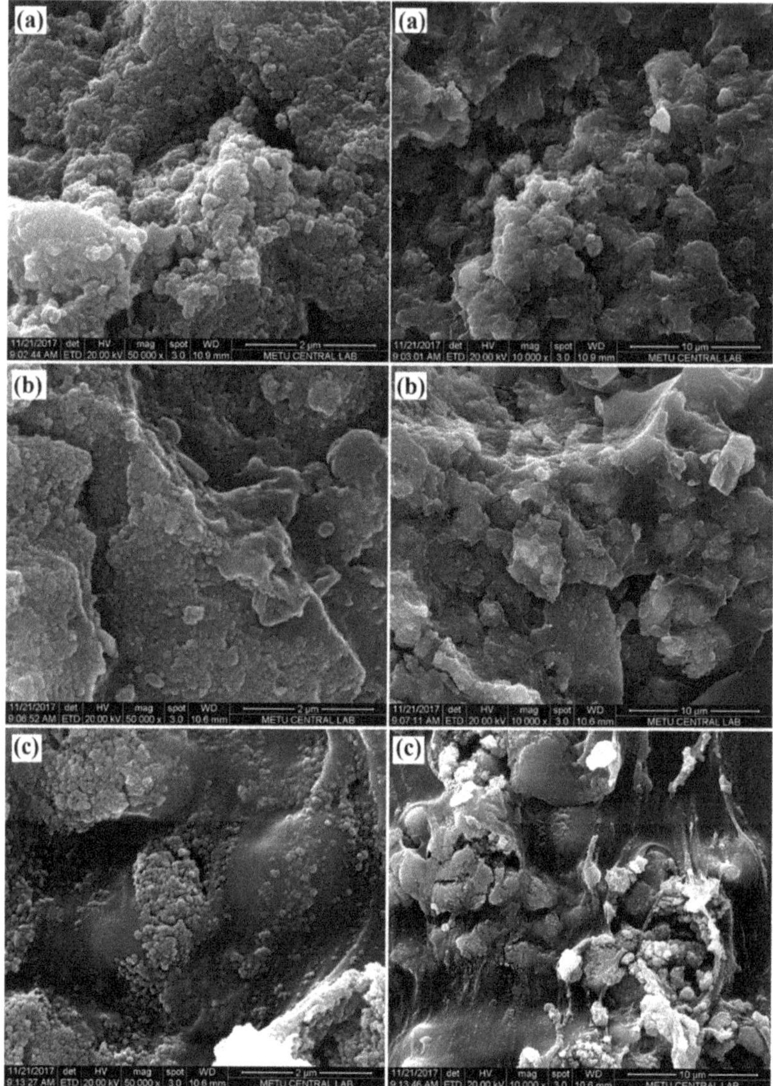

Figure 3. Scanning electron microscopy (SEM) images of (**a**) PVB/TiO$_2$ (5 wt.%) composite, (**b**) PVB/TiO$_2$ (10 wt.%) composite and (**c**) PVB/TiO$_2$ (15 wt.%) composite.

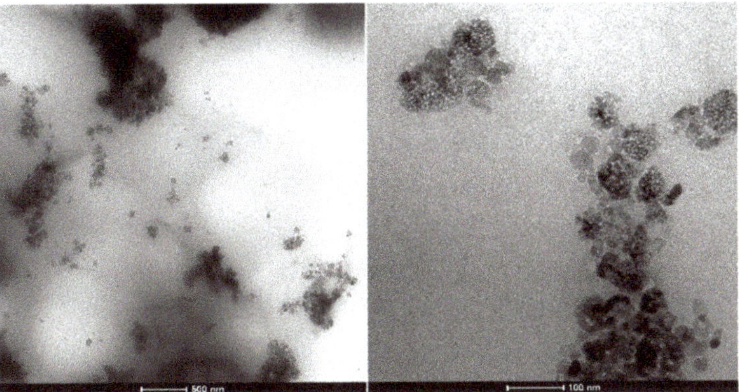

Figure 4. Transmission electron microscopy (TEM) images of PVB/TiO$_2$ (10 wt.%) composite.

Figure 5 presents the solid-phase photocatalytic degradation tendency of the pure polymer and the composites under air atmosphere. The weight loss rate of the composites was higher than that of PVB. The weight loss of the pure polymer and the composites increased continuously with UV irradiation. Pure PVB resulted in insignificant weight loss, almost 2 wt.%, after 240 h under UV irradiation, which could be ascribed to the boron containing chain structure of the pure polymer [4]. TGA results also supported the stated thought. Boron containing chain structure might be the reason for enhanced degradation stability and thermal stability. For the composites with 5 wt.%, 10 wt.% and 15 wt.% of TiO$_2$ nanoparticles, the weight loss values were 8.8 wt.%, 11.2 wt.%, 17.9 wt.%, respectively, after 240 h of UV irradiation. Photoinduced weight loss of the composites increased in parallel with TiO$_2$ content, which demonstrated the effectiveness of the photocatalyst. It was thought that the strong chain structure of PVB lead to the low weight loss values in the composites. UV-Vis spectroscopy was performed to investigate the optical property of TiO$_2$ and the composite with 10 wt.% of TiO$_2$ nanoparticles. According to Figure 6, TiO$_2$ absorbed the majority of the incoming light between 250 and 350 nm. On the other hand, the composite absorbed the majority of the incoming light below 400 nm. UV-Vis spectrum of pure TiO$_2$ exhibits a characteristic absorption band at around 322 nm, which was attributed to the characteristic Ti–O–Ti stretching vibrations [16].

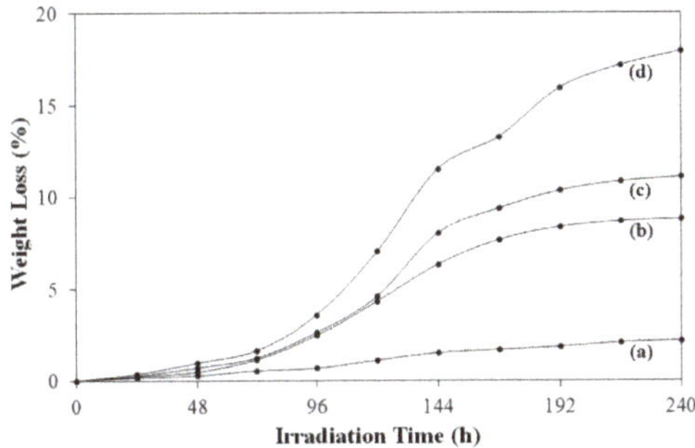

Figure 5. Weight loss of (**a**) pure PVB, (**b**) PVB/TiO$_2$ (5 wt.%) composite, (**c**) PVB/TiO$_2$ (10 wt.%) composite and (**d**) PVB/TiO$_2$ (15 wt.%) composite under UV irradiation as the function of time in air.

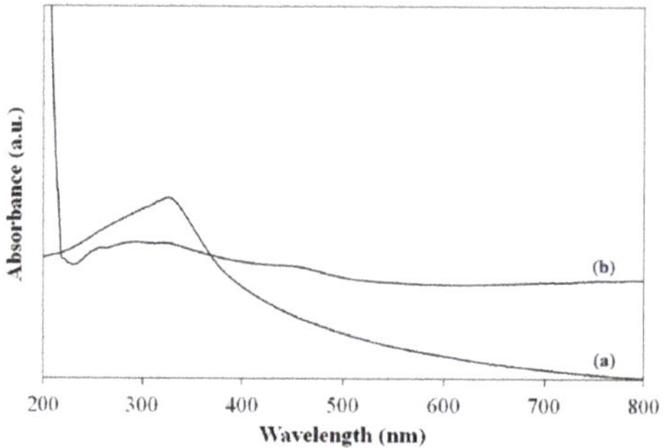

Figure 6. UV-Vis absorbance spectrum of (**a**) TiO$_2$ and (**b**) PVB/TiO$_2$ (10 wt.%).

3. Materials and Methods

Polyvinyl alcohol (PVA) with the molecular weight between 89,000 and 98,000 g/mol, boric acid and titanium dioxide (TiO$_2$, anatase, <25 nm) were obtained from Sigma-Aldrich (Munich, Germany). Pure PVB, which did not contain the photocatalyst nanoparticles, was synthesized through the condensation reaction of PVA and boric acid according to the procedure given in the literature [4]. In detail, 4.0 g of PVA was dissolved in 100 mL of distilled water and the solution was heated up to 80 °C under stirring. At the same time, 4.0 g of boric acid was dissolved in 100 mL of distilled water and the solution was kept under stirring at room temperature. Afterward, the boric acid solution was fed into the polyvinyl alcohol solution. The mixture was maintained at 80 °C under stirring for half an hour, which resulted in the formation of PVB in gel form. PVB in solid form was obtained after drying the gel polymer in an oven at 120 °C [4]. PVB/TiO$_2$ composites, including 5, 10, and 15 wt.% of TiO$_2$ nanoparticles, were synthesized with the same procedure followed to prepare pure PVB. Different from the given synthesis procedure, the photocatalyst nanoparticles were fed into the PVA solution prior to mixing with the boric acid solution.

Fourier transform infrared (FTIR) spectra of pure polymer and the composite, containing 10 wt.% of TiO$_2$ nanoparticles, were recorded on a Thermo Scientific FTIR Nicolet 380 (Nicolet Thermo Corporation, Edina, MN, USA) in the wavenumber range between 600 and 4000 cm^{-1}. The thermogravimetric analyses (TGA) of the pure polymer and the composite, including 10 wt.% of TiO$_2$, were performed with a Setaram Labsys TGA/DTA thermogravimetric analyzer (Setaram Instrumentation, Ankara, Turkey) under nitrogen atmosphere at the heating rate 5 °C/min. The morphology of the pure polymer and PVB/TiO$_2$ composites, including 5, 10, and 15 wt.% of TiO$_2$, respectively, were studied in a QUANTA 400F model field emission scanning electron microscope (FE-SEM) (Thermo Fisher, Hillsboro, OR, USA). A FEI-Tecnai G^2 Spirit Biotwin model conventional transmission electron microscope (CTEM) (Thermo Fisher, Hillsboro, OR, USA) was used for transmission electron microscopy (TEM) analysis of the composite, containing 10 wt.% of TiO$_2$. For this purpose, PVB/TiO$_2$ composite sample was grinded into powder form. Then, the powder sample was dispersed in ethanol and the dispersion was dropped on carbon coated copper grids. The solid-phase photocatalytic degradation of the pure polymer and the composite samples (1.0 g) was carried out in the ambient air using a 30 W UV lamp (254 nm, Philips, Istanbul, Turkey). All samples were weighed before and after UV irradiation to evaluate the weight loss of PVB through the solid-phase photocatalytic degradation. Each photocatalytic degradation experiment used a triplicate set of samples. The UV–Vis absorption spectrum of TiO$_2$ and PVB/TiO$_2$ composite, containing 10 wt.% of TiO$_2$ nanoparticles, were carried out by a Genesys 10S spectrophotometer (Thermo Fisher, Hillsboro, OR, USA) in the wavelength of 200–800 nm.

4. Conclusions

PVB/TiO$_2$ polymer composites were synthesized through the condensation reaction of polyvinyl alcohol and boric acid in the presence of TiO$_2$ nanoparticles. FTIR analysis verified PVB synthesis. TGA results revealed the improvement in thermal stability of PVB with TiO$_2$ contribution. According to SEM and TEM analyses, TiO$_2$ nanoparticles in aggregate structure illustrated good dispersion in PVB matrix. Adding TiO$_2$ nanoparticles in PVB matrix enhanced the solid-phase photocatalytic degradation of the polymer matrix under UV light irradiation.

Funding: This research received no external funding.

Acknowledgments: I thank Ozcan Koysuren for comments that greatly improved the manuscript.

Conflicts of Interest: The author declares no conflict of interest.

References

1. Xing, M.; Qiu, B.; Li, X.; Zhangin, J. TiO$_2$/Graphene Composites with Excellent Performance in Photocatalysis. In *Nanostructured Photocatalysts Advanced Functional Materials*; Yamashita, H., Li, H., Eds.; Springer International Publishing: Zurich, Switzerland, 2016; pp. 23–68. ISBN 978-3-319-26079-2.

2. Zan, L.; Fa, W.; Wang, S. Novel Photodegradable Low-Density Polyethylene-TiO$_2$ Nanocomposite Film. *Environ. Sci. Technol.* **2006**, *40*, 1681–1685. [CrossRef] [PubMed]

3. He, C.H.; Gong, J. The preparation of PVA–Pt/TiO$_2$ composite nanofiber aggregate and the photocatalytic degradation of solid-phase polyvinyl alcohol. *Polym. Degrad. STable* **2003**, *81*, 117–124. [CrossRef]

4. Koysuren, O.; Koysuren, H.N. Photocatalytic activity of polyvinyl borate/titanium dioxide composites for UV light degradation of organic pollutants. *J. Macromol. Sci. A* **2018**, *55*, 401–407. [CrossRef]

5. Lee, S.Y.; Park, S.J. TiO$_2$ photocatalyst for water treatment applications. *J. Ind. Eng. Chem.* **2013**, *19*, 1761–1769. [CrossRef]

6. Fa, W.; Guo, L.; Wang, J.; Guo, R.; Zheng, Z.; Yang, F. Solid-phase photocatalytic degradation of polystyrene with TiO$_2$/Fe(St)$_3$ as catalyst. *J. Appl. Polym. Sci.* **2013**, *128*, 2618–2622. [CrossRef]

7. Zan, L.; Tian, L.; Liu, Z.; Peng, Z. A new polystyrene–TiO$_2$ nanocomposite film and its photocatalytic degradation. *Appl. Catal. A-Gen.* **2004**, *264*, 237–242. [CrossRef]

8. Zhang, L.; Liu, P.; Su, Z. Preparation of PANI–TiO$_2$ nanocomposites and their solid-phase photocatalytic degradation. *Polym. Degrad. Stab.* **2006**, *91*, 2213–2219. [CrossRef]

9. Yang, C.; Gong, C.; Peng, T.; Deng, K.; Zan, L. High photocatalytic degradation activity of the polyvinyl chloride (PVC)-vitamin C (VC)-TiO$_2$ nano-composite film. *J. Hazard. Mater.* **2010**, *178*, 152–156. [CrossRef] [PubMed]

10. Cho, S.; Choi, W. Solid-phase photocatalytic degradation of PVC-TiO$_2$ polymer composites. *J. Photochem. Photobiol. A* **2001**, *143*, 221–228. [CrossRef]

11. Koysuren, O.; Koysuren, H.N. Photocatalytic activities of poly (methyl methacrylate)/titanium dioxide nanofiber mat. *J. Macromol. Sci. A* **2018**, *54*, 80–84. [CrossRef]

12. Geng, S.; Shah, F.U.; Liu, P.; Antzutkin, O.N.; Oksman, K. Plasticizing and crosslinking effects of borate additives on the structure and properties of poly (vinyl acetate). *RSC Adv.* **2017**, *7*, 7483–7491. [CrossRef]

13. Lawrence, M.B.; Desa, J.A.E.; Aswal, V.K.; Rai, R. Properties of poly (vinyl alcohol)-borax gel doped with neodymium and praseodymium. *Bull. Mater. Sci.* **2014**, *37*, 301–307. [CrossRef]

14. Yanase, I.; Ogaware, R.; Kobayashi, H. Synthesis of boron carbide powder from polyvinyl borate precursor. *Mater. Lett.* **2009**, *63*, 91–93. [CrossRef]

15. Riaz, N.; Bustam, M.A.; Chong, F.K.; Man, Z.B.; Khan, M.S.; Shariff, A.M. Photocatalytic Degradation of DIPA Using Bimetallic Cu-Ni/TiO$_2$ Photocatalyst under Visible Light Irradiation. *Sci. World J.* **2014**, *2014*, 1–8. [CrossRef]

16. Sarmah, S.; Kumar, A. Photocatalytic activity of polyaniline-TiO$_2$ nanocomposites. *Indian J. Phys.* **2011**, *85*, 713–726. [CrossRef]

Article

The Synergistic Effect of Pyridinic Nitrogen and Graphitic Nitrogen of Nitrogen-Doped Graphene Quantum Dots for Enhanced TiO$_2$ Nanocomposites' Photocatalytic Performance

Fei Li [1], Ming Li [2], Yi Luo [1], Ming Li [1,*], Xinyu Li [1], Jiye Zhang [3] and Liang Wang [2,*]

[1] College of Science, Guilin University of Technology, Guilin 541004, China; m17753101516@163.com (F.L.); adam2513@163.com (Y.L.); lixinyu5260@163.com (X.L.)

[2] Institute of Nanochemistry and Nanobiology, School of Environmental and Chemical Engineering, Shanghai University, Shanghai 200444, China; limingdobest@163.com

[3] School of Materials Science and Engineering, Shanghai University, Shanghai 200444, China; jychang@shu.edu.cn

* Correspondence: liming928@163.com (M.L.); wangl@shu.edu.cn (L.W.); Tel.: +86-773-369-6613 (M.L.); +86-021-661-3526 (L.W.)

Received: 9 September 2018; Accepted: 30 September 2018; Published: 4 October 2018

Abstract: In this study, nitrogen-doped graphene quantum dots (N-GQDs) and a TiO$_2$ nanocomposite were synthesized using a simple hydrothermal route. Ammonia water was used as a nitrogen source to prepare the N-GQDs. When optically characterized by UV-vis, N-GQDs reveal stronger absorption peaks in the range of ultraviolet (UV) light than graphene quantum dots (GQDs). In comparison with GQDs/TiO$_2$ and pure TiO$_2$, the N-GQDs/TiO$_2$ have significantly improved photocatalytic performance. In particular, it was found that, when the added amount of ammonia water was 50 mL, the content of pyridinic N and graphitic N were as high as 22.47% and 31.44%, respectively. Most important, the photocatalytic activity of N-GQDs/TiO$_2$-50 was about 95% after 12 min. The results illustrated that pyridinic N and graphitic N play a significant role in photocatalytic performance.

Keywords: N-doped graphene quantum dots; TiO$_2$; photocatalytic performance; pyridinic N; graphitic N

---

1. Introduction

In recent decades, increasing environmental pollution has attracted more and more attention, especially the discharge of dye wastewater from factories. It is therefore appropriate to find an effective, low-cost and pollution-free replacement for traditionally problematic energy production. Photocatalysis could be one of the most effective measures to solve the problems of energy shortage and environmental pollution [1–4]. In many semiconductor metal oxide materials, for example, titanium dioxide (TiO$_2$) is extensively used as a photocatalyst [5–7], due to its beneficial characteristics. It is inexpensive, non-poisonous, and has excellent chemical and physical stability [5,8]. Although it has so many superior properties, use of TiO$_2$ as a photocatalyst is limited by some disadvantages in practical application, such as a wide band gap (3.2 eV) and a high electron-hole recombination rate, which leads to low photocatalytic efficiency [3,9–11]. To perfect the photocatalytic activity of TiO$_2$, various measures were utilized, such as many ions being doped into the lattice of TiO$_2$ [12], sensitization via absorbed molecules [13–15], compound with other materials [16,17], and the surface being coated with other cocatalysts possessing excellent performance [18–21]. Among the methods mentioned above, surface loading with other cocatalysts is relatively facile and effective in enhancing the photocatalytic activity of TiO$_2$. Although some auxiliary catalysts can improve the photocatalytic performance of

TiO$_2$, for instance Pt, Au and Ag, their high cost limits their application [22,23]. Therefore, it would be significant to find highly efficient, simple and eco-friendly cocatalysts which enhance the photocatalytic performance of TiO$_2$.

Graphene quantum dots (GQDs) are a novel kind of 0D carbon nanomaterial with dimensions below 10 nm. In addition to all the properties of graphene, GQDs also have unique edge effects and quantum confinement [24]. GQDs are widely used in various fields due to their excellent physical and chemical properties. They are used in photovoltaic devices [25], catalysis [26–30], drug delivery [31], and cell imaging [32–35]. GQDs are environmentally friendly materials with strong anti-chemical corrosion and anti-ultraviolet (UV) irradiation capabilities. Pure GQDs display low catalytic activity due to their high exciton binding energy [36]. In many past studies, GQDs as auxiliary catalysts effectively improved the photocatalytic performance of TiO$_2$ [37], and some reports showed that the doped GQDs displayed excellent effects on improving the photocatalytic performance of TiO$_2$, for example, when nitrogen [38], sulfur [3] and nitrogen and sulfur co-doped [39]. However, there is little work on the effect of different N-bonding structure for the photocatalytic performance of nitrogen-doped GQDs (N-GQDs).

In our study, N-GQDs with different N contents were synthesized by a facile hydrothermal stratagem using different volumes of ammonia water and GQDs. N-GQDs were attached tightly to the surface of TiO$_2$ with a facile hydrothermal method. The photocatalytic performance of N-GQDs/TiO$_2$ was tested by introducing methyl orange (MO). A possible mechanism for improving photocatalytic performance was also investigated and analyzed by comparing the photocatalytic effect of N-GQDs/TiO$_2$ and pure TiO$_2$. All the results showed that N-GQDs effectively improved the photocatalytic performance of TiO$_2$, in which pyridinic N and graphitic N play a decisive role. This work may provide a new perspective for the future study of complexes based on N-GQDs.

2. Results and Discussion

Unless otherwise specified, the N-GQDs-50 with the best optical performance was selected for various characterizations.

2.1. Morphology and Structural Characterization of GQDs and N-GQDs

Figure 1a–d displays the transmission electron microscopy (TEM) and high resolution transmission electron microscopy (HRTEM) images of GQDs and N-GQDs-50. Figure 1a,b shows the TEM images of GQDs and N-GQDs-50, and the size distribution is homogeneous. The size of GQDs and N-GQDs-50 ranged from 2–16 nm and 1–5 nm, the average diameter was 8.66 nm and 3.12 nm respectively (insert in Figure 1a,b). The reduction in the size of N-GQDs-50 was likely to be due to further decomposition of GQDs during the subsequent hydrothermal reaction (refer to the experiment for details). The HRTEM images show that the plane lattice spacing of GQDs and N-GQDs-50 was 0.21 nm, which is similar to the in-plane lattice spacing of graphite (002) [40,41]. Figure 1e shows a TEM image of TiO$_2$ nanoparticles with the thin film. After the hydrothermal reaction, the N-GQDs were compounded on the surface of TiO$_2$. With the oxygen-containing functional groups of N-GQDs, the hydroxyl functional groups of TiO$_2$ may be able to construct functional and relatively stable composites.

XRD was employed to determine the crystalline structure of GQDs and N-GQDs-50. It was clear from the pattern (Figure 2a) of pure TiO$_2$ that there were two types of TiO$_2$, namely anatase and rutile. Peaks at $2\theta = 25.6°$, $37.18°$, $48.25°$, $54.02°$, $55.24°$ and $62.7°$ represented (101), (004), (200), (105), (211) and (116) planes of anatase. Others at $41.44°$ and $56.82°$, represented (110) and (114) of rutile, which identified with P25. The XRD pattern of N-GQDs-50 showed the peaks of N-GQDs/TiO$_2$-50 were the same as that of TiO$_2$, indicating that the structure of TiO$_2$ was not affected by N-GQDs. FT-IR spectroscopy can also characterize samples. As shown in Figure 2b, broad absorption bands at 480–700 cm^{-1} were associated with stretching vibrations of Ti-O-Ti and Ti-O-C. The peak at 1380 cm^{-1} was related to nitrate ion and the peak at 1633 cm^{-1} was due to δH$_2$O vibration of the water molecule [42]. The figures of FT-IR indicated that N-GQDs was successfully coupled with TiO$_2$.

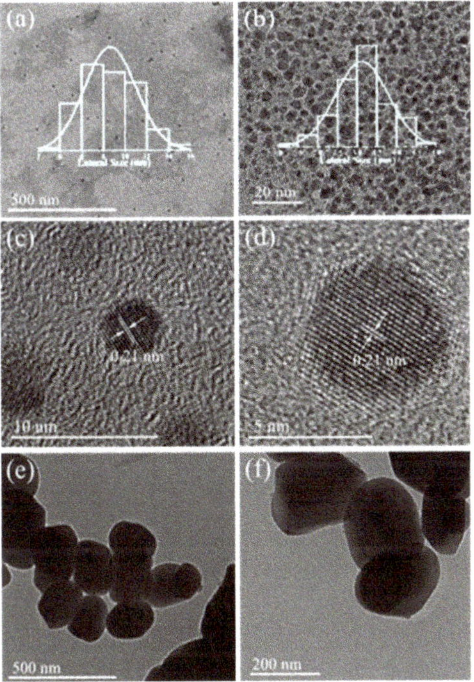

Figure 1. TEM images of GQDs (**a**), N-GQDs-50 (**b**), TiO$_2$ (**e**) and N-GQDs/TiO$_2$-50 (**f**), insets are of corresponding lateral size distribution. HRTEM images of GQDs (**c**) and N-GQDs-50 (**d**).

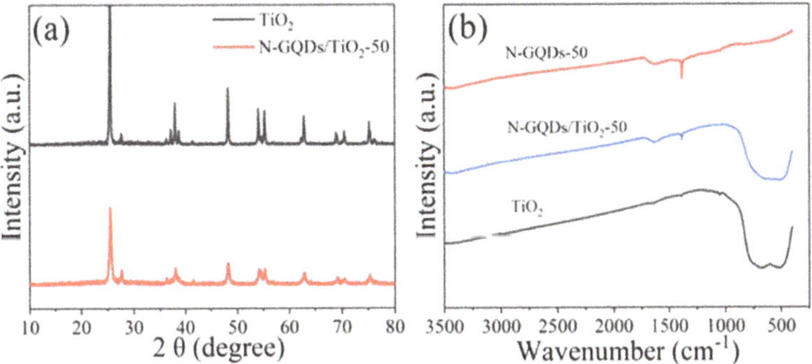

Figure 2. XRD pattern (**a**) of pure TiO$_2$ and N-GQDs/TiO$_2$-50; FT-IR spectra (**b**) of pure TiO$_2$, N-GQDs-50 and N-GQDs/TiO$_2$-50 composites.

To further investigate the composition of GQDs and N-GQDs, XPS (X-ray photoelectron spectroscopy) measurement was employed. Figure 3a shows the full XPS spectra of GQDs, N-GQDs-50 and N-GQDs-100. Peaks can be seen at approximately 284 eV (C 1s), 399.08 eV (N 1s) and 531.08 eV (O 1s) in all the samples. Compared to GQDs, the intensity of N 1s peak N-GQDs-50 and N-GQDs-100 was relatively enhanced, indicating that the N was successfully doped into the GQDs through the hydrothermal reaction with ammonia water. The results in Table 1 further show that the N content of N-GQDs-50 and N-GQDs-100 was higher than GQDs, and the content of N-GQDs-50 was the highest, reaching 10.64%, also indicating that N-GQDs were synthesized successfully. The high-resolution spectrum of N 1s region of N-GQDs-50 and N-GQDs-100 was divided into three peaks at 398.9 eV (pyridinic N), 399.6 eV (pyrrolic N) and 401.5 eV (graphitic N) [25,43].

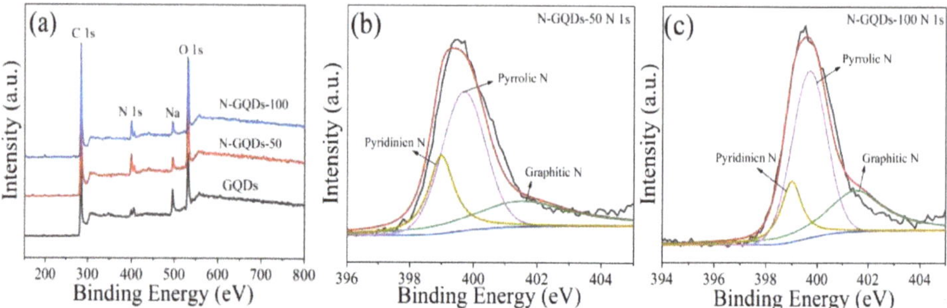

Figure 3. The full XPS spectra (**a**) of GQDs, N-GQDs-50 and N-GQDs-100. High-resolution N 1s spectrum of N-GQDs-50 (**b**) and N-GQDs-100 (**c**).

Table 1. The atomic percent (%) GQDs, N-GQDs-50 and N-GQDs-100 from XPS data.

Samples	C (at%)	O (at%)	N (at%)
GQDs	73.51	18.82	7.67
N-GQDs-50	71.41	17.95	10.64
N-GQDs-100	70.36	19.22	10.42

2.2. Optical Properties

Optical properties were used to characterize the physical nature of carbon-based materials. As shown in Figure 4, the optical absorption ability of GQDs (0.07 mg/mL) (Figure 4a) and N-GQDs-50 (0.07 mg/mL) (Figure 4b) was investigated by UV-vis spectrometer. GQDs displayed a wide absorption peak at 400–500 nm, which was similar to previous studies [44,45]. Compared to GQDs, N-GQDs-50 was also detected as having a strong absorption peak in the UV region at approximately 344 nm. Obviously, differences between GQDs and N-GQDs-50 in UV-vis spectra indicated that GQDs doped with N atom resulted in a strong absorption peak in the UV range.

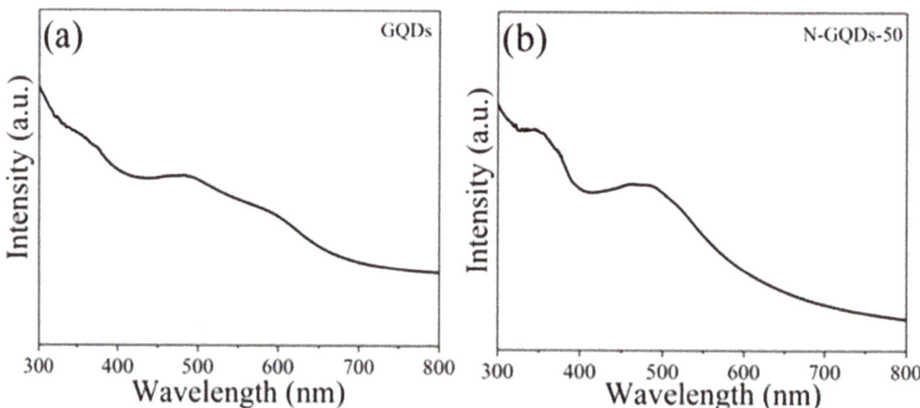

Figure 4. The UV-vis absorption spectra of the GQDs (**a**) and N-GQDs-50 (**b**).

According to the results of UV-vis absorption spectrum, the excitation wavelength of 310–390 nm was chosen for photoluminescent (PL) measurement. In this experiment, both GQDs and

N-GQDs-50 displayed excitation-independent PL behaviors, which were contrary to previous work on carbon-based fluorescent materials [41]. The emission peak of GQDs (Figure 5a) was approximately 540 nm and N-GQDs-50 (Figure 5b) was observed at approximately 520 nm. The excitation wavelength of 350 nm, with an emission peak of N-GQDs had blue shifted 20 nm compared to the GQDs in Figure 5c, which was most likely due to the reduction in size of N-GQDs-50 (see inserts of Figure 1a,b). This phenomenon is consistent with the trend observed by other quantum dots due to the quantum confinement effect at smaller particle size [46]. Figure 5d shows the PLE spectrum of N-GQDs-50 under the emission wavelength at 520 nm. One of the peaks was observed at 388 nm, which was in accordance with the PL results.

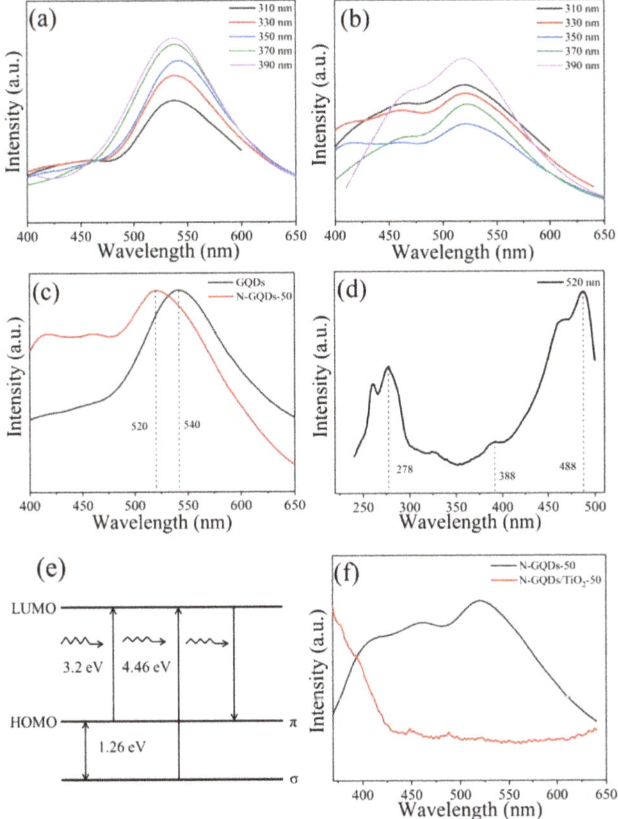

Figure 5. The PL spectra of GQDs (**a**) and N-GQDs-50 (**b**) under excitation with 310–390 nm. (**c**) Comparison of GQDs with N-GQDs under irradiation with 360 nm wavelength. (**d**) PLE spectra of N-QDs-50 when fixing emission wavelength at 520 nm. (**e**) Typical electronic transitions of triple carbenes in the optical spectrum of N-GQDs. (**f**) Comparison of N-GQDs-50 with N-GQDs/TiO$_2$-50 under irradiation with 330 nm wavelength.

The photoluminescence excitation (PLE) spectra showed that luminescence from N-GQDs could correspond to transitions at 278 (4.46 eV) and 388 (3.2 eV) nm, which could have been the result of transition between the σ and π orbital (HOMO) to the lowest unoccupied molecular orbital (LUMO), as shown in Figure 5e. Fluorescence performance was improved by the combination with photo-generated carriers. However, one way to improve photocatalytic performance was to inhibit the recombination of photo-generated carriers so that they could react with organic pollutants on the surface of photocatalysts. Thus, Figure 5f shows that N-GQDs/TiO$_2$-50 possesses excellent photocatalytic activity. The multiplicity of carbine ground-state was connected with energy differences (δE) between the σ and π orbital. According to previous reports, δE should be less than 1.5 eV [47]. In our study, δE of N-GQDs was 1.26 eV, which demonstrated that δE was within the theoretical value.

2.3. Photocatalytic Activity and Possible Mechanism for Improving Photocatalytic Activity

The concentration C/C$_0$ of undegraded MO was used to indicate photocatalytic performance of different catalysts. MO without a catalyst degrades differently under UV light in Figure 6a,b. Pure TiO$_2$ nanoparticles displayed fine photocatalytic activity by UV irradiation, and the degradation rate of MO reached approximately 57% within 12 min. The photocatalytic performance of GQDs/TiO$_2$ was higher than pure TiO$_2$, which reached about 65%. Although the photocatalytic performance of GQDs/TiO$_2$ was preferable to that of pure TiO$_2$, the effect was not satisfactory. MO degradation by N-GQDs/TiO$_2$ was much higher than that of other catalysts. In particular, the degradation of MO by N-GQDs/TiO$_2$-50 reached 95% within 12 min, indicating that the content of graphitic N played a significant function on photocatalytic activity. As shown in Table 2, with the increase of ammonia water content, the content of pyrrolic N obviously increased, while the content of pyridinic N and graphitic N were lessened. The content of pyridinic N and graphitic N of N-GQDs-50 were higher than that of N-GQDs-100, up to 22.47% and 31.4%, respectively.

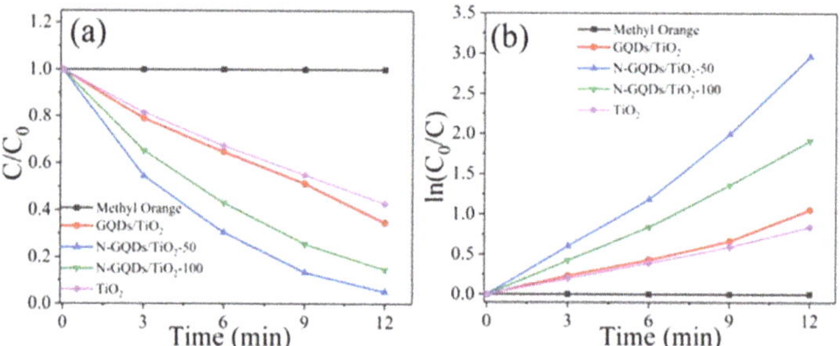

Figure 6. (**a**) The relationship between the concentration of undegraded MO and illumination time for different photocatalysts. (**b**) The relationship between the logarithm of C/C$_0$ and irradiation time of different photocatalysts.

Table 2. The content (%) of doped-N and the different N species of N-GQDs-50 and N-GQDs-100.

Samples	N (at%)	Pyridinic N (at%)	Pyrrolic N (at%)	Graphitic N (at%)
N-GQDs-50	10.64	22.47	44.09	31.44
N-GQDs-100	10.42	16.38	53.03	30.59

To further study the ability and stability of photocatalytic MO degradation by N-GQDs/TiO$_2$ composites, the cyclic stability experiment of photocatalytic degradation of MO by N-GQDs/TiO$_2$-50 was investigated (Figure 7a). After five cycles, the N-GQDs/TiO$_2$-50 was yet to show a good photocatalytic effect. As shown in Figure 7b, the photocatalytic performance of N-GQDs/TiO$_2$-50 and pure TiO$_2$ was slightly reduced, but the photocatalytic activity was still excellent, reaching over 90%. Their results show that the photocatalytic activity of TiO$_2$ could be greatly enhanced by modifying TiO$_2$ with effective methods.

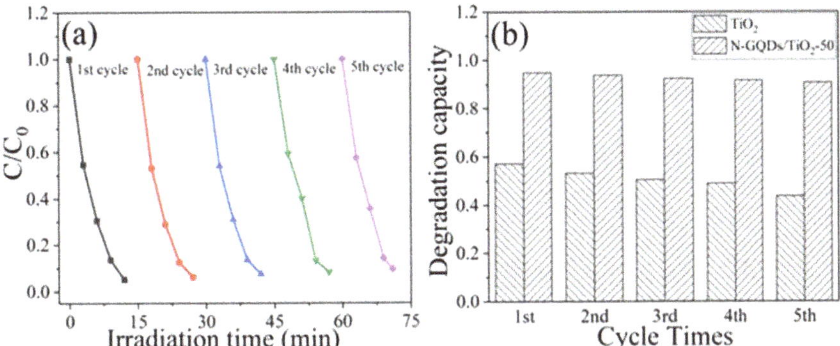

Figure 7. (**a**) Recycle stability of the photocatalytic decomposition of MO by N-GQDs/TiO$_2$-50. (**b**) Repetitive photocatalytic decomposition of MO for TiO$_2$ and N-GQDs/TiO$_2$-50 photocatalysts.

Scheme 1 explains the probable mechanism of degradation of MO by N-QGDs/TiO$_2$ composites. As an n-type semiconductor, TiO$_2$ was able to create electron-hole pairs [44]. Under UV irradiated light, the electron of TiO$_2$ transferred from the valence band to the conduction band to form an electro-hole. Electro-holes reacted with absorbed O$_2$/OH- to produce ·O$_2$/·OH so as to degrade MO. N-GQDs attached to the surface of TiO$_2$, absorbing UV light and raising the excitation of electrons. The excited electrons then transferred to the conduction band of TiO$_2$. With TiO$_2$ as the base of catalytic reaction, the N-GQDs as an unexceptionable electron migration area on the surface of TiO$_2$ could effectively and rapidly transmit photogenerated electrons, inhibiting the fast binding of photogenerated electron-hole pairs, and thus greatly improved the catalytic efficiency of TiO$_2$. Compared with GQDs, N-GQDs showed a strong absorption peak in the UV region. Therefore, N-GQDs/TiO$_2$ displayed strong photocatalytic activity by UV light. The oxygen in MO also combined with electros on N-GQDs to generate ·O$_2^-$, which may have played a significant role in photocatalytic activity [6]. As the main N-binding configuration, pyridine N only existed at the edge of the GQDs, which could be used as the oxygen-reduction active site to enhance the activity of a catalyst. Furthermore, the graphitic N was the electron transfer site [43]. Thus, N-GQDs/TiO$_2$ displayed good photocatalytic performance, and the pyridinic N and graphitic N played a significant position in photocatalytic performance.

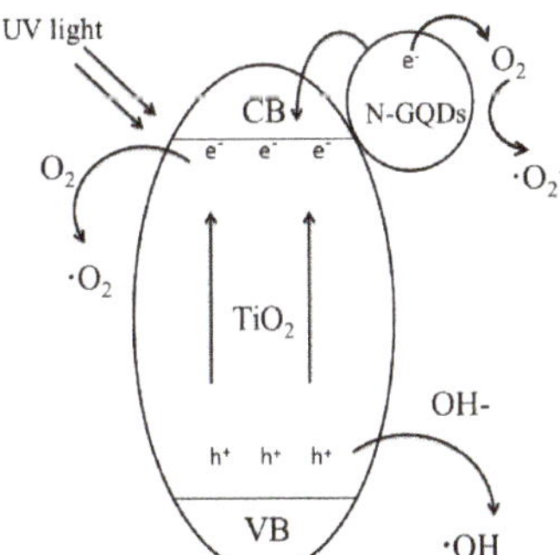

Scheme 1. The possible photocatalyst mechanism of N-GQDs/TiO$_2$ under UV light.

3. Experimental Section

3.1. Chemicals

All reagents were not processed further. Pyrene (C$_6$H$_6$), anhydrous alcohol (C$_2$H$_5$OH), sodium hydroxide (NaOH), nitric acid (HNO$_3$) ammonia water, titanium dioxide (TiO$_2$) were purchased from reagent agent (manufacturer, city, country). Deionized water was used in all the experimental processes. All the chemicals were purchased from shanghai, China.

3.2. Preparation of N-GQDs

GQDs were prepared using a simple hydrothermal method [48]. Then N-GQDs samples (N-GQDs-50, N-GQDs-100, -numbers represent the volume of the added ammonia water) were synthesized by a simple hydrothermal method. Briefly, 0.1 g GQDs was dispersed in 50 mL H_2O, and added to different volumes of ammonia water (50 mL and 100 mL). The mixed homogeneous solution was transferred into a Teflon-lined steel autoclave and then heated at 180 °C for 12 h. After cooling to room temperature, the obtained solution was filtered with a 0.22 μm filter membrane, and the filtered solution was dialyzed for 24 h using a 3500 Da dialysis bag to remove excess ions. Finally, the obtained N-GQDs were dried at 70 °C in air for the subsequent experiment.

3.3. Preparation of N-GQDs/TiO$_2$

N-GQDs/TiO_2 composites were synthesized by a simple hydrothermal method. Weighed 0.4 g Degussa P25 TiO_2 was dispersed into 200 mL N-GQDs aqueous solution (0.2 mg/mL) by ultrasonication (500 W, 40 kHz) for 30 min. Then the solution was transferred into Teflon-lined steel autoclave and then heated at 180 °C for 24 h. After cooling to room temperature, the product was washed three times by centrifugation with deionized water and anhydrous alcohol, then the collected sediment was dried at 70 °C in air to obtain the N-GQDs/TiO_2-50 composite. The GQDs/TiO_2 and N-GQDs/TiO_2-100 composites were also synthesized under the same conditions for comparison.

3.4. Photocatalytic Activity Measurements

The photocatalytic performance of the obtained samples was explored by degrading MO in quartz tubes at the UV light irradiation of a 600 W mercury lamp, and the photocatalytic experiment was carried out at room temperature. The distance between solution and lamp was 10 cm. The experimental procedure is as described: 50 mg samples were dissolved in 50 mL MO (5 mg/L). The resulting solution was roughened without treatment for an hour to achieve adsorption and desorption equilibrium between the catalyst and MO. Then the solution was put under the mercury lamp for illumination with magnetic stirring and 4 mL solution was removed every 3 min to a centrifuge for 5 min (8000 rpm) to remove catalyst particles. The concentration of MO after centrifugation was measured by a UV/vis/near infrared (NIR) spectrometer. The expression formula of degradation rate of MO is $\frac{C_0-C}{C_0}$. The concentration of undegraded MO can be expressed as C/C_0. In this study, C represents the concentration of MO after irradiation, C_0 represents the original concentration of MO before irradiation.

3.5. Characterization

Morphology of samples was measured by transmission electron microscopy (TEM, HT7700, Hitachi, Tokyo, Japan), and X-ray photoelectron spectroscopy (XPS) data were characterized by an ESCALAB 250Xi electron spectrometer (ThermoFisher Scientific, Waltham, MA, USA) with Al Kα Radiation (1486.6 eV). X-ray diffraction (XRD) patterns of samples were recorded within 5–80° (2θ) using a Rigaku D/MAX 2550 diffractometer (Rigaku, Tokyo, Japan) carried out at 40 kV and 100 mA. Fourier transform infrared spectroscopy (FT-IR) was investigated using a Perkin-Elmer spectrum. The UV-vis absorption spectrum was measured by using a UV/vis/NIR spectrometer (Perkin-Elmer, Lambda 750, PerkinElmer, Shelton, CT, USA). The photoluminescent (PL) and photoluminescence excitation (PLE) spectra (Carry Eclipse Fluorescence Spectrophotometer, Agilent Technologies Ltd., Cheadle, UK) were carried out using a fluorescence spectrophotometer.

4. Conclusion

In this study, we synthesized N-GQDs/TiO$_2$ composites by two facile hydrothermal methods. The results show that N-GQDs/TiO$_2$ exhibit excellent photocatalytic performance, and the ability to degrade MO for cyclic stability. In addition, the photocatalytic activity of N-GQDs/TiO$_2$ is associated with the content of graphitic N and the higher content, the better of photocatalytic activity. In particular, it was found that when the amount of ammonia water added was 50 mL and the contents of pyridinic N and graphitic N were as high as 22.47% and 31.44%, respectively. The photocatalytic performance of N-GQDs/TiO$_2$-50 reached about 95% in 12 min. This accomplishment may provide a new perspective for the future study of composites based on N-GQDs.

Author Contributions: Correspondence author, M.L. and L.W.; Data curation, F.L.; investigation, X.L. and J.Z.; methodology, F.L. and M.L. and Y.L.; supervision, M.L. (correspondence author) and L.W.; writing (original draft), F.L.; writing (review and editing), L.W.

Acknowledgments: This work was financially supported by the National Natural Science Foundation of China (Nos. 11764011, 21671129, 51472241, 21571124) and Natural Science Foundation of Guangxi Province (No. 2016GXNSFAA380008, 2017GXNSFBA198216), P. R. China.

Conflicts of Interest: The authors declare no competing financial interest.

References

1. Kumar, S.G.; Devi, L.G. Review on modified TiO$_2$ photocatalysis under UV/visible light: selected results and related mechanisms on interfacial charge carrier transfer dynamics. *J. Phys. Chem. Coruña* **2011**, *115*, 13211–13241. [CrossRef] [PubMed]

2. Cheng, H.; Fuku, K.; Kuwahara, Y.; Mori, K.; Yamashita, H. Harnessing single-active plasmonic nanostructures for enhanced photocatalysis under visible light. *J. Mater. Chem. Coruña* **2015**, *3*, 5244–5258. [CrossRef]

3. Luo, Y.; Li, M.; Hu, G.; Tang, T.; Wen, J.; Li, X.; Wang, L. Enhanced photocatalytic activity of sulfur-doped graphene quantum dots decorated with TiO$_2$ nanocomposites. *Mater. Res. Bull.* **2018**, *97*, 428–435. [CrossRef]

4. Bhatia, S.; Verma, N. Photocatalytic activity of ZnO nanoparticles with optimization of defects. *Mater. Res. Bull.* **2017**, *95*, 468–476. [CrossRef]

5. Wang, X.; Kafizas, A.; Li, X.; Moniz, S.J.A.; Reardon, P.J.T.; Tang, J.; Parkin, I.P.; Durrant, J.R. Transient absorption spectroscopy of anatase and rutile: The impact of morphology and phase on photocatalytic activity. *J. Phys. Chem. C* **2015**, *119*, 10439–10447. [CrossRef]

6. Konstantinou, I.K.; Albanis, T.A. TiO$_2$-assisted photocatalytic degradation of azo dyes in aqueous solution: kinetic and mechanistic investigations: A review. *Appl. Catal. B Environ.* **2004**, *49*, 1–14. [CrossRef]

7. Ruggieri, F.; Antonio D'Archivio, A.; Fanelli, M.; Santucci, S. Photocatalytic degradation of linuron in aqueous suspensions of TiO$_2$. *RSC Adv.* **2011**, *1*, 611–618. [CrossRef]

8. Wei, H.; Wang, L.; Li, Z.; Ni, S.; Zhao, Q. Synthesis and photocatalytic activity of one-dimensional CdS@TiO$_2$ core-shell heterostructures. *Nano-Micro Lett.* **2011**, *3*, 6–11. [CrossRef]

9. Ruggieri, F.; Di Camillo, D. Electrospun Cu-, W- and Fe-doped TiO$_2$ nanofibres for photocatalytic degradation of rhodamine 6G. *J. Nanopart. Res.* **2013**, *15*, 1982. [CrossRef]

10. Sawunyama, P.; Yasumori, A.; Okada, K. The nature of multilayered TiO$_2$-based photocatalytic films prepared by a sol-gel process. *Mater. Res. Bull.* **1998**, *33*, 795–801. [CrossRef]

11. Mori, R.; Takahashi, M.; Yoko, T. 2D spinodal phase-separated TiO$_2$ films prepared by sol–gel process and photocatalytic activity. *Mater. Res. Bull.* **2004**, *39*, 2137–2143. [CrossRef]

12. Asahi, R.; Morikawa, T.; Ohwaki, T.; Aoki, K.; Taga, T. Visible-light photocatalysis in nitrogen-doped titanium oxides. *Science* **2001**, *293*, 269–271. [CrossRef] [PubMed]

13. Mori, K.; Kawashima, M.; Yamashita, H. Visible-light-enhanced Suzuki-miyaura coupling reaction by cooperative photocatalysis with an Ru-Pd bimetallic complex. *Chem. Commun.* **2014**, *50*, 14501–14503. [CrossRef] [PubMed]

14. Yuan, Y.-P.; Yin, L.-S.; Cao, S.-W.; Xu, G.-S.; Li, C.-H.; Xue, C. Improving photocatalytic hydrogen production of metal-organic framework UiO-66 octahedrons by dye-sensitization. *Appl. Catal. B Environ.* **2015**, *168–169*, 572–576. [CrossRef]

15. Wang, P.; Wang, J.; Ming, T.; Wang, X.; Yu, H.; Yu, J.; Wang, Y.; Lei, M. Dye-sensitization-induced visible-light reduction of graphene oxide for the enhanced TiO_2 photocatalytic performance. *ACS Appl. Mater. Interfaces* **2013**, *5*, 2924–2929. [CrossRef] [PubMed]

16. Sadhu, S.; Poddar, P. Template-free fabrication of highly-oriented single-crystalline 1D-rutile TiO_2- MWCNT composite for enhanced photoelectrochemical activity. *J. Phys. Chem. C* **2014**, *118*, 19363–19373. [CrossRef]

17. Wan, L.; Long, M.; Zhou, D.; Zhang, L.; Cai, W. Preparation and characterization of freestanding hierarchical porous TiO_2 monolith modified with graphene oxide. *Nano-Micro Lett.* **2012**, *4*, 90–97. [CrossRef]

18. Yu, H.; Liu, R.; Wang, X.; Wang, P.; Yu, J. Enhanced visible-light photocatalytic activity of Bi_2WO_6 nanoparticles by Ag_2O cocatalyst. *Appl. Catal. B Environ.* **2012**, *111–112*, 326–333. [CrossRef]

19. Wang, P.; Xia, Y.; Wu, P.; Wang, X.; Yu, H.; Yu, J. Cu(II) as a general cocatalyst for improved visible-light photocatalytic performance of photosensitive ag-based compounds. *J. Phys. Chem. C* **2014**, *118*, 8891–8898. [CrossRef]

20. Gu, Y.; Xing, M.; Zhang, J. Synthesis and photocatalytic activity of graphene based doped TiO_2 nanocomposites. *Appl. Surf. Sci.* **2014**, *319*, 8–15. [CrossRef]

21. Aleksandrzak, M.; Adamski, P.; Kukułka, W.; Zielinska, B.; Mijowska, E. Effect of graphene thickness on photocatalytic activity of TiO_2-graphene nanocomposites. *Appl. Surf. Sci.* **2015**, *331*, 193–199. [CrossRef]

22. Seh, Z.W.; Liu, S.; Low, M.; Zhang, S.-Y.; Liu, Z.; Mlayah, A.; Han, M.-Y. Janus Au-TiO_2 Photocatalysts with Strong Localization of Plasmonic Near-Fields for Efficient Visible-Light Hydrogen Generation. *Adv. Mater.* **2012**, *24*, 2310–2314. [CrossRef] [PubMed]

23. Zheng, Z.; Huang, B.; Qin, X.; Zhang, X.; Dai, Y.; Whangbo, M.-H. Facile in situ synthesis of visible-light plasmonic photocatalysts M@TiO_2 (M = Au, Pt, Ag) and evaluation of their photocatalytic oxidation of benzene to phenol. *J. Mater. Chem.* **2011**, *21*, 9079–9087. [CrossRef]

24. Wang, L.; Wu, B.; Li, W.; Wang, S.; Li, Z.; Li, M.; Pan, D.; Wu, M. Amphiphilic graphene quantum dots as self-targeted fluorescence probes for cell nucleus imaging. *Adv. Biosyst.* **2018**, *2*, 1700191. [CrossRef]

25. Pan, D.; Jiao, J.; Li, Z.; Guo, Y.; Feng, C.; Liu, Y.; Wang, L.; Wu, M. Efficient separation of electron-hole pairs in graphene quantum dots by TiO_2 heterojunctions for dye degradation. *ACS Sustain. Chem. Eng.* **2015**, *3*, 2405–2413.

26. Fei, H.; Ye, R.; Ye, G.; Gong, Y.; Peng, Z.; Fan, X.; Samuel, E.L.G.; Ajayan, P.M.; Tour, J.M. Boron- and nitrogen-doped graphene quantum dots/graphene hybrid nanoplatelets as efficient electrocatalysts for oxygen reduction. *ACS Nano* **2014**, *8*, 10837–10843. [CrossRef] [PubMed]

27. Tang, L.; Wang, J.; Jia, C.; Lv, G.; Xu, G.; Li, W.; Wang, L.; Zhang, J.; Wu, M. Simulated solar driven catalytic degradation of psychiatric drug carbamazepine with binary $BiVO_4$ heterostructures sensitized by graphene quantum dots. *Appl. Catal. B Environ.* **2017**, *205*, 587–596. [CrossRef]

28. Yeh, T.-F.; Teng, C.-Y.; Chen, S.-J.; Teng, H. Nitrogen-doped graphene oxide quantum dots as photocatalysts for overall water-splitting under visible light illumination. *Adv. Mater.* **2014**, *26*, 3297–3303. [CrossRef] [PubMed]

29. Li, F.; Sun, L.; Luo, Y.; Li, M.; Xu, Y.; Hu, G.; Li, X.; Wang, L. Effect of thiophene S on the enhanced ORR electrocatalytic performance of sulfur-doped graphene quantum dot/reduced graphene oxide nanocomposites. *RSC Adv.* **2018**, *8*, 19635–19641. [CrossRef]

30. Sun, L.; Luo, Y.; Li, M.; Hu, G.; Xu, Y.; Tang, T.; Weng, J.; Li, X.; Wang, L. Role of pyridinic-N for nitrogen-doped graphene quantum dots in oxygen reaction reduction. *J. Colloid Interface Sci.* **2017**, *508*, 154–158. [CrossRef] [PubMed]

31. Dong, J.; Wang, K.; Sun, L.; Sun, B.; Yang, M.; Chen, H.; Wang, Y.; Sun, J.; Dong, L. Application of graphene quantum dots for simultaneous fluorescence imaging and tumor-targeted drug delivery. *Sens. Actuators B Chem.* **2018**, *256*, 616–623. [CrossRef]

32. Wang, L.; Li, W.; Wu, B.; Li, Z.; Pan, D.; Wu, M. Room-temperature synthesis of graphene quantum dots via electron-beam irradiation and their application in cell imaging. *Chem. Eng. J.* **2017**, *309*, 374–380. [CrossRef]

33. Wang, L.; Wu, B.; Li, W.; Li, Z.; Zhan, J.; Geng, B.; Wang, S.; Pan, D.; Wu, M. Industrial production of ultra-stable sulfonated graphene quantum dots for Golgi apparatus imaging. *J. Mater. Chem. B* **2017**, *5*, 5355–5361. [CrossRef]

34. Wang, L.; Li, W.; Li, M.; Su, Q.; Li, Z.; Pan, D.; Wu, M. Ultrastable amine, sulfo cofunctionalized graphene quantum dots with high two-photon fluorescence for cellular imaging. *ACS Sustain. Chem. Eng.* **2018**, *6*, 4711–4716. [CrossRef]

35. Li, W.; Li, M.; Liu, Y.; Pan, D.; Li, Z.; Wang, L.; Wu, M. Three-minute ultrarapid microwave-assisted synthesis of bright fluorescent graphene quantum dots for live cell staining and white LEDs. *ACS Appl. Nano Mater.* **2018**, *1*, 1623–1630. [CrossRef]

36. Li, L.; Yan, X. Colloidal graphene quantum dots. *J. Phys. Chem. Lett.* **2010**, *1*, 2572–2576. [CrossRef]

37. Rajender, G.; Kumar, J.; Giri, P.K. Interfacial charge transfer in oxygen deficient TiO_2-graphene quantum dot hybrid and its influence on the enhanced visible light photocatalysis. *Appl. Catal. B Environ.* **2018**, *224*, 960–972. [CrossRef]

38. Safardoust-Hojaghan, H.; Salavati-Niasari, M. Degradation of methylene blue as a pollutant with N-doped graphene quantum dot/titanium dioxide nanocomposite. *J. Clean. Prod.* **2017**, *148*, 31–36. [CrossRef]

39. Tian, H.; Shen, K.; Hu, X.; Qiao, L.; Zheng, W. N, S co-doped graphene quantum dots-graphene-TiO_2 nanotubes composite with enhanced photocatalytic activity. J. *Alloys Compd.* **2017**, *691*, 369–377. [CrossRef]

40. Zhang, B.-X.; Gao, H.; Li, X.-L. Synthesis and optical properties of nitrogen and sulfur co-doped graphene quantum dots. *New J. Chem.* **2014**, *38*, 4615–4621. [CrossRef]

41. Wang, Y.; Zhang, L.; Liang, R.-P.; Bai, J.-M.; Qiu, J.-D. Using graphene quantum dots as photoluminescent probes for protein kinase sensing. *Anal. Chem.* **2013**, *85*, 9148–9155. [CrossRef] [PubMed]

42. Taibi, M.; Ammar, S.; Jouini, N.; Fiévet, F.; Molinié, P.; Drillon, M. Layered nickel hydroxide salts: Synthesis, characterization and magnetic behaviour in relation to the basal spacing. *J. Mater. Chem.* **2002**, *12*, 3238–3244. [CrossRef]

43. Xu, Y.; Mo, Y.; Tian, J.; Wang, P.; Yu, H.; Yu, J. The synergistic effect of graphitic N and pyrrolic N for the enhance photocatalytic performance of nitrogen-doped graphene/TiO_2 nanocomposites. *Appl. Catal. B Environ.* **2016**, *181*, 810–817. [CrossRef]

44. Dong, Y.; Chen, C.; Zheng, X.; Gao, L.; Cui, Z.; Yang, H.; Guo, C.; Chi, Y.; Li, C.M. One-step and high yield simultaneous preparation of single- and multi-layer graphene quantum dots from CX-72 carbon black. *J. Mater. Chem.* **2012**, *22*, 8764–8766. [CrossRef]

45. Ye, R.; Xiang, C.; Lin, J.; Peng, Z.; Huang, K.; Yan, Z.; Cook, N.P.; Samuel, E.L.G.; Hwang, C.-C.; Ruan, G.; et al. Coal as an abundant source of graphene quantum dots. *Nat. Commun.* **2013**, *4*, 2943. [CrossRef] [PubMed]

46. Melnikov, D.V.; Chelikowsky, J.R. Quantum confinement in phosphorus-doped silicon nanocrystals. *Phys. Rev. Lett.* **2004**, *92*, 046802. [CrossRef] [PubMed]

47. Peng, J.; Gao, W.; Gupta, B.K.; Liu, Z.; Aburto, R.-R.; Ge, L.; Song, L.; Alemany, L.B.; Zhan, X.; Gao, G.; et al. Graphene quantum dots derived from carbon fibers. *Nano Lett.* **2012**, *12*, 844–849. [CrossRef] [PubMed]

48. Wang, L.; Wang, Y.; Xu, T.; Liao, H.; Yao, C.; Liu, Y.; Li, Z.; Chen, Z.; Pan, D.; Sun, L.; Wu, M. Gram-scale synthesis of single-crystalline graphene quantum dots with superior optical properties. *Nat. Commun.* **2014**, *5*, 5357. [CrossRef] [PubMed]

Communication

Improving Interfacial Charge-Transfer Transitions in Nb-Doped TiO$_2$ Electrodes with 7,7,8,8-Tetracyanoquinodimethane

Reo Eguchi [1,2,*], Yuya Takekuma [1,2], Tsuyoshi Ochiai [2,3,4] and Morio Nagata [1,*]

[1] Graduate School of Engineering, Tokyo University of Science, 12-1, Ichigayafunagawara, Shinjuku-ku, Tokyo 162-0826, Japan; tytakekuma@gmail.com

[2] Photocatalyst Group, Research and Development Department, Local Independent Administrative Agency Kanagawa Institute of industrial Science and TEChnology (KISTEC), 407 East Wing, Innovation Center Building, KSP, 3-2-1 Sakado, Takatsu-ku, Kawasaki, Kanagawa 213-0012, Japan; pg-ochiai@newkast.or.jp

[3] Materials Analysis Group, Kawasaki Technical Support Department, KISTEC, Ground Floor East Wing, Innovation Center Building, KSP, 3-2-1 Sakado, Takatsu-ku, Kawasaki, Kanagawa 213-0012, Japan

[4] Photocatalysis International Research Center, Tokyo University of Science, 2641 Yamazaki, Noda, Chiba 278-8510, Japan

* Correspondence: rikadai2018@gmail.com (R.E.); nagata@ci.kagu.tus.ac.jp (M.N.); Tel.: +81-3-5228-8311 (R.E. & M.N.); Fax: +81-3-5261-4631 (R.E. & M.N.)

Received: 25 July 2018; Accepted: 28 August 2018; Published: 30 August 2018

Abstract: Interfacial charge-transfer (ICT) transitions involved in charge-separation mechanisms are expected to enable efficient photovoltaic conversions through one-step charge-separation processes. With this in mind, the charge-transfer complex fabricated from TiO$_2$ nanoparticles and 7,7,8,8-tetracyanoquinodimethane (TCNQ) has been applied to dye-sensitized solar cells. However, rapid carrier recombination from the conduction band of TiO$_2$ to the highest occupied molecular orbital (HOMO) of TCNQ remains a major issue for this complex. In this study, to inhibit surface-complex recombinations, we prepared Nb-doped TiO$_2$ nanoparticles with different atomic ratios for enhanced electron transport. To investigate the effects of doping on electron injection through ICT transitions, these materials were examined as photoelectrodes. When TiO$_2$ was doped with 1.5 mol % Nb, the Fermi level of the TiO$_2$ electrode shifted toward the conduction band minimum, which improved electron back-contact toward the HOMO of TCNQ. The enhancement in electron transport led to increases in both short circuit current and open circuit voltage, resulting in a slight (1.1% to 1.3%) improvement in photovoltaic conversion efficiency compared to undoped TiO$_2$. Such control of electron transport within the photoelectrode is attributed to improvements in electron injection through ICT transitions.

Keywords: photovoltaic conversion; interfacial charge-transfer transition; 7,7,8,8-tetracyanoquinodimethane; Nb-doped TiO$_2$

---

1. Introduction

Interfacial charge-transfer (ICT) transitions between inorganic semiconductors and π-conjugated organic compounds are characteristic electronic transitions that enable direct photoinduced charge separation. Due to this feature, ICT transitions are applicable to photovoltaic conversions [1]. To date, dicyanomethylene-based compounds, such as tetracyanoethylene (TCNE) and 7,7,8,8-tetracyanoquinodimethane (TCNQ), form surface complexes with TiO$_2$ that have been reported to absorb visible light due to ICT transitions from the π-conjugated system to the conduction band of TiO$_2$ [2–4]. Although photovoltaic effects due to direct electron injection have been investigated using these surface complexes as photoanodes in photoelectrochemical cells, the photovoltaic conversion efficiencies under AM1.5 illumination are quite low (under 2%) when compared to those of dye-sensitized solar cells (DSSCs) [5].

To date, photovoltaic conversions and electron injections between surface complexes, such as bis(dicyanomethylene) compounds (TCNX) [TCNE, TCNQ, and 11,11,12,1 2-tetracyanonaphtho-2, 6-quinodimethane (TCNAQ)] and TiO_2, have been studied theoretically by Fujisawa et al. using density functional theory (DFT) on the basis of Marcus theory, which revealed that the structure and formation mechanism of the surface complex need to be considered to control interfacial electronic transitions and carrier recombinations by adjusting the electron affinity of TCNX [6–8]. In addition, they also demonstrated that carrier recombinations from the conduction band of TiO_2 to the highest occupied molecular orbital (HOMO) levels of these compounds occur quite rapidly, which, as geminate recombinations, are more severe than in DSSCs. Hence, the rapid carrier recombinations of surface complexes hinder the use of ICT transitions in photovoltaic conversion [9] and is the most significant problem faced.

To overcome rapid electron recombination, the properties of surface complexes require further investigation through different approaches that include investigating the interactions between TCNX and modified-TiO_2 photoanodes. Zaban et al. studied the suppression of electron recombination in surface complexes aided by a thin $SrTiO_3$ coating layer on TiO_2 that improved electron injection and electron transport [10]. On the other hand, many researchers reported that the TiO_2 photoanode in a DSSC is one of the most important components affecting photovoltaic performance because it acts as the support for dye molecules as well as the electron-transport region. Furthermore, TiO_2 can, in principal, be n-type-doped to enhance charge collection and electron-transport efficiency within the TiO_2 layer [11].

Among n-type-doped TiO_2 systems, Nb-doped TiO_2 photoanodes have been studied to improve electron conductance and injection because Nb has one more electron than Ti (IV) [12–18]. Recently, Lin et al. reported the influence of TiO_2 doped with Group V-b metal atoms on the photovoltaic performance of dye-sensitized solar cells [18]. Although V, Nb, and Ta belong to the same group (V-b) and have one more electron than Ti, DSSCs based on Nb-doped TiO_2 showed the best photovoltaic performance as a result of the creation of donor levels, which increased the concentration of the carriers. According to this study, the charge transport and conductivity for Nb-doped TiO_2 were superior to those of V- and Ta-doped TiO_2. Therefore, Nb is a superior doping element.

In this study, to inhibit rapid electron recombinations in surface complexes, we examined the effects of Nb-doped TiO_2 electrodes with TCNQ on photovoltaic performance. The effects of different amounts of Nb-doping on the photovoltaic properties of surface complexes were evaluated by means of incident photon-to-current efficiency (IPCE) spectroscopy and by acquiring current-density–voltage (J-V) curves. TiO_2 doped with 1.5 mol % Nb exhibited improved Jsc and Voc values, resulting in a 17% improvement in photoconversion efficiency compared to undoped TiO_2.

2. Results and Discussion

2.1. Structural Characterization of Nb-Doped TiO_2

Nb-doped TiO_2 nanoparticles were formed by a hydrothermal method. Figure 1 displays the X-ray diffraction (XRD) patterns of undoped and Nb-doped TiO_2 samples with varying Nb contents. The anatase and rutile phases of TiO_2 are distinct, and the intensity of the peak corresponding to the anatase phase increased with increasing Nb content in the sample, while that of the rutile phase decreased (Figure 1a). Hence, doping the TiO_2 lattice with Nb was observed to enhance the growth of the anatase TiO_2 phase while hindering the formation of the rutile TiO_2 phase [19]. The average crystallite size of each phase was calculated by the Scherrer-equation [20]:

$$D = \frac{k\lambda}{B \cos \theta} \tag{1}$$

where D is the crystallite size, k is a dimensionless shape factor that has a typical value of 0.94, λ is the Cu-Kα X-ray wavelength (1.5406 Å), θ is the Bragg angle in degrees, and B is the full-width-at-half-maximum (FWHM) of the peak. The anatase and rutile crystallinities were determined from their (101) and (110) peak heights, respectively; crystallite size are listed in Table S1. Incorporation of the Nb dopant into the TiO$_2$ structure typically decreased crystallite growth; indeed, the crystallite size of the anatase phase slowly decreased from 10.1 nm to 8.8 nm with increased doping, led by the effect of Nb on nucleation following doping into the TiO$_2$ lattice. A similar effect of different oxide additives on the TiO$_2$ phase transformation was reported by Yanagida et al. [19]. In addition, the diffraction peaks shifted to lower θ values with increasing Nb content as a result of the larger radius of Nb^{5+} (0.64 Å) compared to that of Ti^{4+} (0.61 Å) [15], in accordance with the Bragg equation: $\lambda = 2d\sin\theta$ (Figure 1b).

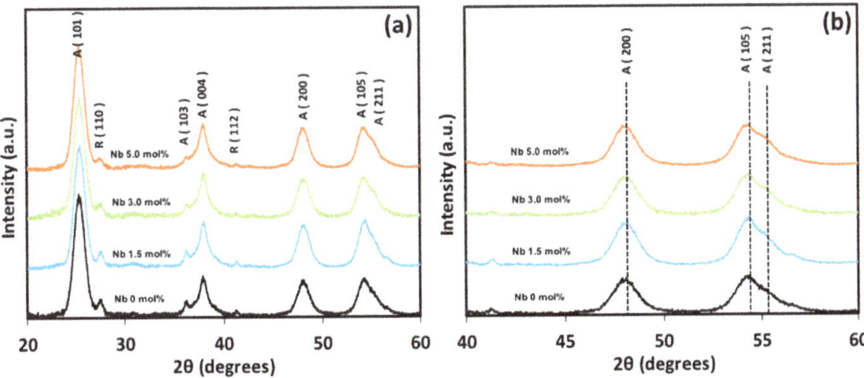

Figure 1. (**a**) X-ray diffraction (XRD) patterns of TiO$_2$ with varying Nb contents (A: Anatase, R: Rutile). (**b**) XRD patterns between 2θ values of 40° and 60°.

Figure 2 displays the X-ray photoelectron spectroscopy (XPS) spectra of the undoped and Nb-doped TiO$_2$ samples with varying Nb contents. Nb 3d$_{3/2}$ and 3d$_{5/2}$ peaks are evident in the spectra of the Nb-doped TiO$_2$ samples, and their intensities increased with increasing Nb content (Figure 2a). The Nb 3d$_{3/2}$ and 3d$_{5/2}$ peaks are located at binding energies of 209 and 206 eV, respectively. Nb-doping also caused the Ti 2p$_{1/2}$ and 2p$_{3/2}$ peaks to shift slightly toward higher binding energies, which is attributable to the higher electronegativity of Nb (1.6) compared to that of Ti (1.54) (Figure 2b) [15]. The peaks in the O 1s spectra correspond to Ti^{4+}–O bonds; the positions of these peaks show similar trends to those observed for Ti due to increases in both lattice oxygen and Nb^{5+}–O bonds in these samples. Figure 2d displays a double-band structure, with the main peak (29.4 eV) related to O 1s electron binding in TiO$_2$ and the other peak, at a higher binding energy (531 eV), attributed to OH groups on the surfaces of these samples. Typically, OH groups are beneficial for TCNQ anchoring [21].

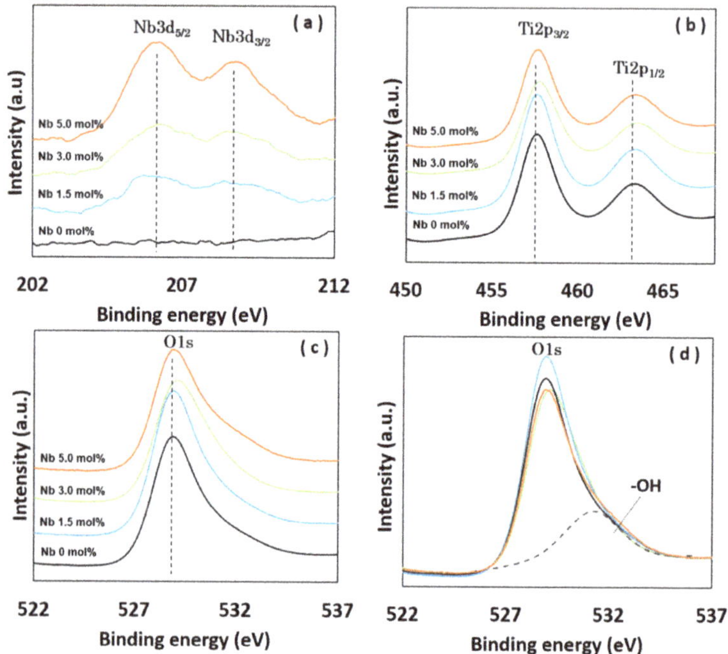

Figure 2. High-resolution (**a**) Nb 3d, (**b**) Ti 2p, and (**c**) O 1s X-ray photoelectron spectroscopy (XPS) spectra of TiO_2 with varying Nb contents. (**d**) O 1s spectral fitting, revealing the presence of peaks related to OH groups.

2.2. Optical Properties of Nb-Doped TiO_2

Examination of the electronic structures of the Nb-doped TiO_2 electrodes is essential to understand the ICT transition from the highest occupied molecular orbital (HOMO) of TCNQ to the conduction band of TiO_2. XPS and calculated Eg values for the valence band can provide electronic-potential information. The reflectance spectra displayed in Figure S1a enable the bandgaps of the samples to be determined. Eg values were calculated after converting the reflectance data into the equivalent absorption coefficients using Equation (1), as shown in Figure S1b and listed in Table 1. Eg increased from 3.06 eV to 3.14 eV as the Nb content was increased from 0 mol % to 5.0 mol %, which is ascribable to strong hybridization between the Ti 3d and Nb 3d states that forms a d-type conduction band [22]. The observed increase in bandgap with increasing Nb content can be also explained by the decrease in the amount of rutile phase TiO_2 formed (Table S1); indeed, the bandgap of anatase is 3.2 eV, which is larger than that of rutile (3.0 eV).

Table 1. Optical bandgap values for TiO_2 samples with different Nb contents.

Nb Content (mol %)	0	1.5	3.0	5.0
E_g (eV)	3.06	3.07	3.11	3.14
$E_{CBM}-E_F$ (eV)	0.46	0.37	0.41	0.64

Figure 3 shows the photoemission spectra of TiO_2 samples with different Nb contents, in which the binding energies are referenced against the Fermi energy level (E_F). Two peaks are evident in each photoemission spectrum, one centered at 6.4 eV and the other at 8.2 eV. These peaks are attributed to emissions from π- and σ-type O 2p orbitals, respectively. The positions of the valence band maxima

(VBMs) were directly determined from the photoemission spectra by linear extrapolation of the onsets of the valence-band emissions [12]. In the case of the undoped TiO_2 sample, the VBM was found to be located 2.6 eV below the Fermi level. With the optical bandgap determined to be 3.06 eV, we deduce that E_F is 0.46 eV lower than the energy of the conduction band minimum (E_{CBM}). The calculated positions of the conduction band minimum (CBM) with respect to the Fermi level are summarized in Table 1. The energy of the CBM of the 1.5 mol % Nb-doped TiO_2 was found to be 0.37 eV higher than that of the Fermi level, while that of the undoped TiO_2 was 0.46 eV higher (Table 1). The CBM of the sample containing 1.5 mol % Nb was therefore 0.09 eV lower in energy than that of undoped TiO_2. The observed lowering of the CBM of the sample containing 1.5 mol % and 3.0 mol % Nb due to donor levels induced by these dopants [18] is attributable to higher electron transitions between the Nb-doped TiO_2 electrode and the anchoring TCNQ. On the other hand, the CBM for the sample containing 5.0 mol % Nb exhibited a large offset between its CBM energy and E_F, compared to the undoped TiO_2, due to intra-band transport [15].

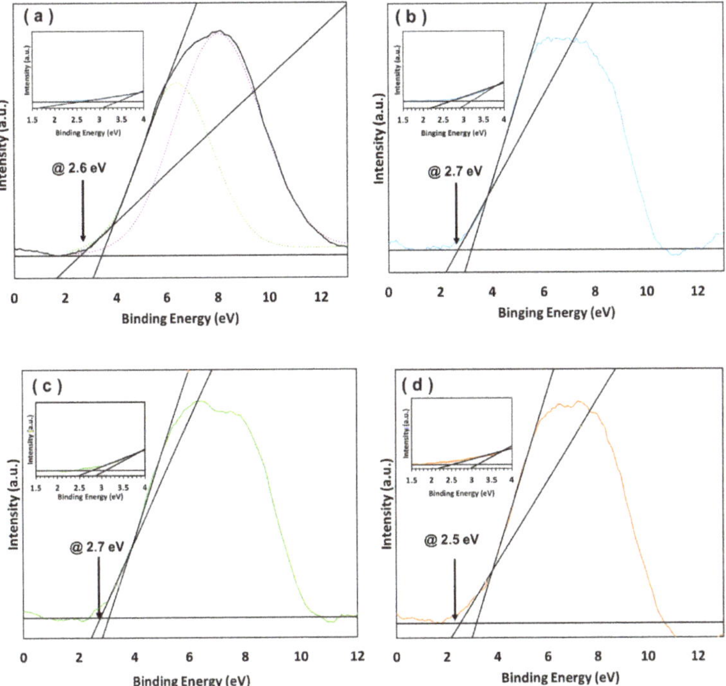

Figure 3. XPS spectra and valence-band maxima of the Nb-doped TiO_2 with Nb contents of (**a**) 0, (**b**) 1.5, (**c**) 3.0 mol %, and (**d**) 5.0 mol %. The Gaussian fitted electron emissions from the π (green) and σ (purple) O 2p orbitals are also shown in (**a**).

2.3. Photovoltaic Performance of Interfacial Charge-Transfer (ICT) Photoconversion Devices

The photovoltaic performance of ICT-transition devices based on the undoped and Nb-doped TiO_2 electrodes with varying Nb contents under solar illumination (AM 1.5 G, 100 mW/cm^2) are displayed in Figure 4a, with the photovoltaic performance parameters summarized in Table 2. Jsc was observed to increase from 4.5 mA/cm^2 to 5.7 mA/cm^2 at an Nb content of 3.0 mol %, which is attributable to improved electronic transitions between the nanoparticles and the anchoring TCNQ associated with the negative CBM-energy shift (Table 1). On the other hand, Voc increased slightly, from 0.40 V to 0.41 V, as the dopant content was increased to 1.5 mol % due to the increased gap between E_F and the redox potential of the I^-/I^{3-} couple. This is also attributable to Nb-doped TiO_2-surface passivation, according to electrochemical impedance spectroscopy (EIS) (Nyquist plots, Figure S2) [23,24]. However, increased Nb doping has a negative influence on photovoltaic behavior at levels over 5.0 mol %; indeed, Voc decreased from 0.40 V to 0.36 V, which is ascribable to a detrimental conduction-band-edge effect. Indeed, the XRD result shows the decrease in rutile crystallization in 5.0 mol % Nb-doped TiO_2 (Table S1) led to defects in crystallization, and affected electron recombination such the degradation of V_{oc} parameter.

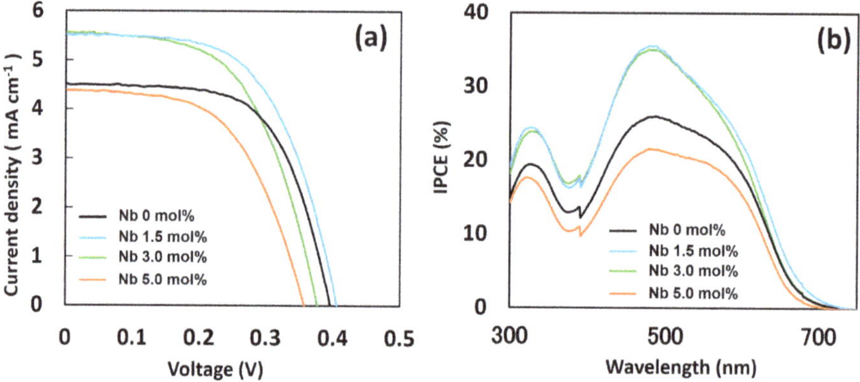

Figure 4. (**a**) Photocurrent-density–voltage curves and (**b**) incident photon-to-current conversion efficiency spectra of Interfacial charge-transfer (ICT) photoconversion devices based on the undoped and Nb-doped TiO_2 electrodes prepared in this study.

Table 2. Photovoltaic-performance parameters of Interfacial charge-transfer (ICT) photoconversion devices based on the undoped and Nb-doped TiO_2 electrodes prepared in this study.

Sample	Jsc (mA/cm^2)	Voc (V)	FF (%)	H (%)
Nb 0 mol %	4.5	0.40	63	1.1
Nb 1.5 mol %	5.5	0.41	59	1.3
Nb 3.0 mol %	5.7	0.38	57	1.2
Nb 5.0 mol %	4.4	0.36	56	0.87

Figure 4b reveals that the device containing the 1.5 mol % Nb-doped TiO_2 electrode exhibits enhanced IPCE compared to that of the undoped electrode. Indeed, the absorption edge was observed to shift from 730 nm to 750 nm. Nb has been reported to create oxygen vacancies in TiO_2 that act as active sites, resulting in a photoresponse red shift [25]. However, the lowest IPCE was obtained in the device containing the 5.0 mol % Nb-doped TiO_2 electrode, as revealed by the J-V curves.

According to previous studies on the fabrication of DSSCs based on TiO_2 doped with Nb at concentrations up to 5.0 mol %, Feng et al. and Huang et al. reported that the highest photovoltaic efficiency was observed for Nb concentrations of 2.0 mol % and 5.0 mol %, respectively [15,16]. However, our results demonstrated that Nb-doping of 1.5 mol % produced the highest efficiency. Considering the behavior of Jsc with doping, we can conclude that the optimum Nb concentration for reaching the best efficiency is in the range of 1.5–3.0 mol %. In addition, the error bar of photovoltaic efficiency in this study is shown in Figure S3.

3. Experimental Section

3.1. Preparation of Nb-Doped TiO$_2$

The Nb-doped TiO$_2$ nanoparticles were synthesized following the procedure described by Nikolay et al. [12]. These syntheses used Ti and Nb precursors and hydrothermal processes. To prepare Nb-doped TiO$_2$ (0, 1.5, 3.0, or 5.0 mol %) nanoparticles, 16.4 mL of titanium tetraisopropoxide (Kanto Chemical Co., Tokyo, Japan) was mixed with 0, 60, 120, or 200 μL of niobium ethoxide (Wako Co., Tokyo, Japan), respectively, after which 2.64 mL of acetic acid (Kanto Chemical Co., Tokyo, Japan) was added under stirring condition with a Teflon stirrer blade for 15 min at room temperature. The mixture was dropped into 68.4 mL of deionized water while stirred at about 800 rpm. After stirring for 1 h, 2.35 mL of 65% nitric acid (Kanto Chemical Co., Tokyo, Japan) was added to the solution. The solution temperature was increased to 80 °C over 40 min and then held at 80 °C for 80 min under reflux conditions with intensive stirring. The nanoparticles were hydrothermally grown using the prepared colloidal solution in a Teflon-lined mini-autoclave at 180 °C for 12 h, after which 0.52 mL of 65% nitric acid was added to the colloidal solution, followed by ultrasonication with stirring for 1 h. The prepared mixture was finally washed three times with ethanol by centrifugation.

3.2. Cell Fabrication

The Nb-doped TiO$_2$ pastes were synthesized following the procedure reported by Ito et al. [26]. The final screen-printing pastes correspond to 18 wt % TiO$_2$, 9 wt % ethyl cellulose and 73 wt % terpineol. Two kinds of pure ethyl cellulose (EC) powders, i.e., EC (10 mPas, Kanto Chemical Co., Tokyo, Japan) and EC (45 mPas, Kanto Chemical Co., Tokyo, Japan) were dissolved before usage in ethanol to yield 10 wt % solutions. Then, 0.325 g of EC (10 mPas) and 0.175 g of EC (45 mPas) of these 10 wt % ethanolic mixtures were added to a round-bottomed rotovap flask containing 1 g pure TiO$_2$ (obtained from a previously prepared precipitate) and 4.05 g of terpineol (Kanto Chemical Co., Tokyo, Japan), and diluted with approximately 100 ml of ethanol. This mixture was then ultrasonicated for 48 h. Ethanol and water were removed from these TiO$_2$/ethyl cellulose solutions using a rotary-evaporator. The final formulations of the pastes were made with a three-roll mill (Exakt, Nagase Screen Printing Research Co., Aichi, Japan). The Nb-doped TiO$_2$ electrodes were fabricated by screen-printing pastes onto glass substrates coated with transparent conducting F-doped SnO$_2$ (FTO), Nippon Sheet Glass Co., Tokyo, Japan) with a sheet resistance of 10 Ω sq^{-1}, followed by sintering at 500 °C for 30 min. The thickness of the TiO$_2$ electrode was set to 8 μm.

The TiO$_2$ electrodes were immersed in a 1 mM solution of TCNQ (Tokyo Kasei Kogyo Co., Tokyo, Japan) in acetonitrile at 60 °C for 24 h. Photovoltaic cells were fabricated using the TCNQ-treated TiO$_2$ electrode (active area: 4 mm × 4 mm), a Pt-sputtered FTO glass counter electrode (Geomatec Co., Kanagawa, Japan), an I$^-$/I$_3^-$ redox couple electrolyte (1 M LiI (Sigma-Aldrich Co., St. Louis, MO, USA) and 0.025 M I$_2$ (Kanto Chemical Co.) in acetonitrile), and a spacer film (thickness: 30 μm).

3.3. Characterization

Incident photon-to-current efficiency (IPCE) spectra were acquired using a Hypermonolight system (M10, Bunkoukeiki Co., Tokyo, Japan) with a calibrated silicon photodiode (Bunkoukeiki Co., Tokyo, Japan). Current-density–voltage (J-V) curves were recorded using a potentiostat (1287A potentiostat/galvanostat, Ametek Co., St. Berwyn, PA, USA) under 100 mW/cm^2 AM 1.5 G simulated sunlight produced by a solar simulator (Yamashita Denso Co., Tokyo, Japan). Electrochemical impedance spectroscopy (EIS) was performed using a potentiostat equipped with calculation software (1255B frequency-response analyzer, Ametek Co, Tokyo, Japan). The thicknesses of the films were measured using a surface roughness profilometer (SURFCOM1440D, Accretech Co., Tokyo, Japan). Crystal structures were determined by X-ray diffraction (XRD, Ultima X-ray diffractometer, Rigaku Co., Yamanashi, Japan). The electronic structures and chemical states of the TiO$_2$ electrodes were investigated by X-ray photoelectron spectroscopy (XPS) (JPS-9010MC,

Nihondensi Co., Tochigi, Japan). The binding energies were calibrated against the C 1s peak at 284.60 eV. For all XPS measurements, undoped TiO_2 and Nb-doped TiO_2 were deposited on carbon sheets. UV-vis spectroscopy was performed using a U-3900H spectrometer (Hitachi Co., Tokyo, Japan) in reflectance mode, and the spectra were analyzed using the Kubelka-Munk formalism to convert reflectance into the equivalent absorption coefficient, α_{KM} [27–29]:

$$\alpha_{KM} = \frac{(1 - R_\infty)^2}{2R_\infty} \tag{2}$$

where: $R\infty$ is the reflectance of an infinitely thick sample with respect to the reference at each wavelength.

4. Conclusions

In this study, we examined the abilities of Nb-doped TiO_2 electrodes with 7,7,8,8-tetracyanoquinodimethane (TCNQ) to inhibit electron recombinations in surface-complexes. The 1.5 mol % Nb-doped TiO_2 electrode exhibited improved photovoltaic performance and superior short circuit current and open circuit voltage, resulting in 1.3% photoconversion efficiency, which is 17% higher than that of the undoped photoelectrode. This improvement is ascribed to enhanced electron injection resulting from a shift in the Fermi level of the TiO_2 electrode toward the conduction band minimum, and the effect of passivation, as revealed by Electrochemical impedance spectroscopy. As expected, these experimental data for Nb-doped TiO_2 with TCNQ reveal that semiconductor modification can be used to achieve efficient photovoltaic conversion through Interfacial charge-transfer transitions by suppressing surface-complex carrier recombinations and improving electron transport.

Supplementary Materials: The following are available online at http://www.mdpi.com/2073-4344/8/9/367/s1. Table S1: Crystallite sizes of TiO_2 samples with various Nb contents; Table S2: Fitted EIS spectra of TiO_2 samples with varying Nb contents. Figure S1: (a) Reflectance spectra and (b–e) optical bandgaps of TiO_2 samples with varying Nb contents; Figure S2: Electrochemical impedance spectra (Nyquist plots) of the undoped and Nb-doped TiO_2 electrodes; Figure S3: Error bars for Conversion efficiency of DSSCs employing different Nb dopants.

Author Contributions: R.E., Y.T., T.O., and M.N. participated in the study design and conducted the experiments. Data were collected and analyzed by R.E. and Y.T. The manuscript was written by R.E., M.N. and T.O. provided valuable input and advice regarding the manuscript.

Funding: This research received no external funding.

Conflicts of Interest: The authors declare no conflicts of interest.

References

1. Fujisawa, J.-I. Large Impact of Reorganization Energy on Photovoltaic Conversion Due to Interfacial Charge-Transfer Transitions. *Phys. Chem. Chem. Phys.* **2015**, *17*, 12228–12237. [CrossRef] [PubMed]
2. Xagas, A.P.; Bernard, M.C.; Hugot-Le Goff, A.; Spyrellis, N.; Loizos, Z.; Falaras, P. Surface Modification and Photosensitisation of TiO_2 Nanocrystalline Films with Ascorbic Acid. *J. Photochem. Photobiol. A* **2000**, *132*, 115–120. [CrossRef]
3. Tae, E.L.; Lee, S.H.; Lee, J.K.; Yoo, S.S.; Kang, E.J.; Yoon, K.B. A Strategy to Increase the Efficiency of the Dye-Sensitized TiO_2 Solar Cells Operated by Photoexcitation of Dye-to-TiO_2 Charge-Transfer Bands. *J. Phys. Chem. B* **2005**, *109*, 22513–22522. [CrossRef] [PubMed]
4. Manzhos, S.; Jono, R.; Yamashita, K.; Fujisawa, J.-I.; Nagata, M.; Segawa, H. Study of Interfacial Charge Transfer Bands and Electron Recombination in the Surface Complexes of TCNE, TCNQ, and TCNAQ with TiO_2. *J. Phys. Chem. C* **2011**, *115*, 21487–21493. [CrossRef]
5. Manzhos, S.; Fujisawa, J.-I.; Segawa, H.; Yamashita, K. Isotopic Substitution as a Strategy to Control Non-Adiabatic Dynamics in Photoelectrochemical Cells: Surface Complexes between TiO_2 and Dicyanomethylene Compounds. *J. Appl. Phys.* **2012**, *51*, 10NE03. [CrossRef]

6. Fujisawa, J.-I.; Nagata, M.; Hanaya, M. Charge-Transfer Complex Versus [Sigma]-Complex Formed between TiO$_2$ and Bis (Dicyanomethylene) Electron Acceptors. *Phys. Chem. Chem. Phys.* **2015**, *17*, 27343–27356. [CrossRef] [PubMed]

7. Michinobu, T.; Satoh, N.; Cai, J.; Li, Y.; Han, L. Novel Design of Organic Donor-Acceptor Dyes without Carboxylic Acid Anchoring Groups for Dye-Sensitized Solar Cells. *J. Mater. Chem. C* **2014**, *2*, 3367–3372. [CrossRef]

8. Fujisawa, J.-I.; Hanaya, M. Electronic Structures of TiO$_2$–TCNE, –TCNQ, and –2,6-TCNAQ Surface Complexes Studied by Ionization Potential Measurements and DFT Calculations: Mechanism of the Shift of Interfacial Charge-Transfer Bands. *Chem. Phys. Lett.* **2016**, *653*, 11–16. [CrossRef]

9. Wang, Y.; Hang, K.; Anderson, N.A.; Lian, T. Comparison of Electron Transfer Dynamics in Molecule-to-Nanoparticle and Intramolecular Charge Transfer Complexes. *J. Phys. Chem. B* **2003**, *107*, 9434–9440. [CrossRef]

10. Hod, I.; Shalom, M.; Tachan, Z.; Rühle, S.; Zaban, A. SrTiO$_3$ Recombination-Inhibiting Barrier Layer for Type II Dye-Sensitized Solar Cells. *J. Phys. Chem. C* **2010**, *114*, 10015–10018. [CrossRef]

11. Zhang, S.; Yang, X.; Numata, Y.; Han, L. Highly Efficient Dye-Sensitized Solar Cells: Progress and Future Challenges. *Energy Environ. Sci.* **2013**, *6*, 1443–1464. [CrossRef]

12. Nikolay, T.; Larina, L.; Shevaleevskiy, O.; Ahn, B.A. Electronic Structure Study of Lightly Nb-Doped TiO$_2$ Electrode for Dye-Sensitized Solar Cells. *Energy Environ. Sci.* **2011**, *4*, 1480–1486. [CrossRef]

13. Ghartavol, H.M.; Mohammadi, M.R.; Afshar, A.; Chau-Nan Hong, F.; Jeng, Y.-R. Efficient Dye-Sensitized Solar Cells Based on CNT-Derived TiO$_2$ Nanotubes and Nb-Doped TiO$_2$ Nanoparticles. *RSC Adv.* **2016**, *6*, 101737–101744. [CrossRef]

14. Liu, W.; Wang, H.-G.; Wang, X.; Zhang, M.; Guo, M. Titanium Mesh Supported TiO$_2$ Nanowire Arrays/Nb-Doped TiO$_2$ Nanoparticles for Fully Flexible Dye-Sensitized Solar Cells with Improved Photovoltaic Properties. *J. Mater. Chem. C* **2016**, *4*, 11118–11128. [CrossRef]

15. Su, H.; Huang, Y.-T.; Chang, Y.-H.; Zhai, P.; Hau, N.Y.; Cheung, P.C.H.; Yeh, W.-T.; Wei, N.T.-C.; Feng, S.-P. The Synthesis of Nb-Doped TiO$_2$ Nanoparticles for Improved-Performance Dye Sensitized Solar Cells. *Electrochim. Acta* **2015**, *182*, 230–237. [CrossRef]

16. Lü, X.; Mou, M.X.; Wu, J.; Zhang, D.; Zhang, L.; Huang, F.; Xu, F.; Huang, S. Improved-Performance Dye-Sensitized Solar Cells Using Nb-Doped TiO$_2$ Electrodes: Efficient Electron Injection and Transfer. *Adv. Funct. Mater.* **2010**, *20*, 509–515. [CrossRef]

17. Kim, S.G.; Ju, M.J.; Choi, I.T.; Choi, W.S.; Choi, H.-J.; Baek, J.-B.; Kim, H.K. Nb-Doped TiO$_2$ Nanoparticles for Organic Dye-Sensitized Solar Cells. *RSC Adv.* **2013**, *3*, 16380–16386. [CrossRef]

18. Liu, J.; Duan, Y.; Zhou, X.; Lin, Y. Influence of V$_B$ Group Doped TiO$_2$ on Photovoltaic Performance of Dye-Sensitized Solar Cells. *Appl. Surf. Sci.* **2013**, *277*, 231–236. [CrossRef]

19. Hishita, S.; Mutoh, I.; Koumoto, K.; Yanagida, H. Inhibition Mechanism of the Anatase-Rutile Phase Transformation by Rare Earth Oxides. *Ceram. Int.* **1983**, *9*, 61 67. [CrossRef]

20. Cullity, B.D. Elements of X-Ray Diffraction. *Am. J. Phys.* **1957**, *25*, 394–395. [CrossRef]

21. Jono, R.; Fujisawa, J.-I.; Segawa, H.; Yamashita, K. Theoretical Study of the Surface Complex between TiO$_2$ and TCNQ Showing Interfacial Charge-Transfer Transitions. *J. Phys. Chem. Lett.* **2011**, *2*, 1167–1170. [CrossRef] [PubMed]

22. Taro, H.; Hideyuki, K.; Koichi, Y.; Hiroyuki, N.; Yutaka, F.; Shoichiro, N.; Naoomi, Y.; Akira, C.; Hiroshi, K.; Masaharu, O.; et al. Electronic Band Structure of Transparent Conductor: Nb-Doped Anatase TiO$_2$. *Appl. Phys. Express* **2008**, *1*, 111203.

23. Bisquert, J. Theory of the Impedance of Electron Diffusion and Recombination in a Thin Layer. *J. Phys. Chem. B* **2002**, *106*, 325–333. [CrossRef]

24. Zhang, Z.; Zakeeruddin, S.M.; O'Regan, B.C.; Humphry-Baker, R.; Grätzel, M. Influence of 4-Guanidinobutyric Acid as Coadsorbent in Reducing Recombination in Dye-Sensitized Solar Cells. *J. Phys. Chem. B* **2005**, *109*, 21818–21824. [CrossRef] [PubMed]

25. Karvinen, S. The Effects of Trace Elements on the Crystal Properties of TiO$_2$. *Solid State Sci.* **2003**, *5*, 811–819. [CrossRef]

26. Ito, S.; Murakami, T.N.; Comte, P.; Liska, P.; Grätzel, C.; Nazeeruddin, M.K.; Grätzel, M. Fabrication of Thin Film Dye Sensitized Solar Cells with Solar to Electric Power Conversion Efficiency over 10%. *Thin Solid Films* **2008**, *516*, 4613–4619. [CrossRef]

27. Kubelka, P. New Contributions to the Optics of Intensely Light-Scattering Materials. Part I. *J. Opt. Soc. Am.* **1948**, *38*, 448–457. [CrossRef]

28. Yang, L.; Kruse, B. Revised Kubelka–Munk Theory. I. Theory and Application. *J. Opt. Soc. Am. A* **2004**, *21*, 1933–1941. [CrossRef]

29. Yang, L.; Miklavcic, S.J. Revised Kubelka–Munk Theory. III. A General Theory of Light Propagation in Scattering and Absorptive Media. *J. Opt. Soc. Am. A* **2005**, *22*, 1866–1873. [CrossRef]

 *catalysts*

Article

Photocatalytic Antibacterial Effectiveness of Cu-Doped TiO$_2$ Thin Film Prepared via the Peroxo Sol-Gel Method

Benjawan Moongraksathum [1], Jun-Ya Shang [1] and Yu-Wen Chen [1,2,*]

[1] Department of Chemical and Materials Engineering, National Central University, Jhong-Li 32001, Taiwan; bmoongraksathum@gmail.com (B.M.); alice19940430@gmail.com (J.-Y.S.)

[2] Department of Chemistry, Tomsk State University, 36 Lenin Prospekt, Tomsk 634050, Russia

* Correspondence: ywchen@cc.ncu.edu.tw; Tel.: +886-3422-7151 (ext. 34203)

Received: 23 July 2018; Accepted: 20 August 2018; Published: 27 August 2018

Abstract: Cu-doped titanium dioxide thin films (Cu/TiO$_2$) were prepared on glass substrate via peroxo sol-gel method and dip-coating process with no subsequent calcination process for the degradation of organic dye and use as an antibacterial agent. The as-prepared materials were characterised using transmission electron microscopy (TEM), X-ray diffraction (XRD), scanning electron microscopy (SEM) and X-ray photoelectron spectroscopy (XPS). For photocatalytic degradation of methylene blue in water, the samples were subjected to Ultraviolet C (UVC) and visible light irradiation. Degraded methylene blue concentration was measured using UV-Vis spectrophotometer. The antibacterial activities of the samples were tested against the gram-negative bacteria Escherichia coli (ATCC25922). Copper species were present in the form of CuO on the surface of modified TiO$_2$ particles, which was confirmed using TEM and XPS. The optimal observed Cu/TiO$_2$ weight ratio of 0.5 represents the highest photocatalytic activities under both UVC and visible light irradiation. Moreover, the same composition remarkably exhibited high antibacterial effectiveness against *E. coli* after illumination with ultraviolet A. The presence of CuO on TiO$_2$ significantly enhanced photocatalytic activities. Therefore, active Cu-doped TiO$_2$ can be used as a multipurpose coating material.

Keywords: antibacterial; copper oxide; photocatalyst; titanium dioxide; thin film; visible light

---

1. Introduction

Photocatalysis has garnered plenty of attention from the scientific community in recent decades, resulting in various commercialized products having photocatalytic functions. Among photocatalysts, titanium dioxide (TiO$_2$) has received the greatest interest because of its remarkable stability and non-toxicity. Modification of TiO$_2$ has been extensively studied to improve its physical and chemical properties and to overcome the limitation of TiO$_2$ in photocatalytic processes. Modified TiO$_2$ has been deployed both environmentally and hygienically, including in the photocatalytic decomposition of organic pollutants [1–5], in self-cleaning materials [6–9], and as an antibacterial agent of photo-induced photocatalytic reactions [10–13].

The activity of TiO$_2$ nanoparticles is due to the oxidative stress and/or the production of reactive oxygen species (ROS), including hydroxyl radical (OH$^\bullet$) and hydrogen peroxide (H$_2$O$_2$) under UV light irradiation; therefore, TiO$_2$ is used as an antimicrobial agent. The produced ROS can cause cell membrane damage, cell cycle cessation, DNA damage and lipid peroxidation in microorganisms via direct contact between cells and nanoparticles, thereby resulting in cell death [14–16].

Several transition metals are toxic to various microbial pathogens. In addition to considerable commercialization in this field, this finding has led to widespread research on the use of such materials as practical antimicrobial agents. Among several transition metals, copper has gained considerable attention, both as dispersions and, in the case of elemental copper and as alloys [10,17,18]. For use as antibacterial agents, copper and its compounds achieve antibacterial activity by the accumulation of copper ions within cells, eventually causing degradation of cell membranes [19–21].

Using titanium tetrachloride (TiCl$_4$) as a precursor and H$_2$O$_2$ as a peptizing agent, previous studies have proposed a peroxo sol-gel method for the synthesis of neutral TiO$_2$ sol [22–25]. The advantages of this method include not needing a calcination process to obtain the anatase structure of TiO$_2$, as well as the ability to use H$_2$O$_2$ as an oxidizing agent. It results in the formation of TiO$_2$ nanoparticles dispersed in neutral, stable, and transparent sol.

The purpose of this study was to prepare Cu-doped TiO$_2$ thin films using the peroxo sol-gel method, thus determining how the addition of Cu to TiO$_2$ influenced its antibacterial effectiveness, as well as the photocatalytic degradation of methylene blue (MB) aqueous solution under either UVC or visible light irradiation.

2. Results and Discussion

2.1. Characteristic of Cu-Doped TiO$_2$ Particles

The peroxo sol-gel method was used to prepare Cu-doped TiO$_2$ sol by direct addition of the precursor of copper (Cu(NO$_3$)$_2$·3H$_2$O) during the heating of TiO$_2$ sol. To obtain powder nanoparticles, the as-prepared sols were further dried at 70 °C for several days. The X-ray diffraction (XRD) patterns of samples are shown in Figure 1. The diffraction peaks of the TiO$_2$ and a series of Cu-modified TiO$_2$ were located at the same positions and showed a similar pattern. The diffraction peaks located at $2\theta = 25.31°, 37.80°, 48.05°, 53.89°, 55.06°, 62.69°, 68.76°$, and $75.03°$ corresponded to the anatase phase of (101), (004), (200), (105), (211), (204), (116), and (215), respectively (JCPDS 21-1272). Furthermore, no additional peaks of copper oxide or other forms were found, implying that copper oxides were either highly dispersed with little TiO$_2$ particles or the amount of Cu dopant was below the detection level of the technique. The size of the crystallite was calculated by the Scherrer equation [26,27]:

$$L = \frac{0.9\lambda}{\beta \cos\theta} \tag{1}$$

where λ is the X-ray wavelength (0.1540 nm), β is the full width at half maximum (FWHM), θ is the diffraction angle, and L is the average crystallite size. The results are listed in Table 1. The presence of copper slightly decreased the crystallite size of TiO$_2$, which was attributed to the inhibition of titania condensation and the crystallization in the Cu-doped system [21].

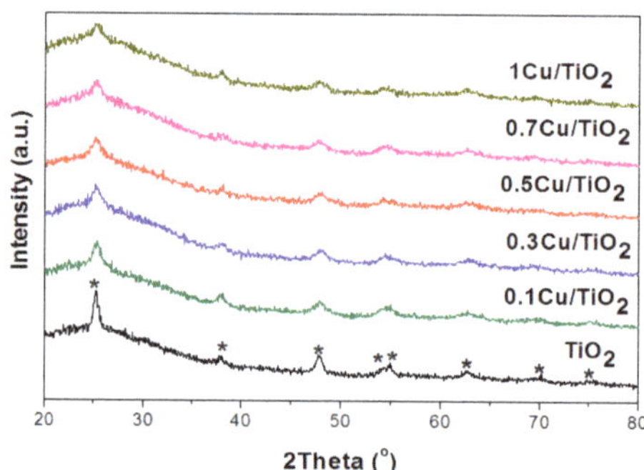

Figure 1. X-ray diffraction patterns of TiO$_2$ and a series of Cu-doped TiO$_2$ powder (* = anatase).

Table 1. Crystallite sizes of TiO_2 and a series of Cu-doped TiO_2.

Sample	Weight Ratio of Cu:TiO_2	Crystallite Size (nm)
TiO_2	0:100	14.96
0.1Cu/TiO_2	0.1:100	9.12
0.3Cu/TiO_2	0.3:100	9.40
0.5Cu/TiO_2	0.5:100	9.21
0.7Cu/TiO_2	0.7:100	9.60
1Cu/TiO_2	1:100	10.16

Morphology of the as-prepared samples was investigated using TEM and HRTEM. Figure 2 depicts TEM images of TiO_2 and 0.5Cu/TiO_2 particles. The morphology of the TiO_2 particles prepared via the peroxo sol-gel method is best described as an elliptical shape with particle size of 40–60 nm and 15–30 nm for long and short axes, respectively [3,7,12,23–25]. In Figure 2b, HRTEM image displayed the lattice fringe of TiO_2 of 0.327 nm, corresponding to the anatase (101) plane. Figure 2c shows 0.5Cu/TiO_2 particles having the particle size ranging from 20 to 30 nm and 5 to 10 nm for long and short axes, respectively. In addition, magnified view clearly identified some small copper nanoparticles ($\leq$4nm) deposited on the surface of the TiO_2 nanoparticles (see Figure 2c). Therefore, elliptical anatase TiO_2 nanoparticles can be synthesised via the peroxo sol-gel method without a subsequent annealing process, and the presence of copper could decrease the particles size and crystallite size of TiO_2, attributed to phase deterioration of the TiO_2 anatase [21].

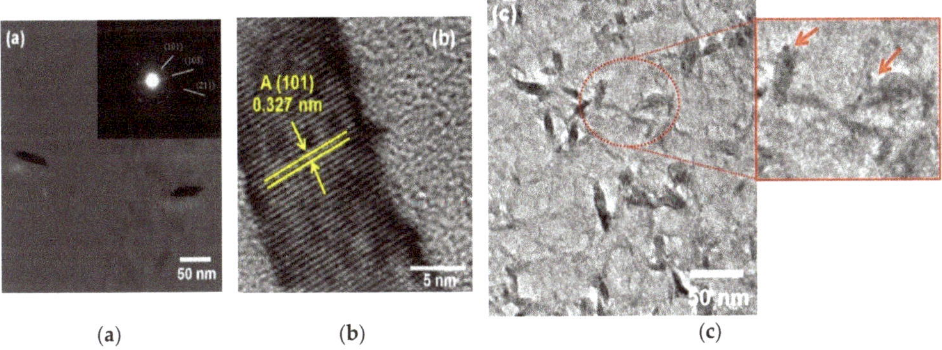

(a)   (b)   (c)

Figure 2. TiO_2 particle: (**a**) TEM image and (**b**) HRTEM micrograph representing lattice fringe attributed to the (101) plane of titania; (**c**) TEM image and magnified view of 0.5Cu/TiO_2 particle.

2.2. Characterization of Cu-Doped TiO_2 Thin Film

XPS was performed to investigate the electronic state of each element in TiO_2 and Cu-doped TiO_2 thin films. Figure 3a–c shows the XPS spectra of Ti 2p, O 1s and Cu 2p in the TiO_2 and 0.5Cu/TiO_2 thin films. The peaks of Ti $2p_{1/2}$ and Ti $2p_{3/2}$ in pure TiO_2 were located at 464.6 and 458.9 eV, respectively, corresponding to the tetravalent state (Ti^{4+}) [28]. The characteristic peaks of Ti 2p did not change in the presence of copper, indicating that the cations in the Cu-doped TiO_2 film are all in the Ti^{4+} state. Figure 3b shows the XPS spectra of the O 1s region for TiO_2 and 0.5Cu/TiO_2 films. The binding state of O 1s region of TiO_2 was deconvoluted into two peaks centred at 530.7 eV and 531.7 eV, which were ascribed to lattice oxide ions in TiO_2 and hydroxyl groups on the surface, respectively (see Figure 3b and Table 2) [22]. The characteristic Cu $2p_{3/2}$ and Cu $2p_{1/2}$ peaks were observed at 934.9 eV and 954.9 eV, respectively, which were consistent with those of the Cu^{2+} cations [29]. The presence of Cu ions could capture the photogenerated carriers to accelerate the separation of charge carriers, subsequently transferring them to the surface of TiO_2 thin film, resulting in the improvement in the photocatalytic activity of the Cu-doped TiO_2 [30].

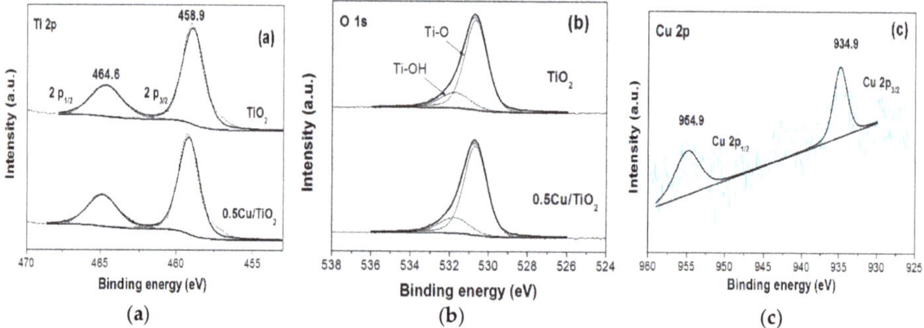

Figure 3. X-ray photoelectron spectroscopy (XPS) spectra of (a) Ti 2p, (b) O1s of TiO_2 and $0.5Cu/TiO_2$, and (c) Cu 2p of $0.5Cu/TiO_2$.

Table 2. O1s XPS data and the fraction of total area of TiO_2 and $0.5Cu/TiO_2$ thin films.

Sample	Lattice O^{2-}		Ti-OH	
	BE (eV)	Fraction (%)	BE (eV)	Fraction (%)
TiO_2	530.7	81.48	531.7	18.52
$0.5Cu/TiO_2$	530.6	73.72	531.6	26.28

Wettability measurements were performed using a customized in-house contact angle meter. To measure the water contact angle (WCA), a 5 µL DI water drop was dripped on the films. The TiO_2 film prepared via the peroxo sol-gel method showed hydrophilicity with an average WCA of 6.4°. After doping with copper, the average WCA of the TiO_2 dramatically increased from 6.4° to 35.9° (see Figure 4). In general, elemental copper exhibits super hydrophilicity, whereas copper oxide (e.g., CuO and Cu_2O) shows hydrophobicity. This finding confirms the presence of CuO (Cu^{2+}) in the Cu-doped TiO_2 films, which was in accordance with the XPS [31,32].

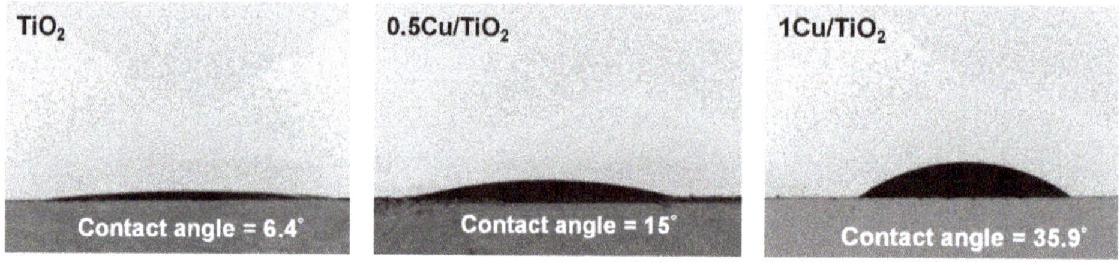

Figure 4. Water contact angle of TiO_2 and Cu-doped TiO_2 films.

2.3. Photocatalytic Degradation of MB Aqueous Solution

Prior to the photocatalytic activity, all samples were immersed into the set-up reactor and kept in the dark for 1 h to attain equilibrium adsorption of MB. The photocatalytic degradation of MB in water under UVC and visible light irradiation is shown in Figure 5. The highest photocatalytic activity under both UVC and visible irradiation was shown for the modified TiO_2 with the weight ratio $Cu:TiO_2$ = 0.5:100 ($0.5Cu/TiO_2$). The amount of copper beyond a certain loading decrease the photocatalytic activity of TiO_2 due to the Cu light absorption [33]. Furthermore, larger doping of copper resulted in CuO acting as a recombination centre [34].

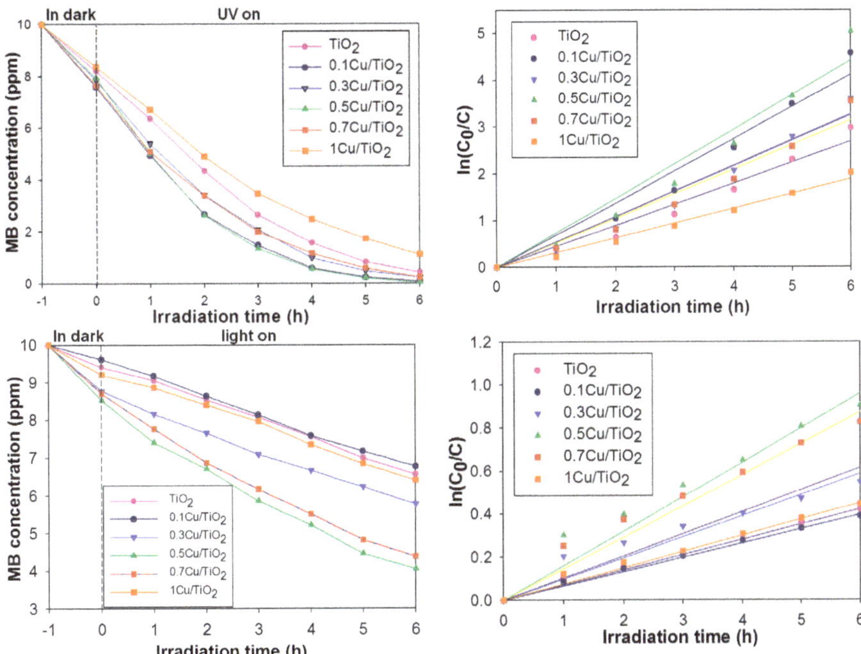

Figure 5. Photocatalytic activities of TiO$_2$ and Cu–doped TiO$_2$ films under UV light illumination (**top**) and visible light illumination (**bottom**).

The photodegradation of MB was fitted to the Langmuir–Hinshelwood model. The slope of ln (C$_0$/C) plotted versus irradiation time (h) indicates the reaction rate constant of the sample. Under UVC and visible light irradiation, the photocatalytic activity was the highest for 0.5Cu/TiO$_2$, with a rate constant of 0.737 h^{-1} and 0.160 h^{-1}, respectively (Table 3). Higher photocatalytic activity was attributed to the photoelectron transfer of the conduction band (CB) of TiO$_2$ to CuO, leaving the hole on TiO$_2$ to take part in oxidation reaction. In other words, under UV irradiation, the presence of CuO can slow down the recombination of electron/hole pairs in TiO$_2$, as shown in the following equations [34,35].

$$Cu^{2+} + e^- CB \rightarrow Cu^+ \tag{2}$$

$$Cu^+ + (O_2,\ H_2O_2\ \text{or other ROS}) \rightarrow Cu^{2+} + e^- \tag{3}$$

CuO nanoparticles deposited on the TiO$_2$ surface received the photogenerated electrons from the CB of TiO$_2$ to form Cu$^+$ ion as shown in Equation (2), and Cu$^+$ ions could be re-oxidised to Cu^{2+} by the ROS species present in the surrounding media (3).

Table 3. Rate constant of the reaction from pseudo-first order kinetics under visible light irradiation.

Samples	Rate Constant (k, h^{-1})	
	Under UV light	Under Visible Light
TiO$_2$	0.449	0.070
0.1Cu/TiO$_2$	0.687	0.066
0.3Cu/TiO$_2$	0.546	0.098
0.5Cu/TiO$_2$	0.737	0.160
0.7Cu/TiO$_2$	0.524	0.146
1Cu/TiO$_2$	0.317	0.076

2.4. Photocatalytic Antibacterial Effectiveness of Cu-doped TiO$_2$ Thin Film

Under irradiation with UVA light, the antibacterial activity of the samples was tested against *Escherichia coli* (ATCC25922). *E. coli* is present as a normal intestinal flora and is commonly found in contaminated drinking water. After a UVA radiation of 3 h at a low UVA intensity of 33 µW/cm^2, the 0.5Cu/TiO$_2$ coating showed high antibacterial effectiveness of >99% when tested against *E. coli*. In contrast, TiO$_2$ displayed an average effectiveness of 61.20% when tested against *E. coli* (see Figure 6). Antibacterial activities of Cu-doped TiO$_2$ could be attributed to the production of ROS species (e.g., $O_2{}^-$, OH$^\bullet$, and H$_2$O$_2$) using TiO$_2$, as well as CuO nanoparticles that trigger oxidative stress and cell damage in bacteria [36]. In addition, released Cu ions (Cu^{2+}) increased intracellular ROS in bacteria using the following pathway [17,21]:

$$Cu^+ + H_2O_2 \rightarrow Cu^{2+} + OH^- + OH\bullet \tag{4}$$

H$_2$O$_2$ is a byproduct of normal metabolism of oxygen in bacterial cells. The accumulation of ROS dramatically increased, eventually causing cell death [10,17–21,36]. Therefore, Cu-doped TiO$_2$ showed antibacterial effectiveness, even though a low-intensity of UVA light source was applied.

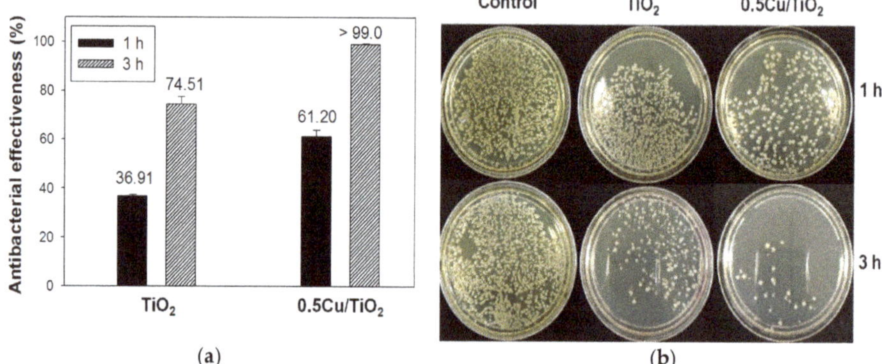

Figure 6. Antibacterial effectiveness (%) against *E. coli* following periods of illumination with 33 µW/cm^2, UVA radiation of 1 and 3 h (**a**), and counts of viable *E. coli* after incubation (1:10^3 dilution) (**b**).

3. Materials and Methods

3.1. Materials

TiCl$_4$ (purity > 99.9%) and H$_2$O$_2$ (30% in water) were purchased from Showa Chemicals Industry, Ltd. (Tokyo, Japan). NH$_4$OH was purchased from Merck Co. (Kenilworth, NJ, USA). Copper (II) nitrate trihydrate (Cu(NO$_3$)$_2$·3H$_2$O) was purchased from Sigma-Aldrich (St. Louis., MO, USA). Distilled water was used throughout the experiments.

3.2. Preparation of TiO$_2$ and Cu-Doped TiO$_2$ Sol

The typical procedure of preparing TiO$_2$ and Cu/TiO$_2$ sols went as follows: Three milliliters of TiCl$_4$ were added dropwise into 150 mL 1 M HCl aqueous solution (under magnetic stirring) and kept in an ice bath so as to maintain the temperature at 0 °C for 30 min. An aqueous solution of 1 M NH$_4$OH was then added dropwise to form the white hydrated titanium oxide gel Ti(OH)$_4$. The pH of the solution was adjusted to 8.0 by adding the required amount of ammonia solution. After aging and stirring for 30 min, a Ti(OH)$_4$ cake was filtered and washed with distilled water until no chloride ions were detected (QUANTOFIX®). An amount of the as-prepared gel was re-dispersed in distilled water under magnetic stirring to form a milky solution. An aqueous solution of H_2O_2 was added dropwise to the solution under vigorous stirring for 1 h. The resultant solution was heated at 95 °C for 4 h under magnetic stirring.

For preparation of Cu-doped TiO_2 sol, the copper precursor was added into the TiO_2 sol during heating at 95 °C. The solid content of the TiO_2 in the solution was 1.0 wt %; the molar ratio of H_2O_2:TiO_2 was 4:1; and the weight ratio of Cu:TiO_2 was: 0.1:100, 0.3:100, 0.5:100, 0.7:100, and 1:100. These ratios were denoted as 0.1Cu/TiO_2, 0.3Cu/TiO_2, 0.5Cu/TiO_2, 0.7Cu/TiO_2, and 1Cu/TiO_2, respectively.

3.3. Preparation of Films

The sols were aged at room temperature for 8 h before deposition on glass. The films were prepared using the dip-coating method. Soda lime glass was used as the substrate. The total coating surface area of each glass substrate was 40 cm^2. The glass substrate was cleaned using a commercial dishwashing detergent. Subsequently, it was ultrasonicated for 30 min in 1 M NH_4OH solution, thoroughly rinsed with distilled water and oven-dried at 60 °C. The glass substrate was vertically dipped into the as-prepared sol with a withdrawal speed of 30 cm/min for 7 times. The thickness of the films was 250–300 nm, as measured by SEM.

3.4. Characterization of Cu-Doped TiO$_2$ Particles and Thin Films

All samples were air dried at 80 °C for 1 h before further characterisation. The XRD patterns of the samples were determined using a Siemens D500 powder diffractometer (Siemens, Westborough, MA, USA) with Cu Kα source (λ = 1.5405 Å). The morphology and structure of the samples were investigated using transmission electron microscopy (TEM) on a JEM-2000 EX II (JEOL, Tokyo, Japan) operated at an accelerating voltage of 120 kV and high resolution TEM (HRTEM) on a JEOL JEM-2010 (JEOL, Tokyo, Japan) operated at 200 kV. The lattice spacing of the samples was measured using Gatan Digital Micrograph software. The chemical composition and chemical state of the samples were determined using X-Ray photoelectron spectroscopy (XPS) with a Thermo VG Scientific Sigma Prob spectrometer (Thermo Fisher Scientific, Logan, UT, USA). The XPS spectra were collected using Al K$_\alpha$ radiation at a voltage of 20 kV and current of 30 mA. The binding energy (BE) was calibrated using contaminant carbon (C_{1S} = 284.6 eV). The peaks of each spectrum were organised using XPSPEAK software (Thermo Fisher Scientific, Logan, UT, USA); Shirley type background and 30:70 Lorentzian/Gaussian peak shape were adopted during the deconvolution. The thickness of the films was measured using scanning electron microscopy (SEM) (Hitachi-3000, Tokyo, Japan).

3.5. Photocatalytic Degradation of MB Aqueous Solution

The photocatalytic activity of the samples was determined by inducing the decomposition of MB under irradiation with either UVC or visible light. An aqueous solution of MB (40 mL) with a concentration of 10 mg/L was loaded onto a quartz glass plate. The samples were horizontally immersed into the MB solution. Before the photocatalytic activity was measured, the reactor was kept in the dark under magnetic stirring for 1 h to achieve the saturation adsorption of MB on the coatings. The catalysts were irradiated by either two 9 W UVC lamps (wavelength = 254 nm, TUV PL-L 18W/4P 1CT/25, Philips) or two 18 W compact fluorescent lamps (1200 lumen, Philips (Pro UV Lamps Ltd., Bucks, UK)) equipped with UV cut-off filters with a cut-off wavelength of 410 nm. The distance from the lamp to the surface of the solution was 15 cm, and the concentration of the aqueous solution of MB was determined at intervals of 1 h using UV-vis spectrophotometer (JASCO V-630 (Japan Spectroscopic Company, Tsukuba, Japan)). The wavelength selected for the measurements was 663 nm, which is the characteristic maximum absorption wavelength of MB.

3.6. Study of Antibacterial Activity

The test method of the coatings against *E. coli*'s (ATCC25922) antibacterial activity was modified from the certificate "JIS Z 2801: 2000 (E)—Antimicrobial products-Test for antimicrobial activity and efficacy" and "TN-050—Standard on nano anti-bacterial coating." The strains were grown on tryptic soy agar and diluted to 5.5×10^6–6.0×10^6 cells/mL using distilled water. The bacterial concentrations were measured from the optical density reading at 600 nm (OD_{600}). The as-prepared sols were deposited on a 5 cm $\times$ 5 cm glass substrate using the dip-coating technique and a catalyst content of 0.2 mg/cm^2. Bare glass substrates were used as controls. Before testing, all samples were sterilized and activated using two 9 W UVC light (λ = 254 nm, TUV PL-L 18W/4P 1CT/25, Philips (Pro UV Lamps Ltd., Bucks, UK)) for 1 h. In order to quantitatively evaluate the antibacterial activity of the coatings, 0.4 mL of bacterial suspension was added onto each coating. Next, the test pieces were covered with 4 $\times$ 4 cm of adhesive film (with a transparency of >85% at 340–380 nm). Covered with the adhesive film, the test piece was placed into a Petri dish and exposed to two UVA lamps (λ = 365 nm, PL-L 36W/09/4P, Philips (Pro UV Lamps Ltd., Bucks, UK)) for either 1 h or 3 h. The irradiance of the UVA intensity was measured at 33 μW/cm^2, using a UV light meter (model UV-340A, Lutron (Lutron Electronic Enterprise Co., Taipei, Taiwan)). After UVA irradiation, the bacterial suspension on each coating was washed off and diluted with phosphate buffer saline (PBS). Bacterial colony-forming units (CFUs) were enumerated by plating serial dilutions (1:10–1:10^5). The number of surviving bacterial colonies was counted (CFU/mL) after incubation at 37 °C for 24 h. The experiments were repeated three times for each sample type; therefore, three parallel CFU values were obtained for each type of sample.

The antibacterial effectiveness was calculated according to the following equation [12,25]:

$$C(\%) = \left(\frac{A - B}{A} \right) \times 100\% \tag{5}$$

where C represents antibacterial effectiveness, A is the average number of colonies formed in the blank control group (CFU/mL), and B is the average number of colonies formed in the experimental group (CFU/mL).

4. Conclusions

0.5Cu-doped TiO$_2$ nanoparticles can be successfully prepared via the peroxo sol-gel method without needing further calcination. The CuO nanoparticles, having a particle size of <4 nm, were deposited on the TiO$_2$ surface. The photocatalytic activity was the highest for 0.5Cu/TiO$_2$, with a rate constant of 0.737 h^{-1} and 0.160 h^{-1} under UVC and visible light irradiation, respectively. Moreover, the 0.5Cu/TiO$_2$ coating showed high antibacterial effectiveness of >99% against *E. coli* after illumination with 33 μW/cm^2 UVA radiation for 3 h. Therefore, the presence of CuO significantly enhanced the photocatalytic activity as well as antibacterial effect of TiO$_2$. Therefore, the Cu-coped TiO$_2$ materials prepared via the peroxo sol-gel method can be an alternative and promising solution to increasing environmental contamination.

Author Contributions: B.M. and Y.-W.C. designed the experiments; B.M. and J.-Y.S. performed the experiments, analyzed the data and contributed material characterization and analysis; B.M. wrote the paper and Y.-W.C. supervised the project.

Funding: This research was funded by MOST 107-0205-2511.

Acknowledgments: This research was supported by the Ministry of Science and Technology, Taiwan.

Conflicts of Interest: The authors declare no conflicts of interest.

References

1. Vinodgopal, K.; Wynkoop, D.; Kamat, P. Environmental photochemistry on semiconductor surfaces: Photosensitized degradation of a textile azo dye, acid orange 7, on TiO$_2$ particles using visible light. *Environ. Sci. Technol.* **1996**, *30*, 1660–1666. [CrossRef]

2. Mohamed, M.; Al-Esaimi, M. Characterization, adsorption and photocatalytic activity of vanadium-doped TiO$_2$ and sulfated TiO$_2$ (rutile) catalysts: Degradation of methylene blue dye. *J. Mol. Catal. A Chem.* **2006**, *255*, 53–61. [CrossRef]

3. Moongraksathum, B.; Chen, Y.W. CeO$_2$–TiO$_2$ mixed oxide thin films with enhanced photocatalytic degradation of organic pollutants. *J. Sol-Gel Sci. Technol.* **2017**, *82*, 772–782. [CrossRef]

4. Chong, M.; Jin, B.; Chow, C.; Saint, C. Recent developments in photocatalytic water treatment technology: A. review. *Water Res.* **2010**, *44*, 2997–3027. [CrossRef] [PubMed]

5. Fujishima, A.; Zhang, X.; Tryk, D. TiO$_2$ photocatalysis and related surface phenomena. *Surf. Sci. Rep.* **2008**, *63*, 515–582. [CrossRef]

6. Guan, K. Relationship between photocatalytic activity, hydrophilicity and self-cleaning effect of TiO$_2$/SiO$_2$ films. *Surf. Coat. Technol.* **2005**, *191*, 155–160. [CrossRef]

7. Moongraksathum, B.; Chen, Y.W. Preparation and characterization of SiO$_2$–TiO$_2$ neutral sol by peroxo sol-gel method and its application on photocatalytic degradation. *J. Sol-Gel Sci. Technol.* **2016**, *77*, 288–297. [CrossRef]

8. Sakai, N.; Fujishima, A.; Watanabe, T.; Hashimoto, K. Quantitative evaluation of the photoinduced hydrophilic conversion properties of TiO$_2$ thin film surfaces by the reciprocal of contact angle. *J. Phys. Chem. B* **2003**, *107*, 1028–1035. [CrossRef]

9. Watanabe, T.; Fukayama, S.; Miyauchi, M.; Fujishima, A.; Hashimoto, K. Photocatalytic activity and photo-induced wettability conversion of TiO$_2$ thin film prepared by sol-gel process on a soda-lime glass. *J. Sol-Gel Sci. Technol.* **2000**, *19*, 71–76. [CrossRef]

10. Sunada, K.; Watanabe, T.; Hashimoto, K. Bactericidal activity of copper-deposited TiO$_2$ thin film under weak UV light illumination. *Environ. Sci. Technol.* **2003**, *37*, 4785–4789. [CrossRef] [PubMed]

11. Pelaez, M.; Nolan, N.; Pillai, S.; Seery, M.; Falaras, P.; Kontos, A.; Dunlop, P.; Hamilton, J.; Byrne, J.; O'Shea, K.; et al. A review on the visible light active titanium dioxide photocatalysts for environmental applications. *Appl. Catal. B* **2012**, *125*, 331–349. [CrossRef]

12. Moongraksathum, B.; Chen, Y.W. Anatase TiO$_2$ co-doped with silver and ceria for antibacterial application. *Catal. Today* **2018**, *310*, 68–74. [CrossRef]

13. Verdier, T.; Coutand, M.; Bertron, A.; Roques, C. Antibacterial activity of TiO$_2$ photocatalyst alone or in coatings on *E. coli*: The influence of methodological aspects. *Coatings* **2014**, *4*, 670–686. [CrossRef]

14. Maness, P.; Smolinski, S.; Blake, D.M.; Huang, Z.; Wolfrum, E.J.; Jacoby, W.A. Bactericidal activity of photocatalytic TiO$_2$ reaction: Toward an understanding of its killing mechanism. *Appl. Environ. Microb.* **1999**, *65*, 4094–4098.

15. Castro, C.; Sanjines, R.; Pulgarin, C.; Osorio, P.; Giraldo, S.A.; Kiwi, J. Structure-reactivity relations of the Cu-cotton sputtered layers during *E. coli* inactivation in the dark and under light. *J. Photochem. Photobiol. A* **2010**, *216*, 295–302. [CrossRef]

16. Richardson, S.D.; Thruston, A.D.; Collette, T.W.; Ireland, J.C. Identification of TiO$_2$/UV disinfection byproducts in drinking water. *Environ. Sci. Technol.* **1996**, *30*, 3327–3334. [CrossRef]

17. Leyland, N.; Podporska-Carroll, J.; Browne, J.; Hinder, S.; Quilty, B.; Pillai, S. Highly Efficient F, Cu doped TiO$_2$ anti-bacterial visible light active photocatalytic coatings to combat hospital-acquired infections. *Sci. Rep.* **2016**, *6*, 24770. [CrossRef] [PubMed]

18. Litter, M. Heterogeneous photocatalysis Transition metal ions in photocatalytic systems. *Appl. Catal. B Environ.* **1999**, *23*, 89–114. [CrossRef]

19. Espirito Santo, C.; Quaranta, D.; Grass, G. Antimicrobial metallic copper surfaces kill Staphylococcus haemolyticus via membrane damage. *MicrobiologyOpen* **2012**, *1*, 46–52. [CrossRef] [PubMed]

20. Grass, G.; Rensing, C.; Solioz, M. Metallic copper as an antimicrobial surface. *Appl. Environ. Microbiol.* **2011**, *77*, 1541–1546. [CrossRef] [PubMed]

21. Rtimi, S.; Pulgarin, C.; Kiwi, J. Recent developments in accelerated antibacterial inactivation on 2D Cu-Titania surfaces under indoor visible light. *Coatings* **2017**, *7*, 20. [CrossRef]

22. Sasirekha, N.; Rajesh, B.; Chen, Y.W. Synthesis of TiO$_2$ sol in a neutral solution using TiCl$_4$ as a precursor and H$_2$O$_2$ as an oxidizing agent. *Thin Solid Films* **2009**, *518*, 43–48. [CrossRef]

23. Chen, Y.W.; Chang, J.Y.; Moongraksathum, B. Preparation of vanadium–doped titanium dioxide neutral sol and its photocatalytic applications under UV light irradiation. *J. Taiwan Inst. Chem. Eng.* **2015**, *52*, 140–146. [CrossRef]

24. Moongraksathum, B.; Hsu, P.T.; Chen, Y.W. Photocatalytic activity of ascorbic acid-modified TiO_2 sol prepared by the peroxo sol-gel method. *J. Sol-Gel Sci. Technol.* **2016**, *78*, 647–659. [CrossRef]

25. Moongraksathum, B.; Chien, M.Y.; Chen, Y.W. Antiviral and antibacterial effects of silver-doped TiO_2 prepared by the peroxo sol–gel method. *J. Nanosci. Nanotechnol..* Accepted.

26. Monshi, A.; Foroughi, M.R.; Monshi, M.R. Modified Scherrer equation to estimate more accurately nano-crystallite size using XRD. *World J. Nano Sci. Eng.* **2012**, *2*, 154–160. [CrossRef]

27. Alexander, L.; Klug, H.P. Determination of crystallite size with the X–ray spectrometer. *J. Appl. Phys.* **1950**, *21*, 137–142. [CrossRef]

28. Dake, L.S.; Lad, R.J. Electronic and chemical interactions at aluminum/TiO_2 (110) interfaces. *Surf. Sci.* **1993**, *289*, 297–306. [CrossRef]

29. Su, J.; Li, Z.; Zhang, Y.; Wei, Y.; Wang, X. N-Doped and Cu-doped TiO_2-B nanowires with enhanced photoelectrochemical activity. *RSC Adv.* **2016**, *6*, 16177–16182. [CrossRef]

30. Wang, S.; Meng, K.; Zhao, L.; Jiang, Q.; Lian, J. Superhydrophilic Cu-doped TiO_2 thin film for solar-driven photocatalysis. *Ceram. Int.* **2014**, *40*, 5107–5110. [CrossRef]

31. Eshaghi, A.; Eshaghi, A. Preparation and hydrophilicity of TiO_2 sol–gel derived nanocomposite films modified with copper loaded TiO_2 nanoparticles. *Mater. Res. Bull.* **2011**, *46*, 2342–2345. [CrossRef]

32. Xu, Y.; Li, J.A.; Yao, L.F.; Li, L.H.; Yang, P.; Huang, N. Preparation and characterization of Cu-doped TiO_2 thin films and effects on platelets adhesion. *Surf. Coat. Technol.* **2015**, *261*, 436–441. [CrossRef]

33. Behnajady, M.; Shokri, M.; Taba, H.; Modirshahla, N. Photocatalytic activity of Cu doped TiO_2 nanoparticles and comparison of two main doping procedures. *Micro Nano Lett.* **2013**, *8*, 345–348. [CrossRef]

34. Moniz, S.J.A.; Tang, J. Charge transfer and photocatalytic activity in CuO/TiO_2 nanoparticle heterojunctions synthesised through a rapid, one-pot, microwave solvothermal route. *ChemCatChem* **2015**, *7*, 1659–1667. [CrossRef]

35. Janczarek, M.; Kowalska, E. On the origin of enhanced photocatalytic activity of copper-modified titania in the oxidative reaction systems. *Catalysts* **2017**, *7*, 317. [CrossRef]

36. Applerot, G.; Lellouche, J.; Lipovsky, A.; Nitzan, Y.; Lubart, R.; Gedanken, A.; Banin, E. Understanding the antibacterial mechanism of CuO nanoparticles: Revealing the route of induced oxidative stress. *Small* **2012**, *8*, 3326–3337. [CrossRef] [PubMed]

Article

Modification to L-H Kinetics Model and Its Application in the Investigation on Photodegradation of Gaseous Benzene by Nitrogen-Doped TiO$_2$

Peng Sun, Jun Zhang, Wenxiu Liu, Qi Wang and Wenbin Cao *

School of Materials Science and Engineering, University of Science and Technology Beijing, Beijing 100083, China; ustbsunpeng@163.com (P.S.); zhangjunustb@foxmail.com (J.Z.); liuwenxiu@outlook.com (W.L.); wangqi15@ustb.edu.cn (Q.W.)
* Correspondence: wbcao@ustb.edu.cn; Tel.: +86-010-6233-2457

Received: 11 July 2018; Accepted: 6 August 2018; Published: 9 August 2018

Abstract: In this paper, the Langmuir-Hinshelwood (L-H) model has been used to investigate the kinetics of photodegradation of gaseous benzene by nitrogen-doped TiO$_2$ (N-TiO$_2$) at 25 °C under visible light irradiation. Experimental results show that the photoreaction coefficient k_{pm} increased from 3.992×10^{-6} mol·kg^{-1}·s^{-1} to 11.55×10^{-6} mol·kg^{-1}·s^{-1} along with increasing illumination intensity. However, the adsorption equilibrium constant K_L decreased from 1139 to 597 m^3·mol^{-1} when the illumination intensity increased from 36.7×10^4 lx to 75.1×10^4 lx, whereas it was 2761 m^3·mol^{-1} in the absence of light. This is contrary to the fact that K_L should be a constant if the temperature was fixed. This phenomenon can be attributed to the breaking of the adsorption-desorption equilibrium by photocatalytically decomposition. To compensate for the disequilibrium of the adsorption-desorption process, photoreaction coefficient k_{pm} was introduced to the expression of K_L and the compensation form was denoted as K_m. K_L is an indicator of the adsorption capacity of TiO$_2$ while K_m is only an indicator of the coverage ratio of TiO$_2$ surface. The modified L-H model has been experimentally verified so it is expected to be used to predict the kinetics of the photocatalytic degradation of gaseous benzene.

Keywords: modified L-H model; N-TiO$_2$; photocatalytic degradation; benzene

1. Introduction

Gaseous benzene released from paints, artificial panel or furniture is threatening to human health, particularly for children. However, the gaseous benzene in indoor air is difficult remedy with traditional methods due to its low concentration (ppm or ppb level) [1–3]. However, TiO$_2$ can decompose gaseous benzene under ultraviolet light irradiation, thus it has attracted growing attention [4–8]. In fact, the photodegradation of gaseous benzene by TiO$_2$ photocatalyst is a heterogeneous reaction occurring at a gas-solid interface, and the reaction rate is strongly affected by the environmental factors, particularly illumination intensity [9–11]. So the kinetic study of photocatalytic reaction is important for revealing the effect of these factors on the photocatalytic reaction rate.

The heterogeneous reaction includes two consecutive steps. Firstly, the reactants are adsorbed on the surface of the photocatalysts and secondly, the photocatalytic reaction commences. Generally, the adsorption rate is slower than the photocatalytic reaction rate. So the overall photocatalytic reaction rate is mainly dominated by the adsorption rate. Furthermore, the adsorption rate can be equivalently expressed using the coverage ratio of the adsorbed reactants on the surface of the photocatalysts [12–15]. So the photocatalytic reaction rate r can be expressed as Equation (1) [16–18], which is widely known as the original L-H model.

$$r = -\frac{dc}{dt} = k_p\theta \tag{1}$$

where c is the concentration of the reactant, t is the photocatalytic reaction time, θ is the coverage ratio of pollutants on the TiO$_2$ surface, k_p is photoreaction coefficient.

According to Langmuir adsorption theory, the coverage ratio is related to adsorption capacity and the concentration of the reactant. K_L was defined as adsorption equilibrium constant to measure the adsorption capacity of TiO$_2$ and coverage ratio θ can be expressed as Equation (2) according to adsorption theory [19].

$$\theta = \frac{K_L c}{1 + K_L c} \tag{2}$$

Input θ from Equation (2) to Equation (1), the photoreaction coefficient r can be expressed as Equation (3) [20–22],

$$r = k_p \frac{K_L c}{1 + K_L c} \tag{3}$$

Equation (3) is the much known expression of L-H model and has been widely used in investigating the kinetics of photocatalytic reactions. Lin et al. [23] studied the photocatalytic degradation pathway of dimethyl sulfide. They used original and derivative L-H models to study the kinetics under different temperatures and found that temperature can enhance photocatalytic activity. Dhada et al. [24] investigated the photocatalytic degradation of benzene by TiO$_2$ under sunshine and UV light. They found that UV light can promote photocatalytic reaction than visible light due to its higher energy of the photons. Cheng et al. [25] studied the photocatalytic degradation of benzene. They found that higher temperature, illumination intensity and humidity can promote the reaction rate greatly.

The works mentioned above are focused in revealing the effect of environmental factors such as illumination intensity, the amount of the photocatalyst and some processing parameters on the photodegradation ratio. However, the effect of illumination intensity on adsorption equilibrium coefficient of gaseous pollutant was neglected in most articles. In liquid phase photocatalysis, some authors have reported their research on the effect of the illumination intensity on both the photoreaction coefficient and the adsorption coefficient [26–29]. Du [30] found that the value of the adsorption coefficient calculated from the L-H model was illumination intensity-dependent in photodegradation of liquid dimethyl phthalate (DMP).

Coincidentally, it has also been found that the adsorption coefficient has been affected by light intensity in the gaseous photocatalytic reactions [31,32]. Brosillon [31] studied the kinetic model of photocatalytic degradation of butyric acid, and they found that the adsorption coefficient K_R can be expressed as Equation (4)

$$K_R = \frac{(k_{rLH} C_{Rads0} + k_{d1} + k'_{d2} I) K}{k_{d1} + k'_{d2} I} \tag{4}$$

where k_{rLH} is the reaction rate of the reaction between ·OH and reactants, k_{d1}, k'_{d2} is the decomposition rate of ·OH in the routes of ·OH $\rightarrow$ OH$^-$ + h$^+$ and ·OH + e$^-$ $\rightarrow$ OH$^-$, I is the light intensity, K is the adsorption constant without light irradiation. Their results indicate that the adsorption coefficient in gas photocatalytic reaction is a function of light intensity, which is not reasonable as the adsorption coefficient should be a constant under a fixed temperature. And, the parameter kr_{LH}, k_{d1}, k'_{d2} are difficult to calculate as the concentration of ·OH is difficult to accurately measure [33] during the process of photocatalytic degradation of benzene and its concentration changes during the progression of the photocatalytic reaction. So this model is not applicable to predict the concentration of the reactant at different reaction times under different illumination intensities. He [32] investigated the degradation of benzene by mesoporous TiO$_2$ and also found that the adsorption coefficient could be affected by light intensity. They attributed it to the decrease of available active sites as the increased photo-induced radicals will occupy more of the active sites under higher illumination intensity. However, the effect of photocatalytic decomposition of the adsorbed benzene by the increased radicals on the adsorption coefficient was not considered. So, it's necessary to accurately describe the relationship between the adsorption coefficient and the illumination intensity in gaseous photocatalytic reactions.

In the present work, the effect of illumination intensity on photoreaction coefficient k_{pm} and adsorption equilibrium coefficient has been studied under a constant 25 °C. Photoreaction coefficient was introduced as the modification to K_L and the compensation K_m was used to replace K_L in the original L-H model. The modified L-H model can reveal the interaction between the adsorption, desorption and photo-oxidation process. The results showed that the K_m and k_{pm} can be obtained under different illumination intensity at 25 °C, thus the concentration at different reaction times can be predicted.

2. Results and Discussion

2.1. Characterization of the N-TiO$_2$ Photocatalysts

The N-TiO$_2$ catalysts were characterized by X-ray diffraction (XRD), Transmission electron microscopy (TEM), UV-Vis spectra (UV-Vis) and X-ray photoelectron spectroscopy (XPS) and the results were illustrated in Figure 1. Figure 1a shows the XRD patterns of N-TiO$_2$. It is clear that all the diffraction peaks were indexed to that of anatase TiO$_2$ (JCPDS no. 21-1272). The crystal size calculated by Scherrer's Equation was also around 10.2 nm. Figure 1b shows the morphology of the N-TiO$_2$ powders. It can be found that the prepared sample was composed of spherical TiO$_2$ and the size was ranged from 9 to 12 nm, which is in consistent with the calculated result. The light absorption spectrum was measured by UV-Vis spectrum and was shown in Figure 1c. It is well known that the bandgap of pristine anatase is 3.2 eV, while the light absorption has been extended into the ranged of 400 to 600 nm of as-prepared N-TiO$_2$. And its bandgap energy was 2.9 eV shown in the inset of Figure 1c calculated by using the method in other works [34,35]. The chemical state of N1s was also investigated by XPS and the result was shown in Figure 1d. Only one peak located at 399.9 eV can be found, which can be attributed to the interstitial doping of nitrogen into TiO$_2$ lattice with Ti–O–N bond [36].

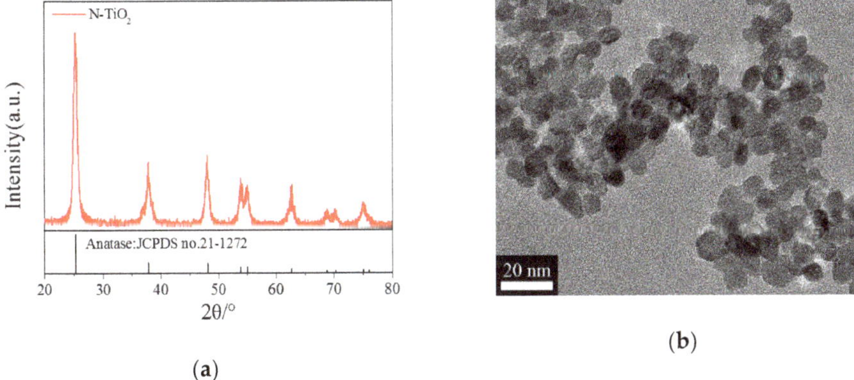

(a)

(b)

Figure 1. *Cont.*

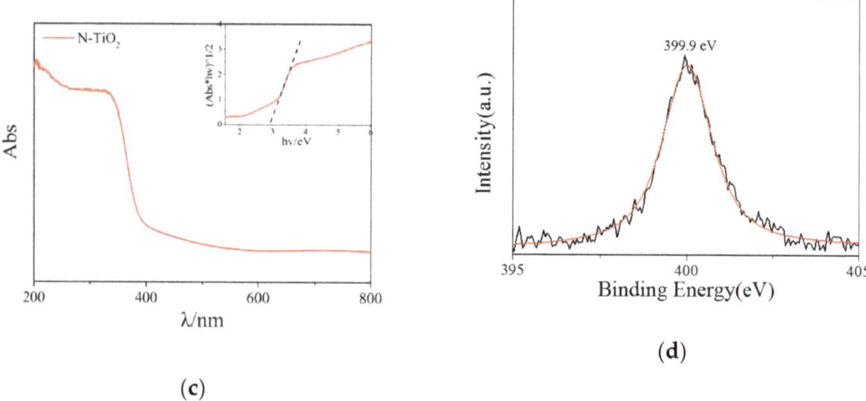

Figure 1. Characterization of N-TiO$_2$ catalysts (**a**) XRD patterns, (**b**) TEM image, (**c**) UV-Vis spectrum, (**d**) N1s binding energy peak.

2.2. Kinetic Study of Photocatalytic Degradation of Benzene under Different Illumination Intensity

Figure 2 shows the variation of benzene concentration with photocatalytic degradation time under different illumination intensities. It shows that the concentration of benzene remained almost unchanged during the first hour without light irradiation, indicating that adsorption and desorption processes of benzene on TiO$_2$ surface have reached equilibrium, thus the decrease of benzene after illumination can be ascribed to the photocatalytic degradation process. When it was illuminated for 4 h under different illumination intensity of 36.7×10^4, 46.9×10^4, 61.7×10^4 and 75.1×10^4 lx, the removal ratio of benzene was 72.1%, 84%, 90% and 92.4%, respectively. The removal ratio increased dramatically under higher illumination intensity, indicating that illumination intensity can promote the photocatalytic degradation performance.

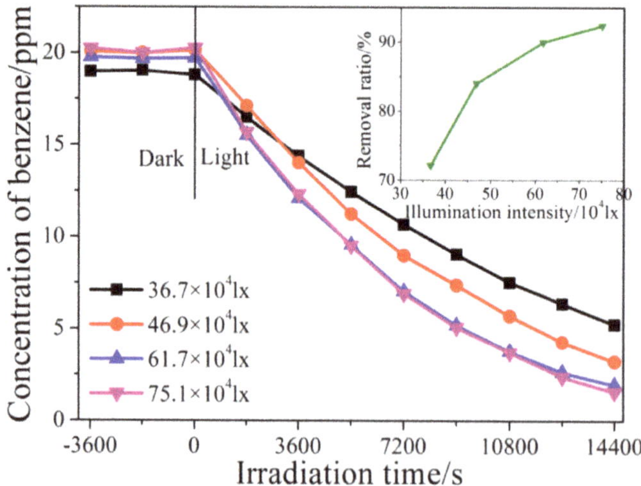

Figure 2. Variation of benzene concentration vs. photocatalytic degradation time under different illumination intensity.

During the photocatalytic degradation process, the amount of degraded benzene per unit time can be calculated by Equation (5).

$$\Delta n = rV = -\frac{dc}{dt}V \tag{5}$$

where Δn is the amount of degraded benzene per unit time, r is the photocatalytic degradation rate, V is the volume of the reactor, c is concentration of benzene and t is photocatalytic degradation time. The detailed form of r is shown by original L-H model in Equation (3) [32,37], so after inputting r from Equation (3) to Equation (5), we can get

$$\Delta n = -\frac{dc}{dt}V = k_p \frac{K_L c}{1 + K_L c} \tag{6}$$

In Equation (6), k_p is the photoreaction coefficient of the whole reaction system and is related to the mass of the catalysts. So the photocatalytic degradation rate coefficient per unit mass k_{pm} can be expressed in Equation (7)

$$k_{pm} = \frac{k_p}{m} \tag{7}$$

Input k_{pm} from Equation (7) into Equation (6), then we can get

$$-\frac{dc}{dt}V = mk_{pm}\frac{K_L c}{1 + K_L c} \tag{8}$$

So the relationship between dc and dt can be expressed in Equation (9)

$$-\frac{V}{mk_{pm}}\frac{1 + K_L c}{K_L c}dc = dt \tag{9}$$

The relationship between c and t can be obtained after making integration to Equation (9), that is

$$-\frac{V}{k_{pm}m}\int_{c_0}^{c}\frac{1 + K_L c}{K_L c}dc = \int_{0}^{t} dt \tag{10}$$

The result of Equation is

$$t = \frac{V}{mk_{pm}}\left[(c_0 - c) + \frac{1}{K_L}(\ln c_0 - \ln c)\right] \tag{11}$$

After rearranging in terms of $1/(c_0 - c)$, the linear form of Equation (11) is obtained.

$$\frac{\ln(c_0/c)}{c_0 - c} = \frac{m}{V}k_{pm}K_L\frac{t}{c_0 - c} - K_L \tag{12}$$

In Equation (12), it can be found that $\ln(c_0/c)/(c_0 - c)$ and $t/(c_0 - c)$ is a linear relationship, and the slope and intercept of the line is $mk_{pm}K_L/V$ and K_L respectively.

Figure 3 shows the plots of $\ln(c_0/c)/(c_0 - c)$ vs. $t/(c_0 - c)$ under different illumination intensity. According to the obtained slopes and intercepts, the values of k_{pm} and K_L were calculated and summarized in Table 1. And the standard deviation R^2 for each case were also listed in Table 1.

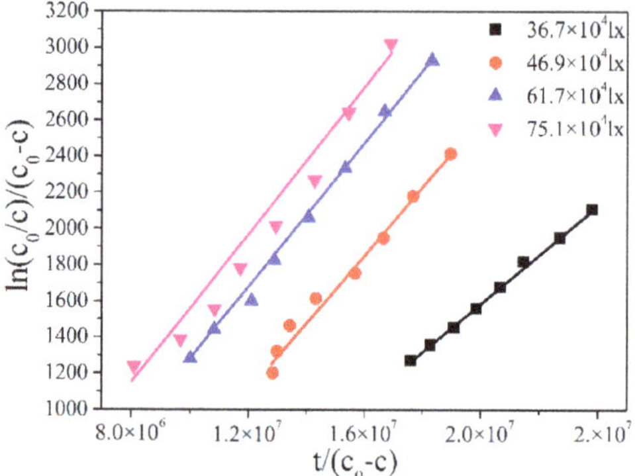

Figure 3. Plots of $\ln(c_0/c)/(c_0 - c)$ vs. $t/(c_0 - c)$ under different illumination intensity. (solid points: experimental results; solid line(curve): fitted results).

Table 1. Calculated K_{pm} and K_L under different illumination intensity using the original L-H model.

Illumination Intensity/10^4 lx	$k_{pm}/$ 10^{-6} mol·kg^{-1}·s^{-1}	$K_L/$ m^3·mol^{-1}	R^2
36.7	3.992	1139	0.9981
46.9	5.731	1064	0.9847
61.7	8.589	791	0.9961
75.1	11.55	597	0.9674

It can be seen from Table 1 that k_{pm} was calculated as 3.992×10^{-6}, 5.371×10^{-6}, 8.589×10^{-6}, 11.55×10^{-6} mol·kg^{-1}·s^{-1} corresponding to the illumination intensity of 36.7×10^4, 46.9×10^4, 61.7×10^4, and 75.1×10^4 lx, respectively. And k_{pm} increased greatly with increases in illumination intensity, which means that the photodegradation rate of benzene can be significantly promoted by increasing the illumination intensity in our experiment conditions. It's reasonable that the increased illumination intensity means more photon irradiated on TiO$_2$ surface, that can produce more ·OH, which is the main radical in photocatalytic reaction. According to other works [32], photoreaction rate coefficient k_{pm} depends on illumination intensity in a power law

$$k_{pm} = \alpha I^n \tag{13}$$

The value of intensity coefficient α and exponent n was 2.24×10^{-14} and 1.482 obtained by using the results in Table 1.

And the value of adsorption constant K_L was decreased from 1139 m^3·mol^{-1} to 597 m^3·mol^{-1} when the illumination intensity was increased from 36.7×10^4 lx to 75.1×10^4 lx. That is, K_L varied with the variation of the illumination intensity. However, the adsorption constant K_L is related to the temperature and should be a constant as the temperature of the reactor was carefully maintained at 25 °C according to Langmuir adsorption theory. So the obtained results are inconsistent with the basic fact that the K_L should be kept unchanged if the temperature was fixed for a certain adsorption-desorption balance, which shows that original L-H model cannot be used to describe the photocatalysis processes accurately.

Generally, it is widely recognized that the photocatalytic degradation of gaseous chemicals mainly includes two steps, gas adsorption on the surface of the photocatalyst and photodegradation. After the gas chemicals were adsorbed on the surface of the photocatalyst, certain amount of the adsorbed molecules were decomposed by photocatalytic degradation.

However, the original L-H model only considers the adsorption and desorption equilibrium of the gas molecules on the surface of the photocatalyst. So the amount of the adsorbed benzene molecules Δn_a and lost desorbed benzene molecules Δn_d of N-TiO$_2$ surface per unit time can be defined as Equation (14) and Equation (15) respectively [38].

$$\Delta n_a = k_a c(1-\theta)S \tag{14}$$

$$\Delta n_d = k_d \theta S \tag{15}$$

where k_a and k_d is adsorption and desorption constant of benzene and is all thermodynamic constant.

When adsorption and desorption process reach equilibrium, there is $\Delta n_a = \Delta n_d$, and the detailed form is shown in Equation (16).

$$k_a c(1-\theta)S = k_d \theta S \tag{16}$$

So coverage ratio θ and adsorption equilibrium constant K_L can be obtained [19]

$$\theta = \frac{k_a c}{k_d + k_a c} = \frac{\frac{k_a}{k_d}c}{1 + \frac{k_a}{k_d}c} \tag{17}$$

$$K_L = \frac{k_a}{k_d} \tag{18}$$

K_L is thermodynamically constant due to k_a and k_d being thermodynamic constants, and is an indication of adsorption ability of the catalysts. While in photocatalytic reaction, the degradation process would cause the decrease of benzene on TiO$_2$ surface, which is equivalent to the increase in the desorption rate of benzene molecules. So the equilibrium between adsorption and desorption process would be broken. However, adsorption equilibrium constant K_L is only related to k_a and k_d in Equation (18), which make it impossible to reveal the effect of degradation process on the equilibrium. Therefore, original L-H model based on Langmuir adsorption theory is not entirely suitable for the photocatalytic degradation of benzene and necessary modification should be applied to original L-H model for better understanding kinetics of the photocatalysis process.

2.3. Modication to the L-H Model and Kinetic Results under Different Illumination Intensity

In the photocatalytic reaction, there are three processes: Adsorption, desorption and the photocatalytic degradation process. The photocatalytic degradation process will cause decrease of benzene molecules on interface, so the amount of lost benzene molecules Δn_b is the sum of desorbed and photocatalytic degraded benzene molecules per unit time.

$$\Delta n_b = k_d \theta S + k_{pm}\theta S \tag{19}$$

Combing Equation (13) and (18), the coverage ratio θ becomes

$$\theta = \frac{k_a c}{k_d + k_{pm} + k_a c} = \frac{\frac{k_a}{k_d + k_{pm}}c}{1 + \frac{k_a}{k_d + k_{pm}}c} = \frac{K_m c}{1 + K_m c} \tag{20}$$

$$K_m = \frac{k_a}{k_d + k_{pm}} = \frac{k_d}{k_d + k_{pm}}\frac{k_a}{k_d} = \frac{k_d}{k_d + k_{pm}}K_L \tag{21}$$

$k_a/(k_d + k_{pm})$ can be defined as coverage coefficient K_m in Equation (21). The coverage coefficient K_m is a function of k_a, k_d and k_{pm}, so K_m is not thermodynamic constant due to k_{pm} is photodynamic. The value of K_m is equal to that of K_L while there is no light due to k_{pm} is zero without irradiation. And the value of k_{pm} will increase greatly under high illumination intensity, thus will result in a decrease of K_m, which is in accordance with the experimental results in Table 1.

The original L-H model can be modified by using K_m to replace K_L in Equation (13) and (14) there is

$$t = \frac{V}{mk_{pm}}\left[(c_0 - c) + \frac{1}{K_m}(\ln c_0 - \ln c)\right] \tag{22}$$

$$\frac{\ln(c_0/c)}{c_0 - c} = \frac{m}{V}k_{pm}K_m t \frac{1}{c_0 - c} - K_m \tag{23}$$

The expression form of Equation (23) is similar to that of original L-H model except coverage coefficient K_m and equilibrium coefficient K_L. K_L in original L-H model is an indicator of adsorption capacity of TiO$_2$, while K_m is the indicator of the amount of benzene on TiO$_2$ surface. The parameters k_{pm} and K_m can be obtained through the plots of $\ln(c_0/c)/(c_0 - c)$ vs. $t/(c_0 - c)$ which were shown in Figure 4a and the results were listed in Table 2. And after taking reciprocal on both sides of Equation (21), the linear relationship exists between $1/K_m$ and k_{pm} can be found in Equation (24) and was shown in Figure 4b. Then the values of k_a, k_d and K_L can also be obtained and summarized in Table 2. The value of K_L in modified L-H model is 2629 m$^3\cdot$mol^{-1} under different illumination intensity at 25 °C, which is consistent with Langmuir adsorption theory. The value of k_a and k_d is constant in a given temperature at 25 °C and the relationship of k_{pm} and I is revealed in Equation (14), thus K_m under different illumination intensity can be obtained by Equation (21). Therefore, the concentration c at different photocatalytic reaction time t under different illumination intensity I can be predicted from Equation (22)

$$\frac{1}{K_m} = \frac{k_{pm}}{k_a} + \frac{k_d}{k_a} \tag{24}$$

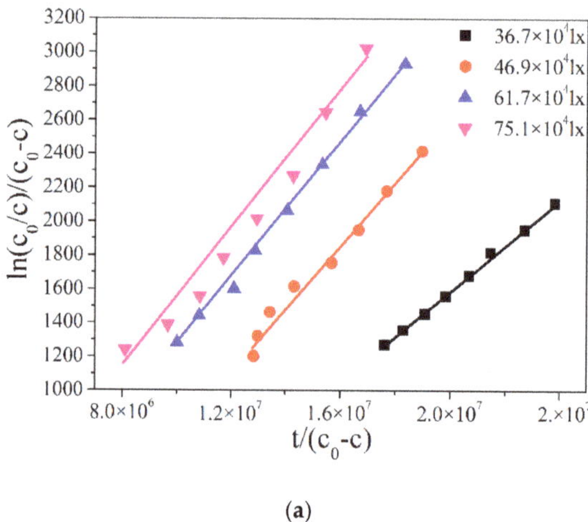

(a)

Figure 4. *Cont.*

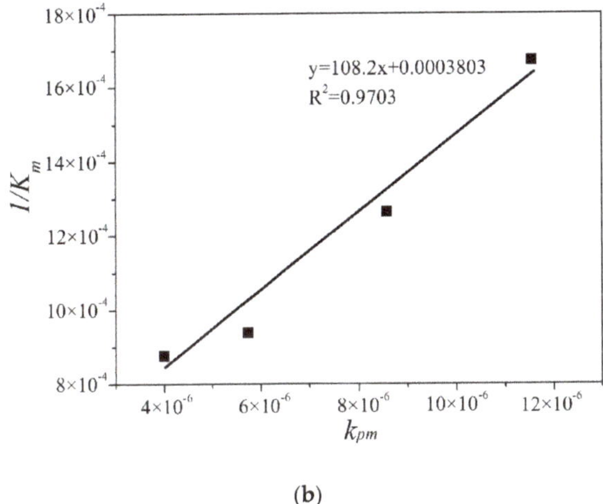

(**b**)

Figure 4. The relationship of the kinetic parameters in modified L-H model (**a**) The linear of $\ln(c_0/c)/(c_0 - c)$ vs. $t/(c_0 - c)$, (**b**) The linear of $1/K_m$ vs. k_{pm} (solid points: experimental results; solid line(curve): Fitted results).

Table 2. Results of modified L-H model under different illumination intensity.

Illumination Intensity/10^4 lx	Photoreaction Coefficient $k_{pm}/10^{-6}$ mol·kg^{-1}·s^{-1}	Coverage Coefficient K_m/m^3·mol^{-1}	Adsorption Constant k_a/m^3·kg^{-1}·$^{-1}$	Desorption Constant k_d/mol·kg^{-1}·s^{-1}	Adsorption Equilibrium Constant K_L/m^3·mol^{-1}
36.7	3.992	1139			
46.9	5.731	1064	9.242×10^{-3}	3.514×10^{-6}	2629
61.7	8.589	791			
75.1	11.55	597			

2.4. The Adsorption Equilibrium Constant K_L Obtained by Using Adsorption Theory

In fact, the adsorption equilibrium constant K_L is thermodynamically constant and can be used to evaluate the adsorption ability. In Langmuir adsorption theory, the adsorption equilibrium constant K_L without light irradiation can be obtained as follow [39–41]:

$$\frac{c_0}{(c_T - c_0)V} = \frac{c_0}{c_m V} + \frac{1}{K_L c_m V} \tag{25}$$

where c_T is total concentration of benzene filled into the reactor, c_0 is initial concentration of gaseous benzene after adsorption equilibrium, c_m is the maximum concentration that can be adsorbed by N-TiO$_2$. It is obvious that there is a linear relationship between $c_0/(c_T - c_0)V$ and c_0 in Equation (25). By filling different volume of benzene into reactor, c_T and c_0 can be measured after adsorption equilibrium and were summarized in Table 3. The plot of $c_0/(c_T - c_0)V$ vs. $c_0/c_m V$ was shown in Figure 5. The slope and intercept of the linear is $1/c_m V$ and $1/K_L c_m V$, respectively. The value of K_L was 2761 m^3·mol^{-1}, which is an indicator of the adsorption ability of benzene of N-TiO$_2$ at 25 °C.

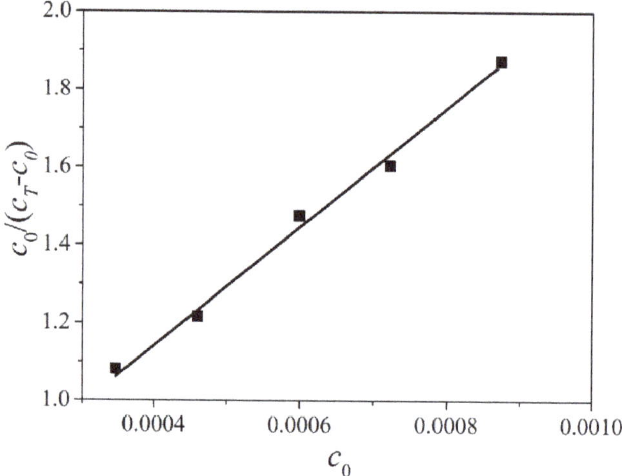

Figure 5. Linear relationship between $c_0/(c_T - c_0)$ and c_0 (solid points: experimental results; solid line(curve): fitted results).

Table 3. Concentration of benzene before and after adsorption equilibrium at 25 °C.

Total Concentration Filled into the Reactor c_t/ppm	Initial Concentration after Adsorption Equilibrium c_0/ppm
15	7.79
18.75	10.29
22.5	13.41
26.25	16.17
30	19.56

2.5. Verification of the Modified L-H Model

To verify the modified L-H model, the photodegradation of benzene under the illumination intensity of 23.8×10^4 lx was carried out by fixing other conditions except the initial concentration of benzene was 14.81 ppm. In this case, the calculated k_{pm} and K_m is 2.101×10^{-6} mol·kg^{-1}·s^{-1} mol and 1645 m^3·mol^{-1} respectively. By inputting the values of k_{pm} and K_m into Equation (22), the predicted concentration variation of benzene vs. irradiation time was obtained, which is shown in Figure 6 (denoted with the black solid line). The experimentally measured concentration of the benzene was denoted with red solid squares in Figure 6. It is clearly seen that the theoretical prediction shows very good agreement with the experimental results. So the modified L-H model can be used to predict benzene concentration under different illumination intensities at a constant temperature.

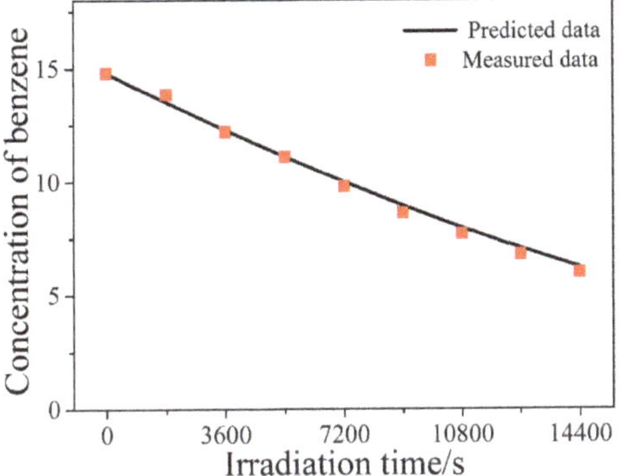

Figure 6. The predicted and measured concentration of benzene vs. time.

3. Materials and Methods

3.1. Preparation and Characterization of Samples

Nanocrystalline N-TiO$_2$ powders were prepared by hydrothermal method following the route used in our previous work [42]. The phase of the nano powders was determined by X-ray diffraction (XRD) with Cu Kα source in the 2θ ranging from 20 to 80°. The morphology of N-TiO$_2$ was characterized by Transmission electron microscopy (TEM, Hitachi, Jeol 200CX, Tokyo, Japan). UV-Vis spectra of the as-prepared sample was measured by Pgeneral UV-1901 instrument. The valence state of N was characterized by X-ray photoelectron spectroscopy (XPS, Thermo Fisher Scientic, Escalab 250, Waltham, MA, USA). Then the N-TiO$_2$ catalysts were dispersed into alcohol with ultrasonic wave of 50 kHz by an ultrasonicato (S6103, Aladdin, Shanghai, China) for two hours. After dispersing, the suspension was spray-coated on the surface of a SiO$_2$ glass substrate (5 cm × 5 cm) and the amount of coated N-TiO$_2$ catalysts was 30 mg. The N-TiO$_2$ coated glass was dried in air under 60 °C for 2 h.

3.2. Photocatalytic Reaction System

The schematic setup of the photocatalytic reaction system is illustrated in Figure 7. The cylindrical reactor with 15 cm in height and 10 cm in diameter was made of 316 L stainless steel. The temperature of the reactor were maintained at 25 °C by a bath circulator. A xenon lamp with a cut-off filter of 420 nm was used as the visible light illumination source. The illumination intensity could be adjusted at the range of 0 to 80 × 10^4 lx. A quartz window was mounted on the reactor for light irradiation. A gas chromatography (GC-2014, Shimadzu, Kyoto, Japan) was connected to the reactor to measure the concentrations of charged benzene in the reactor. The gas chromatography was equipped with Rtx-wax capillary column (Shimadzu) with 60 m in length, 0.53 mm in internal diameter and 1.0 μm in thickness.

3.3. Photocatalytic Reaction Procedures

The N-TiO$_2$ loaded glass was put into the photocatalytic reaction chamber. After a leakage check, the reactor was pumped to a vacuum of 0.1 atmosphere pressure, then the reactor was irradiated for 24 h under 254 nm ultraviolet light to clean the possible pollutants that may be adsorbed on the surface of the photocatalysts and the reactor as well. After a certain volume of benzene was charged/flushed into the reactor, clean air (N$_2$:O$_2$ = 80%:20%) was flushed into the reactor until the inner pressure was balanced with the atmospheric pressure. The concentration of benzene was set at 30 ppm as much as possible. Then the reactor was kept in dark for 60 min to reach the balance of adsorption-desorption. After that, the xenon lamp was turned on to make the irradiation through the quartz window, while the illumination intensity was adjusted at 36.7 × 10^4, 46.9 × 10^4, 61.7 × 10^4 and 75.1 × 10^4 lx by adjusting the distance between the light source and the sample. The concentration of the benzene in the reactor was measured and recorded every 30 min. The temperature of the reactor was maintained at 25 °C by a bath circulator.

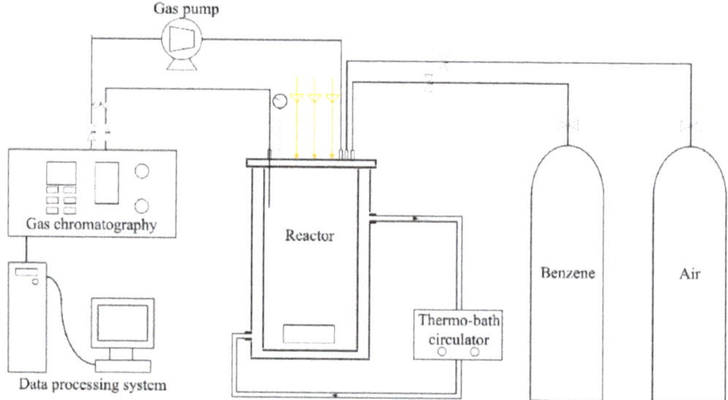

Figure 7. Schematic illustration of the photocatalytic reaction.

4. Conclusions

The L-H model has been used to investigate the kinetics of photodegradation of gaseous benzene by N-TiO$_2$ at 25 °C under visible light irradiation. Experimental data indicates that the adsorption equilibrium constant K_L calculated according to the L-H model decreased from 1139 to 597 m^3·mol^{-1} when the illumination intensity was increased from 36.7×10^4 lx to 75.1×10^4 lx, whereas it was 2761 m^3·mol^{-1} when in absence of light. This is contrary to the fact that K_L should be a constant if the reaction temperature was fixed. The benzene molecules adsorbed on the surface of the N-TiO$_2$ were dynamically photodegraded by the photocatalyst and thus the equilibrium of adsorption-desorption was broken would account for that. Photoreaction coefficient k_{pm} was introduced in the L-H model to compensate the disequilibrium of the adsorption-desorption caused by photodecomposition. Experiment result shows that k_{pm} is proportional to the light intensity $I^{1.482}$. As a result, the new parameter K_m ($k_a/(k_d + k_{pm})$) is closely related to the light intensity. Therefore, the concentration variation of benzene c vs irradiation time t under different light intensity I can be predicted.

Author Contributions: In this paper, P.S., J.Z. and W.L. designed the experiments; P.S., J.Z. and W.L. performed the experiments; P.S., J.Z., W.L. and Q.W. analyzed the data; the manuscript was written by P.S. and edited by W.C.

Funding: This research was funded by the National Key Research and Development Plan of China [Grant Nos. 2016YFC0700901, 2016YFC0700607].

Acknowledgments: This work is financially supported by the National Key Research and Development Plan of China [Grant Nos. 2016YFC0700901, 2016YFC0700607].

Conflicts of Interest: The authors declare no conflict of interest.

References

1. Sui, H.; Zhang, T.; Cui, J.; Li, X.; Crittenden, J.; Li, X.; He, L. Novel off-gas treatment technology to remove volatile organic compounds with high concentration. *Ind. Eng. Chem. Res.* **2016**, *55*, 2594–2603. [CrossRef]
2. Jiang, N.; Hui, C.-X.; Li, J.; Lu, N.; Shang, K.-F.; Wu, Y.; Mizuno, A. Improved performance of parallel surface/packed-bed discharge reactor for indoor VOCs decomposition: Optimization of the reactor structure. *J. Phys. D Appl. Phys.* **2015**, *48*, 40. [CrossRef]
3. Ye, C.Z.; Ariya, P.A. Co-adsorption of gaseous benzene, toluene, ethylbenzene, m-xylene (btex) and SO$_2$ on recyclable Fe$_3$O$_4$ nanoparticles at 0–101% relative humidities. *J. Environ. Sci.* **2015**, *31*, 164–174. [CrossRef] [PubMed]

4. Zeng, L.; Lu, Z.; Li, M.; Yang, J.; Song, W.; Zeng, D.; Xie, C. A modular calcination method to prepare modified N-doped TiO$_2$ nanoparticle with high photocatalytic activity. *Appl. Catal. B Environ.* **2016**, *183*, 308–316. [CrossRef]

5. Ren, L.; Mao, M.; Li, Y.; Lan, L.; Zhang, Z.; Zhao, X. Novel photothermocatalytic synergetic effect leads to high catalytic activity and excellent durability of anatase TiO$_2$ nanosheets with dominant {001} facets for benzene abatement. *Appl. Catal. B Environ.* **2016**, *198*, 303–310. [CrossRef]

6. Yadav, H.M.; Kim, J.-S. Solvothermal synthesis of anatase TiO$_2$-graphene oxide nanocomposites and their photocatalytic performance. *J. Alloys Compd.* **2016**, *688*, 123–129. [CrossRef]

7. Fujimoto, T.M.; Ponczek, M.; Rochetto, U.L.; Landers, R.; Tomaz, E. Photocatalytic oxidation of selected gas-phase VOCs using UV light, TiO$_2$, and TiO$_2$/Pd. *Environ. Sci. Pollut. Res.* **2017**, *24*, 6390–6396. [CrossRef] [PubMed]

8. Wongaree, M.; Chiarakorn, S.; Chuangchote, S.; Sagawa, T. Photocatalytic performance of electrospun CNT/TiO$_2$ nanofibers in a simulated air purifier under visible light irradiation. *Environ. Sci. Pollut. Res.* **2016**, *23*, 21395–21406. [CrossRef] [PubMed]

9. Sabbaghi, S.; Mohammadi, M.; Ebadi, H. Photocatalytic degradation of benzene wastewater using PANI-TiO$_2$ nanocomposite under UV and solar light radiation. *J. Environ. Eng.* **2016**, *142*, 05015003. [CrossRef]

10. Fang, J.; Chen, Z.; Zheng, Q.; Li, D. Photocatalytic decomposition of benzene enhanced by the heating effect of light: Improving solar energy utilization with photothermocatalytic synergy. *Catal. Sci. Technol.* **2017**, *7*, 3303–3311. [CrossRef]

11. Lan, L.; Li, Y.; Zeng, M.; Mao, M.; Ren, L.; Yang, Y.; Liu, H.; Yun, L.; Zhao, X. Efficient UV-Vis-Infrared light-driven catalytic abatement of benzene on amorphous manganese oxide supported on anatase TiO$_2$ nanosheet with dominant {001} facets promoted by a photothermocatalytic synergetic effect. *Appl. Catal. B Environ.* **2017**, *203*, 494–504. [CrossRef]

12. Melián, E.P.; Díaz, O.G.; Araña, J.; Rodríguez, J.M.D.; Rendón, E.T.; Melián, J.A.H. Kinetics and adsorption comparative study on the photocatalytic degradation of o-, m- and p-cresol. *Catal. Today* **2007**, *129*, 256–262. [CrossRef]

13. Chen, M.; Bao, C.; Cun, T.; Huang, Q. One-pot synthesis of ZnO/oligoaniline nanocomposites with improved removal of organic dyes in water: Effect of adsorption on photocatalytic degradation. *Mater. Res. Bull.* **2017**, *95*, 459–467. [CrossRef]

14. Zhi, Y.; Li, Y.; Zhang, Q.; Wang, H. Zno nanoparticles immobilized on flaky layered double hydroxides as photocatalysts with enhanced adsorptivity for removal of acid red g. *Langmuir* **2010**, *26*, 15546–15553. [CrossRef] [PubMed]

15. Dong, W.; Lee, C.W.; Lu, X.; Sun, Y.; Hua, W.; Zhuang, G.; Zhang, S.; Chen, J.; Hou, H.; Zhao, D. Synchronous role of coupled adsorption and photocatalytic oxidation on ordered mesoporous anatase TiO2-SiO2 nanocomposites generating excellent degradation activity of rhb dye. *Appl. Catal. B Environ.* **2010**, *95*, 197–207. [CrossRef]

16. Kim, S.B.; Hong, S.C. Kinetic study for photocatalytic degradation of volatile organic compounds in air using thin film TiO$_2$ photocatalyst. *Appl. Catal. B Environ.* **2002**, *35*, 305–315. [CrossRef]

17. Golshan, M.; Zare, M.; Goudarzi, G.; Abtahi, M.; Babaei, A.A. Fe$_3$O$_4$@hap-enhanced photocatalytic degradation of Acid Red73 in aqueous suspension: Optimization, kinetic, and mechanism studies. *Mater. Res. Bull.* **2017**, *91*, 59–67. [CrossRef]

18. Deng, X.-Q.; Liu, J.-L.; Li, X.-S.; Zhu, B.; Zhu, X.; Zhu, A.-M. Kinetic study on visible-light photocatalytic removal of formaldehyde from air over plasmonic Au/TiO$_2$. *Catal. Today* **2017**, *281*, 630–635. [CrossRef]

19. Langmuir, I. The constitution and fundamental properties of solids and liquids. Part I. Solids. *J. Am. Chem. Soc.* **1916**, *38*, 2221–2295. [CrossRef]

20. Liu, P.; Yu, X.; Wang, F.; Zhang, W.; Yang, L.; Liu, Y. Degradation of formaldehyde and benzene by TiO$_2$ photocatalytic cement based materials. *J. Wuhan Univ. Technol.-Mater. Sci. Ed.* **2017**, *32*, 391–396. [CrossRef]

21. Yuzawa, H.; Aoki, M.; Otake, K.; Hattori, T.; Itoh, H.; Yoshida, H. Reaction mechanism of aromatic ring hydroxylation by water over platinum-loaded titanium oxide photocatalyst. *J. Phys. Chem. C* **2012**, *116*, 25376–25387. [CrossRef]

22. Einaga, H.; Mochiduki, K.; Teraoka, Y. Photocatalytic oxidation processes for toluene oxidation over TiO$_2$ catalysts. *Catalysts* **2013**, *3*, 219. [CrossRef]

23. Lin, Y.-H.; Hsueh, H.-T.; Chang, C.-W.; Chu, H. The visible light-driven photodegradation of dimethyl sulfide on S-doped TiO$_2$: Characterization, kinetics, and reaction pathways. *Appl. Catal. B Environ.* **2016**, *199*, 1–10. [CrossRef]

24. Dhada, I.; Nagar, P.K.; Sharma, M. Photo-catalytic oxidation of individual and mixture of benzene, toluene and p-xylene. *Int. J. Environ. Sci. Technol.* **2016**, *13*, 39–46. [CrossRef]

25. Cheng, L.; Kang, Y.; Li, G. Effect factors of benzene adsorption and degradation by nano-TiO$_2$ immobilized on diatomite. *J. Nanomaterials* **2012**, *2012*, 6. [CrossRef]

26. Ollis, D.F. Kinetics of liquid phase photocatalyzed reactions: An illuminating approach. *J. Phys. Chem. B* **2005**, *109*, 2439–2444. [CrossRef] [PubMed]

27. Xu, Y.; Langford, C.H. Variation of langmuir adsorption constant determined for TiO$_2$-photocatalyzed degradation of acetophenone under different light intensity. *J. Photochem. Photobiol. A Chem.* **2000**, *133*, 67–71. [CrossRef]

28. Giovannetti, R.; Rommozzi, E.; D'Amato, C.; Zannotti, M. Kinetic model for simultaneous adsorption/ photodegradation process of alizarin red s in water solution by nano-TiO$_2$ under visible light. *Catalysts* **2016**, *6*, 84. [CrossRef]

29. Silva, C.G.; Faria, J.L. Effect of key operational parameters on the photocatalytic oxidation of phenol by nanocrystalline sol-gel TiO$_2$ under uv irradiation. *J. Mol. Catal. A Chem.* **2009**, *305*, 147–154. [CrossRef]

30. Du, E.; Zhang, Y.X.; Zheng, L. Photocatalytic degradation of dimethyl phthalate in aqueous TiO$_2$ suspension: A modified langmuir–hinshelwood model. *React. Kinet. Catal. Lett.* **2009**, *97*, 83–90. [CrossRef]

31. Brosillon, S.; Lhomme, L.; Vallet, C.; Bouzaza, A.; Wolbert, D. Gas phase photocatalysis and liquid phase photocatalysis: Interdependence and influence of substrate concentration and photon flow on degradation reaction kinetics. *Appl. Catal. B Environ.* **2008**, *78*, 232–241. [CrossRef]

32. He, F.; Li, J.; Li, T.; Li, G. Solvothermal synthesis of mesoporous TiO$_2$: The effect of morphology, size and calcination progress on photocatalytic activity in the degradation of gaseous benzene. *Chem. Eng. J.* **2014**, *237*, 312–321. [CrossRef]

33. Soltani, T.; Lee, B.-K. Novel and facile synthesis of Ba-doped BiFeO$_3$ nanoparticles and enhancement of their magnetic and photocatalytic activities for complete degradation of benzene in aqueous solution. *J. Hazard. Mater.* **2016**, *316*, 122–133. [CrossRef] [PubMed]

34. Chen, S.-H.; Hsiao, Y.-C.; Chiu, Y.-J.; Tseng, Y.-H. A simple route in fabricating carbon-modified titania films with glucose and their visible-light-responsive photocatalytic activity. *Catalysts* **2018**, *8*, 178. [CrossRef]

35. Li, C.X.; Jin, H.Z.; Yang, Z.Z.; Yang, X.; Dong, Q.Z.; Li, T.T. Preparation and photocatalytic properties of mesoporous RGO/TiO$_2$ composites. *J. Inorg. Mater.* **2017**, *32*, 357–364.

36. Xu, J.; Liu, Q.; Lin, S.; Cao, W. One-step synthesis of nanocrystalline N-doped TiO$_2$ powders and their photocatalytic activity under visible light irradiation. *Res. Chem. Intermed.* **2013**, *39*, 1655–1664. [CrossRef]

37. Wang, J.; Ruan, H.; Li, W.; Li, D.; Hu, Y.; Chen, J.; Shao, Y.; Zheng, Y. Highly efficient oxidation of gaseous benzene on novel Ag$_3$VO$_4$/TiO$_2$ nanocomposite photocatalysts under visible and simulated solar light irradiation. *J. Phys. Chem. C* **2012**, *116*, 13935–13943. [CrossRef]

38. Cong, Y.; Zhang, J.; Chen, F.; Anpo, M.; He, D. Preparation, photocatalytic activity, and mechanism of nano-TiO$_2$ co-doped with nitrogen and iron (iii). *J. Phys. Chem. C* **2007**, *111*, 10618–10623. [CrossRef]

39. Foo, K.Y.; Hameed, B.H. Insights into the modeling of adsorption isotherm systems. *Chem. Eng. J.* **2010**, *156*, 2–10. [CrossRef]

40. Wei, D.; Li, S.; Fang, L.; Zhang, Y. Effect of environmental factors on enhanced adsorption and photocatalytic regeneration of molecular imprinted TiO$_2$ polymers for fluoroquinolones. *Environ. Sci. Pollut. Res.* **2018**, *25*, 6729–6738. [CrossRef] [PubMed]

41. Li, Z.; Kim, J.K.; Chaudhari, V.; Mayadevi, S.; Campos, L.C. Degradation of metaldehyde in water by nanoparticle catalysts and powdered activated carbon. *Environ. Sci. Pollut. Res.* **2017**, *24*, 17861–17873. [CrossRef] [PubMed]

42. Xu, J.; Sun, P.; Zhang, X.; Jiang, P.; Cao, W.; Chen, P.; Jin, H. Synthesis of N-doped TiO$_2$ with different nitrogen concentrations by mild hydrothermal method. *Mater. Manuf. Processes* **2014**, *29*, 1162–1167. [CrossRef]

MDPI

St. Alban-Anlage 66

4052 Basel

Switzerland

Tel. +41 61 683 77 34

Fax +41 61 302 89 18

www.mdpi.com

Catalysts Editorial Office

E-mail: catalysts@mdpi.com

www.mdpi.com/journal/catalysts